Purchased with funds from
mouse Genetics Grant (MRC- MA-1062)
1980.

HISTOPATHOLOGIC TECHNIC AND PRACTICAL HISTOCHEMISTRY

HISTOPATHOLOGIC TECHNIC AND PRACTICAL HISTOCHEMISTRY

FOURTH EDITION

R. D. Lillie, M.D.

Research Professor of Pathology
Louisiana State University Medical Center, New Orleans

Harold M. Fullmer, D.D.S.

Director of Institute of Dental Research
Professor of Pathology
University of Alabama Medical Center, Birmingham

McGraw-Hill Book Company

A Blakiston Publication

New York   St. Louis   San Francisco   Auckland
Düsseldorf   Johannesburg   Kuala Lumpur   London
Mexico   Montreal   New Delhi   Panama   Paris
São Paulo   Singapore   Sydney   Tokyo   Toronto

**HISTOPATHOLOGIC
TECHNIC AND
PRACTICAL
HISTOCHEMISTRY**

Copyright © 1976, 1965 by McGraw-Hill, Inc. All rights reserved.
Copyright 1954 by McGraw-Hill, Inc. All rights reserved.
Formerly published under the title of HISTOPATHOLOGIC TECHNIC, copyright 1948
by McGraw-Hill, Inc. All rights reserved.
Printed in the United States of America. No part of this publication may be reproduced,
stored in a retrieval system, or transmitted, in any form or by any means, electronic,
mechanical, photocopying, recording, or otherwise, without the prior written permission
of the publisher.

234567890 KPKP 783210987

This book was set in Times Roman. The editors were J. Dereck Jeffers and Anne T.
Vinnicombe; the designer was Jo Jones; the production supervisor was Dennis J. Conroy.
The drawings were done by Oxford Illustrators Limited.
Kingsport Press, Inc., was printer and binder.

Library of Congress Cataloging in Publication Data

Lillie, Ralph Dougall, date
    Histopathologic technic and practical histochemistry.
    First published in 1947 under title: Histopathologic technic.
    "A Blakiston publication."
    Bibliography: p.
    Includes index.
    1. Histology, Pathological—Laboratory manuals.
2. Histochemistry—Laboratory manuals. I. Fullmer, Harold M., joint author.
II. Title. [DNLM: 1. Histocytochemistry—Laboratory manuals. 2. Histological
technics. 3. Pathology—Laboratory manuals. QZ25 L729h]
RB37.L65 1976   616.07'583   75-34296
ISBN 0-07-037862-2

" ὅταν δὲ ἔλθῃ ἐκεῖνος, τὸ πνεῦμα τῆς ἀληθείας,
ὁδηγήσει ὑμᾶς εἰς την ἀλήθειαν πᾶσαν." Ιω. xvi–13
δίδασκε ἡμᾶς, κύριε, γνῶναι ταύτην ἀλήθειαν.

"When he comes, the Spirit of Truth, he will
drive you on the path into all truth."
John XVI:13. Teach us, Lord, to recognize
that truth. (Translation according to
Dr. R. D. Lillie)

NOTICE

Medicine is an ever-changing science. As new research and clinical experience broaden our
knowledge, changes in treatment and drug therapy are required. The editors and the
publisher of this work have made every effort to ensure that the drug dosage schedules
herein are accurate and in accord with the standards accepted at the time of publication.
Readers are advised, however, to check the product information sheet included in the
package of each drug they plan to administer to be certain that changes have not been made
in the recommended dose or in the contraindications for administration. This recommendation
is of particular importance in regard to new or infrequently used drugs.

CONTENTS

Foreword   xi

Preface   xiii

General References   xv

1 Microscopy   1

Light; the microscope; illumination; oblique and dark-field illumination; polariscopy; fluorescence microscopy; absorption spectroscopy; phase microscopy; micrometry; mechanical stages.

2 Equipment   17

Microtome knives; paraffin oven; heating equipment; cryostat; staining equipment; cleaning of slides and covers; labeling; specimen bottles; solvent table; reagents.

3 Fixation   25

General procedures; aldehydes; fixative buffers; tannic acid; Ruthenium red; lanthanum; dimethyl suberimidate; alcohols; acetone; lead; mercury; chromates; picric acid; osmium tetroxide; pyroantimonate; heat; fresh tissues.

4 Sectioning   69

Freehand sectioning; tissue slicers; frozen sections; ultrathin sections; freeze substitution; embedding for light microscopy with polyvinyl alcohol, Carbowax, gelatin, agar, paraffin, ester wax, and celloidin: embedding for electron microscopy with Aquon, Durcupan, glycol methacrylate, 2-hydroxypropyl methacrylate, cross-linked polyampholytes, urea formaldehyde, dimethylol urea, methacrylates, Vestopal, Rigolac, epoxy resins, Araldite maraglas 655, and miscellaneous other embedding media.

5 General Staining and Mounting Procedures   105

Dewaxing, hydration, and dehydration methods; mounting methods; handling celloidin-embedded sections; various mounting media and refractive indices; formulas of various water-soluble mounting media and diluting fluids.

6 Stains and Fluorochromes   123

Nuclear stains: mordant dyes, basic aniline dyes, fluorochrome stains; plasma stains; fat stains; diazonium salts; tetrazolium salts; iodine; reactive dyes (Procion); table of solubilities of various common dyes.

7 Nuclei, Nucleic Acids, General Oversight Methods   165

Structures of DNA, nucleic acid digestions; Feulgen, ferrocene carbohydrazide, methyl green-pyronin, and other basic dye stains for nuclei, stains for ribonucleic acid; isoelectric point; methods for nucleohistone; squash preparations for chromosomes; azure-eosin methods; iron and alum hematozylins; other mordant nuclear stains; anthocyanins and anthocyanidins.

8 Chemical End Groups   217

Ethylenes; peroxides; alcohols; hydroxyl and amine blockade; glycols; methods for SH; azo coupling reaction; acylation blockade; methods for tyrosine, histidine, histamine, tryptophan, chromaffin, noradrenalin, enterochromaffin, 5-hydroxytryptamine; the Argentaffin stain; carotid body tumors; methods for amines and blockade of amines; methods for arginine and citrulline; methods for fibrin and fibrinoid; digestion methods for trypsin, pepsin, and elastase; antibody localization; aldehyde and carbonyl reactions; methods for acid groups.

9 Cytoplasmic Granules and Organelles   327

Mitochondria; Golgi; granules of salivary gland, pancreas, pancreatic islets, hypophysis, testis, epididymis, peptic glands, Paneth cells, renal juxtaglomerular granules, keratin, and keratohyalin.

10 Enzymes   357

Tissue preparation; principles of azo dye methods; lysosomes; methods for phosphatases, adenylcyclase, nucleoside phosphorylase, acetyltransferase, phosphoglucomutase, glycerol 3-phosphate dehydrogenase, creatine kinase, disaccharases, aldolase, hexokinase, L-hexonate dehydrogenase, ketose reductase, β-D-galactosidase, α-D-galactosidase, α-D-glucosidase, β-D-glucosidase, amylophosphorylase, UDPG glycogen transferase, N-acetyl-β-glucosaminidase, β-glucuronidase, aryl sulfatase, nonspecific esterases, cholinesterase, choline acetyltransferase, proteases, cathepsin B1, cathepsin D, tyrosine aminotransferase, substrate-film enzyme methods, gelatin protease, DNase, RNase, histonase, cellulase, cysteine desulfurase, carbonic anhydrase, urease, uricase, peroxidases, G-Nadi-oxidase, cytochrome oxidase, peroxisomes, catalase, tyrosinase, dopamine-β-hydroxylase, dopa decarboxylase, phenylethanolamine-N-methyltransferase, monoamine oxidase, dehydrogenases, creatine phosphokinase.

11 Endogenous Pigments   485

Heme pigments; acid hematins; biliverdin, bilirubin and hematoidin; iron-containing pigments; melanosis and pseudomelanosis pigments of intestine; ochronosis; lipofuscins; ceroid; other combined pigment stains; melanins.

12 Metals, Anions, Exogenous Pigments   529

Carotenes; vitamin A; riboflavin; vitamin C; pneumonyssus pigment; exogenous pigments; hematoxylin as a reagent for metals; methods for iron, aluminum, asbestos, titanium, beryllium, beryllium oxide, barium, strontium, calcium, calcium oxalate, sodium, magnesium, silver, copper, cobalt, nickel, lead, arsenic, antimony, bismuth, gold, mercury, manganese, potassium, lithium, thallium, zinc, uranium, thorium, sulfates, phosphates.

13 Lipids   559

Classification; birefringent lipids; differential solubility; fixation; staining methods with oil-soluble dyes; methods for phosphoglycerides, plasmalogens, cerebrosides, sulfatides, gangliosides, cholesterol, fatty acids; reactions for ethylene groups, myelin, carbonyl lipids; acid fastness; electron microscopic histochemical methods; lipid autoradiography; quantitative lipid histochemistry.

14 Polysaccharides, Mucins   611

Classification of mucosubstances; periodic acid Schiff method and effects of various blockades; aldehyde-hydrazine-formazan formation; periodic acid-salicyl-hydrazide method; methods for glycogen for light and electron microscopy; dextran; chitins; cellulose, starch; lignin, inulin, pectin; mucin, cartilage; metachromasia; alcian blue; mixed, low, and high iron diamine stains; Hale stain; concanavalin A-iron dextran stain; mucicarmine; stains for sulfomucins; aldehyde fuchsin; phosphotungstic acid; sialomucins and sialidase; hyaluronidase digestions; amyloid stains; stains for mast cells.

15 Connective Tissue Fibers and Membranes   679

Collagen; reticulum; basement membranes; silver methods; phosphotungstic acid-hematoxylin methods; acid aniline dye methods; picric acid mixtures; hydrochloric acid methods; allochrome stain; elastic tissue stains; oxytalan fiber stains.

16 Smear Preparations, Bacteria, Protozoa, and Other Parasites   719

Smear preparations; Papanicolaou-type stains; Gram stain; Brown-Brenn stain; stains for fungi; stains for diphtheria, influenza, encephalitozoa, and toxoplasma organisms; acid-fast stains; stains for bacterial spores; stains for gram-negative bacteria and Rickettsiae; Giemsa-type stains; stains for leukocyte granules, platelets, reticulocytes, and various oxyphil inclusion bodies; methods for spirochetes and trichinae.

17 Glia and Nerve Cells and Fibers   765

Methods for oligodendrocytes, glia cells, astrocytes; silver and other methods for cells and their processes; stains for retina and other ocular structures.

18 Hard Tissues; Decalcification   787

Tissue preparations; decalcifying methods; ground sections of bones and teeth; stains for bone canaliculi, dentinal tubules, osteoid; labeling of bones and teeth; stains for bone cells and tissues; stains for cartilage.

19 Various Special Procedures   809

Vascular injections; corrosion methods; principles of autoradiography; contact, liquid emulsion, stripping film, and Baserga's two emulsion method for ^{3}H and ^{14}C autoradiographic methods; autoradiographic methods for electron microscopy; microradiography; quantitative microchemical methods; zymograms; scanning electron microscopy; freeze etching; negative staining; electron probe microanalyzer; laser microprobe analysis; principles of fluorescence microscopy; immunofluorescence: preparation of antisera, methods of conjugation of antisera, purification of conjugates, preparation of tissues for immunofluorescence, immunohistochemical detection of cell surface receptors, immunofluorescence quantitation, ferritin-labeled antibody methods; immunoenzyme methods: conjugation with glutaraldehyde, unlabeled antibody enzyme method; microincineration.

20 Buffers and Buffer Tables, Normal Acids and Alkalies   869

Index   891

FOREWORD

This book has been written to give a systematic treatment of recent advances in histopathologic technic. In it I have endeavored to include workable methods on unusual subjects as well as those in frequent use. No attempt has been made to attain encyclopedic scope and many methods which have appeared in previous works have been excluded as obsolete or impractical.

I have endeavored to find methods depending for constancy of results on controllable factors such as time, temperature, hydrogen ion concentration, and concentration of reagents, rather than on the skill of the individual technician. An attempt has been made to avoid methods depending on special primary fixations of tissue. Particularly I have tried to include methods which will work well after routine formalin fixation, and to choose those which may be completed with a few steps and in a short time, having in mind the demands of time on the pathologist.

I have made many modifications of older methods with these purposes in mind. My greatest regret is that in the interests of completeness I have had to include some methods with which I have had no personal experience, and others which I have tried and regard as not fully satisfactory.

I am indebted to my associates, Drs. L. L. Ashburn, K. M. Endicott, B. Highman, R. C. Dunn, and L. R. Hershberger, for some special methods, for advice, and for criticism; to Mrs. C. Jones, Miss D. Plotka, Mr. R. R. Reed, Mr. R. Faulkner, Miss A. Laskey, and Miss J. Greco for their aid in testing and devising many of the special methods included.

I also acknowledge my indebtedness to my predecessors and colleagues from whose works I have borrowed freely. Such of these borrowings as have been taken directly from their original publications are usually so cited in the text, but many have been taken, often in modified form, from other laboratory manuals. These texts are usually cited simply by the author's name, except that in the case of Ehrlich's "Encyklopädie" I have often cited the contributor's name. This last text I have often preferred as a source of those older methods which are still used in unmodified form. The following texts listed in "General References" have been thus used, as well as earlier editions of some of them.

R. D. Lillie

PREFACE

This edition represents a major revision. Most chapters have been greatly enlarged. Dr. Fullmer has assisted with the enormous task of bringing together in one text a broad coverage of cytologic, histopathologic, histochemical, and cytochemical methods employed in current laboratories throughout the world today. We have attempted to provide the methods in a manner that a moderately qualified technician could employ. In addition to results, the theory, mechanism of action, problems that may be encountered, and limitations of interpretations of the staining reactions are provided when known.

We are particularly indebted to Drs. W. Fishman, F. B. Johnson, and S. S. Spicer for reviewing pertinent chapters of this revision. We are indebted further to Drs. M. Hartley, Philip Pizzolato, and J. Mestecky for advice and counsel; to Dr. C. A. McCallum, Dean, School of Dentistry, University of Alabama in Birmingham, and Dr. J. F. Volker, President, University of Alabama in Birmingham, for moral and administrative support; to Mr. Clifton Link, Jr., Mrs. Patricia Donaldson, and Miss Annie Jo Narkates for technical support and Mrs. Charlotte Hughes and Mrs. Dena Darby for secretarial services.

The new material has been the special task in some chapters of one of us, in other chapters of the other, but we have both taken active part in the revisions of all, and there has been remarkably little dispute about the new material. This is probably due to our having written several joint research papers in our National Institutes of Health days. We have both been more interested in practicalities and results than in pure theory. This calls to mind a story about the old master Carl Weigert told by the late Dr. S. P. Kramer, who spent his two years in Germany under Weigert and Virchow about 1890 to 1892. Professor Weigert demonstrated one of his new staining methods to the class, and at the end of the lecture a few students came to the podium to ask questions. One young man asked, "But Professor, what is the theory of the method?" and Dr. Weigert snorted, "Ach! *Die Theorie*! You should ask my nephew Paul [Ehrlich] about that!"

This is not to state that we have not inquired diligently into the probable chemical mechanisms, not only of new histochemical procedures, but also of many traditional staining methods. These inquiries have added greatly to our understanding of stains and reactions.

Specific page references have been eliminated from this edition. Many topics come up in more than one place in the text. We suggest that the reader consult the index for complete listings of pages where information is given on any subject.

This has been a large task that neither of us could have done alone in the time it has taken, and we hope that our effort will prove useful to pathologists, histologists, and technicians alike, as well as to others who may find matters of interest in it.

The very modest list of 17 "General References" published in the first edition of this book grew to 30 in the second edition and to 46 in the third. With the appearance of more general texts in histochemistry as well as histology and enzymology, and with the necessity of entering to some extent the field of electron microscopy, this list has grown to 64 titles. The same system of referring in the text to this list simply by author's name, year, and perhaps a page reference has been followed.

R. D. Lillie
Harold M. Fullmer

GENERAL REFERENCES

Adams, C. W. M.: "Neurohistochemistry," Elsevier, Amsterdam, New York, 1965.

Altmann, R.: "Die Elementorganismen und ihre Bezielungen zu den Zellen," Veit, Leipzig, 1890.

American Registry of Pathology: "Manual of Histologic Staining Methods of the Armed Forces Institute of Pathology," 3d ed., edited by L. Luna, McGraw-Hill, New York, 1968.

Anderson, C. A.: An Introduction to the Electron Probe Microanalyzer and Its Application to Biochemistry, in D. Glick (ed.), "Methods of Biochemical Analysis," vol. 15, p. 147, Interscience, New York, 1967.

Armed Forces Institute of Pathology (AFIP): "Manual of Special Staining Technics," edited by M. F. Gridley, Washington, 1953; "Manual of Histologic and Special Staining Technics," 2d ed., edited by L. P. Ambrogi, McGraw-Hill, New York, 1960; 3d ed., edited by L. Luna, McGraw-Hill, New York, 1968 (see American Registry of Pathology, above).

Baker, J. R.: "Principles of Biological Microtechnique," Methuen, London, and Wiley, New York, 1958.

Barka, T., and Anderson, P. J.: "Histochemistry: Theory, Practice and Bibliography," Hoeber-Harper, New York, 1963.

"Colour Index," 2d ed., Society of Dyers and Colourists, Bradford, Yorkshire, England, and American Association of Textile Chemists and Colorists, Lowell, Mass., 1956; Supplement, 1963. See also Rowe.

Conn, H. J.: "Biological Stains," 4th, 5th, and 6th eds., Biotech, Geneva, N.Y., 1940, 1946, 1953.

———: "Biological Stains," 7th ed., Williams & Wilkins, Baltimore, 1961.

——— and Darrow, M. A.: "Staining Procedures," Biotech, Geneva, N.Y., 1943–1945, 1947–1949.

———, ———, and Emmel, V. M.: "Staining Procedures," 2d ed., Williams & Wilkins, Baltimore, 1960.

Cowdry, E. V.: "Microscopic Technique in Biology and Medicine," Williams & Wilkins, Baltimore, 1943, 1948.

Davenport, H. A.: "Histological and Histochemical Technics," Saunders, Philadelphia, 1960.

Ehrlich, P., et al.: "Encyklopädie der microskopischen Technik," Urban & Schwarzenberg, Berlin, 1903.

Feigl, F.: "Qualitative Analysis by Spot Tests," 3d ed., Elsevier, Amsterdam, 1946.

———: "Chemistry of Specific Selective and Sensitive Reactions," Academic, New York, 1949.

———: "Spot Tests in Organic Analysis," 5th ed., Elsevier, Amsterdam, 1958.

———: "Spot Tests in Inorganic Analysis," 5th ed., Elsevier, Amsterdam, 1958.

Ganter, P., and Jolles, G.: "Histochimie Normale et Pathologique," Gauthier-Villars, Paris, 1970.

Gatenby, J. B., and Beams, H. W.: Bolles Lee's "The Microtomist's Vade Mecum," Blakiston, Philadelphia (McGraw-Hill, New York), 1950.

Glick, D.: "Techniques of Histo- and Cytochemistry," Interscience, New York, 1949.

Gomori, G.: "Microscopic Histochemistry: Principles and Practice," University of Chicago Press, 1952.

Gray, P.: "The Microtomist's Formulary and Guide," Blakiston, Philadelphia (McGraw-Hill, New York), 1954.

Harms, H.: "Handbuch der Farbstoffe für die Mikroskopie," Leverkusen, Staufen Verlag, 1965.

Hayat, M. A.: "Principles and Technics of Electron Microscopy," vols. I and II, Van Nostrand, New York, 1970, 1972.

Hickinbottom, W. J.: "Reactions of Organic Compounds," 2d ed., Longmans, London, 1948; 3d ed., 1957.

Hueck, W.: Pigmentstudien, *Zieglers Beitr. Pathol. Anat.*, **54:** 68, 1912.

———: Die pathologische Pigmentierung, in Kraehl-Marchands (ed.), "Handbuch der allgemeinen Pathologie," 1921.

Ingram, M. J., and Hogben, A. M.: Procedures for the Study of Biological Soft Tissue with the Electron Microprobe, in "Developments in Applied Spectroscopy," edited by W. K. Baer, A. J. Perkins, and E. L. Grove, vol. 6, p. 43, Plenum, New York, 1968.

Jensen, W. A.: "Botanical Histochemistry, Principles and Practice," Freeman, San Francisco, 1962.

Jones, R. M.: "McClung's Handbook of Microscopical Technique," Hoeber-, Harper, New York, 1950.

Karrer, P.: "Organic Chemistry," 4th ed., Elsevier Publishing Company, New York and Amsterdam, 1950.

Klein, G.: Allgemeine und spezielle Methodik der Histochemie, in T. Peterfi (ed.), "Methodik der wissenschaftlichen Biologie I, Springer, Berlin, 1928.

Krajian, A. A., and Gradwohl, R. B. H.: "Histopathological Technic," 2d ed., Mosby, St. Louis, 1952.

Kraus, Gerlach, and Schweinburg: "Lyssa bei Mensch und Tier," Urban & Schwarzenberg, Berlin, 1926.

Lange, N. A.: "Handbook of Chemistry," 9th and 10th eds., Blakiston, Philadelphia (McGraw-Hill, New York), 1956, 1961; rev. 10th ed., 1967.

Langeron, M.: "Précis de microscopie," 5th ed., Masson, Paris, 1934; 7th ed., 1949.

Lee, A. B.: "The Microtomist's Vade-Mecum," 10th ed., edited by J. B. Gatenby and T. S. Painter, Blakiston, Philadelphia (McGraw-Hill, New York), 1937.

Lillie, R. D.: "Histopathologic Technic," Blakiston, Philadelphia for McGraw-Hill, New York, 1948.

———: "Histopathologic Technic and Practical Histochemistry," 2d ed., Blakiston, Philadelphia (McGraw-Hill, New York), 1954; 3d ed., Blakiston, Philadelphia (McGraw-Hill, New York), 1965.

——— (ed.): "Conn's Biological Stains," 8th ed., Williams & Wilkins, Baltimore, 1969.

Lison, L.: "Histochimie animale," Gauthier-Villars, Paris, 1936.

———: "Histochimie et cytochimie animales," 2d ed., Gauthier Villars, Paris, 1953; 3d ed., 1960.

Long, C.: "Biochemists' Handbook," Van Nostrand, New York, 1961.

McClung, C. E.: "Handbook of Microscopical Technique," Hoeber-Harper, New York, 1929.

McLean, R. C., and Cook, W. R. I.: "Plant Science Formulae," Macmillan, New York, 1941.

McManus, J. F. A., and Mowry, R. W.: "Staining Methods, Histological and Histochemical," Hoeber-Harper, New York, 1960.

Mallory, F. B.: "Pathological Technic," Saunders, Philadelphia, 1938.

Mellors, R. C. (ed.): "Analytical Cytology," 2d ed., Blakiston, Philadelphia (McGraw-Hill, New York), 1959.

Nairn, R. C.: "Fluorescent Protein Tracing," 3d ed., Williams & Wilkins, Baltimore, 1969.

Pearse, A. G. E.: "Histochemistry:

Theoretical and Applied," Little, Brown, Boston, 1953; 2d ed., 1960.

————: "Histochemistry, Theoretical and Applied," 3d ed., vol. 1, Churchill, London, 1968; vol. 2, Churchill Livingstone, Edinburgh, London, 1972.

Ramón y Cajal, S.: "Histology," translated by M. Fernan-Nunez, Wood, Baltimore, 1933.

Ranvier, L.: "Traité Technique d'Histologie," Libraire F. Savy, Paris, 1875.

Romeis, B.: "Taschenbuch der mikroskopischen Technik," 13th ed., Oldenbourg, Munich, 1932.

————: "Mikroskopische Technik," Leibnitz, Munich, 1948.

Roth, L. J., and Stumpf, W. E.: "Autoradiography of Diffusible Substances," Academic, New York, 1969.

Roulet, F.: "Methoden der pathologischen Histologie," Springer-Verlag, Vienna, 1948.

Rowe, F.,: "Colour Index," 1st ed. and Supplement, 1924 and 1928. See also "Colour Index."

Schmorl, G.: "Die pathologisch-histologischen Untersuchungsmethoden," 15th ed., Vogel, Leipzig, 1928.

Stecher, P. G. (ed.): "The Merck Index of Chemicals and Drugs," 7th ed., Merck, Rahway, N.J., 1960; 8th ed., 1968.

Sternberger, L. A.: "Immunocytochemistry," Prentice-Hall, Englewood Cliffs, N.J., 1974.

Thompson, S. W.: "Selected Histochemical and Histopathological Methods," Charles C. Thomas, Springfield, Ill., 1966.

Weast, R. C.: "CRC Handbook of Chemistry and Physics," 50th ed., Chemical Rubber Co., Cleveland, 1969.

1

MICROSCOPY

We do not propose to enter into any theoretical discussion of the optics concerned in the use of the compound microscope. Rather, the purpose of this chapter is to bring in certain practical points in the use of the microscope in which we have found it necessary to instruct technicians and physicians in training in pathology.

LIGHT

The advice in older manuals about the necessity for north windows for microscopic work, the avoidance of direct sunlight, and the preferability of a white cloud as a source of illumination is still applicable for the monocular microscope. However, daylight seldom gives adequate lighting for binocular microscopes or for more than low powers; hence some form of artificial lighting generally is necessary for microscopic work. Such lighting has the further advantage of not being subject to variations in the weather.

A tungsten-filament electric lamp gives satisfactory illumination for most purposes. The slightly yellowish color of the light can be corrected by insertion of a thin blue-glass disk into the microscope substage, by the use of blue-glass daylight bulbs, or by the interposition of a water filter containing a weak solution of copper sulfate to which sufficient ammonia water has been added to change its color from green to blue. In form this water filter may be a cell with flat parallel sides such as the microscope lamp manufacturers often supply, or it may be spherical, such as a 500-ml Florence flask. The latter serves also as a converging lens.

Filament images, which give rise to uneven illumination of the microscopic field, are avoidable by the use of ground-glass disks placed in the microscope substage, by the use of frosted- or milky-glass bulbs, or, in more elaborate lamps, by the use of a homogeneous light source large enough to fill the field completely, such as the 6-V ribbon-filament lamp.

In all lamps which do not possess a focusing or condensing device to produce parallel or converging light rays, it must be borne in mind that the intensity of the illumination is inversely proportional to the square of the distance of the light source from the object. The same law applies when a ground-glass disk is inserted in the path of a parallel or converging light beam, since this ground-glass surface acts as though it were the light source and the available illumination diminishes with the distance from the disk to the object.

With the larger microscope lamps, which employ lens systems to focus the light accurately on the condenser of the microscope, it is advisable to mount both lamp and microscope in permanent positions on a baseboard, so that once proper optical alignment is established, it need not be disturbed. In this case it is desirable to have a cloth bag or some form of rigid cover to place over the microscope when it is not in use, in order to protect it from dust. The old-fashioned bell jar functioned well in this respect, but it was heavy and breakable. A cylinder of cellophane or similar transparent plastic, of sufficient diameter and height to cover the microscope readily and with a handle on top, makes a very satisfactory lightweight, transparent substitute which is not readily broken.

The baseboard on which the microscope and lamp are mounted may be made of sufficient thickness to support the microscope at a level such that the eyepieces are at the most convenient height for the individual observer. Among seven workers in one laboratory the most convenient height of the eyepieces above a table 76 cm (30 in) high varied from 33 to 40 cm (13 to $15\frac{3}{4}$ in). A swivel chair with adjustable height may also be used to bring the user's eyes to the approximate level of the eyepieces. Larger lamps with large light sources and focused beams are needed for critical work at high magnifications, for dark-field work, and, principally, for photomicrography. Of late years many of the microscope manufacturers have been furnishing integrated lighting systems attached to the microscope, usually in the substage area. Description of these systems and statements as to the type of illumination furnished are to be found in the manufacturer's catalogs. Some of them include devices for polarized light and dark-field illumination and for substitution of phase-contrast optics.

As with other projection-lamp bulbs, it is often profitable to purchase bulbs of somewhat higher voltage rating than specified by the lamp manufacturer and to operate the lamp at a slightly lower voltage than the stated rating of the bulbs employed. When used this way, the bulbs last much longer, and the difference in color value of the light is scarcely appreciable.

THE MICROSCOPE

For the average worker, the use and care of the microscope are adequately described in the booklets furnished by the manufacturers. Only a few practical points will be discussed here. As Schmorl aptly states, the microscope should be obtained from a reputable manufacturer, and from personal experience we would recommend that when possible the manufacturer's plant be in the same country as the user. Necessary repairs and adjustments are greatly expedited if it is not necessary to send instruments or lenses out of one's own country.

When practical, it is preferable to have a binocular microscope with inclined ocular tubes, so that wet mounts may be studied without standing up over the instrument. The binocular instrument furthermore lessens the fatigue of prolonged use, as compared with the monocular. By training both eyes to observe, it also guards against incapacity during temporary losses of the use of one eye.

Either achromatic or apochromatic objectives may be selected. The former are corrected for two colors only and are considerably cheaper. They give quite satisfactory service for almost all visual work. The latter are corrected for three colors and are preferable for photomicrography, especially in color photomicrography.

The Abbé test plate is a glass slide with a thin film of metallic leaf through which a series of parallel lines has been scored so as to leave clear lines bounded by narrow, opaque, metallic bands with jagged edges. This film is covered by a long, narrow coverslip which varies progressively in thickness from about 90 μ at one end to 230 μ at the other. At the side of this coverslip are graduations indicating the approximate cover-glass thickness at any point.

This test plate is used for testing objectives for chromatic aberration, for spherical aberration, for sharpness of definition, and for flatness of field. A complete substage with a device for oblique illumination is needed. Low-power objectives should be tested between

150 and 200 μ equivalent cover-glass thickness. Number 1 cover glasses average about 150 μ, Number 2 about 210 μ in thickness. Test 4-mm apochromats with correction collars at at least two points, with corresponding adjustment of the correction collar. Immersion objectives should be tested immersed in their proper immersion fluids.

With oblique illumination, achromatic objectives give relatively broad fringes of complementary colors on the edges of the metallic strips. With apochromats, these fringes are narrower—often almost inappreciable.

When sharply focused with the condenser centered and properly focused, a good objective should continue to give sharply defined points on the edges of the metallic bands when the illumination is decentered across the direction of the bands. Similar performance should be obtained in the central and peripheral portions of the field.

Relative flatness of field can be judged by the amount of focusing necessary to give sharp definition respectively in the center and at the periphery of the field. It should be borne in mind that lenses with the greatest resolving power in the center of the field ordinarily do not give so flat a field as some others inferior in resolving power. This property of flatness of field is more important with lower powers and for photographic purposes.

Resolving power may be tested on various test slides. For instance, the diatom *Pleurosigma angulatum* at 250 × should show three distinct striation systems. One runs perpendicular to the median rib; the other two cross obliquely at an angle of about 58°. At higher magnification the striae appear as material between rounded globules which is dark at high and low focus, bright at normal focus. The wing scales of *Epinephele janira* ♀ show longitudinal striation at 40 ×. Between these striae a fine cross striation is seen at 150 ×. At 800 to 1000 × the longitudinal striae are doubly contoured and contain round granules. (The material for this paragraph was derived from Romeis. For a fuller account consult Langeron.)

The objectives to be selected for a microscope naturally vary widely with the purpose to which each is to be put. For general pathology the following seem the most desirable: An achromat of about 3 to 6 × initial magnification is almost essential for general views of sections. Achromats or apochromats of 10 and 20 × (16 and 8 mm) are needed for more detailed study. A 31 × (5.5 mm) achromat has proved quite useful in practice. It is similar in performance to the English $\frac{1}{4}$-in objective. We have found dry achromats of 45 × and even 60 × very useful on occasion, especially when the use of oil is inconvenient. An immersion objective of 60 or 90 to 100 × initial magnification (3 or 2 mm) is required for very high magnification. These last are available in three grades: achromatic, fluorite, and apochromatic, in ascending grades of performance and cost. The second will serve almost all purposes; the last are somewhat better for photography and maximum resolution. The 4-mm apochromat is in practice rather unsatisfactory because of the necessity for adjustment of the correction collar for variations in thickness of coverslips and of film of the mounting medium. An oil-immersion objective of 4 mm (40 to 45 ×) has been found very useful for differential cell counts of leukocytes in thin blood films, because of the larger field afforded. This magnification is still adequate for identification of ordinary blood leukocytes, but for marrow films a 2- or 3-mm (90 or 60 ×) objective is required.

Among eyepieces the 7.5, 10, and 12.5 × seem the most useful. Visual impairment in the user may be partly compensated for with 15 or 20 × oculars. In selecting oculars it should be remembered that objectives do not give effective magnifications of over 1000 times their numerical aperture (NA). Hence a 60 × objective with NA 1.40 can be used with 15 × oculars, giving about 900 diameters final magnification, but a 90 ×, NA 1.30 objective will accept only a 12.5 × eyepiece, giving 1125 diameters, or, perhaps better, a 10 × eyepiece, yielding 900 diameters. Attempts to obtain higher magnifications by use of higher oculars result in blurring of detail.

For apochromatic objectives, compensating eyepieces should be used. For achromats, the Huygenian type is satisfactory. For fluorite objectives, eyepieces should be either compensating or of an intermediate grade designated as hyperplane or planoscopic. These last can also be used with achromatic objectives and even with apochromats, though they are not recommended for the latter.

Generally two eyepieces separated by 5× are adequate. For the binocular instruments, only matched pairs should be used, at a constant interpupillary distance which one may determine by trial for oneself. Both should be brought to focus on some individual detail in a microscopic field by means of the focusing collar on one of the ocular tubes.

In regard to the question of parallel or converging ocular tubes, both have their defenders, and either seems to be satisfactory to the individual observer who has become accustomed to it. Changing from one to the other is difficult. Note that the parallel design has been adopted by three of the four manufacturers whose microscopes have been commonly used in the United States.

Condensers are commonly used in the substage of microscopes to bring to bear on the object a sufficient amount of light at an adequate angular aperture to illuminate the field adequately. For work with ordinary transmitted light, the usual Abbé condenser serves for routine work with achromatic objectives. For apochromatic and fluorite objectives an aplanatic or achromatic condenser is necessary whose numerical aperture should approximate the highest numerical aperture of the objectives likely to be used.

ILLUMINATION

To obtain the best results with any suitable combination of optical components, it is necessary to relate them to one another in a definitely prescribed way. This principle is generally understood with respect to the focusing on the object of the parts above the microscope stage, and it is hardly less important with respect to the components below the stage. The necessary adjustments are (1) alignment of the optical axes of the condenser and illuminator with that of the objective and ocular, (2) focusing of the substage and illuminator condenser, and (3) regulation of the iris diaphragms of the system. If these procedures are not carried out correctly, the most elaborate equipment is no better than the plainest and is frequently worse.

Dim and uneven illumination is the most frequent consequence of faulty axial alignment. To avoid this, align the equipment as described in the following paragraphs:

Focus the microscope on a slide placed on the stage; then move the slide to obtain an empty field. Remove the ocular, and observe the back lens of the objective through a pinhole eyepiece. Close the condenser diaphragm until it begins to restrict the lighted disk seen. Move the condenser by means of its centering screws until the restriction is evenly distributed around the edge of the disk. The condenser is now centered. (The pinhole eyepiece required for this operation can be made by making a needle hole through the center of a cardboard cap fitted over the tube after the eyepiece is removed.)

Before aligning the illuminator with the microscope, the centration of the light source itself with respect to the illuminator condenser should be checked. If the illuminator is aimed horizontally at a nearby wall and the image of the lamp filament is focused upon it, the center of the filament image should be at a level with the center of the illuminator condenser. If the housing has a reflector behind the filament, the direct and reflected images of the filament should fall on top of one another, except that with filaments composed of multiple parallel coils, the two images should be displaced just enough to alternate the coils in the combined image.

To align the illuminator with the microscope, place it in a convenient position in front of the microscope and aim it roughly at the flat side of the substage mirror. Close the illuminator diaphragm, and with the microscope still focused as for centering the condenser, move the mirror and adjust the height of the substage condenser until the aperture in the illuminator diaphragm is seen sharply focused in the center of the field. Place a piece of white paper in the substage filter holder. Examine the image of the lamp filament projected on it. Adjust the lamp position to center the image on the condenser axis. The image should be large enough to cover the lower lens of the condenser. Remove the paper, and again center the image of the diaphragm in the visual field by moving the mirror. Replace the paper, and should it be necessary to adjust the lamp position again, repeat the mirror adjustment also. The optical system will be in axial alignment. In addition, after this procedure the substage and illuminator are also correctly focused for critical microscopy. (This method of aligning the lamp is essentially that of Galbraith, *Q. J. Microsc. Sci.*, **96**: 515, 1955.)

The arrangement described, with illuminator condenser imaged in the plane of the object, is often referred to as "Köhler illumination" and is distinguished from "critical illumination," in which the light source is imaged in the plane of the object. These terms are unfortunate, because both methods are suitable for critical work; furthermore, Köhler did not discover the method referred to here but did devise another quite different method, also named after him, for evenly illuminating very large fields, such as are required with very low-power objectives.

"Critical illumination," in the sense of imaging the actual incandescent light source in the object plane, is rarely used, because it requires a large and uniformly luminous filament. Using this arrangement with standard equipment, even the ribbon filament can fill only the restricted fields of higher-power objectives. However, since for visual microscopy it is necessary to reduce the intensity and correct the color of the light from the usual incandescent sources, it is usual to place a blue ground-glass diffuser in front of the lamp condenser. This diffuser in effect becomes the light source, and since, if the procedure recommended above is used, it is imaged by the condenser in the object plane of the microscope, the arrangement becomes the equivalent of "critical illumination." When a diffuser *is* inserted in this way in the system described above, the lamp condenser should be moved as close to the light as possible, as this results in the most uniform illumination of the ground glass. It is important to secure axial alignment of the system whether a diffuser is to be used or not, because even the diffused light is preferentially radiated in the axial direction. Some observers feel that a diffuser detracts from the quality of the illumination.

Finally, before actual observation is begun, the lamp and substage diaphragms should be adjusted. It is especially important that correct use be made of the latter, as it controls the numerical aperture of the condenser. In theory any objective can achieve its maximum resolution only if it is used with a condenser adjusted to equal it in numeral aperture; if the condenser diaphragm is opened too wide, light which the objective cannot use will pass, and some of it, reaching the objective as stray reflections and refractions, will degrade the visual image; if the condenser aperture is reduced too far, resolution will be sacrificed, and excessive diffraction may also degrade the image. In practice the most satisfactory image seems to be obtained if the substage diaphragm is closed slightly more than theory requires. At this point a slight restriction of the lighted area may be seen by examining the back lens of the objective with the ocular removed. The adjustment of the lamp, or field, diaphragm is a relatively minor matter, but for critical work, if it is closed so as just to avoid encroaching on the visual field, it may improve the image slightly by excluding a small amount of stray light.

In view of the foregoing considerations it should not be necessary to warn against controlling the brightness of the image by closing the condenser diaphragm. Brightness can be controlled satisfactorily by equipping the lamp with a rheostat, by inserting neutral filters in the light path, or by inserting two sheets of Polaroid film in the light path and varying their relative alignment. So long as both Polaroids are on the same side of the object, the birefringence of collagen, striated muscle, or other anisotropic tissue components will not interfere with the dimming effect.

The practice of controlling image brightness by lowering the condenser is also to be discouraged. It not only throws the condenser out of proper focus but also reduces its effective numerical aperture, thereby sacrificing resolution and introducing diffraction effects. Lowering the condenser to illuminate the whole field of a low-power objective may be justified on grounds of expedience if it is being used simply to locate objects for critical study at higher magnifications, but for really adequate lighting with such objectives other measures are required. Some microscopes provide auxiliary condenser lens which may be swung into the light path to enlarge the image of the light source; others achieve the same result by the use of a variable-focus "split" condenser. Removal of the top element of the condenser is also satisfactory; a swinging top element is a built-in feature of some condensers.

It was assumed above that every microscopist knows when the objective of his microscope is focused. Unfortunately this is not the same as saying that every microscopist knows how to focus his objective. One simple precaution ensures against damage to slides and equipment while focusing: *focus up!* After placing a slide on the microscope stage, lower the tube slowly by means of the coarse adjustment while watching the space between slide and objective. When it is certain that the objective is below its focused position, look into the microscope, and raise the tube slowly until the specimen appears. When using immersion objectives, the tube should be lowered before the immersion medium is applied, the objective being rotated to the side to permit application of the medium after the tube has been brought to a suitable level. When the objectives in a nosepiece have been adjusted to be parfocal, this precaution is necessary only when bringing the object into focus for the first time. However, when relying on this feature, one should be on guard against thick coverslips or inverted slides, either of which will interfere with the use of the higher powers.

We are indebted to Dr. James B. Longley, now at the University of Louisville, for contributing the foregoing section on illumination. We have little to add to it.

About focusing of the several objectives, try if possible to have them adjusted so as to be parfocal. It is especially valuable to be able to focus sharply with a 20, 30, or 45 × dry lens, center the area desired for examination, swing the first lens out of position, apply a drop of immersion oil, swing the immersion lens into position, and find the field sufficiently visible to permit sharp focusing with a small movement of the fine adjustment.

It helps also to learn on your own microscope which way to focus on changing objectives. Many objectives have incorporated devices for altering their focus, so that they can be rendered parfocal. Learn from the local representative of your microscope manufacturer how to do this, or have him do it for you. Unstained sections sometimes present special problems in finding the focal plane. In such cases the plane can be found with phase equipment, or by use of a mounting medium of quite low refractive index, so that the unstained section is readily visible grossly. Or one may place a small dot of india ink or black slide-marking ink on the lower surface of the cover glass, to one side of the specimen in the course of mounting. Two ink lines drawn transversely and longitudinally on the surface of smear preparations which are to be studied with oil immersion without coverslips serve the same purpose of locating the focal plane.

Generally the microscope can be used without eyeglasses except when the wearer's optical defect includes a considerable degree of astigmatism or is extreme in grade. A mere 10-diopter spherical correction is readily adjusted for by focusing. So-called high-point oculars are essential for persons who find it necessary to wear their spectacles while working at the microscope. These oculars are now readily available. Or individual caps carrying small lenses of the worker's eyeglass prescription may be procured from the microscope manufacturers and placed over the microscope eyepieces.

Objectives, eyepieces, and condensers should be cleaned by breathing on the glass and wiping with the lens paper made for that purpose. Immersion oil should be cleaned off daily at least or when the use of that objective is finished for the time being. To do this it is well to dampen a spot on a piece of lens paper with a drop of xylene and wipe first with this damp spot, then with the remaining dry lens paper in a single movement.

For *immersion objectives* it is preferable to use one of the nondrying oils now made for the purpose. These are available in high and low viscosities and may be blended according to the particular need of the user. Low viscosities are better when rapid motion of the slide is to be used with short working distances and when fresh wet preparations are being studied under a cover glass. The most practical oil for routine use may be an equal volume mixture of the two grades. On one occasion it was found necessary to make immersion oil of a very low viscosity by mixing approximately 4 volumes of light mineral oil with 1 volume of α-bromonaphthalene. If it is necessary to make such mixtures, they should be checked for index of refraction with a refractometer if possible. The index of refraction should be between n_D 1.515 and 1.520.

If the refractometer is not available, the proper index of refraction may be approximated by immersing a white glass slide (or a glass rod of the proper refractive index) in the mineral oil and then gradually adding α-bromonaphthalene until the glass can no longer be seen through the oil.

If one persists in the use of thickened cedar oil for immersion, the utmost care should be taken to have it cleaned off the immersion lenses at least daily. Dry balsam or cedar oil is probably best removed by carefully chipping off the outer portion with a knife, avoiding contact with either metal or glass, and then removing the remainder with lens paper or a soft cloth moistened with benzene or xylene. Alcohol should be avoided, since it softens the cement in which the front lenses of many older immersion objectives were mounted.

Immersion oil is conveniently removed from cover glasses covering fresh resinous or glycerol-gelatin or similar mounts by first inverting the slide on a blotter and pressing down lightly. This absorbs the bulk of the oil. The small residue is readily removed by gently wiping on a blotter wet with xylene or by dragging a piece of lens paper moistened with xylene across the soiled area, steadying the cover glass with a finger on one corner if necessary.

Blood films may be cleaned of cedar oil by dipping repeatedly in a jar of xylene or by dropping xylene slowly on the slanted slide just above the oil drop until it runs off the edge of the film. The nondrying mineral oils may be removed in the same manner; but it is not necessary to remove them from blood films, since they apparently do not cause fading of Romanovsky stains, as did cedar oil. In fact, it appears that this type of oil may even act to better preserve Giemsa and Wright stains.

When the utmost resolution is required with lenses of a numerical aperture above 1.0 and for dark-field work, a drop of oil is also placed between condenser and slide. Usually this procedure is unnecessary and (if much movement of the slide is required) both troublesome and messy.

If after immersion contact has been made and focus has been attained a grayish haze appears moving in from one side to obscure the image, a tiny air bubble in the immersion fluid is probably responsible. The trouble is remedied by swinging the objective out laterally to break the immersion contact, and then back in again without changing the focus.

OBLIQUE ILLUMINATION

Oblique illumination is used to make more prominent the lineal detail transverse to the plane of oblique lighting. It is obtained by excluding light from all but one side of the undersurface of the condenser. A device for this purpose is included in research-model microscopes, and a similar device may be improvised for student microscopes by sliding a piece of cardboard gradually across the undersurface of the condenser from the desired side. This type of lighting materially increases the resolving power of the objective in the plane of the oblique illumination.

DARK-FIELD ILLUMINATION

Dark-field illumination is a device of essentially similar nature, but it provides oblique illumination from all sides while excluding directly transmitted light. Particulate matter in a fluid medium is thus caused to glow against a dark background, and particles materially smaller than can be discerned by direct transmitted light are thus rendered visible.

Dark-field illumination can be simply achieved for low magnifications by inserting beneath the condenser a disk or stop which shuts off light from the center of the condenser while admitting it to most of the periphery. For immersion work a special dark-field condenser is needed. An immersion objective with a numerical aperture somewhat below 1.0 is required. The usual immersion objective of NA 1.25 to 1.40 may be used if a funnel stop designed for the purpose is inserted into the upper side of the objective before screwing it into the nosepiece. More convenient are immersion objectives with normal numerical apertures which are equipped with an iris diaphragm for reducing the aperture. For routine use, the objectives specially built for dark-field work are desirable. Also very convenient are dark-field illumination units which may be attached directly to the dark-field condenser. This eliminates the troublesome centering of illumination.

Both condenser and objective must be immersed. Otherwise the desired extremely oblique rays are lost by reflection from the undersurfaces of the slide and the objective.

We have on one occasion observed a sample of one of the modified mineral oil type of immersion oils which, although perfectly satisfactory for bright-field work, gave unsatisfactory dark-field illumination when used for immersion of the condenser. Focusing on the oil film revealed large numbers of minute bright particles. These were not removed by filtration. A mixture of 4 parts heavy mineral oil and 1 part α-bromonaphthalene proved free of these particles and was quite satisfactory.

POLARISCOPY

The use of polarized light in histology is particularly valuable in the detection and identification of certain crystalline substances and of certain lipids, and in the study of such tissues as cross-striated muscle, collagen, and myelinated nervous tissue. Polarized light may be obtained by disks of polaroid material, by interposition of series of obliquely placed thin glass plates such as cover glasses or by the use of a Nicol prism. The polarizing device may be set at any convenient place between the light source and the object under study. In research microscopes a place is usually provided for a polarizer just beneath the condenser. For observation, a second polarizing device is required, and this is usually placed over the microscope ocular. This is called the *analyzer*.

The first Nicol prism, or polarizer, permits passage of light vibrating in a single plane. When the second prism, or analyzer, is in the optically identical position, the light which passed through the polarizer passes also through the analyzer, and the field of the micro-scope is bright. Now if one of the prisms is rotated 90° on its long axis in relation to the other, light which passed through the first prism is unable to pass through the second, and the field is dark. In this dark field substances that rotate the plane of polarized light are seen to glow. Such substances are referred to as *anisotropic*, or *doubly refractile*. When the slide is rotated about the axis of the light beam, the bright material fades and brightens 4 times during a complete rotation. By inserting the mica disks supplied with polarizing outfits, color changes are substituted for the lightening and darkening. For this purpose a rotating circular stage carefully centered is almost essential, though it is possible to achieve the same results by simultaneously rotating both polarizer and analyzer in opposite directions with thumb and forefinger. This is fairly easily done with a little practice, so that the overall darkness of the field is maintained.

In using polarizing equipment in an inclined monocular-tube microscope, the polarizer must be adjusted so that its polarizing effect is added to (not neutralized by) the polarizing effect of reflection through a single prism. The optimum point is that at which the reddish cast of the darkened field changes abruptly to greenish. At that point the field is at max-imum darkness. This effect may be completely obviated by placing the analyzer just above the objectives and below the inclined monocular or binocular body prism system. In this case rotation of the plane of polarized light and lightening and darkening of the field are accom-plished by placing the polarizer in the rotating substage mount.

To center a rotating mechanical stage, place on the stage the centering slide provided for that purpose, a stage micrometer, or any other section. With the eye select some con-spicuous detail near the center of the field and commence rotating the stage. If the selected object describes an arc, manipulate the stage centering screws so as to bring the apparent center of rotation to near the middle of the field. Then select some object on the edge of the field and rotate the stage. The selected object should follow around the edge of the field. Make the necessary slight adjustment with the stage centering screws until it does.

When using polarized light, it is necessary to employ appropriate preliminary technical procedures for the preservation of the material that one desires to study. Fats and lipids must be studied in material that has not at any time been subjected to the action of fat solvents. The usual fat-stain preparations from frozen sections are usable, though it may sometimes be desirable to have an unstained frozen section as well. For myelin, frozen sections are also necessary, while for striated muscle and crystals such as silica, either paraffin or frozen sections will serve. It is to be noted that a common doubly refractile material in paraffin sections is the so-called formaldehyde pigment, acid formaldehyde hematin, which is formed from hemoglobin by acid formaldehyde during fixation. The bire-fringence of collagen, well discerned in such mounting media as cellulose caprate and the resins, may be completely lost in preparations mounted in glycerol gelatin.

After treatment with acetic Zenker and certain other acid aqueous fixatives, red cor-puscles may become doubly refractile, lightening and darkening like crystals on rotation of the stage.

Skeletal muscle, when correctly oriented to the plane of polarized light, gives sharply contrasting light and dark bands. At a 45° rotation these bands may be quite inconspicuous; hence rotation is necessary. Either stained or unstained, frozen or paraffin sections may be utilized. Cardiac muscle often gives only a barely discernible diffuse glow without distinct bright and dark bands, and no appreciable change is noted on rotation. Sometimes, however, distinct dark and light bands, brightening and darkening on rotation, are discernible, and on occasion it appears that insufficient illumination is the cause of the failure to show bright and dark bands. Use of azure-eosin–stained sections of skeletal or cardiac muscle gives bright bands which are nearly white or bluish white and deep pink in alternating quadrants of rotation. Lint or cellulose fibers accidentally included in the mounting medium often glow brilliantly, but above the section plane.

Smooth muscle, especially when sections are mounted in a medium of low refractive index, such as xylene–cellulose caprate, often exhibits a distinct birefringence, more diffuse than that of striated or cardiac muscle and generally less brilliant than that of the collagen fibers in the same section.

This property of collagen and of hair can be very valuable in finding the focal plane of unstained sections with inconspicuous or sparsely distributed colored histochemical reactions. It has further enabled the decision as to whether a given histochemical procedure blocked the staining reactivity of collagen or solubilized and destroyed it.

FLUORESCENCE MICROSCOPY

This procedure is employed for the demonstration of substances which of themselves possess the property of emitting light of longer wavelength (in the visible range) when excited by light of a shorter wavelength. Also, objects stained with certain fluorescent dyes may be demonstrated by this means. The latter demonstration in its application to the auramine staining of tubercle bacilli is the commonest use of the procedure.

Usually ultraviolet light of 350- to 400-mμ wavelength is employed as the exciting light. For this purpose we have used a General Electric Mazda 100-W A4H mercury-vapor lamp with the General Electric autotransformer for 4H lamps. This gives an initial voltage of 245 to establish the arc and then drops automatically to a lower voltage. This lamp should be shielded with an adequately ventilated lamp housing and a wooden screen so placed that the light of the lamp falls only on the substage mirror of the microscope. Stray light can be very disconcerting, both in photographic work and in occasioning cutaneous or corneal burns. Lempert (*Lancet*, **247**: 818, 1944) used the Mercra lamp of the British supplier Thomson Houston, Ltd., Crown House, Aldwych, London. The glass bulb of this lamp is made of a dark glass which cuts off most of the visible light—only dark red is seen—and 95% of the light is of the 364-mμ band of the mercury arc. With the General Electric lamp that we have used, it is necessary to interpose filters between the lamp and the condenser to screen out most of the visible light. Most of these filters also pass a certain amount of far-red and infrared light. Their transmission bands in the ultraviolet vary in width, position, and total transmittancy. The width and position of the band are more important when a carbon arc lamp is used as a light source than with the mercury-vapor lamp. The latter emits a bright line at 365 mμ, with no other bright lines in the nearby portion of the spectrum.

Suitable filters are Corning's red-purple Corex A 9863, maximum 81% transmittance at about 320 to 350 mμ, but transmitting over 10% between 264 and 402; red-purple ultra 5970, 86% at about 360 to 370 mμ, over 10% from 318 to 407; HR red-purple ultra 5874, 66% at about 362 to 368 mμ, over 10% from 322 to 398; red ultra 5840, 56% at about 356 to 363, over 10% from 320 to 386; and violet ultra 5860, 28% at about 360, over 10% from 340 to 379. The last is preferred for selectivity but is of rather low intensity. The HR red-purple ultra 5874 is probably the best among the higher-intensity group, or perhaps the red ultra 5840.

Most of these filters transmit a considerable amount of light at the far-red end of the visible spectrum. Insertion of a thin glass water cell containing 5 to 10% acidulated copper sulfate solution is recommended to cut off this red light.

These mercury-vapor lamps, both the British and the American, require perhaps a minute to warm up after they are turned on, and when the current is turned off, one must let them cool perhaps 10 min before they can be turned on again. The lamp is brought as close to the substage mirror as possible—Lempert specified $4\frac{1}{2}$ in (114 mm)—and no converging lens system is used. The glass lenses of such lens systems would take out too much light. However, if a converging light is desired to obtain greater intensity or more critical illumination, a Florence flask or a double watch-glass lens filled with distilled water could be used, or a concave mirror placed behind the lamp.

A condenser is necessary for adequate illumination, but for light in the 350- to 380-mμ spectral region it need not be of quartz, though greater intensity may thus be obtained. Similarly, quartz slides transmit more ultraviolet light at this wavelength than do glass slides. Popper (*Arch. Pathol.* **31:** 766, 1941) has used quartz slides and a glass condenser; Endicott in Lillie's laboratory conversely used a quartz condenser and glass slides. The latter procedure allows greater transmission of ultraviolet light because of the shorter light path through glass. If a quartz condenser is available, it is much more convenient to be able to use ordinary glass slides. Lempert used both glass slides and glass condenser. Metcalf and Patton (*Stain Technol.*, **19:** 11, 1944) found glass slides and condensers satisfactory. Note also that the Abbé type is satisfactory: the aplanatic is better because of the better spherical correction and greater concentration of light; the achromatic type occasions too much light loss. A removable dark-field disk about 166 mm in diameter should be inserted in the substage below the condenser. This gives a luminous object in a perfectly dark field. The cardioid condenser was not so satisfactory. Of the filters mentioned, only the violet ultra 5860 gave a black background without the dark-field disk.

The ordinary glass objectives and eyepieces are used, and a yellow filter is employed over the eyepiece to screen out ultraviolet light and avoid ocular damage. Such a filter should absorb as fully as possible the 340- to 420-mμ region of the spectrum. For photography it is necessary to absorb also the red end of the spectrum, which most of the ultraviolet passing filters also transmit in addition to the 365- to 366-mμ mercury line. Or one may use a film which is insensitive to red light.

As a yellow filter Lempert (*Lancet*, **247:** 818, 1944) first used tartrazine-stained gelatin filters, and later obtained "suitable" glass filters which did not fluoresce in ultraviolet light. Corning's Noviol O 3060 is the nearest to colorless of the ultraviolet cutoff filters that we have used and is the best for judging the color of fluorescences. However, it transmits some far-violet light below 400 mμ and can give headaches on continued use. The same company's Noviol A 3389 is a quite pale yellow which transmits very little light below 425 mμ; and the next, Noviol B 368, gives little below 450 but is a deeper yellow.

All these filters give some diffuse fluorescence which occasions no difficulty visually but is quite disturbing for photographic purposes, producing a diffuse fogging of films.

Paraffin and balsam and cedar oil fluoresce of themselves; the first must be removed from sections, and the other two must be avoided as mounting media. According to Metcalf and Patton (*Stain Technol.*, **19:** 11, 1944) a xylene or toluene solution of isobutyl methacrylate is a satisfactory mounting medium, at least for temporary mounts. Preparations can be mounted in liquid petrolatum or in glycerol. Popper (*Arch. Pathol.*, **31:** 766, 1941) found an objectionable fluorescence in glycerol in his vitamin A studies. The low-fluorescence modified mineral oils used as immersion fluids possess a higher index of refraction than unmodified heavy mineral oil (n_D 1.515 versus 1.483) and like it do not fade the sensitive Romanovsky stains. These oils and glycerol can also be used as immersion fluids for smears, though Lempert preferred to examine auramine-stained smears with dry lenses.

In the fixation of tissue which is to be studied by fluorescence microscopy, formalin, alcoholic, and acetic formalins and an acetic variant of Regaud's fluid have been used; but

Metcalf and Patton warn against fluids containing heavy metals (except zinc), chlorides, bromides, iodides, or nitro compounds. Picric acid is the only one of the last group commonly used in fixing fluids.

NATIVE FLUORESCENCE. A number of substances possess natural fluorescence, including carotenes and vitamin A, chlorophyll, porphyrins, ceroid, riboflavin, atabrine, and a number of alkaloids. These possess each its own fluorescence colors and spectra. The latter may be observed with Amici prisms or with the Jelley microspectroscope.

According to Popper (*Arch. Pathol.*, **31**: 766, 1941) vitamin A yields a green fluorescence which fades fairly promptly: small quantities perhaps in 10 s, normal amounts in the liver in about 45 s. Glycerol gave a disturbing fluorescence of its own, and so sections were studied in water (oil dissolves vitamin A). Malaria pigment, hemosiderin, and bile pigment did not fluoresce. Hepatic lipofuscin gave a brown fluorescence which was stable under continued ultraviolet irradiation. After alcohol extraction the lipofuscin fluorescence changed to red, while the extract gave the green labile vitamin A fluorescence. Similar brown-red fluorescence was given by adrenal lipofuscin. Brown fluorescence was given by a brown acetone-soluble pigment in the testicular Leydig cells, and there was an alcohol-insoluble lipofuscin with brown-red fluorescence in the testicular germinal epithelium, as well as brown-fluorescing lipid droplets, and, in involuting testes, granules of ultraviolet-stable, bright yellow fluorescent material. Lutein cells yield a faint brown fluorescence after the green vitamin A fluorescence has faded. Heart-muscle lipofuscin gives red fluorescence. Yellow fluorescent granules are seen in the epithelium of sweat glands and of the prostate. Amyloid gives a dim blue fluorescence. Ceroid, according to Endicott (*Am. J. Pathol.*, **20**: 44, 1944) and to Popper, yields golden brown fluorescence in paraffin sections; in frozen sections, greenish yellow slowly changing to yellowish white. This is given by both paraffin sections and frozen sections in water, dry, or in paraffin, but not in xylene or xylene Clarite (*Am. J. Pathol.*, **20**: 149, 1944; *Arch. Pathol.*, **37**: 161, 1944). Riboflavin (vitamin B$_2$) gives a green fluorescence like that of vitamin A, but is quickly reduced and rendered nonfluorescent by saturated aqueous sodium hydrosulfite (Na$_2$S$_2$O$_4$·2H$_2$O) solutions. Thiochrome, the oxidation product of thiamine (vitamin B$_1$), has a bluish fluorescence. Chlorophyll, used sometimes as a fat stain, gives a fiery red, quickly fading fluorescence. Atabrine gives an intense green fluorescence; prontosil, a red; penicillin, a green; sodium salicylate, a blue.

Heating of preparations to 170 to 200°C for 3 to 5 min alters the natural blue and green fluorescences of muscle and of certain drugs to give contrasting colors [Helander, *Nature (Lond.)*, **155**: 109, 1945]. Thus, heating to 170°C for 5 min gives yellow sulfathiazole and blue muscle fluorescence. Heating of sulfanilamide, sulfapyridine, papaverine, or inulin to 200°C for 3 min gives yellow drug and blue muscle fluorescence.

Enterochromaffin cells exhibit a golden yellow fluorescence in formaldehyde-fixed tissues (Erös, *Zentralbl. Allg. Pathol.*, **54**: 385, 1932). This property has been urged in support of the identification of the enterochromaffin substance with 5-hydroxytryptamine, whose formaldehyde condensation product gives a similar fluorescence. (See Lillie, *J. Histochem. Cytochem.*, **9**: 44, 1961, for tabulation of fluorescence and other reactions of certain phenols and phenylamines in formaldehyde-fixed serum models.)

According to du Buy and Showacre (*Science*, **133**: 196, 1961), tetracycline added to tissue cultures at 10 to 20 μg/ml induces intense yellow fluorescence sharply localized in mitochondria. Injection of mice with 100 μg/g body weight induced similar localization in liver, spleen, and brain mitochondria in sucrose centrifugates and in fresh frozen sections. The treatment of fresh frozen sections or cell suspensions with tetracycline in Ringer's or Locke's solution at 1 to 2 mg/100 ml might be a useful procedure for demonstration of mitochondria.

Tetracycline given in vivo may also be used to localize newly deposited calcium salts in bone and cartilage (Milch, Tobie, and Robinson, *J. Histochem. Cytochem.*, **9**: 261, 1961).

ABSORPTION SPECTROSCOPY OF TISSUES

Two rather dissimilar technics have been applied. In the one, undispersed light is passed through the condenser, object, and objective-ocular magnifying system, and then through a dispersing prism or grating system so located as to receive the light from only a small portion of the microscope field. The resulting band spectra are photographed and compared with similar spectra taken from unoccupied areas in the object plane. Of necessity this method lacks precise cytologic localization, but it gives a general view of the total spectrum of the tissue component studied.

In the second technic, the light source is a monochromatic beam proceeding from a monochromator through the usual quartz or glass optics of the microscope. The results may be registered qualitatively by photomicrographs taken at selected wavelengths or quantitatively by means of electrophotometric cells placed in the projection plane to receive light only from the area under study. Since the actual size of the sensitive surface of the photoelectric cell is constant in any given apparatus, variation in object area is achieved through changes in the lens system and the bellows extension. Again it is necessary to compare the light intensity received through the structure under study with that received (preferably simultaneously) through unoccupied areas of the object field. And again only the average absorption received over the total sensitive surface of the photometric cell can be recorded.

For spectroscopic studies with ultraviolet light between 200 and about 350 mμ, quartz condensers, objectives, and oculars are required. Objectives may be corrected for a wavelength between 500 and 550 mμ (green) as well as for the ultraviolet range desired, so that the microscope may be focused visually, and the monochromator then shifted to the desired wavelength. Or fluorescent screens may be employed for focusing, and materials selected which emit fluorescence in the visible spectrum when excited by the required ultraviolet bands.

A considerable amount of attention has been paid to the photometric estimation of the deoxyribonucleic acid (DNA) content of individual nuclei, usually after application of a carefully standardized Feulgen procedure. The subject is discussed *in extenso* by P. M. B. Walker and B. M. Richards in Chapter 4 of Brachet and Mirsky, "The Cell," pp. 91–138, Academic, New York, 1959, and by C. Leuchtenberger, Cytochemistry of Nucleic Acids, in P. M. B. Walker's "New Approaches to Cell Biology," Academic, New York, 1960. Pollister and Ornstein's Chapter 1 in R. C. Mellors' "Analytical Cytology," McGraw-Hill, New York, 1955, is classic in this field and is brought up to date in the second edition of R. C. Mellors' "Analytical Cytology," McGraw-Hill, New York, 1959. Birge reports a new photometer assembly which may be of interest; see *J. Histochem. Cytochem.*, **7:** 395, 1959. Papers by Kasten, ibid., **4:** 462, 1956, and **5:** 398, 1957; Barka, ibid., **6:** 197, 1958; Di Stefano et al., ibid., **7:** 83, 1959, and *Endocrinology*, **67:** 458, 1955; and Goldstein, *J. Histochem. Cytochem.*, **2:** 274, 1954, may be consulted for details and further references.

The reflecting microscope used by Mellors (*Science*, **111:** 627, 1950) would appear to lend itself not only to the studies in the ultraviolet range for which he used it but also for work in the infrared and visible ranges. Since it is without the chromatic aberrations inherent in refracting systems, objects are in focus at the same objective plane throughout the utilizable spectrum.

Mellors used Kodak 103-0 UV spectroscopic plates for high sensitivity and speed in photography of living tissue cultures, and Kodak 1372 (35-mm film) for high resolution with diminished speed and sensitivity. Corning Vycor No. 791 ultraviolet transmitting cover glasses were used. The tissue culture fragment was enclosed in an aqueous film between two such cover glasses. Mellors washed the nutrient medium off the hanging-drop culture with a suitable Locke type of saline solution before enclosing it with the second cover glass for photography. Focusing was done with the green mercury line and exposure to the 253.7, 265.2, 275.3, 280.4, 312.6, and 334.1 lines kept to the minimum required for photographic exposure.

PHASE MICROSCOPY

The refractive index, n_D, is a constant, characteristic of each transparent substance, and represents the ratio of the velocity of (sodium) light in a vacuum to the velocity of light in the substance in question. It is the ratio of the sine of the angle of incidence to the sine of the angle of refraction in the medium.

It follows that when two transparent substances of the same or nearly the same refractive index abut each other, there is no refraction at the interface, or very little, and the two substances are not visually distinguishable in the unstained state.

In the case of the relation of the refractive indices of mounting or examination media to the index of the included tissues, it is possible to vary greatly the refractive index of the including medium by suitable selection of its constituents. This problem is considered in more detail in Chap. 5.

In the case of interfaces between the various transparent substances within tissues, small differences in refractive index may be exaggerated by advancing or retarding by $\frac{1}{4}$ wavelength the phase of the diffracted rays relative to that of the direct rays passing through an object and forming an image at the rear focal plane of the objective. This phase change is accomplished by depositing under high vacuum thermal evaporation on a phase plate or directly on an objective lens, usually in an annular pattern, a film of glass of sufficient thickness to alter the phase of green light by $\frac{1}{4}$ wavelength.

In order to compensate for the greater intensity of the undiffracted light, a metal absorbing film is deposited in the area of the phase-plate annulus to equalize the intensity of the diffracted and undiffracted light.

An annular aperture diaphragm is placed in the front or lower focal plane of the substage condenser and is illuminated by the Köhler method to form a secondary source of light. In accord with the principle of Köhler illumination, the lighted circle of the annular aperture forms in the rear focal plane of the objective an image which should coincide accurately with the phase altering annulus. Thus the oblique light which passes directly through the object without diffraction passes also through the altering annulus of the phase plate, while the diffracted light which has been deviated from its direct path by encountering interfaces of differing refractive indices passes through the remaining area of the objective phase plate without the advancement or retardation of $\frac{1}{4}$ wavelength produced by the phase-plate annulus. When these diffracted rays are brought to focus along with the undiffracted light passing through the same object area, they differ in phase by $\frac{1}{4}$ wavelength and consequently interfere. The positive phase plate accelerates the undiffracted light, and the interference produces dark areas in a bright background; the negative phase plate retards the undiffracted light, and the interference pattern gives brighter spots on a darker background.

Specially designed objectives covering the usual range of magnifications are available for positive and negative phase contrast, and annular aperture substage diaphragms are provided for each objective.

A telescope ocular is provided for insertion in place of the usual observation oculars for use during accurate centering of the image of the substage annulus on the objective or phase-plate annulus.

The method of phase microscopy appears to offer its greatest advantages in the observation of surviving cells in warm stage preparations, tissue cultures, and the like.

MICROMETRY

In order to measure the absolute size of microscopic objects, one uses either projection or an eyepiece micrometer. The projection method is convenient when many measurements are to be made in the same field. This is accomplished by first projecting the image of the stage-micrometer ruling on the ground-glass screen of a bellows camera and then adjusting the magnification by varying the bellows length so that a simple magnification factor is reached. The rulings of the usual stage micrometer are at $10\text{-}\mu$ intervals and cover a total space of 1 mm. The image on the ground glass is readily measured with a millimeter rule. If the total 1-mm ruling covers a space of 10 cm on the screen, it is evident that one has a magnification of 100 diameters.

After the projection system has thus been set up at an appropriate magnification, the desired field is substituted for the stage micrometer and photographed or measured directly on the ground-glass screen with a millimeter rule. Division of the measurements in millimeters by the magnification factor yields the true measurements.

More often the virtual image of a tiny scale engraved on a clear-glass disk is used as a standard of comparison. This disk is inserted into the desired eyepiece by unscrewing its top lens, inserting the disk on the shelf within the eyepiece so that the engraved figures are erect and not reversed, and replacing the top lens of the eyepiece. The scale on this disk is brought directly over or beside the object to be measured, and the number of divisions on the scale is noted.

The value of each single division of the eyepiece micrometer scale, as well as that of the whole 100 or 200 such divisions comprised in the total ruling, in terms of absolute measurement, should be determined in advance by direct comparison of the eyepiece micrometer scale with that of the stage micrometer. It is necessary to record these values for each objective that is used with the microscope, and if the micrometer disk is used with different eyepieces, for each eyepiece as well.

Each worker should possess one of these eyepiece micrometer disks and should have a card beside his microscope carrying the value of the eyepiece micrometer divisions in microns for each lens combination to be used. One stage micrometer should suffice for a group, as this is used only occasionally. If no stage micrometer is available, the rulings of a good hemocytometer may be used instead. The small squares in the central area are $\frac{1}{400}$ mm^2, and hence measure $\frac{1}{20}$ mm, or 50 μ, each way. The group of 25 small squares measures 250 μ each way, as do the 16 subdivisions of the outer 8 squares of 1 mm^2 each.

MECHANICAL STAGES

Although it is quite possible to explore a section adequately by freehand movement of the slide on a plain stage, one can never be sure that every area has been seen by this method, especially when higher magnifications are being used. For systematic search of a preparation, the mechanical stage is invaluable. Generally the types which form an integral part of the microscope stage are superior to the detachable-accessory type, since the former are less apt to present lost motion and do not give the distressing variation in vernier readings which is contingent on removing and replacing a stage of the detachable-accessory type.

Periodically the mechanical stage should be thoroughly cleaned with xylene or benzene or other appropriate solvent and then regreased with petrolatum. This facilitates free movement and reduces wear.

Verniers are usually placed on mechanical stages and serve to permit the reading of tenths of a millimeter. The small vernier scale presents 10 divisions covering a space of 9 mm.

When the 0 on the vernier scale is directly opposite one of the markings on the stage scale, read the whole number of millimeters indicated. When the 0 on the vernier scale lies between two markings on the stage scale, read the whole number of millimeters next lower than the point opposite the 0 on the vernier, and then count the number of vernier divisions to the point where one of these is directly opposite one of the divisions on the stage scale. The number of vernier divisions thus counted is the number of tenths of a millimeter to be added to the whole number of millimeters just below the 0 point on the vernier.

2

EQUIPMENT

MICROTOME KNIVES

These should be of the type recommended by the manufacturer for freezing, paraffin, and celloidin microtomes, respectively. Commonly these knives are provided with detachable backs for use during sharpening. The purpose of these backs is to ensure the correct bevel in the edge of the knife. Handles are also generally provided which can be screwed into the base of the knife.

The process of sharpening can be divided into three stages: the removal of gross nicks in the edge on a fine Carborundum stone or on a glass plate with emery powder; the honing proper on a fine water hone or on a glass plate with diamantine powder; and the stropping on a leather strop mounted on a board or on a glass plate with rouge powder. When these procedures are done by hand, in the first two the edge of the knife is moved forward, and the blade is drawn toward the operator in a diagonal stroke. In the stropping step the back of the knife is drawn forward and toward the operator, with the edge following.

When several technicians can be served, it is a true economy to purchase a machine-operated knife sharpener, even though the price appears to be high. Sharpening by hand takes an average of more than an hour daily for a histologic technician, and even $200 or $300 can soon be saved in technicians' time. Further, the machine sharpening is more uniform and will grind the full length of the knife evenly rather than produce the familiar concave edge of the hand-sharpened knife.

Devices for holding safety-razor blades are quite valuable for paraffin and frozen sections, since these blades are readily replaced when they become dull and they cut nearly as good sections on most material as microtome knives. Mallory recommended them particularly for partially calcified material. Many workers find them unsatisfactory when section thicknesses less than 10 μ are required.

PARAFFIN OVEN

An incubator of bacteriologic pattern that can be regulated to between 55 and 60°C is quite satisfactory. Incubators with water jackets maintain a more even temperature. Sufficient space should be available for paraffin to be filtered within the oven. Beakers holding between 30 and 50 ml are useful containers for the paraffin infiltration of individual specimens, and a 2-l beaker forms a good reservoir for the stock of melted paraffin. A vacuum chamber installed within the oven or a specially built vacuum oven is useful for quick infiltration of tissues, particularly of lungs. Such a vacuum chamber can be improvised from a domestic pressure cooker of appropriate size. The services of a plumber are required to make the necessary connections.

We have made an even simpler model by fitting a porcelain electric light socket to the inside of the lid of the pressure cooker and setting the cooker itself in a wooden box packed around the cooker with sawdust covered with plaster of paris. The temperature was regulated by changing the wattages of the electric bulbs until one was found which kept the temperature at 55 to 60°C.

The small paraffin units on the Technicon tissue changer are also quite satisfactory. They expedite infiltration by a continuous gentle rotation which brings fresh paraffin continually to the surface of the tissue and removes the solvent as it diffuses out.

For field use we have improvised a paraffin oven by using a flat tin for a paraffin container, a larger tall tin with a hole cut in the side at the bottom so that the paraffin container could be inserted for part of its length, and an electric light bulb set in the lid of the tall tin as a heat source. The position of the paraffin container, partly under the light, should be so adjusted that paraffin remains unmelted at its outer end. Tissues for infiltration are placed in the melted paraffin immediately adjacent to the remaining unmelted paraffin.

For work where the preservation of undenatured proteins (such as enzymes) is important, the vacuum-type unit is highly recommended.

MECHANICAL TISSUE CHANGERS

Electrically operated devices are now available which will transfer tissues from one fluid to another at prescribed intervals. Tissues may thus be carried through a succession of fixing, washing, dehydrating, and clearing baths, and even through paraffin infiltration if desired. One such device not only transfers from one fluid to another by the clock but also continually agitates the fluid by rotation of the specimen carrier in it, so that diffusion and fluid interchange are materially accelerated. Such machines can be set for a schedule of 24 or 48 h or longer by cutting clock disks appropriately. By substitution of a shorter-period clock and a slide carrier, the device can be altered to a routine slide stainer, carrying slides through deparaffinization, hydration, staining, washing, counterstaining, dehydration, and clearing. At present there is a limit of 12 steps, but the usual routine technics can be adapted readily to such a schedule.

To remove paraffin from metal embedding molds, Technicon tissue carriers, etc. (including the nylon-plastic carriers furnished by the Technicon Company), boil them for 5 to 10 min completely immersed in a tall metal vessel containing about 10 to 12 g (a level tablespoonful, or 16 ml) of powdered Oakite, Calgonite, or other technical sodium phosphate detergent in about a liter of water. Then cool until the paraffin can be removed as a solid cake; rinse and dry. The greasy film left by xylene cleaning is absent with this method, and the danger of working with an inflammable solvent is eliminated (Peers, *Am. J. Clin. Pathol.*, **21:** 794, 1951).

OTHER INCUBATORS

Besides the paraffin oven, which is kept regulated to a temperature 2 to 5°C above the melting point of the paraffin used, it is often necessary to maintain other temperatures over fairly long periods. Small thermostatically controlled incubators that can be regulated at 37°C and at 45 to 50°C are valuable for enzyme digestions, chemical extractions, metallic impregnations, enzyme localization technics, and other methods.

At least one such incubator should be available, preferably two, if such incubation technics are in frequent use.

WATER BATH

Of even greater value than incubators for maintenance of temperature during various staining and incubation procedures is a thermostatically regulated serologic water bath. The contents of Coplin jars quite promptly reach a temperature within 1 or 2°C of the surrounding bath, and temperature maintenance seems to be more steady than in incubators. It is valuable to have a well-fitting but easily removable cover to aid in maintaining temperatures above 60°C and to retard evaporation of the bath.

EXTRACTION WITH HOT SOLVENTS

A reflux extraction apparatus for treatment of slides with boiling acetone, ether, xylene, and the like is readily improvised by selecting a beaker large enough to contain a metal or glass slide rack and a Florence flask of such diameter that it will fit the top of the beaker quite closely but will not quite go down into it. The Florence flask is fitted with a two-hole rubber stopper, and a current of cold water is run through the Florence flask while the contents of the subjacent beaker are kept gently boiling on an electric hot plate. We have boiled acetone in such an apparatus for several hours without having to replenish it.

MICRO REACTION CHAMBER FOR VISUAL OBSERVATION

A simple reaction chamber for on-slide histochemical tests may be improvised by applying thin 22×22 cover glasses to the slide toward each end from the observation area, fastening them down with petrolatum jelly or, for a more permanent chamber, with xylene cellulose caprate. The section, smear, or suspension material is deposited on the slide between the two coverslips with a drop of water or other suitable aqueous fluid. Then a long coverslip is applied, resting on the two small coverslips and fastened down by capillary attraction or with petrolatum.

As usual for such observations, fluid is sucked out from one side by filter paper and reagent is fed in on the other with a small medicine dropper while the significant features of the preparation are under continued microscopic observation.

With No. 1 coverslips the depth of the chamber is about 0.15 mm, the width is that of the superimposed coverslip, and the length is the distance between the edges of the small coverslips. The reaction volume is thus readily computed, say 80 to 100 μl.

The contents of the chamber may be recovered for microchemical assay after completion of visual observation.

CRYOSTAT

Cryostats have undergone a considerable number of modifications since Linderström-Lang's first model was introduced in 1938, and a fair number of differing models have been marketed commercially. Many of these are quite satisfactory and permit the preparation of coherent frozen sections of fresh unfixed tissues at temperatures ranging from -2 to around -20°C. Most of the models have enclosed sectioning chambers which are observed through glass windows and which require insertion of the operator's hands into heavy fur-lined gloves to manipulate microtome, sections, and slides.

Chang (*Am. J. Clin. Pathol.*, **35:** 14, 1961; *J. Histochem. Cytochem.*, **9:** 208, 1961, and **8:** 310, 1960) has described two models, now commercially available, in which the axle of the microtome wheel passes through the wall and is turned by a wheel crank on the outside. The top is open, permitting ready access to and viewing of the interior without frosting. With this ready access to the interior, manual control of sections with the traditional camel's hair brush of paraffin technics becomes possible, and the mechanical antirolling devices, which so often become clogged, are unnecessary and have been eliminated.

In the 1961 model the expansion valve is inside the cold chamber. Thermostatic control of cold-chamber temperature has been introduced, and cooling coils are included in the four lateral walls. The hand wheel readily disengages, permitting easy removal of the microtome for cleaning and oiling. Many other models by several companies are now commercially available. For oiling, Berg (*J. Histochem. Cytochem.*, **8:** 310, 1960) suggests Dow Corning No. 510 silicone lubricant as satisfactory down to −50°C. The whole assembly weighs less than 100 kg (200 lb) and should be readily moved from one laboratory to another. Cryostats manufactured in the United States usually have rotary microtomes with small tolerances subject to cold. When cooled to operating conditions, the microtome is sometimes very stiff and difficult to operate. The simpler Cambridge Rocking microtome and the Slee Retracting Rocker do not suffer from this disadvantage.

INSTRUMENTS

Instruments should include surgeon's knives, scissors, thumb forceps, scissors-type forceps (both with and without mouse tooth tips), a hacksaw with spare blades for cutting bone specimens, section lifters of flat metal for celloidin sections, dissecting needles for handling frozen sections, thin double-edged brain knives, safety razor blades, and some glass syringes for injecting. A slab of paraffin affords a convenient surface on which to cut tissue blocks for later embedding. Its surface is readily cleaned and smoothed from time to time by flaming or by melting and recasting.

STAINING DISHES AND CARRIERS

For unattached "loose," or "free-floating," frozen sections, small, flat covered glass dishes are desirable. In these a few milliliters of stain or enzymatic substrate will suffice. Loose celloidin sections can be similarly handled.

For paraffin sections and attached cryostat, frozen, Carbowax, gelatin, or celloidin sections, slotted containers or carriers are used. Glass Coplin jars with 5 to 15 slots are valuable, particularly for corrosive solutions and reagents. The five-slot Coplin jar is also available with a screw cap which is fairly gas-tight. These are used for treatments with hot solutions of volatile reagents. The ordinary Coplin jar is quite valuable for staining methods in which it is desired to use a relatively small (30 to 40 ml) quantity of stain or reagent. See Table 2-1 for fluid requirements for a small Coplin jar with one, five, or nine slides in it.

In using five-slot vertical Coplin jars for staining it is essential to see that the two slides occupying the end slots face toward the center of the jar. The construction of these jars is such that the end slides may be so closely applied to the glass ends that little or no stain or reagent is present into the slide-jar interval.

Recently there have been developed 12-dish assemblies of oblong plastic vessels accommodating plastic staining racks carrying 25 or 50 slides.[1] We have used the slide

---

[1] Available from Scientific Products, Dallas, Tex.

TABLE 2-1   VOLUME OF FLUID AND
HEIGHT OF FLUID LEVEL
RELATED TO NUMBER OF
SLIDES IN AVERAGE FIVE-
SLOT COPLIN JAR

Volume of fluid, ml	Height of fluid, mm		
	1 slide	5 slides	9 slides
15	15	20	25
17	20	25	32
20	25	33	40
25	33	42	50
30	42	50	65
35	50	62	75

size with much satisfaction. A 12-dish assembly occupies 66 × 11.4 cm (26 × 4.5 in) of table space and stands 108 mm (4.25 in) high. Three-dish units are also available, which we use for xylene for deparaffinization and for clearing. They are readily moved to the mounting area for application of resin and coverslips.

The plastic slide-carrying racks carry slides standing on end, with clear space beyond each end slide sufficient to supply an adequate volume of stain or reagent to the other surfaces of the end slides. With fully loaded slide racks the staining dishes require about 200 ml fluid. A space of 58.4 mm (3.2 in) is allowed for the 25 slides, or with the end wells 63.5 mm (3.4 in), over 2 mm (0.1 in) per slide.

It is sometimes important to know in advance what volume of an unstable or expensive reagent will be required to immerse the sections on slides. Table 2-1 gives a rough indication of requirements for a Coplin jar with one, five, or nine fairly thin slides.

For many procedures slides are stained face up with small amounts of reagent. For this purpose parallel glass rods joined together at one end at about a 5-cm (2-in) interval are convenient. These should rest over a trough or large bowl or sink to catch spilled and discarded reagents. For this purpose we have used for years large copper trays with a drain at one corner to empty into a small sink, a bucket, or other convenient receptacle. These trays measure 25 × 50 × 5 cm (10 × 20 × 2 in). Fastened in each, near one corner, is a smaller and deeper container—say 15 × 10 cm and 10 cm deep (6 × 4 × 4 in)—in which running water is received. This serves for washing slides in running water. On one edge is cut a shallow V-shaped depression to allow overflow of water. It is convenient to place this V on the side opposite the drain so that the overflowing water will traverse most of the area of the large pan before reaching the outlet.

Very convenient for dispensing small quantities of stain and reagents on individual slides are dropping bottles which have glass stoppers with slots on each side to match slots cut on the inside of the bottle neck. A 90° turn closes the bottle tightly, and the stopper is equipped with an overhanging point opposite one of the slots, from which the drops fall.

Serologic pipettes of 1-, 5-, and 10-ml capacities are invaluable for dispensing measured quantities of stains and reagents on individual slides. Rubber bulb pipettes or medicine droppers of 1- to 2-ml capacity with straight or curved tips can also be very useful for application of small amounts of reagents or stains. A hemocytometer can be used as a reaction chamber; the sections are mounted on an ordinary cover glass or even a slide, which is then inverted over the fluid-containing area of the chamber. Reagents are then introduced from one side with a medicine dropper, syringe, or pipette, filling the space 0.1 mm deep by capillarity. Fluid may simultaneously be drawn off from the other side with blotting paper or a pipette or dropper tip inserted in one of the moat areas on one side while fluid is introduced in the other moat area on the other side of the slide.

Graduated cylinders, both stoppered and open, with capacities of 10, 25, 50, and 100 ml should be available, as well as a few larger open cylinders, with capacities of say 250, 500, and 1000 ml.

A balance of perhaps 1- to 2-mg sensitivity and 100-g capacity serves well for most stains and reagents in the histologic laboratory. It is well to have access also to quantitative analytic balances and to scales of larger capacity.

A pair of 50-ml burettes with burette stand may sometimes be useful.

In *heating equipment*, bunsen burners, a one-burner gas plate for heating large amounts of water, an electric hot plate with covered elements for heating flammable fluids, thermostatically controllable warm plates which may be set at 35 to 40°C for spreading paraffin sections or at 70 to 80°C for heating to steaming during staining are accessories which will prove their value.

However, we have done very well with a long copper plate which can be heated at one end with an alcohol lamp.

CLEANING OF SLIDES AND COVERSLIPS

New slides and coverslips may often be satisfactorily cleaned by simply immersing them in alcohol and carefully polishing them with a soft cloth. Usually we have found it preferable to wash them in warm soapy water first and rinse them in several changes of warm water before putting them in alcohol for polishing. In this procedure, if a 1% aqueous solution of acetic acid is substituted for the alcohol, slides take a brilliant polish. Some writers recommend acid alcohol from which to polish slides.

In cleaning used slides on which blood films or unstained paraffin sections remain, slides should first be boiled in an aqueous solution of sodium carbonate or trisodium phosphate or a commercial detergent powder of similar nature. Remnants of sections are then readily wiped off, as is glass-marking ink. The usual soapy water, water, alcohol, or acetic acid sequence follows.

LABELING

Labeling of slides before staining is a necessity in a pathology laboratory. The traditional diamond pencil can be used, but slides so marked are quite prone to breakage during and after staining. For many years now we have used a black glass-marking ink which resists all ordinary reagents and even serves in place of paper labels when the slides are filed after examination. Such inks are removed by alkali, especially when fresh.

Paraffin blocks are readily labeled by affixing with a hot iron on one side a small paper label inscribed with india ink.

Specimen bottles can be conveniently labeled by inserting within the bottle a label first written with india ink on paper and then immersed briefly in smoking-hot paraffin. The paper should promptly lose all its contained air and water as bubbles and should have a translucent appearance. Opaque-looking paper labels are not adequately waterproofed and tend to go to pieces. Similar labels may often profitably be prepared and affixed with a hot iron to the outside of reagent bottles, especially those containing reagents which tend to destroy ordinary gummed labels.

TABLE 2-2 PHYSICAL CHARACTERISTICS OF SOLVENTS

Solvent	mp[a]	bp[b]	n_D	Pounds av. per liter	Kilograms per liter
Acetone	−95	56.5	1.3591	1.747	0.793
Alcohol 100%	−112	78.4	1.3610	1.748	0.789
Alcohol 95%	. . . .	78.1		1.796	0.8115
Amylbenzene sec	. . . .	187/189	1.4894[b]	1.897	0.860[b]
Aniline	−6.2	184.4	1.5683	2.252	1.022
Benzene	5.5	80.1	1.5017	1.938	0.879
Benzyl benzoate	21	323/324	1.5685	2.473	1.122
n-Butyl alcohol	−79.9	117	1.3991	1.786	0.810
sec-Butyl alcohol	−114.7	99.5	1.3968	1.781	0.808
Isobutyl alcohol	−108.0	108.1	1.3924	1.768	0.802
tert-Butyl alcohol	25.6	82.6	1.3878	1.717	0.779
Carbon bisulfide	−108.6	46.3	1.6276	2.786	1.263
Carbon tetrachloride	−22.6	76.8	1.4630	3.515	1.595
Cedar oil, thin	. . . .	168/237	1.5030[b]	2.043	0.927
Chloroform	−63.5	61.2	1.4457	3.283	1.489
p-Cymene	−73.5	176/177	1.4866[b]	1.890	0.857
Diethylbenzene[d]	−32/−84	181/184	1.4957[b]	1.876	0.851[b]
Diethylene glycol	−10.5	244.8	1.4475	2.467	1.118
1,4-Dioxane	9.5/10.5	101	1.4221	2.278	1.033
Ether USP solvent	−116	34.6	1.3497	1.576	0.708
Gasoline 100 octane	−107.4	99.2	1.4040[e]	1.584	0.718[b]
Glycerol 95%	17.9	290	1.4660[b]	2.762	1.252
Glycol	−15.6	197.4	1.4318	2.455	1.113
Methyl alcohol, synth.	−97.8	64.7	1.3288	1.746	0.792
Methyl benzoate	−12.5	198/199	1.5144	2.397	1.087
Methyl salicylate	−8.3	222.2	1.5377	2.606	1.182
Paraffin 56°	56.5	324[c]	1.4262[f]	1.714	0.777[b]
Petroleum ether 20–40	−130/−160	22/38	1.3554[g]	1.367	0.620[b]
Petroleum ether 30–65	ca. −120	30/65	1.3754[h]	1.394	0.632[b]
Petroleum ether 30–80	. . . .	30/80	1.3876[i]	1.480	0.670[b]
n-Propyl alcohol	−127	97.8	1.3854	1.772	0.804
Isopropyl alcohol	−85.8	82.5	1.3776	1.738	0.789
Toluene	−95	110.8	1.4955	1.921	0.866
Trichloroethylene	−73	87.2	1.4777	3.232	1.466
Trimethylbenzene	−25/−45	165/176	1.4931[b]	1.898	0.861[b]
Water	0.0	100.0	1.3330	2.205	0.9982
Xylene	13/−47	138/144	1.4966[b]	1.903	0.863[b]

[a] Slant bar indicates melting and boiling range of compounds in the commercial mixtures. To judge by its melting point, the paraffin seems to be chiefly pentacosane.

[b] Melting and boiling points, densities, and refractive indices determined by Greco, Pathology Laboratory, National Institutes of Health. (Other melting points, etc., are from N. A. Lange, "Handbook of Chemistry," 10th ed., McGraw-Hill, New York, 1967.)

[c] Boiling point of tetracosane.

[d] These are technical grades and are homolog mixtures. Melting and boiling ranges are given.

[e] Approximate n_D of octanes of the same density.

[f] Refractive index of pentacosane at 80°.

[g] n_D for mixture of pentane and isopentane.

[h] n_D of hexane.

[i] n_D of heptane.

To find cost per liter, multiply the price per pint by 2.113, or the price per gallon by 0.264, or the price per pound by the pounds per liter factor given above, or the price per kilogram by the density in kilograms per liter.

SPECIMEN BOTTLES

Specimen bottles, both for storage of specimens and for their collection, fixation, dehydration, and clearing, should be procured in a variety of sizes ranging from perhaps 25 ml or 1 oz, up to brain jars containing some 4 l for the fixation of whole human brains. They should be wide-mouthed and furnished with covers. In the case of storage bottles the covers should fit closely to prevent evaporation. We have found that 25-ml bottles are most often required for storage, 50- and 100-ml bottles for fixation and dehydration procedures. The bottles used commercially for mayonnaise are quite convenient. They have a paraffined-paper inside seal and a screw cap which may be removed by a quarter turn. These are available in quart, pint, half-pint, and quarter-pint sizes, or about 960-, 480-, 240-, and 120-ml capacity.

REAGENTS

Reagents should be purchased from reputable manufacturers under appropriate specifications for the purpose for which they are to be used. There is no point in paying for reagent grades when ordinary technical grades are perfectly satisfactory.

Dyes are unstable organic chemicals and should be bought in such quantities as are likely to be used up within 2 or 3 years. It is good practice to stamp the date of receipt on each bottle of dye purchased. Unless the worker is prepared to test his or her own dyes for quality, certification should be required by the Biological Stain Commission.

Inorganic chemicals of stable nature may be bought in larger quantities, especially if considerable amounts are apt to be required suddenly.

Solvents are often interchangeable one with another as far as results are concerned, and cost may be a material item in the selection of a dehydrating agent or a paraffin solvent. Here prices per unit weight are deceiving, since these solvents are used by volume. Carbon tetrachloride, for example, is nearly twice as heavy per unit volume as toluene, and at 17 cents/lb is actually the same price as the latter at 30 cents/lb. For convenience a table of conversion factors for common solvents is appended (Table 2-2). The relative prices per liter, given in previous editions, have been omitted. These prices are readily calculated from the factors given.

3
FIXATION

The greatest handicap to the histologist working in pathology is improper preservation of material. Several factors enter into this, among which are delay in fixation, postmortem decomposition, drying of tissues, inadequate quantity of fixing fluid, poor penetration of fixing fluid, fixing fluids improper for the material, prolongation of fixation beyond the proper interval, improper storage fluids for prolonged storage, and poor dehydration, embedding, and sectioning technics.

Tissues should be fixed as promptly as possible after cessation of circulation. Autopsies should be made as soon after death as possible, and when this is not immediately possible, prompt refrigeration is of material advantage. Surgical specimens should preferably be fixed as soon as removed. The practice of keeping tissues unfixed until an operation is concluded often results in a distorting dehydration of surface layers, and the practice of keeping them in physiologic saline solution at operating-room temperatures permits autolysis to progress often to a confusing extent. Animal autopsies are preferably made on animals killed or dying immediately before dissection. Complete evisceration by a trained attendant and fixation of the entire visceral mass is a procedure preferable to storage on ice, as far as histologic detail is concerned; but when bacteriologic investigation is an essential part of the autopsy, prompt refrigeration should be the rule when immediate autopsy is not possible.

Slow freezing of unfixed tissue at temperatures near the freezing point of water is to be scrupulously avoided; relatively enormous ice-crystal artifacts are produced. Repeated freezing and thawing disrupts cell organelles, releases enzymes, and produces diffusion of solubilizable constituents.

The practice of immediately embalming human bodies before autopsy does avoid autolytic changes and gives more uniform histologic preservation. It prevents, however, the use of a number of special fixing procedures such as the chromaffin reaction or the application of fat-solvent fixatives to remove lipids completely. Some fixations can be satisfactorily applied after embalming, others cannot. These considerations are entirely aside from the distortion of gross pictures and prevention of microbiologic investigation. Many enzymes are inactivated; some resist brief formalin fixation quite well.

Tissue blocks should be cut of such thickness that the fixing fluid readily penetrates throughout in a reasonably short period of time. This time varies with the fixative, and inversely with the fixation temperature. While low temperatures retard fixation, they also stop autolytic changes, so that the best fixation with many fixatives may be attained by prolonged fixation at temperatures approaching the freezing point of the mixture. With mixed fixatives, it must be recalled that the rates of penetration may vary for the different constituents. For example, mixtures of acetic acid and mercuric chloride, such as acetic sublimate, susa solution, or Zenker's fixative, may show virtually pure acetic acid effect in the deeper parts of the block, with dissolution of albuminous and zymogen granules, lysis of erythrocytes, and the like, while in the outer zone these structures are preserved. Lillie et al. (*Am. J. Clin. Pathol.*, **59:** 374, 1973) reported that these lytic changes could be avoided in fixatives containing 5 to 6% acetic acid by adding an equimolar amount of sodium chloride or nitrate to the fixative. Sodium chloride succeeded well in acetic formol, acetic sublimate, Zenker's fixative, and the like; sodium nitrate was used with lead and silver solutions because of the low solubility of the chlorides of those metals.

The volume of fixing fluid employed should be 15 or 20 times that of the tissue to be fixed. The length and breadth of blocks to be fixed should be such that they are not bent or folded by the container in which tissue is fixed. It is well to open hollow viscera or fill them with fixing fluid. Lungs of small animals may be conveniently fixed by filling them by intratracheal injection, taking care to allow the fluid to escape freely around the injecting needle so as to avoid overdistension. This procedure may also be applied to small areas of human or large-animal lungs.

Fixation by intravascular perfusion must be preceded by washing out blood with an indifferent fluid such as Ringer's or Locke's solution or 0.85% saline solution (Table 5-3). Here it is important so to regulate the injection pressure that it does not exceed the blood pressure. Otherwise, fixation artifacts are produced. This method has the disadvantage that the blood content of the vessels is lost. Further, the method is not possible when postmortem clotting of the blood in the vessels has occurred. Further still, the process of perfusion limits the histologist and histochemist to the use of a single primary fixative, thus preventing comparison of differing fixative effects on adjacent blocks of the same tissue from the same animal.

In the event that the fluid around fixed tissues has evaporated and tissues have become dry, Sandison (*J. Clin. Pathol.*, **19:** 522, 1966) advises rehydration of the tissues in 96% ethanol, 30 volumes; 1% aqueous formaldehyde, 50 volumes; 5% aqueous sodium carbonate, 20 volumes, for 1 day (or longer until tissues become pliable). Double embedding is advised subsequently.

From the point of view of the histologist, the practice of hardening an entire human brain without perfusion, by immersion in dilute formaldehyde solution or other fixative before dissection, can only be condemned. It seems preferable, when topographic study is contemplated and perfusion cannot be done, to divide the brainstem first by a single transverse section just anterior to the oculomotor roots and the anterior margins of the anterior colliculi, thus separating the cerebrum from the mid- and hindbrain. Then make a series of transverse sections through the brainstem and attached cerebellum at 5 to 10-mm intervals, leaving part of the meninges unsevered so as to keep the slices in sequence. Then separate the two cerebral hemispheres by a sagittal section, perhaps better slightly to one side of the median plane, or through the third ventricle. Then on the sagittal surface identify dorsal and ventral points through which sections should pass so that the brain sections will agree in plane with one of the standard cross or frontal section atlases. Make the first section through these points perpendicular to the sagittal surface. Then section the rest of the brain at perhaps 10-mm intervals, cutting parallel to the first plane. This greatly facilitates identification of various areas for the antomist or pathologist to whom the human brain is an occasional object of study. The use of comparative multiple fixations is also permitted.

For these sections a long, thin-bladed knife is essential. We find an 8-l (2-gal) brain jar suitable for thus fixing a human brain.

In general, solid viscera should be cut in slices perpendicular to the surface, in such a way as to expose their anatomic structure to best advantage. For instance, kidney sections should show cortex, medulla, pelvis, and pyramid. Adrenal sections should show cortex and medulla. Sections of tumors should show adjoining tissue sufficient to identify the blocks anatomically and to give the relation of the expanding tumor margin to preexistent tissue. Abscess walls should also show adjacent, relatively uninvolved tissue. The margin as well as the center of a pneumonic focus is often instructive. Digits and other skin-covered objects should be opened so as to admit fixative to the significant areas; skin is almost waterproof. Small bones should be largely stripped free of muscle if their marrow is of interest, or they may be opened with a fine-toothed saw.

Friedenwald ("The Pathology of the Eye," Macmillan, New York, 1929) condemned the practice of freezing the eye and bisecting it before fixation because of the damage resulting from the formation of ice crystals and because of immediate collapse of the tissues and wrinkling when they are placed in fixative after freezing; yet we have had excellent results from the very rapid freezing obtained by immersion in petroleum ether with solid carbon dioxide, followed by immediate axial or paralenticular anteroposterior section and immersion in fixatives while still frozen. Even with Carnoy's fluid the rods and cones remain fully expanded and clearly delineated, and the retina generally remains in contact with the choroid. A sharp razor blade must be used, and care must be taken not to fracture the frozen tissues.

The changes effected in tissues by various fixatives, particularly at the molecular level, are largely unknown. Table 3-1 illustrates the wide variation in just one feature of membranes (size) induced by a few common fixatives. In practice, it is probably prudent in many cases to fix tissues in two different fixatives if critical measurements and assessments are to be made.

Squier et al. (*J. Oral Pathol.*, **2**: 136, 1973) conducted a quantitative study of the effects of several fixatives with differing osmolalities on the size of intercellular spaces observed at the electron microscopic level in normal and burned oral epithelium of rats. The size of intercellular spaces varied widely in relation to the fixative used and the degree of inflammation. For valid comparisons the same fixative must be employed for both normal and inflamed tissues.

Medawar (*J. R. Microsc. Soc.*, **61**: 46, 1941) studied the rate of penetration of various fixatives into a blood clot. He concluded that fixatives neither retard nor facilitate their

TABLE 3-1   SUMMARY OF MEMBRANE MEASUREMENTS*

Fixation	Plasma membrane	Mitochondria	Ribosome-coated endoplasmic reticulum	Vesicotubules	Synaptic vesicles
Permanganate	51 ± 7 (23)	41 ± 5 (82)	44 ± 4 (32)	51 ± 7 (189)	
Osmium tetroxide	54 ± 7 (104)	41 ± 7 (22)	46 ± 6 (12)	53 ± 6 (113)	
Glutaraldehyde	69 ± 4 (31)	45 ± 3 (97)	45 ± 4 (25)	65 ± 7 (232)	
Acrolein	62 ± 7 (132)	50 ± 10 (51)	47 ± 7 (26)	61 ± 7 (97)	
Yamamoto	52 ± 2	44 ± 3	46 ± 2		51 ± 2

* Measurements given in angstrom units. The numbers in parentheses are the number of measurements.

Source: Lillibridge, *J. Ultrastruct. Res.*, **23**: 243, 1968.

own entry into the coagulum and that the distance penetrated by fixatives is directly proportional to the square root of the time of fixation. Among the fixatives tested, the following are listed in order from fastest to slowest in penetrability into a blood clot: 4.5% acetic acid, 10% formalin (nonneutralized), saturated aqueous mercuric chloride, 1% chromic acid (anhydride), 1% osmium tetroxide, saturated aqueous picric acid, absolute ethanol, 1% aqueous uranium nitrate, and 1% tannic acid.

Ganote and Moses (*Lab. Invest.*, **18:** 740, 1968) studied "light cells and dark cells" in rat parenchymal liver at the electron microscopic level and concluded that light and dark cells can be produced as an artifact of immersion fixation. Light and dark cells were not observed in livers perfused and fixed under optimal conditions.

Gil and Weibel (*J. Ultrastruct. Res.*, **25:** 331, 1968) recommend the use of an osmometer in the preparation of fixatives. Adjustment of the osmolality to slight hypertonicity (280 to 320) is recommended (Warshawsky and Moore, *J. Histochem. Cytochem.*, **15:** 542, 1967; Gordon et al., *Exp. Cell Res.*, **36:** 440, 1963). Cells shrink in hypertonic and swell in hypotonic media.

Osmolarity is a relatively new term in relation to fixatives and has been used particularly in reference to fixatives used for electron microscopy. It is a measure of the number of moles and ions of a solute in a stated volume of solvent. This number is directly related to the osmotic pressure of the solution. In a solution containing 1 mmol of an un-ionizable solute the milliosmolarity is 1. A solution of 1 mmol sodium chloride, if completely ionized to the constituent Na^+ and Cl^- ions, gives 2 as the milliosmolarity. With a millimole of Na_2SO_4 or $CaCl_2$, each dissociating to three ions, the solution milliosmolarity is 3; $MgSO_4$, with two divalent ions, gives a 2-milliosmolar solution; CrF_3, dissociating to three fluoride and one chromic ion, gives a 4-milliosmolar solution. With large molecules, as of proteins, a milliosmometer is required, but for simple solutions of metal salts and the like the value is sufficiently accurately determined from the molecular weight; in incompletely ionized solutions the values will be lower, and consultation of ionization-constant tables may be needed.

With very thin tissue slices and rapidly coagulating fluids, milliosmolarity is of relatively less importance than with thicker blocks and more slowly denaturing fluids. Avoidance of osmotic deformation of structure is particularly important for electron microscopy.

The foregoing on osmolarity recalls the efforts of histologists of the latter nineteenth century to balance ingredients which occasioned tissue swelling, such as acetic acid, with those which caused shrinkage, as alcohol does, to achieve a fluid which would maintain original tissue volume while accomplishing the desired hardening. This probably was the purpose of the sodium sulfate in Müller's fluid. From this it was carried over into the Orth and Zenker derivatives of Müller's fluid. Later, when pathologists such as Mallory could discern no important function for it, it was often omitted.

It also later appeared that the complex fluids so devised interfered with histochemical study by their varied and complex action on various tissue elements. The tendency thus veered toward simplification or even omission of chemical fixation.

One must conclude that the demands of perfect morphologic preservation and of unrestricted histochemical reactivity may be irreconcilable and that a choice may have to be made in accordance with the primary purpose for which the tissue is to be used. Rapid processing for prompt pathologic diagnosis does not agree with the slow processing for more precise preservation of morphologic detail or for special histochemical purposes.

FIXING FLUIDS FOR ELECTRON MICROSCOPY AND ENZYME HISTOCHEMISTRY

Aldehyde Fixatives

_None of the aldehyde fixatives (glyoxal, formaldehyde, glutaraldehyde, hydroxyadipaldehyde, crotonaldehyde, methacrolein, pyruvic aldehyde, acetaldehyde, malonaldehyde, malialdehyde, and succinaldehyde) meet all requirements of an ideal general fixative. Methacrolein and crotonaldehyde are very active cross-linking agents. They combine with amine, hydroxy, sulfhydryl, and many active hydrogens. Generally, the more active the aldehyde, the greater the denaturation and inactivation of enzyme activity. Dialdehydes such as glyoxal, glutaraldehyde, hydroxyadipaldehyde, and pyruvic aldehyde are frequently sufficiently reactive to allow some fixation with preservation of cytologic structure and still allow enzyme activity. Cross-linking activity is greater in aldehyde fixatives containing a double bond and in those with dialdehydes.

The amount of cross-linking with the dialdehydes in small pieces of tissues immersed in large volumes of fixative usually employed for histology and electron miscroscopy is open to serious question. Fixed tissues exhibit large amounts of unreacted aldehyde, amine, guanidyl, and other functional groups indicating the presence of one free aldehyde not used for cross-linkage. The situation differs radically from the leather-tanning process whereby a large amount of protein is reacted with a small amount of dialdehyde.

No one procedure can be used for all aldehyde fixatives. The fixatives penetrate at different rates. Among the aldehyde fixatives, acrolein penetrates most rapidly, formaldehyde penetrates more slowly, glutaraldehyde still more slowly, and hydroxyadipaldehyde is the slowest (Sabatini et al., _J. Histochem. Cytochem._, **12:** 57, 1964). The type of fixation to be employed will also vary according to the type of experiment being conducted and the information desired. The concentration of aldehyde fixative may vary from 2 to 12%. The fixative should be buffered to near physiologic pH except for glyoxal and pyruvic aldehyde. Aldehyde fixatives that are too alkaline may manifest reduction of activity due to a Cannizzaro-type reaction. But Cannizzaro reactions occur chiefly with aromatic, not aliphatic, aldehydes. Phosphate, carbonate, and cacodylate buffers are generally satisfactory. Hannibal and Nachlas (_J. Biophys. Biochem. Cytol._, **5:** 279, 1959) note that the addition of certain salts to formalin may sometimes increase enzyme activity.

Purity of aldehydes may be a problem. For example, if the pH of a cold stock solution (25%) of glutaraldehyde falls below 3.5, impurities may be the cause. Buffering such a solution to 7.4 will not result in good fixation of tissues.

Hopwood (_Histochemie_, **11:** 289, 1967) used Sephadex G10 gel filtration to purify various samples of glutaraldehyde. Two fractions which absorbed at 235 and 280 mμ were detected. Only the fraction at 280 mμ had aldehyde activity.

Except for acrolein or glutaraldehyde, sucrose may be added to the aldehyde fixatives to raise the osmolality resulting in an enhancement of fine-structure preservation.

Generally, fine structural detail is obtained best when tissues are fixed in aldehydes at 1 to 4°C rather than at 24°C. For routine studies, blocks not to exceed 1 cm^3 may be fixed in aldehydes for periods of 1 day to several weeks (Sabatini et al., _J. Histochem. Cytochem._, **12:** 57, 1964). However, for enzyme histochemical studies, the use of blocks 1 to 2 mm in thickness is advocated. Fixation at 4°C for 1 to 2 h is recommended except for hydroxyadipaldehyde, which requires 2 to 6 h.

Following fixation, blocks may be washed in cold buffer or sucrose for several hours, for example overnight. At this point, aldehyde-fixed tissues may be stored in cold buffer for days or even months. One may wish to take advantage of this opportunity to conduct histochemical or other studies employing light or electron microscopy. At this stage, the tissues may even be shipped from one laboratory to another.

Aldehyde-fixed, buffer-sucrose–washed tissues may be subsequently stained with uranyl acetate (0.5%, 2.5 h). Nucleic acids are thereby revealed more definitively. Cisternae of endoplasmic reticulum are also better visualized. Lead staining of aldehyde-fixed material results in intense staining of ribosomes and nucleoli.

Lipids are not very reactive with aldehyde fixatives. For this reason, lipid-containing membraneous structures in aldehyde-fixed cells are less adequately delineated than are those fixed in osmium tetroxide. However, the ability of aldehyde fixation to immobilize acetone-insoluble lipids has been useful histochemically (Lillie and Geer, *Am. J. Pathol.*, **47:** 965, 1965). The degree of preservation at the fine structural level achieved by aldehyde fixation compares favorably with that achieved by freeze-drying, freeze substitution, and cryofixation (Sabatini et al., *J. Histochem. Cytochem.*, **12:** 57, 1964; Bullivant, *J. Biophys. Biochem. Cytol.*, **8:** 639, 1960; Seno and Yoshizawa, *J. Biophys. Biochem. Cytol.*, **8:** 617, 1960; Sjostrand and Baker, *J. Ultrastr. Res.*, **1:** 239, 1958).

Osmium tetroxide fixation of aldehyde-fixed tissues sometimes retains structures not revealed by fixation with osmium tetroxide alone. Dales (*Proc. Natl. Acad. Sci. U.S.A.*, **50:** 268, 1963) was able to preserve spindle fibers of the mitotic apparatus in cultured cells fixed first in glutaraldehyde.

Several types of enzyme activity are preserved after aldehyde fixations. Examples are the carboxylic acid esterases, several types of phosphatases (alkaline and acid phosphatases; adenosine mono-, di-, and triphosphatases; glucose 6-phosphatase; and pyrophosphatase), aminopeptidases, and several oxidative enzymes.

Turchini and Malet (*J. Histochem. Cytochem.*, **13:** 405, 1965) supply a note regarding preservation of enzymes for histochemical assay. Small blocks of tissues were placed in 50% glycerol in water and frozen in a freezer set at $-20°C$. Excellent histochemical activity and localization of alkaline phosphatase, ATPase, nucleotidase, monoamine oxidase, and succinic dehydrogenase were obtained 9 months later.

BUFFER SYSTEMS

The buffer used in enzymic reactions is not a matter solely of the proper pH range. In metal ion capture reactions, account must be taken of the effect of the buffer anion on the solubility of the capturing metal ion. In azo dye technics it is reported that certain buffer systems are better than others for specific enzymic reactions. Hopsu and Glenner (*J. Histochem. Cytochem.*, **12:** 674, 1964) reported much greater renal esterase activity in guinea pig kidney homogenates with phosphate and acetate buffers than with tris maleate of the same pH and concentration. Hence buffers should be changed only after making a direct comparison of the two systems.

FIXATIVES FOR ROUTINE PATHOLOGY

FORMALDEHYDE.   The most widely used fixing agent for pathologic histology is formaldehyde. It not only is used as the sole or principal active agent in fixing fluids, but it also enters into many fixing mixtures. Formaldehyde is ordinarily available as a 37% (w/w) solution of the gas in water. This is equivalent 40% w/v. They contain also 10 to 15% methanol as a stabilizer and, when fresh, only traces of formic acid. The commercial solutions have been called *formol* or *formalin* ever since Blum's (*Z. Wiss. Mikrosk.*, **10:** 314, 1893) first report, and hence these are no longer trademarks. Formol seems more common in European usage, formalin in American. In description of fixatives these fluids are regarded as 100% formol, or formalin, and the usual 10% formol contains correspondingly 4% w/v formaldehyde. The term *formal* refers to dimethoxymethane, $(CH_3O)_2CH_2$; *methanal* (HCHO) is a correct synonym for formaldehyde.

The specific gravity of formol (solution of formaldehyde USP) is somewhat variably reported: 1.081 to 1.085 in the *The Merck Index* (8th ed., 1968), 1.111 in our 1948 table (Table 3-2), 1.109 in one manufacturer's catalog, and 1.084 in another. The figures vary probably with the amount of added methanol. Currently some catalogs are offering 55 to 60% solutions with an unspecified amount of butanol as the stabilizer in place of methanol.

Some workers are now dissolving paraformaldehyde, 4 g/100 ml for example, in distilled water and depolymerizing by heating to 80 or 90°C, claiming superior results. A 1 M solution of formaldehyde calls for 3 g of formaldehyde (HCHO) per 100 ml, whether derived from formol or paraformaldehyde.

Solutions of formaldehyde diluted with distilled water are commonly acid, owing to the presence of small amounts of formic acid either as an impurity remaining during manufacture or as a result of oxidation of part of the formaldehyde. For certain silver impregnation technics this natural acidity may be desirable, but for azure-eosin methods and for study of possibly iron-containing pigments, it is essential to correct it. A common practice is to shake the diluted formaldehyde solution with calcium carbonate and store it over a layer of this salt. This gives only approximate neutrality; Romeis cites pH levels of 6.3 to 6.5. Others have used magnesium carbonate, attaining pH levels of about 7.5. With either of these methods, formaldehyde solution drawn from the storage reservoir and used for fixation very promptly becomes more acid as the tissue is fixed. Levels of pH 5.7 to 6 are not uncommon after calcium carbonate treatment.

This shift in pH is avoided by using a soluble buffer in the dilute formaldehyde solution used for fixation. Addition of 4 g monohydrated acid sodium phosphate (or of the anhydrous acid potassium phosphate) and 6.5 g anhydrous disodium phosphate per l gives approximately pH 7 and a total salt content of the two sodium salts of about 1%, dry weight.

Substitution of this fluid for fixation in certain toxicologic studies resulted in a definite increase in frequency of demonstrability of ferric iron in blood pigments, and it almost entirely prevents formation of the so-called formalin pigment.

Fixation in formaldehyde is influenced by the concentration of the reagent and by temperature, just as are other chemical reactions. A 4.1 to 4.5% formaldehyde solution (10% formalin) fixes *adequately* in 48 h at 20 to 25°C (68 to 77°F), in 24 h at 35°C (95°F), and an 8 to 9% formaldehyde solution (20% formalin) hardens in 3 h at 55°C. However, autolysis is also hastened by higher temperatures, so that better fixation is attained with longer exposures at lower temperatures; some writers have recommended fixation at 0 to 5°C. The use of higher formaldehyde concentrations also tends to overharden outer tissue layers

TABLE 3-2 FORMALDEHYDE CONTENT OF SOLUTIONS

% by weight	Specific gravity	Grams per 100 ml	Molarity
40	1.124	44.96	14.97
39	1.120	43.67	14.54
38	1.116	42.39	14.12
37	1.111	41.12	13.69
36	1.107	39.86	13.27
35	1.103	38.605	12.86

and to affect staining adversely, especially with azure-eosinates. With alcoholic formaldehyde fixations the above times may be reduced by 50%.

The use of temperatures above 60°C involves the factors of heat coagulation and of loss of formaldehyde through volatilization. Small pieces of tissue up to 5 mm thick may be hardened throughout by 2 min boiling in 0.85% sodium chloride solution alone, and an egg may be boiled hard in 10 min.

Substitution of alcohol as the diluent of formaldehyde solution results in faster fixation, greater hardening, loss of fats and lipids, better preservation of glycogen, poorer preservation of iron-bearing pigments, and sometimes partial lysis of red corpuscles and loss of eosinophil leukocyte granules.

The foregoing fixation times are traditional. In recent work, where it was desired to minimize the harman condensation of tryptophan or to preserve dopa oxidase activity, we have used times of 1 to 2 h in calcium acetate formalin with very satisfactory results. The enzyme reaction is carried out immediately after fixation and washing of very thin slices; for the indole reactions, hardening in alcohol is prolonged to 1 or 2 days to give better cutting consistency.

Addition of 0.85% sodium chloride to 4% formaldehyde solutions is strongly recommended by some writers. On most material little difference is discerned.

Ramón y Cajal's ammonium bromide formalin, "formol ammonium bromide," is often prescribed for central nervous system tissues, especially for silver impregnation technics. It is a strongly acid mixture containing free hydrobromic acid and methenamine as well as NH_4 and Br ions and at least 4.2% unreacted formaldehyde. On mixing of the formaldehyde (pH 4 to 7; Ramón y Cajal prescribed "neutral") with the ammonium bromide (pH 5 to 5.2) the pH of the mixture promptly falls below 2.0, and readings of 1.4 have been obtained on unused stock mixture. During fixation the pH rises slightly (1.7 after 3 weeks). The fluid is not recommended for histochemical studies. Red corpuscles are lysed, nuclei become directly Schiff-positive from a Feulgen hydrolysis during fixation, and it is probable that hemosiderin iron is at least partly extracted.

Baker's (1%) $CaCl_2$ (10%) formalin and his (1%) $CaCl_2$, (1%) $CdCl_2$, (10%) formalin (Q. J. Microsc. Sci., **85**: 1, 1944) were recommended especially for the fixation and preservation of phospholipids in tissues. Pearse (1960) recommends a variant containing 1.1% $CaCl_2$ and 15% formalin (2 M HCHO). To combine the effect of the calcium ion and of buffering we have substituted 2% calcium acetate (monohydrate) for the $CaCl_2$ of Baker's formula, attaining an approximate pH 7 and about equal resistance with the phosphate formula to pH displacement by deliberate addition of formic acid.

In regard to these calcium–formalin solutions, Baker's $CaCl_2$ formula is 90.2 mM with respect to Ca^{2+}, and Lillie's acetate formula is 113.5 mM. Baker's requires adjustment to neutrality by shaking with $CaCO_3$ or by adjustment with NaOH; in the acetate formula the calcium salt itself acts as an automatic buffer with a similar capacity for formic acid to the phosphate-buffered formalin.

Formaldehyde alone is a rather soft fixative and often does not harden certain cytoplasmic structures adequately for paraffin embedding. Brush borders of renal epithelium are often frayed, and radial striation is partially obscured in paraffin sections, while at least the latter feature is plainly discernible in frozen sections of the same material. Also, cross striations of heart muscle are better defined in frozen than in paraffin sections of material fixed with formaldehyde. Pyramidal cells in the cerebral cortex often are surrounded by clear spaces because of shrinkage in paraffin sections.

To remedy this soft fixation one may substitute for or add to formaldehyde various other reagents such as mercuric salts, chromic acid and its salts, osmium tetroxide, picric acid, alcohol, or various other less commonly used reagents. Likewise, treatment with chromate, picric acid, and mercuric chloride solutions may be used after a previous formalin fixation.

FORMULAS FOR FORMALDEHYDE SOLUTIONS

1. 10% **FORMALIN**

Concentrated formaldehyde solution (37–40%)	100 ml
Tap water	900 ml

2. **FORMOL SALINE SOLUTION**

37–40% formaldehyde solution	100 ml
Sodium chloride	8.5 g
Tap or distilled water	900 ml

3. NEUTRAL 10% FORMALIN

37–40% formaldehyde solution	100 ml
Water	900 ml
Calcium or magnesium carbonate to excess	

4. NEUTRAL BUFFERED FORMALDEHYDE SOLUTION (pH 7)

37–40% formaldehyde solution	100 ml
Water	900 ml
Acid sodium phosphate, monohydrate	4 g
Anhydrous disodium phosphate	6.5 g

5. CALCIUM ACETATE–FORMALIN

37–40% formaldehyde solution	100 ml
Distilled water	900 ml
Calcium acetate (monohydrate)	20 g

6. SUCROSE FORMALIN.   In an effort to avoid intrafixation diffusion or disruption of cell organelles, sucrose solutions have been recommended. We have used (*J. Histochem. Cytochem.*, **8:** 182, 1960) ice cold 30% (0.88 *M*) sucrose for 30 min followed by fixation in formalin 10 ml, calcium acetate monohydrate 2 g, sucrose 30 g, distilled water to make 100 ml for 18 h at 2 to 5°C. Material is then washed 4 h in running water at 20 to 30°C to remove sucrose, and dehydrated over a 48-h period in 70, 80, 95, and 100% alcohols, infiltrated 18 h in 1% collodion (Parlodion, celloidin) in ether alcohol, cleared, and hardened in two changes of chloroform (30 min each), and infiltrated for 2 h in hard paraffin (57°C) under reduced pressure (15 mm mercury). This procedure permitted the preparation of serial sections of rodent intestine at 4 μ.

It is essential that the sucrose be thoroughly washed out before dehydration. Cutting consistency is seriously impaired if sucrose is not removed.

Holt (*Exp. Cell Res. Suppl.*, **7:** 1, 1959) prescribed a gum sucrose solution now used widely for enzyme preservation consisting of 0.88 *M* sucrose in aqueous 1% gum acacia. For preparation, 2 g dry gum acacia and 60 g dry sucrose are added to distilled water to make 200 ml. The solution is stored at 4°C.

De Lellis and Fishman (*Histochemie*, **13:** 1, 1968) noted that erythrocytes stained with the PAS (periodic acid Schiff) method in tissues preserved in Holt's gum sucrose. Further studies revealed the PAS reactivity was due to adherence of gum acacia to erythrocytes. The amount of adherence to other tissues was not investigated. On this basis, caution should be exercised in interpretations of PAS-stained tissues that have been processed through Holt's gum sucrose solution.

7. BAKER'S FORMOL-CALCIUM FOR LIPIDS

	Formol calcium	Calcium cadmium formol	Pearse's variant	
CaCl$_2$ (anhydrous)	1 g	1 g	1.3%	85 ml
CdCl$_2$ (anhydrous)		1 g		
40% formaldehyde	10 ml	10 ml		15 ml
Distilled water	90 ml	90 ml		

8. RAMÓN Y CAJAL'S FORMOL–AMMONIUM BROMIDE, FAB

	Ramón y Cajal	Davenport	Conn and Darrow
37–40% formaldehyde solution	140 ml	120 ml	150 ml
Ammonium bromide	20 g	20 g	20 g
Distilled water	1000 ml	1000 ml	850 ml

9. ALCOHOLIC FORMOL

Formol 37–40% HCHO	10 ml
Distilled water	10 ml
Absolute ethanol	80 ml

10. TELLYESNICZKY'S ACETIC ACID–ALCOHOL–FORMALIN

	Tellyesniczky	Fekete	Opie & Lavin	Lillie's AAF	Bodian
37–40% formaldehyde	5 ml	10 ml	5 ml	10 ml	5 ml
Glacial acetic acid	5 ml	5 ml	5 ml	5 ml	5 ml
Alcohol:					
Concentration	70%	70%	80%	95–100%	50%
Amount	100 ml	100 ml	90 ml	85 ml	100 ml

Source: Tellyesniczky, *Arch. miks. Anat.*, **52**: 202, 1898); Fekete, *Am. J. Pathol.*, **14**: 557, 1938; Opie and Lavin, *J. Exp. Med.*, **84**: 107, 1946; Lillie, *Anat. Rec.*, **103**: 611, 1949; Bodian, *Anat. Rec.*, **69**: 153, 1937.

Alcoholic formalin and acetic acid–alcohol–formalin are excellent for glycogen preservation and are good cytoplasmic fixatives when ribonucleic acid (RNA) digestion tests are to be performed. In the latter case it is well to fix at 0 to 5°C or lower for 24 h only. Tellyesniczky recommended a similar fluid to the last, containing 50 ml each of formalin and acetic acid and 1000 ml alcohol. This fluid is to be distinguished from the same author's acetic dichromate formula. Both are often referred to as *Tellyesniczky's fluid*. Mallory, Langeron, and Romeis noted only the acetic dichromate fluid, Cowdry gave only the acetic acid–alcohol–formalin, and Lee gave both.

Traditionally, material fixed in aqueous formalin solutions is stored indefinitely in the same fluid. For this purpose the buffered solution is superior to the unbuffered, even if solid calcium carbonate is included in the storage bottle. On storage in formalin, gradual decrease in basophilia of cytoplasm and nuclei occurs, as well as a progressive loss in reactivity of myelin to Weigert's iron hematoxylin method. Lipids undergo not too well understood alterations.

Buffered formalin retards the loss of basophilia, but storage in 70% alcohol or in 10 to 20% diethylene glycol in water appears to be better. Ethylene glycol probably can be used in the same way, but has not been tested so exhaustively.

After alcoholic formalin fixation, storage in 70% alcohol would appear to be superior to leaving the tissue in the formalin solution.

GLUTARALDEHYDE.   Seligsburger and Sadlier (*J. Am. Leather Chem. Assoc.*, **52:** 2, 1957) and Fein and Filachione (*J. Am. Leather Chem. Assoc.*, **52:** 17, 1957) used glutaraldehyde as a tanning agent for preparation of leather. A few years later Sabatini et al. (*J. Cell Biol.*, **17:** 19, 1963) compared glutaraldehyde with several other aldehydes for preservation of enzyme activity and structure at the electron microscopic level. Subsequently, many authors have employed glutaraldehyde in various concentrations, with and without additional fixatives or other additives, and at various temperatures. Several of these are listed in Table 3-3. The number of fixatives recommended is reminiscent of the number of silver-impregnation methods available to the prospective user. It is important for the user to recognize that no one knows the effectiveness or accuracy of any fixative. Fixation is a process of which we have very little understanding. That good reproducibility occurs after employment of certain methods with a fixative is not evidence of accuracy, preservation of the original structure, or effectiveness of the fixation.

According to Hopwood (*Histochem. J.*, **4:** 267, 1972) glutaraldehyde exists in solution both as the monomer and its mono- and dihydrated forms and as a series of polymers formed by aldol condensations

$$OHC-(CH_2)_3-CHO \rightarrow OHC-(CH_2)_3-CH{=}\underset{\underset{CHO}{|}}{C}-(CH_2)_2CHO$$

ranging up to products with six-membered rings. As with formaldehyde the —CHO aldehyde group may be hydrated to a methylene glycol radical $-CH\underset{\diagdown OH}{\overset{\diagup OH}{}}$, which may well be the reactive form.

Ultraviolet spectroscopy reveals peaks at 235 mμ, representing the polymer, and at 280 mμ, and the proportion of the latter increases with rise of temperature.

The specific gravity of 50% glutaraldehyde is about 1.1161; it remains liquid down to −21°C but when stirred starts to freeze at once. The melting point is about −15.5°C.

PURIFICATION   Fractionation on a Sephadex column will remove the polymers. Distillation at normal or reduced pressure can be used to isolate the monomer. This eliminates the ultraviolet absorption peak at 235 mμ, leaving only the 280-mμ peak. Filtration on activated charcoal also removes the polymers. Evans (University Microfilms, Ann Arbor, Mich.,

TABLE 3-3 ALDEHYDE FIXATIVES

Reference	%*	Buffer	pH	Other fixatives	Other additives	Comments
Sabatini et al, *J. Cell Biol.,* **17:** 19, 1963	2.5	0.1 *M* phosphate, 4 or 24°C	7.4	0	0	900 mOsm
Karnovsky, *J. Cell Biol.,* **27:** 137A, 1965	2.5	Cacodylate	7.2	2.5% formaldehyde	0	
Chambers et al, *Arch. Pathol.,* **85:** 18, 1968	5.0	Cacodylate	7.2	4.0% formaldehyde	0	
	2.5	0.1 *M* phosphate	7.4	0	0	
Gordon et al, *Exp. Cell Res.,* **31:** 440, 1963	2.0	0.05 *M* cacodylate	7.2	0	0	280–320 mOsm
Trump and Bulger, *Lab. Invest.,* **15:** 368, 1966		0.0625 *M sym*-collidine	6.8–7.1	0.05% OsO$_4$	0	875–925 mOsm
Williams and Luft, *J. Ultrastr. Res.,* **25:** 271, 1968	1.2	0.1 *M* phosphate	7.0	1% [tris(1-aziridinyl) phosphine oxide]†	0	
Behnke and Zelander, *J. Ultrastr. Res.,* **31:** 428, 1970	2.0	0.1 *M* cacodylate	6.5	0	1% Alcian blue	370 mOsm
Wachstein and Besen, *J. Histochem. Cytochem.,* **11:** 447, 1963	0.5	0.5 *M* tris maleate	7.2	0	0	
Futaesaku et al, *Proc. Int. Cong. Histochem. Cytochem.,* **4:** 155, 1972; Rodewald and Karnovsky, *J. Cell Biol.,* **60:** 423, 1974	1.0	0.1 *M* phosphate	7.3	1% tannic acid	0	
Silverman and Glick, *J. Cell Biol.,* **40:** 761, 1969	1.0	0.067 *M* phosphate	7.4	0	Sucrose added to make 300 mOsm	
Kalt and Tandler, *J. Ultrastr. Res.,* **36:** 633, 1971	3.0	0.1 *M* cacodylate	7.0	2% formaldehyde 1% acrolein	2.5% dimethylsulfoxide and 0.001 *M* CaCl$_2$	
Luft, *Anat. Rec.,* **171:** 369, 1971	1.2	0.66 *M* cacodylate	7.3	0	Ruthenium red, 550 ppm	
Flickinger, *Z. Zellforsch. Mikrosk. Anat.,* **78:** 92, 1967	2.5	0.1 *M* cacodylate	7.2	2% formaldehyde	0.05% CaCl$_2$	
Scallen and Dietert, *J. Cell Biol.,* **40:** 802, 1969	2.5	0.1 *M* cacodylate	7.2	2% formaldehyde	0.05% CaCl$_2$ and 0.2% digitonin	Digitonin is used to preserve cholesterol
Overton, *J. Cell Biol.,* **38:** 447, 1968	2.5	Cacodylate	7.2	0	1% lanthanum nitrate	

* Glutaraldehyde.
† Alkylating agent.

70: 634, 1969) treated this with anhydrous $CaCl_2$, extracted into ether, distilled off the ether, and collected the fraction distilling at 15 torr between 82.5 and 84°C. The glassy solid forming in distillation residues reverts to the monomer on heating.

ASSAY. The most satisfactory assay appears to be an iodine titration of the bisulfite compound $RCHO \cdot NaHSO_3 + 2I_2 + 3Na_2CO_3 \rightarrow RCOONa + Na_2SO_4 + 4NaI + H_2O + 3CO_2$.

STORAGE. Glutaraldehyde, 50%, buffered to pH 5, loses only about 2.5% in titer in 11-month storage at 1°C. At pH 8 polymerization with separation of a white precipitate occurs rapidly.

FIXING QUALITIES. Hopwood was mainly concerned with its qualities in fixing for electron microscopy. Hypotonic solutions perform poorly; slightly hypertonic, well; and more hypertonic solutions tend to produce shrinkage and extracellular spaces. Dextran and polyvinyl-pyrrolidone improve this feature when added. Formaldehyde glutaraldehyde mixtures are used and are thought to interreact, possibly, to form a 1,3-cyclohexanedione.

Aldehydes, including glutaraldehyde, react with osmium tetroxide to form corresponding acids and osmium black, and aldehyde prefixation hinders osmium tetroxide penetration.

Hopwood has almost nothing to say about histochemistry of fixed tissue as observed under optical microscopy other than noting the artifactual aldehyde reaction. He seems convinced that this fixative produces extensive protein cross bonding by its two aldehyde groups. In view of the extensive presence of an aldehyde reaction at much the same sites as are demonstrated by anionic dyes and by sequence ferrous ion hematoxylin which Lillie et al. (*J. Histochem. Cytochem.*, **20:** 116, 1972) have shown to be assignable largely to lysyl and arginyl residues, it would seem that not much cross bonding occurs by utilization of the second aldehyde residue of this and probably other dialdehydes under the conditions used in tissue fixation, since a large proportion of these —CHO residues remain free and reactive.

We have noted elsewhere in this book (Chap. 8) that glutaraldehyde is superior to formaldehyde for the demonstration of noradrenaline by the diazosulfanilic azure A technic, as reported by Lillie et al. (*J. Histochem. Cytochem.*, **21:** 448, 1973). Coupland and Hopwood (*J. Anat.*, **100:** 227, 1966) found glutaraldehyde more effective in preserving noradrenaline for iodate and osmic reactions and noted that adrenaline, in contrast to noradrenaline, was not precipitated and retained in tissue by it. We have already noted the inferior preservation of the azo coupling reaction of erythrocytes in our adrenal study as compared with that seen in formaldehyde-fixed tissue and suspect there may be many other reaction alterations assignable to this fixative. It has been little used in microscopic histochemistry and staining procedures aside from electron microscopic technics.

Artifacts may be produced by cross-linking of free amino acids to the fixed tissue. Hopwood does not explain this, but the many free aldehyde residues present in glutaraldehyde-fixed tissue should be capable of binding amines.

Hopwood (*J. Anat.*, **101:** 83, 1967) and Jahn (*Histochemie*, **29:** 298, 1971) have commented on the direct Schiff reaction produced by glutaraldehyde fixation. Lillie suggested this verbally at the 1964 International Congress for Histochemistry and Cytochemistry, and it is discussed later in this chapter, as it was in the third edition (1965) of this book.

End groups react with glutaraldehyde in the following decreasing order: NH_2, peptide, guanidine, NHR, and OH. Glycine, serine, and proline reacted most at pH 6 to 7. Lysine and arginine reactivity increased with rising pH.

Rost and Ewen (*Histochem. J.*, **3:** 207, 1971) reported formation with glutaraldehyde vapor of fluorescent compounds with catecholamines, tryptamines, histamines, and other arylethylamines.

Hopwood (*J. Anat.*, **101**: 685, 1967) gives a table of amount of preservations of enzyme activities after glutaraldehyde fixations of stated duration. Glutaraldehyde is most active in binding ε-amino groups (lysine, diaminophimelic acid). It polymerizes and immobilizes proteins. Immune bodies retain much specific reactivity. Glutaraldehyde lowers the isoelectric points of proteins. Phosphatidylserines and phosphatidylethanolamines are better immobilized by glutaraldehyde than by formaldehyde. (Hopwood's account seems limited to electron microscopic effects. The immobilization of myelin by formaldehyde was known long before Roozemond and l'Hermite and Issael in 1969.)

Glutaraldehyde preserves glycogen slightly better than formaldehyde. Ethanol is still best.

As a fixative glutaraldehyde penetrates rather slowly, and perfusion is advocated by some workers. Hypotonic solutions tend to fix poorly; hypertonic solutions perform better, though they may induce shrinkage and large extracellular spaces. Dextrans and polyvinylpyrrolidone improve this feature.

Mixtures of glutaraldehyde and formaldehyde prepared from paraldehyde are used; it is suggested that a 1,3-cyclohexanedione may form from condensation of the two aldehydes.

Glutaraldehyde and other aldehydes react with osmium tetroxide to form the corresponding acids and osmium black. Apparently glutaraldehyde and formaldehyde prefixations hinder the penetration of osmium tetroxide, the former more. The concept of cross-linking by glutaraldehyde seems firmly established in current thinking.

ARTIFACTS.   Free amino acids bind to protein much more with glutaraldehyde than with formaldehyde. Hopwood's text is not clear, but it seems to be indicated that the free amino acid is bound by already fixed protein. (This is chemically intelligible in view of the extensive presence of attached reactive aldehyde in glutaraldehyde-fixed tissue.)

Jahn (*Histochemie*, **39**: 298, 1971) warned that false-positive reactions in PAS and other aldehyde reactions were given by glutaraldehyde-fixed tissues. Hopwood (*J. Anat.*, **101**: 83, 1967) also commented on this, and it is discussed later in this chapter.

Permeability of membranes to water and variously to ions is partly preserved by glutaraldehyde, and volume changes in certain morphologic components can still occur after glutaraldehyde fixations—e. g., acromers of retinal rods. Diffusion of enzymes tends to be prevented.

A solution of buffered 5% glutaraldehyde (10 ml 50% glutaraldehyde and 90 ml 0.1 M phosphate buffer at pH 7) has been very satisfactory in studies of the azo coupling reaction of noradrenaline islets in the adrenal medulla (Lillie et al., *J. Histochem. Cytochem.*, **21**: 448, 1973). Application of borohydride reduction before staining does away with the unwanted aldehyde effect.

Glutaraldehyde formaldehyde mixtures have been favored for electron microscopy by some writers, but a mixture of even 1% glutaraldehyde in 5% formaldehyde was found by Baccarini and Powell (*Stain Technol.*, **48**: 77, 1973) to adversely affect Nauta-Gygax silver impregnations for axons, by rendering background too dense. Since formaldehyde is commonly used as the reducing agent in Bielschowski and Maresch argyrophil methods, this action of the unused aldehyde group widely found with dialdehyde fixations should be expected. Whether the adverse effect would be remedied by destroying the unused second aldehyde with borohydride remains to be determined. When native tissue aldehydes as in young elastin or in lipofuscins are to be studied, acrolein and all the dialdehydes should be avoided as fixatives.

During the past few years the aldehyde fixatives have been widely used either alone or in combination with other fixatives for subsequent electron microscopic investigations. Also, several aldehydes have been used particularly at 4°C to provide limited fixation for enzyme histochemical studies. Unfortunately, a uniform pattern of enzyme conservation and tissue preservation does not appear to exist. Each enzyme is peculiar, and at this time one cannot without trial predict the effect of any aldehyde fixative on enzyme activity. Janigan (*Lab. Invest.*, **13**: 1038, 1964) studied the effects of aldehyde fixation on retention of β-glucuronidase, β-galactosidase, N-acetyl-β-glucosaminidase, and β-glucosidase in several types of rat organs. After 7-h fixation at 2°C, only 18 to 27% of β-glucuronidase activity could be retained in rat liver fixed in glutaraldehyde; 56 to 62% was retained in tissues fixed in 4% formaldehyde; and 72 to 75% was retained in tissues fixed in 12.5% hydroxyadipaldehyde. Under identical conditions, approximately 64 to 70% was retained in rat liver fixed with 10% glyoxal. For β-D-glucosidase, 82% was retained in hydroxyadipaldehyde-fixed, 56% in formalin-fixed, and 32% in glutaraldehyde-fixed kidney tissues.

In further studies Janigan (*J. Histochem. Cytochem.*, **13**: 476, 1965) noted that more enzyme activity was preserved by aldehydes that provided "poorer" fixation (glyoxal, hydroxy-adipaldehyde) and least enzyme activity by aldehydes that are "better" fixatives (acrolein, glutaraldehyde, and crotonaldehyde). An intermediate state was noted with formaldehyde.

Silverman and Glick (*J. Cell Biol.*, **40**: 761, 1969) fixed several tissues including rat liver and cardiac muscle overnight at 24°C in 1% glutaraldehyde in 0.067 M potassium phosphate buffer at pH 7.4 with sucrose added to make 300 $\pm$ 10 mOsm (milliosmoles). After washing in phosphate-buffered sucrose, blocks were stained with 5% phosphotungstic acid in 6.25% Na_2SO_4, washed in 2% ammonium acetate brought to pH 2 by addition of formic acid, dehydrated, and embedded in Epon. Sections were examined by electron microscopy. In other experiments, stoichiometry of the reaction was determined between selected proteins and phosphotungstic acid using several biochemical methods. Under certain conditions Silverman and Glick believed the stain density reflected the concentration of protein based upon the quantitative reaction of phosphotungstic acid with positively charged groups, although the stoichiometry of the reaction varied with the kind of protein. In tissue sections mitochondrial matrix, cisternae of endoplasmic reticulum, and the Z band of muscle were stained intensely.

Ericsson and Biberfeld (*Lab. Invest.*, **17**: 281, 1967) studied the rate of fixation of rat and mouse liver with 1, 3, or 6% glutaraldehyde or 3% paraformaldehyde in 0.1 M phosphate buffer at pH 7.4. Ericsson and Biberfeld recommend the use of 3% glutaraldehyde for tissue blocks no thicker than 0.5 mm. Assessments were made on the basis of preservation of tissues as observed at the ultrastructural level.

Glutaraldehyde–osmium tetroxide double fixation also appears to have been initially reported by Sabatini et al. (*J. Cell Biol.*, **17**: 19, 1963). In studies of lung tissues Gil and Weibel (*J. Ultrastr. Res.*, **25**: 331, 1968) noted the appearance of artifactual spherical electron-dense granules on alveolar surfaces and on erythrocytes *only* if glutaraldehyde-fixed tissues were postfixed with osmium tetroxide in *phosphate* buffer. Gil and Weibel now use *sym*-collidine buffer; however, bicarbonate, cacodylate, and Veronal acetate were satisfactory. Gil and Weibel recommend adjustment of all buffers to an osmolality corresponding to blood plasma using an osmometer. Hypertonic solutions were observed to result in shrinkage of cells and uneven cell surfaces. Hypotonic solutions result in cytoplasmic swelling, ruptured membranes, interstitial fluid accumulation, and detachment of cells from basement membranes.

Trump and Bulger (*Lab. Invest.*, **15**: 368, 1966) advocate simultaneous fixation in osmium tetroxide and glutaraldehyde. The fixative is prepared by mixing 2 volumes of 4% osmium tetroxide, 1 volume of approximately 50% glutaraldehyde, and 5 volumes of 0.1 M *sym*-collidine buffer. The osmolarity ranges from 875 to 925 mOsm/1 (milliosmoles per liter). The pH at the time of preparation at 24°C is 7.0 to 7.1; by 1 h, the solution becomes reddish brown with a pH of 6.8; after 6 h, the solution is black and the pH is 6.6.

Tissues (1 mm thick) are immersed in the fixative for 1 h at 4°C immediately after preparation of the fixative. Trump and Bulger state that many cytologic features are intermediate in appearance between those observed with either fixative alone.

Futaesaku et al. (*Proc. Int. Cong. Histochem. Cytochem.*, **4**: 155, 1972) and Rodewald and Karnovsky (*J. Cell Biol.*, **60**: 423, 1974) highly recommend a tannic acid–glutaraldehyde perfusion fixative. The fixative comprises 1% tannic acid and 1% glutaraldehyde in 0.1 M phosphate buffer at pH 7.3. In studies of the kidney, Rodewald and Karnovsky perfused rats for 15 min; tissues were excised and fixed for an additional 2 h in the same fixative at 24°C. After two rinses (15 min each) in the phosphate buffer containing 0.1 M sucrose at 0°C, tissues were postfixed in 2% osmium tetroxide in phosphate buffer at 0°C, dehydrated, and embedded in Epon. A previously unattainable preservation of detailed morphologic features of the epithelial slit diaphragm of the rat glomerulus was obtained by Rodewald and Karnovsky. Tannic acid appears to enhance high contrast to the diaphragm and other extracellular tissue components.

Luft (*Anat. Rec.*, **171**: 369, 1971) conducted a lengthy study of the localization of ruthenium red in tissues immersed in a fixative containing this reagent. Luft recommends the following procedure:

1. Fix for 1 h (24 or 4°C) in the following mixture:

3.6% glutaraldehyde (biologic grade) or 2% acrolein	0.5 ml
0.2 M cacodylate buffer at pH 7.3	0.5 ml
Ruthenium red (1500 ppm in distilled water)	0.5 ml

2. Rinse tissues in two or three changes of 0.15 M cacodylate buffer over a 10-min period.
3. Fix tissues again for 3 h at 24°C in the following mixture:

5% osmium tetroxide in distilled water	0.5 ml
0.2 M cacodylate buffer at pH 7.3	0.5 ml
Ruthenium red (1500 ppm in distilled water)	0.5 ml

4. Rinse briefly in the buffer above, dehydrate through ethanols, and embed in epoxy resin.

RESULTS: Luft states that ruthenium red probably stains acidic mucopolysaccharides, and is generally excluded by cell membranes, thus permitting tracing their tortuous path at times. Cell surfaces are nicely stained. He notes that increased density sometimes observed in lipid droplets and in membranes of mitochondria may indicate ruthenium red–staining sites of fatty acids and acidic phospholipids.

Hirsch and Fedorko (*J. Cell Biol.*, **38**: 615, 1968) used the Trump and Bulger simultaneous glutaraldehyde-osmium fixative followed by postfixation in 0.25% uranyl acetate in 0.1 M acetate buffer at pH 6.3 in an ultrastructural study of human leukocytes. Endoplasmic reticulum and ribosomes are stated to be well delineated.

Williams and Luft (*J. Ultrastr. Res.*, **25**: 271, 1968) suggest that the addition of an alkylating agent [tris(1-aziridinyl)phosphine oxide] to a final concentration of 1% in 1.2% glutaraldehyde in 0.1 M phosphate buffer at pH 7 results (after postosmication) in a better preservation of detail in filaments, cytoplasmic background, greater density of granules and mitochondria, etc., than with glutaraldehyde alone (after postosmication). Attention should be called to the fact that the alkylating agent is a nitrogen mustard which may be carcinogenic. The agent is highly water-soluble and nearly nonvolatile. The fixative effect may be related to a preservation of polysaccharides or their derivatives, according to Williams and Luft. Direct data for this suggestion were not provided.

Hashimoto (*J. Invest. Dermatol.*, **57**: 17, 1971) added 1% lanthanum nitrate to 2.5% glutaraldehyde in cacodylate buffer at pH 7.2 and fixed skin 1 to 3 days at 4°C. Lanthanum penetrated into the basal cell interspaces, including half-desmosomes and desmosomes. Tight junctions prevented lanthanum entry. Intestinal brush border also stains according to Overton (*J. Cell Biol.*, **38**: 447, 1968).

Behnke and Zelander (*J. Ultrastr. Res.*, **31**: 428, 1970) recommend the addition of Alcian blue to perfusion and immersion mixtures of glutaraldehyde to enhance ultrastructural detail. After perfusion of rats with 2% glutaraldehyde in 0.1 M cacodylate buffer at pH 6.5 (370 mOsm) for 1 to 2 min, perfusion is continued for 20 min with the same fixative containing Alcian blue at 1% concentration. Tissue samples from perfused animals are then immersed in the glutaraldehyde–Alcian blue fixative for 1 to 2 h at 24°C, postfixed (without an intermediate rinse) in 1% osmium tetroxide in 0.1 M cacodylate buffer at pH 6.5, and embedded as usual at 24°C. Sections could also be stained with uranyl acetate and lead citrate.

Alcian blue is osmiophilic. An increased density of cell coats was observed which could be augmented by uranyl acetate and lead citrate staining. By light microscopy, tissue components stained bluish green with Alcian blue, and postosmication did not alter this appearance even though tissue blocks appeared black. Perfusion enhanced penetration of Alcian blue into the tissues. Little penetration was observed in tissues immersed in the stain-fixative alone.

Chambers et al. (*Arch. Pathol.*, **85**: 18, 1968) find that glutaraldehyde is an excellent fixative for use in routine histopathology. In relation to formalin, dye uptake is generally increased with many stains. Differentiation of connective tissues is said to be particularly good. Chambers et al. used glutaraldehyde that had impurities removed by activated charcoal. The fixative solution advocated is 1 to 2 ml 25% glutaraldehyde added to 10 ml 0.1 M phosphate buffer at pH 7.4. Tissue blocks no greater than 2 × 2 × 0.4 cm are immersed in the fixative for 24 h or more, then paraffin-embedded, sectioned, and stained as usual.

Kalt and Tandler (*J. Ultrastr. Res.*, **36**: 633, 1971) obtained excellent preservation of amphibian embryos at the ultrastructural level using a mixture of 3% glutaraldehyde, 2% formaldehyde, 1% acrolein, and 2.5% dimethyl sulfoxide in 0.1 M cacodylate buffer at pH 7 containing 0.001 M $CaCl_2$. Tissues were postfixed in osmium tetroxide.

Schultz and Case (*J. Cell Biol.*, **38**: 633, 1968) note that neuronal microtubules are very sensitive to the type of perfusate used. Bicarbonate buffer was noted to cause a complete disintegration of neuronal microtubules. Acrolein alone and uncombined with any other aldehyde in phosphate buffer at 320 mOsm caused a similar loss.

Hassell and Hand (*J. Histochem. Cytochem.*, **22**: 223, 1974) conducted comparative studies of dimethyl malonimidate (DMM), dimethyl adipimate (DMA), and dimethyl suberimidate (DMS) to glutaraldehyde fixation in relation to several pH levels and buffer types at the light and electron microscopic levels. Diimidoesters are bifunctional reagents that react with the ε-amino groups of lysine forming intermolecular cross-links. Minimal alterations of properties of proteins are believed to occur. Using water-insolubility of proteins as an assay, DMS (16 to 20 mg/ml with 0.02 M Ca^{2+} in 0.15 M tris–HCl buffer at pH 9.5) insolubilized 92.1% of liver proteins, and only 3.3% of proteins were detectable in the fixative solution. The smaller diimidoesters were less effective in rendering proteins insoluble. The diimidoesters have the following general formula:

$$\overset{\overset{+}{Cl^-H_2N}}{\underset{H_3CO-C}{\|}}-(CH_2)_n-\overset{\overset{+}{NH_2Cl^-}}{\underset{C-OCH_3}{\|}}$$

For DMM, $n = 1$; for DMA, $n = 4$; for DMS, $n = 6$. The fixatives are employed as buffered aqueous solutions which are unstable. Fine structure appears to be best preserved with the use of DMS. Hassell and Hand advise the following method of preparation whereby the constituents are added in the order listed to make 10 ml of solution: 7.8 ml distilled water, 1.2 ml 1.0 N NaOH, 182 mg tris base, and 160 to 200 mg DMS. The pH is adjusted to 9.5 with either HCl or NaOH, and 1.0 ml 0.02 M CaCl$_2$ is added dropwise. The fixative must be used immediately, due to its instability. Dutton et al. (*Biochem. Biophys. Res. Commun.*, **23**: 730, 1966) have concluded that diimidoesters have a half-life of 4 h. Maximum fixation (protein insolubilization) was obtained in blocks 1 to 2 mm thick in $2\frac{1}{2}$ h at 25°C in slowly agitated solutions. Fixation was not enhanced in solutions with increased concentrations of DMS. Excellent preservation of tissues was observed at both light and electron microscopic levels. Mitochondria appeared more dense after DMS fixation as compared to glutaraldehyde.

In another study (*J. Histochem. Cytochem.*, **22**: 299, 1974) Dutton et al. noted excellent preservation of glycogen as observed by light and electron microscopy and as determined by biochemical assays. A special advantage of DMS over glutaraldehyde in studies of glycogen is the absence of a positive Schiff reaction in the background due to the unreacted aldehyde of glutaraldehyde. Preliminary studies of the use of diimidoesters as fixatives for immunologic studies were conducted by McLean and Singer (*Proc. Natl. Acad. Sci. U.S.A.*, **65**: 122, 1970).

Tice and Barrnett (*J. Histochem. Cytochem.*, **10**: 754, 1962) used tissue blocks fixed with hydroxyadipaldehyde, $C_4H_7(OH)(CHO)_2$ at 12.5% in 50 mM tris maleate buffer of unstated pH, containing 0.4 M (10.7%) sucrose, for 1 to 2 h at 4°C, using blocks of about 5 mm^3 volume. This fixation is followed by several hours' washing in tris maleate sucrose solution. The fixation is said not to inhibit glucose 6-phosphatase completely. Such material was subjected as frozen sections or as very small blocks to the enzyme reaction and was then postfixed in buffered osmium tetroxide for electron microscopy.

All these fixatives are said to give good morphologic fixation, to be usable for at least some enzyme histochemistry, to be adaptable to buffered osmium tetroxide postfixation for electron micrography. Hydroxyadipaldehyde has been used for electron microscopic studies of glucose 6-phosphatase.

Workers in enzyme electron micrography add 5.3 to 10.7% sucrose (0.2 to 0.4 M). Glutaraldehyde is used at 4 to 5.6% in pH 7.4 phosphate or cacodylate as above. Wachstein and Besen (*J. Histochem. Cytochem.*, **11**: 447, 1963) fixed kidney at 4°C for electron microscopic localization of adenosine triphosphatase in 0.5% glutaraldehyde at pH 7.2 (tris maleate buffer). After the phosphatase reactions had been performed on 40-μ frozen sections, the sections were postfixed in buffered osmium tetroxide as usual.

ACROLEIN. Acrolein, $CH_2\!=\!CH\!-\!CHO$, and glutaraldehyde, $OCH\!-\!CH_2CH_2CH_2CHO$, were reported by Sabatini, Bensch, and Barrnett (*J. Cell Biol.*, **17**: 19, 1963) as good fixatives when electron microscopy was to follow, and Tice and Barrnett (*J. Histochem. Cytochem.*, **10**: 754, 1962) reported similar good morphologic fixation with *hydroxyadipaldehyde*, $C_4H_7OH\!-\!(CHO)_2$.

Acrolein (bp 52.5°C) is lachrymatory and toxic and should be handled in a chemical fume hood. It is freely soluble in water (40%) and melts at −87.7°C. Aqueous ‹10% solutions have been used, with added 1% CaCl$_2$ for lipids, adjusted to neutrality with CaCO$_3$ or buffered with 0.1 M phosphates or cacodylate to pH 7.4.

Van Duijn (*J. Histochem. Cytochem.*, **9**: 234, 1961) described a widespread positive Schiff reaction of tissues after 5% acrolein–95% alcohol treatment (15 min) of formalin- or Carnoy-mixture-fixed material in paraffin sections. This is ascribed to ethylene condensation with SH, NH$_2$, and imidazole groups, and to phospholipids. Apparently guanidyl, hydroxyl, and peptide bond groups are unreactive. The reactivity of indoles and phenols remains in doubt.

These condensations leave the aldehyde group of acrolein free to react with Schiff and other aldehyde reagents.

Norton, Gelfand, and Brotz (*J. Histochem. Cytochem.*, **10**: 575, 1962) used a 10% solution of freshly distilled acrolein in 1% aqueous $CaCl_2$ to which 0.05% hydroquinone was added before using. The pH varied from 6.4 to 6.8. The solution was best prepared fresh but could be kept 3 to 4 days at 4°C.

This fixative produced, even in a few hours, a definite and rather large decrease in extractable brain lipids (20 to 30% in 24 to 48 h). Decreases in extractable lipid with Baker's Ca-Cd formalin were much less (about 10% at 24 h). Like formaldehyde, acrolein destroyed reactivity of plasmalogen. Norton et al. have noted the widespread positive Schiff reaction of tissue proteins and the binding of acrolein to lipids to produce an artifactual lipid aldehyde.

When these reports are considered, it seems evident that acrolein, at least, should not be used as a general histologic or histochemical fixative, especially when any aldehyde localization technics, such as the Feulgen or the periodic acid Schiff method or ammoniacal silver methods, are involved.

Since the three fixatives are all aldehydes, it is to be expected that all would bind with protein amino and phenol groups, indoles, catecholamines, etc., in much the same way as formaldehyde.

In introducing dialdehyde and acrolein fixations, the expectation was that they would serve to fix proteins more firmly by establishing cross-linkages in which the two aldehyde groupings, or the aldehyde and ethylene groups of acrolein, would both be utilized to combine with available amino groups in neighboring protein chains. This cross-linkage undoubtedly occurs in the tanning of leather. But here a relatively large amount of protein is present to react with a carefully regulated dosage of the aldehyde reagent, whereas in tissue fixation, especially on the electron micrography scale, relatively minute amounts of tissue are exposed to relatively huge amounts of acrolein or dialdehyde, and the expectation would be that the reactive tissue sites would be combined with a reagent molecule whose second reactive group would be left free because of preoccupation of all available tissue sites. The general aldehyde reactivity of tissue reported by Van Duijn after reaction with acrolein supports this view. So do current findings on glutaraldehyde-fixed tissue.

Aldehyde attachment of acrolein, with the known reactivity of osmium tetroxide for ethylenic double bonds, should give rise to some interesting osmium artifacts. And with the dialdehydes, aldehyde reactivity is readily demonstrable wherever a molecule of dialdehyde is attached in a Schiff base union or substituted methylene bridge.

Nevertheless these fixatives do have real value for the specific purposes for which they have been recommended.

CHLORAL HYDRATE.   While this substance was often added in small amounts to fixatives for marine invertebrates as a stupefying agent, its use as a primary aldehyde fixative has been rare. We have used it on icteric livers in 22% concentration in 0.1 M phosphate buffer at pH 7. It turns icteric liver green in a few hours, like formaldehyde and other aldehydes. A 25% solution in 50% alcohol was used by Fawcett and Selby (*J. Biophys. Biochem. Cytol.*, **4**: 63, 1958) to fix nervous tissue destined for Ungewitter urea silver carbonate impregnation. These concentrations are comparable to commonly used formol concentrations: 10% is 1.33 M; 22% chloral hydrate is 1.33 M.

DIETHYLPYROCARBONATE (DEPC).   Pearse et al. (*Histochem. J.*, **6**: 347, 1974) praise this as a vapor-phase fixative for immunofluorescence studies on polypeptide hormones. The following method is prescribed:

Small bits of tissues are quenched at -158°C, freeze-dried, and incubated in a sealed

TABLE 3-4 DEPC* VAPOR FIXATION FOR VARIOUS POLYPEPTIDES

Antiserum to	Antigen	Tissue	Species	DEPC	Heat alone
Gastrin	Synthetic human (2–17)	Antrum	Man	+ +	−
Caerulein	Synthetic	Intestine	Quail, chick	+ +	−
Secretin	Pure, porcine	Duodenum	Dog, quail	+ +	−
Enteroglucagon	90 % pure porcine	Intestine	Dog, quail, chick	+ +	±
Insulin	90 % pure porcine	Pancreas, insulinomas	Man, dog	+ +	+
Glucagon	90 % pure porcine	Pancreas	Man, dog	+ +	±
Gastric inhibitory peptide	Highly purified porcine	Intestine	Man, dog	+ +	+
Calcitonin	Purified dog-fish	Ultimobranchial gland	Quail, chick	+ +	+ +
ACTH	Purified, human	Anterior hypophysis	Sheep	+ +	+ +
Growth hormone	Purified, human	Anterior hypophysis	Sheep	+ +	+ +

* DEPC = diethylpyrocarbonate.
Source: Pearse et al., *Histochem. J.*, **6**: 347, 1974.

vessel containing diethylpyrocarbonate (available from Polysciences, Warrington, Pennsylvania 18976) for 3 h at 55°C. Control blocks are treated in the same fashion with omission of the diethylpyrocarbonate. Control and fixed tissue blocks are then vacuum-embedded in paraffin at 58°C, sectioned at 5 μ, and mounted from water on albuminized slides. Sections are dried overnight at 37°C.

For immunofluorescent staining, the indirect (sandwich) type such as that developed by Coons et al. (*J. Exp. Med.*, **102**: 49, 1955) is advocated using fluorescein-labeled goat anti-rabbit IgG serum as the second layer. As controls, (1) normal rabbit serum, (2) the second layer alone, or (3) absorption of the first layer serum with pure antigen were used when possible.

Table 3-4 reveals details on the efficacy of preservation of antigenic sites. It is noteworthy that fixation was required for the detection of antigens in some cases. In all cases, preservation of cytologic structures was far better in tissues fixed with diethylpyrocarbonate rather than with heat (55°C) alone.

Diethylpyrocarbonate is assumed to react with primary amines in two steps. Pearse et al. suggest the following may occur with an ε-amino group of lysine.

$$\text{RNH} + \underset{\underset{CO-OC_2H_5}{|}}{\overset{\overset{CO-OC_2H_5}{|}}{O}} \xrightarrow{[-C_2H_5OH]} \text{RNHCO} \overset{OC_2H_5}{\underset{O}{<}}$$

After conversion to its *N*-carboxyderivative, reaction with a carboxyl of an adjacent molecule resulting in an amide with elimination of CO_2 and ethanol is postulated.

$$\text{RNHCO} \overset{\overset{CO-OC_2H_5}{|}}{\underset{O}{}} + \text{R'COOH} \longrightarrow \text{RNHCOR'} + 2CO_2 + C_2H_5OH$$

Starke (*Adv. Protein Chem.*, **24:** 261, 1970) provides a review of the use of diethylpyrocarbonate as a protein reagent. Wolf (*Eur. J. Biochem.*, **13:** 519, 1970) also noted that the reagent formed intermolecular peptide-like bonds that involved carboxyl groups and bizarre polymeric proteins insoluble in 8 *M* urea, 6 *M* guanidine, 70% formic acid, trifluoracetic acid, dimethyl formamide, and 5% dodecyl sulfate.

Alcohol Fixatives

Alone, alcohol has only limited application as a fixing fluid. Absolute alcohol is often used for preservation of glycogen, for fixation of pigments and, in many species, for tissue mast cells, and for fixation of blood and tissue films and smears. For the last purpose methyl alcohol is usually preferred and is effective in 80 to 100% concentration. Lower percentages hemolyze red cells and inadequately preserve leukocytes. Preservation of hemosiderin is less adequate with ethyl alcohol fixation than with buffered formaldehyde solutions. In 60 to 80% concentration at low temperature (-20 to $-25°C$) alcohol is a useful fixative for preserving certain proteins and enzymes in a relatively undenatured state. Such tissue should be rapidly frozen with liquid nitrogen or solid carbon dioxide mixtures, to avoid ice-crystal artifacts.

In fixation of tissue, absolute, or 100%, alcohol alone penetrates rather slowly. In combination with formalin at a 9 : 1 ratio it fixes adequately in 18 to 24 h or less. In the Carnoy mixtures fixation may be completed in 1 to 2 h. Thin (3 to 5 mm) slices of tissue should be used. With Carnoy's mixture (formula 2, below), fix for 1 to 2 h. This mixture is the one usually referred to as *Carnoy* in pathology usage and is the only fluid given under Carnoy's name alone by Langeron, Romeis, Schmorl, Mallory, and Cowdry. Fixation at about 0°C for 18 to 24 h is sometimes advantageous. Wash in two changes of 100% alcohol of 30 to 60 min each, and store in thin cedar oil. For prolonged storage one may transfer the slices into liquid petrolatum or embed them in paraffin.

Or if immediate embedding is desired, carry the slices from the second 100% alcohol into a mixture of equal volumes of 100% alcohol and benzene for 20 min, and thence through two 30-min changes of benzene into paraffin.

Carnoy fixatives hemolyze, occasion a considerable shrinkage, and dissolve acid-soluble cell granules and pigments. They do not decalcify appreciably in the 2 to 3 h required for fixation. They give excellent nuclear fixation. Nissl granules are well preserved, and many cytoplasmic structures are adequately fixed. Myelin is lost. Glycogen is apparently better preserved than with 100% alcohol, and we prefer Carnoy for this purpose. Alcoholic 10% formalin, with or without 5% of glacial acetic acid, is perhaps superior to Carnoy for glycogen. Fix 12 to 24 h or longer.

Newcomer's (*Science*, **118:** 161, 1953) fluid is recommended as a substitute for Carnoy's fluid for the fixation of chromosomes. It is said to preserve Feulgen stainability of chromatin better than chloroform-containing fluids.

FORMULAS

1. CLARKE (CARNOY A) (*Philos. Trans. R. Soc. Lond.*, **141:** 607, 1851)

100% alcohol	75 ml
Glacial acetic acid	25 ml

Among cytologists this fluid is usually meant when the term *Carnoy* is used.

2. CARNOY (*Cellule,* **3:** 276, 1887)

100 % alcohol	60 ml
Chloroform	30 ml
Glacial acetic acid	10 ml

This fluid is the one commonly referred to as *Carnoy's fluid* in pathologic practice. It is sometimes called *Carnoy II.*

For both the foregoing, fix thin slices (3 to 5 mm) 4 to 5 h at 25°C or, perhaps better, 18 to 24 h at 3 to 5°C.

Thilander (*Anat. Anz.,* **115:** 89, 1964) found Carnoy's fluid the fixative of choice in the preservation of intravitally injected trypan blue.

Puchtler et al. (*Histochemie,* **16:** 361, 1968) believe Carnoy's fluid is the fixative of choice for studies of fibrous proteins and carbohydrates at the light microscopic level.

According to Murgatroyd (*J. R. Microsc. Soc.,* **88:** 133, 1968), Carnoy's fluid, Helly's fixative, and 10% neutral formalin are adequate fixatives for the Feulgen stain. Tissues fixed in Heidenhain's susa fixative gave lower readings.

Arrighi et al. (*Exp. Cell Res.,* **50:** 47, 1968) were able to isolate DNA (deoxyribonucleic acid) from tissues fixed with various fixatives containing methanol, ethanol, glacial acetic acid, 2-propanol, and acetone, whereas no DNA could be recovered from tissues well fixed in formalin.

Puchtler et al. (*Histochemie,* **21:** 97, 1970) substituted methanol for ethanol in Carnoy's fixative and called the resulting fluid *methacarn.* Very little shrinkage of tissues was observed, and myofibrils were particularly well delineated. Superior preservation of fibrous proteins is claimed, while globular proteins, as in usual Carnoy, and DNA undergo considerable rearrangement. However, good preservation of nuclear structure is seen with light microscopy. The precise effect of alteration of helical and globular proteins in electron microscopy is still under investigation.

3. NEWCOMER'S CARNOY SUBSTITUTE (*Science,* **118:** 161, 1953)

Isopropanol	60 ml
Propionic acid	30 ml
Acetone	10 ml
Dioxane	10 ml

Fix 12 to 24 h at 25°C, and store at 3°C in a fresh portion of the same fluid. Wash and dehydrate with isopropanol.

Engfeldt and Hjertquist (*Acta Pathol. Microbiol. Scand.,* **71:** 219, 1967) noted that 0.5% cetylpyridinium chloride in 10% formalin (nonneutralized or neutralized with $CaCO_3$) did not extract significant amounts of S-labeled chondroitin sulfate from epiphyseal plates of long bones of normal and rachitic rats. Likewise, chondroitin sulfate was not lost in specimens fixed in absolute methanol or in Carnoy's fluid. However, acid mucopolysaccharide was rapidly lost from methanol-fixed sections during staining. Szirmai (*J. Histochem. Cytochem.,* **11:** 24, 1963) and Hale [*Nature (Lond.),* **157:** 802, 1946] also recommend Carnoy's fluid for preservation of acid mucopolysaccharides. It is important to recognize that cetylpyridinium chloride blocks anionic groups and the acid mucopolysaccharides cannot be identified with basic dyes such as azure A.

Allen and Porter (*Immunology,* **18:** 799, 1970) conjugated fluorochromes to IgA, IgG, or IgM to localize antigens in porcine intestinal tissues. Methanol (20 min, 24°C) was the fixative of choice for the demonstration of IgM and IgA. For unknown reasons, methanol fixation was best for IgA in cells of the lamina propria, but either acetone (10 min, 24°C) or ethanol (70%, 30 min, 24°C) provided better fixation for the demonstration of IgA in intestinal crypt epithelium. Fixation of IgG was unsatisfactory.

4. METHANOL PLUS CHLOROFORM. When it is desired to test the lipid nature of some tissue component, fix in a 2 : 1 mixture, changing to 1 : 1 on the second change, preferably at 60°C, in screw-capped bottles (paraffin oven) for 12 to 48 or more h, changing fluid daily. Clear in two changes of chloroform and embed in paraffin.

We now routinely use this fixation schedule: until 5 or 6 P.M. in 2 : 1 methanol chloroform at 60°C; a second bath of the same mixture overnight at 20 to 25°C; 1 : 1 mixture for 9 to 11 h at 60°C; overnight in 1 : 2 methanol–chloroform at 20 to 25°C; from 8 A.M. to 3 P.M. at 60°C in the 1 : 2 mixture; then two 1-h changes of absolute alcohol plus ether (1 : 1) at 20 to 25°C; and 1 to 3 days in 1% celloidin in alcohol ether. Clear in two changes of chloroform, thus hardening the celloidin, and embed in paraffin as usual. We have cut serial sections of human cerebellum at 2 μ by this procedure.

Acetone

Acetone was long used as a rapid fixative for brain tissue in rabies diagnosis and for a time was the fixative of choice for preservation of phosphatases and lipases. In these latter technics 24-h fixation at 0 to 5°C was prescribed. Latterly the use of fresh frozen sections, freeze-dry technics, freeze substitution methods, and ice-cold "formol calcium" has been preferred for these purposes.

Acetone has recently come into favor as a solvent for certain metallic salts for use at low temperatures in freeze substitution technics for tissue blocks.

Chang and Hori (*J. Histochem. Cytochem.*, **9:** 292, 1961) in their *section freeze substitution technic* prepare cryostat sections by quick-freezing in liquid nitrogen, section at about −15°C in their open cryostat, and gently further flatten the sections with a camel's hair brush. The still frozen sections are then transferred to prechilled anhydrous acetone and fixed and dehydrated in a closed vial buried in solid carbon dioxide for 12 h or more, up to 4 weeks. The dehydrated sections are then mounted by floating them in chilled acetone over dry ice and picking them up on a chemically clean cover glass which has been predipped in acetone; a needle is used to guide the sections. The cover glasses with adherent sections are then immersed in cold (−70°C) 1% collodion in 30 : 30 : 40 ether-alcohol-acetone. After a few seconds they are drained and dried at 25°C and kept at room temperature until stained. Storage at +25°C for 7 days or at −10°C for 2 months was reported as giving no loss of enzyme activity. This procedure was recommended especially for the study of water-diffusible enzymes.

For mounting without collodionization, the acetone containing the sections is allowed to warm gradually in an empty Dewar flask or in the freezing area of the refrigerator, and then to room temperature. Sections are then transferred to 95% alcohol for 2 min, floated onto coverslips, flattened, and carried into the staining procedure. This variant is recommended for nonenzymatic stains and reactions.

At 3°C acetone dissolves out fats at once; at −70°C much fat may remain demonstrable by Sudan stains after acetone freeze substitution (Hitzeman, *J. Histochem. Cytochem.*, **11:** 62, 1963).

Certain lipid components, galactolipids and phospholipids, are rendered insoluble in fat solvents by an adequate fixation with formaldehyde or glutaraldehyde. They are acetone-insoluble, and this fact may be used in their histochemical identification: Triplicate blocks are fixed (1) in buffered aqueous 10% formol for 24 h; (2) in three changes of absolute acetone for a total of 24 h and then postfixed 24 h in buffered aqueous 10% formol; and (3) in three changes of absolute acetone and thence directly to ligroin or toluene and paraffin. The first two are routinely dehydrated with alcohol, cleared, and embedded in paraffin. Acetone-insoluble lipids are preserved differentially in block 2, nondifferentially in block 1, and lost in block 3. (Lillie and Geer, *Am. J. Pathol.*, **47**: 965, 1965.)

Kaplow and Burstone [*Nature (Lond.)*, **200**: 690, 1963] successfully demonstrated alkaline and acid phosphatases, esterases, peroxidase, and dopa oxidase in blood films fixed 20 to 30 s at 24°C in 60% acetone in citrate buffer at pH 4.2.

Goland et al. (*Stain Technol.*, **42**: 41, 1967) and Zerlotti [*Nature (Lond.)*, **214**: 1304, 1967] advise the following fixative particularly for preservation of carbohydrates and muco-substances:

1. Immerse 2-mm-thick specimens in 0.5% (0.025 *M*) cyanuric chloride in anhydrous methanol containing 1% (0.1 *M*) *N*-methyl morpholine at 23 to 27°C for 18 to 24 h.
2. Wash in methanol; clear in xylene.
3. Embed in paraffin and section.

Lead

Holmgren (*Z. Wiss. Mikrosk.*, **55**: 419, 1938) introduced the use of 4% basic lead acetate as a fixative which preserved tissue mast cells in a number of species in which alcohol fixation had previously been required for this purpose. Sylvén (*Exp. Cell. Res.*, **1**: 582, 1950), Bunting (*Arch. Pathol.*, **49**: 590, 1950) used basic lead acetate for connective tissue mucins. Holmgren and Sylvén both prescribed fixation for 24 h in lead subacetate followed by 1 day in 10% formol.

Three basic lead acetates are listed by Lange: $Pb_2(CO_2CH_3)_3OH$, $Pb(CO_2CH_3)_2 \cdot Pb(OH)_2 \cdot H_2O$, and $Pb(CO_2CH_3)_2 \cdot 2Pb(OH)_2$. All are quite soluble in water, less so in alcohol. The first of them, also called *lead subacetate*, is soluble about 6% in water and slightly soluble in alcohol. It is probable that this was the salt used by Holmgren and Sylvén; we have used it successfully. Both neutral and basic acetates take up CO_2 from the air and form insoluble lead carbonate. This can be dissolved by cautious addition of acetic acid until just clear, or filtered out. Lead nitrate is very soluble in water (38.8% at 0°C, 56.5% at 20°C, 66% at 30°C), but solubility in alcohol is low: 1% in 95% alcohol, 8% in 75%.

Dissolve 8 g lead nitrate in 15 to 20 ml distilled water and add 95 to 100% alcohol gradually until turbidity appears; clear with a little more water and then more alcohol, proceeding thus until a total volume of 100 ml is reached. Formalin, if desired, is included as part of the water.

LILLIE'S ALCOHOLIC LEAD NITRATE–FORMALIN

Lead nitrate	8 g
40% formaldehyde	10 ml
Water ⎱ see text above	11 ml
95% alcohol ⎰	79 ml

Fix 24 h at 25 to 30°C, or longer at 0 to 5°C (2 to 3 days). At −75°C 10 to 14 days may be needed, and probably only about 1% lead nitrate can be used.

Similar solutions in water are more easily prepared and are also quite effective, and good fixation is obtained with the alcoholic solution from which the formalin is replaced by water.

The pH of 4% lead subacetate is 3.8 to 4.2; that of 8% lead nitrate solutions lies between 3.6 and 4. Formaldehyde has little effect on the pH level. With the addition of 2% sodium acetate ($NaCO_2CH_3 \cdot 3H_2O$) the pH of 8% aqueous $Pb(NO_3)_2$ rises to 5.15.

Recently Lillie et al. (*J. Histochem. Cytochem.*, **21**: 441, 1973, and unpublished studies) have used neutral lead acetate and nitrate at 0.2 M concentration with and without inclusion of 10% formol, alone and with added sodium acetate or acetic acid. Some of these mixtures have proved quite satisfactory for "water-soluble" mast cell granules (hog, guinea pig, cat). Generally 4- to 6-h fixation is adequate, and as with sublimate fixations tissues are then transferred for a day or two to 80% alcohol. The formol variants are perhaps more satisfactory for mast cells.

Neutral lead acetate (sugar of lead) at 0.2 M (7.6% $PbAc_2 \cdot 3H_2O$) (called Pb-3) or with included 10% formol (Pb-8) were the variants most used. Pb-15, 0.2 M (6.6%) lead nitrate with 6% acetic acid and 8.5% $NaNO_3$ added to prevent hemolysis, is an interesting example of a formol-free heavy-metal fixative of lower cost than acetic sublimate.

Lead citrate has found popular usage in electron microscopy recently. Watson (*J. Biophys. Biochem. Cytol.*, **4**: 727, 1958) and later Dalton and Zeigel (*J. Biophys. Biochem. Cytol.*, **7**: 409, 1960) used aqueous solutions of lead salts as stains for thin sections to enhance scatter of electrons and thereby increase contrast of electron photomicrographs. Although saturated solutions of lead hydroxide used by Watson stained sections with more contrast than did either lead acetate or monobasic lead acetate, insoluble lead carbonate formed rapidly upon exposure to air, resulting in troublesome precipitates.

The reagent is made as follows: Add 1.33 g lead nitrate [$Pb(NO_3)_2$] and 1.76 g sodium citrate [$Na_3(C_6H_5O_7) \cdot 2H_2O$] to 30 ml distilled water in a 50-ml volumetric flask. Shake vigorously for 1 min, then intermittently for 30 min. Add 8 ml 1.0 N NaOH, dilute to 50 ml, and mix by inversion of the flask. The stain is generally used as it is, or it may be diluted. The pH is 12 ± 0.1. If faint turbidity develops, it may be removed by centrifugation. The solution is stable for 6 months if stored in tightly stoppered polyethylene bottles.

For staining, a drop of the solution is applied to an embedded section for 5 to 10 min (Epon) or 15 to 30 min (Maraglas or Araldite); however, optimal staining time will vary depending upon the fixative employed previously. Glutaraldehyde- and osmium-fixed tissues require less time. Excessive stain can be removed by application of 0.01 to 0.02 N NaOH.

Citrate apparently forms stable complexes with cationic "basic" lead salts (log $Ka = 6.5$). Reynolds (*J. Cell Biol.*, **17**: 208, 1963) speculates that lead in lead citrate is capable of binding to cysteine, *o*-phosphate, and pyrophosphate. EDTA prevents staining. Glycogen stains intensely and ribosomes are less stained.

Venable and Coggeshall (*J. Cell Biol.*, **25**: 407, 1965) offer an easier method of preparation of lead citrate: Simply add 10 to 40 mg lead citrate and 0.1 ml 10 N NaOH to 10 ml distilled water. Shake the vial vigorously until all lead is dissolved. Store the mixture in a plastic bottle tightly closed to avoid lead carbonate formation.

Bradbury and Stoward (*J. Microsc. (Oxf.)*, **83**: 467, 1964) stained sections with lead citrate according to the method of Reynolds (*J. Cell Biol.*, **17**: 208, 1963) after formazans had formed, indicating enzyme activity. They state that the lead staining is very selective and results in marked contrast with specific identification of enzyme activity.

Strictly speaking the lead hydroxides $Pb(OH)_2$ and $2PbO \cdot Pb(OH)_2$ are almost completely insoluble in water, 0.016 and 0.014%, respectively (Lange). On the basis of its mode of preparation the reagent appears to be sodium plumbite. This can be conveniently prepared by cautious addition of a strong (50%) lead nitrate solution to a volume of 0.1 N NaOH, shaking between each addition to dissolve the white precipitate until with the last drop a faint opalescence or permanent turbidity is attained. This solution has been used by Lillie as a histochemical reagent for cystine and cysteine in hard keratins. The reaction dates back to Salkowski (*Physiologisches Praktikum*, p. 94, 1893). The solution must be protected from air, to prevent lead carbonate formation.

Mercuric Salts

Many of the older formulas specify saturated sublimate. Taking the German laboratory temperature of about 1900 to be 15°C, this means about 5.5%. A convenient standard here would be about 5.6%, or 0.2 M. This facilitates comparison with other heavy metals.

Increasing the mercury content above this level is of questionable value. In view of the present high price of mercury salts and the problem of environmental pollution, used solutions should be poured into a special container and the mercury precipitated as oxide or carbonate, filtered out, and recovered as chloride.

Usually the salt used is mercuric chloride ($HgCl_2$), and this most often in saturated (5 to 7%) solution in water. Mercuric chloride rapidly hardens the outer layers of tissues into a quite hard, white mass. It penetrates poorly after the first 3 or 4 mm; consequently, tissue should be cut in thin slices, not more than 5 mm in thickness. It hardens cytoplasm well, preserves its affinity for acid dyes better than formaldehyde, and increases the affinity of cellular, bacterial, and rickettsial chromatin for basic aniline dyes less than formalin. Nuclear chromatin tends to appear in finer particles with mercuric chloride than with formaldehyde. Because of the relatively increased affinity of cytoplasm for acid dyes, the differences in basophilia and oxyphilia of regenerating, mature, and necrosing cells are less well shown than with formaldehyde. Cytoplasmic structure of renal epithelium, fibrin, connective tissue, cross striae of muscle, and other features may be better shown.

Because of its poor penetration and the shrinkage of tissues produced by it, mercuric chloride is seldom used alone. It is combined with acetic and other acids, with chromates, with formaldehyde, with alcohol, and with various mixtures.

Except for the susa mixtures, material fixed in mercuric chloride fixatives generally requires treatment with iodine to remove granular black precipitates which are distributed throughout the tissues. This may be done by soaking blocks of unembedded tissue in 70% alcohol containing enough tincture of iodine to color it a fairly deep reddish brown or port-wine color (about 0.5% iodine). It is necessary to inspect daily and add a few more drops of tincture of iodine to restore the color, until the alcohol is no longer decolorized.

However, in pathologic practice (particularly when conservation of time is important) it is customary to embed and section without iodizing and to remove the precipitates from the deparaffinized sections before staining. Sections are treated 5 to 10 min in 0.3 to 0.5% iodine in 70 to 80% alcohol (a 1:15 or 1:20 dilution of the USP Tincture of Iodine), rinsed in water, decolorized 1 to 5 minutes in 5% sodium thiosulfate, and washed 5 to 10 minutes in running water before staining.

Lee preferred the practice of iodization before embedding and stated that serious artifacts were produced during embedding if the mercury was not first removed. Mallory condemned this process, preferring to iodize after sectioning, and stated that prolonged iodization of blocks impaired the staining quality of cells. We have used both methods and can see little important difference. Generally we prefer to iodize after sectioning, because of the time factor.

After mercuric chloride fixation the excess fixative should be washed out, preferably with 70 to 80% alcohol, except when combined with chromates. In the latter case washing overnight in running water is recommended. Storage in 70% alcohol is usually recommended.

On long storage in alcohol the material becomes quite hard and brittle. Hence it seems better to complete the dehydration of material not wanted at once and either clear and embed it in paraffin or store it in thin cedar oil. This fluid keeps the tissue reasonably soft and seems to improve its quality for sectioning.

FORMULAS

1. SATURATED AQUEOUS MERCURIC CHLORIDE.   The usual formulas are based on the late-nineteenth-century German laboratory temperature of about 15°C, at which temperature solubility is about 5.6%. The solubility curve is almost a straight line from 10 to 40°C, actual values at 20°C and 30°C being only 0.1% less than thus calculated (Table 3-5).

    This solution is quite acid (pH 3). Spuler (Ehrlich's "Encyklopädie") recommended addition of 0.5 to 0.75% sodium chloride, thereby increasing the solubility of mercuric chloride to about 9%. Aqueous mercuric acetate (5%) is also acid (pH 3.2).

2. ACETIC SUBLIMATE.   Lee added 1% acetic acid to saturated mercuric chloride solution; we have used 2% acetic acid, which gives a pH of 2.3, and 5% acetic acid in 5% mercuric chloride. Losses of acetic acid–soluble proteins are somewhat increased, and sharpness of chromatin staining is improved.

3. BUFFERED SUBLIMATE (B-4).   Latterly we often use a formaldehyde-free variant of the B-5 fixative, containing 5.4 or 6 g mercuric chloride and 1.25 g anhydrous sodium acetate per 100 ml. The pH is about 6, and the fluid serves for fixation of Paneth, pancreatic, and other zymogen granules, when for histochemical reasons it is desired to avoid formaldehyde. Enterochromaffin does not appear to be preserved, but the fluid has been shown to oxidize enterochromaffin in formaldehyde-fixed sections to a quinonoid form, from which it can be reduced by $Na_2S_2O_4$. Mercurial precipitate is sparse or absent, and iodination can be omitted.

4. SATURATED ALCOHOLIC MERCURIC CHLORIDE (33% at 25°C in 99% alcohol, Lange). This solution fixed tissues very rapidly. Much weaker solutions are used currently in freeze substitution technics; Feder and Sidman used a 1% solution at −75°C for blocks not over 3 mm thick which were first frozen in 3 : 1 propane-isopentane at −170°C (using liquid nitrogen as the cooling agent). Fixation took a week, longer for thicker blocks, when done in a dry-ice chest. On completion of the fixation exposure, the fluid is allowed to warm to room temperature, and dehydration and embedding are completed as usual.

    A similar solution in acetone can probably be used, but it must be freshly prepared and promptly chilled to −75°C; at 25°C mercuric chloride and acetone react in a few hours to produce a copious black precipitate.

5. SCHAUDINN'S MERCURIC CHLORIDE ALCOHOL ("sublimate alcohol"), cited from Giemsa (*Dtsche. Med. Wochenschr.*, **35**: 1751, 1909):

Saturated aqueous mercuric chloride	2 parts
Alcohol (absolute)	1 part

Fix 24 h, replace fixative with fresh solution, and continue fixation another 24 h. Then transfer to 70% alcohol.

TABLE 3-5   MERCURIC CHLORIDE CONTENT OF SATURATED SOLUTIONS*

0°C	5°C	10°C	15°C	20°C	25°C	30°C	35°C	40°C
3.6	4.2	4.8	5.7	6.5	7.4	8.3	9.3	10.2

* Italicized figures are interpolated.

We consider the cytoplasmic fixation achieved with this fluid inferior to that achieved with Zenker's fluid. Giemsa recommended it for wet smear fixation, and the fixative has often been attributed to him.

5a. HUBER'S FLUID

Mercuric chloride	5 g
Trichloroacetic acid	15 g
95% alcohol	100 ml

This fluid is said to give no mercurial precipitate and is recommended by Peters (*Stain Technol.*, **33:** 47, 1958) for nerve material destined for his protein-silver nerve impregnation method. The customary iodine treatment is omitted for this purpose.

6. OHLMACHER'S FLUID (Ehrlich's "Encyklopädie") (a), FLUID OF CARNOY AND LEBRUN (Cowdry, McClung, Lee) (b):

100% alcohol	(a) 32 ml	(b) 15 ml
Chloroform	(a) 6 ml	(b) 15 ml
Glacial acetic acid	(a) 2 ml	(b) 15 ml

These may be kept as stock solution. Before using, add 8 g mercuric chloride to the above 40-ml mixture of Ohlmacher's fluid. Lebrun's is saturated with mercuric chloride. Add 4 g, which is an excess. These mixtures fix very rapidly. Ohlmacher's fluid penetrates about 1 mm in the first 10 to 15 min, 2.5 mm in 2 to 3 h. Cut blocks quite thin before fixation. Handle while in the fluid with instruments previously dipped in hot paraffin to avoid contact of the mercurial solution with metal. Wash in two changes of 100% alcohol, about 1 h each; clear in cedar oil and store surplus tissue in this fluid. Ohlmacher's fluid gives comparatively little mercurial precipitate, and there is little difference in appearance between material fixed 30 min and that fixed 150 min. It preserves glycogen poorly in comparison with Carnoy.

FORMALDEHYDE SUBLIMATE FORMULAS

7. HEIDENHAIN'S SUSA FLUID (*Z. Wiss. Mikrosk.*, **33:** 232, 1916; Romeis) consisted of 4.5 g mercuric chloride, 0.5 g sodium chloride, 20 ml 40% formaldehyde, 80 ml distilled water, 4 ml glacial acetic acid, 2 g trichloroacetic acid. A 12-h fixation is followed by washing in 95% alcohol. Mercurial precipitates are said not to be formed. Glycogen was not preserved in the few trials we made of this fluid.

8. ACETIC MERCURIC CHLORIDE–FORMALIN (pH 1.9 to 2):

Mercuric chloride	6 g
Glacial acetic acid	5 ml
40% formaldehyde solution	10 ml
Distilled water	85 ml

Add the acetic acid and formalin to the mercuric chloride solution at time of using.

9. SODIUM ACETATE–MERCURIC CHLORIDE–FORMALIN ("B-5") (pH 5.8 to 6):

Distilled water	90 ml
Mercuric chloride	6 g
Sodium acetate (anhydrous)	1.25 g
40% formaldehyde added at time of using	10 ml

If the salt available is the trihydrate $NaCO_2CH_3 \cdot 3H_2O$, use 2.074 g. Calcium acetate as monohydrate, 2 g, has been substituted, but it may be necessary to filter to remove the slight turbidity resulting probably from the presence of a small amount of calcium hydroxide in the salt.

Fixation with these mercuric chloride–formalin mixtures appears to be adequate in 12 to 24 h, but even rodent livers are not overhardened in 6 to 7 days. Tissues should be transferred directly to, and stored in, 70 or 80% alcohol.

10. ROMEIS' FLUID. This fluid was highly recommended by Puchtler and Sweat (*Histochemie*, **4:** 197, 1964) both as a primary fixative and for postmordanting after other fixations in the Heidenhain iron hematoxylin technic. As used by them, the fluid was prepared with:

Saturated aqueous mercuric chloride	50 ml
5% trichloroacetic acid	40 ml
Formol (37% formaldehyde)	10 ml

The formol should be added at the time of using. The mixture is quite acid (pH 2.1). Fixation intervals of 12 to 24 h for thin blocks are probably adequate. Puchtler and Sweat treated sections 15 to 16 h at 60°C and then removed mercury precipitates with iodine and thiosulfate as usual.

11. FORMOL SUBLIMATE. 5.43 g mercuric chloride, 10 ml formol, 90 ml distilled water, pH about 2.9. Prepare fresh at time of using. Or dissolve mercuric chloride in 10% formol, which may yield a somewhat lower pH.

12. GLUTARALDEHYDE SUBLIMATE. 5 g mercuric chloride, 10 ml 50% glutaraldehyde, 90 ml distilled water. Add glutaraldehyde at time of using.

Spuler (Ehrlich's "Encyklopädie") recommended for blood a strong mercuric chloride solution to which he added 1% acetic acid and 10% formalin. W. H. Cox (*Anat. Hefte*, **10:** 99, 1898) employed for nerve cell (Nissl) granules a mixture of 30 ml saturated aqueous mercuric chloride solution, 10 ml formalin, and 5 ml glacial acetic acid. Stowell (*Arch. Pathol.*, **46:** 164, 1948) recommended Stieve's fluid in studies of liver regeneration. This consists of 76 ml saturated aqueous mercuric chloride solution, 20 ml formalin, and 4 ml glacial acetic acid. Dawson and Friedgood (*Stain Technol.*, **13:** 17, 1938), in studies of the anterior lobe of the hypophysis, used a mixture of 90 ml physiologic saline solution saturated with mercuric chloride (10% or more?) and 10 ml formalin. We have used mixtures of 10 ml formalin, 5 ml glacial acetic acid, 6 g mercuric chloride, and 85 ml water and of 10 ml formalin, 2 g calcium or sodium acetate, 6 g mercuric chloride, and 90 ml water. G. Brecher in Lillie's laboratory found the last mixture (B-5) excellent for differentiation of pancreatic islet cells.

Unbuffered mercuric chloride–formalin mixtures are quite acid (about pH 2.9); the acetate formulas are much less so (pH 5.9–6, 5.85 for sodium and calcium formulas); but the acetic formula is slightly more acid (pH 1.9). The stock solutions appear to be quite stable, but the formaldehyde mixtures are unstable and should be freshly prepared. Formaldehyde reduces mercuric chloride to calomel and metallic mercury, evident as a white to gray or black-flecked precipitate.

Mann (*Z. Wiss. Mikrosk.*, **11:** 479, 1894) gives two mercuric chloride formulas which are still sometimes used. His *osmic-sublimate* is composed of equal volumes of 1% aqueous osmium tetroxide and of 0.75% aqueous sodium chloride solution saturated with mercuric chloride. In his *tannin-picrosublimate* he adds 1 g each of picric acid and of tannin to 100 ml of Heidenhain's 0.75% aqueous sodium chloride solution saturated with mercuric chloride. These fluids were originally used for the study of nerve cells. Lee as early as 1896 changed the solvent to physiologic salt solution, and Cowdry prescribes 0.85% sodium chloride.

Sublimate Dichromate Fluids

Both the original *acetic sublimate bichromate* fluid of K. Zenker (1894) and the formalin variants introduced by Spuler (Ehrlich's "Encyklopädie," 1903) and Helly (*Z. Wiss. Mikrosk.*, **20:** 413, 1904) have enjoyed a wide popularity and have undergone many, usually minor, modifications. The original Zenker's fluid was made by saturating Müller's fluid with mercuric chloride and then adding, just before use, 5% glacial acetic acid. This precaution was necessary in 1894 because acetic acid was then recovered from the pyroligneous acid obtained from the destructive distillation of wood, which was composed of nearly equal amounts of methanol, acetone, and acetic acid. Each of the major constituents, as prepared in commercially pure form, was often considerably contaminated with the other two. Zenker's fluid made with modern reagent-grade acetic acid is quite stable over prolonged periods, even at elevated temperatures (3 months at 60°C). At 100°C acetic acid reduces potassium dichromate slowly.

The formaldehyde and formic acid mixtures start to darken in some hours and become quite dark brown and turbid in a day or two. Hence it is necessary with all these mixtures to defer addition of formalin and of formic acid to the moment of using.

The most common variant from the above formulas is the omission of sodium sulfate. With higher laboratory temperatures than the traditional German 15°C, some workers have added 60 or even 70 g mercuric chloride. It is questionable that the additional sublimate is necessary or even valuable. It probably only adds to the cost of the fluid. In view of the quite adequate mercurial fixation achieved with fluids containing only 1 to 3% mercuric chloride, even 5% is probably more than necessary.

Fixation times vary from 4 to 24 h; longer periods are said to produce overhardening and are not generally recommended. However, we have seen rodent livers fixed 6 to 7 days in Spuler's formalin variant which cut well and gave quite satisfactory histologic staining. If very thin (2 to 3 mm) blocks are cut, 2 to 3 h may well suffice. The premixed acetic Zenker's fluid can be placed on the Technicon and used repeatedly, being changed perhaps weekly to avoid excessive dilution and contamination by blood and serum.

In all cases blocks should be thin; hollow organs should be opened. All Zenker fluids penetrate solid tissue slowly, and the acidic and formaldehyde components penetrate faster than the metallic salts, so that in overthick blocks one soon reaches a zone, 5 to 10 mm deep and below, where the fixation effect appears to be predominantly that of acetic acid or formalin.

ACIDITY. The acetic Zenker fluids are generally quite acid (pH 2.3), and this acidity is not affected by the presence of Na_2SO_4. The formalin variants are a little more acid (pH 3.4) than corresponding fluids without mercuric chloride (pH 3.8–3.95), but they are less acid than unbuffered sublimate without dichromate (pH 3).

The formic acid variant, attributed to M. J. Guthrie in the ninth edition of Lee's "Microtomists' Vade-Mecum" (1928), is an active decalcifying agent and was redevised for that purpose (1943) in our laboratory. Otherwise, fixation effects are similar to those of Zenker's fluid. Formic acid reduces mercuric chloride ($HgCl_2$) to calomel (mercurous chloride, or $HgCl$), so that calomel precipitation starts in about 3 h. Potassium dichromate solution without mercuric chloride changes color from orange to dark red-brown in about 48 h, but without precipitation.

Bencosme (*Arch. Pathol.*, **33:** 87, 1952) prescribed 24 h or *more* washing in running water. Embryonic and very young animal tissues were then dehydrated in successive 2-h changes of 40, 50, 60, 70% alcohol; 3 h in 90% alcohol; and 10 h in 95% alcohol; the last two containing 0.25% iodine. Dehydration was completed with 3- and 5-h changes of 100% alcohol. Tissues from older animals were transferred directly from the wash water to 0.25%

TABLE 3-6 COMPOSITION OF ZENKER AND ZENKER FORMOL FLUIDS

	Zenker (Gray)	Zenker (Enzyk)	Zenker (AFIP)	Zenker (Lillie)	Spuler (Enzyk)	Helly	Bencosme	Guthrie (Lee)	Zenker base	Lillie et al. 1973
Distilled water, ml	960	1000	1000	950	1000	1000	800	1000	1000	950
Potassium dichromate, g	20	25	25	25	25	25	20	25	25	36
Sodium sulfate, g	10	10	10		10	10	8	10	10	60 (NaCl)
Mercuric chloride, g	49.2	50	50	50	50	50	40	50	50	54
Glacial acetic acid, ml				50						50
Add at time of use:										
Glacial acetic acid, ml	50	50	52							
Formic acid, 90% ml								50		
40% (w/v) formaldehyde, ml					100	50	200			

Source: P. Gray, "The Microtomist's Formulery and Guide," McGraw-Hill, New York, 1954; A. B. Lee, "The Microtomists' Vade-Mecum," 10th ed., edited by J. B. Gatenby and T. S. Printer, Blakiston (McGraw-Hill), Philadelphia, 1937; Lillie et al., *Am. J. Clin. Pathol.*, **59:** 374, 1973.

iodine in 95% alcohol for 24 h, followed by four changes of 100% alcohol, the first two of which were used, the last two fresh.

Embryonic tissues were cleared through 1 : 1 and 3 : 1 toluene–100% alcohol mixtures (2 h each), and four changes of toluene (3, 10, 2, and 4 h); adult tissues, directly from 100% alcohol in four changes of toluene (8, 16, 8, 16 h). Used toluene was employed for the first two changes; fresh toluene was used for the last two changes in both cases.

Bencosme infiltrated with a mixture of 9 parts 50 to 52°C paraffin and 1 part yellow beeswax. Embryonic tissues were infiltrated 2 h at room temperature in equal parts of toluene and paraffin and were then transferred to the melted beeswax paraffin at 55°C. With adult tissues this step was omitted, and the tissues were infiltrated directly in beeswax paraffin. Three ovens were required, and tissues were kept in them for 8 h, 16 h, and 1 to 5 days, respectively. It is claimed that the toluene is not adequately eliminated unless three separate ovens are employed.

We cite this schedule as an example of the elaborateness which cytologists can achieve. Because of the high concentration of formalin in this mixture, we would advise very thin blocks. Otherwise considerable differences of fixation of surface and inner zones are apt to be observed. The solvent toluene can be eliminated from the tissue far more expeditiously, and probably more effectively, by infiltration in vacuo (10 mm mercury) for 30 to 60 min. The process can be hastened still more by substituting a low-boiling petroleum ether for the toluene. The latter is also less toxic than toluene.

SALT ZENKER FLUID. In 1973 Lillie, Pizzolato, and Vacca (*Am. J. Clin. Pathol.*, **59**: 374, 1973) proposed a new variant of Zenker's fluid on the basis that it had been observed that addition of 1 M NaNO$_3$ or NaCl prevented hemolysis by 1 N acetic acid in deamination controls (Lillie, *Blood*, **7**: 1042, 1952). Addition of 5.8 or 6% NaCl prevented hemolysis by

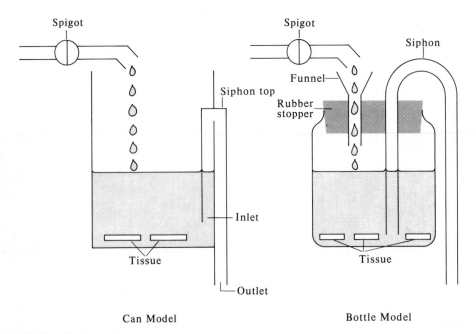

Can Model   Bottle Model

Fig. 3-1 Siphon washer.

various fixative mixtures containing the often used 5% acetic acid. In Zenker's fluid addition of 6% NaCl not only prevented lysis of erythrocytes in the center of the block but also preserved perfectly the acetic acid–sensitive pancreatic zymogen granules. For comparison with other metallic fixatives $HgCl_2$ content was set at 0.2 M or 5.4%. Because of a long experience with Kose's variant of Orth's fluid, $K_2Cr_2O_7$ content was set at 3%. The sodium sulfate was replaced by 6% NaCl; the acetic acid was left at the standard 5%.

A 5- to 6-h fixation at 24°C was used; sometimes 18 to 24 h at 5°C was substituted for convenience.

For all these Zenker fluid variants, wash overnight in running water, or in 2 to 3 changes of water, 1 to 2 h each on the Technicon, then store in 70% alcohol, or proceed at once with dehydration, clearing, and embedding.

For washing of tissue blocks, which is required for the foregoing chromate mercury fixations as well as for the following dichromate methods, the use of a siphon washer (Fig. 3-1) has proved quite convenient. Operating on the same principle as commercial pipette washers, the siphon washer probably removes excess chromate from tissues more rapidly and more surely than the traditional continuous-flow washing, especially when the water inlet tube is not taken all the way to the bottom of the washing jar. We have seen a stratum of deep yellow dichromate solution in the bottom of a bottle, surrounding the tissue blocks, after overnight washing.

Chromates and Chromic Acid

The most commonly used primary chromate fixations are Orth's and Möller's (Regaud's) mixtures. We have used Orth's fluid extensively for routine work. It is equal to formaldehyde for the study of early degenerative processes and necrosis. It is perhaps superior to formaldehyde for demonstrating rickettsiae and bacteria. Chromaffin takes its characteristic brown tone, and its basophilia is well brought out with Orth's fluid. Myelin is better preserved, and the pericellular shrinkage seen about the pyramidal cells after the use of formalin is avoided. Hemosiderin is less well preserved with Orth's fluid than with buffered neutral formalin.

The *fixation period* for Orth's fluid should be between 36 and 72 h. Human tissues are usually given 48 h; brain is perhaps better with 72 h, though 48 h is adequate. Mouse tissues appear to be overhardened in 48 h, and 24 h is often adequate. Rat and guinea pig tissues should be fixed longer, perhaps 36 to 48 h. After fixation, tissues are washed overnight in running water. The traditional storage fluid is 70% alcohol, but tissues may be kept at least a year in 10% formalin, so that material for fat stains can continue to be available. Glycogen is well shown by the Bauer method after Orth's fluid fixation.

Romeis' acetic variant of Orth's fluid gives similar pictures, except that partial hemolysis occurs, more iron pigment is lost, and nuclear chromatin is more sharply defined. This variant was used routinely by Ophüls for many years. He commonly fixed in it for 3 to 4 days. Tellyesniczky used an acetic dichromate mixture without formalin. This mixture appears to be stable.

Lillie employed Kose's variant of Orth's fluid, which contains 3 g potassium dichromate in place of 2.5 g/100 ml, and found that it gives greater hardening in the same period of time, without appreciable change in the overall picture. This fluid, like Zenker's, renders ribonucleic acid (RNA) relatively highly resistant to specific enzyme digestion. This effect is less evident when 2% calcium or sodium acetate is added to Kose's fluid or to Spuler's formalin Zenker mixture.

Möller's or Regaud's fluid is similar in general effect to Orth's but hardens more rapidly

TABLE 3-7 BUFFERED DICHROMATE SOLUTIONS

1 N HCl, ml	0.2 M KH$_2$PO$_4$, ml	0.2 M Na$_2$HPO$_4$, ml	0.2 M Acetic, ml	0.2 M NaAcet, ml	6% K$_2$Cr$_2$O$_7$, ml	pH
26	24				50	1.46
15	35				50	2.01
7	43				50	2.52
?6	47 4				50	3.00
			46.5	3.5	50	3.51
			40	10	50	4.00
			31	19	50	4.50
			13	37	50	5.00
	45.5	4.5			50	5.01
				50	50	5.52
	37.0	13.0			50	5.50
	28.0	22.0			50	5.98
		50			50	6.27

and to a harder final consistency. With this 20% formalin dichromate mixture, as with aqueous 20% formalin and with Bouin fluids containing 20 to 25% formalin, an outer zone of hyperoxyphilia of tissues is produced. In this zone nuclear staining is apt to be poor, and cytoplasm will stain more strongly with acid dyes there than in the center of the block.

Sometimes it is desirable to modify the naturally quite acid pH level (3.8 to 3.95) of potassium dichromate solutions. For example, iron-containing pigments are better preserved at higher pH levels. Lower pH levels seem necessary to elicit the chromaffin reaction of the enterochromaffin cells. Addition of 5% of acetic acid to a 3% potassium dichromate solution lowers the pH to 2.9. This solution is stable for months, but the further addition of formaldehyde quite promptly causes the usual reduction changes: first darkening to dark brown, then to brownish green. Hence formalin should not be added to acidified dichromate mixtures until the time of using. Formulas for the buffering of dichromate solutions appear in Table 3-7. Adding 10 parts 40% formaldehyde to 90 parts 3% dichromate produces little change of pH level: perhaps a rise of 0.05 below pH 4; none above. The upper pH levels do not darken on addition of formaldehyde, but at pH 6.3 some dark green precipitate is formed in 18 to 24 h, the solution remaining clear chromate yellow and smelling strongly of formaldehyde. The more acid pH 5.5 acetate solution shows no precipitation.

FORMULAS

1. MÜLLER'S FLUID (Ehrlich's "Encyklopädie"). Dissolve 2.5 g potassium dichromate and 1 g sodium sulfate in 100 ml distilled water.
2. ORTH'S FLUID (*Berl. Klin. Wochenschr.*, **33**: 273, 1896). To 100 ml Müller's fluid add 10 ml 37 to 40% formaldehyde solution at the time of using. We regularly use 2.5% potassium dichromate in place of Müller's fluid; Mallory used 2 to 2.5%. Romeis and Schmorl reduced the Müller's fluid to 90 ml. Fix 1 to 4 days, usually 2. Wash overnight in running water.
3. ROMEIS' AND OPHÜLS' VARIANTS. Romeis used 85 ml 3% potassium dichromate, 10 ml formalin, and 5 ml glacial acetic acid. Ophüls used 100 ml Müller's fluid; we use 100 ml 2.5% potassium dichromate in place of Romeis' 85 ml 3%. The acid and formalin were added immediately before using. The mixture darkens rapidly. Fix 6 to 24 h according to Romeis, 2 to 4 days according to Ophüls. Wash overnight in running water.

4. CIACCIO'S FLUID. This appears to have been a somewhat variable mixture of potassium dichromate solution, formalin, and a little formic acid or acetic acid, with or without mercuric chloride. Ciaccio changed it according to need from time to time, and it is sometimes uncertain which mixture he used in any given study.

Reference	Potassium dichromate, g	40% HCHO, ml	Acetic acid	Formic acid, gtt	Water, ml
Anat. Anz., **23:** 401, 1903	5	10		3–4*	100
	3	10		3–4	100
Zieglers Beitr., **50:** 317, 1909	4	20		10–15	80
	4	20	10–15† gtt		80
Huber, *Mikroskopie*, **12:** 91, 1957	5	20	1 ml		80

* 0.15–0.22 ml.
† 0.25–0.4 ml.

Ciaccio (and Huber) usually fixed for 3 days and postchromed in 3 to 5% potassium dichromate for a further 3 to 7 days. We usually employed Kose's fluid without added acid and postchromed in 3 to 5% potassium dichromate for intervals of 4 to 10 days, according to the material being studied and the time available. The stronger solution required the shorter time.

Huber (*Mikroskopie*, **12:** 91, 1957) also cites *Sanfelice's fluid* as composed of 64 ml 1% chromic acid to which are added at time of using 4 ml glacial acetic acid and 32 ml 40% formaldehyde. He fixed 24 h at (European) room temperature and washed 1 to 2 days in running water.

5. KOSE'S FLUID (Ehrlich's "Encyklopädie"; *Sitzb. Dtsch. Naturwiss. Vereins "Lotos," Prag*, **46(N.F. 18):** 224, 1898). To 90 ml 3% potassium dichromate add 10 ml 37–40% formaldehyde solution at time of using (pH 3.6–3.9). Fix 1 to 3 days. Wash 16 h in running water.

6. BUFFERED ORTH'S FLUID. Dissolve 25 or 30 g potassium dichromate and 16.2 g sodium acetate crystals in 1000 ml water. To 90 ml of this fluid add 10 ml 40% formaldehyde solution at time of using (pH 5.5). Fix 1 to 3 days. Wash 16 h in running water.

7. TELLYESNICZKY'S ACETIC DICHROMATE (*Arch. mikrosk. Anat.*, **52:** 202, 1898). To 100 ml 3% potassium dichromate add 5 ml glacial acetic acid at time of using. (The mixture keeps well for months after mixing.) Fix 1 to 2 days. If simultaneous decalcification is desired, substitute 5 ml 90% formic acid for the acetic. This latter mixture must be fresh; it darkens in some hours. A 6% acetic acid, changed daily for 2 days, followed by daily changes of 6% acetic acid to a negative oxalate test should go well.

8. MÖLLER'S, OR REGAUD'S, FLUID. To 80 ml 3% potassium dichromate add 20 ml 40% formaldehyde solution at time of using. Fix 1 to 2 days. Möller (*Z. Wiss. Zool.*, **66:** 69, 1899) prescribed 1-day fixation, followed by 3 to 4 days in 3% potassium dichromate. Regaud (*Arch. Anat. Microsc. Morphol. Exp.*, **11:** 291, 1910) recommended 4 days, and followed it with 8 days' further chromation in 3% potassium dichromate for mitochondria. This seems excessive for general purposes. In our experience the prolonged chromation interferes with nuclear staining. However, certain methods for Golgi substance and for mitochondria call for the long chromate treatment. Ehrlich's "Encyklopädie" (1903) and Regaud cite this fluid as Möller's; Mallory, Langeron, Romeis, Conn and Darrow, Lee, and Cowdry all call it Regaud's.

9. KOLMER'S FLUID. For fixation of eyes, Walls (*Stain Technol.*, **13**: 69, 1938) recommends Kolmer's fluid. Mix just before using.

5% aqueous potassium dichromate solution	20 ml
10% formalin	20 ml
Glacial acetic acid	5 ml
50% trichloroacetic acid	5 ml
Saturated aqueous (10%) uranyl acetate	5 ml

Fix 24 h, wash 6 h in water or 25% alcohol, followed by 6-h baths in ascending alcohols: 35, 50, 70, 95, 100%.

10. HELD'S FLUID. To 100 ml 3% potassium dichromate solution add 4 ml 40% formaldehyde and 5 ml glacial acetic acid, and allow to ripen for some weeks until the solution is green. In this fluid the formaldehyde is partly oxidized to formic acid; the chromic acid is partly reduced to chromic salts; and the pH rises from an initial level of about 3 to about 3.8 (Romeis).

Baker (*Q. J. Microsc. Sci.*, **106**: 15, 1965) noted severe distortion and disruption of cells at the electron microscopic level in tissues fixed in either 1.5 to 2.5% potassium dichromate or in Müller's fluid for 1 to 20 days.

McCrae (*J. R. Microsc. Soc.*, **87**: 457, 1967) fixed mouse pancreas in 5% potassium dichromate in distilled water or 0.75% NaCl for 3 to 24 h at either 2 or 24°C, embedded tissues in Araldite, and examined them at the ultrastructural level. Although some earlier light microscopists have recommended dichromate as an excellent fixative for at least certain tissues, McCrae could find no advantages for chrome fixation at the electron microscopic level. Cytoplasmic and nuclear structures could scarcely be recognized. In Lillie's judgment even 5% potassium dichromate, without additive, gave quite inferior cellular fixation in studies of rabbit gastric enterochromaffin cells. Repeating R. Heidenhain's original demonstration (*Arch. Mikrosk. Anat.*, **6**: 368, 1870) he used a 30-day fixing period (unpublished data). In the 1860s and 1870s potassium dichromate did have advantages over alcohol and chromic acid.

Blackstad et al. (*Histochemie*, **36**: 247, 1973) obtained x-ray diffraction patterns of precipitates of tissues treated with potassium dichromate and mercurous or mercuric nitrate for identification of the Golgi complex. Mercurous chromate, Hg_2CrO_4, was identified after mercurous nitrate treatment, and mercuric oxide chromate, $Hg_3O_2CrO_4$, was observed after mercuric nitrate treatment.

Picric Acid

Picric acid gives a quite hard fixation which is excellent for many purposes. Aniline blue methods of the Mallory type, with its Heidenhain and Masson variants, do well after this fixation. Unless very thorough washing of sections is done, it interferes with azure-eosin stains. It penetrates well and fixes small objects rapidly.

Feder and Sidman (*J. Histochem. Cytochem.*, **6**: 401, 1958) recommended 5% picric acid in alcohol in their freeze substitution method. Fix 3-mm blocks 7 days at −75°C; wash in alcohol at +25°C, and embed in paraffin as usual.

The most popular picric acid fixative is Bouin's fluid. Lee recommended not more than 18 h fixation; Masson, up to 3 days; Cowdry, 24 h; and we have found 1 to 2 days satisfactory. Gray condemns it.

This fluid contains 75 ml 1.2% (saturated) aqueous picric acid solution, 25 ml formalin (40% formaldehyde), and 5 ml glacial acetic acid (Ehrlich's "Encyklopädie").

It is quite common practice to add an excess of picric acid crystals to the stock bottle of Bouin's fluid. The resultant fluid should be somewhat more acid. Allen's B-15 modification contains in addition 2 g urea and 1.5 g chromic acid, while in his PFA3 modification the formalin is reduced to 15 ml, the acetic acid increased to 10 ml, and 1 g urea is added (Lee). We have found that the superficial hyperoxyphilia seen in Bouin's fluid–fixed tissues when 20 to 25% formalin is included may be obviated by decreasing the formalin content to 10%. On substitution of 90 to 95% formic acid for the glacial acetic acid the fluid will decalcify small bones during 24 h fixation. This PFF (picric acid–formaldehyde–formic acid) fluid (*J. Techn. Methods*, **24**: 35, 1944) contains 85 ml 1.2% picric acid, 10 ml 40% formaldehyde, and 5 ml 90 to 95% formic acid.

Bouin's fixatives generally lake red cells. They remove the demonstrable ferric iron from blood pigments. They render RNA relatively resistant to RNase digestion. Here the inclusion of 6% sodium chloride could be tried.

Lee recommends extraction with 50 and 70% alcohol until most of the yellow color is removed. Masson (*J. Techn. Methods*, **12**: 75, 1929) transferred tissues from the fixative to water, using enough to cover them and leaving them in it until ready to embed them. We find that alcohol extraction of blocks may take many days with daily changes of alcohol, and we prefer to embed and section after no more than 2 to 3 days in 70 to 80% alcohol. It is far easier to wash the excess picric acid out of the sections after removal of paraffin. Surplus tissue is to be stored in 70% alcohol.

Two alcoholic picric acid fluids which have enjoyed some popularity as glycogen fixatives are Rossman's and Gendre's fluids.

Rossman's fluid (*Am. J. Anat.*, **66**: 342, 1940) consists of 90 volumes of 100% alcohol saturated (about 9%) with picric acid and 10 volumes of neutralized commercial formalin. Fix for 12 to 24 h, and wash for several days in 95% alcohol.

Gendre's fluid (*Bull. Histol. Appliq. Physiol.*, **14**: 262, 1937) is composed of 80 volumes of 90% (by weight, equals 95% by volume) alcohol saturated with picric acid, 15 volumes of 40% formaldehyde solution, and 5 volumes of glacial acetic acid. Fix 1 to 4 h in this mixture and wash in several changes of 80% alcohol. We find that 4 h at 25°C gives excellent fixation of glycogen in liver tissue. Two changes each of 80, 95, and 100% alcohol give adequate washing and dehydration.

Vom Rath's (*Anat. Anz.*, **11**: 286, 1895) picrosublimate acetic fluid is sometimes attributed to Tellyesniczky (*Arch. Mikrosk. Anat.*, **52**: 202, 1898). This was a mixture of equal volumes of mercuric chloride saturated in warm sodium chloride solution and of filtered cold saturated picric acid solution. Vom Rath added either 0.5% or 1% glacial acetic acid to the mixture. Tellyesniczky used the latter proportion and condemned the mixture both for resting nuclei and for cytoplasm.

The *Bouin-Hollande fluid* is recommended by Hartz (*Am. J. Clin. Pathol.*, **17**: 750, 1947) for fixation and gross demonstration of calcification. Bone, calcified necrotic tissue, and calcified fat necrosis appear dark green on a pale yellowish green background. In 100 ml distilled water dissolve successively without heating 2.5 g copper acetate, 4 g picric acid, 10 ml 40% formaldehyde, and 1.5 ml glacial acetic acid. The fluid keeps well and will decalcify small bones in a week or less.

Solcia's GPA (*Stain Technol.*, **43**: 257, 1968) consists of 25% glutaraldehyde, 25 ml; saturated aqueous picric acid, 75 ml; and glacial acetic acid, 1 ml, or sodium acetate, 1 g. Fix 24 h and transfer to several changes of 70% alcohol for 24 to 48 h.

Stefanini, de Martino, and Zambori [*Nature* (*Lond.*), **216**: 173, 1967] recommend PAF, a neutral buffered sodium picrate 5% formol. A 0.15 M phosphate buffer of pH 7.3 is present. To 150 ml of twice-filtered saturated aqueous picric acid (about 1.92 g anhydrous trinitrophenol) add 20 g paraformaldehyde, neutralize with 335 mg NaOH in 1 N solution added dropwise with agitation, heat to 60°C until a clear solution is obtained, let cool, and add 3.31 g $NaH_2PO_4 \cdot H_2O$ and 18.0 g Na_2HPO_4 (anhydrous) and enough distilled water to make 1000 ml.

This PAF fluid is quite stable at room temperature and in daylight, its pH is about 7.3, its osmolarity 900 mOsm. It shows no deterioration in 1 year.

Semen (2 drops) is coagulated at once in 20 ml PAF, the coarse clots are removed, the suspension of spermatozoa is centrifuged for 10 min at 1000 r/min, and the pellet consists mostly of spermatozoa. The supernatant is decanted, and the pellet is washed for 15 min in several changes of phosphate buffer, postfixed for 15 min in 10 ml 1% osmium tetroxide, rapidly dehydrated, broken into small fragments, and embedded in Epon 812.

Osmium Tetroxide

Osmium tetroxide ("osmic acid") mixtures are little used in general diagnostic pathology because of their high cost, poor penetration, and interference with various staining methods. Since the development of modern frozen section methods for fat and degenerating myelin, they have less general value than formerly. Osmium tetroxide is especially valuable for fixation of cytoplasmic structures; but nuclei are poorly stainable, and when its action is prolonged, unsaturated fats reduce it to form black masses.

The vapor of 1 to 2% aqueous solution may be used for the fixation of blood and tissue films. This requires 30 to 60 s or more according to the thickness of the film. Slides are placed face down over a small flat dish containing a thin layer of the solution.

1. *Flemming's strong solution* (Ehrlich's "Encyklopädie" and all modern texts) contained 20 ml 2% osmium tetroxide, 75 ml 1% chromic acid, and 5 ml glacial acetic acid. Fix for 1 to 3 days; wash for 6 to 24 h in running water; store in 80% alcohol.

2. *Hermann's fluid* (Ehrlich's "Encyklopädie") substituted 75 ml 1% platinum chloride for the chromic acid in Flemming's solution, with the same 20 ml 2% osmium tetroxide and 5 ml glacial acetic acid. It was used in the same manner.

3. *Marchi's fluid* consists of 1 part 1% osmium tetroxide and 2 parts Müller's fluid. This composition is that given in all modern texts. Its main use is for demonstration of degenerating myelin.

4. *Mann's fluid.* See description earlier in text.

5. *Buffered osmium tetroxide* is generally recommended for preparation of material for electron microscopy. Palade (*J. Exp. Med.*, **95:** 285, 1952) introduced this fluid, fixing for 1 to 4 h in 1% osmium tetroxide in pH 7.4 Veronal acetate buffer. Rhodin (*Exp. Cell Res.*, **8:** 572, 1955) made this fluid isotonic in order to prevent artifactual cell swelling.

    Barrnett and Palade (*J. Histochem. Cytochem.*, **6:** 1, 1958) were using 1% osmium tetroxide in pH 7.6 Veronal acetate buffer containing enough sucrose to raise the osmolar concentration to 0.44 M.

    Rhodin (*Int. Rev. Cytol.*, **7:** 485, 1958) suggested also 1% osmium tetroxide in Veronal acetate with 3% dextran. McGill and Geer at L.S.U. University used 1% osmium tetroxide in pH 7.4 Veronal acetate buffer containing 4.5% sucrose.

6. *Feder and Sidman* (*J. Histochem. Cytochem.*, **6:** 401, 1958) used 1% osmium tetroxide made up in acetone prechilled to −75°C in their freeze substitution technic. Thin blocks, 3 mm or less, are fixed 6 to 7 days at −75°C and washed in 2 to 3 changes of acetone at the same temperature. Although prefreezing in propane-isopentane at −170°C was

prescribed, the usual isopentane at $-150°C$ can probably be used, and even material frozen in solid carbon dioxide–acetone mixture would probably be quite satisfactory, except for electron microscopy. Osmium tetroxide is rapidly reduced by acetone at room temperatures.

Only a few years ago osmium tetroxide (OsO_4) mixtures were rarely used in general histopathologic practice. Today OsO_4 is widely used. The increase in popularity is due to the increased use of OsO_4 as a fixative and stain for electron microscopic investigations. OsO_4 (mw 254.2) boils at $131°C$, melts at $41°C$, and is soluble to the extent of 7.24% in distilled water of pH 7 at $25°C$. It is miscible with saturated lipids, paraffin oil, and organic solvents such as benzene and carbon tetrachloride.

OsO_4 penetrates slowly into tissues; furthermore, the rate of penetration is reduced by as much as 40% in rat liver by the addition of $0.15 M$ sucrose to solutions of $1\% OsO_4$. OsO_4 has been known to add to double bonds (Criegee, *Ann. Chem.*, **75**: 522, 1936). In a series of papers, Korn (*Biochim. Biophys. Acta*, **116**: 317, 1966; *J. Cell Biol.*, **34**: 627, 1967; *Science*, **153**: 3743, 1966) and Korn and Weisman (*Biochim. Biophys. Acta*, **116**: 309, 1966) reacted methyl oleate with OsO_4 at $0°C$ in water and isolated the product bis(methyl-9,10-dihydroxystearate)osmate quantitatively as determined by thin-layer and gas-liquid chromatography, visible and infrared spectra, nuclear magnetic resonance, and elemental analysis. The product is formulated below:

$$H_3C-(CH_2)_7-\overset{\overset{\displaystyle H}{|}}{C}-\overset{\overset{\displaystyle H}{|}}{C}-(CH_2)_7-COOCH_3$$
$$\overset{|}{O}\diagdown\;\diagup\overset{|}{O}$$
$$Os=O$$
$$\diagup\;\diagdown$$
$$\overset{|}{O}\quad\overset{|}{O}$$
$$H_3C-(CH_2)_7-\underset{\underset{\displaystyle H}{|}}{C}-\underset{\underset{\displaystyle H}{|}}{C}-(CH_2)_7-COOCH_3$$

Korn is of the opinion that osmium-fixed membranes observed by electron microscopy reveal nothing about the original molecular structure of the membrane.

Bahr (*Exp. Cell Res.*, **7**: 457, 1954) obtained no evidence to indicate that OsO_4 reacts with DNA or RNA; however, evidence was obtained to indicate oxidation of sucrose to oxalic acid. OsO_4 is also capable of oxidizing glycogen with the formation of aldehyde groups. Substantial loss of protein occurs during fixation in OsO_4 (1% at pH 7.3 to 7.5 in either $0.14 M$ sodium acetate or Veronal buffer). Dallam (*J. Histochem. Cytochem.*, **5**: 178, 1957) noted the loss of as much as 22% protein during fixation in OsO_4 and another 12% during dehydration in ethanol. Luft and Wood (*J. Cell Biol.*, **19**: 16A, 1963) also noted an 8% loss of protein during OsO_4 fixation and another 4% loss during dehydration. Bahr (*Exp. Cell Res.*, **9**: 277, 1955) noted a loss of liver protein in excess of 50% during a 4-h period of OsO_4 fixation. A loss of lipid also occurs. Morgan and Huber (*J. Cell Biol.*, **32**: 757, 1967) calculated a 39% loss of phospholipids in lung tissues during fixation in a glutaraldehyde–osmium tetroxide.

Seligman et al. (*J. Histochem. Cytochem.*, **16**: 87, 1968) prepared osmium coordination compounds with a variety of acidic or basic ligands to produce water-soluble brown-black monomers and occasional polymers bearing multiple basic or acidic groups. For light microscopy, results were obtained similar to those observed for the usual anionic or cationic dyes. When sections of glutaraldehyde-fixed, Araldite-embedded tissues were incubated in aqueous solutions of the dyes for 3 to 18 h, good electron microscopic contrast was observed.

Seligman et al. (*J. Cell Biol.*, **30:** 424, 1966), Sternberger et al. (*J. Histochem. Cytochem.*, **11:** 48, 1963, and **14:** 711, 1966), and Hanker et al. (*Science*, **152:** 1631, 1966) stained tissues for light and electron microscopy by bridging metals with multidentate ligands. Certain osmiophilic reagents such as thiocarbohydrazide also have an affinity for metals bound to tissues. For example, osmium may be bridged to osmium tetroxide used for fixation. This is called the *OTO method* and results in enhanced staining observed by electron microscopy. Membranes are particularly well stained. Seligman et al. suggest the following may occur (Figs. 3-2 and 3-3):

Litman and Barrnett (*J. Ultrastruct. Res.*, **38:** 63, 1972) conclude that osmium in the Os(VIII) state is probably responsible, at least in part, for membrane density observed by

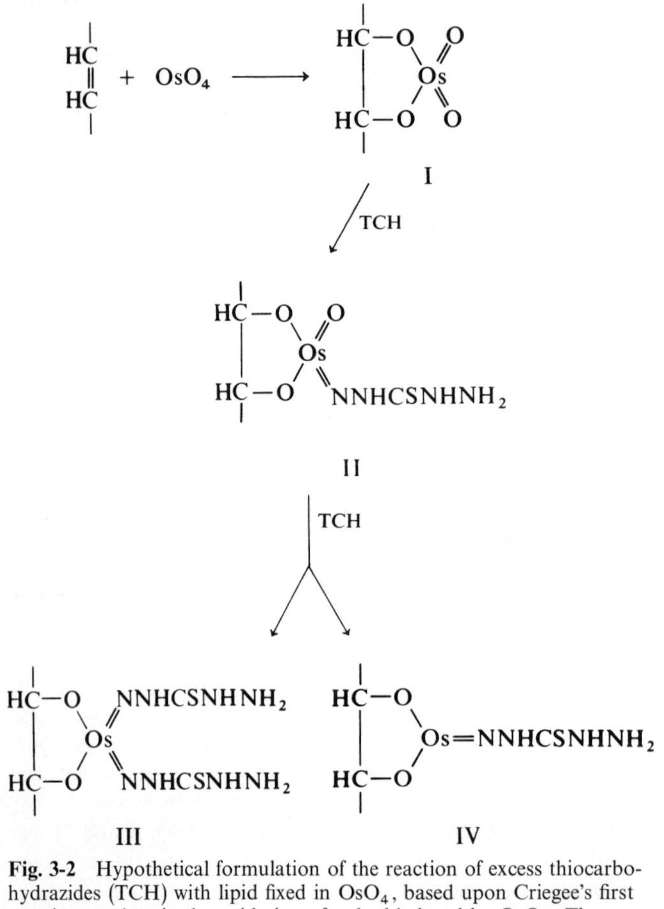

Fig. 3-2 Hypothetical formulation of the reaction of excess thiocarbohydrazides (TCH) with lipid fixed in OsO_4, based upon Criegee's first reaction product in the oxidation of a double bond by OsO_4. The hypothetical reaction products II, III, and IV are formulated to reveal their capability of reacting and binding further OsO_4, the last step in the OTO method. (*By permission of Seligman et al., J. Cell Biol., 30:* 424, 1766).

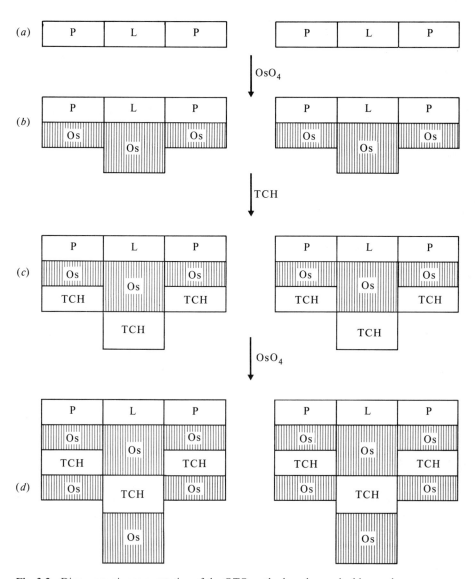

Fig. 3-3 Diagrammatic representation of the OTO method used on a double membrane (Fig. 3-3a) composed of protein (P) and lipid (L). In Fig. 3-3b, the difference in the degree of osmication of the two components (P and L) in osmium tetroxide is represented by a factor of 1 : 2, although the actual ratio would be nearer to 1 : 10. In Fig. 3-3c, the attachment of TCH to components of the membrane fixed in osmium tetroxide is not quantitated. In Fig. 3-3d, the enhancement, most marked with lipid, of bridging osmium through TCH to the tissue-bond osmium is demonstrated. (*By permission of Seligman et al., J. Cell Biol.*, **30:** 424, 1966.)

electron microscopy in OsO_4–fixed tissues. Litman and Barrnett further propose that the Os(VIII) is probably hydrogen-bound in its native state as OsO_4 to protein and to aliphatic side chains of lipid in membranes. Considerable electron density of membranes was removed by $K_4Fe(CN)_6$ in acetic acid. The reagent reacted only with Os(VIII) and *not* with lower oxidation states of osmium to give an insoluble crystalline precipitate $K_2FeOs(CN)_6 \cdot Os(VI)$ esters were unreactive with acetic ferrocyanide. Acetic acid alone decreased membrane density of fixed sections by simple protonation but did not remove osmium from osmate esters.

Hopwood (*Histochemie*, **24**: 50, 1970) investigated the reactions between osmium tetroxide and glutaraldehyde or formaldehyde. They were found to react together to form intermediate products which break down further to form osmium black. Glutaraldehyde reacted with OsO_4 much more rapidly than did formaldehyde.

Fixation has a profound influence on the appearance of tissues at both the light and electron microscopic levels. Landon (*J. Cell Sci.*, **6**: 257, 1970) notes that the fine structure of the Z disks of rat skeletal muscle varies both with the fixatives employed (OsO_4 versus glutaraldehyde) and to a lesser extent with the muscle fiber type.

Davies and Spencer advocate 1% OsO_4 in Veronal-acetate buffer, at pH 6.2, containing 0.24 M sucrose and 0.01 M $CaCl_2$. Palade (*J. Exp. Med.*, **95**: 285, 1952) advised 1% OsO_4 in 0.028 M Veronal-acetate buffer, at pH 7.4, containing 0.14 M sucrose. Tooze (*J. Cell Biol.*, **22**: 551, 1964) noted that 0.01 M $CaCl_2$ in the Davies and Spencer OsO_4 fixative completely prevented loss of hemoglobin from red cells.

Ledingham and Simpson (*Stain Technol.*, **45**: 255, 1970) note intensification of osmium staining of tissues fixed in 4% glutaraldehyde, postfixed in 2% OsO_4, and subsequently treated with 0.8 to 1% solution of *p*-phenylenediamine in 70% ethanol either in the block—prior to embedding in paraffin or Epon—or, in the case of Epon-embedded material, after sectioning for light microscopy.

Sevier and Munger (*Anat. Rec.*, **162**: 43, 1968) noted that 1- to 3-μ cut sections from glutaraldehyde- or formalin-fixed tissues postfixed in osmium and embedded in plastic will stain more intensely with hematoxylin or acid and basic dyes if the embedded sections are first oxidized for 1 to 4 h at 24°C in 5% Oxone (E. I. du Pont de Nemours & Co., Electrochemicals Dept., Wilmington, Delaware).

PROTEIN AND LIPID LOSSES.   Dallam (*J. Histochem. Cytochem.*, **5**: 178, 1957) details losses of proteins and lipids that occur during osmium tetroxide fixation, graduated alcohol fixation, and embedding (Tables 3-8 and 3-9). Although losses of lipids may possibly be reduced by avoidance of fat solvents, it is important to remember that embedding media, such as Epon, are strong lipid solvents (Idelman, *J. Microsc. (Oxf.)*, **3**: 715, 1964).

Pyroantimonate

Tandler and Kierszenbaum (*J. Cell Biol.*, **50**: 830, 1971) recommend perfusion of rats with a saturated aqueous solution of potassium pyroantimonate *alone* (prepared by boiling the salt in deionized or 2 times glass-distilled water, cooling rapidly to 24°C, and centrifugation; pH about 9.2) as a fixative. Although the perfusate does not contain another fixative, thin slices of rat kidney are postfixed in formaldehyde–potassium pyroantimonate, washed, and postosmicated. Thin sections failed to stain subsequently with uranyl acetate or lead citrate. Membranes, mitochondria, and nuclei are said to be particularly well preserved.

TABLE 3-8 ESTIMATION OF PROTEIN LOST DURING PREPARATION OF TISSUE FRAGMENTS, MITOCHONDRIA, AND MICROSOMES FOR ELECTRON MICROSCOPIC EXAMINATION

Sample	Sample no.	Total protein, mg	Losses in reagents (N × 6.25) Osmic-buffer	Graduated alcohol	Methacrylate	Total % protein removed
Heart fragments	1	51.0	5.31	None	None	10.42
	2	28.1	3.75	None	None	13.39
	3	72.8	10.3			14.2
	4	84.1	8.54			10.2
Kidney fragments	1	76.2	8.98			11.7
	2	105.0	17.7			16.9
Liver mitochon-	1	114.8	21.4	12.31		29.4
dria	2	95.1	16.2	11.50		29.2
	3	81.1	20.3	8.50	0.75	36.5
	4	87.8	18.8	8.25	0.94	31.8
Liver microsomes	1	192.3	21.7	4.75		13.8
	2	184.5	22.5	5.06		14.9

Source: Dallam, *J. Histochem. Cytochem.*, **5:** 178, 1957.

Heat

We have occasionally fixed small objects, up to 5 to 10 mm in thickness, by boiling them in physiologic (0.85 to 0.90%) sodium chloride solution for 2 to 3 min. This method may be of value where it is desired to avoid the introduction of alien chemical substances. Some shrinkage is produced. The method has the further advantage of speed. Frozen sections may be completed in a matter of minutes after removal of tissue from the body.

It is possible that formaldehyde solutions employed in this way have some additional insolubilizing effect on tissue proteins. Boiling in saline solution is worthy of further trial in the study of basic proteins and terminal amine groups by histochemical methods.

It is well known that boiling water dissolves starch and glycogen (*Bull. Int. A. Med. Museums*, **27:** 23–61, 1947, especially p. 34). Fixation in boiling distilled water destroys red corpuscles but, if brief, leaves a good deal of glycogen in liver cells, and preserves nuclear and cytoplasmic detail quite well. Collagen is not greatly altered.

TABLE 3-9 ESTIMATION OF LIPID LOST DURING DEHYDRATION OF MICROSOMES AND MITOCHONDRIA

Particulate	Mitochondria*				Microsomes †	
Sample number	1	2	3	4	1	2
Protein, mg	115	95	81	88	189	192
Dry weight of sample	160	130	120	125	362	376
Phosphorus extracted, mg	0.5	0.4	0.4	0.4	0.3	0.4
Phospholipid (P × 25) extracted, percent of total lipid	27.7	30	28	29	5	7

* Dry weight and total lipid calculated on the basis that 70% of mitochondria is protein and 30% lipid (Lindberg and Ernster, 1954).

† Dry weight and total lipid calculated on the basis that microsomes contain 63 mg RNA (ribonucleic acid) phosphorus per mg nitrogen and 43% lipid (Lindberg and Ernster, 1954).

Source: Dallam, *J. Histochem. Cytochem.*, **5:** 178, 1957.

UNFIXED TISSUES

This was of course the original type of material studied by the first histologists and histo-chemists. With the advent of fixation and sectioning procedures this material was little studied for a long period except for the observation of motility of cells and cell organs, of the progress of the mitotic process, of phagocytosis and pinocytosis, and of other phenomena requiring the observation of living or surviving cells.

With the rise in interest in the last two decades in the precise localization in tissues of various enzymes, preparation methods have developed in which denaturation, extraction, and diffusion of enzymes are minimized.

In some instances simple air desiccation of impression smears and of fresh frozen sections gives just enough protein denaturation to prevent diffusion and extraction without seriously imparing activity. The in vacuo desiccation on the Altmann-Gersh freeze-dry procedure seems to have a similar effect. Freezing itself is deleterious in some instances; in these cases the enzyme demonstration reaction has been done by perfusion of intact tissue and by immersion of thin unfrozen slices, and fixation follows the reaction.

Several manufacturers now furnish effective equipment for vacuum desiccation of tissues previously frozen in isopentane or butane with liquid nitrogen. These devices usually operate at 10^{-5} or 10^{-6} mm mercury pressure, and their effectiveness appears to depend on the shortness of the path from the object to a condensing surface kept at a considerably lower temperature to remove water vapor from the system. The object is usually maintained at -30 to $-70°C$, desiccation being faster at the higher temperature, at which also ice-crystal growth is more likely to occur. Ice-crystal size seems to have its greatest importance in electron microscopy.

When fully desiccated, tissues may be directly infiltrated with paraffin or Carbowax without breaking the vacuum. The embedding wax should be defoamed before being used by melting in vacuo.

About 1954, we attempted to utilize the same principle as in organic distillations, of circulating inert gas through the desiccation chamber and a condensing chamber under lower temperature, using only partial vacuum. Mechanical difficulties prevented our success with this process. We believe it has been successfully used by others since.

Kulenkampff (*Z. Wiss. Mikrosk.*, **62**: 427, 1955) prefers propane chilled with liquid nitrogen as the freezing agent, claiming temperatures of about $-190°C$ and hence finer ice-crystal size.

Much work on unfixed tissue is now done by direct sectioning by a cold knife procedure, such as the Adamstone-Taylor and cryostat methods (see also under individual enzyme methods).

4

SECTIONING

FREEHAND SECTIONING

Freehand sectioning with a sharp razor is sometimes employed in making sections, either of unfixed or of fixed tissue, thin enough to be translucent. This method was employed by Terry (*J. Techn. Methods*, **12:** 127, 1929) for his rapid-diagnosis method. Following the preparation of the thin sections they are placed upon a glass slide, and a drop of dilute stain is carefully placed on one surface and rinsed off after a few seconds. With practice it is readily possible to regulate this staining interval so that only one or two surface layers of cells are stained. For this purpose Terry recommended a neutralized polychrome methylene blue made by boiling with alkali for variable periods and then neutralizing. This solution is commercially available from at least one American manufacturer, but 1% solutions of thionin, azure A, or toluidine blue will serve. After rinsing, cover with a cover glass and examine at once. Preparations are temporary.

Aside from this procedure, almost all sectioning is done with microtomes. With these instruments sections may be prepared which are much thinner and more uniform than those prepared by freehand methods. Microtomy requires a firmer consistency than is present in raw or fixed tissue. This consistency is attained by freezing or by infiltration with embedding masses in a fluid state, followed by solidification of the embedding medium. The commonly used embedding media are paraffin and other waxes, soaps, gelatin, agar, Carbowaxes and polyethylene glycols, polyvinyl alcohol, celloidin, and other nitrocelluloses, and for electron microscopy, methyl and butyl methacrylate resins which are polymerized after infiltration.

TISSUE SLICERS

Smith and Farquhar (*Sci. Instr. News RCA*, **10:** 13, 1965) developed an instrument and method to cut fresh nonfixed, nonfrozen sections. The commercial apparatus (Sorval TC-2) has a chopping action. Uniform sections may be cut reliably as thin as 20 μ. The instrument is very useful, particularly for enzyme histochemistry. Sections may be incubated in media for the detection of various enzymes. If the detection system contains an electron-opaque product, sections can be subsequently processed for electron microscopy. Ice crystals and certain fixation artifacts are thereby avoided. Examples of the use of the apparatus are studies by Novikoff et al. (*J. Cell. Biol.*, **29:** 525, 1966), Seligman et al. (*J. Cell. Biol.*, **34:** 786, 1967), Karnovsky (*J. Cell. Biol.*, **35:** 213, 1967), Rutenburg et al. (*J. Histochem. Cytochem.*, **17:** 517, 1969), Fahimi (*J. Cell. Biol.*, **43:** 275, 1969), and Smith (*J. Histochem. Cytochem.*, **18:** 590, 1970).

Another instrument available is the Oxford Vibratome (available from Oxford Laboratories, 107 North Bayshore Boulevard, San Mateo, California 94401). Blade vibration is used rather than a chopping action. The instrument is semiautomatic and permits cutting and collecting fixed or nonfixed sections in varying or alternating thickness (e.g., from 5- to 20-μ sections). The sectioning process is tedious and slow; however, sections as thin as 8 μ may be cut from nonfixed, nonfrozen tissues.

Shnitka et al. (*Lab. Pract.*, **17**: 918, 1968) describe the construction of a simple mechanical tissue chopper fabricated from a domestic electric buzzer capable of chopping tissues 0.1 to 0.4 mm in diameter.

FROZEN SECTIONS

The freezing microtome is used for preparation of sections for rapid diagnosis, for the study of fatty and lipid substances which would be lost if paraffin or nitrocellulose methods were employed, and for many metallic impregnation methods.

Place a few drops of water on the object holder of the freezing microtome, lay a 3- to 5-mm slice of tissue in this water, freeze rapidly until the surface of the tissue appears dry, wait a few seconds, and then try cutting a section at a time at short intervals until satisfactory coherent sections at 10 to 15 μ are obtained. The sections as cut remain on the edge of the microtome knife. They are conveniently transferred to small dishes of water by first moistening the outer side of the left little finger with water, and then wiping the sections off the knife edge with the moistened area and dipping the finger with sections into the dish of water. The sections float off, first as shreds when the block is too hard, then as coherent sections which fold freely in the water as they float. When this point is reached, quickly cut a dozen or more sections, removing three or four at a time to the water. Store sections from each case in an individual small covered dish with a small slip of paper bearing the case number in the water with the sections. The number can be written in pencil; but if prolonged storage is contemplated, write labels with india ink and then dip them in smoking-hot paraffin. For prolonged storage use 10% diethylene glycol by preference; otherwise, 5 to 10% formalin. Vials holding 10 to 15 ml are convenient for this purpose. The india ink label is put inside with the sections.

In selecting sections for staining, complete sections which fold and unfold freely in water are to be chosen. Sections which float rigidly are too thick. For rapid diagnosis such frozen sections may be stained in toluidine blue, thionin, or azure A. Mallory recommended 0.5% thionin or toluidine blue in 20% alcohol for 30 to 60 s. For fixed tissue the addition of 0.5 to 1% of glacial acetic acid sharpens the stain and makes it more selective for nuclei. Sections may be stained thus by immersion in a small amount of stain in a watch glass, then fished out with a needle after the required time, rinsed in a large container of water, and floated onto a slide. Bring the edge of the section on the slanted slide carefully to just above the surface of the water. Tease the remainder of the section half floating in the water into a reasonably flat position, and gradually raise the slide out of the water. If any portion of the section is not flat, dip that side again under the water so as to float it smooth, and again withdraw. Blot dry with hard, smooth filter paper; dehydrate with a few drops of acetone; clear with acetone and xylene, then with xylene; and mount in synthetic resin.

Some prefer first to float sections onto slides as above, then blot them down firmly with hard filter paper, immerse them briefly in 0.5 to 2% collodion solution in ether and 100% alcohol, drain for 30 to 60 s, and harden in chloroform, 80% alcohol, or water before staining. This method is to be preferred if sections are to be heated or treated with alkaline solutions. Such sections may then be handled as are paraffin or attached nitrocellulose or celloidin sections.

Sections stained for demonstration of fats with oil-soluble dyes are to be floated out on slides and blotted flat as above, but all dehydrating agents or fat solvents are to be avoided. Instead, mount directly in a gum arabic, gelatin glycerol, syrup, or other aqueous mounting medium; or temporarily in pure glycerol, water, or other indifferent fluid. Gelatin media require melting before use, and air bubbles in gelatin may be very tenacious. They may be degassed by placing the bottle in a vacuum paraffin oven and turning on vacuum to about 350 to 400 mm mercury (15 in). Care must be taken not to boil the gelatin solution. Water boils at about 124 mm at 55°C, 149 at 60°C. These pressures correspond to 636 mm = 25.2 in and 611 mm = 23.9 in on the usual vacuum gauge. Gum arabic (acacia) media are fluid at room temperatures of 25 to 15°C, and most of them dry hard. Fructose syrup will dry hard if the humidity is not too high but, like Arlex gelatin, will remain sticky for some time in hot, humid weather. Such syrup mounts may be sealed by painting the edges of the cover glasses with a polystyrene or other resin solution; cellulose caprate is excellent.

Adamstone-Taylor Cold Knife Technic for Frozen Sections

In this procedure (Adamstone and Taylor, *Stain Technol.*, **23:** 109, 1948) fresh tissue is quickly frozen in slices of 2 mm thickness or less, either on the freezing block of the microtome or by immersion in liquid nitrogen or in isopentane chilled with liquid nitrogen or in petroleum ether containing chunks of solid carbon dioxide. By using one of the latter procedures a number of blocks may be quickly frozen and then stored in dry ice at −75°C or in a deepfreeze compartment at −25°C until sectioning becomes convenient. Since the essence of the technic is the avoidance of thawing until sections are finally attached to slides, it is necessary to chill the knife of the microtome by fastening to it on each side of the cutting area blocks of solid carbon dioxide. Use Scotch tape or thin strips of sheet metal cut from scrap copper or tin plate.

When the knife is chilled, the tissue blocks are placed on the freezing head and frozen to the block. As the sections are cut, a small camel's hair brush is used to hold them flat as they come onto the knife blade. Then Adamstone and Taylor used a small, hollow scoop containing dry-ice chips to transfer the flat sections to slides and pressed them down. As the sections soften but before complete melting, they must be immersed in the chosen fixative. If the objective is microincineration for demonstration of soluble salts, quick heating to coagulate tissue protein would seem preferable to exposure to the solvent action of any liquid fixative.

Adamstone and Taylor warn that the procedure is much more difficult in a warm or humid atmosphere. The whole process might be much easier (though less comfortable to the operator) if it could be carried out in a refrigerator room at a temperature somewhat below 0°C.

For this purpose the use of a cryostat is suggested. In this type of apparatus the sections can be kept for a time in the frozen state, without the necessity of immediate processing required by the open-air cold knife method.

Use of the cold knife procedure on formalin-fixed tissue may permit preparation of thinner and more coherent sections of such tissues as heart muscle.

Ultrathin Frozen Sections for Electron Microscopy

Bernhard and coworkers [*J. R. Microsc. Soc.*, **3:** 579, 1964; *Ann. Biol. Clin. (Paris)*, **4:** 5, 1965; *J. Cell Biol.*, **34:** 757, 1967; *J. Cell. Biol.*, **49:** 731, 1971] have written a series of papers regarding optimal methods of preparation of frozen sections for electron microscopy. Initially, Bernhard and coworkers used a Porter-Blum MT-1 microtome mounted in an ordinary deepfreeze. Subsequently, Ivan Sorvall Inc., Norwalk, Connecticut, developed a cryokit used with the MT-2 microtome. Hudson and Marshall (*J. R. Microsc. Soc.*, **91:** 105, 1970) developed a similar system. Dollhopf and coworkers (*Mikroskopie*, **25:** 17, 1969) conducted studies leading to the development of the apparatus now available commercially from Reichert Optische Werke AG, Vienna. The LKB apparatus was developed following studies by Persson (*Proc. 7th Int. Cong. Electron Microscopists*, vol. 1, p. 349, 1970) and Appleton (LKB Instruments, Rockville, Maryland). The objective of Bernhard and coworkers was to obtain a reasonable number of aldehyde-fixed high-quality sections useful for a variety of cytochemical studies. The following suggestions are taken from the Bernhard papers:

1. Fresh tissues are cut into 1- to 1.5-mm^3 pieces and fixed for 1 h at 25°C in 25% glutaraldehyde in 0.2 M cacodylate buffer at pH 7.2. (In studies employing the enzyme-labeled antibody method, Bernhard sometimes used 4% depolymerized paraformaldehyde in cacodylate buffer at pH 7.2 for 24 h at 4°C.)

2. Embedding is strongly recommended by Bernhard, although he regards it as not absolutely essential. Tissue blocks are immersed in a stirred 10 to 20% aqueous gelatin solution at 37°C for 5 to 30 min and then cooled. Improved sections are believed to be obtained in specimens sectioned 2 to 3 days after embedding.

3. Gelatin-embedded blocks are placed in 30% glycerol for 5 to 15 min. Next, glycerol-soaked blocks are placed on the object holder at 24°C and then frozen in liquid nitrogen. Treatment with glycerol tends to minimize ice-crystal formation. If isopentane cooled with liquid nitrogen is used, fewer gas bubbles around the specimen tend to form.

4. Blocks are trimmed with a cooled razor blade during observation with a dissecting microscope. The specimen is kept at -70°C, mounted in the microtome. Cutting sections at high speed was once considered essential for high-quality sections. High-quality sections now can be cut with the commercially available instruments without high speed.

5. Sections can be picked up one by one from the edge of the knife with a thin steel needle or a tiny brush and placed directly on a Formvar-coated grid (Belden Mfg. Co., Chicago). Sections are flattened to the grid by pressure applied to the end of a highly polished copper rod (Christensen, in "Autoradiography of Diffusible Substances," p. 355, Academic, New York, 1969). Sections may also be picked up in a trough of 50% dimethylsulfoxide (DMSO) or 50% glycerol in the usual fashion with a Marinozzi ring (*J. Ultrastruct. Res.*, **10:** 433, 1964). Sections are then transferred to distilled water and allowed to float in the plastic rings until used for cytochemical analysis.

    In order to avoid precipitates of reaction products on copper grids, it is important to keep the sections floating in plastic rings throughout the reactions.

    Drying of sections must be prevented until all steps of the staining reaction are completed.

6. Either negative or positive staining may be used. For negative staining, either 2% phosphotungstic acid or 4% silicotungstate (best) for 15 s at 37°C may be employed. A disadvantage is that sections appear homogeneously gray without preferential contrast. For positive stains, 0.5% uranyl acetate in distilled water for 1 min followed by lead citrate for 5 s (up to 1 min) is recommended.

7. Sometimes shrinkage artifacts that may occur around certain organelles appear to be minimized by rinsing grids in either 2.5 to 5% polyethylene glycol (Carbowax *M* 600) or 5 to 10% glycerol for 1 to 3 s. Sections are then air-dried. If sections are treated with glycerol or Carbowax, it is important to recognize that this step results in the loss of some lead and uranium. To allow for this, lead citrate staining should be increased to 1 min and uranyl acetate staining increased to 2 to 5 min in step 5. Bernhard notes that improved image contrast has not been achieved in sections freeze-dried at $-50°C$ at this step.

Increased fixation time results in harder tissues that cut with greater ease, and structure preservation is better. Of course, activity of some enzymes will be reduced. Bernhard believes that fixation time of 1 h gives the best average results for many tissues.

Freeze Substitution

Simpson (*Anat. Rec.*, **80:** 173, 1941) is credited with initiating the technic of freeze substitution in light microscopy. He tried several solvents including ethanol, ether, chloroform, and methyl Cellosolve at temperatures between $-40°C$ and $-78°C$. Feder and Sidman (*J. Histochem. Cytochem.*, **6:** 401, 1958; *J. Biophys. Biochem. Cytol.*, **4:** 503, 1958) prescribe a method, detailed below, which is sometimes used currently:

Thin (1- to 3-mm) slices of tissues are attached to bits of aluminum foil and quenched in 3 : 1 propane-isopentane at $-170°C$, transferred to the substitution fluid (1% OsO_4 in acetone, 1% $HgCl_2$ in ethanol, and 5% picric acid in ethanol) at $-75°C$ in a dry-ice chest for a week or more. The fixative fluid gradually substitutes for the tissue fluids. After substitution with $HgCl_2$ and picric acid, tissues are brought to 24°C, embedded, sectioned, and stained in the conventional manner. OsO_4-fixed tissues are washed with acetone at $-75°C$ prior to conventional treatment thereafter.

Fernández-Moran (*Ann. N.Y. Acad. Sci.*, **85:** 689, 1960), Bullivant (*J. Biophys. Biochem. Cytol.*, **8:** 639, 1960; *Lab. Invest.*, **14:** 1178, 1965), Rebhun [*J. Biophys. Biochem. Cytol.*, **9:** 785, 1961; *Fed. Proc.* **24** (*Suppl.* 15): S-217, 1965], and van Harreveld et al. (*Anat. Rec.*, **149:** 381, 1964; *J. Cell. Biol.*, **25:** 117, 1965) studied freeze substitution methods as applied to electron microscopy. Fernández-Moran and Bullivant employed increasing concentrations of glycerol. Bullivant substituted in ethanol, infiltrated with butyl methyl methacrylate, and embedded in Durcupan followed by Epon. Fernández-Moran substituted in acetone–ethanol–ethyl chloride mixtures containing either 1 to 2% gold or platinum chloride or OsO_4 for several weeks.

Pease (*J. Ultrastruct. Res.*, **21:** 75, 1967) employed 70% eutectic ethylene glycol and pure propylene glycol as substituting media for dehydration of frozen tissues subsequently examined by electron microscopy. These studies are successors to his previous report (*J. Ultrastruct. Res.*, **14:** 356, 1966) describing the use of ethylene glycol, glycerol, or sugar syrup as an "inert" physical solvent as a replacement for water in tissue processing. Pease makes several practical points worthy of observance:

1. Thin slices (0.5 mm) of tissue are quenched in 70% ethylene glycol in Hank's solution cooled to $-77°C$ in a bath of dry ice–acetone or ethanol. Pease could find no advantage in using freezing agents at lower temperatures such as liquid nitrogen. An advantage of using 70% ethylene glycol with this method is that it is also used as the substitution fluid at $-50°C$. Pease also notes that pure propylene glycol may also be used effectively as a substitution fluid at $-40°C$.

Sometimes tissues were preglycerinated (washed in 6 to 8% glycerol for 2 to 5 min and then 12 to 15% glycerol for 3 to 8 min at 25°C) prior to quenching; however, Pease is uncertain if the glycerol treatment aids in the preservation of structure.

2. It is important to use relatively large volumes of substitution fluid in relation to the size of the tissue specimen. This permits removal of tissue water without substantial dilution of the substitution fluid, and temperature fluctuations are fewer during substitution manipulations. Substitution is enhanced in specimens rotated slowly (2 r/min) in the freezer. Pease notes that substitution is completed in a few hours in thin specimens. Progress can be followed visually. Tissues become transparent. If islands of ice remain, they appear opaque. If tissue slices are too thick, air-filled clefts and bubbles may form with refractile properties difficult to distinguish from ice. Pease recommends that substitution be continued overnight despite appearances indicating completion within a few hours.

3. Using successive decantings over a 10- to 30-min period, 70% ethylene glycol is substituted for 100% ethylene glycol (freezing point −13°C). Pease notes that it may be desirable here to substitute with 100% propylene glycol (freezing point −40°C) in order to minimize changes of temperature which could promote ice-crystal formation. After substitution in pure glycol is complete, tissues may be brought to 25°C without damage.

4. Pease obtained excellent preservation of tissues substituted in 70% ethylene glycol containing 5% glutaraldehyde or formaldehyde. Osmium tetroxide (2% in 70% ethylene glycol) was used without success. Blackening was not observed in tissues substituted below 0°C, and membranes were observed in negative contrast.

   In tissues substituted with an included fixative, tissues are warmed to −10°C during the 100% ethylene glycol step (step 3).

5. Glycol-substituted tissues are readily infiltrated and embedded in hydroxypropyl methacrylate without an intermediate reagent. Pease employs a partially prepolymerized plastic containing benzoyl chloride as catalyst, and 5% divinyl benzene is added subsequently as a cross-linking agent (*J. Ultrastruct. Res.*, **14**: 356, 1966). Disadvantages of the use of this embedding material are difficulty in sectioning and a coarse-grained pattern which prevents resolution at high magnification.

Results from thinnest parts of sections are comparable to those achieved in plastic-embedded specimens. Ultrastructural preservation of comparably treated unfixed tissues was poor due to mechanical compression of the tissues, disintegration of sections in the trough, excessive folding of sections, and lack of contrast.

Embedment of tissues in gelatin appears to be essential in loose tissues such as lymph nodes; however, it may not be necessary in tissues such as liver, kidney, or pancreas.

Quick Freezing

Moline and Glenner (*J. Histochem. Cytochem.*, **12**: 777, 1964) have noted that the rate of cooling with liquid nitrogen may be doubled if tissues to be cooled are first coated with a fine film of talcum powder (Fig. 4-1). Rapid homogeneous freezing tends to reduce fracturing of tissues during cooling.

Komender et al. (*Experientia*, **21**: 249, 1965) compared losses of DNA and RNA from rat liver in tissues freeze-substituted in methanol at room temperatures. In methanol-substituted tissues, no loss of RNA was detected, whereas a 17% loss of DNA was obtained. Furthermore, a 16% loss of dry tissue mass was observed in methanol-substituted tissues that were subsequently extracted with water.

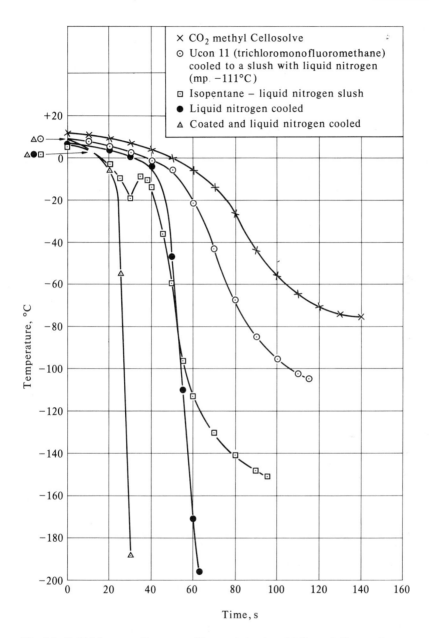

Fig. 4-1 Rabbit heart cooling curves. (*By permission from Moline and Glenner, J. Histochem. Cytochem.*, **12**: 777, 1969.)

EMBEDDING METHODS

The other commonly used sectioning methods require infiltration with embedding masses. Of these the polyvinyl alcohol, Carbowax, gelatin, agar, and soap masses are soluble in water, and tissue may be infiltrated directly. Nitrocellulose is soluble in a mixture of 100% alcohol and ether, paraffin in various fat solvents; and tissues must first be freed of water (dehydration) and brought into the appropriate solvent before infiltration.

FEDER'S POLYVINYL ALCOHOL EMBEDDING. Feder's method (*J. Histochem. Cytochem.*, **7:** 292, 1959; **10:** 341, 1962), as modified by Wachstein, is as follows: Dissolve 200 g polyvinyl alcohol in 600 ml water, add 200 g glycerol, and boil 30 min. Fix tissue in suitable aqueous fixative, block, wash well in water, and transfer directly to the polyvinyl alcohol mixture. Infiltrate at 25°C using three graded water mixtures of, say, 1 : 2, 2 : 1, and undiluted, of 24 h each. Embed in small waxed paper containers, and harden at icebox temperature (0 to 5°C) for enzyme studies (about 3 to 5 weeks), at room temperature for other purposes. Perhaps a week will be required at 25°C; at 37°C 3 days might suffice. Shrinkage is considerable in Feder's original method, in which free evaporation was permitted. This should be minimized in Wachstein's 5°C variant.

Enzymes and lipids, including birefringent lipids, are said to be well preserved.

At 25°C and 0 to 30% humidity (use desiccator with P_2O_5 if necessary) the mass dries to a firm, tough, transparent block which is cemented to wooden or metal block carriers and sectioned on a rotary microtome at 1 to 100 μ.

Feder prescribes Gelvatol-130, procurable from the Shawinigan Corp., Springfield, Massachusetts. We have had completely unsuccessful results with another brand.

CARBOWAX (POLYETHYLENE GLYCOLS). These water-soluble waxes conserve many lipids well. Mellors (*J. Natl. Cancer Inst.*, **10:** 1331, 1950) stated that glycols and polyethylene glycols did not dissolve neutral fats but that they did have considerable solvent activity on aromatic compounds such as steroids, especially those of the adrenal cortex. Pearse (1960, p. 65), however, states that cholesterol and cholesterol esters are insoluble both in aqueous solutions and in melted polyethylene glycols. While Carbowax embedding has been considerably used in the study of tissue lipids, specific statements as to the amount of birefringent lipid in Carbowax sections and in control sections are scarce. Wade (*Stain Technol.*, **27:** 71, 1952) specifically notes good preservation of the acid-fast material in leprosy bacilli. Infiltration can be done directly from water or, more rapidly, on frozen dried material (Hack and Blank, *J. Natl. Cancer Inst.*, **12:** 225, 1951). Sections can be cut on the usual paraffin microtome and are said to ribbon better at humidities below 60% and temperatures under 20°C (Pearse, 1960, p. 774), but Zugibe et al. (*J. Histochem. Cytochem.*, **6:** 133, 1958) put the limits at 75% and 27°C and reported satisfactory ribboning at 6 μ routinely and as low as 2 μ. Cutting consistency is improved by preheating Carbowax 4000 to 175°C for about 30 s (Firminger, *J. Natl. Cancer Inst.*, **10:** 1350, 1950).

The practice of first infiltrating with a low-molecular-weight, relatively low-melting-point Carbowax, followed by a second infiltration with the final embedding medium seems as futile to us as the practice of infiltrating first with a low-melting-point "soft" paraffin, and then with the final higher-melting-point harder paraffin for sectioning. Blank's suggestion of infiltrating at a lower temperature in an aqueous solution, followed by in vacuo desiccation, to avoid thermal destruction of enzymes seems eminently sensible.

Most writers infiltrate tissues directly from water in the polyethylene glycol. Blank and McCarthy (*J. Lab. Clin. Med.*, **36:** 776, 1950) used 9 : 1 Carbowax 4000–Carbowax 1500 at 55 to 60°C, allowing 80 min for blocks 2 mm thick, and longer in proportion for thicker blocks. Even 4 days' exposure was said to produce little distortion. Firminger (*J. Natl. Cancer Inst.*, **10:** 1350, 1950) preferred pure Carbowax 4000 for Washington, D.C., summer use (30 to 36°C). McLane (*Stain Technol.*, **26:** 63, 1951) infiltrated at about 52°C in Carbowax 4000 (mp 50°C) and claimed preservation of tetrazolium dehydrogenase activity in plant tissue. However, we have reported inactivation or extraction of M-Nadi oxidase and verdoperoxidase by glycols and melted Carbowax (*J. Histochem. Cytochem.*, **1:** 8, 1953). Pearse (1960, p. 774) records considerable diffusion of alkaline phosphatases, less of acid phosphatase and peroxidase. Wade used a 15:85 mixture of Carbowaxes 1540 and 4000 and infiltrated 6 h, using two changes. Hack's practice of freeze-drying followed by Carbowax infiltration has been adopted by Zugibe et al. (1958) for fresh frozen material. He also dehydrated formalin-fixed tissue in graded Carbowax 1000 solutions (30, 50, 70, and 95% and pure Carbowax) at 1 h each, ending with 2 h in 100% Carbowax at 47°C. Pearse's directions are similar: Place 3- to 4-mm blocks of formalin-fixed tissue in 70% polyethylene glycol 1000, mp about 40°C (?), for 30 min, in 90% for 45 min, and in 100% for 1 h, stirring at intervals. Embed in wax-paper cups, chilling to 4°C. Cut on rotary microtome at 4 to 6 μ, float sections on 40% diethylene glycol, 10% formalin, and 50% water mixture, pick up on gelatin-coated slides, and dry at 37°C for 10 min. Room temperatures are presumed to be 15 to 20°C. For laboratory temperatures of 20 to 25°C a harder wax would be preferred, say a mixture of Carbowax 1000 and 1500; at 25 to 30°C mix Carbowax 1500 and 4000; above 30°C use Carbowax 4000.

Since Carbowax sections dissolve instantly in water with violent diffusion currents, various other flotation liquids have been proposed. Metallic mercury is not suitable because of the scum of oxide particles, Carbowax, and probably albumin which forms on the mercury. Also, particles of metallic mercury adhere to the sections (Firminger, personal communication). Wade (*Stain Technol.*, **27:** 71, 1952) used 5 mg/100 ml Turgitol 7 in distilled water; Blank and McCarthy (*J. Lab. Clin. Med.*, **36:** 776, 1950) used 0.02% each of gelatin and potassium dichromate, boiled together in daylight for 5 min.

Zugibe et al. (*J. Histochem. Cytochem.*, **7:** 101, 1959) make a stock gelatin-dichromate mixture at 0.2% each, boil, cool, and filter. For use take 10 ml stock mixture, 4 ml glycerol, 4 ml formalin, 1 g cetyltrimethylammonium bromide, and distilled water to make 100 ml. Clear this solution by heating to 30°C. Float sections on this, pick up sections on slides, and dry at 42°C.

Giovacchini (*Stain Technol.*, **33:** 248, 1958) similarly smears slides with a gelatin-glycerol adhesive, places dry Carbowax sections on them, and exposes them in a horizontal position to a temperature of 58 to 60°C for 15 min. The mounted sections are then dried at 58 to 60°C for 24 h and are said to adhere well in routine oil red O hemalum stains. This gelatin solution is made by dissolving 15 g gelatin in 55 ml warm water, then adding 50 ml glycerol and 0.5 g phenol.

Reid and Sarantakos (*Stain Technol.*, **41:** 207, 1966) believe they have overcome some of the problems associated with Carbowax used in the usual fashion. A water-insoluble polyvinylacetate resin identified as AYAF (Union Carbide Chemicals Co., 30 East 42d Street, New York, New York 10017) is employed as a 7.5% mixture in Carbowax. The method follows:

1. Formalin-fixed tissues ready to be embedded are immersed in Carbowax (polyethylene glycol) 200 for two changes of 4 h each.
2. Tissues are immersed in a Carbowax 1540 and 4000 (3:1) mixed with a 1:1 mixture of Carbowax 200 for two changes of 6 h each.
3. Tissues are immersed in a Carbowax 1500 and 4000 (3:1) mixture to which the polyvinylacetate resin AYAF has been added to make a 7.5% mixture. Tissues are infiltrated for two changes of 12 h each.
4. Blocks are cast in a plastic mold, kept in a desiccating jar, and sectioned at 24°C or lower.

5. Sections may be cut as thin as 2 μ on a rotary microtome, and floated on water to which trace amounts of Carbowax 1540-4000 had been added. (Fingers are used to rub off Carbowax from a block.)
6. Sections are collected on albuminized slides and dried overnight in an oven at 56 C.
7. Residual resin matrix is removed by immersion of sections in methyl alcohol for 1 to 2 min prior to processing for staining. After most stains, sections are dehydrated and mounted in the usual fashion in a synthetic resin. For sections stained for fat, as with oil red O, sections are mounted directly from tap water in Apopon. Lillie (1965, p. 100) condemns Apopon for this purpose; it promptly bleaches hematoxylin.

Blocks from human autopsy material up to 10 × 10 × 2 mm were processed successfully. The method is not recommended for tissues with large amounts of interstitial fat because of difficulty of infiltration. Tissues will not cut properly if temperatures rise much above 24°C.

GELATIN. Zwemer (*Anat. Rec.*, **57**: 41, 1933) recommends washing tissues 4 h or more in water, followed by 24-h infiltration at 37°C in 5% gelatin and 12 to 16 h more in 10% gelatin. The tissues are then oriented and embedded in 10% gelatin, which is allowed to harden by cooling at 0 to 5°C.

Blocks may then be sectioned by the freezing method and mounted before staining. The gelatin may be removed with warm water. This avoids gelatin-staining artifacts, which can be disturbing, especially with connective tissue stains. We have found this method useful for the sectioning of atheromatous small arteries, such as coronaries. We are indebted to Mrs. Margaret Giglioni for notes on this procedure. She infiltrated directly in 10% gelatin, sectioned in the cryostat at 5 to 7 μ, collected sections directly on slides, and dried them for some hours at 20 to 25°C so that the sections would adhere well. Then gelatin was rinsed off with warm water (35 to 40°C), and slides were dried at 52°C before staining.

Usually it is preferable to harden the gelatin first by a day or more of fixation in 10% formalin. Zwemer prescribed several hours. Blocks may be sectioned on the freezing micro-tome or may be dehydrated and embedded in paraffin.

The advantage of the method lies in making coherent sections of friable or fragmented tissues. Such material as uterine curettings may thus be handled as a single block. The dis-advantage is that the gelatin stains rather strongly with basic aniline dyes and the stained mass may be confusing. We have found the method helpful, on occasion, in outlining individual cells in masses that were otherwise apparently syncytial in nature.

Baker (*Q. J. Microsc. Sci.*, **85**: 1, 1944) infiltrated in 25% gelatin at 37°C for 20 to 24 h. Ordinarily he then cooled, blocked, and hardened the sections in his calcium-cadmium-formalin solutions. If thin sections were required, he infiltrated instead for 30 h in an open container over anhydrous calcium chloride in a vacuum desiccator at 37°C. Concentration was stopped best just before the gelatin sol set to a gel. The solution was then cooled, blocked, and hardened as before.

In both instances sections were cut on the freezing microtome. With the vacuum con-centration technic, 5-μ sections were prepared by Baker.

Nicolas [*Bibl. Anat.* (Paris), 1896, p. 274] virtually perfected the gelatin infiltration method for preparing frozen sections of eyes. Oakley (*J. Pathol. Bacteriol.*, **44**: 365, 1937) fixed first for 4 days in 10% formalin saline solution. He recommended postchroming in Müller's fluid for 6 weeks in the incubator (37°C), or in Perdrau's fluid ($K_2Cr_2O_7$ 5 g, CrF_3 2.5 g, water to 100 ml), or in Perdrau's fluid half strength for 4 days at room temperature. Large eyes are then windowed before being washed for 24 h. After washing them, cut eyes in half and infiltrate at 37°C overnight in 12.5% gelatin and in 25% gelatin for 24 h. Use gelatin not more than twice, and add 1% phenol to prevent mold growth. Embed cut surface down in 25% gelatin, and harden by cooling to not below 0°C. Trim the blocks and harden them 2 to 3 days in 10% formalin. Wash them for 15 min in water before sectioning. Sections of 5 to 10 μ are claimed for this method. Since the gelatin tolerates dehydration poorly, sections should be mounted in glycerol gelatin.

Albrecht (*Stain Technol.*, **29**: 89, 1954) recommends the use of alcoholic gelatin to assure adherence of fresh frozen sections (particularly large frozen sections of brain) to slides. The alcoholic gelatin is also used as a mounting medium for sections that cannot be dehydrated.

Snodgress and Dorsey (*Stain Technol.*, **38**: 149, 1963) recommend the following water-soluble embedding procedure for frozen sections of central nervous system and peripheral nerves. The method uses egg albumen and is stated to be better than gelatin in that less shrinkage and distortion occurs during dehydration and mounting in a resinous medium.

Formalin-fixed tissues are thoroughly washed and immersed in the aqueous albumen solution. (To 20 g powdered egg albumen is added 5 ml distilled water and mixed. Then 25 ml additional distilled water is added and mixed thoroughly.) Specimens that do not require special orientation are placed in a parchment bag or other semipermeable membrane container, which is enclosed and inserted in Lillie's aqueous calcium acetate–neutral formalin solution. For specimens that require special orientation, Snodgress and Dorsey suggest placement of the specimens with the egg albumen in a box constructed of parchment paper which is immersed in the formalin for hardening. Frozen sections are cut as usual. If tissues remain too long on the freezer, they cannot be removed.

1. Wash in distilled water, and soak for 5 min or longer in a mixture of equal parts of 1.5% aqueous gelatin and 80% alcohol. The fluid should be at least 2.5 cm (1 in) deep. If sections are swollen as a result of the preceding staining procedure, the alcoholic mixture will cause shrinkage to approximately normal dimensions.
2. Using a fine brush, tease a section onto a clean slide. Withdraw the slide from the fluid in a strongly tilted position to encourage rapid drainage of excess fluid. Carefully wipe the slide around the section. In a short time, surface fluid will have evaporated evenly from the section and it will appear velvety.
3. Quickly blot the section gently but completely with a strip of clean, smooth filter paper, using a fresh strip for each subsequent section. Blotting is best done by bending the strip down upon the section until the entire surface of the section has been in contact with the paper. Although this is usually adequate, some sections require a rapid second blotting in which gentle pressure is applied with the fingertip.
4. Rapidly immerse the now almost entirely dry section in 95% alcohol, which will congeal the remaining gelatin, anchoring the section to the slide.
5. If sections are to be stained after mounting, it is advisable to further ensure proper adherence by coating the sections with celloidin. To this end, pass the section through absolute alcohol and subsequently immerse it for 5 min in a 0.5% solution of celloidin in equal parts of absolute alcohol and ether. Then remove the slide, drain it, expose it to the air for several seconds, and immerse it in 80% alcohol for 5 min or longer.

Shands (*Stain Technol.*, **43**: 15, 1968) provides the following method for embedment of cells or tissues in suspension:

1. Fix tissues in a conical centrifuge tube.
2. Centrifuge at low speed, and decant supernatant thoroughly.
3. Resuspend tissues in 2% bovine serum albumin in 0.05 M tris buffer pH 7.5, and transfer to a Beckman cellulose tube.

4. Add 4 drops 25% glutaraldehyde per ml serum albumin, mix, and centrifuge in a swinging-bucket-type centrifuge again at low speed. A gel forms in about 5 min as indicated by the material's becoming opaque.
5. Slice the tube and its contents at 1-mm intervals and place on filter paper to absorb excess moisture, cut away the rim, dehydrate the tissues in ethanol, and process them for electron microscopy in the usual fashion.

AGAR. We have occasionally infiltrated tissues from water in melted 2% agar at 55 to 60°C for 2 to 4 h. This mass becomes quite stiff on cooling and may be cut on the freezing microtome. Its value is for holding exudates and friable tissues in place. We have used it for the study of fat content of loose pulmonary alveolar exudates. It does not stain appreciably with the usual dyes. The method is not suitable when specific carbohydrate stains are to be used.

Friedland (*Am. J. Clin. Pathol.*, **21**: 797, 1951) accelerates this procedure by transferring tissues previously fixed briefly in boiling formalin to melted agar and boiling gently for 1 min. The agar-infiltrated tissue is then sectioned on the freezing microtome. This variant is recommended for friable and partially necrotic tissues, to improve coherence.

PARAFFIN. In order to infiltrate tissues with paraffin it is necessary to displace from them the aqueous or alcoholic fixing fluids, or the water or alcohol used to wash these out, and to replace with a fluid miscible with paraffin. This is usually accomplished by diffusion of the contained fluid out of the tissue block into a surrounding bath of another fluid, which in turn diffuses into and replaces the original fluid content.

Since most water-miscible fluids are not paraffin solvents, it is usually necessary to dehydrate first with a water-miscible fluid and then to replace the dehydrating agent with a paraffin solvent which is at the same time miscible with the dehydrating agent used.

One fluid—1,4-diethylene dioxide, or dioxane—is freely miscible with water on the one hand and with melted paraffin on the other. Lee (10th ed.) warns against the cumulative toxicity of this compound and speaks of a lack of warning odor. Actually there is a distinctive odor, but this is no longer noticed after a short time. We recommend the interposition of a bath of equal volumes of dioxane and paraffin at 56°C after dehydration. Mallory recommended the schedule of Graupner and Weissberger, which included successively three changes of dioxane and three of melted paraffin.

The greatest disadvantage of dioxane is the very low tolerance of its mixtures with paraffin solvents or with paraffin for small amounts of water (Table 4-1). A slight excess of water carried into the dioxane-paraffin bath will occasion stratification with a lower dioxane-water layer and an upper dioxane-paraffin layer. To combat this water intolerance

TABLE 4-1 PERCENTAGE OF ADDED WATER REQUIRED TO RENDER TURBID EQUAL-VOLUME MIXTURES OF DEHYDRATING AND CLEARING AGENTS

Dehydrant	Benzene	Xylene	Chloroform	Carbon tetrachloride	Petroleum ether B	100-octane gasoline	Methyl salicylate
Methanol, absolute	6.5–7	3–3.5	17.5–18	4.5–5	1	Immisc.	6.25
Ethyl alcohol, absolute	12–14	7–8	13–14	10–11	4.5	3.5–4	13.25
Isopropanol, 99%	9.5–10	10	7–7.5	7.5–8.5	11–11.5	8.5–9	17
Acetone, technical	1.75–3	1.5–2	2–2.2	1.5–1.6	0.5–1	1.5–1.6	4.75
1,4-Dioxane, pure	1.5–2	1–1.5	1.5–2	1–1.5	1.5	1–1.5	3.75

Tannenberg (*Am. J. Clin. Pathol.*, **19:** 1061, 1949) adopted an automatic siphonage chamber into which dioxane drips at a fairly rapid rate (3.75 l/h) and which drains down to a level just above the specimens by means of a siphon as often as the fluid level rises to the top of the siphon tube, 2 to 3 cm above the tissue. Adequate dehydration of blocks 2 mm thick in 2 h is claimed, and of curettings in 1 h. The used dioxane is again dehydrated with anhydrous $CaCl_2$ or CaO for 18 to 22 h before it is used again. The reuse of the dioxane is required because of its high initial cost (currently about 3 to 4 times that of tax-free 100% alcohol) and the large volume used in this procedure.

Tannenberg then transferred tissues directly to melted paraffin (two changes of 30 to 60 min each). He suggested further shortening of the total time required by use of vacuum infiltration (Tables 4-2 and 4-3). The boiling point of 1,4-dioxane is 101.1°C, and its volatilization at reduced pressure would be much slower than that of some of the more volatile solvents.

The traditional dehydrating agent is ethyl alcohol. It is usual to transfer tissues from water into 70% alcohol, thence to 80 or 85%, then to 95 and 100% (absolute) alcohol. Two changes daily may be made, ending with at least two or three changes covering 2 to 4 h in 100% alcohol. Some workers omit the first step and start directly with 80% alcohol.

For critical cytologic work and for dehydration of embryos, eyes, and other large or delicate objects, longer intervals in each grade of alcohol and a greater number of gradations are used, and it is advisable to start with alcohol as low as 50 or 60%.

By continued mechanical agitation and multiple changes of each grade of alcohol, and by cutting thin blocks for embedding, the intervals may be cut down to 2 or 3 h in each grade of alcohol.

Isopropanol, 99%, is probably the best all-round substitute for ethanol and is procurable at a modest price even when compared with tax-free ethanol. The tolerance of its mixtures with various paraffin solvents for small amounts of water at least equals on the average that of corresponding ethanol mixtures (Table 4-1). As seen in the same table, the tolerance of corresponding acetone and dioxane mixtures for water is low, that of methanol mixtures is intermediate.

According to Hauser (*Mikroskopie*, **7:** 208, 1952), tissues may be infiltrated in paraffin directly from isopropanol or from a warm (50°C) mixture of paraffin and isopropanol.

Nevertheless, in spite of its avidity for atmospheric water vapor and the low tolerance of its fat-solvent mixtures for water, acetone is often a very effective dehydrating agent when rapidity of action is desirable. No fewer than four changes of acetone of 40 min each should be used as a matter of routine. Of these only the fourth need be fresh acetone. For the third change use the acetone which has previously been used once for the last change, for the second use that previously used twice, and for the first that used for three previous changes. The acetone used 4 times can be saved for cleaning paint brushes or for redistillation, or it may be discarded.

Even this process may be expedited by using four changes of fresh acetone for 20 min each.

Following dehydration with alcohol or acetone, blocks should be transferred to a paraffin solvent which is miscible with the dehydrating agent. An intermediate bath composed of equal parts of the dehydrating agent and of the paraffin solvent or even two such baths, the first of a 2 : 1, the second of a 1 : 2 mixture, are recommended, particularly when working in very humid atmospheres.

PARAFFIN SOLVENTS. The best paraffin solvents are benzene, toluene, xylene, petroleum ether, carbon bisulfide, chloroform, carbon tetrachloride, and cedar oil. All these agents are

TABLE 4-2 SCHEDULES* FOR DEHYDRATION, CLEARING, AND PARAFFIN INFILTRATION

I	II	III	IV	V	VI	VII
Water	Water	Water	Water	Water	Water	Water
70% alcohol 16 h		Acetone (used 3 times) 40 min	Acetone (fresh) 20 min	Acetone (used) 40 min	Alcohols or acetones as before	Dioxane 1 h
85% alcohol 8 h	80% alcohol 16–24 h	Acetone (used 2 times) 40 min	Acetone (fresh) 20 min	Acetone (used) 40 min		Dioxane 1 h
95% alcohol 16 h	95% alcohol 16–24 h	Acetone (used 1 time) 40 min	Acetone (fresh) 20 min	Acetone (used) 40 min		Dioxane 1 h
100% alcohol 2 h	100% alcohol 1 h	Acetone (fresh) 40 min	Acetone (fresh) 20 min	Acetone (fresh) 40 min		
	100% alcohol 1 h					
100% alcohol + xylene āā, 1 h		Acetone + xylene āā 40 min		Acetone + xylene āā 40 min	Cedar oil 16 h	
Xylene 30 min	Xylene 30 min	Xylene 40 min	Xylene 20 min	Xylene 30 min	Xylene 30 min	
Xylene 30 min	Xylene 30 min	Xylene 40 min	Xylene 20 min	Xylene 30 min	Xylene 30 min	
				Xylene 30 min		
				Xylene 30 min		
	Paraffin (55°) 30 min		*or*	Paraffin (55°C) in vacuo at 25 mm mercury 15 min		Dioxane + paraffin āā, 1 h
	Paraffin 30 min					Paraffin 15 min
	Paraffin 30 min					Paraffin 45 min
	Paraffin 30 min					Paraffin 2 h
			Embed			

* Schedules I and III are routine alcohol and acetone schedules; II and IV are accelerated schedules; V is a special schedule for fatty tissues; VI is a cedar oil schedule for skin, muscle, uterus, and other difficult objects; and VII is a dioxane schedule.

Labels written on paper with india ink and dipped in smoking-hot paraffin should be carried through the solvents with the tissue and finally attached to the paraffin block.

TABLE 4-3 TECHNICON SCHEDULES FOR FIXATION, DEHYDRATION, AND PARAFFIN INFILTRATION

	Surgical schedules			Autopsy tissue schedules			Peers' brain schedule: formalin 4 days, chromate 3 days; wash 6 h
	Fresh tissues		Prefixed tissues	48-h formalin-chromate fixatives	24-h sublimate-formalin fixatives	24-h alcohol-acetic acid fixatives	
Fast schedule	Alcoholic formalin	Zenker's fluid					
A	B	C	D	E	F	G	H
1. Alcoholic formalin 15 min	Alcoholic formalin 6 h	Zenker's fluid 8 h	80% alc. 2 h	Water 2 h			80% alc. 6 h
2. 95% alc. 15 min	95% alc. 1 h	Water 2 h	80% alc. 1 h	Water 2 h			80% alc. 6 h
3. 99% alc. 15 min	95% alc. 1 h	Water 2 h	95% alc. 1 h	80% alc. 2 h	80% alc. 2 h		95% alc. 6 h
4. Acetone 15 min	95% alc. 1 h	80% alc. 0.5% I_2 1 h	95% alc. 1 h	80% alc. 16 h	80% alc. 16 h		95% alc. 6 h
5. Acetone 15 min	95% alc. 1 h	95% alc. 0.5% I_2 1 h	100% alc. 1 h	95% alc. $CaCO_3$ 1 h	95% alc. $CaCO_3$ 1 h	95% alc. $CaCO_3$ 1 h	100% alc. 6 h
6. Acetone 15 min	95% alc. 1 h	100% alc. 1 h	100% alc. 1 h	95% alc. 1 h	95% alc. 1 h	95% alc. 1 h	100% alc. 6 h
7. Chloroform 15 min	95% alc. 1 h	100% alc. 1 h	Acetone + xylene āā 1 h	95% alc. 2 h	95% alc. 2 h	95% alc. 2 h	Acetone + xylene āā 6 h
8. Chloroform 15 min	99% alc. + xylene āā 1 h	Methyl salicylate 1 h	Xylene 2 h	100% alc. 2 h	100% alc. 2 h	100% alc. 2 h	Xylene 6 h
9. Paraffin 15 min	Xylene 1 h	Xylene 1 h	Cedar oil 2 h	100% alc. 2 h	100% alc. 2 h	100% alc. 2 h	Cedar oil 6 h
10. Paraffin 15 min	Xylene 1 h	Xylene 1 h	Xylene 2 h	Cedar oil 16 h	Cedar oil 16 h	Cedar oil 16 h	Xylene 6 h
11. Paraffin 15 min	Paraffin 1½ h	Paraffin 1½ h	Paraffin 2 h	Gasoline 1 h	Gasoline 1 h	Gasoline 1 h	Paraffin 6 h
12. Paraffin 15 min	Paraffin 1½ h	Paraffin 1½ h	Paraffin 2 h	Gasoline 1 h	Gasoline 1 h	Gasoline 1 h	Paraffin 6 h
Time 3 h	Time 18 h	22 h	18 h	48 h	44 h	26 h	72 h
Embed	Embed	Embed	Embed		Paraffin in vacuo 15–30 min		Embed
				Embed	Embed	Embed	

miscible with paraffin at 56°C. The disagreeable odor and the toxicity of the fumes of carbon disulfide usually operate to exclude it as a paraffin solvent, although it is said to give excellent results (Lee).

As far as we can determine on direct comparison of blocks of the same tissues, chloroform (bp 61°C, vapor pressure 160 to 248 mm at 20 to 30°C) has no definite superiority over the far cheaper and probably less toxic carbon tetrachloride (bp 77°C, vapor pressure 91 to 143 mm at 20 to 30°C). (The lethal dose of carbon tetrachloride is perhaps double that of chloroform in mice.)

Benzene, although its fumes are more toxic, seems at least as good as its more expensive and less volatile homologs toluene and xylene, on the same basis of direct comparison.

For many years Lillie used with great satisfaction a high-test gasoline, designated *white gas* or *aviation gasoline, lead-free*, which is essentially a petroleum ether. (Caution: lead tetraethyl is dangerously toxic.) Ligroin, petroleum naphtha, or the British "white spirit" can be used. These are fluids of the hexane-heptane range.

From the data in Table 4-1, methyl salicylate would seem to have very interesting possibilities as a clearing agent with a relatively high tolerance for water. Its odor is powerful, but it should be useful when anhydrous alcohol is unavailable.

Of the fluids cited, cedar oil, gasoline, and petroleum ethers occasion the least hardening of tissues. Overnight immersion in gasoline and petroleum ether does not render tissues brittle. Chloroform and xylene appear to make tissues more brittle, but this does not interfere especially with sectioning. Thin cedar oil is an excellent dealcoholization or clearing agent and is highly recommended for such objects as human skin, uterus, thick masses of smooth muscle and tendon, and the like, since it appears to improve their consistency for cutting. It requires multiple changes of paraffin to remove it from the tissue. We usually prefer to interpose two or three baths of 20 to 40 min each in gasoline, xylene, or benzene. Tissues may be left in cedar oil for months without harm. After use for clearing for some time, cedar oil may be restored by filtering and then heating to 60°C in vacuo for 30 to 60 min to remove alcohol, acetone, and water.

Occasionally, after use for clearing of material fixed in acetic acid–alcohol fixatives, cedar oil may crystallize into a firm mass melting at around 35°C. A small quantity of this altered oil will cause a whole bottleful to crystallize. Heating to 200°C restores the normal behavior of the oil.

Popham (*Stain Technol.*, **25:** 112, 1950) recommends, instead of heating, the addition of a little (1 : 80) xylene to reliquefy the altered cedar oil.

Except for cedar oil, we recommend at least two changes of the paraffin solvent between the dehydration and the paraffin infiltration.

The use of saturated solutions of paraffin in the solvent has been recommended as an intermediate step between clearing and paraffin infiltration. It is unnecessary for routine work, and we have not used it for years. It may be necessary, however, for such difficult objects as parasitic worms, fleas, and ticks, on account of their chitinous exoskeletons.

PARAFFIN INFILTRATION.   This is done in an oven regulated to a temperature a few degrees above the melting point of the paraffin used. Ordinarily paraffin is quite satisfactory as obtained from the refiners. It may be obtained in approximately 5-kg blocks directly from the refiners in lots of eight blocks at a materially lower cost than when bought in pound blocks from biologic supply houses.

Ribboning consistency is sometimes improved by addition of 10 or 20% beeswax or by addition of 3 to 5% Halowax. Soft, or 52°C, paraffin will dissolve 15% by weight (but not 20%) of this Halowax (mp 115 to 125°C) with a resultant *fall* of the congelation point to about 50.5°C and quite pronounced hardening of the paraffin. Paraffins of even lower melting point (40 to 42°C) are available, and may be similarly hardened. Beeswax makes the mixture more sticky than pure paraffin, so that sections stick together better.

Paraffins recommended for sectioning range in melting point from 45 to over 60°C. For our laboratory conditions Lillie finds 45 and 50°C paraffin too soft. Some workers recommend a mixture of 50 and 55°C paraffin, and some prescribe infiltration first by a soft, then by a hard paraffin. For us this practice simply prolongs the heating period unnecessarily. For ordinary work Lillie found a paraffin of 55 to 56°C melting point satisfactory with laboratory temperature ranging from about 22°C in winter to 30 to 35°C in summer.

The usual practice is to use three or four changes of melted paraffin of 30 to 60 min each, in order to infiltrate tissue thoroughly and rid it of traces of the solvent which might unduly soften the paraffin within the tissue.

Such prolonged heating inevitably shrinks tissues to a considerable extent. For some 15 years Lillie used a vacuum chamber within the oven for infiltration in vacuo. With the use of a volatile solvent, 10 to 30 min infiltration at an absolute pressure of 25 mm mercury is adequate, and pressures as high as 175 to 200 mm can be used ($= -585$ to -560 mm, or 23 to 22 in, of vacuum). Vacuum infiltration furthermore removes air and gases from cavities within the tissues, notably the pulmonary alveoli, and permits their filling with paraffin.

Peers used a 3-day Technicon schedule for whole human brainstems (Table 4-3, schedule H); also for cerebellum, large blocks of cerebral cortex, whole hemispheres of dog, cat, and monkey brains, and the like. Blocks are cut transversely about 6 mm thick after fixation in 10% formalin for some days or even weeks. They are then hardened for 3 days in 2.5% aqueous potassium dichromate solution and washed 6 h in running water. On the Technicon they then pass through twelve 6-h baths for dehydration, clearing, and paraffin infiltration. Sections are floated out on 70% alcohol warmed to 40 to 45°C, dipped out on large slides, immediately blotted down with hard, smooth filter paper, and then dried for 1 to 18 h at 40 to 45°C.

We have included also in Table 4-3 Technicon schedules for fresh and formalin-fixed surgical tissues (A, B, C, D) and for autopsy tissues (E, F, G). Schedules E, F, and G are designed to use the same time disk and solutions. Tissues fixed in chromate fixatives (including Zenker's fluid variants) are started at step 1 on the schedule; mercuric chloride tissues, which are normally transferred directly to 70 or 80% alcohol, start at step 3; and tissues fixed in acetic acid–alcohol fluids, such as Carnoy, acetic acid–alcohol–formalin, etc., start at step 5. The calcium carbonate in the 95% alcohol of step 5 is intended to neutralize acetic acid which is carried over from the fixing fluid.

Latterly Lillie has preferred to complete the dehydration, clearing, and paraffin infiltration by hand when fixation has been done in Clark or Carnoy fluid or in methanol chloroform mixtures. From the Carnoy fluid thin tissue blocks are passed into equal volumes of ethanol and chloroform for 30 to 60 min and thence to two 30- to 60-min changes of chloroform. Tissues from the second (1 : 1) methanol chloroform are cleared directly in the two changes of chloroform.

In working with Zenker's fluid variants it is necessary to use plastic capsules for tissue and to have the metal parts completely protected by first coating them with smoking-hot paraffin. It is probably best to put tissues into the Technicon only after the fixation period is completed.

Incomplete dehydration is a common cause of the shrinkage and hardening of tissues within paraffin blocks after they have been cut from and put away. It is to be avoided by the use of sufficient changes of pure dehydrating agent to dehydrate the tissues completely before clearing and paraffin infiltration.

EMBEDDING. During the process of embedding, the tissue blocks must be oriented so that sections will be cut in the desired plane. Generally it is convenient to place the surface of the block from which sections are desired next to and parallel to the bottom of the embedding container.

A convenient embedding container may be made from a 5×7.5 cm (2×3 in) oblong of fairly heavy (2 to 3 mm thick) sheet metal and two L-shaped cross sections of angle iron about 2 to 3 cm high. Angle iron may be obtained in various sizes from 3 to 5 or more cm arm length, and cut with a hacksaw into desired lengths. The two L-shaped pieces are laid on the metal plate so as to enclose a rectangular or square space. This is then partly filled with paraffin, and the tissue is placed in it in the desired position and pressed firmly against the bottom. The container is then filled with melted paraffin and allowed to cool until a coherent film forms on the surface. It should then be chilled rapidly in cold or preferably ice water. Rapid cooling gives smaller crystal size to the embedding mass and improves cutting consistency, as well as decreasing permeability to water and air. The metal blocks are removed by dropping them sharply on the table, when the three metal pieces and the paraffin block usually separate cleanly.

TRIMMING OF BLOCKS AND PREPARATION FOR SECTIONING. The upper curved surface which is to form the back of the block is then trimmed off flat with a fairly stout knife. The four sides are then cut down square to within 1 or 2 mm of the edge of the tissue block on the cutting surface and sloping outwardly toward the back, so as to form a truncated pyramid. When ribbons are to be cut, it is convenient to cut narrow triangular wedges from the four corners of the block, so that in the ribbon formed by the coherent sections V-shaped notches are present on both sides between the individual sections. When the block is trimmed, the number should be affixed to one side with a hot spatula.

SPECIAL BLOCKING AND TRIMMING PROCEDURE FOR CROSS SECTIONS OF MULTIPLE SMALL TUBULAR STRUCTURES. When it is desired to prepare cross sections of a number of small tubular structures in a definite arrangement in the final section, the following special procedure may be of value. It can be used to arrange in proper sequence several levels of small animal intestine or several small arteries and veins.

On collecting the material, each block is placed successively in the desired order on an ordinary straight pin. For example, with guinea pig intestine we first place on the pin two levels of duodenum, starting from the pyloric end, then two or three levels of jejunum and two or three of ileum so spaced as to represent the whole extent of the small intestine. We have often placed seven or eight segments of intestine 1 to 2 cm in length on a single pin. The pin is left in place throughout the dehydration, clearing, and paraffin or celloidin-paraffin infiltration procedures. When the embedding stage is reached, lay the combined block flat in the bottom of a metal-plate L block assembly which has been partly filled with paraffin. The row of intestinal segments is placed so that the end from which the first sections are to be cut abuts one end of the embedding mold and the head of the pin which still holds them together abuts one of the sides. Press down firmly and chill the block.

When trimming, the original bottom of the block forms the face which will first strike the microtome knife, and the end which the extremities of the intestinal segments abut is the surface from which sections are to be cut. On trimming the two sides the head of the pin is uncovered and the pin pulled out with a slight twisting motion, using forceps if necessary. The original upper surface of the poured block is the side away from the knife in cutting and is trimmed to come fairly close to the intestines at the cutting surface and sloped away to form a broad base at the other extremity of the block, which is to be attached as usual to a metal block carrier.

If animals are starved or fed on soft cooked food, free of cellulose and lignin, for 24 h before killing, the material should section readily. Calcified small blood vessels should be decalcified.

For attachment to the microtome it is necessary to fasten the paraffin blocks to wooden or fiber blocks or to metal object holders. The last are the most convenient, as they may simply be heated to above the melting point of paraffin and then pressed firmly against the back of the paraffin block. The block and object holder are then immersed in cold water. The object holder should be hot enough to melt its way 1 to 2 mm into the back of the block, but no further.

For attachment to wooden or fiber blocks, melt the back surface of the block with a hot metal spatula, press it immediately against the wood or fiber, and immerse it at once in cold water.

For most routine embedding, commercially prepared molds which can be clamped directly into the block clamp of the microtome are very convenient and are preferred by technicians generally. They are not readily adaptable to the preservation of order of multiple cross sections of small animal intestine or small blood vessels. Here the pin procedure with the L blocks outlined above should be preferred.

SECTIONING. The wooden or fiber block or metal object holder is then inserted into the object carrier of the microtome and clamped firmly in place. The object carrier is then oriented so that the surface of the block is parallel to the knife edge. While adjusting the microtome, its motion should be locked to avoid accidental cuts either to the specimen or to the hands of the operator.

One then cuts off paraffin in rather thick sections by operating the microtome in the usual manner, until the tissue is reached. It may be necessary to readjust the orientation of the block at this point, so as to obtain complete sections in the desired plane.

Sections may then be removed individually by a small dampened scalpel, or, by holding the end section with the same implement at a slight tension, a number of sections may be cut in series so as to form a ribbon. This is necessary when serial sections are required. Convenient lengths of ribbon are transferred to a smooth board or cardboard surface and cut into segments of the desired number of sections with a sharp knife. They adhere to the cardboard along the line of cut and are loosened from it by passing the edge of the knife obliquely along the line of adhesion. Sections are then lifted singly or in strips onto the surface of a pan of water heated to about 38°C, or onto water on a 25 × 75 mm glass slide which is then warmed on a metal plate. In preparation of serial sections, a number of rows or sections may thus be arranged in proper order on a single slide. The warming causes the wrinkles in the section to flatten out, and the whole section or ribbon segment elongates during the process. Persistent wrinkles or rolls may be flattened out by stretching the section with a pair of small knives or needles. From the pan the sections are then floated onto clean glass slides and removed from the water; one edge of the section is held in place with a needle or knife. If the sections are floated on slides, the excess water is drained off to one side, the section being held in place as before. When preparing serial sections of several rows to a slide, it is well to place a length of blank paraffin ribbon at each margin, so that sections will not lie under the edge of the cover glass when stained.

It is preferable to use fresh distilled water for floating out sections. If rapid drying is required, one may use 30 % or even 95 % ethyl alcohol instead.

Many workers prefer to spread a minimal quantity of Mayer's albumen-glycerol fixative on the surface of the slide before floating the section onto it. A very small drop is placed on one end of the slide, the previously thoroughly cleansed ball of the little finger is used to spread it over the whole surface of the slide, and any excess is then wiped off on the hypothenar portion (little finger side) of the palm of the hand.

Traditionally (Ehrlich's "Encyklopädie") Mayer's albumen-glycerol is made by thoroughly mixing 1 volume of the whites of perfectly fresh eggs with an equal volume of pure glycerol. The mixture is then filtered through absorbent cotton or relatively coarse filter paper at 55 to 58°C (paraffin oven temperature). A small lump of thymol or camphor is added to prevent growth of molds.

Faulkner and Lillie (*Stain Technol.*, **20**: 99, 1945) have substituted a 5% solution of dried egg white made by shaking gently at intervals for 1 day in 0.5% sodium chloride solution. Avoid frothing when shaking and filter on a Buchner funnel with vacuum. At least 90 to 95% of the solution should be recovered as a clear filtrate. To 50 ml of this filtrate add 50 ml glycerol and 0.5 ml 1 : 10,000 merthiolate (sodium ethylmercurithiosalicylate).

Since 2 or 3 ml distilled water is used to float out a section on a slide thinly coated with this fixative and the constituents of the fixative are readily soluble in water, it is difficult to imagine that any of the "fixative" remains on the slide to fasten a section to it. It seems probable instead that it functions as a surface tension depressant and thereby secures closer capillary adhesion of the section to the slide.

Priman (*Stain Technol.*, **29**: 105, 1954) recommends diluted mammalian blood serum (15 ml plus 5 ml distilled water plus 6 ml 5% formalin; filter through filter paper) and states that the material spreads well, does not stain, and is superior to egg albumen fixatives in adhesiveness. Although blood plasma in vessels often colors red with periodic acid Schiff technics, it is probable that a thin film of a diluted serum would have too low a density to occasion an appreciable staining artifact.

Pappas (*Stain Technol.*, **46**: 121, 1971) advises the use of chrome alum-gelatin (subbing) solution as a general adhesive for paraffin sections. Sections are stated to be resistant to removal by acids and bases including 1 N HCl or H_2SO_4, 5% oxalic and trichloroacetic acids, 1 N NaOH, organic solvents, and hypochlorite. The procedure involves cleaning slides for 12 h in dichromate-H_2SO_4 (the standard solution), washing in running water for 6 h, rinsing in distilled water, draining for 2 or 3 s, dipping in the subbing solution briefly, and drying in a vertical position free of dust. Do not allow slides to dry between the cleaning and coating steps.

The subbing solution is prepared by dissolving 5.0 g USP gelatin in 1 liter of warm distilled water and adding 0.5 g chrome alum [$CrK(SO_4)_2 \cdot 12H_2O$]. After cooling, the solution is filtered through Whatman No. 1 filter paper. The solution keeps for *only* 48 h at 5°C and should be discarded thereafter. If a smaller quantity will serve, the amounts above may be divided by 10, to give a final 100 ml.

Paraffin sections spread on a warm-water bath are floated on subbed slides in the usual manner and dried. The whole procedure seems subject to the same criticism as made above to glycerol albumen in general (Lillie, personal communication).

Cook (*Stain Technol.*, **40**: 321, 1965) tested the efficacy of several adhesives [glycerol-albumen (G. T. Gurr) diluted 1 : 3 with distilled water containing 2 drops (0.1 ml) formalin per 50 ml; starch; amylopectin; gelatin; agar; and undiluted human plasma] for their ability to retain fresh frozen, paraffin-embedded, and double-embedded sections through alkalis. Cook concluded that the best overall adhesive is plasma and that 56°C is the best drying temperature for section retention. He did not compare distilled water or 30 to 95% alcohol.

Stoward et al. (*Histochemie*, **14**: 212, 1968) recommend the use of low-viscosity nitrocellulose as an adhesive for tissue sections that resists the action of sodium hypochlorite. Grease-free slides are immersed for a few seconds in 0.25% solution of low-viscosity nitrocellulose in 50 : 50 ether–absolute ethanol. After slides are dried at 24°C, paraffin-embedded sections are mounted from a water bath (37 to 40°C), dried at 37°C for 2 to 3 h, and dried overnight at 56°C. Sections are then used in the usual fashion.

Moore and Berni (*Stain Technol.*, **37**: 383, 1962) advise the use of Epoxol 9-5 (Swift & Co., Chemicals for Industry Dept., Chicago, Illinois 60604) as a reagent-resistant adhesive for attachment of methacrylate-embedded sections to slides. Approximately 1 ml Epoxol 9-5 is placed on a slide, and 3 drops of catalyst [ethanol borontrifluoride, $(C_2H_5OH)_2 \cdot BF_3$, available from Harshaw Chemicals Co., Cleveland, Ohio 44106] is added and mixed rapidly. The catalyst is used as a 1 : 5 dilution with methanol. The mixture hardens in about 3 min. Sections are rapidly placed on the adhesive and made adherent by rolling a glass rod over the section in contact with the adhesive. The specimen in the section should not be pressed into the adhesive where it will not be subsequently exposed to reagents or staining.

Slides are allowed to remain covered at 24°C for 1 h, rinsed in three changes of methanol (time unspecified), and cured at 70°C for 48 h. After curing, methacrylate embedding medium is removed with 1 : 1 methyl ethyl ketone and chloroform, and slides are drained and air-dried and are now ready for testing and staining. The method is stated to be particularly useful as a medium for holding specimens subjected to strong acids and alkalis.

Ordinarily such paraffin sections are dried overnight before deparaffinizing and staining. With the aid of gentle heat and an air current they may often be dried sufficiently for staining in an hour or so. By using dilute alcohol for floating the sections onto the slides, and heat for drying, Lillie often stained sections without undue losses in 30 min.

Some workers routinely dry at paraffin oven temperature or even at 65°C in the open. Smith (*Stain Technol.*, **37**: 339, 1962) warns against this procedure, noting considerable interstitial shrinkage of cells as a result of it, as well as diminished staining with specific connective tissue stains. We consider it probable that drying at 5 to 10°C above the melting point of the paraffin tends to permeate all structures thoroughly with melted paraffin, thereby rendering it more difficult to remove, enhancing the frequency of Nedzel's paraffin artifacts, and preventing the use of procedures employing undeparaffinized sections for staining. The practice of floating on 30 to 80% alcohol for spreading and smoothing of sections also accelerates drying and can be carried out successfully at room temperatures of 20 to 30°C.

Hard paraffin is more apt to curl when sections are cut than soft, but soft paraffin permits more lateral compression of tissues along the path of the knife. Paraffin blocks may be made harder for cutting by preliminary immersion in ice water, and it may be advisable to chill the microtome knife as well. An electric light bulb near the object carrier of the microtome tends to make the blocks softer.

Sometimes during ribbon sectioning the ribbon acquires a charge of static electricity which causes it to move about violently and to adhere to various adjacent objects. Breathing on the ribbon and block surface may remedy this difficulty.

Tissues containing much blood or yolk (and certain other substances, including bone and cartilage) are apt to be brittle and to shatter under the microtome knife instead of yielding coherent sections. Painting the surface of the block before cutting each section with a 0.5% solution of collodion in absolute alcohol and ether or acetone will prevent their crumbling. The sections smooth out better if floated on 80% alcohol, but if the collodion film is thin enough, this is often not necessary.

Lendrum (*Stain Technol.*, **19**: 143, 1944) recommends that when tough, hard tissues fail to give satisfactory paraffin sections, the paraffin block be soaked face down in a mixture of 9 parts glycerol and 1 part aniline for 1 to 3 days. This probably serves best with incompletely dehydrated tissues, as it seems to make little difference when care has been taken to dehydrate completely before paraffin infiltration.

After sectioning from a paraffin block Lillie found it good practice to dip the cut surface briefly into rather hot (70 to 80°C) paraffin before putting it away. This helps to prevent drying out and shrinkage of perhaps imperfectly dehydrated tissues. With adequate dehydration this practice should not be necessary, but it takes little time and also helps protect tissue from roaches. We have seen unprotected tissue completely eaten out of the paraffin blocks. The paraffin recoating also protects against oxidation.

Paraffin blocks should be stored in a reasonably cool place. We have used small pasteboard boxes containing the blocks of one or more consecutive cases. These are packed consecutively, standing on end, in drawers of slightly greater depth than the greatest diameter of the boxes, with the exposed end bearing the case number. The drawer bears the first and last numbers on its exposed end.

Sainte-Marie (*J. Histochem. Cytochem.*, **10**: 250, 1962) recommends the following embedding method for studies employing immunofluorescence:

1. Remove tissue promptly after death of the animal, cut slices no thicker than 5 mm, and immerse into 95% ethanol cooled to 4°C. Avoid squeezing, tearing, warming, or drying the specimen.
2. After hardening in ethanol, trim specimens with a razor blade so that the specimen does not exceed 2 to 4 mm in thickness. Place them in a perforated capsule or cheesecloth to facilitate handling, and fix them for 15 to 24 h in 95% ethanol at 4°C.
3. Dehydrate them in four changes of precooled absolute ethanol (1 to 2 h each depending on thickness of the specimen) at 4°C.
4. Pass them through three changes of xylene (1 to 2 h each) at 4°C. Tissues may be stored in either the absolute ethanol or xylene for 1 to 2 days at 4°C.
5. Bring the tissues in xylene to room temperature, and pass them through four vessels of melted paraffin (1 to 2 h each) at 56°C. Embed at 56°C and store the blocks at 4°C. It is preferable to cut blocks immediately, because autofluorescence increases with time. However, Sainte-Marie states blocks 2 to 3 months old have proved satisfactory.
6. Carry out sectioning as usual; however, flotation of sections on water at 4°C should be brief to avoid loss of antigens. Place sections on slides without egg albumen and dry them at 37°C for 30 min. Sections may be stored in a desiccator at 2°C for a few days or weeks; however, best results are achieved if they are stained promptly.
7. Sections are best deparaffinized in two xylene baths for 10 to 15 s each with gentle agitation. Then bring them briefly through graded ethanols and into the buffer system to be employed.

ESTER WAX.   Steedman's ester wax (*Q. J. Microsc. Sci.*, **88**: 123, 1947) has been considerably modified to improve flattening and adherence of sections (*Q. J. Microsc. Sci.*, **101**: 459, 1960); a variant that permits sectioning at room temperatures up to 37°C is called *tropical ester wax* (ibid., **101**: 463, 1960). These are formulated as follows:

	Ester wax 1960, g	Tropical ester wax, g
Diethylene glycol distearate	60	60
Glyceryl monostearate	30	30
300-Polyethylene glycol distearate	10	
Triethylene glycol monostearate		10
Total	100	100

Both preparations are obtainable from British Drug Houses, Poole, Dorset, England.

Ester wax made in 1960 melts at about 48°C. About 4 h is required for infiltration; this time may be reduced by stirring. Sections are cut at 17 to 27°C. The knife bevel angle should be about 25°. Sections are flattened on water, either on the slide or in a pan at 45 to 50°C.

Infiltration may be done from 2-ethoxyethanol (Cellosolve), *n*-butanol, ethanol, or xylene, but dewaxing of sections is done with xylene.

With tropical ester wax infiltration is done at about 50°C, and sectioning may be performed at room temperatures of 17 to 37°C. If blocks and microtome are prechilled, sectioning can be done at even higher temperatures.

Both waxes tend to flocculate when kept liquid for some days at 48 to 50°C, but they may be cleared by heating to 70 to 80°C. After this reheating they may again be kept at normal infiltration temperatures for similar periods.

Salazar (*Stain Technol.*, **39:** 13, 1964) and Taleporos (*J. Histochem. Cytochem.*, **22:** 29, 1974) recommend highly the use of diethylene glycol distearate (DGD) as an embedding medium for high-resolution light microscopy. Sections may be cut as thin as 1 μ with a steel knife. The translucent wax is sufficiently hard to serve as an embedding medium for osmium-fixed material, and it is readily removed to permit phase contrast studies. Taleporos recommends the following method:

DGD (mp 47 to 52°C) is obtained from Ruger Chem. Co., 83 Cordier, Irvington, N.J., as white flakes which may be melted at 60°C in the paraffin oven, filtered, and used without further purification.

Fixed and dehydrated tissues are cleared in two changes of xylene or benzene (30 min each) and transferred to a wax-xylene mixture and later to wax. Taleporos obtained adequate infiltration of fixed, dehydrated, 1-mm slices of tissues within 50 min. Infiltrated tissues are placed in a mold (as for paraffin) and cooled. DGD wax contracts markedly during cooling. In shallow embedding dishes, it may be necessary to add wax with a pipette to the center of the specimen to ensure coverage after final setting. Glass storage beakers with DGD will crack when cooled.

Blocks are trimmed with a sharp razor blade, mounted on a chuck with melted DGD, and inserted into either an ultramicrotome with a glass knife and water trough or a rotary microtome with a sharp steel knife with a water trough. *The water trough is essential, and it must be adjusted to wet the inner facet of the very edge of the knife.* The edge must be free of grease.

Sectioning is done without cooling, with a slow movement of the flywheel. Section curling can usually be diminished by raising the trough water level. Sections as thin as 1 μ may be cut with a steel knife and as thin as 0.5 μ with a glass knife with minimal compression. In general, the thinner the section cut, the greater the friction and tendency for distortion. The necessity for a sharp knife without defects cannot be overemphasized.

DGD is brittle. For this reason, sections are carried on a flat surface (such as a shaved flattened swab stick) directly to a clean slide without adhesive. A wire loop used for ultramicrotomy may also be used. Water about the section is blotted, and the section is dried at 24°C. Sections placed in a desiccator at 24°C overnight are very resistant to harsh staining procedures and rarely loosen. The use of hot plates is not advised. Sections may be dewaxed with xylene.

A principal value of DGD wax is its hardness. It is also brittle. To obtain maximum value, it is essential that optimal fixation and dehydration of tissue has occurred prior to embedding, otherwise, maximum occurrence of artifacts is observed. Taleporos does not advise the use of conventional methods for histology and freeze-drying. Taleporos was unable to obtain sections thin enough for routine electron microscopy, although the wax did not melt in the electron beam.

CELLOIDIN AND LOW-VISCOSITY NITROCELLULOSE EMBEDDING. The lower viscosity of low-viscosity nitrocellulose permits more rapid infiltration with solutions of higher concentration. Consequently, harder blocks may be obtained in less time, and thinner sections can be cut.

In pathology this method is utilized principally for eyes and for the study of bones and their surrounding tissues *in situ*, and when it is essential to avoid shrinkage and the creation of artificial spaces. For such tissues a rather slow dehydration and a gradual infiltration are required.

The following schedule can be recommended for eyes:

1. Fix in 10% formalin for at least 48 h.
2. Dehydrate the eye with successive 24-h baths of 35, 50, 65, and 80% alcohol.
3. Open the eye by cutting off the upper and lower portions of the posterior chamber with two cuts passing horizontally respectively through the upper and lower margin of the anterior chamber and including the nerve head and the fovea centralis between them.
4. Return it to 80% alcohol for 3 to 4 h.
5. Then give it a 24-h bath in 95% alcohol.
6. Then immerse it for 24 h in two changes of 100% alcohol.
7. Next give it a 6-h bath in equal volumes of 100% alcohol and ether.
8. Infiltrate 5 to 7 days in 10% nitrocellulose in 100% alcohol and ether, and a similar period in 20% nitrocellulose, in a tightly stoppered container.
9. Embed in a fairly deep, cylindrical glass dish in which the 20% nitrocellulose, sufficient to cover the eye, reaches about halfway up the side. Place under a bell jar and examine daily.
10. When the surface is solid and the deeper portion still somewhat soft, flood with chloroform and let stand 16 to 24 h.
11. Pour off the chloroform, and let dry until the solid nitrocellulose block can easily be dislodged from the dish. Trim the block into octagonal form, leaving 1 to 2 mm of nitrocellulose on all sides. Dip the back surface of the block into 20% nitrocellulose, and mount it on a fiber or wooden block with a scored, or incised, surface. Fasten an identifying number to the side of the block with a little 20% nitrocellulose.
12. Let dry a few minutes, and return the mounted block to chloroform for several hours.
13. Transfer through successive 24-h baths of 3:1 and 1:1 chloroform and cedar oil mixtures to pure cedar oil. Keep in cedar oil at least 1 day before cutting.
14. Drain, and fasten the fiber or wooden block in the object clamp of the celloidin microtome. Cut sections and store them in 80% alcohol.

The technic for decalcified bone is similar. Sections have been cut as thin as 5 μ by this technic. Ordinarily 10-μ sections will serve.

After susa fixation for 12 h, transfer eyes directly to 95% alcohol. Treat them with iodine until the alcohol is no longer decolorized by adding tincture drop by drop with constant agitation. Let the brown solution stand 30 min. Wash out the excess iodine with three or four changes of 95% alcohol. Let the specimens stand overnight in 95% alcohol. Open the eyes as described above. Complete dehydration, infiltration, and embedding are as above.

When sectioning is completed, cut the nitrocellulose or celloidin blocks off the fiber or wooden block and store them in cedar oil or mineral oil. The latter is cheaper and does not dry or evaporate.

Chesterman and Leach (*Q. J. Microsc. Sci.*, **90**: 431, 1949) used three baths of low-viscosity nitrocellulose, at 5, 10, and 20% in ether alcohol mixture, allowing 3 to 5, 1 to 2, and 1 to 5 days, respectively. They then embedded in a solution of 140 g "industrial nitrocellulose damped with 7:3 butyl alcohol. HX.30/50," in 210 ml 100% alcohol and 250 ml ether, adding 5 ml tricresyl phosphate as a plasticizer to prevent cracking and separation from the tissue blocks. This solution is designated "20%." Apparently the nitrocellulose contains 98 g (or 70%) of the principal ingredient, and the 20% is by weight of the final mixture. Blocks are allowed to set in a loosely closed space for 1 to 3 days until a moderately stiff but still flexible gel is formed. They are then hardened for 1 to 3 days in two or more changes of 70% alcohol. Such blocks become hard and may be cut dry on a celloidin or paraffin microtome. Sections of 15 μ from half a cat brain or 5 to 7 μ from blocks of 5 mm^2 are reported.

RAPID NITROCELLULOSE EMBEDDING.   Koneff and Lyons (*Stain Technol.*, **12:** 57, 1937) recommend the following rapid schedule for low-viscosity nitrocellulose embedding. We have not tried this method.

1. Fix 1 h at 56°C or 2 to 5 h at 37°C in neutral 10% formaldehyde (formalin?), Bouin, susa, or Carnoy fluid (ethanol–chloroform–acetic acid, 6 : 3 : 1). Use screw-capped bottles.
2. Wash in several 1-h changes of distilled water, except Carnoy material, which is extracted 2 h at 56°C in three to five changes of 100% alcohol, and then transferred directly to 100% alcohol and ether 1 h at 56°C (step 4).
3. Dehydrate in two 30-min changes each of 70, 80, 95, and 100% alcohol at 56°C. Use iodized 70% alcohol for the first step if fixation was in susa or other mercury solution.
4. Transfer to 100% alcohol and ether 1 h at 56°C.
5. Infiltrate 1 h at 56°C in 10 g low-viscosity nitrocellulose dissolved in 50 ml 100% alcohol and 50 ml ether.
6. Infiltrate overnight in 25 g nitrocellulose in 45 ml 100% alcohol and 55 ml ether.
7. Infiltrate 2 to 3 h at 56°C in 50 g low-viscosity nitrocellulose in 40 ml 100% alcohol and 60 ml ether.
8. Embed on a fiber block in 50% nitrocellulose as usual.
9. Harden the blocks 1 h in two changes of chloroform.
10. Transfer to 80% alcohol, three changes, 20 min each.
11. Section the blocks while wet with 80% alcohol, and store blocks and sections in the same.

We suggest the cedar oil procedure from "Embed" on as an alternative, since this method appears to give thinner sections than sectioning from alcohol.

For squirrel and lizard eyes, Wall (*Stain Technol.*, **13:** 69, 1938) recommends fixation by immersion in Kolmer's fluid for 24 h, followed by successive 6- to 18-h baths in water, 35, 50, 70, and 95% alcohols, and infiltration by a hot nitrocellulose process. For demonstration of the ellipsoids particularly, Heidenhain's iron hematoxylin and phloxine are recommended.

For thinner sections of retina, Wall recommends removal of the lens before embedding. In any case it is necessary to cut a window through the sclera with a sharp razor blade before nitrocellulose infiltration.

NITROCELLULOSE AND PARAFFIN EMBEDDING.   For difficult and fragile objects a combined celloidin or low-viscosity nitrocellulose infiltration and paraffin infiltration and embedding process has been recommended. One infiltrates with the nitrocellulose as usual, clears with chloroform and cedar oil as above, and carries the blocks thence through two changes of benzene or chloroform into paraffin as usual.

Lillie used this process with a variable degree of success. The sections do not flatten well when floated on water, and are likely to curl and separate from the slides. They flatten better if floated on clean slides with 95% alcohol. This softens but does not dissolve the nitrocellulose. If sections still curl, it may be necessary to soften them with ether vapor, blot down with filter paper, and soak in 0.5 to 1% collodion before deparaffinizing. In any case deparaffinize the section, and simultaneously harden the nitrocellulose in chloroform, transferring thence directly to 95% alcohol.

For sectioning bones without decalcification, Arnold (*Science*, **114:** 178, 1951) dehydrates in acetone and infiltrates in a mixture of 55 g air-dried ($\frac{1}{2}$-s) nitrocellulose, 45 g diamyl or dibutyl phthalate, and 65 to 100 g anhydrous acetone for several days. The acetone-dehydrated tissues are placed in a separatory funnel, and the air is evacuated down to 10 mm mercury with a vacuum pump attached to the upper end of the funnel. The lower end is then immersed in the nitrocellulose solution, and the stopcock is opened slowly to allow the nitrocellulose to enter and cover the specimens. When 5 or 10 times the volume of the tissue has entered, the vacuum line is opened, and the pressure is allowed to rise to atmospheric pressure. Tissues are then transferred to closed containers, and infiltration is continued at 58°C under 3 atm pressure. (Acetone boils at 56.5°C; infiltration in an ordinary screw-capped bottle in the paraffin oven should be adequate to prevent evaporation.) Embedding is then completed by filling a paper container to 4 or 5 cm height with the nitrocellulose-plastic solution, orienting the specimen in the bottom and allowing evaporation at 20°C to a height of about 1 cm to produce a puttylike consistency. The specimen is then cut out, reoriented, and cemented to a regular fiber block with the same solution. It is allowed to dry for a week at 25°C until it is quite hard. The hard tissues are cut at 5 to 8 μ, wetting the block and sections in a 1% aqueous aerosol solution. A heavy sliding microtome and a specially hardened knife blade are used.

LILLIE AND HENSON'S SUCROSE FORMALIN AND FORMALIN, CELLOIDIN, PARAFFIN PROCEDURE.   For the preparation of thin serial sections of rodent intestine for the differential study of enterochromaffin and Paneth cell granules we have successfully utilized the following procedures (*J. Histochem. Cytochem.*, **8:** 182, 1960):

1.   Fix either directly in calcium acetate formalin or in a similar formula containing 30% sucrose after a prior 30-min soaking in 30% sucrose at 3°C. Fixation in either formalin solution is done at 3°C for 18 to 24 h.
2.   Wash 4 h in running water to remove sucrose.
3.   Dehydrate over a 48-h period in 70, 80, 95, and 100% alcohol.
4.   Infiltrate 16 to 20 h in 1% celloidin in equal volumes of absolute alcohol and ether. This step may be prolonged to several days if expedient.
5.   Drain briefly on a paper towel or blotter, and harden in two changes of chloroform, 2 h each.
6.   Infiltrate in paraffin containing 5% beeswax for 2 h at 60°C and 15 mm mercury pressure.
7.   Embed and section serially at 4 μ.

In this study sections were stained by azure A–eosin B, ferric ferricyanide, fast garnet GBC salt, the postcoupled benzylidene procedure, and a sequence combining the two last methods.

Excellent cell structure was shown by this procedure.

EMBEDDING MEDIA FOR ELECTRON MICROSCOPY

Water-miscible media theoretically could have many advantages, particularly in the area of enzyme histochemistry. If infiltration can be conducted at a low temperature, enzyme inactivation may be minimized. If the medium is chemically inert, inactivation of enzymes is minimized. It is hoped that the embedding material will not alter reactivity of staining methods to be employed and the medium will possess desirable cutting qualities. Strong dehydrating agents routinely employed for processing tissues for electron microscopy result in substantial modification of tissues. The usual dehydrating agents also remove lipids. It is noted that no embedding medium is available at this writing for the satisfactory preservation and processing of lipids. The reader should be aware that monomers of most water-miscible resins are also lipid solvents unless low temperatures are used. Although several water-miscible embedding media are available and described below, all have the cited disadvantages.

AQUON. Gibbons [*Nature* (*Lond.*), **184**: 375, 1959] prepared a water-miscible fraction from Epon 812. Aquon is highly miscible with water below 15°C, but it is less miscible at higher temperatures. Being an epoxy resin, reactivity with nucleic acids and to a lesser extent with proteins occurs. Aquon polymerizes into a hardened resin without much shrinkage. Although sections tend to soften slightly in water, thick sections may be cut and stained readily for light microscopy. Cosslett (*J. R. Microsc. Soc.*, **79**: 263, 1960) notes that the loss of material from sections is somewhat comparable to that of sections embedded in Araldite.

The stock embedding mixture is composed of Aquon resin, 10 ml, and dodecenyl succinic anhydride (DDSA), 25 ml (available from R. P. Cargille Labs., 117 Liberty Street, New York, New York 10006, and Polysciences Inc., Warrington, Pennsylvania). The stock solution may be kept at 25°C for at least a week without ill effects. The final mixture used for embedment of tissues prepared immediately before use is composed of stock embedding mixture, 10 ml, and benzyldimethylamine (BDMA), 0.1 ml (available from Maumee Chemical Co., 2 Oak Street, Toledo, Ohio).

For processing, washed tissue is dehydrated through increasing concentrations of resin in water at 4°C until tissues are completely dehydrated. Tissues are retained for 4 h in the resin, transferred to dry gelatin capsules in fresh resin, and polymerized for 4 days at 60°C. It is essential to store the hardened blocks in a desiccator; otherwise moisture is absorbed.

DURCUPAN. Available from Polysciences Inc., Warrington, Pennsylvania, and Ciba Chemical Company, Plastics Division, Kimburton, Pennsylvania, this is a water-soluble epoxy resin introduced as an embedding medium by Stäubli [*C. R. Acad. Sci.* (*D*) (*Paris*), **250**: 1137, 1960]. Kushida [*J. Electron Microsc.* (*Tokyo*), **12**: 167, 1963; ibid., **14**: 52, 1965] also used it as a dehydration agent. Durcupan is a resin with relatively low viscosity. It is a strong irritant to skin. The embedding mixture is composed of Durcupan, 5.0 ml; DDSA, 11.5 ml; accelerator 960, 1.0 to 1.2 ml; dibutyl phthalate, 0.2 to 0.4 ml (available from Barrett Division, Allied Chemical and Dye Corporation, New York). Durcupan must be thoroughly dehydrated before hardener and accelerator are added; otherwise, polymerization will not occur. Hayat (1970) recommends dehydration according to the following schedule: Durcupan in distilled water 50% w/v for 30 min; 70% for 30 min; 90% for 30 min; and 100% Durcupan (two changes for 1 h each).

Dehydrated Durcupan-impregnated specimens are placed in gelatin capsules in fresh resin and polymerized for about 24 h at 45°C. Leduc and Bernhard (*J. Biophys. Biochem. Cytol.*, **10**: 437, 1961) advise collection of sections on water, whereas Stäubli [*C. R. Acad. Sci.* (*D*) (*Paris*), **250**: 1137, 1960] recommended collection of sections on acetone. Sections collected on acetone tend to fragment, and those collected on water tend to soften. Thorough curing of the block tends to minimize softening.

Hardened Durcupan tends to shatter when making ultrathin sections. Kushida [*J. Electron Microsc.* (*Tokyo*), **13**: 139, 1964] added a softener, Cardolite NC-513 (available from Minnesota Mining and Manufacturing Co., Irvington Chemical Div., 500 Doremus Avenue, Newark, New Jersey), in varying proportions along with nadic methyl anhydride (NMA) (available from Polysciences Inc., Warrington, Pennsylvania), in attempts to obtain optimal cutting and other qualities. Kushida studied six mixtures containing Durcupan, Cardolite

NC-513, and NMA. All six contained 100 g Durcupan. Mixtures 1, 2, 3, 4, 5, and 6 contained 0, 10, 20, 30, 40, and 50 g of Cardolite NC-513, respectively. Finally, mixtures 1, 2, 3, 4, 5, and 6 contained 124, 131, 139, 147, 155, and 163 g NMA, respectively.

The amine accelerator 2,4,6-tri(diethylaminomethyl)phenol (DMP-30) is added in the proportion of 1.5 to 2.0 ml/100 ml of the above mixture. Cardolite is a resin with one epoxy group per molecule. Mixture 1 is very hard but mixtures 2, 3, and 4 generally permit sectioning at 20°C.

GLYCOL METHACRYLATE. Glycol methacrylate (GMA, 2-hydroxyethyl methacrylate) has the following structure:

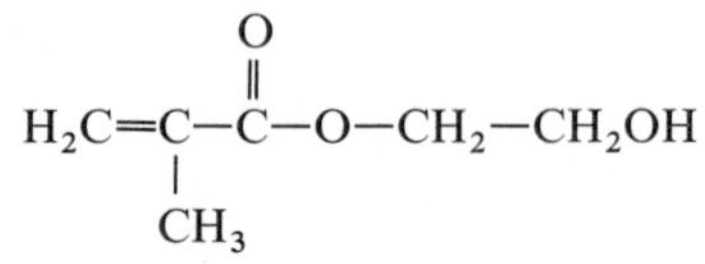

$$H_2C{=}C{-}\overset{\overset{\textstyle O}{\|}}{C}{-}O{-}CH_2{-}CH_2OH$$
$$\underset{\textstyle CH_3}{|}$$

GMA has a density of 1.065 at 20°C, a viscosity of 0.70 poise at 20°C, an index of refraction of 1.4540 at 20°C, and a boiling point of 37°C/3.5 mm mercury. It is miscible with water, methanol, ethanol, and ether. GMA will polymerize with conventional free radical methods using a catalyst and ultraviolet radiation. A photosensitizer such as benzoin may be added to the extent of 0.5% of the monomer to enhance decomposition of the catalyst by ultraviolet light. The degree of polymerization of GMA is a function of time, temperature, and concentration of catalyst. Hardness of a block may be modified by variation of the amount of comonomer or water used. Softer blocks are obtained by using increased amounts of butyl methacrylate or water. Leduc and Bernhard (*J. Ultrastruct. Res.*, **19**: 196, 1967) recommend the use of 2,4-dichlorobenzoyl peroxide (Luperco) as a catalyst because it is readily decomposed to free radicals resulting in reduced time required for polymerization.

Leduc and Bernhard recommend the following procedure for tissues fixed in 1.25 to 2.5% glutaraldehyde in either 0.1 M cacodylate or phosphate buffer pH 7.2 from 15 to 60 min. All procedures including fixation are carried out at 3°C. After tissues are washed from 15 min to overnight in the buffer, they are dehydrated in graded GMA monomer as follows: 80% GMA monomer with 20% distilled water for 20 min; 97% GMA monomer and 3% distilled water, 20 min; unprepolymerized embedding mixture, 20 min; and, for impregnation, prepolymer of embedding mixture overnight. For embedding, tissues are placed in gelatin (not polyethylene) capsules, and fresh prepolymer is added to the top.

The embedding mixture is composed of 7 parts GMA, 3 parts distilled water, 3 parts butyl methacrylate containing 2% Luperco. Capsules are left uncapped for 30 min to permit air bubbles to escape. Then caps are added and adjusted to permit minimal air space in the capsule. For polymerization, capsules are held upright by wire or clear plastic supports which permit passage of maximum ultraviolet light, and subjected to an ultraviolet light (315 to 400 mμ) from a type A lamp (P. W. Allen & Co., London) containing Philips 9-in, 6-W fluorescent tubes having a special actinic blue phosphor (color type 05). The top of the gelatin capsule is 1 cm from the tube, the bottom is 2.5 cm from it.

The time required (1 to 3 days) for polymerization varies with the viscosity of the prepolymer, with the amount of Luperco added, and if other sources of ultraviolet light are used. GMA is readily sectioned with glass or diamond knives. Sections are mounted on Formvar and carbon-coated grids, and may be stained with uranyl acetate and lead citrate in the usual fashion. The unprepolymerized embedding mixture is prepared by adding a small amount of the GMA–butyl methacrylate–water–catalyst mixture in a large Erlenmeyer flask and rapidly bringing the mixture just to the boiling point, after which it is promptly cooled in a large ice-water bath. The viscosity of the prepolymer should approximate thick syrup. If not, alternate heating and cooling should be continued until the viscosity is appropriate. The prepolymer can be stored indefinitely in a deepfreeze and warmed to 3°C just prior to use.

In ordering GMA from a supply house, it is important to request a fresh lot, because methacrylate polymerizes spontaneously. Polysciences Inc. (and perhaps others?) adds 0.2% hydroquinone to prevent polymerization. Rosenberg et al. (*J. Ultrastruct. Res.*, **4**: 298, 1960) recommend 1 h infiltrations of 10, 20, 40, 60, and 80% aqueous GMA with a final overnight infiltration in 97.5% catalyzed GMA at 3°C. Addition of 0.1 to 2.0% ethylene glycol dimethacrylate to the prepolymerized mixture is stated to result in sections that manifest less swelling during aqueous staining procedures.

Ashley and Feder (*Arch. Pathol.*, **81**: 391, 1966) believe that formalin-fixed tissues may be embedded in GMA with no more difficulty than with paraffin. Ashley and Feder believe the preservation of structure is better than with paraffin embedding as determined by stained sections viewed by light microscopy.

Dehydrate fixed tissue blocks not exceeding $2 \times 2 \times 2$ mm in four changes of methyl alcohol (30 min each), and then transfer them to two changes of the plastic monomer (1 h each at 24°C). Blot tissues to remove excess monomer, place them in gelatin capsules No. 00 with a paper label, add monomer mixture to the capsule until it is nearly filled (except for a few millimeters), carefully add a small drop of melted paraffin and a cap to exclude air, and polymerize in a 60°C oven overnight. (Sections may be cut as early as after 6 h.)

Remove the blocks from the oven, remove the gelatin capsule with a razor blade, and file the surface of the block to obtain a 2×2 mm surface. (Sections larger than 2×2 mm cannot be cut easily.) The embedding medium is very hard and brittle when first removed from the 60°C oven. At room temperature the methacrylate is less brittle. Also, if the blocks are very dry, it may be helpful to place the plastic block in an uncovered dry glass jar inside a tightly closed metal can containing 50 ml water for 30 min at 24°C.

Tissues may be sectioned as thin as 1 to 2 μ with a conventional rotary microtome using a steel knife. A single portion of the knife is useful for cutting only 6 to 8 sections. The knife must then be shifted. Single sections are removed from the knife with a jeweler's forceps, floated on a drop of water on a slide, and dried at 80 to 100°C on a hot plate. Sections are stained as usual without removal of embedding medium.

The monomer mixture contains 95 ml GMA (hydroxyethyl methacrylate, available from Rohm & Haas, Philadelphia, Pennsylvania), 5 ml Carbowax 200 (polyethylene glycol 200, Ruger Chem. Co., Irvington-on-Hudson), and 0.15 g catalyst (2,2-azobis[2-methyl]pro-prionitrile, Matheson, Coleman, East Rutherford, New Jersey). If unpurified GMA is used, the amount of catalyst may need to be increased to 0.35 g/100 ml. Use of unpurified GMA may result in a slight stain with toluidine blue. The monomer mixture containing catalyst may be stored at 24°C *in the dark* for several months without apparent deterioration.

Ashford et al. (*J. Histochem. Cytochem.*, **20**: 986, 1972) advocate embedding in GMA for the preservation of fine structure and enzyme activity according to the following method:

1. Fix tissues ($1.5 \times 2 \times 0.5$ mm) for 3 to 12 h in either 3% buffered redistilled glutaraldehyde of pH 7 or 4% paraformaldehyde of pH 7.2 at 0°C.
2. Dehydrate through graded series of GMA (GMA, 95% v/v, polyethylene glycol, 5%;

azobisisobutyronitrile, 0.05 to 0.15% v/v) and water at progressively lower temperatures; 5, 10, 20, 40, 60, and 80% aqueous solutions of GMA monomer mixture at 0°C; and 90 and 95% at −25°C. Leave tissues in each graded series for 3 h except for the 60%, in which case leave them overnight for the sake of convenience.

3. Infiltrate specimens in 100% monomer at −25 to −35°C in a deepfreeze. The time required for dehydration varies with the specimen, but increased viscosity at lower temperatures results in retarded infiltration. Remove gas bubbles by placing the GMA under a vacuum prior to insertion of specimens.

4. Polymerize with a Philips actinic blue 15-W fluorescent lamp (maximal energy at 370 mμ) 7 to 14 cm from the GMA. It is important that the polymerization occur in aluminum containers (such as weighing dishes) refrigerated by an insulated bath of 56% v/v aqueous ethylene glycol maintained at a constant temperature of −45°C. Blow dry nitrogen over the samples to minimize dilution of the GMA by condensation and to aid polymerization, which is inhibited by atmospheric oxygen. Maintain the nitrogen atmosphere by placing Saran Wrap plastic film over the apparatus. Enhancement of polymerization may be achieved by addition of ultraviolet absorbent dyes, such as acriflavine, to the embedding medium. Addition of 0.025% acriflavine results in polymerization within 5 h. Polymerization without the dye requires 36 h.

5. Keep the polymerized blocks in a freezer at −35°C until sectioning.

Ashford et al. reported good preservation of several hydrolases including acid and alkaline phosphatases, β-glucuronidase, aminopeptidase, and esterase. Catalase was also preserved.

Weber (*Histochemistry*, **39**: 155, 1974) was able to demonstrate nonspecific esterase and succinic dehydrogenase in glutaraldehyde or acrolein-fixed and glycol methacrylate–embedded tissues.

2-HYDROXYPROPYL METHACRYLATE.   The next homolog to GMA, 2-hydroxypropyl methacrylate (HPMA), is a water-miscible monomer. Miscibility varies inversely with the temperature. At 20°C, 4 parts HPMA and 1 part water are miscible when vigorously shaken together. If miscibility does not occur at these proportions and temperature, partial polymerization is probably the cause, and the mixture should not be used. Hydroquinone is added to HPMA to prevent premature polymerization during storage; however, it should also be refrigerated.

$$H_2C{=}C{-}\overset{\overset{\displaystyle O}{\|}}{C}{-}O{-}CH_2{-}\underset{\underset{\displaystyle OH}{|}}{CH}{-}CH_3$$
$$\underset{\underset{\displaystyle CH_3}{|}}{}$$

HPMA is readily polymerized by heat or ultraviolet light without removal of the stabilizer by the addition of a catalyst such as benzoyl peroxide, azobisisobutyronitrile, or 2,3-dichloro-benzoyl peroxide. Storage of HPMA should be in the cold and in the dark. Addition of about 10% activated charcoal (v/v) to the monomer with shaking for about an hour until the optical density at 295 mμ shows no further reduction results in the removal of hydroquinone. Both HPMA (nonpolymerized) and prepolymerized HPMA are employed. Twenty percent HPMA with a consistency of free-flowing syrup is converted to the prepolymer by addition of the catalyst azobisisobutyronitrile to a concentration of 0.1% and heating to 120°C with constant stirring. The desired degree of prepolymer formation should occur within 5 min. The degree of prepolymer formation is controlled by occasional rapid cooling on ice. Water is then added to make 97% HPMA.

Leduc and Holt (*J. Cell. Biol.*, **26**: 137, 1965) and Pease (*J. Cell. Biol.*, **27**: 124A, 1965) recommend glutaraldehyde fixation of tissue blocks not to exceed 0.5 mm in any direction. For dehydration, the following schedule is recommended: Tissues are dehydrated with 80% HPMA, 1 h; repeated, 1 h; 97% HPMA, 1 h; repeated, 1 h; prepolymerized HPMA 97%, 1 h. Tissues are then transferred to capsules with prepolymer HPMA, capped to exclude air to a maximum, and heated at 60°C until hard. Air (oxygen) in the capsule will retard polymerization, resulting in a softened surface. If enzyme histochemistry is to follow, ultraviolet radiation in the cold as detailed above for GMA (glycol methacrylate) is advised. The addition of a photosensitizer such as benzoin (0.1%) may reduce the time required for ultraviolet polymerization (usually 12 to 24 h).

CROSS-LINKED POLYAMPHOLYTES (POLYAMPH 10). McLean and Singer (*J. Cell. Biol.*, **20**: 518, 1964) introduced the use of Polyamph 10 (available from Polysciences Inc., Warrington, Pennsylvania) as a water-soluble embedding medium. Polyampholyte is a highly polar vinyl type of polymer containing both positive and negative charges formed by the copolymerization of the anionic monomer methacrylic acid (MA) with the cationic monomer dimethylaminoethyl methacrylate (DMA). Tetramethylene dimethacrylate (TMA) is used as a cross-linking agent, and azodiisobutyronitrile is a catalyst employed for polymerization.

To prepare, MA is distilled at 10 mm mercury pressure, and the fraction which boils off at 60°C is collected and stored at 4°C. Tissues that have been washed with buffer after fixation are dehydrated and infiltrated with a 1:1 mixture of distilled water–combined MA and DMA in the series 10, 20, 40, and 80% for 30 min each at 4°C. Tissues are then transferred to embedding capsules and filled with the following mixture: MA, 20 ml; DMA, 10 ml; TMA, 3.3 ml; and azodiisobutyronitrile, 0.25 ml. In order to avoid substantial heat and polymerization during mixing, MA is added dropwise. Polymerization is accomplished by ultraviolet light exposure for 48 h at 4°C. Sectioning is accomplished by a knife at an acute angle of 38°, and sections are collected on unsupported 1000-mesh grids.

Polyamph-embedded tissues appear to retain protein and nonpolar constituents. Denaturation is stated to be minimal. Spendlove and Singer (*Proc. Natl. Acad. Sci. U.S.A.*, **47**: 14, 1961) employed the medium in studies of bacterial antibodies. Thomson et al. [*Nature (Lond.)*, **215**: 393, 1967] and Singer and McLean (*Lab. Invest.*, **12**: 1002, 1963) used Polyamph in studies of ferritin-antibody conjugates.

UREA FORMALDEHYDE. In the presence of alkali acting as a catalyst, urea and formaldehyde form a clear resin as formulated below according to Steedman ("Section Cutting in Microscopy," Blackwell, London, 1960).

$$
\underset{\text{Urea}}{\begin{array}{c} NH_2 \\ | \\ C{=}O \\ | \\ NH_2 \end{array}} + \xrightarrow{O{=}CH_2} \underset{\text{Monomethylol urea}}{\begin{array}{c} NH{-}CH_2OH \\ | \\ C{=}O \\ | \\ NH_2 \end{array}} + \xrightarrow[O{=}CH_2]{} \underset{\text{Dimethylol urea}}{\begin{array}{c} NH{-}CH_2OH \\ | \\ C{=}O \\ | \\ NH{-}CH_2OH \end{array}}
$$

DIMETHYLOL UREA. The urea-formaldehyde resin is water-soluble, and the use of organic solvents is not required. Casley-Smith (*J. R. Microsc. Soc.*, **87**: 463, 1967) recommended use of the medium for studies of lipids. Disadvantages include substantial shrinkage during polymerization, difficulty in sectioning, loss of lipid during processing, and a tendency for the polymerized resin to fracture with time.

Nir and Pease (*J. Histochem. Cytochem.*, **22**: 1019, 1974) embedded small bits of rat kidney in the highly polar polymerized glutaraldehyde-urea (Pease and Peterson, *J. Ultrastruct.*

Res., **41:** 115 and 133, 1972) and subsequently prepared tissues for the demonstration of cytochrome oxidase using the Graham and Karnovsky diaminobenzidine reagent (DAB). The highly polar embedding material tends to retain lipids; and organic solvents were not employed during processing. Reaction product occurred only between inner and outer mitochondrial membranes. Nir and Pease point out that after the use of osmiophilic reagents such as DAB, it is a common practice to subject the entire block of tissue to OsO_4 in order to achieve insolubility prior to dehydration. However, some extraction of tissue constituents is known to occur because solutions become discolored. In this case, osmication is believed unnecessary if the water-containing glutaraldehyde-urea embedding medium is employed. Organic solvents are not used, and DAB polymers are believed to be visualized in their true *in situ* location.

Recently Stäubli (*J. Cell. Biol.*, **16:** 197, 1963) and Kushida (*J. Electron Microsc. (Tokyo)*, **14:** 52, 1965) have employed the water-miscible resin (Durcupan) as a dehydrating agent which is gradually replaced with a water-immiscible resin (Araldite). Inasmuch as organic solvents are not employed, loss of organic soluble tissue constituents is believed to be minimized.

Stäubli prescribes the following Durcupan-water dehydration sequence carried out at 4°C: 50 : 50 (15 to 30 min); 70 : 30 (15 to 30 min); 90 : 10 (15 to 30 min); and 100% Durcupan (two changes 30 to 60 min each).

After impregnation in Durcupan (Fluka AG, Buchs, Switzerland), embed at 50°C in the following: Araldite I–Durcupan 30 : 70 (1 h); 50 : 50 (1 h); 70 : 30 (several hours or overnight); 100% Araldite I (1 h); and 100% Araldite II (three changes of 30 min each).

Araldite I is Araldite M (10 ml) and hardener 964B (10 ml). Araldite II consists of Araldite I to which is added accelerator 964C (0.4 ml). Tissues are allowed to harden for 24 to 48 h at 50°C. Blocks are cut with no difficulty. If Durcupan remains in the tissues, the blocks will be softened.

METHACRYLATES. Newman et al. (*J. Res. Natl. Bur. Standards A*, **43:** 183, 1949) introduced methacrylates as an embedding medium. Methyl methacrylate has the following structure:

$$CH_2{=}\underset{\underset{\textstyle CH_3}{|}}{C}{-}\overset{\overset{\textstyle O}{\|}}{C}{-}O{-}CH_3$$

Methacrylates are transparent, and monomers are miscible with organic solvents. Hardness of the blocks may be regulated by adjustment of the proportion of *n*-butyl and methyl methacrylate employed. Increased proportions of methyl methacrylate result in harder blocks. A common proportion of 80 : 20 of *n*-butyl–methyl methacrylate is useful for many tissues. Spontaneous polymerization of the resins is reduced by storage in the absence of light at low temperature and the presence of hydroquinone in the resin mixture.

According to Reimer (*Z. Naturforsch [B]*, **14:** 566, 1959) and Cosslett (*J. R. Microsc. Soc.*, **79:** 263, 1960) serious deficiencies of methacrylates as embedding media include a substantial (20%) and uneven shrinkage that occurs during polymerization and an instability and loss (as much as 50%) of resin during electron bombardment. Concomitant with this loss, flow of remaining resin is believed to occur, resulting in distortion of tissue constituents. Tissues fixed with $KMnO_4$ may appear more adversely affected than those fixed with OsO_4.

Because of these disadvantages, methacrylates are not commonly used by electron microscopists today. They may be employed usefully for phase microscopy, however, and sections may be stained by some of the stains employed for light microscopy. Borysko and Sapranauskas (*Johns Hopkins Med. J.*, **95**: 68, 1954) were able to reduce the shrinkage artifact somewhat by the use of prepolymerized resin. Prepolymerized resin results in increased viscosity and retarded infiltration of tissues.

VESTOPAL W. This embedding medium results from esterification of malic anhydride with glycerol or another polyhydric alcohol. Vestopal W polymerizes evenly into a hard block with minimal shrinkage. The resin is relatively stable under electron bombardment, and sections stain easily. Vestopal can be stored for months at 4°C without apparent deterioration; however, the activator (cobalt naphthenate) and the initiator (benzoyl peroxide) are not usable after 1 to 2 months even if stored at 4°C in the dark. Although catalyzed Vestopal keeps only a few hours at 25°C, it may keep for months at 4°C. It is recommended that the final mixture be made up immediately before use as follows: Vestopal, 100 ml; benzoyl peroxide, 1 ml; cobalt naphthenate, 0.5 ml. The Vestopal should be thoroughly mixed with the benzoyl peroxide before the cobalt naphthenate is added. Complete homogenization of the mixture may be obtained after about 40 to 60 min. *An explosion may occur if activator and initiator are mixed before adding to Vestopal!*

Vestopal is miscible with acetone, and tissues are readily infiltrated. Provided tissues are small, they may be successively dehydrated for 5 to 10 min each in 40, 80, 95, 100, and 100% acetone followed by 30 to 40 min each in Vestopal-acetone 1:3, 1:1, 3:1, and terminated in 100% Vestopal for 1 h. Infiltrated tissues are placed in gelatin capsules, the capsules are filled with fresh Vestopal, and polymerization is accomplished in 12 or more hours at 60°C. If tissues are less than 0.5 mm^3 and agitation of the dehydrating and embedding fluids is managed, more thorough embedding occurs more rapidly.

RIGOLAC. Kushida [*J. Electron Microsc. (Tokyo)*, **9**: 113, 1960] developed Rigolac (Riken Goseijushi Co., 3,6-chome, Ginza Chuo-Ku, Tokyo, Japan), a new polyester embedding medium for electron microscopy. Kushida recommends the following schedule: Tissues are dehydrated for 15 to 30 min each in 30, 50, 70, 90, 100, 100, and 100% acetone followed by 1 to 2 h each in an acetone-Rigolac mixture, 1:1, in 100% Rigolac, and again in 100% Rigolac. The Rigolac mixture is as follows:

Rigolac 2004	75 ml
Rigolac 70F	25 ml
Benzoyl peroxide paste	1 g

Tissues are entered into gelatin capsules containing fresh Rigolac mixture. At 55°C, polymerization is accomplished between 18 to 24 h. The density of Rigolac sections exceeds that of methacrylate. Excellent sections may be collected on 10% acetone. The mixture is believed to be better than Vestopal W due to a low viscosity, which permits easy penetration into tissues. The hardness of the blocks can be regulated by varying the proportions of the two Rigolac constituents. Rigolac is insoluble in ethanol.

EPOXY RESINS. Epoxy resins are yellowish condensation products of epichlorhydrin with polyhydroxy compounds. They may be considered polyaryl ethers of glycerol-containing terminal epoxy groups as shown below:

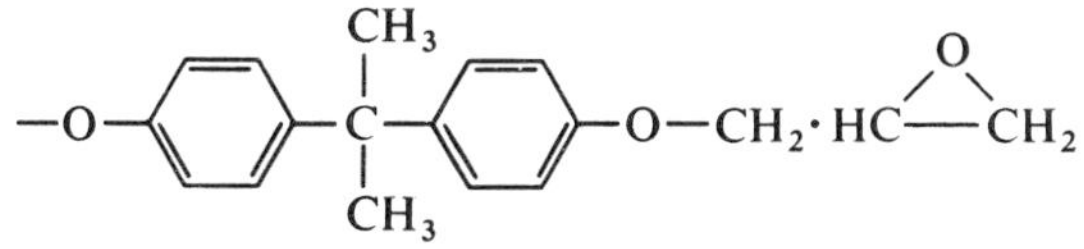

Epoxy resins require cross-linking (curing) agents to develop a hardened tough highly inert solid. Cosslett (*Proc. Europ. Reg. Conf. Electron Microsc.*, **2**: 678, Delft, N.V. Drukkerij Trio, The Hague, 1960) has calculated that approximately 25% of the mass of the section may be lost during electron bombardment; however, flow of the remaining embedding material does not occur as it does with methacrylate (see above). Such stability results in clarity of resolution in epoxy-embedded tissues. Contrast can be further enhanced by the use of heavy metals and by mounting sections on unsupported grids. It is important to remember that epoxy resins cause severe irritation especially upon prolonged or repeated contact. Prompt removal is advised.

ARALDITE.   Among the epoxy resins, Araldite is commonly employed (Ciba Products Corp., Fair Lawn, New Jersey). The Araldites were developed by Maaløe and Birch-Andersen (*6th Symp. Soc. Gen. Microbiol.*, p. 261, 1956), Glavert et al. [*Nature (Lond.)*, **178**: 803, 1956], Glavert and Glavert (*J. Biophys. Biochem. Cytol.*, **4**: 409, 1958), Davis [*Nature (Lond.)*, **183**: 200, 1959], Luft (*J. Biophys. Biochem. Cytol.*, **9**: 409, 1961), and Fink (*J. Biophys. Biochem. Cytol.*, **7**: 27, 1960).

Araldite is a very viscous (3000 to 6000 cP at 23°C) resin that polymerizes with very little shrinkage. Hardness may be regulated by the use of dibutyl pthalate as a plasticizer. Glavert and Glavert (*J. Biophys. Biochem. Cytol.*, **4**: 409, 1958) recommend the following embedding mixture: Araldite, CY212, 10 ml; HY 964 (DDSA—dodecenyl succinic anhydride, available from R. P. Cargille Laboratories, 117 Liberty Street, New York, New York 10006, and from Polysciences Inc., Warrington, Pennsylvania), 10 ml; DY 064 (DMP-30— 2,4,6-tri(dimethylaminomethyl)phenol), 0.5 ml; and dibutyl phthalate, 1 ml. DDSA is a hardener. DMP-30 is an amine accelerator. Block hardness can be modified by variation in the amount of plasticizer (dibutyl phthalate) added.

Luft (*J. Biophys. Biochem. Cytol.*, **9**: 409, 1961) prescribes the following mixture: Araldite 502, 27 ml; DDSA, 23 ml; DMP-30, 0.75 to 1 ml. Araldite 502 contains plasticizer supplied by the manufacturer. Block hardness may be modified by a shift of resin-hardener proportions.

Winborn (*Stain Technol.*, **40**: 227, 1965) prescribes the following mixtures: Araldite 502, 68 ml; DDSA, 19 ml; triallyl cyanurate (TAC, available from Vaughn, Inc., P.O. Box 1495, Memphis, Tennessee 38101), 10 ml; DMP-30, 3 ml. Blocks containing TAC at a concentration of 10% are of medium hardness. Harder blocks have a lesser concentration of TAC. Sections obtained with this mixture stain readily with lead hydroxide. Sections mounted on unsupported grids are stable to electron bombardment, and excellent image contrast is achieved.

Hökfelt (*J. Histochem. Cytochem.*, **13**; 518, 1965) and McNary et al. (*J. Histochem. Cytochem.*, **12**: 216, 1964) note that Araldite is an especially suitable medium in which to embed sections for fluorescence microscopy because it manifests minimal autofluorescence.

EPON.   Epon is a glycerol-based aliphatic epoxy resin. Studies by Kushida [*J. Electron Microsc. (Tokyo)*, **8**: 72, 1959], Fink (*J. Biophys. Biochem. Cytol.*, **7**: 27, 1960), and Luft (*J. Biophys. Biochem. Cytol.*, **9**: 409, 1961) resulted in the development of Epon 812 for use in electron microscopy. In Europe, Epon is known as *Epikote*. Due to its relatively low viscosity (90 to 150 cP) Epon 812 infiltrates tissues more readily than does Araldite. It is very hygroscopic.

For Epon embedding, preparation of three mixtures is recommended: Mixture A is composed of Epon 812, 5 ml, and DDSA, 8 ml. Mixture B is Epon 812, 8 ml, and NMA (nadic methyl anhydride, available from Polysciences Inc., Warrington, Pennsylvania), 7 ml. The final embedding mixture is composed of mixture A, 13 ml; mixture B, 15 ml; and DMP-30, 16 drops.

Thorough mixing of each of the mixtures for at least 20 min is advised. Immediately prior to use, mixture A and mixture B are mixed thoroughly and accelerator DMP-30 (1.5 to 2.0%) is added. In general, a 1 : 1 mixture of A and B will provide a block of suitable hardness. An increased proportion of mixture B will provide a harder block. Darkly colored blocks too brittle to cut may be due to excessive DMP-30 content.

Tissues are immersed successively for 10 min in 30, 60, 90, and 100% (3 times) acetone followed by a 30-min immersion in 1 : 1 acetone–final Epon mixture and in 1 : 3 acetone–final Epon mixture. Tissues are then embedded in the final Epon mixture in dried gelatin capsules and polymerized first for 12 h at 45°C and then 24 h at 60°C.

Hayat (1970) states that a rapid embedding method for Epon 812 provides sections of a quality indistinguishable from those embedded in the routine fashion. Hayat's method follows:

Small pieces of washed, fixed tissues are treated successively for 4 min each in acetone 30, 70, 95, and 100% (2 times), followed by acetone–Epon final mixture 15 min, final Epon mixture 10 min (2 times), followed by embedment in predried gelatin capsules, and heated for 1 h at 60°C. Sections may be cut in blocks cooled to 25°C.

The ratio of anhydride to epoxy content in the embedding mixture influences the quality of sections. Coulter (*J. Ultrastruct. Res.*, **20**: 346, 1967) believes lowering the anhydride-epoxy ratio from 0.7 to 0.6 results in blocks with improved quality for sectioning. Some commercial firms have now standardized the epoxy equivalent of Epon. Epon-embedded sections reveal good contrast under electron microscopic observation, although the granularity of Epon is evident at high magnification. Epon mixtures should be stored in the dark and cold. Precautionary measures should be taken for its hygroscopic nature.

EPON REMOVAL FOR LIGHT MICROSCOPY. Lane and Europa (*J. Histochem. Cytochem.*, **13**: 579, 1965) find that Epon can be removed from Epon-embedded tissues by the use of an ethanol-saturated solution of NaOH. They prescribe as follows: A saturated solution of NaOH in absolute ethanol is prepared and allowed to stand at 24°C a few days until it turns dark brown, whereupon it is ready for use.

1. Completely immerse Epon-embedded sections on slides in the saturated NaOH–ethanol solution for at least 1 h in a *covered* Coplin jar.
2. Remove slides, drain on bibulous paper (do not blot because sections will detach), and place in absolute ethanol for 5 min. Repeat the drainage and ethanol treatments 3 more times. Any water at these stages results in section detachment.
3. Proceed directly from absolute ethanol baths to a concentrated phosphate buffer of pH 7 (Fisher Scientific Co.) for three changes of 5 min each and then to the staining procedure.

MARAGLAS 655. Maraglas is a clear epoxy resin with a viscosity of 500 cP at 25°C in the liquid state. Freeman and Spurlock (*J. Cell. Biol.*, **13**: 437, 1962) first advocated its use for electron microscopy. Maraglas is readily miscible with acetone, styrene, methyl methacrylate, propylene oxide, and DER (Dow epoxy resin) 732; however, it is not miscible with alcohols. Proponents claim many advantages for Maraglas including stability under electron bombardment, greater transparency than Araldite, less granularity than Epon, and low viscosity with ease of embedding and ease of sectioning and of staining.

Spurlock et al. (*J. Cell. Biol.*, **17**: 203, 1963) recommend the following embedding mixture: Maraglas 655, 68 ml; Cardolite NC-513, 20 ml; dibutyl phthalate, 10 ml; and BDMA (benzyldimethylamine, available from Maumee Chemical Co., 2 Oak Street, Toledo, Ohio), 2 ml. Cardolite is available from the Irvington Chemical Div., Minnesota Mining & Mfg. Co., 500 Doremus Avenue, Newark, New Jersey.

Erlandson (*J. Cell. Biol.*, **22**: 704, 1964) modified the mixture advocated by Spurlock et al. He claimed the modified medium infiltrates tissues with greater ease with fewer bubbles formed. The Erlandson mixture is Maraglas 655, 36 ml; DER-732, 8 ml; dibutyl phthlate, 5 ml; and BDMA, 1 ml.

Acetone dehydration of tissues is advised. After impregnation, tissues should remain for 4 h in the degased Maraglas mixture prior to polymerization at 60°C. Softer blocks may be obtained by increasing the DER 732–Maraglas ratio. Winborn (*Stain Technol.*, **40**: 227, 1965) advocates the use of polyethylene instead of gelatin capsules in order to achieve uniform polymerization. Capsules containing unpolymerized resin should also be degased in a vacuum desiccator for 30 min prior to use.

Winborn (*Anat. Rec.*, **148**: 422, 1964) advocates the following mixture to obtain blocks of medium hardness that provide sections with high contrast and excellent thermal stability: Maraglas 655, 48 ml; Cardolite NC-513, 40 ml; TAC (triallyl cyanurate), 10 ml; DMP-30, 2 ml. Winborn states that the medium infiltrates tissues readily and that blocks are easily sectioned. Polymerization is accomplished in 8 h at 65 to 70°C.

OTHER EMBEDDING MEDIA. Kushida [*J, Electron Microsc. (Tokyo)*, **16**: 278, 1967] developed an embedding medium which combined Epon 812 and DER 736. DER 736 is a flexible resin obtained by condensation of polypropylene glycol and epichlorohydrin with the following structure:

$$
\underset{H_2C-CH-CH_2-O}{\overset{O}{\triangle}}\!\!-\!\!O\left[\!\!\begin{array}{c}CH_3\\ |\\ CH-CH_2-O\end{array}\!\!\right]_N\!\!CH_2-HC\!\!-\!\!CH_2
$$

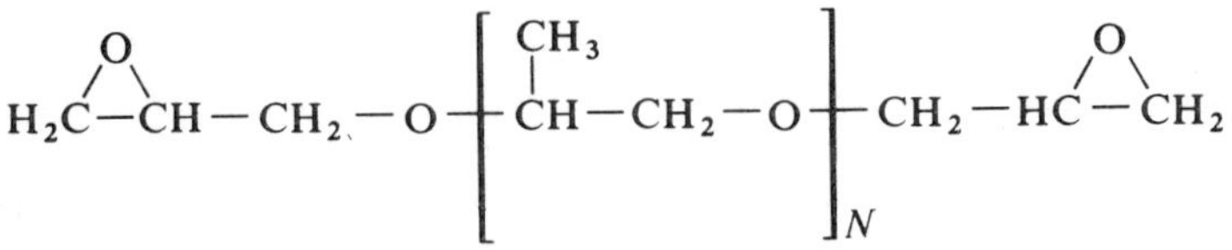

$$N = 4$$

Embedding medium is composed of DER 736, 40 ml; Epon 812, 10 ml; NMA (nadic methyl anhydride), 45 ml; DMP-30, 1.4 to 2 ml. DER 736 is an epoxy resin reactive with all epoxy curing agents. The viscosity is low (30 to 60 cP at 25°C), which permits rapid penetration into tissues.

Spurr (*J. Ultrastruct. Res.*, **26**: 31, 1969) combined DER 736 and ERL 4206 (Union Carbide Corp., 270 Park Avenue, New York, New York 10017), to produce a low-viscosity embedding medium with the following formulation: DER 736, 6 g; ERL 4206, 10 g; NSA (nonenyl succinic anhydride, available from Humphrye Chemical Co., Devine Street, New Haven, Connecticut 06473), 26 g; S-1 (dimethylaminoethanol, an epoxy curing agent available from Pennsalt Chemical Corp., 3 Penn Center, Philadelphia, Pennsylvania 19102), 0.4 g.

According to Trigaux (*Mod. Plastics*, **38**: 147, 1960) ERL 4206 is a cycloaliphatic diepoxide with the following structure:

$$
\begin{array}{c}
\overset{H_2}{C}\\
\diagup\quad\diagdown\\
HC\quad\; CH-HC\overset{O}{\diagdown}CH_2\\
\diagdown O\quad|\qquad|\\
HC\quad CH_2\\
\diagdown C\diagup\\
H_2
\end{array}
$$

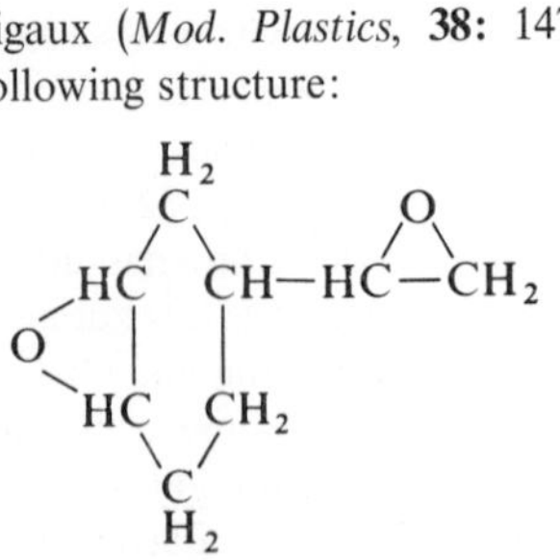

ERL 4206 has a molecular weight of 140.18, has a sp gr of 1.1 at 20°C, and is miscible with ethanol in any proportion. The embedding medium of Spurr has a viscosity of 60 cP at 25°C.

5

GENERAL STAINING AND MOUNTING PROCEDURES

DEWAXING AND HYDRATION BEFORE STAINING

After sectioning, the prepared sections must be brought into appropriate solvents for the staining procedure to be used, and when the staining procedure has been completed, the sections must be brought without loss of the desired stain into media suitable for examination and temporary or permanent preservation.

The technics employed for these purposes vary with the method of sectioning and with staining and mounting methods.

PARAFFIN SECTIONS. These are generally first warmed to just above the melting point of paraffin and then immersed in xylene to dissolve out the paraffin. If immersion in xylene is prolonged for some minutes, heating is unnecessary. A second change of xylene is necessary to prevent carrying paraffin into succeeding reagents. The xylene is then removed from the sections in two successive changes of 100% alcohol. They turn white at this point. Two changes each of 95 and 80% alcohol follow. These reagents should be changed at intervals; the second change of each may be moved over so as to form the first change, and fresh reagents then replace the second changes.

PARAFFIN ARTIFACTS. Nedzel's observations (*Q. J. Microsc. Sci.*, **92**: 343, 1951) indicate that the usual practice of immersing unheated sections in cold xylene may be inadequate to remove paraffin from sections. Birefringent crystals with the same melting point as the paraffin employed may often be found in cell nuclei. These may be stained with oil red O heated to above the melting point of the paraffin. These birefringent crystals are more apt to persist after brief acetone dehydration than after alcohol-xylene sequences. The use of hot xylene at either the deparaffinization or the clearing stage, or the practice of melting the paraffin before immersion in xylene, will tend to eliminate these artifacts; and heating of sections mounted in resins to 60 to 65°C will cause most of them to disappear permanently.

COLLODIONIZATION. *Torn or ragged sections*, or those which tend to become detached from the slides during staining or metallic impregnation procedures, may be attached more securely to the slides by immersion in 0.5 to 1% ether alcohol solution of collodion for 5 or more min. This step is inserted immediately after the 100% alcohol step. After soaking them in collodion, drain the sections for about 1 min, and then harden for 5 min or as much longer as is convenient in 80% alcohol. When it is also necessary to iodize the sections for removal of mercury precipitates, it is advantageous to combine the hardening of the collodion with the iodine treatment by using a 0.5% solution of iodine in 80% (or 70%) alcohol for both purposes. It may be necessary to prolong staining intervals to as much as double the time usual in technics not devised for collodionized paraffin or attached celloidin and nitrocellulose sections. This procedure, like celloidin embedding, tends to diminish shrinkage and creation of artificial spaces during staining. Before acetylation or benzoylation procedures which employ pyridine as the solvent, or methylation in absolute methanol mixtures, collodionization is futile. Pyridine and methanol both dissolve collodion. If later portions of the procedure require collodionization, wash the sections with 100% alcohol after the pyridine or methanol, and then collodionize as usual.

Collodion films prevent the access of enzymes to their substrates in the sections Therefore, if the procedure is necessary for later parts of the technic, dehydrate and collodionize after the completion of digestion.

UNDEPARAFFINIZED SECTIONS. Goetsch, Reynolds, and Bunting (*Proc. Soc. Exp. Biol. Med.*, **80:** 71, 1952) report successful staining of undeparaffinized paraffin sections by direct 18-h immersions in aqueous or alcoholic 3% eosin solutions, Van Gieson's stain, 2% aniline blue, 0.25% light green, 0.05% toluidine blue, 2% methylene blue, 0.5% methyl green, or Bullard's hematoxylin.

Less cytologic distortion than in deparaffinized sections is claimed for this method. After staining, the sections are washed in water or alcohol to remove the excess; they are then dried, deparaffinized, and mounted. The paraffin, unlike collodion, does not prevent access of amylases to glycogen.

In using undeparaffinized sections for staining, avoid heating to near or above the melting point of the paraffin. Even momentary melting forms a virtually impenetrable film. Ordinary staining procedures may require 6 to 12 times as long as in deparaffinized sections; enzyme digestions, 30 to 60 times as long.

IODIZING AND HYDRATION. When material fixed with mercury or lead salts is being handled, substitute a 5-min bath in 0.5% iodine in 80% alcohol for the usual second 80% alcohol step. The usual prescription calls for 0.3 to 0.5% in 70 to 95% alcohol. Mallory used 0.5% in 95% alcohol. After the iodine, rinse the material in water and immerse it for a few seconds in 5% sodium thiosulfate ($Na_2S_2O_3 \cdot 5H_2O$). Mallory used a 5-min bath in 0.5% solution. Then wash 5 min in running water.

However, when the following technic contains an iodine step, as in the Gram and Gram-Weigert methods, preliminary iodization may be omitted on mercury-fixed material.

With material fixed in nonmercurial fixatives the foregoing steps are omitted, and the sections are transferred either directly to the staining solution (if this is in a hydroalcoholic solvent) or to water. However, if both types of fixations—mercurial and nonmercurial—are included in the material being stained, the iodine thiosulfate procedure seems to be harmless to the latter in most instances.

POSTMORDANTING. Often one desires to use some special staining method for which the author has prescribed a fixation method other than that used on the tissue in question. Although in many instances the methods will work quite as well after other fixations than that prescribed, it is still often desirable to have the prescribed fixation or to be able to modify the tissues by some pretreatment in lieu of the prescribed fixation.

In lieu of Bouin or Zenker fixations for Masson trichrome methods, we have found that mordanting for 2 min in a saturated (6 to 8%) alcoholic picric acid solution in place of Bouin's fluid, or with saturated aqueous mercuric chloride solution for 5 min in place of Zenker's fluid, will serve (*Stain Technol.*, **15:** 17, 1940).

Peers (*Arch. Pathol.*, **32:** 446, 1941) reported that 3-h mordanting in saturated aqueous mercuric chloride solution at 58°C was required for formalin-fixed material in place of the primary Zenker fixation prescribed by Mallory for his phosphotungstic acid hematoxylin stain. Earle, however, was able to stain successfully after fixation in buffered neutral formalin (personal communication).

We have long used a 48-h mordanting of brain tissue in 2.5% potassium dichromate after primary formalin fixation, before dehydration and embedding, in lieu of a primary Orth fixation, to prevent pericellular shrinkage during embedding. This procedure does not preserve the characteristic staining of chromaffin in the adrenal.

In place of primary fixation in Ramón y Cajal's formalin–ammonium bromide for metallic impregnation methods on brain tissue, Globus (*Arch. Neurol. Psychiat.*, **18:** 263, 1927) prescribes soaking frozen sections of old formalin-fixed material in a 1 : 10 dilution of 28% ammonia water for 24 h, then rinsing rapidly in two changes of distilled water and immersing for 2 to 4 h in a 1 : 10 dilution of 40% hydrobromic acid. Then rinse in two changes of 1 : 2000 dilution (Conn and Darrow) of 28% ammonia water and proceed.

Arcadi (*Stain Technol.*, **23:** 77, 1948) modifies the foregoing procedure by placing the sections in concentrated (28%) ammonia water and then washing slowly with a 1-mm stream of water for 24 h, repeating this procedure for a second 24-h period and ending the ammonia treatment with a 7-min bath in concentrated ammonia water. From this, the sections are transferred directly to a $\frac{1}{20}$ dilution of 40% hydrobromic acid and incubated at 38°C for 1 h. This is followed by four washes of 4 min each in distilled water. Arcadi used this method for preparation of old formalin material of monkey brains for a modified Del Río Hortega–Penfield method for oligodendroglia.

DEHYDRATION

After staining, paraffin sections are usually dehydrated with two changes of 95% alcohol followed by two of 100% alcohol. Then follows a mixture of equal volumes of 100% alcohol and xylene, followed by two changes of xylene. In place of alcohol two to three changes of acetone or of 99% isopropyl alcohol may be used, and these are followed by an equal-volume mixture of the dehydrating agent and xylene, and then two changes of xylene as before. Some workers have used toluene in place of xylene as a clearing agent, but there is more hazard of drying, with more evaporation from the clearing jars, and it is at least no better.

The choice of a dehydrating agent depends on several factors: Alcohol extracts methylene blue and other thiazin dyes from the sections; acetone does not. Isopropyl and, even better, *tert*-butyl alcohol also fail to extract thiazins appreciably. Acetone dissolves collodion films from sections; isopropyl alcohol does not; and ethyl alcohol does partly. Ethyl alcohol aids in the differentiation of certain stains.

MOUNTING

PARAFFIN SECTIONS STAINED WITH OIL-SOLUBLE DYES. For the demonstration of certain lipid substances which are not lost in the dehydration, clearing, embedding, deparaffinization, and hydration procedures, these sections must be mounted in media which are not fat (and dye) solvents. For this purpose it is sufficient to drain off the excess water and mount in gum syrup, glychrogel, glycerol, gelatin, or the like.

MOUNTING IN RESINOUS MEDIA. After sections stained by ordinary methods have been cleared in xylene, they are mounted preferably in some resinous medium. Formerly xylene and chloroform solutions of Canada balsam or gum dammar, thickened cedar oil, euparal, and the like were preferred as mounting media. Canada balsam is useful for hematoxylin-eosin stains. For aniline blue and acid fuchsin connective tissue stains Curtis (*Arch. Méd. Exp. Anat. Pathol.*, **17:** 603, 1905) first prescribed acidifying with salicylic acid. A quantity of xylene balsam of thin syrupy consistency is saturated with salicylic acid crystals, filtered through filter paper and mixed with an equal quantity of xylene balsam. The use of the fully saturated solution as a mounting medium is apt to produce visible crystals in the preparations.

The rationale of the Curtis procedure was that acid solvents tend to fix acid dyes such as Curtis used for differential collagen stains to their position in tissue. It would seem that addition of 0.1% glacial acetic acid to the balsam would serve as well and avoid the sometimes annoying separation of salicylic acid crystals in the mounted preparation. Glacial acetic acid is freely miscible with xylene.

All the above resins are unsatisfactory for Romanovsky stains, causing progressive fading of the blue component. The best medium for preservation of these stains is heavy mineral oil, but preparations must be sealed with a gelatin, pyroxylin, or other suitable cement, and they inevitably leak. However, preparations so mounted, though messy, were readily cleaned and showed good stain preservation after 20 years. Langeron suggests Apáthy's gum syrup as a cement.

Actual *mounting* of resin and gum syrup mounts is accomplished in this way: Select cover glasses of appropriate size to cover the section. We use 22×22, 22×32 and 22×44 mm sizes. In the center of a square cover a moderate-sized drop of thin syrupy xylene resin or gum syrup is deposited; with long covers, 2 drops. The cover glasses lie flat on a paper towel or blotter on the table. As many as six cover glasses may be thus laid out at one time for xylene resin mounts; but with gum syrup mounts of frozen sections, one at a time is better. The stained section is taken from xylene or water, as the case may be, drained, and placed face down with one long edge of the slide in contact with the blotter adjacent to the edge of the selected cover glass and with the other edge perhaps 1 cm up from the table. The slide is then gradually rotated downward with the edge in contact with the blotter as an axis until the section comes in contact with the drop of xylene resin or gum syrup. The drop spreads to the edges of the cover glass, picking it up from the table surface. If air bubbles chance to be included under the cover glass, they may be coaxed out by gentle pressure on the cover glass with a needle point, or by slightly raising one edge of the cover glass. Sometimes it is necessary to add more mounting medium. One then blots the two lateral edges of the cover glass, holding the slide face down at an angle of about 30° to the table top, to remove excess mounting medium. Slides are then laid flat, face up, on trays in a warm place to dry.

Do not wipe the edges of the cover glass with a xylene wet pledget of gauze. This often dissolves a little resin and spreads it over the previously clean upper surface of the cover glass. When this dries, it forms an optically uneven, streaky surface, which seriously interferes with microscopy and is difficult to remove.

When cellulose tricaprate dries, any excess is extruded at the edges of the cover glass as small, firm beads which can readily be cut off with a small knife or with the thumbnail.

In mounting large sections for topographic study of whole human breasts, cerebral hemispheres, bones, embryos, and the like, air spaces are apt to form between the base and cover glasses after mounting has been accomplished with apparent success. The use of thicker resin solutions, warmed to lower the viscosity, is suggested to avoid aspiration of air after mounting. The polystyrenes are unsuitable for this purpose because of their very high viscosity at relatively low solution concentrations. Natural balsam, the relatively neutral ester gums, and some of the cycloparaffin and terpene resins are suggested; if warmed somewhat, they can be used at 60 to 80% concentration in xylene.

Opaque or cloudy areas in mounted stained sections are often due to incomplete dehydration. Microscopic examination reveals numerous fine droplets of water in and above the section. Such areas tend to show severe fading of stains. The remedy is simple:

Remove the coverslip, wash off the synthetic resin or balsam with xylene, and again dehydrate with the appropriate agent (100 % alcohol, acetone, or isopropyl alcohol); again clear through the appropriate xylene mixture and two or three changes of fresh xylene, and remount.

Hollander and Frost [*Acta Cytol. (Baltimore)*, **15**: 419, 1971] report that addition of 2,6-di-*tert*-butyl-*p*-cresol to Permount sufficient to make a 1 % solution (v/v) results in reduction of fading of stains and cracking and crazing of the mounting medium which normally occurs over a period of years. The cresol is an antioxidant.

RESTAINING.   If fading has already occurred, it may be necessary to restain. In this case, after soaking off the cover glass in xylene (and it may take a day or so to loosen it), pass the section through 100 % alcohol and succeeding reagents just as with a fresh section.

The same technic may be employed when only a stained section is available and some other staining procedure is desired than that originally employed. In this case one may need to decolorize the previous stain with acid alcohol or perhaps weak ammonia alcohol, then wash thoroughly with water, and proceed with the new technic. Ordinarily sections previously stained with iron hematoxylin are not suitable for the iron reaction with potassium ferrocyanide, since the iron lake reacts. Here iron may be removed by 1 to 2 h in 5 % $Na_2S_2O_4$ (dithionite).

Where there were two sections of the same block on a slide, we sometimes restained one of them by another method, leaving the other covered by a cover glass during the new staining procedure.

CELLOIDIN SECTIONS

NITROCELLULOSE SECTIONS.   These sections may be affixed to slides either before or after staining. They are floated out in a vessel of appropriate size onto the surface of clean slides and manipulated on the slanted, partly immersed slide until they lie smooth. They are then blotted firmly with filter paper.

If the section is not already stained, one should tilt an open bottle of ether so that the heavy vapors can be seen descending from the mouth of the bottle onto the surface of the slide. This softens the nitrocellulose so that it adheres firmly to the slide. Or one may dip the slide into 0.5 % nitrocellulose in ether alcohol mixture for a few seconds, drain, wipe the back of the slide clean, and harden the film by immersing the slide in chloroform for 5 to 10 min. Sections are then carried into successive baths of 95 % alcohol, 80 % alcohol (0.5 % iodine in 70 % alcohol 10 to 15 min, 5 % sodium thiosulfate 5 to 10 min for mercury fixations only), and water, or directly into stain if this is in a hydroalcoholic solution. After staining, wash free of excess stain as prescribed in the method used.

LOOSE NITROCELLULOSE SECTIONS.   Some prefer to stain celloidin and nitrocellulose sections before attaching them to slides. This is particularly convenient when large numbers of sections from the same block are to be prepared for class use.

In this case sections are transferred on the slightly curved spatulas known as *section lifters*, perhaps with the aid of a needle, from the 80 % alcohol in which they are stored, through 70 and 50 % alcohols and water, or through 0.5 % iodine in 70 % alcohol if it is necessary to remove mercury precipitates, and thence to sodium thiosulfate solution and to water, or directly into alcoholic staining solutions.

DEHYDRATION AND CLEARING OF NITROCELLULOSE SECTIONS.   After staining, loose sections are dehydrated and cleared by transfer through dishes of the same successive reagents used for attached nitrocellulose sections. Attached sections are handled much as are paraffin sections, and the same technics apply to collodionized paraffin sections.

The more resistant stains may be dehydrated with 95 % alcohol, cleared in Weigert's

(Ehrlich's "Encyklopädie") "carbol-xylene," which is a mixture of 1 volume of melted phenol crystals with 3 volumes of xylene, washed in two to four changes of xylene, and mounted as described for paraffin sections in Depex, Permount, balsam, or other suitable resin.

DESTAINING OF NITROCELLULOSE.   If the staining of the nitrocellulose with basic aniline dyes is objectionable, it may be removed in 2% alcoholic rosin solution, and dehydration and clearing may be completed as above or by one of the following methods.

CLEARING OF SECTIONS STAINED BY EASILY EXTRACTED DYES.   For basic aniline dyes which are extracted by his carbol-xylene, Weigert (Ehrlich's "Encyklopädie") suggested clearing by repeated application of xylene and blotting with filter paper between applications.

Many writers have used various essential oils for clearing, among which the best seems to have been the Cretan origanum oil. Sections stained with basic aniline dyes, especially with azure-eosinates, may be dehydrated with two or three changes of 99% isopropyl or *tert*-butyl alcohol, which do not dissolve nitrocellulose and do not extract the azure either from the section or from the surrounding nitrocellulose, and then passed through a mixture of equal parts of the alcohol and xylene and two or three changes of xylene. Or they may be differentiated with a 2% solution of colophonium (pine rosin) in 95% alcohol until the nitrocellulose is colorless, and then dehydrated and cleared by the isopropyl alcohol–xylene sequence just described. Or after the rosin-alcohol one may blot and flood with xylene alternately for 2 or 3 times until the sections are clear.

FROZEN SECTIONS

FAT STAINS.   Frozen sections stained for fats are floated out smooth on clean slides. One brings one edge of a section in contact with the partially immersed, obliquely held slide, raises the slide gradually so that the section settles smoothly on the surface of the glass, smoothing out folds by varying the angle of immersion so that the folded portion is floated loose while the rest remains above water on the slide. Needle manipulation may occasionally be necessary, especially with irregular or torn sections. When the section is satisfactorily smoothed out, it is drained and blotted with smooth, hard filter paper by running the tip of one finger along the slide on top of the paper while one end of the paper and slide are firmly held down with the other thumb. It is then mounted in a suitable aqueous medium. Glycerol may be used for temporary mounts. These may be sealed by carefully drying the border zone of the slide and cover glass and painting with xylene or toluene balsam or with a pyroxylin cement. With a small drop of glycerol which does not quite reach the edge of the cover glass, edging with polystyrene, as below, should afford a fairly permanent mount.

POLYSTYRENE EDGING.   To seal sticky glycerol gelatin or Arlex gelatin mounts, we often dip a glass rod in 20% polystyrene in xylene and run it along each edge of the cover glass. It dries to nonstickiness in an hour and can be packed away in direct contact with other slides the next morning, without fear that the slides will stick together. Xylene–cellulose caprate serves even better.

LANOLIN-ROSIN.   This medium was recommended by Romeis as the best of the sealing media for aqueous mounts. Dry 20 g anhydrous lanolin in a hot porcelain evaporating dish, stirring it for 15 or 20 min. Then add, bit by bit, 80 g rosin, stirring the while until a clear, light brown fluid results. It is best to heat it on a closed electric hot plate to avoid the chance of fire. Pour it out into small molds and let it harden. Heat a glass rod or metal spatula until it is quite hot, melt off a few drops from a block of the lanolin-rosin, and fix the corners of the cover glass in place. Then apply more hot, melted rosin to complete the sealing.

For permanent mounts, media containing gum arabic or gelatin are usually employed. The first dry hard after a time; the second have to be melted for use, and they set on cooling. Formulas for several usually successful media are described later in this chapter and tabulated in Table 5-2.

In tropical and subtropical climates sealing is needed also for gum syrup.

MOUNTING MEDIA

The material on mounting media is derived from Lillie et al. (*Stain Technol.*, **25:** 1, 1950 and **28:** 57, 1953) and Greco (*Stain Technol.*, **25:** 11, 1950) as later amplified and supplemented by Lillie (1954, 1965, and *Stain Technol.*, **30:** 133, 1955).

Resinous media are composed of a solid natural or synthetic resin dissolved in a suitable solvent to lower the viscosity to a point where the solution will readily enter tissue interstices, flow between slide and cover glass to fill the space completely, and quickly release entrapped air bubbles. The solvent (either that in which a natural resin is dissolved as it comes from its source or a suitable added solvent, usually an aromatic hydrocarbon but sometimes an alcohol or a chlorinated hydrocarbon) must be sufficiently volatile to allow fairly prompt drying of the resin to a hard state and at the same time not so volatile as to dry prematurely during mounting.

It has been observed that highly volatile solvents, such as benzene and to a lesser extent toluene, are likely to produce air spaces under cover glasses because of their excessive evaporation. This fault may be assigned also to low concentration of resin in the solvent. However, higher-boiling solvents, such as xylene (bp 138 to 144°C), trimethylbenzene (bp 165 to 176°C), and diethylbenzene (bp 181 to 184°C) are less likely to give rise to air bubbles. Indeed, trimethylbenzene solution mounts remain liquid internally for a long time.

The rate of drying and the tendency of some mounting media to aspirate air into the mounting space have importance in relation to the conditions of study. Prompt drying to nonstickiness and to nondisplaceability of cover glasses is of great importance to the surgical pathologist; the late aspiration of air bubbles poses problems only if it is necessary to refer back to the same section. On the other hand, with 50- to 100-μ sections and with mounts of membranes, eggs, and small embryos, time is of less consequence, and it becomes highly important that air should not enter the preparations.

The Refractive Indices of Mounting Media

These indices have been much discussed by microscopists, some of whom strongly advocate media with low indices, whereas others insist on high. Without going into theoretical optics, the practical effect seems to be that in media of refractive indices ranging from 1.44 to 1.50 much detail is apparent microscopically, even in unstained and uncolored objects, by reason of the difference in average refractive index of tissues (1.530 to 1.540) from that of these mounting media. As the index of refraction of the mounting medium approaches that of the tissue, the latter becomes more and more transparent, and unstained objects may be extremely difficult or quite impossible to discern. We have seen partly faded Nissl preparations in which we had great difficulty in even finding the section, either grossly or microscopically. When the critical range (about 1.535) is passed, unstained objects again become evident by reason of their now lower refractive index, and media with refractive indices above 1.60 may be quite useful.

Much stress has been laid by some writers, notably Groat, on the refractive index of the dry resin as opposed to that of the solution used for mounting. Groat agreed with Lillie (personal correspondence, 1950) that both indices are important. Obviously the refractive index of a resin solution will lie between that of the solvent and that of the dry resin; and as the solvent evaporates, the refractive index of the medium surrounding the specimen gradually approaches that of the dry resin.

It appears from the foregoing discussion that no single mounting medium will serve all purposes equally well. For this reason data regarding refractive indices of mounting solutions and of dry resins are presented in Table 5-1.

Natural Resins

Balsams are deep amber or yellow resins composed to a considerable extent of terpenes and their carboxylic acids such as abietic acid (pine rosin) and levopimaric acid (French fir), both of which are said to contain two carbon double bonds (Karrer).

CANADA BALSAM. This resin is derived from the liquid rosin of the Canadian fir *Abies balsamea*. Its acid number ranges from 88 to 106; its saponification number, from 105 to 116 (Lange). When prepared by heating with water until the evolution of steam ceases, it solidifies to a clear yellow, hard, brittle resin which is freely soluble in xylene, toluene, benzene, chloroform, etc. Xylene solutions of 60 to 65% by weight of the hard resin correspond in consistency to the syrupy solutions commonly recommended. The resin sets slowly and takes many months to dry hard enough so that slides will not stick together if warmed to summer room temperatures (30 to 35°C).

Hematoxylin and eosin stains are well preserved in Canada balsam, though there is a gradual differentiation of the eosin so that after a year or two muscle, connective tissue fibers, cytoplasms, and oxyphil inclusion bodies present a considerably greater difference in intensity of staining than was evident when the preparations were fresh. We have found renal intranuclear inclusion bodies much more conspicuous in such aged preparations than by most other methods. Canada balsam is superior to many of the synthetic media for preservation of the cobalt sulfide deposits in alkaline phosphatase preparations. Basic aniline dyes are poorly preserved, and Prussian blue fades fairly soon. Acid fuchsin in Van Gieson stains fades, but this fading may be retarded by half saturating the balsam with salicylic acid. This procedure Lillie has modified from Curtis (*Arch. Méd. Exp. Anat. Pathol.*, **17**: 603, 1905). Its rationale has been commented on in the discussion earlier in this chapter (see Mounting in Resinous Media).

Natural syrupy Oregon fir balsam possesses a refractive index of 1.5271. On being heated to 200°C it does not boil, but with an air stream it loses about 18% of its weight and solidifies on cooling. Its refractive index then reads 1.5407.

This resin sets slowly and takes months to dry hard. It fades basic aniline dyes, bleaches Prussian blue promptly, and conserves cobalt sulfide well. Thus its properties are closely similar to those of the usual xylene solutions of Canada balsam.

DAMMAR. This is a common resin used in the varnish and lacquer industries, derived from various East Indian trees of the genus *Shorea*. It softens at 75°C and melts at 100°C. Its acid and saponification numbers are 35 and 39, respectively (Lange), and it contains unsaturated compounds (iodine number 64 to 112). It dissolves in aromatic hydrocarbons, chloroform, ether, and the like. It is commonly used as a xylene solution of about 60 to 75% resin content by weight.

Since dammar is often dirty and its 60 to 75% solutions are virtually impossible to filter because of their viscosity, the usual practice is to dissolve it in a much larger volume of benzene or xylene, filter, and then evaporate down to the required viscosity or to the predetermined weight of solution. Evaporation of xylene is greatly expedited by passing a fairly rapid air stream over the surface of the solution while heating it on a closed electric hot plate under a chemical hood. The xylene is recovered by use of a condenser.

ROSIN, OR COLOPHONY.   Obtained from various pines, rosin is composed largely of abietic acid. It is a dark brown, brittle resin with a melting point of 120 to 135°C. Its acid number is 155 to 175; saponification number, 167 to 194; iodine number, 80 to 220. This rosin is sometimes used as a 1% alcoholic solution for acid differentiation of azure stains, occasionally as the xylene solution for mounting, but most often as a constituent of varnishes and cements.

CEDAR OIL.   When used as a mounting medium, cedar oil is usually concentrated to the point where its refractive index has risen from the initial 1.503 to about 1.5150. On being heated in an oil bath, the native solvent boils off at 168 to 187°C, and the oil loses about 52% of its initial weight as the thin cedar oil for clearing. The refractive index of the residual resin is 1.5262, and the resin is solid at room temperatures.

Cedar oil mounts set slowly and take months to dry hard. Basic aniline dyes fade gradually. Wolbach used this action to differentiate his Giemsa stain for rickettsiae. Prussian blue fades completely in this medium, but cobalt sulfide is well preserved.

Semisynthetic Mixtures

EUPARAL.   This is composed of gum sandarac dissolved in a mixture of eucalyptol and paraldehyde with a liquid mixture of camphor and phenyl salicylate of unstated proportions and a refractive index of 1.53576, according to Gilson (*Cellule*, **23:** 425, 1906). Gum sandarac contains 85% sandaracolic acid and 10% callitrolic acid. Its acid number is 140 to 154; saponification number, 142 to 174; iodine number, 66 to 160 (Lange). These characteristics of acidity and unsaturation place this mixture in the same general group as the natural resins. The carbonyl group of the paraldehyde offers a possible additional reducing agent.

The refractive index of euparal is usually given as 1.483. At 20°C we got 1.4776. On concentration in partial vacuum at 60°C it rose to 1.5174, and the resin solidified on cooling. Romeis states that the solid resin has a refractive index of 1.535.

Euparal sometimes causes discoloration and fading of hematoxylin stains (Lee). The green, or *vert*, variant contains a small amount of a copper salt to prevent this action. Basic aniline dyes and Romanovsky stains are fairly well preserved, though not so well as in some synthetic resins. Prussian blue is reduced and bleaches. Cobalt sulfide is well preserved.

DIAPHANE.   This is composed of a juniper gum base with certain natural and synthetic phenols, according to its manufacturers, the Will Corporation. No formula has been published. The refractive index of diaphane and of green diaphane is quoted as 1.483; at 20°C we found 1.4777 and 1.4792. On concentration in vacuo at 60°C, diaphane lost 59.6% of its original weight and solidified on cooling. Its refractive index was 1.5486.

Like euparal vert, green diaphane contains a little copper to intensify and conserve hematoxylin stains. Basic aniline dyes and Romanovsky stains are fairly well preserved, Prussian blue is decolorized, and cobalt sulfide is excellently conserved. Some fading of fuchsin and of acid fuchsin is observed, which we are inclined to attribute to reduction to leuco dyes rather than to acid action.

Ester Gums

Attempts at neutralizing the acid of the natural balsams with small amounts of sodium or calcium carbonate in the cold have been largely futile, since the resins themselves consist largely of carboxylic acids.

Esterification of abietic acid from pine rosin with glycols or glycerol has resulted in the formation of essentially neutral esters, called generally *ester gums* and used widely in the varnish industry. The class of these which has proved most useful in microscopy is the neutral or low-acid class, with acid numbers below 8. Clear, amber to light brown resins result, soluble in aromatic hydrocarbons to 70 or 80% by weight at syrupy viscosity.

After mounting in 75% ester gum in xylene, cover glasses become fairly immovable in an hour or two; but the resin remains rather soft, and slides will stick together if packed back to face, unless dried for some weeks.

Stains with basic aniline dyes and Romanovsky stains are well preserved. Prussian blue is reduced and decolorized. Cobalt sulfide is well preserved. Fuchsin and acid fuchsin tend to decolorize.

Synthetic Resins

Those now in use include styrene polymers (Monsanto's Lustron L 2020 and Lustrex L-15, Zirkle's L-15, Gurr's Depex, Distrene and others); Groat's styrene plus isobutyl methacrylate copolymer; synthetic terpene resins, especially β-pinene polymers (Harleco synthetic resin HSR, Fisher's Permount, and the Will Corporation's Bioloid); a coumarone plus indene copolymer (Technicon resin), a coumarone mixture with other resins (Gurr's Medium), a maleic polymer with plasticizer (Gurr's Xam), the naphthalene or cycloparaffin polymers (Clarite and Clarite X, introduced by Groat and now unavailable, and the old Fisher Permount, now replaced by a new β-pinene polymer sold under the same name); some entirely secret proprietary preparations such as Rhenohistol of the Rheinpreussen Chemical Works, Homberg-Niederrhein, Germany; and Mahady's Micromount. According to the Farbenfabriken Bayer, Leverkusen-Bayerwerk, Germany (letters, 1951) Caedax is a mixture of a chlorinated aromatic hydrocarbon and a cyclohexanone dissolved in xylene. Usual solvents and concentrations by weight, nature of synthetic resin, and refractive index are presented in Table 5-1.

DIATOM MEDIA.   Brief mention is made also of a few media of very high refractive index, such as Hyrax, n_D 1.82248 (Hanna, *J. R. Microsc. Soc.*, **50**: 424, 1930); Flemming's Naphrax, n_D 1.76 to 1.80 (*J. R. Microsc. Soc.*, **63**: 34, 1943); and two media of the George T. Gurr Co.: Clearax, a diphenyl resin of 66°C mp and n_D 1.666 (for the solid resin) furnished as a chloroform solution of n_D 1.602, and Refrax, a naphthalene formaldehyde polymer, similar to Naphrax, of 60°C mp and n_D 1.78 (as solid resin) furnished as a xylene solution containing plasticizer (Flemming used dibutyl phthalate in Naphrax), n_D of solution 1.598. These media are employed particularly by students of diatoms.

GENERAL PROPERTIES OF SYNTHETIC RESINS.   In general these synthetic resins are quite neutral, and basic aniline dyes are well preserved in them. The chlorinated aromatic hydrocarbon of the usually neutral Caedax may occasionally (in contact with water?) break down and become highly acid, and Mahady's Micromount appears to be acid from the start. Otherwise the resins differ importantly in refractive index (see discussion earlier in this chapter), in rate of setting and drying to a nonadhesive state, and in degree of residual unsaturation. The more unsaturated resins, and the ketonic resins including the natural resins and ester gums, the terpenes and β-pinene resins, the coumarones and coumarone-indene resins, though often quite satisfactory for stains where oxidation-reduction potential is not involved, appear to reduce Prussian blue to the greenish white ferrous ferrocyanide, but they preserve cobalt sulfide preparations relatively well. Conversely, the more oxidized or saturated resins preserve Prussian blue well but allow fading of cobalt sulfide preparations. There are also important differences in solubility in solvents, in viscosity of solutions, in rate of drying, and in tendency to form air bubbles in mounts. Plasticizers are added to combat this last tendency, but they tend to retard drying to nonstickiness.

TABLE 5-1   REFRACTIVE INDICES OF RESINOUS MOUNTING MEDIA IN SOLUTION AND DRY

Resin	Class	Solvent and % resin	Refractive index Solution	Refractive index Solid
Canada balsam	Dried natural resin	Xylene, 60%	1.5232	1.5447 c
Oregon fir balsam	Liquid natural resin	Turpentines, ca. 82%	1.5251	1.5407 o
Gum dammar	Dried natural resin	Xylene, 60%	1.5317	1.5589 c
Cedar oil	Liquid natural resin	Turpentines, ca. 48%	1.5030 1.5151*	1.5262 o
Euparal	Semisynthetic resin mixture	(See text)	1.4776	1.5174 o 1.535 R
Diaphane	Semisynthetic resin mixture	(See text)	1.4777 1.4792	1.5486 o g
Ester gum (5 samples)	Glycol resin acid esters	Xylene, 75%	⎰1.5352 ⎱1.5379	1.5516 c 1.5552 c
Harleco HSR	β-Pinene polymer	Xylene, 60%	1.5202	1.5390 c
Fisher Permount	β-Pinene polymer	Toluene, 60%	1.5144	1.5286 c
Willco Bioloid	β-Pinene polymer	Xylene, 60%	1.5272	1.5505 c
Technicon Resin	Coumarone indene polymer	Benzene + xylene, 60%	1.5649	1.6205 c
Gurr's Medium	Coumarone resin mixture	Cineol, ca. 77%	1.5310	1.5574 o
Gurr's Pale Medium		Cineol, ca. 77%	1.5082	1.5296 o
Cellulose caprate	Cellulose tricaprate neutral ester	Xylene, 50%	1.4860	1.4734 LH
Gurr's Xam	Maleic polymer plasticized	Xylene, ca. 77%	1.5219	1.5401 o
Clarite	Cycloparaffin polymer	Toluene or xylene, 60%		1.544 G
Clarite X	Cycloparaffin polymer	Xylene, 60%	1.5352	1.5647 c
Fisher's old Permount	Cycloparaffin polymer	Toluene, 62.5% Xylene, 60%	1.5172 1.5170	1.5376 c
Caedax	Cyclohexanone and chlorinated diphenyl resin	Xylene, 82%	1.6306	1.6724 c
Rhenohistol	Ketone + HCHO condensation	Xylene, 60%	1.520	1.533 c
Hyrax	Naphthalene resin (secret)	Aromatic hydrocarbon		1.8225 H
Naphrax	Naphthalene HCHO polymer	Xylene + dibutyl phthalate		1.76– 1.80 F
Gurr's Clearax	Diphenyl resin	Chloroform	1.602	1.666 a
Gurr's Refrax	Naphthalene HCHO polymer	Xylene + plasticizer	1.598	1.780 a
Gurr's Depex	Polystyrene	Xylene + plasticizer	1.5228	1.6 a
Mahady's Micromount		Xylene (?), ca. 43%	1.4918	1.4839 c
Groat's Copolymer	Styrene + isobutyl methacrylate	Toluene, 45%	1.5193	1.5500 c
Lillie's polystyrene (Monsanto L–2020)	Styrene polymer	Diethylbenzene, 20%	1.5150	1.6193 c

* Immersion oil.

Source: c, calculated from solution data; o, observed after concentration of commercial solution; a, as advertised by manufacturer; R, according to B. Romeis, "Taschenbuch der mikroskopischen Technik," 13th ed., Oldenburg, Munich, 1932; H, according to Hanna (*J. R. Microsc. Soc.*, **50:** 424, 1930); F, according to Flemming (*J. R. Microsc. Soc.*, **63:** 34, 1943); G, according to Groat (*Anat. Rec.*, **74:** 1, 1939); g, determined by Greco in Lillie's laboratory for the second edition of this book.

Some fading of the acid fuchsin component of Van Gieson stains is observed with the Bioloid resin, euparal, diaphane, and Canada balsam and to a lesser extent with the Micromount, Xam, and the ester gums. There is moderate fading of the azure component of azure-eosin stains with diaphane and Xam, and severe fading with Micromount, which bleaches even alum hematoxylin. The eosin component of this stain fades somewhat in euparal and Bioloid, and occasionally in polystyrenes.

POLYSTYRENE MEDIA. These are usually employed in aromatic hydrocarbon solvents. Their viscosity is too great to permit much over 20 to 25% concentration of resin. Consequently, they set rapidly but tend to form large air spaces under cover glasses. This tendency is combated either by use of a higher-boiling-point solvent, such as diethylbenzene, trimethylbenzene, p-cymene, or a mixture of equal volumes of amylbenzene and xylene, or by addition of orthocresyl phosphate or dibutyl phthalate as plasticizers. Excessive amounts of plasticizer must be avoided, as they retard setting and hardening. Addition of 5 ml dibutyl phthalate to 70 ml xylene and 25 g polystyrene seems adequate to prevent air aspiration, and does not greatly retard setting; 10 ml dibutyl phthalate and 65 ml xylene did considerably delay setting of the mount.

We strongly recommend 20 to 25% solutions of Monsanto's Lustron L-2020 in diethylbenzene,[1] trimethylbenzene,[1] or a mixture of secondary amylbenzene and xylene in equal volumes. Some workers find the odor of the amylbenzene disagreeable. These fluids are easy to mount in: the cover glasses are immovable in an hour, and the slides may be packed tightly together without sticking by the following morning.

Gurr's Depex resin is quite satisfactory. Kirkpatrick and Lendrum's DPX contained 20 g polystyrene, 15 ml tricresylphosphate, and 80 ml xylene. We have had excellent results also with solutions of 5 or 10 ml dibutyl phthalate, 70 or 65 ml xylene, and 25 g polystyrene. The lower quantity of plasticizer permits faster setting and is usually adequate to prevent bubbles in thin sections. For thick sections Zirkle uses 20 g Lustrex 15, 20 ml dimethoxytetraethylene glycol, and 60 ml xylene. The medium sets slowly but forms no bubbles even with 50- to 100-μ sections. Ollet (*J. Pathol. Bacteriol.*, **63:** 166, 1951) also recommends Lustron L-2020, of which he dissolves 100 g in 50 ml dibutyl phthalate and 300 ml monochlorobenzene. Xylene is omitted because it "leads to fading of the Gram's stain."

Polystyrenes contain virtually no titratable acid and little or no residual unsaturated material. Consequently they are excellent for conservation of stains with basic aniline dyes, hematoxylin, Van Gieson's stain, and Mallory aniline blue variants. The Prussian blue and Turnbull's blue of the Perls, the Tirmann-Schmelzer, and the ferric ferricyanide reduction reactions are well preserved. Cobalt sulfide gradually disappears, presumably by oxidation, but may be restored by demounting and reimmersion in ammonium sulfide.

CELLULOSE CAPRATE (TRICAPRATE OR TRIDECANOATE). Introduced as a mounting medium in 1955 (*Stain Technol.*, **30:** 133), this is a pale yellow resin of low refractive index (1.4743 when dry), which yields solutions of satisfactory viscosity for mounting at 50 g resin: 50 g xylene (n_D 1.4860). Adhesion to glass is excellent, coverslips become immovable in an hour, and preparations are nonsticky in less than a day. Excess droplets of resin which form at edges of coverslips are readily cut off with a knife or thumbnail. Unstained sections remain readily visible; nuclei, cell granules, muscle striations, and the like are readily perceived. Azure-eosin and routine hematoxylin stains are well preserved. Prussian blue is well preserved; cobalt sulfide bleaches somewhat in a month and completely in 15 weeks. Many histochemical reactions are well preserved.

---

[1] Obtainable from the Eastman Chemical Company, Rochester, N.Y.

Chemically it is a neutral ester of a straight-chain saturated fatty acid, capric or decanoic acid $C_9H_{19}COOH$, 3 mol per repeating hexose unit of cellulose, and possesses no reducing capacity. Lillie now uses this medium for all histochemical work where dehydration and clearing in nonpolar solvents are permissible.

HISTOCLAD. This is one of the newer media which has achieved some popularity. It adheres well to glass; its melting point is 120°C; its refractive index when dry is about 1.54, and that of the syrupy xylene solution is lower. It is a relatively saturated resin with a low iodine number. The chemical nature of the resin remains unrevealed. It is said to be neutral and nonreactive and to remain almost colorless on aging. It is available from Clay-Adams, Inc., 141 East 25th Street, New York, New York 10010.

POLYVINYL ACETATE. Burstone (*J. Histochem. Cytochem.*, **5**: 196, 1957) recommended polyvinyl acetate,[1] 20% in 80% alcohol, as a mounting medium in esterase and phosphatase technics where azo dyes are formed by liberated α and AS naphthols in the presence of stable diazonium salts. Stains with Schiff reagent, celestin blue, alum and chrome alum hematoxylins, and aldehyde fuchsin are well preserved, and cobalt sulfide does not fade. The initial refractive index of the alcoholic solution is very low (1.3865) but is said to rise as the solvents evaporate. Drying to immovability of coverslips occurs in an hour, and preparations are nonsticky in 3 to 4 hours.

Mineral Oil

This medium is unexcelled for preservation of azure-eosin, Van Gieson, and similar stains. Lillie used it for several years before the introduction of Clarite, for mounting azure-eosin stains, and sections 20 years old still showed excellent preservation of color. The oil does not dry, however, and preparations sealed with nitrocellulose cements often leaked badly. Nevertheless, stains are well preserved and preparations may be remounted after brief soaking in acetone to dissolve the cement. Refractive indices are low—1.460 to 1.483—the heavier, more viscous oils possessing the higher indices. Modified mineral oils, sold for immersion oil, may be obtained which possess refractive indices about 1.518 to 1.520, or nearly that of crown glass. We have used such oils only for temporary mounts, but it is to be noted that Giemsa-stained blood films, put away without cleaning after use of such immersion oils, do not exhibit the fading in oil-wet areas that is usually seen in films where cedar oil has dried after use.

AQUEOUS MOUNTING MEDIA

Permanent mounting media of this type fall perhaps into four general classes: simple syrups, gum arabic media, glycerol gelatins, and acid media of the lactophenol type. Of these the last type is used principally in botany and insect histology. They do not in general conserve stains well, especially nuclear stains. Abopon,[2] recommended by Lieb (*Am. J. Clin. Pathol.*, **17**: 413, 1947) for mounting crystal violet stains of amyloid, is excellent for *this* purpose, and for acetic orcein stains, which are conserved for years (Hrushovetz and Harder, *Stain Technol.*, **37**: 307, 1962, who also found it useful for Giemsa stains). However, it promptly bleaches alum hematoxylin stains. Gum arabic and glycerol gelatin media often cause diffusion of basic aniline dyes into the medium ("bleeding"). This bleeding occurs both in acid gum arabic media and in the neutral glycerol gelatins. It is not prevented

---

[1] Shawinigan Corp., Springfield, Mass.
[2] Valnor Corp., Brooklyn, N.Y.

by addition of small quantities of potassium acetate that suffice to raise the pH to 6.5 or higher. It is prevented by high salt concentrations, even though the medium remains acid. It does not occur in strong sucrose, fructose, or D-sorbitol syrups, though the pH of fructose may be as low as 4. Addition of large amounts of these sugars to glycerol gelatin, gelatin, or gum arabic media prevents bleeding.

About 20% by weight of potassium acetate or about 60% of one of the sugars suffices to prevent bleeding of crystal violet stains for amyloid.

Refractive indices of permanent aqueous mounting media are generally low (1.41 to 1.43). The highest levels, 1.49 to 1.50, are attained with media containing large proportions of sugars, notably fructose and D-sorbitol.

Gum arabic media are generally acid (pH 3.5 to 4.2). The amount of potassium acetate needed to raise the pH to 6.5 or 7 is more than $\frac{1}{10}$ and less than $\frac{1}{2}$ the weight of gum arabic. Glycerol or sugar, sometimes both, is added to aqueous gum arabic solutions to raise the refractive index or to retard overdrying.

Amann's Viscol was apparently a mixture of phenol, gum arabic, and glycerol, according to Dahl (*Stain Technol.*, **26:** 97, 1951), who gave a substitute formula for Amann's secret preparation. Formulas for gum arabic media are given below.

The commercially prepared media Clearcol[1] and Viscol[2] are quite acid (pH 1.5 and 2.9, respectively), and their refractive indices are low (1.4039 and 1.4167). They set promptly after mounting and are not sticky. Fat stains are well preserved, but crystal violet stains of amyloid bleed badly. The Paragon[3] mountant is less strongly acid (pH 5.6) but affects stains similarly. Its refractive index is 1.4241. Alum hematoxylin counterstains faded in 1 month in Viscol but were fairly well preserved in the other two media.

Addition of 20% potassium or sodium acetate to these media might be worth trying, for the better conservation of hematoxylin and other counterstains.

Syrups often serve well as temporary mounting media, but they remain wet, sticky, and more or less fluid in moist climates, and they furnish excellent nourishment for molds. Sucrose, glucose, invert sugar, Karo, and maltose syrups crystallize around and under cover glasses after a time. A fructose syrup available around 1900 (Ehrlich's "Encyklopädie") did not crystallize and had a refractive index of 1.500. These properties are nearly duplicated by the modern crude sorbitol syrup Arlex. The commercial corn syrup Karo has had some vogue as a temporary mounting medium.

Addition of 10% gelatin to syrups is enough to render them solid at 20 to 25°C. We have used an Arlex gelatin thus prepared with some success, but it tends to be sticky in hot weather and is probably unsuitable for tropical use. Increasing the gelatin content to 15% could be helpful.

The glycerol gelatin media (formulas below) require melting for use, but they set firmly in a few days. Their refractive indices are low (1.41 to 1.42). They are tenacious of air bubbles but can be degassed by exposure to 350 to 400 mm vacuum while melted in a few minutes.

FORMULAS

FARRANTS' GLYCEROL–GUM ARABIC.   Dissolve 50 g gum arabic (acacia) in 50 ml warm distilled water. Add 1 g arsenic trioxide and 50 ml glycerol. The viscosity is rather high. The index of refraction is 1.43600 at 20°C. Addition of 1 g potassium acetate is recommended

---

[1] H. W. Clark Co., Melrose, Mass.

[2] Drogueries Réunies, Lausanne, Switzerland.

[3] Paragon C. & C. Co., New York, N.Y.

if a relatively neutral medium is desired. The As_2O_3 can be replaced by 15 mg merthiolate, 0.1 ml cresol, or 100 mg thymol.

Ehrlich, in his "Encyklopädie" (1903), under glycogen, recommended for conservation of iodine stains a mounting medium made by adding gum arabic to a weak Lugol solution to attain a suitable viscosity. Glycogen stains were conserved for at least 10 years. Since Takeuchi has found a 1 : 2 : 600 iodine(KI)-glycerol-water adequate as a temporary mounting medium (Chap. 6), we suggest adding about 200 mg iodine dissolved in 400 mg potassium iodide in 0.5 ml distilled water to 100 ml Farrants' formula (above) to simulate Ehrlich's extempore medium. This medium should serve for permanent mounts for iodine stains of glycogen, starch, cellulose, and amyloid.

GUM SYRUP. Modified from Apáthy by Lillie and Ashburn (*Arch. Pathol.*, **36**: 432, 1943), the formula for gum syrup is as follows: Dissolve 50 g acacia (gum arabic) and 50 g cane sugar in 100 ml distilled water by frequent shaking at 55 to 60°C. Restore volume with distilled water. Add 15 mg merthiolate (sodium ethylmercurithiosalicylate) or 100 mg thymol as a preservative. Place in vacuum chamber for a few minutes while warm to remove air bubbles. Highman (*Arch. Pathol.*, **41**: 559, 1946) adds 50 g potassium acetate or 10 g sodium chloride to this formula, to prevent bleeding of crystal violet stains for amyloid.

DAHL'S VISCOL. This replaces Amann's Viscol (*Stain Technol.*, **26**: 97, 1951). Dissolve 80 g gum arabic in 40 g (32 ml) glycerol and 90 g water; then add 20 g phenol.

Various sugars may be added in varying amounts to gum arabic media. Their effect is to raise the refractive index and to diminish the setting quality of the mountant.

KAISER'S GLYCEROL GELATIN AS MODIFIED BY MALLORY. Soak 40 g gelatin 2 h in 210 ml distilled water. Add 250 ml glycerol and 5 ml melted phenol. Heat gently, stirring constantly for 10 to 15 minutes until the mixture is smooth. Store in refrigerator at 0 to 5°C and melt as needed. Mallory notes a deleterious effect of the phenol on alum hematoxylin stains. The substitution of 50 mg merthiolate (sodium ethylmercurithiosalicylate) or 100 mg thymol for the phenol is suggested. Some workers call this glycerogel.

GLYCHROGEL. Zwemer (*Anat. Rec.*, **57**: 41, 1933) used the following: Dissolve 0.2 g chrome alum $(KCr(SO_4)_2 \cdot 12H_2O)$ in 30 ml distilled water by heating. Dissove 3 g granulated gelatin in 50 ml distilled water by heating. Add 20 ml glycerol to the still warm gelatin solution and mix thoroughly; then add the warm chrome alum solution, mix thoroughly, and filter in a 37°C incubator. Add a crystal of thymol or camphor as a preservative, or 10 mg merthiolate (sodium ethylmercurithiosalicylate). According to Wotton and Zwemer (*Stain Technol.*, **10**: 21, 1935), this medium possesses quite a high index of refraction (1.75) after drying a week. This compares with 1.46 for glycerol, 1.47 for glycerin jelly, and 1.54 for Canada balsam. On occasion we have had great difficulty with the filtration of this medium, and see no great advantage in it.

FRUCTOSE (LEVULOSE) SYRUP. This sugar formerly occurred in commerce only as a syrup of about 1.500 refractive index (Ehrlich's "Encyklopädie"), which fact probably accounts for its introduction as a mounting medium. Mallory directs: Dissolve 30 g fructose (levulose) in 20 ml distilled water by gentle heat. This was for use in Boston.

In our experience in Washington, D.C., this concentration is too low. A 70 to 75% solution has a more suitable viscosity and a higher refractive index (60% fructose, 1.43892; 70%, 1.46011; 75%, 1.4762; 80%, 1.4906 at 20°C).

Fructose syrups do not crystallize in the mounts, and when sealed can well serve as permanent mounting media. In dry climates sealing may be unnecessary, since the preparations become quite hard. Fructose is much higher in cost.

Most sugars crystallize badly after a time, spoiling the mounts. A 70% sucrose syrup has suitable viscosity for mounting and a refractive index of 1.46468. It serves well for a

TABLE 5-2 COMPOSITION AND PROPERTIES OF WATER-MISCIBLE MOUNTING MEDIA

	Gum arabic, g	Gelatin, g	Glycerol, g	Sugar or syrup, g	Potassium acetate, g	Merthiolate, mg	Other ingredients, g	Water, g	Refractive index	pH	Bleeding methyl violet
Apáthy gum syrup (LA)	50			Sucrose 50		15		100	1.4170	4.1	+
Apáthy gum syrup (H-a)	50			Sucrose 50	50	15	NaCl 10	100	1.4266	6.8	−
Apáthy gum syrup (H-b)	50			Sucrose 50	50	15		100	1.4252	4.0	−
Apáthy gum syrup (L-fr)	50			Fructose 50	50	15		100	1.4228	6.7	+
Glycerol gum syrup (L)	20		20	Sucrose 20	20	5		20	1.4600	7.1	−
Farrants glycerol gum arabic	50		50				As_2O_3 1	50	1.4360	4.4	+++
Farrants (KAc)	50		50		50	15		50	1.4404	7.2	−
Viscol (Dahl)	40		20				Phenol 10	45	1.4167	2.9	+++
Kaiser glycerol gelatin (M)		8	50				Phenol 1	42	1.4164	6.9	+++
Kaiser glycerol gelatin (F)		8	50		10		Phenol 1	42	1.4197	6.7	+++
Kaiser glycerol gelatin (L)		8	50			10		42	1.4130	7.0	+++
Kaiser sucrose glycerol gelatin		8	8	Sucrose 50	1	10		33	1.4519	6.6	−
Arlex gelatin		10		Arlex 89	1			*	1.4936	6.0	−
Zwemer glychrogel		3	20				Chrome alum 0.2	80	1.75 Z	6.0	−

* Water included in the Arlex sorbitol syrup.

Sources: *LA*, Lillie and Ashburn; *H-a* and *H-b*, Highman; *L*, Lillie; KAc, potassium acetate: M. F. B. Mallory. "Pathologic Technic," Saunders. Philadelphia. 1938: *F.* Friedenwald. Table derived in part from Lillie, Windle, and Zirkle, *Stain Technol.*, **25**: 1, 1950 and Lillie, Zirkle, and Greco, *Stain Technol.*, **28**: 57, 1953.

temporary mountant but crystallizes in a month or so. The refractive index of 70% glucose is 1.4614; of 80% maltose, 1.4512; of 75% fructose, 1.4762; of white Karo, 1.4799; and of commercial Arlex D-sorbitol syrup, 1.4860. Glucose, invert sugar, and maltose syrups crystallize in a month after mounting. Karo is somewhat acid (pH 5.8) and crystallizes after a time with or without addition of 1% potassium acetate. The Karo and maltose syrups set hard, but both crystallize. Arlex D-sorbitol syrup did not crystallize and had the highest refractive index, and it was found that addition of 10% gelatin produced a medium which set well and preserved fat and amyloid stains well.

ARLEX GELATIN OF LILLIE AND GRECO. Heat 89 g Arlex D-sorbitol syrup in a boiling water bath, add 10 g gelatin, and stir until dissolved. Add 1 g potassium acetate and 10 mg merthiolate. The pH is 6 or higher; the refractive index is 1.4936. As noted earlier, preparations set promptly but remain sticky for some time, especially in moist climates.

FERNANDO'S DEXTRIN SUCROSE SODIUM CHLORIDE. This solution is recommended for amyloid stains with crystal violet.

TABLE 5-3 FORMULAS OF INDIFFERENT DILUTING FLUIDS

1. Physiologic saline solution
Sodium chloride	8.5 g, usually 9.0
Distilled water	1000 ml

   Sterilize in autoclave.

2. Ringer's solution and Locke variant
Sodium chloride	8.5 g	8.5 g
Potassium chloride	250 mg	420 mg
Calcium chloride	300 mg	250 mg
Sodium bicarbonate	(200 mg)*	200 mg
Distilled water	1000 ml	1000 ml

   Sterilize by filtration with Berkefeld.
   Add calcium chloride last. Make fresh.

3. Locke-Lewis solution
Sodium chloride	8.5 g
Potassium chloride	420 mg
Sodium bicarbonate*	200 mg
Glucose	100–250 mg
Calcium chloride	250 mg
Distilled water	1000 ml

   Add calcium chloride last.
   Sterilize by Berkefeld filtration.
   Make fresh as needed.

4. Tyrode solution pH 7.5–7.8
Sodium chloride	8.0 g
Potassium chloride	200 mg
Calcium chloride	200 mg
Magnesium chloride	100 mg
Sodium acid phosphate	50 mg
Sodium bicarbonate	1 g
Glucose	1 g
Distilled water	1000 ml

   Add salts to water in order given.
   Sterilize by Berkefeld filtration.

5. Pannett and Compton's buffered
   salinet† solution pH 7.5
Sodium chloride	6.4 g
Potassium chloride	366 mg
Calcium chloride	160 mg
Distilled water	960 ml

   Boil or autoclave, cool, and add 40 ml of the following:
   Monosodium phosphate monohydrate, 12.5 mg
   Disodium phosphate, anhydrous, 67.6 mg
   Distilled water, 40 ml, which has been similarly sterilized
   The last approximately 5.7 ml of stock 0.1 M phosphate buffer, pH 7.5, diluted with 34.3 ml water (Table 20-12).

---

* Sodium bicarbonate is often omitted from Ringer's solution. It may be added to the Ringer-Locke solution and to the Locke-Lewis solution after the remaining constituents have been boiled to sterilize. The bicarbonate decomposes at 80°C in solution; hence it should not be added until solutions are below that temperature.

† In the Pannett-Compton solution, the chlorides and phosphates must be autoclaved separately and mixed after cooling.

Sources: The formulas are quoted as follows: no. 1, traditional at National Institute of Health (most authors give 0.9%); no. 2, both formulas are as in A. B. Lee, "The Microtomist's Vade-Mecum," 10th ed., edited by J. B. Gatenby and T. S. Painter, Blakiston, Philadelphia, 1937; B. Romeis, "Taschenbuch der mikroskopischen Technik," 13th ed., Oldenburg, Munich, 1932; and E. V. Cowdry, "Microscopic Technique in Biology and Medicine," Williams & Wilkins, Baltimore, 1943, 1948; no. 3, as in Romeis and Cowdry; no. 4, as in Cowdry, Lee, F. B. Mallory, "Pathological Technic," Saunders, Philadelphia, 1938, and Romeis; no. 5, emended from Lee.

Dissolve 16.7 g sucrose, 16.7 g dextrin, and 10 g sodium chloride in 100 ml distilled water, heating and stirring until clear. Add 10 mg sodium merthiolate (ethylmercurithiosalicylate). Store in tightly closed bottle. It is acid (pH 3.75) and has a high refractive index, n_D 1.54 (*J. Inst. Sci. Technol.*, vol. 7, no. 2, 1961).

Substitution of 2.33 g sodium acetate ($NaCO_2CH_3 \cdot 3H_2O$) for 1 g of the sodium chloride should raise the pH of the mixture to near 7, without altering the molar salt concentration.

Abopon[1] is now available as a thick viscous mass or as crystals. It should be diluted with 0.2 M phosphate buffer, pH 7, to a manageable consistency but should still be saturated, with a few crystals in the bottom when cool. Hrushovetz and Harder (*Stain Technol.*, **37:** 307, 1962) direct as follows (we have supplied amounts):

Heat 60 ml Abopon (gum or crystals) on water bath to 60 to 70°C in a beaker. Add gradually about 20 ml 0.2 M phosphate buffer, pH 7 (8 ml 0.2 M NaH_2PO_4 plus 17 ml 0.2 M Na_2HPO_4 should be about right), rotating slowly to mix, avoiding entrapment of air bubbles, until all crystals have dissolved. Store in about 40-ml amounts in small dropper bottles. A few crystals should form in the bottom; if they do not, add a few from the stock supply. If, as occasionally happens, microscopic debris is seen in the mounting medium, heat to 80°C on water bath, and filter while hot through Whatman No. 1 paper.

POLYVINYL ALCOHOL.   This has been used also by Burstone where it was necessary to avoid strong alcohol (verbal communication). It would seem possible to use for mounting sections the glycerol-water solution of the type used by Feder as an embedding medium, perhaps as recently modified by Masek and Birns (*J. Histochem. Cytochem.*, **9:** 634, 1961): 20 g polyvinyl alcohol, 3 g glycerol, and 45 ml distilled water. The solution solidifies after a time but may be reliquefied by melting in a water bath; it then remains liquid again for a period. See Feder (*J. Histochem. Cytochem.*, **10:** 341, 1962).

MEDIA FOR SURVIVING CELLS

Many procedures exist for the examination of tissues, blood, or their constituent cells or products in the fresh state or in aqueous media. For the study of surviving cells, protozoa, and bacteria, a drop of tissue—perhaps diluted in serum or some indifferent fluid (see Table 5-3 for formulas) such as physiologic ("normal") saline solution, Ringer's fluid, or Locke's fluid—is placed on a clean slide and at once covered with a cover glass. The edges of the cover glass may then be covered with petrolatum to prevent evaporation. Various reagents may be introduced by placing a few drops on the slide at one margin of the cover glass and drawing it into the observed space by applying filter paper to the opposite side of the cover glass.

Some procedures for the observation of living cells demand the use of a warm stage. Warm stages may be procured in various designs from the instrument makers. W. R. Earle (personal communication) found the most practical procedure was to enclose his microscope within a box with hand holes and holes for the eyepieces, and to keep the whole chamber warmed up by using the heat of the microscope lamp or other heat source to maintain the desired temperature, which he controlled with a thermometer. For extended investigations and prolonged observation, thermostatic control is necessary.

---

[1] Valnor Corp., Brooklyn, N.Y.

6

STAINS AND
FLUOROCHROMES

The examination of unstained material often has considerable value. However, it is usually restricted to the study of surviving cells and tissues, examinations with polarized light, fluorescence microscopy, phase microscopy, ultraviolet photography, microincineration procedures, and the study of pigments. In a sense, histochemical blocking and control procedures also often yield essentially unstained preparations.

The purpose of staining is to make more evident various tissue and cell constituents and extrinsic materials. Some stains are strictly solution phenomena, such as the staining of neutral fats with oil-soluble dyes; others are strictly chemical, such as the formation of Prussian blue by hemosiderin in the presence of acid and ferrocyanide; others depend on the presence of mordants, as hematoxylin depends on the presence of ferric or aluminum or other metallic salts; in others a so-called adsorption phenomenon may be responsible.

Generally, sequences or combinations of stains are employed to render two or more tissue elements conspicuous in contrasting colors. No process has been devised which will show all tissue elements to the best advantage, but procedures exist which will stain differentially as many as four elements, such as the elastin-collagen procedures which show cell nuclei and cytoplasmic structures as well as elastic and collagen fibers. Some procedures are best adapted to the general study of cell nuclei and cytoplasms, and these are often employed as general oversight stains for primary examination of tissues.

Certain stain solutions are employed in a variety of procedures. To prevent duplication it seems well to describe these solutions first. Others, used only in certain single or related procedures, are best described in connection with the special methods in question. Simple solutions of dyes in single solvents need no special description and will be referred to as such in the procedures concerned, though they may be used in a variety of procedures.

The dyes used in staining may be classed in various ways: according to origin, as natural and synthetic; according to chemical class, as azo, triphenylmethane, fluorane, quinoneimid; or according to their chief use in microscopy.

Origin is of little significance, and certain dyes, such as the indigos and orcein, may be derived either naturally or synthetically. Each of the chemical classes contains dyes used for diverse purposes, and chemically dissimilar dyes may be used for the same purpose. For example, the disazo dye Sudan IV and the anthraquinone dye coccinel red impart nearly identical colors to neutral fats; hematoxylin and certain oxazin dyes are good iron mordant nuclear stains. Two of the best collagen stains are the triphenylmethane dye methyl blue and the disazo dye amido black 10B.

For practical purposes a classification based on use seems best and will be followed in this work.

Dyes are generally complex organic chemicals whose behavior depends to a variable extent on their precise chemical constitution. They may often be mixtures of homologs varying by differences of one or more methyl, ethyl, or phenyl groups, by the number of sulfonic acid or carboxylic acid radicals included, or by the degree of oxidation or reduction. Many are quite stable in the dry state and in solution; others alter spontaneously in solution or even in the dry state. These alterations are generally accelerated by heat and light, and it is well to store dyestuffs in a cool, dark place.

On purchasing dyes—particularly unusual ones or those whose names closely resemble those of other, perhaps dissimilar, dyes—it is well to specify the "Colour Index" (abbreviated C.I.) number of the dye in question; or if it has no C.I. number, its chemical constitution or the precise designation given by the manufacturer, *including the manufacturer's name.* Do not omit the group of letters and/or numerals which follow the name.

The third edition of the "Colour Index," issued in five volumes in 1971, by the Society of Dyers and Colourists, Bradford, Yorkshire, aided by the American Association of Textile Chemists and Colorists, list nearly 8000 synthetic and natural dyestuffs and pigments, a 30% increase since the 1963 supplement. It gives many physical characteristics, solubility data, chemical constitution, often the method of manufacture, and a list of the various synonyms applied to each dye.

In regard to nomenclature of dyes, we have generally followed that adopted in the Biological Stain Commission's publication "Conn's Biological Stains," eighth edition, 1969 (R. D. Lillie, editor), though in some instances we have preferred the original names from the "Colour Index." The manuscript in preparation for the ninth edition of "Conn's Biological Stains" has also been available.

The Biological Stain Commission is a nonprofit corporation whose trustees are for the most part representatives of the various American scientific societies in those fields which use stains. The function of the commission is to test stains and to supply and disseminate information regarding their constitution, behavior, and uses.

Because of the variability of performance of various lots of dyestuffs, it is well to purchase only dyestuffs which have been tested for the purpose for which they are to be used. In this country such testing is done on many common dyes employed in biologic staining procedures by the Biological Stain Commission in accord with the tests given in the latest edition of "Conn's Biological Stains," and dyes bearing the commission's certificate are generally reliable for the purposes specified by the commission, provided they have not decomposed since the date of certification. Dates of certification may be determined approximately by consulting the periodic lists of certifications published by the Stain Commission in "Stain Technology." It is good practice to write the date of receipt on each bottle of dye purchased, using a nonfading ink.

Statements regarding the identity and chemistry of dyestuffs are derived generally from "Conn's Biological Stains," from the "Colour Index," from the literature, and from our own files, and generally are made without specific reference. For further details the first two references above are recommended. Structural formulas will be given in this book only for dyes not so formulated by Lillie (1969) in "Conn's Biological Stains."

NUCLEAR STAINS

The general procedures for staining of cell nuclei fall into two main classes: (1) those done with basic ("cationic") dyes and depending on the presence of deoxyribonucleic and ribonucleic acids (DNA and RNA) to form dye salt–type unions; (2) those done with sequence or combination procedures using a di- or trivalent metal ion mordant and a dyestuff, usually *o*-diphenolic in nature, capable of forming a chelate complex with the metal ion, which in turn is bound to tissue groups which are not necessarily acid in nature. The latter group often functions quite well for nuclear demonstration in material from which the nucleic acids have been removed by acid extraction, as in decalcified tissue. Hence these mordant stains, particularly the aluminum "lakes" (chelate complexes) of hematoxylin, are more widely useful in general pathology than the more precise and specific basic aniline dyes.

The term *basophilic* is properly applied to those tissue substances of acid nature which color readily with basic aniline dyes, such as the nucleic acids, the sulfated polysaccharides, the sialic and uronic acid polysaccharides, and proteins containing an excess of carboxylic acid over amino groups. Although the metal-mordant dyes often color many of the same morphologic elements, they also color under some circumstances substances lacking acidic groups, such as neutral mucopolysaccharides, and under others lipids such as myelins and even basic (*oxyphil, acidophil,* or *eosinophil*) substances such as keratins, keratohyalin, trichohyalin, pituitary alpha cells, and eosinophil leukocyte granules. Hence it is better not to use the term *basophil* to indicate stainability with alum hematoxylin. The old term *siderophilia* to indicate stainability with iron hematoxylin was at least logical, and a term *metallophilia* would probably be more generally significant of the tissue characteristics inducing staining with these metal-mordant complexes. The term *sudanophilia* is similarly used to denote stainability with oil-soluble dyes such as Sudan III and IV, oil red O, Sudan black, et al.

Mordant Dyes

Generally the dyes classified by the "Colour Index" as mordant dyes exhibit one of several chemical end groups: (1) an azo group with hydroxyl groups ortho to it on both sides, (2) one or more salicylic acid residues, (3) orthophenolsulfonic acids, (4) *o*-carboxyquinone, (5) chromotropic acid groups, (6) catechol groups and *o*-hydroxysemiquinone groups, and (7) 1-hydroxy anthraquinones. Some dyes may possess two such groupings; one dye, pseudopurpurin, C.I. 58220, is 1,2,4-trihydroxy-3-carboxyanthroquinone, qualifying in the second, sixth, and seventh groups.

A number of trisazo dyes with salicyl groups are classed as direct dyes 30005, 30105, 30110, 30115, 30120, 30135, 30140, 30150, 30155, 30160, 30165, but these are often afterchromed with $CuSO_4$. Some direct disazo dyes show group units 29130, 29128, 29125, 29120, 29115. These also are after being treated with Cu salts.

Metal mordanting is an extremely ancient practice in leather and textile dyeing. Madder, which was the first source of alizarin, was in use in ancient Egypt. Herodotus records ("History," lib IV: 189) that the Libyan women dyed their goatskin cloaks with it. More definitely Pliny ("*Historia Naturalis*," lib XXXIV: 26) recorded the reaction of shoemakers' black (native ferrous sulfate) with oak gall extract as giving a black color. This points to the practice of the Roman shoemakers of producing black tannin-cured leather with the iron salt to blacken it.

This last reaction was applied by Link in 1807 to the microscopic demonstration of tannin in plant cells. In a sense this is the first application of mordant dyeing in histology. Otherwise, except for Raspail's (*Ann. Sci. Nat.,* **6**: 224, 1825) use of iodine for the histochemical recognition of developing starch granules in plant tissues, the oldest histologic stains were done with carmine, which is an aluminum lake of carminic acid prepared by treating cochineal extracts with alum (Göppert and Cohn, *Bot. Zeitung,* **7**: 665, 1849; Corti, *Z. wiss. Zool.,* **3**: 109, 1851; Hartig, *Bot. Z.,* **12**: 574, 1854). Gerlach (*Mikroskopische*

Studien aus dem Gebiet der menschlichen Morphologie, Erlangen, 1858) actually brought carmine staining into general use. Boehmer (*Aerztl. Intelligenzbl.,* **12:** 539, 1865) introduced a practical means of using hematoxylin, as its aluminum lake. At the same time he reported briefly that it stained after potassium dichromate and copper sulfate mordants.

Mordant dyeing in the broad sense falls into three categories: (1) mordant dyeing in the strict sense is dyeing in which preparations are first soaked, or "mordanted," in a solution of a metal salt and then transferred, with or without washing, to a solution of a mordant dye for an appropriate interval and then brought, with or without regressive differentiation, into an appropriate mounting medium. (2) In metachrome dyeing, the metal salt and the dye each in a suitable solvent are mixed, and a usually more deeply colored solution of a dye lake is formed which serves as the coloring agent. In these dye lakes solutions basic and acidic dye–metal chelate complexes are formed, often more than one and of both types. The proportions are influenced by the relative amounts of dye and metal salt employed. High proportions of Fe(III), for example, with a lesser amount of hematoxylin yield a lake mixture which is a more selective nuclear stain and on electrophoretic analysis gives a larger proportion of cationic components. (3) The third class, referred to in textile dyeing as *afterchroming,* has had only limited use in histology. The elder Heidenhain (Rudolf) (*Arch. Mikrosk. Anat.,* **24:** 468, 1885) used it when he first soaked preparations in hematoxylin and then developed color in them by treatment with potassium dichromate (black) or potash alum (blue). Kattine (*Stain Technol.,* **37:** 193, 1962) blackened and stabilized

TABLE 6-1   MORDANT DYES USED AS STAINS*

"Colour Index" no. 1st ed.	3d ed.	Class color no., "Colour Index," 3d ed., Part I	Stain class	Common name	Nuclear stains Al^{3+}	Cr^{3+}	Fe^{3+}	Textile mordant colors	Calcified tissue
1027	58000	MR-11	M	Alizarin	R			Al, Cr, Sn, Fe: R	R
1037	58205		M	Purpurin	R				R
1239	75490	Nat R-4	M	Carmine, carminic acid	R			Zn, Hg, Al, Sn: R; Cr: P; Ba: V; U: G; Pb, Cu: FR	
1243	75280	Nat R-24	M	Brazilin	R	CR	GN	Al, Sn: R; Cu: RF; Fe: BG– RF	
1246	75290	Nat N-1	M	Hematoxylin	B	CB	BN	Cu; BG; Sn: PR	
883	51030	MB-10	M	Gallocyanin		B	BN		
894	51045	MB-45	M	Gallamin blue		B	BN		
900	51050	MB-14	M	Celestin blue B		B	BN		
1062	58605	MB-32	M	Anthracene blue SWR		BN		Cr, VN	

Color abbreviations: P, Tyrian purple; R, red; O, orange; Y, yellow; G, green; B, blue; V, spectral violet; F, brown (fuscus); C, Gray (canus); N, black (niger). Combination of two symbols indicates an intermediate color, the second somewhat stronger.

* No proper names included among the dye names in this table and no capitalization is required. The country Brazil was named after the dyestuff, which was known to Chaucer.

TABLE 6-2 OTHER MORDANT DYES POSSIBLY USABLE AS STAINS*

"Colour Index" no. 1st ed.	3d ed.	Class color no., "Colour Index," 3d ed., Part I	Stain class	Common name	Textile mordant colors	COONa SO₃Na group
1063	58610	MB-23	M	Alizarin cyanin BBS	Cr: PN	
1064	58615		M	Alizarin cyanin R	Cr, Al: V, P	
1065	58620		M	Alizarin cyanin black G	Cr: BN	
1066	67410		M	Alizarin blue	Cr: BV	
1067	67415	MB-27	M	Alizarin blue S	Cr: BV	
1068	67405		M	Alizarin green S	Cr: G	
1069	67425		M	Alizarin black P	Cr: N	
1070	67430		M	Alizarin black S	Cr: N	
1040	58255	MR-2	M	Anthrapurpurin	Cr, Al: R	
	58260	MR-2	M	Acid anthrapurpurin	Cr, Al: R	SO₃Na
1045	58500	MV-26	M	Quinalizarin	Cr: V	
722	43820	MB-3	M	Chromoxane cyanin R chromoxane	Cr: VB	COONa 2
720	43830	MB-1	M	Pure blue B	Cr, Ba: B; Al, Be: B	COONa 2

* No proper names are included among the above dye names and no capitalization is required.
Note: Color abbreviations are the same as in Table 6-1.

an alum hematoxylin stain for use before Van Gieson–type procedures by postchelating in a mixture of phospholungstic and phosphomolybdic acids. Pearse's (*J. Histochem. Cytochem.*, **5**: 515, 1957) use of cobalt to stabilize and blacken the MTT tetrazole is another example, and Lillie et al. (*Stain Technol.*, **43**: 203, 1968) postchelated orcein-elastin stains with Fe(III) or Cu(II), converting the familiar brown-purple to black.

Two important sequence mordant hematoxylin methods were developed. The first, dichromate to hematoxylin by Weigert (*Fortschr. Med.*, **2**: 190, 1884), with the copper acetate mordanting added by Weigert (1885), gave rise through the sequence copper hematoxylin of Benda (*Arch. Anat. Physiol., Physiol. Abth.*, p. 186, 1886) to the latter's first sequence iron hematoxylin technic (*Arch. Anat. Physiol., Physiol. Abth.*, p. 562, 1886; *Arch. Mikrosk. Anat.*, **30**: 49, 1887) and to the fully developed sequence iron hematoxylin method of M. Heidenhain (*Festsch. A. Koelliker 50 Anniv. Doctorate*, 1892). Shortly thereafter Weigert (*Encyklopädie*, 1903; *Z. Weiss. Mikrosk.*, **21**: 1, 1904) developed premixed (metachrome) iron hematoxylin.

The mordant dye group includes hematoxylin, brazilin, carmine and carminic acid, alizarin, purpurin, anthracene blue SWR, gallein, gallocyanin, gallamin blue, and celestin blue B. All these dyes possess one or more *o*-diphenol groups; carminic acid also presents a salicylic acid grouping, which should also serve for a metal chelation site.

According to Lillie (1969) alizarin and purpurin form scarlet lakes with aluminum and have occasionally been used as nuclear stains. Their principal use, along with alizarin red S (the monosulfonate of alizarin), is as histochemical reagents for calcium.

Carmine, or, more properly, carminic acid, and hematoxylin have been widely used as aluminum lakes for the staining of nuclei, neutral and acid mucins, and, in the case of carmine, glycogen as well. Hematoxylin has also been widely used with Fe³⁺ in sequence mordant methods, in regressive differentiation lake methods, and in progressive mordant stains. Brazilin has been used similarly to hematoxylin but much less extensively.

Phosphotungstic, phosphomolybdic, and molybdic acids have also been employed in combined solutions with hematoxylin. The three oxazin dyes celestin blue B, gallamin blue, and gallocyanin have been used as iron lakes for nuclear staining in place of hematoxylin. Gallocyanin with chrome alum in acid solution is an excellent nuclear and tigroid stain, and anthracene blue SWR has been used as an aluminum lake in place of hematoxylin.

Although it is usually stated that hematoxylin, brazilin, and carmine do not function as stains without metal mordants, it has been shown, first by Clara (*Z. Zellforsch. Mikrosk. Anat.*, **22**: 318, 1935) for enterochromaffin, that hematoxylin and several other dyes with catechol groups slowly react, in the absence of mordants, to form deeply colored complexes. This reaction has been extended to trichohyalin, keratohyalin, eosinophil leukocyte granules, and some elastic membranes (*J. Histochem. Cytochem.*, **4**: 318, 1956). Lillie et al. (*Acta Histochem.* (*Jena*), **42**: 204, 1974) left the elastin reaction unexplained, assigned the enterochromaffin reaction to the presence of a small amount of Fe(II) in the cells, and related the other sites to an arginine condensation reaction.

Two acid dyes of the triphenylmethane group are classed in the "Colour Index" as mordant blues (Nos. 1 and 3). C.I. 43830, chromoxane pure blue B, usually called *pure blue B* with various trade name prefixes, was used by Pearse (*Acta Histochem.* (*Jena*), **4**: 95, 1957) as a reagent for the demonstration of Be^{2+} and Al^{3+}, under the trade name Solochrome Azurine BS. Pearse used mordant blue No. 3, C.I. 43820, chromoxane cyanin R, commonly called *cyanin R* or *chrome cyanin R* with various trade prefixes, as a stain for nuclei (blue) and various cytoplasmic structures (red) under the trade name of Solochrome Cyanine R. (See "Conn's Biological Stains," pp. 206–207.)

Closely allied to those mordant dyes is a group of reagents also yielding strong color reactions with metallic ions by the formation of chelate complexes. The principal use of these reagents, however, is in the demonstration of metallic ions in the tissues. Some of them are included in the "Colour Index" as mordant dyes.

Resorcin green (C.I. 10000, 2,4-dinitrosoresorcinol) has been used for the demonstration of iron in hemosiderin. A dye designated as naphthochrome green B (Clayton Aniline Co.) has been used by Denz for the demonstration of beryllium. This dye seems to correspond most closely to C.I. 44530, mordant green No. 31, naphthochrome green G (Lillie, 1969, p. 227).

Dithizone (diphenylthiocarbazone: Eastman 3092, mw 256.34) is used to produce a red color with zinc in tissue. Purpurin (C.I. 59205), alizarin red S (C.I. 58005), anthrapurpurin or alizarin SX or A (C.I. 58255), and calcium red (Kernechtrot, nuclear fast red, an aminoanthroquinone sodium sulfonate, C.I. 60760) are similarly used for demonstration of calcium carbonate and phosphate deposits, but they do not show the oxalate. Dipotassium rhodizonate (Eastman 2942) or the sodium salt is used for the demonstration of strontium and barium, to which it gives red colors, and of lead, which it colors brown. Aurin tricarboxylic acid (C.I. 43810, chrome violet CG) colors aluminum salts dark red and yields violet with chromic acetate in textile dyeing.

In the foregoing group Kernechtrot, nuclear fast red, needs some special discussion because of the confusion which has arisen from the application of these two names not only to the sulfonated aminoanthroquinone dye supplied as calcium red but also to a totally unrelated basic azin dye related to neutral red and also to the stable diazonium salt fast red B, *Kernechtrotsalz B*, C.I. 37125. Neither of the first two is entered in the "Colour Index" or its 1963 "Supplement" or in the seventh edition of Conn's "Biological Stains"; however, they have been identified in the eighth edition of "Conn's Biological Stains" and in the third edition of the "Colour Index" (C.I. 60760).

Dithiooxamide or rubeanic acid (Eastman 4394) gives a greenish black with Cu^{2+}, blue

TABLE 6-3 CHELATE REAGENTS FOR METAL IONS*

"Colour Index" no.		"Colour Index," 3d ed., Part I, class color no.	Stain class	Name	Color(s)	Metallic ions†	SO₃Na residues
1st ed.	3d ed.						
1	10000		M	Resorcin green	G	Fe(II)	
	44530	MG-31	M	Naphthochrome green G	G	Be	
			M	Dithizone	R	Zn	
1057	58205		M	Purpurin	RP	Ca	
1034	58005	MR-3	M	Alizarin red S	R	Ca	1
1040	58255	MR-2	M	Anthrapurpurin	RP	Ca	
	60760		M	Calcium red or nuclear fast red	R	Ca	
			M	Rhodizonate (K or Na)	R, R, F	Sr, Ba, Pb	
	43810		M	Chrome violet CG	R+	Al	
			M	Rubeanic acid	GN, BN, YF	Cu, Ni, Co	
			M	Diethyldithiocarbamate	YF, F	Cu, Mn	
2	10005	MG-4	M	Naphthol green Y	R, GN, FG, F	Co, Fe(II), Ce, Cu	
1027	58000	MR-11	M	Alizarin	R	Ca	
720	43830	MB-1	M	Pure blue B	B	Be, Al	
			M	Zincon	B	Zn, Cu	
					BG	Co	
					G	Mg	

* No proper names are included among the above dye names and no capitalization is required.

† The metal ions are all divalent except Al(III).

Abbreviations: M, mordant; R, red; Y, yellow; F, brown; G, green; B, blue; P, purple; N, black; R+, dark red.

violet with Ni²⁺, and yellowish brown with Co²⁺. Sodium diethyldithiocarbamate (Eastman 2635) also reacts with Cu²⁺ to yield a yellow-brown color.

Naphthol green Y (C.I. 10005, 1-nitroso-2-naphthol) reacts with cobalt (red), iron (green to black), chromium (olive), and copper (brown).

Basic Aniline Dyes

The *basic azo dyes* Janus green B (C.I. 11050) and the Bismarck browns (C.I. 21000, 21010) are valuable nuclear stains in contrast staining for fats, as they possess the valuable property of relative permanence and lack of diffusion in aqueous syrup media. Janus green B seems also to excel methylene blue as a nuclear counterstain for the acid-fast method for tubercle bacilli in tissues; but for permanence alum hematoxylin is to be preferred for this purpose. The Bismarck browns possess a metachromasia in yellower tone, and stain mucus and cartilage well. Janus green B, Janus black I (a mixture of Janus green B and some brown dye), and the related Janus blue are widely used in the vital staining of mitochondria.

Basic azo dyes are tabulated also in Table 6-4 following the thiazins. It will be noted that two Janus greens are listed, one made by coupling diazotized phenosafranin into dimethylaniline, the other using diethylsafranin (C.I. 50206). The latter (C.I. 11050) appears to be the true composition of Janus green B, rather than the former (C.I. 11045), which was that in the first edition of the "Colour Index."

Similarly, tolusafranin (C.I. 50240) and diethyltolusafranin diazotized and coupled into β-naphthol yield C.I. 12210, basic blue 16, indazole blue R, indoine blue R or 3B, a textile

TABLE 6-4 BASIC ANILINE DYES

"Colour Index," 1st ed.	3d ed.	"Colour Index," 3d ed., Part I, class color no.	Name*	Color Ortho-chromatic	Color Meta-chromatic	Molecular weight†	prim-NH₂ ($prim$-NH_2)	sec-NHR (sec-NHR)	tert-NR₂ ($tert$-NR_2)	Solubility Water	Solubility Alcohol
										(g/100 ml)	(g/100 ml)
Azins											
840	50200		Phenosafranin	R		322.800	2			Sol.	Sol.
843	50205	BV-5	Methylene violet RR	PR		350.854	1		1	Sol.	Very sol.
847	50206		Diethylphenosafranin	P		378.858	1		1	Sol.	Sol.
	50225		Heliotrope B, 2B	V		435.017			2	Sol.	Sol.
841	50240	BR-2	Safranin T, A, O	R	YO	350.854	2			5.45	3.41
825	50040	BR-5	Neutral red	R		288.783	1		1	Sol.	Sol.
Pyronins											
740	45000		Acridine red 3B	R		274.753		2		Sol.	Sol.
739	45005		Pyronin G, Y	R		302.807			2	8.96	0.6
741	45010		Pyronin B	PR		358.915			2	Sol.	Sol.
Oxazins											
(877)	51010		Brilliant cresyl blue	B	PR	332.836	2		1	Sol.	
			Cresyl fast violet	V	R	339.028	1		1	0.38	0.25
913	51180	BB-12	Nile blue (sulfate)	B		732.865	1		1	Sol.	Sol.
Thiazins											
920	52000		Thionin	V	R	263.751	2			0.25	0.25
	52002		Azure C	BV	R	277.778	1	1		Sol.	Sol.
	52005		Azure A	VB	PR	291.805	1		1	Sol.	Sol.
	52010		Azure B	B	V?	305.832		1	1	Sol.	Sol.
922	52015	BB-9	Methylene blue	GB		319.859			2	3.5	1.5
924	52020	BG-5	Methylene green	G	BV	364.856			2	1.5	0.1
927	52030	BB-24	New methylene blue N	B	PR	347.913		2		13.3	1.6
	52041		Methylene violet Bernthsen	V		256.329			1	Insol.	Sol.
925	52040	BB-17	Toluidine blue	B	PR	305.832	1		1	3.8	0.5

Basic azo dyes

No.	C.I. No.	Code	Name	Color		Mol. wt.†				Sol.	Sol.
135	11050		Janus green B	G		511.080			2	Sol.	Sol.
	12210	BB-16	Indazole blue	B		465.915			1	Sol.	Sol.
	12211		Janus blue G	B		562.120	1			Sol.	Sol.
331	21000		Bismarck brown Y	YF	Y	419.320	4			1.3	1.1
332	21010		Bismarck brown R	RF		461.401	4			1.1	1.0

Triphenyl- and diphenylnaphthylmethanes

No.	C.I. No.	Code	Name	Color		Mol. wt.†				Sol.	Sol.
657	42000	BG-4	Malachite green	G		364.922			2	Sol.	Very sol.
676	42500	BR-9	Pararosanilin chloride	R		223.828				0.26	5.98
	42500		Pararosanilin acetate	R		347.820	3			4.15	13.63
677	42510	BV-14	Rosanilin, fuchsin (anhyd.)	R		337.855	3			0.39	8.16
	42510	BV-14	Rosanilin, fuchsin (cryst 4 H₂O)	R		409.916					
678	42520	BV-2	New fuchsin	R		365.910	3			1.13	3.20
680	42535	BV-1	Methyl violet	P	R	379.937	1, 0			2.93	15.21
681	42555	BV-3	Crystal violet	V	PR	407.991			3	1.68	13.87
684	42585	BB-20	Methyl green	G		458.479			2‡	Sol.	Insol.
685	42500		Ethyl green	G		517.017			2‡	Sol.	
686	42556		Iodine green	G	R	472.506			2‡	Sol.	
689	42775	SB-3	Spirit blue	B		Variable	2, 3			Insol.	1.1
690	42563	BB-8	Victoria blue 4R	B		520.123			3	3.23	20.49
728	44040	BB-11	Victoria blue R	B		458.091	1		2	Sol. hot	Very sol.
729	44045	BB-26	Victoria blue B	B		506.096	1		2	Sol.	Sol.
731	44085	BB-15	Night blue	B		576.232	1		2	Sol.	Readily sol.

Phthalocyanines (composition not yet completely revealed)

No.	C.I. No.	Code	Name	Color		Mol. wt.†				Sol.	Sol.
	74240	Ingram	Alcian Blue 8GX						4(?)	Sol.	Sol.
		B1	National Fast Blue 8XM							Sol.	Sol.
		SB-37	Luxol Fast Blue AR								
		SB-34	Luxol Fast Blue G								
		SB-38	Luxol Fast Blue MBSN								
808			Pinacyanole	R		480.401			2 ring	Sol.	Sol.

* Among the above dye names the words Nile, Bernthsen, Janus, Bismarck, and Victoria are proper names and require capitalization. The Luxol dyes, Alcian Blue 8GX, and National Fast Blue 8XM, bear trade names and are of partly or entirely secret composition. Their names are capitalized as above. The Luxol dyes are properly classed as fat stains. AR and G are polyazo, not phthalocyanins.

† Molecular weights differ from those given in Lillie, "Conn's Biological Stains," 1969; they have been recomputed on the 1966 atomic weights: ^{12}carbon = 12.00000.

‡ 1—NR$_3$ group present.

Note: B, blue; G, green; P, Tyrian purple; R, red; V, spectral violet; Y, yellow.

dye, and Janus blue B, respectively. The constitution of C.I. 135 in the first edition of the "Colour Index" was apparently incorrect for Janus blue.

Janus green and Janus blue have been used principally for supravital staining of mitochondria, a process which now seems to be related to the oxidative enzymes localized in these bodies. They are also useful as basic dyes for counterstains.

The *safranins*, safranin T, A, or O (C.I. 50240), phenosafranin (C.I. 50200, safranin B extra) and its N-N-dimethyl and -tetraethyl derivatives, methylene violet RRA (C.I. 50205), and heliotrope B, 2B or amethyst violet (C.I. 50225), have all been used as nuclear stains. The first (C.I. 50240) has been preferred for this purpose and as a counterstain for gram-negative organisms and tissue cells in the Gram stain. After ferrocyanide and particularly ferricyanide technics, safranin O produces dark red crystalline precipitates which are difficult to prevent or remove. Hence the fuchsins are preferred in this case. Safranin O exhibits a strong yellow orange metachromasia which apparently demonstrates more of the mucins than does azure A or thionin.

Safranin O and methylene violet RRA have been used as diazotizable amines to produce fresh diazonium salts which demonstrate enterochromaffin in blue-black or very dark violet on a red to red-purple background stain of protein aromatic amino acid residues. This stain is useful both for enterochromaffin and for protein studies. Safranin O has been the chief dye used for this purpose. We have also used methylene violet RRA, and phenosafranin should be usable, but heliotrope B (C.I. 50225) has both its amine groups fully ethylated and would not be diazotizable. Molecular weights are 322.807 for phenosafranin, 350.861 for safranin O and methylene violet RRA, and 435.023 for heliotrope B. Diethylphenosafranin is used chemically as the diazonium base in the synthesis of Janus green and Janus black.

The principal use of the azin dye neutral red (C.I. 50040) is in vital staining, in tigroid staining, and in such neutral stains as Twort's light green neutral red. It can be a useful indicator, changing from red to yellow at pH 6.8 to 8.0.

Three weakly basic red dyes of the xanthene group are the pyronins Y and B (C.I. 45005 and 45010) and rhodamine B (C.I. 45170). We have occasionally used these as red nuclear stains but find other dyes better, notably the fuchsins and the safranins.

All three have been used as basic stains for cytoplasm, basophil granules, and bacteria in contrast to a nuclear stain with methyl green. The pyronins are used for this purpose in the Unna-Pappenheim-Saathof methods and are also employed as counterstains in the α-naphthol oxidase methods. Rhodamine B has also been employed in the supravital staining of mitochondria, and as a fat stain in aqueous solution for fluorescence microscopy.

The *oxazin dyes*, brilliant cresyl blue (C.I. 51010) and cresyl-fast violet, or *Cresylechtviolett*, and its American substitute, cresyl violet acetate, are used, the one chiefly for supravital staining, the other for Nissl staining and for its metachromatic properties. Gallamin blue, gallocyanin, and celestin blue have been discussed earlier in this chapter. The remaining commonly used member of this group is Nile blue (C.I. 51180). This dye has been used both as an ordinary basic dye and as a special reagent for the staining of fatty acids (blue) and neutral fats (red). This use is discussed at length in Chap. 8.

The new dye Darrow red (Powers et al., *Stain Technol.*, **35**: 19, 1960) has been recommended as a red dye giving a precision of Nissl staining similar to that of cresyl-fast violet. Both are oxazin dyes of similar constitution (Lillie, 1969, pp. 289–290).

The *thiazins*, thionin (C.I. 52000) and its mono-, di-, tri-, and tetramethyl derivatives azure C, azure A (C.I. 52005), azure B (C.I. 52010), and methylene blue (C.I. 52015), and its relatives methylene green (C.I. 52020), toluidine blue (C.I. 52040), and new methylene blue (C.I. 52030), are valuable nuclear stains and exhibit in varying measure the property of metachromasia, or of staining cartilage matrix, mucin, and the granules of mast cells in a more violet or redder tone than they do nuclei. Of these, new methylene blue is the reddest, then thionin, azure C, azure A, toluidine blue, azure B, and methylene blue in order.

The middle three members of the series afford perhaps the greatest color contrast between nuclei and cytoplasm on the one hand and cartilage and mucus on the other.

They are much used as bacterial stains and (in combination with dyes of the eosin group) as general tissue stains, especially for blood and blood-forming tissues.

The higher homologs, methylene blue and azure B, readily undergo oxidation with loss of methyl groups and evolution of formaldehyde, and give rise to lower homologs and to certain deaminized oxidation products which are relatively insoluble in water and have been thought to contribute to the polychromasia of blood stains. The substance Bernthsen's methylene violet appears to be a mixture of these. This oxidation of methylene blue to azures and methylene violets is known as *polychroming*, and mixtures of methylene blue with these products are called *polychrome methylene blue*.

This process of polychroming occurs freely in alkaline solutions without added oxidants, and is expedited by rise in pH level above 8 and by heat. It is inhibited in acid solution. It may be induced by deliberate oxidation in acid solution by addition of chromic acid, but in this instance only azures appear to be formed, and the amount of alteration is strictly proportional to the amount of available oxygen furnished. Thus, use of 250 mg potassium dichromate per g 88 % methylene blue produces a product which is spectroscopically and tinctorially chiefly azure B, while double that amount produces chiefly azure A. The reactions are thus: $K_2Cr_2O_7$ furnishes 3O, Cr_2O_3 and K_2O, the metallic oxides being promptly converted into salts by the excess of acid present. One mol methylene blue $C_{16}H_{18}N_3SCl$ plus one O yields one mol azure B $C_{15}H_{16}N_3SCl$ plus one mol formaldehyde HCHO.

A similar decomposition of methylene blue and the higher azures appears to go on even in the dry dyes and occurs readily also in methanol solutions of their eosinates and more slowly in glycerol methanol solutions. Also in these solutions the change is accelerated by higher temperatures and by presence of alkali and is retarded by acid. Particularly annoying is its occurrence during the drying of eosinate precipitates. Here thorough washing with distilled water, rigid adherence to temperatures below 40°C for drying, and the use of alcohol or vacuum to accelerate drying are of help.

The *triphenylmethane basic dyes*, pararosanilin (C.I. 42500) as chloride, or the more soluble acetate fuchsin or rosanilin (C.I. 42510), and new fuchsin (C.I. 42520), are widely used for demonstration of acid-fast bacteria in sputum, exudates, and tissue sections, as a component of the Weigert iron resorcin fuchsin and the Fullmer iron orcinol new fuchsin elastic tissue stains, as a red nuclear stain for use in Prussian and Turnbull blue reactions, and in variants of the Gram stain for bacteria, in its sulfite leuco form in the Schiff reagent for aldehydes as used in the Feulgen nucleal and plasmal reactions and in the Bauer and McManus polysaccharide methods, as well as in direct reactions for native aldehydes.

Crystal violet (C.I. 42555, *N*-hexamethylpararosaniline) and methyl violet (C.I. 42535, the related mixture of somewhat redder lower homologs) are used extensively in the Gram stain for bacteria in exudates and tissues and the Gram-Weigert fibrin stain, as stains for amyloid, which exhibits an alcohol labile red metachromasia to these dyes, and as the basic component of certain neutral stains. Iodine green (C.I. 42556) is also sometimes used for amyloid.

Methyl green (C.I. 42585) and ethyl green (C.I. 42590), methyl and ethyl addition products to crystal violet, are used for nuclear chromatin (deoxyribonucleic acid) in the methyl green pyronin and related procedures. C.I. 42590 was originally named methyl green and is the dye furnished by at least some manufacturers.

Spirit blue (C.I. 42775), or alcohol-soluble aniline blue, is used in alcoholic solutions as a stain for nuclei, cartilage, and other acidic substances; it has also been used in hydro-alcoholic solutions as a fat stain.

Among these dyes pararosanilin (C.I. 42500) has also been used as a base for diazotization, and the resultant diazonium salt hexazoniumpararosaniline has been used for the capture of naphthol in enzyme localization work.

The Alcian dyes of Imperial Chemical Industries (ICI) and a group of similar dyes produced by Farbenfabriken Bayer A.G., (FBy) are basic dyes in which the chromogen is the phthalocyanin nucleus and the basic side chain is partially identified (Alcian Blue 8GX, C.I. 74240; Phthalogen Brilliant Blue 1F3G, C.I. 74160) or quite unrevealed (Alcian Blues 2GX, 5GX, 7GX; Alcian Greens 2GX and 3BX; Phthalogen Blue 1B, Phthalogen Brilliant Blue 1F3GM, Phthalogen Blue-Black IVM, and Phthalogen Brilliant Green 1FFB). Alcian Yellow GX is formulated as a complex monoazo dye by Ravetto (*Riv. Istochim. Nor. Patol.*, **19**: 257, 1968):

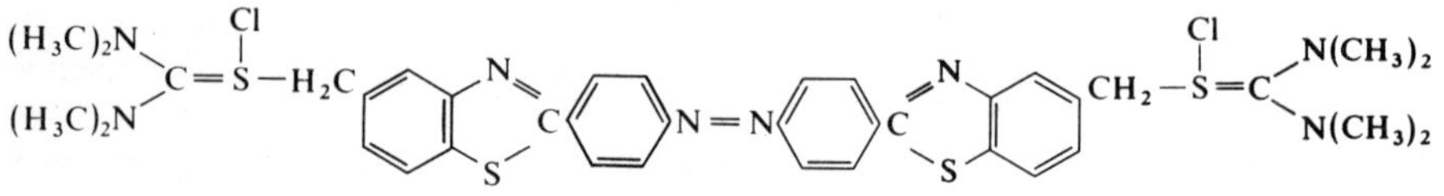

The reader is referred to the eighth edition of "Conn's Biological Stains," edited by Lillie, and to the third edition of the "Colour Index" for further information regarding constitution of these dyes. Mowry has noted that Imperial Chemical Industries has changed the composition of Alcian Blue 8GX at least once, to render it more soluble.

Of these dyes, one, Alcian Blue 8GX, has been widely used as a stain for mucopolysaccharides, and some workers have claimed histochemical specificity for it. Such claims appear to rest on about the same kind of ground as the pine splinter test for indole, since the constitution of the reagent is unrevealed and is subject to change at its manufacturer's convenience.

Mowry and Emmel (*J. Histochem. Cytochem.*, **14**: 799, 1966) have added National Fast Blue 8XM to the list of useful phthalocyanin dyes in mucin staining. The basic end group on rings 2 and 3 is

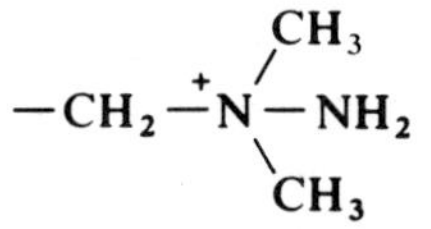

The dye is less soluble than Alcian Blue 8GX, and lower concentrations are used.

The Astra blue of the Germans appears to be used similarly to Alcian Blue 8GX and to have similar "specificity" for acid mucopolysaccharides. The 1963 supplement to the "Colour Index," though it does not formulate the dye in Part II, reveals in Part I that Astra Blue 4R is a triarylmethane dye, not improbably similar to the known dark blue Astracyanine B, C.I. 42705, which is a triaminoditolylphenylmethane basic dye that could well prove equally useful.

Fluorochrome Stains

These stains comprise a number of acid and basic dyestuffs, mostly yellow or orange, which fluoresce more or less intensely in near-ultraviolet (and violet) light. Most of them function as ordinary acids and bases to color oxyphil and basophil (basic and acid) tissue elements in the same manner as the acid and basic dyes used with visible light. Three, fluorescein, rhodamine B, and sulforhodamine B, have been used more or less extensively to condense with free protein amino groups, thereby "tagging" them so that these proteins may be localized by their green or red fluorescence. Fluorescein and rhodamine B have been used as isocyanates or, preferably, as the more stable and commercially available isothiocyanates. With sulforhodamine B, a sulfamido condensation (Chadwick et al., *Lancet*, **1**:412, 1958) has been used. This dye has usually been referred to by one of its trade names, often without noting the manufacturer's name: Lissamine Rhodamine B200, Imperial Chemical Industries Ltd., and Acid Rhodamine B, National Aniline Division, Allied Chemical and Dye Corporation.[1]

A few basic (amino) dyes and two acid dyes have been used as fat stains, by reason of their relatively higher oil solubility. Since they do contain amino or acid groups, their specificity for fat staining is open to question. Acetyl or benzoyl esters might be prepared, as has been done with oil red O, Sudan IV, and Sudan black B (*J. Histochem. Cytochem.*, **1**: 8, 1953).

Of the basic acridine fluorochromes, one, acridine orange, C.I. 46005, has received special attention. It was said for a time to permit discrimination between living and dead cells; now it is found that with an appropriate technic it gives a green fluorescence to deoxyribonucleic acid and a red fluorescence to ribonucleic acid. The free base, C.I. 46005B, is said to be selectively very soluble in stearic acid ("Colour Index," Part I, Solvent Orange 15).

In regard to magdala red, the "Colour Index" relates the basic red No. 6, C.I. 50375, contains as first product of synthesis about 92 to 94% rhodindine (*a*), a monoamino trinaphthosafranin, from which 6 to 8% of true magdala red (*b*), the diamino homolog, is extracted with boiling water. It is believed that this latter fraction is meant by the term *Magdalarot echt*. We do not know whether rhodindine has ever been used intentionally as a stain, and it would appear to be relatively insoluble in water.

Thiazol yellow G, C.I. 19540, has been supplied as a biologic stain also under the synonyms *titan yellow* and *Clayton yellow*.

The alkaloid berberin and the yellow drugs rivanol and atabrine (quinacrine) are included with the dyes in Table 6-5.

Kasten [*Nature (Lond.)*, **184**: 1797, 1959] has used a number of fluorescent basic primary amine dyes to form Schiff reagents, chiefly for use in the Feulgen reaction. These include auramine O (C.I. 41000), acridine yellow G (C.I. 46025), acriflavine (C.I. 46000), coriphosphine O (C.I. 46020), phosphine (C.I. 46045), flavophosphine N or benzoflavine (C.I. 46065), neutral red (C.I. 50040), phenosafranin (C.I. 50200), safranin O (C.I. 50240), and rhodamine 3G (C.I. 45210).

The antibiotic tetracycline apparently functions as a mordant dye to demonstrate newly deposited calcium in vivo, after the traditional manner of madder.

Rubalcava (*Biochemistry*, **8**: 2742, 1969) noted that 1-anilinonaphthalene-8-sulfonate, a fluorescent probe, does not fluoresce in water, but fluoresces strongly in organic solvents when bound to certain native proteins such as hemoglobin-free erythrocyte membranes of rabbits. The binding is sensitive to cations in the suspending medium.

Benjaminson et al. (*Science*, **160**: 1359, 1968) conjugated fluorescein isothiocyanate to deoxyribonuclease. The conjugate was applied to washed cells grown in tissue culture, a coverslip was applied, and the material was viewed immediately with the fluorescence microscope. Specific fluorescence of nuclei was observed which gradually diminished due to enzyme activity.

---

[1] The National Aniline Division's Biological Stains department was sold in 1972 to Matheson, Coleman & Bell, Inc. and is consolidated with their Biological Stains department.

TABLE 6-5 FLUOROCHROME DYES AND DRUGS USED IN HISTOCHEMISTRY AND HISTOLOGY

"Colour Index" no. 1st ed.	3d ed.	"Colour Index," 3d ed., Part I, class color no.	Stain class*	Preferred name	Molecular weight	−SO₃Na (−CO₂Na)	Amino groups NH₂	NHR	NR₂	Color Ortho-chromatic	Ultra-violet fluorochrome	Solubility,† gm/100 ml Water	Alcohol
655		BY-2	BF	Auramine O	303.838		1		2	Y	Y	Very sol. hot; sol.	Sol.
749	45170	BV-10	BF, OSF	Rhodamine B	479.024	1 CO₂H			2	R	YO, R	Sol.	Sol.
753	45210	BR-3	BF	Rhodamine 3G	436.943				1	R	R	Sol.	Sol.
790	46000		BF	Acriflavine	296.202		1			Y	Y	Very sol.	Sol.
788	46005	BO-14	BF	Acridine orange	381.822		2		2	OR, G	OR, G	Sol.	Sol.
787	46020	BY-7	BF	Coriphosphine	287.795		1		1	Y	Y	Sol.	Sol.
785	46025		BF	Acridine yellow	273.768		2			Y	G	Sol.	Sol.
793	46045	BO-15	BF, OSF	Phosphine	329.840		2			YO	YG	Sol.	Sol.
791	46065		BF	Benzoflavine	349.867		2			Y	YG	Sol.	Sol.
815	49050	BY-1	BF	Thioflavine TCN	457.967				1 ring; 1	GY	YG	Sol.	Sol.
	50375a	BR-6	BF	Rhodindine	472.482		1			R	O	Very slightly sol.	Sol.
857	50375b	BR-6	BF, OSF	Magdala red (echt)‡	804.824		2			R	O	Sol. hot	Sol.
	75160	NY-18	BF	Berberine sulfate	508.921				1 ring		Y	1.0	Slightly sol.
			BF	Quinacrine-2HCl-2H₂O (atabrine)	338.346			1	1	Y	GY	2.9	
			BF	Rivanol			2					6.5	
			MF	Tetracycline 3H₂O	498.4712		1 amide		1	Y	Y	Sol.	
225	14780	DR-45	AF, OSF	Thiazine red R	599.575	2				R	V	Very sol.	Sol.
813	19540	DY-9	AF, OSF	Thiazole yellow G	645.729	2		1		Y	B	Sol.	Sol.
370	22120	DR-28	A, AF	Congo red	696.676	2	2			FR	R	Sol.	Sol.
692	42685	AV-19	AC, AF	Acid fuchsin	585.545	3	3			R	R	45.0	3.0
748	45100	AR-52	AF	Sulforhodamine B	580.659	2			2	R	R	Sol.	Sol.
766	45350	AY-73	AF	Fluorescein diNa	376.229	1 CO₂Na				Y	G	50.0	7.0
768	45380	AR-87	A, AF	Eosin Y diNa	691.863	1 CO₂Na				R	Y	44.0	2.0
812	49000	DY-59	AF	Primuline	475.536	1	1			Y±	B, BV	0.25	0.03
816	49010	DY-7	AF	Thioflavine S	Uncertain	+	1			Y	G, BG	Very sol.	Sol.

* Abbreviations: BF, basic fluorochrome; AF, acid fluorochrome; Y, yellow; G, green; O, orange; R, red; B, blue; V, violet; F, brown (fuscus); OSF, oil-soluble fluorochrome; MF, mordant fluorochrome.

† See also Table 4-12.

‡ Although Magdala would seem to have been a proper name, H. J. Conn ("Biological Stains," 7th ed. Williams & Wilkins, Baltimore, 1961) writes it ir lower case. The other dye names are all uncapitalized.

PLASMA STAINS

For the most part these are sulfonic and carboxylic acids which combine more or less firmly with tissue bases, mainly proteins containing an excess of the basic amino acids arginine, lysine, hydroxylysine, and histidine over acidic amino acids. A few nitro and nitroso dyes react similarly. These are tabulated in Table 6-6.

A few among these merit special mention. The nitro dye picric acid, formulated as 2,4,6-trinitrophenol and as its yellower tautomer (Lillie, 1969, p. 60), serves both as an acid and as a yellow contrast stain in collagen methods of the Van Gieson type. It is used also as a simple plasma stain in contrast to hematoxylin and to various basic dyes. In the latter usage, either the acid or its ammonium salt may be used; ammonium picrate is ineffective in Van Gieson and similar stains. If ammonium picrate is not available, it may be made by adding ammonium hydroxide solution gradually to boiling water containing 10 to 15% picric acid, until the steam a minute after the last addition is alkaline to moist litmus or nitrazine paper. Then continue boiling until the steam is no longer alkaline.

Most of the simple plasma stains included in Table 6-6 need no special discussion at this point. Merits of certain dyes as "specific protein stains" will be discussed in Chap. 8.

A few acid dyes which exhibit strong fluorescence in violet and near-ultraviolet light have been presented as fluorochrome dyes (Table 6-5), with the usage symbol AF.

In Nocht-type Romanovsky stains, eosin B performs equally well with eosin Y and gives more brilliant red colors. For azure-eosin and Giemsa substitute staining on most pathologic material we have long abandoned eosin Y in favor of eosin B. It is likewise our experience that for tissue staining the extempore mixtures of the aqueous solutions of the azures and the two eosins B and Y are to be preferred to the prepared eosinates in methanol or methanol glycerol stock solutions. The erythrosins and phloxines have not given satisfactory results in neutral staining, probably because of relative insolubility of their azure salts, but they do better in sequence procedures such as Mallory's phloxine methylene blue and alum hematoxylin eosin technics.

Naphthalene black was recommended by Bower and Chadwick as a selective dark blue stain for Paneth cell and eosinophil leukocyte granules, with greenish blue erythrocytes. The stain follows a PAS (periodic acid Schiff) reaction: 0.5% dye in 50% propylene glycol 15 min; differentiate with 0.01% lithium carbonate. Unfortunately Bower and Chadwick omitted the following letters from the dye name and also left out the source.

Naphthalene black 10B, $C_{22}H_{15}N_6O_6SNa$, mw 514.455, is listed by E. Gurr (1960, p. 280)

as soluble in water (2%), alcohol (2.75%), Cellosolve (3.25%), glycol (7.25%), xylene (nil). He gives naphthalene black 12B as a synonym. In G. Gurr's catalog naphthalene black 10B is amido black 10B, C.I. 20470. Naphthalene black TS is blue-black NSF, C.I. 20480, and is also called naphthalene black 12R. Naphthalene black D or 12 BR, C.I. 20500, is soluble in water, slightly soluble in acetone, and insoluble in benzene.

TABLE 6-6 ACID DYES USED PRINCIPALLY AS PLASMA STAINS

"Colour Index" no. 1st ed.	3d ed.	Name*	Stain class	"Colour Index," 3d ed., Part I, class color no.	Molecular weight	Color	SO$_3$Na	CO$_2$Na and other acids	NH$_2$	NHR	NR$_2$	Solubility, g/100 ml Water	Alcohol
7	10305	Picric acid	A†	AY-1	229.107	Y		1 NO$_2$H				1.18	8.96
10	10316	Naphthol yellow S	A		358.196	Y	1				(2 NO$_2$)	8	Slight
	10316	Flavianic acid	A		314.232	Y	1				(2 NO$_2$)	Very sol.	Very sol.
					Monoazo dyes								
27	16230	Orange G	A	AO-10	452.375	O	2					10.86	0.22
29	16570	Chromotrope 2R	A	AR-29	468.374	R	2					19.30	0.17
79	16150	Ponceau 2R	A	AR-26	480.429	R	2					Sol.	Very slight
88	16180	Bordeaux R	A	AR-17	502.435	R	2					Sol.	Slight
138	13065	Metanil yellow	A	AY-36	375.384	Y	1			1		5.36	1.45
153	16540	Azofuchsin G	A	AR-21	463.474	R	2					Sol.	Very slight
154	16535	Azofuchsin S	A		438.394	R	1					Sol.	Sol.
225	14780	Thiazine red R	A	DR-45	599.575	R	2				Thiazole ring	Very sol.	Sol.
640	19140	Tartrazine	A	AY-23	534.369	Y	2	1			1 pyrazole ring	Very sol.	Slight
					Disazo and polyazo dyes								
252	27290	Brilliant crocein	A	AR-73	556.487	R	2					5.04	0.06
280	26905	Biebrich scarlet	A, AF	AR-66	556.487	R	2					Sol.	Slight
370	22120	Congo red	A	DR-28	696.676	R	2		2			Sol.	Sol.
375	22145	Congo Corinth G	A	DR-10	647.659	R	2		1			Sol.	Slight
454	23510	Brilliant purpurin R	A	DR-15	826.775	F	3		2			Sol.	Mod.
520	24400	Benzo pure blue	A	DB-15	992.816	B	4					6	0.5
581	30235	Chlorazol black E	A	DN-38	781.742	N	4		3			Sol.	Mod.
					Triphenylmethane acid dyes								
696	42571	Fast acid violet 10 B	A	AB-13	643.762	V	2				3	Very sol.	Slight
698	42650	Formyl violet S4B	A	AV-17	775.969	PV	2				3	Very sol.	Very sol.
699	42576	Eriocyanine A	A	AB-34	705.834	VB	2				3	Sol.	Sol.
712	42051	Patent blue V·$\frac{1}{2}$Ca$^+$	Redox A	AB-3	579.725	B	2				2	Very sol.	Slightly sol.

No.	C.I. No.	Generic	Class	Common name	M.W.	Color	(1)	(2)	(3)	Sol. in water	Sol. in alcohol
715	43535	Redox	A	Cyanol FF	554.621	B	2	2		Sol.	Very sol.
716	43530	Redox	A	Ketone blue 4BN, "Cyanol"	602.666	B	2	2		Sol.	Sol.
Quinolines											
801	47005	AY-3	A	Quinoline yellow	Uncertain	Y	+(2?)			Sol.	
802	47010	AY-2	A	Quinoline yellow	Uncertain	Y	+			Sol.	
Fluorane-xanthene dyes											
768	45380	AR-87	A	Eosin Y	697.936	OR	1			44.2	2.18
771	45400	AR-91	A	Eosin B	624.066	R	1			39.11	0.75
773	45430	AR-51	A	Erythrosin B	879.865	R	1			11.1	1.87
778	45410	AR-92	A	Phloxine B	829.644	PR	1			Sol.	Sol.
779	45440	AR-94	A	Rose Bengal	1017.645	PR	1			36.25	7.53
Azins											
828	50085	AR-101	A	Azocarmine G	579.590	R	2	1		Sl. sol.	
829	50090	AR-103	A	Azocarmine B	681.634	R	3	1		Very sol.	
Acid dyes used both as plasma stains and as collagen fiber stains											
246	20470	AN-1	AC	Amido black 10B (naphthol blue-black)	616.499	BN	2	1		Sol.	
282	27195	AR-112	AC	Ponceau S	760.575	R	4			Sol.	
670	42095	AG-5	AC	Light green SF	792.863	G	3		2	20.35	0.82
	42053	FG-3	AC	Fast green FCF	808.862	G	3		2	16.04	0.35
692	42685	AV-19	AC	Acid fuchsin	585.545	R	3	3		Very sol.	Insol.
706	42780	AB-93	AC	Methyl blue	799.814	B	3			Very sol.	
707	42755	Mixture	AC	Aniline blue	737.742	B	3	3		Sol.	
707	42755	AB-22	AC	"Water blue I"	737.742	B	3			Sol.	
737	44090	AG-50	AC	Wool green S	576.627	G	2	1	2	Sol.	Slight
758	45190	AV-9	AC	Violamine R		PR	1	1		Sol.	
1180	73015	AB-74	AC	Indigocarmine	274.235	B	2	1	2	1.68	0.01
805	50420	AN-2	AC	Nigrosin ws	Unknown	N	Unknown	+	2	Insol.	Insol.

* Among the above dye names the words **Biebrich**, **Congo**, **Corinth**, and **Bengal** are proper names and require capitalization. i = indole N.

† Abbreviations: A, acid dye; AC, acid dye; collagen stain colors: R, red; O, orange; Y, yellow; G, green; B, blue; V, violet; F, brown; P, red purple; N, black.

Note: Before 1955 aniline blue was a mixture of methyl blue and water blue I. The name has now been transferred to C.I. 42765 and replaces the prototype named water blue I (Conn's Biological Stains, 9th edition). The identity of the old naphthol blue black is uncertain.

Source: H. J. Conn, "Biological Stains," 6th ed, Biotech, Geneva, N.Y., 1953.

Naphthalene black 12 BR is another synonym for amido black 10B, C.I. 20470.

These are the naphthalene blacks found in dye catalogs and the "Colour Index." Puchtler et al. [*J. Microsc. (Oxf.)*, **89**: 329, 1968] in a combined PAS technic used sulfone cyanin GR ex (C.I. 26400, acid blue 120) to stain myofibrillar elements blue. The dye is a water-soluble, alcohol-soluble acid disazo dye, $C_{33}H_{22}N_5O_6S_2Na_2$, mw 695.689:

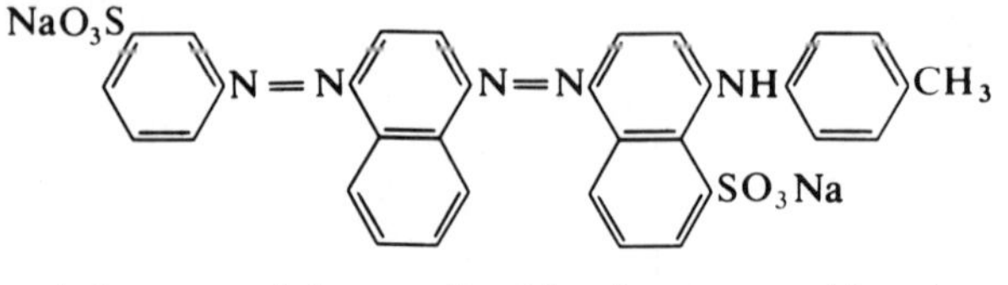

Aqueous solutions are violet, turning blue in strong acid and garnet red in alkali; alcoholic solutions are deep blue.

COLLAGEN STAINS.   A considerable group of dyes, included in Table 6-6, can be used both as simple plasma stains and for the specific purpose of staining collagen fibers which appear to be only weakly basic, in technics involving concomitant or sequential use of phosphomolybdic, phosphotungstic, picric, and other acids. These uses are taken up in Chap. 15, and the dual usage of these dyes as acid dyes (A) and collagen fiber stains is noted in the table by the symbol AC.

Among this group ponceau S requires some special discussion. This dye was used by Curtis (*Anat. Pathol. (Paris)*, **17**: 603, 1905), who called it "ponceau S extra." In 1945 (*J. Tech. Meth.*, **25**: 1) Lillie identified this dye as C.I. 282, formulated in Conn's "Biological Stains," sixth edition, as a monosulfonated disazo dye. Lillie (1969, pp. 106–107) adopted the tetrasulfonate formula.

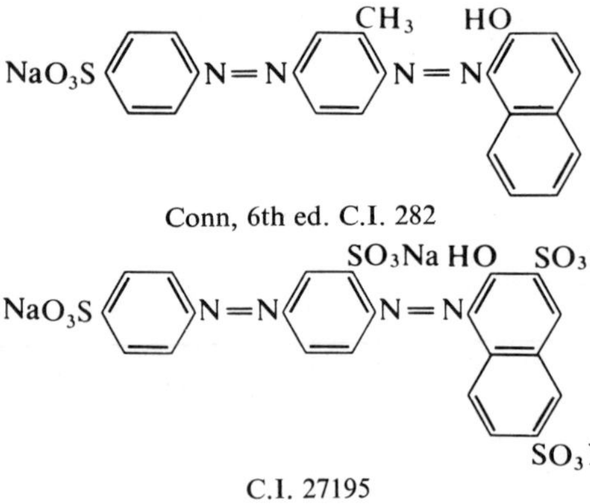

Conn, 6th ed. C.I. 282

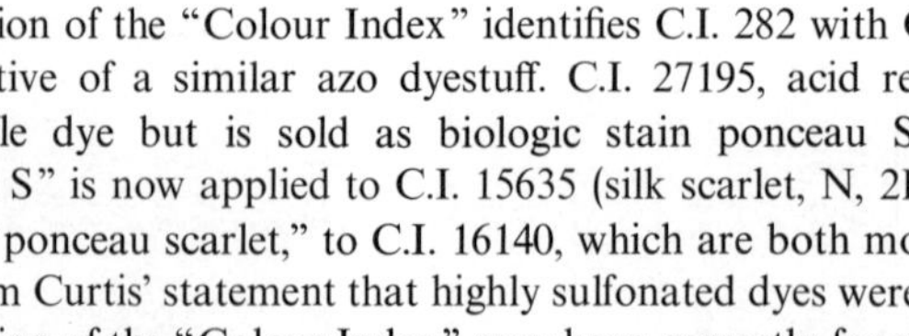

C.I. 27195

But the second edition of the "Colour Index" identifies C.I. 282 with C.I. 27195, which is a tetrasulfonic derivative of a similar azo dyestuff. C.I. 27195, acid red 112, is apparently obsolete as a textile dye but is sold as biologic stain ponceau S or Java scarlet R. The term "ponceau S" is now applied to C.I. 15635 (silk scarlet, N, 2R, S); "ponceau SX," to C.I. 14695; and "ponceau scarlet," to C.I. 16140, which are both mono- and disulfonated monoazo dyes. From Curtis' statement that highly sulfonated dyes were superior, we suspect that the second edition of the "Colour Index" may have correctly formulated the dye which Curtis used.

Water-soluble nigrosin (C.I. 50420) and water-soluble indulin (C.I. 50405), in addition to their use as negative background stains for microorganisms and their use for differentiating basic fuchsin spore stains according to Dorner (*Lait*, **6:** 8, 1926) and the occasional use of water-soluble nigrosin in picronigrosin for collagen, have recently come into use as selective elastin stains from alkaline solutions, working better on animal and young human elastin (Lillie et al., *Acta Histochem.*, **51:** 109, 1974).

Spirit-soluble nigrosin (C.I. 50415, solvent black 5) and nigrosin base (C.I. 50415B, solvent black 7) have interesting properties in the staining of lipofuscins, myelin, and elastin (Lillie et al., *Acta Histochem.*, **51:** 109, 1974).

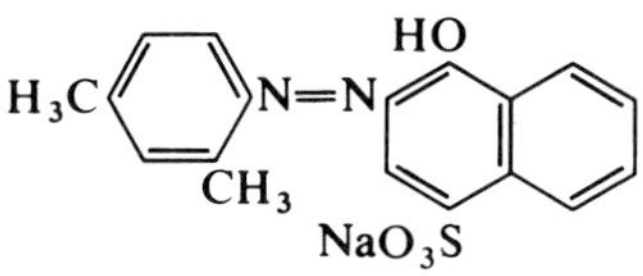

$C_{18}H_{15}N_2O_4SNa$, mw 378.385
C.I. 14695

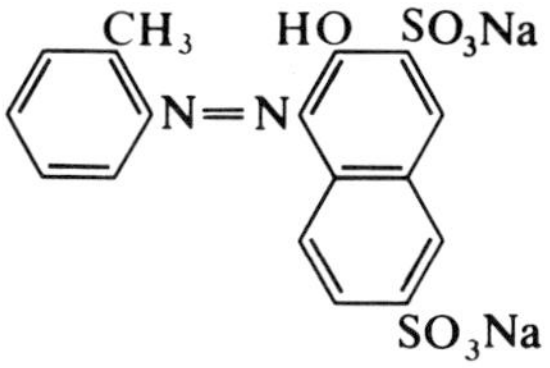

$C_{17}H_{12}N_2O_7S_2Na_2$
mw 466.402
C.I. 16140

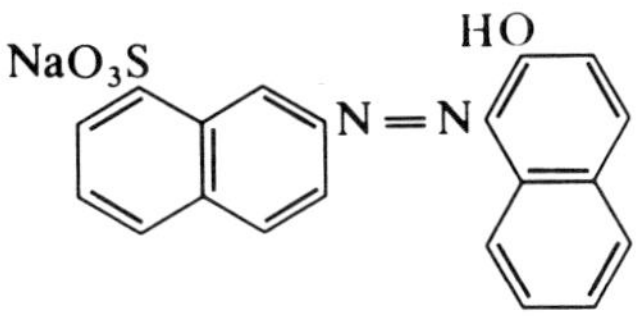

$C_{20}H_{13}N_2O_4SNa$
mw 400.391

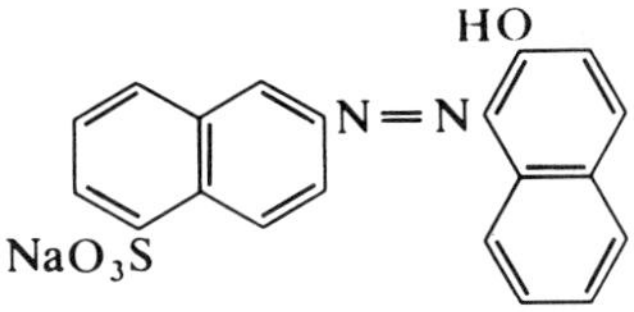

C.I. 15635

Aniline blue, used in these methods, was, according to Conn, a somewhat variable mixture of methyl blue, C.I. 42780, and water blue I, C.I. 42755. The latter has not been used specifically as such, although it would probably function in the same way as the mixture. Methyl blue is now being furnished under the designation *aniline blue* by one manufacturer. We have interchanged the two dyes successfully in Mallory, Curtis, and Mann technics and have no doubt that methyl blue can be used in place of aniline blue WS with no appreciable difference in results.

Another dye in the above group, naphthol blue-black, also deserves special mention. Under its original German name, *Amidoschwarz 10B*, it has had considerable use in Europe and among chemical chromatographers as a "specific protein stain." The name *amido black 10B* is the one used in this book.

A small group of sulfonated triphenylmethane dyes, acid fuchsin, patent blue V, and cyanol FF, produce leuco dyes on reduction with zinc and acid; these leucos are used for the demonstration of hemoglobin and are included in Table 6-6 as "redox" dyes. Methyl blue behaves similarly in the hemoglobin peroxidase technic.

Members of a further group of acid disazo dyes have been used as *supravital stains* for phagocytic and pinocytic activity. The biologists who have used them have retained the traditional names dianil blue RR or 2R, trypan red, and trypan blue for these three dyes but have renamed some of the others. Vital red was "diamine purpurin 3B." Trypan blue is itself more commonly referred to as "blue 3B" or "direct blue 3B," with or without various brand prefixes.

Vital new red is formulated by Conn (6th ed.) as the disulfonic derivative of the same base as C.I. 25375, which is a tetrasulfonate and agrees almost exactly, except for the position of the two naphthyl sulfonic groups, with E. Gurr's formulation of vital new red.

TABLE 6-7 ACID AMINO DYES USED AS SUPRAVITAL STAINS*

"Colour Index" no.		"Colour Index," 3d ed., Part I, class color no.	Stain class	Name†	Molecular weight	Color	SO_3Na	CO_2Na and other acid	Amino groups			Solubility in 100 parts	
1st ed.	3d ed.								NH_2	NHR	NR_2	Water	Alcohol
438	22850	(DR)	SVA	Trypan red	1002.808	R	5		2			3.5	0
456	23570	DR-34	SVA	Vital red	826.793	R	3		2			3.75	0.17
	23860	DB-53	SVA	Evans blue	960.817	B	4		2			7	0.7
465	23690	DB-25	SVA	Dianil blue 2R	841.743	B	3					5	0.01
477	23850	DB-14	SVA	Trypan blue	960.817	B	4		2			1	0.02
			SVA	Vital new red (Conn, 6th ed.)	782.770	R	2			4		Sol.	
			SVA	Vital new red (E. Gurr)	986.858	R	4			4		1.3	0
25375		DR-49	SVA?	Benzo light eosin BL	986.858	R	4			4		Sol.	Slight

* Abbreviations: SVA, supravital acid; colors: R, red; B, blue.
† Among the above dye names only the word Evans is a proper name and must always be capitalized.

It is probable that C.I. 25375, benzo eosin LB or benzo light eosin LB would serve. Evans blue is identified with direct pure blue BF (Fran), which is C.I. 23860.

The class referred to in textile dyeing as *direct cotton dyes* is finding increasing application in the staining of linear proteins, apparently by hydrogen bonding rather than by anionic salt unions as with ordinary cytoplasmic structures. Many dyes of this class are formed by tetrazo coupling of benzidine, tolidine, and di-*o*-anisidine into various usually sulfonated and often amino naphthols. Their elongated structures and multiple azo, hydroxyl (and amino) groups along the chain offer multiple H-bonding sites for linear proteins. Puchtler has used a number of these, structurally resembling Congo red in her modification of the Bennhold and Highman technics for amyloid.

The Congo red family of direct disazo dyes used for direct dyeing of cotton and applied histologically in variants of the Bennhold technic for staining amyloid includes the familiar Congo red, C.I. 22120, itself; Congo Corinth G, C.I. 22145 (Erie garnet B) (E. Gurr's Congo Corinth R is probably the same dye, though he gives an isomeric formula); and benzo fast violet or diamine violet R, C.I. 22570. Benzo blue BB, C.I. 22610, has not been reported as an amyloid stain, but should serve. Manchester blue was formulated by E. Gurr:

$$C_{33}H_{23}N_6O_{11}S_3Na_3 \quad mw \ 844.746$$

Trypan red, C.I. 22850, stains elastin and might be usable for amyloid. Sirius light blue GL, C.I. 23160 (used as a collagen stain), might serve for amyloid. Benzopurpurin 4B, C.I. 23500 (used as indicator and as plasma stain), is a direct cotton dye. Brilliant purpurin R, C.I. 123510, has served as amyloid stain as well as plasma stain. Vital red, C.I. 23570, is used as vital stain and for amyloid. We omit azo blue (By), C.I. 23680, and azo blue, C.I. 23685, both direct cotton dyes because of their poor light-fastness. Dianil blue 2R, C.I. 23690, another direct cotton dye, has been used only as a supravital stain. Brilliant Congo blue RRW, C.I. 23745, another direct cotton dye not much used in histology, is probably not useful for collagen or amyloid because of its precipitability by HCl and NaOH. Trypan blue, C.I. 23850, has been used for amyloid and as a vital stain. Evans blue, C.I. direct blue 53, is used chiefly as a vital stain. Diamine blue 3R, C.I. 24020, is another direct cotton dye sometimes used in histology in Europe, as is also the direct cotton dye Chicago blue RW, C.I. 24280. Dianil blue G, C.I. 24340, a direct cotton dye whose light-fastness is much improved by $CuSO_4$ afterchroming, has had some use as an elastin stain (without $CuSO_4$). It is formulated:

C.I. 24340, direct blue 10
$$C_{34}H_{22}N_4O_{18}S_4Na_4, \quad mw \ 944.786$$

Enianil azurine J, C.I. 24360, direct blue 152, is highly recommended by Horobin and James (*Histochemie*, **22**: 324, 1970) as a red elastin stain from pH 9 solution (see also Lillie et al., *Acta Histochem.*, **44**: 163, 215, 1972). It is formulated:

$$\text{H}_2\text{N} \quad \overset{\text{OH}}{\underset{\text{SO}_3\text{Na}}{\bigcirc\bigcirc}} \text{N}=\text{N} \overset{\text{H}_3\text{CO}}{\bigcirc} \bigcirc \overset{\text{OCH}_3}{\underset{\text{NaO}_3\text{S}}{\bigcirc}} \text{N}=\text{N} \overset{\text{HO}}{\underset{}{\bigcirc\bigcirc}} \text{NH}_2$$

$C_{39}H_{26}N_6O_{10}S_2Na_2$, mw 788.722

It is soluble in water and sparingly soluble in alkaline water; its light-fastness is poor but is improved to excellent by $CuSO_4$ afterchroming (dark blue); its isomer, enianil violet ND, C.I. 24370, has similar properties of staining and of copper afterchroming.

$$C_{37}H_{22}N_6O_{10}S_2Na_2 \text{, mw 788.722 (direct violet 37)}$$

Benzo sky blue, C.I. 24400, and Chicago blue 6B, C.I. 24410, are further direct cotton dyes sometimes used in histology; their light-fastness is improved by $CuSO_4$ afterchroming.

Orcein, whether prepared from *Roccella* and *Lecanora* lichens or from the air oxidation of orcinol in the presence of ammonia or by a Wurster synthesis with H_2O_2, NH_3, and orcinol, is composed of some 14 chromatographically separable fractions, including 7-amino- and 7-hydroxy-2-phenoxazones and their derivatives (Musso, *Planta Med.*, **8**: 432, 1966), of which the 7-amino-phenoxazones are probably the constituents which act as active elastin stains (*Histochemie*, **19**: 1, 1969).

There was for a time some preference among chromosome cytologists for natural orcein. The synthetic form now appears to be equally effective for chromosome cytology as well as for elective elastin staining. The simpler dye resorcein, made from resorcinol and ammonia by the Wurster peroxide synthesis by Lillie et al. [*Acta Histochem. (Jena) Suppl.*, **IX**:625, 1971] and resorcin blue from air oxidation of resorcinol in the presence of NH_3, are also effective elastin stains, giving darker colors than orcein.

FAT STAINS

The oldest member of this group, osmium tetroxide, is not a dye at all but an unstable oxide which is reduced to a black substance, by unsaturated fats and fatty acids, by eleidin, and by other substances. Osmium tetroxide, commonly called osmic acid, is now finding increasing use in histology as a fixative for electron microscopy, in addition to its histochemical usage in the demonstration of unsaturated fats and other reducing substances such as eleidin, melanin, or catecholamines. For demonstration of degenerating myelin we now prefer frozen section methods which permit combination of normal myelin methods with fat stains of contrasting color.

The first synthetic aniline dye to be used as a fat stain was quinoline blue (C.I. 806, 1st ed.). Ranvier (1875) used it in hydroalcoholic solution as a general stain and noted that in glycerol mounts nuclei soon faded and myelin sheaths of nerves and fat in course of absorption through intestinal epithelium were selectively colored blue. It fades badly in light and is not used as a textile dye; it is included in the list of pH indicators by Lillie (1969).

The commonly used oil-soluble dyes fall into two main groups: (1) basic arylamines with very low water-solubility, such as Sudan black B and Sudan red VII B, also called Fettrot VII B, among the azo dyes, the basic triphenylmethane dye spirit blue, and the seven aminoanthraquinone dyes included in Table 6-8, and (2) β-naphthols such as the original disazo dyes, Sudans III and IV and their close relatives oil red O and oil red 4B or EGN and Sudan IV BA (which may well be a synonym for oil red 4B) and a small group of monoazo dyes, Sudan II, Sudan brown, Sudan R, Sudan red, and a few others. We have included with them in the table permanent red R, which has the interesting trade names Blazing Red, Flaming Red, and Fire Red, and has been used occasionally as a fat stain (Putt, *Lab. Invest.*, **5**: 377, 1956).

Of the three acylated dyes tested, which do not form stable unextractable pigments with neutrophil leukocyte granules, as their parent dyes do, one, acetyl Sudan black, is commercially available from National Aniline Division, which also prepared for Lillie a benzoylated oil red O. Oil red 4B or EGN (*Stain Technol.*, **19**: 55, 1944) was identified for Lillie by the National Aniline Division as *p*-xylylazo-*o*-tolylazo-β-naphthol, C.I. 26120; and oil red O as *p*-xylylazo-*p*-xylylazo-β-naphthol, which is now included under C.I. 26125.

Oil red 4B   Oil red O

Several other excellent dyes for fat staining by our supersaturated isopropanol technic were mentioned in the same report: oil brown D (NAC), Sudan brown 5B (G), Sudan red 4B (G), and Sudan Corinth B (G). Sudan red 4BA (G) is stated to be similar to C.I. 26105, Sudan IV; Sudan Corinth 3B (G) is identified as solvent red 40, an azo dye of unstated composition, and oil brown D (W) is solvent brown 7, an unidentified monoazo dye.

Salthouse (*J. Histochem. Cytochem.*, **13**: 133, 1965) formulated Luxol Fast Blue G. It is the diarylguanidine salt of the acid trisazo dye Sirius Light Blue G, C.I. 34200.

Sirius Light Blue G C.I. 34200 $C_{102}H_{97}N_{19}O_{13}S_4$, mw 1925.280

Diarylguanidine

It is insoluble in water, soluble in gelatin solution and in diethylene glycol, very soluble in alcohols, 0.9% in methanol, soluble in Cellosolve, slightly soluble in acetone. It is used in methanol for collagen and elastin, in other alcohols for myelin staining.

In fluorescence microscopy several acid and basic dyes have been used in simple aqueous solution of fat stains. Of these, phosphine, a basic acridine dyestuff, seems to be one of the

TABLE 6-8 OIL-SOLUBLE DYES USED AS FAT STAINS*

"Colour Index" no. 1st ed.	"Colour Index" no. 3d ed.	"Colour Index," 3d ed., Part I, class color no.	Stain class	Name†	Molecular weight	Amino groups NH₂	NHR	NR₂	OH	Solubility in 100 parts Water	Alcohol	Glycol	Xylene	Color
				Monoazo dyes										
73	12140	SO-7	OS	Sudan II	276.341				1β	Insol.	0.28	1.25	3.0	Y
81	12020	SF-5	OS	Sudan brown	298.347				1α	Insol.	Sol.		Sol.	F
113	12150	SR-1	OS	Sudan R, Sudan red B	278.313				1β	Insol.	Sol.		Sol.	R
	12085	PR-4	OS	Permanent red R	327.729				1β	Insol.	Sol.		Sol.	R
		SF-9	OS	Oil brown D						Insol.	Sol.		Sol.	F
		SR-40	OS	Sudan Corinth 3B						Insol.	Sol.		Sol.	R
	12155	SR-17	OS	Sudan red (E. Gurr)	292.340				1β	Insol.	1	1.25	1.25	R
				Disazo dyes										
248	26100	SR-23	OS	Sudan III	352.399				1β	Insol.	0.25	3.15	2.25	OR
258	26105	SR-24	OS	Sudan IV	380.455				1β	Insol.	0.5	2.5	3.5	R
Similar	26120	SR-26	OS	Oil red 4B, EGN	394.480				1β	Insol.	Sol.		Sol.	R
	26125	SR-27	OS	Oil red O	408.507				1β	Insol.	Sol.	2.5	3.5	R
	26050	SR-19	OS	Sudan red VII B	379.468		1β			Insol.	4.25	4	10.45	R
	26150	SN-3	OS	Sudan black B	456.555		2			Insol.	0.25	1.0	2.5	BN, GN

Acylated dyes

C.I. No.	Class	Dye	Mol. wt.	Group					Color
26105 (Ac)	OS	Acetyl Sudan IV	422.491	Ester	Insol.	Sol.			R
26150 (Ac)	OS	Acetyl Sudan black	540.630	Amide	Insol.	Sol.			BN GN
26125 (Bz)	OS	Benzoyl oil red O	512.617	Ester	Insol.	Sol.			R

Triphenylmethane dyes

C.I. No.	Class	Dye	Mol. wt.						Color
689 42775 SB-3	OS, B	Spirit blue	490.053 / mixture / 552.125 (1 : 2 : 3)		Insol.	1.5	4.0		B

Aminoanthroquinone dyes

C.I. No.	Class	Dye	Mol. wt.	n					Color
61100 V-1	OS	Sudan violet R	238.248	2	Insol.	3.2	4.0	5.0	V
61520 Disp.	OS	Sudan blue GN	342.401	2	Insol.	0.05			
61525 SB-63	OS	Sudan blue G	342.401	2	Insol.	Sl. sol.		Sol.	
62545 SB-11	OS	Sudan green BB	434.499	2	Insol.	3.2	4.85	1.6	G
Sol.	OS	Oil blue N	293.369	1	Insol.	Sol.		Sol.	B
61555 SB-14	OS	Caryinel red	438.575	2	Insol.	Sol.		Sol.	R
	OS	Coccinel red			Insol.	Sol.		Sol.	R

Fluorochrome dyes

C.I. No.	Class	Dye	Mol. wt.						Color
	OSF	3,4-Benzpyrene	252.319		Very slight	Sol.?		Sol.	fBW

* Abbreviations: OS, oil-soluble dye; OSF, oil-soluble fluorochrome; B, basic dye. Colors: R, red; O, orange; Y, yellow; F, brown; G, green; B, blue; N, black; fBW, blue-white fluorescence. Sudan red 4BA is "similar" to Sudan IV, C.I. 26105.
† Among the above dye names only the words Sudan and Corinth are proper names and must be capitalized.

TABLE 6-9 STABILIZED DIAZONIUM SALTS USED IN HISTOLOGIC STAINING PROCEDURES*

"Colour Index" no., 3d ed.	"Colour Index" Azoic Diazo no.	Preferred name†	Colors developed			Usual stabilizers	Molecular‡ weight of amine	Coupling rate				Synonyms
			β-Naphthol AS naphthol	α-Naphthol	Entero-chromaffin			ICI	N	B	L	
37245		Fast black B	N		RF-N	ZnCl$_2$	199.3				f	Black B, BS
37190	38	Fast black K	BN		N	ZnCl$_2$, also Zn free	302.3				f	Black K, NK
37235	48	Fast blue B	B+	FN	FP	ZnCl$_2$	244.3	s	f	s	m	Blue B, BNS, diazo blue B
37175	20	Fast blue BB	VB-VN		OR	ZnCl$_2$	300.4	s			m	Diazo blue BB, blue 2B, DB
37155	24	Fast blue RR	PR-VB+	N	R±	ZnCl$_2$	272.3	s		s	s	Blue RR
37255	35	Fast blue VB	VB+		O	$\overline{C}$l	263.3	s		s	s	Varamine blue B, BA, BD, BN blue V, fast blue BM, MB
37135	1	Fast Bordeaux GP	PR	P	O+	ZnCl$_2$	168.2	f		f		Bordeaux GP
37020		Fast brown RR	F			ZnCl$_2$	177.0	f		f		Brown RR
37200	21	Fast brown V	F			Not stated	320.7		s			Fast brown VA
37160	43	Fast Corinth LB	PR			ZnCl$_2$	276.7			f		Corinth LB, diazo Corinth LB
37220	39	Fast Corinth V	PR-FR			ZnCl$_2$	300.3					Corinth V
37195	51	Fast dark blue R	VBN	RF	BN	ZnCl$_2$	371.2	s		f		Navy blue RN
37210	4	Fast garnet GBC	R		OR+	$\overline{S}O_4H$	225.3	f	m	f	f	Garnet GBC
37215	27	Fast garnet GC	FR			$\overline{C}$l	225.3	f	f		f	Fast garnet AC, diazo garnet GC, garnet GC, AC, GCD
37025	6	Fast orange GR	RO		OR	ZnCl$_2$	138.1	f	f			Fast orange GR

							MW‡				Synonyms†
37275	36	Fast red AL	R		F	ZnCl₂	223.3	f	f		Diazo red AL, Red AL, ALS 1-Aminoanthraquinone diazo
37125	5	Fast red B	R+		OR+	Disulfonaph-thalene	168.2	f	f	f	Red B, red V Fast red 5NA, BN
37035	37	Fast red GG	R		OR+	BF₃, SŌ₄H, etc.	138.1	f	f	f	Nitrazol CF, nitrosamine red Para red, red 2G, GG, 2J
37110	8	Fast red GL	R+		PR	Disulfonaph-thalene	152.2	f	f		Diazo red G, red G
37040	9	Fast red 3GL	R		FR-OR+	ZnCl₂	172.6	f	f	f	Red 3GK, 3GL, 3G Diazo red or fast red 3GL
37150	42	Fast red ITR	R	RF		ZnCl₂	258.3	mf			Red ITR Diazo red RC, red RC
37120	10	Fast red RC	R			ZnCl₂	157.6	f			Fast red 4GA
37100	14, 34	Fast red RL	PR		R+	BF₃	152.2 HCl	f	f	f	Red RL, fast red NRL Fast red 5CT, TRN
37085	11	Fast red TR	R+	F		Disulfonaph-thalene C̄l	178.1	f			Red TR, TA, TRS Red violet LB
37010	3	Fast red violet LB Fast scarlet GG	PR OR		OR+	ZnCl₂	260.7 162.0	f f	f		Scarlet 2G, GG Fast scarlet GGS, GGN
37130	13	Fast scarlet R	R	RF		ZnCl₂	168.2	f	f		Diazo scarlet R, Scarlet R, fast scarlet 4NA
37165	41	Fast violet B	BV		OF	ZnCl₂	256.3	s			
37265		α-Naphthylamine	R	R	R	Disulfonaph-thalene	143.2				
37260	45	Fast black G	N, FN		FN			s			

* Abbreviations: Colors—R, red; O. orange; Y, yellow; G, green; B, blue; V, violet; P, purple (redder than violet); F, brown; N, black. Coupling rates: s, slow; m, medium; f, fast. N, Nachlas; B, Burstone; L, Lillie; "ICI, Imperial Chemical Industries Manual." β-Naphthol and naphthol colors are from the "Colour Index"; enterochromaffin, mostly from Lillie; α-naphthol colors, from various sources.

† Among the "preferred names" only the words Bordeaux and Corinth are proper names and require capitalization. The synonyms are further subject to the usual rules covering the trademark names; the names used in this table are those specified by the Journal of Histochemistry and Cytochemistry, 7: 281, 1959. Certain new additions will also appear in the ninth edition of "Conn's Biological Stains."

‡ Molecular weights were calculated by R. D. Lillie, 1974.

best. Also used are methylene blue, rhodamine B, magdala red (echt), all basic dyes, and the acid dyes titan yellow or thiazol yellow G (C.I. 19540) and thiazine red R (C.I. 14780) (see Table 6-5).

STABILIZED DIAZONIUM SALTS

We have included most of the commercially available stabilized diazonium salts in Table 6-9, which gives the third edition "Colour Index" numbers and the azoic diazo numbers, the molecular weights of the undiazotized bases, the colors given with naphthols and with enterochromaffin, the nature of the usual stabilizer or anion, available data as to coupling rates, and the commoner synonyms. A few fresh diazonium salts are employed.

Amine contents of the stabilized salts vary from 18% to a stated 100%, and they vary also among manufacturers for the same salt; 20% is a common level. Some data on amine content are included in "Conn's Biological Stains," eighth ed., 1969. Ordinarily, when knowledge of this subject is important, it may be obtained from the manufacturers.

It is necessary to take care that the base is not furnished under the same name. When ordering, always specify which is required, the base or the salt. In using the *base* it is necessary to diazotize with nitrous acid according to the usual chemical procedure for the specific diazonium salt in question; the *salt* is ready for immediate use. Diazonium salts possess limited stability. With only occasional use, performance should be checked on a known object.

THE TETRAZOLES

These are colorless, fairly readily water-soluble compounds of the general structure given below, which yield insoluble, usually highly colored pigments on reduction and are used in histochemistry as hydrogen acceptors to localize results of various enzymatic oxidations (dehydrogenations).

The tetrazoles are divided into mono- and ditetrazoles; in the latter, two tetrazole rings share a biphenylene group as one of the substituents. Purely for the sake of uniformity we have formulated all ditetrazoles with the biphenylene group at 2,2'. See Table 6-10.

Nitro blue tetrazolium is written by Tsou et al. (*J. Am. Chem. Soc.*, **78:** 6139, 1956) as (III) 2,2'-dinitrophenyl-5,5'-diphenyl-3,3'-(3,3'-dimethoxy-4,4'-biphenylene)tetrazolium chloride but is drawn with the methoxy groups ortho to the diphenylene bond. We follow Conn and Pearse in the position of the methoxy groups, and have formulated the diphenyl at 2,2' to agree with neotetrazolium and blue tetrazolium, recognizing that the 2,2' and 3,3' diphenylene formulations are tautomeric to each other.

Nitro NT is accordingly 2,2'-biphenylene-3,3'-di-*p*-nitrophenyl-5,5'-diphenylditetrazolium chloride. Note that nitro NT is distinguished from NNT.

Tetranitro NT is 2,2'-biphenylene-3,3',5,5'-tetra-*p*-nitrophenyl-ditetrazolium chloride, and tetranitro BT is the corresponding 2,2'-di-*o*-anisylene compound.

MTT is 3-(4,5-dimethylthiazolyl-2)-2,5-diphenyl tetrazolium chloride. Pearse's considerations as to formation of two 5-membered chelate rings require that the charged N be at 3 rather than 2. He does not account for the second valency bond of the Co ion (*J. Histochem. Cytochem.*, **5:** 515, 1957).

TABLE 6-10 MONO- AND DITETRAZOLES WITH SHORT NAME AND COMMON ABBREVIATION*

Usual abbreviation	Short name	Chemical description
TTC, TPT	Triphenyl tetrazolium	2,3,5-Triphenyltetrazolium chloride
NT	Neotetrazolium	2,2'-Biphenylene-3,3',5,5'-tetraphenylditetrazolium chloride
	M and B 1767	2,5-Biphenyl-3-*p*-styrylphenyltetrazolium chloride
INT	Iodonitrotetrazolium	2-*p*-Iodophenyl-3-*p*-nitrophenyl-5-phenyltetrazolium chloride
BT	Blue tetrazolium	2,2'-Di-*o*-anisylene-3,3',5,5'-tetraphenylditetrazolium chloride
Nitro-BT	Nitro blue tetrazolium	2,2'-Di-*o*-anisylene-3,3'-di-*p*-nitrophenyl-5,5'-diphenylditetrazolium chloride
Nitro NT	Nitroneotetrazolium	2,2'-Biphenylene-3,3'-*p*-nitrophenyl-5,5'-diphenylditetrazolium chloride
Tetranitro NT	Tetranitroneotetrazolium	2,2'-Biphenylene-3,3',5,5'-tetra-*p*-nitrophenylditetrazolium chloride
Tetranitro BT	Tetranitro blue tetrazolium	2,2'-Di-*o*-anisylene-3,3',5,5'-tetra-*p*-nitrophenylditetrazolium chloride
MTT	....	3-(4,5-Dimethylthiazolyl-2)-2,5-diphenyltetrazolium chloride
TV	Tetrazolium violet	2,5-Diphenyl-3-α-naphthyltetrazolium chloride
NNT	*m*-Nitroneotetrazolium	2,2'-Biphenylene-3,3'-diphenyl-5,5'-*m*-nitrophenylditetrazolium chloride

* This list of names of tetrazoles contains no proper names, and no capitals are required. The abbreviations are capitalized as given.

Tetrazolium violet, TV, is 2,5-diphenyl-3-α-naphthyl-tetrazolium chloride (Glenner et al., *J. Histochem. Cytochem.*, **5**: 591, 1957; Pearson, ibid., **6**: 112, 1958). Pearson highly recommends *m*-nitroneotetrazolium, NNT, 2,2'-diphenyl-5,5'-di-*m*-nitrophenyl-3,3'-*p-p*-diphenylene, which we write as the tautomer to agree with the other ditetrazole formulas. The formazan is very finely crystalline and insoluble in alcohol. An 80% alcohol extraction is recommended to remove fat and prevent later crystal growth in the aqueous mounts.

At this writing, three ditetrazoles seem about equal for fineness of crystal size, rapidity

Formulation of tetrazoles

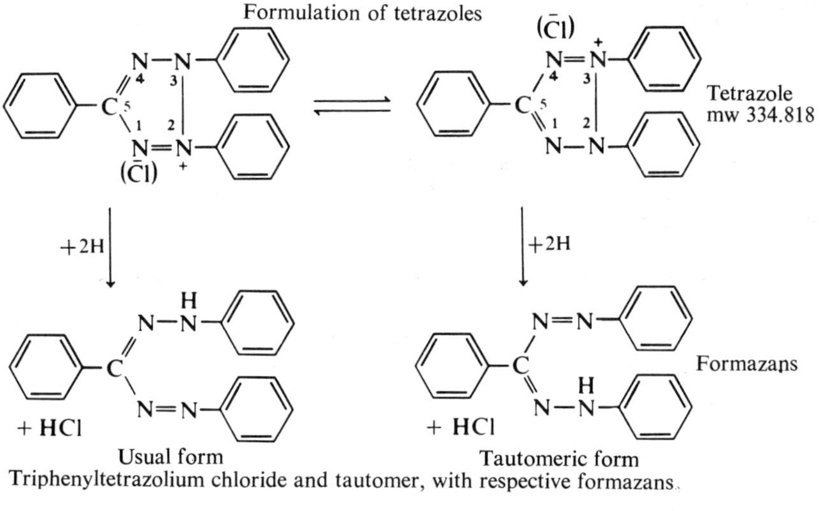

Tetrazole mw 334.818

Formazans

Usual form   Tautomeric form

Triphenyltetrazolium chloride and tautomer, with respective formazans.

of reaction, and insolubility in fats: nitro-BT, tetranitro BT, and Pearson's 5,5'-*m*-nitroneotetrazolium (NNT). This last is to be distinguished from the nitro NT of Tsou et al.

(*J. Am. Chem. Soc.*, **78**: 6139, 1956), which is a 3,3′-di-*p*-nitrophenyl neotetrazolium, is partly fat-soluble, and forms mixed fine and coarse crystals. Seligman's group preferred the blue of the nitro BT to the red-brown of the otherwise equal tetranitro BT. Of the three, the first two have been made commercially available.

The thiazolyl monotetrazole MTT is preferred for electron micrography because of its metal chelation capability, but it is difficult to obtain, even in England.

Iodonitrotetrazolium (INT) seems to be preferred for microchemical assay of extracts by colorimetry at 494 mμ.

IODINE

One of the oldest of all stains,[1] used by Raspail (*Ann. Sci. Nat.*, **6**: 224, 384, 1825) and Caventou (*Ann. chim.*, **31**: 358, 1826) in the study of the structure of the starch granule, the element iodine is still widely used in a variety of tests, which are detailed in the appropriate sections. It gives reactions with amyloid, cellulose, chitin, starch, carotenes, and glycogen. It is used as a reagent to alter crystal and methyl violet so that they are retained by certain bacteria and fibrin. It may serve as an oxidizing agent. It is widely used for the removal of mercurial fixation artifacts.

It is used in the form of dilute alcoholic solutions and in various modifications of Lugol's solution. The term *Lugol's solution* has been quite variously used by different writers, as shown in Table 6-11. Many writers simply refer to it by name, without hint as to which formula was used.

DYE SOLUBILITIES

Table 6-12 gives solubilities of most commonly used dyes, from various sources.

REACTIVE DYES

The new class of reactive dyes was introduced into cellulose dyeing in 1956 by L. D. Rattree and W. E. Stephen of Imperial Chemical Industries, Ltd. (ICI). In the first group of these cyanuric chloride

is combined by use of one of its chlorine atoms with amino dyes of several chemical classes and colors. For cyanuric chloride the 1971 "Colour Index" lists 46 monoazo, 2 dis- and polyazo, 4 phthalocyanin, and 4 aminoanthraquinone dyes, and the list is growing. The dichlorotriazinyl-substituted dyes readily react in the cold by utilizing the second chlorine under alkaline conditions to accomplish a Schotten-Baumann reaction with part of the cellulose hydroxyls. The dye is thus covalently bonded to the tissue and firmly adherent. The third chlorine is less readily replaced and requires heat for reaction. Such dyes contain a phenol, an aliphatic or aromatic amine, or an alkoxide. The first class is called Procion M (for cold dyeing) and Procion H (for hot dyeing) by ICI.

---

[1] Although Robert Hooke (*Micrographia*, Royal Society, London, 1665) notes the use of tinctures of "Logwood and Cocheneel" in tingeing liquors, in his observation of dyed hairs with the microscope he does not identify the dyes used.

According to H. J. Conn's *History of Staining* (Biotech, Geneva, N.Y., 1948) Leeuwenhoek (letter of August 24, 1714, to the Royal Society, published in 1719) used saffron to stain muscle because it was too transparent otherwise. Hooke's study of dyed hair and wool in 1660 was the earliest microscopic observation of stained animal tissue elements (*Micrographia*, Royal Society, London, 1660).

TABLE 6-11   COMPOSITION OF LUGOL'S IODINE SOLUTION

Formula*	Iodine	Potassium iodide	Distilled water
Lugol's "rubefacient solution," 1830	1	2	12
USP 1870 to XIV, AMA editors	5	10	100
Lee, 1890–1937; Cowdry, 1943, 1948	4	6	100
Mosse in Ehrlich's "Encyklopädie"; Schmorl, 1907, 1929; Roulet, 1948; Lillie, 1948; Langeron's "double," 1934, 1949†	1	2	100
Langeron, 1934, 1949	1	2	200
Romeis, 1932–1948; Roulet, 1948; Gomori‡	1	2	300

* For all these formulas, we recommend that the potassium iodide be first dissolved in 1 to 2 times its weight of water. The iodine then dissolves easily in this concentrated solution. When solution is complete, add the rest of the prescribed volume of water.

† Commonly called Weigert's or Gram-Weigert's iodine.

‡ Commonly called Gram's iodine.

Sources: E. V. Cowdry, "Microscopic Technique in Biology and Medicine," Williams & Wilkins, Baltimore, 1943, 1948; P. Ehrlich, "Encyklopädie der mikroskopischen Technik," Urben & Schwarzenberg, Berlin, 1903; Roulet, "Methoden des Patholgischem, Histologie," Springer-Verlag, Vienna, 1948; R. D. Lillie, "Histopathologic Technic," Blakiston, Philadelphia, 1948; M. Langerm, "Precis de microscopie," 5th ed., Masson, Paris, 1934, 1949; B. Romeis, "Taschenbuch der Mikroskopische Technik," 13th ed., Oldenberg, Munich, 1932; G. Gomori, "Microscopic Histochemistry: Principles and Practice," University of Chicago Press, 1952; A. B. Lee, "The Microtomists' Vade-Mecum," Blakiston (McGraw-Hill), Philadelphia, 1890, 1937.

When cotton dyed with a dichlorotriazinyl monoazo dye is treated with dithionite, it is cleaved to two colorless amino compounds, one of which remains covalently bonded to the fiber and can be then diazotized and coupled with a phenol or an aryl amine to form a new covalently bonded dye. If dichlorotriazinyl dyes are used without alkali, they are readily washed out, but may be fixed by an alkali posttreatment.

Reactive dyes for wool (ICI's Procilan, Farbwerke Hoecht's (FH) Remalan dyes) and nylon (ICI's Procinyl dyes) differ chemically and operate in neutral or slightly acid dye baths. Apparently amino and perhaps amide groups in the fiber are utilized.

There are 79 reactive yellows, 60 reactive oranges, 115 reactive reds, 22 reactive violets, 101 reactive blues, 18 reactive greens, 22 reactive browns, and 32 reactive blacks given in Part I of the "Colour Index," mostly of only partly revealed constitution. For each of these reactive dyes indication can be given in Part I of the "Colour Index" of the chemical class of the dye used, of the reactive system employed, and of the Part II "Colour Index" (C.I.) number. The last was revealed for only 19 of the 449 dyes given in Part I. The dye class is usually stated (440 dyes), the reactive class is given for 121 dyes, and Part II C.I. numbers are furnished for 10 red dyes, 1 violet, 6 blue, 1 brown, and 1 black—19 dyes in all.

Of the 19 dyes, 17 are derivatives of cyamuric chloride, 6 being dichlorotriazinyl and 11 monochlorotriazinyl compounds. The other two are respectively trichloropyrimidinyl and vinyl sulfonyl compounds. Five dyes are used for nylon, silk, and wool as well as cellulose (three monochloro- and two dichlorotriazinyl), five for cellulose only (1 vinyl sulfonyl, four monochlorotriazinyl), and the rest for wool, silk, or nylon as well as cellulose.

From the foregoing data it would be expected that mono- and dichlorotriazinyl dyes would bind to tissue polysaccharides such as glycogen, starch, and various mucins, to collagen, and to basic proteins including basic nucleoprotein as well as the carbohydrate moieties of nucleic acids. It is probable that technics may be devised to restrict the reactivity and render binding more specific. Use for premarking of experimentally introduced proteins and carbohydrates seems feasible. In vivo marking of specific tissues with the less reactive monochlorotriazinyls is being used. For further details consult the individual dye accounts.

TABLE 6-12 STAIN SOLUBILITIES FOR DYES IN CURRENT AND PAST COMMON USES

C.I. no.	Preferred name	"Colour Index" and other sources					Holmes		Gurr				
		Water	Alcohol	Acetone	Benzene	Ligroin	Water	Alcohol	Water	Alcohol	Cellosolve	Glycol	Xylene
58000	Alizarin	Sol	Sol. h.[CI]	Sol[CI]	Sl. sol[CI]							10.0	0.125
67451	Alizarin blue S		Insol										
58005	Alizarin red S	7.69[BS]	0.15[BS]	Sl. sol.[CI]			0.40	0.57	0.5	2.0	3.75	7.69	0.15
14030	Alizarin yellow R						25.84	0.04	0.0	1.0	8.0	3.25	0.85
14029	Alizarol yellow CW						7.20	0.01	65	0.25	5.0	1.0	0.005
16185	Amaranth								50	0.0	4.5	4.75	0.01
42755	Aniline blue WS	Sol.[DR]	Sl. sol.[DR]	Sl. sol.[CI]					1.0	0.75	1.0	7.8	0.0
58605	Anthracene blue SWR	Sl. sol.[L]	Sl. sol.[CI]				0.74	4.49	1.0	4.0	1.25	4.0	0.0
41000	Auramine O	0.7[DR]	4.5[DR]				0.0	0.33	0.1	0.55	3.0	1.75	0.05
10360	Aurantia	Sol.[CI]	Sol.[CI]									1.0	0.25
43810	Aurine tricarboxylic acid	Sol.[CI]	Insol.[CI]										
29250	Azidine fast scarlet GGS (Marshal red)	Sol.[CI]	Sol.[CI]						3.0	0.1	2.75	1.75	0.003
29255	Azidine fast scarlet 4GS	Sol.[CI]	Sol.[CI]										
29260	Azidine fast scarlet 7BS	Sol.[CI]	Sol.[CI]										
13090	Azo acid yellow	V. sol.[CI]					2.17	0.81					
50090	Azocarmine B	Sl. sol.[CI]	V. sl. sol.[CI]						2.0	0.05	3.75	3.5	0.01
50085	Azocarmine G	Sol.[CI]	Sl. sol.[DR]						1.0	0.1	3.75	4.5	0.01
16540	Azofuchsin G	3[DR]	Sl. sol.[DR]						10	0.0	1.6	3.5	0.0
18050	Azophloxine	Sol.[DR]	Sol.[DR]										
52005	Azure A	Sol.[CI]	Sl. sol.[CI]										
52010	Azure B	Sol.[DR]	Sol.[DR]										
52002	Azure C	Sol.[DR]	Sol.[DR]										
24200	Benzo fast scarlet 4BFA	Sol.[CI]	Sol.										
23500	Benzopurpurin 4B	Sol.[CI]	Insol.[CI]	Insol.[CI]	Insol.[CI]	Insol.[CI]	13.31	0.0	1.0	0.25	2.8	7.25	0.1
24400	Benzo sky blue, 3,4-Benzpyrene	V. sl. sol.[M] S.O[M]	Sl. sol.[M] 1.0[M]		Sol.[M]				6.0	0.5	2.5	5.5	0.0
75160	Berberin	Sol.[CI]	Sl. sol.[CI]	Sl. sol.[M]	Sl. sol.[M]								
26905	Biebrich scarlet	1.3[DR]	1.1[DR]	Insol.[CI]	Insol.[CI]		1.36	0.05	5.0	0.25	1.5	1.5	0.0
21000	Bismarck brown			Insol.[CI]	Insol.[CI]		1.10	1.08	1.5	3.0	3.0	7.0	0.05
21010	Bismarck brown R	V. sol.[CI]	Sol.[CI]	Sl. sol.[CI]	Insol.[CI]			0.98	1.5	3.0	4.5	3.0	0.0
27255	Blue-black B	Sol.[CI]	Sl. sol.[CI]										
16180	Bordeaux R	Sol.[CI,M]	V. sol.[M]	Insol.[CI]	Insol.[CI]		3.83	0.19					
75280	Brazilin	Sol.[DR]	Sl. sol.[DR]										
51010	Brilliant cresyl blue	Sol.[DR]	Sol.[L]						10.0	10.0	9.5	15.0	0.0
75470	Carminic acid	Insol.	Sol.[CI]		Insol.[M]				3.0	2.6	2.25	8.0	9.5
	Carycinel red	Sol.[L]							0.0	0.2	0.85	4.0	0.0
51050	Celestine blue B	Sol.[CI]							2.0	1.5	2.25	6.5	0.005
	Chloranilic acid	Sol.[L]											
29065	Chlorantine fast red 6BLL	Sol.[CI]	Insol.[CI]	Insol.[CI]	Insol.[CI]				1.0	0.45	3.0	3.5	0.0

C.I. No.	Name												
30235	Chlorazol black E	Sol.[CI]	Sol.[CI]				19.50	0.17	19.0	0.5	3.75	7.75	0.0
16570	Chromotrope 2R	19[DR]	0.15[DR]				0.23	0.99	12.0	2.5	8.25	11.0	0.6
43820	Chromoxane cyanin R	Sol.[DR]	Sol.[DR]				0.86	2.21	0.1	1.5	5.3	15.0	0.01
43830	Chromoxane pure blue B		V. sl. sol.[CI]						5.5	4.75	6.0	9.5	0.005
11320	Chrysoidin R			Insol.[CI]	Insol.[CI]								
11270	Chrysoidin Y			V. sl. sol.[CI]									
22145	Coccinel red	Sol. h.[CI]	Sol.[L]						5.0	0.1	4.5	6.0	0.0
22120	Congo Corinth G	Insol.	Sl. sol.[CI]						5.0	0.75	1.75	2.5	0.0
45510	Congo red	Sol.[CI]	0.2[DR]					0.18	1.0	0.02	0.75	2.5	0.0
46020	Coerulein S	Sol.[DR]	Sol. h.[CI]						4.75	0.6	1.6	3.25	0.0
	Coriphosphine O	Sl. sol.[CI]	Sol.[CI]						9.5	6.0	9.0	1.50	
	Cresyl fast violet	Sol.[CI]	Sl. sol.[DR]										
	Cresyl violet acetate	Sol.[DR]											
27160	Crocein scarlet	0.13[SP]	Sol.[CI]				0.38	0.25	5.0	2.25	5.0	5.5	1.2
16250	Crystal ponceau 6R	Sol.[CI]	0.5[DR]				0.80	0.05	9.0	8.75	7.5	7.0	0.0
42555	Crystal violet	3.0[DR]	13[DR]				1.68	13.87	1.0	1.5	1.5	2.0	0.0
806	Cyanin	1.7[DR]	Sol.[BS]										
43535	Cyanol FF	Sol.[BS]	V. sol.[CI]				13.8	0.44	2.0	0.0	7.35	3.5	0.0
42530	Dahlia	Sol.[CI]	Insol.[CI]						6.0	7.5	8.5	7.5	0.0
	Darrow red	Sol.[CI]											
23565	Deltapurpurin 5B	Sol.[SP]	Sl. sol.[CI]										
24340	Dianil blue G	Sol.[CI]	Insol.[CI]	Insol.[CI]	Insol.[CI]								
	Dithizone	Sol.[CI]	Sl. sol.[M]										
24360	Enianil azurin J	Insol.[M]	Sol.[L]										
45400	Eosin B	Sol.[L]	V. sol.[M]				39.11	0.75	10.6	3.0	6.75	7.0	0.0
45380	Eosin (Y, G)	V. sol.[M]	2[DR]				44.20	2.18	44.0	2.0	25.0	27.5	0.0
14940	Erika B	44[DR]					0.64	0.17					
42576	Eriocyanin A	Sol.[CI]	Sol.[CI]										
45430	Erythrosin B	11[DR]	2[DR]				11.10	1.87	10.0	5.0	7.0	8.25	0.0
45386	Ethyl eosin	Sl. sol. h.[CI]	Sl. sol.[CI]				0.03	1.13	0.0	1.0	2.25	1.0	0.07
37190	Fast black K salt	4.0[BS]											
37235	Fast blue B salt	10.0[BS]											
37175	Fast blue BB salt	4.0[BS]											
37155	Fast blue RR salt	Sol.[DR]											
37220	Fast Corinth V salt	Sol.											
37210	Fast garnet GBC salt	5.0[BS]	5[CI]										
11160	Fast garnet GBC base	0.0[CI]	0.35[DR]	8[CI]	5-8[CI]	0.5[CI]							
42053	Fast green FCF	16[DR]					16.04	0.35	S4.0	S9.0	S8.0	S625	S0.0
37145	Fast red B salt	20.0[BS]											
37035	Fast red GG salt	10.0[BS]											
37150	Fast red ITR salt	6.0[BS]											
37120	Fast red RC salt	16.0[BS]											
37085	Fast red TR	20.0[BS]											
	Fast red violet LB	Sol.											

TABLE 6-12 *(continued)*

C.I. no.	Preferred name	"Colour Index" and other sources					Holmes		Gurr				
		Water	Alcohol	Acetone	Benzene	Ligroin	Water	Alcohol	Water	Alcohol	Cellosolve	Glycol	Xylene
10316	Flavianic acid	V. sol.^{NL}	V. sol.^{NL}				0.08	2.21	0.01	2.25	9.0	5.25	1.0
46065	Flavophosphine N	Sol.^{CI}	Sol.^{CI}				50.20	7.19	50.0	7.0	5.0	8.25	0.0
45350	Fluorescein	Sol.^{CI}	Sol.^{CI}										
45350	Fluorescein Na	Sol.	Sol.										
	Fluorescein isothiocyanate								1.0	8.0	4.0	8.5	0.0
42500–	Fuchsin	0.4^{DR}	8^{DR}										
42510													
51045	Gallamine blue	Sol.^{CI}							0.5	0.5	1.25	3.5	0.0
45445	Gallein	Sol. h.^{CI}	Sol. h.^{CI}						0.2	1.5	2.8	6.0	0.0
45445	Gallein Na	Sol.^{CI}											
51030	Gallocyanin	Insol.^{DR}	Sl. sol.^{DR}						0.5	1.25	4.0	4.5	0.0
42533,	Gentian violet	Sol.^{CI}	Sol.^{CI}						9.0	8.75	9.0	8.5	0.01
42555	Glyoxal-bis(2-hydroxyanil)		Sol.^{L}				0.01	0.0155					
13025	Helianthin												
60760	Helio fast rubin BBL	0.25^{BS}							10.0	10.0	9.5	10.0	0.0
75290	Hematoxylin	1.5^{DR}	30+^{DR}						0.0	0.75	1.0	1.0	0.008
73000	Indigo	Sol.^{CI}	Insol.^{CI}				1.68	0.61	1.3	0.0	4.0	2.85	0.0
73015	Indigocarmine	Sol.^{CI}	Sl. Sol.^{CI}						0.0	1.5	10.0	3.5	2.0
50400	Indulin	Sol.^{CI}	Sol.^{CI}										
42536	Iodine green	Sol.^{CI}	Sol.^{CI}										
11825	Janus black	Sol.	Sol.^{CI}										
12211	Janus blue	Sol.^{CI}					5.18	1.12					
11030	Janus green B	0.25^{BS}	Sol.^{CI}										
60760	Kernechtrot	Sol.^{BS}							10.0	10.0	8.0	8.25	0.0
	Lacmoid	Sol.^{BS}	Sol.^{BS}						5.0	5.0	6.5	7.25	
42095	Light green SF	20^{DR}	0.8^{DR}	Sl. sol.^{BS}			20.35	0.82	20.0	4.0	6.25	12.0	0.0
	Luxol Fast Blue G	Insol.^{BS}	V. sol.^{BS}										
	Luxol Fast Blue MBS	V. sl. sol.^{DR}	Sol.^{DR}							3.0	4.8	1.25	0.0
	Luxol Fast Yellow TN (SY47)		V. sol.^{CI}	Sl. sol.^{CI}									
58000	Madder						See Alizarin						
50375–b	Magdala red echt	Sol. h.^{CI}	Sol.^{CI}				7.60	7.52	10.0	8.5	5.5	7.0	0.0
42000	Malachite green	Sol.^{CI}	V. sol.^{CI}				4.37	0.16	1.0	0.0	3.0	2.25	0.005
	Martius yellow Na												
10315	Martius yellow	4.5^{DR}	0.15^{DR}				0.5	1.90					

13065	Metanil yellow	Sol.[CI]	Sol.[CI]	Sl. sol.[CI]	Sol.[CI]		5.36	1.45	5.0	1.5	3.0	15.0	0.0
74360	Methasol fast blue	V. sl. sol.[DR]	Sol.[DR]										
42780	Methyl blue	V. sol.[CI]	Sl. sol.[DR]						40.0	0.0	1.75	8.0	0.0
42585	Methyl green	Sol.[DR]	Insol.[DR]						8.0	3.0	1.8	3.5	0.0
13025	Methyl orange						0.52	0.68	0.2	0.2	1.0	2.75	0.02
42535	Methyl violet	3[DR]	15[DR]				2.93	13.21	9.0	8.25	9.0	8.5	0.01
52015	Methylene blue	3.5[DR]	1.5[DR]				3.55	1.48	9.5	6.0	6.5	10.0	0.0
52020	Methylene green	Sol.[CI]	Sl. sol.[CI]				1.46	0.12	1.5	0.1	5.15	8.5	0.3
52041	Methylene violet (Bernthsen)	Sl. sol.[BS]	Sol.[BS]						1.0	3.3	3.65	4.0	0.4
50205	Methylene violet RRA	Sol.[CI]	V. sol.[CI]				0.69	3.18	7.0	5.5	7.0	6.75	0.0
75660	Morin	0.025 alk. sol.[M]	V. sol.[M]										
44630	Naphthochrome green B	Sl. sol.	Sol.										
20470	Naphthol blue black	Sol.[CI]	Sol.[CI]	Sl. sol.[CI]					2.5	3.0	3.15	7.25	0.0
10020	Naphthol green B	Sol.[CI]							10.0	0.0	2.75	10.0	0.0
10316	Naphthol yellow S	Sol.[CI]	Sl. sol.[CI]				8.96	0.025	12.5	0.65	1.0	10.0	0.0
	β-Naphthoquinone-4-SO_3Na	Sol.[M]	Sl. sol.[M]	Sol.[M]	Insol.[M]	Insol.[M]							
26300	Naphthylamine black D	Sol.[CI]	Sol.[CI]										
15511	Narcein						10.02	0.06					
50040	Neutral red	5.5[DR]	2.5[DR]				5.64	2.45	4.0	1.8	3.75	3.0	0.0
42520	New fuchsin	V. sol.[CI]	V. sol.[CI]				1.13	3.26					
52030	New methylene blue	Sol.[CI]	Sol.[CI]				13.32	1.65					
44085	Night blue	Sol.[CI]	V. sol.[CI]						2.25	2.25	2.5	8.0	0.0
50420	Nigrosin	Sol.[CI]	Sol.[CI]						10.0	0.0	1.0	11.0	0.0
51180	Nile blue	V. sol. h.[CI]	V. sol.[CI]						6.0	5.0	5.0	5.0	0.25
	Nitro blue tetrazolium	Sol.[BS]											
60760	Nuclear fast red	0.25[BS]											
26125	Oil red O	Insol.[DR]	0.5[DR]	Sol.[CI]	V. sol.[CI]				0.0	0.5	2.75	2.5	2.5
14600	Orange I						5.17	0.61	2.4	0.5	6.15	4.5	0.008
15510	Orange II	Sol.[CI]	Sol.[CI]				11.37	0.15	3.0	0.15	2.5	8.0	0.0
13080	Orange IV						0.16	0.20	0.7	7.5	5.5	4.0	0.07
16230	Orange G	10[DR]	0.2[DR]				10.36	0.22	8.0	0.22	1.7	3.0	0.0
1242	Orcein	Sol.[BS]	Sol.[BS]						2.0	4.2	3.75	6.0	0.0
42500	Pararosanilin acetate						4.15	13.63					
42500	Pararosanilin chloride	Sol. h.[CI]	V. sol.[CI]				0.26	5.93	1.0	8.0	9.0	8.5	0.0
42051	Patent blue V	V. sol.[CI]	Sl. sol.[CI]						10.0	3.3	3.25	6.25	0.0
	Phenazine methosulfate	Sol.[BS]	Sol.[BS]										
50200	Phenosafranin	Sol.[CI]	V. sol.[CI]						6.5	5.25	7.0	6.5	0.0
45410	Phloxine B	50[DR]	9[DR]				50.90	9.02	10.5	5.0	9.0	14.5	0.0
46045	Phsphine	Sol.[DR]	Sol.[DR]						0.6	0.85	2.0	7.5	0.0
10305	Picric acip	1.2[DR]	8.0[DR]				1.18	8.96	1.2	9.0	15.0	5.0	10.0

TABLE 6-12 (continued)

C.I. no.	Preferred name	"Colour Index" and other sources					Holmes		Gurr				
		Water	Alcohol	Acetone	Benzene	Ligroin	Water	Alcohol	Water	Alcohol	Cellosolve	Glycol	Xylene
808	Pinacyanol	Sol.[M]	Sol.[M]				1.75	0.21	Sol.				
16100	Ponceau 2G	6[DR]	0.1[DR]										
16150	Ponceau 2R	Sol.[CI]	Sl. sol.[CI]				0.41	0.06	5.0	0.1	1.5	2.8	0.0
27305	Ponceau 5R (MLB)						12.98	0.01					
16290	Ponceau 6R	1.2[DR]	Insol.[DR]						1.35	1.2	1.5	4.65	0.0
27195	Ponceau S	Sol.[CI]							0.35	0.03	0.75	2.0	0.0
49000	Primulin	h. alum. sol.[CI]											
58205	Purpurin								0.0	0.7	2.15	1.0	0.08
45010	Pyronin B	Sol.[CI]	Sl. sol.[CI]				0.07	1.08	10.0	0.5	4.5	3.0	0.0
45005	Pyronin G, Y	9[DR]	Sl. sol.[CI]				8.96	0.60	9.0	0.5	1.85	4.35	0.0
	Quinacrine	2.86[M]	0.6[DR]		Insol.[M]								
806	Quinoline blue	Sol.[BS]	Sl. sol.[M]						1.0	1.5	1.5	2.0	0.0
	Resorcin fuchsin	Insol.[L]	Sol.[BS]										
10000	Resorcin green	Sl. sol.[BS]	Sol.[L]	Sl sol.[CI]									
45170	Rhodamine B	0.8[DR]	1.5[DR]				0.78	1.47	2.0	1.75	5.5	7.0	0.02
45210	Rhodamine 3G	Sol.[CI]	Sol.[CI]						1.35	5.75	5.6	7.0	0.001
	Rhodizonate Na	Sol.[M]	Insol.[M]										
	Rivanol	6.67[M]	0.9[M]										
15620	Roccellin	0.4[DR]	8[DR]				1.67	0.42	1.25	2.5	2.5	2.0	0.35
42510	Rosanilin	Sol.[CI]	V. sol.[CI]				0.39	8.16					
45440	Rose Bengal	Sol.[CI]					36.25	7.53	30.0	7.0	5.0	7.0	0.005
42510	Rosein lake acetate	Sl. sol.[M]											
	Rubeanic acid	Sol.[BS]	Sol.[M]										
75100	Saffron	Sl. sol.[BS]											
50240	Safranin O	5.5[DR]	3.5[DR]				5.45	3.41	Sol.	3.5	5.0	3.5	0.0
27925	Sirius light blue F3R	Sol.[CI]	Sol.[CI]	Insol.[CI]	Insol.[CI]				4.5	2.35	12.5	3.0	0.0
34200	Sirius light blue G	Sol.[CI]	Sol.[CI]	Insol.[CI]	Insol.[CI]				5.0	1.5	3.0	4.0	0.0
28160	Sirius red 4B	Sol.[DR]	Sl. sol.[DR]	Insol.[CI]	Insol.[CI]								
42775	Spirit blue	Insol.[CI]	Sol.[CI]				0.0	1.10					
12055	Sudan I (Sudan orange R)	Insol.[CI]	2.5[CI]	0.15–1[CI]	0.5–2[CI]	0.65[CI]	0.0	0.37	0.0	0.25	9.5	3.6	15.0
12140	Sudan II	0.0[CI]	0.01–0.05[CI]	1–8[CI]	0.02–3[CI]	0.02–1[CI]	0.0	0.39	0.0	0.25	4.15	1.25	3.0
26100	Sudan III	0.0[CI]	0.08–1[CI]	0.15–3[CI]	0.01–5[CI]	0.03–1[CI]	0.0	0.15	0.0	0.25	3.0	3.25	2.25
26105	Sudan IV	0.0[CI]	1.13[DR]	Sol.?	Sol.?			0.09	0.0	0.5	2.25	2.5	3.5
26150	Sudan blue BZL	0.0[PR]	Sol.?	Sol.?	Sol.?				0.0	0.25	9.0	1.0	2.5
	Sudan black B	Insol.?		0.9–3[CI]	0.12–0.3[CI]	0.3			0.0	4.0	7.0	3.75	5.0
12020	Sudan brown	0.0[CI]	0.2–1[CI]						0.0	2.0	60	1.0	1.5
26050	Sudan red 7B	0.0[CI]	0.08–1[CI]	2.5–3.5[CI]	3–10[CI]				0.0	4.35	4.0	4.0	10.45

CI No.	Name										
26400	Sulfone cyanin GR	Sol.[CI]	Sol.[CI]	Insol.[CI]				4.5	5.0	10.0	10.0	0.0
19140	Tartrazine	11 v. sol.[M] 1.09[M]	0.1[DR] 0.79[M]	Insol.[CI] 0.025[M]	0.027[M]			6.0	0.0	2.5	8.0	0.0
	Tetracycline	V. sol.[CI]	V. sol.[CI]									
49540	Thiazol yellow G	Sol.[DR]	Sol.[DR] 0.25[DR]	Insol.[CI]		2.11	3.17	1.5	1.0	6.0	3.8	0.0
49005	Thioflavine TCN	0.25[DR]	0.25[DR]			0.25	0.25	2.0	1.0	1.25	8.75	0.0
52000	Thionin	3.8[DR]	0.5[DR]			3.82	0.57	1.0	1.0	10.0	3.5	0.0
52040	Toluidine blue					0.37	0.10	3.25	1.75	3.5	5.5	0.0
14270	Tropaeolin O	Sol.[CI]	Insol.[CI] Sol.[L]	Insol.[CI]				3.0	3.0	2.75	9.25	0.25
23850	Trypan blue	Sol.[CI]	Sol.[L]					1.0	0.02	2.10	7.15	0.0
22850	Trypan red	Sol.[CI]	Sol.[CI]					2.5	0.0	0.85	14.5	0.0
44045	Victoria blue B	0.5[DR]	4[DR]			0.54	3.98	4.2	8.25	6.75	6.25	0.0
44040	Victoria blue R	Sol.[CI]	V. sol.[CI]			3.23	20.49	0.6	4.25	9.0	2.5	0.0
42563	Victoria blue 4R	Sol.[CI]	V. sol.[CI]			1.65	1.18	3.0	20.0	7.0	7.0	0.0
10310	Victoria yellow											
45190	Violamine R	Sol.[CI]	Sol.[CI]	Sl. sol.[CI]				4.5	3.25	4.0	11.5	0.0
23570	Vital red	Sol.[CI]	Sl. sol.[CI]					3.75	0.17	4.0	2.3	0.0
42755	Water blue I	Sol.[CI]	Sl. sol.[CI]					50	0.0	4.5	7.8	0.0
44090	Wool green S	Sol.[CI]	Sol.[CI]					4.0	0.0	4.25	7.25	0.0
13015	Wool yellow G							10.0	1.5	1.25	4.0	0.0
	Zincon	Sol.[L]				18.40	0.24					

Source: Data are derived from the "Colour Index" (2d ed., Society of Dyers and Colourists, Bradford, Yorkshire, England, and American Association of Textile Chemists and Colorists, Lowell, Mass., 1956; Supplement, 1963) (CI), from the table of Disbrey and Rack ("Histological Laboratory Methods," E. & S. Livingstone Ltd., Edinburgh and London, 1970) (DR), from "The Merck Index" (8th ed., P. G. Stecher (ed.), Merck, Rahway, N.J., 1968) (M), from Lillie (3d ed., 1969, and his laboratory notes) (L), from "Conn's Biological Stains," 8th ed. (1968), and earlier (7th ed., Williams & Wilkins, Baltimore, 1961; 4th, 5th, and 6th eds., Biotech, Geneva, N.Y., 1940, 1946, 1953) (BS), from "Staining Procedures (Conn and Darrow, 3d ed., Clark, 1973) (SP), from "Handbook of Chemistry," (N. H. Lange, 9th and 10th eds., McGraw-Hill, New York, 1956, 1961; rev. 10th ed., 1967) (NL), from Holmes (*Stain Technol.*, 2:68, 1927, 3:12, 1928, and 4:73, 1929), and from "Gurr's Encyclopaedia of Microcopic Stains," 1960.

The dyes chosen for inclusion in the table include most of those actually used by Lillie (1965), Mallory ("Pathological Technic," Saunders, Philadelphia, 1938), Lee ("The Microtomist's Vade-Mecum," 9th ed., Blakiston, Philadelphia, 1928, and Gatenby and Cowdry (Gatenby and Beams (eds.), Bolles Lee's "The Microtomists' Vade Mecum," Blakiston, Philadelphia, 1950).

Sources of data under "Colour Index" and Other Sources were indicated by superscript abbreviations corresponding to those given in parentheses in this footnote.

The list is not meant to be encyclopedic; a number of dyes at first included in this list were omitted because no solubility data were found in the sources listed here.

TABLE 6-13 NAMES, "COLOUR INDEX" NUMBERS, SYNONYMS, CLASS, REACTIVE GROUP, AND TEXTILE DYEING WITH REACTIVE DYES

C.I. numbers		Names from "Colour Index," Vol. 5, 1972	Dye class†	Reactive group‡	Textile dyeing§
Part I*	Part II				
RR8	17908	Procion Scarlet M-G; Mikacion Scarlet MGS	Azo	Cl-2	CW
RR9	17910	Procion Scarlet H-3G; Cibacron Scarlet 2G	Azo	Cl-1	C
RR7	17912	Procion Rubine H-2B; Cibacron Rubine RE	Azo	Cl-1	CNS
RN1	17916	Procion Black H-G; Cibacron Black FBG-A	Azo	Cl-1	CNS
RR6	17965	Procion Rubine MB; Mikacion Rubine MB	Azo	Cl-2	CNSW
RR4	18105	Procion Brilliant Red H-7B; Cibacron Brilliant Red 3BA	Azo	Cl-1	CSW
RR17	18155	Drimarene Red Z-2B(S); Reactional Red 2B (Gy)	Azo	TCP	CW
RR12	18156	Cibacron Brilliant Red BA	Azo	Cl-1	CW
RV2	18157	Procion Violet H-2R; Cibacron Violet 2RA, 2RF	Azo	Cl-1	CNSW
RR1	18158	Procion Brilliant Red M-2B; Mikacion Brilliant Red 2BS	Azo	Cl-2	CNSW
RR3	18159	Procion Brilliant Red H-3B; Ostazin Brilliant Red H-3B	Azo	Cl-1	CN
RB88	18205	Helaktyn Red F-4BAN	Azo	Cl-2	CNW
RB81	18245	Helaktyn Blue F-2R	Azo	Cl-1	CW
RF1	26440	Procion Orange Brown HG; Cibacron Brown 3GR-A	Disazo	Cl-1	C
RB19	61200	Remazol Brilliant Blue R; Remalan Brilliant Blue R(FH); Ostazin Brilliant Blue VR; (Chem); Diamira Brilliant Blue R (MCI); Primazin Brilliant Blue RL (BASF); Sunfix Brilliant Blue R (NSK)	Anthraquinone	VS	CW
RB4	61205	Procion Brilliant Blue M-R; Mikacion Brilliant Blue RS Ostazin Brilliant Blue SR	Anthraquinone	Cl-2	CNSW
RB5	61210	Procion Brilliant Blue H-GR; Cibacron Brilliant Blue FBR-P; Ostazin Brilliant Blue H-BR	Anthraquinone	Cl-1	CW
RB2	61211	Procion Blue H-B; Cibacron Blue F-3GA	Anthraquinone	Cl-1	CNSW
RB7	74460	Procion Turquoise H-G; Cibacron Turquoise Blue GE	Phthalocyanin	Cl-1	C

* In the Part I C.I. numbers, R stands for reactive, the second letter R = red, N = black (*niger*), V = violet, B = blue, F = brown (*fuscus*).

† Azo stands for monoazo.

‡ Cl-2 = dichlorotriazinyl; Cl-1 = monochlorotriazinyl; TCP = trichloropyrimidinyl; VS = vinyl sulfonyl.

§ C stands for cellulose or cotton, W for wool, N for nylon, S for silk.

$C_{19}H_9N_6O_{11}S_3Na_3Cl_2$, mw 733.385
Procion Rubine MB, C.I. 17965
C.I. reactive red 6; synonym, Mikacion Rubine MB

$C_{32}H_{18}N_8O_{14}S_4Na_4Cl$, mw 994.214
Procion Brilliant Red H-7B, C.I. 18105
C.I. reactive red 4; synonym, Cibacron Brilliant Red 3BA

$C_{19}H_9N_5O_{10}S_3Na_3Cl$, mw 738.830
Drimarene Red Z-2B(S), C.I. 18155
C.I. reactive red 17; synonym, Reactional Red 2B (Gy)

$C_{19}H_{11}N_7O_{10}S_3Na_3Cl$, mw 697.955
Cibacron Brilliant Red BA, C.I. 18156
C.I. reactive red 12

$C_{20}H_{12}N_5O_8S_2Na_2Cl_2$, mw 631.371
Procion Scarlet M-G, C.I. 17908
C.I. reactive red 8; synonym, Mikacion Scarlet MGS

$C_{27}H_{18}N_8O_{11}S_3Na_3Cl$, mw 831.106
Procion Scarlet H-3G, C.I. 17910
C.I. reactive red 9; synonym, Cibacron Scarlet 2G

$C_{27}H_{16}N_8O_9S_2Na_2ClCu$, mw 805.593
Procion Rubine H-2B, C.I. 17912
C.I. reactive red 7; synonym, Cibacron Rubine RE

$C_{23}H_{10}N_8O_{10}S_2Na_2Cl + Cr \cdot Co \cdot C_6H_6O_3$
mw 703.944+ probably around 951.5.
Procion Black H-G, C.I. 17916
C.I. reactive black 1; synonym, Cibacron Black FBG-A

$C_{20}H_{11}N_6O_{10}S_3Na_3Cl_2$, mw 731.412
Helaktyn Red F-4BAN, C.I. 18205
C.I. reactive red 88

R = mixture C_2H_2O- and $-NH_2$
Procion Violet H-2R, C.I. 18157
C.I. reactive violet 2; synonym, Cibacron Violet 2RA, 2RF

$C_{19}H_9N_6O_{10}S_3Na_3Cl_2$, mw 717.385
Procion Brilliant Red M-2B, C.I. 18158
C.I. reactive red 1; synonym, Mikacron Brilliant Red 2BS

$C_{25}H_{14}N_7O_{10}S_3Na_3Cl_2$, mw 808.999
Helaktyn Blue F-2R, C.I. 18245
C.I. reactive blue 81

$C_{25}H_{14}N_7O_{10}S_3Na_3Cl$, mw 773.046
Procion Brilliant Red H-3B, C.I. 18159
C.I. reactive red 3; synonym, Ostazin Brilliant Red H-3B

$C_{31}H_{18}N_9O_9S_3Na_3Cl$, mw 861.158
Procion Orange Brown HG, C.I. 26440
C.I. reactive brown 1; synonym, Cibacron Brown 3GR-A

$C_{32}H_{16}N_2O_{12}S_3Na_2$, mw 642.551
Remazol Brilliant Blue R, C.I. 61200
C.I. reactive blue 19; synonyms: Remalan Brilliant Blue R (FH),
Ostazin Brilliant Blue VR (chem), Diamira Brilliant Blue R (MCI),
Primazin Brilliant Blue RL (BASF), Sumifix Brilliant Blue R (NSK)

$C_{29}H_{20}N_2O_{11}S_3Cl$, mw 774.163
Procion Brilliant Blue H-GR, C.I. 61210
C.I. reactive blue 5; synonyms, Cibacron Brilliant Blue FBR-P
Ostazin Brilliant Blue H-BR

$C_{29}H_{20}N_7O_{11}S_3Cl$, mw 774.174
Procion Blue H-B,[1] C.I. 61211
C.I. reactive blue 2; synonym, Cibacron Blue F-3GA

$C_{23}H_{14}N_6O_8S_2Cl_2$, mw 637.437
Procion Brilliant Blue M-R, C.I. 61205
C.I. reactive blue 4; synonyms, Mikacion Brilliant Blue RS,
Ostazin Brilliant Blue SR

Copper Phthalocyanin

Procion Turquoise H-G, C.I. 74460
C.I. reactive blue 7; synonym, Cibacron Turquoise Blue GE

[1] Used by Marshall (*Stain Technol.*, **41**: 68, 1966) for mucin stains.

Reports on the use of reactive dyes in histology of pathology have been rare. Goland's papers (*J. Histochem. Cytochem.*, **11:** 757, 1963; *Stain Technol.*, **42:** 41, 1967) were mainly concerned with the action of cyanuric chloride and its dye derivatives as cross-linking tissue fixatives, and the dyes which he mentions are now difficult to identify. Marshall (*Stain Technol.*, **41:** 68, 1966) reported on the use of two dichloro- and two monochloro-triazinyl dyes, reactive reds 1 and 3, C.I. 18158 and 18159, and reactive blues 2 and 4, C.I. 61211 and 61205, finding that they stained mucopolysaccharides with open —OH groups, but not those which were sulfated. Henry (*Stain Technol.*, **43:** 297, 1968) reported in vivo staining of mucins and collagen from feeding or injection of Procion Navy Blue M-3RS (C.I. reactive blue 9) whose composition is still secret as of 1971. Collagen remained blue in the living injected rodents for several months.

Mucins, mucoproteins, and collagen are all heavily hydroxylated tissues and would be expected to react with reagents which possess the same type of reactive capacity as benzoylchloride.

Sivitz et al. (*J. Histochem. Cytochem.*, **21:** 87, 1973) reported that Procion Yellow M-4RS [C.I. reactive orange 14, constitution not revealed, a dichlorotriazinyl monoazo (pyrazolone) compound] did not interfere with development of the formaldehyde fluorescence of dopamine or noradrenaline, and was itself fluorescent, containing four chromatographic species.

7

NUCLEI, NUCLEIC ACIDS, GENERAL OVERSIGHT STAINS

Nuclear chromatin, which is the substance demonstrated when nuclei are stained in histologic sections, is composed of the two nucleoproteins. Deoxyribonucleoprotein, which occurs only in cell nuclei, is the major component of the chromosomes, replicates itself during the intermitotic-mitotic cycle, and also serves in the synthesis of ribonucleoprotein, found in chromosomes, nucleoli, and cytoplasmic granules. This second nucleoprotein is present in considerable amount wherever active synthesis of other proteins is being carried out.

Deoxyribonucleic acid (abbreviated DNA or formerly PNA in English, ADN in French, and DNS in German) is composed of units of 2-deoxyribose condensed in successive units with the purines adenine and guanine and the pyrimidines thymine and cytosine. See Figs. 7-1 and 7-2. Small amounts of 5-methylcytosine may be present as well, and in some micro-organisms, notably *Mycobacterium tuberculosis*, this is an important constituent. The purines are linked to the sugars by a condensation of the imidazole-imide nitrogen condensing with the sugar-aldehyde group with elimination of water as a type of Schiff base. The pyrimidines bind by a similar condensation of one of the imide groups present in the ketol tautomer. These nucleoside units are linked by 3,5-pentose ester linkages by moles of phosphoric acid, whose third acid group serves to bind the nucleic acid to a basic protein. Ribonucleic acid (RNA) is similarly constituted, except that the sugar is D-ribose and the pyrimidine uracil replaces the thymine of the deoxyribonucleic acid. In actively growing or protein-synthesizing cells at least, there is fairly rapid turnover of purines, pyrimidines, phosphorus, and sugars. This turnover has been utilized for selective radioautographic labeling of deoxyribonucleic acid with tritium-tagged thymine, or of ribonucleic acid with ^{3}H-uracil. ^{14}C-labeled adenine, ^{35}P-labeled phosphate, and other labeled compounds have also been used.

Deoxyribose, when hydrolyzed from its purine (and pyrimidine) linkages, is promptly Schiff-positive, like other 2-deoxyaldoses. Ribose does not react promptly with Schiff reagent, nor do glucose, mannose, and other ordinary aldoses. This reactivity forms the basis of the Feulgen reaction, which is regarded as highly specific for deoxyribonucleic acid.

The phosphoric acid residues, which are ordinarily bound to nuclear histones by salt linkages, are apparently readily available for the binding of basic (cationic) aniline dyes, and since phosphoric acid is a fairly strong acid, nuclei stain with azure A, methylene blue, safranin, and the like, even from quite weak solutions, at pH levels as low as 2, and with stronger solutions down to pH 1. Only sulfate esters of mucopolysaccharides and oxidatively created cysteic acid residues stain at lower pH levels. It is characteristic of salt-linkage basic dye staining that stained preparations are again readily decolorized by exposure to dilute mineral acids in aqueous or alcoholic solution, and by certain salt solutions.

165

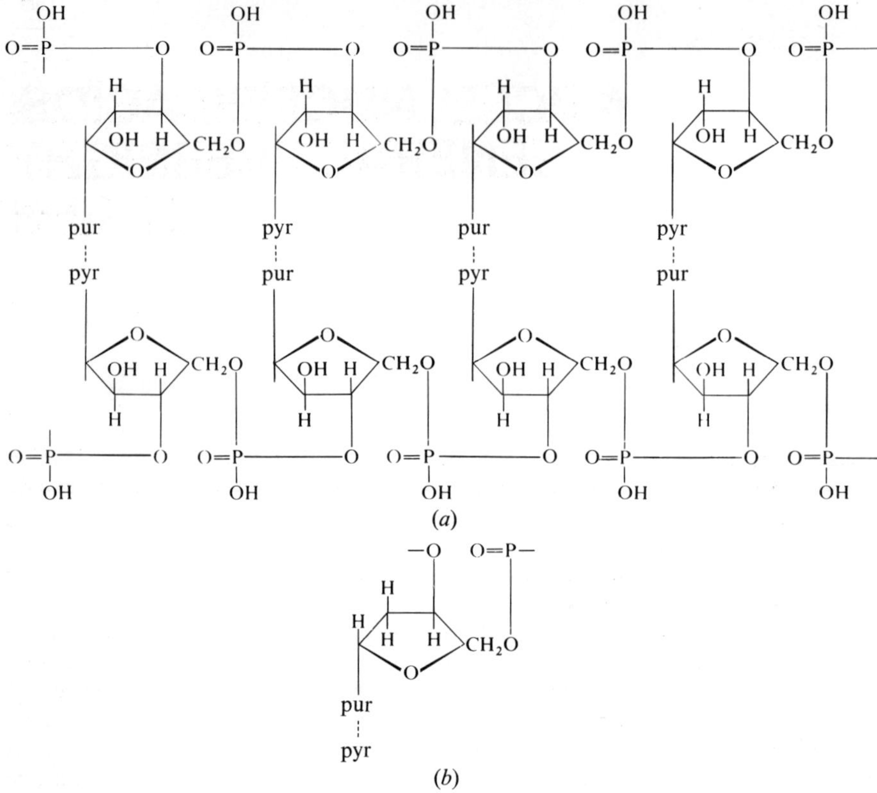

Fig. 7-1 (a) Segment of double helix of deoxyribonucleic acid. Dashed lines indicate interchain purine-pyrimidine hydrogen bonding. H^+ indicates phosphoric acid residues for bonding to basic proteins or cationic dyes. (b) Larger-scale ring for detail.

Certain basic dyes with free NH_2 groups, notably rosanilin and pararosanilin, are capable of irreversibly staining nuclei from warm (60°C) alcoholic (70%) hydrochloric acid (0.1 to 0.2 N). The mechanism of this staining has not been conclusively worked out, but it is believed to result from hydrolysis of the purine (and pyrimidine) bonds with the aldoses and condensation of the liberated aldehydes with the dye amino groups to form Schiff bases of the secondary amine or "diphenamine" type (*J. Histochem. Cytochem.*, **10**: 303, 1962). This coloration is acid-fast and probably is the type occurring in those acid-fast stains for tubercle bacilli which result in failure of nuclear decolorization. Since this diphenamine condensation with aldehyde and the hydrolytic liberation of aldehyde itself both occur more rapidly in acid than in neutral solutions, it follows that for differential staining of acid-fast organisms and lipids, strongly acid dye solutions probably should be avoided.

Mordant dyeing with alum hematoxylin, posthydrolytic staining with pure dilute hematoxylin, and azo coupling of nuclei (Pearse) persist after salt-type basic dye staining and the Feulgen reaction have been abolished by a 1-h or more hydrolysis at 60°C in 1 N hydrochloric acid; hence these stainings are to be assigned to the basic proteins or histone fraction of the nucleus. Staining of nuclei by acid ("anionic") dyes, with or without prior

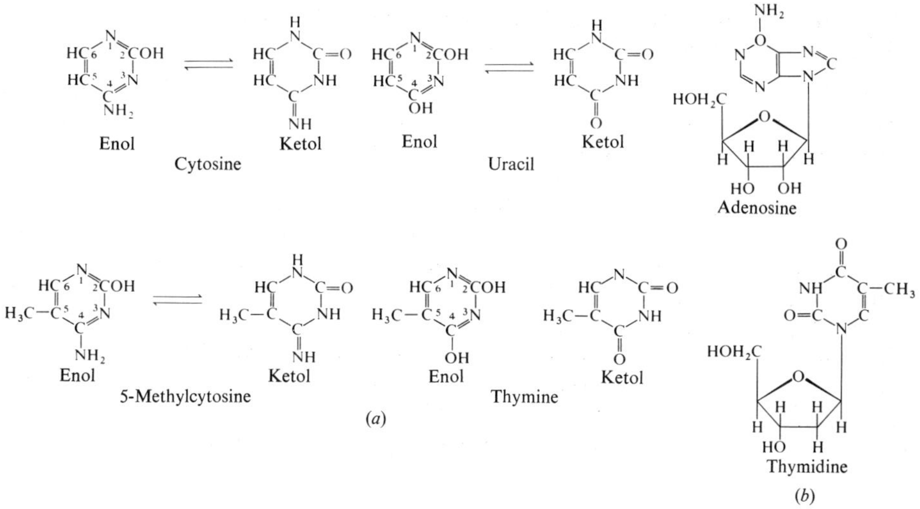

Fig. 7-2 The pyrimidines are given in enol and ketol forms. The purines consist of a pyrimidine ring fused at 4 and 5 with the 4 and 5 positions of an imidazole ring. Two nucleosides, adenosine and thymidine, are shown to illustrate the sugar linkage.

methylation to esterify the free phosphoric acid residues, also depends on histone, and specific arginine reactions can be similarly employed to demonstrate the histone. The last is well shown in spermatozoa.

In practice it is often desirable, for morphologic detail, to combine a nuclear stain with a second stain designed to demonstrate some other tissue element or elements, such as basic proteins, connective tissue fibers, mucopolysaccharides, or other elements, and technics are given below exemplifying this practice. In addition nuclear stains are often added when the primary objective is the demonstration of other tissue constituents. Such technics appear in other chapters.

NUCLEIC ACIDS

The Caspersson school has used extensively the ultraviolet absorption of the nucleic acids at 260 mμ as a means of localizing these substances. Caspersson (*J. R. Microsc. Soc.*, **60:** 8, 1940) used a rather complicated apparatus for photometric estimation of ultraviolet absorption in quite small field areas. The estimates obtained by photography were quantitatively cruder, though qualitatively more accurate. In its applicability to living cells the method has certain special values. Mellors (*Science*, **111:** 627, 1950) found that living tissue cultures would tolerate as many as 30 photographic exposures at 260 mμ before being killed. This 260-mμ absorption band is attributed to the presence of the pyrimidine ring in both purines and pyrimidines of the nucleosides. The apparatus is expensive, whether for ultraviolet photography alone or for photometry as well, and seems adapted to highly specialized studies. For most purposes the histochemical procedures for the identification of nucleic acids are easier, more flexible, and more differential.

Enzymatic Digestion and Acid Extraction Procedures for Selective Removal of the Nucleic Acids

DEOXYRIBONUCLEASE DIGESTION. Kurnick (*Stain Technol.*, **27**: 233, 1952) used a preparation of DNase (obtained from the Worthington Laboratories, Freehold, New Jersey) at a concentration of 2 mg/100 ml, in 0.01 M tris buffer, pH 7.6 (Gomori's, diluted 1 : 5 with distilled water). On material fixed in 80% alcohol or Clarke-Carnoy fluid, digestion periods of 2 h at 37°C or 24 h at 21°C were recommended. Controls treated with buffer alone for the same period and unextracted controls should be used. Kurnick used his variant of the methyl green pyronin technic as the demonstration method; but see the method of Love and Rabotti, described below.

The definitive test for the histochemical identification of ribonucleic acid is its digestion by the enzyme ribonuclease. This is ordinarily prepared from pancreas. Brachet used a boiled acid extract of pancreas. Ribonuclease is thermostable in acid. Most workers have preferred the crystalline substance isolated by the method of M. Kunitz (*J. Gen. Physiol.*, **24**: 15, 1941) and now available commercially. The commercial enzyme has proved active in our hands at a 1 : 100,000,000 dilution. It functions well in 0.8% saline solution buffered with phosphates to pH 6. We prefer this pH level to the 6.75–7 levels of other workers because of the decreased section losses at the more acid level. We have found that commercial barley malt diastase also contains a thermostable ribonuclease (Lillie, *Anat. Rec.*, **103**: 611, 1949).

This digestion, at least in the case of the malt diastase ribonuclease, is largely inhibited by fixation in Kose's, Orth's, and Möller's dichromate formalin fluids, less so by Zenker's fluid fixation. With formalin fixation the digestibility is influenced both by the duration of fixation and by the diluent. Longer aqueous formalin fixations make tigroid and ribonucleic acids generally more resistant. Brief alcoholic formalin fixations facilitate this ribonuclease activity of the diastase. Still better is the Clarke-Carnoy fluid. Bouin's fluid fixation almost entirely prevents barley malt ribonuclease digestion of ribonucleic acids, but Gendre's alcoholic picroformalin acetic fixation allows fairly prompt digestion. Pancreatic ribonuclease does not appear to affect the basophilia of cartilage matrix or nucleus pulposus, in contrast to the crude barley malt diastase.

TECHNIC OF RIBONUCLEASE DIGESTION OF TIGROID AND CYTOPLASMIC RIBONUCLEIC ACID

1. Fix in Clarke-Carnoy fluid at 0°C or in acetic alcohol formalin for 18 to 24 h. Wash in 95% alcohol over calcium carbonate. Complete the dehydration, dealcoholization, and paraffin infiltration as usual.

2. Bring paraffin sections through xylene and alcohols to water. Do *not* collodionize. Collodion films prevent enzyme action almost completely.

3. Immerse for 1 h or more at 37°C (or 50°C for more drastic action; the enzyme is destroyed at 60°C) in a buffered solution. Use simultaneous digestion controls in the solvent without enzyme.

    The solvent contains 8 g sodium chloride, 0.28 g anhydrous disodium phosphate, and 1.97 g sodium acid phosphate monohydrate per 1 to give pH 6. This pH level has less detergent effect on sections than pH 7 and allows effective enzyme action. Commercial ribonuclease, obtained from Armour in 100-mg lots, is quite effective on cytoplasmic ribonucleic acid and on tigroid at 1 : 1,000,000 but is ineffective on glycogen at 1 : 5000. A concentration of 1 : 100,000 seems appropriate for testing purposes. The ribonuclease activity of malt diastase is evident to about 1 : 3000 dilution; its glycogenolytic action, to about 1 : 100,000. Hence a 1 : 1000 dilution is used for both purposes as a matter of routine.

4. Rinse in water and counterstain by azure-eosin, by thionin, or by the other procedures which effectively demonstrate the tissue elements under consideration in undigested controls.

For precise work, after determination of the fact of digestibility under the conditions of the test, it is desirable to determine the minimum enzyme concentration which is effective in a set period, such as 1 h at 37°C. This figure may be used for rough comparison with other sets of conditions or other objects.

Some enzyme preparations may contain also small amounts of a deoxyribonuclease (or protease?), whose action is manifest after 8- to 16-h digestion periods. This activity (as well as amylase activity) may be selectively destroyed by adding 0.57 ml glacial acetic acid per 100 ml solution in distilled water (0.1 M) and boiling for 5 min. Then cool and neutralize with 1 N sodium hydroxide to about pH 6.2, using 1 to 2 drops 1% aurin (rosolic acid) or nitrazine as an indicator, or adjusting with the pH meter. Since the solution now contains approximately 0.1 M sodium acetate (0.82%), addition of sodium chloride is unnecessary and has been found deleterious under some circumstances.

Leuchtenberger and Lund (*Exp. Cell Res.*, **2**: 150, 1951) record that the keratohyalin granules of dog epidermis are Feulgen negative and lose their basophilia to toluidine blue or azure A on digestion with ribonuclease. With neutral stains these granules are oxyphil (arginine content).

As noted later, human saliva also exhibits ribonuclease activity. Its action on tigroid was noted by Szent-Györgyi in 1931 [*Nature (Lond.)*, **128**: 761], who interpreted the action as indicating a polysaccharide nature of the tigroid.

For nuclease digestions Love and Rabotti (*J. Histochem. Cytochem.*, **11**: 603, 1963) used crystallized deoxyribonuclease (Worthington) at 1 mg/100 ml and ribonuclease (Worthington, 2 times or 5 times crystallized) at 10 mg/100 ml, both in 0.02 M tris buffer adjusted to pH 7.3 with 1 N HCl, containing 45 mM MgCl$_2$ and 5 mM CaCl$_2$. Digestions were for 2 h at 37°C. For the differential effects of digestion in much weaker solutions, 0.6 to 6 μg/100 ml deoxyribonuclease and 6 μg/100 ml ribonuclease may be used.

The above concentration was the same as that recommended for Armour ribonuclease in the 1954 edition of this book, noting that it was still effective at 100 μg/100 ml but not noting the effect of a further decimal dilution to 10 μg/100 ml. In 1953 the effects recorded in 1963 by Love and Rabotti would have been regarded simply as incomplete digestion, so it appears that the additional crystallizations have operated to purify the effect, but not especially to increase the potency of the enzyme.

TABLE 7-1 SOLUTION FORMULAS FOR NUCLEASES ACCORDING TO LOVE AND RABOTTI

Deoxyribonuclease (crystallized)	0.4 mg		1 mg	
Ribonuclease (2× or 5× crystallized)		4 mg		10 mg
0.2 M magnesium chloride (4.066% MgCl$_2$·6H$_2$O)	9 ml	9 ml	22.5 ml	22.5 ml
0.2 M calcium chloride (2.22% CaCl$_2$)	1 ml	1 ml	2.5 ml	2.5 ml
0.1 M tris buffer pH 7.3	8 ml	8 ml	20 ml	20 ml
Distilled water	22 ml	22 ml	55 ml	55 ml
Total volume	40 ml	40 ml	100 ml	100 ml

PERCHLORIC ACID EXTRACTION

It has long been known that the acids used in decalcification tend to destroy cytoplasmic basophilia and to impair nuclear staining. A number of workers have used perchloric acid solution for the differential extraction of ribonucleic acid: Ogur and Rosen (*Fed. Proc.*, **8:** 234, 1949), Seshachar (*Science*, **110:** 659, 1949), Sulkin and Kuntz (*Proc. Soc. Exp. Biol. Med.*, **73:** 413, 1950), Di Stefano (*Science*, **115:** 316, 1952).

Koenig and Stahlecker (*J. Natl. Cancer Inst.*, **12:** 237, 1951) report differential removal of ribonucleic acid from nerve cells by 10% aqueous perchloric acid in 15 min at 37°C, 2 h at 25°C, 6 h at 23°C, 12 h at 20°C, and failure to extract in 4 days at 4°C. Liver cell ribonucleic acid was easier to remove, requiring 18 h at 4°C and only 1 h at 23°C. As with ribonuclease, prolonged formalin fixation greatly retards solution of ribonucleic acid in perchloric acid.

Extraction in hot perchloric acid (90°C) also removes deoxyribonucleic acid. Most other proteins are dissolved only to a minor extent; so the method is fairly differential.

It may be noted that after brief alcohol fixation, cytoplasmic basophilia of blood lymphocytes (in smears) may be removed by extraction for 1 h at 58°C in distilled water or in 0.9% sodium chloride solution, and is impaired by even 10 min extraction at this temperature.

Although the specificity of the extraction has not been determined, it is noted that brief treatment with alcoholic potassium hydroxide (1% in 80% alcohol, 20 min) differentially destroys cytoplasmic basophilia of formaldehyde-fixed tissue, leaving nuclear staining unimpaired.

As noted before, other acids share this property of removing ribonucleic acid. Fisher (*Stain Technol.*, **28:** 9, 1953) showed that 2 M dilutions of nitric, hydrochloric, or sulfuric acid at 0 to 5°C would differentially remove cytoplasmic basophilia of pancreas, various epithelia, gastric chief cells, and Nissl substance of nerve cells in 16 to 24 h. Extraction with stronger acid at the same temperature or with the same concentrations at 25°C or 37°C was less differential; a good deal of nuclear basophilia was extracted as well.

These extractions induce sufficient hydrolysis of deoxyribonucleic acid to yield a Feulgen reaction with sulfurous acid leucofuchsin but under the recommended conditions do not appreciably impair nuclear staining with azure A, even when the extraction at 0 to 5°C is prolonged to 32 h.

The metachromasia of mast cells and of pyloric gland mucin, as well as the periodic acid Schiff reaction, were unimpaired by these extractions.

Atkinson (*Science*, **116:** 303, 1952) showed the parallelism in action between normal perchloric and hydrochloric acids, and pointed out that both reagents also extracted mucus, even at 5°C. Only toluidine blue stains were mentioned. The basophilia of mast cells and of cartilage was almost as resistant to these extractions as was that of cell nuclei.

Kasten (*Stain Technol.*, **40:** 127, 1965) fixed KB human carcinoma culture films 15 to 30 min in Clarke-Carnoy 3 : 1 ethanol–acetic acid and extracted with 10% $HClO_4$ at 4°C for 1, 3, 6, 9, 12, 24, and 30 h. Controls were kept in water 30 h.

Acridine orange, chrome alum gallocyanin, Feulgen, and fluorochrome Feulgen stains were used. Cytoplasmic ribonucleic acid was removed completely in 6 h; nucleolar required 12 h for complete removal. Early in the extraction large intranuclear granules appeared which were basophilic to azure B, stained by chrome alum gallocyanin, and were Feulgen-positive by leucofuchsin and fluorochrome methods. The same granules stained for protein by mercuric bromphenol blue and by alkaline Biebrich scarlet. At 24 h there was 10 to 20% loss of deoxyribonucleic acid by Feulgen photometry. A progressive loss of nuclear and cytoplasmic but not nucleolar protein occurred, and concurrently large protein granules appeared in cytoplasm. Glycogen was gradually lost and redistributed in the cells (periodic acid–fluorochrome Schiff). Lipids were unaffected (Sudan black B).

At the conclusion of their graded extraction period in $HClO_4$ coverslip films were transferred to distilled water to complete the 30 h. It is not specifically stated that films were washed before transfer to the distilled water. The acridine orange technic gave green deoxyribonucleic and red ribonucleic acid fluorescence.

Di Stefano (*J. Natl. Cancer Inst.*, **12**: 238, 1951) stated that 12-h $HClO_4$ extraction removed all purine bases from deoxyribonucleic acid, thus creating optimal Feulgen hydrolysis. He used Carnoy material. Gössner's results and Kasten's agree.

The conclusion is that crystalline ribonuclease is the reagent of choice for selective removal of ribonucleic acid.

SPECIFIC STAINING METHODS

Deoxyribonucleic (Thymonucleic) Acid

The Feulgen reaction is now generally considered specific for deoxyribonucleic acid. It depends on acid hydrolysis of the nucleic acid, which by liberating first the purines (and then the pyrimidines) from the deoxyribose phosphoric acid complexes leaves a reactive aldehyde group free on the latter. Although a similar liberation undoubtedly also occurs with ribonucleic acid, it is an observed fact that normal aldoses such as glucose, ribose, xylose, lyxose, and the like do not form the deep red-purple color complex with Schiff's sulfite leucofuchsin reagent, except perhaps very slowly, and that 2-deoxy aldoses, as well as other aldose sugars in which C-2 does not present a hydroxyl group, react promptly. Ketone also reacts, but more weakly and slowly. Hence it would appear that the application of Schiff reagent to sections should not be unduly prolonged.

Hydrochloric acid hydrolysis of nucleic acids is not instantaneous. It is observed that longer hydrolysis may give a more intense reaction, and that further prolongation of the treatment beyond the optimum weakens and finally completely destroys it. This last is sometimes observed when bone is treated with mineral acids for an unduly long period. For routine tissues hydrolysis with normal hydrochloric acid for 10 min gives a more intense reaction than it does in 5 min at 60°C. One minute baths in 1 N HCl at 25°C preceding and following the 60°C bath are frequently prescribed. We fail to discern any advantage in these steps and have omitted them for some time as a matter of routine.

The optimum hydrolysis time for a maximal Feulgen reaction varies considerably with the object under study and the method of fixation employed. For the best results the time should be determined experimentally for each object studied.

The reaction of the hydrolyzed nuclear material with the Schiff reagent occurs quite promptly and appears to be as strong in 2 min as it is with the traditional 2-h treatment. Use of the shorter period would appear to lessen the chance of confusing the slower reactions of ketones and normal aldoses.

The precise fuchsin content and method of manufacture of the Schiff reagent appear to have little influence. The same Schiff reagent as used for other purposes seems quite satisfactory for the Feulgen test.

THE FEULGEN REACTION. This reaction for formalin-fixed material is as follows:

1. Bring paraffin sections through xylene and alcohols to water as usual (with the usual iodine thiosulfate sequence for removal of mercurial precipitates if required).
2. Place in normal hydrochloric acid (preheated) at 60°C for 10 to 15 min (15 min with formol fixation).

3. Immerse in Schiff reagent for 10 min.
4. Wash in three successive baths, for 2 min each, of about 0.05 M (0.5%) $Na_2S_2O_5$ (sodium metabisulfite). The sulfite baths should be discarded daily.
5. Wash 5 min in running water.
6. Counterstain a few seconds in 0.01% fast green FCF in 95% alcohol. The stain does not wash out in alcohol, but if it is too intense, it may be removed promptly in water. Many workers prefer to omit this step.
7. Complete the dehydration with 100% alcohol; clear through one change of alcohol and xylene (50:50) and two of xylene. Mount in polystyrene, ester gum, Permount, HSR, or other synthetic resin or in balsam.

RESULTS: Nuclear chromatin is a deep red-purple. The chromatin of plasmodia, sarcosporidia, toxoplasmata, histoplasmata, and some yeasts is Feulgen-positive, though often paler red than that of host cells. Feulgen-positive globules are seen in *Cryptococcus neoformans* in some specimens and not in others. Similarly, small red bodies are seen in endosporulating coccidioides. No definite Feulgen-positive material was found in vegetative forms of this fungus or of *Haplosporangium parvum*. Mold mycelia and gram-positive and gram-negative bacteria generally fail to stain. Typhus rickettsiae in yolk sac material are Feulgen-negative; and *Pasteurella tularensis* and *Chlamydia psittaci* bodies usually are Feulgen-negative, though we have seen occasional clusters of each stain red. Trophonuclei of *Trypanosoma cruzi* in its leishmania forms in heart muscle stain as deep purplish red rings. In some clusters blepharoplasts are unstained; in others they appear as deeper red-purple rods. In vegetative trypanosomes in the blood (*T. brucei, T. equiperdum*) trophonuclei may stain with difficulty, and blepharoplasts may fail to stain. In encephalitozoa, chromatin varies from vague, poorly stained pink granules to fairly definite, oval and round, small red-pink vesicles. Many coccidia (*Eimeria*) lack Feulgen-positive material, but some young intraepithelial forms present more or less numerous small red rings apparently outlining the refractile granules seen in these organisms. Tigroid, mast cell' granules, cartilage matrix, and mucus are Feulgen-negative (Lillie, *J. Lab. Clin. Med.*, **32**: 76, 1947).

Glutaraldehyde fixation is not suitable; glutaraldehyde, acrolein, and other dialdehyde fixations should be avoided for the Feulgen reaction, because of the widespread amine-bound positive Schiff reaction due to the unused free aldehyde group. If necessary, sections thus fixed may be first reduced in borohydride (0.2% $NaBH_4$ or 0.25% KBH_4 in 1% Na_2HPO_4) for 2 to 5 min (Lillie and Pizzolato, *Stain Technol.*, **47**: 13, 1972).

Though elastic fibers in arteries and elastic ligaments and laminae of some species are often Schiff-positive after the Feulgen hydrochloric acid hydrolysis (especially in rodents), the same fibers tend to be more strongly Schiff-positive when the acid treatment is omitted; and the reaction of elastica in Feulgen stains may be prevented by interposing a 30-min bath in 5% phenylhydrazine or 2 min in 0.2% KBH_4 in 1% Na_2HPO_4 before the acid hydrolysis step.

Kasten [*Nature (Lond.)*, **184**: 1797, 1959] recommends 15-min 60°C hydrolysis in 1 N HCl for formalin-fixed tissues, using his fluorochrome-Schiff reagents, but he often prefers room temperature hydrolysis, 6 min in 6 N HCl.

Spicer (*Stain Technol.*, **36**: 337, 1961) utilizes the fact that Bouin fixation renders ribonucleic acid insoluble and at the same time hydrolyzes deoxyribonucleic acid to the aldehyde state to give a two-color differentiation of the two nucleic acids.

The exact structure of the deoxyribonucleic acid–Schiff's reagent complex is still disputed. Hardonk and Van Duijn (*J. Histochem. Cytochem.*, **12**: 533, 1964) synthesized cellulose films to which amino groups, sulfhydryl groups, or deoxyribonucleic acid were bound covalently. Absorption spectra were compared after HCl hydrolysis in specimens treated with pararosanilin followed by a sulfite rinse or with an HCl and water rinse or with Schiff's reagent followed by sulfite and water. Hardonk and Van Duijn conclude that the substituted group of pararosanilin with deoxyribonucleic acid is $-NH-CHR-SO_3H$. Their data are in concurrence with previous conclusions reached by Rumpf (*Ann. Chim.*, **3** (11): 327, 1935) and by Hörmann et al. (*Justus Liebigs Ann. Chem.*, **616**: 125, 1958). Their data do not support the $-NH \cdot SO_2 \cdot CHR \cdot OH$ group postulated by Wieland and Scheuing (*Ber. Dtsch. Chem. Ges.*, **54**: 2527, 1921).

Butcher (*Histochemie*, **13**: 263, 1968) hydrolyzed sections of tissues in 1 *N* HCl at 60°C and measured the release of products which absorbed at 265 mμ (nucleic acids?). Butcher suggests that the nucleic acid content of sections may be used as a unit of comparison for quantitative histochemistry. Ribonucleic acid appeared to be removed within 4 min and deoxyribonucleic acid after a further 30-min hydrolysis.

Rasch and Rasch (*J. Histochem. Cytochem.*, **21**: 1053, 1973) developed a mathematical model to explain the kinetics of deoxyribonucleic acid hydrolysis. Duijndam et al. (*J. Histochem. Cytochem.*, **21**: 723, 1973) also developed a mathematical model to distinguish separate components in cytochemically stained macromolecules by an analysis of the kinetics of the staining and destaining processes. The analysis is performed by a procedure of computerized curve fitting between experimentally determined and theoretically expected absorbance curves. In further studies, Duijndam et al. (*J. Histochem. Cytochem.*, **21**: 729, 1973) applied the method to a kinetic analysis of the staining and destaining process of the Feulgen method in polyacrylamide model films. They noted that the Schiff-reactive complex is formed via a colorless but ultraviolet-absorbing intermediate. Analysis of destaining with 1 *N* HCl revealed two components in the complex. The components differ only slightly in spectral characteristics, but significantly in stability toward acid.

Lamm et al. (*J. Cell Biol.*, **27**: 313, 1965) and Feder and Wolf (*J. Cell Biol.*, **27**: 328, 1965) noted that ribonucleic acid stained metachromatically and deoxyribonucleic acid stained orthochromatically in acrolein-fixed polyester wax–embedded tissue sections stained with toluidine blue (C.I. 52040). Thin (less than 5 mm) slices of tissues were fixed in 10% acrolein in 0.1 *M* phosphate buffer of pH 7.2 in Tyrode's balanced salt solution for either 1 h at 25°C or overnight at 0°C. Polyester-embedded sections were dewaxed, hydrated, and stained for 8 min in aqueous toluidine blue in 0.02 *M* benzoate buffer pH 4.2, dehydrated in tertiary butyl alcohol for 5 min, cleared in xylene, and mounted in synthetic resin. The contrast between metachromatic ribonucleic acid and orthochromatic deoxyribonucleic acid was less in formalin-fixed tissues.

Klein and Szirmai (*Biochim. Biophys. Acta*, **72**: 48, 1963) conducted spectrophotometric studies of the interaction of azure A with either deoxyribonucleic acid or deoxyribonucleoprotein. They detected a quantitative stoichiometric binding of azure A by deoxyribonucleic acid corresponding to a 1 : 1 molar ratio of deoxyribonucleic acid phosphate to the bound dye. The corresponding ratio for azure A with deoxyribonucleoprotein was 1 : 2, suggesting that only about one-half of the phosphate groups are available to the dye.

In an electron microscopic study, Albersheim and Killias (*J. Cell Biol.*, **17**: 93, 1963) presented evidence to indicate that bismuth combines in vitro with the phosphate of nucleic acids in a manner similar to its reaction with inorganic phosphate. Protein (albumin) was unreactive with bismuth under identical conditions. Dividing cells of onion root tips were employed as the tissue test object.

Chan-Curtis et al. (*J. Histochem. Cytochem.*, **18**: 609, 1970) synthesized an acriflavine-phosphotungstate complex which was used to stain deoxyribonucleic acid at the electron microscopy level. Further work needs to be done before the method can be recommended for routine studies.

Beck et al. (*Exp. Cell Res.*, **39**: 292, 1965) employed one of the newer immunohistochemical methods at the electron microscopy level for the identification of deoxyribonucleic acid–histone using specific serum from individuals with autoimmune disease. Rat liver nuclei isolated by centrifugation in 0.25 M sucrose containing 0.002 M $CaCl_2$ and 0.001 M $MgCl_2$ and purified by sedimentation through dense sucrose are suspended for 30 min in 8 volumes of a 1 : 4 dilution of human serum (from a patient with autoimmune disease) in 0.15 M NaCl containing 10% 0.15 M phosphate buffer of pH 7.2. After washing 3 times in buffered saline solution to remove unattached proteins, nuclei are collected by centrifugation and treated for 30 min at 24°C with 6 volumes of rabbit anti-human γ-globulin previously conjugated to ferritin. Unattached ferritin is removed by washing in 20 to 30 volumes of buffered saline solution (3 times). Nuclei are recovered by centrifugation, fixed, dehydrated, embedded, and examined by electron microscopy.

RESULTS: Electron-dense granules were observed throughout the nucleus. Nearly all granules were related to chromatin. Nuclei treated with normal human serum revealed a few scattered granules.

FERROCENE CARBOHYDRAZIDE FEULGEN SUBSTITUTE. Allen and Perrin (*J. Histochem. Cytochem.*, **22**: 919, 1974) provide the following method for the demonstration of nuclei at both light and electron microscopic levels. The reagent, ferrocenylmethyl carboxyhydrazide, was synthesized by those authors.

Tissues are fixed for 3 h at 4°C in Karnovsky's formaldehyde-glutaraldehyde fixative buffered to pH 7.4 with 0.1 M sodium cacodylate buffer, washed overnight in the same buffer at 4°C, postfixed in 2% OsO_4 for 3 h at 4°C in the same buffer, dehydrated in ethanols, and embedded in Durcupan.

Thin sections mounted on gold grids were hydrolyzed with 1 N HCl at 60°C for 6 to 10 min (depending upon the degree and kind of fixation), rinsed in cold 1 N HCl, rinsed in distilled water, and stained with 2% ethanolic ferrocenylmethyl carboxyhydrazide for 1 h at 60°C. The buffer for the hydrazide is prepared by adding 15.2 ml 1 M acetic acid and 10 ml 1 M KOH and distilled water to make 1 l. The hydrazide is made up as a 2% solution in the 1 : 1 buffer–absolute ethanol mixture followed by readjustment of the buffer to pH 5. Slight warming may be necessary. The solution should be clear and golden yellow.

For electron microscopy, sections are rinsed and examined without further treatment. For light microscopy, sections mounted on glass slides are carried through the same procedure as detailed above and then stained with 2% silver nitrate for 1 h at 25°C in the dark. Sections are then rinsed in distilled water, dehydrated, and mounted in synthetic medium.

Nuclei stain black for light microscopy. Chromatin appeared dense by electron microscopy. Failure of staining occurred in sections with omission of acid hydrolysis and in sections blocked with *m*-aminophenol after hydrolysis.

SPICER'S FEULGEN–AZURE A SEQUENCE STAIN FOR NUCLEIC ACIDS

1. Fix fresh tissues 18 to 24 h in Bouin's fluid at 25°C.
2. Dehydrate, clear, infiltrate, and embed in paraffin as usual. Section shortly before staining; cut sections deteriorate on storage in the exposed state, perhaps because of air oxidation or hydration of aldehyde residues.
3. Deparaffinize and hydrate as usual; wash 5 min in running water to remove remaining free picric acid.
4. Immerse in Schiff reagent (2 g fuchsin per 100 ml), 10 min.
5. Wash in 0.5% $Na_2S_2O_5$, three changes 2, 2, and 2 min each; wash 10 min in running water.
6. Stain 30 min in 0.02% azure A or methylene blue in pH 3 or 3.5 McIlvaine citric acid disodium phosphate. It is presumed that other appropriate buffers may be substituted.

    Take 0.8 ml 1% azure A, 0.35 ml 0.2 M Na_2HPO_4, 1.65 ml 0.1 M citric acid, and distilled water to make 40 ml for the pH 3 mixture, 0.6 and 1.4 for the pH 3.5 mixture. The lower pH level gives less dense thiazin dye staining.
7. Dehydrate in acetones or *tert*-butyl alcohol, clear, and mount in synthetic resin.

RESULTS: Nuclear deoxyribonucleic acid is purplish red; ribonucleic acid, blue. Metachromatic staining is more with azure A than with methylene blue. Carboxylic acids may be blocked from methylene blue staining by 1-h 60°C methylation in 0.1 N HCl methanol, but most mucins do not stain with methylene blue after Bouin fixation. At least at pH levels below 4.5, acidic protein basophilia is not a complicating factor.

Bryan (*Q. J. Microsc. Sci.*, **105**: 363, 1964), using Clarke-Carnoy-fixed insect gonads, substitutes a 6- to 8-h, 40°C incubation in an ammoniacal silver prepared with 2% silver nitrate, following the usual 12-min 60°C 1 N HCl hydrolysis. Excess silver oxide is removed by a 3- to 5-min bath in dilute ammonia water after the silvering. Distilled water washes precede and follow the silver bath. The procedure has been adapted (ibid., p. 367) to osmium-fixed tissue for electron microscopy. Glutaraldehyde fixation is unsuitable because of the widespread aldehyde reactivity induced by dialdehyde fixation.

Korson (*J. Histochem. Cytochem.*, **12**: 875, 1964) advocates a modified Feulgen stain whereby fresh frozen or neutral formalin-fixed sections are hydrolyzed with 1 M citric acid for 30 min at 60°C. Sections are thereafter treated with methenamine silver (5% silver nitrate in 3% aqueous methenamine) for 60 min at 60°C.

AZO COUPLING REACTION. In nuclei this varies considerably with the diazo used and with the fixative employed. With diazosafranin at pH 8 in formol or glutaraldehyde fixation nuclei are scarcely more deeply stained than cytoplasm. With buffered sublimate fixation cytoplasms stain more weakly, and nuclear chromatin stands out in dark red to reddish black. With the diazosulfanilic acid, pH 1 azure A procedure (DAS-AzA) and formol or glutaraldehyde, nuclei stand out in moderate pure green with lighter yellowish green to greenish yellow cytoplasm. This is not cationic dye binding; methylation does not interfere, and the coupling seems assignable to basic nucleoprotein aromatic amino acids, at least in part to histidine.

Mitchell (*Br. J. Exp. Pathol.*, **23**: 285, 1942) reported that adenine and guanine still gave orange colors and uracil an intense crimson after benzoylation on coupling with *p*-diazobenzene sulfonic acid in vitro, and he and Danielli (*Symp. Soc. Exp. Biol.*, **1**: 101, 1947) considered that the relatively benzoylation- and acetylation-resistant nuclear staining by the coupled tetrazonium method might be assignable to purines and pyrimidines in the nucleic acids. Gomori (1952, p. 77) got negative results on ribonucleic acid as well as guanine, adenine, uracil, xanthine, and uric acid, using Coujard models in gelatin, and Pearse (1953, p. 102) obtained no reaction with crystalline deoxyribonucleic acid. Pearse found that the reaction persisted after nucleic acid extraction. Barnard and Danielli [*Nature* (*Lond.*), **178**: 1450, 1956] assigned the benzoyl chloride–resistant tetrazonium reaction to histidine in the basic nuclear protein.

In a consideration of the possible role of the purines adenine and guanine and the pyrimidines cytosine, uracil, and thymine in the azo coupling reaction of cell nuclei we encountered Mitchell's (*Br. J. Exp. Pathol.*, **23**: 285, 1942) statement that adenine and guanine still gave orange colors in vitro after benzoylation and uracil an intense crimson. Gomori (1952) got no reactions on Coujard gelatin models with adenine, guanine, uracil, xanthine, uric acid, and ribonucleic acid. Since it was desirable to see how the in vitro reactions of these substances compared in intensity and pH range with those of the aromatic amino acids recently reported by Lillie et al. (*J. Histochem. Cytochem.*, **21**: 1973), kits of purines and pyrimidines were obtained from Sigma Chemical Co. (St. Louis, Missouri) and Calbiochem Co. (Monsey, New York) including adenine, guanine, xanthine, hypoxanthine, cytosine, 5-methylcytosine, uracil, and thymine.

As in the previous series, 5-mg amounts were dissolved or suspended in 4.5 ml water or alcohol, with 5 ml buffer at about 0.2 M at pH levels graded from 2 by 1 pH steps to 7 and 0.5 pH steps to 10.5 and then 0.5 ml 0.1 M freshly diazotized p-nitroaniline added. Only light yellow colors were seen at low pH levels; slight orange tints appeared in the middle range and deepened to quite deep reds at pH 9.5 to 10.5: cytosine, orange-red to red at pH 9.5 to 10.5; 5-methylcytosine, orange-red above pH 8; thymine, orange to red at pH 10 to 10.5; uracil, yellow-orange colors above pH 4; adenine, orange above pH 8; guanine, orange to red at pH 9 to 10.5; xanthine, orange to red at pH 8 to 10.5; hypoxanthine, only light orange-yellow above pH 8.5.

In accordance with Burian's report (*Ber. Dtsch. Chem. Ges.*, **37**: 696, 1904) we expected those purines with an unsubstituted NH group on the imidazole ring and open C-8 (the C-2 of the imidazole ring) to give positive reactions.

In the pyrimidines according to Day and Mosher (*J. Franklin Inst.*, **255**: 455, 1955) 2-hydroxypyrimidines with an open 5 position readily form 5-azo pyrimidines. This requires an enol (lactim) configuration of at least positions 1 and 2. Johnson and Clapp (*J. Biol. Chem.*, **5**: 163, 1908) found that thymine and 5-methylcytosine both formed red azo compounds with p-diazobenzene sulfonic acid, which they could not isolate, and found further that pyrimidines in which N substituents were present at 3 or at 1 and 3 did not azo-couple. For this azo coupling the ketol (lactam) configuration is required, at least for positions 3 and 4, to furnish the necessary directing NH group, and coupling occurs in the unsubstituted 6 position.

With the NH group of the purine imidazole ring occupied by the nucleoside sugar linkage the purines lack the open NH group required by the azo coupling on the imidazole ring. In the pyrimidines the sugar linkage is formulated to involve the N-1 group of a lactam (ketol) form by White, Handler, and Smith (*Principles of Biochemistry*, 4th ed., McGraw-Hill, New York, 1968, p. 185), and the 3 NH is still available to direct azo coupling. Azo coupling at 5 would be inhibited in uracil and cytosine, since the 1,2 grouping is in ketol form and there is no directing OH group. Hence we may conclude that pyrimidine azo coupling is theoretically possible in nucleosides, though it does not appear to have been conclusively demonstrated.

Turchini (*Trav. Mem. Soc. Chim. Biol.*, **25**: 1329, 1943) reported a method for the two nucleic acids which depends on specific colors produced by deoxyribose and ribose with 9-methyl- or 9-phenyl-2,3,7-trihydroxy-6-fluorone. Deoxyribose gives a blue-violet color;

ribose gives a yellowish rose in spot tests; purines, pyrimidines, and phosphoric acid give no significant color. Backlar and Alexander (*Stain Technol.*, **27**: 147, 1952) reported a workable modification. At that time we had difficulty in obtaining a satisfactory sample of the reagent, and there appears to have been little subsequent use of the method. Recent attempts (1974) to obtain the 9-phenyl derivatives for another purpose have been unsuccessful. For details the reader is referred to the second edition of this work and to the reports cited above.

BASIC ANILINE DYES. Staining 5 to 30 min in 0.1 to 1% solutions of safranin O (C.I. 50240), Janus green B (C.I. 11050), thionin (C.I. 52000), toluidine blue O (C.I. 52040), azure A, methyl or ethyl green (C.I. 42585, 42590) in 0.5 to 1% acetic acid solution in water or, even better, in 0.01 N HCl, will give quite selective chromatin staining even without much further differentiation.

Many prefer the practice of staining in solutions of similar strength in distilled water (or 20% alcohol to prevent growth of molds) for 5 to 20 min, followed by 1-min differentiation in 5% to 5 min in 1% acetic acid. Mitotic nuclei are rendered more conspicuous when these differentiations are carried to a point where resting nuclei remain only rather lightly stained. This corresponds to a point where the sections appear only lightly colored on gross inspection and when the differentiating fluid ceases to extract color clouds.

Many writers use far stronger solutions for nuclear staining than the 0.1 to 1% aqueous solutions noted above, alleging superior results. Saturated alcoholic and aqueous solutions and mixtures of them are recommended in the case of safranin; and one widely used formula, that of Babes (*Z. Wiss. Mikrosk.*, **4**: 470, 1887), recommended saturating 2% aniline water with (5 to 6%) safranin O (C.I. 50240) by heating to 60 to 80°C. This solution keeps only 1 to 2 months. It stains instantaneously. Differentiation in alcohol gives good staining of both resting and dividing nuclei, and use of 0.1 to 0.5% hydrochloric acid alcohol decolorizes resting nuclei more rapidly than chromosomes.

Material fixed with osmium tetroxide methods or subjected to prolonged chromation is more difficult to stain, and more prolonged exposures, higher pH levels, and stronger dye solutions are necessary.

Methyl Green Pyronin

This method had a revival of popularity in the 1950s. When correctly used on properly fixed material, it differentiates specifically between deoxyribonucleic acid (green) and ribonucleic acid (red).

Its behavior is much influenced by prior fixation procedures. Best results have been obtained after neutral and acid alcoholic fixatives, especially the Clarke and Carnoy fluids (including alcohol formalin and acetic alcohol formalin), and after the aqueous dichromate formaldehyde fixatives. Aqueous formalin and Zenker's fluid give inferior results: sections stain either all red or all blue.

The proportions of methyl green (C.I. 42585) or ethyl green (C.I. 42590) and pyronin Y (C.I. 45005) or pyronin B (C.I. 45010) vary greatly in various formulas.

The amount of phenol in the mixture should be small. More than 3% seems definitely deleterious. Some seems to be beneficial, though we have had good results without any.

The presence of glycerol is probably chiefly valuable in retarding evaporation when slides are stained flat with small volumes of solution. Alcohol seems unnecessary.

With fresh samples of methyl green and pyronin, it may be necessary, since satisfactory pyronins have been difficult to prepare, to vary the proportions. Make 2% aqueous solutions of each dye, mix in the desired proportions, and add an equal volume of a mixture of 4 g phenol, 56 ml water, and 40 ml glycerol. The phenol is 2% in the final mixture and may be reduced to 2 or 1 g if it is desired to try a 1% or 0.5% level.

The quality of the pyronin used also has considerable bearing on the success of the stain. Kasten (*Stain Technol.*, **37**: 277, 1962) has reported the presence of a primary amine containing dye in many samples of both pyronin Y and pyronin B and of the lower homolog symmetrical dimethyldiamino acridine red, C.I. 45000. Kasten (*Stain Technol.*, **37**: 265, 1962) related the continuing difficulties with American pyronins and post-World War II German pyronins which were actually rhodamines and good performance of the old Grübler dye and of the British E. Gurr and G. T. Gurr dyes, and the good results of the German return to the original pyronin G synthesis in the 1950s. These recent German samples are free of primary amine contaminants.

The differential staining has been thought to depend on the relative molecular weights of the two nucleic acids.

The technic is simple:

1. Deparaffinize and hydrate paraffin sections as usual.
2. Stain 20 min in methyl green pyronin.
3. Wash in cold, recently boiled distilled water.
4. Dehydrate with acetone; clear through acetone plus xylene (50 : 50) and two changes of xylene; mount in synthetic resin.
4a. Or: blot dry, differentiate a few seconds in equal volumes of triethyl phosphate and xylene, clear in xylene, mount in synthetic resin (Rosa, *Stain Technol.*, **25**: 165, 1950).

RESULTS: Nuclei are blue-green; bacteria and basophil cytoplasm, red.

Rosa's procedure is said to preserve the methyl green better than the usual alcohol dehydration. We have used acetone for the same reason.

Some workers—notably Kurnick, and Trevan and Sharrock, among others—insist on the purification of methyl green by extraction of the aqueous solution with chloroform[1] until no more methyl violet appears in the chloroform solution. Trevan and Sharrock use 2% aqueous methyl green, extracting it in a separatory funnel with chloroform, and state that the extracted solution remains free of further extractable violet for 6 months.

Improved results are claimed in a variant by Elias (*Stain Technol.*, **44**: 201, 1969). He purified 200 ml 0.5% methyl green in pH 4.1 Walpole acetate buffer by shaking the solution with chloroform in a separatory funnel until no more methyl violet was extracted and then dissolved 0.4 g pyronin Y in the methyl green solution. The solution was kept at 4°C and filtered before using. The staining was as above, but washing in water was brief and at 0°C (ice in the water), and dehydration was done in ethanol: in *n*-butanol, 1 : 3, or in *n*-butanol-*tert*-butanol, 19 : 1, or, best, in *tert*-butanol alone, and again best at 0 to 4°C. The low-temperature washes appear to be the critical factors in the improved final staining. Elias used methyl green, C.I. 42585, molecular weight 458.488, and pyronin Y (GS), C.I. 45005, molecular weight 303.844, both from Chroma, and considered that the smaller molecular weight permitted the penetration of pyronin into the denser ribonucleic acid, discounting any chemical difference between the two dyes since both are basic. In pyronin the single positive charge is on a dialkyl quinoneimine N; in methyl green there is a similar dialkyl imine and in addition the strongly basic trialkylaryl ammonium halide radical, making this a much more strongly basic dye. If it were solely a question of density, the denser chromosomes should take more red than resting nuclei.

---

[1] Crystal violet is soluble 5.14% in chloroform (Donaldson, unpublished data).

TABLE 7-2 CONSTITUTION OF METHYL GREEN PYRONIN FORMULA CALCULATED TO APPROXIMATELY 100 ml VOLUME

Formula[a]	Methyl green, mg	Pyronin, mg	Alcohol, ml	Glycerol, ml	Phenol, ml	Water	"Carbolwater" %	"Carbolwater" ml	Total
Saathoff, 1905[b,c]	150	500	5	20	1.5	73.5	2	75	100
Saathoff-Conn[d,e]	800	200	4	16	1.6	78.4	2	80	100
Unna[b,c,f,g]	150	250	2.5	20	0.39	77.11	0.5	77.5	100
Ramón y Cajal[h]	150	250	2.5	20	0.5	78			101
Weil[i]	300	700				100			100
Masson[i]	2000	2000			5	100			105
Slides and Downey[j]	500	500	2.5	20	0.5	99.5	0.5	100	122.5
Slides and Downey reduced to 100	408	408	2.04	16.3	0.41	81.3	0.5	81.7	100
Lillie[k]	700	300		20	0–2	80–78			100
Trevan and Sharrock[l]	36	157.6				100			100+

65 mg orange G

[a] Pappenheim's original prescription was inexact (Romeis, Unna, in Ehrlich's "Encyklopädie," 1903) and is omitted.

[b] Saathoff, "Die Methylgrün Pyronin Methode für elektive Färbung der Bakterien im Schnitt," Dtsch. Med. Wochenschr., 32: 2047, 1905.

[c] F. B. Mallory: "Pathological Technique," Saunders, Philadelphia, 1938.

[d] H. J. Conn: "Biological Stains," 4th ed, Biotech, Geneva, N.Y., 1940.

[e] H. J. Conn and M. A. Darrow: "Staining Procedures," Biotech, Geneva, N.Y., 1947.

[f] B. Romeis: "Taschenbuch der mikroskopischen Technik," 13th ed, Oldenburg, Munich and Berlin, 1932.

[g] G. Schmorl: "Die pathologisch histologischen Untersuchungsmethoden," 15th ed., Vogel, Leipzig, 1928.

[h] S. Ramón y Cajal: "Histology," translated by Fernán Núñez, Wood, Baltimore, 1933.

[i] M. Langeron: "Précis de microscopie," 5th ed., Gauthier-Villars, Paris, 1934.

[j] E. V. Cowdry: "Laboratory Technique in Biology and Medicine," Williams & Wilkins, Baltimore, 1948.

[k] R. D. Lillie: "Histopathologic Technic and Practical Histochemistry," 2d ed., McGraw-Hill, New York, 1954.

[l] D. J. Trevan and A. Sharrock: J. Pathol. Bateriol. 63: 326, 1951.

Specificity of this method is better with acetic acid–alcohol fixatives; formol gives inferior results. Ribonucleic acid is well differentiated from nuclei in iron hematoxylin–safranin sequence staining, but here it seems probable that nuclear histone rather than (or in addition to) the deoxyribonucleic acid may be responsible for the metachrome iron hematoxylin stain.

Kurnick noted that pyronin was considerably extracted from cytoplasm by dehydration and by the differentiating agents used for the removal of methyl green from background structures. Therefore he converted the method into a sequence procedure and used the pyronin as a saturated or one-tenth saturated acetone solution after dehydration and differentiation of the methyl green, thus avoiding irregular extraction of the pyronin by the acetone.

KURNICK'S METHYL GREEN PYRONIN
(*Stain Technol.*, **27**: 233, 1952)

1. Fix in Carnoy, cold 80% alcohol, cold acetone, neutral formalin; embed and section in paraffin as usual; or use frozen dried material embedded in paraffin directly.
2. Deparaffinize and hydrate as usual.
3. Stain 6 min in repurified 0.2% methyl green in water or in pH 4.2 10 mM acetate buffer. (The solution should be extracted in a separatory funnel by shaking it with successive changes of chloroform until no more color is extracted.)

    Scott (*Histochemie*, **9**: 30, 1967) studied the specificity of the methyl green–pyronin staining for nucleic acids. Greatest selectivity of staining was obtained in sections stained with 0.15% w/v methyl green and 0.25% pyronin in 0.05 M sodium acetate buffer pH 5.6 containing 2.0 M MgCl$_2$. Scott observed that nonnucleotide staining is suppressed by addition of a salt such as MgCl$_2$. The presence of MgCl$_2$ appears to enhance selective uptake of methyl green by deoxyribonucleic acid and pyronin by ribonucleic acid. The optimal level of salt concentration may vary from one tissue to another.
4. Blot dry and dehydrate with two changes of *n*-butyl (or *tert*-butyl) alcohol.
5. Stain 30 to 90 s in acetone freshly saturated with pyronin B. For more delicate staining dilute 1 part of the saturated solution with 9 parts acetone, and prolong the time somewhat.
6. Clear directly in cedar oil, wash in xylene, and mount in terpene resin (Bioloid, Permount, HSR).

RESULTS: Blue-green chromatin is observed, red nucleoli, pink to red cytoplasm, green cartilage, purplish pink bone, and brown erythrocytes are seen.

For formalin-fixed tissue Trevan and Sharrock used a weaker formula than most others (Table 7-2) and added orange G as an acid cytoplasmic stain, selecting a slightly bluish pyronin which did not precipitate with orange G. Two stock solutions were kept: one of 72 mg chloroform-extracted methyl green and 315 mg pyronin in 100 ml distilled water; the other of 130 mg orange G in 100 ml 0.2 M acetate buffer of pH 4.8. Working mixtures of equal volumes of the two stock solutions keep for about 2 weeks. They direct: Deparaffinize paraffin sections of 3 to 5 μ; hydrate; stain 10 min to 24 h; wash about 7 s in flowing distilled water; blot; dehydrate with acetone; clear through acetone xylene and two changes of xylene; and mount in polystyrene. Nuclei and mucin are green to blue-green; tigroid and basophil cytoplasms, magenta; nucleoli, pink; oxyphil cytoplasmic components, orange; mast cell granules, usually dark red-brown.

Lillie had no great faith in the specificity of this method, especially on formalin-fixed tissue. It was often capricious and difficult to control. Such methods as the acid–iron hematoxylin–safranin sequence stain seem to yield a similar differentiation with far greater certainty and less limitation as to fixing methods. However, this variant is probably not demonstrating deoxyribonucleic acid with the iron hematoxylin, but rather the basic nuclear proteins in contrast to the safranin-stained ribonucleic acid and acid mucins, heparin, etc.

For assaying the effect of ribonuclease digestions, we find the azure A–eosin B technic superior. The pink of the basic protein left after removal of ribonucleic acid contrasts much better with the blue than does the remaining pink of pyronin staining with the red of undigested preparations.

Roque et al. (*Exp. Mol. Pathol.*, **4**: 266, 1965) advise the following selective method for staining deoxyribonucleic acid and ribonucleic acid. Tissues (2-mm slices) fixed for 3 h in 4% formaldehyde or 2.5% glutaraldehyde in 1% sodium acetate are recommended. Fair results are obtained in tissues fixed with Carnoy's, Zenker's, or Helly's fluids.

1. Deparaffinize sections and bring them to water.
2. Stain for 30 min at 40°C in the following mixture:

Methyl green (C.I. 42585)	0.1 g
Thionin (C.I. 52000)	0.0165 g
Citrate buffer (0.02 M, pH 5.8)	100 ml

The thionin is dissolved in water, methyl green and buffer are added, and the mixture is shaken and filtered. Heating to 60°C may be needed to bring all the thionin into solution. The stain does not keep for more than 2 or 3 days.

3. Rinse sections in distilled water.
4. Dehydrate in three changes of an alcohol mixture (80 : 20 of *tert*-butanol–absolute ethanol) for $\frac{1}{2}$, 3, and 3 min, respectively.
5. Rinse in absolute ethanol, clear in xylene, and mount in synthetic medium.

   Sections should be observed under the yellowish white light of a tungsten bulb without the blue daylight filter; otherwise the red metachromasia of ribonucleic acid cannot be seen.

RESULTS: Nuclei are green and ribonucleic acid is metachromatically red. The stain may be employed in conjunction with a digestion with ribonuclease or deoxyribonuclease. If fixation in formaldehyde or glutaraldehyde is prolonged, nuclei tend to stain more blue, and ribonucleic acid tends to stain more bluish violet.

Semmel and Huppert (*Arch. Biochem. Biophys.*, **108**: 158, 1964) note that ribonucleic acid may react with different basic dyes (toluidine blue, acridine orange, pyronin) to shift the absorption spectrum to a longer (negative metachromasy) or a shorter (positive meta-chromasy) wavelength. The direction of the metachromatic shift is dependent on both the relative concentration of the polymer to the dye and on the secondary structure and chain length of the polymer. In studies with synthetic ribonucleic acid polymers, Semmel and Huppert noted that the extent of positive metachromasy was dependent upon the length of the polymer chains.

Shea (*J. Histochem. Cytochem.*, **18**: 143, 1970) observed stoichiometric binding of azure B in relation to concentration of ribonucleic acid in gelatin films. Good agreement was obtained in comparative estimates of nucleolar ribonucleic acid in mouse Purkinje cell nuclei.

ACRIDINE ORANGE. Von Bertalanffy and Bickis (*J. Histochem. Cytochem.*, **4**: 481, 1956)[1] used acridine orange 0.1% in distilled water, further diluted with buffered Ringer's

---

[1] See 67 papers in bibliography.

solution, on impression smears and cold knife fresh frozen sections of liver at 15 μ. Clarke-Carnoy acetic alcohol fixation was used for the colored pictures, and exposures of 10 to 15 min in 1 : 5000 or 1 : 10,000 acridine orange in pH 6 Ringer's solution were employed. Cytoplasmic ribonucleic acid and mast cell granules colored with red fluorescence; nuclei colored with bright green. Green nuclear fluorochrome staining and red mast cell coloration still occurred at 1 : 100,000 dilution; at 1 : 1000 overstaining in red dominated even with 1-min exposure. Carnoy's alcohol–chloroform–acetic acid and absolute alcohol gave similar preparations to acetic alcohol.

Formalin fixation largely prevented and Bouin fixation completely prevented differential staining of ribonucleic and deoxyribonucleic acid sites, giving only green fluorescence with weak solutions and red with strong.

CYTOPHOTOMETRY OF RIBONUCLEIC ACID.   For this purpose Ritter, Di Stefano, and Farah (*J. Histochem. Cytochem.*, **9:** 97, 1961) used 10-μ paraffin sections of tissues fixed in neutral 10% formalin. Sections were deparaffinized and hydrated as usual and stained 2 h at 45°C in 0.1% cresyl violet (Chroma Gesell.) which was adjusted to pH 4.2 with 0.1 N hydrochloric acid. To remove nonspecific staining, sections were then extracted from 30 min to 16 h in absolute ethanol. After 16 h no further dye extraction occurred up to 24 h, and the bound dye showed strict proportionality to section thickness when measured photometrically at 585 mμ. This absorption peak did not shift with changes in dye concentration as it does with toluidine blue, azure A, and gallocyanin. Digestion with ribonuclease (5 mg/100 ml, 20 min, 40°C, pH 6) completely prevented cytoplasmic cresyl violet staining. Nuclei are also stained by cresyl violet. Preparations mounted in Permount have not appreciably faded or lost in photometric density in 14 months.

Since the cresyl violet also stains nuclear deoxyribonucleic acid, photometric studies require the use of adjacent serial sections, the one digested with ribonuclease and the other not.

TIGROID, NISSL, SUBSTANCE.   This substance occurs as angular and elongate particles of varying size in the cytoplasm of nerve cells. It stains deeply with thionin, azure A, methylene blue, toluidine blue, and the like; less conspicuously with safranin, fuchsin, or neutral red; more differentially from acid solutions or with acid differentiation; poorly or not at all with alum hematoxylin.

Generally, alcohol fixation is prescribed, formalin serves well, and quite good pictures may be obtained with the azure-eosin method on fresh material fixed in Orth's fluid, or in formalin followed by chromate treatment before embedding.

Gersh and Bodian (*Biol. Symp.*, **10:** 161, 1943) stained nerve cells overnight in filtered 1% toluidine blue, differentiated in alcohol and mounted in balsam. We have found a 30-s stain in 0.1% aqueous thionin (buffered to pH 4 with acetates) followed by the acetone, xylene, resin sequence quite effective for most of the tissues noted above, but generally prefer the azure A–eosin B technic because of the more definitive color change from blue to pink when the digestion tests are performed.

Popa and Repanovici (*Biochim. Biophys. Acta*, **182:** 158, 1969) employed agarose gel electrophoresis to study ribosomal ribonucleic acid–toluidine blue complexes.

ISOELECTRIC POINT

While the influence of addition of acetic acid in tending to restrict basic dye and complex staining to nuclei was noted by a number of nineteenth-century workers, no systematic study of the influence of pH on staining with acid and basic dyes seems to have been made until Pischinger's (*Z. Zellforsch. Mikrosk. Anat.*, **3:** 169, 1926) study of the influence of the isoelectric point on stainability with acid and basic dyes. Giemsa (*Zentralbl. Bakteriol.* [*Orig.*], **37**: 308, 1904) had used a few drops of acetic acid in water for differentiating too blue blood films stained with his azure II–eosin blood stain, and when in 1931 we accidentally had a predifferentiated Maximow azure-eosin stain and traced it to an unusually acid distilled water, we published our first paper on buffered azure-eosin staining (Lillie and Pasternack, *Arch. Pathol.*, **14**: 515, 1932). We were not aware of Pischinger's study at the time and did not realize for some years that this technic could be used for determination of isoelectric points.

Highman (*Stain Technol.*, **20**: 85, 1945) used buffered thiazin dyes at graded low pH levels for demonstration of various mucins.

Dempsey (*Anat. Rec.*, **98**: 417, 1947) used the dye uptake at various pH levels for the characterization of the basophilic substances in tissues. He stained 24 h at 25°C in 0.5 mM methylene blue (0.1595% of the pure dye, or 0.18% at 87.5% dye content; 0.2% should do about the same). He measured the dye uptake photometrically, obtaining quite characteristic "pH signatures" for each substance.

It should be pointed out, however, that these "pH signatures," like the isoelectric points which govern acidophilia and basophilia toward azure-eosin stains, are considerably altered by the fixative employed. With brief formaldehyde or Orth fixation, nuclei and tigroid still take up thionin at pH 1.5 and melanin as low as at pH 0.8; after long storage in 10% formalin there is no staining of nuclei at pH 2 and only feeble blue at pH 3. At the latter level melanin colors green; at pH 2 it remains almost pure brown.

The following data are based on material fixed 48 h only in buffered 10% formalin or Orth's fluid. Dempsey used Zenker fixation of unstated duration.

Staining for 1 to 30 min in 0.05% thionin in 0.01 M acetate buffer of pH 3.5, or even in 0.01 M acetic acid (pH about 3.4), gives good staining of nuclei, tigroid, and strongly basophil cytoplasm, and metachromatic staining of mast cells and cartilage. Oxyphil material, muscle, and connective tissue remain unstained, and red corpuscles appear in light yellow. In a similar buffer of pH 4.11, muscle and connective tissues appear faintly greenish, and at pH 5.1 they take a fair grade of light blue-green. Phosphate buffers from pH 5 to 1.5 precipitate thionin almost quantitatively but do not throw down azure A or toluidine blue.

At higher pH levels (6 to 7.5) and at higher dilutions (10 mg/l = 27.5 μM of 88% dye-content methylene blue, or similar concentrations of azure A, thionin, and toluidine blue) and overnight staining, connective tissue, keratin, muscle, and oxyphil materials in general assume a rather diffuse and light green; red corpuscles—a darker, somewhat olive green; basophil cytoplasms—blue-violet; nuclei—fairly deep blue, greenish blue, or lighter bluish green; and cartilage matrix and mast cell granules (at least with azure A, thionin, and toluidine blue)—a brilliant purplish red. Epithelial mucins may fail to stain.

The Windle (Conn and Darrow) and Laskey frozen section methods using thionin also at about pH 4 were essentially similar. The one stained at about pH 3.5 or at 4 following 0.2% acetic acid differentiation, alcohol dehydration, and xylene; the other used frozen sections of old or fresh formol material stained at pH 4, dehydrated in alcohol, and cleared with 1 : 4 carbol-xylene.

Grimley (*Stain Technol.*, **39**: 229, 1964) provides the following tribasic stain for plastic-embedded sections prepared for electron microscopy:

Thin sections of OsO_4-fixed tissues embedded in plastic are attached to slides by heating to 70°C for 1 min. Sections are stained in a Coplin jar for 2 to 5 min in the ethanolic dye mixture or 10 to 30 min in the aqueous dye mixture at 24°C. The ethanolic dye solution is composed of 8% toluidine blue O (C.I. 52040) and 8% malachite green (C.I. 42000). These are saturated. The aqueous dyes are also saturated with both dyes at 4% (w/v). The

TABLE 7-3 BUFFER TABLE FOR STAINING WITH ACID AND BASIC DYES AT GRADED pH LEVELS*

pH	2 N HCl	1 M KH₂PO₄	Distilled H₂O	Dye solution
1.0	10.3	19.4	9.3	1.0
1.5	8.1	23.8	7.1	1.0
	0.2 N HCl	0.1 M KH₂PO₄		
2.0	8.0	24.0	7.0	1.0
2.5	4.4	31.2	3.4	1.0
3.0	2.0	36.0	1.0	1.0
	1 N CH₃COOH	1 M CH₃COONa		
3.5	7.5	0.5	31.0	1.0
4.0	6.5	1.5	31.0	1.0
4.5	4.6	3.4	31.0	1.0
5.0	2.2	5.8	31.0	1.0
	0.2 M KH₂PO₄	0.2 M Na₂HPO₄		
5.5	17.6	2.4	19.0	1.0
6.0	16.8	3.2	19.0	1.0

pH	0.2 M KH₂PO₄	0.2 M Na₂HPO₄	Distilled H₂O	Dye solution
6.5	12.7	7.3	19.0	1.0
7.0	6.8	13.2	19.0	1.0
7.5	2.8	17.2	19.0	1.0
	0.2 N HCl	0.2 M Sodium Veronal		
8.0	5.7	14.3	19.0	1.0
8.5	3.3	16.7	19.0	1.0
9.0	1.3	18.7	19.0	1.0
	1 N HCl	1 M K₃PO₄		
9.5	9.8	16	13.2	1.0
10.0	9.4	16	13.6	1.0
10.5	8.8	16	14.2	1.0
11.0	7.6	16	16.4	1.0
11.5	4.8	16	18.2	1.0

*The stock dye solution at 2% will yield a 1 : 2000 (0.05%) final dilution; at 0.8% the final dilution is 1 : 5000 (0.02%). Biebrich scarlet has been used at 1 : 10,000 final dilution from about pH 3 to 11.5; eosin B does well at 0.05% from pH 4 to 11.5; staining is weakened from depression of dissociation at pH 3. Azure A and thionin have been used at levels of pH 1 to 8; they are prone to precipitate in alkaline solutions. Thionin precipitates in HCl plus KH₂PO₄ buffers; azure A does not.

For azure A–eosin B at pH 3.5 to 9.0: Replace 13 ml water with 5 ml acetone and 4 ml each 0.1% azure A and eosin B, making 13 ml deducted from 31 or 19 ml water.

1 to 5% H_2SO_4 can be used as solvent for lower pH levels. The pH level is determined on the completed solution with a pH meter.

ethanolic dye solution was generally preferred except that some epoxy resins (particularly Epon) stained. Sections embedded in Epon were stained with the aqueous solution. To improve contrast, epoxy-embedded sections may be mordanted in 2% $KMnO_4$ in acetone for 1 min at 24°C. Sections may be counterstained in 4% aqueous basic fuchsin for 1 to 30 min depending upon the type of embedding medium. The least time is required for specimens embedded in methacrylate.

RESULTS: Nuclei are blue to violet, erythrocytes and mitochondria are green, collagen and elastic tissues are magenta, and mucin and cartilage are bright red.

Estable-Puig et al. (*J. Neuropathol. Exp. Neurol.*, **24:** 531, 1965) highly recommend the following general overall staining method for osmium-fixed plastic-embedded tissues for light and phase microscopy. The method employs the strong reducing agent *p*-phenylenediamine. Estable-Puig et al. speculate that osmium tetroxide in tissues may oxidize the reagent, resulting in highly colored oxidation products.

1. Cut sections 0.5 to 2 μ thick on a Porter-Blum ultrathin microtome with glass knives, float them onto glass slides, and allow them to dry at room temperature.
2. Place slides in a Coplin jar containing a freshly filtered 1% solution of *p*-phenylenediamine. The tissue becomes visible as brownish black spots within 15 min to 1 h depending upon the thickness of the section and the osmium content of the tissue.
3. Rinse slides in distilled water for 2 min.
4. Dehydrate in two changes of 95% or absolute ethanol. The alcohol clears the sections of a nonspecific background stain that is sometimes seen if an old solution of *p*-phenylenediamine is used, but it does not alter or reduce the intensity of staining of osmiophilic material.
5. Allow the slides to dry, and then mount them under a coverslip with cedar oil, Canada balsam, or Permount.

Crystals of what appear to be oxidative products of *p*-phenylenediamine may cover the sections if old or unfiltered solutions are used. A quick dip into weak acetic acid solution before step 3 can save the slide. The use of xylene is to be avoided, because it will soften the plastics and cause wrinkling of the sections.

NUCLEOHISTONE METHODS

Alfert and Geschwind (*Proc. Natl. Acad. Sci.*, **39:** 991, 1953) recommend an alkaline fast green method for the histochemical identification of basic nuclear protein. Sections of paraffin-embedded tissues fixed for 3 to 6 h in Lillie's calcium acetate neutral formalin are deparaffinized, brought to water, and hydrolyzed for 15 min in 5% aqueous trichloroacetic acid at 100°C. Sections are washed in three changes of 70% ethanol (10 min each), rinsed in distilled water, stained for 30 min at 24°C in 0.1% aqueous fast green FCF (C.I. 42053) adjusted to pH 8 to 8.1 with NaOH, rinsed in distilled water for 5 min, dehydrated, and mounted in synthetic resin.

RESULTS: Basic nuclear proteins stain green.

Noeske (*Histochem. J.*, **5:** 303, 1973) conducted cytophotometric assays of bone marrow smears stained with fast green FCF according to the Alfert and Geschwind method for histone. Very little dye uptake could be observed in immature cells, whereas high values were obtained in nuclei of mature polymorphonuclear leukocytes. Noeske concludes that the stain uptake may be related more to the functional state of the nucleus than to the amount of histone present.

Garcia (*J. Histochem. Cytochem.*, **17**: 475, 1969) conducted quantitative analyses of fast green FCF uptake with nuclear histone using Alfert and Geschwind's method. Dye uptake increased as much as 50% in cells exposed to hypotonic media. Garcia concludes that although histone binding is similar to the Feulgen with respect to changes in the physical state of the DNA-protein (deoxyribonucleic acid–protein) complex, quantitation of fast green–histone is a less reliable assessment of these changes.

Desai and Foley (*J. Histochem. Cytochem.*, **22**: 40, 1974) used the Alfert and Geschwind method for basic nuclear proteins combined with the toluidine blue pH 9 method of Smetana and Busch (*Cancer Res.*, **26**: 331, 1966) for acidic nuclear protein in a quantitative study of cells of human lymphocytic leukemia. Normal and leukemic cells did not differ in DNA–basic nuclear protein or in DNA–acidic nuclear protein ratios.

FULLMER'S HEMATOXYLIN METHOD FOR BASIC NUCLEAR PROTEIN.
Fullmer (*J. Histochem. Cytochem.*, **10**: 502, 1962) has modified Clara's hematoxylin reaction to use it as a quite specific stain for basic nuclear protein. The reaction is prevented by deamination, acetylation or prolongation of acid hydrolysis beyond 18 h. Lillie noted nuclear staining during an investigation of the catecholic reactions of keratohyalin and trichohyalin (*J. Histochem. Cytochem.*, **4**: 818, 1956). The technic follows:

1. Fix 24 h in neutral formalin, prepare paraffin sections, deparaffinize, collodionize, and hydrate as usual.
2. Hydrolyze 1 h at 60°C in preheated 1 N hydrochloric acid.
3. Rinse in distilled water.
4. Stain 1 h at 37°C in preheated 0.1% hematoxylin in pH 8.14 tris buffer.
5. Wash in distilled water, dehydrate, clear, and mount in synthetic resin.

RESULTS: Nuclei are dark blue; background, unstained.

It is to be noted that azure A staining and the Feulgen reaction have disappeared with 20- to 45-min hydrolysis, and that this reaction is maximal at 1 to 2 h, and still moderately strong at 6 h. The hematoxylin reaction of keratohyalin, eosinophils, and enterochromaffin may resist prior acid hydrolysis for some time, but these tissue elements are readily distinguished. Lillie et al. [*Acta Histochem. (Jena)*, **49**: 204, 1974] have shown this reaction to be blocked by alkaline benzil, thus relating it to the arginine content of the nucleoprotein.

Winkelman and Bradley (*Biochim. Biophys. Acta*, **126**: 536, 1966) conducted spectroscopic titrations of histone and Biebrich scarlet and concluded that the anionic dye binds and stacks to polycations in a fashion analogous to cationic dyes on polyanions.

Ringertz (*J. Histochem. Cytochem.*, **16**: 440, 1968) recommends staining of histone with the fluorescent dye dansyl chloride (dimethylaminonaphthalene sulfonyl chloride) after DNA extraction with 1 N HCl.

Dansyl chloride $C_{12}H_{12}NO_2SCl$, mw 269.752

Ringertz stained 20 μg polypeptide and other substances in spots on filter paper with 0.05 to 1% dansyl solutions in 0.035 M borate buffer of pH 8.3 for 120 min. After rinsing to remove excess stain, dansyl was eluted with 1.5 ml 0.003 M NaOH at 22°C. Absorbance was measured at 245 mμ in 1-cm cuvettes. Further work needs to be done before application to routine histochemistry is to be recommended.

ACETIC ORCEIN. This is a very precise simultaneous fixing and staining procedure for chromosomes in crushed, smeared, or suspended cells. Natural orcein was preferred for a time; now synthetic orcein made by air oxidation of orcinol in the presence of ammonia has taken its place, and application of the Wurster H_2O_2 ammonia synthesis as devised for resorcein by Lillie et al. [*Acta Histochem. (Jena) Suppl.*, **9**: 625, 1971] has produced excellent synthetic orcein (F. Green of Matheson, Coleman, Bell, letter, 1972). Lacmoid, similarly synthesized from resorcinol, has also had wide use in this chromosome technic [Darlington and LaCour, *The Staining of Chromosomes*, G. Allen, London, 1942; Koller, *Nature (Lond.)*, **149**: 193, 1942; O'Mara, *Stain Technol.*, **23**: 201, 1948]. Spirit-soluble nigrosin (C.I. 50415) was used similarly by Von Rosen [*Nature (Lond.)*, **150**: 121, 1947] and Pares [*Nature (Lond.)*, **172**: 1151, 1953].

The technics are essentially similar. Small tissue fragments, centrifuged cell suspensions, impression smears, and the like are immersed fresh in 1% orcein or lacmoid in 45% acetic acid for 3 to 5 min to 48 h. Tissue is then gently crushed in the acetic orcein under a coverslip to attain a single cell layer. The slides are then stood on end in an alcohol vapor chamber for 48 h to fix on the prealbuminized slide and transferred to alcohol until the coverslip falls off or is easily removed. Suspensions are similarly handled; impression smears can be immersed directly in alcohol. Slides may then be briefly counterstained in 0.017 to 0.02% fast green FCF in absolute alcohol, until they appear faintly greenish. After being counterstained in absolute alcohol sections are mounted in diaphane or euparal, or cleared in xylene and mounted in synthetic resin or balsam. The above technic is a composite from Conn and Darrow, and Dalton's method in the third edition of this book.

This is a very precise red chromatin stain. Chromosomes are very conspicuous. Cytoplasm is pale green, without appreciable difference in tone between apparently surviving and definitely necrotic cells. Herein the method seems to us inferior to the azure-eosin methods, but for chromosome identification studies it is unexcelled.

Hrushovetz and Harder (*Stain Technol.*, **37**: 307, 1962) make permanent preparations from cover-glass tissue cultures. Fix 10 min in Clarke-Carnoy, let dry thoroughly in air, and stain 15 min in a moist chamber (45% acetic acid vapor) at 37°C in filtered (natural) orcein, 2% in 45% acetic acid, floating each cover glass face down on a drop of stain on a clean glass plate or slide. Wash in 45% acetic acid to remove excess stain, and in several changes of distilled water. Mount in Abopon.

Stains thus mounted have remained unimpaired as long as 7 years.

IN VIVO STAINING OF NUCLEI. DeBruyn and coworkers (*Exp. Cell Res.*, **4**: 174, 1953) report that nuclei of living cells may be stained with fluorescent diaminoacridine dyes. The affinity appears to be specific for intranuclear nucleic acids. Proflavine hydrochloride is administered intravenously to mice in doses of 25 mg/kg. Fresh frozen sections are observed under long-wave ultraviolet for fluorescence.

On fixed material these dyes show only the same specificity as other basic dyes.

Tjio and Whang (*Stain Technol.*, **37**: 17, 1972) provide the following squash and air-dried preparations for chromosomes. An advantage of the first method is that preparations are made directly from bone marrow without prior in vitro culture or in vivo colchicine administration.

SQUASH PREPARATION

1. Aspirate 0.5 ml sternal, etc., bone marrow, and add 2 to 3 ml 0.85% NaCl in 66 mM phosphate buffer containing 1 μg/ml colchicine or diacetylmethylcolchicine. Free

of blood clots, change the solution, and incubate at 24°C for 1 to 2 h. Two changes may be needed to free clots.

2. For mammals, transfer for 20 min, at 24°C, to hypotonic sodium citrate (1% in distilled water); for birds, use the NaCl-phosphate buffer (without colchicine) diluted with distilled water (1 : 4); for amphibia, use Ringer's solution diluted with distilled water (1 : 1).

3. Transfer to a watch glass containing the following: Stock solution A is made up of orcein (1st ed. C.I. 1242), 2 g; hot glacial acetic acid, 45 ml; and distilled water, 55 ml. For use at this step, dilute 9 : 1 with 1 N HCl and heat gently in the watch glass, but *do not boil.*

4. A piece of marrow is transferred to a slide; stock solution A (step 3) is added; a coverslip is applied and tapped gently and repeatedly with a blunt pencil or needle; excess fluid is removed from the edges of the coverslip by blotting; and the coverslip is sealed with a waterproof cement (Krönig cement, Fisher Scientific Co., R.R. 23, Philadelphia, Pennsylvania; xylene-cellulose caprate should serve). If slides are stored in a cold place, they may be kept for about 3 months. By freeze-drying, they may be kept permanently (Schultz et al., *Science*, **110:** 5, 1949).

AIR-DRIED PREPARATIONS

1. Proceed as for squashes, except that the second change of solution A may be omitted if the marrow pieces are very small.

2. Centrifuge at room temperature at 400 r/min for 4 to 5 min. Remove supernatant, and add 2 to 3 ml 1% sodium citrate in distilled water. Shake to loosen cells and leave for 0.5 h.

3. Centrifuge and remove the supernatant. Add 5 ml Clarke-Carnoy fluid (alcohol–acetic acid, 3 : 1); resuspend by shaking and leave for 2 to 5 min. Repeat this procedure.

4. Centrifuge and remove all the supernatant, add a drop of fresh fixative, and agitate to resuspend the cells.

5. With a Wintrobe or similar pipette put 1- to 2-m droplets of the cell suspension on chemically clean slides. Blow on each droplet gently as soon as it is in place to assist in spreading and drying. Leave to dry thoroughly.

6. Stain 10 to 20 min at 24°C in a staining jar of the orcein staining solution (stock solution A) in step 3 of the squash method.

7. Dehydrate in graded alcohols. Clear in two changes of absolute alcohol–xylene (1 : 1) to avoid excessive extraction of stain, and two changes of xylene, and mount in a synthetic resin.

Staining can also be performed with the Feulgen reaction, Wright's or Giemsa stain, or crystal violet.

Advantages of the air-dried preparation are that mechanical distortions and ruptures of cells are minimized. Flattening of cells is more uniform, and photographs are better.

Ford and Woollam (*Stain Technol.*, **38:** 271, 1963) provide a technic which combines the colchicine–hypotonic citrate sequence of Ford and Hamerton (*Stain Technol.*, **31:** 247, 1956), the method of Tjio and Whang (*Stain Technol.*, **37:** 17, 1962) for obtaining mitoses in vitro without prior culture, and the air-drying technic of Rothfels and Siminovitch (*Stain Technol.*, **33:** 73, 1958) with modifications.

Pregnant female mice are used. The fourteenth day of gestation is used; day 1 is the day the vaginal plug is observed. The animal is given an IP injection of 0.3 ml of 0.025% solution of Colcemid (Ciba). The animal is killed 1 h later; the liver from a killed fetus is removed and placed in 5 ml 0.1% Colcemid in 0.85% NaCl buffered with 0.0066 M

phosphate at pH 7. The liver is fragmented by repeated aspirations and expulsions from a fine nozzled pipette and allowed to set for $1\frac{1}{4}$ to $1\frac{1}{2}$ h at 25°C.

1. Centrifuge liver cell suspension for 5 min at 400 r/min, and remove supernatant.
2. Add 5 ml 1% sodium citrate, and let stand for 15 to 20 min and no longer.
3. Centrifuge as before and discard supernatant. If centrifugation is too rapid, cell destruction will occur.
4. Fix cells in acetic acid–alcohol (1 : 3) for 30 min 4°C.
5. Centrifuge as before, remove supernatant, and add 1 to 2 ml 45% acetic acid (time not specified).
6. Centrifuge as before, discard supernatant, and add sufficient 45% acetic acid to make slides (0.5 ml for six slides).
7. Using a fine-nozzled pipette, add the suspension in drops on the slides maintained at 54°C on a hot plate. Two rows of 4 drops each are most suitable for 3×1 slides. Each slide dries rapidly to produce a series of concentric rings in and around which numerous mitotic figures may be observed. Rapid dehydration aids in good spreading.
8. After drying, stain slides with lactic acid–acetic acid–orcein for at least 30 min at 25°C in 2% orcein in a 1 : 1 solution of 70% aqueous lactic acid in glacial acetic acid.
9. Wash off excess stain with three changes of 45% acetic acid, and allow slides to dry in air.
10. Examine under an oil-immersion lens. Permanent preparations may be made by dehydration and mounting in the usual fashion.

Caspersson et al. (*Chromosoma*, **30:** 215, 1970) advocate the following method for the selective identification of both plant and mammalian metaphase chromosomes. The highly fluorescent alkylating agent Atabrine (quinacrine) mustard effects selective binding, presumably to guanine residues in DNA. Intercalation of the Atabrine group into the double helix is also believed to occur. Human chromosomes 3, 13–15, and Y manifest particularly strong fluorescence with the reagent. Caspersson et al. suggest that because of peculiar staining patterns this method should be of special value in the identification of human chromosome groups 4–5 and 6–12. Individual patterns were also observed in groups 13–15, 17–18, and 21–22.

Ethanol-fixed slides are brought to aqueous buffer (MacIlvaine's disodium phosphate–citric acid, pH 7) and stained in the above buffer containing 50 μg/ml Atabrine mustard dihydrochloride for 20 min at 20°C. Slides are then washed in the buffer and sealed in the buffer. Specimens are examined by fluorescence microscopy. The ninth edition of "Conn's Biological Stains" provides the following structure for Atabrine mustard.

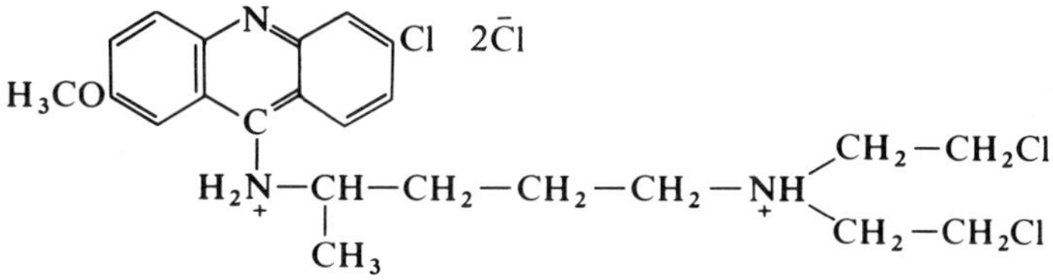

Atabrine mustard $C_{23}H_{30}N_3OCl_5$, mw 541.780

Laberge and Gagne (*Johns Hopkins Med. J.*, **128:** 79, 1971) employed the Atabrine mustard method of Caspersson (*Chromosoma*, **30:** 215, 1970) and confirmed that the distal segment of the long arms of the human Y chromosome fluoresces brightly to permit selective identification.

George [*Nature* (*Lond.*), **226:** 80, 1970; *Stain Technol.*, **46:** 44, 1971] and Pearson et al. [*Nature* (*Lond.*), **226:** 78, 1970] also used Atabrine mustard to selectively fluorochrome the heterochromatin of Y chromosomes.

Snow (*Stain Technol.*, **38**: 9, 1963) advises the use of alcoholic acid carmine as a stain for chromosomes in squash preparations. The stain is made by boiling gently 4 g carmine (C.I. 75470) in 15 ml distilled water containing 1 ml concentrated HCl. After cooling, 95 ml of 85% ethanol is added, and the solution is filtered. Squashes are prepared as usual, and fixed tissues are observed until staining is sufficient. Tissues are mounted in 45% acetic acid. Snow indicates that chromosomes are stained intensely with very lightly stained background.

De Martino et al. (*Histochemie*, **5**: 78, 1965) stained chromosomes with a silver-hexamethylenetetramine solution after hydrolysis for 5 min at 60°C in 1 N HCl.

Studzinski (*Stain Technol.*, **42**: 301, 1967) notes a remarkable affinity of nucleoli and chromosomes for zinc after chromic acid oxidation.

1. Rinse touch preparations or cultured cells in 0.14 M NaCl, and fix in Lillie's acetic acid–alcohol formalin for 10 min.
2. Oxidize in 5% aqueous chromic acid for 30 min at 25°C.
3. Wash in running water 1 min.
4. Immerse sections in 2 mM zinc acetate in 0.14 M veronal-acetate buffer pH 6.5 for 30 min at 37°C.
5. Treat sections for 10 min at 25°C in dithizone. The dithizone (diphenylthiocarbazone) is made up as a 0.1% solution in acetone at least 24 h prior to use. The staining solution prepared immediately before use is composed of dithizone solution and distilled water (1.5 : 1).
6. Rinse in acetone-water (1 : 1) for 5 s.
7. Mount in an aqueous medium such as Kaiser's glycerogel.

RESULTS: Selective red staining of interphase cell nucleoli and chromosomes undergoing mitosis is observed. Selective staining is not observed in paraffin-embedded or formalin-fixed frozen cut sections.

Guard (*Am. J. Clin. Pathol.*, **32**: 145, 1959) recommends both the following staining methods for the sex chromatin (Barr body). The methods employ fast green as a differentiating agent to displace Biebrich scarlet from nuclear chromatin except the Barr body. The first method requires periodic microscopic observation to determine the stage at which differentiation should be terminated. In method 2 sections are treated overnight without need for checking.

METHOD 1

1. Fix smears in 95% ethanol for 10 min at 24°C.
2. Immerse in 70% ethanol 2 min 24°C.
3. Stain in the following mixture for 2 min at 24°C:

Biebrich scarlet (C.I. 26905)	1 g
Phosphotungstic acid	0.3 g
Glacial acetic acid	5.0 ml
50% ethanol	100 ml

4. Rinse in 50% ethanol.
5. Differentiate in the fast green mixture for 1 to 4 h as needed:

Fast green FCF (C.I. 42053)	0.5 g
Phosphomolybdic acid	0.3 g
Phosphotungstic acid	0.3 g
Glacial acetic acid	5.0 ml
50% ethanol	100 ml

During this step, check the differentiation under a microscope at hourly intervals. When all the cells reveal green cytoplasm and the vesicular nuclei are also green, usually in about 4 h, the reaction is complete. The pyknotic nuclei, however, will not be differentiated and will reveal bright red color.

6. Immerse in 50% ethanol for 5 min.
7. Dehydrate and mount.

RESULTS: Sex chromatin (Barr body), pyknotic nuclei, and nuclei of leukocytes are red; all else stains green.

METHOD 2 Hematoxylin–Biebrich scarlet–fast green (H-BS-FG)

1. From 95% alcohol transfer to 70% alcohol for 2 min.
2. Stain in dilute hematoxylin for 15 s only, or dip the slides in the solution just 10 times.

Harris hematoxylin	0.50 ml
50% ethyl alcohol	100.00 ml

3. Without rinsing, stain in Biebrich scarlet mixture for 2 min as in method 1.
4. Rinse in 50% alcohol.
5. Differentiate in the fast green FCF mixture of method 1 by leaving the slides in the solution for 18 to 24 h. The process here does not require microscopic checking.
6. Dehydrate, clear, and mount.

RESULTS: Sex chromatin is red; nuclei are polychromatic; nucleoplasm and cytoplasm are green. It is important to restrict hematoxylin staining to 15 s; otherwise, overstaining will occur, and detection of sex chromatin will be obscured.

Pansegrau and Peterson (*Am. J. Clin. Pathol.*, **41**: 266, 1964) later modified Guard stain method 1 at step 2 so that the staining solution contains 50% ethanol.

Hollander et al. [*Acta Cytol.* (*Baltimore*), **15**: 452, 1971] recommend the following method for reliable and consistent demonstration of the Y chromosome (Barr body). Atabrine (quinacrine hydrochloride) or Atabrine mustard may be employed.

Smears (*not* dried) are fixed for 1 h or longer in 95% ethanol, placed in absolute methanol for 3 min, and then for 3 min each in graded ethanols to water. Smears are stained for 5 min or longer in 0.5% aqueous Atabrine, rinsed in distilled water for two changes of 3 min each, placed in 0.01 M citric acid–phosphate buffer pH 5.5 for 3 min, transferred to 0.01 M phosphate buffer at pH 7.4 for two changes of 3 min each, mounted in 0.01 M phosphate buffer at pH 7.4, and sealed with nail polish or cellulose caprate. Smears are examined with a fluorescence microscope using dark-field illumination. Incident illumination may also be employed.

RESULTS: Well-preserved male interphase nuclei contain a single well-defined, randomly located fluorescent spot within the nucleus. In metaphase plates, a brightly fluorescent area is located at the distal end of the long arm of the Y chromosome, and fainter fluorescence is noted on other chromosomes. Prolonged exposure of stains to buffers at pH 5.5 caused loss of the chromosome stain.

Latt (*Am. J. Hum. Genet.*, **25**: 45a, 1973; *Proc. Natl. Acad. Sci. USA*, **70**: 3395, 1973; *J. Histochem. Cytochem.*, **22**: 478, 1974; *J. Cell Biol.*, **62**: 546, 1974) noted that the fluorescence of 33258 Hoechst is partially quenched in chromosomes with incorporated 5-bromodeoxyuridine, thereby permitting a microfluorometric analysis. Chromosomes from cells which have replicated twice in medium containing 5-bromodeoxyuridine exhibit one brightly and one dully fluorescent chromatid indicating incorporation of 5-bromodeoxyuridine into one or two chains of chromatid DNA, respectively. Sister chromatid exchanges, evident as sharply demarcated reciprocal alterations in fluorescence in chromosomes, can be located in relation to quinacrine banding patterns. 33258 Hoechst was a gift from Dr. H. Loewe, Hoechst A/G, Frankfurt, Germany.

Dreskin and Mayall (*J. Histochem. Cytochem.*, **22:** 120, 1974) monitored thermal denaturation of chromatin *in situ* in single cells using absorption cytophotometry. Cells were stained with gallocyanin chrome alum. Above 70°C a progressive decrease of stain uptake was observed up to 90°C, at which point the dye uptake was approximately one-half that observed in cells maintained at 22°C. Between 70 and 90°C chromosomes also became susceptible to a specific endonuclease that digests single-stranded DNA only.

Dyer (*Stain Technol.*, **38:** 85, 1963) advises the following stain for chromosomes in plant tissues:

1. Fix tissues in Carnoy's fluid for 5 min.
2. Hydrolyze for 5 min at 60°C in 1 *N* HCl.
3. Tap out in lacto-propionic acid–orcein (2 g orcein is added to 100 ml of 1 : 1 mixture of propionic acid and lactic acid and filtered).
4. Squash under a coverslip and observe.

Dyer notes that if the coverslip has been coated with egg albumen prior to use, permanent preparations can be made by removal of the coverslips with attached cells during immersion in 45% acetic acid. Preparations are then dehydrated rapidly through absolute ethanol and mounted in euparal (Flatters and Garnett, Manchester, England). Probably other resinous media of about n_D 1.48 will serve (Table 5-1).

Sharma and Chatterji (*J. Histochem. Cytochem.*, **12:** 266, 1964) stained the root tips of *Allium cepa* for tryptophan and tyrosine. They noted a uniform stain for tryptophan throughout the length of the chromosome, whereas tyrosine was present in specialized euchromatic regions and absent in primary and secondary constriction regions containing heterochromatin. *p*-Dimethylaminobenzaldehyde procedures and modified Millon reactions were used.

Freytag (*Stain Technol.*, **39:** 167, 1964) states that separation and staining of orchid chromosomes is enhanced if living root tips are soaked for 1 h in distilled water; placed in a solution of concentrated HCl for 10 min; 50% ethanol, 10 min; and in a commercial bleach (Clorox, Purex, or Hy-Pro), 5 min; v/v for about 10 min, or until the tips are soft. Specimens may then be stained in the usual fashion.

HUMASON-LUSHBAUGH PINACYANOLE FOR FROZEN SECTIONS

1. Mount fresh or fixed frozen section on a slide, and drain (we suggest blotting down firmly with hard filter paper; cryostat sections should be melted and allowed to dry briefly to ensure adhesion).
2. Flood with 0.5% pinacyanole in 70% alcohol 6 to 10 s.
3. Here Humason and Lushbaugh float and wash in water; we suggest draining, blotting firmly, and washing in distilled water.
4. Remount, blot off excess water, cover with glycerol, glycerol gelatin, or Abopon. Mounts are temporary, and leaching of dyes occurs. Alcohol removes most of the stain. Highman-Apáthy gum syrup might be suitable for longer conservation.

RESULTS: Chromatin is blue to violet; muscle, violet to purple; connective tissue, pink; elastin, dark violet; hemosiderin, orange; plasma cell granuloplasm, red; amyloid, carmine red; red cells and leukocyte granules, unstained; neutral fat, unstained to faint blue; lipids, violet to purple.

Kodama et al. (*Biochim. Biophys. Acta*, **129:** 638, 1966) made flow-dichroism measurements to study the interaction of pinacyanol with DNA.

AZURE-EOSIN METHODS

Excellent as the hematoxylin-eosin stains are for permanence, we prefer for routine use procedures of the azure-eosin type. The latter give definitely good staining of bacteria and rickettsiae in tissues, and they demonstrate tissue mast cells, which are often not evident with hematoxylin.

Both azure A–eosin Y and azure A–eosin B stains at pH 4 to 4.5 on formol-fixed material yield blue nuclear DNA and cytoplasmic RNA (ribonucleic acid) and Nissl substance with pale pink collagen, pink to red muscle, and orange-red erythrocytes. By varying the pH level between 3 and 9 it is found that at some point various tissue elements will fail to take up either the acid or the basic dye. This point is generally quite sharply defined: raising the stain solution pH by 0.5 unit will cause the object to stain pale blue; lowering it by 0.5 unit makes the same object pink. With the native yellow color of the erythrocyte the colors will be yellow-green just above the pK, yellow-orange just below.

The pK levels of individual proteins can be altered by fixatives such as formaldehyde, alcohol, and heavy metals or by deamination or methylation and can be readily determined by this procedure. (See Table 7-3.) They act as sensitive indicators of early necrobiotic changes in cells, in that the normal light blue staining of cytoplasm is replaced abruptly by a brilliant pink. Regenerating liver cells are blue, in contrast with the normal, more or less violet tone. With the buffer procedure now in use it is possible to stain as many as 100 slides at a time by a routine technic without the troublesome individual differentiation formerly required, and to attain a great uniformity of results even on miscellaneous formalin material sent in by mail from various places.

These azure-eosin procedures comprise sequence stains such as Mallory's eosin or phloxine "methylene blue," and neutral stains employing azures and eosin in extempore mixtures or in preformed azure-eosinates in glycerol and methanol stock solutions. In the neutral stain group, such methods as Maximow's and Wolbach's were carried out at approximate neutrality and the excess of azure subsequently differentiated out by means of alcohol, dilute acid, "colophonium alcohol," or sunlight. We have preferred to regulate the red blue balance of the final stain by variation of the pH of the stain mixture by means of buffers. This last procedure gives very constant results on material fixed in formalin, dichromate, and mercury fixatives, with some adjustment of pH for the fixative employed. It is less suitable for picric acid and absolute alcohol–acetic acid fixations. With this method, more acid pH levels give sharper and more selective chromatin staining, denser eosin staining of muscle and red cells, and less cytoplasmic basophilia. Less acid, higher pH levels give denser nuclei, pink to blue-gray muscle, orange erythrocytes, and increased cytoplasmic basophilia. Results are more or less similar for the following methods and are given in detail later in this chapter for azure-eosin methods in general, to avoid duplication.

MALLORY'S EOSIN OR PHLOXINE METHYLENE BLUE. For Mallory's eosin or phloxine methylene blue method, Zenker fixation was prescribed. Salt Zenker preserves erythrocytes and zymogen granules better. Sections were deparaffinized and hydrated in the usual manner, treated with 0.5% iodine in alcohol for 5 to 10 min and with 0.5% sodium thiosulfate for 5 min (5% removes iodine at once), and washed in water.

1. Stain 1 h or longer at 52 to 55°C in 2.5% eosin Y or, preferably, phloxine B (C.I. 45410).
2. Cool, decant, and rinse carefully in water.
3. Stain 5 to 20 min in a mixture of equal volumes of 1% azure II in distilled water and of 1% methylene blue in 1% borax solution. This is filtered just before use and poured onto and off the slides several times to ensure even staining.
4. Then place slides in the water and decolorize individually in 0.2 to 0.5% colophonium alcohol. Mallory recommended adding 2 to 5 ml stock 10% solution in absolute to 100 ml 95% alcohol. With formalin-fixed material 3 to 10% colophonium alcohol is recommended for differentiation. (Mallory's "colophonium" is apparently commercial dry pine rosin, which is largely abietic acid.) Slides are kept in constant motion during differentiation.
5. When the general background becomes pink and nuclei are still blue (microscopic control), dehydrate in three changes of 100% alcohol.
6. Pass through one change of equal parts 100% alcohol and xylene and two changes of xylene. Mount in synthetic resin. (Mallory, of course, used balsam.)

Mallory preferred phloxine B (C.I. 45410) to eosin Y (C.I. 45380) in this technic because of the deeper reds achieved. Borax methylene blue is essentially an alkaline methylene blue which gradually alters to azures and methylene violets, changing faster when kept warm than when cold. Azure II was a mixture of equal parts of methylene blue and of azure I, which was originally azure B, or trimethylthionin. However, the usual azure furnished in the United States from 1925 to 1940 was azure A, or asymmetrical dimethylthionin, a more violet dye than azure B, and it seems probable that this dye was meant by Mallory. The borax methylene blue varies in composition with age, and may contain all three azures—A, B, and C—as well as methylene blue and methylene violets.

MAXIMOW'S HEMATOXYLIN—AZURE II—EOSIN.   Fixation in Spuler's (Maximow's —Z. Wiss. Mikrosk., **26:** 177, 1909; Mallory) variant of Zenker formalin was prescribed. Originally Maximow removed mercury precipitates with iodine alcohol before embedding and insisted on celloidin (nitrocellulose) embedding. He attached the sections to slides with ether vapor before staining. Lillie successfully used paraffin sections and the usual iodine and thiosulfate treatment after sectioning, and has found that the method works quite well on Orth- and formalin-fixed material.

1. Stain sections with alum hematoxylin as usual. Lillie used a 5-min stain in an acid, iodate-ripened solution of full (0.5%) strength. Mallory recommended 24 h in very dilute Delafield's hematoxylin (1 or 2 drops per 100 ml water). Wash in water.
2. Stain 18 to 24 h in azure II eosin: Dilute 5 ml 1:1000 eosin Y with 40 ml distilled water, and add 5 ml 1:1000 azure II. Mallory recommended water buffered to pH 6.8 to 7 with Sörensen's phosphate mixture, but see under Lillie's methods, below.
3. Differentiate sections individually in 95% alcohol until gross blue clouds cease to come out into the alcohol, and red cells and collagen are pink.
4. Pass through two changes of 100% alcohol, 100% alcohol and xylene, and two changes of xylene. Mount in synthetic resin (Maximow used dammar; Lillie had considerable fading in dammer in 2 to 3 months).

RESULTS: Nuclei stain blue; basophil leukocyte and mast cell granules, purple to violet; cartilage, purple; red corpuscles, pink; cytoplasm, blue to pink; secretion granules and eosinophil granules, pink.

GIEMSA STAIN. In place of the extempore mixture of azure II and eosin, a 1 : 50 dilution of *Giemsa's blood stain* may be used with identical results. *Wolbach* (personal communications; Mallory) recommended addition of 2 to 4 drops (0.1 to 0.25 ml) 0.5% sodium bicarbonate to a 1 : 40 dilution of Giemsa's stain. After staining he differentiated in weak colophonium alcohol, dehydrated in 100% alcohol, cleared in xylene, and mounted in cedar oil. Exposure to diffuse daylight or even sunlight after mounting produced a further differentiation which was valuable for rickettsiae. Wolbach prescribed fixation in Möller's (Regaud's) fluid rather than Zenker's.

The azure-eosinate method of Lillie and Pasternack was derived from a staining accident with the Maximow method, in which the use of acid distilled water obviated the necessity for the usual differentiation. Any commercial azure eosinate may be used, or eosinates may be prepared with the yellowish eosin Y or the deeper red eosin B (C.I. 45400) from methylene blue, azure B, toluidine blue, azure A, azure C, thionin, or new methylene blue. Nuclei vary from pure blue at the methylene blue end of the series to purple at the new methylene blue end; mast cell granules, mucin, and cartilage matrix, from blue-violet to purplish red (the *metachromatic* colors). Contrast is greatest between the normal and metachromatic colors in the middle of the series; hence we usually prefer azure A or the often nearly identical azure C. It makes little difference which is used. Commercial samples labeled with either name usually contain more or less of the other, so that on spectroscopic examination some samples of azure A have been found to be more nearly azure C than some other samples labeled "azure C," and vice versa. Other zinc-free thiazin dyes may be substituted in the same amounts, and the zinc double salts may be used as well, if appropriate allowance is made for their lower dye content. The phloxines, rose Bengal, and the erythrosins do not yield satisfactory thiazin salts for use in this technic. The eosinates are made up as 1% stock solutions in equal volumes of glycerol and chemically pure methyl or 100% ethyl alcohol.

Since about 1943 Lillie has preferred to use, in place of the prepared eosinates, aqueous solutions of the eosin and the thiazin dye preferred, mixed at the time of using (*J. Techn. Methods*, **24:** 43, 1944). The same dyes may be used. The following technic may be carried out in Coplin jars or in the Technicon staining racks:

LILLIE'S AZURE A–EOSIN B

1. Bring paraffin sections to water as usual, using the 0.5% iodine, 5% sodium thiosulfate sequence for material fixed with mercuric chloride.
2. Stain 1 h in:

	Coplin jar	Technicon
Azure A	4 ml 0.1%	6 ml 1%
Eosin B	4 ml 0.1%	6 ml 1%
0.2 *M* acetic acid	1.7 ml	34 ml
0.2 *M* sodium acetate	0.3 ml	6 ml
Chemically pure acetone	5 ml	90 ml
Distilled water	25 ml	580 ml

The larger quantities prescribed for the Technicon schedule are allowed to stand 1 h after mixing and are then filtered. The mixture is used repeatedly and discarded at the end of each week. The smaller, Coplin jar quantities are used at once and discarded after use, although we have, on occasion, successfully stained a second group of slides just after the first.

3.   Dehydrate in two or three changes of acetone and clear in a 50 : 50 acetone xylene mixture and two changes of xylene. Mount in synthetic resin. Depex, Permount, HSR, polystyrene, cellulose tricaprate, Groat's copolymer, and ester gum are satisfactory.

Because of the poor keeping quality of aqueous citric acid solution Lillie replaced the McIlvaine buffer with the Walpole series. For pH 4, use 1.7 ml 0.2 M acetic acid and 0.3 ml 0.2 M sodium acetate; for pH 4.5, use 1.25 and 0.75 ml; and for pH 5, use 0.7 and 1.3 ml, respectively. Further variations may be made from the buffer tables. Use the phosphate series below pH 3 and Sörensen's phosphates from pH 5.5 to 8. With surgical material and fresh animal tissue fixed in formalin, pH 4 is about optimal; autopsy tissues may stain better at 4.5. Zenker fixations require pH 4.5; Bouin and Carnoy fixations require pH 5 or 5.5; heavily chromated tissue may require pH 6 or 7. Lead acetate–formalin takes pH 5 to 5.5.

The optimal pH for staining with azure A–eosin B after fixation with buffered mercuric chloride (B-4, B-5) is at about 5 to 5.5. Omission of the customary iodine thiosulfate sequence before staining raises this level by about 0.5 pH. After formalin fixation, iodination has a less marked effect and perhaps acts to depress the optimal staining pH slightly. The effect of omission of the iodine thiosulfate sequence on pure acid dye staining is greater and rather irregular, so that it seems that the excess acid dye uptake may be due to absorbed Hg^{2+} rather than to protein iodination. In the mixture one may substitute 0.5 ml 1% azure-eosinate in glycerol and methanol for the separate aqueous solutions of azure and eosin (10 ml for the Technicon schedule). With Giemsa stain use 0.6 and 12 ml, respectively.

PHTHALATE BUFFER.   Substitution of 10 ml 0.2 M sodium (3.76%) or potassium (4.16%) hydrogen phthalate for the other buffer and part of the water in the above mixture gives a final buffer concentration of 0.05 M and pH 4, which is quite satisfactory for formalin-fixed surgical and experimental tissues.

Nitrocellulose sections and collodionized sections require 2 to $2\frac{1}{2}$ h staining, or 2 or 3 times as much dye. If it is necessary to preserve the collodion coat, substitute isopropanol for acetone in the dehydration and clearing schedule.

RESULTS WITH AZURE-EOSIN METHODS.   The following results apply generally to the foregoing azure-eosin technics: Blue nuclei, bacteria, rickettsiae, tigroid, and ribonucleoprotein generally; blue-violet mast cell and basophil leukocyte granules; reddish violet cartilage matrix; dark blue calcium deposits; light blue to violet or lavender cytoplasm of surviving cells; bright pink cytoplasm of necrosing and necrotic cells and muscle fibers (Zenker's degeneration or necrosis); pink to red secretion granules in pancreatic and salivary gland acini and Paneth cells; pink cytoplasm of gastric gland parietal cells, and blue cytoplasm of chief cells; yellowish green to green chromaffin after chromate fixation only; pink to red eosinophil and pseudoeosinophil granules; pink keratin, amyloid, fibrin, muscle cytoplasm, thyroid colloid, nuclear and cytoplasmic oxyphil inclusion bodies, and bone matrix; orange-red erythrocytes and hemoglobin. Various hyaline degeneration products in liver cells, in Hassall corpuscles of the thymus, in the follicles of the spleen, and elsewhere also stain pink. The mucins vary in color from unstained through pale greenish blues to fairly deep blue-violet. Variation of buffer level toward the acid side increases precision of nuclear staining and decreases diffuse basophilia. Conversely, variation to the alkaline side increases the amount of blue in various elements. James Freeman in our laboratory notes that raising the pH of the azure A–eosin B stain to 6 makes it more selective for hypophyseal alpha cell granules. Erythrocytes and alpha granules still stained red in formalin-fixed human material.

At about pH 5 Orth- and formalin-fixed material will present pale gray-blue smooth muscle, and at higher pH levels perhaps only eosinophil granules and erythrocytes remain red. Conversely, mast cell granules, cartilage matrix, and lymphoid cell cytoplasm retain their blue or violet staining at lower pH levels than do most other elements. After certain alcoholic fixations the ellipsoids of retinal rods stain brilliantly in red.

The azure-eosin methods are also widely useful in the identification of the cells of inflammatory exudates in the tissues. Lymphocytes and plasma cells present the characteristic basophilic (blue) cytoplasm (broader and eccentric in the latter) and the round nucleus with relatively coarse, deeply stained chromatin granules, perhaps radially placed. The cytoplasm of monocytes, histiocytes, or macrophages is broader and less basophilic, and may enclose phagocytosed material. The nucleus of this cell type is characteristically larger than that of the lymphocyte; is round, oval, or indented in form; and is relatively pale in appearance (*leptochromatic*), with quite finely divided chromatin. The nucleus of *Entamoeba histolytica* is much smaller, vesicular, and as pale as that of the macrophage group or paler, and contains a small nucleolus. Its cytoplasm is similar to that of the macrophage group, and may contain phagocytosed erythrocytes. The granules of the polymorphonuclear leukocyte are less well shown in section material than in Giemsa stained smears. They are much more conspicuous in some species, notably rabbits and guinea pigs, than in others.

Leishmania and the leishmania forms of *Trypanosoma cruzi* in heart muscle, skeletal muscle, and skin are well shown, with deep blue rod-shaped blepharoplasts and lighter blue rounded trophonuclei in relatively lightly basophilic cytoplasm. Sarcosporidia, encephalitozoa, and toxoplasmata all appear as fairly conspicuous bodies with blue chromatin in the pink backgrounds of striated muscle and central nervous system tissues. *Bartonella bacilliformis* in human liver is well shown by azure-eosin methods, notably Wolbach's Giemsa variant. The buffered azure-eosin technics should serve with this material. Plasmodia present deep blue chromatin and lighter blue cytoplasm. In human spleen they may sometimes be more easily discerned with an iron hematoxylin technic.

Young and Smith (*J. R. Microsc. Soc.*, **82:** 233, 1964) employed Euchrysine GGNX (C.I. 46040) as a stain for fixed and living blood cells, bone marrow, cultures of rabbit corneal epithelium and endothelium, trypanosomes, and sections of rabbit pancreas and stomach. Nuclei emitted a green fluorescence, nucleoli gave an orange fluorescence, and the nuclear sex chromatin body manifested a yellow fluorescence. Cytoplasmic granules containing RNA or acid mucopolysaccharide manifested a red fluorescence, particularly at pH 7.4. Sections fixed with alcoholic fixatives are stained for 5 min in a 1 : 10,000 solution at pH 7.4. Sections fixed in formalin required a 10-min staining period. Sections are then washed 3 to 10 min in running water, blotted, air-dried, mounted in DPX, and examined by fluorescence microscopy.

MORDANT STAINS FOR NUCLEI AND OTHER STRUCTURES

These include stains with hematoxylin; brazilin; carmine; the oxazin dyes gallocyanin, gallamin blue, and celestin blue; and anthracene blue and other catecholic substances combined with or in sequence after treatments with salts of iron, chromium, copper, aluminum, tin, and other chelate-forming metals and with phosphotungstic, phosphomolybdic, and molybdic acids.

Iron Hematoxylins

These fall into overstaining and regressive differentiation methods, divided into sequence (mordant) methods and combined solution (metachrome) methods, and into progressive methods with acid or an excess of iron salt to prevent overstaining.

The sequence methods depend on a preliminary mordanting with an aqueous solution of a ferric salt, staining with an aqueous solution of hematoxylin until black, and then differentiating with an acid or a ferric salt solution until the desired grade of differentiation is evident on microscopic examination. These methods are widely used by cytologists and protozoologists for the study of chromatin, chromosomes, spindles, centrosomes, mitochondria, and other cytologic details. They are too cumbersome for routine use in pathology because of the excessive amount of time required for each slide.

Hematoxylin solutions for these technics are variously prescribed as fresh or ripened. Ripening is usually accomplished by allowing the solution to stand at 15 to 25°C for weeks or months. Some of the proposed ripening technics arouse skepticism, such as the proposal to ripen by bubbling air through the solution for a few hours (Rawlins and Takahashi, *Stain Technol.*, **22**: 99, 1947). Although it is possible that bubbling air through an alkaline aqueous solution would serve to oxidize catechol residues of hematoxylin, Rawlins and Takahashi do not say whether an aqueous or an alcoholic solution was used. Acid solutions are more resistant to oxidation.

Recent experience indicates that air bubbling requires some 3 to 4 weeks to ripen alum hematoxylin, and that even pure oxygen takes some days (Palmer and Lillie, *Histochemie*, **5**: 44, 1965). Moreover hematoxylin as it comes from the manufacturer contains a variable proportion of hematein, sometimes containing as much as 100% ("Colour Index"). Also, some technics encourage the carryover of a little iron-alum solution into the hematoxylin bath. In this case the Fe^{3+} ion itself serves as an effective oxidant.

Short exposures to mordant and to hematoxylin, and fresh hematoxylin solutions tend to yield blue-black stains; longer exposures and aged hematoxylins yield black or even brownish black stains.

Benda's method, two of M. Heidenhain's, a Masson-Regaud variant, and an iron chloride method of Mallory's are presented in Table 7-4.

The myelin methods, which generally require alkaline differentiation, are discussed under lipids in Chap. 13.

Regressive iron hematoxylin methods are still much used for the identification of intestinal *Entamoebae* in smear preparations. For this purpose a preliminary 15-min fixation of fresh, still moist films in Schaudinn's mercuric chloride alcohol is prescribed. Then treat with 0.5% iodine in 70% alcohol for 3 min, and decolorize with 95% alcohol, or for 1 to 2 min in 5% sodium thiosulfate. Wash in water and stain as usual with Mallory's, Heidenhain's, or other technics.

We see no especially good reason why one of the premixed "neutral" or acid iron hematoxylins should not be used for this purpose. We have often used Weigert's acid iron chloride hematoxylin with picrofuchsin in the study and identification of *Trypanosoma cruzi* in heart muscle. The blepharoplasts appear as intensely black, rod-shaped structures; the trophonuclei, as rounded, gray bodies. *Entamoeba histolytica* is also well shown in sections by this method, and plasmodia and sarcosporidia are readily identified. However, such parasites are often more easily found when azure-eosin methods are employed.

Some elaborate schedules for staining *E. histolytica* in tissue have been prepared. Lillie cited one of these, Goldman's (*Am. J. Clin. Pathol.*, **21**: 198, 1951), in the second edition of this book. However, Lillie has found the usual Weigert iron chloride hematoxylin, Van Gieson picrofuchsin sequence quite satisfactory for demonstration of *Entamoeba* in tissues, and far less time-consuming.

More generally useful in pathology are the premixed ferric salt hematoxylins. These fall into two groups: those mixed with an optimal amount of iron for dense staining and used

TABLE 7-4 COMPARATIVE SCHEDULES FOR SEQUENCE IRON HEMATOXYLIN STAINING

	Benda	Heidenhain	Heidenhain	Masson-Regaud variant	Mallory
Mordant solution	Liq. Fer. Sulph. Oxid. PG in 50–33% dilution	1.5–4% iron alum in water	2.5% iron alum	5% iron alum	5% ferric chloride
Time and temperature	15–20°C, 24 h	15–20°C, ½–3 h	15–20°C, 3–12 h	45–50°C, 5–10 min	25°C, 1 h or more
Washing	Dist., then tap water	Water	Water	Thorough, water	3 changes water
Hematoxylin	1% aq.	0.5% aq.	0.5% aq.	1%	0.5% aq.
Ripened?	No	No	Yes, 4–6 weeks	No	No; fresh
Staining time and temperature	15–20°C, until black	15–20°C, 30 min	15–20°C, 24–36 h	45–50°C, 5–10 min	25°C, 1 h or more
Differentiating agent	5–30% acetic, or weaker acetic, or 1:30 Liq. Fer. Sulph. Oxid. PG	1.5–4% iron alum	2.5% iron alum	⅔ saturated (6%) alcoholic picric acid	0.25% ferric chloride
		See note	See note		

Note: Liquor Ferri Sulphurici Oxidati PG (*Pharmacopeoia Germanica*) consisted of 80 g iron sulfate, 40 g water, 15 g sulfuric acid, and 18 g nitric acid, and contained 10% metallic iron by weight. Iron alum is $Fe(SO_4)_3 \cdot (NH_4)_2SO_4 \cdot 24H_2O$. 15 to 20°C is taken as average German laboratory temperature, 25°C as American. Iron alum solutions should be freshly prepared from clean violet crystals. Saturated alcoholic picric acid solution is approximately 9% according to H. J. Conn, "Biological Stains," 7th ed., Williams & Wilkins, Baltimore, 1961. Heidenhain directs periodic transfer to water for microscopic examination during differentiation.

Sources: Benda's formula and the second formula of Heidenhain above are from P. Ehrlich, "Encyclopädie der mikroskopischen Technik," Urban & Schwarzenberg, Berlin, 1903; the first formula of Heidenhain is from A. B. Lee, "The Microtomist's Vade-Mecum," 10th ed, edited by J. B. Gatenby and T. S. Painter, Blakiston, Philadelphia, 1937; the Regaud formula is from Masson (*J. Techn. Methods*, **12**: 75, 1929); and Mallory's is from F. B. Mallory, "Pathological Technic," Saunders, Philadelphia, 1938.

for regressive staining (principally in the myelin and related methods) and those made more or less selective for nuclei by addition of acid or by an excess of ferric salt.

About 0.5 grams metallic iron in the ferric state is required for each gram of hematoxylin for maximal staining (Lillie and Earle, *Am. J. Pathol.*, **15**: 765, 1939). This corresponds to 3.9 ml of the official solution of iron chloride, to 2.419 g $FeCl_3 \cdot 6H_2O$, or to 4.32 g iron alum. Doubling the quantity of ferric iron in the mixed solution prevents overstaining.

The two commonest mixed iron hematoxylins used for regressive staining are Weigert's and Weil's. Both prescribed the mixture of an equal volume of an aged and "ripened" 1% alcoholic hematoxylin with an aqueous solution of a ferric salt. Weigert used 4 ml official (*Pharmacopoeia Germanica*, or PG) iron chloride solution and 96 ml water. A 2.5% (w/v) solution of ferric chloride crystals may be substituted for the dilute official solution (4 ml USP or PG solution of iron chloride = 2.481 g $FeCl_3 \cdot 6H_2O$). Weil used 4% iron alum. With Weil's hematoxylin Lillie found a hematoxylin solution only 1 to 5 days old superior to the solution several months old prescribed by Weil. Also it is readily available at all times. Inasmuch as ferric salts oxidize hematoxylin promptly, it is difficult to see the necessity for previously oxidized hematoxylin.

Satisfactory formulas for progressive staining of nuclei without overstaining are those of Janssens and of Weigert. Both stain nuclei in 3 to 5 min and do not overstain in 10 to 30 min, and color effects are the same.

Janssens (*Cellule*, **14**: 203, 1898) dissolved 2 g hematoxylin in 60 ml methanol, and 20 g iron alum in 200 ml water, mixed the two solutions, and added 60 ml glycerol. The solution becomes blue-black at once, then purplish, and in about an hour brownish black. It remains usable for perhaps 4 to 5 weeks.

Weigert's (*Z. Wiss. Mikrosk.*, **21**: 1, 1904) solution is less stable, lasting only 2 to 3 weeks, and if large numbers of sections are being stained, it may have to be made fresh twice weekly. It is the easiest to prepare, requiring only one weighing and no glycerol. Mix 100 ml fresh 1% hematoxylin in 95% alcohol with 95 ml distilled water, 1 ml concentrated hydrochloric acid, and 4 ml of the official (USP XI = PG) solution of iron chloride (2.48 g $FeCl_3 \cdot 6H_2O$ may be substituted). This hematoxylin undergoes the same color changes as Janssens' formula, but even more rapidly.

Since the official iron chloride solution has disappeared from the *United States Pharmacopeia*, it has become relatively difficult to obtain, and it now seems indicated to replace it with a simple aqueous solution of reagent-grade ferric chloride. The pharmacopeial

TABLE 7-5 EQUIVALENCE OF IRON MORDANT SALTS AND SOLUTIONS

Salts usually prescribed	mmol Fe	Metallic iron, mg	$FeCl_3 \cdot 6H_2O$, mg	$Fe_2(SO_4)_3 \cdot 9H_2O$, mg	Iron alum, mg	Official solution iron chloride mm³	Official solution iron chloride mg
Fe, 1 mmol	1.0	55.85	270.32	281.02	482.21	390.55	558.5
Iron alum, 4 g*	8.3	463.3	2242.5	2337	4000	3240	4633
Sol. iron chloride, 4 ml†	9.2	512.6	2481	2579	4941	4000	5126
Sol. iron chloride, 4 g	7.2	400	1936	2013	3544	2747	4000
Iron alum, 1 g	2.1	115.8	560.6	583	1000	809.5	1158
Sol. iron chloride, 1 ml	2.3	128.15	620	645	1235	1000	1281.5
Sol. iron chloride, 1 g	1.8	100	484	503	863.4	699.3	1000

* Iron alum is $(NH_4)_2SO_4 \cdot Fe_2(SO_4)_3 \cdot 24H_2O$.
† Sol. iron chloride = Liquor Ferri Chloridi PG and USP I–IX.

solution contained 10% by weight of metallic iron and had a specific gravity of 1.230 to 1.283. Thus it was 2.29 M ferric chloride, or 37.2% (w/v). However, the Weigert aqueous solution can be further simplified by using 1.5 g anhydrous $FeCl_3$ ($=2.5$ g $FeCl_3 \cdot 6H_2O$), 1 ml concentrated HCl (37.5%), and 99 ml distilled water (92.5 mM). A 0.1 M $FeCl_3$ solution in 0.1 N HCl would probably serve.

HANSEN'S FERROHEMATEIN. F. C. C. Hansen's iron hematoxylin (Z. *Wiss. Mikrosk.*, **22:** 45, 1905) was conceived of as ferrous hematoxylin lake, according to a reaction $Fe_2(SO_4)_3 + C_{16}H_{14}O_6$ (hematoxylin $\rightarrow 2FeSO_4 + C_{16}H_{12}O_6$(hematein) $+ H_2SO_4$, or 1 mol iron alum $(NH_4)_2Fe_2(SO_4)_4 \cdot 24H_2O) + 1$ mol hematoxylin gives 2 mols ferrous sulfate and 1 mol each of hematein, ammonium sulfate, and sulfuric acid. Hansen's ferrohematein corresponds exactly to this formulation, making the assumption that hematoxylin is a pure substance, uncontaminated by hematein.

Dissolve 10 g iron alum and 1.4 g ammonium sulfate in 150 ml distilled water and pour into a solution of 1.6 g hematoxylin in 75 ml distilled water. Heat and boil not more than 30 to 60 s. Cool, and keep in a nearly full bottle. Stain 1 to 10 min and differentiate in 2 to 3% sulfuric or 0.5 to 1% acetic acid; or for pure nuclear staining add $\frac{1}{3}$ to $\frac{1}{2}$ volume of 1% sulfuric acid to 1 volume of stain (from Romeis).

The official solution of ferric sulfate, used by Benda in his sequence iron hematoxylin method, also contained 10% iron by weight, according to both the 1884 German pharmacopeia (*Pharmacopoeia Germanica*) and "The United States Dispensatory." Its specific gravity of 1.43 indicates a content of 14.3 g iron or 71.9 g ferric sulfate crystals, $Fe_2(SO_4)_3 \cdot 9H_2O$ per 100 ml.

The common iron alum $(NH_4)_2SO_4 \cdot Fe_2(SO_4)_3 \cdot 24H_2O$ used in histology contains 11.58% metallic iron.

Since the active agent in each of these three preparations is ferric iron, they are readily interchangeable in equivalent quantities. The Weigert 2.5% $FeCl_3 \cdot 6H_2O$ should be suitable.

Doubling the HCl content of the iron solution and reducing the amount of hematoxylin render the Weigert formula more selective for nuclei, and addition of ferrous sulfate stabilizes it so that with occasional to moderate use the solution remains active for several weeks.

STABILIZED IRON CHLORIDE HEMATOXYLIN (Weigert-Lillie)

Distilled water	298 ml
Concentrated hydrochloric acid (37.5%)	2 ml
Ferric chloride crystals $FeCl_3 \cdot 6H_2O$	2.5 g
Ferrous sulfate crystals $FeSO_4 \cdot 7H_2O$	4.5 g
Fresh 1% alcoholic solution of hematoxylin, added when salts are dissolved	100 ml

RESULTS: The solution turns blue-black at once and is usable within a few minutes. Over some weeks the color slowly shifts toward brown. If the solution is much used, it may be necessary to replace it sooner because of depletion of dye.

A further, quite stable variant of this formula is our *ferrous sulfate hematoxylin* (*Anat. Rec.*, **159:** 368, 1967), which gave relatively pure nuclear staining, adequate in 2 min, not overstained in 30 min. It can be used 1 h after mixing, but is better after 24 h and can be used repeatedly for several weeks. It probably should be replaced after 2 months.

Two solutions are prepared: 100 ml fresh 1% alcoholic hematoxylin, and the aqueous iron solution of 98 ml distilled water, 65 g ferrous sulfate ($FeSO_4 \cdot 7H_2O$), 0.5 g ferric chloride ($FeCl_3 \cdot 6H_2O$), and 2 ml concentrated (12 N) HCl. On mixing, a blue-black solution is produced.

During the 1973 hematoxylin shortage crisis Lillie, Pizzolato, and Donaldson (*Stain Technol.*, **48:** 350, 1973) substituted gallein (C.I. 45445) for hematoxylin in a modified Weigert-Weil acid-iron formula. Gallein is with difficulty soluble in water (0.2%), rather slowly in alcohol (1.5%), and readily in glycol (6.0%). These solubilities are from Gurr (1960).

Dissolve 500 mg gallein in 10 ml glycol. Add 40 ml absolute alcohol. Dissolve 2 g iron alum $[NH_4Fe(SO_4)_2] \cdot 12H_2O$, violet crystals only, in 49 ml distilled water, and add 1 ml concentrated hydrochloric acid. Mix the two solutions. The mixture at once presents a blue-black color and is ready for use. Alone it stains adequately in a few minutes. For use in nuclear stain in an acidified Van Gieson procedure, stain 20 min, wash in water, and stain 4 to 5 min in 100 mg acid fuchsin, 100 ml 1.25% picric acid, and 0.25 ml concentrated HCl. Dehydrate and differentiate in two changes of 95% alcohol, then absolute alcohol and xylene, two changes each. Mount in synthetic resin.

RESULTS: Collagen is red; nuclei, black; muscle and erythrocytes, yellow; and cytoplasms, grayish yellow. The amount of hydrochloric acid in the iron gallein regulates the specificity of nuclear staining. In cytoplasms 0.5 ml gives much gray; as much as 1.5 ml can be used, but 1 ml gives about the best contrast on formol-fixed tissue.

The iron gallein is being used successfully for amebas in Schaudim-fixed fecal smears (Dees and Abadie, *J. Parasitol.*, **60:** 1036, 1974). Many of the older formulas specify saturated sublimate. Taking German laboratory temperature of about 1900 to be 15°C, this means about 5.5%. A convenient standard here would be 5.43%, or 0.2 *M*. This facilitates comparison with other heavy metals.

Increasing the mercury content above this level is of questionable value. In view of the present high price of mercury salts and the problem of environmental pollution, used solutions should be poured into a special container and the mercury precipitated as oxide or carbonate, filtered out, and recovered as chloride.

IRON "TONING" OF ALUM HEMATOXYLIN STAINS.   When a preparation stained in alum hematoxylin is rinsed in distilled water and then immersed in 1.5% $FeCl_3$ (2.5% $FeCl_3 \cdot 6H_2O$) for 30 s, nuclear staining is changed from blue to black. Use of alum hematoxylin with picric acid plasma stains or with picric acid mixtures of the Van Gieson type yields a deep red-brown nuclear stain. Introduction of a 30-s 1.5% ferric chloride bath after the hematoxylin step changes the nuclear coloration to black. See also the Kattine phosphotungstic phosphomolybdic postmordanting.

SUBSTITUTES FOR HEMATOXYLIN.   For iron hematoxylin formulas the best substitute so far encountered (1974) is gallein, C.I. 45445, a xanthene dye closely related to the eosins, produced by condensing phthalic anhydride with pyrogallol. It carries one catechol group and an *o*-hydroxysemiquinone group, just as does hematoxylin, but on a quite different structure.

Iron gallein is prepared by dissolving 1 g gallein in 20 ml glycol and 80 ml ethanol and mixing with an iron solution of 4 g iron alum, 98 ml distilled water, and 2 ml concentrated HCl. The mixture is at once blue-black and ready for use. Lillie et al. (*Stain Technol.*, **48:** 348, 1973) prescribe for a Van Gieson stain:

1. Dewax paraffin sections and bring them to water as usual.
2. Stain 20 min in iron gallein; wash in distilled water.
3. Stain 4 to 5 min in Lillie's Van Gieson stain variant: acid fuchsin, 100 mg; saturated aqueous picric acid, 100 ml; concentrated HCl, 0.25 ml.
4. Dehydrate and differentiate in two changes of 95% alcohol.
5. Pass through absolute alcohol and xylene. Mount in synthetic resin.

RESULTS: Black nuclei are seen on the usual Van Gieson background.

Dees and Abadie (*J. Parasitol.*, **60:** 1036, 1974) find the above iron gallein a good substitute for iron hematoxylin for *Entamoeba*, *Balantidium*, and other parasites. They used 1.6 ml concentrated HCl for *Entamoeba histolytica*, 2.5 ml for *Balantidium coli* trophozoites and *Onhnocerca volvulus*, with a Van Gieson counterstain and a 20-min stain for the last two. In fecal smears 3 ml HCl was used for *Entamoeba*, and an eosin counterstain.

OTHER HEMATOXYLIN SUBSTITUTES.   Other dyestuffs can replace hematoxylin in forming iron and aluminum mordant stains. Besides the closely related carmine and brazilin (C.I. 75280), there are gallamin blue (C.I. 51045), gallocyanin (C.I. 51030), celestin blue B (C.I. 51050), and anthracene blue SWR (C.I. 58605). All these dyestuffs possess *o*-diphenolic groupings or quinonoid oxidation products thereof. It is worthy of note that good gray to black nuclear stains may be obtained by immersing sections in fresh mixtures of ferric chloride solution with aqueous solutions of catechol, pyrogallol, or hydroxyhydroquinone (1,2-, 1,2,3- and 1,2,4-hydroxybenzene, respectively) but not with phenol, resorcinol, hydroquinone, or phloroglucinol, which do not possess *o*-dihydroxy groupings. The iron *o*-diphenol mixtures darken quite rapidly and are probably unstable.

METAL UPTAKE REACTIONS WITH DILUTE HEMATOXYLIN DEMONSTRATION. Dilute hematoxylin may be used to demonstrate uptake of various metals by tissue elements. Controls with hematoxylin alone should always be compared, and further controls exposed to solvent alone at the same pH as that determined for the metal solution under test are required.

We have used 1- to 10-μM solutions of various metal salts in this way, incubating sections 1 h at 40°C, washing thoroughly in double distilled water, and then reacting for 1 and 6 h in 0.01% hematoxylin in 0.01 M pH 7 phosphate buffer.

$FeCl_3$, CrF_3, $CuSO_4$, and $ZnCl_2$ give marked enhancement of the 1-h staining of keratohyalin and trichohyalin, with blue-black colors for Fe^{3+} and Zn^{2+}, greener tones with Cr^{3+} and Cu^{2+}. Decrease in the metal concentration decreases the effect; the metals may be removed by extraction with 1 N HCl and then replaced by reexposure, indicating that it is probably a true tissue chelation of the metal. Cell nuclei require higher metal concentration, at least 1 mM, and their staining is enhanced by prolongation of the time in hematoxylin to 6 h.

Substitution of pyrogallol for hematoxylin in the above technic resulted in brown colors of keratohyalin and trichohyalin only, best with Fe^{2+} and Cr^{3+}. Dopa and catechol gave, respectively, feeble and negative results.

Wigglesworth reports some interesting results regarding iron uptake by fixed tissue sections, with particular reference to iron, which apparently goes into a masked state and becomes nonreactive to ferrocyanide. This iron is unmasked by ammonium sulfide. It has been shown that nucleic acid solutions precipitate from saturated urea solution when a little ferric chloride is added, and that the iron thus precipitated is nonreactive to thiocyanate. He attributes iron uptake to mono- and diesters of phosphoric acid, which form firm, nonionized unions, or to free carboxyl groups. The stability of the complexes formed is attributed to chelation.

WIGGLESWORTH'S IRON SULFIDE TECHNIC (*Q. J. Microsc. Sci.*, **93:** 105, 1952)

1. Fix in Carnoy, Bouin, susa, Zenker, or Helly. Prepare paraffin sections, deparaffinize, and hydrate as usual.
2. Immerse 1 min in saturated iron alum solution.
3. Wash well in running water.
4. Immerse 15 to 20 s or more in dilute ammonium sulfide.
5. Rinse, dehydrate, and mount in balsam.

RESULTS: Nuclei are blue-black; chromosomes, intense blue-black; cytoplasm, gray or brown; fibrous tissue, pale chocolate; basal layer of epidermis, blue; erythrocytes, pale blue-gray.

Staining may be intensified by transferring after step 4 to potassium ferricyanide solution, which forms Turnbull's blue. Then sections may be again put through steps 2 to 4. They acquire additional iron sulfide in the process without removal of the ferrous ferricyanide, and thus the intensity of the reaction is deepened.

Wigglesworth interprets the reaction as demonstrating free carboxyl and phosphoric acid groups. He finds that iron uptake is greatly reduced by 2 or 3 days of methylation in 0.1 N HCl in methanol.

Maximum uptake of iron by nucleic acid occurs at pH 1.5; by proteins, at pH 4 or higher. For further interesting details Wigglesworth's article should be consulted.

The recent studies of Lillie and Pizzolato (*J. Histochem. Cytochem.*, **20**: 116, 1972) relative to the selective uptake of Fe(III) and Fe(II) in sequence hematoxylin technics by tissue amino (lysyl) and guanidyl (arginyl) groups are presented at some length in Chap. 8. The methylation relied upon by Wigglesworth to identify the iron uptake sites as phosphoric and carboxyl groups was shown to be amine alkylation, and the deamination practiced by him was completely inadequate for that purpose.

Alum Hematoxylin

Aluminum salt complexes of hematoxylin are usually made with ammonium aluminum sulfate or ammonia alum and are commonly referred to as alum hematoxylins. Occasionally the potassium or sodium alum is substituted with essentially equivalent results. Since the oxidation product of hematoxylin, hematein, is required for staining and the aluminum salts, unlike ferric salts, are not oxidants in themselves, it is necessary to oxidize, or "ripen," alum hematoxylin for use. Hematein is formed gradually by bubbling air through hematoxylin solutions (3 to 4 weeks may be needed for adequate staining) and by some weeks of exposure of solutions in open containers, and also by exposure of the solid dye in opened bottles in moist climates. Certain chemical oxidants such as peroxides, iodates, permanganates, perchlorates, mercuric oxide, and ferric salts ripen hematoxylin at once, though some of these agents require heat as well.

The selectivity of alum hematoxylin for nuclei is increased by the presence of an excess of aluminum salts or, better, by the presence of acids in the solution. Many workers prefer to overstain with an unacidified alum hematoxylin and then differentiate with acid or acid alcohol or other differentiating agents.

Hematoxylin in aqueous 10% ammonium alum gives pH readings of about 2.95; inclusion of 2% acetic acid in the solution brings it to pH 2.34; inclusion of 4% acetic acid, to pH 2.22.

As recorded in the "Colour Index," the amount of hematein present in various batches of hematoxylin as manufactured is variable, the product sometimes being nearly all hematein. We have seen individual lots of alum hematoxylin ripened by the "natural process" which performed adequately in as little as 8 to 10 days; 8 to 10 weeks is, of course, more usual, though Schmorl recorded an 8- to 10-day ripening period for Boehmer's hematoxylin. Occasional lots of hematoxylin are encountered which are severely overoxidized by the standard amount of sodium iodate. In one such instance 1 g $NaIO_3$ was used to 6 g hematoxylin. This is below Mayer's proportion, though more than Lillie recommended in 1965–1972.

Adequately ripened alum hematoxylin solutions, diluted in distilled water to an equivalence of about 10 mg hematoxylin per l, present an absorption spectrum characterized by a rather broad peak with maximum at 554 to 560 mμ and a secondary low shoulder at 430 mμ. The optical density at 560 is generally 3 to 4 times that at 430. On overoxidation, whether by age and long exposure to air or by overdose of a chemical oxidant, the solution changes its color from purple through red to orange or even yellow-brown, and the 430 peak now becomes the dominant absorption maximum spectroscopically. The 560 peak may become quite inapparent.

On the basis of the studies of Palmer and Lillie (*Histochemie*, **5**: 44, 1965) Lillie recommends with iodate ripening that one try about 40 mg $NaIO_3$ for each gram of hematoxylin used and forthwith test the batch for staining, allowing graded intervals of, say, 2, 6, 20, and 60 min. If satisfactory staining is attained only with the longest interval, add another 10 mg $NaIO_3$ per g hematoxylin and test again. In this way overoxidation can be avoided, and a hematoxylin can be prepared which stains adequately in 2 to 6 min and which will improve for a time with use. Such hematoxylins are apt to have a longer useful life than those in which maximum ripening or perhaps slight overripening is attained at once.

It is hoped that definite spectrophotometric density standards for adequate ripening can be developed. The indications are good that such will be the case, but a considerable period of comparison of spectrophotometric and practical staining data on various batches of dye will be necessary first.

Other Oxidants

Though the traditional amounts, 177 mg potassium permanganate, 200 mg sodium iodate, and 500 mg mercuric oxide per g hematoxylin, work well in practice, recent work indicates that considerably smaller quantities of some of these oxidants may be used with good immediate ripening and probably a longer useful life of the solution. Quantities recommended per gram hematoxylin are sodium iodate, $NaIO_3$, 40 to 100 mg; potassium permanganate, $KMnO_4$, 175 mg; potassium periodate, KIO_4, 50 mg; mercuric oxide, HgO, 100 mg. Of these the first three are used cold, the fourth is boiled. $KMnO_4$ and KIO_4 dosages need restudy.

Considering these properties, it makes little difference how the hematein is formed. Solutions vary in vigor of staining according to the amount of dye used in their preparation, and to their stage in oxidation or overoxidation, and in selectivity on progressive staining according to their acidity. The function of alcohol seems to be principally preservative against molds; that of glycerol, to stabilize the solution against overoxidation. In fact, deliberately overoxidized 30% glycerol solutions may regain their staining capacity to a considerable extent on standing.

Table 7-6 gives a number of the more usual formulas on a per liter basis. Some methods prescribe dissolving the hematoxylin first in alcohol or water and oxidizing either by exposure to air or by addition of an oxidizing agent. On trial it seems to make no difference whether the alum, glycerol, water, acetic acid, and other constituents are added to the hematoxylin solution before or after oxidation, and the method of oxidation is similarly unimportant. Alcohol and glycerol may themselves consume chemical oxidants.

Delafield's formula is cited from Mayer in Ehrlich's "Encyklopädie." We have set the amount of alum at 60 g, which is a slight excess over the amount demanded by the 1 : 11 solution described as saturated by Mayer. Mallory called for a 15% alum solution and used ethyl alcohol throughout. The latter recommendation seems logical.

Both Mallory's and Schmorl's formulas are essentially modifications of the vague original Boehmer formula ("Encyklopädie") and are cited from the texts of the two authors.

Mayer's formula (*Z. Wiss. Mikrosk.*, **20**: 409, 1904) is cited in essentially the form given by nearly all texts subsequent to the "Encyklopädie." Lillie's modification (*Stain Technol.*,

TABLE 7-6 FORMULAS FOR 1 | ALUM HEMATOXYLIN*

Ingredients	Mayer hemalum	Mallory aqueous	Harris	Boehmer-Schmorl	Apáthy	Lillie-Mayer	Delafield	Ehrlich acid	Bullard
Hematoxylin	1	2.5	5	5	3.3	5	6.4	6.4	8
Ammonium alum	50	50	100	100	30	50	60	40+	60
Ethyl alcohol			50	60	233		200†	322	333
Glycerol	1000	1000			333	300	160	322	330
Water			1000	1000	423	700	640	322	334
Chemical ripening agent	0.2 NaIO₃‡	None	2.5HgO	None	None	0.2-0.4 NaIO₃‡	None	None	8HgO
Time and temperature	15°C, inst.	25°C, 10 days 0.44KMnO₄ 25°C, inst.	100°C, few min	15°C, 8 days Air bubbling 2-6 wk, 23°C	15°C, 6 wk	25°C, inst.	25°C, 6-8 wk	15°C, 6-8 wk 0.64NaIO₃ 15°C. inst.	100°C, few min
Alternate ripening Time and temperature									
Added Acid: Normal method	None	None	None	None	Acetic 10	Acetic 20 *omit B, c ox*	None	Acetic 32	Acetic 34
Variant method	1 citric or 20 acetic		40 acetic						
Stated life	2-3 mo	2-3 mo	Months to years	Months to years		Months to years	Years	Years	Indef.
Preservative	50 chloral hydrate	2.5 thymol			0.33 acid salicylic				

*In general it is prescribed that glycerol and acid be added after ripening is well along. In case of iodate ripening the sequence of combination of ingredients appears to make no difference. 25°C is adopted as average American room temperature, 15°C as German. Quantities are in grams for solids, in milliliters for liquids. "inst." = instantaneous. Acetic is glacial acetic acid.

† Originally 40 ml ethyl and 160 ml methyl alcohol.

‡ See text on iodate ripening.

Sources: Mayer, Z. Wiss. Mikrosk., **20**: 409, 1904; Mallory, modified from P. Ehrlich's "Encyklopädie der mikroskopischen Technik," Urban & Schwarzenberg, Berlin, 1903 and F. B. Mallory, "Pathological Technic," Saunders, Philadelphia, 1938; Harris, from Ehrlich's "Encyklopädie"; Boehmer-Schmorl, modified from Boehmer's formula in Ehrlich's "Encyklopädie" and from G. Schmorl, "Die pathologisch-histologischen Untersuchungsmethoden," 15th ed., Vogel, Leipzig, 1928; Apáthy, modified from A. B. Lee, "The Microtomist's Vade-Mecum," 10th ed., edited by J. B. Gatenby and T. S. Painter, Blakiston, Philadelphia, 1937; Lillie-Mayer, from Lillie, Stain Technol., **16**: 1, 1941, **17**: 89, 1942: Delafield, from Mayer, in Ehrlich's "Encyklopädie"; Ehrlich, from his "Encyklopädie"; Bullard, in MacCallum, "Laboratory Methods of the U.S. Army," 2d ed., p. 69, Lea & Febiger, Philadelphia, 1919.

16: 1, 1941; **17:** 89, 1942) resembles Mayer's glycerol hemalum in its solvent but contains 0.5% hematoxylin in place of the 0.4% hematein of that mixture (Ehrlich's "Encyklopädie"). Langeron cites a similar, weaker formula with 0.1% hematoxylin, 0.02% sodium iodate, 20% glycerol, and 5% alum, which was devised by Carazzi (*Z. Wiss. Mikrosk.*, **28:** 273, 1911).

Clark, Reed, and Brown (*Stain Technol.*, **48:** 189, 1973) report a modification of Cole's (ibid., **18:** 125, 1943) iodine-ripened hematoxylin which is apparently effective, easily prepared, and economical of dye: Mix 2 ml 1% alcoholic iodine, 1 ml 10% alcoholic hematoxylin, and 100 ml 1.2% aqueous potassium alum. Stopper and place in 60°C oven overnight. No note is made of how often such a solution may be reused.

Apáthy's formula is cited from Lee in modified form. The (1%) hematoxylin was first ripened in 70% alcohol, and this was then mixed with equal volumes of glycerol and 3% acetic, 9% alum solution.

Harris' formula and Ehrlich's mixture are cited from the "Encyklopädie." Lillie has set the amount of alum in the latter at 4%, which is more than sufficient for saturation at 30°C. Iodate ripening of Ehrlich's formula was suggested by Mallory, and we have used it successfully.

Bullard's hematoxylin is the most concentrated formula that we have encountered. Dr. Bullard told Lillie (April 1949) that the formula was devised for the relatively crude, dark-colored American hematoxylin available in 1915–1918. It was originally published by MacCallum in "Laboratory Methods of the U.S. Army." Bullard directs: Dissolve 8 g hematoxylin in 144 ml 50% alcohol, add 16 ml glacial acetic acid and a (heated) solution of 20 g ammonia alum in 250 ml water. Heat to boiling and add slowly (to avoid frothing) 8 g red mercuric oxide. Cool quickly, filter, and add 275 ml 95% alcohol, 330 ml glycerol, 18 ml glacial acetic acid, and 40 g ammonia alum.

If spontaneous ripening of hematoxylin solutions is preferred to one of the chemical means, it is necessary to anticipate one's needs by several months and keep a supply ahead in various stages of ripening.

Palmer and Lillie (*Histochemie*, **5:** 44, 1965) conducted spectroscopic studies of ripening and overripening products of hematoxylin.

Marshall and Horobin (*Histochem. J.*, **4:** 493, 1972) employed thin-layer chromatography and visible, ultraviolet, nuclear magnetic resonance and infrared spectroscopy to oxidation products of hematoxylin. Hematein is the only reaction product if atmospheric oxygen is the oxidant. However, oxidation with H_2O_2, $KMnO_4$, or $NaIO_3$ resulted in the formation of oxyhematein. Oxyhematein acted as an acid dye and did not complex with aluminum ions under staining conditions. At this writing the precise structure of oxyhematein has not been determined to our knowledge.

Gill, Frost, and Millis (in press) recommend a formula made in 730 ml distilled water, 250 ml ethylene glycol, 2 g hematoxylin, 200 mg sodium iodate, 17.5 g aluminum sulfate, $Al_2(SO_4)_3 \cdot 18H_2O$, and 20 ml glacial acetic acid. The equivalent amount of ammonium alum is 23.537 g, of potassium alum 24.915 g. This formula keeps well, does not form the metallic scum which is so troublesome with Harris' formula, and requires periodic filtration only to remove dislodged tissue fragments. A 2-min stain is found adequate for Papanicolaou smears, 3 to 4 min for sections when fresh, 2 min after a week's use.

For testing new samples of hematoxylin we now use a similar formula, without the glycol, since only the immediate testing results are sought. We use 500 mg hematoxylin, 96 ml distilled water, 2 g ammonium alum, 20 mg sodium iodate in that order, standing overnight, taking 0.1 ml for spectroscopy at 10 to 20 mg/l, and then adding 4 ml glacial acetic acid for staining. Part of the sample is diluted with 4% acetic acid, 2% ammonium alum to give dye concentrations of 1, 2, and 3 g/l as well as the stock 5 g/l test solution. A good hematoxylin sample gives somewhat suboptimal staining with 1 g/l in 5 min, and optimal with 2, 3, and 5 g/l.

We prefer the solutions containing glycerol because of their greater stability and those containing 0.5% or higher of hematoxylin because of their greater vigor of staining and capacity for staining larger numbers of sections before exhaustion. Acidified hematoxylins are preferable for routine nuclear staining, but the nonacidified ones have valuable properties as well. Alum percentages generally approach saturation and are lower in the partly alcoholic and glycerol solutions than in the aqueous. The alum prescribed is always ammonium aluminum sulfate, $(NH_4)_2SO_4 \cdot Al_2(SO_4)_3 \cdot 24H_2O$. Since this is a salt of strong acid with relatively weak bases, alum hematoxylin solutions present pH levels of about 2.6 to 2.8 even without added acid.

STAINING TECHNICS

Alum Hematoxylin and Eosin Y

For permanence with reasonable differentiation of nuclei and cytoplasm the preferred method is still hematoxylin and eosin. Many variants have been proposed. Some workers use acid hematoxylins; others differentiate with acid after staining with an unacidified or partly neutralized hematoxylin. Some use strong eosin solutions before the hematoxylin; some mix the hematoxylin and eosin solutions; others use aqueous or alcoholic solutions of eosin after the hematoxylin; and some insist on the color acid of eosin Y (C.I. 45380) dissolved in alcohol. Others prefer other dyes to eosin Y, such as eosin B (C.I. 45400), erythrosin B (C.I. 45430), or phloxine B (C.I. 45410). Eosin B is a slightly bluish pink, erythrosin B is quite similar, phloxine B is a deeper red, while eosin Y is a yellowish pink. The technic follows:

THE ACID HEMATOXYLIN VARIANT

1. Bring sections to water.
2. Stain 2 to 5 or more min in an acid alum hematoxylin. Zenker material may need as much as an hour. Sections do not overstain.
3. Wash thoroughly in tap water until sections are blue. Some workers add a few drops of ammonium hydroxide to 500 ml water. Some use a weak (1%) sodium acetate, or sodium bicarbonate or disodium phosphate solution.
4. Stain 1 min in 0.5% aqueous eosin Y.
5. Rinse in water.
6. Dehydrate in two changes each of 95% and 100% alcohol.
7. Pass through 100% alcohol plus xylene (50 : 50) mixture into two changes of xylene. Mount in a synthetic resin (or balsam).

RESULTS: Nuclei are blue; cartilage and calcium deposits, dark blue; cytoplasm, muscle, and other structures, varying shades of pink; mucin, often light blue; keratohyalin, often dark blue.

THE UNACIDIFIED HEMATOXYLIN VARIANT

1. Bring sections to water.
2. Stain sections 5 min or more in an unacidified alum hematoxylin. The sections are overstained.
3. Differentiate in acid alcohol until red (concentrated hydrochloric acid, 1 ml; 70% alcohol, 99 ml).

4. Wash thoroughly in dechlorinated tap water or 1% sodium acetate or bicarbonate or disodium phosphate until blue.
5. Counterstain 1 min in 0.5% eosin Y.
6. Rinse in distilled water.
7. Dehydrate with two changes each of 95% and 100% alcohol.
8. Pass through 100% alcohol plus xylene, then two changes of xylene.
9. Mount in Permount or balsam.

RESULTS: With this method results differ only slightly from the preceding method. Myelin and similar substances may retain some blue if hematoxylin staining has been prolonged and formalin or a chromate fixation has been used.

   Substitution of certain acid azo dyes for the eosins may give pleasing effects. Orange G (C.I. 16230) gives orange to yellow tones; chromotrope 2R (C.I. 16570), pink and orange-red tones; Bordeaux red (C.I. 16180), somewhat redder tone; Biebrich scarlet (C.I. 26905), varying pink to scarlet tones. Lee recommended this last in 1% solution as the best plasma stain he had tried.

   Bedrick (*Stain Technol.*, **39**: 33, 1964) recommends the use of an acid textile dye, milling red SWB (acid red 114, C.I. 23635) as a 1% solution in 1% potassium alum following hematoxylin as a particularly useful stain for intercellular bridges in the spinous layer of epithelial cells. The staining method is the same as for eosin.

   Pool (*Stain Techol.*, **44**: 75, 1969) employs the following hematoxylin and eosin method for osmium-fixed, Epon-embedded tissues. Sections are handled horizontally on a rack.

1. Cover sections (0.5 to 1 μ) with 1 to 2 drops of peroxide–sulfuric acid (30% H_2O_2 is diluted 1 : 2 with distilled water, and the pH is adjusted to 3.2 with either 0.1 or 0.01 N H_2SO_4). Sections are treated for 1 min and drained.
2. Rinse and dry sections on a hot plate for 1 min at 50°C.
3. Cover sections with 1 or 2 drops of recently filtered Harris' hematoxylin. Stain at 50°C for 1 to 2 min, or longer if necessary.
4. Remove the slide from the hot plate, and immediately wipe away excess hematoxylin. Rinse the section with 4 or 5 drops of distilled water. Drain and dry 1 min, observe microscopically, and repeat steps 3 and 4 if staining is not adequate.
5. Allow the slide to cool, and place a drop of ammoniated 70% alcohol (0.2 to 0.3% of 28% NH_4OH) on the section. A bluing effect can be seen within 3 to 5 s without microscopic aid. Drain and dry for 1 min on the hot plate; then remove and cool.
6. Place 1 or 2 drops of 5% aqueous eosin on the section. Stain for only 30 to 45 s. Rinse with distilled water, drain, dry, and make a microscopic check. If staining is inadequate, repeat this step. Use heat if necessary; however, avoid overstaining, which cannot be satisfactorily corrected.
7. Dry the preparation and place in xylene (Coplin jar) 1 to 2 min. Mount in a synthetic resin.

RESULTS: Nuclei are blue; cytoplasm is pink; nuclear and cell boundaries are well defined; Epon is clear.

Chrome Alum Hematoxylin

C. C. Hansen's chrome alum hematoxylin (*dioxyhämatein*) was recommended as a good lightfast and acid-resistant nuclear stain, superior to Boehmer's alum hematoxylin for most purposes (*Z. Wiss. Mikrosk.*, **22**: 45, 1905).

Dissolve 10 g chrome alum in 250 ml distilled water, and boil until the solution is green. Dissolve 1 g hematoxylin in 15 ml hot distilled water, and add to the chrome alum solution, mixing well. Add 5 ml 10% sulfuric acid (w/v; 2 N should serve) and then dropwise with swirling 552 mg potassium dichromate in 20 ml distilled water. Boil the mixture 1 to 2 min to complete formation of the lake; then cool. Filter before each use. The solution keeps well and stains nuclei blue-black in 30 s to 5 or more min, but does not readily overstain. The stain is resistant to fading. Plasma staining may be enhanced by decreasing the sulfuric acid to 1 to 3 ml, or by addition of ammonia or Na_2CO_3.

Stain sections 1 h (20 min to 16 h), rinse in 0.36 N HCl, wash in running water until blue, counterstain as desired, dehydrate in alcohols, clear, and mount.

RESULTS: Nuclei are well stained, also in Van Gieson preparations. The solution gives excellent stains of the tigroid granules of nerve cells, resembling chrome alum gallocyanin in this respect.

Berube, Powers, and Clark (*Stain Technol.*, **40:** 165, 1965) have prepared a stable chrome alum hematoxylin lake which may be isolated and stored in the dry state and made up for staining in 0.33 N HCl as desired. The solutions are stable for only a few weeks, depositing a black film on the walls of the container.

The lake is made as follows: Mix 10 g hematoxylin, 10 g NaOH, and 70 g potassium chrome alum. Dissolve in 600 ml distilled water. Boil gently 20 min. The solution becomes deep blue. Cool and filter, allowing filtrate to fall directly into a large container containing 3500 ml absolute alcohol, and stirring frequently to prevent aggregation of the precipitate, which forms in alcohol into a viscous mass. Immediately filter the alcohol mixture on a medium-porosity, fritted-glass funnel with vacuum. Divide precipitate into small masses with spatula, and dry at room temperature. Grind to a fine powder, and store in tightly capped glass container.

We suspect that humid climates would demand modification of the final drying stage so as to employ a dehumidified or predried airstream.

For staining dissolve 3 g of the powder in HCl of 0.33 to 0.375 N concentration. Berube, Powers, and Clark treat "*M*/3" and "3%" as equivalent. Obviously a 3% dilution of concentrated HCl is meant, not 3% HCl w/w, which is 0.828 N according to Lange. The stain gradually forms a black deposit on the walls of the staining bottle in 3 to 4 weeks. Filter when this occurs, and add more dry dye if the stain becomes too weak.

Other Mordant Dyes

Neumeyer (*Zentralbl. Allg. Pathol.*, **84:** 109, 1948) recommends as a substitute for alum hematoxylin a 0.1% solution of anthracene blue SWR extra of Badische aniline (C.I. 58605, 1,2,4,5,6,8-hexahydroxyanthraquinone) in 5% aqueous aluminum sulfate solution. Boil the fresh solution 8 min, filter while hot, and add 1 ml 40% formaldehyde per l. The solution is quite stable.

NEUMEYER'S ANTHRACENE BLUE SWR AND EOSIN TECHNIC

1. Hydrate paraffin sections (10 μ) of formalin-fixed tissue as usual.
2. Stain 10 to 30 min in the anthracene blue SWR solution.
3. Differentiate 30 s to 2 min in 2% concentrated HCl in alcohol. Nuclei resist longer differentiation quite well.
4. Wash 15 min in running water.
5. Counterstain in 0.5% eosin $\frac{1}{2}$ to 2 min, or other suitable method.
6. Pass through alcohols, xylene alcohol, and xylenes. Mount in synthetic resin.

RESULTS: Nuclei are blue-black; muscle is violet; connective tissue, bright red; cytoplasm of glycogen-laden cells, blue-violet. *Note:* a fairly fresh dye sample is required.

PEARSE'S PHOSPHORIC ACID CYANIN R METHOD FOR NUCLEI AND VARIOUS CYTOPLASMIC BODIES. Chromoxane cyanin R, C.I. 43820, usually called cyanin R or chrome cyanin R, with various brand name prefixes, is soluble in hot and cold water and in alcohol. Pearse's staining method [*Acta Histochem. (Jena)*, **4**: 95, 1957] is simple.

Deparaffinize and hydrate as usual. Stain 5 to 20 min in 0.2% cyanin R in 1% phosphoric acid at 18 to 25°C. Wash briefly in hot water (3 to 5 min) until color of section changes from orange through red and begins to turn bluish. Dehydrate rapidly in graded alcohols (70%, 85%, 95%, absolute), clear in xylene, and mount in suitable synthetic medium.

RESULTS: Nuclei are blue; collagen and reticulum, unstained or faint pink; eosinophil granules, Russell bodies, enterochromaffin cell granules, keratin, muscle, fibrin, fibrinoid, erythrocytes, myelin, nucleoli, spermatozoon heads, hypophyseal alpha cell granules, red; Paneth granules, pink. Reactions of other zymogen granules were not recorded by Pearse. $Al(OH)_3$ is figured as stained rose-red.

Iron lakes of the oxazin dyes are recommended by Proescher and Arkush (*Stain Technol.*, **3**: 28, 1929) as selective nuclear stains. The lakes are prepared by boiling for 3 to 5 min 0.5% solutions of celestin blue B (C.I. 51050), gallamin blue (C.I. 51045), and gallocyanin (C.I. 51030) in 5% ferric ammonium alum. The first of the three dyes was preferred. Nuclei stain blue-black in 3 to 5 min, leaving cytoplasm unstained. Van Gieson counterstains change the nuclear color to green.

Our recent experience with oxazin mordant dyes as alum hematoxylin substitutes indicates that celestin blue–aluminum lake tends to form a gel which is quite unsuitable for staining; that the chrome alum lakes are much too slow for routine diagnostic staining, though they give excellent deep blue nuclear stains with 24- to 48-h exposures; that the iron lake of gallocyanin, besides requiring boiling to be solubilized, gives too purple a color to contrast well with eosin; and that celestin blue B is sufficiently soluble to readily form the iron lake on simple mixing of the solutions.

Stains with this iron lake, followed by eosin, or used in place of Harris' hematoxylin in the Papanicolaou method, give results which may be difficult to distinguish from those of an alum hematoxylin (Lillie et al., *Am. J. Clin. Pathol.*, **60**: 817, 1973).

1. Dissolve 500 mg celestin blue B in 50 ml distilled water, and mix with 50 ml 4% iron alum. The solution turns blue-black on mixing and is ready to use.
2. Stain 5 min in iron alum–celestin blue B. Staining 1 min gives too light a stain; 2 min may be sufficient for some workers; preparations are not overstained in 30 min.
3. Wash in several changes of distilled water.
4. Dehydrate in 95% alcohol.
5. Counterstain according to preference 30 to 60 s in 0.1% eosin Y or B in 95% alcohol.
6. Rinse in alcohol, dehydrate, and clear through absolute alcohol and xylene; mount in synthetic resin.

RESULTS: Nuclei are blue; background is pink to red.

The same dye mixture has proved useful as a nuclear stain with oil red O fat stains (Catalano and Lillie, *Stain Technol.*, **48**: 354, 1973). A 3- to 5-min nuclear stain follows a 30-min oil red O stain.

A gallocyanin–chrome alum method has been highly recommended by Einarson (*Am. J. Pathol.*, **8**: 295, 1932) for the staining of Nissl granules, and is widely used for this purpose, especially by Scandinavian workers.

EINARSON'S METHOD. Dissolve 10 g chrome alum $KCr(SO_4)_2 \cdot 12\ H_2O$ in 200 ml distilled water, add 300 mg gallocyanin, and heat slowly to boiling. Boil with frequent shaking for 15 to 25 min, cool slowly, and filter. This solution has a pH of about 1.84. Sections of material fixed in alcohol, 20% formalin, 10% alcoholic formalin, sublimate alcohol, or a solution of 10 ml formalin, 40 ml 95% alcohol, and 50 ml 6% aqueous mercuric chloride solution stain well in 24 h, but acetic Zenker material requires a longer time. Either paraffin or nitrocellulose sections may be used. These are brought to water in the usual way before staining, stained 24 to 48 h, dehydrated as usual with alcohols, cleared with xylene, and mounted in balsam (or synthetic resin). This stain is not extracted by alcohol.

Although Stenram (*Exp. Cell Res.*, **4**: 483, 1953) and Terner and Clark (*Stain Technol.*, **35**: 167 & 305, 1960) have challenged the specificity of Einarson's gallocyanin stain for nucleic acids, the method has gained wide popularity. Berube et al. (*Stain Technol.*, **41**: 73, 1966) recommend the following modification to enhance stain specificity. Formalin-fixed, paraffin-embedded sections may be used.

Mix 150 mg gallocyanin (C.I. 51030) and 15 g chrome alum in 100 ml distilled water, boil 10 to 20 min, cool, filter, wash the precipitate, and add distilled water to make 100 ml. The filtrate may be used as a stain; however, a chelate (believed to be the actual stain) may be separated from the filtrate. Bring the filtrate to pH 8 to 8.5 by addition of ammonia, filter with suction through a medium-porosity fritted-glass funnel, and wash the precipitate with a small amount of anhydrous ether. (The dye is insoluble in ether.) The remaining precipitate is the dye. The stain is used as a 3% solution in 1 N H_2SO_4.

Shires (*Stain Technol.*, **42**: 67, 1967) employed RNase (ribonuclease), DNase (deoxyribonuclease), perchloric acid extractions, methylation, and deamination blockades on L-929 fibroblasts. He concluded that DNA and RNA stained with Einarson's (*Acta Pathol. Microbiol. Scand.*, **28**: 82, 1967) gallocyanin chrome alum method.

Kiefer et al. (*Histochemie*, **14**: 65, 1968) noted that nuclei manifested a 30% increase in staining with methylene blue (Deitch, *J. Histochem. Cytochem.*, **12**: 451, 1964) after acetylation, whereas the intensity of staining with Einarson's gallocyanin chrome alum remained constant. Kiefer et al. postulate that differences in dye uptake by the two methods are due to the influence of peculiar steric and spatial relationships of the two dyes.

Haddad [*Acta Anat. (Basel)*, **70**: 260, 1968] conducted an extensive study of the effects of several fixatives (Carnoy, 1 h; 80% ethanol, 24 h; Bouin, 6 h; Zenker, 6 h; a mixture of 80% ethanol, formalin, glacial acetic acid in the proportions 85 : 10 : 5; 10% buffered formalin pH 5.5, 24 h; and Heidenhain's susa, 4 h) on the cytoplasmic material in neurons stainable with Einarson's gallocyanin chrome alum method. The effects of RNase digestion on the cytoplasmic stainable material was also studied. RNase was able to remove all stainable cytoplasmic material in sections fixed with either Carnoy's or 80% ethanol. Stainable material remained in sections fixed with the other fixatives. Haddad concludes that Einarson's gallocyanin chrome alum method stains only RNA in the cytoplasm of neurons.

Vejlsted (*J. Histochem. Cytochem.*, **17**: 545, 1969) studied the degree of fading of brain sections stained with Einarson's gallocyanin after storage in the dark, in sunlight, or in a lighted room or mounted in several types of synthetic media. Irrespective of the mounting media and irrespective of storage in light or dark, the intensity of gallocyanin stain remained constant for approximately 3 weeks and then declined. A new stable level was reached in about 4 months at about two-fifths to three-fifths the original intensity, depending upon the tissue type.

For quantitative cytophotometry of the nucleic acids Sandritter, Kiefer, and Rick (*Histochemie*, **3**: 315, 1963) direct as follows:

Boil 150 mg gallocyanin and 5 g chrome alum in 100 ml distilled water for 10 min. Filter and restore volume to 100 ml. Readjust pH to 1.64, adding HCl or NaOH. Use only freshly prepared solution. Stain 48 h, and wash 30 min in running water. Dehydrate in 70, 95, and 100% alcohol for 5 min each. Mount in n_D 1.55 medium from Cargille, New York.

The method is used on impression smears from fresh-cut surfaces of fresh organs air-dried for 24 h and then fixed for 10 min in Carnoy ethanol–chloroform–acetic acid in a 6 : 3 : 1 mixture. For DNA determination, predigest films 2 h at 50°C in Worthington RNase (0.1 mg/ml, pH 6.5 Sörensen buffer).

Photometry was done with a 2-mμ band in a Zeiss PMQ II spectrophotometer. About 2400 cells were measured with an NA 0.35 condenser, NA 1.25 100 X immersion objective, and 10 X ocular. The measurement area diameter was 0.56 μ, and 10 to 20 measurements per cell were made. The cell area was planimetrically determined on a photograph of the cell. Results were recorded in "work units," equal to extinction multiplied by surface. Reproducibility was $\pm 3\%$. Measurements were made outside the absorption maximum at 497 mμ with a 10 mμ $\Delta\lambda$ filter, since readings at maximum (575 mμ) were too high on most objects.

STOICHIOMETRY. Absorption curves in tissue were determined with a second spectrophotometer, Zeiss PMQ II, where the cuvette box was replaced by a microscope with a 0.35 NA condenser, 100 X Neofluar NA 1.25 objective, and 10 X ocular. The measured area was 0.5 μ in diameter. A 10-mμ monochromator light band was used. The stand was equipped with a fine-movement mechanical stage from Zeiss giving a movement regulation of 0.1 μ of the measurement point I_n and blank area I_o.

The Fluka dye sample used was chromatographically pure, containing 1% ash, 11.5% H_2O of crystallination, plus 87% dye. After removal of 2.68% NaCl plus 9.5% H_2O, elementary analysis was as follows:

		C	H	N	Cl		O
mw 336.73	computed	53.48	3.89	8.32	10.53	76.22	23.78
$C_{15}H_{12}N_2O_5 \cdot HCl$	found	54	3.89	8.18	10.82	76.89	23.11

A glacial acetic acid solution 3 days old did yield three components as in the work of Terner and Clark (*Stain Technol.*, **35:** 167, 1960). In HCOOH-CH$_3$COOH-H$_2$O, 15 : 10 : 75, electrophoresis gave a single broad band going slowly to the negative pole.

Gallocyanin has an absorption maximum at 530 mμ (Conn gives 636 mμ for gallocyanin); the chromium lake, a 15 X stronger one at 575 mμ (Sandritter, Kiefer, and Rick, *Histochemie*, **3:** 315, 1963).

For combined cytophotometry of DNA and RNA in the same section Kiefer, Zeller, and Sandritter (*Histochemie*, **20:** 1, 1969) fix tissue impression smears with 95% alcohol, stain according to their usual procedure (Sandritter, Kiefer, and Rick, *Histochemie*, **3:** 315, 1963) in chrome alum–gallocyanin for 48 h, hydrolyze 12 min in 1 N HCl, and then react with a coriphosphine O–Schiff reagent for presumably 60 min, wash as usual in HCl bisulfite and in water, dehydrate with alcohols, clear in xylene, and mount in synthetic resin. Stepwise photometry after the chrome alum gallocyanin step and again after the Feulgen hydrolysis indicates how much of the original D550 is lost, and since gallocyanin uptake by DNA is proportional to coriphosphine uptake, the computation of the amount assignable to RNA is readily determined. Only about half of the RNA–gallocyanin Q complex is lost in the 12-min hydrolysis as compared with essentially total loss of uncombined RNA in a 4-min HCl hydrolysis.

PHENOCYANIN TC (C.I. 51145). Phenocyanin Fe(II)–eosin is a new technic which gives a satisfactory replacement for the traditional hematoxylin-eosin (Lillie et al., *Am. J. Clin. Pathol.*, **63:** 576, 1975). Preparations of formol-fixed tissue are dewaxed and hydrated as usual.

1. Stain 5 to 10 min in 1% phenocyanin TC in 20% ethylene glycol or 20 min in 0.1% phenocyanin TC in 1% concentrated HCl in water or in distilled water.
2. Rinse in distilled water.
3. Afterchrome 5 to 10 min in 0.28% (10 mM) ferrous sulfate (FeSO$_4$·7H$_2$O) to convert the rather light and easily extracted blue-green stain to a highly insoluble dark blue.
4. Wash in distilled water; dehydrate in 95% alcohol.
5. Counterstain 30 s in 0.1% eosin (Y or B) in 95% alcohol.
6. Dehydrate in 95% and absolute alcohols, clear in xylenes, and mount in synthetic resin.

RESULTS: Nuclei are dark blue; bacteria, small yeasts, Nissl substance, pancreatic acinar, and gastric chief cell cytoplasms, blue to deep blue; cartilage, mast cell granules, certain acid mucins, dark blue-violet; axons, violet; other cytoplasms, smooth and striated muscle, and collagen, pink to red with eosin.

In its good staining of bacteria, yeasts and other RNA sites as well as of cartilage, mast cell granules, and certain acid mucins, the stain, as done above, seems superior to the usual alum hematoxylin technics.

Since the stain reacts strongly with bisulfite aldehyde sites (sulfonic acids?), it gives a dark blue partly replacing the expected red-purple when used as a nuclear stain after the periodic acid Schiff (PAS) reaction. When used before the PAS reaction, the nuclear stain is considerably weakened, probably by bisulfite reduction and removal of iron. Hence, for the PAS reaction, the use of one of the iron gallein formulas is suggested as the nuclear stain if iron hematoxylin must be replaced.

Nuclear staining is prevented by selective extraction of nucleic acids in 1 N HCl 60°C or 10% perchloric acid for 5 min at 95°C, leaving the mucopolysaccharide staining intact.

A simple phenocyanin stain is readily washed out by acid-alcohol, ammonia-alcohol, pyridine, glycol, chloroform plus methanol, 36% urea, etc. A 10-min afterchroming in 10 mM copper sulfate moderately increased resistance to these solvents, and ferrous sulfate afterchroming was almost completely effective in preventing extraction. Potassium oxalate (2 M), which is known to remove even hemosiderin iron from sections, moderately impaired the iron phenocyanin stain, being less effective on simple and copper phenocyanin.

C.I. 51145, Phenocyanin TC
$C_{21}H_{15}N_3O_6Cl$, mw 426.8165

Carmine (C.I. 75470)

Formerly the most important dyestuff in the histologist's armamentarium, carmine is less and less used today. We have used it as a nuclear stain in such procedures as Weigert's fibrin stain and occasionally in other methods but now no longer use it in any procedure except the

Best carmine stain for glycogen, for Mayer's mucicarmine, and for coloring injection masses red. Formulas for these will be given under the special methods.

Orth's formula (Mayer in Ehrlich's "Encyklopädie") called for 2.5 to 5 g carmine, dissolved by boiling 15 min in 100 ml saturated aqueous lithium carbonate solution. After cooling, 1 g thymol was added as a preservative. Sections were stained 2 to 5 min and differentiated in a 1 : 100 dilution of concentrated hydrochloric acid in 70% alcohol.

In Grenacher's formula Mayer (Ehrlich's "Encyklopädie) recommended 2 g carmine dissolved by boiling for 1 h in 100 ml 5% ammonium alum, restoring water to maintain volume. After cooling Mallory added 1 g thymol. Grenacher originally used 0.5 to 1% carmine and 1 to 5% alum.

Mayer's (Ehrlich's "Encyklopädie") alcoholic carmine ("paracarmine") may be used when it is desired to avoid aqueous solutions. Dissolve 1 g carminic acid, 0.5 g aluminum chloride, and 4 g calcium chloride (crystals?) in 100 ml 70% alcohol. Filter before using. Stain sections 15 to 30 min, and wash with 70% alcohol, adding 2.5% glacial acetic acid if a more purely nuclear stain is desired.

MAYER'S CARMALUM (MALLORY). Dissolve carminic acid, 1 g, and ammonium (or potassium) alum, 10 g, in 200 ml distilled water, warming if necessary; cool and add 0.2 g salicylic acid or other suitable preservative, such as merthiolate. See also Mucicarmine, in Chap. 14.

CHROME ALUM CARMINE. Fyg (*Z. Wiss. Mikrosk.*, **45:** 442, 1928) recommended highly a chrome alum carmine which gives a dark blue nuclear stain that combines well with either eosin or Van Gieson stain.

Boil 1 g carmine 15 min in 6% chrome alum solution, cool, and filter. Stain in this 30 min to 12 h. It does not overstain. It can be differentiated slowly with 1% (0.12 N) HCl in 70% alcohol, or more rapidly with Weigert's borax ferricyanide mixture. Wash in water, and counterstain 5 min in Van Gieson stain or 30 to 60 s in 0.1% alcoholic eosin. Dehydrate in alcohols, clear in xylene, and mount in balsam or synthetic resin. Results compare favorably with those of alum hematoxylins.

Anthocyanins, Anthocyanidins

A group of flower and fruit colors, anthocyanins, are generally glucosides of glucose or cellobiose with anthocyanidins. These have the general formulation

Cyanidin

Delphinin, C.I. 75190
(R represents the glucoside linkage.)

existing when isolated from the glucoside usually as chlorides or sometimes picrates. They present a varying amount of hydroxy and often methoxy substitution. Thus cyanidin is 3,5,7,3′,4′-pentahydroxyanthocyanidin, syringidin (malvidin) is 3,5,7,4′-tetrahydroxy-3′,5′-dimethoxyanthocyanidin, delphinidin is 3,5,7,3′,4′,5′-hexahydroxypyrylium chloride.

Those used in histology, chiefly as nuclear stains, are chiefly myrtillus or myrtillin and sambucin.

Myrtillus is the juice of the bilberry or European whortleberry, the *Heidelbeere* of German writers (*Vaccinium myrtillus*). It was used by Lavdowsky (*Arch. Mikrosk. Anat.*, **23:** 506, 1884). Pappenheim (*Virchows Arch. Pathol. Anat.*, **157:** 19, 1899), Lipp (*Mikroskopie*, **6:** 32, 1951), and Erdmann (*Mikrokosmos*, **40:** 151, 1951). According to Harms myrtillus is a mixture of delphinidin and syringidin.

Sambucin, the other commonly used product, is the juice of the black elderberry (*Sambucus niger*), the *Hollunderbeere* of the German writers. Novelli (*Experientia*, **9:** 152, 1953) called it *sambucyanin*. Harms identifies it as a mixture of cyanidin mono- and diglucosides. It has been used as a nuclear stain by Claudius (*Zentralbl. Bakteriol.* **5:** 579, 1899), Kappers (*Z. Wiss. Mikrosk.*, **28:** 416, 1911), and Gruber (*Mikroskopie*, **4:** 187, 1949).

These stains have been used as the fresh juices, as fermented and boiled juice, as exhaustive alcohol extracts, and as lead acetate precipitates freed from lead by H_2S. Dry products are prepared by low-temperature vacuum desiccation and are quite stable. These are generally used at 1% levels in water, slightly acidified. Alum addition has appeared beneficial with myrtillus, but is not spoken of with sambucin. This probably relates to the 3′,4′,5′-trihydroxy structure of delphinidin. Myrtillin shows some enhancement and stabilization of color with alum. Iron, chromium, and other metallic mordants do not appear to have been tried. The dyes apparently function as cationic dyes from somewhat acid aqueous solutions. Some writers have commented on their availability as alum hematoxylin substitutes.

One of the earliest nuclear stains was observed by Hartig in 1850 with phytolacca juice (Gertz, *Z. Wiss. Mikrosk.*, **33:** 7, 1916). Other juices tried have been less satisfactory as far as the reports go. They include blackberry (Claudius), black currant (H. Fol, "Lehrbuch der Vergleichenden Mikroskopische Anatomie," 1884, p. 183, and also 2d ed., 1896), phytolacca (Hartig, cited by Gertz), blackberries and dahlia leaves (Fol, from Kappers), *Sambucus ebulis* (Novelli), phytolacca (Novelli, *Experientia*, **9:** 224, 1953), wild strawberry juice (Kappers).

Clerocyanidin from *Clerodendron trichotomum* was reported by Novelli (*Bull. Microsc. Appl.*, **4:** 6, 1954) to stain collagen from picric acid mixtures. Harms indicated a similar effect with red beet juice.

8

CHEMICAL END GROUPS

ETHYLENES

Ethylenic linkages exist principally in histochemistry in relation to the unsaturated fatty acids and in certain long-chain aliphatic alcohols which are found in cutaneous fats such as hair oil or sebaceous secretions.

There are four principal histochemical methods for the demonstration of ethylenic double bonds: the oldest empirically discovered reaction with osmium tetroxide (Schultze, *Arch. Mikrosk. Anat.*, **1**: 299, 1865); the demonstration of aldehyde residues with Schiff reagent or other reagents after oxidation with peracetic or performic acid (Lillie, *Stain Technol.*, **27**: 37, 1952); the new bromination-silver technics depending on the introduction of bromine into double bonds and its subsequent removal by silver nitrate to form insoluble, photosensitive silver bromide which is then reduced to metallic silver *in situ* (Mukherji et al., *J. Histochem. Cytochem.*, **8**: 189, 1960; Norton et al., ibid., **10**: 83, 1962); and the fourth, oxidation with $K_2Cr_2O_7$ at pH 3.5 and 3°C, followed by acetic hematoxylin to yield a black lake, proposed by Lillie (*Histochemie*, **20**: 338, 1969) in explaining the mechanism of the Smith-Dietrich reaction. This reaction was observed by Boehmer (*Aertzl. Intell. Bl. München*, **12**: 539, 1865) and utilized by Weigert (*Fortschr. Med.*, **2**: 190, 1884) in the myelin method.

With most lipids osmium tetroxide adds quantitatively to double bonds in the proportion of 1 molecule osmium tetroxide to each C=C residue (Kahn, Riemersma, and Booij, *J. Histochem. Cytochem.*, **9**: 560, 1961) for oleic acid and simple oleates, 2 mol osmium tetroxide to 1 glycerol trioleate, but the reaction is modified in choline and ethanolamine (colamine) phosphatides (lecithins and cephalins). Generally the first reaction product is an osmic ester

$$
\begin{array}{c}
\mathrm{H-\overset{|}{C}} \\
\mathrm{\underset{|}{\overset{\|}{C}}-H}
\end{array}
+
\begin{array}{c}
\mathrm{O} \quad \mathrm{O} \\
\diagup \mathrm{Os} \diagdown \\
\mathrm{O} \quad \mathrm{O}
\end{array}
\longrightarrow
\begin{array}{c}
\mathrm{H-\overset{|}{C}-O} \quad \mathrm{O} \\
\diagup \mathrm{Os} \diagdown \\
\mathrm{H-\overset{|}{C}-O} \quad \mathrm{O}
\end{array}
\xrightarrow[+2H_2O]{}
\begin{array}{c}
\mathrm{H-\overset{|}{C}-OH} \\
\mathrm{H-\overset{|}{C}-OH}
\end{array}
+
\begin{array}{c}
\mathrm{HO} \quad \mathrm{O} \\
\diagup \mathrm{Os} \diagdown \\
\mathrm{HO} \quad \mathrm{O}
\end{array}
$$

which then hydrolyzes to a glycol and a lower hydroxy acid of osmium (Criegee, *Liebigs Ann. Chem.*, **522**: 75, 1936; **550**: 99, 1942).

Osmium tetroxide also blackens other substances: enterochromaffin, melanin, adrenaline and noradrenaline, probably also dopamine, eleidin, tannin, and other strongly reducing substances. When osmium tetroxide treatment is preceded by treatment with more selective oxidants, such as dichromates, the blackening is restricted to substances, such as triolein, which resist oxidation by the potassium dichromate bath. This is the basis of the Marchi procedure for degenerating myelin. During the earlier part of this century, following the introduction of the Sudan dyes, osmium tetroxide passed into almost total disuse in the study of fats, except for some very special purposes, such as the demonstration of the Golgi lipids. With its introduction as a nearly ideal morphologic fixative for electron microscopy it has received renewed attention, not only for fixation but also for histochemical study.

OSMIUM TETROXIDE METHOD.   Cut frozen, gelatin or polyvinyl alcohol sections of formalin-fixed material.

1. Soak part of the sections in 2.5 % potassium dichromate for 2 days, then 1 or 2 days more in a mixture of 6 ml 2.5 % potassium dichromate and 3 ml 1 % osmium tetroxide. Wash in water and mount in gum syrup. Degenerating myelin and neutral fats are blackened.
2. Treat another group overnight in 0.25 % osmium tetroxide, and wash in distilled water.
2a. Mount in glycerol gelatin or aqueous polyvinyl alcohol glycerol.
   Or:
2b. Dehydrate and mount in balsam or synthetic resin.
   In critical studies, multiple sections and graded time intervals should be used, and appropriate controls should be applied. Also, preparations may be mounted in osmium tetroxide with a cover glass sealed with petroleum jelly (Vaseline) and observed repeatedly during the reaction period.
3. Treat a third group of sections 24 h in 1 % osmium tetroxide, wash in water 6 to 12 h, soak several hours in absolute alcohol to obtain the secondary staining of fats. Wash in water and mount in gum syrup. This procedure is supposed to demonstrate saturated neutral fats as well as unsaturated, but probably it actually demonstrates those lipids which are easily oxidized by potassium dichromate and are dissolved in the neutral fats. The alcohol undoubtedly removes unaltered alcohol-soluble lipid and may serve to reduce further the osmic ester or lower acid.

Material treated in the block with osmium tetroxide, alone or with chromates, can be embedded in paraffin or nitrocellulose. For paraffin embedding, clearing with cedar oil or briefly with chloroform is preferred, and chloroform balsam is said to be better for mounting.

The Marchi (*Riv. Sper. Freniatr.*, **12**: 50, 1886) method was based on observations that pretreatment with dichromate prevented osmic reduction by normal myelin but not that of degenerating myelin and neutral fats. Swank and Davenport (*Stain Technol.*, **9**: 11, 1934; **10**: 87, 1935) found that potassium chlorate gave better results than potassium dichromate in the Marchi method, and this variation has been generally adopted. Adams (1968) links the differential reaction to the readier penetration of hydrophilic phospholipids and sphingolipids by chlorate and dichromate while osmium tetroxide also readily penetrates hydrophobic unsaturated fats, fatty acids, and cholesterol esters.

For the Swank-Davenport method anesthetize the animal at a suitable interval after injury (10 days for rabbits, 14 to 29 days for cats and monkeys) and perfuse at once with a fluid containing 70 g $MgSO_4 \cdot 7H_2O$ and 25 g $K_2Cr_2O_7$ per 1 distilled water. Human material obtained 10 to 30 days after the injury may be similarly perfused from local arteries. Then remove the material with as little manipulation as possible and place it for 2 to 7 days in calcium acetate–formol. Cut thin blocks (3 mm), and immerse for 5 to 10 days in the Poirier et al. (*Stain Technol.*, **29**: 71, 1954) fluid: 0.5 % OsO_4, 11 ml; 1 % $KClO_3$, 16 ml; 10 % HAc, 3 ml; formol, 3 ml; distilled water, 67, ml (total 100 ml). According to Smith et al. (*J. Neurol. Neurosurg. Psychiat.*, **19**: 62, 1956) human material can yield significant results up to 15 months after injury and even after 1 to 8 years' storage in formol–saline solution.

After the osmication, wash overnight in running water, dehydrate in alcohols, and embed and section preferably in celloidin or low-viscosity nitrocellulose. If facilities for celloidin sections are not available, infiltrate 4 to 7 days in 1 to 2% celloidin, harden blocks in two changes of chloroform for 30 min each, and infiltrate and embed in paraffin as usual. Dewax paraffin sections, bring to water, counterstain in thionin, azure A, or safranin O at pH 4.5 (acetate buffer 0.1 M) for 20 to 30 min, and dehydrate with 95% alcohol and absolute butanol or isopropanol to avoid softening celloidin. Mount in synthetic resin.

For some purposes Lillie preferred a frozen section method with a modified Weigert myelin stain combined with Sudan II (C.I. 21401) for the degenerating myelin.

In a recent exploration of the mechanism of the chromation hematoxylin method or Smith-Dietrich-Baker procedure, Lillie (*Histochemie*, **20**: 338, 1969) found that by reducing chromation temperature to near 0°C and prolonging the time to some 4 to 6 weeks a very selective myelin, erythrocyte staining with acetic hematoxylin was achieved, and this was completely prevented by a 10-min bromination in 0.1 N Br in 2% aqueous KBr. The Cr deposited from CrO_3 and $K_2Cr_2O_7$ mordanting reacts with hematoxylin in 1% acetic acid; that taken up from chrome alum, CrF_3, or $CrAc_3$ requires a pH level of 6 to 7 in the hematoxylin solution. When the Cr(VI) deposit (1) is reduced by mild reducing agents (2), it retains greater stability to extraction agents than does Cr(III). It is deduced that the form bound from Cr(VI) mordanting is actually Cr(IV) as described by Ogata and Ogata (*Beitr. Pathol.*, **71**: 376, 1923) in the chromaffin reaction. Lillie concluded that the Cr(IV) was present as a metal ester (2) formed across the double bond as with Mn and Os. This could be reduced by mild reduction to a still stable Cr(III) ester (3), might dehydrate to a perhaps less reactive form (4), and might form a fairly stable hematoxylin chelate (5).

With chromation at 60°C, as in the Baker procedure, bromination only moderately attenuates the reaction, and acetylation and nitrous acid deamination also impair it moderately, indicating that amino and probably also hydroxyl groups participate in Cr(IV) binding. The acetic hematoxylin picture after deamination and 60°C chromation is morphologically and topochemically similar to that of cold chromation.

PERACETIC AND PERFORMIC ACID SCHIFF PROCEDURES. Certain oxidants, such as performic and peracetic acids, and potassium permanganate attack ethylenes to form epoxides, glycols, peroxides, and aldehydes. Chemical studies have shown about 92% recovery of glycolic fatty acids from peracetic oxidation of unsaturated fatty acid (Findley et al., *J. Am. Chem. Soc.*, **67**: 412, 1945; and Swern et al., ibid., **67**: 1786, 1945). Experiments in our own laboratory gave on oxidation of 0.1 mol allyl alcohol with an excess of peracetic acid and distillation of the reaction mixture, a recovery of about 8 mmol formaldehyde, as determined by iodine titration of the bisulfite compound; this agrees perfectly with the works of Swern, Findley, and Scanlan.

These oxidation technics, reported by Pearse (*Q. J. Microsc. Sci.*, **92**: 393, 1951) and Lillie (*Stain Technol.*, **27**: 37, 1952) essentially simultaneously,[1] depend on the production of aldehyde by performic or peracetic acid and the demonstration of that production by Schiff's sulfite leucofuchsin reagent according to the reactions:

$$CH_3-(CH_2)_x-CH=CH-(CH_2)_y-COOH + 2CH_3CO_3H \longrightarrow$$

$$CH_3-(CH_2)_x-\underset{\underset{O-O}{|\quad\quad|}}{CH-CH}-(CH_2)_yCOOH \longrightarrow CH_3-(CH_2)_xCHO$$

$$+ OCH-(CH_2)_y-COOH \text{ by cleavage of the peroxide}$$

Permanganate, by virtue of its ability to also cleave the α-glycols formed by oxidation of the ethylene bond, with the formation of aldehyde, also gives a positive Schiff reaction which might well be more intense by virtue of the possibility of essentially total conversion of the double bond to aldehyde residues, but it also oxidizes aldehyde to carboxylic and carbonic acid residues. Prior bromination (1 h at 25°C in 1 ml bromine to 39 ml carbon tetrachloride) completely blocks the reaction, whereas acetylation does not. After the oxidation step, interposition of blockades with sulfite, aniline, phenylhydrazine, or semicarbazide prevents the lipids from coloring with Schiff reagent when the usual 10-min exposure is used. Long exposures (2 to 3 h) may overcome sulfite blockades and brief (30 min) phenylhydrazine blockades; and the aldehydes react with the usual red-purple color. Prolonged phenylhydrazine and aniline hydrochloride blockades (4 to 72 h) appear to be permanently effective. Elleder and Lojda (*Histochemie*, **24**: 7, 1970) in a study of the plasmal reaction seemed inclined to attribute the reaction entirely to plasmal, apparently ignoring or denying the occurrence of the reaction where plasmalogen and other soluble lipids have been extracted and no plasmal reaction can be elicited with $HgCl_2$. Lillie commented on this paper at greater length (*Histochemie*, **25**: 191, 1971).

The action of performic acid and of peracetic acid is quite comparable, and both reagents are readily prepared. According to Greenspan (*J. Am. Chem. Soc.*, **68**: 907, 1946) performic acid is formed in adequate concentration in 30 to 60 min, but deteriorates virtually to inactivity by the next day. Peracetic acid takes 2 to 3 days to reach maximum concentration; it may be kept several weeks. We have used the same jar repeatedly. Hence the performic acid is preferable for an occasional test, and the peracetic is better for frequent routine use. The following directions for manufacture of the reagents are derived from Greenspan:

PERFORMIC ACID REAGENT. To 8 ml 90% formic acid add 31 ml 30% hydrogen peroxide and 0.22 ml concentrated sulfuric acid. Keep at or below 25°C. About 4.7% performic acid (HCO_3H) is formed within 2 h, and the solution deteriorates after a few more hours. Make fresh daily.

PERACETIC ACID REAGENT. To 95.6 ml glacial acetic acid add 259 ml 30% hydrogen peroxide and 2.2 ml concentrated sulfuric acid. Let stand 1 to 3 days. Add 40 mg disodium phosphate as stabilizer. Store in the refrigerator at 0 to 5°C. We have kept such solutions for months. A single Coplin jar of this reagent may be used for 8 to 10 groups of nine slides before it is discarded; but a positive control should be included, at least in the later groups.

---

[1] *Q. J. Microsc. Sci.*, vol. 92, no. 4, actually appeared about 2 weeks later than *Stain Technol.*, vol. 27, no. 1.

Using the foregoing formulas, a maximum concentration of about 4.7% performic acid is reached in 2 h, and a maximum of 8.6% peracetic acid in 80 to 96 h. If 90% hydrogen peroxide is used, concentrations of peracetic acid around 45% are reached. Since there are indications that this high a concentration of peracetic acid may destroy aldehyde (Findlay's letter, 1954), its routine use is not recommended.

LILLIE'S PERACETIC OR PERFORMIC ACID SCHIFF REACTION

1. Fix and section in accordance with the solubility requirements of the lipid under study: for ceroid, any fixation and paraffin or frozen sections; for lipofuscin pigments, routine formalin and paraffin sections; for retina and myelin, aqueous formaldehyde or, better, formaldehyde dichromate sequence fixations and paraffin sections.
2. Bring sections to water as usual.
3. Oxidize 2 h in peracetic acid reagent or 90 min in performic acid reagent.
4. Wash 10 min in running water.
5. Immerse in Schiff reagent for 10 min.
6. Wash in three changes of 0.5% sodium bisulfite or metabisulfite, 2, 2, and 2 min.
7. Wash 10 min in running water.
8. If desired, counterstain 1 to 2 min in Weigert's acid–iron hematoxylin, wash 4 min in running water, and stain 1 min in saturated aqueous picric acid solution, or use other suitable counterstain.
9. Dehydrate, clear, and mount by the alcohol, alcohol plus xylene, xylene, polystyrene sequence, or if required because of the solubility of the lipids concerned, wash in water and mount in glycerol gelatin or polyvinyl alcohol.

RESULTS: Ceroid, retinal rod acromere lipid, and many lipofuscin pigments react with a red-purple color. Myelin sometimes reacts quite well, especially in frozen sections or in well-chromated paraffin material.

If no counterstain is employed, nuclei are also colored red by the concurrent Feulgen reaction. The latter may be induced by corresponding mixtures of formic or acetic acid with a little sulfuric acid and distilled water replacing the hydrogen peroxide solution. The red of nuclei is almost completely suppressed by iron or alum hematoxylin counterstains.

Hair cortex also colors red-purple, a reaction which Pearse (*Q. J. Microsc. Sci.*, **92:** 393, 1951) attributed to cystine, although he was not able to repeat the reaction with cystine in vitro. Like that of ceroid, this reaction is prevented by prior bromination and is not reversed by even prolonged extraction in fat solvents, hot or cold. Like ceroid, hair cortex is also acid-fast by Ziehl-Neelsen technics; but, unlike ceroid, it is not sudanophilic in paraffin sections. The Schiff reaction is best prevented by sulfite blockade.

Pearse makes peracetic acid by mixing 5 ml 30% hydrogen peroxide with 20 ml acetic anhydride and notes that it is less active than his performic acid mixture. Apparently he used it immediately, without allowing the 24- to 48-h interval which Greenspan recommends for reaction to form peracetic acid. His performic acid reagent is made by adding 4 ml 30% hydrogen peroxide to 40 ml 98% formic acid and letting the mixture react for 1 h before using it.

Greenspan used about 1.57 mol of hydrogen peroxide per mol acetic acid, a proportion designed to give a maximum yield of peracid. Pearse's prescription calls for only 0.2 mol of hydrogen peroxide per mol acetic anhydride. Thus the latter contains much less water and much more acetic acid in proportion than does Greenspan's mixture.

With performic acid, Greenspan's directions call for about 1.55 mol peroxide per mol formic acid; Pearse's, for only 0.038 mol, again providing less water and a far greater excess of formic over performic acid.

This reaction is controlled by omission of the hydrogen peroxide from the performic or peracetic acid reagent, substituting the same amount of water.

BROMINATION BLOCKADES AND THE BROMINE-SILVER PROCEDURES. Prior chlorination or bromination, but not iodination, prevents the foregoing reaction. Iodine chloride and bromide should also serve but have not been used histochemically. For insoluble lipids, bromine in carbon tetrachloride 2.5% by volume is usually employed; for substances soluble in carbon tetrachloride the preferred mixture now seems to be a solution in aqueous potassium bromide (Norton et al., *J. Histochem. Cytochem.*, **10**: 83, 1962), rather than bromine-water or gaseous bromine. Chlorine and bromine and strong halogen acids react with ethylene groups to form halogen-substituted saturated compounds:

$$RCH{=}CH{-}R'{-}COOH + Br_2 \longrightarrow RCHBr{-}CHBr{-}R'COOH$$

and

$$RCH{=}CH{-}R'COOH + HI \longrightarrow RCHI{-}CH_2R'{-}COOH$$

It must be noted that with prolonged exposures such polysaccharides as starch and glycogen are also brominated or chlorinated (3 to 6 h in 2.5% bromine in carbon tetrachloride).

Since prolonged exposure of aliphatic halogen compounds to silver nitrate results in the formation of silver halides, bromination has been made the basis of two bromine-silver methods for localization of ethylenic residues, that of Mukherji et al. (*J. Histochem. Cytochem.*, **8**: 189, 1960), which used bromine vapor and preferred an ammoniacal to a neutral silver nitrate solution, and that of Norton et al. (*J. Histochem. Cytochem.*, **10**: 83, 1962), which brominated in a weak potassium bromide–bromine solution and silvered with strongly acid silver nitrate to avoid nonspecific silver reduction. Norton and coworkers did not specifically note the negative silver reactions of melanin and enterochromaffin, but it is known that the argentaffin reactions of these substances are prevented by adequate prior oxidation and by bromination and iodination specifically (*J. Histochem. Cytochem.*, **5**: 325, 1957, for melanin; ibid., **9**: 184, 1961, for enterochromaffin). Whether the 1-min bromination of this technic is adequate for this purpose remains undetermined. Mukherji and coworkers have noted a positive reaction of sterols.

The reactions involved are as follows:

$$RCHBr{-}CHBr{-}R'{-}COOH + 2AgNO_3 \longrightarrow$$
$$RCH(NO_3){-}CH(NO_3){-}R'{-}COOH + 2AgBr$$

$$AgBr + H \longrightarrow Ag + HBr \text{ (light or photographic developer)}$$

BROMINATION-SILVER TECHNIC OF NORTON et al. FOR UNSATURATED FATS. Fix 18 h or more in Baker's calcium-cadmium-formol saturated with calcium carbonate. Wash blocks 3 to 5 h in running water. Cut 10-μ frozen sections and collect in distilled water.

1. Immerse 1 min in 0.1 N bromine in 2% potassium bromide (add 1 ml bromine to 388 ml 2% potassium bromide) and rinse in distilled water.
2. Immerse 1 min in 1% sodium bisulfite.
3. Wash in seven changes of distilled water.
4. Treat 18 to 22 h in 1% silver nitrate in 1 N nitric acid in the dark.
5. Wash in seven changes of distilled water.

6. Develop in Eastman Dektol diluted one-half in distilled water, for 10 min.
7. Wash well in water, mount in glycerol gelatin or glycerol polyvinyl alcohol. Reactive sites are brown to black. Suitable counterstains can be tried if desired.

The reaction is controlled as follows:

1. By omission of free bromine from the 2% potassium bromide of step 1.
2. By 1- to 2-h oxidation in performic acid before step 1. Wash in seven changes of distilled water.
3. By removal of silver bromide by immersion in 5% sodium thiosulfate for 5 min at step 5. To accomplish the removal of silver bromide in this way steps 4 and 5 and the thiosulfate extraction must be done in the photographic darkroom (Pizzolato, *J. Histochem. Cytochem.*, **10**: 102, 1962). A 1% KCN extraction in the presence of air will, as usual, remove the silver, either before or after development. About 20 min should serve.

The reacting lipids may also be removed by extraction with appropriate solvents. In this case it must be remembered that Baker's calcium-cadmium-formol was designed to insolubilize phospholipids and that after formalin fixation many such lipids remain demonstrable in paraffin sections with Sudan and hematoxylin procedures.

The bromine solution is stable for some weeks at least.

With chromation at 60°C, as in the Baker procedure, bromination only moderately attenuates the reaction; acetylation and nitrous acid deamination also impair it moderately, indicating that amino and probably also hydroxyl groups participate in Cr(IV) binding. The acetic hematoxylin picture after deamination and 60°C chromation is morphologically and topochemically similar to that of cold chromatin.

PEROXIDES

Peroxides are formed in the course of air oxidation of unsaturated fats and give rise to aldehyde cleavage. The same process apparently occurs to some extent in the course of performic and peracetic acid oxidation of olefins, although in this case the principal products are epoxides and glycol groups.

Sehrt (*Munch. Med. Wochenschr.*, **74**: 139, 1927) and Lison (*Bull. Soc. Chim. Biol.*, **18**: 185, 1936) showed that the Winkler-Schultze, or "M-Nadi," indophenol blue synthesis from α-naphthol and *p*-amino-*N,N*-dimethylaniline could be mediated nonenzymatically by partially oxidized unsaturated fats. Lillie found that part of the lipofuscin pigment of the human ovary and rat liver ceroid in choline deficiency cirrhosis give this reaction under relatively anaerobic conditions, and the reaction is enhanced or engendered anew by prior peracetic acid oxidation, thus supporting the view that the aldehyde produced by this oxidant from ethylenes goes through a peroxide stage. The following directions are from Lillie and Burtner (*J. Histochem. Cytochem.*, **1**: 8, 1953):

THE WINKLER-SCHULTZE REACTION APPLIED TO FATTY ACID PEROXIDES

1. Paraffin or frozen sections are brought to distilled water as usual. Dissolve 576 mg (4.02 mmol) α-naphthol in 6 ml 1 N sodium hydroxide, and dilute to 50 ml with distilled water. Dissolve 691 mg (4.00 mmol) *p*-amino-*N,N*-dimethylaniline (as *p*-dimethylphenylenediamine) in 50 ml distilled water. Mix, filter, and use at once.
2. Immerse sections in mixture for 3 min, or flood them with it.
3. Decant, and rinse in distilled water.

4. Immerse 3 min in dilute Lugol's solution (0.2% iodine, 0.4% potassium iodide). Wash 20 min in lithium carbonate, 5 mg/100 ml distilled water. Mount in glycerol gelatin. Faint to moderate blue or green indicates presence of peroxides. The browner pigments give a greener color.

Granados and Dam (*Acta Pathol. Microbiol. Scand.*, **27**: 591, 1950) noted that the earlier, lighter yellow stages of vitamin E deficiency pigment gave the leucodichloroindophenol reaction for peroxides, and Glavind et al. (*Experientia*, **5**: 34, 1949) reported the following method devised by that group for the histochemical localization:

THE HEMIN-CATALYZED LEUCODICHLOROINDOPHENOL PEROXIDE REACTION OF GLAVIND et al.

SOLUTION A

1. Prepare the stable *stock solution of hemin* (obtained from Hoffman, LaRoche) by dissolving 40 mg in 10 ml pyridine and 20 ml glacial acetic acid. This keeps for a long time.
2. Prepare leuco-2,4-dichlorophenolindophenol (identified by Gibbs as 3,5-dichloro-4,4'-dihydroxydiphenylamine) by the method of Gibbs, Cohen, and Cannan (*Public Health Rep.*, **40**: 699, 1925) or by reducing a solution of 2,6-dichlorophenolindophenol in 50% alcohol with the stoichiometric amount of ascorbic acid. Precipitate as sodium salt with NaCl. Purify by redissolving in alcohol and reprecipitating with water, repeating several times to eliminate traces of ascorbic acid.

Gibbs and coworkers made the indophenol by condensing 2,6-dichloroquinonechloroimid[1] with phenol in alkaline solution and salting out with NaCl and purifying by repeated aqueous re-solution and salting out, and then reduced the product with K_2S and recrystallized from an acid solution containing H_2S to prevent reoxidation (ibid., **39**: 381, 1924).
SOLUTION B. Dissolve 25 mg leuco-2,4-dichlorophenolindophenol (3,5-dichloro-4,4'-dihydroxyphenylamine) in 3.5 ml absolute alcohol. Mix with 5 ml distilled water. Use at once.

Float frozen sections onto slides, and blot dry. Add 0.74 ml solution A to solution B, mix, and pour onto slides. Let stand 3 to 5 min, wash thoroughly with distilled water, and mount in glycerin gelatin or Farrants' gum arabic.
RESULTS: Peroxide sites are shown in red. Since the retention of the red dye locally depends on its oil solubility in the fats, preparations are not dehydrated. The blotting before staining is important, since an excess of water will precipitate the hemin and prevent the reaction.

ALCOHOLS

As Lison aptly states, this grouping has had little histochemical interest. Undoubtedly alcohol groups are among those rendered strongly metachromatically basophil by sulfation and phosphorylation, but these reactions do not permit their distinction from α-glycols and perhaps phenols as well. Hence detailed comparison of results of these procedures with those of procedures demonstrating α-glycols and phenols more specifically would be necessary to localize alcohol sites which do not fall into the other categories.

Actually the class of compounds which should be considered as nonglycolic aliphatic alcohols is considerably larger and more important than Lison seemed to consider, if one

---

[1] Available from K and K Laboratories, Jamaica, N.Y.

includes in it, as seems proper, the nonperiodate reactive acetylaminopolysaccharides which are not already sulfated. We hesitate to include the Landing-Hall (*J. Histochem. Cytochem.*, **4**: 41, 1956) phosphorylation reaction, since it now appears that phosphorylation of tissues engenders, besides basophilia, a positive Schiff reaction in many structures such as elastin, collagen, pituitary chromophils, gastric parietal cells, Paneth cells, erythrocytes, muscle, and glycogen. Moreover Landing and Hall considered that phosphoryl chloride reacted not only on OH groups but also on SH and primary and secondary amines (Haust and Landing, *J. Histochem. Cytochem.*, **9**: 548, 1961). The Schiff reaction was tentatively attributed to $\equiv$PO or $=$POCl groups. The former seems improbable, since $\equiv$PO appears in normal phosphates, which are not Schiff-positive.

Hence the sulfation technic appears to be the method of choice for creation of sulfate ester residues on alcohol sites. Mowry's technic (*J. Histochem. Cytochem.*, **6**: 82, 1958; McManus and Mowry, p. 142) appears to be convenient for this purpose.

MOWRY'S ETHER SULFURIC SULFATION METHOD

1. Pack cracked ice around a Coplin jar containing 25 ml anhydrous ethyl ether.
2. Add gradually 25 ml concentrated sulfuric acid. The reaction is exothermic. When the mixture is again cold, cover the Coplin jar with a cover that has its edges coated with Vaseline (petroleum jelly; petrolatum USP).
3. Deparaffinize sections with xylene; wash in absolute alcohol and in ether.
4. Sulfate in the ether sulfuric acid mixture for 5 min (or more, if required).
5. Wash in alcohol and water.
6. Stain one pair of sulfated and unsulfated control sections in 0.01% toluidine blue in 3% acetic acid (pH about 2.5) for 5 to 30 min. Dehydrate in acetone, melted *tert*-butanol, or ethanol; clear in xylene, and mount in Permount or cellulose caprate.
7. Stain another pair of sulfated and unsulfated control sections by the periodic acid Schiff method.

Structures which are metachromatically stained by toluidine blue in sulfated tissue but not in unsulfated controls and which are negative to the periodic acid Schiff reaction without sulfation are to be considered as probably containing alcohols and to be periodate-negative polysaccharides. The adequacy of the sulfation may be appraised by the completeness of prevention of the periodic acid Schiff reaction. Mowry, as well as others, recorded strong orthochromatic blue staining of collagen with the sulfation toluidine blue sequence, and metachromatic staining of glycogen, mucins, reticulum, basement membranes, smooth muscle stroma, and, using alcoholic toluidine blue, also dextran. But apparently collagen fibers in mucin or cartilage do not stain. Alcian blue staining after sulfation was largely limited to collagen. Mowry has speculated that the positive orthochromatic reaction of collagen might be due to its hydroxyproline content.

Later (*J. Histochem. Cytochem.*, **12**: 821, 1964) it was found in a study of sulfuric acid catalysis of acetic anhydride 1 : 3 acetic acid acetylation that in histochemical usage such reagents resulted in cessation of the acylation reaction in the sulfation stage, even with very minute amounts of sulfuric acid (1 : 12,500, v/v). Reduction of the acetic anhydride to 5% by volume was tolerated; 2.5% was marginal when 1 : 400 dilutions of sulfuric acid were used. Low-pH azure A staining was induced at 1 : 62,500; blocking of Van Gieson collagen staining occurred at 1 : 2500.

Since such mixtures appear to sulfate rapidly and as fully as the older mixtures containing much larger amounts of sulfuric acid and since they engender inappreciable amounts of heat on mixing, their use seems practical, at least for inducing azure A staining at sulfation sites and for blocking the periodic acid Schiff reaction more completely than the pyridine acetylation mixtures; it appears that they may be useful for these and some other purposes. As with sulfation with the Mowry and Spicer (*J. Histochem. Cytochem.*, **8:** 18, 1960) mixtures, the sulfation effect is readily abolished by methanolysis in methanol–hydrochloric acid or sulfuric acid mixtures and is relatively resistant to alkali saponification.

SPICER'S MIXTURE. A mixture of 10 ml concentrated sulfuric acid plus 30 ml glacial acetic acid gives quite adequate sulfation of gastrointestinal mucins, basement membranes, reticulum, and collagen with a 3-min immersion, as demonstrated by a following 20-min stain with 0.05% azure A in 0.01 N hydrochloric acid. Smooth muscle, cytoplasm, and other normally eosinophilic structures are not stained.

Weaker similar sulfation mixtures, containing 7.5 and 2.5% sulfuric acid by volume, are also effective but require longer sulfation intervals, i.e., 15 to 30 min for the 7.5% and 1 h for the 2.5%.

The glacial acetic–acetic anhydride–sulfuric acid mixtures very promptly confer an intense basophilia toward azure A at pH 1 (0.125 N hydrochloric acid), not only on mucins, starch, glycogen, basement membranes, reticulum, and collagen, but also on smooth muscle and cytoplasm as well. The periodic acid Schiff reaction is blocked, as is staining with acid dyes and with both components of the fast green Van Gieson connective tissue stain. The blockade is broken effectively by 1- to 2-h methylation at 60°C, and in the case of the Van Gieson variant also partially by alcoholic ammonia or 1% potassium hydroxide in 70% alcohol (20 min).

It appears to be indicated that this reagent sulfates NH_2 groups as well as OH groups, but further studies are needed on this point.

Blockade of Hydroxyl and Amine Groups

Benzoylation and acetylation are used to render NH_2 and OH groups nonreactive by esterifying them. In general in unfixed proteins esterification of amines is accomplished more quickly and at lower temperatures. Starch, glycogen, and cellulose are the most difficult objects to render nonreactive to the periodic acid Schiff reaction. With these three hexose polysaccharides, individual particles may remain strongly reactive after fairly prolonged treatment while most of the same material demonstrable in controls has disappeared. Cartilage matrix and thyroid colloid are also strongly resistant to acetylation and benzoylation. In cartilage the pericellular areas are more resistant than the intercellular. Thyroid colloid, in benzoylation, presents strongly reactive areas adjacent to negative areas.

Cartilage matrix is almost completely acetylated after 18 to 24 h at 25°C in 40% acetic anhydride plus pyridine mixture, but 6 h at 58°C in 2.5% benzoyl chloride plus pyridine was not quite adequate. Although most of the glycogen is acetylated in similar periods, a few granules may remain. Benzoylation for 6 h at 58°C completely abolished the reactivity of glycogen. Two hours was nearly sufficient, but 24 h at 25°C was not quite adequate. Epithelial mucins are moderately resistant to acetylation, 6 h at 25°C being adequate for most of them, but are quite readily benzoylated, virtually completely so in 2 h at 25°C in 10% benzoyl chloride plus pyridine. Connective tissues, mast cell granules, cuticular borders, retinal rod acromere lipid, ocular membranes, amyloid, hyalin droplets, plasma, fibrin, and the like are acetylated and benzoylated quite promptly. Benzoylation for 20 h in 10% benzoyl chloride plus pyridine mixture is used by Danielli to prevent the reaction of histidine, tryptophan, and tyrosine with diazonium salts.

Stronger mixtures of benzoyl chloride with pyridine give rise to considerable amounts of crystalline precipitate, which redissolves when sections are brought into alcohol. These precipitates are suspected of causing localized areas of reaction failure. Weaker mixtures—

2.5% or 5% benzoyl chloride by volume in anhydrous pyridine—seem about as effective in blocking the periodic acid Schiff reaction as stronger, crystal-producing mixtures.

Benzoylation at 58°C for 6 h appears to destroy the capacity of nuclei to stain with iron hematoxylin. Hence, either a lower temperature (37 or 25°C) for a longer period should be employed, or exposure to 58°C should be limited to 2 or 3 h. Similarly, acetylation in 40% acetic anhydride in pyridine for 6 h at 58°C is about as effective as a 24-h treatment at 25°C. Neither can be relied upon to acetylate consistently all the glycogen, starch, and cellulose, but cartilage matrix appears to acetylate almost completely in 6 h at 58°C or in 18 to 24 h at 25°C.

Technics used in our laboratory are as follows: Bring sections to 100% alcohol as usual. Dip them in pyridine and transfer them to the reagent solution.

ACETYLATION. Use 16 ml acetic anhydride plus 24 ml anhydrous pyridine. Incubate 1 to 24 h at 25°C in accordance with desired effect, or $\frac{1}{2}$ to 6 h at 58°C. The longer intervals are necessary to approach complete acetylation of glycogen, starch, cellulose, and cartilage.

BENZOYLATION. Use 2 ml benzoyl chloride plus 38 ml anhydrous pyridine. Incubate 1 to 24 h at 25°C or $\frac{1}{2}$ to 6 h at 58°C. Glycogen, starch, cellulose, thyroid colloid, and cartilage require the most drastic treatments. After either treatment wash with two changes each of 100, 95, and 80% alcohol.

RESULTS: A 2-h acetylation blocks the periodic acid Schiff reaction of collagen, basement membranes, reticulum, adrenal lipofuscin, colonic "melanin," and renal brush borders, but not that of glycogen, starch, cellulose, cornea, vitreous, lens capsule, Descemet's membrane, gastrointestinal mucins, cartilage, or (usually) adrenal chromaffin. With 6 h of acetylation, starch, cellulose, glycogen, gastric mucin, cartilage, and sometimes chromaffin remain reactive. With 16 h, cartilage matrix remains faintly reactive, but this is lost at 18 to 20 h.

According to Shackleford (letter, December 13, 1961), phthalic anhydride condenses with polysaccharide sites in the same manner as acetic anhydride and thereby induces intense reactivity to toluidine blue as well as enhanced reactivity to colloidal iron and Alcian blue. It did not, under the conditions employed, apparently impair reactivity to the periodic acid Schiff reaction. This would appear to be another example of the superior efficiency of positively chromogenic reactions over those of related nature which tend to prevent another chromogenic reaction.

SHACKLEFORD'S PHTHALIC ANHYDRIDE TOLUIDINE BLUE METHOD

1. Deparaffinize and hydrate sections as usual.
2. Immerse sections 15 to 20 min in saturated phthalic anhydride pyridine heated to 58°C in a water bath in the chemical hood, heating controls similarly in pure pyridine.
3. Wash in graded alcohols to water.
4. Stain 30 min in 0.01% toluidine blue in 0.2 M acetate buffer pH 2.6 or 3.6.
5. Rinse in buffer and dehydrate quickly in *tert*-butanol or acetone, clear in xylene, and mount in synthetic resin.

Neutral and acid polysaccharides are largely or completely unstained in controls and exhibit strong orthochromatic basophilia after the acylation. As with sulfation, the presence of vicinal glycol or amino alcohol groups would not seem to be required.

Phthalic anhydride   Saccharide hydroxyl   Phthalyl saccharide ester with free carboxyl

Hence, saccharides with 1 to 3 glycosidic linkages, two acetylamino-3-hydroxy groups, and the like should react as well as glycols. It seems not improbable that phenols would also react, but the observations reported do not especially suggest reaction of protein amino groups.

TOSYLATION. In 1961 (*J. Histochem. Cytochem.*, **9:** 184) Lillie reported that tosylation prevented azo coupling, Clara hematoxylin, ferric ferricyanide reduction, and ferrous ion uptake reactions of enterochromaffin. Further studies, not pertinent to that report, indicated that tosyl chloride (*p*-toluenesulfonyl chloride) in dry pyridine and in 87.5% pyridine saturated with borax was ineffective, but usable results were obtained with the 50% acetone borax solution suggested in Noller's text. Studies with the periodic acid Schiff reaction showed that the greatest blocking effect was obtained with an 8-h interval, the blockade being less effective both at 2 to 4 h and at 16 to 32 h. Since the previously reported azo coupling and hematoxylin reactions of enterochromaffin disappear at 2 to 4 h tosylation and reappear when the tosylation time is prolonged to 32 h, it is thought that the alkaline solution hydrolyzes the sulfonyl esters formed earlier in the process. Similarly, the incomplete blocking of eosin staining seems greatest at 4 h, diminishing again thereafter. The dilute hematoxylin staining of rat arterial elastic tissue and proventricular keratohyalin was unimpaired by any of the tosylation intervals tried (2, 4, 8, 16, 32 h).

TOSYLATION TECHNIC. *Reagent:* Dissolve 1 g *p*-toluenesulfonyl chloride in 20 ml acetone. Dissolve 1 g borax ($Na_2B_4O_7 \cdot 10H_2O$) in 20 ml warm distilled water. Mix and use at once. Replacement with fresh solution at 4-h intervals is suggested for longer tosylation intervals.

Deparaffinize sections with xylene, wash in two changes of acetone, and immerse in tosylation reagent for the desired interval. A 4- to 8-h interval seems indicated, shorter for phenols, longer for amines and glycols. Wash in water and carry through the demonstration procedure. Controls are exposed to 2.5% borax in 50% acetone for the same period.

SAPONIFICATION. This was employed histochemically by McManus and Cason to reliberate glycols for reaction with periodic acid after acetylation. It was believed that only acetyl esters would be thus hydrolyzed, and not the amides formed by acetylation of amine groups (McManus and Mowry, p. 105). McManus' original saponification procedure, a 45-min exposure to 0.1 N potassium hydroxide in water, although chemically effective, often also totally removed the sections from the slides. In 1951, Lillie introduced a 20% dilution of concentrated ammonia water (28% NH_3) with absolute alcohol as an effective deacetylating reagent (*Stain Technol.*, **26:** 123, 1951), which, however, required 24-h exposures to unblock acetylated glycols effectively. A 0.1 N NaOH in 85% alcohol had effectively removed sections from slides in 1 h.

Later we tried much shorter intervals in alcoholic potassium hydroxide and found that 15- to 30-min intervals in 1% potassium hydroxide in 70 or 80% alcohol were usually chemically adequate and should not be exceeded if section losses were to be avoided. The 70% alcohol solution is perhaps a little more effective, and 20 min is usually enough.

However, a 20- to 30-min extraction with 1% potassium hydroxide in 70 to 80% alcohol, with or without prior periodic acid oxidation or acetylation, more or less completely and promptly destroys the cytoplasmic basophilia of gastric chief cells, pancreatic acinar cells, Lieberkühn crypt cells, and the like, so that in azure-eosin stains the nuclei now appear deep blue on a pale purple to pink background. This would appear to be another procedure for the selective extraction of ribonucleic acid. For periodic acid Schiff, azo coupling, and other reactions the alcoholic potassium hydroxide deacetylation procedure has been generally satisfactory. Although pancreas fixed in aqueous formaldehyde or in acetic alcohol formalin loses its cytoplasmic basophilia on 10- to 20-min extraction by 1% potassium hydroxide in 70% alcohol, the same tissue remains strongly basophil after a 24-h treatment with 5.6%

ammonia in 80% alcohol (1 volume 28% ammonia plus 4 volumes absolute alcohol). Hence, when preservation of ribonucleic acid is important, this slower saponification procedure may be employed.

This 1% potassium hydroxide, 70% alcohol saponification procedure has successfully restored the capacity of acetylated carbohydrates to react to the periodic acid Schiff reaction, of phenols to azo-couple, of catechols such as reduced melanin to reduce acid silver nitrate, of basic proteins to stain with acid dyes, and, after methylation, of sulfonic, phosphoric, and carboxylic acid residues to stain with basic dyes.

With the ammonia alcohol procedure deacetylation for shorter periods (4 to 6 h, 37°C) after the periodic acid reaction reveals as Schiff-positive or partly so such tissues as ocular lens capsule and Descemet's membrane, cornea, vitreous humor, glycogen, gastric and intestinal mucins, cartilage, and chromaffin. Longer treatment intensifies these reactions. Overnight treatments are required to deacetylate most collagen, vascular basement membranes, cuticular and brush borders of epithelial cells, and some pigments.

The acromere substance of the retinal rods is one of the easiest to acetylate (40 min in acetic anhydride plus pyridine mixture) and probably the most difficult to deacetylate, requiring the full 24 h in the 80% alcohol–20% ammonia water mixture. Even 48 h in 80% alcohol–20% ammonia may be inadequate to deacetylate this material.

GLYCOLS

In histochemistry the most widely used reaction for carbohydrates is the periodic acid Schiff sulfite leucofuchsin reaction, usually abbreviated PAS (French, APS; German, PSS or PJS from *Perjodsäure* Schiff). It is based on the Malaprade reaction, in which 1,2-glycols undergo oxidative cleavage to form aldehyde. In the Nicolet-Shinn reaction one of the hydroxyls may be replaced by a primary or secondary amino group, but apparently not by an amide group, since saccharides bearing an acetylamino group adjacent to the other available ring hydroxyl are unreactive to periodic acid. The isolated amino acids serine and threonine are attacked by periodic acid, but when the α-amino group is used in a peptide bond, as is usually the case with amino acids when protein-bound, what is essentially an amide group is now adjacent to the hydroxyl, and the grouping is unavailable for periodic acid oxidation.

It must not be assumed that this is the only oxidative action of which periodic acid is capable, but in other cases demonstrable aldehyde is not known to be produced, and a histochemically demonstrable aldehyde reaction which has resulted from relatively brief periodic acid oxidation at 15 to 35°C may generally be assumed to denote the presence of 1,2-glycol or amino alcohol groupings. If demonstrable sudanophilia is absent at the site of aldehyde reaction, the reaction is generally assigned to the presence of a polysaccharide. If the reaction is accompanied by sudanophilia, the presence of partially oxidized unsaturated fats or of glycolipids is to be considered.

Other oxidants have been used besides periodic acid. Lead tetraacetate and sodium bismuthate apparently attack much the same groupings as periodic acid. Chromic acid and potassium permanganate also produce aldehyde from polysaccharides, but they further attack and destroy aldehydes, probably converting them first to carboxyl groups and then to carbon dioxide. Further, these two oxidants attack ethylenic double bonds. This is discussed further under that topic.

The first glycol-to-aldehyde oxidative reaction used in histochemistry was the Bauer reaction (*Z. Mikrosk. Anat. Forsch.*, **33**: 143, 1933), which produced a positive Schiff aldehyde reaction in glycogen, starch, cellulose, and various mucopolysaccharides. Because of the further oxidation of aldehyde by chromic acid, less density of aldehyde reaction is achieved, and structures with fewer glycol groups are not demonstrated, such as basement membranes and collagenous and reticular fibrils. Prolonged exposure to chromic acid or potassium permanganate completely destroys aldehydes, both those which they themselves have produced and those previously formed by other agencies; native aldehyde or that produced by less drastic oxidants or hydrolzing agents (Lillie, *Stain Technol.*, **26**: 123, 1951). The potassium permanganate glycol-to-aldehyde reaction in histochemistry was reported independently by Rossman (*Carnegie Contrib. Embryol.*, **30**: 97, 1942), Casella (*Anat. Anz.*, **93**: 289, 1942), and Lillie (*Bull. Int. Assoc. Med. Mus.*, **27**: 23, 1947), the existence of the war conditions at the time preventing mutual access to the publications.

The most often used of the reactions, the periodic acid Schiff, was discovered independently by Hotchkiss in 1945, though his report was delayed 3 years (*Arch. Biochem. Biophys.*, **16**: 131, 1948), by McManus a year later [*Nature (Lond.*), **158**: 202, 1946), and perhaps slightly later by Lillie (*J. Lab. Clin. Med.*, **32**: 910, 1947; *Bull. Int. Assoc. Med. Mus.*, **27**: 23, 1947), who did not report his findings promptly.

McManus was attempting to demonstrate serine, threonine, and hydroxylysine residues in tissue, not realizing at the time, as most of us did not, that the amide nitrogen of the peptide bond did not constitute one of the groupings susceptible to the Nicolet-Shinn reaction. His first report called attention particularly to the reaction of mucins. Lillie's first reports dealt chiefly with reticulum and glycogen, respectively, and came about in the course of an exploration of the mechanism of the Bauer glycogen stain as a result of a verbal suggestion by C. S. Hudson (Jackson and Hudson, *J. Am. Chem. Soc.*, **59**: 2049, 1937) that we try periodate.

Early technics used the alkaline sodium salt Na_3IO_5 and the potassium salt KIO_4 as well as the acid H_5IO_6, depending on commercial availability at the time. Our first attempt to demonstrate glycogen with Na_3IO_5 failed because we had not then learned to acidify the solution. But addition of 0.5% nitric acid to the 1% Na_3IO_5 brought the pH down to 1.6, and the reaction succeeded.

We have tabulated the solutions used in Table 8-1 giving reagent used, concentration in grams per 100 ml and in molarity, pH of solution, presence of buffer or alcohol, and recommended exposure time.

It has been reported by Dempsey, Singer, and Wislocki (*Stain Technol.*, **25**: 73, 1950) that a 1-h oxidation at 37°C of Zenker-fixed, paraffin-embedded tissue in 1% aqueous periodic acid (H_5IO_6) induced very pronounced basophilia in sites containing high cystine concentrations. They attributed the reaction to cysteic acid. In our hands oxidation of skin for 2 h at 60°C in 1% periodic acid gave only a very moderate basophilia of distal hair cortex, as compared with the intense reaction obtained by 10-min oxidation with peracetic acid. An 18-h fixation at 24°C in 1% periodic acid–10% formalin, although the usual structures were rendered Schiff-positive, did not induce basophilia of keratin or other structures.

Hence it would appear that the cystine → cysteic acid oxidation by periodic acid probably does not occur to any appreciable extent at temperatures of 25°C and below, but that their reaction must be considered when temperatures above 35°C are used in the periodic acid oxidation step.

Periodic acid solutions apparently may be used for quite large numbers of sections and over several weeks without becoming exhausted. Nevertheless it is good practice to check a standard test section for adequacy of staining from time to time, for example, the completeness of staining of the fine reticular or sarcolemmal structure in human smooth muscle. Since 0.02 M periodic acid for 10 min appears adequate, it is evident that there is a large margin of safety in the usual solutions.

TABLE 8-1 PERIODIC ACID OXIDIZING BATHS

Author and date	Source HIO₄	g/100 ml	M	Solvent	pH	Time, min	Temperature, °C	Reducing bath
McManus, 1946, 1960	H_5IO_6	0.5	0.022	Distilled water	2.1	5	Room	No
Lillie, 1947	Na_3IO_5	1.0	0.036	0.5% HNO_3 aq.	1.6	10	22–32	No
Hotchkiss, 1948 (A)	H_5IO_6	0.8	0.035	Aq. 0.02 M NaAc	2.5	5	Room	Yes
Hotchkiss, 1948 (B)	H_5IO_6	0.8	0.035	70% alcohol 0.02 M NaAc	2.4	5	Room	Yes
Lillie, 1950	KIO_4	0.69	0.030	0.3% HNO_3 aq.	1.9	10	24	No
Mowry, 1952	H_5IO_6	1.0	0.044	90% alcohol	1.5	120	24	No
Lillie, 1953–1961 ...	H_5IO_6	1.0	0.044	Distilled water	1.95 1.81	10	25	No
Pearse, 1960, p. 831	H_5IO_6	1.0	0.044	0.02 M NaAc		5	18–22	Yes
Pearse, 1960, p. 832	H_5IO_6	0.5	0.022	Distilled water	2.1	2–5	18–22	No
	H_5IO_6	1.14	0.05	Distilled water	1.9	10	24	No

Hotchkiss' recommendation of an alcoholic solution was based on the then current belief that glycogen in fixed tissues was readily soluble in water. Since a 16-h exposure to boiled aqueous diastase solution does not appreciably diminish the amount of demonstrable glycogen in liver tissue (*Anat. Rec.*, **103**: 635, 1949), this precaution seems unnecessary for this purpose.

Hotchkiss' reducing rinse (KI, 1 g; $Na_2S_2O_3 \cdot 5H_2O$, 1 g; water, 20 ml; alcohol, 30 ml; 2 N HCl, 0.5 ml), in McManus' experience as well as ours, appears to block partially the coloration of collagen and reticulum, much as does a deliberately interposed bisulfite blockade step, and is not recommended as a routine procedure.

The usual method of demonstrating the aldehyde produced by the various oxidants is with Schiff's reagent. We have used usually the "cold Schiff" solution with 1 g rosanilin or pararosanilin per 100 ml. We have tried also both hot traditional and "cold Schiff" variants containing 0.5 g fuchsin per 100 ml and find the staining less intense and less complete in the usual 10-min staining interval than with the stronger variant. Longer staining intervals enhance the staining with the 0.5 % fuchsin variant, but there is a tendency also to overcome sulfite blockades with the longer-interval stains.

General technics and detailed results of the periodic acid Schiff and the Bauer and Casella reactions as they relate to carbohydrates are presented in Chap. 14.

Positive HIO_4 Schiff reactions (without information as to Bauer and Casella tests) are recorded for eosinophil leukocytes in the monkey but not in man; for the vitamin E deficiency ceroid pigment of the rat uterus; for the ceroid pigment of the mouse testis and adrenal; for Kurloff bodies in guinea pigs; for goblet cells in the pancreatic ducts; for certain nuclear inclusion bodies in the human vas deferens; for acrosomes and head caps of intra-testicular spermatozoa; for egg albumen and gelatin; for cytoplasmic granules in the rat lacrimal gland; and for a nonlipid component of the Golgi apparatus. Bauer and periodic Schiff-positive materials have been recorded in schistosome egg shells and in opercula of the eggs of *Capillaria* (*Hepaticola*) *hepatica*.

The reactions of blood cells are presented elsewhere in this book.

Keratin and keratohyalin—of pharynx and esophagus generally, of the proventriculus of muridae, and of epidermis and hair cortex—are not stained; but, curiously, hair cortex reacts strongly to the periodic acid Schiff procedure after bromination.

The positive reactions of arterial elastica and of elastic laminae and ligaments which are seen after periodic oxidation, especially in rodents, occur also when the oxidation step is omitted. They may be prevented by interposing a 30-min bath in 5 % aqueous phenylhydrazine hydrochloride before the periodic acid step. Wash 5 to 10 min in running water after the phenylhydrazine. KBH_4, 0.25 %, in 0.1 M Na_2HPO_4 for 2 to 5 min is more convenient. Lignin of plant tissues, often seen in intestinal contents, is also directly Schiff-positive, without prior oxidative treatment (Jensen).

The widespread direct Schiff reaction occurring at amine and other sites after dialdehyde fixations (glutaraldehyde, hydroxyadipaldehyde, glyoxal, and also acrolein) can be quite confusing when Feulgen or periodic acid Schiff or other specific aldehyde technics are being used. A 2- to 5-min bath in 0.2 to 0.5 % Na or KBH_4 in 0.1 M Na_2HPO_4, followed by water washing, will eliminate the artifactual aldehyde reaction. The reaction of native aldehydes such as that of lysinal in young elastin is also eliminated, and for this tissue dialdehyde fixations should be avoided or controlled by other parallel fixations.

Crippa (*Boll. Soc. Ital. Biol. Sper.*, **27**: 599, 1951) reports on the use of lead tetraacetate as an oxidant in place of chromic or periodic acid in a procedure utilizing the oxidation-induced aldehyde reaction for the demonstration of mucin and mucoid substances.

The reagent, which is stable for an indefinite period, is a 1 % solution of recrystallized lead tetraacetate in glacial acetic acid. The solution is clear and colorless. The technic follows:

CRIPPA'S LEAD TETRAACETATE METHOD FOR MUCINS AND MUCOIDS

1. Fix as usual. Embed in paraffin, section, and deparaffinize in xylene.
2. Wash in 50 : 50 xylene–glacial acetic acid and in glacial acetic acid.
3. Immerse for 30 to 60 min at room temperature in 1% lead tetraacetate in glacial acetic acid.
4. Wash in three changes of glacial acetic acid to remove lead acetates.
5. Pass through xylene–glacial acetic acid 50 : 50 mixture, xylene, and descending alcohols to water (or direct through 50% acetic acid to water).
6. Immerse in Schiff reagent 15 to 20 min.
7. Wash in three changes of 0.5% sodium bisulfite, 1, 2, and 2 min.
8. Wash in running water 10 min.
9. Counterstain if desired, as with the periodic acid Schiff method; or dehydrate, clear, and mount directly in balsam or synthetic resin.

RESULTS: Aldehyde deposits engendered by lead tetraacetate oxidation show in red-purple.

If desired, steps 7 to 13 of Gomori's methenamine silver method for glycogen and mucins may be substituted for steps 6 to 9 above. The mucins then appear in black.

McManus (February 1952) wrote us that lead tetraacetate renders Schiff-positive essentially the same structures as does periodic acid. Jordan and McManus used a fresh mixture of 1 part of a strong to saturated solution of lead tetraacetate in glacial acetic acid with 3 parts of a mixture of 20 g anhydrous sodium acetate, 25 ml glacial acetic acid, and 50 ml water. They used a shorter oxidation time, 30 to 120 s. Lhotka (*Stain Technol.*, **27**: 213, 1952) used 5 g potassium acetate dissolved in 100 ml glacial acetic acid saturated with lead tetraacetate, and the same short oxidation times as McManus. Leblond and coworkers (*Stain Technol.*, **27**: 277, 1952) observed that glycogen was not stained by room temperature oxidation but did react vigorously at higher temperatures. Our experience with the Lhotka reagent indicated prompt reaction of glycogen. Shimizu and Kumamoto (*Stain Technol.*, **27**: 97, 1952) used a solution similar to McManus':

1. Deparaffinize and hydrate as usual.
2. Wash in molar sodium acetate solution for 5 min (13.6% $NaCO_2CH_3 \cdot 3H_2O$).
3. Oxidize 10 min in lead tetraacetate, 1 g, dissolved in 30 ml glacial acetic acid plus 70 ml 46.5% (saturated) sodium acetate solution.
4. Wash 5 min in molar sodium acetate solution.
5. Wash 10 min in running water.
6. Immerse in Schiff reagent for 15 min.
7. Wash in three changes of sulfite solution, 2 min each.
8. Wash 10 min in running water.
9. Counterstain, dehydrate, clear, and mount as in periodic acid Schiff methods.

RESULTS: *vic*-Glycols show in red-purple.

Although Shimizu and Kumamoto prescribed the original Schiff reagent and sulfite rinse of Feulgen and Rossenbeck, undoubtedly the more recent variants will give fully satisfactory results.

It is to be noted that lead tetraacetate, in *warm* acetic acid solution, also reacts with olefins to form glycol diacetates and unsaturated acetoxy compounds (Hickinbottom, 1948, p. 32).

Staple [*Nature (Lond.)*, **176**: 1, 25, 1955] recorded that inclusion of 0.01 *M* boric acid (61.8 mg/100 ml) in a 1% lead tetraacetate solution in glacial acetic acid would prevent the subsequent Schiff reaction of *cis*-glycols but not that of adjacent *trans*-hydroxyls in model experiments with galactogen and glycogen. The reaction period in lead tetraacetate was 4 h at 23°C. The reaction of intestinal goblet cell and laryngeal mucous gland mucins was inhibited in the presence of boric acid.

Lhotka (*Stain Technol.*, 27: 259, 1952) recommended the use of sodium bismuthate in aqueous phosphoric acid solution as another reagent for the cleavage of *vic*-glycols and α-hydroxycarboxylic acids to carbonyl. Organic solvents can also be used. Overoxidation is said not to occur. The bismuthate solution in phosphoric acid is unstable and must be used at once.

LHOTKA'S SODIUM BISMUTHATE SCHIFF PROCEDURE FOR *vic*-GLYCOLS AND α-HYDROXYCARBOXYLIC ACIDS

1. Section formalin-fixed tissue in paraffin, deparaffinize, and hydrate sections as usual.
2. Suspend 1 g sodium bismuthate in 100 ml 20% phosphoric acid solution. The suspension is bright orange-yellow and soon turns brown by precipitation of bismuth pentoxide. Oxidize sections 3 min in the fresh suspension.
3. Wash in running water for 1 min.
4. Rinse 15 s in normal hydrochloric acid to remove bismuth pentoxide from sections.
5. Rinse in distilled water.
6. Immerse in Schiff reagent 10 min.
7. Wash in 0.5% $Na_2S_2O_5$ or SO_2 water, three changes, 2 min each.
8. Wash in running water 5 min (10 min is better).
9. Counterstain with alum hematoxylin or as desired.
10. Dehydrate, clear, and mount by alcohol, xylene, synthetic resin sequence.

RESULTS: Pictures are similar to those obtained with periodic acid or lead tetraacetate. Sites of *vic*-glycols and α-hydroxycarboxylic acids in red-purple.

SULFUR AMINO ACIDS

The methods for demonstration of the sulfur amino acids cystine, cysteine, and methionine in protein combination probably all depend on the initial presence of or the formation of the sulfhydryl radical in the course of the demonstration reaction. Methods purporting to distinguish protein-bound disulfide (—SS—) and sulfhydryl (—SH) are to be accepted with great caution. Air oxidation of paraffin sections while they are still undeparaffinized, and even of unsectioned tissue within a previously sectioned paraffin block, can in some days or weeks render negative the previously demonstrable sensitive ferric ferricyanide reaction of the first sections cut from the same block. On the other hand, alkaline demonstration reagents tend to reduce disulfide to sulfhydryl. This occurs in alkaline cyanide solutions without obvious reducing groups (KCN in the presence of atmospheric oxygen converts the reduced metallic silver back to the relatively oxidized double cyanide), as well as in alkaline tetrazolium solutions where the tetrazole itself is a reducing agent. The old sodium plumbite reagent contains no obvious reducing group, yet it reacts with distal hair cortex to form lead sulfide. Perhaps some such reaction goes on in strong alkali as $RSSR + 2H_2O \rightarrow 2RSH + H_2O_2$, and the peroxide in turn acts as a reducing agent, as it does with ferric ferricyanide, releasing 2H and liberating molecular oxygen (O_2).

From these considerations it would appear that when it is desired to distinguish SH from SS groupings, every precaution should be taken to protect the sensitive SH groups from oxidant fixatives, from exposure to air, and from mercaptide-forming heavy-metal fixatives. Here, if the solubilities of the proteins under study permit, such fixatives as the

Carnoy acetic acid–alcohol (1 : 3) and acetic acid–chloroform–alcohol (1 : 3 : 6) are to be recommended. They should be used cold, or even by a freeze substitution process, in tightly closed containers.

Paraffin infiltration is perhaps preferably done in vacuo to reduce the heating interval to a minimum. We have compared the chloroform–Carnoy at 5°C with the often recommended 5 to 10% trichloroacetic acid in 80% alcohol on keratinization of hair in skin of newborn animals and found demonstration of SH by ferric ferricyanide after Carnoy fixation at least equal to, if not better than, after trichloroacetic acid, and general morphologic preservation seems distinctly better. For such proteins as are dissolved in acetic acid–alcohol fixatives, chloroform–methanol may be tried, or freeze-drying, or 1 M NaCl may be incorporated in aqueous acetic acid mixtures (*Am. J. Clin. Pathol.*, **59**: 374, 1973).

Formaldehyde fixation is distinctly inimical to satisfactory —SH demonstration. The acid and neutral reagents would seem preferable to alkaline. We would suggest ferric ferricyanide with appropriate blockade controls, and the mercaptide reagents as more specific though perhaps less sensitive reagents. Barrnett and Seligman's (*J. Natl. Cancer Inst.*, **13**: 215, 1952; **14**: 769, 1953) dihydroxydinaphthyl disulfide (commonly called DDD) and Pearse's (*J. Histochem. Cytochem.*, **1**: 460, 1953) tetrazolium method both employ alkaline solutions and are apt to show some alkali cleavage of SS in their application, thereby giving misleading results (Findlay, *J. Histochem. Cytochem.*, **3**: 331, 1955). Hyde (ibid., **9**: 640, 1961) noted that trichloroacetic acid–fixed material may prove relatively resistant to thioglycolate reduction, thus decreasing the amount of demonstrable SS.

When total protein sulfur demonstration is desired, these latter methods, following an appropriate reduction, give excellent results. The presence of reducing agents during or after the Barrnett-Seligman reaction is to be avoided; the SS bond on whose formation the reaction depends is itself susceptible to cleavage.

The Ferric Ferricyanide Reduction Test

This test was introduced into histochemistry by Golodetz and Unna (*Monatsh. Prakt. Dermatol.*, **48**: 149, 1909), for the demonstration of reduction sites in tissue. At first it was attributed to tyrosine, but all Unna's later papers refer simply to *Reduktionsorte*. Ferric ions are reduced to ferrous and in the presence of ferricyanides in the solution yield insoluble Turnbull's blue.

Lillie and Donaldson (*Histochem. J.*, **6**: 30, 1974) showed that ferricyanide acted as an oxidizing reagent in alkaline solutions but not below pH 7, while ferric chloride readily oxidized tissue sulfhydryls, adrenal noradrenaline, and enterochromaffin at pH 2.5, the approximate usual pH of ferric ferricyanide mixtures. They therefore concluded that reduction of the ferric ion occurred and the liberated Fe(II) reacted with the unreduced ferricyanide to form Turnbull's blue. Most authors had previously considered that a reduction of ferricyanide occurred and formed Prussian blue, following the dictum of Golodetz and Unna.

The method was cited by Schmorl for the identification of lipofuscin pigments. It is found applicable to the pigment in phagocytes and parenchyma cells of the adrenal cortex, to a granular pigment found in phagocytes in the ovary adjacent to involuting corpora lutea, to the so-called melanosis pigment of the human appendix and colon, and to a homologous pigment in the intestine of the guinea pig (*Anat. Rec.*, **108**: 239, 1950).

The method also demonstrates chromaffin of the adrenal medulla in material appropriately fixed in dichromate fixatives (Lillie, *Anat. Rec.*, **108**: 239, 1950).

Gomori (*Arch. Pathol.*, **45**: 48, 1948) and Lillie independently discovered that the enterochromaffin or argentaffin substance of the basal granular cells of the gastrointestinal mucosa gives this reaction. Laskey and Greco (*Arch. Pathol.*, **46**: 83, 1948) compared it with the Masson silver method and found it nearly as efficient and much briefer.

Chèvremont and Frederic [*Arch. Biol.* (*Liege*), **54**: 589, 1943], apparently unaware of the previous use of this reaction for lipofuscins, utilized it for the demonstration of fixed sulfhydryl groups. They recognized, however, that other unidentified substances might react, and required the use of specific sulfhydryl blocking reagents for confirmation of the identification. They used mercuric chloride—Yao (*Q. J. Microsc. Sci.*, **90**: 1401, 1949) prescribed 1 h in 6% aqueous solution—or 1% alcoholic chloropicrin solution or 4% aqueous monoiodoacetic acid.

Fisher and Lillie have noted that thyroid colloid colors quite deep blue by this method.

Mast cells of rats and mice often conspicuously reduce ferric ferricyanide, coloring bright blue diffusely, with small rounded clear areas representing the granules.

Cutaneous keratin, as noted by Golodetz and Unna (*Monatsh. Prakt. Dermatol.*, **48**: 149, 1909), and keratohyalin granules, as noted by Chèvremont and Frederic [*Arch. Biol.* (*Liege*), **54**: 589, 1943], and the corresponding structures in the stratified epithelium of the forestomach of rats and mice color deep blue by this reaction. Mercurial fixations do not inhibit the reaction in the rodent forestomach, and incubation at 37°C for 16 h in 5% mercuric chloride does not stop it in skin sections. Chèvremont and Frederic disagree with us and with Golodetz and Unna and deny the reaction of cutaneous keratin. A 20-min treatment with 2% CrO_3 destroys the reactivity, but 18-h treatments with $K_2Cr_2O_7$ (0.1 M) and HIO_4 (0.03 M) do not.

IN VITRO TESTS WITH FERRIC FERRICYANIDE. The histochemical reaction mixture is promptly reduced by ascorbic, oxalic, and uric acids; by phenols (phenol, resorcinol, hydroquinone, pyrogallol, *o*-cresol, α-naphthol, β-naphthol, adrenaline, L-tyrosine, tyramine, thyroxine); by indoles (indole, skatole, tryptophan, tryptamine, nitrosoindole, nitrosotryptophan); by arylamines (aniline, α-naphthylamine, benzidine, diphenylamine); by thiols (ethyl mercaptan, mercaptoethanol, thiophenol, thio-2-naphthol, rubeanic acid, glutathione, cysteine, DL-methionine); by hydrazines (phenylhydrazine, semicarbazide, aminoguanidine); by stannous chloride, nascent hydrogen, carbon disulfide; by inorganic sulfides, sulfite, dithionite, and thiosulfate; by glyoxal, benzaldehyde, and benzyl alcohol; by nitrous acid; by hydrogen peroxide, lecithin, cod liver oil, and linseed oil.

Pyrocatechol gives a green-black precipitate with $FeCl_3$; *p*-dimethylaminoaniline and salicylic acid, a deep purple; the aminophenols (*o*-2,4-diaminophenol), red to brown; nitrosoresorcinol and 1-nitroso-2-naphthol, brown; *p*-phenylenediamine, dark brown; Na_2S, NaSH, $(NH_4)_2S$ and H_2S, greenish black; but premixed with potassium ferricyanide and allowed to stand for a minute or so, they all promptly yield blue or green with ferric chloride.

Other substances tested took an hour or more, more often overnight, to produce blue or green, and probably need not be considered as demonstrable by the histochemical reaction when the reaction time is restricted to 10 min. These included isopropanol, formaldehyde, acetaldehyde, acrylic acid, tartaric acid, citric acid, cholesterol, and fructose, which gave blue or green in an hour but no precipitate; piperidine and pyridine, which gave brown precipitates with ferric chloride and whose premixed ferricyanide solutions which had stood for 1 h or more gave Prussian blue reactions with ferric chloride; and a larger group giving no reaction at 1 h: DL-serine, DL-lysine, histidine, histamine, DL-valine, L-glutamic acid, L-arginine, DL-aspartic acid, DL-threonine, L-proline, L-cystine, DL-alanine, DL-phenylalanine, methanol, ethanol, *n*-butanol, glycol, diethylene glycol, glycerol, 1,4-dioxane, allyl alcohol, acetone, formic acid, glucose, urea, guanine, caffeine, uracil, thiouracil, inositol, chloral hydrate, mercury bichloride, hydrogen chloride, and distilled water.

FIXATION OF MATERIAL. For demonstration of the melanins, neuromelanin, melanosis, and lipofuscin pigments routine formalin fixation is satisfactory. We have had good results on the adrenal pigment after alcoholic formalin, acetic alcohol formalin, Carnoy, aqueous and alcoholic lead nitrate, and mercuric chloride–formalin mixtures, and a variety of other non-chromate-containing fixatives. Potassium dichromate fixations generally weaken or abolish the reactivity of these pigments.

For the argentaffin or enterochromaffin substance, routine fixations of 1 to 15 days in aqueous 10% formalin or 5% glutaraldehyde are satisfactory. Acetic acid-formalin fixation is unsatisfactory. Alcoholic fixations of all sorts remove the argentaffin substance. It is preserved poorly or not at all with aqueous fixatives lacking aldehyde. Dichromate formaldehyde fixations weaken the reaction and give a green color in place of clear blue. Formalin storage of 8 months to years prevents the reaction.

For chromaffin, fixation in potassium dichromate–formaldehyde mixtures without added acid is preferred. The Möller and Kose mixtures give good results. For other reactions other fixations serve better.

For sulfhydryl Chèvremont and Frederic prescribed a 4- to 18-h fixation in saline formalin or Bouin's fluid and avoided prolonged treatment with melted paraffin or used frozen sections. They stated that longer than 48-h exposure to formaldehyde would reduce or abolish the reactivity of sulfhydryl groups.

Although sulfhydryls form stable nonreactive mercaptals on reaction with aldehydes, strong hydrogen chloride or zinc chloride is required as catalyst. Only thiol acids react directly without catalysts. Hence formalin fixation can be tolerated if not unduly prolonged.

We find Carnoy's acetic acid–chloroform–alcohol mixture and a mixture of 5 g trichloro-acetic acid, 75 ml alcohol, and 20 ml water—both at 5°C—superior to formalin for preservation of sulfhydryl in the basal layers of the epidermis and the hair follicle cells. Use of sublimate formalin with sodium acetate inhibits the reaction of these structures, but not that of keratin. We use a 30-min vacuum infiltration in paraffin for skin. It appears to be necessary to work up this material promptly after preparation. Paraffin sections which have been saved for some weeks after cutting have utterly failed to give the reaction at sulfhydryl sites, though those which were reacted at once from the same blocks gave good results.

For thyroid colloid the use of alcoholic fixatives seems preferable to that of aqueous formalin. The use of dichromate fixatives appears to be contraindicated, both because they tend to oxidize reducing groups and because they appear to engender a more or less diffuse, moderate blue-green coloration of collagen, epithelia, and the like.

THE REAGENT. Golodetz and Unna, Staemmler, Schmorl, Laskey and Greco, and Lillie (1950) all prescribed a 5-min bath in a freshly prepared mixture of equal volumes of 1% ferric chloride and freshly prepared 1% potassium ferricyanide. The mixture should remain clear and greenish brown. Chèvremont and Frederic prescribed a mixture of 1 volume of fresh 0.1% potassium ferricyanide solution and 3 volumes of 1% ferric sulfate solution; they noted that the mixture was acid (pH 2.4) and that it was stable for about 2 h in daylight and longer in the dark. A 20- to 25-min immersion in this solution was required for paraffin sections.

We found this highly disproportionate mixture—about 50 mol ferric salt to 1 mol ferricyanide—advantageous in giving a much clearer background than the Golodetz-Unna mixture, which gives only 2 mol ferric salt to 1 mol ferricyanide, but the concentration of ferricyanide is so low that prolonged exposures are necessary; and even then the demonstration of enterochromaffin cells is uncertain, and the reactivity of some of the pigments is dubious.

After considerable experimentation we (*J. Histochem. Cytochem.*, **1**: 87, 1953) have settled on 30 ml of 1 % ferric chloride, 4 ml fresh 1 % potassium ferricyanide [$K_3Fe(CN)_6$], and 6 ml distilled water. The final mixture thus contains 0.75 % $FeCl_3$ and 0.1 % potassium ferricyanide, a molar proportion of about 15 : 1. A 10-min bath in this solution appears adequate.

THE FERRIC FERRICYANIDE REDUCTION TEST TECHNIC

1. Take 5 to 10 μ paraffin sections of appropriately fixed tissue to water as usual, interposing blockade procedures if desired.
2. Immerse 10 min in 4 ml 1 % potassium ferricyanide [$K_3Fe(CN)_6$] plus 30 ml 1 % ferric chloride ($FeCl_3 \cdot 6H_2O$) plus 6 ml distilled water at 25°C.
3. Wash in 1 % acetic acid (counterstains are not advised).
4. Dehydrate, clear, and mount in cellulose caprate through alcohols and xylene.

RESULTS: Reducing sites are dark blue; background, pale greens. Note that natively present ferrous salts give a Turnbull's blue reaction with $K_3Fe(CN)_6$ alone, as well as with this reagent.

With some tissue elements the reaction can be done in sequence by first reacting with $FeCl_3$, washing, and then applying (neutral or) acid $K_3Fe(CN)_6$. While $K_3Fe(CN)_6$ is itself an oxidant in alkaline solutions, in the above two tests the tissue oxidation and ferricyanide reaction are both done in acid solutions.

To demonstrate SS groups, first reduce a duplicate preparation 10 min in fresh 10 % sodium thioglycolate. Dithionite $Na_2S_2O_4$, hydrosulfide NaSH, or monothioglycol $HOCH_2CH_2SH$ can be used for the same purpose.

Perhaps better for SS groups are the peracetic azure A or Alcian blue methods of Lillie (given later in this chapter) and Adams and Sloper (*J. Endocrinol.*, **13**: 221, 1956), which depend on peracid cleavage of cystine and cysteine to cysteic acid. Bromine can also be used as the oxidant.

THE NITROPRUSSIDE REACTION.

Glick cited only the nitroprusside technics, which so far are applicable only to frozen sections of fresh unfixed tissue. The technic was cited by Giroud and Bulliard (*Protoplasma*, **19**: 381, 1933) and modified by Glick. It is no longer used in histochemistry because of poor localization, and is therefore omitted.

MERCURY ORANGE.

Bennett's method (*Anat. Rec.*, **110**: 231, 1951) uses a red, mercury-containing azo dye, *p*-chloromercuriphenyl-azo-β-naphthol, to form insoluble mercaptide linkages with sulfhydryl groups in tissue. The reaction appears to be specific. It does not occur with the corresponding mercury-free azo dyestuff. It is prevented by agents which oxidize: 25.3 mg/100 ml iodine in propanol (1 mM), molar hydrogen peroxide plus ferric chloride mixture in isopropanol or 1 mM in water; or alkylate sulfhydryl: 1.85 % iodoacetamide or 1.86 % iodoacetic acid (0.1 M) in *n*-propanol. It is prevented also by organic mercurials which form mercaptides with sulfhydryl: 1 mM *n*-propanol solutions of phenylmercuric chloride (31.3 mg/100 ml), tolyl mercuric chloride (32.7 mg/100 ml), or methyl mercuric iodide (34.5 mg/100 ml). Tissues and sections are brought through appropriate solvents to pure *n*-propanol, then immersed for several hours in the propanol solutions of the blocking reagents, washed in several changes of propanol, and then transferred, along with control, unblocked, positive material of the same general character (i.e., sections with sections, smears with smears, teased tissue with teased tissue), to the sulfhydryl reagent.

Mescon and Flesch (*J. Invest. Dermatol.*, **18**: 261, 1952) have further adapted Bennett's method for frozen sections of unfixed and paraffin sections of formalin-fixed material.

Unfixed material is sectioned by the Adamstone-Taylor procedure or in the cryostat, mounted on slides, and at once immersed in the sulfhydryl reagent.

Routine paraffin sections of formalin-fixed tissue are deparaffinized and brought to 80% alcohol as usual. Then stain, dehydrate, clear, and mount as for the frozen sections.

Bennett fixed muscle tissue in 5% trichloroacetic acid, washed in distilled water, dehydrated in graded alcohols, and transferred to propanol or butanol. Alternatively, the tissue was quickly frozen in isopentane cooled with liquid nitrogen, and dehydrated while still frozen by several changes of *n*-propanol or *n*-butanol at −20 to −25°C for 10 to 12 days. Muscle tissue was then teased out to isolate single fibers and small groups, which were stained whole. He also used paraffin and nitrocellulose sections, which were brought to *n*-propanol or *n*-butanol for reaction.

The reagent is *p*-chloromercuriphenyl-azo-β-naphthol. This is dissolved according to Bennett at 12.5 μM (6 mg/l) in butanol or at 5.6 μM (3 mg/l) in *n*-propanol; or according to Mescon and Flesch by dissolving 3 mg in 100 ml 100% ethyl alcohol at room temperature, and then adding 25 ml distilled water to reduce the alcohol concentration to 80%. Mescon's reagent is stored at 4°C and keeps fairly well.

BENNETT'S METHOD FOR SULFHYDRYL, WITH MODIFICATIONS FROM MESCON AND FLESCH

1. Deparaffinize paraffin sections and bring to 80% alcohol. Transfer Adamstone-Taylor frozen sections and nitrocellulose sections directly, and ordinary frozen sections after blotting on slides, to the Mescon-Flesch reagent.
2. Immerse in the Mescon-Flesch reagent for 1 to 3 h.
3. Dehydrate, clear, and mount through 95 and 100% alcohol, alcohol plus xylene, two changes of xylene in a synthetic resin.
1a. Bring teased muscle preparations, tissue fragments to *n*-propanol, or nitrocellulose sections to isopropanol, which does not dissolve the nitrocellulose.
2a. Immerse several hours or overnight, according to size and thickness of material, in Bennett's propanol reagent solution.
3a. Wash 2 or 3 h in *n*-propanol.
4a. Clear through propanol plus xylene, in two changes of xylene, and mount in synthetic resin.

RESULTS: Sites of sulfhydryl are colored red. If phenyl-azo-β-naphthol is used instead of its mercury derivative, there should be no reaction.

Interpose blockade reactions after step 1. The reagents are described earlier. It is suggested that one of the mercaptide reagents be used as well as an alkylating or an oxidizing reagent; or, better, one of each of the three types. The blockade reagent should be used for 4 to 18 h and should be followed by a 2- to 6-h bath in the same solvent to remove excess blocking agent. Then transfer to 80% alcohol or to *n*-propanol and proceed with step 2 or 2a.

The color is not extracted by toluene, benzene, 70% alcohol, 50% propanol, 0.1 N NaOH, 0.1 N CaCl$_2$ (about 0.55%), 0.1 N NH$_4$OH (about 0.68% dilution of 28% ammonia water), or 1 M NaCl (5.85%) on overnight exposure. The color is extracted by overnight immersion in 1 mM propanol solutions of mercaptans: β-mercaptoethanol, cysteine, 2,3-dimercaptopropanol (BAL), glutathione, and thioglycolic acid (mw 78, 121, 124, 307, and 92, respectively).

Hence if extraction tests are required, interpose after step 2 or 2a, hydrating as usual for the use of aqueous reagents, washing first with propanol for use of the propanol reagents, and dehydrating for use of toluene, benzene, or other similar solvent. After the prescribed exposure, wash out the extracting agent with the same solvent, dehydrate if necessary, clear, and mount as usual.

MERCAPTIDE METHOD OF LILLIE AND GLENNER FOR SULFHYDRYL.   After our 1957 introduction of a glacial acetic acid–azo coupling step to intensify the color of the p-dimethylaminobenzaldehyde reaction with indoles, especially protein tryptophan, it seemed desirable to demonstrate that other p-N,N-dimethylaminophenyl compounds with tissue would also undergo an ortho azo coupling reaction in glacial acetic acid. To this end Glenner synthesized p-N,N-dimethylaminophenylmercuriacetate according to Whitmore ("Organic Compounds of Mercury," American Chemical Society Monograph, New York, 1921).

Dark red azo coupling with diazotized safranin was demonstrated at sulfhydryl sites, and was completely decolorized by 2,3-dimercaptopropanol in propanol. The procedure has since proved useful from time to time, and repetition of the synthesis has posed no important difficulties on several batches. Hence we include the method here. However, the reagent proves to have a quite limited shelf life.

The reagent p-N,N-dimethylaminophenylmercuric acetate is not presently available commercially: Dissolve 32 g mercuric acetate in 100 ml distilled water [the calculated amounts are 31.87 g $Hg(CO_2CH_3)_2$ and 12.18 g (12.74 ml) dimethylaniline]. A slight excess of the Hg salt is used deliberately, and the excess is maintained throughout the process to prevent formation of the diaryl compound. Since addition of alcohol alone to warm or room-temperature (24°C) 32% $HgAc_2$ gives rise to a copious yellow precipitate, cool to 3°C and add 12.5 ml N,N-dimethylaniline in 100 ml absolute alcohol also prechilled to 3°C. As the cold alcoholic dimethylaniline solution is poured into the aqueous mercuric acetate, a yellow color and a little precipitate appear, but the latter redissolves on shaking. When about half the alcoholic solution has been added, a white precipitate starts to form, and as the addition is completed, it fills the entire fluid and sets into a firm mass with just a little yellow fluid remaining. The flask is then stoppered with a clean rubber stopper and shaken violently. This soon breaks up the clot into a fairly thick suspension of white precipitate in a colorless fluid. This is allowed to stand overnight at 3°C and then filtered out on paper and washed with distilled water until no more reaction is given with sulfides. The moist precipitate is then transferred to a vacuum desiccator and dried for 3 to 6 days at 25°C. Yield: calculated, 37.32 g; actual, 30.7 g.

Cowden and Curtis (*Histochemie*, **22**: 247, 1970) reported on the use of a number of fluorescent mercurials both as fluorochromes and as color reagents for demonstration of sulfhydryl groups. Of the four dyes tested—mercury orange, 4-(p-dimethylaminobenzene azo)phenylmercuric acetate (available from Sigma Chemical Co., St. Louis, Missouri), fluorescein mercuric acetate, and merbromin (Mercurochrome 220)—the last was the best for direct color reaction, the third gave the strongest fluorescence, the second gave too weak a fluorescence to be useful. The fluorochromes were used at 20 mg/100 ml N,N-dimethylformamide, and adequate fluorochroming was accomplished in 1 h. Merbromin from commercial sources and Mercurochrome from E. Gurr were first dissolved, 20 mg in 0.5 ml deionized water, and then diluted to 100 ml with N,N-dimethylformamide. The use of a nonpolar solvent largely prevented anionic dye staining. The usual blocking agents prevented the fluorochrome and direct staining; stains were extracted by 24-h exposures to 1 mM thioglycolic acid in n-propanol, but not by propanol alone.

MERBROMIN STAIN FOR SULFHYDRYL (SH) GROUPS.   Use Carnoy A–fixed material embedded in paraffin, sectioned, and dewaxed in xylene. If not used at once, sections are dipped in melted paraffin to protect them from air oxidation. The excess wax is cut off with a razor blade from around the section, and the remainder removed in three changes of xylene.

1. Stain in 20 mg/100 ml *N,N*-dimethylformamide 1 h for fluorescence, 48 h for visible staining. For commercial 2% merbromin, dilute 1 ml (= 20 mg) with 99 ml solvent.
2. Wash twice in dimethylformamide for 3 min for each wash.
3. Dehydrate with absolute alcohol, two changes; xylene, two or three changes. Mount in Harleco Fluorescence or, for visible light, any good synthetic resin. For blockade reactions introduce these blockades before step 1; for the SH extraction test introduce them after step 2, transfer from alcohol, and return through alcohol to stain on clearing and mounting, respectively.

Allen and Perrin (*J. Histochem. Cytochem.*, **22**: 919, 1974) provide the following method for the histochemical identification of SH at light and electron microscopic levels: The reagent 1-chloromercuriferrocene (available from K & K Laboratories, Plainview, New York) was employed. They used wool as a test object. Samples of wool were reduced in 0.3 *M* benzylthiol in 20% *N*-propanol for 2 days, washed in 20% *N*-propanol, washed in several changes of dimethylformamide, immersed in 20 ml 0.1 *M* solution of 1-chloromercuri- ferrocene in dimethylformamide saturated with sodium sulfite for 4 days with occasional shaking, washed with dimethylformamide until washings were colorless, and embedded in Spurr epoxy resin. The sulfite is added to the 1-chloromercuriferrocene to act as an oxygen scavenger and to reduce any remaining S—S. The mercury treatment was omitted in controls.

For electron microscopy, ultrathin sections are mounted on stainless steel grids and examined without further treatment. For light microscopy, 1-μ epoxy resin sections are treated as indicated above; however, after the rinse following the mercury treatment, sections are stained with 2% silver nitrate in the dark for 1 h at 25°C, rinsed, dehydrated, and mounted in synthetic medium. Sites of SH stain black; the structure of 1-chloromercuriferrocene is indicated below:

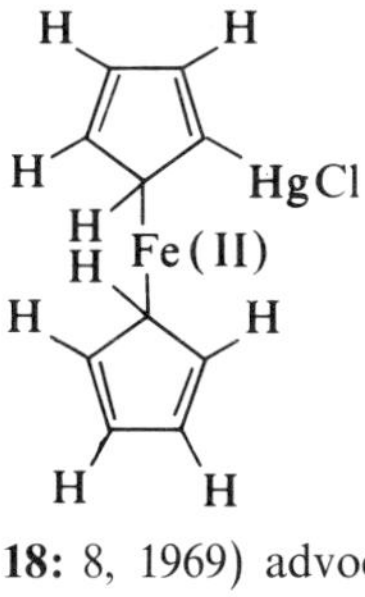

Grillo and Grillo (*Histochemie*, **18**: 8, 1969) advocate the following fluorescent histo- chemical method for sulfhydryl groups:

1. Fresh frozen sections or deparaffinized sections previously fixed in either acetone or ethanol are immersed in fluorescein mercuric acetate (0.01 μM to 1 m*M* in either 0.1 *N* NaOH or dimethyl formamide for 3 to 5 s). Formalin-fixed tissues may not be used because of formaldehyde-induced fluorescence.
2. Sections are rinsed briefly in absolute ethanol (if tissues were subjected to dimethyl formamide) or in distilled water (if tissues were subjected to NaOH).
3. Sections are dehydrated and mounted as usual.
4. Tissues are examined under ultraviolet light for selective fluorescence in regions containing SH groups. The mercurial reagent is obtained from NBC, Cleveland, Ohio. Grillo and Grillo state that staining was greatly diminished or abolished in sections blocked with 0.1 *M* iodoacetic acid for 24 h. The distribution of stain in tissues is reputed to be analogous to that obtained using the Barrnett-Seligman DDD method.

Engel and Zerlotti (*J. Histochem. Cytochem.*, **12:** 156, 1964) recommend the following method for the histochemical detection of SH groups. Freeze-dried paraffin-embedded sections are recommended:

1. Deparaffinize sections in xylol or petroleum ether. (Embedding may be avoided if sections are cut on a cryostat and then freeze-dried.)
2. Denature sections overnight in absolute ethanol at 25°C.
3. Transfer to 95% ethanol for 2 min.
4. Stain with the azomercurial reagent for 30 min at 37°C. The reagent is 4-(*p*-dimethyl-aminobenzene azo)phenylmercuric acetate. Prepare the incubation solution by adding 3 mg of the reagent to 40 ml of alcoholic glycine buffer (mix 1 *M* glycine and 0.4 *M* sodium acetate in a 1 : 1 solution, and adjust the pH to 9 with 1 *N* sodium hydroxide). Add 4 ml glycine buffer to 36 ml of 80% ethanol.
5. Transfer through two changes of 95% ethanol to remove excess reagent.
6. Complete dehydration in absolute ethanol for 5 min.
7. Clear in xylol for 5 min and mount in synthetic resin.

RESULTS: Sites of SH stain light yellow. Engel and Zerlotti explain that the light yellow may not necessarily limit the value of the method in optical microscopy. The dye has a relatively high molar extinction coefficient (25,000). Use of a strong light source in conjunction with an interference filter that transmits at 430 mμ permits good transmission and resolution even at high magnification. In all studies of SH groups, care must be taken to avoid air oxidation. Paraffin-embedded blocks should be sealed with hot (90 to 110°C) paraffin. Both blocks and sections should be kept in an evacuated desiccator containing Drierite.

Mundkur (*Exp. Cell Res.*, **34:** 155, 1964) employed the mercury orange reagent to localize SH groups in yeast cells at the ultrastructural level. Ribosomes are particularly rich in SH.

Sippel (*J. Histochem. Cytochem.*, **17:** 428, 1969) synthesized several maleimides *N*-substituted with aryl substituents on azo or anthraquinone dyes during attempts to develop a reagent of sufficient color to act as a direct stain for SH groups. A blue maleimide AQM-7, 1-(*p*-maleimidoanilino)-4-methylaminoanthraquinone, was one of the more successful re-agents. Light green staining was obtained in sites containing SH groups. Staining was nearly abolished in sections blocked with SH reagents.

Sippel (*Histochem. J.*, **5:** 413, 1973) later advocated the following method for the histo-chemical detection of SH. Formalin (preferably) or Carnoy-fixed, paraffin-embedded tissues may be used.

1. Deparaffinize sections, bring them to ethanol, collodionize if necessary, and rinse in 50% 2-propanol.
2. Immerse sections for 30 to 60 min at 22°C in *N*-(4-aminophenyl)maleimide (5 m*M*) obtainable from Koch-Light Laboratories, Colnbrook, Bucks, England. To prepare, dissolve 38 mg maleimide in 20 ml 2-propanol, and add 20 ml 0.05 *M* sodium phosphate buffer at pH 6. The undissolved material, if any, need not be removed by filtration.
3. Rinse sections in 50% ethanol and then in 0.1 *N* HCl.
4. Diazotize sections for 2 min (add 20 ml 0.1 *M* NaNO$_2$ to 20 ml 0.2 *N* HCl).
5. Rinse sections in 0.1 *N* HCl; then immerse sections in 0.5% sulfamic acid for 30 s. (The sulfamic acid is used to destroy excess HNO$_2$ and may be unnecessary).
6. Couple for 2 min in 0.5% 1-naphthylamine in 0.1 *N* HCl.
7. Rinse in two changes of 0.1 *N* HCl.

8. Repeat step 4.
9. Couple for 5 min in fresh 0.5% chromotropic acid (disodium hydrate) in 0.1 M sodium acetate buffer at pH 4.5.
10. Wash thoroughly in water; counterstain if desired for 5 min in either 0.5% safranin O (C.I. 50240) or neutral acriflavine (C.I. 46000) in 1% acetic acid.
11. Differentiate in 1% acetic acid, dehydrate in ethanols or acetone, and mount in synthetic medium.

RESULTS: Sites of SH stain blue. Complete absence of stain was noted in sections blocked with either methyl mercuric iodide or methyl mercuric chloride. The mercury compounds were dissolved with 25% (v/v) N,N-dimethylformamide in 0.05 M sodium phosphate buffer at pH 7. Sections were treated for 30 min at 22°C prior to exposure to the specific reagent. Substantial but less effective inhibition was observed in sections blocked with 5 mM N-ethylmaleimide (30 min at 22°C).

Barrnett and Seligman utilize 2,2'-dihydroxy-6,6'-dinaphthyl disulfide as a reagent for both —SH and —SS— groups.

BARRNETT AND SELIGMAN'S DIHYDROXYDINAPHTHYL DISULFIDE (DDD) METHOD FOR SULFHYDRYL AND DISULFIDE (*J. Natl. Cancer Inst.*, **13**: 215, 1952; **14**: 769, 1953).

1. Fix 24 h in 80% alcohol containing 1% trichloroacetic acid.
2. Dehydrate, embed in paraffin, and section at 5 to 10 μ. Use minimal amount of albumen fixative, or preferably none.
3. Deparaffinize, collodionize with 0.5% collodion, and hydrate through alcohols as usual.
4. To demonstrate —SS— groups as well as —SH groups, insert here a 2- to 4-h bath at 50°C in 0.2 to 0.5 M thioglycolic acid (1.8 to 4.6%), adjusted with sodium hydroxide to pH 8.
5. Dissolve 25 mg 2,2'-dihydroxy-6,6'-dinaphthyl disulfide in 15 ml 100% alcohol, and add 35 ml Michaelis' pH 8.5 Veronal sodium–HCl buffer. Stain up to nine slides in a Coplin jar at 50°C for 1 h.
6. Remove the Coplin jar from the water bath and cool for 10 min at room temperature.
7. Rinse briefly in distilled water.
8. Wash 10 min in 0.01% acetic acid (pH 4 to 4.5).
9. Dehydrate with graded alcohols, pass through 100% alcohol plus ether 50:50, and extract with ether for 5 min to remove excess reagent and reaction by-products.
10. Rehydrate through graded alcohols to distilled water.
11. Stain 2 min in a fresh solution of 50-mg fast blue B in 50 ml 0.1 M Sörensen's phosphate buffer of pH 7.4.
12. Wash 2 min in running water.
13. Mount in glycerol gelatin, or dehydrate with acetone, clear in xylene, and mount in synthetic resin (Permount, HSR, polystyrene, etc.).

RESULTS: Reaction sites show pink, red, and blue-red to blue with increasing intensity of the reaction.

Barrnett wrote us that interposition of an iodine treatment after step 3 completely blocks the sulfhydryl reaction, thus: Immerse for 4 h at 25°C in 1.5 mM iodine (378 mg/l) containing a trace of KI at pH 3.2.

Blocking is also accomplished by a 4-h bath at 37°C in 0.1 M ethyl maleimide (1.25%), buffered with Sörensen's phosphates to pH 7.4, or by a 20-h bath at 37°C in 0.1 M (Na) iodoacetate (about 2%) at pH 8.

Incubation in 0.03 M glutathione (0.92%) at pH 8.5 for 3 h at 50°C between step 10 and step 11 completely prevents the development of color.

Barrnett formerly used at step 4 a $1\frac{1}{2}$-h bath at 50°C in 10% ammonium sulfide. Step 4 as given above represents his 1953 procedure (personal communication). We have found ammonium sulfide ineffective in rendering distal hair cortex reactive to the ferric ferricyanide test and have had similar failures with sodium dithionite, sulfite, and thiosulfate. Stannous chloride not only failed to render hair cortex reactive but occasioned a diffuse deposit of Turnbull's blue over most other structures in the sections.

Pearse notes that for opening disulfide groups relatively brief treatment with alkaline potassium cyanide is preferable because of the destruction of sulfhydryl by alkali. We have had rather irregular and generally unsatisfactory results with potassium cyanide and prefer 10% sodium thioglycolate adjusted with sodium hydroxide to pH 9.5. This reagent appears to render hair cortex strongly reactive to the ferric ferricyanide method on a 10-min exposure at room temperature (25°C). Longer exposures at higher temperatures appear simply to increase section losses without any appreciable gain in reactivity. This reagent works well after Carnoy, hot methanol chloroform, formalin, and the like. After Spuler's Zenker formol, apparently only sulfhydryl groups become demonstrable, and prolonged dichromate fixations decrease or abolish reactivity. This last finding suggests that the mercurial binding of sulfhydryl groups may operate to protect them against dichromate oxidation.

Unalkalinized 10% sodium thioglycolate solution (pH 5.5) may also be effective in opening disulfide bonds, but it apparently requires heat (60°C) and longer exposures. Its use may be desirable when it is necessary to avoid alkali.

Foraker and Wingo (letter, August 31, 1954) use a solution containing 5 g sodium sulfite (Na_2SO_3) and 10 g sodium acetate per 100 ml, treating sections for 1 h at 22 to 30°C. This is said to reduce —SS— groups to —SH without unblocking —SH groups previously alkylated by iodoacetate. In our hands, using the ferric ferricyanide method, this has failed to demonstrate —SS— in distal hair cortex.

Kekule and Linnermann used iodine to convert mercaptans to disulfides in 1862 (Hickinbottom, p. 131); hence this method of blocking —SH groups prior to trying to localize —SS— groups would seem inappropriate, as the iodine-created —SS— groups might be difficult to distinguish from those occurring naturally.

Recent experience indicates that sodium thioglycolate may deteriorate in dry form, so that this reagent may also fail to open —SS— bonds. Also 5% potassium cyanide (pH 10.8) for 10 min at 25°C may fail to convert hair cystine to sulfhydryl. Good results have been attained with a 10-min exposure at 25°C to 5% sodium sulfide (Na_2S). This solution is strongly alkaline (pH 11), and longer exposures engender undue section losses. It appears, however, that this reagent may overcome iodoacetate blockade.

Barrnett and Seligman (*Science*, **116:** 323, 1952) noted in rat tissues strong reactions in hair cortex at the zone of keratinization, in pancreatic acinar cells, in intestinal smooth muscle, in Hassall's corpuscles of the thymus, in lens, and in Purkinje cells of the cerebellum; moderate reactions of hair cortex near the root, of epithelia of epidermis, hair sheaths, sebaceous glands, and intestinal villi, of Paneth cells and stroma cells of villi, and of striated and smooth muscle of the skin. Intense staining of vascular elastica occurred, but this was not prevented by iodine and hence is not a sulfhydryl reaction.

The sodium plumbite reaction for reactive sulfur in hair dates back to 1893 (Salkowski's *Physiologisches Praktikum*, p. 94). Unna (*Biochemie der Haut*, Fischer, Jena, 1913) directs: Dissolve 0.5 g lead acetate in 10 ml water and add 10% sodium hydroxide until the lead hydroxide at first precipitated is redissolved as sodium plumbite Na_2PbO_2. Treat celloidin sections of alcohol-fixed skin (plantar) with this reagent until the keratin is well blackened. This method demonstrates hard and soft keratin and sites chemically containing cystine, as well as the more easily reactive cysteine and glutathione.

Methenamine silver impregnation at 60°C, carried to the point of blackening collagen, demonstrates the keratinization zone of hair cortex quite well, leaving hair bulb cytoplasm pink and distal hair cortex brown to yellow. When the blackening of collagen is prevented by prior acetylation, keratinizing hair cortex still blackens, and when deacetylation with 1% potassium hydroxide in 80% alcohol (20 min) is practiced, distal hair cortex also blackens, indicating cleavage of SS groups. That the reaction in hair cortex is due to SH groups appears highly probable.

CYSTEIC ACID METHODS. Another group of methods for cystine is based on the fact that peracetic and performic acids convert cystine quantitatively to cysteic acid. Bromine in carbon tetrachloride solution (1 : 39 ml, 1 h) and aqueous potassium permanganate solution (0.5%, 30 min, followed by 5 min in 1% oxalic acid or bisulfite) are somewhat less effective, and periodic acid, which was recommended by Dempsey, Singer, and Wislocki (*Stain Technol.*, **25**: 73, 1950), has been ineffective in our hands when elevated temperatures are not employed. In accordance with the strongly acid character of cysteic acid, keratin thus oxidized takes up thionin even from a solution in 0.1 N hydrochloric acid. In unoxidized skin sections only the mast cells stained with this solution.

THE PERACETIC ACID–AZURE–EOSIN METHOD FOR HAIR KERATIN

1. Paraffin sections of skin fixed in the usual fixatives are deparaffinized and hydrated through xylene and descending alcohols as usual.
2. Oxidize for 10 to 120 min in peracetic acid mixture prepared 1 to 10 days previously.
3. Wash 5 min in running water.
4. Stain 1 h in azure A–eosin B or for 30 min to overnight in thionin, azure A, or methylene blue in 0.125 N HCl (pH 1.0) or in 0.1 M sodium citrate–HCl buffer of pH 2, 3, or 4. The dye concentration should be 0.05% for $\frac{1}{2}$-h staining and may be as low as 10 mg/l for overnight.
5. Dehydrate with acetone, and clear in acetone plus xylene and two changes of xylene. Mount in synthetic resin.

RESULTS: Hair cortex and cuticle are dark violet to black in dermal and free portions, grading to greenish blue toward the root in hypodermal portion; stratum corneum, moderate violet; melanin granules, dark green to dark violet; nuclei, blue-violet; keratohyalin, red to purple; muscle, pink; connective tissue fibers, pale pink; cartilage matrix, violet to purple; mast cell granules, dark purple or dark violet.

Methylation in 0.1 N HCl methanol at 60°C for 24 h, done after the peracetic acid step, abolishes all basophilia except that of the keratins and melanin. The basophilia of hair cortex, engendered by peracetic acid, is destroyed by methylation at 60°C for about 3 to 6 days. Even 7-day methylation at 60°C is without effect on the basophilia of hair cortex when applied before the peracetic acid oxidation. Since methylation acts to destroy basophilia by esterifying open acid groups, and no cysteic acid is present until after oxidation of the disulfide, this behavior is to be expected.

Demethylation in 1% potassium hydroxide in 70% alcohol for 20 to 30 min restores basophilia at pH 1, both in hair cortex cysteic acid and in melanin, but not in mast cells or cartilage.

Adams and Sloper (*J. Endocrinol.*, **13**: 221, 1956) use a cysteic acid method for the demonstration of cystine in the neurosecretory substance of the hypothalamus and hypophyseal stalk. This method depends on the highly acid character of cysteic acid as produced by performic (or peracetic) acid oxidation. They used Pearse's formula for preparation of performic acid and 1% Alcian Blue 8GS in 2 *N* sulfuric acid.

PERFORMIC ACID ALCIAN BLUE METHOD OF ADAMS AND SLOPER FOR DISULFIDES

1. Fix in 4% formaldehyde, embed in paraffin as usual, section at 15 μ or 7 μ, deparaffinize, and hydrate as usual.
2. Oxidize 5 min in Pearse's performic acid (make fresh daily), and wash 5 min in distilled water.
    If the section is loosening from the slide at this point, heat it at 60°C until it is just dry, to secure readherence.
3. Stain 1 h in 1% Alcian Blue 8GS in 2 *N* H_2SO_4 (pH 0.2) (5.4 ml 98% H_2SO_4 plus H_2O to make 100 ml).
4. Wash 5 min in running water, dehydrate in alcohols, clear in xylene, and mount in balsam or Depex.

RESULTS: Neurosecretory substance and hair keratin are bright blue; stratum corneum is unstained. Staining of ground substance in collagen and cerebral corpora amylacea was reported also without performic acid oxidation. Nuclei stain faintly or not at all after performic acid.

The more stable peracetic acid serves as well, and other strongly basic dyes can be substituted for Alcian Blue 8GS with equal or superior results.

AROMATIC COMPOUNDS

The Azo Coupling Reaction

Aside from its use for localizing phenols and naphthols liberated from their esters forming substrates for various enzyme actions this reaction has application to the localization of the aromatic amino acids occurring in various proteins; to the demonstration of phenolic and heterocyclic substances in enterochromaffin cells, in mast cells, in adrenal medulla and various neurons, and in poison and salivary glands of lower vertebrates and mollusks; for demonstration of phenolic substances involved in tanning of arthropod exoskeletons; and for localization of various aromatic substances in botanic histology. The last was the first histochemical use of the reaction of Raciborski (*Bull. Acad. Sci. Cracovie. Clin. Sci. Natl.*, p. 353, 1906).

Protein reactions are generally assigned to the aromatic amino acids tyrosine with a phenol group, histidine with its imidazole group, and tryptophan with its indole group. It appears probable that the strong positive reaction reported in parts of the adrenal medulla by Lillie et al. (*J. Histochem. Cytochem.*, **21**: 448, 1973) and of granules in certain neurons are assignable to noradrenaline, which possesses a catechol group. Lillie et al. (*J. Histochem. Cytochem.*, **21**: 455, 1973) also report that the reaction of enterochromaffin now appears to be due to an as yet unidentified catechol with perhaps minor participation of 5-hydroxytryptamine. The reaction obtainable in mast cells by applying *p*-bromodiazobenzene in alkaline solution after the Falck-Torp freeze-dry, hot formaldehyde gas procedure is assigned to histamine by Lagunoff, Phillips, and Benditt (*J. Histochem. Cytochem.*, **9**: 534,

1961); the diazosafranin reaction of rodent and dog mast cells occurring after heavy metal as well as aldehyde fixations at pH levels of 3 to 9 is assigned to 5-hydroxytryptamine by Lillie et al. (*J. Histochem. Cytochem.*, **21**: 441, 1973). The reaction of bilirubin also occurring over the wide pH range of 3 to 9 with various diazos (Lillie and Pizzolato, *J. Histochem. Cytochem.*, **17**: 738, 1969) is assigned to the two central pyrroles resulting from the central cleavage of the tetrapyrrole chain during coupling.

Both Mitchell (*Br. J. Exp. Pathol.*, **23**: 277, 1942) and Danielli (*Symp. Soc. Exp. Biol.*, **1**: 101, 1947) considered reactions by purines and pyrimidines and prescribed methods for their histochemical identification, using blockade procedures and the coupled tetrazonium reaction, proposed by Clara and Canal (*Z. Zellforsch. Mikrosk. Anat.*, **15**: 801, 1932). Burian (*Ber. Dtsch. Chem. Ges.*, **37**: 696, 1904) explored the reactions of imidazoles and purines to diazonium salts. Intensely colored products were formed with xanthine, hypoxanthine, adenine, guanine, and theophylline, but 7-methylpurines such as theobromine and caffeine did not react. Also C-8 substitution blocked the reaction. Burian interprets the reaction of imidazole and the purines as triazene formation on the imide N rather than as true azo coupling on the μ carbon directed by the imide N ortho to it, as we think of imidazole azo coupling today.

Of the three pyrimidines involved in the nucleic acids all possess in their enol forms an essentially aromatic ring with a phenolic hydroxyl at 2; in thymine the para position 5 is occupied by a methyl group, inhibiting azo coupling; in cytosine and uracil position 5 is open and coupling should be possible.

The purines consist essentially of a 1,3-diazene ring sharing C-4 and C-5 with an imidazole ring, and when C-8 (C-2 of the imidazole ring) is unsubstituted and the imide group is free, it has similar azo coupling capacity to the imidazoles. This situation exists with both adenine and guanine when free. But since this imide group is involved in a Schiff's base union with the aldose carbon of ribose or deoxyribose, azo coupling would not be expected. The pyrimidines must exist in essentially nonaromatic ketol form, with open imide groups to bind to the nucleic acid sugars; hence their enol hydroxyls are not available to direct azo coupling. For these reasons it is considered improbable that the purines and pyrimidines of the nucleic acids participate to any noteworthy extent in protein azo coupling reactions. The moderately strong reaction of nuclei in the diazosulfanilic acid–pH 1 azure A (DAS-AzA) reaction seems assignable to basic protein (histone) histidine; it is prevented by Zn, Fe(II), and other chelations.

The *coupled tetrazonium reaction* has been the principal azo coupling procedure used in the study of proteins until the 1960s. Clara and Canal (*Z. Zellforsch. Mikrosk. Anat.*, **15**: 801, 1932) preferred a two-step diazotization of benzidine, coupling with tissue after diazotizing the first NH_2 group, then washing and diazotizing the second NH_2 group, and coupling with resorcinol or a naphthol to enhance the rather light color given to enterochromaffin by diazobenzidine, but reported also a tetrazo procedure which was adopted by Mitchell, Danielli, and later workers. Pearse (1953) replaced the acid washings prescribed by Danielli to destroy triazenes with simple rinses in pH 9.2 buffer before the second naphthol (H acid) enhancement coupling, and later writers followed this practice.

Burstone (*J. Histochem. Cytochem.*, **3**: 32, 1955) found that the acid washes prevented the enhancement effect of the second coupling, while pH 9.2 rinses did not. He further reported some quite unspecific color effects, as reactions on cellulose fibers, protein, and pyroxylin films. Howard and Wild (*Biochem. J.*, **65**: 651, 1957) reported that a large number of the amino acids formed compounds with diazonium salts though their report was in agreement with Pauly (*Hoppe Seylers Z. Physiol. Chem.*, **42**: 508, 1904) that most of them

TABLE 8-2 SOME CHARACTERISTICS OF COMMON DIAZONIUM SALTS

Name	"Colour Index" no.	Amine, mw	Color			
			Entero-chromaffin	β-Naphthol	Naphthol AS	Protein
Safranin	50240	350.8	Blue-black	Blue	Dark blue	Red
Fast red GG, p-Nitroaniline	37035	138.1	Red	Red	Red	Yellow
Fast red B, p-Nitro-o-anisidine	37125	168.2	Red	Red	Red	Pale yellow
Sulfanilic acid		173.2	Red	Red	Red	Pale yellow
Fast garnet GBC o-aminoazotoluene	37210	225.3	Red	Red	Red	Pale yellow
Fast blue B, di-o-anisidine	37235	244.3	Red-brown	Blue	Blue	Pale yellow
Benzidine		184.2	Red-brown	Blue	Blue	Pale yellow
Fast black K salt (base not available)	37190	302.3	Blue-black	Blue-black	Blue-black	Red-orange

were uncolored. Red compounds were formed with tyrosine and histidine, the original colorless tryptophan triazene rearranged to form a deeply colored carbon azo compound. However, the triazenes formed by tetrazotized benzidine readily couple with the fortifying naphthol (H acid or S acid) by their second azo group, and this rendered the tetrazonium procedure simply a general reaction, quite unspecific for proteins. In this view later writers concurred: Barka and Anderson (pp. 59–61) and Pearse (1960, pp. 91–92; 1968, pp. 127–128). Lison (1960) was less definite but doubled the value of the reaction. We essentially agree with Lison.

Similar data were obtained for amino acids by Lillie et al. (*J. Histochem. Cytochem.*, **21**: 441, 1973). Strong red to brownish red were obtained at pH 7 to 9 with tyrosine, histidine, histamine, tryptophan; histidine gave orange at pH 6, orange-yellow at 5, none below; tryptophan gave decreasing orange to yellow from pH 6 to 3. 5-Hydroxytryptophan and 5-hydroxytryptanine gave red from pH 3 to 9, a little stronger from pH 4 up. Bilirubin gave a good red at pH 3 to 9, slightly stronger above 7; biliverdin did not react. Pyrrole gave orange from pH 3 to 9.

Fortunately diazonium salts are now available which yield sufficiently deep color at presumed tyrosine and histidine sites in tissue on direct alkaline and even "acid" azo coupling to render enhancement by Clara's procedure unnecessary. Diazosafranin and fast black K both yielded red "background"colors which became quite dark in such materials as keratin, some zymogen granules, melanosis pigment, erythrocytes, and muscle. These colors resist prolonged extraction in 0.1 N HCl in water or alcohol, whereas safranin staining is promptly extracted from non-acid-fast substances. The color is promptly bleached by sodium dithionite ($Na_2S_2O_4$), which cleaves the azo bond. It is believed that the following technics will give a truer picture of the azo coupling reaction of tissue proteins than the coupled tetrazonium method.

NORMAL METHOD FOR PREPARATION OF FRESHLY DIAZOTIZED AMINES

REACTIONS

$$O_2N-C_6H_4-NH_2 + NaNO_2 + 3HCl \longrightarrow$$

$$\overset{+}{O_2NC_6H_4N} \equiv N-\overset{-}{Cl} + NaCl + H_2O + HCl$$

$$O_2N-C_6H_4-\overset{+}{N} \equiv N-\overset{-}{Cl} + HCl + 2NaOH \longrightarrow$$

$$O_2NC_6H_4N = N\overset{-}{O}H + 2NaCl + H_2O$$

Dissolve or suspend 1 mmol of the amine in 3 ml 1 N HCl and 6 ml distilled water. (The diazonium salts are generally more soluble than the bases.) Chill to 2 to 3°C in the freezing compartment of a refrigerator or by placing the container in an ice bath. Add 1 ml 1 N sodium nitrite (NaNO$_2$) ($= 1$ mmol, or 69 mg), and return the container to the refrigerator or ice bath. Diazotize 20 to 30 min.

This technic allows 1 mmol HCl to neutralize the amine, 1 mmol to convert the NaNO$_2$ to HNO$_2$, and 1 mmol to furnish the necessary acidity for the reaction. Hence, in transferring the fresh diazotate to an alkaline buffer it is regarded as existing in 0.2 N HCl, and corresponding allowance is made in the mixing of Veronal or tris buffers. Electrometric measurement of the final solutions has confirmed the validity of this assumption. The 10 ml of fresh diazotate is 100 mM in concentration, and dilutions to the desired concentration are made accordingly.

Some arylamines may give difficulties in preparation of the acid diazo because of their low solubility in water. Usually this problem is handled simply by treating the ice-cold suspension of the amine with the prescribed amount of NaNO$_2$ with continued agitation to maintain the suspension. The diazonium chlorides are usually sufficiently more soluble than the amines to yield good solutions.

The problem may also be handled by converting the arylamine to its hydrochloride by wetting a millimolar amount with 1 to 4 ml concentrated HCl and allowing it to react for 20 to 30 min before diluting. After adding the prescribed amount of water, the preparation may be heated to complete the solution of the amine hydrochloride. If this procedure is adopted, as in the case of sulfanilic acid, it may be necessary to increase the amount of NaNO$_2$ above the theoretical 1 mmol, to compensate for the evolution of brown gaseous oxides of nitrogen which may occur on addition of the NaNO$_2$ to the ice-cold solution.

Some of the amines, such as p-nitroaniline, are quite insoluble in water (0.08%, *The Merck Index*, 1960) but much more soluble in alcohol (4%, ibid.). Here the arylamine may be conveniently first treated, 1 mmol in 3 ml 1 N HCl made with absolute alcohol instead of water. Slight warming may be used to assist solution. The final 27% alcohol in the 0.1 M acid diazo becomes insignificant when dilution to 1 : 10 mM is made for alkaline azo coupling.

AZO COUPLING REACTION.   FRESH DIAZOTATES.   For 5 mM solution take 2 ml of fresh diazotate plus 2 ml distilled water plus enough additional 0.1 N HCl to make the prescribed quantity in the Michaelis table (Table 8-3) plus the required amount of Veronal sodium for the pH level desired. For pH 8.5, use 2.1 (6.1 − 4.0) ml 0.1 N HCl plus 33.9 ml 0.1 M Veronal sodium. The buffer solutions are precooled to 3°C. For pH 8, the values would be 6.3 (10.3 − 4.0) ml 0.1 N HCl plus 29.7 ml 0.1 M Veronal sodium. An approximate pH of 7.8 is attained with 0.1 M Na$_2$HPO$_4$ as the diluent.

TABLE 8-3 REAGENTS FOR AZO COUPLING AT pH 3, 4, 8, AND 8.5

Reagents	Fast black K salt 20% amine				Freshly diazotized safranin			
	pH levels				pH levels			
	3	4	8	8.5	3	4	8	8.5
0.1 N HCl, ml	29.4	26.4	11.4	6.1	25.4	22.4	10.6	5.3
0.1 M Na citrate	10.6	13.6			10.6	13.6		
0.1 M Na Veronal, ml			28.6	33.9			28.6	33.9
Stable diazo, mg	300	300	60	60				
Fresh diazo 0.1 M, ml					2	2	0.4	0.4
Distilled water, ml					2	2	0.4	0.4
Time at 3°C	1–2 h	1 h	2–5 min	2–5 min	1–2 h	1 h	2–5 min	2–5 min

Note: For stable diazonium salts the appropriate acetate, borate, or tris buffers may be substituted, if desired. The Sörenson phosphate series is useful at pH 5 to 7.

For stable diazotates, weigh out the calculated quantity to furnish the desired molarity. For example, with a sample of fast red GG containing the equivalent of 20% of the primary amine, mw 138, 138 mg would be adequate for 200 ml 1 mM solution; so for 40 ml one would use 27.6 mg; for 2 mM, 55.2 mg; and for 5 mM, 138 mg in 40 ml (see Table 8-4).

For fast black K we find 1 mM (60 mg/40 ml) and 1 to 5 min often quite adequate at pH 8 or 8.5, but for some special purposes use 2 mM for 2 min. The same values hold for freshly diazotized safranin.

For acid azo coupling at pH 3 or 4, a higher concentration, 5 mM, seems adequate, and longer times are needed—30 min to 2 h. This requires 300 mg/40 ml for fast black K and 2 ml/40 ml for freshly diazotized 0.1 M safranin; 0.1 M acetate buffers are used as the diluent for the stable diazotates; the HCl citrate series at 0.1 M can be used for the fresh diazo, with the equivalent of the amount of diazotate deducted from the prescribed amount of 0.1 N HCl and with the fresh diazotate considered as 0.2 N HCl. It is often convenient to use 0.1 M NaH_2PO_4 as the diluent (pH about 3.2).

TECHNIC. Use thin paraffin sections, preferably 3 to 5 μ.

TABLE 8-4 MOLARITY OF SOLUTIONS OF SOME COMMON STABLE DIAZOTATES

Diazo	mw amine	Stated % base	mg/40 ml	
			1 mM	5 mM
Fast garnet GBC	225.3	18	49.5	247.5
Fast red GG	138.4	20	27.6	138.4
Fast red B	108.2	20	37.6	168.2
Fast blue B	249.3	20	48.8	244.3

1. Bring through two changes of xylene and 100, 95, and 80% alcohols, two changes each, to distilled water.
2. Chill buffers to 3°C in refrigerator or with ice bath. Chill Coplin jar.

In using stable diazotates, compute the amount to be employed on the manufacturer's stated percentage of primary amine represented by the preparation used. Usually this is around 20%. Molarities are computed on this basis. Some supposedly pure diazonium chlorides are supplied. In this instance molarities are computed on the formula weight of the diazonium chloride, not on that of the undiazotized base.

Thus on the basis of 20% amine, a 5-mM solution requires the equivalent of the molecular weight of the base (amine) in milligrams of the stable salt to 40 ml of buffered solvent (20% of 1 mmol in $\frac{1}{25}$ l).

Care should be taken that the stable diazo employed in enzyme demonstration reactions does not contain metal salts inhibitory to the enzyme. Stable diazotates are often zinc chloride double salts; boron trifluoride is also often used. Aluminum salts are often present. It is also needful that the "stable" diazo be reasonably fresh. They deteriorate in a year or so, unless kept cold and in the dark.

Standard Fresh Diazo Method with *p*-Nitroaniline, 5 m*M*, pH 8

PREPARATION OF FRESH DIAZOTATE. Dissolve 138 mg *p*-nitroaniline (1 mmol) in 3 ml 1 N HCl plus 6 ml distilled water, warming if necessary. Cool to 3°C and add 1 ml 1 M sodium nitrite (1 mmol = 69 mg). Let stand 20 min at 3°C to diazotize.

Meanwhile bring paraffin sections through xylene and alcohols to distilled water as usual.

Mix and chill Michaelis buffer (to make pH 8 finally) 0.1 M Veronal sodium, 28.6 ml; 0.1 N HCl, 7.4 ml; distilled water, 2 ml. Then add 2 ml fresh diazotate and pour over slides in prechilled Coplin jar. Let stand 2 min at 0 to 5°C. Wash in three changes of 0.1 N HCl 5 min, 5 min, 5 min; distilled water, two changes, 1 min, 1 min; and counterstain in azure A for 15 min (1% azure A, 2 ml; 0.1 N HCl, 4 ml; distilled water, 34 ml; pH 2). Rinse in distilled water, acetone three times, acetone plus xylene, xylenes, and mount in cellulose caprate.

If preliminary procedures are to be applied to the sections, start the diazotization when the sections are due to be removed from the last step of the preliminary procedure. Sections can be held 10 to 15 min in distilled water, or for some hours in 80% alcohol, unless contraindications are specifically noted.

DIAZOSAFRANIN. For the *p*-nitroaniline in the above technic, substitute 1 mmol safranin O, C.I. 50240, 351 mg, diethylsafranin, C.I. 50206, 378 mg, or dimethylsafranin, C.I. 50205, 351 mg, with appropriate allowances for dye content. These solutions in 3 ml 1 N HCl plus 6 ml water are reasonably stable and may be made up in larger quantities, if desired. For use mix 9 ml with 1 ml 1 N NaNO$_2$. If kept cold the nitrite solution is also stable for a week or more. With the diazosafranins omit the counterstain.

The acid diazo mixture is not very stable and should be used within 2 h of mixing for best results. It is probable that internal acid coupling occurs on longer standing, since only one of the two amino groups of the safranin can be diazotized in acid aqueous solution, and therefore the dye residue remains capable of acid (amine) coupling with diazonium groups. On overnight storage at 3°C the acid diazo mixture changes from blue to dark red and gives only feeble coloration of argentaffin cells on alkalinization. Addition of 1 ml acid diazo mixture to 40 ml 0.1 M disodium phosphate solution brings the reaction to pH 7.7, which is sufficiently alkaline for phenol coupling, and not alkaline enough to remove the sections from the slides.

ACID DIAZOSAFRANIN REACTION FOR BILIRUBIN AND RODENT MAST CELLS. When diluted 2 : 38 with 0.1 M NaH_2PO_4 diazosafranin has about a pH of 3.2 and stains selectively red bile casts, hematoidin (*J. Histochem. Cytochem.*, **17**: 738, 1969) and rat, mouse, and certain other mast cells which have a relatively high 5-hydroxytryptamine content (ibid., **21**: 441, 1973). In this report we demonstrated in vitro azo coupling of 5-hydroxytryptophan and 5-hydroxytryptamine (5-HT) to furnish deep red with *p*-nitrodiazobenzene over a pH range of 3 to 9. Periodic acid oxidation of 5-HT in vitro only slightly retarded but did not appreciably weaken its azo coupling capacity and in sections of the same preoxidation did not prevent the diazosafranin reaction at pH 3 or 8 of rat mast cells. Azo coupling of adrenal medulla and of duodenal enterochromaffin was prevented by HIO_4 oxidation and restored by $Na_2S_2O_4$ reduction. In the absence of the bile pigments and hematoidin on formol- or glutaraldehyde-fixed tissue this acid azo coupling with diazosafranin would appear to be a fairly selective method for 5-HT.

In vitro, alkaline azo coupling reactions occasion no noteworthy color changes with catechol, anisole, phloroglucinol, *o*-aminophenol, amidol, uracil, guanine, phenylalanine, or cysteine. Dark brown precipitates are produced with phenol, *o*-cresol, and *p*-cresol; a very dark red-brown precipitate occurs with resorcinol; red-purple to dark red-purple or purple precipitates occur with hydroquinone, pyrogallol, and tyrosine; and dark blue-violet precipitates, with *α*- and *β*-naphthols.

Model experiments on formalin-fixed, phenol-impregnated, gelatin-paper strips gave coupling colors with diazosafranin as follows: pink to colorless with the blank gelatin control, with phenol, catechol, hydroquinone, pyrogallol, phloroglucinol, *o*-aminophenol, 2,4-diaminophenol, tyrosine, tyramine, 3,4-dihydroxyphenylalanine, adrenaline, noradrenaline, indole, tryptophan, tryptamine, cysteine, guanine, *m*- and *p*-aminophenol. Pink to purple colors were obtained with *o*- and *p*-cresols and 5-hydroxytryptamine; Gomori has reported blue with the last, but we have obtained definite blues only with resorcinol and with *α*- and *β*-naphthols.

For testing the reactions of various phenolic substances in comparison with argentaffin granules, Gomori followed the suggestion of Coujard and used the following technic: Substances to be tested are dissolved 0.25 to 1 at 1% in a mixture of equal parts of serum and 5% gelatin. Marks are written with a clean steel pen dipped in a slightly warmed solution on a slide; more than 25 different substances can be tested on one slide. The slides are dried and subsequently exposed to formaldehyde fumes in a closed jar for several hours; the excess formaldehyde is driven off in the paraffin oven. The dry slides can be stored for weeks. For the test the slides are run through xylene and alcohols to distilled water and stained just as tissue sections would be. It should be noted that substances susceptible to air oxidation will probably be in the oxidized state. Quinones do not azo-couple.

In 1966 Lillie, Palmer, and Gutiérrez (*J. Histochem. Cytochem.*, **14**: 421, 1966) introduced the idea of first azo-coupling with *p*-sulfodiazobenzene, washing thoroughly with acid to destroy triazenes, and then demonstrating the sulfonic acid groups now bound to tissue at azo coupling sites with dilute solutions of cationic dyes at or below pH 1. The method had the defect of also staining cartilage, mast cells, and a few of the more strongly sulfated mucopolysaccharides by the cationic dye. In most instances this is not especially confusing, and in some cases we have done away with this cationic dye staining by a short (2- to 4-h) 60°C 0.1 M methylsulfuric acid–methanol methylation.

The method has proved useful for the selective demonstration of histidine residues by Lillie and Donaldson's (*J. Histochem. Cytochem.*, **20**: 929, 1972) nitration blockade method [derived from Brunswik's (*Hoppe Seylers Z. Physiol. Chem.*, **127**: 268, 1923) xanthoproteic-diazo sequence], for the demonstration of bile casts and hematoidin using 0.01% solutions in 0.12 N HCl of safranin O, rosanilin, parosanilin, azocarmine base (rhodindine), and pyronin B to produce red in bile pigments, for the demonstration of noradrenaline cells in the adrenal medulla and cytoplasmic granules in various neurons.

The original technic has been revised by first dissolving the HCl–sulfanilic acid paste in hot water and then cooling to 3°C for diazotization. The nitrite dose has been increased to allow for our subtropical deterioration of $NaNO_2$, and a urea addition at the end of acid diazotization has been added to destroy excess HNO_2. A 30-mg urea addition suffices for 70 mg $NaNO_2$.

$$CO(NH_2)_2 + 2HNO_2 \longrightarrow CO_2 + 2N_2 + 3H_2O$$

THE DIAZOSULFANILIC ACID–pH 1 AZURE A SEQUENCE (DAS-AzA) METHOD OF 1972 (*J. Histochem. Cytochem.*, **21**: 448, 1973)

1. Treat 195 mg sulfanilic acid monohydrate 20 min with 2 ml concentrated HCl.
2. Dilute the paste with 30 ml distilled water, and heat to 80 to 90°C with swirling to reach a clear solution.
3. Cool in refrigerator or ice bath to about 3°C.
4. Add 200 mg granular or stick $NaNO_2$ in 1 ml distilled water. A little brown gas escapes, and the solution turns slightly yellowish. Keep at 3°C with periodic agitation for 20 min to diazotize.
5. Add 80 to 90 mg urea and mix thoroughly. Some gas may escape (N_2).
6. Add gradually 12 to 15 ml 16% sodium carbonate, waiting for gas evolution to subside after each 3-ml addition. The solution turns distinctly yellow, and the last 1-ml addition provokes no further gas evolution. If the solution does not turn yellow at this point, start over again, since some error has been made and no staining will result.
7. From sections previously brought through alcohols to ice-cold distilled water, pour off water and pour in alkaline diazo. Couple 2 to 3 min.
8. Wash in three changes 0.1 N HCl, for 5 min each. In critical studies one can use three changes of 0.24 N HCl in 70% alcohol for 10 min each or even overnight in the final change.
9. Transfer directly to 0.02 to 0.05% azure A in 0.12 N HCl (filtered before use to remove undissolved dye particles), and stain for 20 min. [0.12 N HCl = 1 ml concentrated (12 N) HCl + 99 ml distilled water.]
10. Rinse in distilled water 2 times for 10 to 15 s each, dehydrate with reagent-grade acetone, clear in xylene, mount in synthetic resin. With blocking procedures xylene–cellulose caprate is useful because unstained areas are still visible at n_D 1.482.

RESULTS: Erythrocytes are green to black after formol, green to brown after glutaraldehyde fixation; nuclei and elastin, green; cytoplasms and muscle, yellowish green; collagen, greenish yellow to yellow; noradrenaline cells of the adrenal medulla, dark blue to greenish blue.

We have occasionally used the Gibbs reaction with 2,6-dichloro- or dibromoquinone-chloroimid (*Am. J. Pathol.*, **36**: 623, 1960) but have found it rather temperamental and unreliable. The reagent condenses with phenols with an open para (or ortho?) position to yield indophenol dyes. J. C. Dacre (*Anal. Chem.*, **43**: 589, 1971) reports that many 4-alkoxyphenols also produce blue and green indophenols with the Gibbs reagent.

GIBBS' INDOPHENOL REACTION FOR PHENOLS (from Pearse, 1953, p. 478)

1. Deparaffinize and hydrate thin (5 μ) paraffin sections of formalin-fixed material.
2. Immerse 15 min in 0.1 % Gibbs' reagent in pH 9.2 Veronal buffer.
3. Wash 5 min in running water, counterstain in 0.5 % safranin or 1 % neutral red, differentiate and dehydrate in 70 % and absolute alcohol, clear in xylene, and mount in synthetic resin.

RESULTS: When successful, the method demonstrates enterochromaffin granules in dark gray-blue.

Acylation Blockade

The azo coupling reaction may be blocked by acylation and restored by saponification procedures. For prevention of the azo coupling reaction of phenols the alcohol–acetic anhydride method proves to be superior to pyridine solutions of either benzoyl chloride or acetic anhydride (*J. Histochem. Cytochem.*, **12**: 522, 1964).

ALCOHOLIC ACETYLATION FOR BLOCKING THE AZO COUPLING REACTION OF PHENOLS

1. Bring thin paraffin sections through xylenes to absolute alcohol.
2. Acetylate 1 h in equal volumes of acetic anhydride and absolute alcohol.
3. Wash in 80 % alcohol and 10 min in running water.
4. Azo-couple acetylated and unacetylated control sections in fast black K, diazosafranin, or diazotized *p*-nitroaniline as usual, wash in 0.1 N HCl and in water, dehydrate, and mount.

Tyrosine

THE MILLON REACTION.   Cowdry, Romeis, and Glick all recommend the method of Bensley and Gersh for protein-containing tyrosine.

Millon reagent: Saturate 500 ml 4 % aqueous nitric acid with mercuric nitrate crystals. Filter, and to 40 ml filtrate add 0.2 ml 70 % nitric acid and 140 mg sodium nitrite.

Prepare paraffin sections of frozen dried material or material fixed in an anhydrous fixative. Mount on slides by floating on 95 % alcohol, or without flotation. Deparaffinize, dip briefly in 100 % alcohol, blot, and immerse directly in Millon reagent. Remove sections at intervals, inspecting for the presence of rose or red color. The action reaches its maximum in about 3 h. Rinse in 1 % nitric acid, dehydrate with 100 % alcohol (dropper bottle or several changes), clear in xylene, and mount in balsam.

RESULTS: Tyrosine is rose red. Glick regards the reaction as specific.

Serra (*Stain Technol.*, **21**: 5, 1946, emended) recommends the following modification of Millon's reaction as essentially specific for tyrosine.

STOCK REAGENT.   Dissolve 7.5 g mercuric sulfate, 5.5 g mercuric chloride, and 7 g anhydrous sodium sulfate in 85 ml distilled water. Add 12.5 g (6.8 ml) concentrated sulfuric acid (sp gr 1.836) and when cool dilute to 100 ml.

PROCEDURE.   Immerse sections—loose, on cover glasses, or on slides—in stock reagent for 30 min at 60°C in covered vessel. Then cool to room temperature by immersion of container in cold water and let stand for 10 min. Add an equal volume of distilled water and a few drops of molar (6.9 %) $NaNO_2$ solution. A red develops, reaching a maximum in 3 min. Mount in glycerol, and seal with lanolin rosin, polystyrene, or cellulose caprate.

THE PROTEIN DIAZOTIZATION COUPLING REACTION FOR THE DEMONSTRA-
TION OF TYROSINE (Lillie, *J. Histochem. Cytochem.*, **5**: 528, 1957; Glenner and Lillie,
ibid., **7**: 416, 1959). That tyrosine-bearing proteins develop yellow on treatment with nitrous
acid has been known since 1888 (Richard, *Bull. Soc. Chim. Ind.*; Mulhouse, ibid., p. 75), and it
was noted at the same time that the yellow of wool and silk turned brown on exposure
to light. The procedure was developed by Morel and Sisley to yield various azoic colors
on coupling in the dark with various phenols and naphthols (*Bull. Soc. Chim.*, **41**: 1217, 1927).
The reaction is formulated thus:

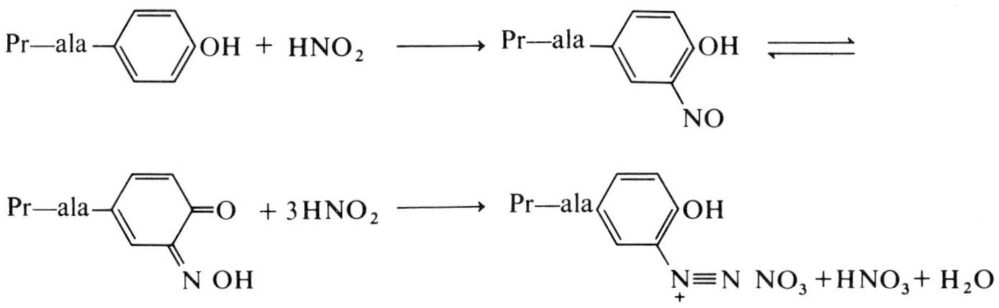

Our own experiments confirmed the necessity for cold and strict darkness, and determined
the optimal diazotization time to be about 16 h at 3°C, and S acid (8-amino-1-naphthol-5-
sulfonic acid) and H acid (8-amino-1-naphthol-3,6-disulfonic acid) to be the most appropriate
coupling agents. Urea is added to the coupling reagent to decompose excess HNO_2. A dark
chamber is improvised by inverting over the Coplin jar containing the slides an empty tin
can of sufficient height (10 cm) and diameter to cover it completely.

The fixation method must be such as to conserve the protein under particular study.
Neutral aqueous formalin has proved suitable for most objects. The technic follows:

1. Prepare paraffin or frozen sections and attach to slides as usual. Bring to distilled water.
2. Diazotize 16 h at 3°C in strict darkness with 1 N $NaNO_2$ in 1 N acetic acid: Dissolve
   2 g $NaNO_2$ in 28 ml distilled water and add 1.7 ml glacial acetic acid. (Do not mix the
   dry salt with the glacial acetic acid; this produces an immediate evolution of brown
   gaseous oxides of nitrogen.)
3. To 50 ml 70% alcohol add successively 500 mg S acid, 500 mg potassium hydroxide,
   and 1 g urea or 500 mg ammonium sulfamate. Dissolve and chill to 3°C.

   Wash sections in four changes of distilled water at 3°C, 5 s each, and introduce into
   the cold alkaline coupling reagent for 1 h. This we usually do also in the dark, though
   Glenner noted that diazotized tyrosine was much less photosensitive in alkali.
4. Wash in three changes 0.1 N HCl, 5 min each, and 10 min in running water.
5. Counterstain, if desired, with alum hematoxylin. We usually omit this step.
6. Dehydrate in alcohols, clear in xylene, and mount in xylene, cellulose caprate, or other
   resin.

RESULTS: Tyrosine sites are red-purple to pink. Hair cortex, soft keratin, elastic laminae,
the inner sheath cells and medulla of the root zone of hairs show the strongest reactions;
moderate reactions are seen in zymogen granules of salivary and gastric glands, pancreas
and Paneth cells, smooth and striated muscle, some nuclei and cytoplasms, thyroid and
hypophyseal colloid. Eosinophil leukocytes, enterochromaffin cells, and mast cells are not
identified.

In step 3 Glenner substituted 500 mg ammonium sulfamate, which consumes 2 mol HNO_2

per mol, equaling the 2 mol consumed by urea. Urea (mw 60) is thus nearly twice as effective as ammonium sulfamate (mw 114) on a weight-for-weight basis and currently costs about half as much. We have adhered to the use of urea in this step. The quantity can probably be safely reduced to 250 mg, since the carryover of HNO_2 in the washed sections must be infinitesimal.

Geyer [*Acta Histochem.* (*Jena*), **13**: 355, 1962] notes that the Morel-Sisley reaction can be accelerated and intensified by a 1- to 6-h pretreatment with 10% iodine in alcohol. The sections are then washed in alcohols and water and nitrosated at 3 to 5°C for 1 h only, using 1 N $NaNO_2$–1 N acetic acid, strictly in the dark.

Pretreatment with tetranitromethane, 0.1 ml in 10 ml pyridine and 20 ml 0.1 N HCl for 6 h at East German laboratory temperature, and washing in alcohol or acetone (2 times, 5 min each) and water (10 min) will completely prevent the Morel-Sisley and Millon reactions for tyrosine and moderately impair the xanthydrol and Adams reactions for tryptophan [Geyer, *Acta Histochem.* (*Jena*), **13**: 357, 1962]. We (*J. Histochem. Cytochem.*, **20**: 929, 1972) found two successive 1-h treatments with Geyer's reagent effective in preventing the azo coupling reaction of tyrosine.

Histidine

A number of procedures purporting to be specific for histidine have been reported. Two of these using the coupled tetrazonium reaction as demonstration reagent and relying on alkaline iodinations to block tyrosine reactivity are mentioned—those of Landing and Hall (*Stain Technol.*, **31**: 197, 1956) and Bachmann and Seitz (*Histochemie*, **2**: 307, 1961)—and are passed over because the coupled tetrazonium method is now considered as depending at least in part on triazene combinations with nonaromatic amino acid residues, and because iodination does not prevent azo coupling of tyrosine. Diiodotyrosine readily azo-couples in vitro, iodine being ejected.

Brunswick (*Hoppe Seylers Z. Physiol. Chem.*, **127**: 268, 1923) reported a method depending on nitration by application of the xanthoproteic reaction to block tyrosine and demonstration by *p*-diazobenzenesulfonic acid. His experimental proof was convincing, and the method was accepted by Klein in 1928 and Serra (*Stain Technol.*, **21**: 5, 1946). Thick sections were required to give adequate colors with the diazosulfanilic acid, and apparently the method has been used only on plant tissues and smears of animal tissue because of the destructiveness of strong HNO_3.

Lillie and Donaldson (*J. Histochem. Cytochem.*, **20**: 929, 1972) overcame these difficulties by nitration in an HNO_3–Ac_2O–HAc 4 : 1 : 5 mixture or in tetranitromethane in pyridine and water. It was shown in vitro that 3-nitrotyrosine did azo-couple but with reduced color and that 3,5-dinitrotyrosine does not appreciably react with fresh alkaline *p*-nitrodiazobenzene, which gives good reds with histidine, histamine, tryptophan, and tyrosine in the alkaline range. The diazosulfanilic acid–pH 1 azure A (DAS-AzA) sequence was used as giving the strongest color reactions to protein, and as permitting after the azo coupling step prolonged acid treatments to eliminate triazene bonding. The function of the pH 1 azure A in the method is simply that of a cationic dye at low pH for the demonstration of the sulfonic acid residues of the azo-bound sulfanilic acid.

TECHNIC. Bring paraffin sections through xylene and absolute alcohol to glacial acetic acid. Nitrate 4 h in 20 ml concentrated HNO_3, 25 ml glacial acetic acid, and 5 ml acetic anhydride. Wash 10 min in several changes of water and stain by the DAS-AzA method. Alternatively nitrate in three successive baths of 1 h each in fresh mixtures of 15 ml 1% tetranitromethane (TNM) in pyridine and 30 ml 0.1 N HCl, wash 10 min in several changes of distilled water, and stain by the DAS-AzA method as before. After the azure A step, dehydrate with three changes of fresh acetone, clear in xylene, and mount in synthetic resin. RESULTS: Nitration prevents the reaction of hair medulla and of adult human vascular elastin as well as that of young animals completely. Sections should be compared both with DAS-AzA stains of unnitrated controls and with simple pH 1 azure A stains for natively sulfated and sulfonated structures.

The TNM technic is perhaps easier, but the HNO_3 method has the advantage of ready reagent availability.

Histamine

Lagunoff, Phillips, and Benditt (*J. Histochem. Cytochem.*, **9:** 534, 1961) direct as follows for the demonstration of histamine in mast cells:

1. Freeze in isopentane chilled with liquid nitrogen; desiccate at -30 to $-40°C$ in vacuo over P_2O_5.
2. When dry, transfer to desiccator containing paraformaldehyde and calcium chloride, seal on lid with petrolatum, partially evacuate, and heat in paraffin oven at 55 to 60°C for 4 to 96 h.
3. Embed in paraffin; section at 5 to 10 μ.

Paraffin sections are subjected to azo coupling in 2.5% Na_2CO_3, with freshly diazotized *p*-bromoaniline: 3 ml 5% in 1 N HCl plus 3 ml 5% $NaNO_2$ for 1 min; add 30 ml ice-cold 2.5% Na_2CO_3; expose sections to the alkaline solution 10 to 15 min at 0°C.

Fast garnet GBC and fast black K were also used in 0.5% solution in pH 7.8 borate buffer, but they gave weaker reactions. It may be pointed out that their fresh diazotate contains a large (2.46 times) excess of nitrite, the unused part of which is carried over into the coupling stage.

The mast cells also give the ferric ferricyanide reduction test, reduce hot methenamine silver, and present a yellow-green fluorescence in near-ultraviolet. A negative tryptophan reaction by the Adams method, a negative chromaffin reaction, and a positive diazotization coupling tyrosine reaction were also observed.

Lagunoff et al. speculate on the possible concurrent presence of tyramine, dopa, or dopamine to account for the positive tyrosine test, but they did not include this test in their model experiments.

Lagunoff et al. quoted Frankel and Zellmer (*Biochem. Z.*, **110:** 234, 1920) concerning the condensation of formaldehyde with histamine to form imidazoleisopiperidine, which we formulate below for both histamine tautomers.

In the presence of a positive tyrosine (or tyramine) reaction at the same site we do not believe that this application of the Pauly reaction can be assigned surely to the presence of histamine, though the idea naturally is attractive.

THE o-PHTHALALDEHYDE FLUORESCENCE METHOD FOR HISTAMINE. When first introduced as a histochemical procedure by Juhlin and Shelley (*J. Histochem. Cytochem.*, **14**: 525, 1966), it was applicable only to fresh tissue as cell suspensions, tissue spreads, and fresh frozen sections. Shelley, Ohman, and Parnes (*J. Histochem. Cytochem.*, **16**: 433, 1968) extended the method to freeze-dried–paraffin sections, and Enerbäck (*J. Histochem. Cytochem.*, **17**: 757, 1969) found brief Carnoy fixation, alcohol dehydration, and vacuum or routine paraffin embedding quite satisfactory. The bright yellow fluorescence appears to be quite specific for histamine. Shelley et al. reported model experiments on 31 reactants: bright yellow only with histamine, bright blue with reduced glutathione, green with DL-β-dopa, faint orange-brown with noradrenaline and procaine, faint blue-green with adrenaline, faint yellow with histidine and taurine, and no fluorescence with DL-alanine, ammonia, DL-arginine, DL-aspartic acid monohydrate, DL-citrulline, L-cysteine HCl, dopamine, DL-glutamic acid, glutamine, glycine, heparin, L-hydroxyproline, 3-hydroxytyramine HCl, imidazole, DL-isoleucine, DL-leucine, DL-methionine, DL-norvaline, and DL-valine. Juhlin added glucosamine and galactosamine to the unreactive list.

The Juhlin-Shelley technics for cell suspensions, fresh tissue, omental spreads, frozen and frozen dried tissue involved 1- to 3-min treatments with pH 11 Tyrode solution, blotting and treatment with 1% o-phthalaldehyde (OPT) in p-xylene with direct observation in that fluid, or transfer after 3 min to tetrahydrofurfuryl alcohol (2-furfurylmethanol). The reaction must be studied within 10 to 15 min.

The Shelley et al. freeze-dried–paraffin section method and the Enerbäck Carnoy-paraffin section method permit use of adjacent paraffin sections for other procedures, and permit better morphologic localization. Enerbäck fixed small blocks (4 × 4 × 4 mm or less) in Carnoy for 2 h at 25°C, then for 3 h in ethanol, 1 h in ethyl benzoate, and paraffin for 30 min in 100 mm mercury vacuum, or routinely without vacuum. With both section methods flatten the sections on slides without water flotation, heat to 60°C to melt paraffin, place in 100% humidity chamber, and flood with 1% OPT-xylene for 2 min or with 1% OPT-ethylbenzene for 4 min. Pour off the reagent, mount in tetrahydrofurfuryl alcohol, and study at once under ultraviolet light on fluorescence microscope.

Juhlin and Shelley used a 3650-Å mercury-vapor lamp; Enerbäck employed a filter excluding light below 4100 Å and spoke of a blue background.

While anhydrous solvents–benzene, toluene, xylene, ethylbenzene, chlorobenzene—are used for the OPT reaction, it is insisted by all that a small amount of water is necessary for the reaction. But water flotation of paraffin sections destroys the fluorescence. The maintenance of histamine *in situ* in the Enerbäck technic, despite the ready solubility of free histamine in alcohol and Carnoy, is ascribed to protein bonding and prompt protein denaturation by Carnoy.

The OPT procedure appears more specific than the Lagunoff et al. azo method and does not require the freeze-dried formaldehyde treatment of that procedure.

Ehinger et al. (*Exp. Cell Res.*, **47**: 116, 1967; *Biochem. Pharmacol.*, **17**: 1997, 1968) describe a histochemical method for the demonstration of histamine. Freeze-dried paraffin-embedded tissues are sectioned, deparaffinized with xylene, and placed in a closed vessel (about a 500-ml size) containing a few milligrams of OPT freshly crystallized from heptane which has previously been heated in an oven at 100°C for 10 min to promote vaporization. Slides are inserted, the vessel is closed, and the sections are treated at 24°C for 90 s. Sections are removed, exposed to steam from boiling water for 5 s, dried at 80°C in an oven for 5 min, mounted in xylene, and examined at 365 mμ with the mercury line from an HBO 200-W high-pressure lamp. A selective blue light is emitted if dry; a more yellow light is emitted under humid conditions. The specificity of the method has been checked in model systems. Further work needs to be done in tissues.

Tryptophan

The Romieu reaction for proteins (according to Glick, Lison, and Cowdry) indicates the presence of tryptophan. Fix in alcohol, Bouin's fluid, or formalin. Prepare rather thick paraffin or nitrocellulose sections. Deparaffinize and hydrate as usual. Blot. Cover with a few drops of syrupy phosphoric acid, and heat for a few minutes at 56°C (use a paraffin oven). Remove, apply cover glass, and examine at once. Red to violet indicates the presence of tryptophan.

THE VOISENET-FÜRTH REACTION. This is recommended by Serra (*Stain Technol.,* **21:** 5, 1946).

1. Fix in a formaldehyde fixative, or mordant hydrated paraffin sections in 10% formalin for 1 to 5 h.
2. Immerse in 12% sodium silicate solution (sp gr 1.10) for 3 to 5 s. If material is well hardened, this step may be omitted. Drain on filter paper.
3. Transfer directly to fresh Voisenet reagent for 10 to 15 min. *Voisenet reagent:* To 10 ml concentrated HCl add 1 drop 2% aqueous formol and 1 drop 0.5% sodium nitrite (fresh aqueous solution). Mix well. Prepare fresh daily.
4. Mount directly in glycerol and examine at once (the color fades).

RESULTS: Violet indicates presence of tryptophan. The reaction is given by indolic compounds and in proteins is specific for tryptophan.

The *p*-dimethylaminobenzaldehyde methods are less destructive and yield stronger colors and sharper localizations.

POSTCOUPLED BENZYLIDENE REACTION FOR INDOLES (Glenner and Lillie, *J. Histochem. Cytochem.,* **5:** 279, 1957). Fix briefly in buffered (pH 7) 10% formalin. Generally, results have been satisfactory with 3-h fixation; both poor and good results have been obtained with material fixed 24 to 72 h. Prolonged formalin and oxidative fixations have been unsatisfactory. Alcohol hardening overnight or longer, after the brief formalin, is helpful for paraffin sections, and the thin celloidin-paraffin technic seems excellent.

Dehydrate, clear, embed in paraffin, and section at 5 μ.

1. Bring through xylene to 100% alcohol, and dry briefly.
2. Immerse in 1 g *p*-dimethylaminobenzaldehyde in 10 ml concentrated HCl and 30 ml glacial acetic acid, 5 min, at 25°C.
3. Wash in glacial acetic acid, three washes, 30, 60, and 60 s.
4. Add 1 ml freshly diazotized S acid[1] to 40 ml cold (15 to 20°C) glacial acetic acid, mix, and pour into a cold Coplin jar over the sections. Let react 5 min.
5. Wash in two changes of glacial acetic acid.
6. If desired, counterstain 5 min in 0.05% new fuchsin in glacial acetic acid, and wash in two changes of glacial acetic acid, 1 min each.
7. Treat with acetic acid plus xylene, 3 to 4 changes of xylene, cellulose caprate.

---

[1] Safranin, previously used as an alternative, gives false-positive reactions in rat mast cells. These cells color red with diazosafranin in glacial acetic acid alone.

RESULTS: Reaction sites are blue. Stomach: peptic gland zymogen granules are moderately reactive; eosinophil leukocyte granules, strongly reactive; keratin, unreactive. Pancreas: zymogen granules are deep blue; cytoplasm of acini, unstained; duct contents, blue; alpha cells, light blue; beta cells, unstained. Small intestine: Paneth cell granules and mucosal eosinophils are blue; cytoplasm of epithelium of villi, pale gray blue. Artery: elastica is negative. Salivary gland, parotid type, zymogen granules are blue; rat secretory duct type, deep blue secretory granules.

To demonstrate Paneth granules (deep blue) use an animal which has been starved overnight. If such a preparation is subjected to a following fast garnet GBC azo coupling at pH 8.5, the enterochromaffin cells are also demonstrated in a contrasting deep red. This occurs equally in glutaraldehyde-fixed tissue.

THE ROSINDOLE REACTION FOR TRYPTOPHAN AND OTHER INDOLES.

Adams (*J. Clin. Pathol.*, **10:** 56, 1957) and Glenner (*J. Histochem. Cytochem.*, **5:** 297, 1957) almost simultaneously rediscovered the rosindole reaction of E. Fischer (*Ber. Dtsch. Chem. Ges.*, **19:** 2988, 1886) which Mikosch used histochemically in Reichl's variant (*Monatsschr. Chem.*, **11:** 155, 1890), reporting his otherwise unpublished results privately to Reichl. This reaction depends on the same benzylidene condensation with indoles as in the previous method. The blue color is developed, however, not by azo coupling after the condensation in step 2 but by immersion in a 5 ml concentrated HCl, 35 ml glacial acetic acid mixture which is poured onto 500 mg sodium nitrite ($NaNO_2$) in a Coplin jar; slides are transferred from acetic acid (step 3) immediately into this bubbling mixture for about 1 min at 25°C. They are then washed in two changes of glacial acetic acid, acetic acid plus xylene, and 3 to 4 changes of xylene, and mounted in xylene–cellulose caprate.

Results are essentially similar to those of the preceding method but are somewhat less precise, and, particularly with the Adams variant, cited below, there is more damage to tissue structure from the strong acid employed.

THE ADAMS DMAB (DIMETHYLAMINOBENZALDEHYDE) METHOD

1. Sections are brought to absolute alcohol and just dried or collodionized in 0.25% collodion.
2. Sections are transferred directly to 5% DMAB in concentrated HCl (sp gr 1.18) for 1 min.
3. Treat with 1% $NaNO_2$ in concentrated HCl for 1 min. The $NaNO_2$ crystals are dropped into the acid and sections immersed at once, while evolution of brown gas proceeds.
4. Wash for 30 s in water.
5. Rinse in acid alcohol (1% concentrated HCl 70% alcohol).
6. Dehydrate, clear, and mount in synthetic resin.

Results are essentially similar to those of Glenner's rosindole reaction except that morphologic preservation is inferior, though blues are more intense.

In 1972 Previero et al. (*Biochim. Biophys. Acta*, **285:** 269) reported that acyl chlorides in dry trifluoroacetic acid attacked tryptophan and its peptides predominantly at position 2, to a less extent at 1 and 5. The 2 product was an indolyl alkyl ketone which promptly reacted with the α-amino group to create a β-carboline ring giving specific absorption in the ultraviolet and a blue fluorescence. Rost and Pearse (*Histochem., J.*, **5:** 577, 1973) have utilized this reaction for the histochemical localization of high tryptophan concentrations in tissue. Identical localization patterns are reported as with the Adams DMAB (E. Fischer rosindole) reaction.

Rost and Pearse (*Histochem. J.*, **5:** 577, 1973) applied the reaction for tryptophyl units described by Previero et al. (*Biochim. Biophys. Acta*, **285**: 269, 1972) to develop a histo-chemical method. Formalin-fixed or freeze-dried sections may be used. Paraffin- or Araldite-embedded tissues may be used. Paraffin-embedded sections placed in a closed Coplin jar with vapors of acetyl chloride for 15 min at 80°C, freeze-dried dewaxed tissues treated for 15 min at either 22 or 80°C, and resin-embedded sections treated for 30 min at 22°C produced acceptable blue fluorescence with excitation-emission maxima at 370 : 460 mμ as observed by microspectrofluorometry.

Freeze-dried paraffin sections, dried cryostat sections of fresh unfixed tissue, or paraffin sections of glutaraldehyde-fixed material are used but not formol, since the position 2 condensation of formaldehyde to form a norharman ring interferes. These sections are placed vertically and supported by a small piece of glass rod, in a Coplin jar, over 2 to 4 ml acetyl chloride so as to expose the sections to the vapor of the reagent. Undewaxed paraffin and Araldite sections may be so used but require longer exposures, up to 30 min. At 20°C 15-min exposures of naked sections may be needed; at 80°C 5 min is sufficient. A non-fluorescent mounting medium is required, and a lamp giving 365 mμ ultraviolet is used.

Acetyl chloride of course reacts with protein and carbohydrate hydroxyl groups, tyrosine and probably histidine, the catechol groups of the catecholamines, but Rost and Pearse record no confusion with other sources. Further study on possible confusion sites seems to be indicated, but the method should prove a useful alternative to the aldehyde-indole condensation reaction.

Augsten et al. (*Histochemie*, **19**: 44, 1969) conducted quantitative studies of several proteins and peptides containing tryptophan and concluded that the amount of blue pigment obtained with the Adams *p*-dimethylaminobenzaldehyde-nitrite reaction is dependent both upon tryptophan content and the structure of individual proteins.

Håkanson and Sundler (*J. Histochem. Cytochem.*, **19**: 477, 1971) condensed model dipeptides containing NH_2-terminal tryptophan with formaldehyde and evaluated the intensity of fluorescence. The fluorescence yield was improved 20 to 30 times if the formaldehyde condensation reaction was carried out in the presence of ozone.

Xanthydrol condenses with indoles to form purple to violet complexes. In glacial acetic acid the condensation occurs only at position 3, which is, of course, occupied and un-available in tryptophan. However, in the presence of hydrochloric acid in the glacial acetic acid, an oxonium salt of xanthydrol is formed, and this condenses readily with indoles at positions 1, 2, and 3 to form colored complexes. Accordingly we tried xanthydrol in a 3 : 1 mixture of acetic and hydrochloric acids, and obtained strong coloration in violet of pan-creatic zymogen granules (*J. Histochem. Cytochem.*, **5**: 188, 1957).

Paraffin sections of formalin-fixed tissue are deparaffinized in xylene, rinsed in glacial acetic acid, and reacted 5 min in a fresh solution of 1 g xanthydrol in 36 ml glacial acetic acid with which 4 ml concentrated hydrochloric acid is mixed just before using. The prepara-tions are then washed in glacial acetic acid, acetic acid plus xylene, and four changes of xylene and mounted in cellulose caprate.

Preparations are good for 2 to 3 months after mounting but may fade completely in 6 months.

Hršel's [*Acta Histochem. (Jena)*, **4**: 47, 1957] chromic acid followed by phosphomolybdic acid and sequence staining with eosin and light green is said to give a specific red to tryptophan-containing proteins, but Hršel admits some important exceptions in his protein model staining, and he does not make any topochemical comparisons with the *p*-dimethyl-benzaldehyde procedures as published by Lison or with our 1956 method. The method does not seem to have been used to any great extent since. Further exploration of it might be of interest.

Chromaffin

Wildi (*Science*, **113:** 188, 1951) records reduction (in vitro) of neutral silver nitrate by various *o*- and *p*-dihydroxyaryl compounds: pomiferin, isopomiferin, dihydroisopomiferin, 3',4',7,8-tetrahydroxyflavanone, eriodictol, catechol, hydroquinone, tetrabromocatechol, and pyrogallol. Phenol, *m*-dihydroxyaryl compounds, and compounds which formed insoluble silver salts failed to react. Wildi's reactions were performed in about 91% alcohol, with about 0.5% $AgNO_3$. Most reactive substances reacted in 15 min at 20 to 25°C. Some required heating to 60°C for 1 min.

Chromaffin cells share the property of reducing diammine silver salts exhibited by argentaffin cells, and Masson considered the latter as belonging to the chromaffin system. THE OGATA-OGATA SILVER METHOD (*Beitr. Path. Anat.*, **71:** 376, 1923). This method depends directly on this property.

1. Treat fresh tissue for 1 to 2 h in the dark in a 1% dilution of 28% ammonia water (1 ml and 99 ml water).
2. Place blocks in the Bielschowsky-Maresch diammine silver solution, diluted with an equal volume of distilled water, for 3 to 5 h in the dark.
3. Transfer to several changes of 1 : 100 dilution of ammonia water, during 30 min, in the dark.
4. Fix 1 h in 3% sodium thiosulfate in the dark.
5. Wash 1 h in running water.
6. Fix 1 to 2 days in 10% formalin. Cut frozen sections, float onto slides, dehydrate, clear, and mount in balsam or a synthetic resin as usual.

Ogata and Ogata note also the reduction of osmium tetroxide by adrenal medulla cells, but this is overshadowed by the greater blackening of the cortical lipids.

The chromaffin reaction, first reported by Gregor Joesten (*Arch. Heilkd.*, **5:** 97, 1864) and by Henle (*Z. Rat. Med.*, **24:** 143, 1865), is a brown coloration of the cytoplasm of adrenal medulla cells and of the chromaffin cells of various paraganglia in the abdomen. It is shown also in varying measure by the enterochromaffin cells of the stomach (R. Heidenhain, *Arch. Mikrosk. Anat.*, **6:** 368, 1870) and intestine (J. E. Schmidt, *Arch. Mikrosk. Anat.*, **66:** 12, 1905) and in "dopamine" cells of ruminant duodena and capsules of lung and liver (Falck, Hillarp, and Torp, *J. Histochem. Cytochem.*, **7:** 323, 1959).

In the adrenal the reaction is assigned to oxidation products of adrenaline and noradrenaline. Lison formulates the reactions thus:

Adrenaline Adrenochrome Oxoadrenochrome

Melanoid Pigment

Noradrenaline undergoes similar reactions

Iodate oxidation, at first thought to be equivalent to chromate oxidation, is now believed to demonstrate noradrenaline specifically. Hillarp and Hökfelt (*J. Histochem. Cytochem.*, **3:** 1, 1955) prescribed immersion of very thin (0.5 mm) slices of adrenal medulla in 10% KIO_3 for 48 h, followed by fixation for 24 h in 10% formalin and sectioning at 10 to 20 μ on a freezing microtome. The oxidation product does not resist paraffin-embedding procedures, but Hillarp and Hökfelt prescribe a red nuclear counterstain and mounting in Canada balsam. But Carbowax may be used, or 15 to 30 min paraffin infiltration (*J. Histochem. Cytochem.*, **7:** 149, 1959).

During this fixation process adrenaline is converted to sparingly soluble, increasingly red 2-iodoadrenochrome. Hence adrenal medulla cells lacking noradrenaline (guinea pig, rabbit) remain uncolored. The two catecholamines appear in separate cells (rat, hamster). Noradrenaline cells are very conveniently located in the latter species in the periphery of the medulla. Eränkö (*J. Histochem. Cytochem.*, **4:** 11, 1956) used the same procedure and compared the iodate reactive areas with fluorescent areas in the contralateral formalin-fixed hamster adrenal sections (cryostat sections postfixed in Baker's formol-calcium).

Owing to the substitution of the terminal amine group of adrenaline it does not condense so readily with formaldehyde to form the fluorescent tetrahydroisoquinoline derivative, although formalin-fixed serum models of both catecholamines are fluorescent, adrenaline somewhat less than noradrenaline (Lillie, *J. Histochem. Cytochem.*, **9:** 44, 1961). Falck et al. (ibid., **10:** 348, 1962) found much smaller amounts of noradrenaline and dopamine detectable by this method.

There appears to be no good reason for using the potassium salt instead of the commoner sodium iodate ($NaIO_3$), unless one wishes to conduct the reactions at 0 to 5°C. The solubility advantage of KIO_3 is present only at low temperatures, disappearing at about 15°C.

Eränkö (*J. Histochem. Cytochem.*, **5:** 408, 1957) also demonstrated the binding of ^{131}I from radioactive potassium iodate (KIO_3) by noradrenaline cells of the hamster adrenal radioautographically. Iodate-formalin mixtures, on the other hand, did not give specific noradrenaline localization, but rather presented a general medullary radioactivity and also a strong cortical reaction.

TABLE 8-5 SOLUBILITY OF $NaIO_3$ AND KIO_3 AT 0 TO 100°C IN GRAMS PER 100 ml WATER

	Temperature, °C							
	0	10	20	30	40	60	80	100
$NaIO_3$	2.50	5.8*	9.0	12.0*	15.0	21.0	27.0	34.0
KIO_3	4.73	6.4*	8.13	11.2	12.8	18.5	21.8	32.2

* Interpolated.

Source: N. A. Lange, "Handbook of Chemistry," 10th ed., McGraw-Hill, New York, 1961.

The *chromaffin reaction proper*, with potassium dichromate, is most often done with formalin-dichromate mixtures, though the earlier workers, before the introduction of form-aldehyde into histologic technic in the 1890s, used it alone or with sodium sulfate or other salts. We have repeated Heidenhain's experiment with 5% $K_2Cr_2O_7$ and found the chromaffin reaction of rabbit gastric argentaffin cells excellent, but a good deal of cell separation was present, and nuclear staining left much to be desired.

Some evidence exists that the pH of the dichromate solution may be quite important in the production or nonproduction of a satisfactory chromaffin reaction. Hillarp and Hökfelt (*J. Histochem. Cytochem.*, **3:** 1, 1955) found the adrenal chromaffin reaction most intense when a pH level of 5 to 6 was maintained (100 ml 5% $K_2Cr_2O_7$ + 10 ml 5% K_2CrO_4). They preferred to defer formalin fixation until after chromation, as with the iodate method.

The formaldehyde condensation noted above occurs fairly rapidly and prevents the subsequent performance of a satisfactory chromaffin reaction.

Even as little as 1 h of preliminary treatment with formalin may prevent the chromaffin reaction, though the tissues are fixed thereafter for the usual length of time in Orth's fluid. The presence of acid in the dichromate fixatives may also prevent chromaffin from staining. Ophüls' acetic variant of Orth's fluid and Tellyesniczky's acetic dichromate fluid and the acetic Zenker fluid may fail to demonstrate chromaffin in the same material as gave good chromaffin staining after primary Orth fixation. The Spuler-Maximow variant of Zenker's fluid demonstrates chromaffin better than the original acetic Zenker, but not so well as Orth's fluid.

After Orth, Kose, or Möller (Regaud) fixation, chromaffin is manifest, even in frozen sections stained with fat stains and in paraffin sections stained with alum hematoxylin, as a diffuse brown coloration of the cytoplasm of the chromaffin cells. This material is tinged orange by safranin O and green by toluidine blue O, thionin, and azure A. The azure A–eosin B technic stains chromaffin yellowish green to bluish green. Schmorl used a 24-h Giemsa stain in formalin-dichromate fixations. Wiesel's method (from Schmorl) required sequence staining in 1% toluidine blue or aniline blue and by safranin O. The two stains were used for 20 min each, with 5 min of washing between. Sections were differentiated in 95% alcohol until the blue color reappeared, then cleared in carbol xylene and xylene and mounted in balsam. Chromaffin stained green; other cytoplasms, blue; nuclei, red.

The chromaffin reaction of enterochromaffin cells engenders a basophilia manifest by green azure A staining at pH 5 to 8, yellowish green at pH 4, and greenish yellow at pH 3. It is prevented by 8-h 60°C 0.1 N HCl–methanol methylation and restored by 20-min 1% KOH–70% alcohol saponification. It is indicated that a carboxylic acid is formed during the chromaffin reaction.

The *Vulpian* (*C. R. Acad. Sci.* [*D*] (*Paris*), **43:** 663, 1856) *reaction*, according to Cowdry, indicates the presence of adrenaline. Immersion of fresh adrenal tissue in a dilute solution of ferric chloride produces a green coloration of the chromaffin cells of the medulla. According to Karrer, ferric chloride gives green with solutions of catechol. Since adrenaline and noradrenaline are both catechol derivatives, this may be the explanation of the Vulpian reaction.

Ciaccio (*Anat. Anz.*, **23:** 401, 1903) adapted the Vulpian reaction for histologic use. He fixed very thin slices in a 5% solution of ferric chloride in 100% alcohol for about 10 min, transferred them to a mixture of 1 part ammonia with 10 parts 100% alcohol, and followed this treatment by hardening in 100% alcohol. Penetration was poor, only to about 100 μ. The medulla cells and medullary veins contain granules of violet to brown material.

Fixation in aqueous formaldehyde–ferric chloride mixtures does not give the Vulpian reaction or preserve chromaffin.

It would appear that Ciaccio's procedure could well be applied to paraffin sections of frozen dried material.

ANILINE BLUE PLUS ORANGE G. Gomori (*Am. J. Clin. Pathol.*, **10:** 115, 1946) reported on the use of the Mallory-Heidenhain technic for demonstration of the chromaffin cells of the adrenal medulla. Aqueous formalin, Bouin's fluid, and M. Heidenhain's mercuric chloride–formalin are the recommended fixatives. Dichromate fixatives were considered inferior, and alcoholic fixatives were unsuitable.

But in this case the azocarmine G is reduced to 0.05%, and staining is altered to 60 to 90 min at 55 to 60°C only. The aniline alcohol differentiation is done under microscopic control until chromaffin cells stand out in deep pink with pale cortex cells. The phosphotungstic acid bath is reduced to 2 or 3% and to 20 min. Mallory's aniline blue–orange G mixture is used undiluted for 15 to 40 min only. Quinoline yellow (C.I. 47005?) and tartrazine (C.I. 19140) are recommended as substitutes for orange G in Mallory's mixture; they yield a greenish yellow cortex cell stain in place of the dull orange of the original mixture, contrasting better with the purplish red granules of the chromaffin cells.

Three quinoline yellows appear in the "Colour Index": C.I. 47000, 47005, and 47010. The first is insoluble in water; the other two are sulfonic acids. Probably C.I. 47005 is meant.

Alpha cells of pancreatic islets, some cells of the anterior pituitary, neutrophil leukocytes and myelocytes, and enterochromaffin cells also possess granules staining deep purplish red to violet by this technic; hence it appears that this technic demonstrates other cell granules of chromaffin cells, not chromaffin itself.

Wood (*Am. J. Anat.*, **112:** 285, 1963) used a modified Orth fixative made from a Müller fluid made up in 0.2 M pH 4.1 acetate buffer and stained 15 min in a modified Mann stain (equal volumes of 1% eosin Y, 0.5% aniline blue, and 0.2 M pH 4 acetate buffer), differentiated in 95% alcohol, dehydrated and cleared with two changes each of absolute alcohol and xylene, and mounted in Permount. Noradrenaline cells stained yellow, adrenaline cells purplish brown.

RESERPINE DEPLETION. In rats noradrenaline cells were depleted by reserpine at 0.5 mg/kg; adrenaline cells were largely depleted at 1.25 mg/kg, in several daily doses. This use of reserpine was reported previously by Camanini, Lozana, and Molinatti (*Experientia*, **14:** 199, 1958). Eränkö [*Nature (Lond.)*, **175:** 88, 1955] used nicotine at 1 mg/kg to deplete fluorescent noradrenaline islets. Chronic nicotine intoxication (2 to 4 mg daily, $6\frac{1}{2}$ months) produced medullary hyperplasia evident after a two-week rest period, with abundant noradrenaline.

Perhaps the best methods of demonstrating chromaffin substance in paraffin sections after chromate fixations are the ferric ferricyanide reduction test, which colors chromaffin a deep greenish blue, and the periodic acid Schiff reaction, which colors medulla cells a dark grayish red to gray-pink. Both tests should be employed in critical cases.

It has been shown in test tube experiments that the potassium dichromate precipitation products formed with catechol, adrenaline, dopa, and the like promptly reduce ferric ferricyanide mixtures.

Since these reactions occur also in the guinea pig adrenal, which lacks noradrenaline, and are limited to chromate-fixed material, it would appear that they must be reactions of adrenochrome, 3-ketoadrenochrome, or the dimer of 3-ketoadrenochrome postulated by Lison. We have shown in vitro that when an excess of catechol is reacted with periodic acid, no demonstrable aldehyde is produced. But the condition of the histochemical reaction is that of a large excess of periodic acid reacting on adrenochrome or an oxidation polymer thereof. With polymers of the type postulated by Lison, the suggestion made to us by C. S. Hudson when we first encountered the reaction with periodic acid seems much more

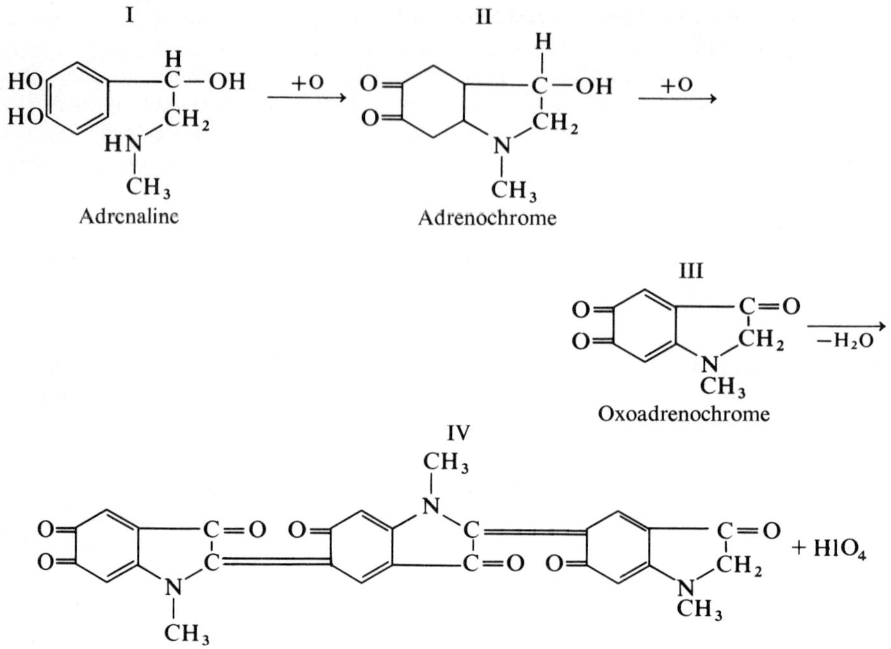

I

II

Adrenaline

Adrenochrome

III

Oxoadrenochrome

IV

Trimer of Lison's "Melanine" type (Lison, 1960, p. 249)

pertinent than with adrenaline or adrenochrome, because of the increased probability that the reaction product would remain insoluble. Hudson proposed that the quinone carbons of adrenochrome would be further oxidized to carbon dioxide and be eliminated and that carbons 4 and 7 would form aldehyde.

GLUTARALDEHYDE–DICHROMATE SILVER METHOD FOR ADRENAL NOR-ADRENALINE CELLS. Jones (*Stain Technol.*, **42**: 1, 1967) prescribes 6- to 24-h 5% buffered glutaraldehyde fixation followed by a 6- to 12-h chromation in 3.5% $K_2Cr_2O_7$, routine washing, dehydration, clearing, paraffin embedding, and sectioning. Sections are brought to distilled water and treated 24 h in diammine silver (Jones said Fontana silver), rinsed in distilled water, toned 5 min in 0.5% gold chloride, "fixed" 5 min in 0.2 M $Na_2S_2O_3$ (5% hypo), washed 5 to 10 min in running water, dehydrated, cleared, and mounted in synthetic resin (Jones used Depex).

RESULTS: Noradrenaline cells appear black, adrenaline cells colorless. Omission of the argentaffin reaction gives yellow noradrenaline cells, unstained adrenaline. Jones used neonatal and adult rats.

The method requires chromation in the block and is apparently not applicable to routine surgical material.

Wood (*J. Histochem. Cytochem.*, **22**: 1060, 1974) employed electron microscopic x-ray analysis to specifically identify chromium in the reaction product developed in noradrenaline-containing cells in the adrenal medulla of cats after glutaraldehyde-fixed tissues were treated with 2.5% potassium dichromate and 2% sodium sulfate in cacodylate buffer of pH 4.1 for 3 h.

Eränkö [*Nature (Lond.)*, **175**: 88, 1955; *Ann. Med. Exp. Biol. Fenn.*, **33**: 378, 1955] described fluorescent islets in the adrenal medulla of the rat, shown by their discharge by nicotine and their correspondence with the iodate-positive islets to be composed of noradrenaline-storing cells. In this study formalin fixation of fresh frozen sections was used. Later enhanced fluorescence was produced and greater sensitivity achieved by Falck and Torp (*Med. Exp.*, **5**: 429, 1961) and Falck et al. (*J. Histochem. Cytochem.*, **10**: 348, 1962). Tissues were freeze-dried and exposed to hot formaldehyde gas and then infiltrated and embedded in paraffin. This technic was applied in vitro to a variety of catecholamines and related compounds. The strongest reactions were obtained with dopamine, dopa, and noradrenaline; metatyrosine, metatyramine, α-methyl dopa gave weaker reactions, then adrenaline, methoxytyrosine, and epinine. No reactions were obtained with phenylalanine, tyrosine, β-phenylethylamine, normetanephrine, hordenine, or 3,4-dihydroxyphenylacetic acid.

The procedure has become quite involved. Pearse (1972, pp. 1102–1108 and 1894) cites the method of Falck and Owman (*Acta Univ. Lund*, sec. II, no. 7, 1965). The formaldehyde gas must be evolved from paraformaldehyde which has been equilibrated over sulfuric acid in an atmosphere of 50 to 70% humidity for 10 days. To achieve 50% relative humidity use sulfuric acid, sp gr 1.34, 586 g/l; for 60% humidity, sp gr 1.29, 503 g/l; for 70%, sp gr 1.29, 428 g/l. These values are equal to 12.4 N, or 45%; 10 N, or 38.5%; and 8.7 N, or 34%.

For freeze-drying Pearse (1968, p. 593) prescribes quenching in isopentane at $-160°C$, propane at $-185°C$, in CCl_2F_2 (Arcton 12) at $-158°C$, or Freon at $-22°C$ ($CHClF_2$), all chilled with liquid nitrogen on strips of copper or aluminum foil or copper or brass plates. Tissue thickness is between 1 and 6 mm. The other dimensions are limited by the size of the drying chamber. Vacuum of 0.01 torr ($=10 \mu$) mercury is prescribed. Phosphorus pentoxide is employed as a desiccant within the drying chamber. The performance of the vacuum chamber should be prechecked while the chamber is empty. Before drying, dislodge frozen tissue from the foil or metal plate with a metal rod and decant excess cooling fluid. Transfer the tissue with cold forceps to a tissue holder in the drying chamber; close the chamber. When pressure falls to 0.5 torr, set machine to desired temperature. For adrenal medulla -45 to $-50°C$ is suggested as optimal by Eränkö (*Histochem. J.*, **4**: 213, 1972); higher temperatures allow diffusion. Desiccation time according to Pearse for a 3-mm block is 4 h 30 min at $-40°C$, 6 h 15 min at $-50°C$. Shorter times may be used at higher temperatures for more routine material, e.g., 3 h 30 min for $-10°C$, 4 h at $-20°$, or $-30°$ for a 3-mm block. Thicker blocks take longer; with thinner blocks the tissue density enters and influences the usually shorter times.

When desiccation is complete, allow the temperature to rise, open the chamber, and transfer the tissue to a paraformaldehyde chamber or for controls directly to paraffin. Eränkö (ibid.) prescribes 30 min at 50°C followed by 1 h at 80°C for catecholamine fluorescence. After fixation Pearse infiltrates in paraffin in vacuo. Routine paraffin infiltration has also been used. Sections for fluorescence are simply mounted without dewaxing in a drop of mineral oil with a cover glass.

Eränkö (*Histochem. J.*, **4**: 213, 1972) placed tissues on copper screen wire mesh, froze them in propane with liquid nitrogen at $-185°C$, transferred the tissues to a vacuum desiccator containing phosphorus pentoxide above the tissue holder, evacuated with a mechanical and diffusion pump, sealed, and stored them at -45 to $-50°C$ for 7 days or more. Shorter periods allowed diffusion artifacts, as did also desiccation temperatures above $-40°$. Then the desiccator was warmed to 50°C, the vacuum was broken, and the tissues were transferred to glass jars containing paraformaldehyde equilibrated with air of 60% relative humidity. The jar was closed and kept at 50°C for 30 min and in an oven at 80°C for 1 h. Then tissues were embedded in paraffin (or in Epon-Araldite for electron microscopy).

Paraffin sections at 3 to 10 μ are mounted directly in mineral oil and examined by fluorescence microscopy. Duplicate sections may be dewaxed in xylene and used for staining procedures.

If liquid nitrogen is not available, freezing in isopentane or 20- to 40°C petroleum ether with pieces of solid carbon dioxide in it should serve.

AZO COUPLING REACTION. Azo coupling with diazosulfanilic acid was reported in 1915 by Aldrich (*J. Am. Chem. Soc.*, **37**: 203, 1915) for adrenaline, tyramine, and histamine. Gomori noted only weak yellow in adrenal medulla with the then available stable diazos. Our experience with *p*-nitrodiazobenzene and other similar diazos was equally unsatisfactory. Diazosafranin and diazomethylene violet gave only rather inconspicuous red and dull violet colors, lighter than erythrocytes (*J. Histochem. Cytochem.*, **21**: 448, 1973). In the same paper Lillie et al. reported greenish blue to dark blue staining of the noradrenaline islets of the adrenal medulla in 10 mammals including human beings with the diazosulfanilic azure A method. Fixation of adrenal with 5% glutaraldehyde in pH 7 phosphate buffer gave better results than buffered 10% formol. The reaction was prevented by a 10- to 30-min exposure to 1% periodic acid, which did not appreciably alter staining of other tissue components. Stainability was restored by 4-h exposures to 5% $Na_2S_2O_3 \cdot 5H_2O$, 5% $Na_2S_2O_4$, or 5% $Na_2S_2O_5$. Iodine, ferric chloride, and dichromate oxidations similarly prevented the noradrenaline (and enterochromaffin) reaction and were similarly reversed by reductions, but these oxidants also adversely affected protein azo coupling. The action of ferric chloride ($FeCl_3$) seemed to be a chelate reaction on histidine; the protein effect was duplicated with 0.1 M $FeSO_4$, and $ZnAc_2$, which did not affect the noradrenaline reaction. Certain nerve cells—the pigmented cells of the substantia nigra and locus coeruleus and some sympathetic ganglion cells—contain more or less numerous, perhaps peripherally located cytoplasmic granules which color dark blue by this DAS-AzA method. These are thought to agree with the formaldehyde fluorescent granules which have been identified as noradrenaline.

Eränkö (*J. Histochem. Cytochem.*, **12**: 487, 1964) freeze-dried small bits of adrenal and treated them with formaldehyde vapor (generated from paraformaldehyde) for 15 min to several hours at 40 to 80°C; then sections were examined for specific fluorescence. One-hour exposure at 50°C produced fluorescence of monoamines. Sections were then photographed, stained for one of several enzymes including nonspecific esterases, acetylcholine, and cholinesterases, and lactic or other tetrazolium reductases, and reexamined, and the same region was rephotographed. Thus, sites of enzyme activity were correlated with localization of monoamines.

Ritzen (*Exp. Cell Res.*, **44**: 505, 1966) dissolved various catecholamines in a protein or sucrose solution which was subsequently reacted with formaldehyde on slides to determine the relative yield of fluorescence. A linear relationship of fluorescence of catecholamine concentration was observed only at very low concentrations of catecholamine.

Carrodi and Jonsson (*J. Histochem. Cytochem.*, **13**: 484, 1965) added dopamine, nor-adrenaline, and 6,7-dihydroxyisoquinoline hydrochloride (1 to 2 mg/ml) in 2% aqueous solutions of bovine serum albumin. Solutions were spread on glass films, allowed to dry, and exposed to gaseous formaldehyde. The fluorescent products of dopamine and nor-adrenaline were noted to have nearly identical activation and emission spectra. However, the 4,6,7-trihydroxy-3,4-dihydroisoquinoline formed from noradrenaline has a labile OH group at the 4 position which can be removed with thionyl chloride vapor at 50°C within 30 to 45 min, resulting in the formation of the fully aromatic 6,7-dihydroxyisoquinoline. The isoquinoline possesses characteristics other than the 3,4-dihydroisoquinolines and is not reduced by sodium borohydride. Thus a distinction between dopamine and noradrenaline may be possible using fluorescent histochemical methods.

Ehringer et al. (*J. Histochem. Cytochem.*, **17**: 351, 1969) modified the Falck and Hillarp method for the fluorescent demonstration of adrenergic nerves. The technic permits a more rapid identification. Essentially, sections (50 to 100 μ) of tissues are sliced with a tissue slicer (Sorvall TC-2), separated in 1 drop of saline solution, and stretched on glass slides during observation under a dissecting microscope. Sections mounted on glass slides are then dried over phosphorus pentoxide in vacuo for 1 h and prepared for fluorescence microscopy as usual according to Falck and Hillarp.

Centrifugal separation methods employed by Ende [*J. Physiol. (Lond.)*, **141**: 183, 1958] and Schumann [*J. Physiol. (Lond.)*, **137**: 318, 1957] as well as the morphological studies by Wood (*Am. J. Anat.*, **112**: 285, 1963), Wood and Barrnett (*Anat. Rec.*, **145**: 301, 1963; *J. Histochem. Cytochem.*, **12**: 197, 1964), and Yates et al. (*Tex. Rep. Biol. Med.*, **20**: 494, 1962) suggest the localization of noradrenaline in large granules and the presence of adrenaline in small granules of adrenomedullary cells. Wood and Barrnett (*J. Histochem. Cytochem.*, **12**: 197, 1964) appear to have successfully distinguished noradrenaline and adrenaline granules at the electron microscopic level utilizing selective chromate oxidation. Freehand slices of hamster adrenal medulla fixed in 0.1 M cacodylate–buffered 6.25% glutaraldehyde, washed, and incubated with 2.5% potassium dichromate in 0.2 M acetate buffer of pH 4.1 containing 1% sodium sulfate for 24 h at 4°C were sectioned at 10 μ on a freezing microtome and examined by light or phase microscopy. For electron microscopy, chromate-treated specimens were trimmed under the dissecting microscope, rinsed in cacodylate buffer, refixed in 1% osmium tetroxide–containing sucrose (Caufield, *Biophys. Biochem. Cytol.*, **3**: 827, 1957), dehydrated, and embedded in Epon. Only noradrenaline-containing granules chromated at pH 4.1 stained densely, whereas adrenaline-containing granules chromated at pH 4.1 failed to stain densely, but stained densely if chromated at pH 6.5. But it has been our experience that guinea pig adrenal which contains chiefly adrenaline and little or no noradrenaline gives an excellent chromaffin reaction after fixation in Orth's or Kosé's fluid in buffers of about pH 3.5.

Hakanson and Sundler (*J. Histochem. Cytochem.*, **22**: 887, 1974) modified the formaldehyde condensation method for the demonstration of catecholamines and indolamines, achieving an improved sensitivity in staining for noradrenaline and dopamine. The modification consisted of treatment of specimens with formaldehyde gas (generated from paraformaldehyde) for 1 h in a closed vessel at 80 or 110°C or for 1 to 5 h in a connected adjacent chamber at 25 or 3°C. For noradrenaline, dopa and dopamine optimal fluorescence was produced by exposure at 25°C, and increased with longer exposures, up to 5 h, the longest interval tried. For adrenaline, the best results were obtained in 1 h at 80 or 110°C. Poorer results at higher temperatures are ascribed to partial oxidative degradation of noradrenaline (dopa and dopamine?). 5-Hydroxytryptamine fluorescence after 1 h at 80°C equaled that at 25°C after 3-h exposure. Tryptamine and tryptophan behaved similarly.

Freeze-dried formaldehyde-exposed adrenals gave differential orange-yellow noradrenaline islet fluorescence with a 3-h 25°C exposure. With the standard 1-h 80°C exposure all medullary cells fluoresced. The adrenaline fluorescence is greenish yellow (emission maximum 470 : 480 mμ, that of noradrenaline 550 : 570 mμ).

Eränkö [*Acta Endocrinol. (Kbh.)*, **18**: 180, 1955] noted that noradrenaline-containing cells became fluorescent after brief fixation in formalin. Subsequently Falck et al. (*J. Histochem. Cytochem.*, **10**: 348, 1962), Falck (*Acta Physiol. Scand. [Suppl.]*, **56**: 197, 1962), Corrodi and Hillarp (*Helv. Chim. Acta*, **46**: 2425, 1963; **47**: 911, 1964), and Björklung et al. (*J. Histochem. Cytochem.*, **16**: 263, 1968; **20**: 56, 1972) have developed and refined the formaldehyde condensation method for the histochemical detection of noradrenaline and dopamine.

Tissues are freeze-dried at $-35°C$ after quenching in propane cooled with liquid nitrogen, subjected to paraformaldehyde gas at 80°C for 1 h in a closed vessel, and embedded in epoxy resin or in paraffin under a vacuum for 10 min at 60°C. Björklund et al. (*J. Histochem. Cytochem.*, **20**: 56, 1972) later noted that a relative humidity of 50 to 70% is useful to minimize the formation of secondary reactions with yellow fluorescent products. Sections cut at 6 to 10 μ are placed on nonfluorescent slides and warmed to the melting point of paraffin, mounted in liquid paraffin, covered with nonfluorescent coverslips, and examined by fluorescence microscopy.

Falck (*Acta Physiol. Scand.* [*Suppl.*], **56**: 197, 1962) noted that the biogenic amines in rat iris could be demonstrated without freeze-drying. Iris tissue could be stretched on a slide, air-dried for 5 to 20 min at 25°C, heated with formaldehyde gas, washed with xylene, and whole-mounted. Falck notes that it is important to process tissues within 10 to 15 min after the animal is sacrificed in order to obtain reproducible results. He noted that the method demonstrates catecholamines and 5-hydroxytryptamine in neurons and 5-hydroxytryptamine in blood platelets and in mast cells of rats and mice.

If spectral recording at neutral pH is desired, sections mounted on coverslips are deparaffinized on a hot plate, and the coverslip is inverted with the section facing the quartz dark-field condenser. Liquid paraffin is used as immersion medium. The section on the quartz slide is mounted in liquid paraffin. For plastic-embedded sections, nonfluorescent immersion oil is used. Excitation and emission spectra are now taken. At neutral pH, the main excitation peak of both fluorophores (from dopamine and noradrenaline) is about 410 mμ.

To distinguish dopamine from noradrenaline, an HCl treatment is prescribed. Liquid paraffin is *incompletely* removed from sections with xylene, and HCl vapor is applied in two steps. Sections are exposed to HCl vapor for 15 to 30 s at 25°C. (Sections on slides are placed in a Coplin jar, 10 ml HCl is added, and the lid is closed. The sections are above the level of the liquid HCl. Plastic-embedded sections are exposed for 2 to 4 min.) Sections are mounted and excitation spectra are recorded. In the second step, sections are again deparaffinized incompletely with xylene, exposed for 2 to 3 min to HCl vapors, mounted, and examined as above. Plastic-embedded sections require 3 to 6 min in HCl vapors.

In the histochemical reactions between catecholamines and formaldehyde Björklund et al. (*J. Histochem. Cytochem.*, **20**: 56, 1972) suggest the formation of fluorophores occurs according to the following formulations:

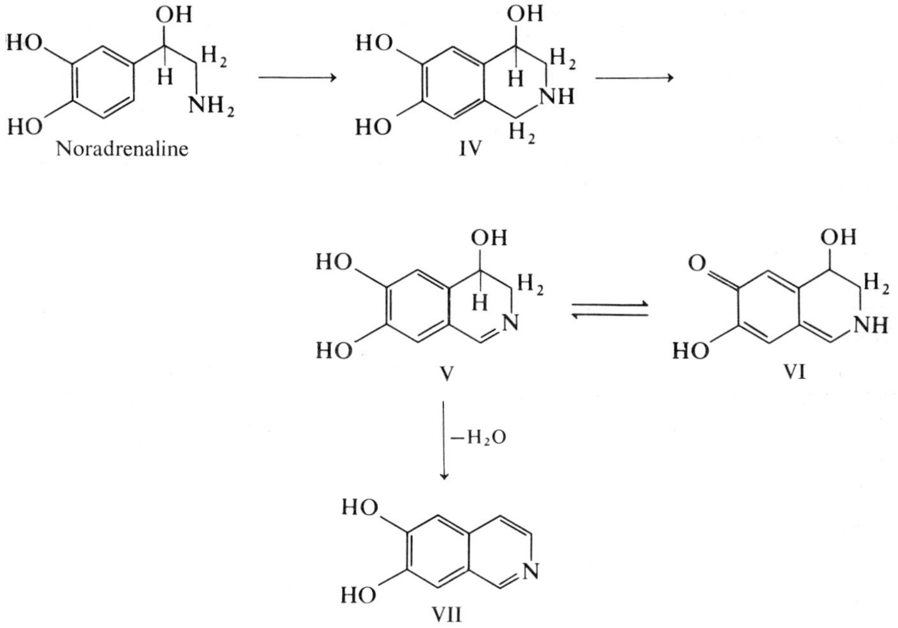

OH
HO
H₂
H
NH₂
HO
Noradrenaline

OH
HO
H₂
H
NH
HO
H₂
IV

OH
HO
H₂
H
N
HO
V

OH
O
H₂
NH
HO
VI

−H₂O

HO
N
HO
VII

The amines are condensed with formaldehyde to form 6,7-dihydroxy-1,2,3,4-tetrahydroiso-quinoline (I) and 4,6,7-trihydroxy-1,2,3,4-tetrahydroisoquinoline (IV). Dehydrogenation occurs in a second step to form the fluorescent 3,4-dihydroderivatives (III) and (VI). At neutral pH, the fluorophores are in their tautomeric quinoidal forms (III) and (VI) with excitation maxima at 410 mμ. A brief exposure to HCl gas promotes conversion to the nonquinoidal forms (II) and (V) with excitation maxima at 370 mμ. Further exposure to HCl is believed to result in transformation of 4,6,7-trihydroxy-3,4-dihydroxyisoquinoline (V) to 6,7-dihydroxyisoquinoline (VII) with an absorption maximum of 320 mμ. The 6,7-dihydroxy-3,4-dihydroisoquinoline (II) cannot react in this fashion.

Björklund et al. note that the HCl vapor treatment permits distinction between dopamine and noradrenaline. They also call attention to the problem that the second HCl vapor treatment results in an increasing overall background tissue fluorescence which causes a reduction in the 370 : 320 mμ ratio. They also recommend the use of coverslips rather than quartz slides, which may manifest an autofluorescence which further obscures the 370 : 320 ratio.

Lindvall and Björklund (*Histochemistry*, **39**: 97, 1974) provide a method for the demonstration of catecholamines using glyoxylic acid–induced fluorescence.

Perfusion of animals with cold (0 to 4°C) glyoxylic acid solution is recommended. A Krebs-Ringer bicarbonate buffer is prepared composed of 6.923 g NaCl, 0.354 g KCl, 0.27 g CaCl₂, 0.162 g KH₂PO₄, 0.294 g MgSO₄·7H₂O, 2.10 g NaHCO₃, and 1.8 g glucose. Distilled water is added to make 1 l. Glyoxylic acid is added to make a 2% solution. The addition of glyoxylic acid causes a reduction in pH, which is returned to 7 by addition of NaOH. The buffer is saturated with a mixture of 95% O₂ and 5% CO₂.

The perfusion is performed at high pressure using 150 ml for adult rats and 10 ml for young rats. In adult rats the ascending aorta is perfused, and in young animals the left ventricle is perfused.

Brains are removed rapidly, cooled in the 0 to 4°C buffer, and sliced at 30 to 35 μ with the Vibratome. The specimens are kept in oxygenated Krebs-Ringer bicarbonate buffer at 4°C during preparation.

With a blunt glass rod, sections are transferred to the neutralized buffered cold 2% glyoxylic acid solution for 3 to 5 min. The sections must be kept below the surface of the liquid.

Sections are mounted on glass slides, dried under a warm air stream provided by a hair drier for 15 min, and then dried overnight over phosphorus pentoxide in a desiccator. Sections may be kept in the desiccator for up to a week without apparent interference with the quality of the fluorescence.

Sections are then exposed to glyoxylic acid vapor in a vacuum. Lindvall and Björklund have constructed a special apparatus for this treatment. In essence, the slides in their racks are prewarmed in an oven for about 3 min. This prevents condensation of glyoxylic acid onto slides when introduced into the reaction chamber. Slides are then placed in the reaction chamber in an oven, and a vacuum is developed. Hot (100°C) glyoxylic acid–saturated air is introduced from an adjacent 1-l chamber at a partial pressure of 300 torr (mm mercury). Hot air from the oven is then allowed into the reaction vessel until atmospheric pressure is reached. The specimens are maintained at 100°C for 2 min and then removed from the vessel.

The hot glyoxylic acid is prepared as follows: Two grams of glyoxylic acid which has been dried in a desiccator over phosphorus pentoxide for 24 h is heated to 100°C for at least 1 h in a specially made 1-l vessel with an arm containing a valve which in turn connects to the arm of the other vessel. After heating for 1 h at 100°C, the hot vapors are admitted to the chamber containing the sections as described above.

Sections are mounted in liquid paraffin and examined with a fluorescence microscope equipped with a Schott BG12 primary lamp filter and Zeiss 41 + 47 as secondary filters. A background fluorescence may develop in sections kept at room temperature. This can be avoided for as long as several months in sections stored at −20°C.

Catecholamines, dopamine, and noradrenaline are believed to be converted to intensely fluorescent 2-carboxymethyldihydroisoquinoline derivatives with glyoxylic acid. Central and peripheral dopamine- and noradrenaline-containing neurons are demonstrated nicely. Entire adrenergic neurons including nonterminal portions of axons and sometimes dendrites become fluorescent.

Lindvall et al. (*Histochemistry*, **39**: 197, 1974) investigated fluorophore formation of 32 different phenylethylamine derivatives with glyoxylic acid. They noted the development of strong fluorescence from 3-hydroxylated and certain 3-methyoxylated primary phenylethylamines, whereas low fluorescence was obtained from secondary, tertiary, and N-acetylated phenylethylamines.

Björklund et al. (*J. Histochem. Cytochem.*, **21**: 263, 1973) propose the reaction scheme below between glyoxylic acid and peptides with NH_2-terminal tryptophan. The peptide (I) reacts with glyoxylic acid (HOOC—CHO) in an acid-catalyzed Pictet-Spengler-type cyclization to form the weakly fluorescent 1,2,3,4-tetrahydro-β-carboline-1-carboxylic acid derivative (II) via a Schiff base. Thereafter either autooxidative decarboxylation to the 3,4-dihydro-β-carboline derivative (III) may occur, or an intramolecularly acid-catalyzed reaction with glyoxylic acid to yield the 2-carboxymethyl-3,4-dihydro-β-carbolinium (IV) may develop. Decarboxylation of (IV) may yield 2-methyl-3,4-dihydro-β-carbolinium (V).

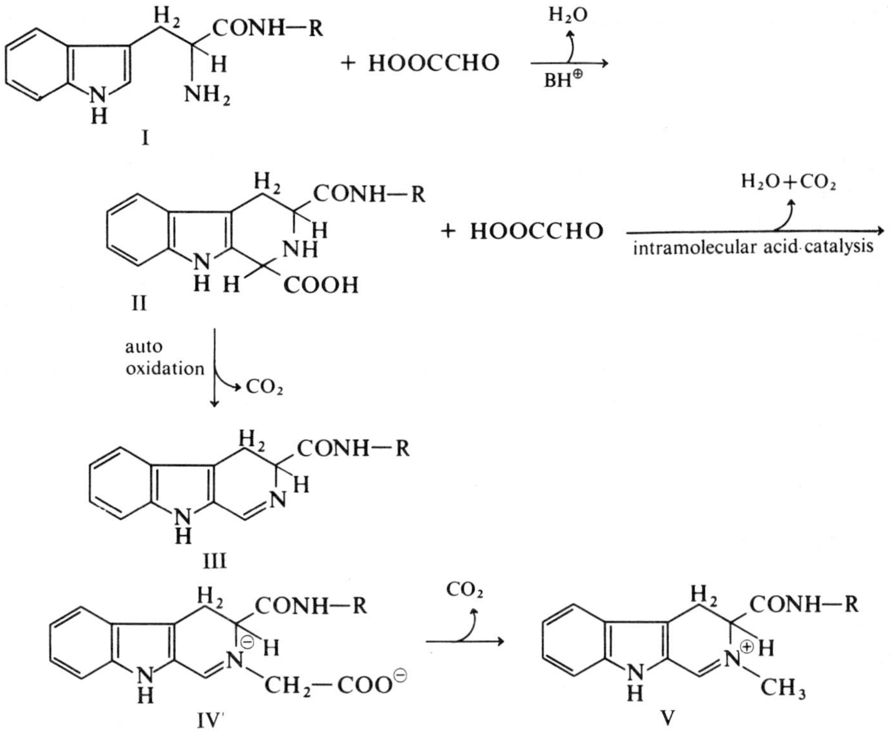

$$H_2O$$

$$\text{I} \quad + \text{HOOCCHO} \xrightarrow[BH^{\oplus}]{}$$

$$\text{II} \quad + \text{HOOCCHO} \xrightarrow[\text{intramolecular acid·catalysis}]{H_2O + CO_2}$$

auto
oxidation $\rightarrow CO_2$

III

IV $\xrightarrow{CO_2}$ V

Enterochromaffin

R. Heidenhain identified yellow chromaffin cells in the gastric mucosa of dog, cat, and rabbit in 1870 (*Arch. Mikrosk. Anat.*, **6**: 368, 1870) using 6% $K_2Cr_2O_7$ for 30 days as the fixing reagent. Grüzner and Menzel (*Pfluegers Arch.*, **20**: 395, 1879) described them as blackened by osmium tetroxide in stomach and duodenum. Nicolas (*Int. Monatsschr. Anat. Physiol.*, **8**: 1, 1891) and Kultschitzky (*Arch. Mikrosk. Anat.*, **49**: 7, 1897) recognized them as containing fine oxyphil granules basal to the nucleus, whence their title of basal granular cells. Schmidt (*Arch. Mikrosk. Anat.*, **66**: 12, 1905) recognized chromaffin cells in the human duodenum. Ciaccio (*Arch. Ital. Anat. Embriol.*, **6**: 482, 1907) attributed their chromaffin reaction to adrenaline. Masson (*C. R. Acad. Sci.* [*D*] (*Paris*), **158**: 59, 1914) described the argentaffin reaction and (*Am. J. Pathol.*, **4**: 181, 1928) considered them the cells of origin of carcinoid tumors of the gastrointestinal tract. Cordier and Lison (*Bull. Histol. Appliq.*, **7**: 140, 1930) discovered the azo coupling reaction and established the phenolic nature of the enterochrom-affin substance. Lison (1936) expressed the then prevailing opinion apparently originated by Masson that the phenolic substance was a catechol with a short, probably para side chain. Clara (*Z. Zellforsch. Mikrosk. Anat.*, **22**: 318, 1935) reported the still unexplained intense reaction with very dilute hematoxylin without mordant and other dyes with a catechol grouping. Jacobs (*J. Pathol. Bacteriol.*, **41**: 9, 1939) proposed a xanthopterin or uropterin as the enterochromaffin substance. Vialli and Erspamer [*Arch. Sci. Biol.* (*Bologna*), **28**: 101, 1942] reported on "enteramine" recovered in acetone extracts of the gastrointestinal mucosa. Gomori (*Arch. Pathol.*, **45**: 48, 1948) reported on the ferric ferricyanide reaction of Golodetz and Unna as demonstrating enterochromaffin cells and records their reaction with Gibbs' 2,6-dichloroquinone chlorocinide to give a blue indophenol.

In 1952 Erspamer and Asero [*Nature* (*Lond.*), **169**: 800, 1952] announced the identification of enteramine as 5-hydroxytryptamine and reported that this was the enterochromaffin substance. Lembeck [*Nature* (*Lond.*), **172**: 910, 1953], Ratzenhofer and Lembeck (*Z. Krebsforsch.*, **60**: 169, 1954), Gebauer, Rumelin, and Becker (*Dtsch. Med. Wochenschr.*, **83**: 620, 636, 642, 1958), Snow et al. (*Lancet*, **2**: 1004, 1955), Hornykiewicz (*Acta Neuroveg.*, **13**: 110, 1956), and Hand, McCormick, and Lumb (*Am. J. Clin. Pathol.*, **30**: 47, 1958) have reported on presence of 80 μg to 2.5 mg/g 5-hydroxytryptamine in carcinoid tumors, and Lillie and Glenner (*Am. J. Pathol.*, **36**: 623, 1960) reported demonstration of gray-blue material basal to tumor cell nuclei with the postcoupled benzylidene reaction ($2\frac{1}{2}$ h in formol, 3 h postmortem).

Blocking of the azo coupling reaction by oxidation with diammine silver and its deblocking by silver extraction (KCN) and $Na_2S_2O_4$ reduction was reported by Lillie and Glenner (ibid.), and Lillie (*J. Histochem. Cytochem.*, **9**: 44, 1961; **9**: 184, 1961) reported further oxidative blocking and dithionite deblocking experiments and concluded that a catechol structure was perhaps more probable than 5-hydroxytryptamine, ruling out the Gomori resorcinol structure.

Reserpine discharge of enterochromaffin cells was reported by Vialli and Quaroni (*Riv. Istochim.*, **2**: 111, 1956) in chickens, by Benditt and Wong (*Am. J. Pathol.*, **32**: 638, 1956; *J. Exp. Med.*, **105**: 509, 1957) in guinea pigs (80% with full regeneration at 4 days), by Marks, Surgen, and Hayes (*Proc. Soc. Exp. Biol. Med.*, **97**: 413, 1958) with only partial discharge, and by Lillie, Weissbach, and Glenner (*J. Histochem. Cytochem.*, **7**: 23, 1959). Argentaffin, ferric ferricyanide, and azo-positive cells responded; argyrophil cells were not decreased. Lillie et al. noted that Udenfriend, Weissbach, and Sjoerdsma's (*Science*, **123**: 669, 1956) assays for 5-hydroxytryptamine had returned to 50% of normal assay figures when cell counts had only reached 10% at 8 days and a higher amount per cell is indicated for stomach and ileum than for duodenum, both in the normal and in the 7- to 8-day figures. Pentilla (*Acta Physiol. Scand.* [*Suppl.*], **69**: 281, 1966) reported similar depletion of azo-positive and argentaffin cells, with no depletion of argyrophil or fluorescent cells.

The golden yellow fluorescence of enterochromaffin cells was first described by Erös (*Zentralbl. Allg. Pathol.*, **54**: 385, 1932) and Hamperl (*Virchows Arch. Pathol. Anat.*, **292**: 1, 1932). Aside from enterochromaffin it has been used to detect 5-hydroxytryptamine in rat and mouse mast cells, where its presence has been clearly established by chemical means on isolated rat mast cells by Benditt, Wong, Arase, and Roeper (*Proc. Soc. Exp. Biol. Med.*, **90**: 303, 1955) and by Udenfriend et al. assays of mouse and dog mastocytomas by Meier (*Proc. Soc. Exp. Biol. Med.*, **100**: 815, 1959). This fluorescence is now generally used for identification of 5-hydroxytryptamine after fixation by the Falck-Torp freeze-drying–hot formaldehyde gas method.

THE INDOLE REACTIONS IN RELATION TO ENTEROCHROMAFFIN AND 5-HYDROXYTRYPTAMINE.   The benzylidene reaction with *p*-dimethylamine benzaldehyde gives strong violet with open alpha or beta positions. This includes among biologic substances free and protein-bound tryptophan, free tryptamine, 5-hydroxytryptamine, pyrogallol, orcinol, resorcinol, carbazole, indole, serotonin, pyrrole, skatole, 2-methylindole, including only those substances from Glenner and Lillie (*J. Histochem. Cytochem.*, **5**: 279, 1957) which gave spot tests with 10 mg to 1 μg. Even 1 mg 1,2,3,4-tetrahydronorharman, which is the reaction product of formaldehyde and tryptamine, did not react.

Barter and Pearse [*Nature* (*Lond.*), **172**: 810, 1953; *J. Pathol. Bacteriol.*, **69**: 25, 1953] explained the nonreactivity of the presumed 5-hydroxytryptamine in enterochromaffin by the formation of a similar β-carboline, 6-hydroxy-1,2,3,4-tetrahydronorharman. Lillie and Glenner (*Am. J. Pathol.*, **36**: 623, 1960) also reported a negative postcoupled benzylidene reaction, and Glenner (*J. Histochem. Cytochem.*, **5**: 297, 1957) reported a negative rosindole reaction. This reaction also depends on the benzylidene condensation and shows much the same reactivities. Lillie's (*J. Histochem. Cytochem.*, **5**: 188, 1957) xanthydrol reaction also fails to demonstrate enterochromaffin cells. Xanthydrol shows some important differences from *p*-dimethylamino-benzaldehyde as a reagent. Strong reactions are obtained with tryptophan, tryptamine, 5-hydroxytryptamine, indole, skatole, pyrrole, carbazole, 2-phenylindole, 2-methylindole, 2-carbethoxytryptophan, 2-phenyltryptophan, and tetrahydronorharman. Isatin and 2-hydroxytryptophan did not react. Here Barter and Pearse's explanation of the nonreactivity of enterochromaffin does not hold, since the β-carboline tetrahydronorharman reacts just as strongly as tryptamine and serotonin.

Solcia, Sampietro, and Vassallio (*J. Histochem. Cytochem.*, **14**: 691, 1966) reported that while no indole reaction was obtained on formol-fixed tissue, glutaraldehyde-fixed guinea pig enterochromaffin cells gave strong reactions by the postcoupled benzylidene and xanthydrol methods. Identification of the enterochromaffin cells rested apparently on a purely morphologic basis; the test of Lillie and Greco-Henson (*J. Histochem. Cytochem.*, **8**: 182, 1960) of combining a red azo reaction with the blue indole reaction was not used. The later papers of Solcia et al. [*Nature* (*Lond.*), **214**: 196, 1967; *Histochemie*, **17**: 273, 1969] added further in vitro tests and further data on the fluorescence reaction. Geyer's [*Acta Histochem.* (*Jena*), **31**: 146, 1968] report, using acetaldehyde, acrolein, and glutaraldehyde fixations as well as formol and adding Adams' (*J. Clin. Pathol.*, **10**: 56, 1957) dimethylaminobenzaldehyde nitrite method to the indole methods fully confirmed Solcia et al. (*J. Histochem. Cytochem.*, **14**: 691, 1966). He also used no histochemical proof of the identity of his enterochromaffin cells.

At the Fourth International Congress for Histochemistry and Cytochemistry in Kyoto, Japan, August 21 to 25, 1972, Lillie et al. announced that rat mast cells coupled with diazosafranin at pH 3 to 8, that 5-hydroxytryptamine and 5-hydroxytryptophan azo-coupled with *p*-nitrodiazobenzene to give prompt strong red colors at pH 3 to 9, that the azo coupling of the mast cells and 5-hydroxytryptophan was not inhibited by preoxidation, that the noradrenaline islets of the adrenal medulla gave a strong dark blue reaction on a light green background with the diazosulfanilic–pH 1 azure A sequence, that this reaction and the azo coupling reaction of *p*-nitrodiazobenzene with guinea pig duodenal enterochromaffin were both inhibited by the same oxidations and restored by $Na_2S_2O_4$ reduction, that noradrenaline gave a deep red with *p*-nitrodiazobenzene at pH 8, that after preoxidation with periodic acid only a pale yellow resulted, that when the oxidized noradrenaline was again reduced with $Na_2S_2O_3$ a strong red reaction again resulted, that when the Lillie and Greco-Henson test was applied to formol- and glutaraldehyde-fixed guinea pig duodenal enterochromaffin, numerous blue indole-positive cells appeared, and when *p*-nitrodiazo-benzene at pH 8 was applied, numerous red enterochromaffin cells appeared between the blue indole cells, not replacing or altering the color of any of them, and that with further reversing the sequence on the test no indole cell replaced or altered the color of the azo-positive cells. From these tests it was concluded that the azo-positive substance in enterochromaffin is a catechol and not 5-hydroxytryptamine, that acid diazosafranin colors 5-hydroxytryptophan depots red as well as bilirubin and hematoidin, and that a useful azo method for adrenal medulla is recorded.

After the detailed publication of these findings by Lillie et al. (*J. Histochem. Cytochem.*, **21**: 441, 448, 455, 1973) Solcia and Buffa (ibid., **22**: 127, 1974) reported that cells morphologically identified as enterochromaffin cells colored blue with the xanthydrol reaction and were then rehydrated and subjected to azo coupling with fast garnet GBC. During this process the xanthydrol staining was lost. They stated that the same cells which had stained by xanthydrol now colored red with the azo reaction. They did not report visual observation of the replacement of the indole reaction by the azo reaction; in fact they could not, since their indole reaction was lost during the preparation for restaining. They did not try doing the more stable azo coupling reaction first and then applying the indole reaction under direct visual observation as Lillie et al. did, nor did they use the more stable postcoupled benzylidene reaction first and watch the red azo-positive cells appear between the blue indole-positive cells.

The Solcia and Buffa presentation is regarded as unconvincing.

The reactions used for the identification of enterochromaffin cells and of the phenol-producing cells of carcinoid tumors fall into two general classes: (1) the metal reduction reactions, comprising the chromaffin reaction, the argentaffin reaction, the ferric ferricyanide reaction, and the seldom used osmium tetroxide reaction; and (2) the specific phenol reactions, including the azo coupling reaction and the Gibbs indophenol condensation reaction but probably not Clara's catechol dye condensation reaction. There are also a few other infrequently used procedures.

The chromaffin reaction was described earlier in this chapter. Apparently the basophilia created during the chromaffin reactions is due to a carboxylic acid. Azure A at pH 1 to 2 failed to stain; at pH 3 a greenish yellow results; at pH 4, a yellowish green; and at 5 to 8, green. Methylation in 0.1 N HCl–methanol required 8 h at 60°C to fully inhibit the green staining with azure A at pH 5; saponification in 1% KOH–70% alcohol for 20 min restores the basophilia.

THE ARGENTAFFIN REACTION; ARGYROPHILIA.   The term *argentaffin* is of Latin origin: *argentum*, silver; *affinis*, neighboring, related by marriage. Argyrophilia is of Greek derivation: *argyros*, silver; *philos*, a friend, or *philein*, to love. Despite their etymologic similarity in meaning, the terms *argentaffin* and *argyrophil* have by convention acquired distinctly different significance. *Argentaffin* means possessing the capacity to reduce silver salts in the dark, without the aid of any added or following reducing agent. Lison's term *argentoreductrice* was more definite but has the same meaning; it never gained general acceptance. *Argyrophil*, on the other hand, indicates that the tissue element so described can be impregnated with silver, but that light or a reducing agent is required to produce the black deposit of metallic silver. The hybrid term *argentophil* was in wide use for a considerable period, with the same meaning. Clara (*Acta Neuroveg.*, **16**: 294, 1957) suggested that it be discarded and replaced by the purely Greek equivalent *argyrophil;* this practice seems logical and has been generally adopted.

Masson reported a block method for demonstration of argentaffin cells. We have abandoned it completely in favor of section methods which allow the comparison of a variety of other procedures on adjacent sections.

We now prefer a simple diammine silver reaction similar to that of Hamperl, done without gold toning or counterstaining. The same diammine silver solution and the same basic technic are used for studies of melanin and lipofuscin pigments. Exposure times vary according to the object under study, and in the characterization of an unfamiliar substance, graded exposure times are used.

DIAMMINE SILVER.   To 2 ml 28% ammonia (sp gr 0.900) add about 35 to 40 ml 5% silver nitrate, fairly rapidly at first, swirling between each addition to dissolve the brown silver oxide. As the clearing interval starts to lengthen, add smaller amounts each time; stop when a faint but definite permanent turbidity is achieved.

GLASSWARE.   Beakers and Coplin jars should be chemically clean. Remove previous silver deposits by gently flowing a small amount of concentrated nitric acid over the entire inner surface. Then wash with several changes of distilled water.

LILLIE'S AND MASSON'S DIAMMINE SILVER METHODS. Deparaffinize and hydrate thin paraffin sections as usual.

1. Wash in distilled water.
2. Expose to diammine silver prepared as above in a covered Coplin jar in the dark 1 to 10 min for melanin, 15 min for neuromelanin, 1 to 2 h for bilirubin and hematoidin, 8 to 24 h for enterochromaffin, lipofuscins, and melanosis coli pigment.
3. Wash in distilled water.
4. For Masson's section method tone 5 min in 2 ml 1% gold chloride, 3 g ammonium thiocyanate, 3 g sodium thiosulfate, 98 ml distilled water (modified from Ramón y Cajal's gold toner). Wash in distilled water. For Lillie's method omit step 4.
5. Fix 1 to 2 min in 5% (0.2 M) sodium thiosulfate.
6. Wash 5 to 10 min in running water.
7. Counterstain 1 min in 0.1% safranin in 1% acetic acid and wash 2 to 3 min in distilled water. Omit this step according to Lillie's method.
8. Dehydrate with alcohols and alcohol plus xylene, clear in xylenes, and mount in synthetic resin or cellulose caprate.

This procedure combines Masson's section method with Lillie's diammine silver. Lison used a simple 0.1% gold chloride, 4 to 6 min for toning in step 4; Jacobson omitted the gold. Masson applied the method to frozen, paraffin, celloidin sections of formol-fixed materials. Collodionization is advisable to prevent section losses with overnight exposures.

Gomori (*Arch. Pathol.*, **45**: 48, 1948) detailed a modification of his methenamine silver method for enterochromaffin, which required 12 to 48 h. Burtner and Lillie devised a much more rapid variant of this procedure which seems adequately selective in practice.

THE GOMORI-BURTNER METHENAMINE SILVER METHOD FOR ARGENTAFFIN CELLS (*Stain Technol.*, **24**: 225, 1949). SOLUTIONS. *Stock methenamine silver solution:* Dissolve 3 g methenamine (hexamethylene tetramine) in 100 ml distilled water. Add 5 ml 5% aqueous silver nitrate. This solution can be stored in a cool, dark place for months.

Working solution: To 30 ml stock methenamine silver solution add 8 ml Holmes' pH 7.8 borate buffer.

GLASSWARE. The Coplin jars should be chemically clean. Previous silver mirror deposits should be removed with concentrated nitric acid.

TECHNIC

1. Treat deparaffinized sections with Weigert's iodine solution, 10 min.
2. Bleach with 5% sodium thiosulfate ($Na_2S_2O_3 \cdot 5H_2O$), 2 min.
3. Wash in running water, 10 min.
4. Rinse in two changes of distilled water.
5. Place sections in Coplin jars containing preheated buffered methenamine silver solution, and put in a 60°C (paraffin) oven for 2 to 3 h. Preheating to 60°C reduces impregnation time by $\frac{1}{2}$ to 1 h and is now our regular practice. To avoid repetitions, take out slides at 2, $2\frac{1}{2}$, and 3 h.
6. Rinse in distilled water.
7. Tone in 0.1% gold chloride ($HAuCl_4$), 10 min.
8. Rinse in distilled water.
9. Fix in 5% aqueous sodium thiosulfate, 2 min.

10. Wash in running water, 5 min.
11. Counterstain with 0.1% safranin O in 0.1% acetic acid, 5 min.
12. Dehydrate with acetone, clear in xylene, and mount as usual.

RESULTS: At 2 to 3 h impregnation is optimal. Argentaffin cells are well blackened and appear in numbers as great as with longer impregnations. Coarse connective tissue of the submucosa sometimes shows a variable amount of blackening at this time. In 2-h incubation, partial blackening of argentaffin cells in reduced numbers is seen. At $1\frac{1}{2}$ h, only lignin is black. At longer intervals than 3 h, besides the connective tissue, the granules of eosinophil leukocytes, nuclei, smooth muscle, and surface epithelium become blackened. By 4 h a silver mirror begins to appear on the sides of the slides and Coplin jars.

Human mast cell granules remain unblackened and brilliantly red with the safranin counterstain after nuclei and reticulum are blackened. Rat mast cells may blacken selectively even before enterochromaffin, or along with it, when preparations are silvered $2\frac{1}{2}$ h at 60°C.

Gomori (*Arch. Pathol.*, **45:** 48, 1948) used Gram's iodine in step 1. In step 2 he used either a bisulfite or a thiosulfate solution of unspecified strength. His borate buffer was slightly different: To 100 ml 0.2 *M* boric acid add 8 or 12 ml 0.2 *M* borax at 40 to 50°C for pH 7.8 or 8.2, respectively. Borax is soluble 3.9%—about 0.1 *M*— at 30°C (Lange). His methenamine silver working solution contained 25 ml of the above stock solution, 25 ml distilled water, and 5 ml borate buffer. He silvered at 37 to 45°C for 12 to 24 or even 48 h.

If preparations appeared oversilvered at step 6, Gomori differentiated in 0.5% sulfuric acid containing 0.1 to 0.2% iron alum until the background was almost clear. He then washed in 0.5 to 1% hydrochloric acid in 70% alcohol, and resumed the above procedure with step 9.

Gomori's regressive differentiation of oversilvered preparations with weak acidified iron alum solution did not give particularly satisfactory results with our variant. With the shorter incubation periods used in this new technic, it is more practical to repeat the procedure with a briefer silvering time on duplicate sections.

If a 10-min bath in 1% ferric chloride or a 1-h bath in 5% chromic acid is introduced at step 3, the silvering of enterochromaffin cells is inhibited, and the method is converted to a more or less selective demonstration technic for collagen and mucin. The time in the methenamine silver is reduced to 2 or $2\frac{1}{2}$ h for this purpose.

Neither treatment prevents the silvering of the enterosiderin pigment of guinea pigs. Ferric chloride prevents the silvering of adrenal lipofuscin, but chromic acid does not.

Effective blackening of collagen, much reticulum, and the mucins of the colonic, pyloric, and duodenal glands and of intestinal goblet cells was obtained by both methods. Ferric chloride afforded partial blackening of gastric surface epithelial mucin; chromic acid, a more complete silvering. The reaction of cellulose and lignin is enhanced and accelerated by both variants.

The chromic acid variant is of course essentially the Gomori silver method for mucin and glycogen.

The ferric ferricyanide reaction, which is probably the most sensitive as well as the easiest reaction for enterochromaffin, has been presented earlier in this chapter. Its value is similar to that of the diammine silver reactions. It was first observed for enterochromaffin about 1947 in Gomori's laboratory (*Arch. Pathol.*, **45:** 48, 1948) and in ours (Laskey and Greco, ibid., **46:** 83, 1948).

Argyrophil cells are found in the gastrointestinal mucosa in much the same localization as argentaffin cells, but in larger numbers and in certain locations where argentaffin cells are not demonstrable, e.g., in normal human and rat stomach mucosa (Hamperl, *Virchows Arch. Pathol. Anat.,* **371**: 482, 1952; Dawson, *Anat. Rec.,* **100**: 319, 1945), and at earlier stages in embryonic development. Argyrophil cells are also reported in pancreatic islets.

The identity of at least part of the argyrophil cells of the intestine with argentaffin cells was shown in comparison microscope studies on adjacent serial sections stained respectively with the Masson-Hamperl method and with a Gros-Bielschowsky variant similar to that in parallel studies by Hamperl (*Virchows Arch. Pathol. Anat.,* **371**: 482, 1952) and by Hellwig (*Z. Zellforsch. Mikrosk. Anat.,* **36**: 546, 1952).

It may well be that the principal difference between the intestinal argentaffin cell, which is also argyrophil, and the argyrophil cell, which is nonargentaffin, lies in the relatively minute quantity of reducing substance present in the latter. In the reserpine-induced discharge of the enterochromaffin substance ferric ferricyanide, azo coupling, Clara hematoxylin, and methenamine silver reactions disappear completely during the peak period of discharge, serotonin assays fall dramatically, but counts of cells demonstrable by Hamperl's Bodian procedure for argyrophil cells remain essentially unaltered (Lillie et al., *J. Histochem. Cytochem.,* **7**: 23, 1959). The traces remaining in the discharged cells are adequate for argyrophil staining.

The argyrophil cells of the rat stomach have been shown to be nonfluorescent in the Falck-Torp technic and store dopamine to give a brilliant green fluorescence after parenteral administration of dopa (Häkanson et al., *Histochemie,* **21**: 189, 1970).

We have replaced Feyrter's standardized Gros-Bielschowsky variant (Lillie, 1965) with the Bodian-Hamperl technic used by Lillie, Weissbach, and Glenner (*J. Histochem. Cytochem.,* **7**: 23, 1959) in their study of the reserpine discharge of guinea pig enterochromaffin. Hamperl (*Virchows Arch. Pathol. Anat.,* **321**: 482, 1952) directs as follows:

1. Deparaffinize sections, bring to distilled water, and incubate 24 h at 37°C in 1 % protargol containing 4 to 6 g metallic copper per 100 ml.
2. Rinse briefly in distilled water.
3. Reduce 5 to 10 min in 1 % hydroquinone, 5 % Na_2SO_3 solution.[1]
4. Wash in distilled water.
5. Immerse 2 to 5 min in 2 % oxalic acid.
6. Wash in distilled water.
7. Dehydrate through graded alcohols, clear in xylene, and mount in balsam.

RESULTS: Argyrophil cells are brownish black to black on a yellow to brown background. Counterstains are not needed.

Gomori further recorded (*Arch. Pathol.,* **45**: 48, 1948) the pale green staining of argentaffin cells by very dilute ferric chloride solution; the production of dark yellowish shades by nitric acid, bromine water, and iodates; a brilliant ruby to purplish or bluish red with the Mallory-Heidenhain stain; and the application of Gibbs' reaction with 2,6-dichloroquinonechloroimid to produce blue indophenol dyes in the presence of phenols as giving a distinct reaction with argentaffin granules.

AZO COUPLING REACTION. This is one of the more specific identification reactions for the enterochromaffin substance. Its use dates back to Cordier and Lison (*Bull. Histol. Appl.,* **7**: 140, 1930). Its sensitivity is generally less than that of the ferric ferricyanide

---

[1] Emended from 10 to 100 ml distilled water. Neither the hydroquinone nor the sulfite amounts prescribed are soluble in 10 ml.

reaction, as determined by cell counts on adjacent serial sections. Freshly diazotized *p*-nitro-aniline (fast red GG) is about as sensitive with brief coupling periods as ferric ferricyanide. Freshly diazotized safranin gives black enterochromaffin cells on a red protein background and has been quite useful. These last two are our preferred methods. Among the stable diazotates fast black K gives similar contrasts, but the background is more orange. Red enterochromaffin cells with pale yellow backgrounds are given by the stable diazos fast red B and fast garnet GBC. For further diazonium salts and their results see Lillie, Henson, and Cason, *J. Histochem. Cytochem.*, **9:** 11, 1961.

We now use Clara's dilute hematoxylin (*Z. Zellforsch. Mikrosk. Anat.*, **22:** 318, 1935) as a 0.01 % solution in pH 6.5 0.01 *M* phosphate buffer (36 ml 0.01 *M* NaH$_2$PO$_4$ + 14 ml 0.01 *M* Na$_2$HPO$_4$ + 0.5 ml 1 % alcoholic hematoxylin, dissolved overnight). The aqueous solution must be freshly prepared and is used once. A 30-h staining period is usually adequate. Sections of formol-fixed tissue are dewaxed and hydrated as usual, stained, and dehydrated directly with alcohols, cleared in xylene, and mounted in xylene, cellulose caprate (n_D 1.482), or similar resin.

RESULTS: Soft keratin is blue-gray; cutaneous eleidin, keratohyalin of skin, and gastrointestinal tract are dark blue; trichohyalin may stain also in alcohol-fixed material but may fail to stain after formalin fixation. Enterochromaffin cells, basal cells in some carcinoid tumors, and eosinophil leukocytes show blue to black granules. Elastic laminae of rodent arteries often color dark blue. Hair cortex may sometimes take a light pink. Hemosiderin and enterosiderin color black; this last coloration is prevented by acid or dithionite desiderization. Copper gives a blue-green.

Lillie et al. [*Acta Histochem. (Jena)*, **49:** 204, 1974] in a study on the several mechanisms involved in the Clara hematoxylin reaction showed a quite general agreement in susceptibility to blockade reactions and stain extraction tests between enterochromaffin and enterosiderin and suggested that this reaction of enterochromaffin cells may be due to their content of a small amount of ferrous iron. The hematoxylin reaction for iron appears to be somewhat more sensitive than the Prussian blue reaction, and the sensitivity of the direct ferricyanide reaction for Fe(II) may well be somewhat less still. The reaction of keratin, keratohyalin, eosinophil leukocyte granules, and Fullmer's basic nucleoprotein are preventable by a benzil blockade at pH 13, which also prevents the 1,2-naphthoquinone-4-sodium sulfonate (NQS) reaction for arginine content of these structures. The reaction of young animal and human vascular elastica is apparently not due to the aldehyde (lysinal) content of developing elastin; it is not prevented by borohydride reduction of the aldehyde.

The Holcenberg-Benditt (*Lab. Invest.*, **10:** 144, 1961) Ninhydrin reaction for indole ethylamines was proposed as a specific reaction for enterochromaffin which would support the 5-hydroxytryptamine (5-HT) theory of the nature of that substance.

The technic was fairly simple, and we were led to try it again recently in studies concerned with the 5-HT content of rat and mouse mast cells despite our earlier (unreported) repeated failures on demonstration of enterochromaffin.

Formol-, glutaraldehyde-, 0.2 *M* lead acetate-, and B-4 sublimate–fixed tissues were subjected to control stains with pH 8.5, 2.5 m*M* *p*-nitrodiazobenzene for 2 min for enterochromaffin cells in pig duodenum and 0.05 % azure A of pH 1 for 30 min for mast cells in rat and mouse tongues and to a Holcenberg-Benditt reaction executed as follows:

1. Sections were flooded 1 or 2 min with 0.2 % Ninhydrin in glacial acetic acid–acetone 1 : 9, decanted, and allowed to dry at 25°C.
2. They were then heated 15 min at 100°C in an oven.
3. They were cleared in xylene and mounted in cellulose caprate.

Despite the presence of numerous typical azo-positive enterochromaffin cells only rarely did we find a dubious pale pink cell in the glutaraldehyde-fixed tissue and none with the other fixatives. The lead and mercury fixations did not preserve demonstrable enterochromaffin but did conserve numerous mast cells in the pig duodenum. Neither those nor the mouse mast cells gave any trace of reaction with Ninhydrin. There was irregular red staining of collagen and muscle with the heavy-metal fixations in both hog and mouse.

These observations lead to the conclusion that the Holcenberg-Benditt reaction cannot be regarded as a reliable practical method for enterochromaffin cells. This is perhaps to be expected now that it has become evident that the major azo-positive substance in entero-chromaffin is a catechol and not 5-HT.

It is possible that Falck-Torp freeze-dried, hot formaldehyde gas fixation and the Ninhydrin sublimation technic proposed as an alternative method by Holcenberg and Benditt may yield more consistently positive results, but these variants seem impractical for the usual diagnostic laboratory.

Carotid Body Tumors

Barroso-Moguel and Costero (*Am. J. Pathol.*, **41**: 389, 1962) report a variant of Del Río Hortega's method for pigments and pigment precursors which they used in a study of silver-reactive cells in carotid body tumors. This method is claimed to give more reliable results, showing more numerous "argentaffin cells" with more copious granulation than other methods.

1. Fix in aqueous 10% formalin and cut frozen sections. Wash in distilled water.
2. Impregnate in 2% silver nitrate until yellowish.
3. Rinse in distilled water.
4. Impregnate in Del Río Hortega's ammoniacal silver carbonate, to which add 3 drops of pyridine for each 10 ml, and heat to 60°C, until sections are light tobacco color.
5. Rinse in distilled water.
6. Tone in 0.2% gold chloride at 60°C for several minutes.
7. Fix in 5% sodium thiosulfate (0.2 M).
8. Wash 10 min in several changes of distilled water, dehydrate in alcohol, clear in xylene, and mount in balsam or Permount.

RESULTS: Silver-positive granules are black; other structures, pale purplish.

In our judgment this is not a histochemical technic for true argentaffin substances. The second silver reagent is used hot and contains pyridine, both of which factors increase "sensitivity" and decrease "specificity." It is possible that methenamine silver when used hot also shares this decreased specificity and enhanced reactivity, but at least some time limits have been placed on that method, and it is recognized that in 5 to 6 h everything is black.

These same cells have, however, been shown to react to a rigidly applied Masson-Hamperl technic (Glenner, *Proc. 59th Annu. Meet. Am. Assoc. Pathol. Bacteriol.*, pp. 18–19, May 2–4, 1962). Glenner identified the argentaffin substance as noradrenaline. Chromaffin and iodate reactions are applicable.

The fluorescence method using the Falck-Torp freeze-dried, hot formaldehyde gas procedure is undoubtedly usable to detect noradrenaline in this organ and its tumors, and the Lillie et al. (*J. Histochem. Cytochem.*, **21**: 455, 1973) diazosulfanilic azure A method for adrenal medulla is also applicable.

AMINES

The Ninhydrin Test

Serra (*Stain Technol.*, **21**: 5, 1946) directs as follows: Deparaffinize and hydrate as usual, or use frozen sections. Flood sections with a mixture of equal volumes of pH 6.98 phosphate buffer and of a freshly prepared 0.4% aqueous solution of Ninhydrin (triketohydrindene hydrate). Place sections on a rack over boiling water and steam for 1 to 2 min. A blue or violet color develops. Drain, mount in pure glycerol, and cement with lanolin rosin cement, cellulose caprate, or polystyrene. Observe at once, since the color fades in a day or so.

Serra states that the reaction is given in adequate intensity by the amino acids except proline and hydroxyproline, and by peptides and proteins. Langeron notes ready diffusion of color.

The Alloxan Reaction

Langeron cites this reaction for amino acids and proteins generally. Alloxan reacts with amine groups to form murexide, or purpurate of ammonium.

Flood sections with 1% alcoholic alloxan solution. Proteins color red or rose. Serra (ibid.) recommends use of a neutral (pH 7) phosphate buffer as solvent and accelerates the reaction by heating over a boiling water bath. On fixed tissues the reaction is weak. The color is diffusible and is given also by nonprotein NH_2 groups and perhaps by SH groups.

The Ninhydrin and Alloxan-Schiff Reactions

According to Yasuma and Ichikawa (*J. Lab. Clin. Med.*, **41**: 296, 1953), the Ninhydrin reaction with α-amino acids may be used for the demonstration of protein with Schiff reagent. Briefly, Ninhydrin decomposes amino acid residues to carbon dioxide, ammonia, and the next lower aldehyde, suffering reduction of one of its own hydroxyl groups in the process. This reduced Ninhydrin then conjugates with 1 mol ammonia and 1 mol Ninhydrin to form a blue-violet compound. This compound is quite diffusible, and its use for protein identification has been unsatisfactory.

However, the aldehyde formed on the oxidized amino acid residues may be quite well demonstrated with Schiff reagent. Thus Longley in our laboratory obtained quite good reactions with muscle protein, collagen, and thyroid colloid. It may be noted that deamination prevents their reaction.

Fixations with 10% formalin, with Zenker variants, or with absolute alcohol are recommended. Paraffin sections at 5 to 10 μ are deparaffinized and brought to 100% alcohol as usual. Then incubate at 5 to 24 h in 0.5% Ninhydrin (or 1% alloxan) in 100% alcohol at 37°C. Wash for a few minutes in running water, immerse for 30 min in Schiff reagent, wash 1, 2, and 2 min in three changes of 0.5% $Na_2S_2O_5$, wash 10 min in running water, dehydrate, and clear, and mount through an alcohol, xylene, synthetic resin series. Counterstains have not been successful in our hands so far.

This chemical explanation given by the original authors is, of course, the reaction of Ninhydrin or alloxan with a free uncombined amino acid and is manifestly inapplicable to amino acids which are incorporated by peptide bonds in protein chains. The precise nature of the reaction is still controversial. Puchtler and Sweat (*J. Histochem. Cytochem.*, **10**: 365, 1962) attacked the validity of the reaction on much the same grounds, pointing out the ready solubility of aldehydes formed from loosely included amino acids and those hydrolyzed from protein by Ninhydrin. Although gelatin is Schiff-positive, nonpretreated collagen is not, and glycine, proline, and hydroxyproline do not yield aldehyde on treatment with Ninhydrin.

Kasten (*J. Histochem. Cytochem.*, **10**: 769, 1962) defends the value of the reaction, admitting the obscurity of the chemical mechanism. He apparently postulates an amino acid with free carboxyl and α-amino groups, bound to protein at its other end. He offered no speculation as to which amino acids might be so attached.

Glenner (*J. Histochem. Cytochem.*, **11**: 285, 1963) points out that oxidative deamination to aldehyde is not limited to α-amino acids, occurring with many other primary aliphatic amines. Hence the ε-amino groups of the lysines and the α-amino groups of γ-glutamyl and β-aspartyl peptides should be available and reactive. This last suggestion accords well with the known blocking of Ninhydrin and alloxan-Schiff reactions by nitrous acid deamination.

The probable reaction is hence a simple oxidative deamination:

$$\begin{array}{ccc} | & & | \\ OC & & OC \\ | & & | \\ HC-(CH_2)_3CH_2NH_2 + O & \longrightarrow & HC-(CH_2)_3CHO + NH_2 \\ | & & | \\ HN & & HN \\ | & & | \end{array}$$

Van Dalen and Van Duijn (*Histochemie*, **26**: 180, 1971) applied the Ninhydrin Schiff method to amino cellulose films. Further studies are needed prior to use as a quantitative procedure.

The blocking of the Ninhydrin Schiff reaction reported by Rappay [*Nature (Lond.)*, **200**: 274, 1963] produced by inclusion of micromolar and higher concentrations of $CuSO_4$ in the alcoholic Ninhydrin solution, but not by pretreatment of tissue with $CuSO_4$ at 0.1 M concentration, remains chemically obscure. We suggest a chelate bonding to the hydrate group of the Ninhydrin itself, which should effectively block its reactivity with protein NH_2 groups.

The Xanthoproteic and Murexide Reactions

Serra (*Stain Technol.*, **21**: 5, 1946) prefers "strong" fixations but uses fresh material as well. Cover sections with concentrated nitric acid for several minutes until they become intensely yellow. Wash in distilled water. Expose to ammonia fumes or immerse in dilute ammonia solution. The color changes to orange. Mount in glycerol after rinsing in water.

The reaction is produced by tyrosine, phenylalanine, or tryptophan, as well as by phenols. Langeron (7th ed., p. 1264) notes in addition that purine bases give violet to purple by this method, the *murexide reaction*. Uric acid is distinguished from guanine by its relative solubility in piperazine hydrate and its relative insolubility in iron alum and mineral acids. A negative test means nothing.

THE METHOD OF WEISS, TSOU, AND SELIGMAN. This method (*J. Histochem. Cytochem.*, **2**: 29, 1954) is said to demonstrate only primary amino groups (II) utilizing the azo-methine condensation (III) of 2,3-hydroxynaphthaldehyde (I), followed by akaline azo coupling with fast blue B (IV). It is doubtful whether the second azo group of the di-*o*-anisidine tetrazonium salt is utilized in formation of the final colored reaction product (V or probably V*a*), since the initial coupling probably uses up all the available tissue naphthol sites.

In this, we dissent from the formulation in the original paper and in the paper of Hopman (*Mikroskopie,* **12:** 1, 1957). Unlike the arylamine condensation with tissue aldehydes, this reaction is said to proceed better in alkaline solution. The lack of reactivity in the acid media is assigned to ring formation by hydrogen bonding (VI).

WEISS, TSOU, SELIGMAN HYDROXYNAPHTHALDEHYDE METHOD FOR TISSUE AMINES

1. Bring paraffin sections through xylene and alcohols to distilled water.
2. Immerse for 1 h at 24°C in 2,3-hydroxynaphthaldehyde,[1] 20 mg, in 20 ml acetone plus 30 ml Veronal HCl buffer of pH 8.5.
3. Wash in distilled water, three changes, for a total of 15 min.
4. Azo-couple 3 to 5 min in 25 mg fast blue B in 50 ml 0.1 *M* Veronal HCl buffer of pH 7.4 at 25°C. The dry diazonium salt is stirred into the buffer at the time of using.

---

[1] Available from Borden-Dajec, Philadelphia, and from K and K, Jamaica, N.Y.

5. Wash 5 min in running water, dehydrate, clear, and mount in synthetic resin through the usual alcohol-xylene sequence.

RESULTS: Reaction sites are red to blue.

Formalin-fixed tissue is only faintly reactive. Best results for nuclear proteins are attained with acetic Zenker fixation; for cytoplasmic structures, with 80% alcohol. Paneth, keratohyalin, and trichohyalin granules are not mentioned in the descriptions. Ribonucleoprotein and deoxyribonucleoprotein sites stain strongly; smooth and striated muscle react strongly.

It is noted, however, that pancreatic zymogen granules reacted strongly in formalin-fixed material and were not conserved with the other fixations. Erythrocytes color brown after formalin, purple-brown after alcohol; they remain unstained after Zenker fixation. Use of salt Zenker would remedy the loss of erythrocytes and acetic acid–soluble zymogen granules. Eosinophil leukocyte granules do not stain. Collagen stains paler than muscle, and elastin is unstained.

Strong staining is noted in the mucin of esophageal glands and in the lumen of the esophagus, in thyroid colloid, in ovarian follicular fluid, in epididymal tubule fluid, and in cartilage matrix, but is not specifically mentioned for gastric or intestinal mucins.

The so-called o-diacetylbenzene method has been much used in Europe since its introduction by Voss (*Z. Mikrosk. Anat. Forsch.*, **49**: 51, 1941) and Dietz (ibid., **51**: 14, 1942). Wartenberg (*Acta Histochem. Cytochem.*, **3**: 145, 1956) considered it a highly specific reagent for primary amines, but the exact mechanism remains unexplained.

Barka and Anderson prescribe fixation in Zenker, Gendre, or absolute alcohol, but state that formol and mercurial fixatives are unsuitable.

Paraffin sections are brought through xylene and alcohols to water as usual.

1. Immerse in pH 8 Veronal buffer for 1 to 2 min.
2. Mix 20 ml 2% diacetylbenzene in 70% alcohol with 20 ml pH 8 Veronal buffer. Immerse sections in freshly mixed reagent 30 to 60 min.
3. Wash in pH 7 Veronal buffer and thoroughly in distilled water.
4. Dehydrate, clear, and mount in balsam or synthetic resin through usual alcohol-xylene sequence.

RESULTS: Reaction sites are red to purple or violet.

The total alcohol content of the reaction mixture may be increased to about 50 to 60% to avoid loss of more water-soluble proteins and peptides. We have not tried this method.

According to Barka and Anderson the reagent was not commercially available in 1962; we have not seen it listed in recent American catalogs, but the para isomer is available from Eastman.

Structurally the reagent as formulated could be described as a dimethylphenyl-o-diketone (I). It would be interesting to try other arylmethyl ketones.

I   o-Diacetylbenzene   II

Rosselet and Ruch (*J. Histochem. Cytochem.*, **16**: 459, 1968) believe dansyl chloride (1-dimethylaminonaphthalene-5-sulfonylchloride) (II) above in 95% ethanol saturated with

$NaHCO_3$ reacts only with α-NH_2 and ε-NH_2 groups. Acetone- or trichloracetic acid–precipitated smears were rinsed in 95% ethanol, reacted with a 0.1% solution of dansyl chloride in 95% ethanol, saturated with $NaHCO_3$ for 4 h at 24°C, rinsed for 30 min in 95% ethanol, and mounted and examined in either ethanol or Fluormount (Gurr). Rosselet and Ruch calculated that 50% of lysine in sperm is reactive and 95% of lysine in rat liver nuclei is dansylated.

Acid Dyes

Protein stains with acid dyes have long been used for the demonstration of the more basic proteins. The effect of solution pH control on the results of staining with neutral stains, particularly with reference to blood films, was reported by Pischinger (*Z. Zellforsch.* [*Mikrosk. Anat*], **3:** 169, 1926), and we applied this procedure to tissue azure-eosin staining in 1932 (*Arch. Pathol.*, **14:** 515), noting at that time the disappearance or overpowering of eosinophilia of erythrocytes and eosinophils at about pH 5.6 on formalin-fixed and Orth-fixed human and experimental material. The effect of mercurial and alcoholic fixations in raising the pH optimum for such stains was reported in 1941 (*Stain Technol.*, **16:** 1): Zenker formol, pH 5; methanol, ethanol, and Carnoy, 6.5; versus pH 4.2 for neutral formalin or Orth's fluid, pH 4.5 for acid formalin, and pH 5.2 for pH 7 buffered glutaraldehyde. The hot chromation of the Elftman procedure (*J. Histochem. Cytochem.*, **2:** 1, 1954) has recently been observed to raise the pH optimum to about pH 7; with buffered $HgCl_2$ (pH 6) and unbuffered 0.2 M lead acetate the pH optimum was about 5.5.

MERCURIC BROMOPHENOL BLUE OF MAZIA, BREWER, AND ALFERT FOR PROTEINS. (Davenport). Thin sections of fixed material, unfixed smears, and cell suspensions are brought into 0.1% bromophenol blue, either in saturated aqueous mercuric chloride solution or preferably in 95% alcohol containing 10% $HgCl_2$. Stain 15 min, blue in pH 6.5 to 7 buffer (0.1 M Na_2HPO_4 should serve), mount and examine in water, or dehydrate rapidly, clear, and mount.

Although binding of the blue dye to basic protein occurred also in the absence of mercury ions, Mazia, Brewer, and Alfert considered that acid proteins reacted by virtue of COOH, SH, and aromatic residues binding the mercury, which in turn took up the dye. Runham (*J. Histochem. Cytochem.*, **9:** 87, 1961) notes that the reaction is completely prevented by deamination and considers that it demonstrates only protein amino groups. It is not clear why the dye should not react also with guanidino groups of arginine, which resist deamination.

Ringertz and Zetterberg (*Exp. Cell Res.*, **42:** 243, 1966) studied the specificity of bromophenol blue reaction for histone. Dye uptake at alkaline pH was found to be influenced not only by the quantity of basic protein present but also by the qualitative properties of the basic proteins, the mode of fixation, the method of deoxyribonucleic acid extraction, and the staining method employed. Deitch's (*Lab. Invest.*, **4:** 324, 1955) procedure with naphthol yellow S, disodium flavianate $C_{10}H_4O_8N_2SNa_2 \cdot 3H_2O$, mw 412.25 (C.I. 10316), was designed for photometric estimation of histone-type proteins in material which could be simultaneously reacted by the Feulgen method for deoxyribonucleic acid. The absorption maxima for naphthol yellow S at 435 mμ and for the Feulgen condensation product at 570 mμ are sufficiently far apart to accomplish this end, and the absorption curves are of such shape that no significant overlapping occurs.

DEITCH'S STANDARD NAPHTHOL YELLOW S TECHNIC. Fix in Clarke-Carnoy acetic alcohol. Embed in paraffin, section, deparaffinize, and hydrate sections as usual.

1. Stain 15 min in 1% naphthol yellow S in 1% acetic acid at 20 to 25°C.

2. Transfer (as many as 10 slides) to Coplin jars or similar staining jars containing 50 ml 1% acetic acid, and differentiate 15 to 24 h. Discard the differentiating fluid when it has been used for 10 slides.

3. Blot and transfer directly to (liquid) tertiary butanol (mp 25.6°C), clear in xylene, and mount in Shillaber's high-refractive-index oil (n_D 1.540).

The use of xylene–cellulose caprate is suggested for preparations which are to be handled extensively, but the high refractive index mountant is required for photometry.

The Biebrich scarlet (C.I. 26905) method of Spicer and Lillie (*Stain Technol.*, **36**: 365, 1961) for basic proteins depends on the same principle of selective dye uptake of certain tissue elements at high pH levels. Certain tissue elements present considerable differences in maximum-staining pH between material fixed in formaldehyde and non-formaldehyde-containing fixatives, and in the critical pH range formaldehyde treatment of sections of nonformalin material inhibits staining. With other tissue elements pH maxima are the same for formalin- and non-formalin-fixed material.

BIEBRICH SCARLET METHOD OF SPICER AND LILLIE FOR BASIC PROTEIN.
For non-formalin-fixed material, buffered mercuric chloride (B-4) was preferred. The usual iodine thiosulfate sequence was prescribed, and sections were stained 30 to 90 min in 0.01% Biebrich scarlet in appropriate buffer solutions. The Laskey glycine-NaOH series was preferred for pH levels above 7.5, the McIlvaine citric acid–sodium phosphate series for lower levels. After staining, sections were dehydrated directly in 95 to 100% alcohol, cleared in xylene, and mounted in synthetic resin.

We have generally preferred a somewhat stronger solution and a shorter staining period; e.g., take 1 ml 1% Biebrich scarlet plus 49 ml buffer and stain 20 min, dehydrate with alcohol or acetone, clear in xylene, and mount, preferably in cellulose caprate, so that unstained tissue elements can be seen.

RESULTS: Tissues that were oxyphil at pH 9.5 after mercuric chloride fixation included seminal vesicle fluid, Kurloff bodies, trichohyalin, keratohyalin and stratum lucidum, Paneth granules, eosinophil leukocytes, elastic laminae, and spermatozoon heads. With primary formalin fixation only elastic laminae and eosinophil leukocyte granules still colored strongly. Posttreatment (6 h) with formaldehyde gave less reduction of oxyphilia, spermatozoon heads, trichohyalin, and Paneth granules remaining oxyphil.

Spicer (*Exp. Cell Res.*, **28**: 480, 1962) noted that most mucous cells contain basic protein which stains with Biebrich scarlet at pH 9.5.

Ahsmann and Van Duijn (*Histochemie*, **12**: 285, 1968) described a method to link amino acids covalently to cellulose films. The α-amino groups are coupled to carboxyl groups of slightly oxidized cellulose. The amino acid–cellulose films may be useful for quantitative studies of side groups of amino acids using a membrane colorimeter.

Deamination and Acylation of Amines

Monné and Slautterback (*Ark. Zool.*, **1**: 455, 1950) report that, with a Mallory-Heidenhain aniline blue stain with azocarmine, oxyphil components of nuclei, chromatin, and myoglobin resist nucleic acid extractions and are probably histones, basic proteins, and protamines. Deamination with Van Slyke's mixture, 6 g $NaNO_2$, 35 ml distilled water, and 6 ml glacial acetic acid for 1 to 12 h at 20 to 25°C, completely inhibits both aniline blue and azocarmine staining. A 12-h bath in 5% chloramine-T at 20 to 25°C has the same effect, while 0.4% Ninhydrin for 12 h at 80°C inhibited aniline blue staining of yolk but not azocarmine staining of chromidia. Lillie (*Blood*, **7**: 1042, 1952) reported that a 10-min 1 N $NaNO_2$–HAc deamination destroyed the oxyphilia of erythrocytes in methanol-fixed,

TABLE 8-6   pH LEVELS OF NITRITE SOLUTIONS

Solution	pH	Solution	pH
Van Slyke's reagent (2.17 N)	3.65	12.5% acetic acid	2.0
1 N NaNO$_2$ in 1 N HCl	0.68	1 N NaNO$_2$ in H$_2$O	6.7
1 N NaNO$_2$ in 1 N acetic acid	3.7	1 N acetic acid (6%)	2.2
1 N NaNO$_3$ in 1 N acetic acid	2.2	1 N NaNO$_3$	5.0

Giemsa-stained human blood films, while the oxyphilia of eosinophil leukocyte granules resisted 2 to 6 h. Later studies (*J. Histochem. Cytochem.*, **6**: 352, 1958) indicated that acetic acid solutions of NaNO$_2$ were more effective than hydrochloric acid, that formol fixation retards but does not prevent deamination.

The formaldehyde effect was more fully explored by Lillie et al. (*Histochemie*, **24**: 156, 1970). 2 N NaNO$_2$-HAc–deaminated alcohol fixed blood films in 2 min. With exposure to alcoholic 10% formol for 10 and 30 min and 1, 2, 4, 8, and 24 h the deamination time required was 2, 10, and 10 min and 1, 6, 8, and 8 h, respectively. Aqueous 10% formol gave 5, 10, 20 min and 2, 6, 8, 12 h for the same exposure times. With 3°C 10% formol fixation of tissue for 2, 4, 6, and 24 h, deamination times were 4, 4, 4, and 8 h, and with 25°C 10% formol for 2, 4, 6, and 24 h, deamination required 4, 6, 10, and 12 h. Fixation times of 1, 3, 7, 14, 21, 28, 42, and 56 days, and 3 and 14 months required 12, 36, 36, 36, 24, 18, 24, 24, 18, and 18 h. The falloff in deamination times with longer fixations may have been due to general impairment in azure-eosin staining with longer formol fixations.

The oxyphilia of eosinophil leukocytes, assignable largely to arginine residues, persists up to 18 h. However, when the technic of graded pH staining with acid dyes is adopted, it is found that the maximum pH at which eosinophil leukocyte granules are stained by the acid dye is depressed from 11 to 7 by 24-h nitrosation (1 N NaNO$_2$–1 N HAc), which indicates a considerable amount of amine destruction by HNO$_2$. It is known that there is a slow loss of a second N in free arginine after prompt destruction of the α-amino group, and it is possible that this accounts for the shift in protein pK during nitrosation.

Nitrosation at 25°C in 1 N or 2 N NaNO$_2$–acetic acid mixtures does not appreciably impair Van Gieson staining (*J. Histochem. Cytochem.*, **6**: 352, 1958).

Acylation for 1 to 2 h in 0.1 sulfuric acid–10 acetic anhydride–30 glacial acetic acid weakens or abolishes both muscle and plasma staining with the fast green of the fast green Van Gieson technic, and the deep purple-red staining of collagen and reticulum by the acid fuchsin component. A 1- to 2-h methylation at 60°C in 0.1 M aged sulfuric acid–methanol mixture fully restores both plasma and collagen staining. Saponification overnight in 5.6% ammonia in 80% alcohol does not unblock the acylation blockade of the Van Gieson stain. However, a 20-min saponification in 1% potassium hydroxide in 70% alcohol does partly, though not fully, restore the fast green Van Gieson staining.

Strong pH 1 azure A staining occurs in smooth muscle and cytoplasms, as well as in the usual mucopolysaccharide sulfation sites, and this also is abolished by methylation but not by saponification as above. Inclusion of 10% propanol, isopropanol, or, somewhat less effectively, *n*-butanol in the above acylation mixture, replacing part of the acetic acid, fairly effectively restricts azure A staining to mucopolysaccharide sites, as with the sulfuric acid–glacial acetic acid mixtures which lack acetic anhydride. Ethyl and methyl alcohols did not prevent amine sulfation under the same conditions.

Periodic acid Schiff staining of gastric surface epithelial and pyloric gland mucin and of intestinal striated borders, goblet cells, and Brunner's gland and colonic gland mucins is similarly prevented by the same acylation and restored by the same methylation, and the blockade resists ammonia alcohol saponification.

It thus appears that sulfation of amines may be accomplished with acetic acid solutions containing both sulfuric acid and at least 1 to 2% acetic anhydride, and this effect is prevented if an excess of a suitable alcohol over the sulfuric acid is present.

SULFATION OF AMINES (Lillie, *J. Histochem. Cytochem.*, **12:** 821, 1964). Cut thin paraffin sections of material fixed in formalin or otherwise. Deparaffinize as usual in two or three changes of xylene and two of absolute alcohol.

1. Rinse in glacial acetic acid.
2. Immerse 3 to 5 min in mixture of 10 ml acetic anhydride, 30 ml glacial acetic acid, and 0.1 ml concentrated sulfuric acid, thoroughly mixed by being poured back and forth once or twice between a beaker and the Coplin jar.
3. Wash 10 min in running water.
4. Stain sulfated sections and unsulfated controls by a suitable acid dye method, such as Biebrich scarlet, the fast green Van Gieson technic, or other method for basic protein.

Further controls may be methylated after sulfation to restore stainability, using 0.1 N hydrochloric acid or 0.1 M sulfuric acid in methanol, 2 h at 60°C. Saponification by the potassium hydroxide or ammonia alcohol technics should be ineffective or only partially effective in overcoming the blockade.

Acetylation with acetic anhydride–pyridine mixtures or with hot acetic anhydride alone is less effective against simple acid dyes such as eosin Y, eosin B, or Biebrich scarlet, but the sulfation-induced basophilia is absent, and such mixtures as that in the azure A–eosin B technic can be used. Attempts to block Van Gieson staining by such procedures are usually unsuccessful. Sulfuric catalysis tends to improve pyridine acetylations and does not appear to induce sulfation basophilia. The oxyphilia to eosin B of smooth and striated muscle, erythrocytes, eosinophils, zymogen granules, and the like, when blocked by acetylation, is readily restored by 20 min in 1% potassium hydroxide in 70% alcohol.

Deitch (*J. Histochem. Cytochem.*, **9:** 477, 1961) complained of difficulties in the prevention of her modified Sakaguchi arginine reaction by acetylation. Although her pyridine reagent, 20% acetic anhydride, 1 h at 25°C, and her acetic acid reagent, 5% acetic anhydride, 1 h at 25°C, would appear to have been inadequate in view of our experience with similar and more concentrated reagents, her 2-h exposure to pure acetic anhydride at 80°C is only slightly less than that of Monné and Slautterback and should have had more effect. Hydrolysis of the acetyl esters in the strongly alkaline Sakaguchi reagent may, of course, have occurred. Here the foregoing amine sulfation technic could well have been more effective.

Urea

LESCHKE'S METHOD. Fixation of tissue in a mixture of equal volumes of saturated mercuric nitrate solution and 2% nitric acid is followed by treatment of sections with ammonium sulfide to convert the mercury-urea compound to black mercuric sulfide. This method is regarded by both Lison and Glick as highly unspecific.

THE XANTHYDROL REACTION. This is quite specific; but the reaction is slow, urea is highly diffusible, and the localization is therefore poor. Gomori recommends Oliver's method (*J. Exp. Med.*, **33:** 177, 1921): Fix small pieces of tissue in a filtered solution of 6 g xanthydrol in 35 ml alcohol and 65 ml glacial acetic acid for 6 to 12 h. Alcohol dehydration, paraffin embedding, and sectioning follow. The xanthydrol urea crystals are demonstrated under polarized light.

We have used an essentially identical fixing fluid by first freezing tissue in solid carbon dioxide and 100% alcohol mixture and then fixing for 14 days at −25°C. At this temperature the alcohol–acetic acid mixture remains liquid, and the tissue melts only as the fixative penetrates. The method needs further study to put it on a good routine basis.

Uric Acid and Urates

Sodium urate is the commonest of the urates occurring in gouty tophi. Its crystals are slightly soluble in cold water, and insoluble in alcohol and ether. Mallory directed fixation in 95% or absolute alcohol. Schmorl recommended Oestreicher's (*Virchows Arch. Pathol. Anat.*, **257**: 614, 1925) preliminary treatment of 6 h in 6% xanthydrol (9-hydroxyxanthene) in glacial acetic acid, followed by 48-h fixation in 100% alcohol. Mallory preferred celloidin; Schmorl, paraffin embedding. Stain sections briefly—Schmorl says 2 min—in hemalum or other unacidified alum hematoxylin. Wash in water, dehydrate, and clear by an alcohol-xylene sequence and mount in balsam or Permount. According to Mallory's technic the crystals are deep blue; by Oestreicher's xanthydrol method, bright yellowish green.

Schultz (*Virchow Arch. Pathol. Anat.*, **280**: 519, 1931) reported a rather complicated procedure which Mallory recommended. We have not tried it. Fix tissues in 100% alcohol. Pass through three changes of acetone of about $1\frac{1}{2}$ h each. Then place in equal volumes of acetone and benzene for 30 min, then two changes of benzene for 30 min each. Embed in paraffin, and section.

1. Deparaffinize and bring to 100% alcohol.
2. Stain 5 min in carmine, keeping slide in motion. The carmine solution is essentially similar to Best's and is prepared thus: Boil 1 g carmine, 2 g ammonium chloride, 0.5 g lithium carbonate in 50 ml distilled water. Cool and add 20 ml 28% ammonia water. Mix 6 ml filtered stock solution with 3 ml 28% ammonia water and 5 ml methyl alcohol for staining.
3. Wash in several changes of 100% alcohol.
4. Stain 30 s, keeping slide in motion, in a half-saturated (about 0.75 to 0.94%) solution of methylene blue in 100% alcohol.
5. Rinse in 100% alcohol.
6. Stain 15 s, keeping slide in motion in sodium sulfate–picric acid mixture: 9 ml saturated aqueous picric acid solution, 1 ml saturated aqueous sodium sulfate solution (nearly 50 g of the anhydrous salt per 100 ml). Filter.
7. Wash in several changes of 100% alcohol.
8. Clear in xylene and mount in balsam.

RESULTS: Nuclei are gray-blue; cytoplasm is yellowish; uric acid crystals, deep blue-green; sodium urate, brilliant green.

Gerard and Cordier [*Arch. Biol.* (*Liege*), **43**: 367, 1932] used Carnoy fixation and observations under polarized light for the demonstration of experimental uric acid deposits.

Uric acid exists in tautomeric trihydroxyl (acid) and (predominant) tricarbonyl (ketonic) forms and is readily oxidizable. It has been found to reduce silver nitrate, on which fact the Gomori and De Galantha methods depended. We have shown that in vitro uric acid and sodium urate reduce ferric ferricyanide promptly and hot (60°C) methenamine silver blackens in a few seconds. Ammoniacal silver blackens in a few seconds with the sodium salt, more slowly with the acid. Even neutral silver nitrate solutions turn brown with the acid in 10 min, and acid silver nitrate (pH 4, 0.1 M) turns brown in a few hours (*J. Histochem. Cytochem.*, **5**: 311, 1957).

Gomori used his methenamine silver solution at pH 9, preheating to 37°C and incubating sections 30 to 60 min to blacken urate crystals. Most other tissue elements remain colorless at this stage. Complete the process as usual.

De Galantha's procedure (*Am. J. Clin. Pathol.*, **5**: 165, 1935) seems unnecessarily complicated, and by its use of a reducing procedure loses in specificity. (See De Galantha's article.)

ARGININE

Arginine is traditionally localized in tissue by variants of Sakaguchi's [*J. Biochem.* (*Tokyo*), **5**: 25, 1925] α-naphthol and (ibid., **31**: 231, 1950) 8-hydroxyquinoline methods. These methods and McLeish's (*Exp. Cell Res.*, **12**: 120, 1957) 2,4-dichloro-1-naphthol variant are regarded as quite specific for guanidyl groups, which in mammalian tissue means arginine. All these depend on hypochlorite or hypobromite oxidation to form an orange to red compound of variable stability and solubility. The nature of this compound has been much discussed and debated. Bhattachandrya considered an indophenol structure and reported recovery of ornithine and a 1,4-naphthoquinone from hydrolystates [*Nature* (*Lond.*), **184**: 153, 1959; *Ann. Biochem. Exp. Med.*, **20**: 93, 1960; *Arch. Biochem. Biophys.*, **77**: 237, 1958].

Lempert and Lempert-Sréter (*Chem. Ber.*, **91**: 796, 1961) formulated the condensation product of guanidine and benzil as forming an imidazole ring. These studies were extended by Itano's group to combinations with cyclohexanedione (Toi et al., *J. Biol. Chem.*, **242**: 1036, 1967) and the fluorochrome 9,10-phenanthrenequinone (Yamada and Itano, *Biochem. Biophys. Acta*, **130**: 538, 1966) as reagents for the specific binding and identification of arginine peptides. Alkaline solutions of benzil, 1,2-cyclohexanedione, glyoxal, and 9,10-phenanthrenequinone are now used by Lillie et al. (*J. Histochem. Cytochem.*, **19**: 487, 1971) as specific blocking reagents for arginine, and Magun and Kelly (*J. Histochem. Cytochem.*, **17**: 821, 1969) introduced 9,10-phenanthrenequinone as a specific histochemical fluorochrome for localization of arginine. Speculating that the Sakaguchi reaction might function by condensation of a 1,2-naphthoquinone with the guanidyl group to form a similar imidazole ring, after oxidation of the 1-naphthol to a 1,2-quinone, Lillie et al. (*J. Histochem. Cytochem.*, **19**: 487, 1971) tried 1,2-naphthoquinone-4-sodium sulfonate (NQS) in barium hydroxide solution following Deitch's (*J. Histochem. Cytochem.*, **9**: 477, 1961) variant of McLeish's method, without NaOCl, and achieved brilliant deep red staining of arginyl sites, blocked the reaction with benzil, cyclohexanedione, phenanthrenequinone, and glyoxal, and showed in vitro specificity. The new reaction regularly gives a greater color depth than the NaOCl and NaOBr methods, preparations can be routinely dehydrated and mounted without special precautions, and their colors remain stable for at least 8 months. It is observed that while the color product is formed equally in vitro in NaOH and in $Ba(OH)_2$ solutions, the former product is water-soluble and remains in the solution and the latter can be centrifuged out. The optimal pH is 13 to 13.4, best about 13.2. If NaOH is used alone, the localizing reaction fails, but if an excess of $BaCl_2$, $SrCl_2$, or even $CaCl_2$ is present, reactions are as good as or better than with $Ba(OH)_2$.

From these findings we deduce that the Sakaguchi reaction proceeded by oxidation of 1-naphthol with NaOCl to 1,2- and 1,4-naphthoquinones, that the 1,2-napthoquinone (III) then cyclizes with the guanidyl group to form an imidazole (IV).

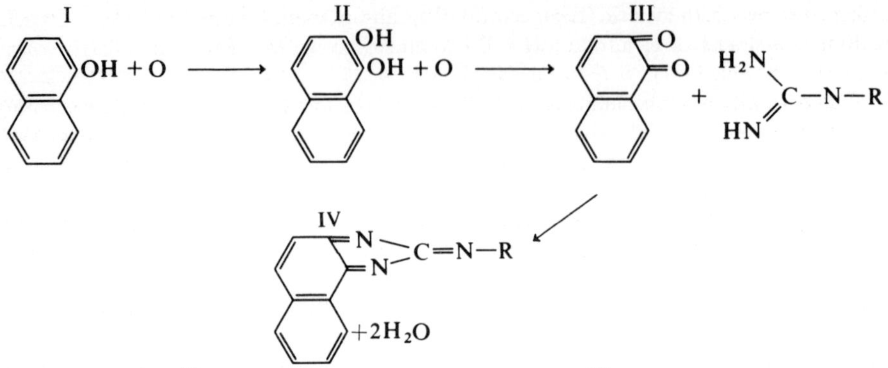

I   II   III

$$\text{[naphthol]}\ \text{OH} + \text{O} \longrightarrow \text{[naphthalenediol]}\ \text{OH} + \text{O} \longrightarrow \text{[naphthoquinone]} + \underset{HN}{\overset{H_2N}{>}}C-N-R$$

IV

$$\text{[naphtho]}\ \underset{N}{\overset{N}{>}}C=N-R$$

$+2H_2O$

V 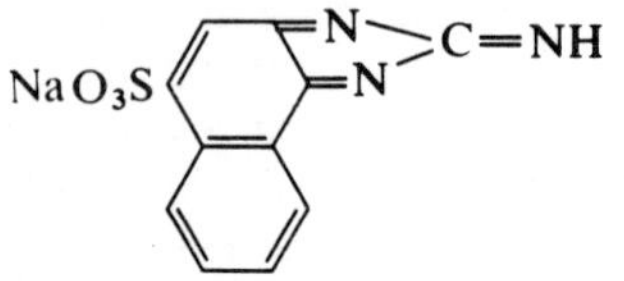   VI

$$NaO_3S\text{[naphtho]}\ \underset{N}{\overset{N}{>}}C=NH \qquad -Ba-O_3S\text{[naphtho]}\ \underset{N}{\overset{N}{>}}C=NH$$

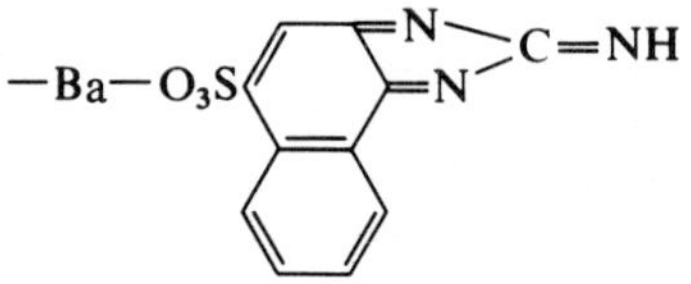

The cleavages noted in V and VI account for Bhattachandrya's finding of ornithine in the reaction product. The 1,4-naphthoquinone remains free and cannot cyclize to form the imidazole. The Na salt (V) is soluble; the Ba salt (VI) is insoluble. Without the sulfonic acid group 1,2-naphthoquinone apparently forms a soluble product, and the Ba has no point of attachment. However, a formula preserving attachment to the protein chain (IV) corresponds better to the resistance of the final NQS reaction product to various extractions. Yamada and Itano formulated the reaction product of the 9,10-phenanthrenequinone reaction with cleavage of the guanidyl residue from the protein chain (VII), formula (VIII).

VII   VIII

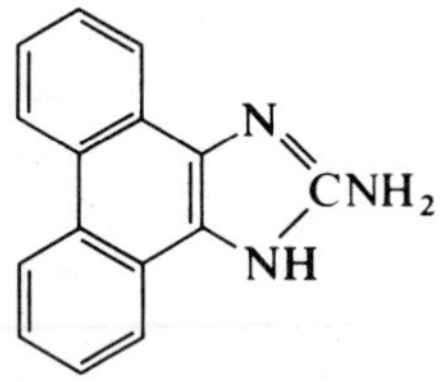

$$\text{[phenanthro]}\ \underset{NH}{\overset{N}{>}}CNH_2 \qquad or \qquad \text{[phenanthro]}\ \underset{NH}{\overset{N}{>}}C-NH-(CH_2)_3-\underset{NH}{\overset{C=O}{CH}}$$

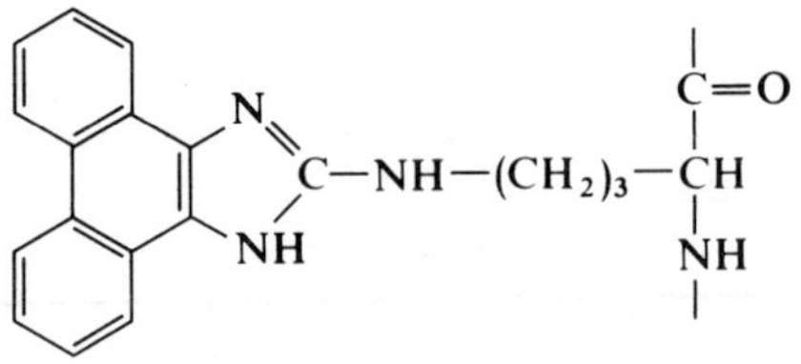

THE 9,10-PHENANTHRENEQUINONE REACTION OF MAGUN AND KELLY. To 40 ml of 0.1 N NaOH in 80% alcohol add 10 ml fresh 1% 9,10-phenanthrenequinone in dimethylformamide. Use at once, since the red solution soon turns yellow and becomes inert.

Bring sections to 80% alcohol, immerse 10 min in the reagent, rinse in 95% alcohol 3 times for 1 min, hydrate, and mount in freshly mixed glycerol–water–1 N NaOH, 9 : 1 : 1. Examine with a mercury HBO 200 lamp, dark-field condenser, UG-3 excitation filter, and Zeiss 470-mμ barrier filter. The preparations remain usable for at least an hour.

Study at once by fluorescence microscopy, using a UG-3 excitation filter and Zeiss 470-mμ barrier filter and, with Hitachi-Perkin-Elmer spectrofluorometer, using right-angle excitation and emission paths. Fluorescence reaches maximum in about 5 min and remains unaltered for at least 60 min. Nucleoli react strongly, nuclear chromatin in proportion to concentration.

Cytoplasms of neurons, liver cells, renal convoluted tubule cells, villus-striated border, collagen, and keratin fluoresce strongly.

Bock and Schluter (*Z. Zellforsch.* [*Mikrosk. Anat.*], **122**: 456, 1971) fixed hypothalamus for 7 days in picroformol (ibid., **87**: 534, 1968), cut 10-μ sections, washed out picric acid in water, and dehydrated and reacted by the Magun-Kelly technic. The neurosecretory substance gave a positive reaction for arginine by this method and also by the following NQS method (Bock, personal communication, December 1971).

We have used 9,10-phenanthrenequinone successfully as a blocking reagent against the NQS reaction at 100 mg in 1 g Ba(OH)$_2$ in 40 ml 50% alcohol, allowing 1-h exposure; 15 min may be adequate.

THE NEW 1,2-NAPTHOQUINONE-4-SODIUM SULFONATE (NQS) METHOD (Lillie, Pizzolato, Dessauer, and Donaldson, *J. Histochem. Cytochem.*, **19**: 487, 1971). This method gives deeper red colors than any of the previous Sakaguchi technics and is easier to perform. The heavy section losses occasioned by present-day commercial Clorox are avoided. Fading of the reaction after mounting is largely eliminated. The reaction product is stable to 72-h Versene, 4-h 0.5 M potassium oxalate, 24-h sodium acetate, and 3-h 0.12 N hydrochloric acid extractions. No suitable counterstain has been found; basic dyes tend to override and obscure the reaction.

The reagent is unstable, and the NQS must be added to the solvent immediately before using. It remains active for about 10 min, and longer exposures are useless. The optimal pH is about 13.2; the reaction fails above 13.6 and below 12.5.

PROCEDURE. Bring sections through xylene and alcohols to water; include the iodine thiosulfate treatment for mercury-fixed tissue. Mix 10 ml 0.4 N NaOH, 10 ml 1 M BaCl$_2 \cdot$2H$_2$O (24.4%), and 20 ml distilled water. Dissolve 100 mg NQS in this mixture and pour over sections. React 10 min (5 min may give quite adequate staining; in 12 to 15 min the solution turns brown and has no further staining capacity). Rinse 10 s in 1% sodium acetate and 10 s in distilled water. Dehydrate in alcohols, clear in xylene, and mount in cellulose caprate.

RESULTS: Distal hair cortex is unstained; keratohyalin, eleidin, trichohyalin, and hair medulla stain deep red or brownish red. Nuclei are pink; cytoplasm, light pink; collagen, pink; muscle, light orange to pink. A 1-h exposure to 2.5% Ba(OH)$_2$ produces a positive reaction of distal hair cortex but removes part of keratohyalin. Sperm heads, Paneth cell granules, some arterial elastin, eosinophil leukocyte granules, and neurosecretory substance of the hypothalamus (Bock, personal note, 1971) also color red. Pretreatments with 0.2 N NaOH or 0.26 N Ba(OH)$_2$ solutions of 1,2-cyclohexanedione or glyoxal, similar alkalized 50% ethanol solutions of benzil or 9,10-phenanthrenequinone prevent the NQS reaction with 30- to 60-min exposures. These reagents are known to condense with arginine to form stable heterocyclic ring compounds (Toi et al., *J. Biol. Chem.*, **242**: 1036, 1967; Yamada and Itano, *Biochim. Biophys. Acta*, **130**: 538, 1966).

With the availability of the simpler and more reliable 9,10-phanthrenequinone fluorochrome method and the more strongly chromogenic β-napthoquinone-4-SO$_3$Na (NQS) technic it is not expected that the capricious hypochlorite and hypobromite procedures will often be used. However, we cite here a *Baker* (*Q. J. Microsc. Sci.*, **88**: 115, 1947) *hypochlorite α-naphthol method.* Baker prescribed 24-h fixation of thin (3 mm) tissue slices in Bouin, Zenker, susa, HgCl$_2$–acetic acid, or formol–saline solution and usual paraffin embedding, sectioning at 6 μ, and attachment of sections with Mayer's albumen glycerol.

1. Dewax in xylene, pass through ethanol, and collodionize (1 to 5 min) in 1% collodion–ether plus alcohol. Drain, harden in 70 to 80% alcohol, wiping back of slide first.

Iodize to remove mercury precipitate if necessary, remove iodine with 5% (0.2 M) hypo, and wash in water.

2. To 2 ml 0.1% NaOH add 0.05 ml (2 drops) 1% α-naphthol in 70% alcohol and 0.2 ml (4 drops) 1% sodium hypochlorite (diluted from usual commercial solutions 5% NaOCl). Drain or blot section and place in a horizontal position face up. Flood with the naphthol-hypochlorite mixture. Let stand 15 min to develop color fully.

3. Drain, blot, and clear in pyridine-chloroform (3 : 1) for 3 min. Baker prescribed mounting in pyridine chloroform, sealing with gold size, and immediate examination. (CCl_4 is less volatile and less toxic).

RESULTS: Sperm heads, keratohyalin, Paneth granules are pink to red. Later writers have used 0.02% aniline in Permount or mineral oil as mounting media (Carver, Brown, and Thomas, *Stain Technol.*, **26:** 89, 1953). Carver, Brown, and Thomas used a sequence technic of 0.3% 8-hydroxyquinoline (oxine) in 30% alcohol for 15 min, transferred directly to 9.4 ml 5% sodium hypochlorite, 15 ml 0.1 N KOH, H_2O to make 100 ml, and then 15 g urea, 15 ml 0.1 N KOH, 70 ml *tert*-butanol, H_2O to make 100 ml, two changes for 10 s and 2 min, dehydration in *tert*-butanol, and clearing in xylene. Deitch (*J. Histochem. Cytochem.*, **9:** 477, 1961) used 5% tributylamine (TBA) in *tert*-butanol to dehydrate and 10% TBA in mineral oil to mount. She allowed 10 min for reaction in freshly filtered saturated (25°C, 4%) Ba(OH$_2$), 25 ml; 1% NaOCl, 5 ml; and 75 mg 2,4-dichloro-1-naphthol in 5 ml *tert*-butanol added just before using.

The other technics all use a vertical slide position in a Coplin jar. Because of frequent section losses by these methods the Baker collodionization horizontal staining may be valuable.

DEAMINATION, FERROUS SULFATE, HEMATOXYLIN SEQUENCE FOR ARGININE.   Fix thin tissue slices 4 to 8 h in buffered mercuric chloride, about 0.2 M HgCl$_2$·0.2 M sodium acetate: The solution is moderately stable. Overnight fixation is usable, perhaps better at 3°C. Transfer directly to 70 or 80% alcohol. Harden 1 to 5 days as convenient. Trim blocks at this stage. Complete dehydration with 80, 95%, absolute alcohol, alcohol plus naphtha or toluene, naphtha or toluene, and embed in paraffin as usual.

Carnoy, methanol chloroform, and 80% alcohol fixations can be used if preservation of zymogen granules and erythrocytes is not desired. Avoid formaldehyde and other aldehydes. They greatly hinder deamination.

1. Bring paraffin sections through xylene, alcohols, iodine, and thiosulfate to remove excess mercury, and wash 10 min in water. Rinse in distilled water.
   Direct variant: Omit steps 2 and 3.

2. Deaminate 2 h in 7 g sodium nitrite (sticks or pellets), 44 ml distilled water, and 6 ml glacial acetic acid added just before using.

3. Wash 10 min in four changes of distilled water.

4. Immerse in 0.01 M ferrous sulfate (278 mg FeSO$_4$·7H$_2$0/100 ml distilled water) for 2 h. The solution must be fresh, and the reagent should not be too old.

5. Rinse in three changes of distilled water, for 5 min each.

6. Use 5 ml 1% alcoholic hematoxylin (fresh to 7 days old) for 2 h, 10 ml 0.1 M phosphate buffer of pH 7, and 35 ml distilled water, freshly mixed. Hematoxylin oxidizes rapidly at this concentration and pH.

7. Since the chlorine and chloramines often present in municipal water supplies can rapidly bleach hematoxylin, transfer directly to alcohols, clear in xylene, and mount in synthetic resin. Cellulose caprate is good when unstained material should be seen and identified (n_D about 1.480).

RESULTS: *Direct variant:* Nuclei, keratin, eosinophils, Paneth granules, myelin, bile pigment, trichohyalin, and keratohyalin are black or blue-black. Erythrocytes are blue-black; muscle, blue to dark blue. Cytoplasm is blue-gray.

Deamination prevents reaction of lysine sites (muscle, erythrocytes, cytoplasm) leaving arginine sites reactive. Bile pigment is bleached by nitrous acid. Myelin retains moderate staining.

Ferrous sulfate in warm humid climates tends to become caked in the bottle, even on local dealers' shelves, and in these circumstances it may not perform satisfactorily as a mordant. In that case Mohr's salt, ferrous ammonium sulfate, is said to be more stable and performs satisfactorily at the same molar concentration. Recently we have resorted to preparation of a fresh 1-M ferrous salt as follows: Place 2 g bright iron wire in a 25-ml graduated cylinder. Add 20 ml freshly boiled distilled water at 24 to 26°C and 4 ml 12 N hydrochloric acid. Then cover with a 2.5- to 3-cm layer of mineral oil. The resulting approximately 1 M ferrous chloride solution is ready for use in 2 to 3 days and remains usable for about 2 weeks. The excess iron remains bright, no brown precipitate should be formed, and the pH of the solution is about 4 to 4.5. Amounts required for use are withdrawn with a graduated pipette, through the oil layer.

The amount of iron required for the 24 ml, 2 N acid (48 mmol) is about 1.34 g (24 meq iron), so the above prescription allows a 50% excess. (See *J. Histochem. Cytochem.*, **20**: 116, 1972; *Stain Technol.*, **46**: 145, 1971.)

Notenboom et al. (*Histochemie*, **9**: 117, 1967) note that the Deitch method for arginine results in a specific fluorescent product that persists for at least 2 months if sections are stored in the dark at 24°C. The Deitch method must be modified to omit aniline in the mounting medium, and a nonfluorescent mounting medium must be employed. It is suggested that tissues are examined using excitation filter BG3 (4 mm) with ultraviolet light–absorbing filters 44 and 50.

Citrulline

Rogers (*J. Histochem. Cytochem.*, **11**: 700, 1963) investigated Fearon's diacetyl monoxime reaction for substituted ureides of the type $RNH-CO-NH_2$ and in particular of citrulline

$$NH_2-CO-NH-CH_2-CH_2-CH_2-CH\begin{array}{l} ^{\displaystyle NH_2} \\ ^{\displaystyle COOH} \end{array}$$

and found it too drastic for histochemical use because of the strong acid and high temperature employed. Our own experiences, reducing the temperature to 60°C and using glacial acetic acid as the menstruum for the hydrochloric acid, were no more successful. However, Rogers obtained quite usable results by condensing with *p*-dimethylaminobenzaldehyde in dilute hydrochloric acid to obtain yellow. A high concentration of sodium chloride is used to prevent swelling.

ROGERS'S METHOD FOR CITRULLINE IN STIMULATED HAIR FOLLICLES OF RAT.   Denude hair from back by plucking 18 days in advance, so that anagenesis stage will be reached. Excise skin from preplucked area and fix 48 h in phosphate buffered formalin. Embed and section in paraffin at 8 μ. Deparaffinize and hydrate as usual.

Make 1% solution of *p*-dimethylaminobenzaldehyde (Ehrlich's reagent) in 4 M sodium chloride 0.5 N hydrochloric acid (Ehrlich's reagent, 1 g; concentrated hydrochloric acid, 4 ml; sodium chloride, 23.38 g; distilled water to make 100 ml).

Place a drop of this reagent on the section, quickly cover with a cover glass, and watch under a microscope for the appearance of yellow in the hair medulla and in the Henle and Huxley layers of the sheath, immediately adjacent to the medullary and lateral areas occupied by the trichohyalin granules. Preparations are not permanent but can be photographed in color.

The trichohyalin granules give a strong red Sakaguchi reaction for arginine and a negative reaction for citrulline. As the hair root is followed toward the keratinization zone, the Sakaguchi reaction for arginine becomes abruptly much weaker, and the Rogers reaction for citrulline conversely becomes stronger.

Biochemically no protein breakdown and resynthesis seem to be involved; there is no liberation of arginyl residues nor uptake of citrulline. Instead the guanyl residues of arginine appear to convert directly to ureide residues of citrulline. An interchange with adjacent glutamic acid residues to form γ-glutamide residues is suggested. This would entail no overall change in charge of the protein and would account well for the relatively strong oxyphilia of both areas.

Rogers writes the reaction thus:

$$
\begin{array}{cc}
\overset{|}{\underset{|}{\text{NH}}} & \overset{|}{\underset{|}{\text{NH}}} \\
\text{CH}-\text{CH}_2-\text{CH}_2-\text{CH}_2-\text{NH}-\overset{\text{O}}{\underset{\text{NH}}{\overset{\|}{\text{C}}}}-\text{NH}_2 \qquad & \underset{\text{HO}}{\overset{\text{O}}{\text{C}}}-\text{CH}_2-\text{CH}_2-\text{CH} \\
\overset{|}{\underset{|}{\text{CO}}} & \overset{|}{\underset{|}{\text{CO}}} \\
\text{Arginyl} & \text{Glutamyl}
\end{array}
$$

$$
\begin{array}{cc}
\overset{|}{\underset{|}{\text{NH}}} & \overset{|}{\underset{|}{\text{NH}}} \\
\text{CH}-\text{CH}_2-\text{CH}_2-\text{CH}_2-\text{NH}-\overset{\text{O}}{\overset{\|}{\text{C}}}-\text{NH}_2 \qquad & \underset{\text{H}_2\text{N}}{\overset{\text{O}}{\text{C}}}-\text{CH}_2-\text{CH}_2-\text{CH} \\
\overset{|}{\underset{|}{\text{CO}}} & \overset{|}{\underset{|}{\text{CO}}} \\
\text{Citrullyl} & \gamma\text{-Glutamide}
\end{array}
$$

Holmes (*J. Histochem. Cytochem.*, **16**: 136, 1968) has adapted the carbamidodiacetyl reaction of Fearon (*Biochem. J.*, **33**: 902, 1939) with diacetyl monooxime (α-nitrosoethyl methyl ketone) to the histochemical demonstration of citrulline.

Holmes gives the reactions: hydrolysis to diacetyl (butandione) which then condenses with ureide such as citrulline.

$$
1 \quad
\begin{array}{c}
\text{CH}_3 \\
| \\
\text{C}=\text{NOH} \\
| \\
\text{C}=\text{O} \\
| \\
\text{CH}_3
\end{array}
\;+\;\text{H}_2\text{O}
\quad\xrightarrow{\text{H}^+}\quad
\begin{array}{c}
\text{CH}_3 \\
| \\
\text{C}=\text{O} \\
| \\
\text{C}=\text{O} \\
| \\
\text{CH}_3
\end{array}
\;+\;\text{NH}_2\text{OH}
$$

Diacetyl oxime   2,3-Butanedione, diacetyl

$$
\begin{array}{c}
\underset{\text{Citrulline}}{
2\ \begin{array}{c}
\mathrm{CH_3} \\ | \\ \mathrm{C}{=}\mathrm{O} \\ | \\ \mathrm{C}{=}\mathrm{O} \\ | \\ \mathrm{CH_3}
\end{array}
+
\begin{array}{c}
\mathrm{NH_2} \\ | \\ \mathrm{C}{=}\mathrm{O} \\ | \\ \mathrm{NH} \\ | \\ (\mathrm{CH_2})_3 \\ | \\ -\mathrm{OC}{-}\mathrm{C}{-}\mathrm{NH}{-} \\ \mathrm{H}
\end{array}
}
\end{array}
\quad \xrightarrow{\ \mathrm{H^+}\ } \quad
\underset{\text{Orange product}}{
\begin{array}{c}
\mathrm{CH_3} \\ | \\ \mathrm{C}{=}\mathrm{N} \\ | \ \ \ \ \searrow \\ \mathrm{HC}{-}\mathrm{N} \ \ \mathrm{C}{=}\mathrm{O} \\ | \\ (\mathrm{CH_2})_3 \\ | \\ -\mathrm{C}{-}\mathrm{C}{-}\mathrm{NH} \\ \ \ \| \ \ \mathrm{H} \\ \ \ \mathrm{O}
\end{array}
}
$$

TECHNIC

1. Paraffin sections of formol-fixed tissue are flattened on water at 45 to 50°C and picked up on slides thinly coated with a silicone rubber cement spread with a wooden applicator. Clear Seal of General Electric Silicone Products Dept. was used (Waterford, New York 12188). Sections adhere at once and cannot be moved, so care must be taken to place them where wanted on the slide.
2. Dry 90 min at 25°C and 30 min more at 60°C.
3. Dewax in xylene; hydrate through graded alcohols.
4. Block tryptophan with performic acid (40 ml 8% formic acid, 4 ml "100 vol" 30% H_2O_2, 0.5 ml concentrated H_2SO_4) 30 min (Pearse, 1960, 1968, 1 h).
5. Counterstain by Prussian blue reaction 15 min, and dip into 3% $FeCl_3$ until blue. Step 5 may be omitted; it may obscure the citrulline reaction.
6. In a 250-ml beaker heat 200 ml 10% HCl (1.2 N) to boiling. Add 3 ml 6% diacetyl-monooxime, and immerse sections in boiling mixture for 10 min.
7. Cool to 25°C and add 8 ml 10% potassium persulfate. Let stand 2 min 15 s.
8. Remove sections, and blot dry with several thicknesses of filter paper. Add a drop of silicone rubber (Dow Corning glass and ceramic adhesive) and pick up coverslip for permanent mounting. The reaction fades in other media. (Dow Corning Corporation, Midland, Michigan.)

RESULTS: Citrulline sites bright orange to red-orange. Omission of the performic acid step allows a Hopkins-Cole red-purple reaction for tryptophan in fibrin, renal casts, prostatic gland contents, cartilage, epidermal keratin, thyroid colloid. (We question part of these as tryptophan sites; Holmes does not mention Paneth cells, eosinophils, pancreatic zymogen, or lens protein.)

PROTEINS

Fibrin

Fibrin is a fibrous protein occurring in various acute inflammatory processes, notably diphtheritic inflammation and lobar pneumonia, and in antemortem and postmortem thrombi.

It is eosinophilic with hematoxylin-eosin and azure-eosin stains. It retains the plasma stains in procedures of the Van Gieson and fast green Van Gieson types and the Masson-Mallory methods. It stains orange in the allochrome method, and red to pink in the periodic acid Schiff method, changing to orange if a picric acid counterstain is used. It retains the crystal violet–iodine complex of the Gram stain when Weigert's aniline xylene differentiation is used, but not when alcohol or acetone is the differentiating agent. Mallory's phosphotungstic acid–hematoxylin is much favored by some workers.

Stains of the Masson type are more brilliant after Bouin, Zenker, or Spuler fixation. With material fixed in formalin, they may be improved by premordanting in saturated alcoholic picric acid for 2 min at 25°C or in saturated aqueous mercuric chloride for an hour at 58°C. Bouin fixation, however, interferes with the Weigert fibrin stain. The allochrome method is quite differential and does well on formalin-fixed tissue.

According to Glynn and Loewi (*J. Pathol. Bacteriol.*, **64:** 329, 1952) fibrin is digested in 3 h at 37°C by 0.1% trypsin in pH 8 phosphate buffer, but fibrinoid and collagen are not.

Localization by fluorescein-labeled specific antiserum is a research possibility, scarcely of interest in diagnostic pathology.

The Weigert Fibrin Stain

The *Gram-Weigert technic* devised by Weigert in 1887 for fibrin and bacteria differs from the Gram technics in that counterstaining, if any, must be done before the Gram reaction is carried out, and in that the differentiating agent is aniline, alone or weakened by admixture with xylene. Numerous variants have been suggested, chiefly in the counterstains used. Weigert, writing for the 1903 "Encyklopädie der mikroskopischen Technik," still recommended a carmine prestain. He stated that it was necessary to treat chromate-fixed material with oxalic acid before staining for fibrin, and suggested also the pretreatment with 0.33% potassium permanganate. This procedure is unnecessary for bacterial staining. On the other hand Weigert also stated that a simple aqueous methyl violet solution was adequate for fibrin staining, while for bacterial staining the aniline methyl or crystal violet solution was to be preferred. Inasmuch as it is often desirable to have both fibrin and bacteria demonstrated in the same preparation, the technic adopted should conform to the above requirements for both.

GRAM-WEIGERT TECHNIC FOR FIBRIN AND BACTERIA.   Bring paraffin sections to water as usual.

1. Stain 20 min in 1% pararosanilin or basic fuchsin in 1% acetic acid (boil, cool, and filter).
2. Wash in three changes distilled water.
3. Stain 5 min in 1% crystal violet in distilled water.
4. Rinse quickly in 1% sodium chloride.
5. Treat 30 s in iodine–potassium iodide–distilled water, 1 : 2 : 100.
6. Blot with filter paper.
7. Decolorize slides lying flat, face up, with 1- to 2-ml portions of equal volumes of aniline and xylene, pouring off several times until no more color comes out into the mixture.
8. Clear in three changes of xylene.
9. Mount in Permount or Canada balsam.

This is nearly the original Weigert method. In step 1 he (Ehrlich's "Encyklopädie") stained with carmine; for step 3 he used aniline methyl violet (10% alcoholic methyl violet, 12 ml; fresh aniline water, filtered, 100 ml). In step 7 when preparing class sets, we have used a succession of three Coplin jars of the aniline-xylene mixture, allowing sections to stand 2 to 3 min in each and following with three or four changes of xylene similarly applied. RESULTS: Nuclei are red; fibrin is violet; gram-positive organisms are blue-black. If a fuchsin nuclear stain has been used and the fixation was with Helly or Orth fluid, red corpuscles may remain a deep red, and nuclei are violet to red. Collagen may retain some violet, but in a lighter shade than fibrin. Keratohyalin granules and hair shafts often retain violet, just as they may retain red in Ziehl-Neelsen stains. Young intraepithelial rabbit coccidia (*Eimeria*) contain numerous blue-black granules between capsule and nucleus by this technic. In older, free forms these granules are more apt to be gram-negative, and with the acetone technic they are gram-negative in all stages. Mucus also may remain violet with the carbol fuchsin 30-min prestain at step 1 but is gram-negative with the acetone technic. Sometimes glycogen is demonstrated by Gram-Weigert methods, as noted by Lubarsch (Ehrlich's "Encyklopädie," 1903). Hypophyseal beta granules are violet.

NOTES ON MODIFICATION OF WEIGERT FIBRIN STAIN

1. Preiodization of mercury-fixed tissue may be omitted; step 5 removes mercury deposits.
2. The use of a preliminary potassium permanganate oxidation–oxalic acid reduction, sometimes prescribed for material fixed in dichromate fixatives, appears to be quite unnecessary.
3. Weigert elastin stain may be interposed before step 1 (Lillie, 1948, p. 209).
4. A periodic acid Schiff sequence may also be interposed at this point.
5. We have sometimes used Weigert's acid iron chloride–hematoxylin for step 1. Violet-black effects may appear and confuse the picture.
6. MacCallum's variant seems unnecessarily complicated; see *J.A.M.A.*, **72**: 193, 1919.
7. Use of the Feulgen reaction at step 1 gives red-purple nuclei, but its effect on bacterial ribonucleoprotein needs further evaluation.
8. Various crystal violet, methyl violet, or gentian violet mixtures have been prescribed in step 3. Timing is varied from 1 to 10 min. Aniline and phenol formulas have been urged for bacteria. For these the Hucker-Conn ammonium oxalate formula seems adequate, and aqueous solutions suffice for fibrin.

Both fibrin and fibrinoid give positive indole reactions according to Pearse; Glenner and Lillie (*J. Histochem. Cytochem.*, **5**: 279, 1957) recorded both positive and negative reactions of fibrin in various species. The protein fibrinogen presents 3.3% tryptophan.

Moe and Abildgaard (*Acta Pathol. Microbiol. Scand.*, **76**: 61, 1969) tested the efficacy of Mallory's phosphotungstic acid–hematoxylin, Masson's trichrome stain, and an immunofluorescence stain for fibrin. Fibrin clots were prepared in vitro. Phosphotungstic acid–hematoxylin stained only those clots to which either calcium chloride or serum colloids had been added. Fibrin stained with the Masson trichrome only if either calcium chloride was added to the clot or if the clot was agitated during clotting. Fibrinogen precipitates stained with phosphotungstic acid–hematoxylin or the trichrome if calcium chloride was added. Immunofluorescence methods revealed specific stains for both fibrinogen and fibrin.

Fibrinoid

Fibrinoid is a homogeneous, refractile, oxyphilic substance occurring in degenerating connective tissue, in term placentas, in rheumatoid nodules, in Aschoff bodies, and in pulmonary alveoli in some prolonged pneumonitides.

With toluidine blue it is often metachromatic, sometimes not; and in some instances this metachromasia is abolished by predigestion with hyaluronidase, in others not. With phosphomolybdic or phosphotungstic acid–aniline blue stains of the Mallory type, it is described sometimes as staining red with the counterstain, sometimes as cyanophil. With phosphotungstic acid–hematoxylin it colors partly blue, partly orange to yellow. Altshuler and Angevine record purple staining with crystal violet. Its uptake of acid and of basic dyes at various pH levels is similar to that of fibrin. According to Altshuler and Angevine, fibrinoid contains considerable amounts of arginine. Pearse records a positive rosindole reaction (tryptophan).

Regarding the Weigert fibrin stain, the literature is vague. Klinge (*Virchows Arch. Pathol. Anat.*, **278**: 486, 1930) spoke of most of the fibrinoid material as giving a positive fibrin reaction. The fibrin reaction in common use at that time was the Weigert fibrin stain (Schmorl; Romeis).

Fibrinoid is conceived of as a precipitation of acid mucopolysaccharide with a basic protein (Altshuler and Angevine, *Am. J. Pathol.*, **25**: 1061, 1949).

Like Altshuler and Angevine, Glynn and Loewi (*J. Pathol. Bacteriol.*, **64**: 329, 1952) report a strongly positive periodic acid Schiff reaction. This reaction is abolished by pectinase digestion and resists tryptic digestion but is weakened by tryptic digestion after denaturation for 18 h in 36 % (6 *M*) urea.

They also report blackening and demonstration of a fibrillar structure by Gomori's reticulum technic. Fibrin did not react by this method. This reaction of fibrinoid was also abolished by pectinase digestion.

For tryptic digestion they used a 0.1 % solution of Armour's crystalline trypsin in pH 8 phosphate buffer for 3 h at 37°C.

In the pectinase digestion a crude preparation from onions was used at 0.4 % in pH 4 acetate buffer, for 3 h at 37°C.

Enzymatic Digestion Tests

Before proceeding to individual tests it seems pertinent to offer the following commentary on this method of identification of tissue components:

1. Since even *purified enzymes* or *crystalline enzymes* are seldom free of all other enzymatic activities, digestion of a given tissue component by a given enzyme preparation does not necessarily identify that tissue component as consisting of the specific substrate of the enzyme employed.
2. When a given enzyme preparation is known to digest a given substrate under specified conditions of fixation and preparation, failure of that enzyme preparation to digest a tissue component under the same specified conditions may be taken to definitely exclude identification of that tissue component with the stated given substrate.
3. For even presumptive identification of a tissue component by reason of its digestion by a given enzyme, it must be shown that the tissue component in question is not destroyed by the enzyme solvent alone, or by the enzyme after inactivation by physical or chemical means, or in the presence of known specific inhibitors of the enzyme action in question.

TRYPTIC DIGESTION

We have used effectively a "1–300" trypsin of the Nutritional Biochemical Company at 0.1 % concentration in 0.01 *M* phosphate buffer of pH 7.6, with or without 0.4 % sodium chloride and 0.1 % sodium fluoride. This trypsin had also quite a strong glycogenolytic activity. Purer crystalline products are now available. On digestion at 37°C, nuclear and cytoplasmic staining is destroyed in as little as 1 h in material fixed in 80 % alcohol, but it persists much longer in material fixed in aqueous formalin for long periods or in chromate fixatives. Collagen and reticulum are quite resistant; basement membranes are somewhat less so (*Lab. Invest.*, **1**: 30, 1952). Sarcolemma swells and stains poorly, displaying prominently the fine connective tissue fibrils traversing its outer surface.

Controls of the same buffer solvent without the enzyme should be employed. After alcohol fixation, distilled water removes ribonucleic acid at 60°C.

PEPTIC DIGESTION

We have used Difco pepsin at 1 : 1000 concentration in 0.1 *N* hydrochloric acid, digesting for varying periods at 37°C. For controls use 0.1 *N* hydrochloric acid without enzyme.

Cytoplasmic staining is abolished in 2- to 4-h digestion after cold alcohol fixation, persisting longer in material fixed in aqueous formalin. Periodic acid Schiff positive basement membranes are digested much more rapidly than the collagen or reticulum which is stained blue by the allochrome method. Nuclei become Schiff-positive in 4 to 6 h and then swell and vacuolate and lose their sharp definition in 18 h. Ocular lens capsule and Descemet's membrane are quite resistant to peptic digestion.

Chromation of tissue almost completely inhibits peptic digestion.

TECHNICS FOR TRYPTIC AND PEPTIC DIGESTION

1. Fix in 80% alcohol, Carnoy, or the like. Alcoholic formalin for 18 h (not more) may be used. Avoid use of prolonged fixation in aqueous formalin and especially in dichromate mixtures. Embed and section in paraffin, not celloidin.
2. Sections may be digested without deparaffinizing according to the suggestion of Goetsch, Reynolds, and Bunting (*Proc. Soc. Exp. Biol. Med.*, **80:** 71, 1952) regarding amylase digestion. We have not tested this. Generally, deparaffinize and hydrate as usual.
3. Immerse control sections in the same solvent as is used for the enzyme: a 0.01 *M* phosphate buffer of pH 7.6 containing 0.1% sodium fluoride and 0.4% sodium chloride for trypsin, and 0.1 *N* hydrochloric acid for pepsin. Leave controls in until the last test sections are taken out. Use 1 : 1000 solutions of trypsin and pepsin in the solvents described above. Digest for 30 min, 1, 2, 4, 6 or 7, and 16 to 18 h, removing test slides at the various times.
4. Wash in water.
5. Counterstain to demonstrate the particular tissue elements under study.
6. Wash and mount in glycerol gelatin; or dehydrate, clear, and mount in synthetic resin.

Compare digested preparations with each other, with undigested controls, and with solvent controls.

We have used 0.1% chymotrypsin and 0.1% papain, both in pH 7.6 phosphate buffer, 1 h at 37°C, in essentially similar procedures (*J. Histochem. Cytochem.*, **7:** 204, 1959).

Collagenase Digestion

Collagenases are specific enzymes that act on native collagen. Unless the enzyme can effect hydrolytic cleavage of the coiled portion of the native collagen molecule, it does not qualify as a specific collagenase.

Several strains of *Clostridium histolyticum* and *Clostridium perfringens* produce specific collagenases [Maschmann, *Biochem. Z.*, **295:** 1, 1937; Bidwell et al., *Biochem. J.*, **42:** 140, 1948; Nagai et al., *J. Biochem.* (*Tokyo*), **51:** 382, 1962]. Certain strains of *Bacteroides melaninogenicus* (Gibbons and MacDonald, *J. Bacteriol.*, **81:** 614, 1961), *Streptomyces madurae* (Rippon, *Biochim. Biophys. Acta*, **159:** 147, 1968), and *Trichophyton schönleini* (Rippon and Lorincz, *J. Invest. Dermatol.*, **43:** 483, 1964) are known to produce specific collagenase.

Vertebrate collagenase was first demonstrated in tadpole tail tissues by Gross and Lapiere (*Proc. Natl. Acad. Sci. U.S.A.*, **48:** 1014, 1962) and by Lapiere and Gross ("Mechanisms of Hard Tissue Destruction," Publication 75, p. 663, AAAS, Washington, D.C., 1963). Since that time collagenase has been shown to be produced by gingivae and skin (both epithelial and connective tissues), synovia, bone, polymorphonuclear leukocytes of peripheral blood, uterus, and macrophages [see reviews by Fullmer in W. Fishman (ed.), "Metabolic Conjugation and Metabolic Hydrolysis," p. 301, Academic, 1970; and by Seifter and Harper in P. Boyer (ed.), "The Enzymes," vol. III, p. 649, Academic, 1971]. Several of the vertebrate collagenases act once only on the collagen molecule, producing one piece three-fourths of the length from the NH_2-terminal end of the molecule, the other one-fourth near the COOH-terminal end. Collagenases derived from microbial sources cleave the molecule in many places. Pure collagenases from vertebrate sources are not available. Vertebrate collagenases become progressively more unstable with increasing purity. Fairly pure collagenase from clostridial sources is now available commercially.

TECHNIC.   Fresh frozen and paraffin-embedded sections are acceptable; however, formalin fixation must be avoided. Collagen fixed thoroughly with formalin resists digestion with collagenase.

Sections are brought to water and incubated with the purest preparations available at a concentration of 100 to 500 $\mu g/ml$ 0.05 M tris–HCl buffer of pH 7.5 containing 1 mM $CaCl_2$ for 15 min to several hours at 37°C depending upon the nature of the collageneous tissue. The degree of collagenolytic activity varies widely from batch to batch. Likewise, the purity of the enzyme preparations varies widely. Substantial nonspecific activity against casein is not uncommon.

Elastase Digestion

Wälchli (*J. Prakt. Chem.*, **17**: 71, 1878) noted that ox pancreas digested ligamentum nuchae. Kühne (*Untersuch. Physiol. Inst. Univ. Heidelberg*, **1**: 219, 1878) observed that impure preparations of trypsin dissolved elastic tissue. Mall (*Rep. Johns Hopkins Hosp.*, **1**: 171, 1896) detected dissolution of elastic fibers by pancreatin whereas collagen fibers persisted. Balo and Banga (*Schweiz. Z. Pathol. Bacteriol.*, **12**: 350, 1949) appear to be the first to have distinguished a specific enzyme elastase partially purified from bovine pancreas. A purified enzyme appears to have been obtained by Lewis et al. (*J. Biol. Chem.*, **222**: 705, 1956) from porcine pancreas. Many elastases from plants and microorganisms have been reported (see Mandl, *Adv. Enzymol.*, **23**: 163, 1961; *Methods Enzymol.*, **5**: 665, 1962). None of the vertebrate elastases have been studied in detail except that from porcine pancreas.

Recrystallized elastase can be obtained as a single band with a sedimentation coefficient $S_{20, w}^{\circ}$ of 2.58 ± 0.02S as determined by Lewis et al. [*J. Biol. Chem.*, **222**: 705, 1956; Hartley and Shotton, in P. Boyer (ed.), "The Enzymes," vol. III, p. 323, Academic, 1971]. Common contaminants of impure elastase preparations are other pancreatic endopeptidases such as trypsin and chymotrypsin. These enzymes will slowly hydrolyze the synthetic substrates of elastase. Elastase is unable to hydrolyze to any significant extent N-benzoyl-L-arginine ethyl ester (BAEE) or N-acetyl-L-tyrosine ethyl ester (ATEE), which are specific substrates for trypsin and chymotrypsin, respectively. For this reason, assay for activity against these substrates is useful to determine contaminating endopeptidase activity.

The existence of a proelastase (a precursor of elastase) was first shown by Grant and Robbins (*Arch. Biochem. Biophys.*, **66**: 396, 1957; *Proc. Soc. Exp. Biol. Med.*, **90**: 264. 1955). Activation of proelastase to elastase is accomplished by tryptic cleavage of a single peptide bond near the NH_2-terminal region of the zymogen proelastase, permitting the enzyme to assume an active conformation.

Elastase is readily soluble in water and dilute salt solutions up to 50 mg/ml from pH 4 to 10.5. Elastase solutions are stable for prolonged periods at an acid range at 2°C. Optimal activity of elastase is at pH 8.8.

Lewis et al. (*J. Biol. Chem.*, **222**: 705, 1956) noted that elastase will digest other proteins. On unoxidized chains of insulin, preferential activity of elastase was noted for peptide bonds on the COOH-terminal side of Ala 14, Val 18, and Gly 23 in the beta chain, and Ala 8 and Ser 12 in the alpha chain (Sampath et al., *Biochem. J.*, **114**: 11, 1969). Klee (*J. Biol. Chem.*, **240**: 2900, 1965) noted that elastase splits the Ser—Ala bond in ribonuclease. Hartley and Shotton (in P. Boyer (ed.), "The Enzymes," vol. III, p. 323, Academic, 1971) view the physiologic activity as a protease with broad specificity that may complement the activity of trypsin and chymotrypsin.

TECHNIC

1. Fresh frozen sections or paraffin-embedded sections fixed in formalin or Carnoy are deparaffinized and brought to water.
2. Sections are incubated at 37°C for a variable period in purified elastase (5 mg/50 ml 0.1 *M* borate buffer of pH 9). The length of time required for digestion of elastic tissue depends upon the size and density of the elastic tissue. Dermal elastic fibers require 1 to 4 h. The degree of digestion of elastic fibers can be assessed after staining with one of the elastic tissue stains. Do not collodionize sections.

As noted above, elastase has a broad spectrum of proteolytic activity. Fullmer (*J. Histochem. Cytochem.*, **8**: 113, 1960) noted removal of mucopolysaccharides, muscle, and epithelial cells in sections digested with commercial preparations of elastase. Elastase digestion of many structures including elastic tissue was greatly increased in sections oxidized with peracetic acid for 30 min at 23°C. Peracetic acid–oxidized human and rat hair was readily digested by elastase.

Fullmer et al. (*J. Dent. Res.*, **48**: 646, 1969) and Lazarus and Fullmer (*J. Invest. Dermatol.*, **52**: 545, 1969) employed elastase to separate epithelial tissues from connective tissues from gingivae and skin respectively. The basement membrane appears to be selectively digested. The technic involves slicing fresh gingiva or skin into very thin strips (less than 1 mm, if possible). After washing thoroughly in Tyrode's solution containing 100 μg streptomycin, 100 units penicillin, and 100 units Mycostatin per ml, tissue bits are incubated in a change of the same solution containing 2 times crystallized elastase (Nutritional Biochemical Corp., Cleveland, Ohio), 1 mg/ml for 1 to 3 h at 37°C. The pH is adjusted to pH 9 with 1 *N* NaOH. The end point of incubation is determined by the ease of removal of epithelium from underlying connective tissue as observed under a dissecting microscope.

Antibody Localization

Coons, Leduc, and Kaplan (*J. Exp. Med.*, **93**: 173, 1951) in their study of localization of fluorescent antibodies, first cut hard-frozen sections, then attached them to gelatin-coated slides in the cryostat, removed them from the cryostat, and melted the section by pressing a finger against the back of the slide under the section. Sections were then dried onto the slides in an air stream at room temperature for an hour and were stored at 4°C overnight. They were then fixed to immobilize the egg albumen and γ-globulin by immersion for 30 min at 37°C in preheated 95% alcohol. For bovine serum albumin, acetone was used at room temperature for 15 min. After fixation the slides were dried in a vertical position in the 37°C incubator for 30 min.

Sections were then reacted with a drop of fluorescein-treated antibody for 30 min at 37°C, washed in buffered saline solution for 10 min, and mounted in buffered glycerol. They were then examined under the fluorescence microscope. After photographing, the cover glass was floated off, and the section was fixed in 10% formalin for 10 min and counterstained with hematoxylin and eosin. The same area was then rephotographed.

Coons had to synthesize not only the isocyanate but also the fluoresceinamine from which it was derived. Moreover, the isocyanate was unstable, though it could be kept for limited periods in acetone solution at very low temperatures. Hence, the method is practically restricted to laboratories with facilities for organic synthesis. Silverstein's (*J. Histochem. Cytochem.*, **5**: 94, 1957) technic with rhodamine B isocyanate was subject to the same difficulties, requiring synthesis of a nitrorhodamine B from 4-nitrophthalic anhydride and *m*-diethylaminophenol reduction to the amine, and conversion to the isocyanate with phosgene.

The introduction of the more stable isothiocyanates, which are said to yield better fluorescent antibodies and other proteins and polysaccharides below, and which have recently been made available commercially, and the introduction of sulforhodamine B, which is readily converted to a sulfonyl chloride to condense with protein amines by sulfamide condensation, should serve to extend the uses of these procedures.

Production of specific antiserums is beyond the scope of the present work. Texts on immunity and original papers should be consulted.

Riggs et al. (*Am. J. Pathol.*, **34**: 1081, 1958) gave detailed directions for the synthesis of fluorescein isothiocyanate and rhodamine B isothiocyanate, proceeding from isomer I of Coons' (*J. Exp. Med.*, **91**: 1, 1950) nitrofluorescein diacetate and from 4-nitrophthalic acid and *N,N*-diethylaminophenol, reducing to amines and converting the amines to isothiocyanates with thiophosgene. It is probable that other fluorescent amines can be similarly converted to isothiocyanates, but since fluorescein and rhodamine B isothiocyanates are reasonably stable and have been made commercially available, it seems probable that they will replace the unstable isocyanates previously used. Fluorescein isothiocyanate has been placed on a certification basis by the Biological Stain Commission, and up to October 1973 nine batches from seven different manufacturers have been certified as giving satisfactory performance.

Chen (*Arch. Biochem. Biophys.*, **133**: 263, 1969) conducted a quantitative study of the fluorescence of a number of dyes [fluorescein isothiocyanate, dansyl chloride, rhodamine B isothiocyanate, anthracene-2-isocyanate, sulforhodamine B sulfonyl chloride (from ICI), and 4-acetamido-4'-isothiocyanatostilbene-2,2'-disulfonic acid] conjugated to γ-globulins as a function of labeling. Quantum yields, excitation and emission spectra, fluorescence decay times, and polarization values were measured. These data may be helpful in making a choice of dye for various applications and provide a quantitative basis for observed fluorescent conjugates.

CONJUGATION PROCEDURE FOR ISOTHIOCYANATES.   Riggs et al. direct as follows: Mix 10 ml 0.85% sodium chloride, 3 ml 0.5 *M* carbonate bicarbonate buffer adjusted to pH 9, and 2 ml acetone. Chill the mixture by immersing the Erlenmeyer flask in acetone dry-ice mixture until ice crystals form. Then add, with stirring, 10 ml diluted globulin fraction of known protein content. Rechill the mixture to the point of ice-crystal formation. Dissolve 50 μg fluorescein or rhodamine B isothiocyanate per mg protein in 1.5 ml acetone and add gradually, with stirring, to the cold protein solution. Continue stirring for 18 h in a cold room at 4°C.

Riggs et al. then removed excess dye by dialysis (but see Lipp, under Separation of Labeled Protein from Dye, below) filtered through a millipore filter, and froze 2-ml quantities in ampules, which were stored in the dry-ice chest until used.

The sulfonyl chloride of sulforhodamine B (C.I. 45100, Acid Rhodamine B NAC, Lissamine Rhodamine B 200 ICI) has also been used for the tagging of proteins, yielding a brilliant orange fluorescence (Chadwick et al., *Lancet*, **1**: 412, 1958). Chadwick et al. direct as follows for preparation of the sulfonyl chloride: Place 1 g sulforhodamine B and 2 g phosphorus pentachloride in a mortar and grind together for 5 min in the chemical hood. Then add 10 ml anhydrous acetone and stir occasionally for 5 min. Then filter the acetone solution of sulforhodamine B sulfonyl chloride. Successful conjugates are obtained up to 18 h from preparation. Conjugates are prepared at 0 to 2°C.

To each 1 ml serum or protein solution of similar concentration add 1 ml 0.85% sodium chloride solution and 1 ml 0.5 *M* carbonate bicarbonate buffer of pH 9. Mix and add 1 ml acetone solution of sulforhodamine B–sulfonyl chloride drop by drop with constant stirring. Continue stirring in the cold for 12 to 18 h. Dialyze against 0.85% sodium chloride in the cold for 6 to 8 days, changing fluid at regular intervals, until no more color appears in the dialysate. The final volume of the conjugate is about three times that of the original mixture; it may be concentrated in an air stream, with or without partial vacuum. The serum conjugates may be stored at 15 to 20°C and appear to be reasonably stable.

SEPARATION OF LABELED PROTEIN FROM DYE.   Lipp (*J. Histochem. Cytochem.*, **9**: 458, 1961) recommends gel filtration for the freeing from uncombined dye of labeled protein. A gel is formed by wetting a special dextran, Sephadex G-25 (Pharmacia, Uppsala, Sweden), with the solvent to be used for elution and packing it to form a column 20 cm high in an ultraviolet-transmitting glass tube 4 to 5 cm in inside diameter. The column is then filled with the buffered (0.85%?) saline solution, and drainage is started. The dye and labeled protein mixture are added when the fluid level descends to the top of the column, and when the fluorescent mixture has entirely entered the column, more eluent is added. Owing to the restricted pore size of the dextran particles, small molecules enter them and are much retarded in passing, while the tagged protein passes by without entering the particles. As the fluorescent protein descends, a nonfluorescent zone appears behind the protein band, separating it from the adsorbed dye. After the fluorescent protein has been collected, the column may be washed free of adsorbed dye with more eluent. Passage of tagged protein through such a column requires 45 to 60 min, in place of the several days of dialysis needed in previous technics.

RECONCENTRATION OF PROTEIN SOLUTION.   Lipp (ibid.) places dry Carbowax 4000 or similar high-polymer polyethylene glycol in a dialysis tube, which is then immersed in a slightly wider column containing the protein solution. The lower end of the dialysis tube is sealed; the upper end opens into a funnel. Salts and water diffuse into the dialysis tube and dissolve the Carbowax, forming a solution of high osmotic pressure which cannot diffuse back through the membrane, thereby removing considerable amounts of water from the protein solution.

See also Curtain (*J. Histochem. Cytochem.*, **9**: 484, 1961) for closely similar procedures.

ALDEHYDE AND CARBONYL REACTIONS

Schiff's Sulfurous Acid Leucofuchsin Reagent

This reagent for carbonyl groups, especially aldehydes, has shown itself extremely well adapted to the demonstration of these groups in tissue sections. It is used in Feulgen's nucleal reaction for the aldehyde group of deoxyribose when freed from purine and pyrimidine bases (nucleotide base plus sugar plus acid) by acid hydrolysis, for the aldehyde unmasked in acetal phosphatides by the action of mercuric chloride (the plasmal reaction), for aldehydes arising from the peroxidation of unsaturated fats, for the demonstration of aldehydes formed by Criegee cleavage reactions of 1,2-glycols and 1,2-aminoalcohols by periodic and chromic acids, potassium permanganate, lead tetraacetate, and sodium bismuthate, for the demonstration of aldehydes resulting from the peracetic and performic acid oxidation of carbon-carbon double bonds, in the Oster amine oxidase method, the Liang nerve ending technic, and for the native lysinal aldehyde arising from oxidation of lysine in the synthesis of the desmosines in developing elastic tissue. This last aldehyde was clearly described and separated from nucleal and plasmal by Feulgen and Voit (*Pfluegers Arch.*, **206**: 389, 1924). It is seen in arterial elastica of young animals and children, absent in the ligmentum nuchae of an old horse, but present in this structure in young oxen (Pizzolato and Lillie, *Arch. Pathol.*, **88**: 581, 1969).

Fuchsin in 0.5 to 1% solution may be reduced with 6 to 8% SO_2 water (sulfurous acid), with 0.5 to 2% sodium bisulfite, metabisulfite, or the corresponding potassium salts ($NaHSO_3$, $Na_2S_2O_5$, $KHSO_3$, $K_2S_2O_5$), with 2.25% sodium thiosulfate ($Na_2S_2O_3 \cdot 5H_2O$), with 0.5% sodium dithionite ($Na_2S_2O_4$) (Alexander, McCarty, and Alexander-Jackson, *Science*, **111**: 13, 1950), or with 1 ml thionyl chloride ($SOCl_2$) per g fuchsin (Barger and DeLamater, *Science*, **103**: 121, 1948). The last three reducing agents have been little used in histology.

The directions of Feulgen called for dissolving the fuchsin in boiling-hot water, filtering while cooling, adding the sulfite, and storing in the dark for a day or two before use. This practice was followed by nearly all writers up to 1950, when we evolved a satisfactory, abbreviated cold procedure, which will be given here.

Coleman (*Stain Technol.*, **13**: 123, 1938) introduced the procedure of completing the bleaching of the reagent with activated charcoal, which removed a variable amount of residual coloring matter not taken out by the sulfite. Barger and DeLamater state that the charcoal treatment may either precede or follow the sulfite bleaching; Longley (*Stain Technol.*, **27**: 161, 1952) showed that charcoal treatment before bleaching removed a good deal of fuchsin as well as the so-called yellow component.

The schema (Fig. 8-1) of the formation and aldehyde reaction of Schiff's sulfurous acid–fuchsin reagent is after Wieland and Scheuing (*Ber. Dtsch. Chem. Ges.*, **54**: 2527, 1921), modified (*Int. Rev. Cytol.*, **10**: 1, 1960) and condensed somewhat from Kasten. In this scheme pararosanilin (I) reacts with sulfurous acid to form the leukosulfonic acid (II), and its free amino groups take up one or two equivalents of sulfur dioxide (SO_2) to form the leuko-sulfinic acid derivatives III and IV. Since, with the usual excess of SO_2 used, product IV is the more probable, its reactions to form monoaldehyde and dialdehyde derivatives (VII and V) are followed. These derivatives are highly unstable and promptly hydrolyze respectively to the colored products VII and VI. In view of the great excess of Schiff reagent over aldehyde in the usual histochemical reaction products VII and VIII probably represent the histochemical products.

Amounts of fuchsin used also vary considerably: the 1 g in 220 ml cited by Lee (9th ed., 1928) from Feulgen (Feulgen and Voit, *Pfluegers Arch.*, **206**: 389, 1924), and others; the 0.25% of Barger and DeLamater (*Science*, **103**: 121, 1948); the 1 g in 120 ml of Bauer (*Z. Mikrosk. Anat. Forsch.*, **33**: 143, 1933), C. Bensley (*Stain Technol.*, **14**: 47, 1939), Cowdry, and others; the 1 g/100 ml of our method, presented here; the 2 g/100 ml which we have occasionally used.

PREPARATION OF SCHIFF REAGENT: TRADITIONAL METHOD

1. Bring 100 ml distilled water to boiling, remove from the flame, and at once dissolve 1 g basic fuchsin (pararosanilin, rosanilin, or new fuchsin).
2. When the solution cools to 60°C, filter and add to the filtrate 2 g sodium or potassium bisulfite or metabisulfite ($NaHSO_3$, $KHSO_3$, $Na_2S_2O_5$, $K_2S_2O_5$) and 20 ml 1 *N* hydrochloric acid.

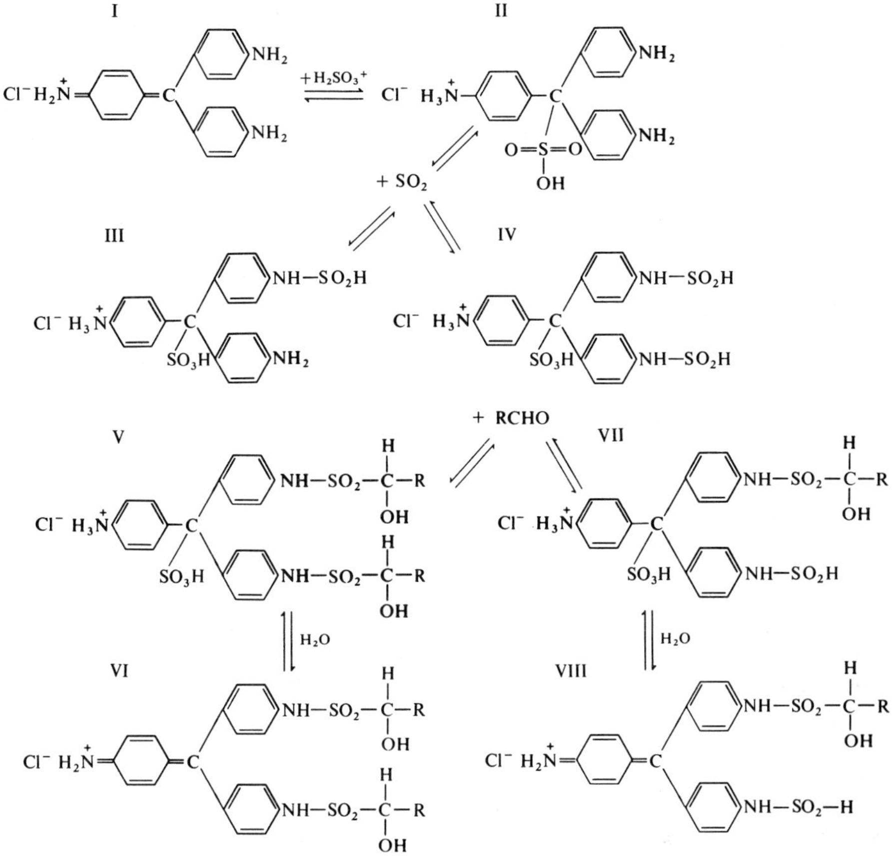

Fig. 8-1 Schema of formation of Schiff reagent from pararosanilin and its reaction with aldehyde to form colored products.

3. Stopper and store solution in the dark at room temperature for 18 to 24 h.
4. Add 300 mg activated charcoal, shake vigorously 1 min, and filter.
5. Store at 0 to 5°C. The solution should be a clear light yellow. Discard it when a pink color develops. We have kept this solution for several weeks.

If thionyl chloride is used in place of sulfite and acid, use 1 ml/g fuchsin. The pH of the final reagent is about 1.25 with this variant.

PREPARATION OF REAGENT BY "COLD SCHIFF" PROCEDURE

1. Weigh out 1 g fuchsin and 1.9 g sodium metabisulfite ($Na_2S_2O_5$). Dissolve in 100 ml 0.15 N hydrochloric acid. For 2 g fuchsin use 0.25 N hydrochloric acid and double the $Na_2S_2O_5$.
2. Shake the solution at intervals or on a mechanical shaker for 2 h. The solution is now clear and yellow to light brown.
3. Add 500 mg *fresh* activated charcoal; shake 1 to 2 min.
4. Filter into graduated cylinder, washing the residue with a little distilled water to restore the original 100-ml volume.

The solution should be water white. If it is yellow, a fresh lot of activated charcoal should be obtained, and the charcoal decolorization should be repeated. Store at 0 to 5°C.

Such solutions have been so stored for 2 months, still remaining colorless. Stronger solutions may form a white precipitate on refrigeration. This precipitate is perhaps less apt to form at lower pH levels. It does not dissolve on warming gently. The pH of this fluid is about 2.2. For dilute Schiff reagent, add 1.9% (0.1 M) sodium metabisulfite in 0.15 N hydrochloric acid to the desired dilution.

The *dithionite (hydrosulfite) solution of Alexander et al.* (*Science*, **111**: 13, 1950) is prepared by dissolving 1 g fuchsin in 200 ml boiling-hot water, filtering while hot, and adding at once 1 g sodium dithionite ($Na_2S_2O_4$) and 500 mg activated charcoal. Shake thoroughly and filter. The finished product is light amber and of unspecified pH. It does not seem to have been used much.

Mowry (*Ann. N.Y. Acad. Sci.*, **106**: 402, 1963) in his combined Alcian blue and Hale–periodic acid Schiff technics prescribes a weaker variant of the cold Schiff reagent, which is also apparently less acid than the above:

He used basic fuchsin, 0.5 g; sodium sulfite (Na_2SO_3), 5 g; concentrated hydrochloric acid, 3 ml; and distilled water to make 200 ml. The amount of fuchsin is 25% that above and the sulfite content is still approximately 0.2 M, but owing to the use of Na_2SO_3, the excess of anions over Na and NH_2 is reduced by about 0.1 N (0.262 versus 0.165 N). Weaker Schiff reagents of this type are quite satisfactory for the demonstration of reactive epithelial mucins by periodic acid methods. For such objects as mucosal reticulum and smooth muscle stroma, the stronger Schiff reagent appears to be necessary. For Feulgen reactions the usual 0.5 g fuchsin per 100 ml seems adequate and should be adhered to for photometric studies. But we see no virtue in boiling the water and filtering it hot before adding the sulfite.

Longley's (*Stain Technol.*, **27**: 161, 1952) conclusions that 0.25% fuchsin is often too weak, that 0.5% is usually adequate, that precipitation of the leuco substance is more apt to occur above pH 3, and that a lower pH than the 2.2 achieved in the 1% cold Schiff is of no special value are in general agreement with the foregoing and with our present position. He further noted that the presence of an excess of SO_2 was not deleterious. Van Duijn (*J. Histochem. Cytochem.*, **4**: 55, 1956) proposed a sulfite leucothionin to give a contrasting color (blue) to Schiff reagent (red) in a combined Feulgen-McManus procedure.

Dissolve 0.5 g thionin in 250 ml distilled water, boil 5 min, cool, restore volume, add 250 ml tertiary butanol (mp 25°C), transfer to stoppered bottle without filtration, add 75 ml 1 N hydrochloric acid, and at once 5 g $Na_2S_2O_5$. Stopper, shake, let stand 24 h at 25°C and then 48 h at 4°C. Filter out a portion of the stock solution for use. After using it, return the solution to the stock bottle and keep it stored at 4°C. The reagent keeps about 6 weeks from initial mixing and should not be used until aged as specified above, i.e., after 1 day at 25°C and 2 days at 4°C.

Reaction time is longer than with Schiff reagent, Van Duijn recommending 15 min for routine work and an hour for quantitative work. The reaction color is deep blue for Feulgen and for McManus reactions. The blue Feulgen color remains stable with a subsequent periodic acid Schiff reaction.

In the Feulgen reaction Dutt (*J. Histochem. Cytochem.*, **18**: 221, 1970) achieves a green nuclear stain by using two Schiff reagents, the one from acridine yellow, the other from brilliant cresyl blue, either in an equal volume mixture or separately in either sequence.

Fluorochrome Schiff reagents have been made by Kasten [*Nature (Lond.)*, **184**: 1797, 1959] by treatment with gaseous SO_2 or thionyl chloride from acridine yellow G, C.I. 46025; acriflavine, C.I. 46000; coriphosphine O, C.I. 46020; flavophosphine N, C.I. 46065; neutral red, C.I. 50040; phenosafranin, C.I. 50200; safranin O, C.I. 50240; rhodamine 3G, C.I. 45210; phosphine, C.I. 46045; and, last and perhaps best, auramine O, C.I. 41000.

These dyes, together with the basic and acid fuchsins and thionin, have one chemical structural group in common: all possess at least one primary arylamine group. Feigl, in earlier editions of his "Spot Tests," spoke of a Schiff reagent made with malachite green which possesses only fully alkylated amino groups. This dye has failed to yield a Schiff reagent in our hands, and we are led to wonder whether Feigl had a partly dealkylated sample. The primary arylamine group appears necessary to form the sulfite complex which serves as a chromogenic aldehyde reagent.

The auramine O Schiff reagent tends to precipitate and is stable for only a few hours, but it gives a very striking intense greenish yellow fluorescence to nuclei in the Feulgen reaction.

Saturate an 0.5% solution of auramine O with sulfur dioxide just before using. Expose hydrolyzed sections (6 N HCl, 6 min, 25°C) to the SO_2 auramine O for 45 min, and wash as usual in 0.1 N bisulfite and 10 min in running water.

The other dyes are similarly prepared by treating solutions with SO_2 or thionyl chloride, using 0.5 to 0.25% dye for Feulgen reagents and 0.1% solutions in periodic acid technics.

Shanklin and Azzam (*Stain Technol.*, **40:** 117, 1965) report that the usual deep red of basement membranes, smooth muscle stroma, striated borders of intestinal epithelia, and brush borders of renal tubule epithelium when stained by the periodic acid Schiff reaction is replaced by blue when observed by phase contrast microscopy. Only a part of the amylase-digestible periodic acid Schiff–positive material in liver cells appears blue in phase microscopy; the rest remains red. Intestinal goblet cell mucus, figured in deep red under bright field, became only a lighter red in phase microscopy. Only part of the renal glomerular stroma appeared blue, the rest red under phase contrast. More densely Feulgen-positive nuclei appeared deep blue with phase microscopy, and the Ninhydrin Schiff reaction also showed areas of blue under phase microscopy. But the pink to red stains in hematoxylin-eosin preparations do not appear in blue with phase microscopy.

Hydrazine Reactions

Albert's plasmalogen reaction with dinitrophenylhydrazine is described in Chap. 13, and Seligman's method with 2-hydroxy-3-naphthoic acid hydrazide, followed by azo coupling with fast blue B and other stable diazonium salts, has also been used as an aldehyde demonstration reagent after periodic acid oxidation, but it seems to offer no striking advantage over the Schiff reagent. Further, the diazonium salt couples to a variable extent with protein-bound aromatic amino acids, usually yielding yellow or orange.

This last objection applies also to our new "black periodic" technic (*Stain Technol.*, **36:** 361, 1961), but here the contrast often afforded by black staining of fine reticulum and basement membranes to light red muscle and gland cell cytoplasm enhances the value of the periodic acid method for the study of these structures.

Schiff's Base, or Diphenamine, Reactions

Arylamines condense readily with tissue aldehydes to form Schiff's bases. Because of the large excess of arylamine present, these are of Eibner's (*Ber. Dtsch. Chem. Ges.*, **30:** 1444, 1897) diphenamine type, as Lillie demonstrated (*J. Histochem. Cytochem.*, **10:** 303, 1962), thus: $RCHO + 2ArNH_2 \rightarrow RCH(NHAr)_2$. These Schiff bases, or, perhaps better, to avoid confusion with the better-known azomethines (RCH = NAr), diphenamine bases are quite resistant to acids and indeed are best formed from acid solutions or, even better, glacial acetic acid.

This diphenamine reaction may well be responsible for the success of Arzac's and Kligman's periodic acid–basic fuchsin methods for the demonstration of glycogen, fungi, etc.; it is definitely the reaction in question in our black periodic and black Bauer methods (*Stain Technol.*, **36**: 361, 1961) and in Spicer's rapid periodic acid–diamine procedure for mucopolysaccharides (*J. Histochem. Cytochem.*, **9**: 368, 1961).

The black periodic procedure is based on the diphenamine Schiff base condensation of *m*-aminophenol in glacial acetic acid, which at this point prevents the aldehyde Schiff reaction quite completely, followed by alkaline azo coupling with fast black K, C.I. 37190. Black and dark brown are to be regarded as specific for aldehydes, since this condensation does not occur with ketones. The black reaction of enterochromaffin cells with fast black K is completely prevented by periodic acid oxidation.

Diphenamines can also be formed with secondary arylamines, at least of the type $ArNHCH_3$ (*J. Histochem. Cytochem.*, **10**: 303, 1962), so that dyes with primary and secondary amino groups of the type specified may be expected, under appropriate conditions, to form stable compounds with tissue aldehydes. Reference to Tables 6-4 and 6-6 will indicate the commonly used basic and acid dyes which might be thus used.

THE BLACK PERIODIC ACID PROCEDURE OF LILLIE, GILMER, AND WELSH (*Stain Technol.*, **36**: 361, 1961). Fix in formalin or other fixative suitable for mucopolysaccharides. Embed and section in paraffin at 3 to 6 μ. Deparaffinize and hydrate through xylene and graded alcohols as usual.

1. Oxidize 10 min in 1% periodic acid (H_5IO_6).
2. Wash 10 min in running water and drain.
3. Dehydrate in one or two changes of glacial acetic acid. Drain.
4. React for 90 min in 11% (1 *M*) *m*-aminophenol in glacial acetic acid. (If protected from addition of water or alcohol, this solution may be reused for 2 to 3 weeks or longer. Test sections treated in it periodically for complete blockade of the Schiff reaction.)
5. Wash 10 minutes in running water. Rinse in distilled water.
6. Azo-couple 2 to 4 min at 3°C in fast black K (3 mg/1 ml) freshly dissolved in 0.1 *M* Michaelis Veronal–HCl buffer at pH 8 which has been precooled. This solution remains usable for about 2 h and may be used for two or three batches of slides in close succession.
7. Wash directly in three changes 0.1 *N* HCl, 5 min each.
8. Wash 10 minutes in running water.
9. Dehydrate in alcohols, clear in xylene, and mount in cellulose caprate, Permount, or other suitable resin.

RESULTS: Cytoplasm is gray-pink to gray-red; nuclei are poorly contrasting in somewhat deeper color; glycogen, starch, cellulose, and lignin in intestinal contents, black; gastric surface epithelial mucin, intestinal goblet cells, striated border of villi, fungus cell walls, epithelial, glandular, and vascular basement membranes, reticulum, rodent elastic tissue, stroma of smooth and striated muscle, black; colonic and some other mucins, black to brown.

Although some measure of success was attained in a diphenamine Schiff base reaction of rosanilin in 0.2 *N* HCl–alcohol or water and the reaction product was resistant to extraction with the same solvents, the foregoing black periodic procedure seems to fill the same purpose better. However, the Schiff base reaction with acid fuchsin still seems to have some theoretical interest, since ordinary acid-base nuclear staining is avoided by the use of acid dyes with open amino groups. We have used aniline blue with some success, but the best dye which we have tried is acid fuchsin.

SCHIFF BASE REACTION WITH ACID FUCHSIN. After the usual periodic acid oxidation (5 to 10 min in 0.5 to 1%) wash well in water and immerse in 1% acid fuchsin in 0.01 N HCl for 18 to 24 h. (Shorter times may well suffice but have not been tested in detail). Wash 5 min in running water, decolorize 10 min (or more) in 1% borax ($Na_2B_4O_7 \cdot 10H_2O$), which leaves control, unoxidized sections pale pink or colorless, except for rodent elastic fibers. Rinse in water, counterstain as desired, e.g., 10 s in 0.1% fast green FCF in 95% alcohol, then dehydrate in 100% alcohol, and alcohol plus xylene, clear in xylene, and mount in synthetic resin.

A number of acid dyes with substituted amino groups failed to give this reaction: fast green FCF, wool green S, patent blue V, cyanol. The azo dye Biebrich scarlet also was unsuccessful.

The borax acetone condensation of m-aminophenol with periodic acid–engendered aldehyde sites is slower and less effective than that in glacial acetic acid. The Schiff reaction of the connective tissue elements, including elastica, disappears promptly (30 min), but that of gastrointestinal mucins persists with only slight attenuation for at least 4 h. Similarly, only the mucins appear in deep purple in the fast black K demonstration procedure with short exposures to the amine aldehyde condensation process. With medium and longer exposures (1, 2, 4 h) mucins become black, elastica is black, and basement membranes, reticulum, and collagen fiber attain a final deep gray-red at 4-h exposure.

These findings support the view that aldehyde concentration is actually lower at the connective tissue sites in that the blocking procedure reduces the Schiff reaction below visibility with relatively brief exposures, and that relatively long exposures are required to build up enough density in the positive azo coupling sequence.

Aldehydes: The Sawicki Reaction

Sawicki et al. (*Anal. Chem.*, **33**: 93, 1961) used 3-methyl-2-benzothiazolone hydrazone hydrochloride to react with aldehyde and then oxidized with ferric chloride to produce a deep blue tetraazopentamethine cyanin dye (Hünig et al., *Angew. Chem.* [*Engl.*], **1**: 640, 1962). Davis and Janis [*Nature (Lond.)*, **210**: 318, 1966] adapted this spot test method for histochemical demonstration of aldehyde. The method detects as little as 0.1 μg formaldehyde and amounts under 10 μg of a considerable number of other aldehydes in spot tests, and according to Davis and Janis has about 100 times the sensitivity of standard Schiff reagent.

THE DAVIS-JANIS TECHNIC FOR TISSUE ALDEHYDES. Thoroughly wash formol-fixed tissue in water, pH 7.2 osmium tetroxide fixation gives the same results. Dewax thin (2 to 3 μ) paraffin sections in three changes of xylene (5 min), carry through absolute alcohol (3 times for 3 min each), 95% alcohol (2 times for 3 min each), to distilled water, and wash thoroughly. Then react 15 min at 0°C or (British) room temperature (20°C?) in a 2 g/100 ml distilled water solution of MBTH=N-methylbenzothiazolone hydrochloride (Aldrich Chemical Co.). Then transfer to the oxidizing bath, 2 g MBTH, 2.5 g ferric chloride, 100 ml distilled water; for 45 min at 0 or 20°C. Rinse quickly in distilled water, three dips in each of two 95% alcohol baths and in three absolute alcohol baths and in xylene, clear 5 min in xylene, mount in Permount.

Examine on the same or the following day; fading occurs to a notable extent in 3 or 4 days. Collagen, basement membranes, and fibrinoid color deep blue-green, cartilage and mucins pale to faint blue, epithelial cells a faint pea green. Amyloid and chitin do not stain.

With prior periodic acid oxidation aldehyde reactions are seen at the usual sites; the collagen reaction is enhanced. Treatment of sections with sodium borohydride reduces aldehydes and prevents the reaction. (NaBH$_4$ is soluble in water 5.5% at 25°C; the reaction time is not given.)

Jahn [*Acta Histochem. (Jena)*, **34:** 387, 1969] reports that a 2 *M* solution of (24.2%) tris(hydroxymethyl)aminomethane in water, 25 or 50% acetic acid or 0.5 *N* hydrochloric acid completely blocks pseudoplasmal and periodic acid–engendered aldehydes with a 15- to 20-min exposure. The aqueous solution is quite alkaline (pH 11). In reagent grade this agent is still moderately expensive (100 g, $5, Eastman 4833, 1971). The practical grade is fairly cheap.

SILVER REACTIONS. Aldehydes reduce silver nitrate rather slowly at pH 6 and scarcely at all at pH 4. Ammoniacal and hot methenamine silver solutions are reduced fairly promptly in vitro (*J. Histochem. Cytochem.*, **5:** 311, 1957). But the situation in sections is less definite. Most periodic acid–engendered aldehyde sites, though not all, reduce hot methenamine silver about as well when the periodic acid step is omitted.

Noteworthy exceptions to the above, in which it clearly appears that aldehydes engendered by periodic acid oxidation are actually reducing methenamine silver, are the lens capsule and Descemet's membrane in the eye (but not glycogen, reticulum, or cornea), and the surface epithelial mucin of the stomach (but not that of its glands or of the intestine, nor the mucosal or muscular reticulum, nor the cuticular border of the villi) (*J. Histochem. Cytochem.*, **2:** 130, 1954).

BLOCKADE METHODS. Blockade methods are chemical procedures which of themselves fail to give recognizable color reactions in tissue elements but which so alter the latter that they fail to give other color reactions. In this way some quite specific chemical procedures can be made to alter a general color reaction or stain so as to make the failure to react specific evidence of the presence of specific chemical groupings.

Thus deamination prevents certain tissue elements from staining with acid aniline dyes, and thereby gives evidence that the usual staining of those elements was due to the presence of amino groups.

BLOCKADE OF ALDEHYDE GROUPS. The use of sulfite, semicarbazide, and phenylhydrazine to combine with aldehydes and ketones, and thereby prevent color reactions with Schiff reagent or Seligman's carbonyl reagent, aids in the identification of aldehydes and ketones. The reaction of sulfite or sulfurous acid with most aldehydes is quite easily reversed by application of mild oxidants such as 3% hydrogen peroxide; 1% iodine or its equivalent as potassium hypoiodite; 1% sodium iodate or its equivalent in iodic acid; 1% potassium chlorate or its equivalent in chloric acid; or 1% ferric chloride. Intervals of 2 to 10 min in these reagents are sufficient to nullify a 2- to 4-h blockade in 0.1 *M* NaHSO$_3$ and to render the Schiff reaction of HIO$_4$–generated aldehydes again fully positive. Even 10-min washing in tap water from a chlorinated municipal water supply suffices to reverse completely the sulfite blockade of the Schiff reaction. None of these oxidants by itself will produce aldehyde either from 1,2-glycols or from ethylenes.

Though a prolonged treatment with bisulfite after the Feulgen hydrochloric acid hydrolysis, the periodic acid glycol oxidation, or the performic or peracetic ethylene oxidation will completely prevent the positive reactions which normally occur in 5 to 10 min in Schiff reagent, a prolonged exposure (2 to 3 h) to the Schiff reagent completely overcomes the sulfite blockade; and during this prolonged bath the more strongly reacting tissue elements recover their positive reactions first.

Similarly a 30-min exposure to 5% phenylhydrazine hydrochloride after periodic or performic acid oxidation completely prevents reaction to a 10-min bath in Schiff reagent, but is overcome by a 3-h exposure. However, treatments with phenylhydrazine prolonged for several hours or overnight render some periodic acid–generated aldehydes permanently Schiff-negative; others are more loosely combined and again become Schiff-positive on 3-h exposure to Schiff reagent, or by interposition of a weak oxidizing agent. But cellulose, mucus, and cartilage remain reactive to a 10-min Schiff bath after 1 h in phenylhydrazine.

Pearse prescribes 2 to 3 h at 60°C in 10% phenylhydrazine in 20% acetic acid. This is undoubtedly effective but necessitates a further control section, heated for the same period in plain 20% acetic acid.

Dimedone we have not used as a blocking agent for carbonyl. It is said to be specific for aldehydes, and some workers recommend it highly. Pearse writes that it is found to block extremely slowly in practice: "It is often difficult to distinguish the reduction in recolorization of leucofuchsin caused by dimedone from that observed in control sections treated with the solvent (acetic-alcohol) only."

Nevertheless, when it does block, the reaction is to be regarded as specific for aldehyde. Failure to block does not exclude aldehyde. The same is true of phenylhydrazine, aniline chloride, and hydroxylamine.

Pearse uses a saturated solution of dimedone (5,5-dimethylcyclohexane-1,3-dione) in 5% acetic acid (in alcohol?) for 1 to 16 h at 60°C, or for 2 to 3 days at 22°C.

Pearse prefers hydroxylamine in aqueous sodium acetate solution (10 g hydroxylamine hydrochloride, 20 g sodium acetate crystals, 40 ml distilled water), and treats sections 1 to 3 h at 22°C. He states that with the solution "condensation with tissue aldehydes of all varieties is rapid and apparently complete."

We have noted some curious failures to block the Schiff reaction after periodic acid with hydroxylamine, either as a 20% solution in 30% sodium acetate or as a 10% solution in pH 4.5 0.2 M acetate buffer. (The pH optimum for the hydroxylamine aldehyde condensation is said to be 4.7.) The Schiff reaction of gastric surface epithelial mucin and gland mucin, of ocular lens capsule and rod acromeres, of cellulose and lignin, of cartilage and nucleus pulposus, and sometimes of muscle glycogen remained positive with 2 and 24 h of hydroxylamine blocking after periodic acid in one series of trials in which molar aniline chloride had been completely successful in 24 h and largely so in 2 h. Only gastric gland mucin gave a slight Schiff coloration. Usually exposure to 5% phenylhydrazine hydrochloride for 24 h completely prevents the Schiff reaction when interposed after periodic, peracetic, or hydrochloric acid, and 2 h is often adequate, but on some objects this agent may also fail to block.

A 1-h exposure to 5% KCN, adjusted with acetic acid to pH 6.7 to 7.3, applied after periodic acid oxidation completely prevents the following Schiff reaction of hepatic, muscle, and cartilage cell glycogens, of cartilage matrix, tracheal and gastric mucins, and liver ceroid. The blockade is broken by a second 10-min exposure to 1% periodic acid (*J. Histochem. Cytochem.*, **4:** 479, 1956). A drop of nitrazine solution or of phenolsulfonphthalein as indicator facilitates the pH adjustment. Neutral red, amber at pH 8, is red at pH 7 in KCN solutions.

ANILINE, *m*-AMINOPHENOL AND *p*-TOLUIDINE. Arylamines present in excess in the surrounding medium react with tissue aldehydes to form diphenamine Schiff bases (*J. Histochem. Cytochem.*, **10:** 303, 1962). For insoluble tissue aldehydes this is conveniently accomplished by immersion of sections in 1 M solutions of the arylamine in glacial acetic acid. At 23 to 25°C aniline takes 2 to 3 h; *m*-aminophenol, 60 to 90 min; *m*-toluidine, 2 to 3 h; *N*-methyl-*m*-toluidine, about 4 h; and *p*-toluidine, 30 to 60 min to completely block the Schiff reaction of periodic acid–oxidized gastric and intestinal mucins.

Of these, all except *p*-toluidine can be used to direct azo coupling with freshly diazotized safranin (5 mM in glacial acetic acid, 18°C, 10 min) in positive demonstration reactions. The use of m-aminophenol for directing alkaline azo coupling has been detailed earlier in this section.

When lipid aldehydes which are soluble in glacial acetic acid are under study, it is necessary to go back to our older technic with aqueous 1 M aniline hydrochloride. It appears probable that the p-toluidine would exhibit similar greater rapidity of action in this solvent also, and that m-aminophenol could be used in a technic similar to the "black periodic" but avoiding fat solvents. However, these alternative procedures have not been tested. m-Aminophenol requires 11% for a 1 M solution, p-toluidine 10.7%, or 14.36% of the commercially available hydrochloride in water. The last has a water-solubility, 25% at 25°C, high enough for a 1.75 M solution.

For aniline chloride in molar solution dissolve 9.3 g (=9.1 ml) aniline in 8.1 ml concentrated HCl (0.1 mol each), shaking during addition (the mixture becomes quite hot); dilute to 100 ml with distilled water. This solution effectively renders Schiff-negative (with a 10-min Schiff bath) the aldehyde formed in the Feulgen, peracetic acid, and periodic acid reactions. Even a 15-min aniline chloride bath is adequate to block the Feulgen reaction. A 6-h exposure to molar aniline chloride or 18 h in 0.1 M prevents the aldehyde reaction of retinal rod acromeres after peracetic acid oxidation. The periodic acid Schiff reaction of glycogen, retinal acromeres, and corneal and scleral collagen is inhibited by 2 h in 1 M or 18 h in 0.1 M aniline chloride. That of epithelial mucin requires 4 h in 1 M or 18 h in 0.4 M aniline chloride. That of the lens capsule and Descemet's membrane is not blocked by 18 h in molar aniline chloride; it requires a 2.5 M concentration for the same period.

As prepared above, aniline hydrochloride has a pH of 1.4 to 1.9. The pH level may be raised to about 4 without obvious decomposition; at about pH 4.5 the solution becomes milky, and at pH 5 it separates into an oily layer and an aqueous phase. The amount of sodium hydroxide required to produce turbidity corresponds to conversion of enough aniline chloride back to aniline to furnish about 1.9% aniline water, just about the usual solubility limit. The advantage of raising the pH level to 4 lies in avoidance of destruction of acid-soluble tissue elements. Aldehyde blockade is perhaps more effective with the acid solution and progresses even more rapidly in glacial acetic acid.

Rappay and Van Duijn (*Stain Technol.*, **40:** 275, 1965) report complete oxidation of aldehydes to carboxyls by chlorous acid treatments of 8 h at 20°C and pH 3.5. Aldehydes produced by periodic and Ninhydrin oxidation, by Feulgen acid hydrolysis, and by the plasmal reaction with mercuric chloride were successfully blocked.

RAPPAY AND VAN DUIJN'S CHLOROUS ACID REAGENT.   Dissolve   2.19   g NaClO$_2$ in 70 ml distilled water. Add 20 ml 5 N acetic acid. Add distilled water to make 100 ml. For the periodic acid and Ninhydrin Schiff reactions and the Feulgen reaction 10-μ cryostat sections were fixed 5 min in neutral 10% formalin; for the plasmal reaction unfixed cryostat sections were carried directly into 1% mercuric chloride for 4 min.

BROMINATION.   Bromine oxidation of aldoses to aldonic acids was probably first reported by Hlasiwetz (*Justus Liebigs Ann. Chem.*, **119:** 281, 1861). Jackson and Hudson (*J. Am. Chem. Soc.*, **60:** 989, 1938) used it to convert erythrose and glyoxal derived from hydrolysis of periodate-oxidized starch to erythronic and oxalic acids. Lillie, Pizzolato, and Donaldson [*Acta Histochem. (Jena)*, **44:** 215, 1972] used a 30-min 1 N bromine–CCl$_4$ treatment to block the aldehyde reaction of elastin. Labella (*J. Histochem. Cytochem.*, **6:** 260, 1958) observed that a liquid aldehyde isolated from an elastase digestion of isolated elastin lost its Schiff reactivity on bromination.

BOROHYDRIDES.   On the other hand, borohydrides promptly reduce aldehydes to the corresponding primary alcohols and ketones to secondary alcohols. Quinones are also reduced according to the chemical literature (Chaikin and Brown, *J. Am. Chem. Soc.*, **71**: 122, 1949). Carboxyls, anhydrides, esters, nitriles, ethylenes, nitro groups, lactones, epoxy groups, lactams, and amines are unreactive. Histochemical reports by Jobst and Horvath (*J. Histochem. Cytochem.*, **9**: 711, 1961), Geyer [*Acta Histochem. (Jena)*, **15**: 1, 1963], and David and Janis [*Nature (Lond.)*, **210**: 318, 1966] used rather long exposures and weak solutions. Lillie and Pizzolato (*Stain Technol.*, **47**: 13, 1972) explored the use of $NaBH_4$ and KBH_4 in water and other solvents at varied concentrations and pH levels. They found aqueous solutions adequate against periodic acid–produced and other aldehydes at 1, 0.5, and 0.2% $NaBH_4$ in 1% Na_2HPO_4 in 1 min; 0.1% took 2 min.

KBH$_4$ is equally effective and in 250-mg pellets keeps better in warm, moist climates. Solutions deteriorate in a few hours. Isopropanol solutions (0.4%, saturated) and 1% pyridine solutions required 10 to 15 min; dioxane and methyl Cellosolve solutions gave no satisfactory results with 1-h exposures. More acid solutions evolve hydrogen copiously and were not used histochemically. Solutions in water and in 1% borax were quite alkaline and required 15 to 20 min for complete blocking. After 7 h 0.625% KBH_4 still blocked completely in 1 min; 1% solutions after 24 h required 15 min to effect complete blocking.

Hence it appears that solutions should be made shortly before using and that presence of a buffer substance to prevent rapid rise of pH enhances the activity.

Thus far we have not worked out conditions for the histochemical reduction of quinones except that they apparently require greater exposures than aldehydes.

ACETYLATION AND BENZOYLATION OF ALDEHYDES. A 24-h acetylation at 23 to 25°C in 40% acetic anhydride–pyridine introduced after periodic acid oxidation of gastrointestinal mucins, glycogen, rodent arterial elastica, etc., more or less completely prevents the following Schiff reaction. The acid azo coupling reaction of aldehydes condensed with *m*-toluidine is similarly prevented, and the "black periodic" reaction is quite severely impaired. Benzoylation in 5% benzoyl chloride pyridine 4, 8, or 24 h at 23 to 25°C is similarly effective. Alcoholic 40 or 25% acetic anhydride is less effective than the pyridine solutions at all time exposures.

Sometimes longer exposures (24 h) are less completely effective than 4- to 8-h exposures. The addition of 0.1 ml sulfuric acid to 10 ml acetic anhydride and 30 ml pyridine appears to increase its effectiveness somewhat and in the presence of the pyridine produces no positive evidence of complicating sulfation of other structures.

The reactions are written $RCHO + 2(CH_3CO)_2O + H_2O \rightarrow RCH(CH_3COO)_2 + 2CH_3COOH$. The acetyl (or benzoyl) esters may be saponified and the aldehydes again liberated in chemical experiments, though the latter, at this writing, has not been tested histochemically. Ketones do not react (Lillie et al., *J. Histochem. Cytochem.*, **14**: 529, 1965).

Mixtures of acetic anhydride–glacial acetic acid, 25 : 75, with 0.2 to 0.25% by volume of concentrated sulfuric acid added are also quite effective, but it appears at this moment debatable whether the blockade is due to acetylation or to sulfation.

ACID GROUPS

The usual acid groupings demonstrable in tissues are the carboxylic acid residues of fatty acids, of carbohydrate uronic acids, and of relatively acid proteins such as pepsinogen; the phosphoric acid residues of the nucleic acids; the sulfuric acid half esters of the amino-polysaccharides such as heparin, chondroitin, and mucoitin sulfuric acids; and the sulfonic acid residues such as are produced by oxidative cleavage in the cystine proteins like keratin, and such as not improbably exist in cutaneous and ocular melanins.

Though each of these groups has other identifying reactions, such as the sudanophilia of the fatty acids, the specific amino acid reactions of certain proteins, the immunochemical fluorescence reactions of various globulins, the Feulgen reaction of deoxyribonucleic acid, the Criegee oxidation cleavage reactions of the glycolic polysaccharides, and the reducing reactions of the melanins, they have in common the property of staining with basic aniline dyes. This property is known generally as *basophilia.*

Basophilia is conveniently studied with simple dilute solutions of basic dyes in buffered aqueous solution, and also with balanced mixtures of acid and basic dyes such as the azure A–eosin B mixture, also in buffered aqueous solutions.

The discriminating factor in both these methods is that sulfuric and sulfonic acid residues take up basic dyes at lower pH levels than do phosphoric acid residues and that these in turn stain at lower pH levels than do various carboxylic acid groups.

Several factors enter into the estimation of the lowest pH at which a given tissue element stains: (1) the pK of tissue acids, which, especially in proteins, is influenced by the fixation employed. For example, erythrocytes are strongly basophil when stained with azure-eosin at pH 8 to 10; after brief methanol fixation they become amphoteric at about pH 7 and strongly oxyphil at about pH 6.5 to 6; with 1- to 3-day formaldehyde fixation this shift occurs at about pH 6 to 5.5; (2) the concentration of the basic dye employed: higher concentrations stain the same acidic elements at lower pH levels; (3) the precise dyestuff employed: azure A at 1 mM concentration stains generally at lower pH levels than methylene blue or thionin, also at 1 mM, and pararosanilin stains at lower levels than crystal violet.

Further it must be recalled that strongly acid solutions hydrolyze nucleic acids to liberate deoxyaldose residues which can in acid solution readily condense with primary and less freely with secondary arylamines to form stable Schiff bases of the secondary or tertiary amine–diphenamine type. The dye aldehyde compounds thus formed are relatively acid-fast, whereas the usual salt union, basic dye plus tissue acid reaction product, is readily decolorized by acids.

Another factor in the estimation is the time of exposure. It would be expected that 18- to 24-h staining ought to yield more intense staining than shorter intervals. This is not regularly the case. On some such trials staining has actually been greater in 30 min to 2 h than in 18 to 24 h. The reason for this paradoxical behavior seems related to the probability that in the longer interval hydrolysis occurs and dye base is adsorbed onto the glass container, with consequent lowering of dye concentration in the solution and redecolorization of stained objects in the tissue. This phenomenon is observed only in quite weak solutions (0.1 to 20 μM).

THE pH SIGNATURE TECHNIC OF DEMPSEY, SINGER, AND WISLOCKI (*Endocrinology*, **38:** 270, 1946; *Anat. Rec.,* **102:** 175, 1948; *Int. Rev. Cytol.,* **1:** 211, 1952). This method used dilute solutions of methylene blue, originally on Zenker-fixed material, for demonstrating differences in basic dye uptake by various tissue elements. Foraker (*J. Histochem. Cytochem.,* **7:** 284, 1960) has applied the method to routine formalin-fixed surgical material. He used 0.5 mM methylene blue in acetate buffers of pH 3, 4, and 5; phosphate buffers of pH 6 and 7; and Veronal buffers of pH 8 and 9; and stained overnight, dehydrated in alcohol, cleared in xylene, and mounted in HSR synthetic resin.

Because of the instability of methylene blue, of its usual contamination by lower homologs and Bernthsen's methylene violet, and of its generally weaker staining power we have usually employed 0.02% azure A (0.685 mM), extended the pH range down to 1 or occasionally to 0.5, and limited the staining time to 30 to 60 min. Because of the partial extraction of thiazin dyes by ethyl alcohol, we have usually dehydrated with acetone. In this we follow the practice reported by Highman from our laboratory in 1946 (*Stain Technol.,* **20:** 84) for buffered toluidine blue staining, rather than the later modifications of the Harvard group. Formalin fixation is used.

BUFFERED AZURE A STAINING.   To 49 ml of the selected buffers add 1 ml 1% aqueous or alcoholic azure A. For pH 1 we usually employ 1 ml 37.5% hydrochloric acid in 99 ml distilled water (0.125 N); for pH 2 and 3, the HCl–KH$_2$PO$_4$ series; for pH 4 and 5, the acetate series; for pH 6 and 7, Sörensen's phosphates; and for pH 8 and 9, Michaelis' Veronal–HCl series.

Acid-base staining with basic dyes should be completely decolorized by a 5-min exposure to 0.2 N HCl in water or 70% alcohol.

RESULTS: At pH 1 only mast cells, some cartilage, cutaneous and ocular melanins, and cysteic acid produced by bromination or performic or peracetic acid oxidation of keratin, etc., should stain. Denser nuclei and mitotic figures appear at about pH 1.5; pancreatic acinar and gastric chief cell cytoplasm, at about pH 2. Epithelial mucins, varying according to species and locality, present minimum pH staining levels of 2 to 4, related probably to their composition, mucoitin sulfates, sialomucins, and uronic acid mucins. Fatty acids require pH 3 to 4, and muscle, collagen, and some mucins become basophilic above pH 4.5 to 5. Erythrocytes acquire basophilia at a somewhat higher level.

Staining with Nile blue constitutes a special case which will be discussed in greater detail below, in relation to fatty acids. When this dye is used in buffered solutions similar to those applied to the thiazin dyes and sections are then dehydrated in alcohol or acetone, results are essentially similar to those with azure A. But when staining is done in 1% sulfuric acid (v/v, pH 0.85), a dark blue stain is conferred on liquid fatty acids, lipofuscin pigments, and peptic gland zymogen granules which can be completely decolorized in a few seconds by acetone in the cases of lipofuscins and pepsinogen or by glycol in the case of soluble fats with a somewhat longer interval, provided that the decolorization is done directly after the acid staining bath. The coloration is conserved when preparations are mounted in aqueous media. If such stained preparations are washed with water after being stained in the sulfuric acid solution, the fat-solvent-extractable stain is converted to a salt type of union, and lipofuscin preparations can then be dehydrated and mounted in resinous media. The stain changes, however, to the same relatively light green seen in preparations stained in azure A at pH 4 and similarly dehydrated.

For further details see Lillie, *J. Histochem. Cytochem.*, **4:** 377, 1956; **6:** 130, 1958.

THE SULFURIC ACID–NILE BLUE TECHNIC FOR FATTY ACIDS (*Stain Technol.*, **31:** 151, 1956). Use frozen sections for soluble fats, paraffin sections for lipofuscins, pepsinogen, and other protein-bound fatty acids. Stain 20 min in 0.05% Nile blue in 99 ml distilled water plus 1 ml concentrated sulfuric acid. For extraction tests transfer frozen sections to diethylene glycol (HOCH$_2$—CH$_2$)$_2$O for 4 h; transfer paraffin sections directly to anhydrous acetone, agitating for 1 min. After acetone, the sections may be cleared in xylene and mounted in cellulose caprate or returned to water and mounted in glycerol gelatin or temporarily in water or glycerol for examination before restaining.

Sections for direct examination after Nile blue and restained sections are washed in several changes of distilled water or in running water of suitable quality and mounted in glycerol gelatin.

It is to be noted that this procedure does not discriminate between free fatty acids and alkaline earth soaps, since the sulfuric acid decomposes the latter to more or less insoluble sulfates and free fatty acids. According to Gomori (p. 205) calcium soaps may be converted to lead soaps by a 10-min incubation at 55 to 60°C in 1 to 2% lead nitrate. The lead soaps are then converted to insoluble brown sulfide by 5- to 10-min exposure to dilute ammonium sulfide. Free fatty acids, being insoluble in water and essentially nonionized, do not react. Calcium soaps are distinguished further from simple fatty acids by their insolubility in alcohol and other fat solvents. However, the lipofuscins are at least partly insoluble in boiling methanol-chloroform and in hot pyridine, retaining their sudanophilia and Nile blue staining capacity and other reactions.

Lead soaps are often soluble in fat solvents; newly formed lead oleate is readily soluble in alcohol, lead stearate dissolves in hot alcohol, the linoleate is soluble in chloroform ("The Merck Index," 7th ed.).

ALKYLATION. Usually the Fraenkel-Conrat (*J. Biol. Chem.*, **161**: 259, 1945) procedure as modified by Fisher and Lillie (*J. Histochem. Cytochem.*, **2**: 81, 1954) for application in histology has been used. This process uses 0.05 to 0.1 N hydrochloric acid in methanol at 25, 37, or 58 to 60°C for varying intervals. The basophilia of nuclei, cartilage, mast cells, mucins, pigments, and proteins generally is abolished at intervals which are fairly constant for each constituent at the temperature used, on comparably fixed material. One disadvantage noted has been a swelling, fragmentation, and loss of birefringence of collagen fibers which occurs in the hot acid-methanol solutions. In about 1960 we tried using sulfuric acid in place of hydrochloric acid, with the purpose of obtaining a more nearly anhydrous methylation reagent and thereby perhaps accelerating methylation and decreasing the collagen destruction which was thought to be hydrolytic in nature. The 0.05 M sulfuric acid solution proved to be at least as effective in methylating as the 0.1 N hydrochloric acid solution. It was found, moreover, that sulfuric acid–methanol solutions of 0.25 to 0.1 M esterified fairly promptly, so that in 1 or 2 weeks no free sulfuric acid was detectable on testing with barium chloride. Molar solutions, however, remained unesterified or quite incompletely esterified even after several months at 25°C.

It was early observed that the methylation procedure promptly abolished the metachromatic basophilia of mast cells and cartilage and that longer intervals were required for lipofuscins, nucleic acids, etc., with still longer ones needed for cysteic acid produced by oxidation cleavage of cystine and for melanin. In 1957 Kantor and Schubert (*J. Am. Chem. Soc.*, **79**: 152) reported that the Fraenkel-Conrat reagent desulfated chondroitin sulfates, liberating methyl hydrogen sulfate, and leaving unsulfated chondroitins.

In 1958 (*J. Histochem. Cytochem.*, **6**: 136) Lillie showed that the methylation blockade of the Nile blue staining of peptic gland zymogen granules was broken by saponification in alcoholic KOH, and in 1959 (*J. Histochem. Cytochem.*, **7**: 123) Spicer and Lillie showed that while the metachromatic basophilia of peracetic acid–oxidized hair cortex was abolished by drastic methylation and restored by saponification as was the orthochromatic staining of proteins and cartilage above pH 4, the metachromatic staining of mast cells and cartilage occurring at pH 2 and below was not restored by saponification, even though only a quite mild methylation was employed. Spicer (*J. Histochem. Cytochem.*, **8**: 18, 1960) later utilized this mild methylation both in combinations with sulfation and alone, and in contrast with more drastic methylation in studies of the reactivities of rodent mucopolysaccharides to azure A at low and medium pH levels and to Alcian blue. This mild methylation apparently attacked only certain strongly acid carboxylic acids.

METHYLATION PROCEDURES; FISHER-LILLIE TECHNIC. For 0.05 N hydrochloric acid take 0.4 ml, for 0.1 N take 0.8 ml concentrated hydrochloric acid and make up to 100 ml with absolute methanol (reagent grade).

1. Deparaffinize sections in xylene, wash in two changes absolute alcohol, drain, and immerse in methylation reagent. For methylation at 37 or 60°C use screw-capped Coplin jars, tightly closed, and check at intervals as a precaution against evaporation (methanol boils at 64.7°C).

2. After expiration of the prescribed interval, wash in descending alcohols and water.
3. Stain with azure A–eosin B, 0.02 to 0.05% azure A at graded pH levels for 30 to 60 min, with the sulfuric Nile blue technic or other method under study.
4. Dehydrate and clear or wash in water and mount as prescribed for the demonstration technic used.

SPICER'S MILD METHYLATION.   Use the 0.1 N hydrochloric acid mixture above and methylate 1 h at 60°C or 4 h at 37°C.

THE 0.1 M METHYLSULFURIC ACID TECHNIC.   Add 2.7 ml concentrated sulfuric acid to about 490 ml absolute methanol, mix thoroughly, and add methanol to make 500 ml. Let stand at 25°C for about a week. Then test every day or two, by dropping 0.5 ml into 1 ml 1% barium chloride. When turbidity is no longer produced, the methylation reagent is ready and is stable for some months. Use enough at a time to cover sections in a Coplin jar. Preheat to 60°C or lower temperature if desired. At 60°C use tightly closed screw-cap Coplin jars to avoid loss of methanol and increased concentration of monomethyl sulfate (bp 188.3 to 188.6°C).

The methylsulfuric acid reagent is similar to the Fraenkel-Conrat-Fisher reagent in its effect on basophilia, or, on some trials, it is somewhat slower. It is noteworthy that rat mast cells seem to resist methylsulfuric acid methylation up to 4 h at 0.1 M concentration.

It has been further observed that while the methanol–hydrochloric acid reagent completely removes iron from enterosiderin and hemosiderin pigments during a 4- to 8-h 60°C methylation exposure, the sulfuric reagent apparently does not attack the iron oxide component of these pigments.

Since the ethyl sulfates of Ba, Sr, Ca, and Pb are known to be soluble in water, it remained uncertain whether the active agent in this reagent was dimethyl sulfate or methylsulfuric acid, $CH_3—O—SO_2—OH$. On testing this reagent with barium carbonate under microscopic observation, it was found that the $BaCO_3$ crystals become smaller and smaller and finally disappear and that small bubbles of gas are formed.

It is evident that methylsulfuric acid is at least an important, if not the sole, methylating constituent of the solution.

THIONYL CHLORIDE.   This reagent is a fuming liquid boiling at 79°C whose vapors are corrosive to skin, eyes, mucous membranes, and clothing, hydrolyzing in water to H_2SO_3 and HCl.

Geyer [*Acta Histochem. (Jena)*, **14**: 284, 1962] reported that this reagent methylated more rapidly and at lower temperatures than the HCl–methanol reagent, and was more nearly anhydrous, with less tendency to protein hydrolysis.

The reagent is freshly prepared before use by gradually flowing 4 ml clear white thionyl chloride into an Erlenmeyer flask containing 100 ml absolute methanol, with the flask tilted away from the operator. Methylation is largely completed in 4 to 6 h at 20°C, nuclei perhaps requiring 24 h. At 56°C with preheated reagent methylation effects are evident in 30 min and complete in 2 h. Toluidine blue stains are the most readily blocked; thereafter Hale colloidal iron and Alcian blue stains are the most readily blocked. The Barrnett-Seligman and the Geyer epoxy ether–COOH reactions are blocked; the PAS (periodic acid Schiff) and the Ninhydrin Schiff reactions require more than 6 h. Fast green and amido black 10B stains and Burstone's *m*-nitrobenzaldehyde reactions are enhanced, suggesting liberation of additional amino groups or perhaps increase in basicity of NH_2 groups by alkylation. Morel-Sisley tyrosine and Adams tryptophan reactions are unaffected. Ethanol may be used in place of methanol, but ethylation is slower than methylation, and higher alcohols are even less effective.

With the Geyer formula 4 ml thionyl chloride at sp gr 1.638 is 6.552 g, or 55 mmol, which would yield 110 mmol HCl according to a reaction

$$SO_2Cl_2 + 2CH_3OH \longrightarrow CH_3-O-SO_2-O-CH_3 + 2HCl$$

In 100 ml methanol 110 mmol would give a 1.1 N solution, or 11 times the concentration in the Fisher-Lillie reagent, besides 0.5 M concentration of dimethyl sulfate. Since Fisher and Lillie found that 0.1 N HCl–methanol functioned about twice as fast as 0.05 N, the greater reaction speed of the Geyer reaction is really less than might be expected.

BORON TRIFLUORIDE.   Coalson and Schwartz [*Acta Histochem.* (*Jena*), **36:** 227, 1970] report that boron trifluoride is an excellent esterifying reagent for tissue acid groups, comparing it with methanol–HCl and thionyl chloride. Carboxyls and sulfate esters are blocked first, then phosphoryl groups of nucleic acids. Methylation was more rapid after Bouin fixation, slower with neutral formol fixation, intermediate with Helly, Carnoy, Zenker, etc.

Boron trifluoride is available as a 14% solution in methanol (boron trifluoride boils at $-101°C$ and is also very soluble in water), 105.7 g/100 ml at 0°C. It has a pungent, suffocating odor, and is corrosive to skin and mucous membranes. It decomposes in water to fluoroboric and boric acids. The solution should be used within 6 months of preparation.

Coalson and Schwartz found a 7% concentration, used at 28°C, effective in abolishing all basophilia in 24 h after formol fixation, 6 h after Bouin fixation. Sections are brought through xylene, ethanol, and methanol to the equal volume dilution of the stock 14% solution in methanol.

Demethylation is accomplished with the usual reagents.

Terner and Clark's (*J. Histochem. Cytochem.*, **8:** 184, 1960) ethyl chloride alkylation was aimed at a specific primary amine reaction and is discussed under that topic.

These methylation technics, particularly the Fisher-Lillie 0.1 N HCl procedure, have been used not only to prevent basic dye staining of nucleic acids, protein and polysaccharide carboxyls, and fatty acids as in lipofuscins, but also in a considerable variety of other procedures, such as the staining by elastin stains of elastins and altered collagen (Fullmer, *J. Histochem. Cytochem.*, **5:** 11, 1957; **6:** 425, 1958), the periodic acid Schiff reaction of glycogen, gastric mucins, thyroid colloid, and connective tissues (Lillie, 1954, p. 163), the prevention of enterochromaffin reactions with diazonium salts and with dilute hematoxylin (Lillie, *J. Histochem. Cytochem.*, **9:** 184, 1961), the prevention of the deamination of proteins (Terner and Clark, *J. Histochem. Cytochem.*, **9:** 184, 1960), and the uptake of Fe(II) and Fe(III) by tissue lysyl and other NH_2 groups (Lillie and Pizzolato, *J. Histochem. Cytochem.*, **20:** 116, 1972) staining in sequence iron hematoxylin technics.

Demethylation of carboxylic, phosphoric, and sulfonic acid residues is essentially similar chemically to the deacylation of acetylated hydroxyls, except that in this instance it is the acyl radicle which remains immobile and demonstrable in the tissue.

The technic reported in 1958 (*J. Histochem. Cytochem.*, **6:** 130) for the restoration of Nile blue staining of pepsinogen granules after methylation has proved applicable in several other situations and is essentially identical to the saponification procedure given under Deamination and Acylation of Amines, thus:

After methylation treat sections 20 to 30 min in 1% potassium hydroxide in 70% alcohol; wash and stain by the selected demonstration procedures.

It is particularly to be observed here that the staining of ribonucleic acid sites is only partially or perhaps not at all restored. This we attribute to alkaline alcoholic solution of the ribonucleic acid. It occurs equally on treatment of unmethylated material with alcoholic KOH solutions. It may be avoided by use of the ammonia alcohol saponification reagent recommended by Lillie, thus:

After methylation treat 24 h in 40 ml absolute alcohol plus 10 ml 28% ammonia, wash, and proceed with selected demonstration procedures.

Or the ribonucleic acid may be rendered insoluble in alcoholic KOH by postfixing sections 24 h at 60°C in 8% aqueous formaldehyde or in 5% $K_2Cr_2O_7$ or 5% mercuric chloride. Picric acid and 5% lead nitrate treatments of similar duration were ineffective.

The same 1% KOH–70% alcohol treatment was found by Lillie and Pizzolato to dealkylate the lysine NH_2 groups so that they would again take up Fe(II) and Fe(III). Blocking of the same sites by HNO_2 deamination was not reversible by the saponification procedures.

The permanganate demethylation reported by Fisher and Lillie has been little used and requires further exploration. Similar restoration has occasionally been observed when other oxidants are applied after methylation, e.g., HNO_2, HIO_4, but these occurrences have not been adequately explored. Possibly some such reaction occurs as the following, which parallels the familiar dealkylation of tertiary amines by oxidants seen in the dealkylation of methylene blue.

$$R-CO-OCH_3 + O \longrightarrow R-CO-OH + CH_2O$$

$$R-N-(CH_3)_2 + O \longrightarrow R-NH-(CH_3) + CH_2O$$

Alkyl halides have occasionally been tried as such in histochemistry for alkylation of primary and secondary amines. Terner and Clark (*J. Histochem. Cytochem.*, **8:** 184, 1960) used ethyl chloride. We have employed methyl iodide and both propyl iodides. The excessive volatility of the first two (bp 12.3 and 42.4°C, respectively) is a serious handicap to their histochemical use. Ethyl iodide (bp 72.4°C), *n*-propyl bromide (bp 71°C), and isopropyl and *n*-propyl iodides (89.5 and 102.5°C) would seem perhaps more convenient.

Before closing the section on methylation and saponification as a means of studying tissue acid radicals it seems necessary to discuss briefly the lactonization hypothesis advanced by Vilter (*Ann. Histochem.*, **13:** 205, 1968). Lactonization of a uronic acid polymer is theoretically possible. A δ-lactone could be formed from C-6 to C-3

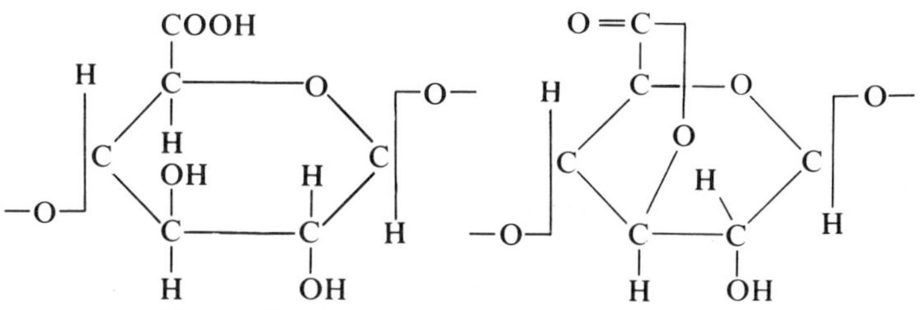

Hexuronic acid δ-Lactone

on a 1,4-polymer. In support of his hypothesis he reports that dry HCl in dry benzene blocks Alcian blue staining of a mucopolysaccharide. The reaction is blocked also in *o*-dichlorobenzene. To support his thesis he claims that the blocking of sulfated mucopolysaccharides is due to formation of internal esters and denies Kantor and Schubert's report of desulfation. Kantor and Schubert (*J. Am. Chem. Soc.*, **79**: 152, 1957) present a solid chemical report with isolation of a methyl ester and a potassium salt of desulfated chondroitin sulfate with elementary analyses. Neither product contains sulfur against calculated and found 5.3 and 5.2% in the original chondroitin potassium sulfate. Vilter's contention here appears absurd and unsupported by analytical data. The effect of methylation in abolishing basophilia is not limited to acid carbohydrates. It occurs in pepsinogen granules. Glutamic and aspartic acids contain no nearby hydroxyls on which lactones could be formed. It occurs in cysteic acid residues produced in hair by peracetic acid albeit quite drastic treatment is required (2 or more days at 60°C in 0.1 *N* HCl–methanol; Spicer and Lillie, *J. Histochem. Cytochem.*, **7**: 23, 1959) and is reversible by a saponification. It alters the Nile blue color in isolated oleic acid from blue to red and produces a thin oil melting at −20°C (methyl oleate, mp −19.9°C; oleic acid, mp +16.3°C: Weast). This oil was readily saponified and regained its blue staining by Nile blue (Lillie, *J. Histochem. Cytochem.*, **4**: 377, 1956). Oleic acid contains no hydroxyl group on which a lactone could be formed.

Sorvari and Stoward (*Histochemie*, **24**: 114, 1970) supported Vilter's thesis, again working only on mucopolysaccharides. They state that if methylated mucopolysaccharides are first reduced with borohydride at 0°C or reacted with hydroxylamine at room temperature, they then fail to regain basophilia to azure A on saponification. This indicates that the lactone CO group is capable of reacting as a carbonyl rather than as part of a carboxyl radical. Vilter's alternative suggestion of internal esterification to nearby hydroxyls on other chains seems untenable, the methyl radicals are present in large excess and would preoccupy available carboxyls before they could make contact with sterically available hydroxyls. Sorvari and Stoward's findings appear to indicate that in carboxylic acid sugars lactonization is the most probable route of methylation blockade. But this does not apply to the methylation of lipofuscins, pepsinogen granules, nucleic acids, sulfated saccharides, or cysteic acid residues. However, one must recall that the attack of borohydrides on lactones is disputed. The three references cited by Sullivan ("Sodium Borohydride," Metal Chemicals Division of Ventron, Beverly, Mass., 1970) were apparently not accepted by Hickinbottom (1957) or Hilgetag (in Weygand (ed.), "Organisch-Chemisches Experimentierkunst," 3d ed., Barth, Leipzig, 1964), both of whom state that lactones are only exceptionally attacked by sodium borohydride. See also Lillie and Jirge, *Histochemistry*, **41**: 241, 1975.

Terner and Lev (*J. Histochem. Cytochem.*, **11**: 804, 1963) used 1.8% Ba(OH)$_2$(0.1 *M*) in ethanol at 0°C for 30 min after a 2-h, 60°C, 0.1 *N* HCl–methanol methylation. Distilled water is presumed as the solvent. Ba(OH)$_2$·H$_2$O is soluble at 25°C; only 233 mg/100 ml 70% alcohol (Lillie and Pizzolato, unpublished data) and the amount prescribed in about a saturated solution in water, at 3 to 4°C. This procedure induced a γ-metachromasia to thionin and toluidine blue in previously β-metachromatic colonic mucins. The metachromasia of this mucin is extinguished below pH 2.5 and is blocked also by acetylation. Sialomucin in mouse sublingual gland also completely loses metachromasia on acetylation. Hydrolysis with 10% acetic acid for 1 h reverses the acetylation blockade.

Richter (in Anselutz and Reindel (eds.), "Organic Chemistry," vol. 1, Elsevier, Amsterdam, 1947) is cited as stating that lactones hydrolyze readily in water. The acetylation blockade resisted aqueous hydrolysis at 60°C and 0.05 *N* HCl at 20°C for several hours; lactones are relatively stable, the acetic anhydride–pyridine blockade was broken only by 1 h at 60°C in 10% HAc, though readily saponified by alkali. Sialic acids can readily form a five-atom lactone ring not involving the α-carbon hydroxyl or the deoxy-β-carbon, but cyclizing with the γ-carbon to form a cyclic ester, a γ-lactone.

$(CHOH)_2-CH_2OH$

Sialic acid

$(CHOH)_2-CH_2OH$

γ-Sialolactone

Uronic acid γ-lactone

The method of Barrnett and Seligman (*J. Histochem. Cytochem.*, **4**: 41, 1956; *J. Biophys. Biochem. Cytol.*, **4**: 169, 1958) of creation of carbonyl groups of carboxylic acids and the demonstration of the ketone by reaction with 2-hydroxy-3-naphthoic acid hydrazide followed by azo coupling with fast blue B was at first thought to selectively demonstrate COOH-terminal protein carboxyls. Karnovsky et al. in a later series of papers (*J. Histochem. Cytochem.*, **7**: 290, 1959; **8**: 326, 1960; *J. Biophys. Biochem. Cytol.*, **8**: 319, 1960; *Histochemie*, **2**: 234, 1961) explained the reaction as largely due to side-chain carboxyls of aspartic and glutamic acid where mixed anhydrides functioned as carbonyls and azolactones formed on the relatively few C-terminal carboxyls. In both instances the CO group attaches directly to one lateral carbon and by an oxygen bridge to the other, so that no true ketone appears to be formed. Methylation blocked the reaction, and we have in preliminary experiments found that saponification deblocked and that deamination was without adverse effect. When a model experiment with commercial cellulose acetate paper gave a strong azo reaction after a 2-h exposure to hydroxynaphthoic acid hydraide, without use (in our hands) of the acetic anhydride step, we decided that the method was of doubtful specificity and made no further study of it. The method was complicated and appears to have had little use. Our procedure followed closely that of Barrnett and Seligman and of Karnovsky and Fasman (*J. Biophys. Biochem. Cytol.*, **8**: 319, 1960): from absolute alcohol to pyridine to acetic anhydride-pyridine, 40 : 10, 1 h at 60°C, then 0.1% 2-hydroxy-3-naphthoic acid hydrazide in acetic acid–alcohol–water, 2 : 20 : 18, 2 h at 25°C, rinse, and azo-couple in fast black K 120 mg/30 ml pH 8.5 barbital HCl for 3 min at 3°C, then usual acid washes, 5, 5, 5 min 0.1 N HCl, then 10-min wash in running water, alcohols, xylene, and mounting in synthetic resin. This gave deep violet keratin; lighter trichohyalin, smooth and striated muscle, and epithelial nuclei; moderate coloration of collagen and cytoplasms; deeper in glycogen sites in muscle and cartilage cells. This was confirmed by the PAS test and salivary digestion. Methylation prevented the reaction; saponification restored it and gave staining of hair cortex, which does not stain directly.

Geyer [*Acta Histochem. (Jena)*, **19**: 73, 1969] noted that acid mucins and carboxyls did not stain by the Barrnett-Seligman procedure or by two methods of his own, the epoxy ether method [*Acta Histochem. (Jena)*, **14**: 1, 1962] and the hydroxamic acid method (ibid., **14**: 67, 1962). The Barrnett-Seligman technic is followed substituting for the acetic anhydride 50 to 100 mg 1-cyclohexyl-3-[2-morpholine-(4)-ethyl]-carbodiimid-*p*-toluol methyl sulfonate, 50 mg 2-hydroxy-3-naphthoic acid hydrazide, 5 ml tetrahydrofuran, 5 ml 70% alcohol, and 10 ml distilled water. Incubate 2 to 3 h at 30°C, then wash 3 times in 50% alcohol for 10 min each time and 0.5 N HCl for 30 min, wash 4 or 5 times in distilled water and in 1% NaHCO₃ 4 or 5 times and distilled water again 4 or 5 times, and azo-couple 2 to 4 min in 0.1% fast blue B.

The results parallel those of the acid anhydride method. The reaction is prevented by methylation by Geyer's thionyl chloride method. It differs from the Barrnett-Seligman technic in giving positive reactions in mast cells and submaxillary gland mucin, but gastric, goblet cell, and sublingual mucins are negative.

Pearse (1968) regarded the specificity as not yet established.

Stoward and Burns (*Histochemie*, **10**: 230, 1967) first concluded that the reddish lilac stain developed in tissue proteins following use of the Barrnett-Seligman method could not be due to the formation of derivatives of protein side-chain carboxyl groups. In later studies, Stoward and Burns (*Histochemie*, **13**: 7, 1968) were impressed by the failure of significant staining of sections treated with aniline (5% v/v in xylene for 5 days at 4°C) or methanol (containing 5% pyridine v/v for 5 days at 4°C) interposed between acetic anhydride incubation and the 2-hydroxy-3-naphthoic acid hydrazide. Stoward and Burns now conclude that the blockades by aniline and methanol confirm the reliability of the stain for the demonstration of protein side-chain carboxyl groups.

Geyer's [*Acta Histochem.* (*Jena*), **14**: 1, 1962] epoxy ether method for protein carboxyls is a histochemical application of the work of Stevens et al. (*J. Am. Chem. Soc.*, **72**: 1758, 1950; **74**: 618, 1952; **75**: 3977, 5975, 1953) showing that epoxy ethers reacted with carboxyls to form ketoesters. The reagent 1-(*p*-biphenyl)-1-methoxyethylene oxide is applied as a 2 to 5% solution in *p*-xylene in which sections are heated under reflux at about 138°C for 30 to 40 h. After this, sections are washed in xylene, brought to 50% alcohol, and reacted successively with 2-hydroxy-3-naphthoic acid hydrazide and fast blue B as in the Barrnett-Seligman reaction. Results are similar to those of that method. Protein carboxyls are said to react but not uronic acids. Lipofuscins are not mentioned, and pepsinogen granules do not appear to have reacted.

The prolonged heating in a flammable solvent at 138° and the lack of ready availability of the reagent have operated against frequent use of the method.

While the Geyer [*Acta Histochem.* (*Jena*), **14**: 67, 1962] hydroxamic acid method for tissue carboxyls follows a fairly simple technic comprising methylation to esterify carboxyls, treatment with alkaline alcoholic hydroxylamine to form carbamic acids, and chelation with Fe(III) to form a red to brown complex, the preparation of each of the reagents for the latter two steps is complicated. The reaction is said to demonstrate the same sites as the mixed anhydride reaction described earlier, and shares its limitation to protein carboxyls and nonreactivity for hexuronic acids. It is noted that gastric chief cells react only weakly in contrast to the Nile blue reaction. The reaction needs further study and simplification of the reagent to make it practical.

Burns and Stoward (*Histochemie*, **26**: 266, 272, 279, 1971) advocate a fluorescent method for the demonstration of COOH-terminal carboxyl groups of proteins. The method is essentially that previously described by Stoward [*J. R. Microsc. Soc.*, **87**: 247, 1967].

1. Sections of deparaffinized Carnoy-fixed tissues are brought to absolute ethanol and then into two changes of acetic anhydride of 2 min each.
2. Incubate sections in a 1:1 mixture (v/v) of pyridine and acetic anhydride for 1 h at 60°C.
3. Rinse in acetic anhydride 2 min at 25°C.

4. Wash in two changes of 95% ethanol (2 min each).
5. Bring sections to water, and immerse in 0.5% solution of salicylhydrazide (Sigma Pharmaceutical Corporation) in 5% aqueous acetic acid for 20 to 40 min at 25°C. This solution should be allowed to equilibrate for 48 h prior to use and discarded after 1 month.
6. Wash briefly in deionized water.
7. Remove excess salicylhydrazide from sections by immersion in a 0.5% fresh solution of trisodium pentacyanoamminoferroate (Koch-Light, Ltd. or Hopkin & Williams, Ltd., England) for 10 s. Increased exposure to this reagent leads to diminished stain.
8. Rinse briefly in deionized water and immerse sections in 1% aqueous zinc acetate for 5 min at 25°C.
9. Wash in deionized water, dehydrate through ethanols and xylol, and mount in synthetic medium.
10. Examine under ultraviolet light at 366 μg with a fluorescence microscope.

IRON METHODS FOR ACID GROUPS. While colloidal iron methods such as the Hale [*Nature (Lond.)*, **157**: 802, 1946] and its various modifications and the Lillie-Mowry iron mannitol methods have had much use in carbohydrate histochemistry, the mode of action of the iron seems to have been poorly understood. The concept that Fe(III) ions themselves bound to carboxyls and hydroxyls was widespread in biochemistry as well as in histochemistry (Baker; Breslow, *J. Biol. Chem.*, **283**: 1332, 1963; Gurd and Wilcox, *Adv. Pract. Chem.*, **11**: 311, 1956; Wigglesworth, *Q. J. Microsc. Sci.*, **90**: 105, 1952; and others). In 1972 Lillie and Pizzolato (*J. Histochem. Cytochem.*, **20**: 116, 1972) reported that Fe(II) and Fe(III) bound selectively to amino (lysine chiefly) and guanidyl (arginine) groups when used in sequence iron hematoxylin technics, stressing the nonreaction of known acid mucins and cartilage, and demonstrating further that carboxyl groups introduced on PAS–aldehyde sites by condensation with 5-aminoisophthalic acid in glacial acetic acid, though strongly basophil to azure A, did not stain by the sequence iron hematoxylins (Vacca and Lillie, *J. Histochem. Cytochem.*, **20**: 79, 1972).

Mayer (*Mitt. Zool. Stat. Neapel.*, **12**: 303 [326, footnote], 1896) reported that sections flooded with ferrous acetate and exposed to air face up in a moist chamber took up brown iron at mucin sites. The iron deposits could be turned blue by HCl and $K_4Fe(CN)_6$ or black with tannin. Here the $FeAc_2$ oxidized to $FeOHAc_2$, which is insoluble. Lillie and Mowry's (*Bull. Int. Assoc. Med. Mus.*, **30**: 91, 1949) NaOH–$FeCl_3$ controls also stained mucins. Mayer's experiment was repeated by Lillie, Vacca, and Pizzolato (*J. Histochem. Cytochem.*, **20**: 159, 1972) with success. A soluble $FeOHCl_2$ solution was prepared by careful oxidation of freshly prepared $FeCl_2$. Impregnation with this $FeOHCl_2$ at 10 mM for 2 h, followed by HCl ferrocyanide treatment, successfully demonstrated tracheal gland mucin, cartilage, duodenal goblet cells, guinea pig Brunner gland mucin, crypt and goblet cell mucins of ileum and colon, rat mast cells, sublingual and submaxillary gland mucins, umbilical cord matrix, nucleus pulposus, and gastric epithelial mucoprotein of several species. The reaction is prevented by a 4-h 60°C 0.1 M CH_3HSO_4–CH_3OH methylation, impaired but not completely blocked by acetylation, and unaffected by deamination. $FeOHCl_2$ is also prepared by adding varying amounts of 1 N NaOH to 1 M $FeCl_3$ in a final concentration of 0.1 M. This solution was used on tissue at 10 mM as with the $FeCl_2$ oxidation product and gave the best results at a 1 : 1 molar ratio NaOH–$FeCl_3$. A 3 : 1 ratio precipitated all Fe as $Fe(OH)_3$. Intermediate ratios were variably effective. The $FeCl_2$ oxidation product is stable at 0.1 M for 4 months, the NaOH–$FeCl_3$ at 1 : 1 for 3 months, and 10 mM solutions may be used repeatedly for 3 to 4 weeks.

It is known from the chemical literature that iron forms a number of complex hydroxy-chloride and chloride ions: $FeOH^{2+}$, $Fe(OH)_2^+$, $Fe_2(OH)_2^{4+}$, $FeCl^{2+}$, $FeCl_6^{3-}$, etc. (Bray and Hershey, *J. Am. Chem. Soc.*, **56**: 1889, 1934; Rabinowitch and Stockmayer, *J. Am. Chem. Soc.*, **64**: 335, 1942). The reactions are simple: $2FeCl_2 + H_2O_2 \longrightarrow 2FeOHCl_2$ and $FeCl_3 + NaOH \longrightarrow FeOHCl_2 + NaCl$. The mechanism of the colloidal ferric hydroxide in about 2 N acetic acid undoubtedly goes through the phase $Fe(OH)_3 \longrightarrow FeOH^{2+} \longrightarrow FeOHAc_2$ (insoluble basic ferric acetate).

9

CYTOPLASMIC GRANULES AND ORGANELLES

MITOCHONDRIA

Supravital Technics

These round, oval, rod-shaped, or filamentous structures are destroyed by such acid fixatives as Zenker's fluid. They may be stained specifically by supravital technics with Janus green B (C.I. 11050), diethylsafranin (C.I. 50206), Janus blue (C.I. 12211), Janus black (C.I. 11825), pinacyanol (1st ed. C.I. 808), rhodamine B (C.I. 45170), and methylene blue (C.I. 52015). Cowdry identified Janus black I as a mixture of a brown dye and Janus green B, on the latter of which its action depends. However, it appears that true Janus black has been used for mitochondria (Conn, 7th ed., p. 68, 1961). Janus blue or indazole blue is similar in its effectiveness to Janus green B. It is diethylsafranin-azo-β-naphthol (C.I. 12210). Pinacyanol is a red basic dye said to be superior to Janus green B. Rhodamine B is a weakly basic red dye.

SUPRAVITAL STAINING OF BLOOD. Cowdry directs: Mix on a clean slide a small drop of blood with a small drop of 1 : 10,000 Janus green B in 0.85% salt solution (1% aqueous solution 1 ml, distilled water 99 ml, sodium chloride 850 mg, or use 99 ml 0.86% sodium chloride solution). Drop on coverslip and let spread. Ring with petroleum jelly. In about 5 to 10 min mitochondria are colored deep bluish green. Similarly small fragments of fresh tissue may be crushed in the salt solution to a thin film.

The technics with methylene blue, Janus (indazole) blue, Janus black, and rhodamine B are basically similar.

Hetherington (*Stain Technol.*, **11:** 153, 1936) recommended pinacyanol (1st ed. C.I. 808)[1] alone or in combination with neutral red. He used a 0.1% solution of pinacyanol in 100% alcohol, diluting 1 : 40 for use with 100% alcohol, and adding, when desired, between the same and double the amount of neutral red (C.I. 50040). Schwind (*Blood*, **5:** 597, 1950) prescribed stock solutions in 100% alcohol–0.1% pinacyanol and 0.4% neutral red, 9 drops of pinacyanol and 30 drops of neutral red in 5 ml 100% alcohol. We suggest 0.3 ml, 1 ml, and 100% alcohol to make 10 ml.

Flame clean slides, flood with one of the dilute dye mixtures, drain in a vertical position, and let dry, using a gentle, warm air stream in hot humid weather. Place on a clean cover glass a small drop of blood (or marrow diluted with or crushed in a drop of serum), using

---

[1] Obtainable from Eastman Distillation Products Inc., Rochester, N.Y.

only enough blood to give a film one cell layer thick and avoid air bubbles. Cover with the stain-coated slide, and seal edges with petroleum jelly. Schwind recommends application of soft petroleum jelly with a 5-ml syringe and a short, 23-gauge hypodermic needle, and warns against moving the cover glass during the process, as this tends to rupture cells.

Mitochondria soon color a deep blue to violet in still living and motile cells. Hetherington's neutral red mixture is reported to give a red color to nuclei, "neutral red granules and vacuoles." Schwind recorded a purplish blue for nuclei.

Staining occurs more rapidly at 37°C, but preparations do not keep so long. Motility ceases somewhat below this temperature. As the colors are feebler than those in fixed preparations, brilliant illumination and a darkened laboratory are recommended.

As pinacyanol is light-sensitive, its solutions must be stored in the dark, and stained films must be kept in the dark. Preparations, even when stored in the cold, do not keep more than 2 weeks (Schwind). However, mitochondria remain deeply stained for many hours, contrasting with the fairly prompt fading in Janus green preparations.

Fixed Tissue Methods

Of more general application are methods using fixed tissue. Regaud's method (*Arch. Anat. Microsc. Morphol. Exp.*, **11**: 291, 1910) requires fixation with Möller's (Regaud's) fluid for 4 days, followed by 8 days (four changes of 2 days each according to Mallory) in 3% potassium dichromate. Wash in running water 24 h. Dehydrate in alcohol, clear in benzene, and embed in paraffin. Cut thin sections (2 to 5 μ). Deparaffinize, hydrate, and mordant 8 to 10 days in 5 to 15% iron alum [$NH_4Fe(SO_4)_2 \cdot 12H_2O$] solution. Cowdry reduces this iron alum step to 24 h in a 5% solution. Wash in water for a few minutes. Stain 24 h in aqueous 1% hematoxylin containing 10% each of alcohol and glycerol. Regaud specified 6 to 8 weeks, Cowdry required only 3 weeks' aging. Differentiate in 5% iron alum solution with microscopic control. Wash 30 min in running water. Dehydrate by the 95, 100% alcohol series. Clear in xylene and mount in balsam. Mitochondria are sharply stained black.

ANILINE–ACID FUCHSIN–METHYL GREEN. Cowdry especially recommends the aniline–acid fuchsin–methyl green method of Bensley (*Am. J. Anat.*, **12**: 297, 1911), with the Möller-Regaud fixation in place of Bensley's acetic acid–osmium tetroxide–potassium dichromate fixation:

Cowdry fixed thin slices for 4 days in Möller's (Regaud's) fluid in a refrigerator at 3°C, changing the fluid daily, and mordanted in 3% potassium dichromate for 8 days, changing every second day. Bensley used 2 ml 4% osmium tetroxide plus 8 ml 2.5% potassium dichromate, to which he added 1 drop glacial acetic acid and fixed small pieces for 24 h. Altmann had used a closely similar mixture: equal volumes of 2% osmium tetroxide and 5% potassium dichromate, warning against presence of any free chromic acid, and fixing 24 h. See Table 9-1.

After any of these fixations wash 4 to 6 h in siphon washer or 6 h to overnight in running water. Dehydrate (Altmann, Cowdry) in graded alcohols 70 to 75, 80 to 85, 90 to 95, and 99 to 100%, allowing 6 to 18 h in each. Clear with benzene or xylene and embed in paraffin. Section at 1 to 5 μ; Altmann insisted that sections over 2 μ gave inferior results; Cowdry specified 3 to 5 μ.

1. Deparaffinize and hydrate through xylene and graded alcohols. Altmann coated slides with a thin chloroform rubber solution, dried by heating, and placed sections on a slide in a drop of 1 part 4% celloidin in acetone to 4 parts alcohol, smoothed them out, and blotted them firmly with filter paper. The paraffin was then melted and the sections deparaffinized and hydrated as usual.

TABLE 9-1   MIXING SCHEDULE FOR ALTMANN'S AND BENSLEY'S OSMIC DICHROMATE FLUIDS

	2% OsO_4	5% $K_2Cr_2O_7$	0.75% CH_3CO_2H	Final content, %		
				OsO_4	$K_2Cr_2O_7$	CH_3CO_2H
Altmann	5	5		1.0	2.5	
Bensley	4	4	2	0.8	2.0	0.15

Note: One drop of glacial acetic acid is about 0.015 ml.

2. Oxidize 1 min in 1% $KMnO_4$, rinse, reduce 1 min in 5% oxalic acid, and wash thoroughly in water. Altmann did not use this step.
3. Blot excess water and cover with Altmann's aniline–acid fuchsin (Altmann used 20% acid fuchsin in cold saturated aniline water; later writers, 1.5 g in 10 ml filtered 5% aniline water, shaken at intervals for 24 h). Altmann applied the stains cold and heated to steaming, Cowdry here continued the staining 6 min as the solution cooled, Bensley and Mallory each prescribed preheating the solution to 60°C and staining 5 min as the solution cooled.
4a. Here Altmann flooded off the acid fuchsin with a mixture of 1 part saturated alcoholic picric acid and 2 parts water, differentiating with more of the same, warmed, under microscopic control. Stop the differentiation with alcohol. Clear in xylene and mount in balsam or dammar. Stains were better preserved in mineral oil.
4b. Later writers at this point washed 1 min in distilled water, stained 1 to 5 s in 1% methyl green or (Bensley) toluidine blue.
5. Wash in 95% alcohol plus xylene, then in two changes of xylene; and mount in balsam, ester gum, Permount, or mineral oil. If too much blue or green is extracted by alcohol, use *tert*-butanol or acetone to dehydrate.

RESULTS: Mitochondria are crimson, zymogen granules red, nuclei according to counterstain green or blue.

Altmann also used a formic mercuric nitrate (stable) fixation, obtaining more brilliant stains. With these fixatives a 10% acid fuchsin in 30% alcohol could be used in place of the aniline mixture.

Saturate 30% HNO_3 (sp gr 1.185) with red mercuric oxide. For use mix 10 ml of this stock solution with 30 ml water and 10 ml glacial acetic acid or 50% formic acid (sp gr 1.12). These fixatives of Altmann's should fix thin pieces adequately in 4 to 8 h. Transfer directly to 70% alcohol after fixation.

Cain (1948) after the acid fuchsin step differentiates in about 0.1% Na_2CO_3 to decolorize cytoplasm. Stop differentiation by dipping briefly in 0.125 N HCl (1% concentrated HCl in water), then wash in distilled water, counterstain briefly in 0.5% aqueous methyl blue, rinse in distilled water, dip in 0.125 N HCl for 3 s, wash in distilled water, dehydrate, clear, and mount by an alcohol, xylene, balsam sequence.

Chang (*Exp. Cell Res.*, **11**: 643, 1956) gives the following modification of the Altmann aniline–acid fuchsin method for mitochondria for use on 1-μ paraffin sections of frozen dried material. The aniline–acid fuchsin contains only 10 g acid fuchsin per 100 ml aniline-water, is acidulated with 2 drops (about 0.03 ml) glacial acetic acid, and is aged a week or more before using. Filter each time before using.

CHANG'S ANILINE–ACID FUCHSIN VARIANT FOR MITOCHONDRIA

1. Float 1-μ paraffin sections of frozen dried material on 3% formalin 1% $CaCl_2$ on albuminized slides, and warm until flattened; drain and dry on slide warmer.
2. Without deparaffinizing, immerse sections 48 h in 5% $K_2Cr_2O_7$, 1% $CaCl_2$.
3. Wash in running water 90 min, dry in air.
4. Deparaffinize and hydrate rapidly through xylene and alcohols.
5. Rinse in distilled water and stain 8 to 15 min in preheated aniline–acid fuchsin at 60°C.
6. Rinse in distilled water and differentiate 2 to 5 min in aniline water. Rinse again in distilled water.
7. Counterstain 1 min in 1% aqueous methyl green.
8. Dehydrate directly in 95 and 100% alcohols, clear with xylene, and mount in Permount.

RESULTS: Mitochondria are bright red; erythrocytes, dark red; collagen, red to purplish; nuclei, faintly green; cytoplasm, faint pink.

The exact staining time must be determined for each type of material, and the required differentiation time is determined by unstated criteria.

Material fixed in Baker's formol-calcium or alcohol freeze substitution has been found to give good results.

PHOSPHOTUNGSTIC ACID HEMATOXYLIN.   Mallory recommended a technic with phosphotungstic acid hematoxylin. Fix 24 h or more in 10% formalin, mordant 2 to 5 days in three or four changes of 5% aqueous ferric chloride solution, wash briefly in water, harden in three or four changes of 80% alcohol for 1 to 2 days, until the alcohol remains clear. Embed in paraffin as usual, and bring paraffin sections to water.

1. Oxidize in 0.25% potassium permanganate, 5 to 10 min.
2. Rinse.
3. Treat with 5% oxalic acid 3 to 5 min.
4. Wash thoroughly in tap water.
5. Stain in Mallory's phosphotungstic acid hematoxylin for 1 to 2 days.
6. Rinse in tap water.
7. Differentiate in 95% alcohol.
8. Dehydrate with 100% alcohol, clear in xylene, and mount in balsam.

RESULTS: Nuclei and mitochondria are stained deep blue; collagen and elastin, reddish; myoglia and neuroglia fibrils, blue.

Takaya (*Stain Technol.*, **42**: 207, 1967) recommends the following stain for mitochondria. Paraffin-embedded specimens fixed in formalin-Zenker or in Möller's (Regaud's) solution are advised.

1. Sections are incubated at 58°C for 3 to 6 h in a Luxol Fast Blue MBS[1] (Chroma 10980) solution prepared by dissolving 1 g of the dye in 1000 ml 95% alcohol and adding 10 ml 5% acetic acid. This solution keeps well but should be filtered before use.
2. Sections are passed through 95% alcohol to remove the excess stain, then washed in distilled water.
3. Differentiate in 0.05% aqueous Li_2CO_3 for 10 s; then immerse sections in 70% ethanol until parts of sections are colorless.

---

[1] Luxol Fast Blue MBS is C.I. solvent blue 38. It is a sulfonated phthalocyanin-dye diarylguanidinium salt, which behaves as an acid dye. Staining is enhanced by methylation and prevented by adequate deamination (Smith and Dunbar, *J. Pathol. Bacteriol.*, **91**: 117, 1966).

4. Wash in distilled water; then stain for 2 to 3 min in 0.5% aqueous phloxine B (C.I. 45410).
5. Differentiate for 2 min in 5% aqueous phosphotungstic acid, wash 5 min in running tap water, differentiate in ethyl alcohol, dehydrate, and mount in synthetic resin.

RESULTS: Mitochondria are green in a pinkish to light green cytoplasm. Nuclei are red and nucleoli are green.

Liisberg [*Acta Anat. (Basel)*, **73:** 496, 1969] recommends the following stain for mitochondria. The stain is prescribed only for paraffin-embedded tissues fixed in neutralized formalin. Liisberg heats the staining solution to boiling to bring the reagents into solution. Filtering is *not* recommended.

Hydrated deparaffinized sections are stained for 10 min at 24°C in a mixture of fast green FCF (C.I. 42053), 2 g; chromotrope 2R (C.I. 16570), 1 g; phosphomolybdic acid, 1 g; distilled water to make 100 ml. Sections are rinsed in four changes of distilled water, immersed in 1% stannous chloride ($SnCl_2$) in 0.1 N HCl for 15 to 30 s, rinsed in four changes of distilled water, dehydrated, and mounted in synthetic resin.

RESULTS: Mitochondria, erythrocytes, and secretion granules of parenchymal cells of the pancreas are red. Background staining is blue.

GOLGI SUBSTANCE

Golgi apparatus was first described as an intricate anastomosing network of cytoplasmic strands and was first demonstrated by Golgi in nerve cells in 1898. It is demonstrable by impregnations with silver, with osmium tetroxide, or both. The technics are fickle and often fail, and considerable experimentation may be necessary to obtain optimal results with any given type of material. For these reasons its study in routine pathologic material would be a matter of considerable difficulty, especially since it requires separate fixation of special tissue blocks.

RAMÓN Y CAJAL'S URANIUM SILVER METHOD. The following technic is recommended by Cowdry for young animals:

1. Fix 8 to 24 h in uranyl nitrate 1 g, formalin 15 ml, distilled water 100 ml.
2. Wash quickly in distilled water.
3. Impregnate 1 to 2 days in 1.5% silver nitrate.
4. Rinse in distilled water.
5. Reduce 12 h in a freshly prepared solution of hydroquinone 2 g, formalin 15 ml, distilled water 100 ml, anhydrous sodium sulfite 150 mg. (*This is Ramón y Cajal's developer.*)
6. Wash in distilled water. Dehydrate, clear, and embed in paraffin, using a short alcohol or acetone schedule.
7. Deparaffinize sections and mount.

DA FANO'S COBALT SILVER METHOD. This method is of more general application. Thus, from Cowdry:

1. Fix in 1 g cobalt nitrate [$Co(NO_3)_2 \cdot 6H_2O$], 100 ml distilled water, 6 to 15 ml formalin for 3 to 18 h. Embryonic tissues require the smaller amounts of formalin. With ordinary adult tissues use 15 ml. Cartilage and small organs, such as those of mice, are fixed adequately in 3 to 4 h; routine tissues, in 6 to 8 h; central nervous tissues, in 8 to 18 h.
2. Rinse in distilled water.

3. Impregnate 1 to 2 days in 1.5% silver nitrate (1% for very small, easily permeable fragments, 2% for fatty and central nervous system tissues).
4. Rinse in distilled water.
5. Cut blocks thinner than 2 mm. Reduce in fresh Ramón y Cajal's developer (see above) for 12 to 24 h.
6. Wash in distilled water 30 min. Dehydrate, clear, and embed in paraffin. For greater permanence, bring sections to water, tone 1 to 2 h in 0.1 to 0.2% gold chloride, wash, and counterstain with alum carmine.

RESULTS: Golgi apparatus is black. This technic is essentially that given by Cowdry.

ELFTMAN'S AOYAMA SILVER METHOD VARIANT FOR GOLGI SUBSTANCE.
Elftman (*Anat. Rec.*, **106**: 381, 1950, and letter, May 5, 1950) recommends an Aoyama Golgi body technic for the selective demonstration of Sertoli cells. Fix mouse testis 18 to 24 h in neutral 10% formalin (carefully decanted from the excess calcium carbonate) to which is added 1% (43.5 mM) of cadmium chloride ($CdCl_2 \cdot 2\frac{1}{2}H_2O$). Rinse in two changes of distilled water. Immerse for 16 h in 1.5% silver nitrate. Rinse quickly in two changes of distilled water. Reduce 5 h in Ramón y Cajal's developer or in an acid fluid containing instead of 2 g just 1 g hydroquinone, 15 ml formalin, 100 ml distilled water, and 150 mg sodium bisulfite ($NaHSO_3$). Wash, dehydrate, clear, and infiltrate and embed in paraffin as usual.

If desired, counterstain by the azure-eosin method or by some other similar Romanovsky technic.

Sertoli cell cytoplasm blackens, as does Golgi material in the spermatogenic cell series.

Elftman comments that both he and Baker find a 5-h reduction more than ample in this and the Da Fano technic. Elftman has had poor success in applying this procedure to paraffin sections. Frozen sections might be tried.

Elftman later (*Stain Technol.*, **27**: 47, 1952) found prior fixation productive of irregularities and introduced gold toning to confer greater resistance to subsequent histochemical procedures.

ELFTMAN'S DIRECT SILVER METHOD FOR GOLGI SUBSTANCE

1. Place small blocks of fresh tissue in 15% formalin containing 2% silver nitrate (pH 4) for 2 h. (Adjust with a drop or two of acetic acid if necessary, or use pH 4 acetate buffer as the solvent.)
2. Rinse 5 s with 15% formalin or distilled water.
3. Immerse in 15% formalin containing 2% hydroquinone for 2 h.
4. Complete fixation by an additional overnight treatment with 15% formalin.
5. Dehydrate with alcohols, clear, and infiltrate and embed in paraffin, and section as usual.
6. Sections may be deparaffinized and mounted at this point for preliminary examination. Otherwise deparaffinize and hydrate as usual.
7. If necessary, reduce density of silver deposit by oxidizing in 0.1% iron alum for a few minutes under microscopic control.
8. Wash in distilled water.
9. Tone in 0.2% acid gold chloride for 10 or 15 min.
10. Rinse in distilled water and remove silver chloride in 2% sodium thiosulfate, 2 min.
11. Wash 5 min in running water.
12. Counterstain as desired: e.g., the periodic acid Schiff procedure, the allochrome method, a Mallory aniline blue variant, the Van Gieson stain.

13. Dehydrate, clear, and mount in synthetic resin (or balsam; a reducing medium seems indicated).

RESULTS: Golgi elements are blackened, but, as Elftman cautions, not all that is black is Golgi material. Omission of the gold toning leaves the silver deposits much more susceptible to oxidative fading, either by subsequently employed histochemical reagents or by the mounting medium.

THE OSMIUM TETROXIDE METHOD. Cowdry recommends especially Ludford's (*J. R. Microsc. Soc.*, **46:** 107, 1926) variants of the osmium tetroxide methods, thus:

1. Fix thin blocks 18 h in Mann's osmic sublimate fluid.
2. Wash 30 min in distilled water.
3. Impregnate 3 days at 30°C in 2% osmium tetroxide; or 1 day in 2% osmium tetroxide at 35°C, 1 day in 1%, and 1 day in 0.5%.
4. Wash 1 day in water at 30 or 35°C.
5. Dehydrate, clear, and embed in paraffin.
6. Cut sections at 3 to 5 μ; deparaffinize and mount.

RESULTS: Golgi apparatus, yolk, and fat are blackened. Other structures remain yellow to brown.

By treatment with turpentine the blackening is gradually removed from yolk and fat, perhaps completely in 10 to 15 min, leaving the Golgi apparatus black (Lee, Baker). If mitochondria are not blackened, they may be counterstained (Ludford) with Altmann's aniline–acid fuchsin, heating to steaming and letting stand for 30 min. Then wash in water and differentiate with 100% alcohol. Or nuclei may be counterstained with safranin, crystal violet, or neutral red for a few minutes in dilute (1 : 1000?) aqueous solutions. Again dehydrate and differentiate with alcohol.

Estable-Puig et al. (*J. Neuropathol. Exp. Neurol.*, **24:** 531, 1965) highly recommend the following general overall staining method for osmium-fixed, plastic-embedded tissues for light and phase microscopy. The method employs the strong reducing agent paraphenylene-diamine. Estable-Puig et al. speculate that osmium tetroxide in tissues may oxidize the reagent, with highly colored oxidation products as a result.

1. Cut sections 0.5 to 2 μ thick on a Porter-Blum ultrathin microtome with glass knives, float them onto glass slides, and allow them to dry at room temperature.
2. Place slides in a Coplin jar containing a freshly filtered 1% solution of paraphenylene-diamine. The tissue becomes visible as brownish black spots within 15 min to 1 h depending upon the thickness of the section and the osmium content of the tissue.
3. Rinse slides in distilled water for 2 min.
4. Dehydrate in two changes of 95% or absolute ethanol. The alcohol clears the sections of a nonspecific background stain that is sometimes seen if an old solution of paraphenylenediamine is used, but it does not alter or reduce the intensity of staining of osmiophilic material.
5. Allow the slides to dry and then mount them under a coverslip with cedar oil, Canada balsam, or Permount.

Crystals of what appear to be oxidative products of paraphenylenediamine may cover the sections if old or unfiltered solutions are used. A quick dip into weak acetic acid solution before step 3 can save the slide. The use of xylene is to be avoided, because it will soften the plastics and cause wrinkling of the sections.

Chu (*Stain Technol.*, **47**: 27, 1972) states that the Golgi can be nicely demonstrated in many cells (including spinal ganglion and epididymis) in fresh frozen sections stained only with 1% aqueous potassium permanganate.

BAKER'S SUDAN BLACK.   Baker (*Q. J. Microsc. Sci.*, **90**: 293, 1949) revised his 1944 method (ibid., **85**: 1, 1944) cited in Lillie (1948) by introducing a chromation after formaldehyde fixation, less vigorous than that of Ciaccio. The method differs from Ciaccio's in that no fat solvent test is applied. Neutral fats are colored in the same way as Golgi bodies; but Baker now states that the latter do not stain with Sudan IV, reversing his previous position.

TECHNIC

1. Fix blocks 2 to 3 mm thick for 1 h in 10% formalin containing 0.7% sodium chloride and kept over marble chips.
2. Transfer directly to a fresh formalin-dichromate mixture containing 88 ml 2.5% $K_2Cr_2O_7$, 7 ml 10% NaCl, and 5 ml neutral formalin (kept over marble chips), and let stand for 5 h.
3. Transfer directly to 5% potassium dichromate for 18 h at room temperature, and then place in 60°C paraffin oven for 24 h.
4. Wash 6 h in running water.
5. Infiltrate at 37°C for 18 h (or any convenient longer period) in 25% gelatin containing 0.2% sodium *p*-hydroxybenzoate.
6. Cool, block, and harden in *formalum*, a solution of 20 ml formalin, 4 g potassium alum, and 80 ml water, for 18 h, or any convenient longer period.
7. Section at 8 to 10 μ on freezing microtome.
8. Transfer section to 70% alcohol.
9. Stain $2\frac{1}{2}$ min (30 s to 4 min) in a saturated solution of Sudan black B. (Boil 0.5 g in 100 ml 70% alcohol under a reflux condenser for 10 min, or saturate by shaking at intervals for several days.)
10. Wash 5 s in 70% alcohol and then 1 min in 50% alcohol.
11. Wash in water, sinking section below surface.
12. Counterstain in Mayer's carmalum (or other alum carmine) for 2 to 4 min.
13. Rinse in *distilled* water and wash in two large changes of water.
14. Float onto slide, blot, and mount in Farrants' glycerol gum arabic. Harden mounts overnight in paraffin oven.

RESULTS: Solid Golgi bodies and outer parts of hollow ones are dark blue; Golgi vacuoles, colorless; cytoplasm, pale gray-blue to colorless; chromatin, red or pink; neutral fats (triglycerides), dark blue.

CHARACTERISTICS OF THE GOLGI SUBSTANCE.   The Smith-Dietrich procedure stains the Golgi substance gray to black if the differentiation procedure is shortened to 8 h.

The Golgi substance is negative to the Windaus and Schultz cholesterol tests. It is removed when fresh tissues are fixed in alcohol or other lipid solvents. After fixation in calcium-formalin it withstands boiling water (30 min), ether (15 h), 1 N hydrochloric acid at 57°C (4 h), boiling 100% alcohol and boiling ether (30 min each), or ether–alcohol–hydrochloric acid at 50°C for 5 h; but not 30 parts xylene and 70 parts glacial acetic acid at 57°C for 15 h. (The last was devised as a myelin solvent.) It resists the usual paraffin embedding technic, thus resembling myelin after formalin fixation.

Elftman reported at the 1953 meeting of the Histochemical Society that the Golgi substance was peracetic acid Schiff–positive. In the discussion which followed Lillie brought out that the periodic acid Schiff reaction of the Golgi substance reported by Gersh (*Arch. Pathol.*, **47**: 99, 1949) and attributed by him to glycoprotein, and the peracetic acid Schiff reaction of this substance were probably both due to unsaturated fatty acids and their oxidation products (*J. Histochem. Cytochem.*, **1**: 387, 1953).

After formalin fixation the phosphatides cephalin, sphingomyelin, and lecithin likewise resist solution in ether, acetone, or 100% alcohol, but may be extracted by pyridine, by a 30 : 70 xylene-acetic mixture, or by a sequence treatment in saturated aqueous sodium oleate or ricinoleate (24 h) followed by 30 min each in boiling 100% alcohol and boiling ether. Cephalin and lecithin are blackened by osmication and withstand decolorization by turpentine; sphingomyelin does not.

Hence the Golgi substance probably consists of cephalin, lecithin, or both, at least in the snail spermatocytes investigated by Baker.

Of recent years quite a number of reports of specific enzyme localization in the Golgi area of various cell systems have appeared.

Cheetham et al. (*J. Cell Biol.*, **44**: 492, 1970) isolated from rat liver a substance they believe to contain Golgi complex material in approximately 80% purity. Contaminating material included endoplasmic reticulum, mitochondria, lysosomes, plasma membranes, and microbodies. Cheetham et al. noted substantial nucleoside di- and triphosphatase activities.

GENERAL ENDOCRINE GRANULE METHOD

According to Solcia et al. (*Stain Technol.*, **43**: 257, 1968), a 3- to 10-h 60°C 0.2 N HCl hydrolysis eliminates basophilia due to the nucleic acids (DNA, RNA) and mucopolysaccharides and induces basophilia to pH 5 with 0.01 to 0.005% azure A or toluidine blue or to 0.02% pseudoisocyanin in distilled water. Thus demonstrated are pancreatic islet alpha and delta cells, enterochromaffin, basophil cells of gastrointestinal mucosa, adrenaline (Ad) cells of the adrenal medulla, thyroid parafollicular (C) cells, and pituitary thyrotropic (Tt) basophils.

The best fixatives were 6% glutaraldehyde at pH 7.2 for 24 h; glutaraldehyde $HgCl_2$ (6 : 6 : 100); 25% glutaraldehyde, 1 volume; saturated aqueous picric acid, 3 volumes plus 1% NaAc or HAc (GPA); formol sublimate (5 to 6% $HgCl_2$–10% formol; and Bouin's fluid with only 1% HAc. Fixation times were omitted in Table I of Solcia et al. Tissues were hardened 1 to 2 days in 70% alcohol. Paraffin sections at 4 to 5 μ floated on a drop of water on a slide and dried in an oven at 40 to 50°C for 24 h.

HYDROLYSIS.   One to five percent H_2SO_4, HAc, and CCl_3CO_2H all worked, but 0.1 to 0.4 N HCl was best, and 0.2 N was taken as standard; 50°C was inadequate, 80°C too destructive, 60 to 70°C best; time required was 3 to 10 h. Optimum hydrolysis times were 3 h for enterochromaffin and endocrine cells of pyloric mucosa in guinea pigs (G cells), alpha and delta cells of the islets of the pancreas, calcitonin-producing cells of the thyroid (C cells), thyrotropic cells of the pituitary, and adrenaline-producing cells of the adrenal. Nuclear staining decreased from a maximum rating of 6 to 5 at 30 min, 4 at 1 h, 2 at 2 h, and 0 at 3 h. Pancreatic acinar cells gave ratings of 6, 4, 3, 1, 0 for the same times; goblet cells, 4, 6, 6, 4, 2, and 0 at 8 h; mast cells, 6, 6, 6, 5, 4, and 2. Thyroid C cells averaged best at 4 h with Bouin and GPA fixatives.

STAIN.   Toluidine blue, 0.01 to 0.005%, gave good stain with metachromasia. Azure A, 0.01 to 0.005%, gave a stronger but less metachromatic effect. Pseudoisocyanin was superior for fluorescence or in monochromatic light at 5780 Å. Alcian blue resisted alcohol dehydration; the other stains lost metachromasia and fluorescence on alcohol dehydration. The best pH level was 5; there was too much background at 6 to 7; basophilia of alpha and delta islet cells; and G, C, and Ad cells lost staining at pH 4.3 to 4.5, enterochromaffin at 4.1 to 4.2, pituitary Tt cells at 3.8. DNA, RNA, myelin, and sialomucin lost staining below pH 3 and lost metachromasia below pH 1.

Addition of NaCl extinguished staining at 0.3 to 0.5 M for alpha and delta cells of islets of the pancreas, endocrine cells of the pyloric mucosa in guinea pigs, calcitonin-producing cells of the thyroid (C cells), adrenotropic and thyrotropic cells of the pituitary submaxillary sialomucin and myelin; enterochromaffin and mast cells required 0.9 to 1 M NaCl to prevent staining with 0.0005% toluidine blue. The addition of 0.3 M $MgCl_2$ to 0.05% solutions of Alcian blue prevented staining of all endocrine cells as well as submaxillary glands; mast cells required 0.9 M $MgCl_2$ to abolish staining.

Argentaffin and argyrophil methods and azo coupling of enterochromaffin can precede the process.

He explains the reaction as due to hot HCl extraction of nucleic acids and mucins. Glutaraldehyde protects the specific hormones better than formol. He attributes reactions to COOH groups of proteins and suggests that adrenaline and enterochromaffin also contain protein.

TECHNIC.   Fix in glutaraldehyde, glutaraldehyde sublimate 5 : 5 : 100, formol sublimate, or Bouin's or Solcia's GPA, for thyroid C cells. Embed in paraffin and cut at 4 to 5 μ. Float on a drop of water on a slide, and dry in an oven at 45 to 50°C for 24 h.

1. Dewax in xylene, and bring through alcohols to water.
2. Hydrolyze in 0.2 N HCl at 60°C for 8 to 10 h with glutaraldehyde fixations.
3. Rinse in three changes of distilled water, and stain in 0.01% toluidine blue at pH 5 for a few minutes, mount in a drop of stain, and examine. Photograph for preservation of results.

Acetone or *tert*-butanol dehydration, xylene, and cellulose caprate can be tried; Solcia et al. say alcohol destroys metachromasia. Azure A in the same pH 5 buffer and 0.01% concentration gives more intense staining but less metachromasia. Pseudoisocyanin is recommended for sensitivity and selectivity but must be studied by fluorescence microscopy. Alcian blue stains will stand routine use of alcohol, xylene, and balsam.

SALIVARY GLAND

The zymogen granules are only moderately oxyphil to neutral stain mixtures. They are dissolved by fixatives containing acetic acid, such as aqueous or alcoholic 5% acetic–10% formalin, Carnoy, Zenker, Tellyesniczky's acetic dichromate, Bouin, and Gendre fluids. Their oxyphilia is better shown after fixation with such fluids as Kose's dichromate formalin or Spuler's Zenker formalin than with neutral aqueous formalin. The azure A–eosin B technic is suggested for this reaction.

Rabinovitch et al. (*Exp. Cell Res.*, **42:** 634, 1966) noted that secretory granules are partially destroyed in tissues fixed initially in OsO_4; however, excellent preservation of granules (two types) was observed in tissues fixed initially in 6.25% glutaraldehyde in 0.075 M cacodylate buffer at pH 7.4 for 4 h and postfixed in OsO_4. The granules resist digestion with 1 : 1000 malt diastase in pH 6 saline solution, which removes the basophilic ribonucleic acid from the basal cytoplasm.

Parotid zymogen granules give a positive Schiff aldehyde reaction after oxidation with periodic acid but not after potassium permanganate or chromic acid oxidation. The periodic Schiff–positive material resists diastase digestion. These reactions have been recorded in human beings, rabbits, rats, and guinea pigs. Positive (blue-violet) Gram-Weigert staining is sometimes noted in rabbits and rats.

Tryptophan reactions with the xanthydrol, the rosindole, and the postcoupled benzylidene methods have been recorded for human, monkey, and mouse serous parotid acini, for submaxillary demilunes in the monkey, and for the eosinophilic granules of mouse submaxillary duct cells (*J. Histochem. Cytochem.*, **5**: 188, 279, 1957). The tyrosine reaction by the diazotization coupling method also appears in the same locations.

Salivary gland mucins are discussed under aminopolysaccharides.

PANCREAS

Like those of the serous salivary gland acini, pancreatic zymogen granules are destroyed by aqueous and alcoholic acetic formalins, and by Carnoy's and Bouin's fluids, and they disappear from all but the most superficial acini on fixation with Tellyesniczky's dichromate, Zenker's and Gendre's fluids. Pancreatic zymogen granules may be protected from solution in aqueous fixatives containing 5 to 6% acetic acid by adding 5 to 6% NaCl to the solution. Mercuric chloride–acetic acid, acetic formol, and Zenker's fluid have been successfully thus modified and erythrocytes and pancreatic zymogen granules well preserved (Lillie et al., *Am. J. Clin. Pathol.*, **59**: 374, 1973). They are well preserved by buffered neutral formalin, by Orth's, Kose's, and Möller's dichromate formalin fluids, and by formalin-Zenker variants. They are also well preserved by buffered neutral mercuric chloride without formaldehyde (B-4). This fixation is useful for the study of amine group reactions.

They stain conspicuously red after neutral formalin and formalin dichromate fixations with azure A–eosin B at pH 4 to 4.5; at pH 5, for Spuler and Helly fixations; and at 5.5 to 6 after neutral sublimate (B-4). They become especially conspicuous if cytoplasmic ribonucleic acid is first removed by ribonuclease digestion or ice-cold mineral acid or alcoholic alkali extraction, or if it is rendered unreactive by methylation.

In contradistinction to salivary zymogen granules, those of the pancreas may stain or remain unstained by the periodic acid leucofuchsin technic. Both conditions have been seen in man and mice; the negative phase has been seen in rat and guinea pig pancreas; but in rabbits we have seen these granules HIO_4 Schiff–positive. They are Bauer and Casella–negative and are usually violet with the Gram-Weigert method.

Pancreatic Islets

Bensley's method with neutral stains was designed especially for differentiation of pancreatic islet cells. The neutral stains were made by precipitating aqueous crystal violet with an approximate stoichiometric proportion of acid fuchsin or orange G, or safranin with an acid violet, which last procedure is presented in detail below. Prolonged staining with a 24-h-old 20% alcohol dilution of the stock 100% ethyl alcohol solution of the crystal violet salts was used. In all three, differentiation is carried out under microscopic control with clove oil and alcohol.

SAFRANIN ACID VIOLET. Bensley's (*Am. J. Anat.*, **12**: 297, 1911) safranin acid violet has been used with striking contrasts. This is a true neutral stain made by precipitating saturated aqueous safranin O (C.I. 50240) by cautiously adding saturated aqueous acid violet. Bensley did not specify which acid violet, but fast acid violet 10B (C.I. 42571) and eriocyanin A (C.I. 42576)—two rather blue violets—and a reddish violet formyl violet S4B (C.I. 42650) readily form coarse granular precipitates with safranin; and all will probably serve. Eriocyanin A is the bluest of the three and resembles the color depicted in Maximow's "Histology," first edition, p. 692, Saunders, Philadelphia, 1930. The eriocyanin A is added until on splashing the mixture up the sides of the flask the color changes, rather abruptly, from red to violet. The precipitate is collected on a filter, dried, and dissolved in 100% alcohol as with Bensley's other neutral stains, and the staining technic follows the same general procedure.

ERIOCYANIN A (GEIGY). Two dyes are marketed under this name, acid blue 34, C.I. 42561, benzylpentamethylpararosanilin disulfonate, and acid blue 75, C.I. 42576, dibenzyltetramethylpararosanilin disulfonate, "patentblau AE." The second of these was identified as the dye best corresponding to Bensley's acid violet and is prescribed in the safranin–eriocyanin A technic in Lillie (1954). In view of the confusion in naming it would perhaps seem wise to abandon the name "eriocyanin A" and revert to the term "patent blue AE," cited as the original name in the "Colour Index." But patent blue AE is now used by two companies to designate C.I. 42090, which is a dibenzyldiethyldiamino-triphenylmethane trisulfonate (ammonium salt). Hence we adhere to the name previously used and include C.I. 42576 in its description.

Bensley specified fixation with his acetic osmic dichromate mixture or with the Spuler-Maximow formalin-Zenker fluid. Sections should be less than 5 μ thick. Dilute the stock alcoholic stain with an equal volume of distilled water, let stand 30 min, filter, and use at once. Stain 5 to 30 min (Bensley), 40 to 50 min (Maurer and Lewis, *J. Exp. Med.*, **36:** 141, 1922), or as long as 2 h (Cowdry, McClung). Blot and dehydrate with acetone; clear in xylene or toluene. Maurer and Lewis passed sections rapidly through 95% alcohol to remove precipitates, dehydrated in 100% alcohol, and cleared in benzene. Then differentiate sections individually under microscopic control in a mixture of 3 volumes clove oil and 1 volume 100% alcohol. Wash thoroughly with xylene (or toluene or benzene) and mount in balsam (in originals).

Nuclei and granules of beta cells of pancreatic islets stain red; granules of alpha cells of pancreatic islets (Bensley) and of hypophysis and erythrocytes stain blue to violet. Hypophyseal chromophobe cells show at most a faint violet stippling.

Maurer and Lewis identified the red-staining cells as alpha cells and the blue as beta, and commented on the basophilia of the so-called acidophil cells of the hypophysis produced by this method. They further stated that the various types of chromophil cells are impossible to differentiate cytologically, except by the specific granule stains. Lillie found that the same cell areas in which nearly all the cells stain pink with azure-eosin stain predominantly blue in the adjacent section with safranin-eriocyanin A.

On trial of the foregoing method it is found that the high alcohol content (50%) of the final staining mixture inhibits staining in covered Coplin jars, just as it does with Wright's blood stain. However, only Maurer and Lewis spoke of staining on the slide as is customary with blood stains.

In order to test our various acid violets to ascertain which was the best for this method, we devised the following method. It seems to work well in practice and entails far less trouble than the traditional Bensley method.

LILLIE'S SAFRANIN O–ERIOCYANIN A METHOD

1. Bring thin paraffin sections through xylenes and alcohols to water as usual, treating mercury-fixed material 5 min with 0.5% iodine in 70% alcohol and 1 min in 5% sodium thiosulfate.
2. Stain 30 min in

1% safranin O (C.I. 50240)	2 ml
1% eriocyanin A (C.I. 42576)	2 ml
0.1 M citric acid	1.3 ml
0.2 M disodium phosphate	0.7 ml
Distilled water	34 ml

3. Rinse in water, dehydrate with acetone, clear in acetone and xylene and two changes of xylene, mount in polystyrene or other synthetic resin.

RESULTS: Nuclei and beta cells are red; erythrocytes and alpha cells, blue. Staining is already fairly good in 10 min and is perhaps slightly better in an hour than in 30 min. A pH 3.8 acetate buffer can probably be used.

If desired the buffer may be omitted and Bensley's 3 : 1 clove oil plus 100% alcohol mixture used for differentiation instead.

Lillie used this method also on hypophyses fixed in neutral formalin as well as with Orth's and Spuler's fluids, with perhaps even more brilliant contrasts. Negri bodies and necrotic nerve cells appear conspicuous in light blue by this method.

Maldonado and San José (*Stain Technol.*, **42:** 11, 1967) prescribe a phloxine-azure-hematoxylin method to differentiate alpha, beta, and delta cells of the islets of Langerhans. Human material must be fresh. Tissues fixed in Heidenhain's susa or in Bouin's fluid without acetic acid are washed in several changes of ethanol until the picric acid is removed. Tissues fixed in susa are transferred to 75% ethanol containing 0.5% iodine, dehydrated, and embedded in paraffin. After deparaffinized and hydrated sections have been subjected to the usual iodine and thiosulfate treatment to remove mercury, they are stained in a 1% aqueous solution of phloxine B, rinsed in distilled water, treated with 3% aqueous phosphotungstic acid for 1 min at 23°C, rinsed in distilled water, stained with azure B (0.05% aqueous for 0.5 min), washed in distilled water, stained with Weigert's iron hematoxylin for 1 min, rinsed in distilled water, dehydrated, and mounted in synthetic medium. The pH of the azure B staining solution was not provided.

RESULTS: Alpha cells stain purple; beta cells stain violet-blue; delta cells stain light blue. The phosphotungstic mordant after phloxine staining aided retention of the dye. Exocrine cells are stained grayish blue with red secretion granules.

Coalson (*Stain Technol.*, **41:** 121, 1966) advocates the following method for insulin-containing beta cells. Tissues may be fixed in any formalin-containing fixative.

1. Bring paraffin-embedded sections to water.
2. Oxidize sections for 30 to 45 s at 23°C in 0.25% $KMnO_4$ in 0.5% H_2SO_4.
3. Rinse in tap water.
4. Bleach in 5% oxalic acid for $\frac{1}{2}$ min.
5. Rinse in running water for 5 min.
6. Wash in distilled water 3 to 5 min 2 times.
7. Stain for 20 min at 23°C in a solution containing 35 mg *N,N'*-diethylpseudocyanin chloride[1] in 100 ml distilled water.
8. Wash in running water.
9. Wash in cold distilled water (4°C) 1 to 3 h, and mount in an aqueous medium such as Kaiser's glycerogel.

RESULTS: Insulin-containing beta cells are stained metachromatically light to dark purple. The background stains orthochromatically. Coalson notes that the stain from beta granules can be removed by running water within 15 to 20 min. The staining solution is stable at refrigerator temperature and can be kept for at least a year. The results for pseudoisocyanin are comparable to those obtained with aldehyde fuchsin.

Levene (*Stain Technol.*, **39:** 39, 1964) prescribes the following stain for alpha granules

---

[1] The dye may be obtained from Farbenfabrik Bayer, Leverkusen-Küppersteg, Alte Landstrasse 56, Germany.

in islets of Langerhans. Tissues may be fixed with 10% formalin, with 10% formalin with 5% acetic acid, and with Bouin's.

1. Bring paraffin-embedded sections to water.
2. Oxidize in 0.3% aqueous $KMnO_4$ for 10 min at 23°C.
3. Wash in running water.
4. Decolorize in 4% aqueous $K_2S_2O_5$ for 5 to 10 s.
5. Wash thoroughly in running water and rinse in distilled water.
6. Mordant in 4% aqueous iron alum for 1 h at 23°C.
7. Rinse briefly in distilled water.
8. Stain in phosphotungstic hematoxylin overnight.
9. Wash in 95% ethanol 10 to 20 s.
10. Dehydrate and mount in synthetic medium.

RESULTS: Alpha granules are blue in a mauve cytoplasmic background. Staining is slightly more discrete in formalin-fixed than in Bouin-fixed tissues. Other oxidants (iodine, peracetic acid, periodic acid, and chromic acid) were ineffective.

Solcia et al. (*Histochemie*, **20**: 116, 1969) used a modification of MacConaill's lead hematoxylin to stain alpha and delta cells of the islets of Langerhans, C cells of the thyroid, enterochromaffin cells, gastric G and X cells, pituitary ACTH and MSH (melanocyte-stimulating hormone) cells, adrenal medullary cells, and chemoreceptive cells of the carotid body.

To prepare the lead hematoxylin, a 5% solution of lead nitrate in distilled water is added to an equal volume of saturated aqueous solution of ammonium acetate. The mixture is filtered, and 2 ml of 40% formaldehyde is added to each 100 ml filtrate. This makes up the stock stabilized lead solution which will keep for several weeks at room temperature.

The stain is made by adding 0.2 g hematoxylin (Haematoxylin Krist, Merck) in 1.5 ml of 95% ethanol to 10 ml of the lead solution and diluting with 10 ml distilled water, stirring. After 30 min the mixture is filtered and diluted to 75 ml by addition of distilled water.

Deparaffinized and hydrated sections are stained for 2 to 3 h at 37°C, 1 to 2 h at 45°C, or 6 to 8 h at 23°C. At 37°C solutions blackened after 4 to 5 h. Reactive structures listed above stain blue-black.

ANILINE BLUE–ORANGE G. Bensley also recommends a variant of Mallory's aniline blue–orange G stain for pancreatic islets, which we requote from Mallory with slight changes: Fix fresh tissue in thin slices with the Spuler-Maximow fluid for 4 to 24 h.

1. Presumably, treat as usual with iodine and sodium thiosulfate.
2. Stain 10 min in Altmann's aniline–acid fuchsin.
3. Wash rapidly in previously boiled and cooled distilled water.
4. Mordant 10 min in 1% aqueous phosphomolybdic acid solution.
5. Drain and stain 1 h or less in 0.5 g aniline blue (C.I. 42755) or methyl blue (C.I. 42780) and 2 g orange G (C.I. 16230), 100 ml distilled water.
6. Drain, and differentiate in 95% alcohol until gross color clouds cease coming off.
7. Dehydrate with 100% alcohol, clear in xylene, and mount in balsam.

RESULTS: Alpha granules are orange-red; beta granules, bluish; acinar cells, bluish violet, sometimes with orange zymogen granules; erythrocytes, red. Bencosme (*Arch. Pathol.*, **53**: 87, 1952) recommends a very elaborate schedule using a Masson trichrome procedure, for which the reader is referred to the Bencosme article.

CHROME ALUM HEMATOXYLIN PHLOXINE. Bell (*Am. J. Pathol.*, **22**: 631, 1946) strongly recommended Gomori's chrome alum hematoxylin–phloxine method. Gomori (*Am. J. Pathol.*, **17**: 395, 1941) directed as follows: Bouin's fluid or formalin-Zenker fixations are preferred. Cut paraffin sections at 2 to 4 μ.

1. Run sections through xylene and alcohols to water.
2. Refix in Bouin's fluid for 12 to 24 h.
3. Wash sections thoroughly in tap water to remove picric acid.
4. Treat sections for about 1 min with a solution containing about 0.3 % each of potassium permanganate and sulfuric acid.
5. Decolorize with a 2 to 5 % solution of sodium bisulfite. Wash.
6. Stain in the following hematoxylin solution under microscopic control until the beta cells stand out deep blue (about 10 to 15 min): Mix equal parts of 1 % aqueous hematoxylin and 3 % chrome alum. Add to each 100 ml of the mixture 2 ml 5 % potassium dichromate and 2 ml 0.5 N sulfuric acid. The mixture is ripe after 48 h and can be used as long as a film with a metallic luster will continue to form on its surface after 1 day's standing in a Coplin jar (about 4 to 8 weeks). Filter before use.
7. Differentiate in 1 % (0.125 N) hydrochloric acid–alcohol for about 1 min.
8. Wash under the tap until the section is a clear blue.
9. Counterstain with 0.5 % aqueous solution of phloxine (B?) for 5 min. Rinse.
10. Immerse in 5 % phosphotungstic acid solution for 1 min.
11. Wash under the tap for 5 min. The section should regain its red color.
12. Differentiate in 95 % alcohol. If the section is too red and the alpha cells do not stand out clearly enough, rinse the section for about 15 to 20 s in 80 % alcohol.
13. Transfer to 100 % alcohol, clear in xylene, and mount in balsam.

RESULTS: Beta cells are blue; alpha cells, red; delta cells, from pink to red and indistinguishable from alpha cells; acinar zymogen granules, red to unstained in pancreas. In hypophysis, alpha cells are pink; beta cells, gray-blue and not readily distinguished from chromophobes; nuclei, red-purple to blue-violet; erythrocytes, deep pink; smooth muscle, pink; collagen, unstained; goblet cell mucin, coarsely granular and dark, slightly greenish blue in color.

Bell found that this method permitted ready distinction of alpha and beta granules in the pancreatic islets even on material fixed as much as 12 h postmortem.

In place of steps 9 to 12 Bencosme (*Arch. Pathol.*, **53**: 87, 1952) substitutes a ponceau–acid fuchsin mixture as used for Masson stains and stains 15 to 45 min, rinses in 1 % acetic acid, and differentiates in 1 % phosphomolybdic acid until alpha and beta cells are clear (5 to 30 min). Rinse in 1 % acetic acid. Dehydrate with 100 % alcohol.

Mitochondria of islet cells are pale red to orange-red, alpha granules are deep red, beta granules are blue-gray to black, and cytoplasm of delta cells is pale gray or gray-orange.

Ferner (*Virchows Arch. Pathol. Anat.*, **319**: 390, 1951) used the Gros-Schultze silver method for the specific silvering of islet alpha cells and of certain cells other than the enterochromaffin cells in the gastric mucosa (dog). The beta granules are not blackened, but Heidenhain-Kultschitzky cells do blacken along with Ferner's "silver cells" in the gastric mucosa. Ferner's silver cells do not react by the Masson argentaffin method (but see Bencosme, below), with methenamine silver, nor with the ferric ferricyanide reduction test, which demonstrates the argentaffin cells of the duct epithelium.

The Gros-Schultze method was originally devised for demonstration of axons and neurofibrils in frozen sections. Material fixed for 10 days to several months in neutral formalin is used. Sections are first impregnated in silver nitrate and then in an ammoniacal silver nitrate solution containing a little excess of ammonia to inhibit silvering of nuclei and connective tissue. Exact prescriptions vary. The original prescription called for formaldehyde treatment between the silver nitrate and the ammoniacal silver steps. Landau's variant for paraffin sections used formaldehyde before the nitrate and after the diammine silver, but not between.

Bencosme (*Arch. Pathol.*, **53:** 96, 1952) prescribes fixation in 10% formalin or in the trichloroacetic acid variant of Bouin's fluid (saturated picric acid–40% formaldehyde–2% trichloroacetic, 75 : 25 : 5). After paraffin embedding he employed one of Masson's argentaffin cell stains, the Laidlaw reticulum method, or Roger's silver method according to Van Campenhout (*Proc. Soc. Exp. Biol. Med.*, **30:** 617, 1933). It appears probable that all these are argyrophil, not argentaffin, procedures.

Hellweg (*Virchows Arch. Pathol. Anat.*, **327:** 502, 1955) recommends a variant of the Bodian Protargol technic for demonstration of islet alpha cells in formalin-fixed material.

1. Float paraffin sections on slides with Ruyter's fluid (8 ml distilled water, 2 ml acetone, 1 drop methyl benzoate) or other *protein-free* fluid. Dry as usual.
2. Deparaffinize, hydrate, and soak 24 h in 40% formaldehyde.
3. Wash in distilled water.
4. Incubate 12 to 24 h at 48°C in the dark in 1 to 2% Protargol (pH 6.9 to 7.9), containing 2 g metallic copper to 50 ml in a Coplin jar.
5. Wash in distilled water.
6. Develop 5 min in 1 g hydroquinone, 5 g sodium sulfite, distilled water 100 ml.
7. Wash in distilled water.
8. Wash 3 min in 2% oxalic acid.
9. Wash in water, dehydrate in alcohols, clear in xylene, mount in balsam.

RESULTS: Pancreatic argyrophil (alpha) cells are brown to black.

Hypophyseal, adrenal medullary, and gastrointestinal argyrophil cells generally require a full 24 h in Protargol.

The Protargol should be Winthrop's Protargol-S or Bayer's "für die Bodian-Methode," and solutions should not be more than a week old. The hydroquinone solution should be made up fresh each time.

Gold toning appeared to be of little value and is omitted. Horse, dog, and mouse material gave fair demonstrations of pancreatic alpha cells. The procedure was unsuccessful in rat and rabbit. Human material was apparently not tested.

BENSLEY'S ANILINE–ACID FUCHSIN–METHYL GREEN.   This method, detailed earlier in this chapter, gives green acinar cells with deeper green nuclei; red zymogen granules, basal filaments, and mitochondria; deep red alpha granules; green beta granules.

The foregoing methods may be used also for hypophysis, parathyroid, and other glands.

Takaya (*J. Histochem. Cytochem.*, **18:** 178, 1970) advises the following fluorescent stain for the specific identification of alpha cells in pancreatic islets of rat, mouse, swine, hamster, rabbit, dog, guinea pig, hen, pigeon, newt, and crucian carp.

1. Fresh frozen sections are cut at 5 μ and mounted on coverslips.
2. One drop of 1% *o*-phthalaldehyde (Calbiochem) dissolved in either *p*-xylene or ethyl-benzene is added to the section, applied to a slide, and examined directly with a fluorescence microscope (Osram mercury lamp 200 W; BW, 4 mm UG1, 2 mm excitation, and K430 ultraviolet-absorbing filters).

RESULTS: Alpha cells only are fluorescent. The nature of the reactive material is unknown. Reactivity with histamine, catecholamine, and tryptamine were ruled out.

Rosenbloom and Rennert (*Stain Technol.*, **45**: 25, 1970) note that insulin in polyacrylamide gels can be stained with aldehyde fuchsin after oxidation with 0.05% $KMnO_4$ in 0.5% v/v H_2SO_4 for 30 min followed by bleaching in 2% oxalic acid.

HYPOPHYSIS

The alpha or acidophil, beta or "basophil" or cyanophil, and chromophobe cells of the anterior lobe of the hypophysis may be distinguished by use of stains of the hematoxylin-eosin type, notably with phloxine according to Mallory, with which the granules of the alpha cells stain pink or red and those of the beta cells stain blue. Alpha granules stain red by the eosin B–azure A technics.

Laqueur (personal communication) finds the Gomori aldehyde fuchsin method useful for anterior hypophysis. Using a Mallory-Heidenhain counterstain with fast green FCF instead of aniline blue, the alpha granules are orange-red; beta granules, violet to purple; delta granules, green to greenish blue; chief cell cytoplasm, pale gray-green; connective tissue, green; elastica, violet; nucleoli and erythrocytes, red.

By the Mallory-Heidenhain and Masson aniline blue technics alpha granules stain red; beta granules, bright blue; delta granules, lighter blue. After mercuric chloride–formalin fixation the Masson method gives dark violet alpha granules.

Laqueur considers the periodic acid Schiff method to be one of the most reliable for beta granules and colloid (both red-purple). Delta granules are not distinguished from beta granules, though delta cells usually contain less periodic acid Schiff–positive material.

METHYL BLUE–EOSIN. This method, introduced by Mann in 1894 for nervous tissues and extensively used for Negri bodies, has been modified in our laboratory by introduction of buffering of the mixture and staining at 60°C. Water-soluble aniline blue, which was formerly a mixture of diphenylrosanilin trisulfonic acid, C.I. 42755, and triphenylpararosanilin trisulfonic acid, C.I. 42780 (first edition, C.I. 707), can be substituted for methyl blue, C.I. 42780; the results are essentially identical. This is not surprising since at least one American company furnished the same dye, C.I. 42780, under both names. The substitution of eosin B (2,8-dinitro-4,6-dibromo-10-(o-carboxyphenyl)-3-hydroxy-7-xanthone, C.I. 45400) for Mann's eosin Y (C.I. 45380) results in deeper red in erythrocytes, hypophyseal alpha cells, etc. The change in proportion of the two 1% dye solutions from the original equal volume mixture to the 4:1 eosin–methyl blue ratio also seems advantageous (Glenner and Lillie, *Stain Technol.*, **32**: 187, 1957).

EOSIN B–METHYL BLUE TECHNIC. Fix in buffered neutral formalin, buffered sublimate formalin (B-5), or Zenker formalin (Spuler or Helly). After the usual paraffin embedding, cut 4- to 6-μ sections. Attach, deparaffinize, and hydrate as usual. Treat mercury-fixed tissues 10 min in 0.5% iodine–70% alcohol and 2 min in 5% thiosulfate; wash 10 min in running water.

1. Stain 1 h, placing slides in solution in Coplin jars at 25°C and transferring at once to a 60°C oven, using the following mixture:

1% aqueous eosin B	8 ml
1% aqueous methyl blue or aniline blue	2 ml
McIlvaine citric acid 11:9 disodium phosphate buffer	2 ml
Distilled water	28 ml

A 0.01 *M* acetate buffer pH 4.5 to 4.6, 30 ml may be substituted for the McIlvaine buffer and water.

2. Wash 5 min in running water, dehydrate through graded (50, 80, 100%) acetones; clear with acetone plus xylene (1 : 1) and two changes of xylene. Mount in Permount or similar resin.

RESULTS: Hypophyseal alpha cell granules are dark red; beta cell granules, dark blue; chromophobes, gray to pink; colloid, red to blue-violet; erythrocytes, orange-red; collagen fibers, blue.

With staining at pH 3.6 hypophyseal eosinophils, erythrocytes, keratinized epithelium, and muscle stain bright pink to red. Hypophyseal chief cells and epithelial and gland cytoplasms stain purplish or grayish pink. Collagen, reticulin, and hypophyseal cyanophils stain blue, and some mucins tinge pale bluish. Nuclei stain black with iron hematoxylin, or red without it. Granules of eosinophil leukocytes stain red or blue or perhaps both in the same cell.

McLETCHIE'S METHOD.   For certain cyanophil granules in the pituitary this method is perhaps worthy of trial. It depends on pretreatment with iodine and with phosphotungstic acid, and utilizes an acid dye. For this method Lendrum (*J. Pathol. Bacteriol.*, **57**: 267, 270, 1945) first prepares the "carbacid fuchsin" thus: Mix 1 g acid fuchsin (C.I. 42685) with 0.4 g melted phenol crystals; cool and dissolve in 10 ml 95% alcohol. Grind 0.5 g starch fine. Add 0.5 g dextrin and grind to a fine powder. Suspend in 100 ml water by grinding with gradual addition of the water. Heat to 80°C, cool, filter, and add it to the acid fuchsin–phenol–alcohol mixture to make a total of 100 ml.

For this technic Lendrum prescribes a formalin–mercuric chloride fixation and an alcohol, chloroform, paraffin embedding sequence. For the staining of basement membranes by this technic, Lendrum prescribes exposing the paraffin sections face down over a shallow dish of formalin at 55 to 60°C for 2 h, before removal of paraffin. (For cell granule staining this step is not necessary.) Sections are then deparaffinized and brought to water as usual. The usual iodine thiosulfate sequence for treatment of material fixed with mercuric chloride is not mentioned by Lendrum, and would not be required for removal of mercury precipitates, since a treatment with iodine is contained in the technic at a later point. The technic is given by Lendrum as follows:

1. Stain 3 to 5 min with alum hematoxylin.
2. Wash briefly in water.
3. Stain 1 to 2 min in 1% fast green FCF in 0.5% acetic acid, varying the time in accordance with the desired intensity of the green cytoplasmic staining.
4. Rinse in water.
5. Treat for 2 min with Lugol's iodine solution (1 : 2 : 100 or 5 : 10 : 100).
6. Wash off with 95% alcohol.
7. Immerse in 2% alcoholic solution of phosphotungstic acid for 2 min.
8. Rinse in water.
9. Stain 2 to 6 min in "carbacid fuchsin." Longer intervals give more staining of collagen and basement membranes. Lendrum prescribes microscopic control but gives no definite criteria at this point.
10. Rinse in water, dehydrate, clear, and mount.

The cell granules of apocrine cells and of pituitary beta cells are stained a blackish red. Even without the formalin vapor treatment there is some staining of collagen and basement membranes, but with that treatment they are stained intensely red. The apocrine cell granules are also iron-positive with ferrocyanide, and the same cells contain fat droplets.

PEARSE'S METHOD. This method (*J. Pathol. Bacteriol.*, **61**: 195, 1949) combined a Hotchkiss alcoholic periodic acid Schiff reaction and an alpha granule stain for hypophysis: Fix in Helly, in Zenker, in half-saturated $HgCl_2$ in formalin–saline solution, or in formalin–saline solution alone. Embed sections in paraffin.

1. Use iodine and thiosulfate sequence to remove Hg precipitates.
2. Bring to 70% alcohol.
3. Treat 5 min in 0.8 g H_5IO_6, 20 ml distilled water, 10 ml 0.2 M sodium acetate solution, and 70 ml ethyl alcohol.
4. Rinse in 70% alcohol, wash 1 min in 1 g KI, 1 g $Na_2S_2O_3 \cdot 5H_2O$, 20 ml H_2O, 30 ml ethyl alcohol, and 0.5 ml 2 N HCl.
5. Rinse in 70% alcohol, and treat 15 to 45 min in fuchsin-sulfite solution (Schiff reagent).
6. Wash in running water, 10 to 30 min.
7. Stain 30 s in 0.5% celestin blue in 5% iron alum and then 30 s or more in Mayer's hemalum (no intervening wash).
8. Differentiate quickly in 2% acid alcohol and blue in water.
9. Optionally stain alpha granules in 2% orange G in 5% phosphotungstic acid for 5 to 10 s (orange II, C.I. 15510, gives deeper color).
10. Wash in running water until a yellow tinge is just visible in acidophil areas (or use microscopic control).
11. Alcohols, alcohol plus xylene, xylenes, polystyrene (DPX).

RESULTS: Colloid of stalk and vesicles, magenta; cyanophil, "basophil" (or beta) granules, dark red; acidophil (or alpha) granules, orange; erythrocytes, orange; nuclei, blue-black.

GRAM REACTION. Pearse also found that the cyanophil cells were gram-positive, and Foster and Wilson (*Q. J. Microsc. Sci.*, **93**: 142, 1952) use a variant of the Gram stain for differential coloration of cyanophil cells, thus:

Hydrate and stain 2 to 3 min in 1% aqueous crystal violet. Rinse in water and cover with Lugol's (Gram's) iodine for 2 to 3 min. Blot, and differentiate in clove oil. Rinse in xylene and mount in Canada balsam. Beta granules are dark violet.

We find that 30 s in Weigert's iodine is quite adequate and presume that the authors meant Gram's solution rather than the strong pharmacopeial Lugol's solution. As usual with oil-differentiated Gram stains, collagen tends to remain violet. Differentiation should be carried to the point where cell nuclei are fully decolorized.

Use of the usual aniline-xylene mixture for differentiation, as in the routine Gram-Weigert method, also gives excellent results. Lillie used a Feulgen nucleal reaction before a Gram-Weigert stain. The red-purple nuclei contrast beautifully with the violet granules of the cyanophil cells.

IRON HEMATOXYLIN. Another method for staining the eosinophil granules of anterior hypophysis cells as well as those of eosinophil leukocytes is a modified Weigert myelin stain, derived from our variant of the Weil method. Tissue should be fixed in an aqueous fixative containing formaldehyde. Chromation is not essential.

LILLIE'S 1958 IRON HEMATOXYLIN FOR MYELIN SHEATHS, EOSINOPHILS, HYPOPHYSEAL ALPHA CELLS, ETC.

SOLUTION A. Mix 2 g $FeCl_3 \cdot 6H_2O$ plus 0.6 g NH_4Cl plus distilled water to make 100 ml. Check pH (it should be about 2).

SOLUTION B. Use 1 g hematoxylin per 100 ml 95 or 100% alcohol. Mix equal volumes of solutions A and B on same day as used. Use formalin-fixed, frozen, paraffin, or celloidin sections.

1. Stain 1 h at 60°C. Rinse in distilled water.
2. Differentiate in 0.5 g borax plus 1.25 g $K_3Fe(CN)_6$ in 100 ml distilled water until nuclei are decolorized but red corpuscles remain brown to black and myelin sheaths are black. For myelin 10 to 15 min should suffice on formalin-fixed material.
3. Wash in distilled water, 3 to 5 min.
4. Counterstain, if desired, 5 min in 0.1% safranin in 1% acetic acid.
5. Rinse in distilled water, alcohols or acetone, xylenes, balsam, or synthetic resin. (Isopropanol preserves celloidin; acetone dissolves it.)

COUPLED TETRAZONIUM METHOD OF LANDING AND HALL FOR ANTERIOR PITUITARY CELLS (*Stain Technol.*, **31**: 193, 1956). Fix in neutral 10% formalin, dehydrate, and embed in paraffin as usual; bring sections to water as usual.

1. Rinse in Veronal-acetate buffer at pH 9.2.
2. Couple in 0.1% fast blue B in Veronal-acetate at pH 9.2 for 15 min at 0°C.
3. Rinse in buffer.
4. Postcouple with 2% H-acid in pH 9.2 buffer, 15 to 30 s.
5. Wash in running water, 2 min.
6. Treat with 1% sodium periodate (Na_3IO_5) in 0.5% HNO_3 (0.8% H_5IO_6 is equivalent), 10 min at 20 to 25°C.
7. Wash in running water, 5 min.
8. Stain in Van Duijn's sulfite leucothionin, 2 h.
9. Alcohols, xylene, synthetic resin.

RESULTS: Cyanophils are blue; acidophils, brown; chromophobes, unstained.

It appears probable that the blue staining of beta cells in this procedure is simply the periodic acid Schiff reaction which these cells and the colloid are known to give. The brown of the alpha cells seems assignable to tyrosine; Glenner found a strong Morel-Sisley diazotization coupling reaction in the alpha cells (*J. Histochem. Cytochem.*, **8**: 138, 1960). (Glenner and Lillie (*J. Histochem. Cytochem.*, **5**: 279, 1957) reported a strong indole (tryptophan) reaction in colloid and beta cells, using the postcoupled benzylidene reaction.

Kerenyi and Taylor's (*Stain Technol.*, **36**: 169, 1961) "Niagara Blue 4B" (benzo sky blue, C.I. 24400) method for differentiating beta cells in bluish purple with a hematoxylin-eosin background and in deep blue with Mayer's hemalum alone seems to have interesting possibilities for routine use. It is applicable to routine autopsy material after formalin fixation, and the benzo sky blue may be applied by restaining routine hematoxylin-eosin sections with satisfactory results.

KERENYI-TAYLOR BENZO SKY BLUE METHOD FOR HYPOPHYSEAL BETA CELLS

1. Deparaffinize and hydrate 5-μ sections of formalin-fixed hypophysis.
2. Stain 2 min in 1% aqueous benzo sky blue.
3. Wash in water 1 min, tap water or several changes of distilled water.
4. Stain 1 min in alum hematoxylin (Harris or formula of similar strength).
5. Differentiate in 0.125 N HCl in 70% alcohol, under microscopic control.
6. Blue in 1% Na_2HPO_4 or 1% sodium acetate, and rinse in water.
7. Insert eosin counterstain at this point, 10 to 15 s.
8. Rinse, dehydrate, clear, and mount in balsam or synthetic resin.

RESULTS: Beta cell granules are deep blue without eosin, bluish purple with it; alpha cell granules, red with eosin, unstained without; chromophobe cytoplasm, pink with eosin, unstained without; nuclei, blue-black.

Halmi (*Stain Technol.*, **27**: 61, 1962) recommends the following useful stain for identification of cells of the rat anterior pituitary. The method is not recommended for use in man. Paraffin-embedded sections previously fixed with Bouin's fluid (with 0.5% trichloroacetic acid substituted for acetic acid) are deparaffinized and brought to water. Sections are oxidized with Lugol's iodine (probably the official 5:10:100) for 30 min at 24°C (alternatively 5% Oxone (Na_2O_2) or peracetic acid, also 30 min 24°C). After iodine, sections are next treated with 5% aqueous sodium thiosulfate to remove iodine. If other oxidants are used, this step is omitted. Sections are next rinsed in distilled water, stained with aldehyde fuchsin (which must be ripened for at least 3 days at 24°C) for 10 min, differentiated in three changes of 95% ethanol (5 min each), stained in Ehrlich's hematoxylin for 3 to 4 min, blued in running water, counterstained for 45 s in a mixture of 0.2% light green SF yellowish (C.I. 42095), 1% orange G (C.I. 16230), and 0.5% phosphotungstic acid in 1% acetic acid, rinsed briefly in 0.2% acetic alcohol, dehydrated, and mounted in synthetic resin.
RESULTS: In pituitaries of rats, beta granules are purple, delta granules are green, and alpha granules are orange.

Adams and Swettenham (*J. Pathol. Bacteriol.*, **75**: 5, 1957) developed a stain very useful for use on cells of the anterior pituitary of man permitting discrimination of mucoid (basophil) cells into two distinct types. Granules in these cells have been designated S (containing cystine and carbohydrate) and R (containing carbohydrate). During the staining procedure performic (or peracetic) acid is employed to oxidize cystine, and the carbohydrate of the S granule is rendered insusceptible to the subsequent PAS (periodic acid Schiff) stain, whereas the carbohydrate of the R granule remains susceptible to the PAS and is identified by the magenta stain. The mechanism of discriminatory carbohydrate reactivity is unknown.

1. Bring paraffin-embedded sections of fixed (formalin-mercury) tissues to water and remove mercury.
2. Oxidize sections with peracetic or performic acid for 5 min at 24°C.
3. Rinse sections in tap water 10 min, 70% ethanol briefly, blot, rinse in tap water, and dry in 60°C oven *until just dry.*
4. Rinse in absolute ethanol and then in tap water for 1 min.
5. Stain in Alcian blue for 1 h at 24°C (dissolve by heating to 70°C a 3% solution of Alcian blue (C.I. 74240) in 2 N H_2SO_4, cool, and filter.
6. Rinse sections in tap water 5 min.
7. Oxidize sections in 0.5% periodic acid for 10 min at 24°C.
8. Rinse in running water 5 min.
9. Immerse in Schiff's reagent for 10 min.
10. Transfer sections to three reducing rinses of 0.5% sodium metabisulfite ($Na_2S_2O_5$) for 2 min each.
11. Wash in running water.
12. Stain in celestin blue 2 to 3 min. [Into 50 ml distilled water add 2.5 g iron alum and allow to stand overnight at 24°C. Add 0.25 g celestin blue B (C.I. 51050) to the alum and boil for 3 min, filter after cooling, and add 7 ml glycerol. (Boiling is quite unnecessary, see Lillie et al., *Am. J. Clin. Pathol.*, **60**: 817, 1973.)]
13. Rinse sections and stain in Mayer's hemalum for 2 to 3 min, differentiate in 1% acid alcohol, and wash in running water.

14. Counterstain in orange G (C.I. 16230) for 10 s (2 g orange G is added to 100 ml 5% aqueous phosphotungstic acid). Let stand for 24 h and use the supernatant.
15. Wash in running water until sections are pale yellow (about 30 s), dehydrate, and mount in synthetic medium.

RESULTS: Pituitary mucoid granules of the R type stain reddish magenta; S-type mucoid granules stain blue; alpha-type granules of acidophils stain orange; and nuclei are stained blue-black. Comparisons of components of cells of the pituitary to various stains is difficult due also in part to species differences. It may be helpful to know that the R granules are the same as the beta granules identified by the PAS reaction in the stain by Wilson and Erzin (*Am. J. Pathol.*, **30:** 891, 1954) provided below and by Pearse's method (*J. Pathol. Bacteriol.*, **64:** 791, 1952). The S-type granules of this stain correspond to the gamma granules of the stain by Wilson and Erzin. Adams and Pearse (*J. Endocrinol.*, **18:** 147, 1959) conducted further studies to conclude that R-type cells are thyrotrophs and S-type cells are gonadotrophs.

The Adams-Swettenham stain is far too complicated to permit an appraisal of the value of the numerous steps.

We have in succession a peracetic–Alcian blue sequence for cystine; a PAS for the beta cell mucoprotein; an iron alum–celestin blue for nuclei; Mayer's hemalum, also for nuclei; and an acid alcohol which decolorizes celestin blue more rapidly than hematoxylin. Besides that, there has been in the celestin blue step an iron alum mordanting before the hematoxylin which would tend to make it black, followed by an orange G alpha granule stain in phosphotungstic acid which also postchromes the hematoxylin.

We are concerned about interactions of several staining procedures with each other.

Wilson and Erzin (*Am. J. Pathol.*, **30:** 891, 1954) recommend the following method for identification of cell types of anterior pituitaries of man and rat.

1. Bring formalin- or mercury-fixed, paraffin-embedded sections to water and remove mercury as usual.
2. Oxidize with 0.5% periodic acid for 5 min.
3. Rinse, and immerse sections in Schiff's reagent for 15 min.
4. Wash in running water for 10 min.
5. Stain in 1% orange G (C.I. 16230) for 10 to 15 s.
6. Mordant for 15 s at 24°C in 5% phosphotungstic acid.
7. Rinse briefly in running water, and stain in 1% aqueous methyl blue for 1 min.
8. Rinse briefly in 1% acetic acid, dehydrate, and mount in synthetic medium.

RESULTS: Beta granules (R-type granules) are reddish magenta, gamma granules (S-type granules) stain purple, and alpha granules stain orange. Use of a blue filter aids in discrimination.

Swope et al. (*J. Histochem. Cytochem.*, **18:** 450, 1970) compared adjacent sections of rat pituitary with the peroxidase-labeled antibody (to luteinizing hormone, follicle-staining hormone, or thyroid-stimulating hormone) to sections stained with the permanganate–Alcian blue–aldehyde fuchsin sequence. Thyroid-stimulating hormone was located in large polygonal cells that frequently possessed long processes and stained with both Alcian blue and aldehyde fuchsin. Follicle-stimulating hormone was located in cells that stained only with aldehyde fuchsin. Luteinizing hormone was identified in cells that did *not* stain with either aldehyde fuchsin or Alcian blue.

Gutiérrez and Sloper (*Histochemie*, **17:** 73, 1969) embedded a commercial preparation of synthetic oxytocin (Parke-Davis) and lysine vasopressin (Sandoz) in agar. Picroformol-fixed sections oxidized with either acid permanganate or peracetic acid stained intensely with pseudoisocyaninchloride.

McGuire and Opel (*Stain Technol.*, **44**: 236, 1969) have noted that Weigert's resorcin-fuchsin elastic tissue stain also stains neurosecretory cells.

TESTIS AND EPIDIDYMIS

The interstitial cells of the testis contain lipids, pigment, crystalloids, and sometimes glycogen. The fats include cholesterol esters at times. The pigment is a lipofuscin and gives a brown fluorescence in ultraviolet light. The crystalloids are rod-shaped bodies with rounded or pointed ends. They are singly refractile, dissolve in a pepsin–hydrochloric acid mixture, swell in 10% potassium hydroxide, and are insoluble in 10% mineral acids and in fat solvents.

Under certain experimental conditions the lipofuscin pigment of the interstitial cells may be greatly increased in amount and may assume the characteristics of the ceroid pigments: acid-fastness, staining with oil-soluble dyes after paraffin embedding, reactivity to the periodic acid and peracetic acid Schiff methods and to the ferric ferricyanide reduction test.

The spermatozoon head is a favorite cytologic object for the study of the arginine-rich protamines (trout) and nucleohistones. Methods for nucleic acids, spindles, centrosomes, and mitochondria are also considerably used for this organ. The Golgi aggregates of the epithelium of the tubules of the epididymis are readily identified, and this organ has been much used in enzyme distribution studies, especially those concerning the relationship of enzymes to Golgi aggregates.

PEPTIC GLANDS

Peptic gland zymogen granules may be demonstrated in the gastric mucosa of rats and rabbits after fixation with such chromate-formalin mixtures as Spuler's, Helly's, Orth's, and Kose's fluids by staining in 0.05 to 0.1% thionin buffered with 0.01 M acetate buffer to pH 4 to 5 for 1 min or more. The granules appear as discrete deep blue or greenish blue bodies when stained at pH 5, contrasting with the deep violet-blue of the basal cytoplasm. At pH 3 they fail to stain.

The granules are destroyed by acetic acid fixatives and are poorly preserved by neutral formalin. Acid alcoholic fixatives destroy them.

The azure A–eosin B technic stains them faintly pink in rats and rabbits, and red in guinea pigs. In guinea pigs the same cells may contain concurrently present retiform periodic acid Schiff–positive material, but the granules themselves were recorded as periodic acid Schiff–negative in the three species mentioned. The eosinophilic granules resist barley malt ribonuclease digestion, which destroys the basophilia of the basal cytoplasm.

When paraffin sections of formalin-fixed gastric mucosa of various mammals are stained by the sulfuric acid–Nile blue procedure for lipofuscins, washed in water, and mounted in glycerol gelatin, the zymogen (pepsinogen) granules of the chief cells stain selectively dark blue, chief cell basal cytoplasm is a lighter blue, and other structures range from blue-green to pale greenish yellow. This staining was prevented by immediate acetone extraction after the staining, as with the lipofuscins, and is also prevented by prior methylation in 0.1 N HCl in methanol, 4 to 6 h at 60°C, and is restored by demethylation (saponification) of the methyl ester by 1% potassium hydroxide in 70% alcohol, 20 min at 25°C. The deacetylation (saponification) reagent recommended by Lillie (1954), 20 ml 28% ammonia plus 80 ml absolute alcohol, 24 h at 25°C, may be substituted, with the advantage that it does not remove cytoplasmic ribonucleic acid, as alcoholic KOH does.

The peptic zymogen granules, however, do not stain with Sudan black, though parietal cell cytoplasm, smooth muscle, and erythrocytes stain faintly to moderately.

Peptic zymogen granules color strongly with the tryptophan methods and moderately by the Morel-Sisley tyrosine method. In some specimens they may be quite large, as big as pancreatic zymogen granules, and give a distinct orange-pink with p-nitrodiazobenzene, a dark red with alkaline (pH 8) diazosafranin, and dark bluish green with diazosulfanilic acid–azure A. Diazosafranin at pH 3 does not stain them.

The above findings agree with the published analyses of pepsin: tyrosine, 8.9%, tryptophan, 2.3%; glutamic acid, 21.6%; and aspartic acid, 10.0%; with the basic amino acids—arginine, 1.6, histidine, 1.5, and lysine, 1.3%—comparatively low in amount, accounting for the strong acidity of this protein (Block, "Amino Acid Handbook," p. 294, Charles C Thomas, Springfield, Ill., 1956).

The usual tests for ribonucleic acid are applicable to the basal cytoplasm of the chief cells.

The oxyphilia of the parietal cells does not extend to the high pH levels of the arginine proteins and is strongly affected by nitrosation, suggesting that it is due largely to lysine (and hydroxylysine) residues.

PANETH CELLS

These cells contain conspicuous eosinophilic granules of variable size. The granules may lie in clear vacuolar spaces in the cytoplasm between the nucleus and the gland lumen. Like the zymogen granules of the pancreas and the salivary glands, these eosinophilic granules are destroyed by acetic acid fixatives such as aqueous and alcoholic acetic formalin, and Bouin's, Gendre's, and Carnoy's fluids. The acetic dichromate fluids, such as Tellyesniczky's and Zenker's, may preserve them entirely, in part, or not at all. They are well preserved by buffered neutral formalin or mercuric chloride–formalin, by Orth's, Kose's, and Möller's dichromate–formalin mixtures, and by Spuler's formalin-Zenker fluid. Salt Zenker will preserve them.

They stain red by Lillie's azure A–eosin B method at pH 3.75 to 4.1. They resist malt diastase (ribonuclease) and pancreatic ribonuclease digestions, which render them more conspicuous by destroying the normal basophilia of the cytoplasm. They are sometimes gram-positive by the Gram-Weigert technic, especially in human beings, sometimes in guinea pigs, usually not in rabbits or rats.

They often stain red-purple by the periodic acid Schiff procedure. Positive reactions have been observed in human beings, mouse, guinea pig, rat, and rabbit, as well as negative reactions in the first three. Application of a picric acid counterstain changes the color of Paneth granules to orange, thereby permitting their ready distinction from the still red-purple mucigen granules.

The oxyphilia remains evident up to pH 10.5 with Biebrich scarlet after neutral mercuric chloride fixation, but it disappears at pH 7 or 8 in formalin material (Spicer and Lillie, *Stain Technol.*, **36**: 365, 1961). Nitrous acid and adequate acetylation each decreases the oxyphilia, both after formaldehyde and mercurial fixation (*J. Histochem. Cytochem.*, **6**: 352, 1958). The Barrnett-Seligman DDD (dihydroxydinaphthyl disulfide) technic demonstrates SH and SS groupings in formalin-fixed mouse Paneth granules (Selzman and Liebelt, *J. Histochem. Cytochem.*, **10**: 106, 1962). The tryptophan content of the granules is perhaps best demonstrated (in blue) by the postcoupled benzylidene technic on material fixed 2 to

3 h only in neutral buffered formalin. Azo coupling with fast garnet GBC or fast red B may be combined with this to demonstrate the enterochromaffin granules in a contrasting (red) color in the same section (*J. Histochem. Cytochem.*, **8**: 182, 1960). The diazotization coupling reaction for tyrosine has been successfully employed after neutral formalin, sublimate formalin (B5), and Kose's dichromate–formalin fixations (*J. Histochem. Cytochem.*, **7**: 416, 1959). Contrasts are poorer than with the tryptophan reaction, because of the moderate reaction of cytoplasm. We have obtained suggestive reactions with the older α-naphthol and 8-quinolinol hypochlorite arginine technics. The NQS (β-naphthoquinone-4-SO_3Na) method has given deep red coloration of Paneth granules (Lillie et al., *J. Histochem. Cytochem.*, **19**: 487, 1971).

Fixation by fixatives which preserve the granules is required; buffered neutral 10% formol is good. Granules are better demonstrable in animals fasted overnight before killing.

Taylor and Flaa (*Arch. Pathol.*, **77**: 278, 1964) were able to detect arginine, SH groups, and tyrosine in Paneth cell granules of formalin-fixed sections of rat tissues.

RENAL JUXTAGLOMERULAR GRANULES

Harada (*Stain Technol.*, **45**: 71, 1970) tested several methods for the selective demonstration of the juxtaglomerular granules in the kidney and advises the following because of its rapid staining selectivity and simplicity. Tissues may be fixed in acid or neutral formalin. $HgCl_2$ may be added to the neutral calcium acetate–formalin for best results.

1. Deparaffinize and bring sections to water.
2. Wash briefly in tap water.
3. Blot and stain in 0.5% crystal violet in 70% ethanol for 1 to 3 min.
4. Wash briefly in running water and blot well. This step is crucial.
5. Differentiate with frequent blotting in a 1 : 1 mixture of aniline–xylene for several minutes until nearly all color is gone.
6. Clear in xylene and mount in a synthetic resin.

RESULTS: Juxtaglomerular granules stain deep purple in a varying faint purple background. Increasing concentrations of ethanol (greater than 50%) in the staining solution favored greater staining of the juxtaglomerular granules and lesser staining of the other renal tissues.

Harada (*Anat. Anz.*, **128**: 431, 1971) modified the above stain by the interposition of an oxidation between steps 2 and 3. A 0.5% $KMnO_4$ and 0.5% H_2SO_4 in a 1 : 1 mixture for 5 min is followed by a bleach in 1% oxalic acid for 2 min, a rinse in running water for 5 min, and the use of crystal violet (0.1%) in either an acid (0.1 N HCl) or alkaline (0.01 to 0.1 N NH_4OH or 0.1 N NaOH containing 2.5% Na_2CO_3) aqueous solution for 1 to 3 min.

RESULTS: After oxidation and acid crystal violet, juxtaglomerular granules and elastic fibers are stained deep violet against a faint purple of other tissues. After oxidation and alkaline crystal violet, juxtaglomerular granules and erythrocytes stain an intense deep violet in contrast to faint purple of other tissues.

Harada (*Stain Technol.*, **46**: 155, 1971) notes that juxtaglomerular granules are selectively stained by several basic dyes, although a few, notably basic fuchsin and crystal violet, are much better. Curiously, toluidine blue, thionin, safranin O, and azure A were poor.

Deparaffinize sections from paraffin-embedded phosphate-buffered 10% formalin, bring them to water, stain in 0.1% basic fuchsin (C.I. 42510) or crystal violet (C.I. 42555), rinse, and immerse in 0.01 N KOH for 1 to 3 min; wash off excess dye with a 1 : 1 mixture of aniline and xylene on slides laid flat, clear, and mount in synthetic resin.

RESULTS: Granules in juxtaglomerular cells stain red with basic fuchsin or blue-violet with crystal violet. The alkali step is used for differentiation, and sections are detached less readily if flat. Granules in juxtaglomerular cells are destroyed in unfixed frozen sections.

Szokol (*Acta Morphol. Acad. Sci. Hung.*, **18**: 283, 1970) stained juxtaglomerular granules with several fluorochromes. The isoelectric point of the granules was determined to be in the vicinity of 3 to 4.4. All dyes were readily extracted in sections incubated in either 2 *N* NaOH or in 1% acetic in 0.2 *N* HCl for 1 h.

Granules in juxtaglomerular cells of mice are stained after intraperitoneal injections of 1 ml 1% solutions of neutral red, brilliant cresyl blue, Nile blue sulfate, or acridine orange. Kidney cells were examined in squash preparations 1 h after injection of the dye [Szokol et al., *Nature* (*Lond.*), **208**: 1331, 1965].

Janigan (*Arch. Pathol.*, **79**: 370, 1965) notes that juxtaglomerular cell granules in man stain with the fluorescent thioflavine T (C.I. 49005). Sections had been formalin-fixed, paraffin-embedded, and stained in a filtered 1% solution of the dye, differentiated in 1% acetic acid, and mounted in Apathy's syrup.

MYOEPITHELIAL CELLS

Puchtler et al. (*Stain Technol.*, **41**: 15, 1966) indicate the following stain is selective for myoepithelial cells. Human lingual glands were used.

1. Tissues must be fixed in Carnoy's No. 2 mixture (ethanol–chloroform–glacial acetic acid, 6 : 3 : 1) and embedded in paraffin.
2. Bring sections to water.
3. Stain in Mayer's acid hemalum 5 to 10 min.
4. Wash in running water.
5. Treat in 5% tannic acid for 10 min.
6. Rinse in three changes of distilled water.
7. Immerse in 1% aqueous phosphomolybdic acid for 10 min.
8. Rinse in three changes of distilled water.
9. Stain in azophloxine GA (C.I. 18058) solution for 5 min (2 g azophloxine is dissolved in methanol–glacial acetic acid, 9 : 1). The solution is allowed to stand 16 to 20 h before use and is *not* filtered.
10. Rinse in two changes of methanol–glacial acetic acid (9 : 1).
11. Dehydrate and clear through absolute ethanol and xylene, and mount in a nonfluorescent mounting medium.

RESULTS: Myoepithelial cells and muscle stain red. Basement membranes stain yellow. Mucous cells are unstained, and serous cells are grayish yellow. Collagen and reticulum are yellow. Use of a mercury vapor lamp, exciter filter BG 12/3 or BG 12/6 mm, dark-field condenser, and ultraviolet-blue absorption filter (GG9/1 mm and OG1/1.5 mm) resulted in yellowish red myoepithelial cells in sharp contrast to a dark background.

HAIR, KERATIN, AND KERATOHYALIN

Hair cortex is often fully or partly gram-positive, especially when Weigert technics are used. Keratohyalin also retains the violet; keratin retains it less. In Ziehl-Neelsen technics hair cortex is quite strongly acid-fast. Hair cortex colors red-purple with Schiff reagent on direct exposure of a few hours. It becomes Schiff-positive with 10-min exposure after peracetic or performic acid oxidation, but not after periodic acid oxidation. Bromination prevents the peracetic acid Schiff reaction but renders the periodic acid Schiff reaction positive, so that even a 10-min Schiff reagent bath yields a positive reaction. Keratohyalin and keratin are periodic acid and peracetic acid Schiff–negative.

Keratohyalin often colors intensely with progressive alum hematoxylin and iron hematoxylin stains, as well as moderately with basic aniline dyes. Leuchtenberger and Lund (*Exp. Cell Res.*, **2**: 150, 1951) recorded the loss of azure A and toluidine blue basophilia on ribonuclease digestion. The granules are Feulgen-negative. With iron hematoxylin myelin procedures such as our "eosinophil myelin" stain, differentiated to the point where cell nuclei are decolorized but erythrocytes are still black, keratohyalin varies from black to red; keratin, from light to dark brown; and hair cortex, from black near the root to blue or light blue-gray distad.

Liisberg (*Acta Anat.*, **69**: 52, 1968) highly recommends the following stain for clear-cut delineation of cornification.

1. Deparaffinize formalin- or Carnoy-fixed, paraffin-embedded sections and bring to water.
2. Stain for 10 min at 24°C in 0.1% toluidine blue in distilled water (pH not given).
3. Rinse in three changes of distilled water.
4. Stain in 0.1% rhodamine B (C.I. 45170) in McIlvaine's buffer of pH 3.6 for 10 min at 24°C.
5. Rinse sections in three changes of distilled water, dehydrate, and mount in synthetic medium.

RESULTS: Cornified areas appear red; nuclei and other basophilic areas are blue or blue-black.

Swift (*J. R. Microsc. Soc.*, **88**: 449, 1968; *Histochemie*, **19**: 88, 1969) describes a technic for the demonstration of cystine at the electron microscopic level. The method employs Gomori's silver-methenamine reagent. As applied to hair follicles in guinea pigs, Swift noted localization of silver deposits confined to cystine-containing keratin and melanin. Distinction is readily made on the basis of morphology. The method is provided below. The user is cautioned about the well-known capricious behavior of silver.

For fixation, Swift recommended the use of glutaraldehyde for 2 to 3 h. In a comparative study, he could detect no difference in specimens fixed in ethanol. Very thin specimens were dehydrated, embedded in Araldite, sectioned, placed on grids, treated with Gomori's silver-methenamine reagent at pH 9.2 for 2 to 3 h at 45°C, and washed with deionized water. To effect clear observation of covalently bound silver, sections were treated with 10% sodium thiosulfate for 5 min, followed by rinsing in deionized water. In some cases sections were also stained with uranyl acetate and lead citrate.

RESULTS: Metallic silver deposits were observed at the electron microscopic level at sites of cystine. The reaction was completely blocked in acid hydrolyzed sections (0.1 N HCl for 1 h at 45°C) subsequently reduced with benzyl thiol (0.3 M benzyl thiol in 20% n-propanol, 180 min, 24°C) and alkylation with iodoacetate.

Kutlik (*Anat. Anz.*, **128**: 314, 1971) uses permanganate hematoxylin for elective staining of keratohyalin. Mix fresh 4 ml 1% aqueous hematoxylin and 4 ml 0.25% $KMnO_4$; dilute with 72 ml distilled water. Bring sections through distilled water to stain, stain 30 min at 20 to 22°C, wash 3 to 5 min in running water but no longer, dehydrate in alcohols, counterstain a few seconds in 0.5% alcoholic rose Bengal or 30 min in alcoholic brilliant yellow, rinse in 95% alcohol, blot dry, clear in xylene, and mount in balsam.

RESULTS: Keratohyalin is dark violet, keratin pale violet, nuclei unstained. Iron deposits light gray, calcium purplish to brownish red.

The mechanism of this stain is not clear. The stain resists alcohol, but washes out with overexposure to water.

According to Unna, eleidin may be differentially stained by a picric acid–nigrosin sequence technic. Romeis directs as follows: (1) Stain 5 min in saturated aqueous picric acid solution; (2) rinse in water; (3) stain 1 min in 1% aqueous nigrosin (C.I. 50420); (4) wash in water, alcohol, oil, balsam.

RESULTS: Eleidin is blue-black; keratin, bright yellow.

Unlike keratohyalin, trichohyalin does not stain with hematoxylin but stains vividly with acid dyes. The azure A–eosin B technic displays these granules in bright red against the lightly to deeply basophilic cytoplasm of the hair follicle.

Near the root zone, hair cortex gives strong sulfhydryl reactions in appropriately fixed material. The ferric ferricyanide reduction test is positive after trichloroacetic alcohol fixation, but negative after mercurial fixations or postmordanting (—SH blockade). Cutaneous keratin and sometimes keratohyalin also color blue with this test, and in these structures the reaction is also positive in our hands after mercurial fixations and blockade treatments.

With the Barrnett-Seligman dihydroxydinaphthyl disulfide (DDD) method for sulfhydryl, rat hair cortex colors blue at the zone of keratinization, and red nearer the root, and remains unstained distad. Epithelial cells in root sheaths, hair bulbs, and stratum germinativum of the epidermis color red; those of the stratum corneum, pink. The reaction of keratohyalin was not noted.

Sodium plumbite, hot methenamine silver, and cystine–cysteic acid cleavage methods, as applied to hair cortex, are recorded under cystine reactions.

In a reinvestigation of the Meirowsky reaction (*J. Histochem. Cytochem.*, **4**: 318, 1956) it was found that keratohyalin and the eleidins of the stratum lucidum and in non-formalin-fixed tissue the trichohyalin granules color intensely on prolonged exposure to dilute solutions of various catechols, among which were recorded catechol, pyrogallol, dopa, adrenaline, gallocyanin, brazilin, and hematoxylin. These reactions occur optimally at about pH 6.5, require some oxygen, and can be carried out in double-distilled water in the complete absence of chelating metals. Competitive inhibition occurred with $Na_2S_2O_4$ and with cysteine; partial inhibition, with Na_2S; no inhibition, with KCN or NaN_3 (range of inhibition, 1 to 10 mM).

With the use of hematoxylin this is essentially Clara's hematoxylin method. For trichohyalin, use methanol-chloroform (50 : 50 or 60 : 30) fixation; for keratohyalin, calcium acetate–formalin also serves quite well.

The reaction is seen also in stratified squamous epithelia of the pharynx, larynx, esophagus, and rodent forestomach.

Cutaneous keratohyalin and that of certain mucosal stratified squamous epithelia has the property of taking up many divalent and trivalent metal ions, which may then be demonstrated by 1- to 2-h stains with neutral 0.1% hematoxylin to yield blue and blue-black, greenish blue with copper, red with thorium and tin. It forms a convenient model for study of color reactions of various metal ions with color-producing reagents. The hematoxylin color reactions are presented in Chap. 12.

The Clara reaction of keratohyalin can be prevented by a 4- to 8-h methylation or by 1 N HCl for 1 h 60°C. After the hot HCl treatment of formol-fixed tissue the keratohyalin granules and eleidin will again take up such metals as Cu(II), Fe(III), Zn(II), or Sn(II) and can then be stained blue-green, blue-black, blue, or red respectively with 0.1% hematoxylin at pH 7, or also at pH 2.5 for Fe(III) or Sn(II). But in chloroform-methanol-fixed skin the hot HCl leaves round clear spaces in place of the keratohyalin granules, and these do not take up Cu, Fe, Sn, or Zn, though the adjacent eleidin does and becomes restainable.

Hair cortex gives strong reactions to the Millon and Morel-Sisley tyrosine methods. Soft keratins of the epidermis and of gastrointestinal stratified squamous epithelia react well by the latter procedure; smooth and striated muscle also colors light red; elastic fibers are a fair red-purple; and epithelial cytoplasm takes pink to light pink.

Faint to slight tryptophan reactions have been recorded for hair keratin, outer root sheath, and Huxley's layer (Glenner and Lillie, *J. Histochem. Cytochem.*, **5**: 279, 1957) using the postcoupled benzylidene reaction. Smooth and striated muscle react moderately.

Azure A–eosin B stains reveal strong basophilia (ribonucleic acid) in active hair follicle cytoplasm, moderate basophilia of the basal and spinocellular layers, very moderate eosinophilia of stratum corneum of the skin, more pronounced eosinophilia in gastrointestinal epithelia, and strong eosinophilia of trichohyalin, keratohyalin, and stratum lucidum and of part of the hair medulla. Spicer and Lillie (*Stain Technol.*, **36**: 365, 1961) have shown that the oxyphilia of lucidum, keratohyalin, and trichohyalin persists up to pH 10.5 after $HgCl_2$ (B-4) fixation though not after formalin, using the dilute Biebrich scarlet technic. This oxyphilia has proved relatively resistant to nitrosation (*J. Histochem. Cytochem.*, **6**: 352, 1958), which suggests that it may be due largely to arginine residues, and moderately positive Sakaguchi reactions have been obtained on these structures with the older α-naphthol, 8-quinolinol, and 2,4-dichloro-1-naphthol (Deitch) NaOCl methods. Keratohyalin and eleidin (lucidum) react strongly for arginine by the 1,2-naphthoquinone-4-sodium sulfonate (NQS) method of Lillie et al. (*J. Histochem. Cytochem.*, **19**: 487, 1971). Trichohyalin also may react strongly, as does hair medulla but usually not hair cortex. Neutral formol, 1 M NaCl-5 to 6% acetic acid–formol, salt Zenker, or other nonhemolyzing zymogen granule–preserving fixatives are required. Demonstration in rodents is more successful if food is withheld 8 to 18 h before killing.

A benzil blockade prevents not only the NQS reaction of keratohyalin and keratins but also their Clara hematoxylin reaction and their high pH oxyphilia [Lillie et al., *Acta Histochem.* (*Jena*), **49**: 204, 1974].

Large amounts of glycogen may be demonstrated in the large clear cells of the outer sheath of active hair follicles, and some is demonstrable in cutaneous striated muscle.

The usual methods for lipids are applicable to the study of the sebaceous glands. According to Montagna ("Structure and Function of the Skin," second edition, Academic, New York, 1962), human sebum contains cholesterol, free fatty acids, squalene, odd-numbered saturated and unsaturated fatty acids (free and esterified), aliphatic alcohols, and little glyceride or phospholipid. But histochemically Montagna speaks of positive Smith-Dietrich and Baker hematein stains, osmiophilia and osmiophobia, birefringence and the Schultz test, and a yellow-orange fluorescence excited by light of 360-mμ wavelength.

10

ENZYMES

When this chapter was nearly completed it was realized that a considerable number of recent writers had stated substrate and inhibitor concentrations only in molarities, often in such form as $6.3 \times 10^{-5} M$, and had almost regularly failed to supply the molecular weight (mw). Some supplied equivalents in milligrams per 100 ml; to them we are grateful. Since molarities are not immediately useful at the bench when weighing and measuring, we have availed ourselves copiously of data in Lange's "Handbook of Chemistry" and "The Merck Index" and a simple computer to supply the missing data in weight in volume (w/v) percentage or milligrams per 100 ml. Conversion of complex molarity expressions to simple decimals in molarity (M), millimolarity (mM), and micromolarity (μM) have been carried out. We did contemplate using nanomolarity (nM) for $10^{-9} M$ as well, but usually the range to micromolarity (μM) sufficed. In the case of simple metallic salts formula weights (mw) are usually available on the bottle.

PREPARATION OF TISSUES FOR ENZYME HISTOCHEMISTRY

Two requisites must be kept in mind when processing tissues for enzyme histochemistry, namely, the preservation of maximum enzyme activity and localization of the reaction product at the site of the enzyme. These criteria are generally impossible to fulfill, and some compromise is generally made. This is particularly true with soluble enzymes where some fixation is required for sufficient retention of the enzyme for demonstration.

Most enzymes are inactivated by heat and by fixatives, thereby precluding the practice of enzyme histochemistry on tissues prepared by routine methods for histology and pathology. Special methods for enzyme histochemistry have been developed for these reasons. In general, the methods prescribe the sparing use of heat and fixatives. Walker and Seligman (*J. Biophys. Biochem. Cytol.*, **9:** 415, 1961) advocate brief exposure of tissues to cold formalin. Chang and Hori (*J. Histochem. Cytochem.*, **9:** 292, 1961) prescribe treatment of frozen sections with cold acetone. Novikoff (*J. Biophys. Biochem. Cytol.*, **9:** 47, 1961) advises the fixation of small bits of tissues in cold Baker's formol-calcium. A few years ago, several authors were convinced that freeze-drying of tissues was essential for the practice of enzyme histochemistry. This fad has passed somewhat. Freeze-drying is a useful and recommended method of tissue preparation for enzyme histochemical studies; however, it is not essential in many cases as formerly implied. When properly used, excellent tissue preparations are achieved. A common disadvantage has been frequent breakdown of apparatus and/or the development of leaks in the system. Improved apparatus is now available.

No single method of tissue preparation can be recommended for the demonstration of all enzymes. Each enzyme is affected differently by every method of tissue preparation and fixative. Acceptable methods of tissue preparation are prescribed individually for each enzyme.

AZO DYE METHODS

Azo dye methods are generally used for the histochemical detection of acid and alkaline phosphatases, nonspecific esterases, β-glucuronidase, β-glucosaminidase, and aminopeptidases. The methods generally employ a substrate molecule containing a component designating the desired enzyme specificity and another portion of the molecule, usually a naphthol or naphthylamine, which is utilized in the formation of an insoluble dye after enzymatic hydrolysis. Nachlas et al. (*J. Histochem. Cytochem.*, **5**: 565, 1957) have written a comprehensive review of problems associated with histochemical enzyme localization. Several of the points provided below relating to characteristics of substrates, reaction products, diazonium salts, and enzyme localization are reviewed in the article by Nachlas et al.

Desirable characteristics of a substrate include (1) capacity to be rapidly hydrolyzed by the enzyme, (2) sufficient solubility in the buffer to permit optimal concentration, (3) that the substrate should not hydrolyze spontaneously during the incubation period, (4) that it should have no special affinity for tissue components, (5) that it should provide the desired enzyme specificity, and (6) that after hydrolysis the liberated reaction product should couple rapidly with a diazonium salt.

Desirable characteristics of the reaction product (usually a naphthol or naphthylamine) include (1) the ability to react rapidly with the diazonium salts in simultaneous coupling procedures, (2) high insolubility in both aqueous and lipid solvents, and (3) sufficient substantivity for protein to permit attachment to the tissues at the site of enzyme activity.

Diazonium salts are employed to couple with the liberated reaction product to produce a chromogen. Desirable characteristics of a diazonium salt include that (1) it is capable of rapid coupling at the desired pH, (2) it is stable under conditions of the assay, (3) it should not produce background staining of tissue sections, (4) it should not inhibit enzyme activity, (5) it produces an intensely colored chromogen when coupled with the desired reaction product, and (6) the chromogen produced should be amorphous (not in needle formation, etc.), be stable (to permit slide storage), and have sufficient substantivity to attach to tissues at the site of enzyme activity.

Diazonium salts may be employed to couple with the liberated reaction product at the time of enzymatic hydrolysis (simultaneous coupling) or later (postcoupling). Advantages of postcoupling include that (1) it eliminates the possibility of inhibition of enzyme activity by the diazonium salt, (2) incubation for enzyme activity can be conducted at optimal conditions of pH, etc., and (3) optimal conditions can be employed for coupling subsequently.

The accuracy of localization of enzyme activity depends upon many factors including (1) the rate of enzyme activity, (2) the solubility and the degree of diffusion of the reaction product, (3) the rate of coupling of the diazonium salt with the reaction product (Nachlas et al., *J. Histochem. Cytochem.*, **7**: 50, 1959), (4) the solubility of the chromogen, and (5) the substantivity of the chromogen.

Many factors influence staining reactions for hydrolytic enzymes. For example, substrates employed in azo dye methods are unnatural. This contrasts with histochemical methods employed for the detection of dehydrogenases whereby the natural substrates can be used. Complex naphtholic substrates may be too highly insoluble, so that an adequate concentration cannot be obtained in the incubation medium. Several factors may require that incubations be conducted at a nonoptimal temperature or a nonoptimal pH, or in the presence of inhibitors such as lead, calcium, or the diazonium salts themselves. The structure of the substrate itself is an important consideration. For example, the rate of hydrolysis of naphthol AS phosphate esters is generally less than that of unsubstituted naphthols; however, AS naphthols are less soluble in incubation media after hydrolysis.

Glenner (*J. Histochem. Cytochem.*, **16**: 519, 1968) evaluated the specificity of enzyme histochemical reactions. He noted that a single substrate can be acted upon by more than one enzyme (Felgenhauer and Glenner, *J. Histochem. Cytochem.*, **14**: 401, 1966; Martin et al., *J. Biol. Chem.*, **234**: 1718, 1959). The localization of enzymes in a histochemical system may vary depending upon the type of tissue processing employed, the type of substrate employed, and the type of precipitation procedure used (Shnitka and Seligman, *J. Histochem. Cytochem.*, **9**: 504, 1961).

Livingston et al. (*Histochemie*, **18**: 48, 1969; **24**: 159, 1970) synthesized the diazonium salts, namely, triphenyl-*p*-aminophenethyl lead and triphenyl-*p*-aminophenyl lead. They recommend the use of the diazonium salts for the demonstration of acid hydrolases in plant and animal tissues. Stated advantages are increased density permitting electron microscopic investigations, increased solubility of the salts, and knowledge of the exact chemical composition of the coupling moiety. The salts are not available commercially at the time of this writing.

PHOSPHATASES

The technics for acid and alkaline phosphatases fall into three classes: (1) the earlier methods, depending on the capture of the liberated phosphate ion by a suitable cation to form an insoluble salt *in situ*, and permitting the use of varied, more or less physiologic phosphate substrates, which are then visualized as lead or cobalt sulfide or similar-colored insoluble products; (2) methods using highly artificial substrates of phosphates of various simple and complex naphthols, which are visualized by simultaneous or later coupling with diazonium salts to form insoluble azo dyes; (3) methods using indoxyl substrates. Acid phosphatases are considered first. Table 10-1 provides details of three methods.

Nonspecific Acid Phosphatase Procedures

TABLE 10-1 SUBSTRATES FOR THE LEAD–GLYCEROPHOSPHATE ACID–PHOSPHATASE METHOD

	Gomori	McDonald	Barka
Glycerophosphate	$\alpha + \beta$ 8 % 3 ml	$\alpha + \beta$ 3.06 % 3 ml	β* 1.25 % 6 ml (40 mM)
Buffer	Acetate	Acetate	Tris maleate
Volume	30 ml	27 ml	6 ml
Molarity	50 mM	50 mM	0.2 M
pH	5, (W: 6)*	4.7	5
Distilled water	....	....	6 ml
Lead nitrate	36 mg in 30 ml buffer	10 mg in 27 ml buffer	20 ml 0.2 %
Final molarity of lead,			
mM	3.5	1	2.4
	Precipitate of excess lead glycerophosphate filtered out		Mix first 3 ingredients and add lead dropwise with shaking

* $C_3H_5(OH)_2PO_4Na_2 5H_2O.$

Sources: Gomori, *Stain Technol.*, **25**: 81, 1950; McDonald, *Q. J. Microsc. Sci.*, **91**: 315, 1950; Barka, *J. Histochem. Cytochem.*, **10**: 741, 1962; and Wachstein, *J. Histochem. Cytochem.*, **6**: 389, 1958.

TECHNIC

1. Fix fresh frozen cryostat-cut sections affixed to slides for 15 to 30 min in neutral 2.5% glutaraldehyde–neutral formalin ($CaCl_2$ or calcium acetate) at 2 to 4°C, in acetone for 15 min at 2 to 4°C, or in acetone at −70°C for several hours.
2. Wash out formalin or acetone with distilled water (80 s).
3. Incubate in substrate at 37°C for 40 min (Barka substrate; 90 min to 24 h for the Gomori substrate; shorter intervals up to 3 h with Wachstein's pH 6 variant).
4. Rinse in distilled water and for 1 min in 2% acetic acid.
5. Rinse in distilled water, and transfer to a 2% dilution of ammonium sulfide for 2 min.
6. Wash, dehydrate, and mount in synthetic resin as usual, inserting a safranin, fuchsin, or methyl green nuclear counterstain if desired. Hematoxylin and eosin may be used.

RESULTS: Sites of acid phosphatase activity are brown.

The acetic acid wash at step 4 was considered important by Gomori in preventing nonspecific lead staining. Acetic acid concentrations of 1 to 3% are used, and timing is not exact; 2% for 1 min represents a compromise. Later writers do not stress this step.

Regulation of the calcium concentration is essential to permit accurate localization of enzyme activity. Elevation of calcium concentration in the medium inhibits localization in nuclei; a decrease in concentration below 0.05 M results in nuclear staining. This is believed to be due to the failure of rapid capture of enzymatically liberated phosphate which diffuses into nuclei. Likewise, if liberation of phosphate is slow due to low levels of enzyme activity (loss of enzyme due to experimental manipulations, denaturation of enzyme by fixatives, heat, embedding procedures, etc.), insufficient phosphate will be released to exceed the solubility product, calcium salts are not formed, and a failure to stain ensues.

Prolonged incubations favor diffuse staining due to low enzyme activity and failure of rapid capture of released phosphate.

Prolonged washing of tissue sections, especially in acid-distilled water, should be avoided. Reaction product may be washed out, especially at an acid pH.

Desmet (*Stain Technol.*, **37**: 373, 1962) cautions that use of the 1 to 2% acetic acid wash in the Gomori acid phosphatase method as prescribed to remove nonenzymatically precipitated lead will probably also remove lead deposited as a result of enzymatic activity.

For acid phosphatase Wachstein has used free-floating frozen sections of material previously briefly fixed in cold (3°C) neutral formalin. Barka (*J. Histochem. Cytochem.*, **10**: 741, 1962) also used frozen sections of material prefixed 1 to 2 days at 3°C in neutral calcium-formalin, but he placed them on slides and dried them 2 to 24 h before reacting. Allen and Slater (*J. Histochem. Cytochem.*, **4**: 110, 1956) used 15-μ frozen sections for alkaline phosphatase after 12-h formalin fixation. The Feder technic, freeze substitution in acetone followed by polyvinyl alcohol embedding, has been used successfully for acid and alkaline phosphatases, 5-nucleotidase, and adenosine triphosphatase (Birns and Masek, *J. Histochem Cytochem.*, **9**: 204, 1961). Fresh frozen cryostat sections are quite useful for the nonspecific alkaline and the various adenosine phosphatases (Burgos et al., *J. Histochem. Cytochem.*, **3**: 103, 1955; Padykula and Herman, *J. Histochem. Cytochem.*, **3**: 161, 170, 1953; Freiman, *J. Histochem. Cytochem.*, **10**: 520, 1962); they have not proved useful for acid phosphatase, but if transferred after attachment on slides into cold formalin (3°C) or acetone (0 or −70°C) for postfixation, they give very acceptable results. This last procedure has given 72% acid and 92% alkaline phosphatase preservation after 12 h in acetone at −70°C and 21% acid and 36% alkaline after 18 h in formalin at 2 to 4°C. 5'-Nucleotidase, adenosine triphosphatase, and glucose 6-phosphatase gave 96, 72, and 22% preservation after acetone and 62, 6, and 0% preservation after formalin (Chang and Hori, *J. Histochem. Cytochem.*, **10**: 592, 1962). Since 15 to 30 min in formalin vapor is sufficient to prevent diffusion of

M-Nadi oxidase in blood smears and 2 h is enough to prevent diffusion of dopa oxidase even in blocks, it would seem that much shorter formalin postfixation would be adequate for cryostat sections, as it is indeed in the case of the phosphopyridine nucleotide–linked dehydrogenases (Hitzeman, *J. Histochem. Cytochem.*, **11**: 62, 1963). Burstone (ibid., **9**: 146, 1961) reported good results from postfixation at 25°C for a total of about 5 min in absolute and descending percentages of diacetone alcohol. [This is a ketone whose members are the methyl radical and a tertiary butyl alcohol group; $(CH_3)_2-COH-CH_2-CO-CH_3$.]

Since the azo coupling technic for phosphatases is essentially limited to acid and alkaline naphthol phosphatases, these technics will be presented first.

Sections are prepared by one of the foregoing procedures. Adamstone-Taylor cold knife sectioning can be done without elaborate equipment, and postfixation of the attached frozen section with cold acetone probably constitutes the most readily applicable and generally useful procedure. Wash out acetone or formalin with distilled water, and transfer to one of the substrates in Table 10-2, incubate for the indicated time, wash, and mount in glycerol gelatin.

Lojda et al. (*Histochemie*, **3**: 428, 1964) studied the reliability of several azo dye methods for the histochemical detection of acid phosphatase. They noted that the phosphates of naphthols AS, AS-D, AS-BO, AS-CL, AS-LC, AS-BS, AS-BI, and AS-TR prepared by either a cold or hot method are complicated mixtures that differ in composition depending upon the method of preparation used. The admixtures used influence the rate of enzymatic activity. Lojda et al. also noted that the degree of inhibitory effect of diazonium salts varies with the concentration of the diazonium salt (although not linearly dependent), the buffer (greater in Veronal-acetate than in acetate), pH (increased inhibition with rise of pH), and temperature (increased with rise of temperature). Variation in inhibition of hexazotized pararosanilin depends upon the source of the pararosanilin. The authors also caution that sharp localization may not indicate accurate localization.

Hoshino and Kobayashi (*J. Histochem. Cytochem.*, **19**: 575, 1971) note that specimens embedded in water-miscible glycol-methacrylate retain acid phosphatase activity. Specifically, small pieces of tissues are fixed in 2% glutaraldehyde with 3% paraformaldehyde in cacodylate buffer of pH 7 for 30 to 60 min at 4°C. After washing overnight in cacodylate buffer, dehydration and embedding according to Leduc and Bernhard (*J. Ultrastruct. Res.*, **19**: 196, 1967) are prescribed. Acid phosphatase may be demonstrated in thin sections by either azo dye or lead methods without removal of embedding medium.

Evans et al. (*J. Histochem. Cytochem.*, **14**: 171, 1966) described a method for the demonstration of acid phosphatase using 5-iodoindoxyl phosphate as substrate. They attribute superiority of the substrate over other indoxyl substrates to a rapid oxidation of 5-iodoindoxyl to 5,5′-diiodoindigo in the acid range. The method is not recommended for lipid-rich tissues, because 5,5′- diiodoindigo forms crystals with lipids.

1. Fix tissues in Baker's formol-calcium for 24 h at 4°C.
2. Place tissues in 30% sucrose (0.88 *M*) containing 1% gum acacia for at least 24 h at 4°C. (Fresh frozen cryostat sections may also be used.)
3. Free-floating or mounted cryostat-cut sections may be incubated in the following medium for 40 to 60 min at 23°C. Highly reactive sites stained within 15 min.

5-Iodoindoxylphosphate	0.01 to 0.1%
Citrate buffer (0.1 *M*), 2 ml, pH 5.4	
Sodium chloride (2 *M*), 5 ml	11.69%
Potassium ferricyanide (0.05 *M*), 0.5 ml	1.65%
Potassium ferrocyanide (0.05 *M*), 0.5 ml	7.11%

Total volume is 10 ml.

TABLE 10-2 SUBSTRATES FOR THE NAPHTHOL AS PHOSPHATE-SIMULTANEOUS AZO COUPLING TECHNIC FOR ACID AND ALKALINE PHOSPHATASES

	Burstone[a] FD	Burstone[b] OC	Burstone[c] FF	Barka[d] AS-TR, BI	Barka[d] α	Burstone[e] FD	Burstone[c]
Fixing and sectioning procedure	Freeze-dried, celloidin paraffin	Freeze-dried, celloidin paraffin	Cryostat postfixed in diacetone alcohol	18 h formalin attached frozen sections	18 h formalin attached frozen sections	Freeze-dried, celloidin paraffin	Cryostat postfixed in diacetone alcohol
Naphthol AS phosphate[f]	BI(TR,LC)	MX	BI(E,CLAN)	TR, BI	α-Naphthyl	MX	E(KB,CL,BA,TR, AN)
Amount	5 mg	5 mg	10 mg	10 mg	20 mg	5 mg	10 mg
Dimethylformamide	0.1–0.25 ml	0.25 ml	0.25 ml	1 ml		0.25 ml	0.25 ml
Distilled water	25 ml	25 ml	25 ml	q.s. 20 ml	q.s. 20 ml	25 ml	25 ml
Buffer	0.2 M acetate	0.2 M acetate	0.2 M acetate	0.15 M Veronal acetate[g]	0.15 M Veronal acetate[g]	0.2 M tris	0.2 M tris
pH	5.2	5.2	5.2	5	6	8.3	8.3
Vol	25 ml	25 ml	25 ml	5 ml	5 ml	25 ml	25 ml
Activator	10% $MnCl_2$ 0.1 ml = 10 mg	10% $MnCl_2$ 0.1 ml = 10 mg	10% $MnCl_2$ 0.1 ml = 10 mg				
Diazonium salt	Fast red violet LB	Fast blue BB	Fast red violet LB or fast blue BB	Freshly diazotized pararosanilin[h]		Fast red violet LB	Fast blue BB or fast red violet LB
Amount	30 mg	30 mg	30 mg			30 mg	30 mg
Final vol	50 ml	50 ml	50 ml	20 ml	20 ml	50 ml	50 ml
Time	30–60 min	45–60 min	30–120 min	20–90 min	10–30 min	30–60 min	10–30 min
Temperature	25°C	25°C	25°C	25°C	25°C	25°C	25°C

[a] J. Histochem. Cytochem., **7**: 39, 1959.
[b] Ibid, **7**: 147, 1959.
[c] Ibid, **9**: 146, 1961.
[d] Ibid, **10**: 741, 1962.
[e] Ibid, **6**: 322, 1958.
[f] Only the qualifying letters of the AS naphthols are given; e.g., for "BI" read "naphthol AS-BI".
[g] Barka's Veronal-acetate contains 9.714 g sodium acetate (3H$_2$O) (=5.85 g anhydrous) and 14.174 g sodium barbiturate in CO$_2$-free, glass-distilled water to make 500 ml.

[h] Hexazonium pararosanilin, fresh diazotate (Davis and Ornstein, J. Histochem. Cytochem., **7**: 297, 1959) was used by Barka as the diazo reagent in his acid phosphatase method. The diazotization is done at about 20 to 25°C, and no diazotization interval is specifically allowed, the process being presumed to be instantaneous. The raw, approximately 2.5 N HCl solution containing 4% pararosanilin chloride without allowance for dye content is mixed with 4% NaNO$_2$ in equal volumes, and the mixture is dumped directly into the alkaline buffered substrate and diluted with water, the pH then being adjusted to 5 or 6, the former for naphthol AS-BI or -TR phosphate, the latter for α-naphthyl phosphate. Excess nitrous acid is not destroyed, though there appears to be about a 50% excess.

4. Rinse sections in distilled water.
5. Mount in an aqueous medium such as Kaiser's glycerogel.

RESULTS: A red-purple microcrystalline or granular stain developed in areas of low enzyme activity. A blue-purple microcrystalline stain developed in sites of high enzyme activity. In overincubated tissues a fine pink diffuse stain developed. The distribution of stain is similar to that obtained using azo dye methods.

Barka (*J. Histochem. Cytochem.*, **12:** 229, 1964) noted lysosomal distribution of acid phosphatase in light and electron microscopic studies of proximal jejunum of mice after oral administration of corn oil.

Parkin et al. (*J. Histochem. Cytochem.*, **12:** 288, 1964) conducted a quantitative histochemical study of acid phosphatase in carcinomatous and normal cells of the prostate of man. A reduction in cancer cells was found.

Kaluza and Burstone (*J. Neuropathol. Exp. Neurol.*, **23:** 477, 1964) used several azo dye methods for acid and alkaline phosphatases in a study of the central nervous system of rabbits and cats.

Using *p*-nitrophenylphosphate as a substrate, Doty and Shofield (*Histochem. J.*, **4:** 245, 1972) noted that only lysosomal staining in osteoclasts occurred in reactions run at pH 7, whereas in reactions run at pH 5 to 6.5 phosphatase activity was evident in osteocytes and osteoblasts as well as osteoclasts.

Lin and Fishman (*J. Histochem. Cytochem.*, **20:** 487, 1972) employed differential centrifugation and chromatographic methods to separate and distinguish different acid phosphatases in lysosomes and on microsomes from BALB/c mouse kidneys.

Bowen [*J. Microsc. (Oxf.)*, **94:** 25, 1971] provides the following method for the histochemical detection of acid phosphatase and *β*-glucuronidase at the electron microscopic level. Bowen suggests that the method may be employed for the detection of other acid hydrolases.

ACID PHOSPHATASE

1. Fix tissue slices 1 mm thick at 4°C for 1 h in 3% glutaraldehyde in 0.1 M cacodylate buffer at pH 7.4.
2. Section tissues at 50 μ.
3. Wash tissues overnight at 4°C in 0.2 M acetate buffer of pH 5.2.
4. Incubate tissues $1\frac{1}{2}$ to 3 h at 18°C in the following medium: Dissolve 5 mg of either naphthol AS-TR phosphate or naphthol AS-BI phosphate in 0.25 ml dimethyl formamide or dimethyl sulfoxide, and add 25 ml distilled water, 25 ml 0.2 M acetate buffer of pH 5.2, and 30 mg *p*-nitrobenzene diazonium tetrafluoroborate (fast red GG; Kodak).
5. Rinse tissues in 0.2 M acetate buffer pH 5.2.
6. Fix tissues in OsO_4 according to Millonig (*J. Appl. Physiol.*, **32:** 1637, 1961).
7. Dehydrate and embed in Araldite.

A postcoupling procedure may be alternately employed. Couple at pH 7 in 0.1 M cacodylate buffer at the above salt concentration. Acid phosphatase is detected in lysosomes and in ribosomal regions.

Goldfisher et al.(*J. Histochem. Cytochem.*, **12:** 72, 1964) reviewed problems associated with the localization of several phosphatases at the ultrastructural level. Numerous problems in interpretation were discussed. Problems discussed included penetration of fixatives, degrees of fixation with various fixatives, and enzyme inhibition. They noted that many capillaries and bile canaliculi will hydrolyze many nucleoside phosphates, whereas renal tubular cells hydrolyze only triphosphates.

McKeever and Balentine (*Lab. Invest.*, **29**: 633, 1973) studied acid phosphatase activity in ventral motor horn cells of rats at the light and electron microscopic levels. Activity was noted in membrane-bound vacuoles, dense bodies, lamellae and cisternae of the Golgi bodies, and the GERL (Golgi related region of smooth endoplasmic reticulum from which lysosomes appear to develop) of neurons and neuroglia.

LYSOSOMES

Lysosome is a term introduced by de Duve (*Biochem. J.*, **60**: 604, 1955). He and coworkers were fractionating cells by ultracentrifugation, whereupon the presence of a group of acid hydrolases was detected in a fraction slightly different from that of microsomes. The enzymes in this group were all hydrolases; all acted at an acid pH; and enzyme activity was not evident until homogenates were frozen and thawed, exposed to detergents, or suspended in hypotonic media. The particles were designated *lysosomes* in recognition of their hydrolytic enzymes. Lysosomes have been found in many tissues of mammals, in invertebrates, and in unicellular organisms (*Biol. Bull.*, **124**: 285, 1963; *Ciba Found. Symp. Lysosomes*, 1963, p. 201).

Lysosomes may autofluoresce (Koenig, *J. Histochem. Cytochem.*, **11**: 556, 1963), take up neutral red (Cohn and Weiner, *J. Exp. Med.*, **118**: 1009, 1963), and take up fluorescent dyes [Allison and Mallucci, *Nature (Lond.)*, **205**: 141, 1965].

Swift (*Fed. Proc.*, **23**: 1026, 1964), Swift and Hrubon (*Fed. Proc.*, **23**: 1023, 1964), and Novikoff (*J. Cell Biol.*, **15**: 140, 1962) studied the formation of lysosomes. Parts of mitochondria, Golgi bodies, and endoplasmic reticulum have been detected in lysosomes. They may sometimes develop in living cells in response to an unfavorable environment. The origin of the hydrolases is unknown.

Movat et al. (*Lab. Invest.*, **29**: 669, 1973) demonstrated a kinin-generating enzyme in lysosomes separated from human polymorphonuclear leukocytes.

Mayahara et al. (*Histochemie*, **11**: 88, 1967) advocate the following *lead citrate* method for the demonstration of alkaline phosphatase at the ultrastructural level.

1. Tissues are fixed in 2% glutaraldehyde or 4% formaldehyde in 0.1 M cacodylate buffer of pH 7.4 containing 8% sucrose for 1 h at 4°C.
2. Thick frozen sections (40 to 60 μ) are incubated for 5 to 20 min at 24°C in the following medium:

	Final concentration, mM
1.4 ml 0.2 M tris HCl buffer pH 8.5	28
2 ml 0.1 M sodium β-glycerophosphate	20
2.6 ml 15 mM MgSO$_4$	3.9
4 ml saturated (about 0.5%) alkaline lead citrate solution, pH *ca.* 10	2

Sucrose may be added to make a final concentration of 8%. The final pH is adjusted, if necessary, to 9.2 to 9.4 with 0.1 N NaOH. The saturated alkaline lead citrate is kept airtight and filtered or centrifuged before use.

3. Following incubation, tissues are refixed in 1% OsO$_4$ for 1 h at 4°C, dehydrated through graded ethanols and propylene oxide, and embedded in Epon. After sectioning, sections may be poststained with uranium and examined after electron microscopy.

For light microscopy, specimens may be taken after step 2, immersed in dilute ammonium sulfide, rinsed, and mounted in a water-soluble mount such as Kaiser's. Mayahara et al. report localization of alkaline phosphatase is comparable to "other methods."

Bonting and Nuki (*Ann. Histochim.*, **8:** 79, 1963) conducted quantitative assays for alkaline phosphatase of various regions of developing molars of hamsters in freeze-dried microdissected samples. Greatest activity was demonstrated in cells of the stratum intermedium.

ALKALINE PHOSPHATASES: THE CALCIUM–COBALT SULFIDE (Ca-CoS) METHOD STAINING PROCEDURE. The general histologic technic for handling sections for specific and nonspecific alkaline phosphatase methods follows essentially the procedure of Gomori:

1. From acetone or formalin, wash sections with distilled water. Paraffin sections from freeze substitution, freeze dry, or the older fixation procedures are deparaffinized, preferably with petroleum ether, and either washed in isopentane, two further changes of petroleum ether in the pentane range, and let dry (preferable) or hydrated through acetones (preferably ice-cold) to water.
2. Introduce into selected substrate, preheated to 37°C, and incubate for prescribed period, perhaps taking out slides at graded intervals.
3. Wash 5 min in two or three changes of distilled water.
4. Immerse 5 min in 2% cobaltous nitrate (chloride or acetate will serve).
5. Wash in four changes of distilled water, 2 min each.
6. Immerse in approximately neutral 1% ammonium or 0.5% sodium sulfide for 1 min (2.5 ml saturated Na_2S—about 20%—plus 0.56 ml glacial acetic acid plus 97 ml distilled water yields a pH of about 6.8 to 7.5. NaSH may be used, if available.)
7. Wash 5 min in running water.
8. Counterstain if desired.
9. Dehydrate, clear, and mount, preferably in Canada balsam or ester gum. Cobalt sulfide keeps better in "reducing" resins.

Note: If preferred, the Von Kóssa silver technic may be substituted for steps 4 to 7.
RESULTS: Sites of phosphatase activity are dark brown to black.

Cobalt sulfide deposits in sections gradually disappear when they are mounted in certain of the synthetic resins such as old Permount, polystyrene, and the Euparal-Diaphane–type mixtures. Preservation is best in natural Canada balsam and in β-pinene or piccolyte resins (Bioloid, Permount, HSR, etc.) and in ester gums. Preparations mounted in Apáthy's gum syrup, in glycerol gelatin, Arlex gelatin, syrup, and the like fade rapidly. Faded preparations may be fully restored by demounting, hydrating, reimmersing in ammonium sulfide, and then remounting as before.

Ammonium sulfide gradually deteriorates on the laboratory shelf. It is desirable to get a fresh bottle about once a year.

Sodium sulfide, in equivalent concentration, or hydrogen sulfide–water may be used instead, and the pH adjusted to near neutrality. Sodium sulfide is deliquescent, and crystals should be drained and dried before being weighed. Or use 2.5 ml saturated aqueous (about 20%) sodium sulfide and about 0.56 ml glacial acetic acid and 97 ml distilled water to give pH 6.8 to 7.5. Addition of a few drops of this solution to a small supply of Apáthy's gum syrup or glycerol gelatin should retard fading in these media.

Table 10-3 gives comparative details on seven methods for alkaline phosphatases. Molnar (*Stain Technol.*, **27**: 221, 1952) suggests replacing the cobalt bath (step 8) with a 2- to 5-min bath in 2% lead nitrate solution and the ammonium sulfide treatment (step 10) with a 40-min bath in 0.5% aqueous sodium or potassium rhodizonate solution. Dark brown lead rhodizonate deposits are formed, and hemosiderin does not react.

Gomori (letter, 1953) found the rhodizonate color no better than the sulfide, and suggested instead the use of a very dilute solution of gallamin blue, which colors lead deposits selectively and intensely. Other counterstains may follow the gallamin blue.

Incubation solutions at high pH favor insolubility of calcium salts formed with consequent deposition on sections at the site of enzyme activity. After fixation, loss of enzyme activity is not significantly affected by dehydration; however, heat employed for paraffin embedding results in considerable loss of enzyme activity. Danielli (*J. Exp. Biol.*, **22**: 110, 1946) estimates as much as 75% of alkaline phosphatase may be inactivated by heat during paraffin embedding of tissues at 58 to 60°C.

Lowering the pH of the incubation medium results in reduction of enzyme activity, reduced capture efficacy, and reduced accuracy of localization. Sections left in dilute solutions of ammonium sulfide for more than the prescribed time periods will manifest drift of the reaction products. Cleland (*Proc. Linnean Soc. New South Wales, vol. 75*, pt. 1 and 2, pp. 36–70, 1950) studied the effects of many factors on the localization of alkaline phosphatases.

Lane (*J. Histochem. Cytochem.*, **13**: 235, 1965) notes that sections containing metal reaction products (cobalt sulfide, calcium phosphates, etc.) may be treated with 0.5% hematein for 1 to 60 min, "blued" in alkaline tap water for 1 to 2 min, and either dehydrated and mounted in a synthetic medium or mounted in an aqueous medium such as Kaiser's. Permanent blue-purple lakes are formed. Hematoxylin 0.1%, 1 h, should serve.

Rufo and Fishman (*J. Histochem. Cytochem.*, **20**: 336, 1972) studied the effects of various inhibitors of alkaline phosphatases. L-Homoarginine inhibited partially purified liver alkaline phosphatase by 97%. The alkaline phosphatase of leukocytes and macrophages of rat intestinal lamina propria was inhibited by 12.5 mM L-homoarginine, but not by L-phenylalanine. Rufo and Fishman report that both L-phenylalanine and L-homoarginine are uncompetitive inhibitors. The effects are reversible and require optimal concentration of substrate. In contrast, competitive inhibitors are most effective at low substrate concentrations (*J. Biol. Chem.*, **247**: 3082, 1972).

Watanabe and Fishman (*J. Histochem. Cytochem.*, **12**: 252, 1964) studied L-phenylalanine as an inhibitor of alkaline phosphatase in rat intestine. Substantial inhibition was observed when β-glycerophosphate, o-carboxyphenylphosphate, and α-naphthyl acid phosphate were employed as substrates. Although substantial inhibition of alkaline phosphatase activity in the striated border of intestinal epithelial cells was observed, rat kidney and leukocyte alkaline phosphatase activity was relatively little affected.

Lin and Fishman (*J. Biol. Chem.*, **247**: 3082, 1972) have concluded that L-homoarginine is an uncompetitive inhibitor of human liver and bone alkaline phosphatases; however, it does not inhibit human intestinal and placental alkaline phosphatases.

Borgers (*J. Histochem. Cytochem.*, **21**: 812, 1973) conducted light and electron microscopic observations of the effects of two inhibitors of alkaline phosphatase—namely, tetramisole [2,3,5,6-tetrahydro-6-phenylimidazo(2,1-β-)thiazole hydrochloride] and R 8231 [6(*m*-bromophenyl)-5,6-dihydroimidazo(2,1-β-)thiazole oxalate]. The inhibitors are available from Janssen Pharmaceutica Research Laboratories, 2340 Bearse, Belgium. Tetramisole is a broad-spectrum anthelmintic. Complete inhibition of alkaline phosphatase was observed in all tissues except intestine at 0.5 and 0.1 mM, respectively. The activities of acid phosphatase, adenosine triphosphatase, thiamine pyrophosphatase, glucose 6-phosphatase, and 5'-nucleotidase were unaffected by the inhibitors.

TABLE 10-3   GLYCEROPHOSPHATE SUBSTRATE MIXTURES FOR ALKALINE PHOSPHATASE

	Gomori[a] Wt vol	Gomori Final mM	Wachstein[b] Wt vol	Wachstein Final mM	Padykula[c] Wt vol	Padykula Final mM	Burgos et al[d] Wt vol	Burgos Final mM	Feigin and Wolf[e] Wt vol	F&W I Final mM	F&W II Final mM	F&W III Final mM	F&W IV Final mM	Allen[f] Wt vol	Allen Final mM	Freiman[g] Wt vol	Freiman Final mM
Fixation Section method	Acetone Paraffin		Acetone Paraffin		Fresh Frozen		Fresh Frozen		Acetone Paraffin					Fresh Frozen		Fresh Frozen	
Specific substrate	α, β	10–20	α, β(?)	3.3	α, β	12	α, β	13.3	β	33	33	33	33	β (5H$_2$O)	12	β (5H$_2$O)	15
Amount	5–10 ml 3%		50 mg		180 mg		10 ml 2%		500 mg					183 mg		230 mg	
Calcium salt	CaCl$_2$	70–90	CaCl$_2$	8	CaCl$_2$	18	CaCl$_2$	4.5	Ca(NO$_3$)$_2$ 4H$_2$O 163.5 mg	13	13	13	0	CaCl$_2$	15	CaCl$_2$	18
Amount	20–15 ml 2%		4 ml 0.1 M		5 ml 2%		5 ml 0.5%							5 ml 0.15 M		5 ml 2%	
Magnesium salt	MgCl$_2$	5	MgSO$_4$	10	None		MgSO$_4$ (7H$_2$O) 1 ml 0.5%	0.41	MgCl$_2$ (6H$_2$O) 0.102 mg 1.63 g	0	10	160	160	None		None	
Amount	0.5 ml 10%		5 ml 0.1 M														
Buffer	Veronal Na	50–100	Ammediol	80	Veronal Na	80(?)[h]	Veronal Na	20	Veronal Na	49	49	49	49	Veronal Na	33	Veronal Na	25
Amount	0.5–1 g		20 ml 0.2 M		10 ml 0.2 M(?)[i]		10 ml 0.2 M		12.5 ml 0.2 M					16.7 ml 0.2 M		12.5 ml 0.2 M	
Buffer pH			8, 9						9.2					9.5(a)		9.4(a)	
Final volume Final pH	50 ml 9–9.4		50 ml 8, 9		50 ml 9.4[i]		50 ml 8.6[i]		50 ml 9.2					50 ml 9.5		50 ml 9.4	
Incubation time Temp	1–4 h 37°C		15–60 min 37°C		5–180 min 37°C		5–20 min 10, 25, 37°C		5–30 min 37°C					10 min 37°C		60 min 37°C	

[a] G. Gomori, "Microscopic Histochemistry," University of Chicago Press, 1952, p. 184.
[b] M. Wachstein and E. Meisel, J. Histochem. Cytochem., 2: 137, 1954; Science, 115: 652, 1952.
[c] H. Padykula and E. Herman, J. Histochem. Cytochem., 3: 161, 170, 1955.
[d] M. H. Burgos, H. W. Deane, and M. L. Karnovsky, J. Histochem. Cytochem., 3: 103, 1955.
[e] I. Feigin and A. Wolf, J. Histochem. Cytochem., 5: 53, 1957.
[f] J. M. Allen, J. Histochem. Cytochem., 9: 681, 1961.
[g] D. G. Freiman, H. Goldman, and N. Kaplan, J. Histochem. Cytochem., 10: 520, 1962.
[h] Probable, not definitely stated in reference.
[i] Adjusted with NaOH or HCl.

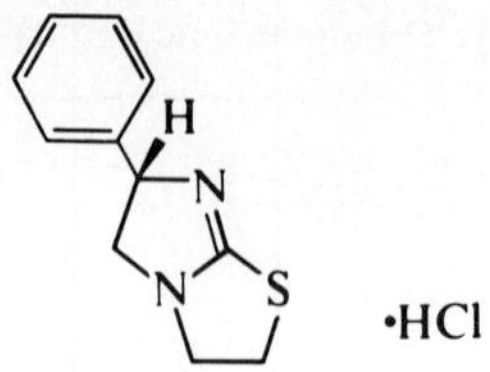

L-Tetramisole or levamisole

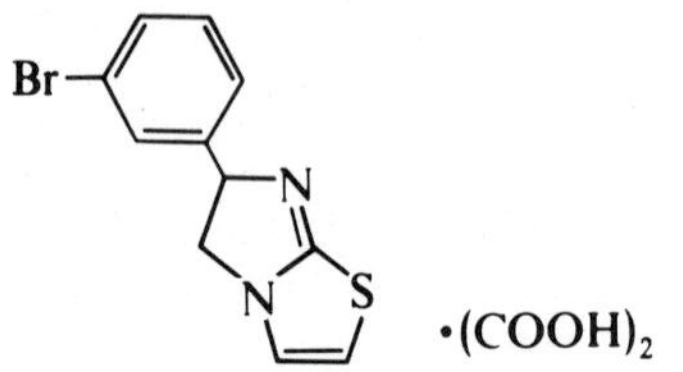

R 8231

Kaplow (*Am. J. Clin. Pathol.*, **39**: 439, 1963) tested several azo dye methods for the demonstration of alkaline phosphatase in leukocytes. He recommends the use of naphthol AS-BI phosphate as substrate, fast violet LB as a diazonium salt, and propanediol buffer of pH 9.7.

Wachstein, Meisel, and Ortiz (*Lab. Invest.*, **11**: 1243, 1962) prescribe overnight fixation at 4°C in 4% formalin with 1% $CaCl_2$ or neutralized 6% formalin without $CaCl_2$ or in Holt's sucrose formalin, and frozen sections at 5 to 8 μ. Cryostat sections are also used of tissue prefrozen in isopentane or petroleum ether at $-70°C$, with or without postfixing in the above $CaCl_2$ formalin for 15 to 180 min. Table 10-4 provides details of Wachstein and Meisel methods for neutral phosphatases.

TABLE 10-4 SUBSTRATES OF WACHSTEIN AND MEISEL FOR SPECIFIC PHOSPHATASES AT NEUTRAL pH LEVELS

	Substrate			
	Na β-GP*	A-5'-P†	ATP‡	K,G-6-P§
Amount	10 ml	20 ml	20 ml	20 ml
Concentration . .	1.25%	0.125%	0.125%	0.125%
Molarity	8.2 mM	1.4 mM	1.0 mM	1.5 mM
Tris maleate 0.2 M	5 ml	20 ml	20 ml	20 ml
pH	7.2	7.2	7.2	6.7
Lead (NO$_3$)$_2$, 2%.	3 ml	3 ml	3 ml	3 ml
Molarity	3.6 mM	3.6 mM	3.6 mM	3.6 mM
MgSO$_4$ 0.1 M . . .	0	5 ml	5 ml	0
Molarity	. . . .	10 mM	10 mM	
Distilled water . . .	32 ml	2 ml	2 ml	7 ml
Total volume	50 ml	50 ml	50 ml	50 ml
Incubation time . .	30–120 min	5–60 min	5–30 min	10–15 min
Temperature	n.s.¶	n.s.	n.s.	32°C

* Na β-GP, sodium β-glycerophosphate, 5H$_2$O, mw 306.126.
† A-5'-P, muscle adenylic acid, adenosine 5'-phosphate, mw 347.23.
‡ ATP, adenosine triphosphate, mw 507.21.
§ K,G-6-P, K$_2$, glucose 6-phosphate, mw 336.33.
¶ n.s., not stated.

Source: From *J. Histochem. Cytochem.*, **5**: 204, 1957.

The substrate for the pH 6 lead sulfide technic contains:

	Amount	Final molarity, mM
Pb $(NO_3)_2$	600 mg	3.3
0.05 M acetate buffer, pH 6	500 ml	45.0
3% sodium glycerophosphate*	50 ml	8.9

* mw 306.126.

Mix, incubate 24 h in 37°C incubator, add 5 ml distilled water, filter. Store in refrigerator at 3°C. The filtered solution is usable for 3 to 4 days only; older solutions, though not turbid, give rise to the nuclear staining artifact. Incubation time was 5 to 120 min.

Substantial controversy exists over the validity of the Wachstein-Meisel method for the histochemical detection of ATPase. Particular concern is expressed with the data which indicates lead-catalyzed hydrolysis of ATP, inhibition of ATPase by lead in the incubation medium, significant nonspecific staining of tissues, and the presence of nucleotide in the lead deposits (Moses et al., *Fed. Proc.*, **23**: 549, 1964; Rosenthal et al., *J. Histochem. Cytochem.*, **14**: 698, 1966; **14**: 702, 1966; **17**: 839, 1969; **18**: 915; 1970, **16**: 530, 1968; and Tice, *J. Histochem. Cytochem.*, **17**: 85, 1969). Inasmuch as the present authors regard the questions still unresolved, the reader is referred to the listed publications in order to familiarize himself with the original data and interpretations therefrom.

Moses and Rosenthal (*J. Histochem. Cytochem.*, **15**: 354, 1967) are convinced that lead-catalyzed ATP occurs in the Wachstein-Meisel mixture to the extent that the validity of the method should be seriously questioned.

Gillis and Page (*J. Cell Sci.*, **2**: 113, 1967) conclude that the lead precipitation methods are unsatisfactory to demonstrate ATPase activity in glycerinated rabbit muscles.

Novikoff has defended the Wachstein-Meisel method (*J. Histochem. Cytochem.*, **15**: 353, 1967; **18**: 916 and 366, 1970). Novikoff (ibid., **18**: 366, 1970) is convinced of the reliability of the Wachstein-Meisel method for the demonstration of the Mg–Ca–ATPase. Ganote et al. (*J. Histochem. Cytochem.*, **17**: 641, 1969) believe it may be reliable.

Tunik (*J. Histochem. Cytochem.*, **19**: 75, 1971) advises caution in the interpretation of the results of the calcium precipitation method for ATPase.

Jacobson and Jørgensen (*J. Histochem. Cytochem.*, **17**: 443, 1969) conclude that the modification of the Wachstein-Meisel incubation mixture detailed below results in an insignificant lead-catalyzed hydrolysis of ATP and probably represents a fairly accurate histochemical demonstration of Mg^{2+}–ATPase and Ca^{2+}–ATPase. They suggest the medium should employ 80 mM tris maleate buffer of pH 7.2 (at 37°C), 3 mM ATP (disodium salt), 3 mM Mg^{2+}, 120 mM Na^+, 30 mM K^+, and 3.6 mM lead nitrate [equivalent to 91.8 mg ATP, 27.9 mg $MgCl_2$, 701.3 mg NaCl, 223.7 mg KCl, and 119.2 mg Pb$(NO_3)_2$ per 100 ml, respectively].

Torak (*J. Histochem. Cytochem.*, **13**: 191, 1965) studied ATPase in rat brain after various fixations.

Severson et al. (*Stain Technol.*, **48**: 221, 1968) studied the distribution of ATPase in demineralized skeletal tissues.

Engel et al. (*J. Histochem. Cytochem.*, **16**: 273, 1968) used the electron probe in conjunction with the ATPase method of Padykula and Herman applied to sections of normal and diseased muscle. Measurements of calcium, phosphate, sodium, iron, and potassium were made. Advantage of the probe is that it can analyze for every element in the periodic table above that of boron. One disadvantage is that resolution does not exceed 2 to 5 μ.

Charnock et al. (*J. Histochem. Cytochem.*, **20**: 1069, 1972) describe a quantitative method for the electron microscopic observation of transport ATPase which employs electron-dense Cs^+ rather than K^+ ions as activator of $(Na^+ + K^+)$–ATPase. The transient formation of an ion-carrier state during the hydrolysis of ATP by erythrocyte ghost membrane $(Na^+ + K^+)$–ATPase was shown by a significant increase in membrane density due to accumulation of Cs^+ ions. The accumulation occurred only in the presence of ouabain.

The method consists of diluting 5 ml packed ghost red cell membranes to 10 ml by addition of the following reagents to permit the final composition of the medium to be 100 mM trisglycyl-glycine buffer of pH 7.6, 80 mM NaCl, 20 mM CsCl, 2 mM MgCl$_2$, 2 mM disodium ATP, and 0.1 mM ouabain, or in milligrams per 100 ml 467, 337, 18.6, 61.2, and 5.8.

Results suggested a uniform distribution of enzyme throughout the entire membrane.

ADENYLCYCLASE

Reik et al. (*Science*, **168**: 382, 1970) and Howell and Whitfield (*J. Histochem. Cytochem.*, **20**: 873, 1972) describe a histochemical method for the detection of *adenylcyclase*. Reik et al. studied the enzyme in rat liver, whereas Howell and Whitfield demonstrated activity in alpha and beta cells of isolated islets of Langerhans from rat pancreas. The procedure depends upon the precipitation of pyrophosphate (PP$_i$) by lead ions catalyzed by adenylcyclase in the reaction

$$ATP \xrightarrow[\text{cyclase}]{\text{adenyl-}} \text{cyclic AMP} + PP_i$$

Howell and Whitfield state that the specificity of the method is greatly enhanced by the substitution of adenylylimidodiphosphate (AMP-PNP) for ATP as substrate.

PROCEDURE

1. Islets of Langerhans cells separated from rat pancreas by the method of Howell and Taylor (*Biochim. Biophys. Acta*, **130**: 519, 1966) are fixed in 1% glutaraldehyde in 0.05 M cacodylate buffer pH 7.4 containing 4.5% glucose for 1 h at 25°C.
2. Wash three times, and store overnight in the above buffer.
3. Incubate tissues in the following medium for 30 to 60 min at 30°C: tris maleate buffer, 80 mM, pH 7.4, containing 8% glucose, 2 mM theophylline, 2 mM magnesium sulfate, 0.5 mM ATP or 0.5 mM 5'-adenylyl-imidodiphosphate (ICN, Ltd., Irvine, California), and 4 mM lead nitrate; or in milligrams per 100 ml, amounts are 360.34 theophylline, 46.48 MgSO$_4 \cdot$7H$_2$O; 15.3 ATP; 21 ADP; 132 Pb(NO$_3$)$_2$.
4. Rinse tissues briefly in tris maleate buffer.
5. Postfix tissues in 1% OsO$_4$ in cacodylate-nitrate-glucose buffer of pH 7.4.
6. Dehydrate in graded ethanols, and embed in an epoxy resin.

RESULTS: Quantitative assays indicated 20 to 40% of adenylcyclase survived glutaraldehyde fixation. The pyrophosphate formed due to adenylcyclase activity was deposited uniformly and only on the outer surface of plasma membranes. The quantity of deposition was markedly increased in tissues incubated in media containing 10 mM fluoride, and lesser increases were noted in tissues incubated in media containing 2 μg/ml glucagon.

Tu and Malhotra (*J. Histochem. Cytochem.*, **21**: 1041, 1973) use the method of Reik et al. (*Science*, **168**: 382, 1970) to demonstrate the presence of adenylcyclase on plasma membranes, nuclear membranes, and mitochondrial membranes of the fungus *Phycomyces blakesleeanus*.

CYCLIC 3′,5′-NUCLEOTIDE PHOSPHODIESTERASE

Shanta et al. (*Histochemie*, **7**: 177, 1966) provide the following method for the histochemical demonstration of *cyclic 3′,5′-nucleotide phosphodiesterase*. The substrate (cyclic 3′,5′-AMP) is acted upon by 3′,5′-AMPase forming 5′-AMP, which is split into adenosine and phosphate by 5′-nucleotidase supplied in the incubation medium. Liberated phosphate is captured by lead.

1. Fresh frozen tissues were sectioned at 10 μ with a cryostat.
2. Sections were dried 3 to 5 min under a fan and subsequently incubated in the following medium for 30 min to 3 h at 37°C:

Cyclic 3′,5′-AMP, 1.44 mM	47 mg/100 ml
Tris maleate buffer, 50 mM	pH 7.62 (the final pH should be 7.5)
Magnesium chloride, 10 mM	95 mg/100 ml
Lead acetate, 2 mM	66 mg/100 ml
Snake venom, 1 mg*	

* Purchased from Sigma Chemical Co., from *Crotalus atrox*.

3. Rinse sections in distilled water 3 times.
4. Place sections in dilute ammonium sulfide 2 min.
5. Wash in distilled water, and mount in a water-soluble medium such as Kaiser's.

Positive results are identified as brownish black stains. Omission of substrate resulted in failure of staining except for background staining. Omission of snake venom resulted in greatly reduced staining. Slight staining is presumed to be due to the presence of endogenous 5′-nucleotidase action on 5′-AMP. Inclusion of theophylline ethylenediamino (0.1 M) in the incubation medium resulted in nearly complete inhibition.

Conjunctiva of the eye, corneal and scleral stromal cells of the eye, proximal and distal tubules of the kidney, glial cells, and cerebral cortex stained strongly. The periphery of red blood cells stained strongly, and liver parenchymal cells stained moderately.

NUCLEOSIDE PHOSPHORYLASE

Rubio et al. (*Am. J. Physiol.*, **222**: 550, 1972) provide the following method for the histochemical localization of *nucleoside phosphorylase* in cardiac tissues:

The following reaction is catalyzed reversibly:

$$\text{Inosine} + \text{inorganic phosphate} \rightleftharpoons \text{hypoxanthine} + \text{ribose 1-phosphate}$$

1. Perfuse rat hearts for 7 min with glutaraldehyde (0.5%) containing dextran (3%), glucose (3%), and dithiothreitol (0.4 mM, 6.17 mg/100 ml at mw 154.25) at pH 7.4.
2. Rinse tissues with tris maleate buffer of pH 7.2 containing 0.2 M sucrose and 0.4 mM dithiothreitol at 0°C.
3. Cut sections in a cryostat at 40 to 50 μ.
4. Rinse tissues for 1 h in the buffer in step 2.
5. Incubate tissues in the following medium for 45 min at 25°C:

Tris maleate buffer, 0.1 M	
Sucrose, 0.2 M	6.8%
CaCl$_2$, 10 mM	111 mg/100 ml
Pb(NO$_3$)$_2$, 2 mM	66 mg/100 ml
Ribose 1-phosphate*, 4 mM	92 mg/100 ml
Hypoxanthine*, 10 mM	136 mg/100 ml

* May be purchased from Calbiochem.

6. Rinse tissues in the above buffer briefly.
7. Postfix tissues in 1% OsO_4 in 0.1 M cacodylate buffer.
8. Rinse with distilled water, dehydrate in ascending ethanols, clear in propylene oxide, and embed in Epon.

RESULTS: Nucleoside phosphorylase activity was detected only in endothelial cells, pericytes, and red cells. Cardiac tissue only was examined. Rubio et al. state that omission of ribose 1-phosphate resulted in absence of staining. No staining was observed in endoplasmic reticulum or mitochondria.

Borgers et al. (*J. Histochem. Cytochem.*, **20:** 1041, 1972) employed the nucleoside phosphorylase method of Rubio et al. (*Am. J. Physiol.*, **222:** 550, 1972) at the electron microscopic level to disclose enzyme activity in the cytoplasmic sap of endothelial cells, adventitial fibroblasts and fibrocytes, nerve axoplasm, and platelets, and in neutrophilic, basophilic, and monocytic leukocytes. Some lymphocytes displayed activity and others did not.

ACETYLTRANSFERASE

Benes et al. (*J. Histochem. Cytochem.*, **20:** 1031, 1972) describe a method for the ultrastructural localization of *acetyltransferases* catalyzing the transfer of acyl groups from palmityl–coenzyme A (CoA) to α-glycerophosphate. The method relies on the formation of a heavy-metal precipitate at the site of CoA release. Enzyme action releases CoA, which is oxidized by potassium ferricyanide, which is thereby reduced to potassium ferrocyanide, which precipitates in the presence of cupric ions.

1. Animals are perfused through the aorta with 1% glutaraldehyde in 0.025 M cacodylate buffer pH 7.4 containing 4.5% glucose.
2. Blocks 2 to 3 mm thick are cut (Benes et al. used rat liver) and washed in cacodylate buffer at 4°C for 3 to 12 h.
3. Slices are cut with a Smith-Farquhar section device.
4. Tissues are incubated in the following medium for 1 h at 37°C:

Potassium ferricyanide (0.5 mM)	16.5 mg/100 ml
Copper sulfate (3.0 mM)	74.9 mg/100 ml
Sodium citrate (5.0 mM)	147 mg (2H_2O)/100 ml
Cacodylate buffer (18.7 mM, pH 7.4)	
Palmityl-CoA (0.18 mM)	18.1 mg/100 ml
L-α-Glycerophosphate (2.5 mM)	43.0 mg/100 ml

5. For electron microscopy, tissue slices are washed 3 times with buffer at 4°C, fixed in 1% osmium tetroxide in cacodylate buffer of pH 7.4 for 1 h, dehydrated through graded ethyl alcohols, cleared in propylene oxide, and embedded in Epon.

RESULTS: Electron-dense deposits were noted in the cisternae of smooth and rough endoplasmic reticulum of rat liver parenchymal cells. Tissues incubated without α-glycerophosphate and palmityl-CoA failed to stain. Omission of α-glycerophosphate alone resulted in slight staining of the above-named structures, possibly due to endogenous α-glycerophosphate or lysophosphatidic acid. Fixation resulted in substantial (as much as 90%) loss of enzyme activity.

GLUCOSE 6-PHOSPHATASE

For glucose 6-phosphatase, unfixed cryostat sections are used in the following substrate, according to the same general technic as given earlier in this chapter for specific and nonspecific alkaline phosphatases.

	Amount, ml	Molarity, mM
0.125% glucose 6-PO$_4$K$_2$*	20	1.5
0.02 M acetate buffer, pH 6	20	8.0
2% Pb(NO$_3$)$_2$	3	3.6
Distilled water	7	
Total	50	

* mw 336.32.

Inhibitors of acid phosphatase were added to the above substrates in the following final concentrations: 5 mM (21 mg/100 ml) NaF; 2 mM (30 mg/100 ml) *d*-tartaric acid. As a preincubation inhibitor, 2.5 mM (84 mg/100 ml) ethylenediaminetetraacetate in pH 6 (50 mM?) acetate buffer, 60 min, to inhibit alkaline phosphatase.

Glucose 6-phosphatase is distinguishable from ordinary nonspecific acid phosphatase by its thermolability and its sensitivity to formaldehyde and to pH levels below 5. It is destroyed in material fixed in cold alcohol or acetone and embedded in paraffin. Tissue may be quickly frozen and stored at $-20°C$ for several days without great loss of activity.

According to Chiquoine (*J. Histochem. Cytochem.*, **1**: 429, 1953) fresh unfixed tissue is frozen and sectioned in the cryostat and dried briefly onto slides.

SUBSTRATE.   Dissolve 250 mg barium glucose 6-phosphate in 10 ml distilled water, and add 0.1 ml 2 N HCl and 120 mg K$_2$SO$_4$. Let stand 2 h with frequent stirring. Centrifuge out BaSO$_4$ and test supernatant for absence of Ba by adding a grain or so more K$_2$SO$_4$. If clear, dilute to 30 ml and adjust to pH 6.7 with normal KOH. For the working substrate dilute 1 volume of this solution with 2 volumes of 0.2% lead nitrate (6 mM). Slight turbidity results. Filter and place 0.5-ml quantities in each of the required number of 3-ml beakers. Coat edges of beakers with petrolatum (USP). Place sections face down over orifice of beaker, invert, and incubate 5 to 15 min at 32°C. Rinse successively in distilled water, in 2% acetic acid and again thoroughly in distilled water. Immerse for 2 min in 1 : 50 ammonium sulfide. Wash 2 to 5 min in running water. Stain 2 min in 0.1% safranin in 1% acetic, or as desired, dehydrate, clear, and mount in balsam. Activity sites are in brown.

Immersion for 10 min in Weigert's iodine or in boiling water abolishes the activity, as does fixation with formaldehyde. Preparations should be compared with duplicates prepared by the glycerophosphate–acid phosphatase method.

Kanamura (*J. Histochem. Cytochem.*, **21**: 1086, 1973) noted progressive loss of glucose 6-phosphatase activity with increasing periods of washing of tissues after fixation in 2% glutaraldehyde. More glucose 6-phosphatase could be detected after a 5-min wash of blocks or sections than without. However, after a 5-min wash in 0.01 M cacodylate buffer containing 0.3 M sucrose at 4°C, a progressive decline in enzyme activity occurred with increasing time of washing.

Manns (*J. Histochem. Cytochem.*, **16**: 819, 1968) studied the effects of fixation, heat, and other factors on the histochemical demonstration of glucose 6-phosphate in rat liver. Small blocks of liver were quenched in liquid nitrogen, sectioned with a cryostat at 10 μ, and mounted on glass slides. Sections were exposed to methanol-free formaldehyde (3, 5, 10, or 15%) in 0.04 M phosphate buffer at 4°C for 2, 5, 10, 30, or 60 min.

There was no visible loss of glucose 6-phosphatase activity in sections fixed in neutral formalin for periods up to 10 min. Enzyme activity was reduced to about 50% after 1 h. With increasing acidity of formalin, increasing damage to enzyme activity was observed. Inactivation was marked at pH 5.4. Also, increases in formaldehyde above 5% were very destructive of enzyme activity.

Total loss of activity followed treatment with water at 90°C for 5 min; 0.1 M acetate buffer of pH 5 at 37°C for 5 min; ethanol, methanol, or acetone at 0°C for 5 min; or the addition of Zn^{2+} (0.01 M), Cu^{2+} (1 mM), cyanide (0.01 M), Mg^{2+} (1.0 M), or Ca^{2+} (1.0 M). Fluoride (0.01 M) caused partial inhibition, while arsenite (0.05 M) or iodoacetate (0.05 M) were without effect. (Amounts per 100 ml are $ZnCl_2$, 136 mg; $CuSO_4 \cdot 5H_2O$, 25 mg; KCN, 65 mg; $MgCl_2$, 9.3%; NaF, 42 mg; Na_2O_3, 850 mg; $NaCO_2CH_2I$, 1.03%).

Manns concludes that previously reported detrimental effects of cold formalin may be due to the presence of methanol in formalin. As much as 13% may be added commercially to prevent polymerization.

Kuhlman (*Proc. Soc. Exp. Biol. Med.*, **85:** 458, 1954) noted that epithelial cells of seminal vesicles and prostate gland would hydrolyze either fructose 6-phosphate or glucose 6-phosphate. He was unable to determine if specific enzymes were responsible for the reactions.

Kanamura (*J. Histochem. Cytochem.*, **19:** 386, 1971) was able to demonstrate glucose 6-phosphatase at light and electron microscopic levels in rat liver after brief glutaraldehyde [2% in 50 mM cacodylate buffer of pH 7 containing 0.25 M (8.55%) sucrose] fixation.

Hori and Yonezawa (*J. Histochem. Cytochem.*, **20:** 804, 1972) advise caution in the interpretation of zymograms on the basis of their studies of rat liver glucose 6-phosphate dehydrogenase isozymes. Isozymes derived from cell sap designated D, F_1, F_2, and F_3 were interconvertible in the presence or absence of mercaptans.

PHOSPHOGLUCOMUTASE

Meijer (*Histochemie*, **8:** 248, 1967) provides the following method for the histochemical demonstration of *phosphoglucomutase*. The enzyme converts the substrate α-D-glucose 1-phosphate to α-D-glucose 6-phosphate. Thereafter, α-D-glucose 6-phosphate is oxidized by exogenous glucose 6-phosphate dehydrogenase to D-glucono-δ-lactone 6-phosphate, whereby endogenous NADPH–tetrazolium reductase reduces nitro BT (nitro blue tetrazolium) to an insoluble formazan.

1. Mount fresh cryostat-cut sections on slides and fix in acetone for 30 min at $-25°C$.
2. Incubate tissues in the following medium for 1 to 2 h at 22°C:

Disodium glucose 1-phosphate	40 mg
NADP	1.2 mg
ATP	2.5 mg
$MgCl_2 \cdot 6H_2O$	12.5 mg
Nitro BT	2 mg
Imidazole buffer, 40 mM, pH 7.4	1.5 ml
Gelatin solution (3%)	3.5 ml
Glucose 6-phosphate dehydrogenase (1 mg/ml)	0.02 ml

Adjust the pH to 7.4.

3. Fix sections in neutral 10% formalin for 30 min.
4. Rinse, and mount in a water-miscible mix such as Kaiser's.

RESULTS: Enzyme activity is manifested by dark blue staining. Meijer noted inhibition of enzyme activity with fluoride (0.21% NaF), 50 mM, and beryllium (0.4% $BeCl_2$), 5 mM. It is noted that localization is that of NADPH–tetrazolium reductase—which is the case with the histochemical demonstration of many dehydrogenases. Efforts to transfer electrons to phenazine methosulfate or menadione and thereafter to the tetrazolium salt were unsuccessful. The method warrants further investigations.

PHOSPHOENOLPYRUVATE PHOSPHATASE

Sandström (*Acta Morphol. Neerl. Scand.*, **8**: 195, 1970–1971) advises the following method for the histochemical detection of *phosphoenolpyruvate phosphatase*.

1. Perfuse animals with either Lillie's neutral calcium acetate–formalin or 2.5% glutaraldehyde and immerse tissues in the same fixative for 1 h at 23°C. Tiny human liver needle biopsies are fixed for 48 h.
2. Embed tissues in polyethylene glycol.
3. Cut sections on a sliding microtome.
4. Transfer sections to tris maleate buffer (80 mM, pH 7.2). The embedding medium is readily dissolved.
5. Incubate free-floating sections in the following medium for 1 h at 37°C:

   Lead nitrate (119 mg/100 ml; 3.6 mM in 50 mM acetate buffer)
   Cobalt (3.4 mM; 44 mg $CaCl_2$/100 ml)
   Phosphoenolpyruvate (8–10 mM; 134–168 mg/100 ml)

$C_3H_5O_6P$, mw 168.0435

   Before use, the pH should be adjusted to 6 and the solution filtered.
6. Rinse tissues in distilled water.
7. Treat sections with dilute (0.5%) ammonium sulfide.
8. Rinse sections with distilled water.
9. Dry sections on albuminized slides, dehydrate, and mount in synthetic medium.

RESULTS: A brownish black stain indicates phosphoenolpyruvate phosphatase activity. Sandström noted selective staining of liver sinusoid endothelial cells.

Sandström states that substrate specificity was determined by using as substrates 10 mM α,β-glycerophosphate, 1 mM adenosine triphosphate, 4 mM nucleoside diphosphate (adenosine, guanosine, cytidine, and uridine diphosphate), 4 mM thiamine diphosphate chloride, 4 mM thiamine monophosphate chloride, 4 mM glucose 6-phosphate, and 4 mM fructose 1,6-diphosphate. Boiled sections (30 min) were also used as controls. No staining was observed except in sections employing phosphoenolpyruvate phosphate as substrate.

Sandström noted that staining occurred only when both cobalt and lead were present in the incubation medium. Further studies of this method appear warranted.

Concentration, mM	mw	Concentration, mg/100 ml	Substrate
10	172.08	172	α,β-glycerophosphate
1	507.21	51	Adenosine triphosphate (ATP)
4	427.22	171	Adenosine diphosphate (ADP)
4	442.14	177	Guanosine diphosphate
4	403.22	161	Cytidine diphosphate
4	404.18	162	Uridine diphosphate (UDP)
4	532.84	213	Thiamine diphosphate chloride
4	451.85	181	Thiamine monophosphate chloride
4	260.14	104	Glucose 6-phosphate
4	340.13	136	Fructose 1,6-diphosphate

GLYCEROL 3-PHOSPHATE DEHYDROGENASE, TRIOSEPHOSPHATE DEHYDROGENASES

Brosemer (*J. Histochem. Cytochem.*, **20:** 266, 1972) used an immunofluorescent histochemical method to demonstrate localization of *glycerol 3-phosphate* and *triosephosphate dehydrogenases* in honeybee flight muscles. His method follows:

1. Section tissues frozen in liquid nitrogen at 4 μ with a cryostat and mount on slides.
2. Fix sections for 10 min at 23°C in 95% ethanol.
3. Wash in phosphate-buffered saline solution (PBS) (0.01 M sodium phosphate in saline solution at pH 7 with an ionic strength of 0.16).
4. Incubate for 15 min at 23°C with goat IgG-PBS buffer (0.8 mg goat γ-globulin per ml PBS).
5. Incubate with rabbit antiglycerol 3-phosphate dehydrogenase or antitriosephosphate dehydrogenase in goat IgG-PBS buffer for 45 min at 37°C.
6. Wash exhaustively with PBS.
7. Incubate with fluorescein-labeled goat antirabbit IgG (1 mg/ml) in goat IgG-PBS buffer for 45 min at 37°C.
8. Wash exhaustively in PBS.
9. Air-dry and mount in bicarbonate-buffered glycerol.
10. Observe for fluorescence.

Brosemer noted that substitution of PBS for goat IgG-PBS buffer did not significantly alter results. He advised use of the latter buffer to decrease the possibility of nonspecific interaction between antibody preparations and the tissues.

For electron microscopy, muscles were fixed in 2.5% glutaraldehyde in 50 mM cacodylate and 0.15 M sucrose at pH 7.3 for 3 h at 3°C. Specimens were postfixed in 1% osmium tetroxide in the same buffer, dehydrated in ethanols, and embedded in Epon. Thin sections were subsequently stained with uranyl acetate and lead citrate.

Both triosephosphate and glycerol 3-phosphate dehydrogenases were located near the Z band cross striations.

Allen (*J. Histochem. Cytochem.*, **9:** 681, 1961) lists a considerable number of organic phosphates which liberate phosphate ion in homogenate studies at pH 9.4, using uniformly a final 12-mM substrate concentration in his substrate as given above: adenosine tri-, di-, and monophosphates, nicotinamide-adenine dinucleotide, nicotinamide-adenine dinucleotide phosphate, thiamine pyrophosphate, 2-deoxyglucose 6-phosphate, fructose 1- and 6-

TABLE 10-5   MOLECULAR WEIGHTS OF PHYSIOLOGIC PHOSPHATES
FOR PHOSPHATASE STUDIES

Phosphate	mw	mg = 12 mM in 50 ml substrates
Adenosine triphosphate	507.21*	304
Adenosine diphosphate (pyrophosphate)	427.22*	256
Adenosine 5'-phosphate (muscle adenylic acid)	347.23*	208
Adenosine 3'-phosphate (yeast adenylic acid)	347.23*	208
Thiamine pyrophosphate (diphosphate)	496.38*	297
Nicotinamide-adenine dinucleotide (NAD)	663.2	398
Nicotinamide-adenine dinucleotide phosphate (NADP)	743.0	445
2-Deoxyglucose 6-phosphate	244.14	146
Fructose 1-phosphate	260.14*	156
Fructose 6-phosphate	260.14	156
Fructose 1,6-diphosphate	340.13*	204
Galactose 1-phosphate	260.14	156
Galactose 6-phosphate	260.14	156
Glucose 1-phosphate (—K$_2$2H$_2$O: 372.35)*	260.14*	156
Glucose 6-phosphate (—K$_2$: 336.32)*	260.14*	156
Glucosamine 6-phosphate	259.16	155
Mannose 1-phosphate	260.14	156
Mannose 6-phosphate	260.14	156
6-Phosphogluconate, Na$_2$(H$_2$: 276.145)	320.12	192
Ribose 5-phosphate	230.12*	138
O-Phosphoserine	171.07	102.6
Phosphocreatine	211.12*	126.6
α-Glycerophosphate Na	216.046*	129.6
β-Glycerophosphate Na 5H$_2$O	306.126	183.5
O-Phosphothreonine	199.107	119
Phosphocholine (phosphorylcholine)	219.61*	131

* P. G. Stecher (ed.), "The Merck Index of Chemicals and Drugs," 7th ed., Merck, Rahway, N. J., 1960.

monophosphates and fructose 1,6-diphosphate, galactose 1- and 6-phosphates, glucose 1- and 6-phosphates, glucosamine 6-phosphate, mannose 1- and 6-phosphates, 6-phosphogluconate, ribose 5-phosphate, O-phosphoserine and -threonine, phosphocholine and phosphocreatinine, as well as the traditional α- and β-glycerophosphates.

Of these, only mannose, glucose, and galactose 1-phosphates, ribose 5-phosphate, phosphocholine, and α-glycerophosphate gave moderate reactions compared with the strong reaction of β-glycerophosphate in the acid phosphatase method. These lists are by no means exhaustive. Allen's acid phosphatase media contained again 12 mM specific substrate, 50 mM acetate buffer of pH 5 or 5.5, and 3.6 mM lead nitrate (the published 0.036 M was a missed typographical error). His glucose 6-phosphatase medium contained in addition 2 mM d-tartrate to inhibit nonspecific acid phosphatase.

Padykula and Herman included adenosine 5'-phosphate and adenosine di- and triphosphates as substrates to be used in their alkaline phosphatase medium at the same 12-mM concentration in the substrate mixture as in Table 10-5 except that they prescribed a 5-mM concentration for the triphosphate.

The addition of d-tartaric acid serves to enhance activity of specific hexose 6-phosphatases, which otherwise react only moderately by acid phosphatase technics, and to inhibit ordinary "nonspecific" acid phosphatases, as demonstrated by the Gomori glycerophosphate-lead procedure.

Thiamine Pyrophosphatase, Thiamine Diphosphatase

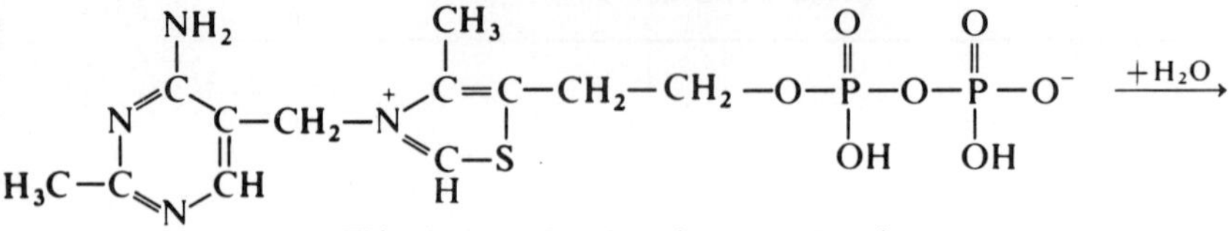

Thiamine pyrophosphate (cocarboxylase)[1]

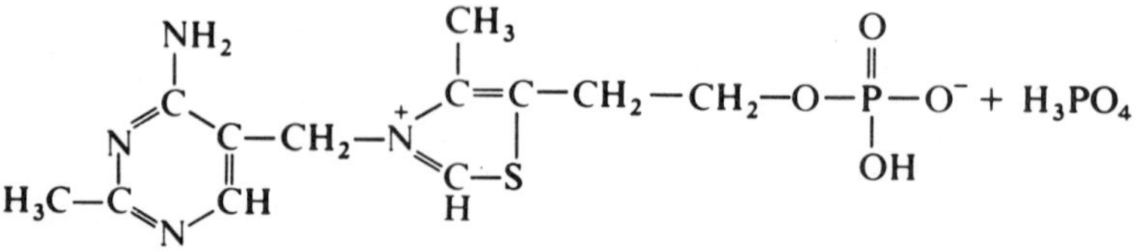

Thiamine orthophosphate

Allen and Slater (*J. Histochem. Cytochem.* **9**: 418, 1961) and Novikoff et al. (ibid., **9**: 459, 1961) refer to this enzyme as *thiamine pyrophosphatase*. Naidoo (ibid., **10**: 580, 1962) uses *thiamine diphosphatase*, in accordance with British usage, but definitely identifies his substrate thiamine diphosphate as cocarboxylase. Hence the terms are to be regarded as synonyms.

Allen and Slater found the enzyme in the well-segregated Golgi zone of the lining epithelium of the mouse epididymis. They used cryostat sections of fresh tissue, attached to slides. Novikoff preferred attached frozen sections either of material fixed 12 to 18 h at 4°C in neutralized formol-calcium or cryostat sections postfixed 5 to 60 min in the same fixative. Acetone postfixation was inferior; formaldehyde improved the intensity of the reaction over that in unfixed frozen sections.

Naidoo used frozen dried paraffin sections of brain, cut at 7 to 15 μ, and carried his sections, dry from the final isopentane, directly into substrate.

Naidoo's method (*J. Histochem. Cytochem.*, **10**: 580, 1962) for "thiamine diphosphatase" (pyrophosphatase): Cut thin (2 mm) slices of brain, freeze in isopentane at $-35°C$, freeze-dry 3 days at $-78°C$, and embed in paraffin (mp 49°C) in vacuo (15 min). Chill blocks, and section at 10 to 15 μ, transferring onto albuminized slides without water flotation. Store blocks at $-10°C$ for less than 7 days; section on same day as incubation.

1. Dewax paraffin sections at 20 to 25°C in 80 to 100°C petroleum ether for 2 min.
2. Wash in two changes isopentane (bp 27 to 31°C). Sections dry at once in air.
3. Thrust quickly into warm substrate with sharp movement to dislodge air bubbles.
4. Incubate in the substrate below at 37°C, removing sections at graded intervals. *Substrate:*

	ml	Final concentration mM
0.1 M maleate buffer pH 7.1	15	50
5 mM stock thiamine pyrophosphate (2.4 mg/1 ml)	3	0.5
0.1 M MgCl₂ (2.03 % MgCl₂·6H₂O)	6	20
0.03 M Pb(NO₃)₂ (or acetate) (993 and 975 mg/100 ml)	1	1.0
Distilled water to make	30	

---

[1] White, Handler, Smith, and Stetten, "Principles of Biochemistry," 2d ed., McGraw-Hill, New York, 1959, p. 345.

TABLE 10-6 SUBSTRATES FOR THIAMINE PYROPHOSPHATASE AND CONTROLS

	Allen and Slater		Naidoo, PbS		Naidoo, Ca—CoS	
	Amount	Molarity, mM	Amount	Molarity, mM	Amount	Molarity, mM
Thiamine pyrophosphate HCl (HCl) mw 460.789	115 mg	5	11.5 mg	0.5	184 mg	8
Thiamine monophosphate HCl (HCl) mw 381.814	95 mg	5	9.5 mg	0.5		
Sodium β-glycerophosphate ($5H_2O$) mw 306.126	76.5 mg	5	7.6 mg	0.5		
Sodium barbital 0.1 M	16.7 ml	33			12.5 ml	25
Tris maleate 0.1 M, pH 7.1			25 ml	50		50
$CaCl_2$ 0.1 M (1.11%) mw 110.99	7.5 ml	15			12.5 ml	25
$MgCl_2$ 0.1 M (2%) ($6H_2O$) mw 203.33			10 ml	20	10 ml	20
$Pb(CO_2CH_3)_2$ 40 mM (1.5%) ($3H_2O$) mw 379.35			1.25 ml	1		1
Distilled water, to make	50 ml		50 ml		50 ml	
Final pH	9.4		6.9			8.7
Incubation time	2–4 min		10–80 min		5–50 min	
Temperature	37°C		37°C		37°C	

Sources: Allen and Slater, *J. Histochem. Cytochem.*, **9**: 418, 1961; Naidoo, *J. Histochem. Cytochem.*, **10**: 580, 1962.

The final pH is 6.8–7.

5. Wash in four changes of distilled water at 0°C for 10 min.
6. Immerse in 1 : 50 dilution of yellow ammonium sulfide for 2 min.
7. Wash in distilled water, dehydrate, clear, and mount in synthetic resin or Canada balsam, as for other PbS stains.

RESULTS: Sites of thiamine pyrophosphatase activity are brown to black.

CONTROLS. Results are negative if thiamine orthophosphate or inorganic diphosphate is substituted in the substrate.

The maleate buffer is made by dissolving 5.804 g dry reagent-grade maleic acid in 400 ml glass-distilled water, adjusting with 1 N NaOH to pH 7.1 and bringing final volume to 500 ml.

Naidoo wrote Lillie (October 25, 1962) that thiamine pyrophosphate (cocarboxylase C) from Roche or Sigma is about 98% pure. Hence, with a molecular weight of 478.81, the above 30-ml substrate mixture contains a total of 7.32 mg of the commercial product, or 15 μmol = 7.18 mg of the pure product.

The further purification of the commercial product was as follows: "Samples used in this work were reprecipitated twice by solution in a little 1 N HCl and the addition of 5 volumes of acetone. Reprecipitation did not lead to any discernible change in the histological results, nor did an independently prepared specimen."

In this series Allen's pH 9.4 preparation and Naidoo's pH 8.6 are visualized as usual by the cobalt sulfide sequence, while with Naidoo's lead-containing substrate, the acid phosphatase–PbS technic is followed.

RESULTS: Reaction sites are shown in black CoS or brown PbS. Glycerophosphate and thiamine O-monophosphate give the usual alkaline phosphatase pattern. Sites which are unreactive to these two substrates but which color with thiamine pyrophosphate are considered specific. In Naidoo's lead technic at pH 6.9 with his low substrate concentration, Mg^{2+} is required, but neither he nor Allen includes it when an alkaline technic with 5 to 8 mM substrate is used.

INHIBITORS. Naidoo notes inhibition by 0.1 mM UO_2^{2+} (acetate) and no effect by 1 mM NaF. Allen and Slater note differential inhibition of glycerophosphatase and thiamine monophosphatase by 4 mM cysteine; the thiamine pyrophosphatase resists this.

Shanthaveerappa and Bourne (*Exp. Cell Res.*, **40**: 292, 1965) employed the thiamine pyrophosphatase method of Novikoff and Goldfisher (*Proc. Natl. Acad. Sci. U.S.A.*, **47**: 802, 1961) in an assessment of the Golgi apparatus in olfactory bulbs.

PHOSPHAMIDASE

In the course of exploration of a variety of substrates in the acid and alkaline phosphatase methods, Gomori found that p-chloroanilidophosphonic acid as substrate gave a quite different distribution picture from that of other substrates when used in the acid range, though the distribution picture was typical of the other alkaline phosphatases when the alkaline to neutral ranges were employed. Gomori (*Proc. Soc. Exp. Biol. Med.*, **69**: 407, 1948) described the method of purification of the crude product made by Otto's synthesis (*Ber. Dtsch. Chem. Ges.*, **28**: 617, 1895) (see also Bredereck and Geyer, *Hoppe Seylers Z. Physiol. Chem.*, **254**: 223, 1938).

GOMORI'S PHOSPHAMIDASE METHOD

SUBSTRATE. Use a 0.1 M stock solution of p-chloroanilidophosphonic acid. Dissolve 21 g in an excess (say 150 ml) of a 10% dilution of 28% ammonia water (the equivalent is about 137 ml). Adjust to pH 8 with dilute (say 10%) acetic acid electrometrically (just colorless with phenolphthalein). Add distilled water to make 1000 ml. Store in refrigerator at 0 to 5°C.

Working substrate is composed of substrate and the following solvent:

Solvent $\begin{cases} \text{MnCl}_2 \ 12.5\% \ (1 \ M) \text{ aqueous solution } 0.4 \text{ ml } (4 \text{ mmol}). \\ \text{Pb(NO}_3)_2 \ 3.31\% \ (0.1 \ M) \text{ aqueous solution } 2.5 \text{ ml } (2.5 \text{ mmol}). \\ \text{Maleate buffer pH } 5.6 \ (5.8\% \text{ maleic acid } 10 \text{ ml } 0.1 \ N \text{ sodium hydroxide } 62 \text{ ml,} \\ \quad \text{distilled water } 28 \text{ ml}) \ 100 \text{ ml } (0.1 \ N). \\ \text{Shake to dissolve initial precipitate.} \end{cases}$

To make working substrate add 4 ml stock 0.1 M p-chloroanilidophosphonic acid. Preheat in paraffin oven 30 min, and filter out excess lead phosphate.

TECHNIC. Fix tissues in acetone, 95% alcohol, or 100% alcohol at 0°C (for 24 h?); dehydrate with acetone or 100% alcohol accordingly. "Clear" with petroleum ether, infiltrate in paraffin at 58 to 60°C in vacuo (10 to 15 mm mercury) for 15 min. Embed and section at 5 to 10 μ.

1. Deparaffinize and hydrate sections as usual. Inactivate a set of control sections by immersion for 10 min in Gram's iodine solution.
2. Immerse test sections and iodine-inactivated controls in preheated, filtered working substrate, and a second set of untreated controls in the preheated, filtered solvent without substrate, for 10 to 24 h at 37°C, arranging slides all facing in the same direction in Coplin jars so tilted that the slides lean about 30° out of perpendicular with faces obliquely downward.
3. Wipe precipitate off the backs of the slides and rinse in distilled water.
4. Wash slide in 0.1 M citrate buffer of pH 5, until the surface of the glass adjacent to the section appears clean.
   Caution: Undertreatment leaves precipitate on the section; overtreatment may remove part of the lead phosphate deposits at sites of activity.
5. Wash 2 to 5 min in running water.
6. Treat in 1% dilution of yellow ammonium sulfide for 1 to 2 min (or 0.5% Na$_2$S buffered to about pH 7).
7. Wash in water; counterstain as desired, 0.1% safranin in 1% acetic for 5 min, or otherwise.
8. Dehydrate in alcohols, clear with xylene, and mount in balsam, ester gum, or a β-pinene resin.

RESULTS: Carcinoma cells and gray substance of brain and cord show much blackening; other tissues show relatively little. Some benign tumor cells color quite heavily; others do not.

Meyer and Weinmann (*J. Histochem. Cytochem.*, **3**: 134, 1955) modified the substrate by dissolving 2.08 g p-chloroanilidophosphoric acid in 100 ml 0.15 N NaOH. This is approximately 0.1 M (mw 206.578). The lead nitrate solution is made at 3.3 mM (110 mg/100 ml) in maleate buffer pH 5.2 containing about 30 mM NaCl (175 mg/100 ml) in place of Gomori's MnCl$_2$. By using the maleate buffer table (Table 20-6) and replacing part of the water by 2.25 ml 0.1 M substrate in a 50-ml mixture, the prescribed concentrations are attained. According to Meyer and Weinmann the alkaline stock 0.1 M substrate is reasonably stable if stored in the cold.

Meyer and Weinmann still used acetone-fixed paraffin sections and incubated slides alone face down in frequently agitated substrate (inclined turntable). They performed the Gomori iodine inactivation on undeparaffinized sections, using the stronger Weigert (1% I_2) solution for 90 min or alternatively 10% HNO_3. The reason for the isolation is that activity transfer from one section to another is observed.

The new substrate and an increase in incubation temperature to 40 to 42°C permitted reduction of exposure time to 1.5 to 4 h. The ammonium sulfide treatment to visualize the precipitated lead phosphate by converting it to sulfide is unchanged and is the same as in the Gomori acid phosphatase methods: 2 min in 2% dilution of yellow ammonium sulfide followed by washing in water.

In a more extended report (*J. Histochem. Cytochem.*, **5**: 354, 1957) they spoke of a pH of 5.15 and of incubation times at 41°C of 1 to 3.5 h, but they still used the same tissue preparation procedure. The iodine inactivation was now done in 20 min after deparaffinizing the control section, still mounted on the same slide with the test sections.

Pearse (1960) quoted the Gomori and Meyer-Weinmann technics.

INORGANIC PHOSPHATASES

Berg, in a series of papers (*J. Histochem. Cytochem.*, **3**: 22, 1955; **4**: 429, 1956; **8**: 85, 92, 1960; *J. Cell Physiol.*, **95**: 435, 1955; *Anal. Chem.*, **30**: 213, 1958) has identified the usual substrate of the inorganic polyphosphatase of fishes and amphibia as a cyclic trimetaphosphate and has demonstrated its occurrence in the intestine of rodents. Trimetaphosphate goes through three steps in degradation to orthophosphate, each apparently requiring its own enzyme.

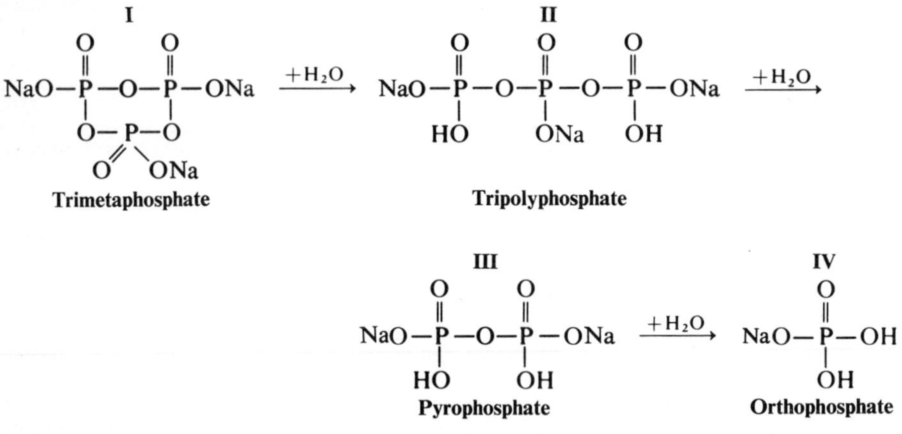

Trimetaphosphate (I) → Tripolyphosphate (II) → Pyrophosphate (III) → Orthophosphate (IV)

REACTIONS. $I + H_2O \rightarrow II$; $II + H_2O \rightarrow III + IV$; $III + H_2O \rightarrow 2IV$.

ENZYMES.   These are trimetaphosphatase, tripolyphosphatase, and pyrophosphatase. Assay studies of rat intestine indicated the presence of about 3 times the amount each of pyrophosphatase and tripolyphosphatase as trimetaphosphatase, so that there would be no deficiency in the second and third enzymes of the degradation series to limit the potentiality of the trimetaphosphatase.

Berg (*J. Histochem. Cytochem.*, **12**: 341, 1964) developed histochemical staining methods for triphosphatases by a *chelate removal method*. In this case, the substrate is the chelate, and the enzyme unmasks bound ions which are precipitated by lead in the incubation mixture. This method permits staining of phosphatases that act on substrates insoluble in Gomori-type mixtures. Differences in these methods are illustrated below.

The following scheme illustrates the Gomori-type method:

$$\text{Substrate and metal salt } (\textit{soluble}) \xrightarrow{\textit{buffer, enzyme}} \text{salt of product and metal } (\textit{insoluble})$$

In this case, the reagents, buffer, and a metal are adjusted to make the product relatively insoluble with the substrate soluble. The procedure cannot be employed with substrates as insoluble as the product. Linear inorganic polyphosphates form insoluble soaps with the usual capture reagents such as lead, cobalt, and calcium. For this reason Berg advances the following method:

$$\begin{array}{l}\text{Substrate-metal chelate} \\ \text{and capturing anion} \\ (\textit{soluble})\end{array} \xrightarrow{\textit{buffer enzyme}} \begin{array}{l}\text{product } (\textit{soluble}) \\ + \text{ salt of metal and} \\ \text{anion } (\textit{insoluble})\end{array}$$

Berg indicates that the key to the method is to use the substrate as the solubilizing agent for another molecule which then is employed to develop a stain. Polyphosphates are sequestering agents that may serve to solubilize divalent cations.

Berg provides the following illustration. Consider the properties of a substrate mixture containing triphosphate and lead. The lead triphosphate chelate is soluble in water in the form of the anion [(I) below] with a pK value near 6 and a partly tridentate structure.

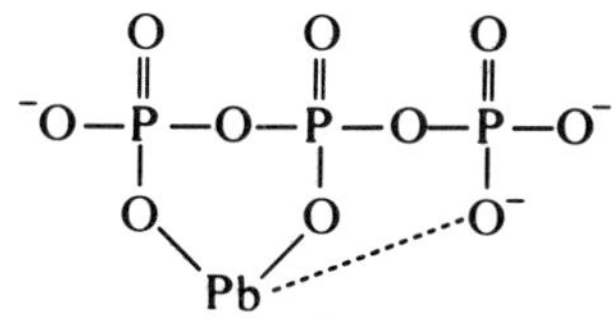

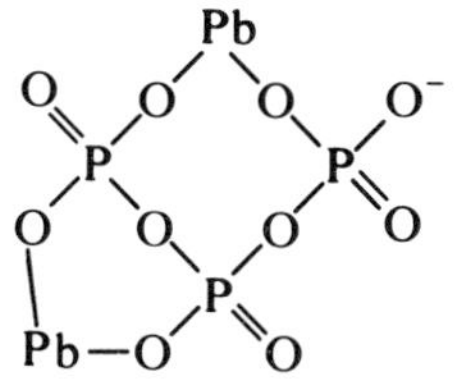

When lead concentrations approach saturation, a bifunctional chelate (II) is found in measurable amounts. The equilibrium at saturation is about 6 : 1 in favor of (I). Chelated lead is bound in non-ionic form and cannot be removed from solution by other anions such as sulfate and phosphate. The lead triphosphate chelate itself is believed to be an anion that forms an insoluble salt.

Enzyme activity on triphosphate results in loss of one of the chelating sites in (I) or (II) above, resulting in the release of a lead ion. Simultaneously hydrolysis produces two precipitating anions (pyrophosphate and orthophosphate) at a site where one existed previously. The solubility product is thereby exceeded, resulting in a precipitate.

BERG METHOD FOR ALKALINE POLYPHOSPHATASE

1. Best results are achieved in paraffin-embedded absolute acetone–fixed (4°C overnight) tissues.
2. Sections are incubated in the following medium at 37°C. Strong staining is observed within 2 to 5 min in sections of rat and mouse duodenum that have been paraffin-embedded.

Sodium triphosphate* (0.096%)	2.6 mM
Sodium barbital (pH 8.7)	20.0 mM
Lead acetate powder to end point (until the solution remains faintly opalescent)	

* mw 367.91.

3. Rinse sections in 2% acetic acid briefly.
4. Rinse sections in distilled water.
5. Treat sections in 2% ammonium sulfide for 2 min.
6. Nuclei may be stained with carmalum, and cytoplasms may be stained with fast green FCF (0.2% in 95% ethanol) if desired.
7. Dehydrate, and mount in synthetic medium.

Alkaline polyphosphatase–reactive sites are stained dark brownish black. Control experiments include those with omission of a "complexing substrate" in an incubation medium or sections with inactivated enzymes.

Media for any histochemical "chelate removal" reaction are prepared in four steps:

1. Mix all ingredients other than the lead salt to the concentrations given in the final formula. This is solution A.
2. Titrate an aliquot of solution A to opalescence with continuous stirring, by adding either lead nitrate powder or a suitably buffered concentrated solution of a lead salt (solution B).
3. Adjust to the desired pH with the acid or base that was used for the buffer.
4. Either filter through a "very fine"–grade filter, or back-titrate with solution A (with continuous stirring). The resulting clear solution is used for incubating sections.

BERG METHOD FOR NEUTRAL TRIPHOSPHATASE

1. Best preservation of enzymatic activity is utilization of freeze-substituted sections of tissues at −17°C with acetone-formaldehyde.
2. Incubate tissues 2 h or longer at 37°C in the following medium:

Sodium triphosphate, 2.6 mM	95.6 mg/100 ml
Sodium barbital, 20 mM	412 mg/100 ml
Na_2HPO_4 and KH_2PO_4 (pH 7), 2 mM	
$Pb(NO_3)_2$ (to opalescence; approximately 3 mM), 0.099%	

3. Rinse sections briefly in 2% acetic acid.
4. Rinse sections in distilled water.
5. Treat sections in 2% ammonium sulfide for 2 min.
6. Nuclei may be stained with carmalum, and cytoplasms may be stained with fast green FCF (0.2% in 95% ethanol) if desired.
7. Dehydrate and mount in synthetic medium.

Neutral polyphosphatase activity is detected by brownish black stain.

Berg no longer uses his method for acid polyphosphatase (personal communication, 1973). In any routine test for pyrophosphatase or triphosphatase activity, Berg recommends use of the alkaline reaction mixtures. Berg recommends use of the neutral triphosphate medium only for testing tissue sites that displayed ATPase activity at neutral but not alkaline pH. Thus discrimination between neutral ATPase and neutral inorganic polyphosphatase is obtained.

Starvation reduced trimetaphosphatase activity in mouse duodenum (Berg, *J. Cell Physiol.*, **56**: 165, 1960). Berg (*J. Exp. Zool.*, **149**: 147, 1962) also studied trimetaphosphatase activity in the developing chick.

CREATINE KINASE

Krasnov (*J. Histochem. Cytochem.*, **21**: 568, 1973) provides the following method for the demonstration of *creatine kinase* in single neurons.

1. The brain region of rabbits containing Deiter's nucleus is quick-frozen at $-150°C$.
2. Sections are cut at $20~\mu$ in a cryostat and lyophilized.
3. Individual neurons are dissected out using a dissecting microscope and weighed with a quartz fiber fishpole balance.
4. Individual weighed neurons are transferred to a microtube of 2 ± 0.1 mm internal diameter and 50 mm in length containing $1~\mu l$ of incubation medium with the following constituent concentrations. Tubes are immersed in ice until incubated.

Imidazole acetate buffer pH 7.4	100 mM
Phosphocreatine	10 mM
Adenosine diphosphate (ADP)	3 mM
Magnesium acetate	5 mM
β-Mercaptoethanol	10 mM
Bovine serum albumin	0.05 %
Triton X-100	0.1 %

Standards consisted of reagents containing 0.6 mM creatine. [Stock solutions of phosphocreatine (100 mM), ADP (100 mM), magnesium acetate (1.0 M), β-mercaptoethanol (500 mM), creatine (50 mM), and imidazole-acetate buffer of pH 7.4 (200 mM) may be stored at $-25°C$. Stock solutions of bovine serum albumin (10%) and Triton X-100 (10%) may be stored at 0 to 4°C.]

5. Microtubes are incubated for 40 min at 38°C.
6. Place in an ice bath and use capillary constriction micropipettes to transfer 0.5 μl of the incubation mixture into 1 ml 0.25% Ninhydrin solution in 40% ethanol in a fluorometer tube. As a positive control, a tube containing 0.5 μl creatine and the Ninhydrin and ethanol is advocated.
7. Under subdued light add 100 μl 4 N KOH.
8. After 10 min, measure fluorescence at 390 mμ in the primary and 495 mμ in the secondary. According to Conn and Davis [*Nature (Lond.)*, **183**: 1053, 1959] the maximum excitation spectrum is at 390 mμ, and the maximum fluorescence spectrum is at 495 mμ.
9. Enzyme activity is expressed as moles of creatine per kilogram of dry tissue per hour.

DISACCHARASES

It appears that the hexosidases demonstrated by hydrolysis of 6-bromo-2-naphthol hexosides differ in distribution and thermostability from biochemically isolated lactase, invertase, and trehalase.

Dahlqvist and Brun (*J. Histochem. Cytochem.*, **10:** 294, 1962) have attempted to localize histochemically disaccharidase activity by demonstrating the liberated glucose with a coupled enzymatic reaction yielding an insoluble formazan.

SUBSTRATES.   Chromogenic substrate is prepared fresh daily:

0.4 *M* phosphate buffer pH 6	40 ml
Glucose oxidase	4 mg
Nitro blue tetrazolium (nitro BT)	10 mg
Phenazine methosulfate	6 mg

SPECIFIC SUGAR SUBSTRATES.   These may be 1 g sucrose; 1.05 g lactose monohydrate, 1.1 g trehalose dihydrate, 1.1 g melibiose dihydrate, or 100 mg glucose, dissolved in distilled water and made up to 10 ml.

WORKING SUBSTRATES.   These consisted of 2 ml specific sugar solution and 18 ml chromogenic substrate. Note: The glucose mixture turns black in less than 1 h.

TISSUE.   Fresh frozen sections are cut at 10 μ by a cold knife process (Adamstone-Taylor or cryostat technic) and attached to slides. The sections are placed in working substrate and incubated 2 to 20 h at about 20°C. When color has developed adequately, fix by immersing slides for 10 min in 5% formaldehyde, rinse 5 min in 15% alcohol, and mount in glycerol gelatin. Counterstain with nuclear fast red or carmine, if desired.

If tetranitro blue tetrazolium is substituted in the chromogenic substrate, the red-brown formazan should contrast well with alum hematoxylin.

REACTIONS.   The disaccharidase releases glucose, which is oxidized by glucose oxidase to gluconic acid. The reduced oxidase reduces the phenazine methosulfate and is itself reoxidized. The reduced phenazine methosulfate reduces the tetrazolium to insoluble formazan and is itself reoxidized.

The phenazine methosulfate is necessary even in conjunction with the active nitro-tetrazoles, in contrast with the case of Krebs cycle dehydrogenases.

The method was successful for demonstration of invertase and trehalase in the brush border zone of the intestinal villus and crypt epithelium. Lactose and melibiose substrates demonstrated no activity.

Maltose and isomaltose could not be used as substrates since they directly reduced the tetrazole. Biochemical assay of the rat intestine used revealed no lactase activity. It is probable that this last should be sought in intestine of nursling animals.

ZYMOHEXASE (ALDOLASE)

Allen and Bourne (*J. Exp. Biol.*, **20:** 61, 1943) developed a method for zymohexase. Zymohexase is an aldolase which converts hexose diphosphate (fructofuranose 1,6-diphosphate) to dihydroxyacetone phosphate and phosphoglyceraldehyde, plus an isomerase, which catalyzes equilibrium between the two products. Inclusion of iodoacetic acid stopped the decomposition at the triose stage. The triose phosphates liberate phosphoric acid in alkaline solution, which is visualized in the usual manner.

This enzyme activity appears to have had no recent attention in histochemistry under the cited nomenclature, and further details are omitted here.

HEXOKINASE AND GLUCOKINASE

Meijer (*Acta Histochem.*, **28:** 286, 1967) provides the following method for the histochemical demonstration of *hexokinase* and *glucokinase*.

1. Cut fresh frozen sections at 7 μ with a cryostat.
2. Fix sections in acetone for 30 min at $-25°C$.
3. Incubate sections in the following medium for 1 to 2 h at 37°C. The medium must be made up fresh.

D-Glucose	30.0 mg
NADP	2.5 mg
ATP	5.5 mg
MgCl$_2$·6H$_2$O	20.0 mg
Nitro BT	2.5 mg
Imidazole buffer 0.04 M, pH 7.5	2.0 ml
0.6 M KCN solution	0.1 ml
Gelatin solution 6%	3.8 ml
Glucose 6-phosphate dehydrogenase solution (Bolhringer, 1 mg/ml)	0.005 ml

The pH is adjusted to 7.5.

The enzyme is soluble. For this reason the incubation mixture contains about 4% gelatin to retard diffusion. Sharpness of localization is also enhanced by acetone fixation. The absence of either ATP or Mg^{2+} results in greatly diminished or absence of staining.

The synthesis of glycogen from blood glucose initially involves the formation of glucose 6-phosphate mediated by hexokinases and glucokinases. In the above method, D-glucose in the incubation mixture is converted to D-glucose 6-phosphate. Glucose 6-phosphate dehydrogenase added to the mixture as well as the endogenous enzyme oxidizes the glucose 6-phosphate to D-gulono-δ-lactone 6-phosphate. During this process, glucose 6-phosphate dehydrogenase also transfers electrons from the substrate to NADP. These electrons are transferred in turn from the reduced coenzyme by endogenous NADPH–tetrazolium oxidoreductase to the final acceptor nitro BT (nitro blue tetrazolium). Nitro BT becomes reduced to form an insoluble dark blue diformazan.

In this case the localization of glucokinase coincides with the localization of endogenous NADPH–tetrazolium reductase. This is the case with many dehydrogenases detected histochemically. Using the incubation conditions described above, Meijer could detect no difference in staining behavior after addition of either D-mannose or D-fructose to the incubation medium.

L-HEXONATE DEHYDROGENASE

Balogh (*J. Histochem. Cytochem.*, **13:** 533, 1965) provides the following method for the histochemical demonstration of L-*hexonate dehydrogenase* activity:

1. Tissues frozen on dry ice are sectioned in a cryostat at $-18°C$ and air-dried at room temperature.
2. Tissues are incubated in the following medium for 30 min at 37°C:

0.2 M tris HCl buffer pH 7.6	4 ml
50% polyvinylpyrrolidone in 0.2 M tris HCl buffer pH 7.6	4 ml
0.02 M sodium gulonate	2 ml
Nitro BT	10 mg
EDTA	20 mg
NADP (nicotinamide-adenine dinucleotide phosphate)	1 mg

The substrate sodium gulonate was prepared by Balogh. Other L-hexonates tested (L-mannonate and L-idonate) gave a histochemical pattern similar to L-gulonate in kidneys in several species. D-Hexonates tested failed to serve as substrates. Balogh noted enzyme activity in the tubular epithelial cells of the renal cortex. He failed to detect activity in sections of tissues from digestive, respiratory, hemopoietic, endocrine, locomotor, and nervous systems.

KETOSE REDUCTASE

Johnson (*J. Histochem. Cytochem.*, **13**: 583, 1965) provides the following method for the histochemical detection of *ketose reductase (sorbitol dehydrogenase)*. The enzyme was demonstrated in the epithelium of the seminal vesicle and coagulating glands of mice. Staining was also obtained in sperm, central hepatic parenchymal cells, and renal proximal tubules.

1. Tissues frozen on dry ice are sectioned at 15 μ in a cryostat.
2. Sections are fixed in acetone for 5 min.
3. Rinse 3 times in tris buffer 300 mM at pH 8.8.
4. Incubate sections in the following medium until the desired staining results (up to 20 min at 37°C):

Tris buffer pH 8.8 (85 mg/ml)	3.0 ml
Nitro BT (1 mg/ml)	2.0 ml
D-Sorbitol (364.0 mg/ml)	0.1 ml
NAD (6 mg/ml)	1.0 ml
EDTA (13 mg/ml)	0.2 ml
KCN (0.23 mg/ml)	0.2 ml
Distilled water	0.5 ml

Sections fixed in a solution containing formalin 10%, ethanol 20%, and tris 600 mM at pH 7.5 required incubation in the above medium up to 45 min at 37°C to acquire satisfactory staining. Most selective localization was achieved in sections treated with both acetone and formalin-ethanol.

β-D-GALACTOSIDASE

The method for this enzyme introduced by Cohen, Tsou, Rutenburg, and Seligman (*J. Biol. Chem.*, **195**: 239, 1952) has been revised by Rutenburg, Rutenburg, Monis, Teague, and Seligman (*J. Histochem. Cytochem.*, **6**: 122, 1958).

It was found that fresh tissue could be stored 10 days at 2 to 4°C without serious loss of activity and that fixation in cold neutral 10% formalin up to 3 days caused no demonstrable impairment. Cryostat sections incubated in saline solution lost enzyme into the solution, which could be demonstrated by its action on substrate solutions, but the sections retained an adequate amount for the histochemical reaction. Hence the following procedure is preferred.

THE 1958 TECHNIC OF RUTENBURG ET AL. FOR β-D-GALACTOSIDASE

1. Fix thin blocks of fresh tissue 16 to 20 h in neutral 10% formalin. (Calcium acetate-formalin or phosphate-buffered formalin should be suitable.) Wash blocks about an hour in several changes of distilled water.
2. Cut frozen sections at 15 μ and transfer directly to substrate. (Cryostat sections attached

to slides, if required for other concurrent studies, may be used directly or after 15-min fixation in calcium-formol.)

3. Incubate 2 to 8 h, usually 4, at 37°C in the following substrate: Dissolve 10 mg 6-bromo-2-naphthyl-β-D-galactopyranoside in 1.5 ml absolute methanol, add 20 ml hot distilled water (70°C), let cool, and add 8.5 ml McIlvaine buffer of pH 4.95 and 10 ml distilled water. (This substrate may be made in 400-ml amounts and kept as long as 6 months at 4°C.)
4. Wash 3 min in three changes of ice-cold distilled water.
5. Postcouple in fast blue B, 1 mg/1 ml 1% sodium bicarbonate (pH 7.5 to 8) for 3 to 5 min.
6. Wash in three changes of ice-cold distilled water, and mount in glycerol gelatin.

RESULTS: Sites of enzymatic activity are deep blue, grading to purple and pink in areas of weaker reaction. Epithelia generally react strongly; muscle, pancreas, lymph nodes react weakly or not at all. Leukocytes stain strongly.

Pearson et al. (*Lab. Invest.*, **12:** 1249, 1963) have developed the following indoxyl method for the histochemical detection of *β-galactosidase* activity.

1. Tissues cut into small blocks not exceeding 2 to 3 mm in thickness are quick-frozen in a dry ice–acetone mixture and sectioned with a cryostat at 6 μ. (Gibson and Fullmer, *J. Periodontol.*, **34:** 470, 1969, retained maximum β-galactosidase activity in mineralized tissues by fixing small blocks of tissues overnight in Lillie's calcium acetate–neutral 10% formalin prior to demineralization with EDTA.)
2. Sections affixed to slides are placed in a mixture of *n*-butanol, acetone, and chloroform (1 : 1 : 1) at -65°C for 30 min.
3. Sections are rinsed in distilled water, and incubated in the following medium for 30 min to 6 h at 37°C depending upon the degree of enzyme activity:

5-Bromo-4-chloro-3-indolyl-β-D-galactoside (4.1 mg/ml dimethyl formamide)	0.25 ml
Acetate buffer (16 mM, pH 5.4)	15.5 ml
NaCl	13.3 mg
Spermidine hydrochloride	4.0 mg

4. Sections are rinsed in distilled water, dehydrated in graded ethanols, and mounted in a synthetic resin.

RESULTS: Sites of enzyme activity are revealed by a very fine blue-green stain. In human beings, greatest activity is noted in cells of the lamina propria of the small intestine, spleen, lymph nodes, adrenals, ovary, thymus, kidney, and liver. Gibson and Fullmer (*J. Periodontol.*, **34:** 470, 1969) noted intense activity in osteoclasts. 1-4-Galactonolactone (mw 178.1) at 3 mM is a strong specific inhibitor. Zn^{2+} at 10 mM was also noted to be inhibitory (136 mg $ZnCl_2$/100 ml).

Gossrau (*Histochemie*, **37:** 89, 1973) notes that β-galactosidase in certain organs of rats and mice is able to hydrolyze naphthol AS-BI substrates, namely, β-galactopyranoside.

α-D-GALACTOSIDASE

Monis et al (*J. Histochem. Cytochem.* **11:** 653, 1963) recommend the following histochemical method for α-D-*galactosidase*.

1. Blocks of tissues not exceeding 3 mm thick are fixed in 10% neutral formalin for 24 h at 4°C.
2. Sections are cut with a cryostat at 6 μ.
3. Free-floating sections are incubated in the following medium for 2 to 4 h at 37°C.

4. The substrate (6-bromo-2-naphthyl-α-D-galactopyranoside) is stored in either dimethyl formamide or dimethylacetamide at 10 mg/0.3 ml.

Phosphocitrate buffer (0.1 M, pH 5)	2.5 ml
Warm distilled water (70°C)	7.5 ml
Stock substrate solution	0.1 ml

The ingredients are soluble in the warm solution. Sections are incubated after cooling to 37°C.

5. After incubation, rinse in distilled water and couple in the following solution for 2 min at 22°C:

Distilled water	10 ml
Fast blue B (tetrazotized diorthoanisidine)	10 mg
Sodium bicarbonate	25 mg

6. Rinse sections and mount in Kaiser's glycerogel.

RESULTS: A purple stain indicates enzyme activity. α-D-Galactosidase is greatly inhibited by fast blue B. For this reason, a simultaneous coupling method could not be employed. Also, use of a postcoupling procedure permited alkaline coupling. Most diazonium salts couple better at an alkaline pH. Monis et al. emphasize the great solubility of the enzyme and that formalin fixation is required to permit any assessment of localization.

α-D-GLUCOSIDASE

Rutenburg, Goldbarg, Rutenburg, and Lang (*J. Histochem. Cytochem.*, **8:** 268, 1960) presented a histochemical technic for the localization of this enzyme based on an earlier colorimetric biochemical study by Goldbarg, Tsou, Rutenburg, Rutenburg, and Seligman (*Arch. Biochem. Biophys.*, **75:** 435, 1958).

Pearse (1960) prescribes free-floating frozen sections of material fixed in cold neutral formalin. Incubate 1 to 2 h at 37°C in 9 ml 25 mM phosphate buffer pH 6.5 plus 1 ml alcohol containing 1 to 1.5 mg 6-bromo-2-naphthyl-α-D-glucopyranoside. Rinse in distilled water and develop color by a 2-min immersion in 10 ml 0.1 M phosphate buffer pH 7.5 to which 10 mg fast blue B is added just before using. Wash in water and mount in glycerol gelatin.

Wolfgram (*J. Histochem. Cytochem.*, **9:** 171, 1961) points out that the localization of glucosidase and galactosidase in brain is in the myelin sheaths and that defatting with chloroform applied to dried cryostat sections completely prevented this localization. However, enzymatically liberated 6-bromo-2-naphthol was recovered from the solution. Immersion of tissue in solutions of 6-bromo-2-naphthol, with fast blue B postcoupling, showed similar myelin localization.

The effect of prior defatting with chloroform on nonnervous tissues in these hexosidase reactions has not yet been reported.

β-D-GLUCOSIDASE

Pearson et al. (*Proc. Soc. Exp. Biol. Med.*, **108:** 619, 1961; *Lab. Invest.*, **12:** 1249, 1963) recommend the following indoxyl method for the histochemical demonstration of *β-D-glucosidase* employing 5-bromo-3-indolyl-β-D-glucoside as substrate.

1. Tissues 2 mm thick are quick-frozen in a mixture of dry ice and acetone, sectioned at 6 μ in a cryostat, affixed to slides, and stored at −25°C until used.

2. Immerse sections in a mixture of *n*-butanol, acetone, and chloroform (1 : 1 : 1 by volume) at −65°C for 30 min to remove lipids.
3. Rinse tissues 3× in distilled water.
4. Incubate tissues for 3 h at 25°C in the following medium:

5-Bromo-3-indolyl-β-D-glucoside (4 mg/ml)	1 ml
Polyvinylpyrrolidone (5 mg/ml)	1 ml
Acetate buffer (0.2 M, pH 5.4)	14 ml

5. Rinse tissues in distilled water, and fix in Lillie's calcium acetate–neutral 10% formalin for 1 min.
6. Counterstain with Mayer's hematoxylin.
7. Dehydrate in graded ethanols, clear in xylol, and mount in synthetic medium.

RESULTS: Greatest enzyme activity was observed in the preputial glands, duodenum, and small intestine. Traces were found in the adrenals, liver, kidney, parotid, pancreas, stomach, spleen, lymph nodes, epididymis, and skin. Enzyme activity was not inhibited by saccharo-1,4-lactone—an inhibitor of β-glucuronidase. Enzyme activity was inhibited by 0.03 M $CuSO_4$ (0.75% $CuSO_4 \cdot 5H_2O$) and by Holt's equimolar 0.05 M ferricyanide-ferrocyanide.

GLYCOGEN SYNTHESIS

TAKEUCHI'S AMYLOPHOSPHORYLASE AND AMYLO-1,4 → 1,6-TRANSGLUCO-SIDASE METHODS. Takeuchi's (*J. Histochem. Cytochem.*, **6**: 208, 1958; **9**: 304, 1961) technics for the demonstration of phosphorylase synthesize first an amylose, coloring blue with iodine, from glucose 1-phosphate and then constitute it by an amylo-1,4→1,6-trans-glucosidase branching enzyme action into the branched polysaccharide glycogen, which colors purplish to reddish brown with iodine.

SUBSTRATE. Dissolve 100 mg glucose 1-phosphate (disodium), 20 mg muscle adenylic acid, and 2 mg glycogen (as a primer, if necessary) in 10 ml acetate buffer of pH 5.8 and 10 ml distilled water (personal communication, 1973).

For inhibition of the branching enzyme action and consequent exclusive production of the straight-chain amylose add 5 ml absolute alcohol or 1 mg mercuric chloride (about 0.1 mM) to the foregoing substrate.

SECTION PROCEDURE

1. Cut frozen sections of fresh, unfixed tissue at 10 to 40 μ by the cold knife (Adamstone-Taylor) or cryostat procedure, and transfer as free-floating frozen sections to the above substrate. Friable tissues may be attached directly to slides and air-dried at 20 to 25°C for a few minutes, but the free-floating procedure is preferable.
2. Incubate 1 to 2 h in the above substrate, using the longer exposure in the presence of the branching enzyme inhibitor.
3. Rinse quickly in 40% alcohol, float free sections onto slides, and dry at 37°C for a few minutes. With preattached sections omit this step.
4. Fix 2 to 4 min in absolute alcohol; dry.
5. Apply control digestion in α- and β-amylases at this point.
6. Immerse directly or after amylase digestion in dilute (1 : 9) Gram's iodine solution (I–KI–HO = 1 : 2 : 3000) for 5 to 10 min, until the iodine coloration—red, brown, purple (or blue for amylose)—appears.
7. Mount directly in Takeuchi's iodine glycerol (Gram's iodine 1 : 9 glycerol), and seal with paraffin or cellulose caprate.

AMYLASE DIGESTION

1. Incubate 30 min to 10 h in 0.5% α-amylase in 4-mM acetate buffer pH 5.7; carry through steps 6 and 7, above.
2. Use 0.5% β-amylase in 1-mM acetate buffer, pH 4.5, rinse, and carry through steps 6 and 7.

RESULTS: Amylose, the result of amylophosphorylase activity alone, is shown in deep blue by the direct procedure and is digested completely by both α- and β-amylase. Glycogen, the result of the sequence action of amylophosphorylase and amylo-1,4 → 1,6-transglucosidase, colors red-brown to purple in the direct procedure, is completely digested by α-amylase, and resists β-amylase. In a later paper Iwamasa and Takeuchi (*Acta Histochem. Cytochem.,* **5**: 57. 1972) indicate that glycogen produced by the action of phosphorylase tends to stain red-brownish in liver, but bluish in muscle with iodine.

In a personal communication (1973) Takeuchi added the following comments: If native glycogen is present, it is not necessary to add glycogen as a primer. Dextran is not needed. The addition of EDTA (1 mM, 0.033%) is sometimes useful (Hori, *Stain Technol.,* **40**: 157, 1965). Addition of polyvinylpyrrolidone (PVP) may reduce enzyme diffusion, but erratic distribution may occur. Eränkö and Palkama (*J. Histochem. Cytochem.,* **9**: 585, 1961) disagree with this. Addition of sodium fluoride (5 mM, 0.02%) is very useful. Addition of 5, 10, or 25% ethanol may be helpful; however, in some tissues a slight inhibition of enzyme activity is noted.

Smith (*J. Histochem. Cytochem.,* **18**: 756, 1970) calls to our attention that two forms of glycogen synthetase, namely, D and I forms, exist. The D form is dependent on glucose 6-phosphate for activity, whereas the I form is active in the absence of glucose 6-phosphate although stimulated by its presence. Using light microscopy, Smith noted variation in the degree of activity of the two forms in clear cells of active and inactive eccrine sweat glands.

Sawyer et al. (*J. Histochem. Cytochem.,* **13**: 605, 1965) report that preservation of stains in sections incubated for the demonstration of phosphorylase according to the method of Takeuchi may be accomplished in the following manner: After incubation, sections are stained in 35% Gram's iodine in distilled water until the color reaction is complete and the background is stained yellow. Then sections are placed in 10% formol, 0.1% chloral hydrate (100 ml 40% formaldehyde, 1 g chloral hydrate, 900 ml H$_2$O, with the pH adjusted to 4.4) for 5 min. Counterstain for 2 min in 0.4 mM methyl green in 1.0 N acetate buffer of pH 4.8. Dehydrate in *tert*-butyl alcohols and mount in synthetic medium. Ehrlich (Ehrlich's "Encyklopädie," 1903) mounted iodine-stained glycogen preparations in iodized Apáthy's gum syrup and reported that stains were stable for 10 years.

Takeuchi and Sasaki (*J. Histochem. Cytochem.,* **18**: 761, 1970) compared polyglucose formed from the action of phosphorylase and that formed by uridine diphosphate glucose glycogen transferase (UDPG glycogen synthetase) using light and electron microscopy. They confirmed earlier findings at the light microscopic level that glycogen formed in vitro by phosphorylase stains blue with iodine whereas that formed in vitro by UDPG-glycogen synthetase stains pale red-purple. In addition, they noted that at the electron microscopic level native glycogen appears in typical particles singly or in groups whereas in the same muscle immersed in a medium containing UDPG, glycogen appeared as beads of varying density that may form a rosary. In muscles incubated in media containing glucose 1-phosphate, newly formed glycogen appears at the electron microscopic level as amorphous particles that stain more faintly with lead.

Miyayama (*Acta Histochem. Cytochem.,* **4**: 87, 1971; ibid., **5**: 1, 1972), Iwamasa (Ibid., **5**: 106, 1972), Takeuchi and Sasaki (*Histochemie,* **23**: 310, 1970), and Takeuchi et al. (*Acta*

Histochem. Cytochem., **3:** 173, 1970) conducted several studies involving the specificity of the Takeuchi method for phosphorylase and for branching enzyme, and the electron microscopic appearance of newly produced (in vitro) glycogen as compared with glycogen existing at the time the specimen was taken.

Eckner (*J. Histochem. Cytochem.*, **19:** 133, 1971) is convinced that one cannot discriminate between new and native glycogen on the basis of Takeuchi's method for phosphorylase.

Guha and Wegmann (*J. Histochem. Cytochem.*, **13:** 148, 1965) localized phosphorylase at the light microscopic level employing uniformly labeled ^{14}C-glucose 1-phosphate in the Takeuchi type of histochemical incubation medium. Sections subsequently processed for autoradiography and examined revealed tissue localization analogous to that achieved by the Takeuchi method.

In studies of phosphorylase and UDPG–glycogen transglycosylase using freeze-dried canine myocardium and liver, Eckner et al. (*Histochemie*, **19:** 340, 1969) were unable to distinguish preexisting and newly formed polysaccharides despite the use of iodine staining, differential digestion with amylases, and acid hydrolysis. In tissues without stainable glycogen before incubation, Eckner et al. were unable to detect glycogen after incubation. When preexisting glycogen was present, Eckner et al. observed less polysaccharide after incubation.

Those using the Takeuchi method are advised to use appropriate controls particularly with respect to interpretation of the iodine-staining reaction. It is perplexing to conceive that an identical product (glycogen) derived from the action of an identical enzyme (phosphorylase) results in a red-brownish stain in liver cells and a bluish stain in muscle.

Roelofs et al. (*Science*, **177:** 795, 1972) employed histochemical methods and were unable to detect phosphorylase activity in skeletal muscles of individuals with McCardle's disease. However, phosphorylase was detectable histochemically in regenerating muscles either in vivo or in vitro.

Dvorak and Cohen (*J. Histochem. Cytochem.*, **13:** 454, 1965) employed the following fluorescent antibody technic for the histochemical detection of *phosphorylase* in skeletal muscle:

1. Tissues are quick-frozen in isopentane cooled to $-60°C$ with dry ice and acetone and sectioned in a cryostat at 4 μ.
2. A drop of the complete medium for the histochemical detection of phosphorylase is placed on the section[1] followed by a drop of conjugated serum against phosphorylase.[2] Tissues are incubated up to 1 h at 37°C.

---

[1] The histochemical medium for phosphorylase is as follows:

α-D-Glucose 1-phosphate	150 mg
Adenosine monophosphate	5 mg
Rabbit liver glycogen	5 mg
Polyvinylpyrrolidone	500 mg dissolved in 12 ml acetate buffer pH 5.9

[2] Conjugated serum was prepared as follows: Twice-crystallized rabbit skeletal muscle α-phosphorylase-(α-1,4-glucan orthophosphate glucosyl transferase) was obtained from Nutritional Biochemical Corporation, Cleveland, Ohio. Adult English shorthair guinea pigs were bled for control serums and then sensitized in the four footpads with 1 to 2 mg enzyme incorporated in complete Freund's adjuvant (total volume 0.2 to 0.4 ml per animal). Six weeks after sensitization, three subcutaneous booster injections of 1 mg α-phosphorylase precipitated in alum (Kabat and Mayer, 1961) were administered on alternate days. Thereafter, intradermal injections of 100 μg enzyme were given weekly. Pooled serums were obtained from sensitized animals by cardiac puncture on alternate weeks, beginning 1 week after the last alum injection. Sensitized animals thus received a total of 4 to 7 mg enzyme.

Collected serums were assayed for antibody activity by the Ouchterlony method as described by Kabat and Mayer ("Experimental Immunochemistry," 2d ed., Charles C Thomas, Springfield, Ill., 1961). Antibody was conjugated to fluorescein isothiocyanate according to the method of Coons and Kaplan (*J. Exp. Med.*, **91:** 1, 1950). Tagged serums were absorbed once or twice with acetone-dried mouse or rabbit liver powder.

3. Sections are washed for 5 min in cold buffered saline solution.
4. Mount in a water-based medium such as Kaiser's.

The distribution of enzyme activity with the above method was analogous to that obtained with the Takeuchi method. Phosphorylase is a soluble enzyme, and cellular localization is favored in sections incubated with the entire histochemical medium for phosphorylase and the polyvinylpyrrolidone prescribed.

Goldberg et al. (*J. Natl. Cancer Inst.*, **13:** 543, 1952) and Hori (*Stain Technol.*, **39:** 275, 1964; **41:** 91, 1966) describe a lead precipitation method for the demonstration of α-*glucan phosphorylase*. The validity of this method awaits further proof. Lindberg and Palkama (*J. Histochem. Cytochem.*, **20:** 331, 1972) noted complete inhibition of phosphorylase activity by lead at concentrations prescribed in the method for the detection of phosphorylase at the electron microscopic level.

Lindberg (*Histochemie*, **36:** 355, 1973) and Lindberg and Palkama (ibid., **38:** 285, 1974) provide the following methods for the light and electron microscopic demonstration of *glycogen phosphorylase*:

LIGHT MICROSCOPIC METHOD OF LINDBERG AND OF LINDBERG AND PALKAMA.

Small slices of tissue such as liver are frozen, sectioned at 20 μ, fixed in freshly prepared 1% paraformaldehyde, buffered with 0.1 M acetate at pH 7.2 for 5 to 10 min at 0°C, rinsed in three changes of acetate buffer for 20 min at 0°C, and preincubated in the following medium for 30 min at 37°C:

0.1 M acetate buffer pH 5.9	10 ml
Absolute ethanol	2 ml
Polyvinylpyrrolidone	900 mg
NaF	180 mg

Sections are then rinsed in three changes of acetate buffer containing the above concentrations of ethanol and polyvinylpyrrolidone and incubated in the following medium for 2 h at 37°C:

0.1 M acetate buffer pH 5.9	10 ml
Absolute ethanol	2 ml
Polyvinylpyrrolidone	900 mg
Dithiothieitol (DTT)	22.2 mg
$FeCl_2 \cdot 4H_2O$	7.14 mg
Glucose 1-phosphate	100 mg

Control sections are incubated in the full media with omission of the substrate glucose 1-phosphate.

Sections are rinsed for 10 min in three changes of distilled water and stained either with Gram's iodine containing 0.3 M sucrose or with dilute (1%) ammonium sulfide, dehydrated, and mounted. Sections stained with iodine must be mounted in an aqueous medium such as Kaiser's glycerogel, preferably iodized (0.1 to 0.03%).

ELECTRON MICROSCOPIC METHOD OF LINDBERG AND LINDBERG AND PALKAMA.

Slices of tissue are quickly fixed in 1% paraformaldehyde buffered with 0.1 M acetate at pH 7.2 for 20 min at 0°C, rinsed in three changes of 0.1 M acetate buffer of pH 7.2 for 20 min at 0°C, preincubated in 0.36 M (180 mg/12 ml) NaF buffered with 0.1 M acetate buffer of pH 5.9 for 15 min at 37°C, rinsed in three changes of acetate buffer of pH 5.9 for 15 min at 0°C, and incubated in the following medium for 1 h at 37°C:

0.1 *M* acetate buffer, pH 5.9	12 ml
DTT	22.2 mg
FeCl$_2$·4H$_2$O	7.14 mg
Glucose 1-phosphate	100 mg

Control sections are incubated without either iron or substrate. After rinsing in three changes of acetate buffer of pH 5.9 for 20 min at 0°C, sections are fixed in 2.5% glutaraldehyde buffered with 0.1 *M* phosphate buffer of pH 7.2 for 90 min at 0°C, rinsed in several changes of phosphate buffer of pH 7.2 for 90 min at 0°C, dehydrated, and embedded in Epon. Iron precipitates were localized in endoplasmic reticulum and in glycogen particles. Glycogen is poorly preserved by the method.

URIDINE DIPHOSPHATE GLUCOSE GLYCOGEN TRANSFERASE

Takeuchi and Glenner (*J. Histochem. Cytochem.*, **8**: 227, 1960; **9**: 304, 1961) reported the histochemical application of this (Leloir, *J. Am. Chem. Soc.*, **79**: 6340, 1957; *Arch. Biochem. Biophys.*, **81**: 508, 1959; *J. Biol. Chem.*, **235**: 919, 1960; Hauk and Brown, *Biochim. Biophys. Acta*, **33**: 556, 1959) pathway of glycogen synthesis.

In this synthesis uridine triphosphate (UTP) reacts with glucose 1-phosphate (G-1-P) to form pyrophosphate (PP) and uridine diphosphate glucose (UDPG), and the latter adds a glycosyl group to preexisting "primer" glycogen in α-1 → 4 linkage. The pathway is demonstrated histochemically by direct use of the intermediate uridine diphosphate glucose (Kornberg, in Long, pp. 533–543).

Takeuchi and Glenner showed that activity was promptly and completely inhibited by 0 to 5°C postfixation of fresh frozen sections (Adamstone-Taylor cold knife) in formalin, more slowly and moderately by alcohol (with some activity remaining at 24 h), slightly by cold acetone, and scarcely at all by drying on slides.

The pH optimum was about 7.6; the presence of glucose 6-phosphate and of Versene in the incubation media enhanced the reaction. With prolonged (10 to 48 h) exposure to substrate glycogen again disappeared, but maximal reactions were attained in 20 to 60 min. The synthesized glycogen colored red-brown with iodine, resisted β-amylase almost completely, and was promptly digested by α-amylase. Addition of as much as 40% alcohol to the substrate still permitted formation of glycogen in distinction from the amylophosphorylase,

TABLE 10-7 URIDINE DIPHOSPHATE GLUCOSE GLYCOGEN TRANSFERASE SUBSTRATES

Substrates	1960	1961	1961 control
Uridine diphosphate glucose	50 mg	50 mg	
Glycogen	10 mg	10 mg	10 mg
Versene	20 mg	20 mg	20 mg
Glucose 6-phosphate	.....	10 mg	10 mg
Distilled water	15 ml	14 ml	14 ml
Dissolve and add 0.2 *M* tris buffer pH 7.4	10 ml	10 ml	10 ml
Absolute alcohol	.....	1 ml	1 ml
Final volume	25 ml	25 ml	25 ml

branching enzyme sequence, where alcohol inhibits the branching enzyme and causes deposits of an amylose which colors blue with iodine and is digested by β-amylase.

PROCEDURE

1. Cut frozen sections of fresh tissue by the Adamstone-Taylor cold knife method.
2. Transfer directly as free-floating sections to substrate. Friable or difficult tissues may be attached to slides directly from the knife and air-dried before incubation. Incubate 1 h at 25°C.
3. Transfer directly to a 1:9 dilution of Gram's iodine $(I-KI-H_2O = 1:2:3000)$ until color appears.
4. Pick up on slides, mount in 90% glycerol–10% Gram's iodine. Seal with cellulose caprate or paraffin. (Use a small drop of glycerol which does not quite fill the space under the cover glass.)

RESULTS: Newly synthesized glycogen is red-brown. No-substrate controls are negative. Preexisting glycogen is apparently lost during the incubation process.

Grillo et al (*J. Histochem. Cytochem.*, **12**: 275, 1964) employed histochemical and chemical methods to determine levels of UDPG glycogen synthetase in skeletal and cardiac muscles of chicks from the third to sixteenth day of incubation.

N-ACETYL-β-GLUCOSAMINIDASE

Enzymes hydrolyzing β-glycosides of N-acetylglucosamine have been isolated from a variety of sources: mammals, molluscs, molds, emulsin.

By use of α-naphthyl-N-acetyl-β-glucosaminide in a simultaneous azo coupling technic, Pugh and Walker (*J. Histochem. Cytochem.*, **9**: 242, 1961) have devised a procedure for the histochemical localization of *N-acetyl-β-glucosaminidase.*

Biochemical tests showed 60 to 80% preservation of the activity of fresh tissue after 20-h fixation in formol–saline solution (1% NaCl) or formol-calcium (1.2% $CaCl_2$) at pH 7 or pH 5.5 formalin containing 0.11 *M* sodium citrate or acetate or 0.1 *M* calcium acetate or in acetone or 80% alcohol. Although the last was best, formol-calcium and calcium acetate–formalin were nearly as good and gave better general histologic preservation.

Fast garnet GBC at 1.5 mg/ml proved to be the most satisfactory diazonium salt: fast blue B, fast red RC, and fast red TR were also tried.

A substrate concentration of 1.8 m*M*, almost saturated in water, proved satisfactory. Increasing the concentration of α-naphthyl-N-acetyl-β-glucosaminide by addition of propylene glycol or methyl Cellosolve to the solvent was of no benefit.

METHOD OF PUGH AND WALKER FOR N-ACETYL-β-GLUCOSAMINIDASE. Fix fresh tissue overnight (16 to 24 h) at 3 to 5°C in formol-calcium (pH 7) or calcium acetate–formalin (pH 5.5). Soaking in gum sucrose before sectioning is recommended but is optional. Friable and fragmentary tissues should be embedded in gelatin.

Cut frozen sections at 5 to 15 μ, and transfer to freshly prepared, nearly saturated (1.8 m*M*), filtered substrate:

α-Naphthyl-N-acetyl-β-glucosaminide, 623.5 μg (1.8 μmol), in 1 ml water
Fast garnet GBC, 1.5 mg, in 0.1 ml water, filtered

Citrate buffer 0.2 *M*, pH 5.5, in 0.1 ml
Final pH about 5.5

Incubate 10 to 60 min at 20°C. Transfer to water, place sections individually in 30% alcohol and then back in water to spread, pick up on clean slides, and mount in glycerol gelatin or Gurr's Hydromount. Counterstaining with neutral celestin blue or 0.5% methyl green in 0.1 *M* pH 4.6 acetate buffer is optional. Locations in renal convoluted tubules, salivary gland and thyroid acini, stomach, intestine, bronchus, uterus, testis, and epididymis, chiefly epithelial, are figured, as well as deposits in solid viscera, spleen, thymus, and ovary.

Strong inhibition is obtained by inclusion of 0.5 to 1 *μM* N-acetylglucosaminolactone in the substrate.

Hayashi (*J. Histochem. Cytochem.*, **13**: 355, 1965) provides the following method for the localization of N-acetyl-*β*-glucosaminidase at the histochemical level.

1. Fix small tissue blocks ($5 \times 5 \times 3$ mm) in Baker's formol-calcium for 24 h at 4°C.
2. Without washing, blot tissues on filter paper to transfer to Holt's hypertonic gum sucrose (0.88 *M* sucrose containing 1% gum acacia) for 24 h at 4°C. Tissues may be kept in gum sucrose for at least a week without enzyme deterioration.
3. Cut sections at 5 to 8 *μ* with a cryostat.
4. Sections are treated with 100% ethanol for 5 min at 0°C and rinsed in distilled water.
5. Free-floating sections are incubated in the following incubation medium for 15 to 30 min at 37°C depending upon the degree of enzyme activity of the tissues. *Stock solutions:* Solution A, pararosanilin: One gram pararosanilin chloride is dissolved in 20 ml distilled water, and 5 ml concentrated HCl is added with gentle warming. Cool, filter, and store at 22°C. Solution B, sodium nitrate: A fresh 4% solution is employed.

   The working solution is prepared by mixing 0.3 ml each of solutions A and B above. Diazotization is completed within a few minutes at 24°C; allow 15 to 20 min at 3 to 5°C.

   Three milligrams of the substrate (naphthol AS-BI–N-acetyl-*β*-glucosaminide) is dissolved in 0.5 ml ethylene glycol–monomethyl ether to which is added 5 ml of 0.1 *M* citrate buffer pH 5.2 and immediately thereafter 0.6 ml of the hexazotized pararosanilin solution prepared as described above. The pH is adjusted to 5.2 with 1 *N* NaOH, diluted to a final volume of 10 ml with distilled water, and filtered through Whatman No. 1 paper. The final concentration of the substrate is 0.5 m*M*, and that of the diazo reagent is 3.6 m*M*.
6. After staining, sections are rinsed in distilled water (two changes) and in 30% ethanol briefly, floated on water again, and mounted on gelatinized slides.
7. Sections may then be fixed in neutral formalin for 5 min, dehydrated in graded ethanols, and mounted in synthetic medium.
8. Alternatively, sections may be fixed, counterstained in 1% methyl green in 0.1 *M* Veronal-acetate buffer of pH 4 for 1 min, dehydrated in graded ethanols, and mounted in synthetic medium.

Hayashi notes that the diazonium salt fast garnet GBC (10 mg) may be substituted for hexazonium pararosanilin.

RESULTS: Sites of enzyme activity are stained bright red. The method appears to be readily reproducible and reliable. Good localization is achieved. The enzyme is inhibited by a specific inhibitor, namely, N-acetylglucosaminolactone at 0.05 m*M* in the incubation medium.

β-GLUCURONIDASE

FERRIC SALT, 8-HYDROXYQUINOLINE METHOD. Fishman and Baker (*J. Histochem. Cytochem.*, **4:** 570, 1956) developed a histochemical stain for the demonstration of *β-glucuronidase* whereby 8-hydroxyquinoline liberated during the course of enzyme activity is precipitated *in situ* as a poorly soluble iron chelate. The iron chelate is subsequently converted into Prussian blue. Although the reliability and specificity of the method have been challenged many times (for example, Janigan and Pearse, *J. Histochem. Cytochem.*, **10:** 719, 1962), Fishman et al. (*J. Histochem. Cytochem.*, **12:** 239, 1964) appear to have appropriately answered all criticisms of the method. Specifically, biochemical methods were employed to demonstrate that the appearance of the staining reaction in tissue sections is dependent upon *β*-glucuronidase activity. Quantitative analytic methods were employed to demonstrate that glucuronic acid was liberated into the incubation medium concomitant with the deposition of ferric hydroxyquinoline as hydrolysis of the substrate proceeded. Fishman et al. further demonstrated the pH dependence of enzyme activity in tissues analogous to that displayed by the purified enzyme. The histochemical staining reaction was inhibited by inclusion in the medium of another specific substrate (naphthol AS-*β*I-*β*-glucosiduronic acid) which does not result in deposition of 8-hydroxyquinoline. Inhibition was also specifically achieved by the inclusion of saccharolactone (1.0 μmol/ml) in the incubation medium. The Fishman-Baker method follows (*J. Histochem. Cytochem.*, **4:** 570, 1956):

SUBSTRATE PREPARATION

1. The substrate (8-hydroxyquinoline glucosiduronic acid) must be prepared biosynthetically. Pure 8-hydroxyquinoline is homogenized with peanut oil and administered into rabbits via stomach tube. Urine (which must be acid) is collected, filtered, adjusted to pH 4 with HCl, and placed in the cold. Green crystals form after 24 h which are removed by filtration and dried. If crystals do not appear, an equal volume of ethanol may be added and crystallization induced by rubbing with a glass rod. Crystals are purified by dissolving them in a minimum of hot water, charcoal adsorption, and cooling. Dried crystals are stable at 23°C.

2. The incubation solution is prepared as follows:

    a. A saturated ferric-8-hydroxyquinoline solution is prepared by dissolving 50 mg 8-hydroxyquinoline in 30 ml 0.1 N acetate buffer of pH 5 by warming in a water bath.

    b. To the above hot solution is added 20 ml hot ferric sulfate solution [1.522 g $Fe_2(SO_4)_3 \cdot 6H_2O$ is added to 100 ml and stirred in a boiling water bath just preceding the point when the solution becomes clear orange]. The ferric sulfate-8-hydroxyquinoline mixture is shaken and heated in a water bath at 37°C for 2 h. The mixture is then centrifuged at 2000 r/min for 15 min and decanted into the substrate mixture described in the next step.

    c. The substrate mixture is prepared by adding 100 mg of 8-hydroxyquinoline glucosiduronic acid to 2 ml 0.1 N acetate buffer of pH 5 in a 125-ml Erlenmeyer flask and warming in a boiling water bath. After the substrate is dissolved, it is mixed with the ferric-8-hydroxyquinoline solution prepared above, centrifuged at 2000 r/min for 15 min, and filtered through Whatman No. 2 paper, and the supernatant is stored in a bottle with a glass stopper. Prior to storage, 1 g finely powdered gum acacia is added to each 100 ml of prepared substrate. The powdered gum is dropped on the surface of the mixture *without stirring* and refrigerated. *Do not shake.* Entrapped air bubbles are tenaciously retained.

TISSUE PREPARATION AND STAINING

1. Tissue blocks not exceeding 2 to 3 mm in thickness are fixed in chloral hydrate–formalin (20 ml 38% formaldehyde, 80 ml distilled water, 100 mg chloral hydrate to which are added marble chips to aid in neutralization) for 17 to 24 h at 8°C (Baker et al., *J. Histochem. Cytochem.*, **6:** 244, 1958).

2. Rinse briefly in distilled water and cut cryostat sections at 10 to 20 μ. (Friable tissues may be embedded in gelatin.)

3. Cut sections are floated on distilled water and transferred to the incubation solution described above. The sections are first incubated in the refrigerator at 8°C for 17 to 24 h and then $\frac{1}{2}$ to 7 h at 37°C depending upon the amount of enzyme in the particular tissue.

4. Use a filter paper to remove the scum from the surface of the substrate, rinse sections, and examine microscopically to determine if incubation is sufficient. Incubate longer if necessary.

5. Place sections in 0.5 M oxalate buffer (sodium oxalate 2.87 g, oxalic acid 0.47 g, distilled water 100 ml) 15 min at 23°C.

6. Wash sections thoroughly in distilled water, blot dry, place on a hot plate at 40°C *briefly* until dry.

7. Rinse in distilled water, and cover sections with 2% aqueous potassium ferrocyanide for 10 s, add 6 drops 1 N HCl, and let stand for 2 min.

8. Wash, and repeat step 7 for 10 min instead of 2.

9. Wash sections thoroughly for 15 min in distilled water.

10. Counterstain in neutral red (1% in distilled water) for 2 min.

11. Wash, dehydrate, and mount in synthetic medium.

Sites of enzyme activity are revealed as Prussian blue. Many tissues exhibit enzyme activity.

An example of the postcoupling method is that of Seligman et al. (*J. Histochem. Cytochem.*, **2:** 209, 1954), who reported the synthesis of the substrate 6-bromo-2-naphthyl-β-D-glycopyruronoside (commercially available from Borden and K and K).

METHOD OF SELIGMAN, TSOU, RUTENBURG, AND COHEN FOR β-GLUCURONIDASE

1. Prepare fresh frozen sections in a cryostat or by the Adamstone-Taylor technic at 6 to 10 μ.

2. Dry on clean slides for a few minutes. Here, according to Wolfgram, extract 1 min in chloroform to prevent lipid staining. Usually this step has been omitted.

3. Dry on slides for a few minutes, fix 10 min at 3 to 5°C in phosphate-buffered 10% formalin (about pH 7), and wash 15 min in cold water.

4. *Substrate solution:* Dissolve 30 mg 6-bromo-2-naphthyl-β-D-glucopyruronoside in 5 ml absolute methyl alcohol, and add 20 ml McIlvaine phosphate citric acid buffer pH 4.95 and 75 ml distilled water. Incubate sections 4 to 6 h at 37°C in this substrate.

5. Rinse 1 min in distilled water and azo-couple 2 min in 0.1% fast blue B in 0.02 M phosphate buffer pH 7.5 at 3 to 5°C.

6. Wash in two changes of cold distilled water and 1 of 0.1% acetic acid, and mount in glycerol gelatin.

RESULTS: Sites of enzyme activity are blue; lipids, red. To avoid this lipid staining in nervous tissues Wolfgram (*J. Histochem. Cytochem.*, **9:** 171, 1961) removed lipid by a 1-min immersion in chloroform, which he stated did not inhibit the enzyme activity but nevertheless totally prevented coloration of central nervous tissues.

Pugh and Walker (*J. Histochem. Cytochem.*, **9**: 105, 1961) substituted naphthol AS-LC glucuronide and *N*-acetyl-β-glucosaminide for demonstration of *β-glucuronidase* and *N-acetyl-β-glucosaminidase*, respectively, in the foregoing technic.

Hayashi et al. and Fishman et al. (*J. Histochem. Cytochem.*, **12**: 293 and 298, 1964) provide the following method for the histochemical detection of β-glucuronidase at the light microscopic level:

1. Fix thin tissue slices (5 × 5 × 3 mm) in formol-calcium (4% w/v formaldehyde and 1% anhydrous calcium chloride) for 24 h at 4°C.
2. Without washing, blot tissues on filter paper, and transfer to Holt's hypertonic 1% gum acacia–30% sucrose medium for 24 h at 4°C. Tissues may remain in this solution for at least a week.
3. Cut frozen sections at 5 μ with a cryostat.
4. Incubate free-floating frozen sections for 30 to 60 min at 37°C in the following medium prepared as directed: The substrate stock solution (solution A) is 0.5 mM naphthol AS-BI glucuronide salt (27.4 mg/100 ml) in 0.2 N acetate buffer pH 5. It is prepared by completely dissolving 28 mg naphthol AS-BI glucuronide (free acid) in 1.2 ml of 0.05 M sodium bicarbonate (420 mg NaHCO$_3$ in 100 ml H$_2$O) and adding 0.2 N acetate buffer (pH 5) to make 100 ml. Kept at 20°C, the solution will be useful for several weeks.

   The dye stock solution (solution B) of Barka and Anderson is prepared as described earlier.

   The tissues are incubated in the following working solution: Mix 0.3 ml each of pararosanilin solution and sodium nitrite solution in a 20-ml beaker. About 1 min later, pour 10 ml of the substrate stock solution into the dye mixture. Adjust pH to 5.2 with 1 N NaOH using a pH meter. Add distilled water to a final volume of 20 ml. Filter through a Whatman No. 1 filter paper. Thus, the final concentration of the substrate is 0.25 mM and that of the diazo reagent is 1.8 mM. The resultant substrate working solution is a slightly yellow, clear solution even after a 30-min incubation.
5. Rinse sections in distilled water and mount on slides.
6. If desired, counterstain in 1% methyl green in 0.1 M Veronal-acetate buffer of pH 4 for 5 min.
7. Rinse thoroughly, dehydrate, and mount in synthetic medium.

Discrete localization of β-glucuronidase is achieved in many tissues. The method is reproducible and reliable. The structure of naphthol AS-BI–β-D-glucuronide follows:

Naphthol AS-BI–β-D-glucuronide

C$_{24}$H$_{22}$NO$_9$Br, mw 548.348

Fishman and Goldman (*J. Histochem. Cytochem.*, **13**: 441, 1965) later developed a postcoupling procedure for the demonstration of β-glucuronidase. The substrate is the same as that employed in the above procedure. After incubation in the substrate solution for the appropriate time interval, sections are washed and coupled with a suitable diazonium salt. Fishman and Goldman indicate that several are suitable, including fast garnet GBC, fast dark blue R, fast blue RR, fast blue 2B, and fast red violet 2B. The salts are made up fresh in a saturated solution in 0.01 M phosphate buffer pH 7.4. Sections are coupled for 2 to 5 min, rinsed in distilled water, and mounted in a water-soluble mount such as Kaiser's.

Fishman and Goldman demonstrated that with the above methods diffusion of enzyme was not a problem, the staining reaction was completely inhibited by saccharolactone (1 mM) and baicalein (trihydroxyflavon-β-D-glucosiduronic acid) at 100 μg/ml; the intensity of the staining reaction was a function of substrate concentration and pH.

Meijer (*Histochemie*, **30:** 31, 1972; ibid., **35:** 165, 1973) and Meijer and Vloedman (ibid., **34:** 127, 1973) firmly applied a semipermeable membrane to fresh frozen sections of tissues in order to limit diffusion of enzymes and thereby effect accurate localization. The authors recommend the use of cellulose dialysis membranes with an average pore radius of 24 Å obtainable from Viking Co., Chicago, Illinois. Excellent results were reported using naphthol substrates for β-glucuronidase and acid phosphatase in the usual fashion. Meijer also recommends the method for studies of lactate dehydrogenase (*Histochemie*, **35:** 165, 1973).

Smith and Fishman (*J. Histochem. Cytochem.*, **17:** 1, 1969) employed p-(acetoxymercuric) aniline diazotate as a reagent for the detection of either acid phosphatase or β-glucuronidase at the light and electron microscopic levels. Their method for *β-glucuronidase* follows. For light microscopy, tissues are fixed for 18 to 20 h at 4°C in 10% formalin in either phosphate- or calcium-containing buffers containing 0.1% chloral hydrate. Tissues may also be fixed in 1.5% glutaraldehyde in phosphate buffer of pH 7.4 containing 1% sucrose.

1. Incubate free-floating sections (15 μ for electron microscopy, 8 μ for light microscopy) in the following medium for 15 to 45 min at 37°C: The medium is prepared as a dilution of a stock solution of the substrate. Add 13.7 mg naphthol AS-BI glucosi-duronic acid in 1 ml 0.05 M NaHCO$_3$. Shake until dissolved, and dilute to 100 ml with 0.1 N acetate buffer of pH 4.5. Stored in the refrigerator, the solution is stable for months. The working solution is prepared by diluting 16 ml stock solution (0.25 mM) to 50 ml with 0.1 N acetate buffer at the desired pH level (for rat liver the optimal pH is 4.5). The reaction may be run from pH 3 to 8. Final concentration of substrate is 0.08 mM.
2. Wash sections in two changes of 7% sucrose in ice-cold distilled water no longer than 15 min.
3. Postcouple for 2 to 3 min at 24°C in the following diazotate: Add 1 g p-(acetoxymercuric) aniline (obtained from Polysciences, Rydal, Pennsylvania 19406, or from Eastman Organic Chem., Rochester, New York) to 25 ml of 50% glacial acetic acid in a 50-ml volumetric flask, and shake at 24°C until dissolved. After solvation, plunge into ice at 2 to 4°C. Use the diazotate the same day.
4. Wash sections at 4°C in Veronal-acetate buffer for three changes of 5 min each.
5. Mount in polyvinylpyrrolidone and examine by light microscopy. Sites of β-glucuronidase are red. The red stain *does not* withstand alcohol or acetone dehydration.
6. For electron microscopic observation, sections are reacted with a 1% solution of thiocarbohydrazide containing 2% sucrose for 30 to 45 min with agitation. (Sections for light microscopy may also be reacted with thiocarbohydrazide for 5 to 10 min, dehydrated, and mounted in synthetic resin without loss of reaction product.)
7. Wash sections in two changes of 7% sucrose 0.1 M Veronal-acetate buffer of pH 7.4 for 15 min at 4°C.
8. Treat sections with 1.5 to 2% OsO$_4$ pH 7.2 to 7.6 for 90 min at 24°C or 45 to 60 min at 35°C.

9.   Dehydrate through ethanols and embed in Araldite, but avoid use of propylene oxide.

RESULTS: At the electron microscopic level, β-glucuronidase was observed in the cisternae of the Golgi, in the endoplasmic reticulum, and within the nuclear envelope of mouse and rat liver parenchymal cells.

Smith and Fishman used the same method for the demonstration of *acid phosphatase*. At step 1, sections are incubated in the medium for 2 h at 37°C. To 10 ml *N,N*-dimethylacetamide is added 100 mg naphthol AS-BI phosphate. The buffer consists of 5 ml Michaelis Veronal-acetate buffer, 7 ml 0.1 *N* HCl, and 13 ml distilled water. The working solution consists of 15 ml buffer, 35 ml distilled water, and 0.6 ml of the substrate stock solution. The pH is adjusted to 5 with 0.1 *N* HCl if necessary.

Jeffree [*J. Microsc. (Oxf.)*, **89:** 55, 1969] also employed naphthol AS-BI–β-D-glucosiduronic acid in a simultaneous coupling procedure for the demonstration of β-glucuronidase. The incubation medium was that employed by Hayashi and Fishman detailed above to which 30 mg of a stable diazonium salt was added. Jeffree concluded that suitable salts were fast Bordeaux OL, fast red violet LB, fast Corinth LB, and hexazonium pararosanilin (as initially prescribed by Hayashi and Fishman).

Bowen [*J. Microsc. (Oxf.)*, **94:** 25, 1971] provides the following simultaneous coupling method for the histochemical demonstration of β-glucuronidase.

1.   Fix tissue slices 1 mm thick at 4°C for 1 h in 3% glutaraldehyde in 0.1 *M* phosphate buffer pH 7.4.
2.   Section tissues at 50 μ.
3.   Wash tissues overnight in 0.1 *M* phosphate buffer pH 7.4.
4.   Rinse briefly in 0.1 *M* acetate buffer pH 4.5.
5.   Incubate tissues in the following medium for $1\frac{1}{2}$ to 3 h. (*Substrate stock solution:* Dissolve 13.7 mg AS-BI glucosiduronic acid in 1 ml 0.05 *M* NaHCO$_3$, and dilute to 100 ml with 0.1 *M* acetate buffer pH 4.5.) The incubation solution is:

Stock substrate solution	16 ml
Acetate buffer, 0.1 *M*, pH 4.5	34 ml
p-Nitrobenzene diazonium tetrafluoborate, 30 mg; fast red GG (C.I. 37035)	

6.   Rinse tissues in acetate buffer 0.1 M pH 4.5.
7.   Fix tissues in OsO$_4$ according to Millonig (*J. Appl. Physiol.*, **32:** 1637, 1961).
8.   Dehydrate and embed in Araldite.

Alternatively, sections may be postcoupled with the diazonium salt at the above concentration at pH 7 in 0.1 *M* cacodylate buffer. β-Glucuronidase was detected in the free ribosomal regions and in lysosomes.

ARYL SULFATASE

Rutenburg, Cohen, and Seligman (*Science*, **116:** 539, 1952) reported a method for aryl sulfatase depending on the decomposition of the sulfuric acid ester of 6-benzoyl-2-naphthol and the coupling of the liberated benzoyl naphthol with fast blue B. Rat tissues may be fixed for some days or even months in cold neutral formalin. From these, frozen sections are prepared as usual. Human and monkey tissues are cut by a cryostat technic, mounted on glass slides directly, and dried in air. With the unfixed tissues it is necessary to incubate in a substrate made up in hypertonic salt solution.

TABLE 10-8 ARYL SULFATASE ACTIVITY IN ORGANS BY SPECIES*

	Human being	Monkey	Mouse	Rat	Rabbit	Guinea pig	Hamster	Dog
Liver ...	+ +	+ +	+ +	+ +	±	±	±	±
Kidney .	+ +	+ +	+ +	+ +	±	±	+ +	±
Pancreas	+ +	+ +	±	+ +	±	±	+ +	±
Adrenal .	±	±	+	+	±	±	+	±

* Compiled from various sources by Lillie.

The substrate is prepared by dissolving 25 mg potassium 6-benzoyl-2-naphthyl sulfate in 80 ml hot 0.85% sodium chloride solution and adding 20 ml 0.5 M pH 6.1 acetate buffer. For unfixed frozen sections the substrate is made hypertonic by the addition of 2.6 g sodium chloride per 100 ml, thus raising the NaCl content to 3.28%. Unfixed frozen sections attached to slides are first immersed in three baths of NaCl solution of 0.85, 1, and 2% concentration, and then placed in hypertonic substrate. Fixed frozen sections are transferred first to a small portion of the normal substrate, and then to the regular incubation bath of 20 ml. Sections from each organ are to be incubated in separate containers to avoid transfer of activity. For organs with high activity incubate 2 to 3 h at 37°C; for other organs, 4 to 16 h in hypertonic substrate.

After incubation, wash formalin-fixed tissues in water and unfixed tissues in descending grades of salt solution (2, 1, and 0.85%); and postcouple 5 min in a freshly prepared cold (4°C) 0.1% solution of fast blue B in 0.05 M phosphate buffer of pH 7.6. Then wash in three changes of cold 0.85% sodium chloride solution and mount in glycerol or glycerol gelatin.

Areas of high activity stain blue; those of lower activity stain purple to red. On storage even at 4°C the blue gradually changes through purple to red, and some diffusion occurs. Activity is largely cytoplasmic.

Roy (*J. Histochem. Cytochem.*, **10:** 106, 1962) quotes the method of Rutenburg, Cohen, and Seligman (*Science*, **116:** 539, 1952) as the most widely used histochemical procedure for this group of enzymes and points out procedures for activity due respectively to aryl sulfatases A, B, and C.

Pearse prescribes a 2- to 8-h 37°C incubation and postcoupling in fast blue B as in the Rutenburg method. Pearse prescribes frozen sections of formalin-fixed tissue, or fresh frozen sections attached to slides. For the hypertonic saline (2.6% NaCl) substrate, Pearse prescribes baths in graded saline solutions before the substrate. Since 56 mM (0.33%) NaCl is adequate for the chloride-dependent sulfatase B, this precaution is unnecessary for the other substrates.

Roy's substrate I (Table 10-9), although he used it for biochemical assay, should give essentially the same demonstration of the two aryl sulfatases as the original method. Use of substrate III increases the activity of sulfatase C and completely suppresses that of A and B. Substrate II permits activity of sulfatase A but suppresses that of sulfatase B, which is chloride-dependent. This distinction can be made only in the absence of sulfatase C, which usually supplies most of the activity.

Roy regards the substrate as satisfactory for aryl sulfatase C, but not for A and B. A 2-h incubation was adequate for assay of C; a period of 16 to 18 h was needed for A and B.

No simultaneous coupling technic has been developed, but the complex naphthol appears to be sufficiently insoluble for use of the postcoupling technic. Hence, after the substrate incubation wash in distilled water, transfer to fast blue B 1 mg/ml in ice cold pH 7.5 to 8 buffer for 5 min, wash in water, and mount in glycerol gelatin.

TABLE 10-9 ARYL SULFATASE SUBSTRATES

| | Pearse | | Roy | | | | | |
| | | | I | | II | | III | |
	Amount mg	Molarity mM	Amount mg	Molarity mM	Amount mg	Molarity mM	Amount mg	Molarity mM
6-Benzoyl-2-naphthyl sulfate........	6.25	0.68	6.25	0.68	6.25	0.68	6.25	0.68
NaCl......	650	495	82	56		100		100
Buffer........	Acetate	100	Acetate	100	Acetate		Tris	
pH	6.1		6.1		6.1		8.0	
Volume........	25 ml		25 ml		25 ml		25 ml	

Sources: Pearse, "Histochemistry: Theoretical and Applied," 2d ed., Lillie, Brown, Boston, 1960; Roy, *J. Histochem. Cytochem.*, **10**: 106, 1962.

Woohsmann and Hartrodt (*Histochemie*, **4**: 336, 1964) advise the following simultaneous azo dye method for the histochemical localization of *aryl sulfatase*.

1. Small blocks of tissues 2 to 4 mm thick are fixed overnight in Baker's formol-calcium at 4°C (not to exceed 24 h).
2. Wash tissues in distilled water briefly, section at 10 μ with a cryostat, collect sections in physiologic saline solution, and mount on slides.
3. Incubate for 1 to 2 h at 37°C in the following medium:

Naphthol AS-BI sulfate (potassium salt)	40–50 mg
NaCl (0.85 %)	20 ml
Acetate buffer (0.2 M pH 6.2 to 7.2)	5 ml
NaCl	650 mg
Fast red salt ITR	25 mg

A slight precipitation of the medium will disappear after slight warming in an incubator and filtration.

4. Wash sections in distilled water, and mount in a water-soluble medium such as Kaiser's glycerogel.

   Highest activity is noted in uterine epithelial cells of pregnant rats, cells of the adrenal and kidney cortex, basal epithelial cells of placenta, and certain pericytes of the central nervous system.

Woohsmann and Hartrodt noted that although the diazonium salt fast red TR could be substituted for fast red ITR, fast red TR tended to form large crystal chromogens, particularly if sections were overincubated. Some background staining is observed irrespective of the diazonium salt employed.

Woohsmann and Hartrodt noted HgCl$_2$ (0.081%, 3 mM), KCN (0.32%, 50 mM), and monoiodoacetic acid (0.93%, 50 mM) are inhibitors. If L-cysteine (0.06%, 5 mM), NaF (0.42%, 10 mM), or NaH$_2$PO$_4$ (0.41%, 3 mM) is added *with* the substrate, inhibition is noted. Sections incubated in HgCl$_2$, KCN, or monoiodoacetic acid in acetate buffer for 1 h at 37°C prior to incubation with the substrate manifest inhibition of the staining reaction.

Enzyme activity was abolished in sections fixed in acetone for 3 h at 5°C. Only slight activity is noted in sections treated in 70% ethanol for 3 h at 5°C.

Gibson and Fullmer (*J. Periodontol.*, **41**: 102, 1970) employed the Woohsmann and Hartrodt method in studies on the distribution of sulfatase in periodontal tissues of rats. They also noted a strong activity in certain pericytes associated with blood vessels.

Wolf et al. (*Proc. Soc. Exp. Biol. Med.*, **124**: 1207, 1967) synthesized 5-bromo-4-chloro-3-indolyl sulfate and developed the indigo method described below for the demonstration of *sulfatase*.

1. Tissues sliced 2 to 4 mm thick were quick-frozen in a glass tube immersed in a flask containing dry ice and acetone.
2. Tissues were embedded in O.C.T. (optimal cutting temperature—Lab-Tek) and sectioned at −20 to −35°C with a cryostat at 6 μ.
3. Sections were fixed briefly in either cold neutral formalin or cold acetone.
4. Sections were incubated for 2 h at 23°C in the following medium:

Tris buffer 0.05 M pH 6.1	14 ml
MgCl$_2$ (to make final concentration 5 mM) (7.6 mg/ml)	1 ml
5-bromo-4-chloro-3-indolyl sulfate, 7.2 mg/ml distilled water	1 ml

RESULTS: Sites of sulfatase activity were stained blue-green. Greatest activity was detected in liver, kidney, pancreas, and adrenal. The reaction was inhibited by sulfite, cyanide, and fluoride (concentrations not given).

ARYL SULFATASE METHOD (Hopsu-Havu et al., *Histochemie*, **8:** 54, 1967) FOR LIGHT AND ELECTRON MICROSCOPY

1. Fix tissues in 5% glutaraldehyde in 0.1 M cacodylate buffer of pH 7.4 at 4°C.
2. Wash tissues for 12 h to 10 days in 0.1 M cacodylate buffer of pH 7.4 containing 7.5% sucrose (0.22 M).
3. Incubate frozen sections (10 μ thick) in the following medium: 160 mg p-nitrocatechol sulfate dissolved in 4 ml distilled water, 12 ml 0.1 M acetate buffer of pH 5.5, and 4 ml 5% barium chloride. The pH is adjusted to 5.5 with 0.2 M acetic acid. For electron microscopy, incubate thin blocks of tissue instead of frozen sections in the above medium.
4. Rinse sections or blocks in several changes of 0.1 M cacodylate buffer pH 7.4 with 7.5% sucrose. Thorough washing is an important step here if electron microscopy is to follow.
5. For light microscopy, immerse sections in 2% ammonium sulfide for 1 to 2 min; dehydrate and mount in synthetic resin. For electron microscopy, small blocks are postfixed in osmium tetroxide in cacodylate buffer of pH 7.4, dehydrated through graded ethanols, embedded in Epon 812, sectioned, and examined.

Comparative biochemical and light and electron microscopic studies revealed aryl sulfatase localization in lysosomes of proximal convoluted tubules of rat kidney. In a follow-up study Kalimo et al. (*Histochemie*, **14:** 123, 1968) advised that tissues must not be stained with lead or uranyl acetate after the barium or lead sulfate precipitates are formed. Such staining results in solubilization of the reaction product and thereby displacement or removal.

Al-Azzawi and Stoward (*Histochemie*, **37:** 187, 1973) note that the activity of aryl sulfatase and acid phosphatase may be greatly reduced in sections of kidneys of rats killed with an overdose of either chloroform or ether.

Sandström and Westman (*Histochemie*, **19:** 181, 1969) recommend embedment of tissue blocks in polyethylene glycol (Carbowax) with a molecular weight of about 1000 after aldehyde fixation. Sections approximately 40 μ in thickness are cut and immersed in the reaction medium for the detection of whatever enzyme is desired. The polyethylene glycol will dissolve into the reaction medium. Sections are then rinsed in buffer 3 times for 5 min each, postfixed in osmium tetroxide for 1 h, embedded in Epon, sectioned, and stained for electron microscopy.

ESTERASES

Histochemically localizable esterases comprise nonspecific esterases (which decompose glyceryl and other esters of short-chain aliphatic acids), lipases (which attack esters of long-chain fatty acids and are found principally in pancreas), and cholinesterases (which hydrolyze fatty acid esters of choline and acetylcholine and are found in motor end organs, neural synapses, nerve cells, and erythrocytes).

Choudhury (*J. Histochem. Cytochem.*, **20:** 507, 1972) obtained many nonspecific esterases from several rat organs in order to study the nature of esterase polymorphism. An increasing susceptibility to organophosphate inhibition was observed to parallel progressive lengthening of the acyl chain of the substrate molecules. Choudhury was able to hybridize an organophosphate-sensitive esterase with a resistant type which yielded a few new esterases.

Choudhury concludes that esterases are built on a subunit structure and exhibit overlapping specificities due to sharing of common subunits. He sees no justification of divisions of esterases into acetyl-, aryl-, or carboxylesterases on the basis of either organophosphate sensitivity or substrate hydrolysis.

Following his work on the phosphatases, Gomori (*Proc. Soc. Exp. Biol. Med.*, **58**: 362, 1945) introduced a method for the demonstration of lipases. In this technic the water-soluble palmitic or stearic acid esters of certain polymer glycols or hexitans are hydrolyzed in the presence of a soluble calcium salt, and the calcium soaps formed *in situ* are converted into lead soaps by treatment with lead nitrate. The lead soaps are converted to brown lead sulfide with ammonium sulfide.

The general procedures for preparation of sections for demonstration of esterases can well be substituted for Gomori's acetone paraffin technic. George (*J. Histochem. Cytochem.*, **11**: 420, 1963) regards Tween 80 or 85 as more specific for lipase than the Tweens prescribed by Gomori. Most later writers who have used the Tween methods simply refer to Gomori's or Pearse's textbooks. Lison (1960) quotes it essentially without change.

The validity and utility of histochemical methods for lipase should be challenged. Pancreatic lipase will hydrolyze short-chain fatty acid esters, namely, formate and butyrate esters of *p*-chlorobenzyl alcohol at rates comparable to those for oleate esters of the same compound. Conversely, esterases are capable of hydrolyzing long-chain fatty acid esters (Brockerhoff, *Biochem. Biophys. Acta*, **212**: 92, 1970; Lake, *Histochem. J.*, **4**: 71, 1972).

THE TWEEN METHOD FOR ESTERASES

1. Prepare frozen sections of fresh unfixed tissue, attach to slides, and postfix 15 min in ice-cold calcium formalin or in ice-cold acetone. Or fix overnight in neutral formol-calcium and cut frozen sections. Or prepare paraffin sections by the freeze-dried, celloidin paraffin or paraffin technic, deparaffinize in isopentane or low-boiling petroleum ether, and let dry. Wash out formalin or acetone with several changes of distilled water. Generally, prefixation, formalin fixation, and frozen sections seem to constitute the preferred procedure.
2. Incubate at 37°C in one of the substrates in Table 10-10.
3. Rinse in distilled water.

TABLE 10-10 TWEEN SUBSTRATES FOR ESTERASES

	Gomori, from Lillie, 1954	Pearse, 1953	Lison, 1960
Tween*	40 or 60, 2%, 3 ml	60 or 80, 5%, 1.2 ml	20, 80, 5%, 0.6 ml
Solvent	30% glycerol, 12 ml	Distilled water, 24.6 ml	Distilled water, 27 ml
CaCl₂	2% CaCl₂, 3 ml	10% CaCl₂, 1.2 ml	2% CaCl₂, 0.9 ml
Buffer	Tris, 0.1 M, 12 ml	Tris, 0.5 M, 3 ml	Tris, 0.05 M, or Veronal 1.5 ml
pH	7.3	7.3	7.2
Final volume	30 ml	30 ml	30 ml
Time	8–24 h	3–12 h	8–24 h
Temperature	37°C	Not stated	37°C

* Tween numbers: 20, laurate; 40, palmitate; 60, stearate; 80, oleate. Hydrolysis of the oleate was claimed to be a specific property of pancreatic lipase.

Sources: R. D. Lillie, "Histopathologic Technic and Practical Histochemistry," 2d ed., Blakiston, Philadelphia, 1954; K. G. E. Pearse, "Histochemistry: Theoretical and Applied," Little, Brown, Boston, 1953; L. Lison, "Histochemie animales et Cytochimie," 3d ed., Gauthier Villars, Paris, 1960.

4. Treat with 2% lead nitrate solution for 10 min.
5. Rinse repeatedly in distilled water.
6. Treat 2 min with a 1 : 100 dilution of light yellow ammonium sulfide in distilled water. (Gomori said 10 drops, about 0.5 ml, in a Coplin jar of water, or about 50 ml.)
7. Wash in running water. Counterstain 3 to 5 min in alum hematoxylin, dehydrate, clear, and mount in synthetic resin. Gomori advises against xylene as a clearing agent and as a solvent for the Clarite resins, alleging fading in xylene media. Instead, use dichlorethylene or ligroin (cf. Timm's lead sulfide method), or since any decoloration of the highly insoluble lead sulfide which might occur would be due probably to oxidation to the almost equally insoluble white lead sulfate, use of natural fir balsam or other reducing resin seems to be indicated.

RESULTS: Sites of lipase activity are evident as dark brown deposits of lead sulfide.

Gomori notes further that treatment for 1 min with Gram's iodine solution or with 5% phenol, or boiling 10 min in water, destroys the enzyme. Addition of 0.2% sodium taurocholate intensifies the action of pancreatic lipase but inhibits that of "all other organs."

Nachlas and Seligman (*J. Biol. Chem.*, **181**: 343, 1949) reported that in the test tube eserine at 3.5 mM (962 mg/l) had only a moderate inhibitory effect both on pancreatic hydrolysis of naphthyl laurate and stearate (lipase) and on hepatic and renal hydrolysis of naphthyl acetate (esterase), especially in human beings. Sodium taurocholate at 0.1 M (0.54%) accelerated the lipase hydrolysis of naphthyl laurate and stearate, and depressed slightly the esterase decomposition of naphthyl acetate. Quinine hydrochloride at 0.05 M (1.894%) inhibited almost completely the pancreatic lipase hydrolysis of naphthyl laurate and stearate; esterase showed a species variable but lesser grade of inhibition or even (in the dog) acceleration. Atoxyl (sodium arsanilate) at 0.1 M (2.39%) and 0.3% (0.07 M) sodium fluoride tended to inhibit the esterase activity of liver, kidney, and pancreas on naphthyl acetate, but they were without effect on the pancreatic lipase hydrolysis of naphthyl laurate and stearate.

In their hands the hydrolysis in vitro of the polyglycol stearic acid esters, such as Gomori used, followed more the organ distribution pattern of esterase than that of true lipase. The activity against naphthyl laurate and stearate was similar in organ distribution to that against olive oil.

Unfortunately naphthyl stearate and laurate are too insoluble to use in the histochemical technics (Seligman et al., *Ann. Surg.*, **130**: 333, 1949), but by including 0.1 M (2.39%) sodium arsanilate (atoxyl) in the naphthyl acetate substrate, they showed that the esterase activity of liver and kidney was inhibited, and they considered the pronounced activity still evident in pancreas as due to lipase.

This finding suggests that the method may actually demonstrate lipase in pancreas and that the activity inhibited by taurocholate may be that of nonspecific esterase, since Seligman et al. (ibid.) showed that the homogenate esterases of other organs were relatively ineffective in hydrolyzing β-naphthyl palmitate and stearate, though pancreatic homogenate enzymes hydrolyzed this substrate readily. Seligman was not able to adapt this palmitate-stearate substrate to histochemical use because of its great insolubility.

THIOACETIC ESTERASE

The cholinesterases of motor end plates (true) and of erythrocytes (pseudo) are reported to hydrolyze noncholine esters, and the acetylcholinesterase of rat brain is active against naphthyl (Ravin, Zacks, and Seligman, *J. Pharmacol. Exp. Ther.*, **107**: 37, 1953) and indoxyl (Barrnett and Seligman, *Science*, **114**: 579, 1951; Pepler and Pearse, *J. Neurochem.*, **1**: 193,

1957) esters. But electrophoretically separated human brain acetylcholinesterase is reported to react with acetylthiocholine substrates and not against thioacetic acid or naphthyl and indoxyl acetates (Barron et al., *J. Histochem. Cytochem.*, **11:** 139, 1963). Thioacetic esterase was concentrated in a zymogram band showing also weak activity against the synthetic protease substrate α-*N*-benzoyl-DL-arginyl-β-naphthylamide (BANA) but none against other esterase substrates or acetylthiocholine iodide.

However, thioacetic acid as substrate has given good demonstrations of acetylcholinesterase sites in rodents (Wachstein et al., *J. Histochem. Cytochem.*, **9:** 325, 1961). Barrnett and Palade (*J. Biophys. Biochem. Cytol.*, **6:** 163, 1959) and Zacks and Blumberg (*J. Histochem. Cytochem.*, **9:** 317, 1961) reported its use for acetylcholinesterase in both human and mouse muscle.

Wachstein's directions for this technic appear a little more complete and adaptable to routine laboratory use than the others cited. It is to be borne in. mind that though usable for demonstration of cholinesterases in human muscle, as well as more generally in rodents, the thioacetic esterase method is not specific for acetylcholinesterase, demonstrating also aliesterases and so-called nonspecific cholinesterase.

THE THIOACETIC ACID ESTERASE METHOD

1. *Section preparation:* Wachstein and others prefer tissue prefixed in neutral formol-calcium for 16 to 24 h. Section on the freezing microtome or use cryostat or cold knife frozen sections of fresh tissue, directly or, preferably, after 15-min fixation at 0 to 5°C in formol-calcium or acetone, or after one of the section freeze substitution technics. Burstone's freeze-dried, paraffin or celloidin paraffin technic is applicable. Petroleum ether (bp 25 to 40°C) should be used to remove paraffin; acetone is needed to remove celloidin; otherwise sections can be dried in air after petroleum ether and placed directly in the substrate.

2. *Substrate* (Wachstein): For solution A, dissolve thioacetic acid, 0.15 ml (= 152.5 mg), in 5 ml distilled water. Adjust to pH 5.5 with about 5 ml 0.1 *N* NaOH. Add 0.2 *M* acetate buffer pH 5.5 to make 100 ml.

    Working substrate (solution B): To 20 ml solution A add 1 ml 0.5% lead nitrate solution drop by drop, with shaking. Centrifuge and filter. Replace clear solution with fresh every 15 to 20 min during incubation, to minimize precipitation. At pH 6 precipitation becomes more troublesome; at pH 5 the intensity of staining is reduced. The final molarity of this substrate for thioacetic acid is about 19 m*M*; of lead, 0.7 m*M* or lower; of acetate buffer, 0.17 *M*.

3. Incubation times of 30 min to 24 h were used, usually 1 h.

4. After incubation wash in two or three changes of distilled water, dehydrate in alcohols, clear in xylene, and mount in Canada balsam or natural fir balsam.

RESULTS: Sites of activity appear in brown. Inhibitor-resistant esterase may be specifically demonstrated by a 1 h 37°C preincubation in 10 μM E-600. Inclusion of 5 mM sodium taurocholate in the substrate is necessary for demonstration of pancreatic lipase. Eserine at 10 μM was used to inhibit acetylcholinesterase activity.

The method apparently has similar capacities for esterase demonstration with the naphthol AS and AS-LC acetate technics. It would appear to offer advantages in electron microscopy in that an electron-dense reaction product is engendered.

Thioacetic acid, perhaps listed under the synonym "thiacetic acid" or "thiolacetic acid," is readily available in practical grade from purveyors of organic chemicals.

Bell and Barrnett (*J. Histochem. Cytochem.*, **13:** 611, 1965) synthesized thiobutyric, thiocaproic, and thiocaprylic acids and developed histochemical methods for the detection of *esterase* at the light and electron microscopic levels. Although distribution of enzyme activity was much the same with all substrates, thioacetic acid could be used in the incubation medium without the addition of an organic solvent (acetone).

1. Fix tissue slices 2 to 3 mm thick for 1 to 4 h in 4% glutaraldehyde in 0.1 M cacodylate buffer of pH 7.2.
2. Cut frozen sections at 10 to 15 μ, and wash sections 1 to 2 h in 0.1 M cacodylate buffer pH 7.2.
3. Rinse sections briefly in cacodylate buffer (0.1 M pH 7.2) and acetone (1 : 1).
4. Incubate in the following medium for 10 to 30 min at 23°C:
   a. To 50 ml 0.1 M cacodylate buffer of pH 7 add 0.05 M acid (0.38 ml thioacetic acid; 0.52 ml thiobutyric acid; 0.66 ml thiocaproic acid or 0.8 ml thiocaprylic), and adjust the pH to 7 with 1 N NaOH. Add sufficient buffer to make 70 ml.
   b. Dissolve (0.01 mol) Pb(NO$_3$)$_2$ (331 mg) in 10 ml 0.1 M cacodylate buffer of pH 7, and add to the substrate solution. A white flocculent precipitate appears. Add 20 ml acetone, 2 to 3 ml at a time, with continual shaking, and adjust the pH to 7. Filter if necessary. The solution should be clear to light yellow.
5. Rinse sections in a 1 : 1 solution of cacodylate buffer, 0.1 M at pH 7, and absolute acetone.
6. Immerse tissues in dilute ammonium sulfide; then rinse in cacodylate buffer.
7. Tissues may be mounted in an aqueous medium such as Kaiser's or dehydrated and mounted in a synthetic medium.

For electron microscopy, proceed as described above through step 6. Sections are then refixed in 1% buffered osmium tetroxide for 2 h, dehydrated through graded ethanols, and embedded in Epon according to Luft (*J. Biophys. Biochem. Cytol.*, **9**: 409, 1961).

Enzyme activity is indicated by brownish black lead sulfide deposits observed by light microscopy and as opaque deposits by electron microscopy. Omission of substrate resulted in absence of lead sulfide deposits. Using thiobutyric as substrate, the reaction is activated by heparin, 1 μg/ml, in the incubation medium, and inhibition is observed at 100 μg/ml. Diethyl-p-nitrophenylphosphate (E-600) at 2.7 μg/100 ml (0.1 μM) completely inhibits the reaction, and diisopropyl fluorophosphate (DFP) at 1.84 mg/100 ml, 0.1 mM is moderately inhibitory and without effect at lower concentrations. With thioacetic as substrate, the reaction in neural tissues is completely prevented by physostigmine (eserine) at 27.5 μg/100 ml, 1 μM, but cytoplasmic reactions in other tissues persist. The cytoplasmic reaction in all tissues is completely inhibited, except in presumed lysosomes, by E-600 at 0.1 μM.

Dermal adipose tissue and pancreatic acinar tissues displayed more enzyme activity with the eight carbon than with two, four, and six carbon substrates; conversely, liver, kidney, and intestine manifested more activity with two and four carbon substrates. Using thioacetic acid as substrate, enzyme activity at the electron microscopic level was observed mostly in rough endoplasmic reticulum and on nuclear membranes. Reaction products were also detected between outer and inner membranes of mitochondria.

ESTERASES: AZO DYE TECHNICS

Nachlas and Seligman (*J. Natl. Cancer Inst.*, **9**: 415, 1949) proposed a nonspecific esterase method utilizing β-naphthyl acetate as substrate and visualizing by simultaneous coupling with α-naphthyl diazonium naphthalene-1,5-disulfonate, which gives a red color; or with the commercial stable tetrazotized diorthoanisidine, fast blue B, which gives a deep blue.

This method has been largely abandoned because of diffusion artifacts; Gomori's α-naphthyl and naphthol AS acetate methods afforded some improvement but have, in their turn, been replaced by more complex naphthyl acetates of the AS series and by the use of improved section preparation procedures.

Following Gomori, Burstone (*J. Histochem. Cytochem.*, **4**: 130, 1956) used α-naphthyl and naphthol AS acetates, propionates, and butyrates on frozen dried paraffin sections, with improvements in localization. The esterase activities were inhibited by 2 m*M* sodium taurocholate, 2 m*M* NaF, 0.5 m*M* diisopropyl fluorophosphate (DFP) and 0.5 m*M* Phemerol, but not by 0.1 m*M* eserine.

Later Burstone synthesized several new naphthol AS acetates; of these, naphthols AS-D, AS-OL, AS-MX, and especially naphthol AS-LC acetates were recommended as suitable substrates for the aliesterase technic as performed on frozen dried paraffin sections.

BURSTONE'S FREEZE-DRIED, ESTERASE METHOD WITH NAPHTHOL AS-LC ACETATE

1. Deparaffinize with petroleum ether or isopentane; hydrate through acetone or by drying isopentane in air and passing section directly into substrate.
2. *Substrate:* Dissolve naphthol AS-LC acetate, 3 mg, in 0.3 ml acetone or dimethyl-formamide; add 15 ml distilled water and 15 ml 0.1 *M* tris buffer pH 7.1, and stir in 15 mg fast garnet GBC or fast blue RR; filter and pour over sections in Coplin jar.
3. Incubate 10 to 30 min at 25°C, or until sufficient red or blue has been developed in sections. A 2-h trial is adequate for the reporting of a negative reaction, as in the presence of specific inhibitors or of lesions inducing loss of esterase activity.
4. Wash thoroughly in distilled water, and mount in glycerol gelatin, polyvinylpyrrolidone, or other suitable medium.

METHOD OF SHNITKA AND SELIGMAN FOR SIMULTANEOUS DEMONSTRA-TION OF INHIBITOR-RESISTANT (A) AND INHIBITOR-SENSITIVE (B) ALIESTERASES (*J. Histochem. Cytochem.*, **9**: 504, 1961).

It was observed that although diisopropyl fluorophosphate inhibition was poorly reversible by treatment with certain oximes, inhibition by NaF or by arsanilate was readily reversed by 90 to 120 min washing of sections in distilled water.

1. Fix fresh tissue in thin blocks (3 mm) 16 to 24 h in neutral formol-calcium at 2 to 5°C; wash in four changes of cold 0.85% NaCl.
2. Transfer to ice-cold 30% sucrose (w/v = 0.88 *M*) containing 0.9% gum acacia (gum arabic), and infiltrate 12 to 24 h at 2 to 4°C. Blot and freeze in petroleum ether or isopentane at −70°C.
3. Transfer blocks to cryostat at −20°C, section at 4 to 8 μ, and transfer sections with needle to surface of 2 *M* NaCl (11.7%) at 5°C.
4. Transfer at once, lifting on spatula, to first substrate. *First substrate:* Pour 10 ml 0.1 *M* phosphate buffer of pH 7.3 containing 3% glycol by volume into small beaker containing 0.1 ml 2.5% solution of naphthol AS acetate in acetone, swirling to mix thoroughly. Add 30 mg sodium fluoride (about 70 m*M*) and 10 mg fast blue BB (C.I. 37175), stir to dissolve, filter, and incubate sections 10 to 20 min. Instead of NaF, sodium arsanilate 0.1 *M* (240 mg) may be used as inhibitor.
5. Wash in three or four changes of distilled water, 30 min each, using 300 ml for each 10 sections, to remove all traces of inhibitor.
6. Transfer to surface of second substrate, made as before, but without inhibitor and using fast red violet LB salt 10 mg in place of fast blue BB (Lillie, 1969, p. 140).
7. Incubate 20 to 35 min at 37°C.
8. Float in distilled water 10 min, pick up on slides, and mount in 90% polyvinylpyrrolidone.

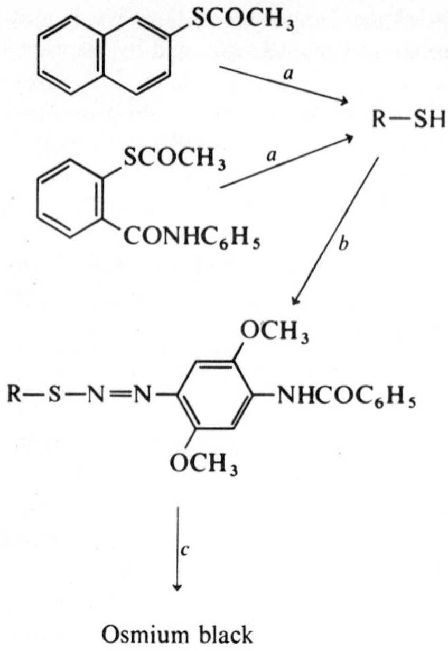

Fig. 10-1 The first substrate shown is NTA and the second substrate is TAB. (*a*) Esterase, (*b*) fast blue BBN, (*c*) osmium tetroxide. (*From Seligman et al., Ann. Histochim.,* **11:** 115, 1966. *By permission of the publisher.*)

Osmium black

RESULTS: Inhibitor-resistant (A) esterases are blue; inhibitor-sensitive (B) esterases, red.

Shnitka and Seligman also combined mitochondrial staining by the reduced diphospho-pyridine nucleotide, nitro blue tetrazolium technic for "diaphorase" in a sequence preceding steps 5 to 8 of the above procedure.

Following Holt's bromoindoxyl acetate technic, which, by reason of its ferroferricyanide redox buffer content, demonstrates only A esterase in rat kidney, by steps 5 to 8 of the above double-stain technic, Shnitka and Seligman also successfully demonstrated the B esterase in red, contrasting with the blue of the bromoindigo method.

Seligman et al. (*Ann. Histochim.,* **11:** 115, 1966) synthesized two substrates, namely, 2-naphthylthiolacetate (NTA) and 2-thiolacetoxybenzanilide (TAB), whereby *aliesterase* may be demonstrated at both acid and alkaline pH levels. The liberated mercaptan is readily captured by S coupling with fast blue BBN to form an osmiophilic diazoether at either pH as indicated in Fig. 10-1.

Seligman et al. prescribe the following procedure:

1. Fix small tissue blocks in Baker's calcium-formol solution for 24 h at 4°C.
2. Rinse tissues 3 times with physiologic saline solution at 4°C, and transfer tissues to 30% sucrose for 24 h at 4°C.
3. Freeze tissues with a dry ice–acetone mixture, and cut sections with a cryostat at 4 μ.
4. Incubate free-floating sections in the following medium first for 10 min (NTA) or 20 min (TAB) at 37°C and then for 60 min at 0°C at either pH 7.4 or 5. The reagents NTA and TAB are available from Polysciences Inc., Warrington, Pennsylvania.

0.1 M phosphate buffer (pH 7.4 or 5)	10 ml
2.5% NTA or TAB in acetone	0.1 ml
Fast blue salt BBN	5 mg

5. Wash sections in distilled water 5 to 10 min.

6. Mount sections on coverslips, place on a Chen rack in a small glass jar containing 0.5 g osmium tetroxide crystals and a few drops of water on the bottom. Osmicate with osmium tetroxide vapor for 5 min (NTA) and 8 min (TAB) by bringing the bottom of the glass jar in contact with a water bath at 50°C.
7. Wash sections in distilled water for 30 min.
8. Dehydrate, and mount in synthetic resin.

For electron microscopy the procedure above is modified slightly. Small blocks of tissues are fixed in either Baker's formol-calcium (24 h, 4°C) or 5% glutaraldehyde containing 0.22 M sucrose in phosphate buffer (0.08 M, pH 7.4) for 4 h at 4°C. Tissues are washed in 0.22 M sucrose for 30 min and frozen sections cut at 25 μ. Sections are incubated in the above media (*except* 0.75 g sucrose is added) for 30 min at 37°C for NTA and for 1 h at 37°C for TAB. The incubation fluid is changed after 30 min.

After incubation, tissues are washed for 30 min in 0.22 M sucrose at 4°C and osmicated as detailed above. Thereafter, tissues are washed in sucrose, dehydrated in graded alcohols, and embedded in Araldite. Seligman et al. note that contrast at the electron microscopic level can be increased if thin sections of the Araldite-embedded material are treated with a 1% solution of thiocarbohydrazide for 1 h at 50°C, washed several times with water at 50°C for 15 min each, and osmicated again for 1 h at 60°C.

ESTERASES: INDOXYL METHODS

Barrnett and Seligman (*Science*, **114**: 379, 1951) introduced indoxyl acetate and butyrate as substrates and described their preparation from sodium indoxyl.

These substances yield indigo on hydrolysis in the presence of air. This pigment is quite insoluble in water and in fats. The method demonstrates nonspecific esterase, lipase, and cholinesterases, which are distinguished by use of the usual inhibitors.

Holt and Withers [*Nature (Lond.)*, **170**: 1012, 1952] substituted 5-bromoindoxyl acetate for the above, using ordinary frozen sections of tissue fixed for 16 h at 4°C in neutral formol–saline solution. Finer-grained deposits were produced, diffusion decreased, and the consistency of performance increased.

Holt reported a series of minor modifications and applications of this technic (see *J. Histochem. Cytochem.*, **4**: 541, 1956, and *Proc. R. Soc.* [*Biol.*], **148**: 520, 1958). Shnitka and Seligman (*J. Histochem. Cytochem.*, **9**: 504, 1961) have slightly modified this procedure and reinterpreted it in view of their own experience with it.

HOLT'S 5-BROMOINDOXYL ACETATE ESTERASE METHOD ACCORDING TO SHNITKA AND SELIGMAN. Sections were prepared by calcium-formol fixation, Holt's gum sucrose infiltration, and cryostat sectioning as quoted in the double esterase method of Shnitka and Seligman.

SUBSTRATE: Mix rapidly 2 ml 0.1 M tris HCl buffer of pH 8.3; 5 ml 2 M NaCl (11.7%); 1 ml ferricyanide-ferrocyanide redox buffer 0.05 M for $K_3Fe(CN)_6$ and for $K_4Fe(CN)_6 \cdot 3H_2O$ [dissolve 1 mmol of each (329 and 422 mg, respectively) in the same 20-ml portion of distilled water]; add 2 ml distilled water. Pour the mixture into a small beaker containing 1.3 mg 5-bromoindoxyl acetate in 0.1 ml acetone, swirling to mix. 5-Bromo-4-chloroindoxyl acetate,[1] 1.5 mg, may be substituted for the 5-bromo derivative. The final concentrations are 0.5 mM for the indoxyl acetate and 5 mM each for $K_3Fe(CN)_6$ and $K_4Fe(CN)_6$. Incubate free-floating sections 30 to 120 min, wash in distilled water, pick up on glass slides,

---

[1] Sigma Chemical Co., St. Louis, Mo.

and mount in polyvinylpyrrolidone, or cautiously dehydrate in alcohols, clear in xylene, and mount in Permount.

RESULTS: Discrete droplet localization of esterases is seen, shown by Shnitka and Seligman to be A esterase, with suppression of B esterases by the redox buffer acting as an inhibitor.

Holt and Hicks (*J. Cell Biol.*, **29**: 361, 1966) demonstrated that an osmiophilic complex can be developed from indoxyl substrates. The indoxyl substrate for esterase introduced by Barrnett and Seligman can be used as an example:

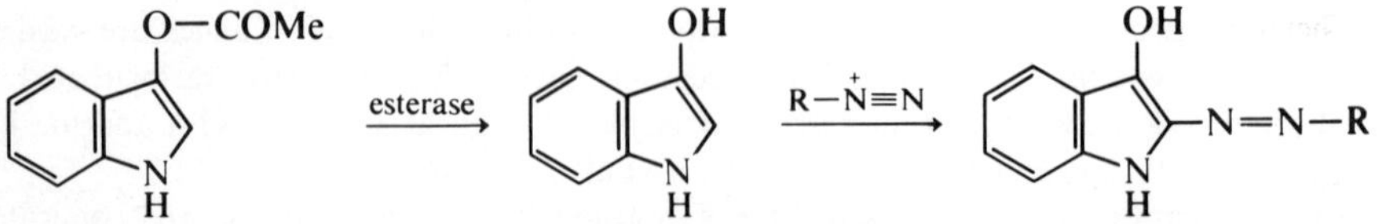

The azo dye (III) is derived from indoxyl (II) after enzymatic hydrolysis of the substrate (I). Holt and Hicks indicate that azo dyes derived from indoxyl have unique chelating propensities in that they are able to form two classes of chelates, namely, metal (M) binding by the azo and enol groups (IV), and by the azo and imino groups (V):

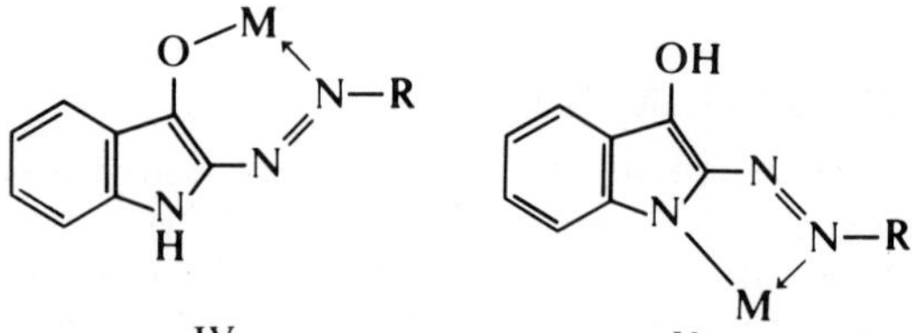

Holt and Hicks disclosed that under suitable conditions hexazotized pararosanilin reacts with unsubstituted indoxyl (II) to form the osmiophilic complex (VI):

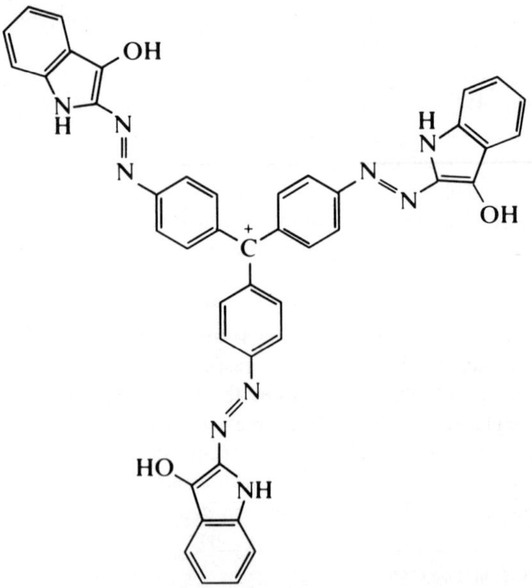

VI

Lillie regards the above formulation as highly improbable. The diazo is applied in 3 times excess after deposit of the indoxyl, and only single indoxyls are likely to bind any one diazonium molecule.

Holt and Hicks claim excellent results with the above method for the demonstration of nonspecific esterase in rat kidney. They prescribe as follows:

1. Fix slices of tissue in 5% glutaraldehyde buffered in 0.067 M cacodylate buffer of pH 7.3 for 4 h at 4°C.
2. Cut frozen sections at 7.5 or 30 μ (light or electron microscopy).
3. Incubate tissues in the Holt medium above with an indoxyl acetate substrate concentration of 1 mM and hexazotized pararosanilin at 1 mM containing citrate buffer of 0.03 M at pH 6.

For electron microscopy, tissues may be rinsed in buffer, postfixed in osmium tetroxide for 1 h, dehydrated, and embedded in Epon.

Li et al. (*J. Histochem. Cytochem.,* **21:** 1, 1973) were able to distinguish nine bands of nonspecific esterase isozymes in a mixed human leukocyte preparation. Cytochemical and polyacrylamide gel electrophoresis methods were employed.

Vladutiu et al. (*J. Histochem. Cytochem.,* **21:** 559, 1973) used both immunofluorescence and immunoperoxidase methods to identify a primate-specific esterase in the perinuclear regions of human and monkey proximal convoluted tubule cells. Both methods were efficacious; however, the peroxidase-labeled preparations are permanent.

Sweetman and Ornstein (*J. Histochem. Cytochem.,* **22:** 327, 1974) employed cationic disk gel electrophoresis to separate several esterases from various types of human leukocytes. On the basis of data derived from the use of several specific substrates and inhibitors (including some new ones) Sweetman and Ornstein concluded that the major esterases in human neutrophilic granules are elastase-like esterases.

CHOLINESTERASE

Gomori (*Proc. Soc. Exp. Biol. Med.,* **68:** 354, 1948) introduced a technic for localization of *cholinesterase,* for which he preferred myristoylcholine as substrate. He prescribed 12- to 24-h fixation in acetone at 0°C, a 1- to 3-h bath at 0°C in equal parts of 100% alcohol and ether, a 12-h infiltration in 4% collodion in alcohol-ether mixture at 0°C followed by two changes of chloroform of 1 h each, and embedding in paraffin (preferably in vacuo for 15 min. Do not heat more than 2 h). Section at 5 to 10 μ.

This procedure should undoubtedly be replaced by one of the later esterase section preparation methods: freeze-dried paraffin, cold neutral formol-calcium 16 to 24 h, and frozen sections, cryostat, or Adamstone-Taylor cold knife sections attached to slides and, perhaps preferably, postfixed 15 min at 3°C in formol-calcium or acetone or by the section freeze substitution method of Chang and Hori.

1. Deparaffinize with petroleum ether and hydrate through acetone or by drying in air; wash formalin- or acetone-fixed sections in distilled water.
2. Incubate 2 to 16 h in solvent or in substrate. Since greater activity should be preserved by the newer fixation methods, the shorter interval should be adequate.
    Solvent:

0.1 M cobaltous acetate = 2.49% of the tetrahydrate	40 ml
0.1 M tris maleate buffer pH 7.6	60 ml
Distilled water	200 ml

Add 1 mg each of $CaCl_2$, $MgCl_2$, and $MnCl_2$.

Specific substrate: To 30 ml solvent add 0.6 ml 0.02 *M* myristoylcholine (0.663%) in distilled water.

Store the stock solvent and the myristoylcholine solutions separately in the refrigerator at 4°C, adding a crystal of camphor to each.

3. Wash 2 min in running water.
4. Immerse 15 min in 0.5% dilution of yellow ammonium sulfide solution or 0.5% sodium sulfide. (The cobalt soaps react rather slowly to form the sulfide.)
5. Wash 2 min in running water.
6. Counterstain 5 min in 0.2% safranin in 1% acetic, if desired.
7. Dehydrate in alcohols, clear in xylene, and mount in Canada balsam, ester gum, or β-pinene resin.

RESULTS: Sites of cholinesterase activity are shown as dark brown; nuclei, red; cytoplasm, etc., in various shades of pink. Addition of Prostigmine bromide (10 μM: 0.5 ml of a 30 mg/100 ml solution to a 50-ml Coplin jar) specifically inhibits the hydrolysis of the substrate.

Nachlas and Seligman (*J. Biol. Chem.*, **181**: 343, 1949) note that cholinesterase is inhibited by 10 μM (2.75 mg/l) of physostigmine (eserine) but that lipase and esterase are only partly inhibited by 3.5 mM (nearly 0.1%).

Gomori noted species differences in localization with myristoylcholine and the palmitic and lauric choline esters. Dog and mouse tissues reacted best with myristic and lauric esters; human and pigeon tissues, with palmitic. Hard, Peterson, and Fox (*J. Neuropathol. Exp. Neurol.*, **10**: 48, 1951) preferred myristoylcholine for their experimental studies on dogs.

Koelle and Friedenwald (*Proc. Soc. Exp. Biol. Med.*, **70**: 617, 1949) substituted acetylthiocholine as substrate in the cholinesterase technic introduced by Gomori (*Proc. Soc. Exp. Biol. Med.*, **68**: 354, 1948), because the choline fatty acid esters originally employed were found to be only very slowly hydrolyzed by brain and purified cholinesterases from erythrocytes and electric organ in eels, in comparison with acetylcholine. Acetylthiocholine was found to hydrolyze even more rapidly than acetylcholine. This substrate is hydrolyzed both by cholinesterase and by nonspecific esterase, and may be blocked by pretreatment of tissue with irreversible cholinesterase inhibitor diisopropyl fluorophosphate at 1 mM (184 mg/l).

KOELLE AND FRIEDENWALD'S TECHNIC FOR CHOLINESTERASE

REAGENTS. Buffer consists of 1 *N* glycine, 50 ml; 1 *N* NaOH, 18 ml; distilled water, 32 ml (pH 9.6). For the solvent, use 0.4 ml buffer, 0.1 *M* copper sulfate (2.5% $CuSO_4 \cdot 5H_2O$), 0.2 ml distilled water, 8.6 ml. Add a trace of copper thiocholine, dispersing thoroughly. Preheat to 37°C for at least 15 min.

SPECIFIC SUBSTRATE: Add 0.8 ml acetylthiocholine solution to 9.2 ml solvent, and filter and use at once.

ACETYLTHIOCHOLINE SOLUTION. Dissolve 14.5 mg acetylthiocholine iodide in 0.75 ml distilled water in centrifuge tube.

Add 0.25 ml 2.5% copper sulfate ($CuSO_4 \cdot 5H_2O$). Centrifuge out cupric iodide and decant.

COPPER THIOCHOLINE. Dissolve acetylthiocholine in copper glycinate solution, and adjust to pH 12 with KOH. Let stand overnight, collect precipitate, and wash free of alkali with distilled water.

TECHNIC

1. Cut frozen sections of fresh unfixed tissue, or make teased preparations of fresh muscle. As with the Gomori technic, later preparation methods may be used.

2. Place control preparations in 1 mM (0.0184%) diisopropyl fluorophosphate in 0.85% sodium chloride solution, and let stand for 30 min, to inactivate.
3. Wash in distilled water.
4. Place untreated preparations and inactivated controls in specific substrate, and incubate at 37°C for 10 to 60 min.
5. Rinse in distilled water saturated with copper thiocholine.
6. Transfer to a 1% dilution of ammonium sulfide solution [Koelle says $(NH_4)_2S$ or 0.5% Na_2S buffered to pH 7] to convert the deposited copper thiocholine to the dark brown, amorphous copper sulfide.
7. Wash in water, float onto slides, blot down.
8. Counterstain if desired, e.g., with 0.2% safranin in 1% acetic for 3 to 5 min, rinse, dehydrate with alcohols or acetone, clear in xylene, and mount in balsam, ester gum, or β-pinene resin.

RESULTS: Sites of enzymatic activity are shown by dark brown deposits of copper sulfide.

In place of the diisopropyl fluorophosphate inhibitor used by Koelle and Friedenwald, it should be possible to use the 10 μM Prostigmine bromide (mw 303) inhibitor of Gomori or the 10 μM physostigmine (mw 275) solution of Nachlas and Seligman. Table 10-11 provides details on a few esterase inhibitors.

Rohlich [*Nature* (*Lond.*), **178**: 1398, 1956] finds that acetylcholinesterase may be demonstrated quite well in muscle infiltrated for 2 to 3 h at 42°C in polyethylene glycol 1000 and sectioned at 5 μ, using a modified Koelle technic. The material must be sectioned promptly, as the activity deteriorates almost completely in storage for about 1 month in the embedded state.

The molecular weights for choline, acetylcholine, butyrylcholine, laurylcholine, and myristylcholine and acetylthiocholine and butyrylthiocholine and sample formulations are presented below:

$$HO{-}CH_2{-}CH_2{-}\overset{+}{N}(CH_3)_3{\cdot}\overset{-}{Cl}$$
Choline chloride

$$HS{-}CH_2{-}CH_2{-}\overset{+}{N}(CH_3)_3{\cdot}\overset{-}{I}$$
Thiocholine iodide

$$CH_3CO{-}O{-}CH_2{-}CH_2{-}\overset{+}{N}(CH_3)_3{\cdot}\overset{-}{Cl}$$
Acetylcholine chloride

$$CH_3CO{-}S{-}CH_2{-}CH_2{-}\overset{+}{N}(CH_3)_3{\cdot}\overset{-}{I}$$
Acetylthiocholine iodide

The butyryl ($CH_3{-}CH_2{-}CH_2{-}CO$), lauryl ($C_{11}H_{23}CO$), and myristyl ($C_{13}H_{27}CO$) groups are substituted for the acetyl (CH_3CO) group in the above cholines and thiocholines.

	mw
Choline chloride	139.632
Acetylcholine chloride	181.670
Butyrylcholine chloride	207.708
Laurylcholine chloride	319.924
Myristylcholine chloride	345.962
Thiocholine iodide	247.151
Acetylthiocholine iodide	289.189
Butyrylthiocholine iodide	315.227

In cases of poisoning by acetylcholinesterase inhibitors, Bergner (*Am. J. Pathol.*, **35**: 807, 1959) compares the grade of inhibition of the Koelle acetylthiocholine CuS method in teased intercostal muscle tissue with that obtained in the same muscle after 30-min immersion in 1 mM solutions of TMB-4 (mw 478.213) or 2-PAM (mw 264.08) (see Bergner's Reactivators, which follow).

TABLE 10-11 ESTERASE INHIBITORS, CHEMICAL NAMES, MOLECULAR WEIGHTS, AND EXAMPLES OF EFFECTIVE MOLARITIES USED HISTOCHEMICALLY

Inhibitor	mw	Acetylcholinesterase	Aliesterase not qualified	Esterases A	Esterases B	Lipase
Physostigmine (eserine) {alkaloid	275.36	$10\ \mu M^{a,b}$	$2.5\ mM^a$	...	$2.5\ mM^a$	$3.5\ mM^a$
{salicylate	413.48					
Prostigmine bromide	303.22	$10\ \mu M^c$		...		
Diisopropyl fluorophosphate (DFP)	184.15	$1\ mM^d;\ 10\ \mu M^b$	$0.5\ mM^e$	...	$0.5\ mM^f$	
Arsanilic acid	217.04	......	$0.1\ M^a$	...	$0.1\ M^f$	
Sodium fluoride (NaF)	41.99	$1\text{--}5\ mM^b$	$2\ mM^e;\ 70\ mM^a$	...	$70\ mM^f$	
Sodium taurocholate	537.68	......	$0.2\%^a$	...	......	$0.1\text{--}0.2\%^{a,c}$ activates
Quinine hydrochloride	378.91		Less effect than on lipase[a]	...	......	$50\ mM^a$
Diethyl-p-nitrophenylphosphate (E-600)	275.20	$0.1\text{--}0.001\ \mu M^b$	......	...	$10\ \mu M^f$	
Phemerol·H_2O	466.11	$0.5\ mM^e$				

Note: A esterase is generally described as resisting those inhibitor levels which inhibit B esterase. Precise levels for its inhibition are not cited.

[a] M. M. Nachlas and A. M. Seligman, J. Biol. Chem., **181**: 392, 1949.
[b] M. Wachstein, E. Meisel, and C. Falcon, J. Histochem. Cytochem., **9**: 325, 1961.
[c] G. Gomori, Proc. Soc. Exp. Biol. Med., **68**: 354, 1948.
[d] G. B. Koelle and J. S. Friedenwald, Proc. Soc. Exp. Biol. Med., **70**: 617, 1949.
[e] M. S. Burstone, J. Histochem. Cytochem., **4**: 130, 1955.
[f] T. K. Shnitka and A. M. Seligman, J. Histochem. Cytochem., **9**: 504, 1961.

Bergner's rat experiments included poisonings by the insecticides Diazinon and Parathion, by TEPP and sarin, and by diisopropyl fluorophosphate (DFP) and tabun. With the first two, considerable postmortem reactivation occurred when the bodies lay at room temperature for 24 h; with the last two there was little or no postmortem spontaneous reactivation.

2-PAM was quite an effective reactivator after Diazinon, Parathion, and sarin, but it was weak after DFP and tabun. TMB-4 gave strong reactivation after DFP, tabun, and sarin.

Although lack of acetylcholinesterase activity of motor end plates in cadavers is not regarded as conclusive evidence of inactivator poisoning, Bergner apparently regards the regeneration of such activity by 1 mM 2-PAM or TMB-4 as important evidence for the existence of such poisoning. However, prolonged postmortem delays before autopsy do permit considerable regeneration of acetylcholinesterase activity after some poisons of this type, but not after all of them. See "The Merck Index" for descriptions of these compounds, except TMB-4 [1,1'-trimethylene-bis(4-formylpyridinium bromide)dioxime].

BERGNER'S CHOLINESTERASE INACTIVATORS

O,O-Diethyl-*O*-(2-isopropyl-4-methylpyrimidyl-6)thiophosphate
(Diazinon, mw 504.36)

O,O-Diethyl-*O*-*p*-nitrophenyl thiophosphate
(Parathion, mw 297.27)

Tetraethyl pyrophosphate
(TEPP, mw 290.198)

Diisopropyl fluorophosphate
(DFP, mw 184.153)

Methylisopropyl fluorophosphate
(Sarin, mw 125.124)

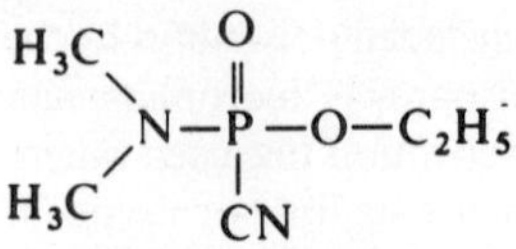

$$\underset{H_3C}{\overset{H_3C}{>}}N-\underset{\underset{CN}{|}}{\overset{\overset{O}{\|}}{P}}-O-C_2H_5^{\cdot}$$

O-Ethyl-*N*,*N*-dimethylcyanophosphamide
(Tabun, mw 162.134)

BERGNER'S REACTIVATORS

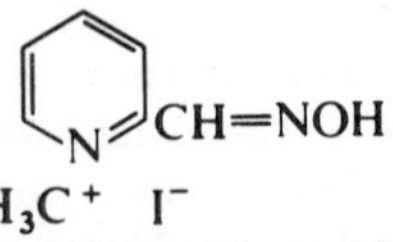

CH=NOH on pyridinium ring

$$H_3C^+ \quad I^-$$

Pyridinium 2-aldoxime methiodide
(2-PAM, mw 264.075)

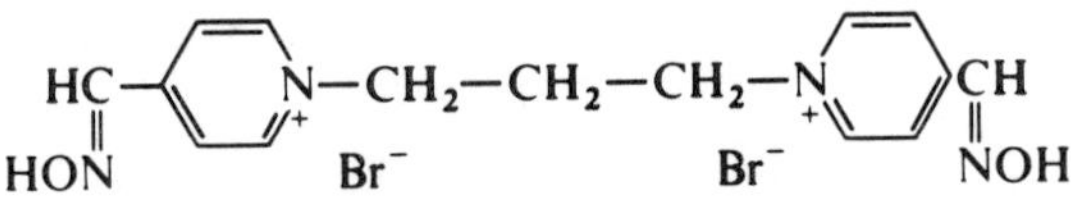

$$HC-\text{〈ring〉}-N-CH_2-CH_2-CH_2-N-\text{〈ring〉}-CH$$
$$\underset{HON}{\|} \qquad Br^- \qquad\qquad Br^- \quad \underset{NOH}{\|}$$

1,1′-Trimethylene-bis(4-formylpyridinium bromide)dioxime
(TMB-4, mw 478.213)

Karnovsky and Roots (*J. Histochem. Cytochem.*, **12**: 219, 1964) recommend the following method for the demonstration of *cholinesterase*. The basis of the method is that ferricyanide is reduced by thiocholine to ferrocyanide which combines with Cu^{2+} ions to form insoluble copper ferrocyanide (Hatchett's Brown). The Cu^{2+} ions in the medium are complexed with citrate to prevent formation of insoluble yellow-green copper ferricyanide.

1. Fix blocks of tissues overnight in 10% formalin containing 1% $CaCl_2$ at 4°C. Tissues may then be stored in Holt's gum sucrose (0.88 M sucrose in 1% gum arabic).
2. Wash blocks in distilled water 5 min, and blot dry.
3. Cut sections in a cryostat and mount on gelatinized slides.
4. Incubate tissues for 15 min to 2 h at 23°C in the following medium: Dissolve 5 mg acetyl- or butyrylthiocholine iodide in 6.5 ml 0.1 M sodium hydrogen maleate buffer of pH 6. Add in order, with stirring, 0.5 ml 0.1 M (2.94%) sodium citrate, 1 ml 30 mM (0.75%) $CuSO_4$, 1 ml distilled water, and 1 ml 5 mM (0.164%) potassium ferricyanide. The clear, greenish incubation solution is stable for hours. One milliliter 27.5% physostigmine sulfate (1 M) may be used as an inhibitor to replace 1 ml distilled water. At pH levels lower than 6, it is necessary to increase the final concentration of citrate to maintain a stable medium, namely, pH 6 (5 mM), pH 5.5 (10 mM), pH 5 (15 mM), and pH 4.5 (20 mM). Karnovsky and Roots recommend incubation of tissues at pH 6 ($Na_3C_6H_5O_7 \cdot 2H_2O$, mw 294.10).

Karnovsky and Roots noted that at pH 6, increasing the concentration of ferricyanide promotes diffuse and nonspecific (e.g., nuclear) staining. SH groups in tissues are not stained.

Bloom and Barrnett (*Ann. N.Y. Acad. Sci.*, **144**: 626, 1967) employed the Karnovsky-Roots method for electron microscopic investigations. Sections cut at 50 μ were incubated at 4 or 18°C for 10 to 90 min. Tissues were then washed in distilled water for 30 to 60 min and embedded in Maraglas. The sections may be postfixed in Veronal-buffered osmium tetroxide.

Douglas [*Acta Histochem. (Jena)*, **24**: 307, 1966] reported his efforts at quantitation of the copper thiocholine method for the histochemical detection of cholinesterase. The method involved constant observation of a slide through a microscope during incubation. A stopwatch was employed to record the time required to form the first perceptible stain.

Koelle et al. (*Ann. N.Y. Acad. Sci.*, **144:** 613, 1967) believe the gold-thiocholine and gold-thioacetic acid methods of Koelle and Gromadzki (*J. Histochem. Cytochem.*, **14:** 443, 1966) are superior for electron microscopic localization. The chief advantage achieved by substitution of gold for copper or lead as capture reagents is that the precipitates formed have fine colloidal dimensions with high electron density, thereby improving localization. Selective inhibition of acetylcholinesterase was achieved with 10 to 30 μM BW 284 [1,5-bis(4-allyl-dimethylammonium phenyl)pentan-3,1-dibromide] and of cholinesterase with 0.01 to 0.1 μM diisopropyl phosphorofluoridate (DFP), mw 184.15, or with 0.03 μM NV-683 (the dimethyl carbonate of 2-hydroxy-5-phenylbenzyltrimethylammonium bromide). Both enzymes were inhibited by 10 μM DFP or physostigmine. The gold-thiocholine method achieves high specificity, whereas the gold–thioacetic acid method is less specific but provides finer localization. The method follows. The following stock solutions were employed:

1. $AuNa_3(S_2O_3)_2 \cdot 2H_2O$ (0.1 M), 100 mg, was dissolved in 1.9 ml H_2O [obtained as 100 mg ampuls (Sanocrysin) from Ferrosan, Medizinalfabrik, Copenhagen].
2. Acetylthiocholine or butyrylthiocholine, 0.05 M, 46 mg acetylthiocholine iodide or 50 mg butyrylthiocholine iodide, is dissolved in 2.05 ml H_2O, and 1.15 ml 0.1 M $AgNO_3$ is added. The solution is shaken to promote aggregation and precipitation of AgI.
3. Physostigmine salicylate and BW 284 are prepared as 1 mM stock solutions and stored frozen until used.

The histochemical procedure for the gold-thiocholine method for light microscopy is as follows:

1. Fix tissue slices in 10 % formalin in 0.028 M maleate buffer of pH 7.4 with 7.5 % sucrose at 4°C for 2 h.
2. Rinse tissues in the cold in the above buffer for 30 min to 2 days.
3. Cut sections in a cryostat at 10 to 20 μ, and allow them to dry on slides.
4. Place sections in a preincubation solution of the following composition for 15 min at 23°C:

$AuNa_3(S_2O_3)_2 \cdot 2H_2O$	0.4 ml
Distilled water	2.9 ml
$NaH_2PO_4 \cdot H_2O$ (6 M)	4.6 ml
K_2HPO_4 (6 M)	2.1 ml

Adjust the final pH to 5.6.
5. Place sections in the incubation mixture for 1.5 to 180 min at 23°C. Prepare the mixture and allow it to stand until it is visibly colloidal (20 to 40 min), at which time filter it through a Whatman No. 3 filter paper immediately prior to use:

$AuNa_3(S_2O_3)_2 \cdot 2H_2O$	0.4 ml
Distilled water	2.1 ml
$Na_2H_2PO_4 \cdot H_2O$ (6 M) 83 %	4.6 ml
K_2HPO_4 (6 M) 104.5 %	2.1 ml

6. Rinse sections in the following solution for 10 min:

Distilled water	3.3 ml
$NaH_2PO_4 \cdot H_2O$ (6 M)	4.6 ml
K_2HPO_4 (6 M)	2.1 ml

7. Immerse sections in acid–alcoholic ammonium sulfide for 10 min at 23°C [3.6 ml $(NH_4)_2S$, 20 ml ethanol, 6 ml glacial acetic acid, enough alcohol to make 30 ml, and 1 drop 0.1 M $AuNa_3(S_2O_3)_2$]. Mix the solution, adjust it to pH 5.5, and filter it with a Whatman No. 3 filter paper.
8. Dehydrate, and mount in synthetic medium.

The histochemical procedure for the gold–thioacetic acid method for light microscopy is as follows:

1. Fix, rinse, and cut sections as with the gold-thiocholine method.
2. Place sections in a preincubation mixture for 15 min at 23°C:

$AuNa_3(S_2O_3)_2 \cdot 2H_2O$	0.3 ml
Distilled water	8.2 ml
$MgCl_2 \cdot 6H_2O$ (1 M) 20.33%	0.32 ml
Maleate buffer (0.703 M)	0.8 ml
Thioacetic acid (0.2 M) 1.522% (0.288 ml dissolved in 3.6 ml 1 N NaOH and diluted with cold freshly boiled distilled water to make 20 ml)	0.3 ml
HCl (1 N)	0.12 ml

Adjust the final pH to 6.2 and filter before use.
3. Immerse sections in the incubation mixture for 1.5 to 180 min at 23°C:

$AuNa_3(S_2O_3) \cdot 2H_2O$	0.3 ml
Distilled water	6.5 ml
$MgCl_2 \cdot 6H_2O$ (1 M) 20.33%	0.32 ml
Maleate buffer (0.703 M)	0.8 ml
Thioacetic acid (as above)	2.0 ml
HCl (1 N)	0.12 ml

Adjust the final pH to 6.2 and filter before use.
4. Transfer slides to 4% formaldehyde in 0.85% NaCl for 10 min at 23°C.
5. Dehydrate, and mount in synthetic medium.

If inhibitors are employed, their inclusion is within the distilled water allotment in preincubation and incubation mixtures. Koelle et al. showed excellent localization in motor end plates and in cervical and ciliary ganglion cells.

Koelle et al. (*J. Histochem. Cytochem.*, **22**: 252, 1974) refined the bis(thioacetoxy)aurate method for electron microscopic localization of acetylcholinesterase and nonspecific cholinesterase previously described by Austin and Berry (*Biochem. J.*, **54**: 695, 1953) and Koelle et al. (*J. Histochem. Cytochem.*, **16**: 754, 1968). The refinements consist of the following: (1) The use of a special formalin–Krebs-Ringer-calcium solution composed of 5 ml formaldehyde plus 50 ml of a solution (0.9% NaCl, 100 ml; 1.15% KCl, 4 ml; 1.22% $CaCl_2$, 1.27 ml; 2.11% KH_2PO_4, 1 ml; 3.13% $MgCl_2 \cdot 6H_2O$, 1 ml; 0.1 M Na_2HPO_4, 40 ml; 1 N HCl, 2 ml; 21 ml distilled water) for perfusion with the fixative at 5 to 10°C, and fixation thereafter of small slices for 3 to 6 h in the same fixative at 4°C, is recommended. Fixed tissues kept at 4°C overnight in the buffered fixative are also satisfactory. For electron microscopy, formalin-fixed small slices embedded in agar are also recommended. (2) To enhance specificity of the staining reaction, immersion of sections in 10^{-3} M iso-OMPA (tetramono-isopropyl pyrophosphortetramide) in the formalin fixative solution for 60 min at 5°C is recommended. Sections are then rinsed in the formalin fixative for 10 min at 5°C, rinsed in two changes (5 min each) of distilled water, air-dried, stored in the refrigerator overnight, and stained the following day. Complete inhibition of cholinesterase activity and no detectable influence of specific acetylcholinesterase activity was observed in iso-OMPA–treated sections as prescribed.

Effective blockage of cholinesterase could also be achieved (except in the central nervous system) in cats injected intravenously with 6 ml 10^{-3} M iso-OMPA in 0.9 % NaCl solution per kg 15 min prior to infusion of the fixative. No detectable loss of acetylcholinesterase activity was observed. Failure of inhibition of cholinesterase activity in the central nervous system is due to the blood-brain barrier.

Koelle et al. had far greater difficulty in selective inhibition of specific cholinesterase. Complete inhibition could be achieved using the reversible inhibitors ambenonium or 1,5-bis-(4-allyldimethylammoniumphenyl)pentan-3,1-dibromide (BW-284C51) with 10-μ sections and thiocholine esters as substrates. However, the usefulness of these inhibitors with the bis(thioacetoxy)aurate[Au(TA)$_2$] method was found to be seriously limited due to their predominant action at the anionic site of acetylcholinesterase (Wilson, *Biochim. Biophys. Acta,* **7:** 520, 1951) and their limited penetrability into cellular and subcellular membranes (Koelle, *J. Pharmacol. Exp. Ther.,* **120:** 488, 1957). Koelle et al. were able to circumvent these problems partially by the use of thin sections of certain tissues such as motor end plates for light or electron microscopy by initial exposure of the tissue blocks to 10^{-3} M ambenonium (unspecified time) and inclusion of 5×10^{-7} M ambenonium in the Au(TA)$_2$ incubation medium.

For cat tissues to be subsequently studied by electron microscopy, Koelle et al. recommend giving the anesthetized artificially respired animal an IV injection of a reversible selective inhibitor of cholinesterase, namely, 10-α-diethylaminopropionyl phenothiazine HCl (Astra 1397) at 100 to 200 μmol/kg followed by an irreversible inhibitor of both acetylcholinesterase and cholinesterase, namely, 2-diethoxyphosphinylthioethyldimethylamine acid oxalate (217 AO) at 0.5 to 1 μmol/kg. Tissues from sacrificed animals thereafter revealed no detectable acetylcholinesterase activity, but some cholinesterase activity was regained.

Koelle et al. provide the amounts and concentrations of reagents in incubation, as well as pre- and postincubation solutions, in Table 10-12. The incubation media are prepared

TABLE 10-12 AMOUNTS AND CONCENTRATIONS OF REAGENTS EMPLOYED PER 10 ml INCUBATION AND PRE-POST SOLUTIONS

Reagents*	Final concentration (M or N)	Solution†	
		Incubation, ml	Pre-post, ml
Au$^+$, 0.1 M	0.005	0.55	
H$_2$O		1.75	3.20
MgCl$_2$, M	0.03	0.30	0.30
HCl, N		0.30	0.40
NaHMal, 1.15 M	0.70	6.10	6.10
I$_2$, 0.200 N		0.50	
TA, 0.200 N	0.01	0.50	

* Au$^+$, Na$_3$Au(S$_2$O$_3$)$_2$·2H$_2$O, final concentration indicated based on amount of Au$^+$ released by I$_2$ in presence of 10 % excess of initial reagent; NaHMal, sodium hydrogen maleate buffer (see text); I$_2$, Lugol's solution (see text); TA, thioacetic acid, triple-distilled from thioacetic acid, practical (Eastman), volume added adjusted according to actual concentration of solution as determined by iodometric titration immediately before use.

† pH of both solutions adjusted to 7 to 7.05 with 1 N HCl or 1 N NaOH.

Source: Koelle et al., *J. Histochem. Cytochem.,* **22:** 252, 1974.

by sequential addition of the following reagents per 10 ml of solution, namely, 0.55 ml 0.1 M $Na_3Au(S_2O_3)_2$ (as Sanocrysin, obtainable from Ferrosan A/S, Copenhagen, Denmark, in 100-mg ampules); 1.75 ml distilled water, 0.30 ml 1 M $MgCl_2$; 0.30 ml 1 N HCl; and 6.10 ml 1.15 M maleate buffer.

At this point, the aliquot to be used for the first hour of incubation is placed in an ice-water bath, and those for hourly replacements are stored in the refrigerator. The thioacetic acid (TA) is titrated in duplicate by adding 0.8 ml and 3 ml cold (4°C) distilled water to a 25-ml chilled Erlenmeyer flask using chilled pipettes. Lugol's solution (0.2 N I_2) is added until it is no longer decolorized (approximately 0.85 ml); the normality of the thioacetic acid solution and the adjustment of the 0.5 ml aliquot to be added are now calculated. Quickly added now are 0.5 ml 0.2 N I_2 and 0.5 ml 0.2 N thioacetic acid (with the volume adjustment, if required). If a black or purple precipitate occurs, the thioacetic acid has not been added sufficiently rapidly to complex with the Au^+ released by I_2, and the solution should be discarded. The pH of properly prepared solutions is then adjusted to 7 to 7.05 with 1 N NaOH or 1 N HCl, and filtered through a Whatman No. 1 filter at 2 to 4°C.

For light microscopy, slides are immersed in the pre-post medium for 30 min at 5°C, transferred to sequential hourly changes of the incubation fluid at 5°C for the desired time, transferred to the pre-post fluid for 30 min at 5°C, rinsed in water, dehydrated, and mounted in synthetic medium.

Immersed slices of tissue or Smith-Farquhar cut sections for electron microscopy are treated similarly except they are kept in the pre-post medium for 1 h before and 1 h after the incubation medium. Thereafter, they are fixed in the formalin fixative, treated with 1% OsO_4 for 2 h at 5°C, dehydrated, and embedded in Epon.

On the basis of the action of selective inhibitors, Koelle et al. ascribe staining of stellate ganglia as due principally to cholinesterase, and staining of motor end plates is due principally to acetylcholinesterase. Koelle et al. also noted staining of lysosomes that was not prevented by inhibitors, whereupon they conclude the staining is due neither to acetylcholinesterase nor to cholinesterase.

Haugaard et al. (*J. Histochem. Cytochem.*, **13**: 566, 1965) showed that aurous ions form a coordinate complex with thioacetic acid containing one metal ion and two thioacetate residues.

Barnard and Rogers (*Ann. N.Y. Acad. Sci.*, **144**: 584, 1967) noted that under suitable conditions ^{3}H-DFP (diisopropyl fluorophosphate) or ^{32}P-DFP is reacted 1 : 1 with a cholinesterase molecule. Nonreacted reagent is fully removed by exchanges and washes. Subsequent autoradiographs disclose localization at end plates. Acetylcholinesterase specifically labeled by DFP (with specificity tested by use of the inhibitor 284C51 from Wellcome Laboratories, and by removal of the DP groups by pyridine-2-aldoxime methiodide) was shown to account for approximately 35% of total enzyme reactivity at mouse skeletal junctions. Approximately 10% of reactivity is due to pseudocholinesterase, and the remainder is nonspecific esterase. The specificity of pseudocholinesterase was tested by reactivity with ethopropazine (mw 312.46) [Lysivane, Parsidol, or 10-(2-diethylamino-1-propyl)phenothiazine hydrochloride] (Warner-Chilcott Labs, Inc.) at 30 μM (937 μg/100 ml). Ethopropazine is a highly selective inhibitor of nonspecific cholinesterase with no inhibition of acetylcholinesterase (Hopsu and Pontinen, *J. Histochem. Cytochem.*, **12**: 853, 1964). Nonspecific esterase selectivity was tested by being unaffected by the above reagents and by physostigmine at 10 μM. Barnard and Rogers calculated that the number of acetylcholinesterase molecules at each end plate varied from 0.4×10^7 to 3×10^7 depending upon the type.

Quantitative electron microscopic autoradiography was employed to disclose that acetylcholinesterase was concentrated over junctional folds in sternocleidomastoid muscles. Very little was detected in axons or in muscle regions. Barnard and Rogers suggest that acetyl-cholinesterase activity is essential for the termination of acetylcholine activity within the folds of the postsynaptic membrane, and that a great surplus of enzyme does not exist within the junction.

Lewis and Shute (*J. Cell Sci.*, **1**: 381, 1966) employed a thiocholine technic in an electron microscopic investigation of the distribution of enzyme activity in cholinergic neurons of the rat.

Namba et al. (*Am. J. Clin. Pathol.*, **47**: 74, 1967) offer the following method for the simultaneous demonstration of *nerve fibers* and *cholinesterase*:

1. Fresh tissues are frozen, sectioned at 30 to 50 μ in a cryostat, and mounted on slides or coverslips.
2. Fix sections for 1 h at 4°C in a formol-calcium solution containing magnesium and cadmium (10 ml formalin, 1 g $CaCl_2$, 0.5 g $MgCl_2 \cdot 6H_2O$, 0.1 g $CdCl_2 \cdot 2\frac{1}{2}H_2O$, and 0.07 M Veronal-acetate buffer of pH 6.45 to make 100 ml).
3. Rinse in distilled water at least 1 h at 4°C.
4. Incubate sections for 60 min at 37°C or at 23°C in the Koelle-Friedenwald medium detailed earlier.
5. Rinse with distilled water.
6. Transfer to 1% solution of ammonium sulfide for 5 min.
7. Rinse in distilled water.
8. Immerse in 0.25% ferricyanide solution for 10 min at 23°C.
9. Rinse in distilled water 5 min 2 times.
10. Place in absolute ethanol 1 h at 23°C.
11. Rinse in distilled water 1 min.
12. Place in a silver nitrate solution for 90 min at 37°C (10 g $AgNO_3$, 0.05 g $CuSO_4 \cdot 5H_2O$, 1 g $CaCO_3$, in 100 ml distilled water).
13. Rinse 5 s in distilled water.
14. Immerse sections in a reducing solution for 10 min at 23°C (1 g hydroquinone, 10 g Na_2SO_3 in 100 ml distilled water).
15. Wash in distilled water twice for 5 min each time.
16. Dehydrate, and mount in synthetic medium.

RESULTS: Sensory, autonomic, and motor nerve axons are stained black. Areas of cholinesterase activity are also stained black. Selective staining was demonstrated in photo-micrographs.

Friedenberg and Seligman (*J. Histochem. Cytochem.*, **20**: 771, 1972) reviewed problems related to specificity, localization, penetration, determination of bound and unbound enzymes, dangers of nonenzymatic reactions, and contrast for electron microscopic observations for several methods utilized for the histochemical and cytochemical detection of acetylcholin-esterase and related enzymes.

Gauguin et al. (*Histochemie*, **34**: 97, 1973) employed the Lewis and Shute method (*J. Cell Sci.*, **1**: 381, 1966) using acetylthiocholine to demonstrate cholinesterase in parafollicular (C cells) of the rat thyroid gland at the light and electron microscopic levels.

CHOLINE ACETYLTRANSFERASE

Burt (*J. Histochem. Cytochem.*, **18**: 408, 1970) proposes the following method for the histo-chemical localization of *choline acetyltransferase* activity. The procedure is based upon the formation of an insoluble lead mercaptide from coenzyme A. Choline acetyltransferase catalyzes the synthesis of acetylcholine from acetyl coenzyme A and choline.

1. Fresh frozen sections of tissues are sectioned at 8 μ and fixed for 5 min in a fixative containing 25 mM HEPES buffer (N-2-hydroxyethylpiperazine-N-2-ethanesulfonic acid) of pH 7, 10% sucrose, and 2% formaldehyde.
2. Wash sections for 15 min with the buffered sucrose solution (omitting the formaldehyde).
3. Incubate in the following medium up to 3 h:

   25 mM HEPES buffer pH 6
   4 mM choline (48.5 mg/100 ml)
   0.2 mM acetyl coenzyme A (16.2 mg/100 ml)
   1.8 mM Pb(NO$_3$)$_2$ (59.6 mg/100 ml)
   0.1 mM 3.83 mg/100 ml phospholine iodide (echothiophate iodide)

4. Wash in three changes of distilled water.
5. Immerse in 2% ammonium sulfide for 2 min.
6. Wash in distilled water.
7. Mount in a water-type mounting medium such as Kaiser's.

Burt noted heavy staining in perikaryon, in nuclei, and on cell surfaces in the form of granules. Less intense staining was noted in neuropil. Sections incubated with Cu^{2+} (an inhibitor of choline acetyltransferase) resulted in heavy staining only in nuclei with light staining of perikaryon. Burt is convinced that the small stained granules on the surface of neurons and motor horn cells represent choline acetyltransferase activity. The granules on the surface of motor horn cells are interpreted as boutons terminaux, which are known to contain choline acetyltransferase activity.

With respect to the specificity of the above method it must be remembered that any enzyme that will hydrolyze acetyl coenzyme A will give a positive reaction. Also, choline acetyltransferase in formalin-fixed tissues under the histochemical conditions specified can catalyze the hydrolysis of acetyl coenzyme A in the absence of the natural acyl acceptor, choline. Burt noted the distribution of staining in tissues did not vary irrespective of the presence or omission of choline in the incubation medium.

Papadimitriou and Van Duijn (*J. Cell Biol.*, **47:** 71, 1970) incubated a polyacrylamide film (containing aspartate aminotransferase in either a mitochondrial or soluble cytoplasmic fraction of liver cells) in a medium containing α-ketoglutarate, L-aspartate, and lead nitrate. The lead oxalacetate formed was later converted to lead sulfide, and absorbance at 520 mμ was determined. Absorbance was determined to be directly proportioned to the concentration of chemically determined oxalacetate in the film.

PROTEASES

Mast Cell Proteases: Chymotrypsin-like Enzyme

Gomori (*J. Histochem. Cytochem.*, **1:** 469, 1953) described in mast cells an enzyme which was quite active against chloroacetyl esters of α-naphthol and naphthol AS. His technic, as presented in Table 10-13, has had to be put together from the above paper and that of Gomori and Chessick (*J. Cell Physiol.*, **41:** 51, 1953) on the use of α-naphthyl and naphthol AS acetates as esterase substrates, to which paper we are indebted for the total volume of the substrate solution and the identity and amount of the diazonium salt used. Gomori noted that the mast cell enzyme resisted 10 μM eserine.

Later Benditt (*Fed. Proc.*, vol. 13, no. 1, abstr. 1646, 1956) showed that naphthol AS chloroacetate was cleaved both by crystalline chymotrypsin and by extracts of isolated rat mast cells. These two in vitro reactions and the histochemical reactions were inhibited by 0.1 mM diisopropyl fluorophosphate. Later papers by Benditt (*Ann. N.Y. Acad. Sci.*, **73:** 204,

TABLE 10-13 TECHNICS AND SUBSTRATES FOR MAST CELL PROTEASE- ESTERASE

	Gomori	Gomori	Benditt and Arase, 1958	Lagunoff and Benditt, 1961	Moloney et al., 1960
Fixation	Formalin	Formalin	Formalin	Cold formalin acacia-sucrose	Methanol, 30 s (smears, imprints)
Section	Paraffin	Paraffin	Paraffin	Frozen sections defatted with methanol	
Substrate	α-Naphthyl chloroacetate	Naphthol AS chloroacetate	Naphthol AS chloroacetate	Naphthol AS phenyl-propionate	Naphthol AS-D chloroacetate
mg = μM	5 mg = 0.9 mM	5 mg = 0.6 mM	0.42 mg (= 50 μM)	0.4 mg (= 40 μM)	25 mg (= 2.9 mM)
Special solvent	Acetone, 0.5 ml	Acetone, 0.5 ml Propylene glycol, 7.5 ml	Acetone, 1.25 ml Propylene glycol, 7.5 ml	Methanol, 10 ml	Acetone, 1.25 ml
Buffer	0.2 M phosphate	0.2 M phosphate	0.1 M PO₄	0.1 M tris	0.1 M Veronal
Volume	2.5 ml	2.5 ml	5 ml	5 ml	12 ml
pH	6.4	6.4	6.4	8	7.4
Water to make final volume	25 ml	25 ml	25 ml	25 ml	25 ml
Diazonium salt, fast garnet GBC	20–50 mg	10–50 mg	10 mg	10 mg	25 mg
Incubation time	5–20 min	5–20 min	5–20 min	15–30 min	30 min
Temperature	20–25°C	20–25°C	20–25°C	20–25°C	20–25°C

Sources: Gomori, *J. Histochem. Cytochem.* **1**: 469, 1953; Gomori and Chessick, *J. Cell Physiol.*, **41**: 51, 1953; Benditt and Arase, *J. Histochem. Cytochem.* **6**: 431, 1958; Lagunoff and Benditt, *Nature (Lond.)*, **192**: 1198, 1961; Moloney et al., *J. Histochem. Cytochem.*, **8**: 200, 1960.

1958) and Benditt and Arase (*J. Histochem. Cytochem.*, **6**: 431, 1958; *J. Exp. Med.*, **110**: 451, 1959) identified further substrates attacked by α-chymotrypsin and isolated (as well as *in situ*) mast cell enzyme, such as *N*-acetyl-L-tryptophan, *N*-acetyl-L-tyrosine, and *N*-acetyl-L-phenylalanine ethyl esters. *p*-Toluenesulfonylarginine methyl ester is cleaved by trypsin but not by chymotrypsin or the mast cell enzyme. Resistance to formaldehyde by chymotrypsin as well as by the mast cell enzyme was noted.

Lagunoff and Benditt [*Nature (Lond.)*, **192**: 1198, 1961], in kinetic studies of hydrolysis of benzoyl, phenylacetyl, phenylpropionyl, and phenylbutyryl naphthol AS esters, using the 310 mμ excitation–515 mμ emission fluorescence of released naphthol AS, reported that hydrolysis was rapid for phenyl propionate, slow for phenyl butyrate, and negligible for phenyl acetate and benzoate. Histochemical studies using simultaneous fast garnet GBC coupling agreed.

Moloney et al. (*J. Histochem. Cytochem.*, **8**: 200, 1960) seemed unaware of Benditt's studies and referred to the enzyme as an esterase, again stressing its presence in neutrophil leukocytes and myelocytes, as well as in mast cells. Superior results were claimed for the substitution of naphthol AS-D chloroacetate (mw 353.813) for Gomori's original preparation.

Moloney and coworkers studied more inhibitors, in reference particularly to neutrophil leukocyte and myelocyte activity, and reported for physostigmine strong inhibition at 1 mM, none at 100 μM; for taurocholate, complete at 10 mM; for arsanilate, complete at 10 mM, none at 1 mM; for diisopropyl fluorophosphate, complete at 10 mM; for NaF, partial at 3 mM; for CuSO$_4$, complete at 1 mM, with no inhibition by 0.1 M Lugol's solution, 3 mM ZnSO$_4$, 0.1% Zephiran, or 3% Versene.

As noted above, none of these substrates was published in complete detail. Some of the data in Gomori's substrates are from Gomori and Chessick (*J. Cell Physiol.*, **41**: 51, 1953); the molarity of the phosphate buffer in Benditt and Arase is set to agree with the estimated concentration of Gomori's. The amount of diazo in Lagunoff and Benditt is supplied from Benditt and Arase, and the amount of tris buffer agrees with that in Burstone's practice, who also worked with Gomori.

Mast Cell Proteases: Trypsin-like Enzyme

Glenner et al. (*J. Histochem. Cytochem.*, **10**: 109, 1962) have noted the presence in the mast cells of dogs and human beings (but not of rabbits, guinea pigs, rats, or mice) of an enzyme hydrolyzing α-*N*-benzoyl-DL-arginyl-β-naphthylamide and certain naphthol AS esters of ε-amino-caproic acid. This enzyme is activated by 0.01 M tetra-*n*-butylammonium iodide, inhibited by 0.1% heparin and tosyl and benzoyl L-arginine methyl esters, in distinction from the mast cell protease of Benditt and Arase, found in all six species noted above and demonstrated by its hydrolysis of naphthol AS chloroacetate.

SUBSTRATE   [*Nature (Lond.)*, **185**: 846, 1960]

α-*N*-Benzoyl-DL-arginine-β-naphthylamide	30 mg
Fast Corinth V	10 mg
Tris maleate buffer 50 mM, pH 7	20 ml

Fast Corinth V may be replaced by fast garnet GBC, but fast blue B completely inhibits the reaction.

The above substrate is rapidly digested by bovine trypsin, but it resists bovine chymotrypsin for 18 h.

Fresh frozen sections directly or after Glenner's acetone chloroform freeze substitution procedure are brought into the above substrate for 20 to 30 min, drained and transferred to 1% copper sulfate (0.04 M) for 10 min, washed, and mounted in glycerol gelatin containing a few drops of 1% copper sulfate.

INHIBITORS. These are 0.1 mM diisopropyl fluorophosphate, 2 mM CuSO$_4$, 2 mM Pb(NO$_3$)$_2$. The enzyme is not inhibited by soybean trypsin inhibitor (1 mg/ml), 1 mM iodoacetate, or 0.5 and 10 mM KCN [mw CuSO$_4 \cdot$5H$_2$O 249.68, DFP 184.15, Pb(NO$_3$)$_2$ 331.20, CH$_2$ICOOH 185.95, KCN 65.12].

γ-GLUTAMYL TRANSPEPTIDASE

Glenner et al. (*J. Histochem. Cytochem.*, **10:** 481, 1962) reported a method for the histochemical localization of *γ-glutamyl transpeptidase*.

Attached fresh frozen sections are used according to Glenner's method. Incubate in the following substrate mixture for 20 min:

N-(γ-L-glutamyl)β-naphthylamide	3 mg in 0.5 ml acetone
Fast garnet GBC	15 mg in 1 ml distilled water
0.1 *M* tris maleate buffer pH 7.2	10 ml
Distilled water	28.5 ml

Sites of activity are shown in red: bile duct epithelium in liver, human pancreatic duct epithelium, pancreatic acini in rat and guinea pig, brush borders of P-2 and D-1 segments of renal tubules in guinea pigs and in P-1 and P-2 segments in rats, testicular germinal epithelium, epithelial border zone in human epididymis and seminal vesicle, human endometrial gland epithelium, rat and guinea pig ova and granulosa cells, some duct cells in salivary glands (rodent), tracheobronchial epithelium (rodent), hypophyseal beta cells (?), thyroid follicle epithelium.

INHIBITORS. There is severe inhibition by Cu^{2+}, Zn^{2+}, Hg^{2+}, and Pb^{2+} at 1 mM, and by 5 mM bromsulphalein and 1 mM bromocresol green.

Without effect were Na$_2$HPO$_4$, MgCl$_2$, MnCl$_2$, CaCl$_2$, KCN, and NaF at 1 mM; diisopropyl fluorophosphate, 0.1 mM (mw 184.15); Versene, 10 mM (mw 336.21 as the disodium salt); sodium pyruvate, 1 mM (mw 110.0); sodium taurocholate, 4 mM (mw 527.67); iodoacetic acid, 1 mM (mw 185.95); acetyltrimethylammonium bromide, 5 mM (mw 182.06); *p*-quinone, 1 mM (mw 108.09); Atabrine, 1 mM (mw 508.93); and *p*-chloromercuribenzoic acid, 5 mM (mw 357.16).

Activation was seen with glycylglycine and L-methionine, inhibition by L-serine and oxidized glutathione, all at 1 mM.

Rutenburg et al. (*J. Histochem. Cytochem.*, **17:** 517, 1969) developed the following method for the light and electron microscopic demonstration of *γ-glutamyl transpeptidase activity*.

FOR LIGHT MICROSCOPY

1. Small tissue blocks were quick-frozen in liquid nitrogen for 10 s. (Tissues may be fixed in 3% glutaraldehyde, but longer incubations are required.)
2. Sections were cut at 4 to 8 μ in a cryostat and dried in air.
3. Tissues are incubated at 25°C until sufficient color develops. Liver, kidney, and pancreas require only 3- to 5-min incubation; seminal vesicle, ovary, and epididymis require 10 min; lung, liver, jejunum, and thymus require 15 min; esophagus, stomach, duodenum, ileum, testis, uterus, spleen, tongue, heart, cerebellum, cerebrum, and skeletal muscle require 45 min.

γ-Glutamyl-4-methoxy-2-naphthylamide (2.5 mg/ml); may be obtained from Cyclo Chemical Corp., Los Angeles	1 ml
Tris buffer (0.1 M) pH 7.4	5 ml
Saline solution (0.85%)	14 ml
Glycylglycine	10 mg
Fast blue BBN (diazotized 4′-amino-2′,5′-diethoxybenzanilide); may be obtained from Polysciences Inc., Paul Valley Industrial Park, Costner Circle, Warrington, Pa. 18976	10 mg

A stock solution of the substrate is prepared by dissolving 25 mg γ-glutamyl-4-methoxy-2-naphthylamide in 0.5 ml dimethyl sulfoxide and 0.5 ml 1 N NaOH and then adding 9 ml distilled water. The solution is stable for 3 days at 4°C.

4. Rinse tissues in 0.85% saline solution for 2 min.
5. Immerse sections in 0.1 M cupric sulfate for 2 min.
6. Rinse in 0.85% saline solution for 2 min. At this point nuclei may be stained with hematoxylin or methyl green (0.2% methyl green in 0.01 M acetate buffer of pH 4.2 for 6 min).
7. Rinse in distilled water, dry, and mount in a water-soluble mounting medium such as Apáthy's or Kaiser's.

FOR ELECTRON MICROSCOPY

1. Tissue blocks 5 × 5 × 3 mm are fixed for 4 to 6 h in 10% formalin buffered with 0.05 M sodium phosphates to pH 7.4 for 4 to 6 h at 4°C.
2. Sections are cut at 30 to 40 μ with the Smith-Farquhar chopper.
3. Wash sections 3 times for 5 min each in 7.5% sucrose in 0.05 M sodium phosphate buffer of pH 7.4.
4. Thick sections are incubated in the following medium for 30 to 40 min at 25°C:

γ-Glutamyl-4-methoxy-2-naphthylamide (1.25 mg/ml)	2 ml
Sodium phosphate buffer of 0.1 M, pH 7.4	5 ml
Distilled water	13 ml
Glycylglycine	10 mg
Fast blue BBN	10 mg
Sucrose	1.5 g

The substrate solution is prepared by dissolving 12.5 mg glutamyl-4-methoxy-2-naphthylamide in 0.25 ml dimethylsulfoxide and 0.25 ml 1 N NaOH and adding sufficient distilled water to make 10 ml.

5. Rinse sections in 7.5% sucrose in 0.1 M (2.5%) cupric sulfate for 2 min.
6. Rinse sections in 7.5% sucrose.
7. Immerse sections in 1% thiocarbohydrazide containing 7.5% sucrose for 15 min at 37°C. (The thiocarbohydrazide may be obtained from Cyclo Chemical Corp.)
8. Rinse 3 times in 7.5% sucrose at 37°C for 5 min each.
9. Place sections on a stainless steel screen in a $\frac{1}{2}$-oz French square bottle containing $\frac{1}{8}$ g OsO_4 with a few drops of water.
10. Close the bottle with a Teflon-lined cover, and heat on a sand bath at 50°C for 15 to 20 min.
11. Dehydrate the blackened sections through graded alcohols, embed in Araldite, section, and examine by electron microscopy.

For light microscopy, enzymatically liberated 4-methoxy-2-naphthylamine couples promptly with the diazonium salt (fast blue BBN) to form an orange-red insoluble azo dye. Chelation with the cupric ion results in the formation of an intense red insoluble complex with an increased affinity for tissue protein. Rutenburg et al. depict the following sequence of events for electron microscopy:

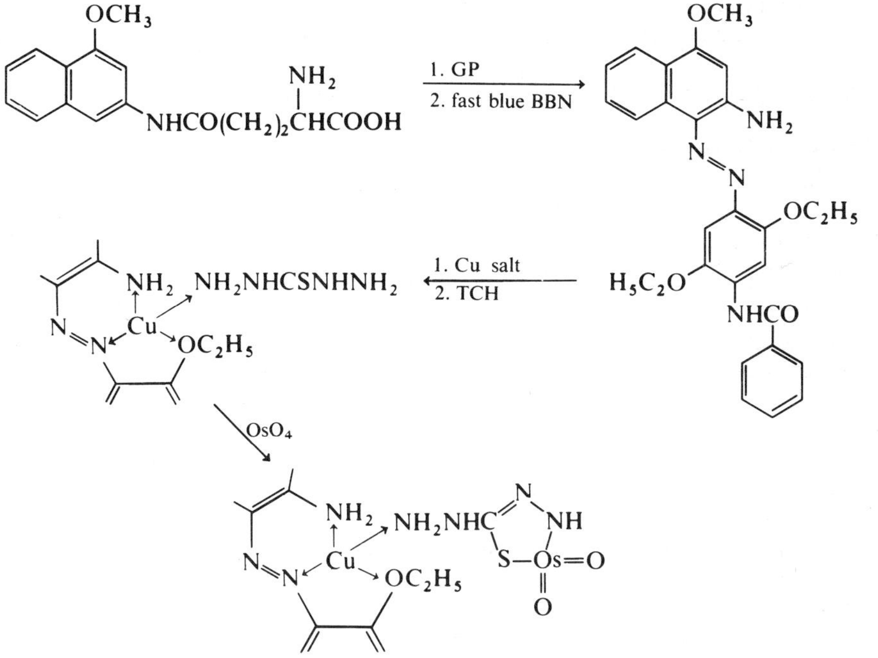

GP indicates γ-glutamyl transpeptidase activity; TCH indicates thiocarbohydrazide. It was necessary to osmicate the copper chelate with the aid of thiocarbohydrazide in order to provide high contrast for the electron beam and to produce an insoluble pigment that would withstand embedment in Araldite.

Rutenburg et al. noted that 3 mM (0.396%) glycylglycine and (0.0445%) L-methionine added to the incubation mixture resulted in 240 and 130% activation, respectively. Cupric sulfate, nickel chloride, or zinc chloride at 1 mM concentration resulted in 100, 100, and 90% inhibition of enzyme activity, respectively.

At the electron microscopic level, γ-glutamyl transpeptidase activity was noted in the endoplasmic reticulum region in the vicinity of zymogen granules of rat acinar pancreatic cells.

CATHEPSIN B1

Sylvén and Snellman (*Histochemie*, **38**: 35, 1974) developed an immunofluorescent method for the histochemical demonstration of *cathepsin B1*. Antiserum was raised in rabbits against pure cathepsin B1. Specific antibodies were isolated and conjugated to fluorescein isothiocyanate (FITC), and the conjugate was purified. Fresh frozen sections postfixed in ethanol-ether (1 : 1) for 30 min at 25°C and fresh frozen sections autolyzed for 24 h at 4°C (to activate lysosomal enzymes) were employed. Conjugates were applied to tissue sections for 2 to 4 h in humidified petri dishes with controls in the usual fashion. Sections examined by fluorescence microscopy revealed cathepsin B1 present particularly in lysosomes and phagocytic vacuoles in many tissues, especially liver.

CATHEPSIN D

Poole et al. (*J. Histochem. Cytochem.*, **20:** 261, 1972) employed a fluorescent immunohistochemical method to demonstrate *cathepsin D* in lysosome-like bodies as well as in a diffuse pattern in the cytoplasm of synovial cells of rabbits with experimental arthritis.

TYROSINE AMINOTRANSFERASE

E. Thompson and G. Tomkins (*J. Cell Biol.*, **49:** 921, 1971) advance the following method for the histochemical demonstration of *tyrosine aminotransferase* (TAT). The reaction catalyzed by the enzyme is L-tyrosine + α-ketoglutarate $\rightleftharpoons$ *sym*-hydroxyphenylpyruvate + L-glutamate. The method is adapted from an assay system for acrylamide gels (Vareriote et al., *J. Biol. Chem.*, **244:** 3618, 1969). The method as applied to cells in tissue culture follows:

1. Remove growth medium from monolayers of cells on glass coverslips, and rinse cells 2 or 3 times by flooding gently with 0.1 M phosphate buffer of pH 7.6 at 4°C containing 0.2 mM (5 mg/100 ml) pyridoxal phosphate[1] and 0.5 mM (0.0073%) α-ketoglutaric acid.[2] (The cofactor and substrate are added to stabilize TAT in cell-free solutions.) Allow the final rinse to remain on the cells for 1 to 2 min, and then remove it by suction.
2. Drain the cells by tipping the coverslips, and allow them to dry at room temperature for 1 to 2 h. (At this stage cellular TAT is stable, and cells may be stored in an ordinary refrigerator—although cell morphologic features are poor.)
3. Cells are incubated in the mixture below at 37°C in the dark in a humidified air–CO_2 incubator for approximately 4 h. Positive staining can usually be observed after 2 h. After 6-h incubation, staining not dependent on TAT occurred in all cells.

	ml
Nitro BT (10 mg/ml H_2O)	1
Phenazine methosulfate 0.4 mg/ml	1
EDTA (0.1 M) (mw Na_2EDTA 336.21)	0.375
Dulbecco's phosphate buffered saline*	2.25
Monoiodotyrosine (0.02 M in Dulbecco's phosphate-buffered saline solution) (mw 307.086)	5
α-Ketoglutaric acid (mw 146.1) (0.3 M in Dulbecco's phosphate-buffered saline solution) adjusted to pH 7 with 7 N NaOH	0.05
Pyridoxyl phosphate 1.3 mM in Dulbecco's phosphate-buffered saline solution (0.4 μg)	0.2

* Phosphate buffered saline (Dulbecco, *J. Exp. Med.*, **99:** 167, 1954).

Component	mg/l	Component	mg/l
NaCl	8000	KH_2PO_4	200
KCl	200	$CaCl_2$ (anhyd.)	100
Na_2HPO_4	1150	$MgCl_2 \cdot 6H_2O$	100

Thompson and Tomkins note that *p*-hydroxyphenylpyruvate reduces the tetrazolium salt when phenazine methosulfate is present. No reaction occurs with omission of the substrate monoiodotyrosine. They indicate that they have not noted clones of cells completely negative for the enzyme. They have noted as much as a tenfold difference.

---

[1] Pyridoxal phosphate mw 247.15.
[2] α-Ketoglutaric acid mw 146.1.

AMINOPEPTIDASES

There has been some dispute as to whether histochemically demonstrable enzymes releasing β-naphthylamine or other arylamines from amide combinations with amino acid carboxyls should be termed *leucine aminopeptidase* or simply *aminopeptidase*. At first it was thought generally that the leucyl naphthylamide gave more specific results. Gomori (*Proc. Soc. Exp. Biol. Med.*, **87**: 559, 1954) worked with an alanyl derivative in 1954 which did not give especially good histochemical results. Burstone and Folk, and Ackerman used both the alanyl and leucyl derivatives; Nachlas and Seligman and coworkers used chiefly the leucyl. Some investigators have thought the results essentially identical. Ackerman thought the alanyl derivative better, and Seligman's school have adhered to use of the leucyl derivative.

A variety of section preparation procedures noted in Table 10-14 and in the text have been used. Glenner's recommendations seem well worthwhile in this respect; if $-70°C$ temperatures are not available, Novikoff's recommendations may be followed, or they may be combined with Glenner's acetone chloroform extractions at 0 to 5°C.

The substrates in Table 10-14 are fairly similar, and incubation times depend in any case on the activity of the particular tissue under study. The temperature is again a matter of convenience. The enzyme activity is greater at 37°C than at 25°C, but so is the rate of decomposition of the diazonium salt.

Molarities have been computed for convenience on the perhaps unwarranted basis that the substrates are reasonably pure substances. The diazonium salts are probably present in considerable excess, and since commercial samples represent 18 to 40% or more of the primary amine, computation of molarities seemed futile for them. The molecular weights may be found in Table 6-9.

Burstone also used fast black K and fast red B in the first report but later generally preferred fast garnet GBC. Nachlas and Seligman prefer fast blue B in almost all enzymatic azo reaction procedures, but they did compare fast garnet GBC in their 1962 paper.

In their first report Nachlas et al. dried their cryostat sections on the slides for 30 min at 37°C; in the later report this drying is not mentioned. Also in the first report they note that colors were at first red and were converted to blue by postchelation in 0.1 M (2.5%) $CuSO_4 \cdot 5H_2O$. This practice is not noted in the later report. Gomori had previously noted diffusion from unfixed frozen sections.

Novikoff (*J. Histochem. Cytochem.*, **8**: 37, 1960), using the Burstone-Folk (1956) substrate, found good preservation of activity in fresh frozen (cryostat) sections which were postfixed in acetone at 2 to 4°C for 15 min. Unfixed sections lost activity by diffusion, and postfixation for 15 min in formol-calcium at 2 to 4°C decreased the activity.

INHIBITORS. Strong inhibition has been reported by Burstone and Folk for 10 mM KCN, 10 mM Versene, 1 mM each of Cu^{2+}, Pb^{2+}, Cd^{2+}, and (Ackerman) Hg^{2+}. Ackerman noted inhibition by 1 mM Zn^{2+} in the presence of 0.1 M $CaCl_2$; he reported activation without the Ca^{2+}. Inhibition by certain diazonium salts was reported by Nachlas et al. (1957) and by Burstone and Weisburger, especially by the borofluorides. Citrate was inhibitory at 10 mM (Nachlas) but not at 5 mM (Burstone and Folk). Ackerman noted activation by Mg^{2+}, Co^{2+}, and Zn^{2+} at 0.1 M. Iodoacetate 1 mM, Mn^{2+} 1 mM, ethyl maleimide 0.1 M, NaF 2 mM, and diisopropyl fluorophosphate 0.5 mM have been reported as without effect. Short methanol or ethanol fixation inactivates, as does aqueous formalin (Ackerman).

Since it appears that the azo dyes formed by azo coupling with the naphthylamine released from amino acid amide linkages in the group of peptidase methods are fat-soluble basic dyes, diffusion occurs in glycerol gelatin or polyvinylpyrrolidone, and crystallization appears in and around lipid droplets, both much to the detriment of the permanency of the

TABLE 10-14   AMINOPEPTIDASES: COMPARISON OF METHODS AND SUBSTRATES

	Section preparation methods			
	Burstone and Folk: Freeze-dried paraffin, petroleum ether, acetone, water	Nachlas et al.: Attached cryostat sections, dried 30 min at 37°C (1957); fresh frozen section (1962)	Burstone and Weisburger: Freeze-dried celloidin-paraffin, petroleum ether, acetone, water; attached fresh frozen section (cold knife)	Ackerman: Blood, marrow films, 2 min at −10°C in 1% OsO_4 in dimethylformamide, wash in water
Amides of aryl-amine	β-Naphthylamine	β-Naphthylamine	3-Aminocarbazole 3-amino-9-ethylcarbazole	β-Naphthylamine
Aminoacid(s)	DL-Alanine L-leucine	L-Leucine	DL-Alanine	Alanine
Amount and final conc. amide .... Special solvent ..	(Alanyl) 5 mg, 0.93 mM (Leucyl) 5 mg, 0.78 mM	10 mg, 1.56 mM H_2O, 1.25 ml	($3\text{-}NH_2$) 2.5 mg, 0.34 mM (9-Ethyl) 2.5 mg, 0.35 mM Acetone, 0.25 ml	10 mg, 1.87 mM
Buffer .... Amount .... Final molarity .. pH	0.2 M tris 5 ml 40 mM 7.1	0.1 M acetate 12.5 ml 50 mM 6.5	0.2 M tris 7.5 ml 60 mM 7.1	0.1 M phosphate 12.5 ml 50 mM 6.7
Activators ...... Final conc..... Diazonium salt ..	.... Fast garnet GBC, 15 mg	1.25 ml 20 mM KCN 1 mM Fast blue B, 12.5 mg	.... Fast garnet GBC, 10 mg	$MgSO_4·7H_2O$, 1 mg 1 mM Fast garnet GBC, 12.5 mg
Additional solvent Final volume....	Water, 20 ml 25 ml	Physiologic saline, 10 ml 25 ml	Water, 17.25 ml 25 ml	Water, 12.5 ml 25 ml
Time range ..... Temperature	15 min–4 h 24–25°C	20 min–4 h 37°C	30 min–18 h 25(37)°C	Up to 4–6 h 20–25°C

Sources: Burstone and Folk. *J. Histochem. Cytochem.* **4**: 217, 1956, and **9**: 332, 1958; Nachlas et al., *J. Histochem. Cytochem.* **5**: 264, 1957, and **10**: 315, 1962; Burstone and Weisburger, *J. Histochem. Cytochem.* **9**: 349, 1961; Ackerman, *J. Histochem. Cytochem.* **8**: 386, 1960.

preparations. The freeze-dried, paraffin technic, with the lipid extraction entailed in deparaffinization with xylene or petroleum ether, eliminated part of the difficulties. To overcome these difficulties Glenner (*J. Histochem. Cytochem.*, **10**: 257, 1962) prescribes for alanyl and leucyl naphthylamidases, mast cell protease, and γ-glutamyl transpeptidase a modified section freeze substitution technic. After freeze substitution for 18 to 24 h in acetone at −70°C, sections are left in pure acetone for 1 h at 4°C, then 1 h each in 2:1 and 1:2 acetone-chloroform and in pure chloroform. They are then transferred to slides, coated with 0.5% celloidin, and dried in air. After incubation in the appropriate substrate for the prescribed interval, wash in 0.9% sodium chloride and in pH 9.2 0.1 *M* Veronal buffer, and mount in the H-A (Highman variant of the medium of Lillie and Ashburn) gum Apáthy. This variant of the Apáthy gum arabic sucrose medium was specifically designed to prevent diffusion of crystal violet in amyloid stains, and has served to keep aminopeptidase preparations unimpaired as long as 8 months. Sealing of the coverslips with cellulose caprate seems superior to pyroxylin cement; we have used both.

Freeze substitution in acetone followed by polyvinyl alcohol or low-melting paraffin may also be used. Naidoo's freeze-dried, low-melting paraffin followed by petroleum ether or isopentane also seems applicable.

OXIDATIVE AMINOCARBAZOLE METHOD OF BURSTONE AND WEISBURGER FOR AMINOPEPTIDASES (*J. Histochem. Cytochem.*, **9**: 712, 1961).   This technic offers a different approach from the foregoing azo coupling methods.

Prepare sections by the freeze-dried, celloidin-paraffin technic or fresh frozen sections in the cryostat or by the Adamstone-Taylor technic. Attach to slides and bring to substrate as usual.

SUBSTRATE

	Final concentration, mM
Dissolve 5 mg DL-alanyl-3-amino-9-ethylcarbazole in 0.5 ml ethanol	0.65
Add 25 ml distilled water and 2 ml 50 mM (1.65%) K$_3$Fe(CN)$_6$	3.5
Add 3 mg 5,6,7,8-tetrahydro-α-naphthylamine in the alcohol	0.7

Burstone comments particularly on the favorable color results when 5,6,7,8-tetrahydro-α-naphthylamine is added to the above substrate, giving a salmon pink which is converted to deep blue by postchelation with copper. The amount is not stated; 3 mg would furnish more than one equivalent for all the carbazole.

Incubate 15 to 60 min at 25°C (Burstone does not specify temperature; this was approximately the temperature of Building 10 at the National Institutes of Health at the time he wrote his 1960 paper there), rinse in distilled water, transfer to 10% aqueous copper sulfate (CuSO$_4$·5H$_2$O = 0.4 *M*) for 1 to 2 h, wash in water, and mount in glycerol gelatin. A counterstain in 0.1% basic fuchsin in 1% acetic acid, 2 to 5 min, and mounting in Highman's 50% potassium acetate–gum Apáthy may be tried if a nuclear stain is needed.

The *nerve cathepsin* of Adams et al. or a similarly located aminopeptidase has been found (Adams and Glenner, *J. Neurochem.*, **9**: 233, 1962) to be well demonstrated by Burstone's aminopeptidase substrates, but biochemically it gives its maximal activity at pH 7 to 7.5 with L-leucyl-β-naphthylamide as substrate. Relative activities at pH 7 were 90% for DL-alanyl-β-naphthylamide, 33% for DL-phenylalanyl-β-naphthylamide, and nil at pH 5, 6, and 7 for benzoyl-DL-arginine-β-naphthylamide, chloroacetyl-α-naphthylamide, and 2-chloroacetyl-3-naphthoic acid anilide (naphthol AS chloroacetate).

AZO DYE TECHNIC OF ADAMS AND GLENNER FOR NERVE AMINOPEPTIDASE.

SUBSTRATE. Dissolve 3 mg L-leucyl-β-naphthylamide hydrochloride in 30 ml 0.1 M tris buffer at pH 7. Preheat substrate to 37°C. Add 15 mg fast garnet GBC just before introducing sections. Incubate fresh frozen 10-μ sections, attached to slides, for 30 to 60 min. Rinse, and mount in glycerol gelatin.

INHIBITORS. These are cysteine, 10 mM; iodoacetate, 10 mM; p-chloromercuribenzoate, 1 mM; KCN, 10 mM; ascorbic acid, 10 mM; diisopropyl fluorophosphate, 10 mM. Preheating to 60, 80, or 100°C for 10 min inactivated nerve aminopeptidase, as did chloroform-methanol and chloroform-acetone extractions. Inactivation was noted when fast red B was used as the diazonium salt.

Niemi and Sylvén (*Histochemie*, **18:** 40, 1969) are convinced that the leucine aminopeptidase stain may be used to recognize cellular injury possibly several hours prior to the time when it may be detected by dye exclusion tests. To detect early cell injury, they advise buffering the incubation mixture to pH 5.5; the use of fast blue B (C.I. 37235) for coupling is advised.

Sylvén and Bois (*Histochemie*, **3:** 341, 1963) compared leucine aminopeptidase activity in fresh cryostat sections of several tissues with enzyme activity of tissue homogenates. They concluded that the degree of staining observed relates more to the number of enzyme sites available to substrate interaction than to total enzyme present in the tissues. Fixation and methods of tissue preparation for histochemical analysis modify the number of reactive sites.

Monis et al. (*J. Histochem. Cytochem.*, **13:** 503, 1965) noted that formaldehyde fixation of tissues results in a greater loss of aminopeptidase activity than does glutaraldehyde. Comparable localization of enzyme activity was achieved using leucyl, methionyl, and alanyl amides with the rapid coupler 4-methoxy-2-naphthylamine.

Meade and Rosalki (*J. Clin. Pathol.*, **17:** 61, 1964) separated leucine aminopeptidase isozymes from human serums, employing cellulose acetate membranes.

Wachsmuth (*Histochemie*, **14:** 282, 1968) reports the demonstration of aminopeptidase using an immunofluorescent method.

Sandström (*Histochemie*, **26:** 40, 1971) has employed the copper chelate obtained in Nachlas' aminopeptidase method for the identification of aminopeptidase in lysosomes of chicken livers at the electron microscopic level.

Wachsmuth and Woodhams (*J. Histochem. Cytochem.*, **21:** 685, 1973) employed the Nachlas et al. (*J. Histochem. Cytochem.*, **5:** 264, 1957) method and a specific fluorescent antibody technic for aminopeptidase to estimate the number of molecules of aminopeptidase in the proximal convoluted tubules of swine kidney. They concluded that 4-μ segments of tubules contained 10^4 molecules of aminopeptidase per μ^2.

SUBSTRATE FILM METHODS OF ENZYME LOCALIZATION

Perhaps inspired by the statement in Wells' "Chemical Pathology" (3d ed., Saunders, Philadelphia, 1918) that smears of blood or pus on a film of fibrin would digest small holes in the fibrin about each leukocyte, there has arisen in recent years a small series of specific enzyme localization methods wherein a cryostat section of unfixed tissue is closely apposed to a film of the specific substrate and the two are allowed to interact for a time and then are separated. The tissue section is then fixed and stained by some appropriate procedure for identification of ordinary morphologic details, and selected areas are photographed. The film is also appropriately fixed and stained for the demonstration of the specific substrate.

The substrate film method has been employed for the histochemical detection of ribonuclease (Daoust, *J. Histochem. Cytochem.*, **8**: 131, 1960, and **14**: 254, 1966), deoxyribonuclease (Daoust, *Exp. Cell Res.*, **12**: 203, 1957, and **24**: 559, 1961), proteases (Adams, *J. Histochem. Cytochem.*, **9**: 469, 1961; Cunningham, *J. Histochem. Cytochem.*, **15**: 292, 1967), fibrinolysin (Kwaam and Astrup, *Lab. Invest.*, **17**: 140, 1967; Todd, *Br. Med. Bull.*, **20**: 210, 1964, and *J. Clin. Pathol.*, **17**: 324, 1964), and amylases (Shear and Pearse, *Exp. Cell Res.*, **32**: 174, 1963; Szemplinska et al., *Acta Biochem. Pol.*, **9**: 239, 1962; Tremblay, *J. Histochem. Cytochem.*, **11**: 202, 1963). Daoust has written a critical evaluation of these methods (*J. Histochem. Cytochem.*, **16**: 540, 1968). The following points must be considered by individuals employing these methods:

The substrate film must be susceptible to enzyme action but remain insoluble. Alterations in the substrate film must be a consequence only of enzyme action. With film mixtures containing a substrate and another constituent such as gelatin, it is important to discriminate between possible action on the substrate and on the gelatin.

False-negative results are sometimes obtained by diffusion of certain stainable constituents from tissues into the film. For example, diffusion of acid mucopolysaccharides from cartilage slices can stain with toluidine blue in preparations designed to demonstrate nucleases.

False-positive reactions may also occur as a consequence of diffusion of certain constituents from tissues into the substrate film, preventing access of the stain to the substrate. In such areas the absence of stain might be interpreted as positive.

Lake [*Nature (Lond.)*, **209**: 521, 1966] calls attention to the fact that trypsin will digest the gelatin prepared for the Daoust methods for the histochemical detection of deoxyribonuclease and ribonuclease, thus permitting a false-positive interpretation of results.

The gelatin-silver protease (and cathepsin) method of Adams and Tuqan (*J. Histochem. Cytochem.*, **9**: 469, 1961) seems the most closely allied to Wells' procedure.

GELATIN-SILVER PROTEASE METHOD OF ADAMS AND TUQAN

1. Expose panchromatic "quarter" plates to daylight for 15 min. Develop, fix in bisulfite hypo, and wash and dry as usual. Cut with diamond point from glass side into pieces of convenient size—e.g., cut a 3 × 5 in (75 × 125 mm) plate into five strips 1 × 3 in (25 × 75 mm).

2. Fix tissues 1 to 3 days at 4°C in 10% formalin, cut frozen sections at 15 μ, apply inactivators if they are required, rinse in distilled water, and immediately float onto the gelatin side of the blackened photographic plate.

3. Allow to just dry, then moisten with (a) 0.15 M phosphate buffer pH 7.6 or (b) 0.15 M acetate buffer pH 5. No free fluid should be left on the section; the buffer should be entirely absorbed by the gelatin, to prevent diffusion.

4. Incubate in a moist chamber saturated with water vapor at 37°C for 30 to 60 min. The slides may be conveniently enclosed in a large petri dish with wet filter paper in the bottom. The petri dish should be preheated to 37°C. Inspect every 5 to 10 min, replenishing buffer if needed.

5. Dry, dehydrate with alcohols, and clear in xylene within 2 min to prevent crazing of gelatin film which may ensue on slow dehydration. Mount in balsam or synthetic resin.

RESULTS: Protease activity is shown by clear areas where gelatin has been digested and loose silver granules have been washed out. Activity is described about pancreatic zymogen cells, intercalated duct cells of rat and mouse submaxillary gland, over Paneth cells, myelin, renal cortex.

Iodoacetate (1%, 1 h) inactivated the myelin cathepsin, but not the pancreatic enzyme; diisopropyl fluorophosphate (1 to 10 mM, 1 h), 6% HgCl$_2$ (1 h) inactivated both. Pretreatment with alcohol or formalin had no effect, but their presence during the reaction inhibited. This iodoacetate-sensitive protease is the myelin cathepsin of Adams and Bayliss (*J. Histochem. Cytochem.*, **9**: 473, 1961).

The autodigestion procedure reported by Lillie and Burtner (*J. Histochem. Cytochem.*, **1**: 8, 1953) is of similar nature, with reliance on the digestibility of alcohol-fixed cytoplasm.

Leukocyte protease has long been recognized from its lytic action on fibrin of exudates (Wells, "Chemical Pathology," 3d ed., Saunders, Philadelphia, 1918, pp. 94–96). It resists formaldehyde for long periods.

Histochemically it may be made evident from the selective cytolysis and later karyolysis of the neutrophil leukocytes which occur in blood films which are first fixed 10 min in 75% alcohol and then exposed to distilled water at 60°C. The enzyme is destroyed by fixation in boiling xylene. Formaldehyde fixation renders cytoplasm and nuclei resistant to digestion and also to trypsin. The autolytic digestion occurs readily after fixation in boiling acetone, benzene, or toluene.

Penn et al. (*J. Histochem. Cytochem.*, **20**: 499, 1972) used autoradiographic film as a gelatin substrate to demonstrate acrosomal proteolytic activity of sperm.

Shear (*J. Histochem. Cytochem.*, **17**: 408, 1969) employed the gelatin-silver substrate film method to demonstrate protease activity in adipose tissues of rats and guinea pigs.

DAOUST'S GELATIN FILM DEOXYRIBONUCLEASE METHOD.   For this method (*Exp. Cell Res.*, **12**: 203, 1957) Daoust directs as follows:

PREPARATION OF SUBSTRATE FILMS.   Mix equal volumes melted 5% gelatin and 0.2% deoxyribonucleic acid by heating on water bath. Spread 0.05 ml of mixture with pipette tip over 25 × 40 mm area; leave horizontal until dry. Fix overnight in neutral 10% formaldehyde (25% formalin). Wash in distilled water and let dry.

PREPARATION OF SLIDES FOR CARRYING SECTIONS.   Coat slides to carry the sections with glycerol gelatin (10 : 7 : 83 water; premelted on water bath). Deposit 0.3 ml on a slide and spread with the pipette tip over a 25 × 40 mm area. Lay slides horizontal for 10 to 15 min to gel.

PREPARATION OF SECTIONS.   Rinse and blot fresh tissue to remove blood, freeze onto cutting stage of cryostat, and section at 15 μ in cryostat at −20°C. Transfer sections quickly in cryostat to glycerol gelatin–coated slides, remove at once from cryostat to avoid freezing of the coating, and lay horizontally on warm plate at 37°C to melt the glycerol gelatin and spread the sections. Then recool to 20 to 25°C and allow glycerol gelatin film to gel again (about 20 min). This procedure is required to prevent adhesion of the section to the substrate film on reseparation of slides after exposure.

INCUBATION.   This is accomplished at 20 to 25°C by apposing the gelated section-bearing slide to the dry substrate film slides and pressing the two firmly together for 5 to 60 min. The two slides are then reseparated, and the section is fixed by immersion in a horizontal position in neutral 10% formalin for 18 to 24 h. The sections are then washed in distilled water, two changes, for 5 min each, and allowed to dry. The substrate film, as soon as separated, is similarly washed and dried. Stain substrate film for 10 min in 0.2% toluidine blue, wash in distilled water, let dry in air, and mount directly in Canada balsam.

Immerse section for 5 min in 25 : 75 acetic acid–alcohol to prevent staining of the gelatin, transfer directly to 0.1% toluidine blue, stain 1 to 2 min, wash in distilled water, dry in air at 20 to 25°C, and mount in balsam.

RESULTS: Areas of deoxyribonuclease activity appear unstained in the substrate film and may be matched with the section by superimposition of photographic images, or by side-to-side comparison. The use of a comparison microscope to match areas could be advantageous. For this purpose the two slides should be accurately apposed to each other during incubation.

It would appear that the technic of tissue staining could be simplified by incorporating a known amount of acetic acid in the toluidine blue, or by buffering it deliberately at pH 3, omitting the acetic acid–alcohol bath. The use of a synthetic resin to prevent the thiazin dye's fading, which occurs in balsam, also seems advisable. Azure A or B would undoubtedly serve as well as toluidine blue.

Daoust did not employ acid dyes after his digestion procedure, to demonstrate that the loss of staining was due to destruction of deoxyribonucleic acid and not simply to proteolytic digestion of the supporting gelatin, as in the Adams-Tuqan method. This same objection applies also to the same author's ribonuclease method.

Likewise no provision is made for exploration of the effects of specific inhibitors, and since there is no pH control in the digestion process, it is quite uncertain .which of the deoxyribonucleases is being shown.

SUBSTRATE FILM METHOD OF DAOUST AND H. AMANO (*J. Histochem. Cytochem.*, **8**: 131, 1960) FOR RIBONUCLEASE (DRAINING TECHNIC). An equal volume mixture of 5% gelatin and of 5 to 10% sodium yeast ribonucleate[1] is melted in a water bath. Apply 3 to 5 drops of the mixture to a clean glass slide, spreading it with the pipette tip. Stand the slide vertically on a paper towel to drain and dry, wiping off excess from the extreme lower end of slide with wet filter paper or gauze. Fix dried film 1 h at 2 to 4°C in neutral 20% formalin. Wash out excess formalin in three changes of distilled water, for 5 min each.

CONTROL.   Exposure of such films to 0.05% ribonuclease (Worthington) impairs toluidine blue staining in 2 h and abolishes it in 8 h.

PROCEDURE.   Cryostat sections are applied to glycerol gelatin–coated slides, reacted, stained, and dried in air according to the same procedure as for Daoust's deoxyribonuclease method.

Daoust and Morais (*J. Histochem. Cytochem.*, **20**: 350, 1972) noted that gelatin films containing soluble ribonucleic acid (RNA) displayed nuclease activity with the same tissue distribution as when the usual RNA is employed. However, the use of polyadenylic acid resulted in a different tissue distribution of enzyme activity. For example, in contrast to RNA films, the lamina propria of the small intestine were negative and muscle layers were positive in polyadenylic acid (poly-A) films. In the ovary, enzyme activities on the two films were just the reverse of each other. The nature of the polyadenylic acid hydrolases is unknown.

Tremblay (*J. Histochem. Cytochem.*, **11**: 202, 1963) has reported an essentially similar procedure for the demonstration of amylase in tissue sections. In this case the substrate is starch, and the demonstration method is the periodic acid Schiff reaction.

TREMBLAY'S STARCH SUBSTRATE FILM METHOD FOR AMYLASE.   Heat a 4% aqueous suspension of Hydrolyzed Starch[2] 15 min in a boiling water bath. Filter while hot through glass wool and gauze. Deposit about 0.5 ml (8 to 12 drops) on a slide, and spread rapidly over a 25 × 40 mm area. Stand upright on filter paper to drain. Wipe excess off lower end, and dry film in air at 20 to 25°C. Fix film 18 to 24 h in methanol–acetic acid–water (5 : 1 : 5) mixture. Wash in three changes of distilled water, 5 min each, and dry in air at 20 to 25°C.

---

[1] Schwartz Laboratories, Inc.
[2] Comnaught Medical Research Laboratories, Toronto, Canada.

PREPARATION OF TISSUE SECTIONS.   This procedure follows that in Daoust's technic for deoxyribonuclease.

INCUBATION.   Appose accurately the two slides bearing, respectively, the tissue section on glycerol gelatin and the substrate film, face to face, and press firmly together. After 1 to 20 min separate films by insertion of a knife edge or razor blade.

Rinse substrate film in distilled water, refix 15 min in the methanol–acetic acid–water mixture, stain by the periodic acid Schiff reaction, wash, dry, and mount in resin.

Wash tissue sections, fix in 4% formaldehyde 18 to 24 h, wash, and stain with 0.2% toluidine blue, as in the deoxyribonuclease technic.

The α-amylase in the control experiments was inhibited by heating 10 min to 100°C or by inclusion of 1 mM $CuSO_4$. Heating slides for inactivation was practiced. It is presumed that inclusion of 1 mM $CuSO_4$ ($\cdot 5H_2O$ = 25 mg/100 ml) in the last wash water applied to the starch film before drying and using would act to inhibit the amylase activity, but Tremblay does not mention such a test.

Substitution of a highly branched polysaccharide, such as glycogen, for the starch in the substrate film should serve to distinguish α- from β-amylase.

Smith and Frommer (*J. Histochem. Cytochem.*, **21:** 189, 1973) register a plea that investigators employ standardized procedures for the localization of amylase activity using the starch-film method. Specifically, they suggest that (1) a pure starch film be used (rather than one mixed with gelatin), (2) sections should be mounted directly on the substrate film, (3) sections should be left permanently on the film (they state this does not result in false-positive or false-negative results), (4) iodine should be used (greater contrast is obtained between positive and negative areas; however, stained sections are not permanent), (5) the concentration of starch in the film should be stated (3% films are good for tissues with low activity; those with high activity may require the use of films containing 5% or more to achieve maximum localization), and (6) differences between fixed and unfixed films were undetectable.

CELLULASE

Sumner (*Histochemie*, **13:** 160, 1968) advocates the following substrate film method for the histochemical detection of *cellulase*. Cellulase hydrolyzes cellulose to oligosaccharides and monosaccharides and is widely distributed in microorganisms, plants, and invertebrates.

FILM PREPARATION.   Clean microscopic slides were dipped into a 10% solution of carboxymethylcellulose (Cellofas B10, ICI Ltd., England) and allowed to dry in an oven at 60°C. Slides were slanted about 10° from the horizontal. Films were then fixed in an acid-ethanol solution (3 parts 1 N HCl, 7 parts absolute ethanol) overnight. Wash in 70% ethanol 15 to 30 min to remove the acid, and allow to dry in air. Cut around the slides and peel the film from the slides. The films may be stored indefinitely.

HISTOCHEMICAL PROCEDURE

1.   Slides are first dipped in chrome-gelatin (chrome alum, 0.05 g; gelatin, 0.5 g; distilled water, 100 ml). Stand the slides up vertically, and allow them to dry at 24°C.
2.   Sections cut at 10 to 15 μ with a cryostat from tissues fixed for 1 h in Baker's formol-calcium are attached to the chrome-gelatin film on slides. Formol-calcium–fixed tissues stored in cold Holt's (*Exp. Cell Res. Suppl.*, **7:** 1, 1959) 1% gum acacia–30% sucrose fluid may also be used.
3.   Cover the section with the carboxymethylcellulose (CMC) film. This is done by floating a CMC film on a water bath and placing a slide bearing the chrome-gelatin coating with the adherent section under the film and raising it out of the bath.

4. Dry the slide in a stream of cold air.
5. Coat the slide with a 0.25% solution of celloidin in absolute ethanol-ether (1 : 1), and allow to dry.
6. Incubate the slide in a moist atmosphere at 37°C. This may be done by inverting the slide over a small petri dish containing water, and covering with a large petri dish. Usually 15- to 60-min incubation is sufficient, but up to 2 h may be needed.
7. Place the slide in 70% ethanol, to stop the reaction, for about 5 min. Then allow the slide to dry in air.
8. Stain in 0.05% aqueous toluidine blue until the film is evenly and sufficiently deeply stained (at least 1 h).
9. Rinse the slide with distilled water, and allow it to dry in air.
10. Soak the slide in xylene for several minutes, and mount in synthetic medium.

RESULTS: The completed specimen is purplish red and weakly metachromatic. Sites of cellulase activity appear as clear or very pale patches in the film.

In the digestive system of several mollusks, cellulase was detected in the crop and stomach and in the lumen and absorptive cells of the digestive gland tubules. Salivary glands and epithelia of crop and stomach were without activity. Sections of control tissues inactivated by boiling water were negative. At step 8 one may wish to substitute the periodic acid Schiff method as a positive control for cellulose.

Pette and Brandau (*Biochem. Biophys. Res. Comm.*, **9**: 367, 1962) and Nolte and Pette (*J. Histochem. Cytochem.*, **20**: 567 and 577, 1972) describe a gel film method for measurement of kinetic enzyme activity in single cells using cryostat-cut sections. The main principle of the method is a comparative activity determination based upon bipositional recording of initial reaction kinetics in two preselected measuring fields in the same tissue section. A modified microscopic photometer with monochromator and variable measuring field diaphragm coupled to a recorder is employed. In determination of dehydrogenase activity, phenazine methosulfate is used, and reduced coenzymes is measured in a 1.5% agarose gel. Although the method will permit comparison of enzyme activity in different parts of a tissue section (for example, in different parts of a hepatic lobule), the quantity of enzyme activity cannot be related to tissue weight, volume, protein, DNA, etc.

ACID DEOXYRIBONUCLEASE (DNASE II)

Aronson et al. (*J. Histochem. Cytochem.*, **6**: 255, 1958) and Vorbrodt (ibid., **9**: 647, 1961) have reported closely related methods for acid deoxyribonuclease (DNase II), based on Gomori's lead sulfide acid phosphatase method. Deoxyribonuclease I is Mg^{2+}-dependent and has a pH optimum of 7 to 8. We have not seen a histochemical method for it. Deoxyribonuclease II functions best at pH 4.5 to 6 and does not require Mg.

SUBSTRATE. Vorbrodt prefers a herring sperm deoxyribonucleic acid obtained from Light, Ltd., which is a relatively low polymer and hydrolyzes more rapidly than calf thymus deoxyribonucleic acid from British, Russian, and American sources. However, he uses 5 to 10 times the amount.

	Final concentration
Deoxyribonucleic acid, herring sperm, 5 mg,	20 mg/100 ml
or calf thymus, 0.5–1 mg	2–4 mg/100 ml
Acid phosphatase, 2.5 mg	10 mg/100 ml
Acetate buffer 0.2 M pH 5.9 }6.2 ml	50 mM
for nuclear staining, pH 5.2	
Lead nitrate 0.1 M (3.31%), 0.5 ml	2 mM
Distilled water to make 25 ml	

Dissolve the substrate and the acid phosphatase in the buffer. Dilute the lead nitrate with 10 to 15 ml distilled water, and add gradually to substrate with shaking. Aronson's substrate contained 2.65 mM Pb and used a pH 5 level. He used 2 mg acid phosphatase for 22.5 ml and 4 mg deoxyribonucleic acid.

SECTION PREPARATION.   Aronson et al. cut fresh frozen sections at 15 μ in a cryostat, attached to slides, and postfixed 5 min in formalin-water-acetone, 10 : 40 : 50, at -10 to $-12°C$ and washed 3 min in 50% acetone and 5 to 15 min in distilled water at 20 to 25°C. Novikoff's substitution of 15 min in formol-calcium at this point could be advantageous. Vorbrodt found 2 min inadequate. Vorbrodt prefixed tissue blocks 12 to 24 h in formol-calcium at 2 to 4°C, cut frozen sections at 15 μ, and affixed to slides. He tried Aronson's procedures, with inferior results.

For routine use Vorbrodt's preferred procedure, overnight fixation and frozen sections, is not only the most practical but apparently the best.

INCUBATION.   Incubate attached frozen sections and formol-calcium–fixed smears 30 min to 4 h (average, about 2 h) at 37°C. Rinse in distilled water, then 1 min in 1% acetic acid, and again in distilled water.

DEVELOPING.   Develop in 2% yellow ammonium sulfide 2 min. Wash in water, counterstain if desired, dehydrate in alcohols, clear in xylene, mount in balsam or synthetic resin.

RESULTS: At pH 5.9 largely lysosomal-type localization is obtained; at 5, one finds largely nuclear localization. This last fact disturbed Aronson, since centrifugate studies had shown cytoplasmic rather than nuclear localization.

INHIBITORS.   Both nuclear and lysosomal localizations are inhibited by 10 mM NaF, 1 mM CuSO$_4$, 5 mM p-chloromercuribenzoic acid, 5 mM Versene. Na$_2$SO$_4$ at 10 mM inhibits lysosomal but not nuclear staining. Substitution of adenosine 3'- or 5'-monophosphate as substrate gives an SO$_4$-insensitive activity; glycerophosphate activity is sensitive to sulfate.

Some question perhaps remains whether the lysosomal reaction is actually deoxyribonuclease, not just acid phosphatase. However, omission of deoxyribonucleic acid does prevent the reaction. (But perhaps glycerophosphate can replace it?)

CYSTEINE DESULFURASE

This enzyme hydrolyzes cysteine to H$_2$S, NH$_3$, and pyruvic acid: HSCH$_2$CH(NH$_2$)COOH + H$_2$O → H$_2$S + NH$_3$ + H$_3$C—CO—COOH.

Jarrett (*J. Histochem. Cytochem.*, **10:** 400, 1962) reports a tetrazolium-formazan method for its localization.

SUBSTRATE.   Mix 20 mg blue tetrazolium (or, preferably, nitro blue tetrazolium) in 10 ml distilled water plus 10 ml 0.1 M phosphate buffer of pH 7.6. Add slowly 64 mg cysteine hydrochloride in 19 ml distilled water. Then add 1 ml 0.1 M MgSO$_4$ and 10 mg pyridoxine 5-phosphate as activator.

PROCEDURE.   Cut fresh frozen tissue in a cryostat at 8 μ, and attach it to coverslip or slides. Immerse in substrate and incubate 18 to 24 h at 37°C. When satisfactory color is attained, rinse in distilled water, fix 10 to 15 min in 10% cold formalin, and mount in an aqueous mountant. Substituting nitro blue tetrazolium permits reduction of the incubation period to 4 h, with sharper localization.

RESULTS: In human skin a strongly positive band is seen at the stratum granulosum.

In some manner, not explained by Jarrett, this reaction is supposed to participate in the cystine cross-linkage of keratin.

INHIBITORS.   Smythe, in Long, p. 496, notes that D-cysteine is inert to the animal enzyme, that pyridoxal phosphate is the coenzyme, and that KCN and As$_2$O$_3$ inhibit.

CARBONIC ANHYDRASE

This enzyme catalyzes the reaction $H_2CO_3 \rightleftharpoons H_2O + CO_2$ over a wide pH range. Its molecular weight has been determined at 30,000, and it contains 1 atom of Zn per molecule.

V. Massey notes in Long, p. 489, marked inhibition by Cu^{2+}, Ag^+, Au^{3+}, Hg^{2+}, Zn^{2+}, and V^{3+}; by oxidants ($KMnO_4$), sulfide, NaN_3, and KCN. Animal but not plant carbonic anhydrase is strongly inhibited by sulfanilamide and other sulfonamides. The plant enzyme is apparently SH-dependent, being inhibited by p-chloromercuribenzoate and similar reagents.

Gomori cited Kurata's method:

1. Fix thin slices in cold acetone for 1 h.
2. Wash briefly in distilled water.
3. Incubate 45 min in a freshly mixed and filtered substrate composed of 50 ml 8% $NaHCO_3$ plus 5 ml 10% $CoCl_2$ or $MnCl_2$.
4. Wash in bicarbonate buffer pH 7.2, dehydrate in alcohols, embed in paraffin, and section.
5. Treat deparaffinized, hydrated sections with 2% dilution of ammonium sulfide to reveal the $CoCO_3$ precipitate as CoS, or with 0.5% periodic acid to convert $MnCO_3$ to MnO_2.

KURATA'S REACTIONS. Carbonic anhydrase accelerates liberation of CO_2 from the $NaHCO_3$ and accelerates precipitation of $CoCO_3$ or $MnCO_3$. This appears highly illogical, since insoluble carbonates should be more rapidly precipitated at higher pH levels, rather than lower, and liberation of CO_2 should decrease the amount of CO_3 ion present.

However, selective staining of erythrocytes and gastric parietal cells was claimed.

Fand et al. (*J. Histochem. Cytochem.*, **7**: 27, 1959), failing to demonstrate carbonic anhydrase by the Kurata method in sites of biochemically demonstrated high activity (endometrium, kidney) and finding the Kurata reaction positive in pancreatic islet beta cells, proceeded to demonstrate that sites yielding the Kurata reaction also reacted with dithizone, including inorganic zinc, insulin, and zinc-insulin models and tissues previously inactivated at 90°C or with KCN or (Braun-Falco) Diamox. Strong reactions to the Kurata method were given also by gelatin Coujard models containing Cd^{2+}, Mn^{2+}, and Ca^{2+}. Fand et al. got strong Kurata staining of renal proximal convoluted tubules in rodents but not in dogs or primates.

In view of Fand's findings, Kurata's method is not to be recommended for localization of carbonic anhydrase.

Pearse, 1960, although agreeing with Fand's conclusions regarding the Kurata method, nevertheless presents Hausler's variant thereof:

SUBSTRATE. Mix 1 ml 0.1 M $CoSO_4$ (1.55% = 2.8% $CoSO_4 \cdot 7H_2O$) and 6 ml 0.1 N H_2SO_4. Just before use, pour in 1 g $NaHCO_3$ dissolved in 50 ml 0.1 M Na_2SO_4 (1.42%), freshly prepared.

METHOD

1. Postfix fresh frozen (cryostat or cold knife) sections 1 h in acetone at 0 to 4°C.
2. Transfer loose-floating sections to substrate at 18 to 20°C for 90 to 120 min. Keep sections floating.
3. Wash for 2 min in distilled water.
4. Treat with dilute (1 : 50) ammonium sulfide for 1 min.
5. Wash, and mount in glycerol gelatin.

RESULTS: Activity sites are black.

INHIBITOR. Carbonic anhydrase is inhibited by 4 mM sodium Diamox (acetazolamide). (The molecular weight of Diamox is given in "The Merck Index" as 222.25; replacing the sulfonamide NH_2 by ONa gives 245.22.)

Like other CoS preparations, these also could be dehydrated and mounted, preferably in Canada balsam, and counterstains may be applied as with alkaline phosphatase technics.

Hansson (*Histochemie*, **11**: 112, 1967) described another method for the histochemical detection of *carbonic anhydrase*. Rosen (*J. Histochem. Cytochem.*, **18**: 668, 1970; **20**: 696, 1972; **20**: 951, 1972; *Histochem. J.*, **4**: 35, 1972) is convinced the method is reliable and specific. Muther (*J. Histochem. Cytochem.*, **20**: 319, 1972) has objected to the method on the ground that (1) CO_2 diffusion is too slow to allow discrimination of enzymatic and nonenzymatic activity, (2) inhibitors are stated to complex with cobalt rather than inhibit enzyme activity, (3) staining or inhibition of staining is stated to occur by reagents shown biochemically not to have or inhibit carbonic anhydrase activity, and (4) no correlation has been demonstrated by the effects of various fixatives employed histochemically and biochemically. Rosen (*J. Histochem. Cytochem.*, **20**: 951, 1972) appears to have answered most of these questions satisfactorily. The reason for the lack of correspondence in the effects of fixatives remains unknown. The Rosen-modified method follows. For a test object, Rosen employed toad or turtle urinary bladder.

1. Tissues are fixed in 3% glutaraldehyde in 0.17 M cacodylate buffer of pH 7.1 for 1 to 2 h.
2. After brief storage in saline solution–soaked gauze, blocks of tissue (1 to 2 cm²) are placed on millipore filter paper (25 μ thick with pore size of 0.45 μ) and incubated (floated on the surface) of the following medium for 20 min at 23°C:

$CuSO_4$	1.75 mM	27.1 mg
H_2SO_4	53 mM	10.6 ml 1 N
$NaHCO_3$	0.157 M	1.319 g
KH_2PO_4	3.5–11.7 mM	47.6–156 mg

Add distilled H_2O to 100 ml.
3. Wash tissues in KH_2PO_4 (0.67 μM pH 5.9).
4. Immerse tissues in a dilute (0.6%) solution of ammonium sulfide for 1 to 2 min at 23°C.
5. Mount tissues in a water mounting medium such as Kaiser's.

For electron microscopic observations, specimens are washed with saline solution after step 4 above, dehydrated, and embedded in Epon 812. Postfixation in osmium could not be utilized because of solubilization of the precipitate.

The reaction is dependent upon the loss of CO_2 from the surface, local hydroxyl accumulation, and precipitation of the cobalt salt. Rosen noted that if tissues on filter paper submerged below the surface, they failed to stain. Incubation time was 12 min with high concentrations of KH_2PO_4 and 30 min at low concentrations. Incubation of specimens with 10 mM acetazolamide resulted in complete inhibition of enzyme activity.

Muther (*J. Histochem. Cytochem.*, **20**: 319, 1972) conducted a critical evaluation of the present methods for the histochemical detection of carbonic anhydrase. He concluded that staining is the consequence of the formation of an unknown compound containing cobalt in a surface layer associated with alkalinization of the medium and independent of enzyme activity.

Churg (*Histochemie*, **36**: 293, 1973) used several tissues and several fixatives to test the reliability of the Häusler (*Histochemie*, **1**: 29, 1958) and Hansson (*Histochemie*, **11**: 112, 1967) methods for carbonic anhydrase and concluded that the localization of stain did not necessarily reveal enzyme activity.

Gay and Mueller (*J. Histochem. Cytochem.*, **21**: 693, 1973) used a labeled specific inhibitor of carbonic anhydrase to achieve accurate localization of the enzyme. Chickens and quail of various ages received 1 to 25 μl ^{3}H-acetazolamide per g body weight. Three and twenty-four hours later 2-mm^3 pieces of various tissues were removed under anesthesia, frozen at $-160°C$, freeze-dried, fixed with OsO_4 vapor, embedded in Epon, sectioned at 1 μ, and covered with NTB-2 emulsion. Free and loosely bound acetazolamide was rapidly cleared from the body within 24 h. No significant label was found in muscle, cartilage, adipose tissue, and nuclei. Carbonic anhydrase activity was found in proximal and distal collecting tubules and mesangium of the kidney; the acinar but not centroacinar cells of the pancreas; the tubular glands and columnar cells of the shell gland; the lining cells but not zymogen cells of the proventriculus; erythrocytes; erythroblasts; heterophils; lymphoid nodules but not fat and interstitial tissue of bone marrow.

Gay and Mueller (*Science*, **183:** 432, 1974) used the above labeled acetazolamide inhibitor method to demonstrate the presence of carbonic anhydrase in osteoclasts and in sites of new bone formation in chickens. The enzyme was not detected in osteocytes and osteoblasts. Distribution of the enzyme was found to be similar to that described by Faleski et al. (*Fed. Proc.*, **32:** 898A, 1973), who used the fluorescent antibody method.

Gay et al. (*J. Histochem. Cytochem.*, **22:** 819, 1974) engendered antiserums to avian erythrocyte carbonic anhydrase in rabbits and purified the α-globulin fraction. Using fluorescein isothiocyanate-conjugated goat antirabbit α-globulins, Gay et al. were able to localize carbonic anhydrase in gastric mucosal lining cells, the columnar and tubular cells of the shell gland, red blood cells, and osteoclasts. Nonspecific staining of bone was observed.

UREASE

Sen (*Indian J. Med. Res.*, **18:** 79, 1930) reported that urease would act in fairly strong alcoholic solution to liberate ammonia and carbon dioxide from urea and made this the basis of a histochemical method. Of the Co, Ca, Ni, Cu, and Pb ions tested for localization of the carbonate formed, the first seemed best. His procedure follows:

Fix animal tissues 1 h in 60% alcohol containing 1% cobalt nitrate. Add an equal volume of 60% alcohol containing 1% urea, and let stand 48 h at room temperature. Dehydrate with alcohols, embed in paraffin or celloidin, section, hydrate sections, and immerse for some minutes in hydrogen sulfide water or dilute sodium sulfide solution. Wash and mount as usual.

Brief trials of this method have not been satisfactory, and we suggest following Glick's (*J. Natl. Cancer Inst.*, **10:** 321, 1949) alternate-section titrimetric procedure to give approximate localization of urease activity. A block of gastric mucosa is frozen quickly at $-25°C$. With a large cork borer cut a cylindrical block from the surface down to muscularis. Cut serial frozen sections horizontally from the surface downward. Reserve alternate sections for staining and histologic evaluation in comparison with the results of titrimetric assay of intervening sections.

URICASE

Graham and Karnovsky (*J. Histochem. Cytochem.*, **13:** 448, 1965) provide the following method for the histochemical detection of *uricase*. Uricase catalyzes the oxidation of urate by O_2, yielding CO_2 and H_2O_2. The enzyme is found in all mammals except human beings and other primates, in other vertebrates except birds and some reptiles, and in most invertebrates except insects and spiders (Keilin, *Biol. Rev.*, **34:** 265, 1959). Significant amounts are found in the liver and kidneys of mammals. De Duve (*Biochim. Biophys. Acta*, **40:** 186, 1960) noted the presence of uricase, D-amino acid oxidase, and catalase in subcellular microbodies of rat liver parenchymal cells.

With this method horseradish peroxidase in the incubation medium catalyzes the oxidation of 3-amino-9-ethylcarbazole by H_2O_2 generated at sites of uricase activity. The 3-amino-9-ethylcarbazole is oxidized to an insoluble red compound. The method follows:

1. Blocks of tissues quick-frozen in isopentane cooled with a dry ice–acetone mixture are cut at 4 to 8 μ with a cryostat and dried in air on coverslips.
2. Fix sections briefly in dry ice–cooled acetone (10 to 15 min).
3. Rinse sections briefly in 0.15 M (0.88%) NaCl.
4. Incubate sections for 30 to 60 min at 37 to 42°C in the following medium: Dissolve 3 mg 3-amino-9-ethylcarbazole (Aldrich Chemical Co., Milwaukee, Wisconsin) in 1 ml *N,N*-dimethyl formamide and add 10 ml 0.05 M tris HCl buffer of pH 8. Shake the mixture thoroughly, let stand for 2 to 3 min, and then filter. Add the following reagents with vigorous shaking: 3 mg sodium urate, 5 mg EDTA, and 3 mg horseradish peroxidase (type II; Sigma).
5. Wash sections in 0.15 M NaCl.
6. Fix sections for 1 to 2 h in 4% formaldehyde.
7. Wash in 0.15 M NaCl.
8. Mount in Kaiser's glycerol jelly (water mount).

Nuclei may be stained with iron celestin blue between steps 7 and 8 (0.5% solutions of celestin blue are boiled for 3 to 5 min in 5% ferric ammonium alum and filtered). Sections are stained for 3 to 5 min.

Uricase could not be detected in formalin-fixed tissues. Omission of peroxidase results in staining of mast cells and polymorphonuclear leukocytes and no staining of hepatic parenchymal cells. The contrary was observed in the presence of exogenous peroxidase. Graham and Karnovsky believe it is likely that in the absence of exogenous peroxidase, H_2O_2 generated during uricase-catalyzed oxidation of urate diffused to sites of endogenous peroxidase activity.

Incubation in the presence of 95% O_2 markedly enhanced the reaction in tissue sites identical to those observed with incubations with lesser oxygen content. Sections either preincubated with the inhibitor 2,6,8-trichloropurine (K & K Laboratories, Plainview, New York) at 0.2% in buffer or with the inhibitor at the same concentration in the incubation mixture resulted in complete inhibition.

In livers of rats and guinea pigs uricase activity appeared as small, discrete granules scattered throughout parenchymal cells. After prolonged incubation, a few scattered granules appeared in distal tubular cells and collecting duct cells of kidneys of rats and guinea pigs. In all cases, the granular distribution differed from that revealed by stains for acid phosphatase.

Yokota (*Histochemie*, **36**: 21, 1973) reported on the localization of urate oxidase using an immunofluorescent method; however, the localization obtained in liver cells varied from that described by Graham and Karnovsky (*J. Histochem. Cytochem.*, **13**: 448, 1965) and by Reddy et al. (*Am. J. Pathol.*, **56**: 351, 1969). Yokota obtained a diffuse cytoplasmic staining, whereas the other workers using other methods obtained a discrete granular pattern.

OXIDASES, PEROXIDASES, DEHYDROGENASES

For practical purposes these fall into three classes, the relatively stable, moderately formaldehyde-resistant hemoglobin peroxidase; the myeloperoxidase or verdoperoxidase—also relatively stable, but somewhat less resistant—of the granular leukocytes and their precursors; and the highly labile tissue oxidases exemplified by cytochrome oxidase (or G-Nadi oxidase of the earlier German writers), succinic dehydrogenase, cysteine desulfurase, and other similar enzymes demonstrable by the tetrazolium-formazan systems.

The methods for their detection may also be classed according to the substances used to yield colored products in the presence of hydrogen peroxide, or without it, and some of these methods may be used for both the stable and the highly labile enzymes.

Hemoglobin Peroxidase

We will cite two general methods for hemoglobin peroxidase: Dunn's modifications of the "zinc leuco" sulfonated triphenylmethane dyes, and one of the nitroprusside benzidine methods.

Lison used first acid fuchsin, and later some patent blue reduced by nascent hydrogen to its leuco state. Dunn (*Stain Technol.*, **21**: 65, 1946) identified the patent blue as patent blue V (C.I. 42051) and later (*Arch. Pathol.*, **41**: 676, 1946) substituted the related dye cyanol FF (C.I. 43535), using identical amounts. The latter dye is slightly more violet but otherwise equivalent, and we prefer the patent blue V.

THE LISON-DUNN LEUCO PATENT BLUE METHOD.   Fix blocks 3 to 5 mm thick in buffered 10% formalin for 24 to not more than 48 h. Prepare paraffin sections at 5 to 8 μ.

Make a stock 1% aqueous solution of patent blue V. To 100 ml of this add 10 g granulated metallic zinc and 2 ml glacial acetic acid and boil until completely decolorized. Stopper well and store. This leuco patent blue solution is stable.

For use filter out 10 ml, and add 2 ml glacial acetic acid and 0.1 ml 30% hydrogen peroxide. This must be freshly mixed.

1.   Bring sections to water.
2.   Stain 3 to 5 min in the leuco patent blue peroxide reagent.
3.   Rinse in water.
4.   Counterstain 30 to 60 s in 0.1% safranin O in 1% acetic acid or in an aqueous carmine.
5.   Dehydrate, clear through a graded alcohol and xylene sequence, and mount in synthetic resin.

RESULTS: Nuclei are red; cytoplasm is light pink; hemoglobin, dark blue-green; oxidase granules, dark blue.

When aniline blue (C.I. 42755) is substituted for patent blue in the foregoing procedure, it decolorizes with zinc and hot acetic acid to a light transparent green. As with patent blue, it also gives a deep blue hemoglobin peroxidase reaction. Mounting is perhaps best in Karo or glycerol gelatin, and preparations may conveniently be examined at once in water. Alcohol tends to extract the color of both the blue dyes on dehydration (Janigan and Lillie, 1963, unpublished).

THE LEPEHNE-PICKWORTH METHOD.   This method of demonstration of cerebral capillary distribution is essentially a benzidine and nitroprusside oxidase method for hemoglobin. The technic follows (emended from Mallory):

Fix brain tissue in 10% formalin for 1 to 3 weeks. Cut frozen sections at 200 to 300 μ (0.2 to 0.3 mm).

1.   Wash sections 30 min in distilled water.
2.   Place sections for 30 min at 37°C in benzidine and nitroprusside reagent: Dissolve 100 mg benzidine in 0.5 ml glacial acetic acid. Add 20 ml distilled water. Dissolve 100 mg sodium nitroprusside in 10 ml distilled water. Mix the two solutions and add 70 ml distilled water. Make this reagent fresh each time. While sections are in this mixture, agitate frequently.
3.   Wash 10 s in distilled water.

4. Place in 0.04 to 0.05% hydrogen peroxide at 37°C for 30 min, shaking frequently. This dilution must be freshly prepared. Use 0.1 ml 30% hydrogen peroxide in 70 ml water.
5. Wash in distilled water.
6. Dehydrate in 70, 95, and 100% alcohol, allowing sections to stand in each until diffusion currents are no longer evident. Place in 100% alcohol and xylene, and then in two or more changes of xylene until clear. Mount in balsam.

RESULTS: Blood cells in the capillaries are colored black; other structures remain pale gray.

Romeis cites the Doherty, Suh, and Alexander (*Arch. Neurol. Psychiat.*, **40:** 158, 1938) variant of the Pickworth method. Step 2 above is shortened to 10 min, step 4 to 20 min at 20°C; but if the capillary net is not yet black, incubate further at 37°C until the background is bleached. The benzidine solution is composed of 0.5 g benzidine dissolved in 50 ml 100% alcohol, mixed with 10 ml 1% aqueous sodium nitroprusside solution (freshly made) and 40 ml distilled water. Deteriorated nitroprusside gives a greenish color on mixing. The peroxide solution contains also 10 ml fresh 1% aqueous sodium nitroprusside, 2 ml glacial acetic acid, 0.5 ml 30% hydrogen peroxide, 50 ml 100% alcohol, and distilled water to make 100 ml.

Myoglobin Peroxidase

Drews and Engel have applied the benzidine peroxidase method to the study of muscle and have successfully demonstrated a peroxidase activity which they assign to myoglobin (*J. Histochem. Cytochem.*, **9:** 206, 1961).

BENZIDINE METHOD OF DREWS AND ENGEL FOR MYOGLOBIN PEROXIDASE

1. Fix fresh muscle 1 to 14 days in formol–saline solution. Surgical material is preferable in human beings, and in animals it may be desirable to perfuse with 0.85% NaCl to remove blood before fixing. In this case the formol–saline solution may follow the saline perfusion.
2. Cut frozen sections at 10 to 75 μ, and place in benzene [*sic*] for 1 min.
3. Stain 5 to 8 min in the modified Van Duijn solution given below.
4. Rinse in 0.85% saline solution, mount in glycerol gelatin, and study at once. The benzidine blue has its usual tendency to fade.

RESULTS: Dark blue microcrystalline deposits are seen at the level of the I bands in striated muscle and in smooth muscle.

MODIFIED VAN DUIJN SOLUTION.   Dissolve 0.5% (100 mg) benzidine base in (20 ml) 0.85% NaCl at 80°C. Cool and filter. To 9 ml add 1 ml saturated (about 40% at 25°C) ammonium chloride solution and 1 drop (about 0.05 ml) 3% hydrogen peroxide solution (freshly diluted from the stock 30% solution). Make benzidine solution fresh every hour.

The postfixation of Villamil and Mancini—5 min in saturated alcoholic picric acid solution —directly after step 3, might act to preserve the color. After picroalcohol, wash in water, dehydrate, clear, and mount in synthetic resin as usual.

MYELOPEROXIDASE

This enzyme has been isolated from leukocytes by Agner (*Acta Physiol. Scand.*, vol. 2 *Suppl.* 8, 1941). It appears to be the enzyme responsible for both the benzidine and naphthol peroxidase reactions and for the Winkler-Schultze–M-Nadi oxidase reaction. Certainly there is a very considerable correspondence in reagents effective in inhibiting the Washburn benzidine nitroprusside peroxidase test and the Winkler-Schultze reaction. Apparently the necessary peroxide for the latter reaction must be added if the reagent is quite fresh or is formed spontaneously in contact with air within the dimethyl-*p*-phenylenediamine and α-naphthol mixture in a fairly short time.

Gomori's statement that leukocyte oxidase is not an enzyme but is fat peroxide ignored Agner's work. It was apparently based on the demonstration by Lison (*Bull. Soc. Chim. Biol.*, **18:** 185, 1936) that fatty peroxides in adrenal cortical substance were capable of completing the indophenol synthesis from dimethyl-*p*-phenylenediamine and α-naphthol without access of air or addition of peroxide, and the considerable correspondence shown by Sehrt (*Munch. Med. Wochenschr.*, **74:** 139, 1927) between the destruction by various reagents of the sudanophilia of leukocytes and of the Winkler-Schultze oxidase reaction.

Lillie and Burtner have shown that a number of reagents promptly destroy myeloperoxidase (and M-Nadi oxidase) while conserving sudanophilia, while certain other reagents conserve the enzyme reactions quite well and destroy sudanophilia. There also appears to be no doubt that the sudanophilia of the granules in neutrophil leukocytes is not due to a fat (*J. Histochem. Cytochem.*, **1:** 8, 1953).

Of all the various technics for leukocyte peroxidase, we generally prefer the Washburn (*J. Lab. Clin. Med.*, **14:** 246, 1928) method. For this we prefer fixation of films for 10 min in 75% alcohol. This has less damaging effect on the enzyme than 100 or 95% alcohol or any of the formaldehyde methods.

THE WASHBURN METHOD (Lillie, 1952, modified from Washburn)

1. Fix air-dried blood or marrow films 10 min in 75% alcohol or in formaldehyde vapor for 30 min as follows: Place a piece of glass rod diagonally in the bottom of a Coplin jar to raise slides a little off the bottom. Add 1 to 2 ml 40% formaldehyde. Put in slides smear end up. Cover jar. After fixing, wash slides in two changes of distilled water.
2. Mix 1 ml 30% aqueous sodium nitroprusside with 99 ml 0.3% benzidine in 95% alcohol. This mixture is said to keep quite well in the cold, though we prefer to keep the benzidine solution alone and add the freshly prepared nitroprusside solution at time of using.

   To the alcoholic benzidine–nitroprusside mixture add an equal volume of freshly diluted 0.04% hydrogen peroxide, diluting the 30% stock solution 0.1 ml in 75 ml distilled water, pipette at once onto slides, and let stand 5 min. To exclude negative reactions, double the time. Allow about 2 ml of the final mixture for each slide.
3. Wash 2 min or more in running tap water.
4. Counterstain 10 min by the "accelerated Giemsa stain."
5. Rinse in water, dry smears, and examine in an immersion oil of the mineral oil type. Dehydrate sections with acetone, clear with xylene, and mount in neutral synthetic resin.

RESULTS: Peroxidase granules in neutrophils are greenish or bluish black; in eosinophils, somewhat greener or browner in tone; nuclei, violet to red-purple; erythrocytes, orange-pink. When reagents which inhibit myeloperoxidase are interposed after step 1, erythrocytes often take a browner tone and may present much granular or crystalline black deposit or long, needle-shaped crystals.

In the foregoing technic Washburn added 0.3 g basic fuchsin to the alcoholic benzidine–nitroprusside solution, and after washing the reaction mixture off, decolorized to faint pink with 95% alcohol, counterstained with Wright's stain, rinsed, dried, and examined.

The following technics include benzidine and hydrogen peroxide as reactive materials, but no nitroprusside.

THE BENZIDINE PEROXIDASE REACTION.   The following technics for the benzidine peroxidase reaction and statements regarding it are a composite derived from Graham (*J. Med. Res.*, **39**: 15, 1918), the texts of Mallory, Romeis, and Schmorl and from Endicott's work in our laboratory. Graham fixed fresh smears in a fresh 10% formalin-alcohol for 1 to 2 min and then washed in water. Loele used frozen sections of formalin-fixed tissue which are collected from the microtome in water. The staining solution is a saturated aqueous or 40% alcohol solution of benzidine to which 0.2 to 0.67 ml 3% hydrogen peroxide per 100 ml is added. Loele considered the aqueous solution more stable and prescribed shaking up 0.5 g benzidine per 100 ml distilled water, filtering, and adding 2 ml 1% hydrogen peroxide to each 100 ml filtrate. The 1% peroxide is a 1 : 30 dilution of the concentrated 30% solution. The Graham-Loele technic follows:

1.   Stain smears 5 to 10 min, sections 3 to 5 min in the benzidine-peroxide reagent.
2.   Wash in water.
3.   Counterstain briefly in 0.1 to 0.5% methylene blue.
4.   Wash in water.
5a. Blot smears, dry, and examine in modified mineral oil of n_D 1.515.
5b. Float sections onto slides, blot, dehydrate with 95 and 100% alcohol, and clear with 100% alcohol and xylene mixture and two changes of xylene. Blotting between changes is necessary to keep sections on slides. Mount in balsam or synthetic resin.

RESULTS: Peroxidase granules are yellow to brown; nuclei, blue.

SATO'S TECHNIC (*Tohoku J. Exp. Med.*, **3**: 7, 1926, cited in Sato, *J. Lab. Clin. Med.*, **13**: 1058, 1928).   In this technic the same benzidine and hydrogen peroxide solution may be used, though usually only 0.07 to 0.13 ml 3% peroxide is added to each 100 ml filtrate, and Sato's benzidine solution was only 0.25%. The technic:

1.   Mordant unfixed fresh air-dried smears in 0.5% aqueous copper sulfate ($CuSO_4 \cdot 5H_2O$) solution for 1 min. Sato used 0.3% copper sulfate in 0.1% acetic acid.
2.   Rinse very quickly in water (Mallory dipped films 3 times in water; Gradwohl simply drained the films).
3.   Stain in the benzidine–hydrogen peroxide for 2 to 8 min.
4.   Wash in water.
5.   Stain 20 s to 2 min in 1% aqueous safranin.
6.   Wash in water, dry, and mount in balsam, cedar oil, or synthetic resin, or examine directly in immersion oil.

Villamil and Mancini (*Rev. Soc. Argent. Biol.*, **23**: 215, 1947; **24**: 337, 1948) described a benzidine-peroxidase technic for the demonstration of labile oxidases in thyroid epithelial cells. Frozen sections of unfixed tissue are required, and as little as 10 to 15 min treatment with alcohol or formalin destroys the enzyme, though a similar exposure to physiologic saline solution is tolerated.

AMMONIUM MOLYBDATE BENZIDINE PEROXIDASE REACTION OF VILLAMIL AND MANCINI

1.   Immediately on excision of human surgical or animal specimen cut frozen sections at 10 to 15 μ.
2.   Immerse sections for 3 to 6 min in 1% ammonium molybdate in 0.9% sodium chloride solution.
3.   Transfer to a saturated benzidine solution in 0.9% sodium chloride solution, to which a

few drops of hydrogen peroxide are added. In about 3 min the sections become an intense blue.

4. Transfer directly to saturated (9%) picric acid solution in 95% alcohol, and fix for 5 min.
5. Wash in water, dehydrate, clear, and mount as usual.

RESULTS: Oxidase granules are dark blue in yellow cytoplasm of gland cells, histiocytes, and adjacent endothelial cells. To distinguish the more stable leukocyte oxidase, some sections should be immersed in 10% formalin for 15 min before step 2.

THE GRAHAM α-NAPHTHOL PYRONIN STAIN (*J. Med. Res.*, **35**: 231, 1916). This stain for oxidase granules prescribes 1- to 2-min fixation of fresh air-dried smears in fresh 10% formalin-alcohol.

1. Wash in water.
2. Stain 4 to 5 min in 1 g α-naphthol dissolved in 100 ml 40% alcohol to which 0.2 ml 3% hydrogen peroxide is added shortly before using.
3. Wash 15 min in running water.
4. Stain 2 min in 0.1 g pyronin (C.I. 45005 = Y or 45010 = B), 4 ml aniline, and 96 ml 40% alcohol.
5. Wash in water.
6. Stain 30 to 60 s in 0.5% aqueous methylene blue.
7. Wash in water, blot, dry, and mount in balsam.

RESULTS: Neutrophil granules giving the oxidase reaction are purplish red; eosinophil granules are larger, lighter red, and more refractile; basophil granules are deep purple; cell nuclei are blue; cytoplasm is pale blue; erythrocytes are greenish yellow to pink.

The above α-naphthol peroxide solution may be used for 4 or 5 days. The aniline pyronin is relatively stable.

Ritter and Oleson (*Arch. Pathol.*, **43**: 330, 1947) carried out the α-naphthol pyronin method on blocks of fixed tissue which were then embedded in paraffin and sectioned. They first fix for 24 h in 10% alcoholic formalin containing 1 ml 0.1 N sodium hydroxide per 100 ml. After washing 10 min in running water the blocks were immersed for 24 h in 100 ml fresh 1% α-naphthol in 40% alcohol, to which 0.2 ml 30% hydrogen peroxide is added at the moment of using. Then, after washing for 10 min in running water they transferred to 0.1% pyronin, 4% aniline, 40% alcohol for 3 to 24 h. After this tissues were dehydrated for 1 h in 80% and 2 h in 95% alcohol and for 2 h in two changes of xylene for 20 min each and embedded in paraffin. Sections were cut at 4 μ and mounted without counterstaining, or they were counterstained with alum hematoxylin or with a rather bluish Romanowsky stain. Eosin stains, if used, should be kept light to avoid confusion with the red peroxidase granules.

They observed that treatment with 100% alcohol before performance of the α-naphthol-peroxide treatment weakens the reaction and that xylene treatment abolishes it.

The Winkler-Schultze reaction and the Nadi reaction depend on the synthesis of indophenol, the indophenol blue of the German writers, from α-naphthol and dimethyl-*p*-phenylenediamine in the presence of air. Lison referred to the reaction as a phenolase reaction.

The Winkler-Schultze method depends on an alkaline solution of α-naphthol; the Nadi method dilutes a 10% alcoholic solution one hundredfold with water, and it utilizes preferably somewhat alkaline buffers during the staining procedure.

THE WINKLER-SCHULTZE REACTION, THE M- AND G-NADI OXIDASES, CYTOCHROME OXIDASE. Person and Fine (*J. Histochem. Cytochem.*, **9**: 190, 197, 1961) used *p*-dimethylaminoaniline (dimethyl-*p*-phenylenediamine) and α-naphthol in studies on G- and M-Nadi oxidases.

For the G-Nadi reaction they used fresh frozen sections; for the M-Nadi reaction, the frozen sections were fixed in 10% formalin for 20 min and washed in four changes of distilled water, for 2 min each.

The substrate was prepared at 10 mM $(CH_3)_2NC_6H_4NH_2 \cdot 2HCl$ and 10 mM α-naphthol dissolved in a minimal amount of alcohol and made up to volume in 0.1 M phosphate buffer.

Positive reactions for M-Nadi oxidase are obtained in parotid and submaxillary glands as well as in cells of the myeloid series.

About 10- to 20-min incubation was required for G-Nadi oxidase, an hour or more for the M-Nadi reaction. This is much slower than the Winkler-Schultze reaction on blood and marrow smears.

As far as we have seen at this date, the p-aminodiphenylamine of Burstone's studies on cytochrome oxidase has not been utilized in the M-Nadi reaction.

Our 1953 technic (*J. Histochem. Cytochem.*, **1**: 8) for the Winkler-Schultze reaction is emended and combined with the Gräff method in accordance with papers of Person et al.

1. Prepare fresh frozen sections by the Adamstone-Taylor cold knife method or in a cryostat, and affix them to clean slides. Prepare blood, marrow, or tissue smears or imprints as usual.
2. For the M-Nadi reaction, fix at once for 20 min in ice-cold 70% alcohol containing 10% formalin. Wash in four changes of distilled water, 2 min each. For the G-Nadi reaction for cytochrome oxidase omit this step.
3. Prepare fresh substrate: Dissolve 21 mg p-amino-N,N-dimethylaniline dihydrochloride and 14.3 mg α-naphthol in 0.1 to 0.2 ml ethanol, and add 0.1 M phosphate buffer of pH 7.6 (Person: 6.4 to 8.1) to make 10 ml. Lay slides face up on staining rack, and deposit 1 to 2 ml on each. Rock gently to cover smear or section. Or for tissue, lay slides flat in a large petri dish and flood, and cover to maintain humidity. Stain films 3 to 10 min (sections, up to 2 h), watching color development. Rinse quickly in distilled water.
4. Preparations may be mounted at this point directly in glycerol gelatin. Immerse in dilute Lugol's solution (0.2% iodine) or in 40% ammonium heptamolybdate $(NH_4)_6 \cdot Mo_7O_{24} \cdot 4H_2O)$ for 3 min. Wash in distilled water [containing a trace of lithium carbonate (1 : 20,000) to blue after iodine].
5. Films may be counterstained with Giemsa as usual, washed, and dried. Mount sections, preferably without counterstain, in glycerol gelatin.

RESULTS: Sites of activity are blue-black to dark green.

Omission of formaldehyde from the 75% alcohol enhances activity in neutrophil leukocytes but permits severe diffusion artifacts.

For G-Nadi oxidase, substitution of Burstone's p-aminodiphenylamine plus 3-amino-9-ethylcarbazole substrate is recommended, and preferably his complete technic should be followed.

LATER TECHNICS FOR CYTOCHROME OXIDASE.   A number of new reagents have been introduced for the indophenol synthesis of the G-Nadi reaction, which is now regarded as a demonstration of cytochrome oxidase activity. (See Table 10-15.)

In 1958 Nachlas, Crawford, Goldstein, and Seligman (*J. Histochem. Cytochem.*, **6**: 445) replaced the p-amino-N,N-dimethylaniline of the Nadi mixture with 4-amino-1-N,N-dimethylnaphthylamine. On oxidation in the presence of α-naphthol this yielded a purple pigment, indonaphthol purple, which gave satisfactory localization and stability.

Nachlas et al. direct: Quickly freeze tissue at $-80°C$, section at 6 to 8 μ in the cryostat at $-20°C$, and thaw sections onto slides in the cryostat. After incubating in the substrate as prescribed, rinse in saline solution, and mount in glycerol gelatin.

Burstone (*J. Histochem. Cytochem.*, **7**: 112, 1959) explored further series of amines and phenols as reagents for the indophenol synthesis, and later (ibid., **8**: 63, 1960) he recommended substrates containing 4-amino-diphenylamine and either 4-amino-4'-methoxydiphenylamine or 9-ethyl-3-aminocarbazole. With these substrates, addition of cytochrome, required with simple amines, could be eliminated. As in the method of Nachlas et al., a small amount of catalase is still added to eliminate confusing action of peroxidases.

Burstone likewise used fresh frozen sections mounted on slides, which could be stored at $-20°C$ for several days. After incubation, slides are fixed 1 h in 10% formalin containing 10% cobaltous acetate (4H$_2$O) or similar (0.4 M?) solutions containing nickel chloride, ferric ammonium sulfate, cadmium sulfate, uranyl nitrate, or lead nitrate. After 5 min washing, sections were mounted in glycerol gelatin or dehydrated, cleared, and mounted in Permount. The metal chelation increased the permanency of the stain. Inhibition of the above reaction is accomplished by inclusion of 1 mM KCN, Na$_2$S, or NaN$_3$ in the medium.

In mixing substrates dissolve the naphthol and mix with buffer, then add catalase, which inhibits peroxidases, and, for rather inactive tissues, the cytochrome c; this may be replaced by the same volume of water with such active tissues as heart, stomach, or brain. Just before using, mix in the freshly prepared arylamine.

The pH 7.4 tris HCl buffer contains 21.1 g tris(hydroxymethyl)aminomethane, 170 ml 1 N HCl, and distilled water to make 1 l.

Montagna and Yun (*J. Histochem. Cytochem.*, **9**: 694, 1961) favored particularly the 8-amino-1,2,3,4-tetrahydroquinoline plus N-phenyl-p-phenylenediamine substrate of Burstone (ibid., **9**: 59, 1961); Baker and Klapper (ibid., **9**: 713, 1961) also used one of Burstone's methods; Tewari and Bourne (ibid., **10**: 42, 619, 1962) employed both the Nachlas and the Burstone procedures.

TABLE 10-15 SUBSTRATES FOR CYTOCHROME OXIDASE

Substrates	Nachlas et al., 1958	Burstone, 1960	Burstone, 1960	Burstone, 1961
Arylamine.............	4-Amino-1-N,N-dimethylnaphthylamine, 16 mg/8 ml	4-Aminodiphenylamine, 6–9 mg	4-Aminodiphenylamine, 6–9 mg	4-Aminodiphenylamine, 6–9 mg
Naphthol	α-Naphthol, 8 mg/6 ml	4-Amino-4'-methoxydiphenylamine, 6–9 mg in alcohol 0.3 ml	9-Ethyl-3-aminocarbazole, 6–9 mg in alcohol 0.3 ml	8-Amino-1,2,3,4-tetrahydroquinoline, 6–9 mg in alcohol 0.3 ml
Cytochrome c.........	30 mg/6 ml			
Catalase	60 μg/2 ml	1.2 mg	1.2 mg	1.2 mg
Buffer	Phosphate 0.1 M	Tris HCl 0.2 M	Tris HCl 0.2 M	Tris HCl 0.2 M
Volume..............	6 ml	9 ml	9 ml	9 ml
pH..................	7.4	7.4	7.4	7.4
Final volume	30 ml	30 ml	30 ml	30 ml
Incubation time........	10–30 min	15–60 min	15–60 min	15–60 min
Temperature..........	25°C	25°C	25°C	25°C

Sources: Nachlas et al., *J. Histochem. Cytochem.*, **6**: 445, 1958; Burstone, *J. Histochem. Cytochem.*, **8**: 63, 1960; Burstone, *J. Histochem. Cytochem.*, **9**: 59, 1961.

Pretreatment with 0.1 % potassium cyanide solution for a few minutes or inclusion of 1 mM in the substrate abolishes the G-Nadi reaction; the Winkler-Schultze reaction is not affected. The alkali α-naphthol solution of that method cannot be used for demonstration of the labile oxidase (even if buffered to the same pH level?). Ordinary formaldehyde fixation prevents the G-Nadi oxidase reaction but not the Winkler-Schultze, though Gräff found tissues fixed in formalin and buffered with phosphates to pH 7.3 to 7.6 for an unstated period suitable for his Nadi oxidase reaction.

Graham et al. (*J. Histochem. Cytochem.*, **13**: 150, 1965) note that one of the problems with the use of benzidine for the detection of peroxidase activity is rapid fading of the blue reaction product. They prescribe the following method for *peroxidase* using 3-amino-9-ethylcarbazole, which forms a red stain under prescribed conditions.

1. Fix small blocks of tissues in 4% formaldehyde in 0.1 M phosphate buffer at pH 7.3 for 18 h at 4°C.
2. Wash tissues for at least 24 h at 4°C in 5 to 30% sucrose solution.
3. Cut sections at 8 μ in a cryostat, and dry sections in air.
4. Incubate sections for 2 to 5 min in the medium prepared in the following manner: Dissolve 2 mg of 3-amino-9-ethylcarbazole (Aldrich) in 0.5 ml N,N-dimethyl formamide; add and mix 9.5 ml 0.05 M acetate buffer of pH 5. Immediately prior to use, add 1 drop (about 0.05 ml) 3% H_2O_2.
5. Wash sections in distilled water.
6. Mount in an aqueous mount such as Kaiser's glycerol jelly.

A red reaction product was observed in sites analogous to those observed using benzidine. No fading of stained sections has been observed by Graham et al. for 1 year. Alcoholic incubation media are unsuitable because of solubility of the reaction product.

Graham and Karnovsky (*J. Histochem. Cytochem.*, **14**: 291, 1966) recommend the following method for the light and electron microscopic localization of peroxidase. They injected horseradish peroxidase into mice and studied the distribution in renal tissues.

1. Slice tissues 3 to 4 mm thick, and fix for 4 to 5 h at 4°C in either 5% glutaraldehyde in 0.1 M phosphate buffer of pH 7.6 or in a 4% formaldehyde and 5% glutaraldehyde solution in 0.1 M cacodylate buffer of pH 7.2 prepared in the following manner: Add 2 g paraformaldehyde to 25 ml distilled water in a flask, and heat to 60 to 70°C, stirring; add 1 M NaOH drop by drop until the solution clears. Cool under flowing tap water, and add 5 ml 50% glutaraldehyde, 20 ml 0.2 M cacodylate buffer of pH 7.2 with 25 mg calcium chloride, and mix thoroughly.
2. Wash tissues overnight in 0.1 M cacodylate buffer of pH 7.2.
3. Cut frozen sections at 40 μ.
4. Incubate sections in a saturated solution of 3,3′-diaminobenzidine (either the free base or the tetrahydrochloride is satisfactory; they may be obtained from Sigma) in 0.05 M tris–HCl buffer of pH 7.6 containing 0.01% H_2O_2. Saturated solutions of the free base require filtration, whereas saturated solutions of the hydrochloride do not.
5. Wash sections in three changes of distilled water.
6. Postfix for 90 min in 1.3% OsO_4 in *sym*-collidine buffered to pH 7.2 and containing 5% sucrose.
7. Dehydrate in graded ethanols, and embed in Epon 812 (Luft, *J. Biophys. Biochem. Cytol.*, **9**: 409, 1961).
8. Sections may be stained with lead (Karnovsky, *J. Biophys. Biochem. Cytol.*, **11**: 729, 1961) although this is optional.

Straus (*J. Histochem. Cytochem.*, **19**: 682, 1971) reports complete suppression of horse-radish peroxidase in sections treated with absolute methanol containing 1% sodium nitro-ferricyanide and 1% acetic acid (15 min at 22°C). This treatment also weakens the antibody reaction. Streefkerk (*J. Histochem. Cytochem.*, **20**: 829, 1972) noted enhanced inhibition of erythrocyte pseudoperoxidase by treatment of unfixed sections with 0.006% H_2O_2 or with 0.0125% H_2O_2 in phosphate-buffered (0.025%) saline solution for 20 min at 23°C after treatment of sections in absolute methanol for 20 min at 23°C.

Hidaka and Udenfriend (*Arch. Biochem. Biophys.*, **140**: 174, 1970) and Straus (*J. Histochem. Cytochem.*, **20**: 949, 1972) note complete inhibition of horseradish peroxidase by phenyl-hydrazine. Straus observed complete inhibition of horseradish peroxidase activity in formalin-fixed sections of lymph nodes subjected to 0.05% phenylhydrazine hydrochloride in 0.05 *M* phosphate buffer of pH 7.1 for 1 h at 37°C, although no inhibition of endogenous peroxidase was detectable.

Novikoff et al. (*J. Histochem. Cytochem.*, **20**: 1006, 1972) modified their earlier (*J. Histochem. Cytochem.*, **17**: 675, 1969) DAB (alkaline-3,3'-diaminobenzidine) method for microperoxisomes by the addition of KCN [0.1 *M* (0.65%), 0.5 ml] to a final concentration of 5 m*M* in the incubation medium. This results in reduction of staining of mitochondria, permitting scanning of sections by light microscopy.

Novikoff et al. (*J. Histochem. Cytochem.*, **21**: 737, 1973) noted that until 1973 micro-peroxisomes had been found in 24 cell types of various tissues.

Novikoff and Goldfisher (*J. Histochem. Cytochem.*, **17**: 675, 1969) prescribe the following method for the demonstration of *peroxisomes* (microbodies) and mitochondria at the light and electron microscopic level employing diaminobenzidine.

DEMONSTRATION OF PEROXISOMES

1. Fix tissues in 3% glutaraldehyde in 0.1 *M* cacodylate buffer of pH 7.4 for 3 to 6 h at 4°C.
2. Cut cryostat sections at 10 to 25 μ, and incubate in the following medium at 37°C:

Diaminobenzidine tetrahydrochloride (Sigma)	20 mg
0.05 *M* 2-amino-2-methyl-1,3-propanediol buffer, pH 10 (Sigma)	9.8 ml
1% hydrogen peroxide (freshly prepared)	0.2 ml

Adjust the pH to 9 if necessary, and filter if a precipitate forms. Inclusion of 0.01 *M* 3-amino-1,2,4-triazole (an inhibitor of catalase) in the incubation medium results in complete inhibition of staining of peroxisomes. Potassium cyanide up to 0.1 *M* does not inhibit staining.

DEMONSTRATION OF MITOCHONDRIA.   Fixed and sectioned tissues as described above are incubated in the following medium at 37°C:

Diaminobenzidine tetrahydrochloride	20 mg
0.05 *M* acetate buffer, pH 5	8.9 ml
0.05 *M* manganese chloride	1.0 ml
Freshly prepared 0.1% hydrogen peroxide	0.1 ml

Adjust the pH to 6 and filter if a precipitate forms.

Novikoff and Goldfisher note that both outer and inner membranes of all mitochondria stain but that the rate of staining varies widely depending upon the cell type. Amino-triazole (0.1 *M* 0.84%) has no effect on mitochondrial staining, whereas complete inhibition of staining occurs with the inclusion of 1 m*M* cyanide in the incubation mixture. Novikoff and Goldfisher speculate that inner membrane staining may be due to cytochrome c and that staining in the outer membrane may be due to cytochrome b_5. Sottocasa et al. [in R. W. Estabrook and M. E. Pullman (eds.), "Methods in Enzymology," vol. 10, p. 448, Academic, New York, 1967] have shown the presence of cytochrome b_5 in outer membranes of mitochondria.

Strum and Karnovsky (*J. Ultrastruct. Res.*, **31**: 323, 1970) described endogenous peroxidase activity in acinar cells of submaxillary glands of rats at the ultrastructural level. Peroxidase activity was noted in perinuclear cisternae, endoplasmic reticulum, and ribosomes.

Baker and Yu (*Am. J. Anat.*, **131**: 55, 1971) employed the Nakane (*J. Histochem. Cytochem.*, **16**: 557, 1968) peroxidase-labeled antibody method to relate thyrotropic cells of the rat pituitary to staining characteristics. Thyrotropic cells were noted to stain with aldehyde fuchsin and with the periodic acid Schiff procedure.

Yamashima and Barka (*J. Histochem. Cytochem.*, **21**: 42, 1973) studied the development of endogenous peroxidase in rat submandibular glands. Peroxidase developed in the cisternae of the rough endoplasmic reticulum and in the nuclear envelopes of acinar cells on the seventeenth day of gestation.

Hand (*J. Histochem. Cytochem.*, **22**: 207, 1974) noted peroxisomes are also evident in striated muscle of rats.

Plapinger et al. (*Histochemie*, **14**: 1, 1968) synthesized a series of *p*-substituted aromatic diamines, some thioureidonaphthols, and some mercaptonaphthols, and studied their cytochemical behavior in the Nadi reaction using fresh frozen sections of rat heart muscle. They found that *N*-benzyl-, *N*-4-methylbenzyl-, and *N*-4-methoxybenzyl-*p*-phenylenediamine were good reagents for the demonstration of cytochrome oxidase activity in the Nadi reaction with light microscopy. Blue indoaniline dyes are produced with 1-naphthols, and self-condensations do not develop. The indoaniline dyes are osmiophilic; however, they have the undesirable behavior of developing a droplet formation. The objectionable droplet conformation was overcome in the use of amine reagents with either a carbonyl or a carboxyl in the 2 position or by the insertion of a bulky substituent in the 5 position of 1-naphthol coupling reagents.

CYTOCHROME OXIDASE

Plapinger et al. prescribe the following method for *cytochrome oxidase*. Their reagents, including *N*,*N*'-bis (*p*-aminophenyl)-1,3-xylylenediamine, are available from Polysciences Inc., Paul Valley Industrial Park, Warrington, Pennsylvania 18976.

1. Small blocks of tissue are either fixed for 1 h at 4°C in 4% depolymerized paraformaldehyde or frozen directly in liquid isopentane cooled with dry ice–acetone at −80°C and sectioned at 6 μ with a cryostat.

2. Sections on clean coverslips are incubated for 20 to 30 min at 23°C depending upon the amount of enzyme in the tissues.

Phosphate buffer (0.1 M pH 7.4)	3 ml
1-Naphthol or substituted 1-naphthol (1 mg/ml)	2 ml
N-benzyl-, *N*-4-methylbenzyl-, or *N*-4-methoxybenzyl-*p*-phenylenediamine (2 mg/ml)	2 ml
Catalase (0.03 mg/ml)	0.5 ml
Cytochrome c	7.0 mg

Plapinger et al. tested several naphthols prepared for the incubation medium by dissolving 10 mg in 0.1 ml ethanol and diethylene glycol dimethyl ether respectively followed by the addition of 10 ml distilled water. The amine solutions were prepared immediately before use.

If *N*,*N*'-bis(*p*-aminophenyl)-1,3-xylylenediamine is substituted in the above incubation mixture, no naphthol is required, inasmuch as it undergoes self-polymerization to form an osmiophilic insoluble brown stain useful also for electron microscopy. Also, adequate staining is achieved with an incubation period of 12 min. For electron microscopy, specimens are osmicated either in 2 to 4% osmium tetroxide for 30 min at 37°C or in vapor at 55°C for 15 min. To avoid excessive background osmium staining, it is helpful to employ a 1% acetic acid–water wash before aqueous washes and osmication. Inclusion of potassium cyanide (0.01 M) or sodium azide (6 μM) in the incubation medium resulted in absence of stain.

Kerpel-Fronus and Hajós (*Histochemie*, **10**: 216, 1967) provide the following method for the histochemical detection of cytochrome oxidase at the electron microscopic level:

1. Rats are perfused 10 to 20 min with 4% formalin in 0.1 M cacodylate buffer of pH 7.4 after the blood has been washed out with a balanced salt solution. Prime fixation time varies with each organ and tissue.
2. Tissues are chopped freehand and washed briefly in 0.2 M tris buffer of pH 7.4.
3. Tissues are incubated 40 to 60 min at 24°C in the following medium:

    10–15 mg *p*-aminodiphenylamine (*N*-phenyl-*p*-phenylenediamine)
    9 ml distilled water
    4 ml 0.2 M tris buffer, pH 7.4
    Sucrose to make 0.44 M final concentration

4. All subsequent steps are carried out at 4°C. Wash tissues for 10 min in a solution containing 4% formalin, 15% (0.44 M) sucrose, and 0.1 M sodium hydrogen maleate buffer of pH 6.
5. Postchelate tissues for 60 min in a solution containing 0.5% copper sulfate, 4% formalin, 15% sucrose in 0.1 M sodium hydrogen maleate buffer of pH 6. This solution must be used immediately.
6. Wash tissues overnight in 0.1 M sodium hydrogen maleate of pH 6 containing 15% sucrose.
7. Postchelate again for 60 min in 0.5% potassium ferrocyanide in 0.1 M sodium hydrogen maleate containing 15% sucrose.
8. Rinse briefly in Millonig buffer of pH 7.4.
9. Fix tissues in 1% osmium tetroxide buffered to pH 7.4.
10. Dehydrate in graded alcohols and embed in Durcupan (Fluka).

Inasmuch as the substrate has only limited ability to penetrate (approximately 100 to 200 μ), only surface layers are useful.

The reaction product appears in droplets almost exclusively within mitochondria. Preservation of tissues is not excellent, and mitochondria are swollen. In the central nervous system, synaptic structures are destroyed beyond recognition.

Weimar and Haraguchi (*J. Histochem. Cytochem.*, **13**: 239, 1965) appear to have detected in healing corneal wounds a cytochrome oxidase inhibited in the usual fashion with cyanide but resistant to azide and formalin. Preincubation of tissues in 1 mM phenyl-hydrazine abolished the staining reaction in liver, but unidentified positive and negative areas were noted in the kidney.

Tsou et al. (*J. Histochem. Cytochem.*, **20**: 741, 1972) report a selective method for the demonstration of cytochrome c at the electron microscopic level. They note that cytochrome c is soluble and readily washed from tissues. A reagent which binds selectively to cytochrome c in mitochondrial membranes needs to be employed. Tsou et al. prepared a metal chelate from the reaction of 4,4′-diamino-2,2′-bipyridyl (DABP) and ferrous chloride which was found suitable.

1. Incubate isolated rat liver mitochondria for 5 min at 23°C in 3.6 mM DABP–ferrous chelate (not available commercially) containing 8.5% sucrose.
2. Add glutaraldehyde (3% in 0.1 M phosphate buffer pH 7.4) drop by drop in the incubation medium.
3. Rinse, dehydrate, and process for electron microscopy with glutaraldehyde fixation only.

RESULTS: Mitochondria incubated with the DABP–ferrous chelate display membranes stained with spacings of 50 to 70 Å between stained regions. Electron-dense particles measuring 34, 43, 51, 60, and 68 Å are observed on the inner surface of outer mitochondrial membranes and cristae. Stained particles are not observed in mitochondria incubated in the absence of chelate.

Seligman et al. (*J. Biol. Chem.*, **38**: 1, 1968), Reith and Schuler (*J. Histochem. Cytochem.*, **20**: 583, 1972), Cammer and Moore (*Biochemistry*, **12**: 2502, 1973), and Roels (*J. Histochem. Cytochem.*, **22**: 442, 1974) clearly demonstrate that diaminobenzidine (DAB) reacts (donates electrons) to the oxidized form of cytochrome c in mitochondria. The reduced form is then reoxidized by cytochrome oxidase and oxygen. Thus, the DAB product localizes cytochrome c and also cytochrome oxidase activity. The reaction was not inhibited by antimycin, which blocks transfer of electrons from cytochrome b to c, and the reaction product was greatly increased after addition of cytochrome c. The use of methods that extract cytochrome c resulted in a substantial reduction or abolition of stain in mitochondria.

Yamashina and Barka (*J. Histochem. Cytochem.*, **20**: 855, 1972) localized peroxidase at the light and electron microscopic levels in developing submandibular glands of normal and isoproterenol-treated rats.

Reddy and Svoboda (*J. Histochem. Cytochem.*, **20**: 793, 1972) studied microbodies in Leydig cell (interstitial) tumors at the light and electron microscopic levels.

Feder (*J. Cell Biol.*, **51**: 339, 1971) prepared a "microperoxidase" from cytochrome c by digestion with pepsin. The heme peptide (mw 1900) was purified by ammonium sulfate precipitation and dialysis. Feder provides the following structural formula:

Microperoxidase

Feder indicates that the utility of microperoxidase exceeds that of horseradish peroxidase because of its comparatively small size. For this reason, microperoxidase will penetrate into extracellular compartments which exclude horseradish peroxidase due to size. Feder provides the following method as applied to mice weighing 20 to 25 g:

1. Unanesthetized mice are injected intravenously with 20 mg myeloperoxidase over a period of 30 to 60 s (10 mg horseradish peroxidase type IV from Sigma may be similarly employed). Overt signs of toxicity were not observed.
2. Mice are sacrificed after 15 to 30 min. Brains are fixed by ventriculocisternal perfusion with Karnovsky's 5% glutaraldehyde, 4% formaldehyde with 0.05% $CaCl_2$ in 0.08 M cacodylate buffer of pH 7.2 (*J. Cell Biol.*, **27**: 137A, 1965), and then immersed in the same fixative for 4 h at 23°C.
3. Tissues are washed for 4 to 16 h at 4°C in 0.2 M cacodylate buffer of pH 7.2.
4. Sections are cut at 50 μ with a Smith-Farquhar tissue chopper (Sorvall TC-2) and stored in 0.05 M tris buffer of pH 7.6 for 30 to 120 min prior to staining.
5. Sections are stained by the method of Karnovsky (*J. Cell Biol.*, **35**: 213, 1967) in a solution containing 0.01% H_2O_2 and 0.05% 3,3'-diaminobenzidine tetrahydrochloride (Sigma) in 0.05 M tris buffer of pH 7.6. Myeloperoxidase-containing specimens were incubated for 3 h.
6. For electron microscopic studies, specimens were rinsed, postfixed in osmium tetroxide, dehydrated, and embedded in Araldite (Ciba) by standard means.

Legg and Wood (*Histochemie*, **22**: 262, 1970) studied catalase synthesis in relation to the formation of microbodies in liver cells in rats and concluded that microbody proliferation is probably not dependent on catalase synthesis, but possibly on the synthesis of non-enzymatic protein.

CATALASE

Hale (*J. R. Microsc. Soc.*, **84**: 323, 1965) provides the following starch film method for the identification of *catalase*. The principle of the technic is that sections are placed in contact with a thin film of starch gel which is then immersed in hydrogen peroxide that saturates the gel except where it is in contact with catalase. At these points of contact the catalase destroys the hydrogen peroxide. Subsequent immersion of the gel in a solution of potassium iodide produces the dark blue iodine-starch reaction in the gel except where the hydrogen peroxide has not entered the gel. This results in a white localization of catalase on a blue background of gel.

1. Fresh frozen sections or sections cut from Holt's formalin-fixed tissues [16 to 24 h at 2 to 5°C in neutral calcium-formalin and then for a variable period in a solution of 0.88 M (30%) sucrose containing 0.9% gum acacia] are satisfactory. Fixed tissues provided better localization.
2. Hydrolyzed starch is prepared in the same manner as used for making blocks for electrophoresis.
3. Thin films on glass slides are made by placing starch on a slide, dropping another slide on top, and flattening.
4. Slides are cooled to room temperature and immersed in water, and the films are floated free.
5. Tissue sections floated on water are collected on slides and dried at 37°C for 30 min.
6. Freshly prepared starch films are floated on top of the tissue sections.
7. Slides with starch films on tissue sections are dried at 37°C for 10 min.
8. Slides are immersed in 0.01 M hydrogen peroxide for 2 h at 4°C.
9. Remove the hydrogen peroxide, and add a 2% solution of potassium iodide. Allow contact for 15 to 120 min at 23°C.
10. Remove the specimens from the iodide solution, gently float off the starch film under

water, mount on a glass slide, and examine under the microscope. One may compare the starch film with the retained tissue section on the slide. Enzyme activity is indicated by pale areas on the starch film.

The basis of the method is that hydrogen peroxide acts as both substrate and electron donor for catalase, whereas it acts only as a substrate for peroxidase; this is shown in the following scheme:

1. $2H_2O_2$ $\longrightarrow$ $O_2 + 2H_2O$ True catalase activity
   Donor Substrate

2. $R—CH_2OH + H_2O_2$ $\longrightarrow$ $R—CHO + 2H_2O$ Peroxidase activity
   Donor   Substrate

Incubation of tissues at 60°C for 24 h prior to conducting the staining reaction resulted in failure to stain. Likewise, incubation of tissues in the presence of 10^{-3} to 10^{-12} M HCN resulted in complete inhibition of the staining reaction. Hale noted that certain cells (leukocytes?) in the kidney were stained blue. He questions whether this represents peroxidase activity, since potassium iodide may act as a substrate for oxidation by peroxidase and thus react to give blue.

STERNBERGER UNLABELED ANTIBODY ENZYME IMMUNOHISTOCHEMICAL METHOD

Sternberger (*Mikroskopie*, **25**: 346, 1969), Sternberger and Cuculus (*J. Histochem. Cytochem.*, **17**: 190, 1969), Sternberger et al. (ibid., **14**: 711, 1966; **18**: 315, 1970) developed an *unlabeled antibody enzyme method* for *immunohistochemistry*. They demonstrated that the localization of antigens in tissues can be identified and the staining reaction intensified by the use of enzymes. Conjugated antibodies need not be employed. The procedure prescribes the sequential application to the tissue of (1) rabbit antiserum to the antigen to be localized, (2) sheep antiserum to rabbit immunoglobulin G (IgG) added in sufficient excess to leave one free combining site after reaction with the rabbit antibody, (3) rabbit antihorseradish peroxidase, which reacts as an antigen with the free combining sites of anti-IgG, (4) horseradish peroxidase, (5) 3,3′-diaminobenzidine tetrahydrochloride and hydrogen peroxide, and (6) osmium tetroxide. The authors note that it is better to employ immunospecifically purified antihorseradish peroxidase rather than total serum IgG, because the portion of serum IgG devoid of antihorseradish peroxidase interfered with step 4 of the above procedure.

Moriarty and Halmi (*J. Histochem. Cytochem.*, **20**: 590, 1972) employed the Sternberger method to identify ACTH-secreting cells of the anterior pituitary glands of rats. The method follows:

1. Tissues are satisfactorily fixed 2 to 4 h at 23°C in 8% paraformaldehyde in 0.1 M phosphate buffer of pH 7.4, 2.5% glutaraldehyde in 0.1 M phosphate buffer of pH 7.4, or paraformaldehyde–picric acid as advocated by Zamboni and De Martino (*J. Cell Biol.*, **35**: 148A, 1967).
2. Tissues are embedded in either methacrylate (75% ethyl, 25% methyl) polymerized at 40°C or Araldite 6005 (Ladd Research Industries, Inc., Burlington, Vermont).
3. Thin sections are cut, mounted on Formvar-coated nickel grids, and stained immediately after sectioning. Methacrylate-embedded sections do not require etching;

however, staining of Araldite-embedded sections may be increased if they are etched by flotation on a drop of 10 % aqueous hydrogen peroxide for 20 min immediately prior to staining.

4. To cover nonspecific sites of protein absorption, grids are first floated on normal goat serum diluted 1 : 30 with 0.5 M tris–HCl buffer diluted 1 : 9 with 0.9 % NaCl at pH 7.6 containing 0.8 % NaCl for 3 min.

5. Rinse in saline 0.5 M tris–HCl buffer of pH 7.6 for 3 min.

6. Place sections on a drop of rabbit anti-ACTH antiserum diluted 1 : 20 or 1 : 100 with the tris–HCl–saline buffer for 3 min.

7. Rinse in saline 0.5 M tris–HCl buffer of pH 7.6 as above for 3 min.

8. Place sections on a drop of goat antirabbit IgG diluted 1 : 10 (or 1 : 5 for Araldite-embedded sections) with the tris–HCl–saline buffer above for 3 min.

9. Rinse sections on a drop of the tris–HCl–saline buffer for 3 min.

10. Place sections on a drop of the horseradish peroxidase–antihorseradish peroxidase (PAP) complex (the antiperoxidase is made in rabbits) diluted 1 : 10 with the tris–HCl–saline buffer above for 3 min. The PAP complex is kept frozen in 0.1-ml amounts and diluted just before use. Frozen-diluted PAP complex keeps no longer than 2 weeks.

11. Rinse section on a drop of the tris–HCl–saline buffer for 3 min.

12. Sections are stained for 3 to 4 min with the 3,3′-diaminobenzidine reagent with hydrogen peroxide prepared as follows: 22 mg 3,3′-diaminobenzidine is placed in a beaker containing 175 ml tris (0.5 M tris brought to pH 7.6 by addition of 1 N HCl) and 1.5 ml 0.3 % hydrogen peroxide. The solution is filtered. The solution must be kept moving during staining. The present authors agitated the incubation fluid with a magnetic stirrer. Kawarai and Nakane (*J. Histochem. Cytochem.*, **18:** 161, 1970) developed a "steady flow" apparatus to effect circulation.

13. Rinse grids in three consecutive changes of distilled water.

14. Stain sections in aqueous 2 % osmium tetroxide for 60 min under a hood. No visible stain occurs without osmication. Osmium serves to stain the reaction product and to counterstain.

Moriarty and Halmi note that two cell types stain for ACTH in the rat pituitary, namely, a distinctive cell type in the anterior lobe and all cells of the intermediate lobe. The normal ACTH cell in the anterior pituitary is small, stellate, or sometimes angular with a few processes extending to vessels and often around other cells. Secretory granules (approximately 200 to 300 mμ in diameter) are normally in a single row near the plasma membrane. Golgi complex generally stained intensely.

TYROSINASE, DOPA-MELANASE

According to presently accepted concepts, the chromophore of melanin consists in considerable measure of a sulfur-bearing protein into which the indolic oxidation product of tyrosine is incorporated. In the formation of this chromophore L-tyrosine goes through a series of oxidative and rearrangement changes, including a probable decarboxylation, to form a product which is probably a quinhydrone. These changes are enzymatically mediated and exhibit a rather long preliminary phase in which L-tyrosine is converted to L-dopa (3,4-dihydroxyphenylalanine). Hence to expedite the reaction L- or DL-dopa is usually used as the substrate; D-dopa is not attacked.

For literature, see Lillie, *J. Histochem. Cytochem.*, **4:** 318, 1956, from which paper the following technic is taken. Lillie has modified the technic slightly from Fitzpatrick and Lerner (*Zoologica*, **35:** 28, 1950) and has elaborated on the description of the results.

1. Quickly freeze skin or other tissue by immersion in petroleum ether containing some fragments of solid carbon dioxide. Slice while frozen parallel to the direction of the hair into slices about 1 mm thick.
2. Fix these slices 1 h in ice-cold calcium acetate–formalin.
3. Wash 1 h in six changes of distilled water.
4. Incubate 24 h at 20 to 25°C in 0.1 % L-(or DL-)dopa in 0.05 M phosphate buffer of pH 7.
5. Wash in water, postfix 1 to 2 days in calcium acetate–formalin, dehydrate, clear, embed in paraffin, and section at 5 μ.
6. Deparaffinize in xylene and mount in cellulose caprate, either directly or after counterstaining as desired. Azure A–eosin B at pH 4 gives good contrasts, coloring melanin and dopa melanin dark green.

RESULTS: With positive reaction the cytoplasm of juxtapapillary and hair matrix cells colors a diffuse brown. Melanin granules tend to be larger than in controls and may form confluent globules. Trichohyalin granules remain uncolored in uncounterstained preparations and appear bright red with azure-eosin. Dendritic melanocytes in the base of the epidermis and margins of the hair follicles appear to be increased in numbers and contain larger confluent globules of dark brown melanin. The normal basophilia of hair follicle and epidermal cytoplasm is largely lost, from extraction of the relatively unfixed ribonucleic acid during the incubation period. Keratohyalin granules inconstantly give a weak brown color; their eosinophilia to azure-eosin is unimpaired. Pigment in more distal hair cortex and medulla remains dark brown and finely granular, as in controls.

INHIBITORS. KCN is a partial inhibitor at 0.1 mM, complete at 1 to 100 mM; NaN$_3$, at 1 mM; Na$_2$S, 0.1 and 1 mM; Na$_2$S$_2$O$_4$, 5 mM. Cysteine, 1 mM, appears to enhance the reaction.

It appears probable that Novikoff's recent practice of postfixing cryostat sections in calcium-formol at 3°C for 15 to 20 min should prove applicable to this reaction. Blocks fixed at -70°C in alcohol and cold acetone–fixed blocks, both embedded and sectioned in paraffin, have proved inert in the dopa reaction.

Use of L-tyrosine at 0.1 % (about 5.5 mM) in the foregoing technic required 24 h at 3°C followed by a second period in fresh substrate of 24 h at 37°C (Fitzpatrick and Lerner, *Science*, **112**: 223, 1950). However, only irradiated skin in the course of active pigment deposit (sun tanning, x-irradiation) gives a positive reaction with tyrosine.

Bloch and Peck (from Mallory) used dopa for the demonstration of M-Nadi oxidase in leukocytes: Fix smears by 20-min exposure to hot formaldehyde vapor, and wash. Then incubate 1 to 2 h in 50 ml 0.1 % L-dopa in 0.85 % NaCl to which 1 ml 0.1 N NaOH is added at time of using. Then wash, dry, and examine. Leukocyte granules are brown.

The color of the granules may be intensified by a 2-h exposure to 2 % silver nitrate, followed by thiosulfate and washing. In either variant an alum hematoxylin counterstain may be used.

DOPAMINE-β-HYDROXYLASE

Dopamine-β-hydroxylase is an enzyme that synthesizes norepinephrine from dopamine. It is the last enzyme in the synthetic pathway to norepinephrine from tyrosine. Dopamine has a broad substrate specificity which includes many phenylethylamine derivatives (Creveling et al., *Biochim. Biophys. Acta*, **64**: 125, 1962), although the preferred substrate is dopamine. Since no other significant enzymatic pathway to norepinephrine is present in mammalian tissues, this enzyme is present at all sites where norepinephrine is synthesized. Dopamine-β-hydroxylase can therefore be used as a specific anatomic marker for norepinephrine synthesis. Dopamine-β-hydroxylase is associated with catecholamine storage vesicles isolated from peripheral tissues (Laduron and Belpaire, *Biochem. Pharmacol.*, **17**: 1127, 1968) and from synaptic vesicles isolated from hypothalamus synaptosomes (Lauder and Austin, *Proc. Aust. Biochem. Soc.*, **4**: 89, 1971).

Immunofluorescent methods have been employed to demonstrate dopamine-β-hydroxylase [Hopwood, *Histochemie*, **13**: 323, 1968; Geffen et al., *J. Physiol.* (*Lond.*), **204**: 593, 1969; Hartmann et al., *Pharmacologist*, **12**: 286, 1970; Hartmann, *J. Histochem. Cytochem.*, **21**: 312, 1973; Frydman and Geffen, *J. Histochem. Cytochem.*, **21**: 166, 1973] at the light and electron microscopic levels. Frydman and Geffen noted depletion of the enzyme in lamb adrenal after administration of reserpine (0.25 mg/kg body weight daily for 7 days). Recovery of dopamine-β-hydroxylase was noted 1 week after cessation of drug administration.

METHOD OF HARTMANN   (*J. Histochem. Cytochem.*, **21**: 312, 1973)

1. Thin slices (3 to 5 mm) of tissues placed on aluminum foil are frozen on a block of dry ice and then sectioned at 10 μ with a cryostat.
2. Sections are placed on slides, dried in air, and fixed 20 to 30 min in chloroform-methanol, 2 : 1, at 0°C. At this stage, fixed slides may be air-dried for 1 h and stored at -90°C.
3. After bringing sections to room temperature, incubate sections for 15 min at 37°C with anti-dopamine-β-hydroxylase serum diluted 1 : 50 to 1 : 70 with phosphate-buffered saline solution containing 0.3% Triton X-100 (Sigma).
4. Rinse sections with phosphate-buffered saline solution with Triton X-100 very gently, and immerse sections in the buffer for two periods of 5 min each.
5. Incubate sections with fluorescein isothiocyanate conjugated to anti-IgG (diluted 0.1 to 0.5 mg protein per ml with 0.3% Triton X-100 in phosphate-buffered saline solution) for 15 min. (Stability of the conjugate is greatly decreased if stored frozen at a protein concentration <2 mg/ml.)
6. After incubation, repeat step 4 above.
7. Mount sections in buffered glycerogel (0.5 *M* carbonate at pH 8.6 and glycerol, 1 : 1).

Hartmann provides lengthy specific details on the preparation of pure antigen, antiserums, and the conjugate. Hartmann concludes that the use of Triton X-100 as indicated above is important in the reduction of nonspecific fluorescence.

Dopamine-β-hydroxylase was detected in neuronal fibers and terminals. Localization is specific for neurons that contain norepinephrine and is not present in neurons that contain dopamine or serotonin. The innervation of small cerebral arteries and arterioles by central noradrenergic neurons was also shown.

Hökfelt et al. (*Histochemie*, **33**: 231, 1973) applied the Coons indirect immunofluorescent method using fluorescein isothiocyanate–conjugated antibodies to demonstrate localization of three catecholamine synthesizing enzymes, namely, *dopa decarboxylase, dopamine β-hydroxylase*, and *phenylethanolamine N-methyltransferase* in rat adrenal tissues. Formulas depicting the action of these enzymes are in Fig. 10-2.

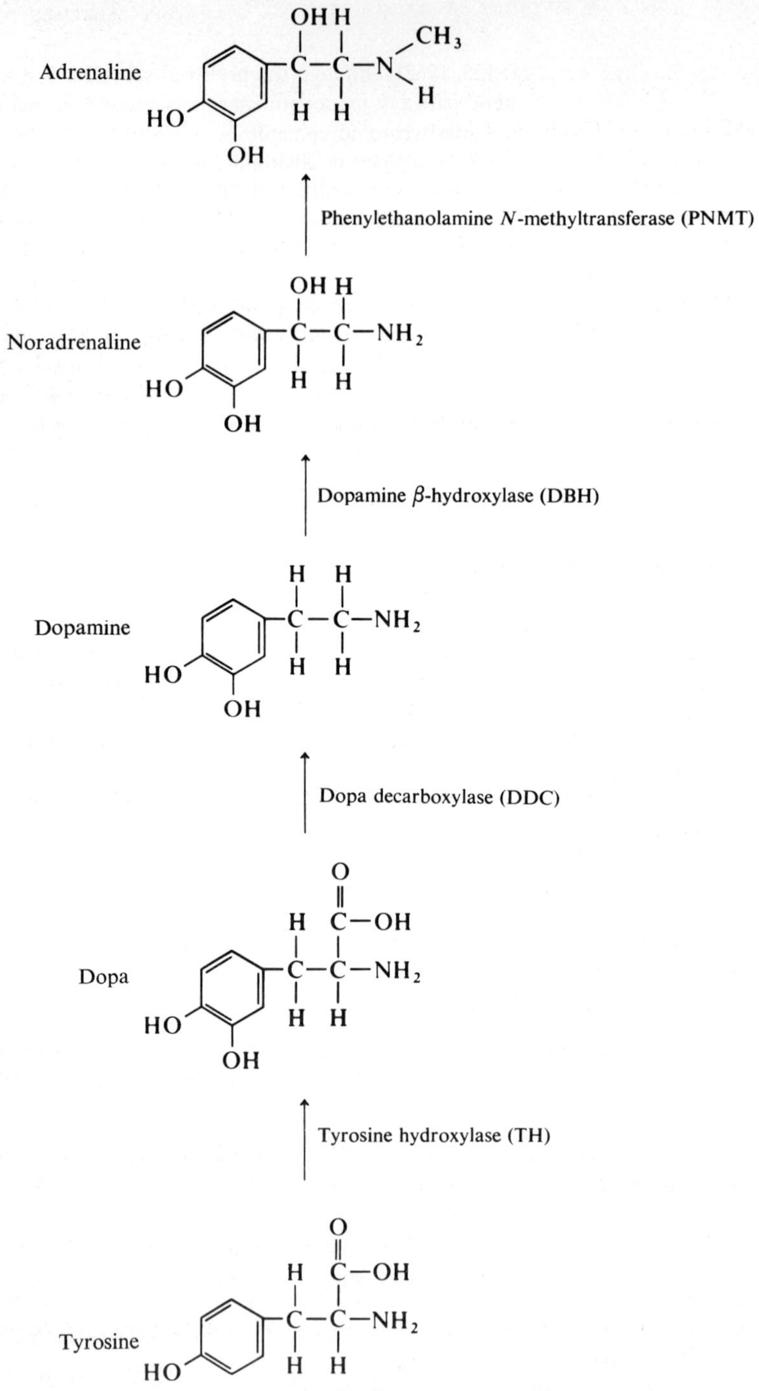

Fig. 10-2 Formulas depicting the action of dopa decarboxylase, dopamine β-hydroxylase, and phenylethanolamine N-methyltransferase. (*From Hökfelt et al., Histochemie,* **33:** 231, 1973. *By permission of the publisher.*)

AMINE OXIDASE AND DECARBOXYLASE

Oster and Schlossman (*J. Cell Physiol.*, **20**: 373, 1942) described a method for the localization of amino acid decarboxylase and amine oxidase in guinea pig kidneys. The decarboxylase first converts the amino acid to the next lower amine, and the amine oxidase converts the amine to the corresponding aldehyde with liberation of ammonia. The aldehyde is then rendered visible by the application of Schiff reagent. Because of the natural occurrence of plasmalogen, which liberates aldehyde (slowly in the presence of acid, more rapidly with mercuric chloride), it was thought necessary to block these aldehydes first with sulfite before proceeding with the specific enzyme technic. In the latter, tyramine was used as a substrate. When the amino acids L-tyrosine or L-tryptophan were used as substrates instead, the same localization was observed, indicating that the active areas (in distal convoluted tubules) possessed both decarboxylase and amine oxidase activities.

TECHNIC

1. Incubate frozen sections of fresh kidney in 2% sodium bisulfite solution for 24 h at 37°C.
2. Wash thoroughly in distilled water.
3. *Plasmal controls:* Take several sections at this point and immerse for 5 min in 1% mercuric chloride solution, wash in water and immerse in Schiff reagent for 15 min, wash in two or three changes of 0.05 *M* sodium bisulfite, wash in water for 5 min, counterstain 1 to 2 min in acetic hemalum, and mount in gum syrup. There should be no red.
    Buffer control: Incubate 24 h in 0.067 *M* pH 7.2 phosphate buffer.
    Enzyme inactivation control: First soak some sections in octyl alcohol for 24 h after the bisulfite treatment, then as follows:
    Experimental: Incubate at 37°C for 24 h in 0.5% tyramine hydrochloride in pH 7.2 0.067 *M* phosphate buffer.
4. Wash in distilled water.
5. Place in Schiff reagent until blue appears, allowing 30 min for buffer and inactivation controls.
6. Pass through three changes, 90 s each, of 0.05 *M* sodium bisulfite.
7. Float out, and mount in potassium acetate–gum syrup. Sections may be counterstained with acetic acid–safranin if desired.

RESULTS: Sites of amine oxidase activity are blue. If L-tyrosine or L-tryptophan is substituted for tyramine in the substrate, the blue indicates the presence of decarboxylase as well as of amine oxidase.

The above technic has been emended from Oster and Schlossman to supply omitted details. The following method, worked out by Glenner et al. in Lillie's laboratory, seems better and has won some measure of acceptance.

Table 10-16 provides details on the Glenner and Yasuda et al. technics for monoamine oxidase.

TECHNIC OF GLENNER ET AL. FOR MONOAMINE OXIDASE.   Collect fresh tissue and freeze; cut frozen sections at 10 to 15 μ by the Adamstone-Taylor method or in a cryostat. Attach to slides.

Immerse sections in preheated substrate, and incubate 30 to 45 min at 37°C. Wash 5 min in running water, fix 24 h in neutral buffered 10% formalin, dehydrate in graded alcohols, clear in xylene, and mount in Permount or similar resin.

Yasuda and Montagna preferred mounting in glycerol gelatin directly after fixation and water washing, but they noted reddish tinging of fat.

TABLE 10-16 MONOAMINE OXIDASE SUBSTRATES AND TECHNICS

| | Glenner et al. | | Yasuda et al. | | | |
| | | | | Concentrations | | |
	Amounts	Concentrations	Amounts	A	B	C
Tryptamine HCl	37.5 mg	6.3 mM	1 ml 0.25 M	8.3 mM	8.3 mM	8.3 mM
Nitro blue tetrazolium	7.5 mg	3.2 mM	7 ml 0.1 %	0.3 mM	0.3 mM	0.3 mM
Phosphate buffer	7.5 ml 0.1 M	25 mM	22 ml 0.1 M	73 mM	73 mM	73 mM
pH	7.6	....	7.6			
Na$_2$SO$_4$	6 mg	1.8 mM				
KCN	....	....	....	....	1 mM	
Marsilid PO$_4$	....	....	....	....	....	100 mM
Final volume	30 ml	....	30 ml			
Incubation time	30–45 min	....	30–60 min			
Temperature.........	37°C	....	37°C			

Sources: Glenner, Burtner, and Brown, *J. Histochem. Cytochem.*, **5**: 591, 1957; Yasuda and Montagna, *J. Histochem. Cytochem.*, **8**: 356, 1960.

Cyanide (1 mM) inhibits cytochrome oxidase but not monoamine oxidase or succinic dehydrogenase, according to Yasuda and Montagna, but Glenner et al. got inhibition at 0.1 mM. Trichloroacetic acid fixation preserves the SH reduction of nitro blue tetrazolium but inactivates monoamine oxidase. Glenner, Burtner, and Brown note inactivation by prior 15-min exposure to 0.1 M phenylhydrazine, Marsilid (1-isonicotinyl-2-isopropylhydrazine), p-chloromercuribenzoate, and ethanol. Heating 20 min at 60°C inactivates. Incorporation of Marsilid, hydroxylamine, semicarbazide, hydrazine, and atabrine at 0.1 M, bisulfite at 1 mM, and cyanide at 0.1 mM into substrate inhibited strongly.

Graham and Karnovsky (*J. Histochem. Cytochem.*, **13**: 605, 1965) provide the following method for the histochemical detection of monoamine oxidase. The method takes advantage of the fact that H$_2$O$_2$ is generated during the monoamine oxidase catalyzed oxidation of tryptamine.

1. Small blocks of tissues (Graham and Karnovsky used rat kidney and liver) are quenched in isopentane cooled with dry ice in acetone.
2. Sections are cut in a cryostat at 8 μ.
3. Immerse in cold acetone for 15 min (for lipid extraction).
4. Wash in 0.9% NaCl.
5. Incubate sections for 20 to 40 min at 37°C in the following medium: Dissolve 2 mg 3-amino-9-ethylcarbazole (Aldrich) in 0.5 ml dimethylformamide. Add 9.5 ml 0.05 M phosphate buffer of pH 7.6. Shake and filter the mixture. To this add 12 mg tryptamine hydrochloride and 10 mg horseradish peroxidase (Sigma), and shake the medium before use.
6. Wash sections in 0.9% NaCl.
7. Fix in 10% formalin for 2 h.
8. Wash in water, and mount in a water medium such as Kaiser's.

Organic solvents remove the reaction product. Sites of monoamine oxidase are detected by a red stain. In kidneys of guinea pigs, greatest activity was detected in the proximal convoluted tubules. Inconsistent activity was detected in rat kidneys. Ordinarily, no reaction product was detected in sites of known peroxidase activity such as in leukocytes, mast cells, and Kupffer cells. If the incubation medium contains insufficient exogenous peroxidase activity, some reaction product is observed at endogenous sites, possibly due to inefficiency of the peroxidatic capture reaction with diffusion of liberated H_2O_2 to sites of endogenous peroxidase activity.

Graham and Karnovsky favor the tetrazolium method of Glenner for routine histochemical use.

Shanthaveerappa and Bourne (*J. Histochem. Cytochem.*, **12**: 281, 1964) described the distribution of monoamine oxidase in the eyes of rabbits.

Hanker et al. (*Histochemie*, **33**: 205, 1973) provide the following method for monoamine oxidase:

1. They advocate perfusion of animals 1 to 4 min with 2% formaldehyde prepared from freshly depolymerized paraformaldehyde, 0.15 M in phosphate buffer pH 7.4 after flushing with physiologic saline.
2. Tissues are cut into small blocks and rinsed for 30 min in several changes of 0.22 M (7.5%) sucrose in 0.1 M phosphate of pH 7.2.
3. For light microscopy, sections are cut at 4 to 10 μ with a cryostat and collected on coverslips.
4. Sections are stained for several hours to overnight in the following medium at 37°C:

Tryptamine hydrochloride	5.5 mg
Sodium acetate (0.06 N, 0.008%)	6.4 ml
Sodium citrate (0.1 M, 2.9%)	0.6 ml
Copper sulfate, dropwise (15 mM, 0.375%)	1.25 ml
Dimethylsulfoxide (DMSO), adjust pH to 6.6–6.8.	1.5 ml
Potassium ferricyanide, dropwise (2.5 mM, 0.085%)	1.25 ml

The final pH should be 6.6 to 6.8. The medium is filtered.

5. Tissues are postfixed in 4% formaldehyde in 0.1 M phosphate buffer pH 7.2 for 1 h.
6. Tissues are rinsed, and then immersed in a freshly prepared solution of 5 mg 3,3'-diaminobenzidine tetrahydrochloride (Sigma) in 10 ml 0.05 M acetate buffer pH 5.6 for 20 min at 25°C.
7. For light microscopy, sections are rinsed in distilled water for three changes of 5 min each. For electron microscopy, the three changes are for 15 min each.
8. For light microscopy, sections are osmicated for 15 min in buffered osmium tetroxide in a closed vessel at 55°C. For electron microscopy, frozen or chopped sections are immersed in buffered 2% osmium tetroxide and heated to 55°C for 15 min in a closed vessel.
9. Sections are dehydrated and mounted in synthetic medium.

Enzyme activity is revealed by black staining. Distribution is stated to be analogous to that obtained with other methods such as Glenner's. Dimethylsulfoxide is stated to facilitate electron transfer and possibly to aid in penetration of reagents. An osmotic effect is also speculated.

Yoo et al. (*J. Histochem. Cytochem.*, **22**: 445, 1974) tested the effects of temperature and concentrations of glutaraldehyde and formaldehyde on inactivation of monoamine oxidase in rat liver. Best preservation of monoamine oxidase as well as structural integrity was observed in tissues fixed in 1.5% formaldehyde or glutaraldehyde in 0.1 *M* potassium phosphate buffer of pH 7.4 at 0°C. Approximately 70% of monoamine oxidase was preserved after 60-min fixation under these conditions.

Shannon et al. (*J. Histochem. Cytochem.*, **22**: 170, 1974) propose the following method for monoamine oxidase for electron microscopic studies: Excised fresh tissues are rinsed for 15 to 30 min in 0.05 *M* phosphate buffer containing 8% sucrose at pH 7 (340 mOsm) at 4°C, fixed for 2 to 5 min in 2% depolymerized paraformaldehyde in 0.1 *M* phosphate buffer of pH 7.4 (690 mOsm) at 4°C, rinsed in the cold phosphate-sucrose buffer for three changes of 5 min each, incubated at 22°C for 30 to 60 min in the following medium modified from Boadle and Bloom (*J. Histochem. Cytochem.*, **17**: 331, 1969):

Tryptamine hydrochloride (12 mg 10 ml 0.05 *M* phosphate buffer pH 7.6)	10 ml
2% sodium sulfate	0.1 ml
Sucrose	0.2 g
BSPT in dimethyl formamide (2.5 mg/0.1 ml)	0.1 ml

The ingredients are added in order with stirring. The medium is filtered and must be used immediately. The final pH is 7.55 at 340 mOsm. BSPT is a tetrazolium salt, 2-(2'-benzothi-azolyl)-5-styryl-3-(4'-phthalhydrazidyl)tetrazolium chloride, available from Polysciences. Controls consist of omission of substrate, omission of the tetrazolium salt, excessive fixation (to inactivate the enzyme), or addition of the following inhibitors prior to or during the incubation mixture at 0.02 *M* concentration: Marsalid (isonicotinic acid-2-isopropyl hydrazide) and tranylcypromine-2 (phenylcyclopropylamine).

Following incubation, tissues are rinsed in three changes (5 min each) of buffer, osmicated with 2% osmium tetroxide for 2 h at 50°C, rinsed once in buffer, dehydrated with ethanol and *N*-butyl glycidyl ether, and embedded in Epon. For light microscopic observations, sections are cut at 4 *μ*. BSPT penetrates poorly into blocks of tissue (particularly in kidney), and for this reason thin sections (500 to 800 Å) are cut at the perimeter of the embedded tissues. Sections are mounted on bare 400-mesh copper grids and examined without further treatment or after staining with alcoholic uranyl acetate and Reynold's lead citrate.

Reaction products were located on endoplasmic reticulum, outer mitochondrial membranes, and nuclear membranes of rat and guinea pig myocardium. It is important not to substitute TC-NBT for BSPT, because the osmiophilia of the tetrazolium salt may result in a staining artifact.

THE TETRAZOLIUM-FORMAZAN REACTION.   Colorless, soluble tetrazolium compounds are converted by the addition of hydrogen into water-insoluble, deeply colored pigments known as formazans. In tissue, this is accomplished by highly labile enzyme systems which are destroyed by the cold alcohol and acetone fixation processes. Tetrazolium salts are employed for the histochemical detection of many oxidative enzymes such as those that follow. Chapter 3 contains a table with a list of tetrazolium salts available at the time the third edition of this book was written. Table 10-17 provides abbreviations,

TABLE 10-17 ABBREVIATIONS, CHEMICAL NAMES, AND MOLECULAR WEIGHTS OF TETRAZOLIUM SALTS

Abbreviation	Chemical name and molecular weight
BSPT	2-(2'-Benzothiazolyl)-5-styryl-3-(4'-phthalhydrazidyl)tetrazolium chloride (mw 501.96)
DS-NBT	2,2'-Di-*p*-nitrophenyl-5,5'-distyryl-3,3'-(3,3'-dimethoxy-4,4'-biphenylene)ditetrazolium chloride (mw 869.73)
Half nitro BT	2-Phenyl-3-(3-methoxy)-4-phenyl-5-(*p*-nitrophenyl)tetrazolium chloride
Half tetranitro BT	[Half TNBT]* 2,5-di(*p*-nitrophenyl)-3-(3-methoxy-4-phenyl)tetrazolium chloride
INST	2,2'-(5,5'-Tetra-*p*-nitrophenyl-3,3'-stilbene)ditetrazolium chloride (mw 873.2)
NT	[Neotetrazolium]* 2,2'-(*p*-diphenylene)-bis(3,5-diphenyl)tetrazolium chloride (mw 667.58)
OsTNST	2,2'-5,5'-Tetra-*p*-nitrophenyl-3,3'-stilbene ditetrazolium chloride osmate (mw 2000.2)
TC-NBT	[Thiocarbamyl nitro BT]* 2,2'-di-*p*-nitrophenyl-5,5'-di-*p*-thiocarbamylphenyl-3,3'-(3,3'-dimethoxy-4,4'-biphenylene) ditetrazolium chloride (mw 935. 83)
YT	[Yellow tetrazolium]* 2,2'-di-(3-nitrophenyl)-5,5'-dimethyl-3,3'-(4,4'-biphenylene)-ditetrazolium chloride (mw 633.5)

* Trivial name

chemical names, and molecular weights of other tetrazolium salts that have become commercially available more recently.

Tetrazolium salts generally have undesirable as well as desirable characteristics. Desirable characteristics for use in staining include (1) solubility in sufficient concentration for histochemical use in appropriate buffer systems, and high insolubility after reduction to a formazan; (2) molecular size that permits penetration through cell membranes; (3) tissue substantivity but not for special cell constituents; (4) lipid solubility; (5) stability to light; (6) no inhibition of enzyme activity, and (7) the formazan should be darkly colored for easy identification and finely granular, not needlelike or crystalline in shape. Large crystals interfere with a determination of enzyme localization.

Eadie et al. (*Histochemie*, **21:** 170, 1970) studied the reduction of tetrazolium salts with respect to suitability for employment in quantitative histochemistry. They noted that the yield of diformazan from nitro BT after chemical reduction, after enzymatic reduction in liver homogenates, and in sections from a "mock" tissue was not in linear proportion to the strength of reducing conditions. Yields from the monoformazans INT and MTT were linear. Eadie et al. suggest that it is essential to calibrate reactions which involve ditetrazolium reduction.

Tsou et al. (*J. Histochem. Cytochem.*, **16:** 487, 1968) synthesized an osmate of 2,2'-5,5'-tetra-*p*-nitrophenyl-3,3'-stilbene ditetrazolium chloride (Os-TNST) which was used successfully to demonstrate an intracristal localization of succinic dehydrogenase in rat heart mitochondria.

Seligman et al. (*J. Histochem. Cytochem.*, **19:** 273, 1971) developed a new ditetrazolium salt, namely, 2,2'-di-*p*-nitrophenyl-5,5'-distyryl-3,3'-(3,3'-dimethoxy-4,4'-biphenylene)ditetrazolium chloride. The new salt (abbreviated DS-NBT) may now be purchased from Polysciences Inc., Paul Valley Industrial Park, Costner Circle, Warrington, Pennsylvania 18976. The structure of the DS-NBT tetrazolium salt is given below (top formula) followed by the formazan structure after reduction (a) and the proposed structure after formation of diosmate esters (b):

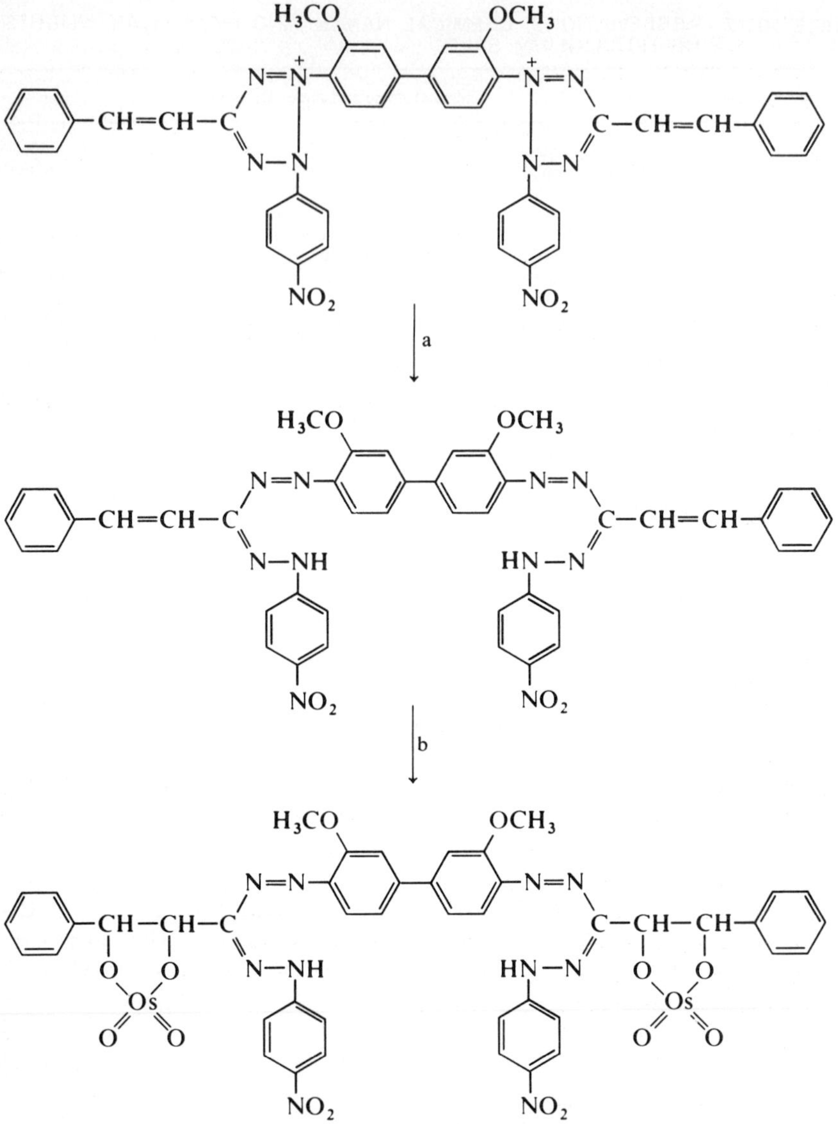

Osmium black

Seligman et al. indicate that DS-NBT may be used with light and electron microscopy. It is reduced more readily than nitro BT, and it is more readily osmicated than TC-NBT.[1] They obtained excellent localization of succinic and lactic dehydrogenase activities using DS-NBT.

Kalina et al. (*J. Histochem. Cytochem.*, **20**: 685, 1972) synthesized several monosmiophilic tetrazolium salts that yielded osmiophilic lipophobic formazans useful for the ultrastructural

---

[1] 2,2'-di-*p*-nitrophenyl-5,5'-di-*p*-thiocarbamylphenyl-3,3'-(3,3'-dimethoxy-4,4'-biphenylene)ditetrazolium chloride.

localization of dehydrogenases. Although all the new formazans were reoxidized to the tetrazolium salts when they were dissolved in dimethylformamide, they were not reoxidized by osmium tetroxide when dissolved in tetrahydrofuran or precipitated by reduction in tissues, whereupon dark complexes formed. One of the tetrazolium salts, namely, 2-(2′-benzothiazolyl)-5-styryl-3-(4′-phthalhydrazidyl)tetrazolium chloride (BSPT), was readily reduced by succinic dehydrogenase, gave a formazan which produced a dark osmium complex relatively rapidly, and demonstrated succinic dehydrogenase activity on mitochondrial membranes of rat myocardial cells. For further particulars on the chemistry of the tetrazolium salts, the reader is referred to Chap. 6.

CITRIC ACID CYCLE DEHYDROGENASES, DIAPHORASES, SUCCINIC DEHYDROGENASE

Semenoff (*Z. Zellforsch. Mikrosk. Anat.*, **22**: 305, 1935) reported a method for demonstration of succinic dehydrogenase depending on the decoloration of methylene blue in the presence of sodium succinate. The quantity of the last was not critical, and increases in concentration accelerated the reaction. The technic follows, purely as a matter of historical interest. It has been replaced by the tetrazolium-formazan methods.

1. Cut frozen sections of fresh unfixed tissue.
2. Mount on clean slides in a few drops of substrate, cover with cover glass, and seal with petrolatum. Observe periodically. The substrate is composed of:

0.05 % methylene blue	2 ml
10 % sodium succinate	1–2 ml
0.067 *M* phosphate buffer of pH 7.6–8 to a total of 10 ml	

3. The preparations are decolorized first in the central area of the cover glass; later, peripherally. Sites of greatest dehydrogenase activity are the first to decolorize. In liver and muscle, decoloration is complete in 30 to 80 min.

Control sections, heated to 60°C for 10 min, fail to decolorize methylene blue. Treatment with potassium cyanide at 10^{-4} or 10^{-5} (M?) completely inhibits the reaction; at 10^{-6} M the reaction is retarded, and at 10^{-7} M no effect is observed. It is presumed that the cyanide is added to the substrate, rather than that pretreatment was practiced. The potassium cyanide inhibition raises the suspicion that cytochrome oxidase was involved.

THE DEHYDROGENASES OF THE TRICARBOXYLIC (KREBS) AND RELATED CYCLES. These cycles are now to some extent demonstrable histochemically by the tetrazolium-formazan reduction reaction, which was previously used only for "endogenous" and succinic dehydrogenases (Table 10-18). The demonstration of several of these enzymes requires (or is facilitated by) the presence as cofactors of nicotinamide adenine dinucleotide and nicotinamide adenine nucleotide phosphate, commonly called NAD[1] and NADP and in their reduced forms NADH and NADPH. Cyanide has been introduced into the substrates to prevent oxidation (= dehydrogenation) by the alternative cytochrome oxidase system (Rosa and Velardo, *J. Histochem. Cytochem.*, **2**: 110, 1954). But with the introduction of faster-reacting

---

[1] Current usage has substituted the following terms for the phosphopyridine nucleotides: nicotinamide adenine dinucleotide (NAD) for diphosphopyridine nucleotide (DPN); dihydronicotinamide adenine dinucleotide (NADH) for reduced diphosphopyridine nucleotide (DPNH): nicotinamide and dihydronicotinamide adenine nucleotide phosphate (NADP, NADPH) for triphosphopyridine nucleotide and its reduced form (TPN, TPNH), respectively. The older terms appear in older papers.

tetrazoles, such as nitroneotetrazolium (Pearson, *J. Histochem. Cytochem.*, **6**: 112, 1958), nitro blue tetrazolium, tetranitro blue, and tetranitroneotetrazoles, the use of cyanide was found to be unnecessary (Nachlas et al., *J. Histochem. Cytochem.*, **5**: 420, 1957). The inclusion in the substrate of minimal amounts of methylene blue, of other thiazin dyes which also form reduced leuco compounds, and particularly of phenazine methosulfate was found by Farber et al. (*J. Histochem. Cytochem.*, **4**: 347, 357, 1956) to accelerate greatly the blue tetrazolium reactions of the dehydrogenases, but at least the methylene blue addition was found to have no accelerating effect in the succinic dehydrogenase reaction with the faster-reacting nitro blue tetrazolium (Nachlas et al., *J. Histochem. Cytochem.*, **5**: 420, 1957; Novikoff, ibid., **8**: 34, 1960).

In 1956 Farber et al. (*J. Histochem. Cytochem.*, **4**: 254, 266, 284) showed that by the use of the coenzymes NAD and NADP, a considerable number of other Krebs cycle substances could be used as substrates for tetrazolium reduction in place of succinate. And the reduced forms of the two nucleotides can themselves serve as substrates. The action of these cofactors was referred to as a diaphorase action, and the histochemical tetrazolium reduction pattern seemed to fall into two groups, according to which of the coenzymes was used. A considerable number of studies by other workers has shown nonidentity of reduction patterns within the two groups, and much current work is being done in this area.

As it became evident from biochemical ultracentrifuge fractionation studies that these enzymes were associated with mitochondria, histochemical studies were increasingly directed toward demonstration of this localization. Two major problems arose: (1) the swelling and distortion of the organelles and possible loss of enzyme by diffusion; (2) false localization on the finer fat droplets of cell cytoplasm.

Some workers, notably Scarpelli and Pearse (*J. Histochem. Cytochem.*, **6**: 369, 1958), have attempted to protect mitochondrial enzymes against swelling and distortion of the mitochondria during the substrate exposure phase by adding 15% sucrose (0.44 *M*) or, later and preferably, 7% polyvinylpyrrolidone (pyrrolidinone of some manufacturers; and PVP of the abbreviators) to the substrate.

Brief fixation procedures, such as that included in the above technic, seem now to be preferred. The formol-calcium fixation, 15 min at 3°C, preserves activity but does not eliminate lipid staining. Acetone for 15 to 30 min at 25°C considerably impairs dehydrogenase activity, but this is largely restored by inclusion of 0.01% Q 10 or menadione in the incubation medium. Acetone, 15 min at 3°C, preserves activity well and prevents the lipid artifact; at −65°C fat remains in the section, but at this temperature the acetone-*n*-butanol-ether mixture of Baker and Klapper (*J. Histochem. Cytochem.*, **9**: 713, 1961) is said to function well, at least for succinic dehydrogenase and NAD diaphorase reactions.

Altmann and Chayen [*Nature (Lond.)*, **207**: 1205, 1965] noted that inclusion of 20% polyvinylpyrrolidone in the incubation medium for the histochemical detection of the dehydrogenases resulted in retention of all nitrogenous material in the tissue sections. Polyvinylpyrrolidone at approximately mw 30,000 was used, and sections were incubated for approximately 20 min. Sections incubated under identical conditions except for omission of polyvinylpyrrolidone lost approximately 44% of their nitrogen content.

It was early observed that neotetrazolium, blue tetrazolium, and the monotetrazoles tended to produce colored formazan deposits in fat droplets, which were explained at least in part by diffusion of imperfectly water-insoluble formazans and their accumulation in fat in the usual manner of Sudan-type staining of lipids. The introduction of dinitro and tetranito ditetrazoles yielding formazans that are quite insoluble in fats and fat solvents permitted removal of lipids after staining and, it was hoped, any formazan they may have contained. Further, the mounting in resinous media should prevent aqueous diffusion of formazan and the crystallization on the surface of fat droplets which occurred slowly after mounting in glycerol and other aqueous media.

The second approach, pretreatment with fat solvents, as at first practiced, resulted in a considerable degree of enzyme inactivation. However, it has been found that a 15-min acetone extraction at 0 to 5°C adequately removed lipids and did not appreciably inactivate. Acetone at dry ice temperature did not adequately remove lipids (Hitzeman, *J. Histochem. Cytochem.*, **11**: 62, 1963). Cold $CaCl_2$–formalin did not prevent the lipid artifact. Wattenberg's (*J. Histochem. Cytochem.*, **8**: 296, 1960) use of coenzyme Q 10 or menadione at 0.01% in the substrate media served to restore essentially all the quantitative succinic dehydrogenase activity loss (two-thirds) ensuing from acetone extraction (15 min, 28°C) of cryostat sections, as assayed with iodonitrotetrazolium. Visual effects on nitro blue tetrazolium and iodonitro-tetrazolium staining were similar. The required amount of quinone was dissolved in acetone and dried on the cover glass on which the reaction was conducted.

Baker and Klapper (*J. Histochem. Cytochem.*, **9**: 713, 1961) successfully eliminated fat staining with nitro blue tetrazolium by a prior 15-min extraction at −65°C in a butanol-ether-acetone mixture. This did not impair the succinic or lactic dehydrogenase reactions or the reduced NAD diaphorase reaction.

To avoid unnecessary duplication, a single technic, basically that of Nachlas, Tsou, de Souza, Cheng, and Seligman (*J. Histochem. Cytochem.*, **5**: 420, 1957), is presented, with modifications derived from papers of Allen, Hitzeman, Baker, Wattenberg, and others. Table 10-19 gives compositions of other substrates for other enzymes of the Krebs cycle and for other tetrazoles. In this table we have given amounts of reagent in milligrams on the basis of a total 30 ml volume and the final concentration of the reagent in the whole volume of substrate expressed as molarities to facilitate comparison with journal references. The latter often present data in confusing mixtures of mols, molarities, and weight and volume units. Table 20-1 giving formulation and molecular weights of reagents has been included for the special benefit of the hard-working technicians who have to translate quantities expressed in molarities back into units of mass and volume.

The following nitro blue tetrazolium technic is a composite, based primarily on the method of Nachlas, Tsou, de Souza, Cheng, and Seligman (*J. Histochem. Cytochem.*, **5**: 420, 1957) with additions and alterations based on papers of Novikoff, Wattenberg, Allen, Chang and Hori, Cogan and Kuwahara, Hitzeman, and others. It is designed also for various of the NAD– and NADP–linked dehydrogenases by substitution of substrates from Table 10-18.

The alternative dimethylthiazolyl-cobalt-formazan technic is offered chiefly as a confirmatory alternative whose reagents are sometimes difficult to obtain. It also seems to be adaptable for various of the Krebs cycle enzymes.

THE TETRAZOLIUM-FORMAZAN METHOD FOR SUCCINIC DEHYDROGENASE

1. Quick-freeze material (3 to 5 mm thick blocks of tissue) by placing in a dry test tube and immersing the tube in dry ice and acetone. Animal material is taken as promptly as possible after killing; human surgical material, as soon after excision as possible; autopsy material, at the beginning of the dissection, as soon as possible after death.

2. Transfer still frozen tissue, after trimming, directly onto freezing head or metal chuck of the microtome, and cut sections with a cold knife, preferably in a cryostat at 10 μ. Collect sections directly from microtome blade onto cold, clean glass slides and cause to adhere by warming back of slide momentarily with the fingertip.

3. Fix sections 15 min in acetone at 2°C. If the Coplin jar with sections is kept in the cryostat, the time may be increased twofold for each 10°C drop in temperature.

4. Originally 15 mg nitro blue tetrazolium (nitro BT) was used; Pearse and, later, Allen

reduced this by half; and Hitzeman in Allen's laboratory now uses 2.5 mg. Dissolve 5 mg nitro blue tetrazolium in 15 ml 0.1 M phosphate buffer of pH 7.6, and add 15 ml 0.1 M sodium succinate. Warm to 37°C, if development of color is not prompt at 25°C.

5. Incubate sections at 25 or 37°C for 10 to 30 min until adequate blue is developed (8 min at 37°C is reported as adequate).
6. Wash for 1 min in distilled water, dehydrate in 30, 50, 70, 85, 95%, and absolute alcohol, for 5 min each, absolute alcohol and xylene, and two changes of xylene. Mount in Permount.

RESULTS: Sites of succinic dehydrogenase activity are blue.

If a red nuclear stain is desired, counterstain a duplicate section after step 5 in 1% safranin O, rinse in water, and dehydrate as in step 6.

CONTROLS

1. After step 3 immerse sections 30 min in 0.01 M iodoacetate (185 mg iodoacetic acid in 100 ml 0.01 N NaOH). Inactivation is complete.
2. Omit succinate from substrate. No staining should result.
3. Heating sections to 80°C for 1 h completely inactivates.

Both the stock succinate solution and the phosphate buffer are stable for some months. Nachlas et al. record successful reuse of the final substrate 4 or 5 times for batches of 15 slides over a period of 2 months. It is recommended that such solutions be kept cold and that they be filtered through a coarse paper to remove tissue fragments between usings. Most workers prefer to use fresh solutions, at least for each day's work.

Hajós and Kerpel-Fronius (*Histochemie*, **23:** 120, 1970) illustrate that unfixed tissues may be incubated for enzyme histochemistry and subsequently prepared for electron microscopy. Good results for succinic dehydrogenase activity and tissue preservation were obtained. The method follows:

1. Small blocks (1 to 2 mm³) of tissue are excised and incubated up to 60 min in media for the demonstration of succinic dehydrogenase (Kerpel-Fronius and Hajós, *Histochemie*, **14:** 343, 1968) or for cytochrome oxidase (Kerpel-Fronius and Hajós, *Histochemie*, **10:** 216, 1967) containing 0.2 M sucrose (6.84%).
2. Immerse blocks in chilled 4% paraformaldehyde for 90 min.
3. Wash for 40 min in two changes of phosphate buffer.
4. Fix for 1 h in 1% osmium tetroxide buffered to pH 7.2.
5. Dehydrate through alcohols and embed in Durcupan (Fluka).

Succinic dehydrogenase activity was restricted to mitochondrial membranes and cristae of rat myocardium. Ultrathin sections may be subsequently stained for contrast with alcoholic uranyl acetate and lead citrate according to Reynolds (*J. Cell Biol.*, **17:** 208, 1963) although this was not employed for the demonstration of succinic dehydrogenase activity.

Kerpel-Fronius and Hajós (*Histochemie*, **14:** 343, 1968) provide the following method for the demonstration of succinic dehydrogenase at the light and electron microscopic levels.

1. For electron microscopy, thin (1 to 2 mm³) slabs of tissues or slices (500 to 800 μ) are cut in a cryostat. For light microscopy, 5- to 10-μ sections are cut in a cryostat.
2. Wash tissues briefly in 0.1 M phosphate buffer of pH 7.6.
3. Incubate tissues for 45 min at 37°C in the following medium. Add the ingredients in order with vigorous shaking.

	ml
0.5 M sodium potassium tartrate dissolved in 0.1 M Sörensen's phosphate buffer, pH 7.6	3
0.3 M CuSO$_4$ (7.5%)	0.35
0.1 M Sörensen's phosphate buffer, pH 7.6	0.8
1.0 M sodium succinate (14%)	0.7
5 mM potassium ferricyanide (0.165%)	0.15

The solution is clear, green, and stable for several hours. The final pH should be between 6.6 and 6.7.

4. For light microscopy, sections are washed briefly in distilled water, fixed in 10% formalin, and mounted in a water-based medium such as Kaiser's. Observation by phase microscopy is recommended inasmuch as contrast is not great. For electron microscopy, tissues are washed briefly in 0.1 M Sörensen's phosphate buffer and fixed for 1 h in osmium tetroxide in Millonig's buffer of pH 7.4, dehydrated through graded alcohols, and embedded in Durcupan (Fluka). Only 50 to 100 μ near the surface of the block contains the reaction product.

Kerpel-Fronius and Hajós initially stained also with Reynold's lead citrate; however, they later discovered that the alkaline lead citrate solution dissolved the copper ferrocyanide reaction product.

They noted reaction product as granules in mitochondrial membranes only. Mitochondria were swollen. Nonspecific copper binding was not observed.

Hanker et al. (*Histochemie*, **33**: 205, 1973) added 13% dimethyl sulfoxide (DMSO) to the Kerpel-Fronius–Hajós incubation mixture, claiming more rapid development of color. They noted DMSO inhibition of glutamic dehydrogenase (with both NAD and NADP) and increased malic and NADPH and NADH dehydrogenases.

Vecsei et al. (*J. Chromatogr.*, **9**: 525, 1962) employed neotetrazolium as a salt to detect succinic dehydrogenase in sections of several organs of the rat. Thereafter, the red monoformazan was quantitatively separated from the blue diformazan by application of paper chromatography.

Seligman et al. (*J. Histochem. Cytochem.*, **19**: 273, 1971) provide the following methods for succinic, lactic, and NADH$_2$-dehydrogenases:

1. For light microscopy, fresh frozen sections are used. For electron microscopy, blocks of skeletal and cardiac muscles 2 × 2 mm are fixed in depolymerized paraformaldehyde (4% in 0.05 M phosphate buffer of pH 7 at 0°C). (For succinic and NADH$_2$-dehydrogenases, 5 min is recommended; for lactic dehydrogenase, 3-h fixation is recommended.)
2. Wash blocks in 7.5% sucrose in 0.05 M phosphate buffer of pH 7 (30 min for succinic and NADH$_2$-dehydrogenases and overnight for lactic).
3. Incubate sections at 30 to 37°C for 20 to 40 min in the following medium for succinic dehydrogenase:

Phosphate buffer pH 7	0.05 M
DS-NBT*	0.5–1 mg/ml
Sodium succinate	0.05 M (1.35%)
Sucrose	2.5%

* 2,2'-di-p-nitrophenyl-5,5'-distyryl-3,3'-(3,3'-dimethoxy-4,4'-biphenylene)ditetrazolium chloride.

For the demonstration of NADH$_2$-dehydrogenase, substitute NADH$_2$ (1 mg/ml) for sodium succinate.

For lactic dehydrogenase demonstration use of the following medium is advised for 20 to 30 min at 30°C:

Tris buffer, pH 7.2	0.05 M
NAD	0.7 mg/ml
DS-NBT	1 mg/ml
Lithium lactate	0.2 M (1.92%)
Phenazine methosulfate	0.3 mg/ml

Seligman et al. used dimethylsulfoxide (DMSO) to aid in the solubilization of DS-NBT. DMSO may also aid in the preservation of fine structure. DS-NBT was first dissolved in DMSO using 0.5 ml for a 10-ml incubation medium with a final concentration of 5%.

Seligman et al. also noted that a 40-s sonication of tissues appeared to facilitate staining of tissues, which permitted a reduction of incubation time to nearly one-half of that prescribed.

One feature which must be kept in mind and controlled at the electron microscopic level is that both the tetrazolium salt and the formazan are osmiophilic. Adherent tetrazolium salt and nonenzymatically reduced DS-NBT will osmicate.

To enhance preservation of tissues for the demonstration of *3-β- and 17-β-hydroxysteroid dehydrogenases* by light microscopy Morse and Heller (*Histochemie*, **35**: 331, 1973) recommend (1) placement of small slices of tissues in 15% Dextran containing 1.5% dimethylsulfoxide in tris maleate buffer at pH 7.4 for 5 min and then freezing the tissue containing the dimethylsulfoxide; (2) soaking tissue sections for 15 min at 34°C in 5% polyvinylpyrrolidone, 5% dimethyl sulfoxide, and 4.3% dimethyl formamide in tris maleate buffer pH 7.4 prior to incubation; (3) the use of KCN in the incubation medium at a final concentration of 5 mM at pH 7.3; and (4) the use of polyvinylpyrrolidone in the incubation medium at a final concentration of 5%.

THE SCARPELLI, HESS, AND PEARSE MTT Co^{2+} METHOD FOR DIAPHORASES. This technic uses 2-(4,5-dimethylthiazolyl-2)-3,5-diphenyl tetrazolium chloride or bromide (MTT) with cobalt chelation (from Pearse, 1960, with further variations from Hitzeman, *J. Histochem. Cytochem.*, **11**: 62, 1963). Because of the high cost of the reagents, the stock solution is prepared in 10-ml quantities which can be kept at 0 to 4°C for 3 to 4 weeks, and the working substrate is prepared in 1-ml amounts and used by applying 1 or 2 drops of the freshly prepared substrate to sections on coverslips.

STOCK SOLUTION

	A, ml	B, ml
0.2 M tris buffer, pH 8	2.5	2.5
MTT, 1 mg/ml	2.5	2.5
Cobaltous chloride, 12%, CoCl$_2 \cdot 6H_2O$ (504 mM)	0.3	0.3
Polyvinylpyrrolidone	0.75	
Distilled water to make total of	10.0	10.0

Adjust pH with the tris buffer to pH 7.2, and bring final volume to 10 ml with distilled water. The cobaltous chloride solution is acid.

Pearse substitutes tris buffer for the original phosphate, thereby avoiding the filtration to remove cobalt phosphates which was required in the original medium.

WORKING SOLUTION. To 1 ml stock solution add 6 mg reduced NAD or NADP. This gives concentrations of about 7 and 6 mM for the two nucleotides respectively.

TECHNIC

1. Quickly freeze fresh tissue with solid carbon dioxide or liquid nitrogen, using small pieces 2 to 4 mm thick. Transfer to cryostat and cut thin frozen sections. The Adamstone-Taylor procedure may be used. Attach sections to cover glasses.
2. Fix for 15 min in acetone at 2 to 4°C or in cold CaCl$_2$–formalin, or proceed directly with step 4.
3. Rinse off formalin with distilled water. Let acetone-fixed sections dry in air.
4. Deposit 0.1 to 0.2 ml (2 to 4 drops) of working substrate solution on section, and incubate unfixed sections for 5 to 30 min at 37°C, using the mixture described above. Acetone-fixed sections are incubated 20 min at 37°C for the NADH diaphorase and 40 min for the NADPH diaphorase.
5. Fix previously unfixed sections for 5 to 30 min in cold formol-calcium. Rinse in distilled water.
6. Counterstain if desired, in a one-tenth dilution of Mayer's (or Grenacher's) carmalum or in chloroform-extracted 0.5% aqueous methyl green. Rinse in distilled water.
7. Mount in glycerol gelatin.

RESULTS: The cobalt-formazan deposit at diaphorase activity sites is black. Tris buffer should be used with the dimethylthiazolyl cobalt tetrazole, usually phosphate with nitro blue tetrazolium.

Pearse lists specifically malic, lactic, α-glycerophosphate, alcohol, glutamic, and isocitric dehydrogenases with NAD and isocitric, glutamic, and malic with the NADP as cofactor. Hitzeman reported specifically on malic, lactic, β-hydroxybutyric, and 3-β-hydroxysteroid dehydrogenases with NAD and isocitric with the NADP derivative.

Other specific dehydrogenases studied in other reports, often with only indicated composition of substrates, include pyruvic (Kuwabara and Cogan, *J. Histochem. Cytochem.*, **8**: 214, 1960) and glyceraldehyde 3-phosphate (Cavazos et al., *J. Histochem. Cytochem.*, **10**: 387, 1962). It is to be presumed that these and other dehydrogenase enzymes not mentioned were or can reasonably be investigated by technics based on those given, employing substrates of the general formulation given in Table 10-18. Polyvinylpyrrolidone is to be included if fresh frozen sections are employed directly; with cold formol-calcium or cold acetone fixation it is apparently unnecessary. The use of cyanide is probably also unnecessary with the newer tetrazoles; its persistence in formulas seems to be a holdover from the neotetrazolium and blue tetrazolium era of enzyme histochemistry.

Hershey et al. (*J. Histochem. Cytochem.*, **11**: 62, 224, 1963) in homogenate studies on mouse breast noted a strong activating effect of Mn^{2+} on isocitric dehydrogenase: 8 μM quadrupled the activity, and 0.1 mM produced a maximal eightfold increase.

In regard to the function of phenazine methosulfate in the reactivity of NAD–linked dehydrogenases, notably lactate and α-glycerophosphate dehydrogenases, Van Wijhe, Blanchaer, et al. (*J. Histochem. Cytochem.*, **11**: 505, 1963) find the activity in white muscle fibers much lower than in red muscle in the absence of this agent, and actually greater when 0.1% phenazine methosulfate (*N*-methylphenazonium methyl sulfate, mw 306) is added to the substrates. The lower activity of white fibers in the absence of phenazine methosulfate thus appears due to limitation in diaphorase activity rather than to lack of dehydrogenases.

Balogh (*J. Histochem. Cytochem.*, **10**: 232, 1962) reported that fresh osseous and dental tissues decalcified in 10% Versene in 0.1 M phosphate buffer at pH 7 with 3-day exposure at 4 to 10°C; magnetic stirring and daily changes of decalcifying fluid could be used for demonstration of enzyme activities. NAD and NADP diaphorases and succinic, lactic, malic, isocitric, and glucose 6-phosphate dehydrogenases were preserved.

TABLE 10-18 GENERAL SUBSTRATE FORMULAS FOR THE NAD– AND NADP–LINKED DEHYDROGENASES OF THE CITRIC ACID CYCLE

| | Pearse, 1960 | | | Hitzeman, 1963 | | Lillie, 1964 | |
	Amounts		Final concentration	Amount	Final concentration	Amount	Final concentration
	For 1 ml	For 30 ml					
Specific substrate........	0.1 ml 1 M	3 ml 1 M	0.1 M	15 ml 0.1 M	50 mM	15 ml 0.1 M	50 mM
Cofactor........	0.1 ml 0.1 M	3 ml 0.1 M	10 mM	0.2 ml 0.1 M	0.67 mM	0.3 ml 0.1 M	1 mM
Cyanide (K, Na)........	0.1 ml 0.1 M	3 ml 0.1 M	10 mM	1.5 ml 0.1 M	5 mM		
Phosphate buffer........	....	....	....	13.5 ml 60 mM	27 mM	9 ml 0.1 M	30 mM
Tris HCl buffer........	0.25 ml 0.1 M	7.5 ml 0.1 M	25 mM				
Nitro BT........	....	....	....	2.5 mg	0.0083%	5 mg	0.017%
MTT........	0.25 mg	7.5 mg	0.025%				
Polyvinylpyrrolidone........	75 μg	2.25 mg	0.0075%				
Distilled water, to make........	1 ml	30 ml	....	30 ml	....	30 ml	

Pearse, "Histochemistry: Theoretical and Applied," 2d ed., Little, Brown, Boston, 1960; Hitzeman, *J. Histochem. Cytochem.*, **11**: 62, 1963; Lillie, "Histopathologic Technic and Practical Histochemistry," 3d ed., McGraw-Hill, New York, 1965.

TABLE 10-19   AMOUNTS IN MILLIGRAMS OF DEHYDROGENASE SUBSTRATES REQUIRED FOR 20, 30, AND 100 ml 50 mM SUBSTRATE SOLUTION

Substrates	Cation	H₂O	mw	For 50 m*M* substrate			H₂O†	mw	For 50 m*M* substrate		
				20 ml, mg	30 ml, mg	100 ml, mg			20 ml, mg	30 ml, mg	100 ml, mg
Ethanol			46.07	46	69	230					
Glucose 6-phosphate	K₂*		336.32	336	504	1682					
Glutamate	Na	1	187.14	187	281	936		147.13	147	221	736
α-Glycerophosphate	Na*		216.05	216	324	1080					
β-Hydroxybutyrate	Na		126.09	126	189	630		104.10	104	156	520
Isocitrate	Na‡	1	276.08	276	404	1380		192.12	192	288	960
Lactate	Na†		112.06	112	168	560		90.08	90	135	450
L-Malate	Na₂*	½	187.07	187	281	935		134.09	134	201	670
Pyruvate	Na		110.05	110	165	550		88.06	88	132	440
6-Phosphogluconate	Ba		409.47								
Succinate	Na₂†	6	270.16	270	405	1351		118.01	118	177	590
NAD			663.4								
NADH	Na₂		763								
NADP	Na	2	801								
NADPH	Na		834								

* P. G. Stecher (ed.), "The Merck Index of Chemicals and Drugs," 7th ed., Merck, Rahway, N.J., 1960.

† N. A. Lange, "Handbook of Chemistry," 10th ed., McGraw-Hill, New York, 1961.

‡ T. Barka and P. J. Anderson, "Histochemistry: Theory, Practice and Bibliography," Hoeber-Harper, New York, 1963.

Note: 6-Phosphogluconic acid is available as the barium salt (mw 409.47) glyceraldehyde 3-phosphate as a bromodioxane addition compound. The chemical manipulations required are not within the scope of this book.

For 20 ml of a 50-m*M* substrate use the molecular weight (formula weight on many bottle labels) in milligrams, dissolved in a convenient amount of the water in the formula. For 30 ml take 1.5 × the milligram molecular weight. When using the above table always check the molecular weight given in the table against the formula weight on the bottle. The amount of water of crystallization may vary, and sodium and potassium salts, though often interchangeable with the free acid, do not have the same molecular weight. Where differences exist, use the formula weight on the bottle and apply the above rule. Smaller amounts of substrate 1, 2, 3, or 10 ml are readily derived by appropriately shifting the decimal point.

Gomori's tellurite reactions for succinic dehydrogenase appear to be only of historical interest, although still occasionally mentioned in review papers.

TELLURITE REACTION FOR DEHYDROGENASES (modified in 1962 from Gomori). Preferably use frozen sections of fresh, unfixed tissue. Postfixation for 15 min in acetone at 4°C, as in Novikoff's procedures with the tetrazolium methods, appears to be indicated. Tissue may be stored at 4°C for several hours without loss of activity. Fixation for 4 h in cold acetone causes only 40 % less activity. The substrate is a 0.1 to 0.05 *M* phosphate buffer of pH 7.3 to 7.6 containing sodium succinate, lactate, etc., in 50 m*M* to 0.2 *M* concentration and 0.1 % potassium tellurite. Other details from the tetrazolium methods should be readily adaptable, using 0.1 % potassium tellurite in lieu of the tetrazoles.

Incubate at 37°C for 20 min to 3 h, inspecting at intervals. Activity sites are shown by deposits of brown to black elementary tellurium. Wash in pH 7 to 7.3 phosphate buffer. Counterstain as desired (avoiding strong acids or alkalies), dehydrate in alcohol, clear in xylene, and mount in balsam. (Tellurium is insoluble in water and in fat solvents, but soluble in strong acids, alkalies, and KCN.)

Brody et al. (*Arch. Pathol.*, **84:** 312, 1967) report delineation of the extent of myocardial infarction in slices of myocardium immersed in nitro BT (5 mg/ml) in 1.0 *M* phosphate buffer of pH 7.4 for 30 min at 37°C. Normal viable myocardium contains endogenous substrates, coenzymes, and necessary dehydrogenases to reduce nitro BT to the blue formazan.

Necrotic and nonviable myocardium does not, and a failure of reduction of nitro BT occurs; thus delineation of the infarct occurs. Brody et al. note that the infarct is less adequately delineated in bodies examined 12 h or more after death.

DIHYDROFOLATE REDUCTASE

Onicescu et al. (*Histochem. J.*, **2**: 289, 1970) propose the following method for *dihydrofolate reductase*. Dihydrofolate reductase catalyzes the reduction of folate to tetrahydrofolate and thereby represents a key role in the intermediary metabolism of one carbon units. The reactions proceed as follows:

$$\text{Folate} + \text{NADPH}_2 \xrightleftharpoons{\text{dihydrofolate reductase}} \text{Dihydrofolate} + \text{NADP}$$

$$\text{Dihydrofolate} + \text{NADPH}_2 \xrightleftharpoons{\text{dihydrofolate reductase}} \text{Tetrahydrofolate} + \text{NADP}$$

1. Fix tissues in one of the aldehydes (formaldehyde, crotonaldehyde, or glutaraldehyde), 20% w/v at 4°C for 30 min.
2. Rinse tissues.
3. Cut sections in a cryostat.
4. Incubate sections until desired staining occurs (time and temperature not provided by Onicescu et al.).

Folate 0.004 M	0.183%, Na folate, mw 463.39
NADP 0.005 M	0.372%
NAD 0.005 M	0.332%
Phosphate buffer (0.1 M pH 7)	
Nitro BT 0.1 M	

5. Rinse sections, dehydrate, and mount in a synthetic medium.

The reliability and specificity of the above method should be checked before use.

TETRAHYDROFOLATE DEHYDROGENASE

Gerzeli and De Piceis Plover (*Histochem. J.*, **4**: 79, 1972) provide the following method for *tetrahydrofolate dehydrogenase*:

1. Fix blood smears in 60% acetone for 30 s. Use unfixed liver sections.
2. Immerse sections in the following incubation medium for 10 min at 37°C:

	mg
Tetrahydrofolic acid (1 mM in 2 mercaptoethanol)	44.5
NADP (1 mM)	74.3
Nitro BT (0.3 mM)	24.5
Phenazine methosulfate (0.82 mM)	25.1
MgCl$_2$ (5 mM)	46.6
NaN$_3$ (10 mM)	65.01
M/15 phosphate buffer, pH 7.4, to make 100 ml	

Oxygen-free nitrogen is bubbled through the medium.
3. Rinse sections and mount in a water medium such as Kaiser's.

Again the reliability and specificity of the above medium should be proved before use. Fahimi and Amarasingham (*J. Cell Biol.*, **22**: 29, 1964) provide the following method

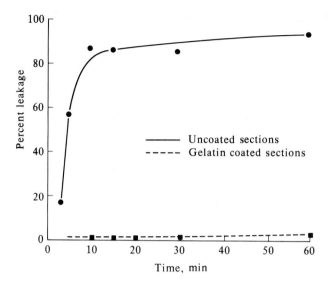

Fig. 10-3 Leakage of LDH from 4-μ-thick sections of white
adductor muscle of the rabbit. (*From Fahimi and Amarasingham,
J. Cell Biol.,* **22:** 29, 1964. *By permission of the publisher.*)

for the histochemical detection of soluble enzymes such as lactic dehydrogenase. They note
that as much as 80% of lactic dehydrogenase may diffuse into the incubation medium within
10 min as indicated in Fig. 10-3.

Fahimi and Amarasingham prescribe the following method for *lactic dehydrogenase*. The
method can obviously be applied for the demonstration of other soluble enzymes.

1. Sections at 4 μ are cut with a cryostat from small blocks quick-frozen in liquid propane-
   isopentane (3 : 1) cooled with liquid nitrogen, mounted on coverslips, and dried under
   a blower for 3 min. Sections may or may not be fixed in cold acetone.
2. The following incubation mixture is prepared:

	μg/ml	
Nitro BT	163.5	0.2 mM
NAD	331.7	0.5 mM
Lithium lactate	9.60	0.1 M
Phenazine methosulfate	61.2	0.2 mM

The above reagents are mixed with an equal volume of 0.2 M tris buffer of pH 7.4 containing 5% purified gelatin.
3. One-fourth milliliter of the above incubation mixture is layered evenly over a slide, and
   the air-dried sections on coverslips are inverted on the mixture. Incubation for 3 to 5 min
   is adequate for very active tissues such as muscle.
4. After incubation, sections on coverslips are separated from slides, washed, dehydrated,
   and mounted in the usual fashion in synthetic medium.

Fahimi and Amarasingham are convinced of the efficacy of the method for the *in situ* retention of enzyme activity. Their data are impressive.

Benitez and Fischer (*J. Histochem. Cytochem.*, **12**: 858, 1964) appear to have improved the method described above with the following suggested modifications:

1. Instead of 5% gelatin, 6.5% is used. This avoids retraction of the film due to drying for at least 1 h.
2. To improve tissue-gel contact and to prevent bubble formation due to trapped air, the gelatin film is placed on a flexible base such as polyvinyl film. Addition of the gel to the film and allowing it to "set" in the dark for at least 15 min prior to application to the section are advised. The flexible base permits easier and more uniform contact.
3. After incubation, the section and coverslip can be readily removed by immersion in warm (45°C) water or by application of forceps.

Van Duijn et al. (*J. Histochem. Cytochem.*, **15**: 631, 1967) provided a theoretical treatment of the influence of slow substrate diffusion on the kinetics of an enzyme reaction in a cytochemical system.

Melnick (*Fed. Proc.*, **24**: 259, 1965) noted that generally greater hydrolase and dehydrogenase activities could be demonstrated in sections cut from specimens frozen rapidly, such as in isopentane cooled with liquid nitrogen, than from specimens cooled slowly, such as at a controlled −1°C per min.

Diculesco et al. (*J. Histochem. Cytochem.*, **12**: 145, 1964) studied several types of dehydrogenases in different kinds of muscles of different species.

Ogata and Mori (*J. Histochem. Cytochem.*, **12**: 183, 1964) studied dehydrogenase activity of muscles of several invertebrates.

Smith (*J. Histochem. Cytochem.*, **12**: 847, 1964) studied several dehydrogenases in rabbit skeletal muscle.

Succinic dehydrogenase, cytochrome oxidase, and NAD diaphorase were identified in synaptic regions of the spinal cord of rats (Nandy and Bourne, *J. Histochem. Cytochem.*, **12**: 188, 1964).

Ogata and Mori (*J. Histochem. Cytochem.*, **12**: 171, 1964) studied the distribution of several oxidative enzymes in different kinds of muscles from human beings, monkeys, cows, swine, dogs, cats, rabbits, guinea pigs, rats, mice, whales, pigeons, lovebirds, chickens, snakes, lizards, geckos, tortoises, frogs, and fish.

Fullmer (*J. Histochem. Cytochem.*, **12**: 210, 1964) studied several dehydrogenases participating in the citric acid cycle, pentose shunt, and fatty acid metabolism. Osteocytes, osteoblasts, and osteoclasts stained for all the dehydrogenases. Enzyme activity was generally greatest in osteoclasts, less in osteoblasts, and least in osteocytes, although variation in enzyme activity was also noted in relation to the location and function of the particular cells.

Nelson and Wakefield (*J. Histochem. Cytochem.*, **21**: 184, 1973) conducted quantitative histochemical assays for glycolytic enzymes, namely, hexokinase, phosphoglucoisomerase, phosphofructokinase, and lactic dehydrogenase in sympathetic ganglia, adrenal medulla, dorsal root ganglia, and the granular layer of the cerebellum of rats.

Engel (*J. Histochem. Cytochem.*, **12**: 46, 1964) noted mitochondrial aggregation in abnormal skeletal muscles from three patients. One patient had major features of myotonia congenita with minor features of paramyotonia congenita. The other two had major features of hypokalemic periodic paralysis with minor features of adynamia episodica hereditaria. Aggregated mitochondria stained for most of the usual dehydrogenase; however, they failed to stain for succinic dehydrogenase and menadione-linked α-glycerophosphate dehydrogenase.

CREATINE PHOSPHOKINASE ISOZYMES

Rosalki [*Nature (Lond.)*, **207**: 414, 1965] recommends the following method for the demonstration of *creatine phosphokinase isozymes* on cellulose acetate membranes:

1. Tissue homogenates, serums, etc., are separated on cellulose acetate membranes in barbitone buffer of pH 8.6 using a constant current of 0.5 mA/cm for $1\frac{1}{4}$ h.
2. Following electrophoretic separation, cellulose acetate strips are cut longitudinally in half. One half is immersed in a full incubation mixture provided below, the other in a control solution consisting of the incubation solution minus the substrate:

Adenosine 5-diphosphate sodium salt	2 mg
Hexokinase (2.8 IU)	10 μl
Glucose 6-phosphate dehydrogenase (1.4 IU)	10 μl
Glucose	2 mg
Magnesium sulfate ($MgSO_4 \cdot 7H_2O$)	7 mg
NADP	3 mg

The solution is divided into equal parts, and 5 mg creatine phosphate is added to the test solution. Immediately prior to use, 80 μl (0.1 %) phenazine methosulfate is added to both test and control solutions.

The strips are incubated in a moist chamber for 1 h at 37°C. Enzyme activity is identified as stained bands.

11

ENDOGENOUS PIGMENTS

THE HEME OR TETRAPORPHIN PIGMENTS

Hemoglobin itself is a relatively basic protein, endowed with or closely associated with peroxidase activities, containing closely bound iron which is usually not reactive for iron ions, and possessing a fairly characteristic absorption spectrum. In hemoglobinurias it is sometimes seen as strongly oxyphilic rhomboid crystals in the lumina of the proximal convoluted renal tubules and as brightly eosinophilic globules in the distal cytoplasm of the tubule epithelial cells. Similar granules are sometimes found filling macrophages in recently hemorrhagic corpora lutea, and they may be identified in the erythrocyte fragmentation that ensues in erythrophagia. This eosinophilia persists at a higher pH level than that of most other tissue elements. In formalin-fixed paraffin sections stained with azure A–eosin B at pH 5.5, smooth muscle, connective tissue, and most cytoplasm stain blue or green; erythrocytes, eosinophil leukocyte granules, and a few other objects retain the eosin stain. This is useful though not highly specific.

Hemoglobin is well stained in erythrocytes by many acid dyes such as the eosins, and often its own yellowish color sufficiently modifies the tinge of red given it by eosin to give it a distinctly different color from that of other cytoplasmic materials. Certain red azo dyes such as azofuchsin G (C.I. 16540) and azofuchsin GN (C.I. 16535) stain erythrocytes deep red and are rather poor plasma stains, while brilliant purpurin R (C.I. 23510) gives good brown cytoplasm and muscle, but rather pale yellowish brown erythrocytes. A mixture of one of the azofuchsins, 1 part of 1 % solution to 4 parts of 1 % brilliant purpurin R, gave red erythrocytes and pinkish brown cytoplasm and muscle. It is probable that with any individual dye samples, varying mixtures would have to be tried to find the most differential one.

Certain basic aniline dyes (notably toluidine blue O and thionin), when used for 30 to 60 s in 1 : 1000 neutral aqueous solution, stain erythrocytes a brilliant green or yellowish green, nuclei deep blue, cytoplasm light blue, cartilage purple, and mast cell granules deep purplish violet.

Alum hematoxylin counterstains of frozen sections stained for fats with oil red O often present deep olive green erythrocytes. When overstained with neutral iron hematoxylin (either in combined solution or sequence technics), hemoglobin is stained black and is among the more difficult substances to decolorize, losing its color only just before myelin in freshly fixed chromated formalin material.

Allied to these stains is the Dunn-Thompson (*Arch. Pathol.*, **39:** 49, 1945) method. It apparently requires neutral formalin fixation, for we have had it behave somewhat erratically on mailed-in formalin material. However, on tissue fixed in neutral buffered formalin it gives quite consistent results. It may also be used on tissue smears provided these are fixed moist in neutral buffered 10% formalin, or in 3% tannic acid in methyl alcohol (3 to 5 min), or in methyl alcohol and subsequently treated with 3% tannic acid–methyl alcohol as above.

THE DUNN-THOMPSON HEMOGLOBIN STAIN.   Smears and sections are brought to water as usual for sections.

1. Stain 15 min in Mallory's aqueous alum hematoxylin. (Probably any other unacidified alum hematoxylin will serve.)
2. Wash in tap water.
3. Mordant 1 min in 4% iron alum.
4. Rinse in tap water.
5. Stain 15 min in a picrofuchsin solution composed of 13 ml 1% acid fuchsin and 87 ml saturated aqueous picric acid solution.
6. Dehydrate and differentiate 3 min in one change or more of 95% alcohol. Dehydrate with 100% alcohol (two changes), and clear in xylene. Mount in synthetic resin, polystyrene, Depex (etc.).

RESULTS: Cytoplasm is brown to yellow; hemoglobin casts, phagocytosed particles, and erythrocytes are emerald green; collagen, red; nuclei, brown to purple to gray-black.

THE OKAJIMA METHOD (*Anat. Rec.,* **11**: 295, 1916).   This is one of the older differential hemoglobin stains. The following is Dunn's modification, worked out in our laboratory: Fix in buffered neutral 10% formalin; embed and section in paraffin. Bring to distilled water as usual.

1. Mordant 1 min in 10% phosphomolybdic acid.
2. Wash in distilled water.
3. Stain 1 h in 9 ml 10% phosphomolybdic acid and 30 ml saturated aqueous (7.69%) alizarin red S (C.I. 58005) (Okajima: 20 min to 20 h).
4. Wash in distilled water.
5. Counterstain if desired with an unacidified alum hematoxylin (3 to 5 min).
6. Wash in water; dehydrate and clear through alcohols and xylene. Mount in synthetic resin.

RESULTS: Hemoglobin is light to dark orange-red; the background is light brown with hematoxylin, orange without hematoxylin.

Of the derivatives of Lison's zinc leuco method for hemoglobin peroxidase, R. Dunn's patent blue V or cyanol (FF) (C.I. 42051 and 43535, respectively) variant is probably the most useful (see Chap. 10).

Photography with violet light, utilizing the strong absorption band at 411 mμ, can be useful for localizing heme pigments photographically. The acid hematins, including formalin pigment, also show this absorption band in vitro.

Liberation of the iron ions from hemoglobin by chemical means consistent with good morphologic preservation of tissue seems to be a matter of some difficulty. We have seen Prussian blue reactions in erythrocytes fixed in deliberately acidified formalin (pH 3.5); Gomori reported liberation of ferric ions by strong peroxide solutions. Neither of these procedures seems to be consistent. Prolonged digestion with 0.1 N HCl in the presence of ferricyanide has failed to produce Turnbull's blue in our hands.

A procedure which might prove of interest would be the production of fluorescent antibodies against specific hemoglobins. We have not heard of this having been tried. It might even be possible so to identify fetal hemoglobin and hemoglobin S. This is pure speculation at this time, though we believe that the specific antibodies have been prepared.

Lipp and Ratzenböck [*Acta Histochem. (Jena)*, **16:** 317, 1963] describe a 9-phenyl-2,3,7-trihydroxy-6-fluorone (Eastman 6346) reaction as a specific stain for erythrocytes, depending on their hemoglobin content. The reagent is prepared fresh weekly. Dissolve 0.5 g in 100 ml 95% alcohol, let stand overnight, add 3 ml 28% ammonia, and mix well. After 1 h filter.

Bring paraffin sections (8 to 10 μ) of formol-fixed tissue to 95% alcohol and stain 25 min. Rinse in 95% alcohol, interpolate nuclear stain with hemalum if desired, blue, wash, dehydrate, and differentiate hemoglobin stain 5 min in absolute alcohol, clear in xylene, and mount in Caedax or other suitable resin. Red corpuscles stain deep violet on a yellowish brown background. Introduction of the blue nuclear stain does not alter the other colors.

In place of the hemalum nuclear stain the Prussian blue reaction can be interpolated. This alters erythrocyte color to orange, contrasting with blue-green hemosiderin.

Bile pigments, unreacted hemosiderin, lipofuscins, etc., retain their native colors.

The mechanism of the reaction has not been explained, but it appears to be specific. Bilirubin and presumably hematoidin do not react. The same reagent reacts in acid alcoholic solution to give violet with slightly hydrolyzed deoxyribonucleic acid according to Turchini's method.

METAL UPTAKE REACTIONS.   Like formol-fixed myelin and eosinophil leukocyte granules, erythrocytes color black with acetic hematoxylin after pH 3.5, 2.5 to 3% $K_2Cr_2O_7$ mordanting; 4 h at 60°C is adequate; 4 to 6 weeks at 0°C serves. The 0°C reaction is blocked by brief bromination. Neutral 0.1% hematoxylin for 2 h at 25°C colors erythrocytes blue-black after a 2-h, 25°C, 10-mM $FeCl_2$ or $FeSO_4$ mordanting, greenish black after a similar 10 mM $CuSO_4$ treatment.

AZO COUPLING REACTION.   Formol-fixed erythrocytes color dark green to black by the diazosulfanilic acid–azure A (pH 1) sequence and dark red with diazosafranin (5 mM, pH 8, 3 min). These stains are largely suppressed by a 2-h mordanting in 0.1 M zinc acetate or ferrous or ferric chloride, or by glutaraldehyde fixation (Lillie, *Histochemie*, **20:** 338, 1969; Lillie and Pizzolato, *J. Histochem. Cytochem.*, **20:** 116, 1972).

Puchtler and Sweat (*Arch. Pathol.*, **75:** 588, 1963) provide the following method for the simultaneous demonstration of hemosiderin (blue) and hemoglobin (red).

1. Formalin pigment must first be removed from formalin-fixed deparaffinized sections by immersion of sections in saturated alcoholic picric acid for 5 min at 24°C. Tissues are then washed until decolorized.
2. Mordant in Zenker-formol overnight.
3. Wash in running water for at least 15 min.
4. Remove mercury pigment with the iodine–sodium thiosulfate as usual; wash under the tap for 5 min.
5. Wash in several changes of distilled water. Avoid any contact with metals.
6. Place in potassium ferrocyanide solution for 30 min (mix equal parts of a fresh 2% solution of potassium ferrocyanide and 2% hydrochloric acid, filter).
7. Wash in three changes of distilled water.
8. Transfer to a 5% aqueous solution of tannic acid for 10 min (make up 2 days prior to use).
9. Rinse in three changes of distilled water.
10. Treat with 1% aqueous phosphomolybdic acid for 10 min.
11. Rinse in three changes of distilled water.

12. Stain in a saturated solution of phloxine B in methanol–glacial acetic acid, 8 : 3 (pH approximately 3.7), for 5 min. Do not filter.
13. Rinse in two changes of methanol–glacial acetic acid, 9 : 1.
14. Dip once in absolute alcohol, clear in xylene, mount in Permount.

RESULTS: Hemosiderin stains dark blue to greenish blue, hemoglobin stains red, elastic fibers stain pink, and all else stains yellow. It is important that tissues be treated at least overnight in Zenker-formol solution.

Puchtler et al. (*Arch. Pathol.*, **78:** 76, 1964) encountered a chromatographic impurity, probably bis-*p*-nitroaniline → H-acid, in amido black 10B which gave a deep greenish blue to hemoglobin and also to collagen and cytoplasm in the Puchtler-Sweat 1962 hemoglobin method, thereby impairing its usefulness in detecting hemoglobin droplets in phagocytes and renal epithelial cells.

ALTERED HEMOGLOBINS.   *Methemoglobin* is colored bright red by the addition of potassium hydroxide in fresh unfixed material. This reaction is less distinct or absent in fixed tissue. Methemoglobin is partly soluble in alcohol. It is colored by certain basic dyes and stains like hemoglobin with acid dyes. Also like hemoglobin it may be stained by myelin methods. It is bleached by hydrogen peroxide. Silver nitrate and osmium tetroxide do not alter its color. It is best identified by spectroscopic study of fresh material.

Sulfhemoglobin, or *sulfmethemoglobin*, is a greenish sulfur methemoglobin compound occasioning the greenish discoloration of the abdominal wall of cadavers. It is distinguishable from other hemoglobin derivatives by spectroscopy. (The foregoing statements concerning the altered hemoglobins are derived from Schmorl and Mallory.)

ACID HEMATINS

Of the acid hematins, three have been studied in tissue to a greater or lesser extent: the all too familiar formalin pigment, malaria pigment, and an HCl hematin which forms in small hemorrhagic lesions in the surface of the gastric mucosa. This last we have seen occasionally in animals in certain intoxications, where the fixation was such as to prevent the formation of formalin pigment.

Formalin pigment is formed on fixation in formaldehyde solutions at pH levels below 5.6, and possibly an alkaline hematin is formed at pH levels above 8. We have been told of the presence of formalin pigment in tissue fixed in sodium acetate–formalin. Phosphate buffering to pH 7 seems to prevent its formation quite reliably.

The acid hematins are microcrystalline dark brown pigments occurring as tiny birefringent needles or rhomboids. They give no iron reactions; they are dissolved by alcoholic picric acid or alcoholic ammonium picrate solutions and by alkalies. They are quite resistant to strong organic and dilute mineral acids, and they resist even concentrated sulfuric acid for some time. Concentrated nitric acid bleaches them in an hour.

Usually the occurrence of the pigment free in vein lumina, coupled with its birefringence and crystal form, is adequate for identification of formalin pigment. The identification of the other two hematins is uncertain in the presence of formalin pigment.

The *formaldehyde pigment*, or *acid formaldehyde hematin*, is formed when acid aqueous solutions of formaldehyde act on blood-rich tissues. It is a dark brown microcrystalline substance which rotates the plane of polarized light. Consequently the individual, minute, rhomboid particles glow and darken alternately with each 90° rotation of the stage when examined with crossed Nicol prisms or good polaroids. The pigment withstands extraction with water, alcohol, acetone, glycols, glycerol, fat solvents, and dilute acids. Formalin pigment is bleached promptly or within an hour by concentrated nitric acid, and partially by 90% formic acid. It is bleached in 30 min by 3% hydrogen peroxide or by 5% chromic anhydride (CrO_3). Sequence treatment with 5% potassium permanganate (but not 0.5%) and 5% oxalic acid removes the pigment, though neither reagent alone appears to bleach it. It withstands extraction by concentrated sulfuric, hydrochloric, phosphoric, and acetic acids. It does not give Prussian blue or Turnbull's blue reactions with ferrocyanide or ferricyanide. It is extracted by treatment with weak alcoholic, aqueous, hydroalcoholic, or water-acetone solutions of sodium, potassium, or ammonium hydroxide, and by ammonia solutions in glycerol and glycols. It is removed at once by saturated alcoholic picric acid solution. It occurs copiously within vascular spaces among apparently intact or laked erythrocytes and also apparently within phagocytes. It seems more probable that the particles within phagocytes are formed from previously phagocytosed erythrocytes or hemoglobin than that the pigment is ingested as such. Spectroscopically it is similar to, but distinct from, hydrochloric acid and acetic acid hematins.

Several methods have been suggested for its removal. Verocay immersed sections for 10 min in 0.01% potassium hydroxide in 80% alcohol and then washed for 5 min in two changes of water. Schmorl warns that this treatment impairs the alcohol fastness of the Gram stain; he finds Kardasewitsch's method harmless in this respect. This method employed a 5-min to 4-h extraction in a 1 to 5% dilution in 70% alcohol of 28% ammonia water. Subsequent washing with water or alcohol is necessary to remove the excess ammonia. The saturated alcoholic picric acid mordanting before Masson stains should remove formalin pigment, especially if prolonged to 5 min. A mixture of 50 ml each of acetone and 3% hydrogen peroxide and 1 ml 28% ammonia water removes the pigment in 5 min or less.

More important than methods of removal of formaldehyde pigment is the fact that it is not formed on fixation with formaldehyde buffered to pH levels above 6. Formalin at pH levels from 3 to 5 forms large quantities of the pigment, with or without obvious lysis of erythrocytes. Fixation in alkaline formalin is said also to produce a formalin pigment. With this we have had no experience. Brief trial of alkaline formalin years ago gave us losses of nucleic acids which we have preferred to avoid.

The foregoing statements on formaldehyde pigment are based largely on studies by Herschberger and Lillie (*Bull. Int. Assoc. Med. Mus.*, **27**: 136, 145, 162, 1947).

Malaria pigment occurs in the parasites (especially the quartan *Plasmodium malariae*), about brain capillaries, and in littoral phagocytes in spleen, liver, bone marrow, and lymph nodes. It is an amorphous, dark brown, granular pigment which does not fluoresce in ultraviolet light but rotates the plane of polarized light, resembling in the latter respect the otherwise quite similar microcrystalline and doubly refractile acid formaldehyde hematin, the so-called formalin pigment. Kósa (*Virchows Arch. Pathol. Anat.*, **258**: 186, 1925) stated that that part of the malaria pigment which was free or in erythrocytes was doubly refractile, while that in phagocytes never was. Like formalin pigment it is soluble in dilute aqueous and alcoholic solutions of sodium, potassium, and ammonium hydroxides; is bleached within an hour by concentrated nitric acid, by hydrogen peroxide, and partially by 90% formic acid; and is insoluble in dilute (5%) aqueous solutions of mineral acids and in concentrated acetic, hydrochloric, phosphoric, and sulfuric acids. On direct experiment it was found that both these pigments remained identifiable in slide preparations for 2 weeks if sections were first blotted dry, treated with the 96.5% sulfuric acid (sp gr 1.84), covered with a cover glass, and sealed with petrolatum. But tissue structure was destroyed (Hershberger, unpublished data).

Schmorl stated that malaria pigment is soluble in 5% alcoholic solutions of sulfuric, nitric, and hydrochloric acids at 40 to 50°C in 1 day or less. It is also soluble in aniline, in pyridine, and in a 4% solution of quinine in chloroform. It is insoluble in fat solvents and is not stained by fat or lipid stains. It is not blackened by osmium tetroxide or silver nitrate. Malaria pigment gives no iron reaction on direct test, but according to M. Kósa (*Virchows Arch. Pathol. Anat.*, **258:** 186, 1925), it may give a Prussian blue reaction with potassium ferrocyanide when sections are first treated for 10 to 12 h with 2% oxalic acid or 1% hydrochloric acid to remove hemosiderin, washed with distilled water, treated for 5 to 10 min with 1% potassium hydroxide, and again washed with distilled water. In our experience the prescribed acid treatment may be inadequate to remove the ferric iron from hemosiderin in formalin-fixed material. Hence we consider the validity of Kósa's statement questionable. Moreover, we have seen artifactual ferrocyanide staining of smooth muscle, liver cell cytoplasm, and other structures after prolonged exposure to oxalic acid used for extraction of iron from hemosiderin, where no iron was seen before. Moreover, we have shown that smooth muscle and other lysine- and arginine-rich sites readily take up Fe(II) and Fe(III) from dilute aqueous solution (Lillie and Pizzolato, *J. Histochem. Cytochem.*, **20:** 116, 1972).

Brown (*Exp. Med.*, **13:** 290, 1911) called malaria pigment *hematin;* Mallory used this term for a dark brown amorphous pigment, not more closely described, which occurred in old extravasations of blood. Neither of these usages should be confused with the usual meaning of an acid- or alkali-soluble hemoglobin derivative with a characteristic absorption spectrum.

ACID HEMATIN.   Sometimes similar microcrystalline brown pigment is found in material fixed in neutral buffered formalin. This occurs in somewhat autolyzed spleens in the surface adjacent to the stomach, and is also seen in perfectly fresh material fixed in neutral formalin in focal hemorrhages of the gastric mucosa. This pigment would appear to be a hydrochloric acid hematin. We have observed this gastric pigment chiefly in experimental toxicologic material fixed in phosphate-buffered formalin, and cannot say whether it would have appeared with other fixations. However, similar pigment may be formed in spleen tissue fixed in 1% acetic or formic acid in 89% alcohol or acetone, without formaldehyde.

BILIVERDIN, BILIRUBIN, HEMATOIDIN

In the destruction of hemoglobin from broken down erythrocytes, the iron-containing tetraporphin heme is first separated from the protein globin and then undergoes oxidative scission, or opening, of the ring between two pyrrole ring carbons, the one vinylated, the other methylated. The resulting iron-containing compound (or compounds) is known as *verdohemin*, which on loss of iron to ferritin or new hemoglobin molecules becomes biliverdin. Biliverdin is formed particularly in littoral cells of the bone marrow and spleen, and passes through the circulation to the liver. When the ring opening precedes the porphin-globin cleavage, the resultant intermediate protein is verdoglobin. This undergoes hydrolysis and iron loss to form biliverdin again. In the liver, biliverdin is normally reduced to bilirubin; on secretion into the intestine it undergoes further reduction to urobilinogen (stercobilinogen) and is later oxidized to urobilin (stercobilin), the yellow-brown pigment of feces and urine. In bilirubin and biliverdin the two vinyl groups remain; in urobilinogen, urobilin, and mesobilirubinogen they are hydrogenated to ethyl groups.

Bilirubin further forms a mono- or diglucuronide by esterification of one or both of the propionic acid residues with the C-1 hydroxyl of glucuronic acid. These, especially the diglucuronide, are water-soluble, and they azo-couple to give the direct van den Bergh reaction. Alcohol solubilizes unconjugated bilirubin, thus making it available for the indirect reaction.

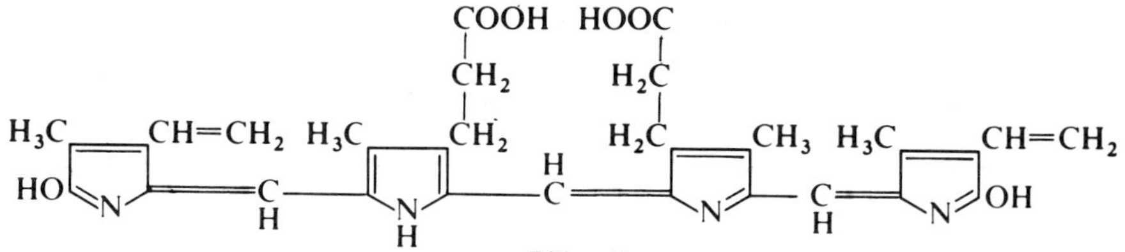

Biliverdin

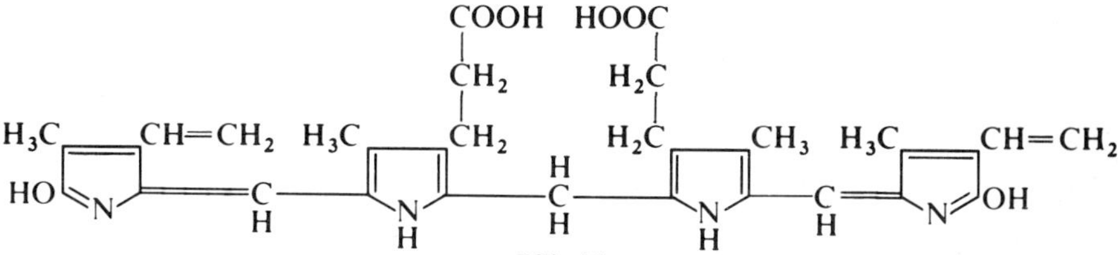

Bilirubin

Hematoidin is a golden yellow to orange or reddish brown crystalline or amorphous pigment occurring particularly within hemorrhagic infarcts and aging hemorrhages. Rich identified it with bilirubin; Virchow clearly distinguished the two. Of the two reports which identify the two pigments, the hematoidin in the case of Fischer and Reindel (*Hoppe Seylers Z. Physiol. Chem.*, **127**: 299, 1923) was derived from an echinococcus cyst of the liver; that in Rich's (*Johns Hopkins Med. J.*, **36**: 225, 1925) case came from an omental cyst, removed surgically, which had apparently been well encapsulated and unconnected with liver or bile passages. These reports would be more convincing had the hematoidin come from a source more widely separated anatomically from the liver.

These pigments were originally characterized by their Gmelin reaction to concentrated sulfuric or nitric acid, in which a succession of colors develops: brownish red, red, purple, violet, blue, and green. This is often performed by adding a crystal or so of sodium nitrite, enough to make 1 to 2% HNO_2, to concentrated nitric acid. The test is often unsatisfactory but should be repeated on two or three slides before it is regarded as negative.

THE GMELIN REACTION. Gmelin first presented this method for bile pigments before the Heidelberg society of physicians and scientists in May 1824 (Tiedemann and Gmelin, "Die Verdauung nach Versuchen," K. Groos, Heidelberg, 1826). Gmelin used a nitric acid of a sp gr of 1.25, about 40%, probably contaminated with HNO_2, and recorded a series of color changes: green, blue, violet, lavender, rose, and finally yellowish brown.

In its modern application Pearse (1960, pp. 658, 921) used a few drops of an equal-volume mixture of ethanol and concentrated nitric acid, watching the reaction under a cover glass. Cruicks (*J. Clin. Pathol.*, **55**: 116, 1971) warns that these reagents react violently, boiling spontaneously in an open vessel and perhaps exploding in a closed one. The mixture should be made of 1 ml of each reagent at the time of using. Undoubtedly some HNO_2 is formed.

Lillie (1954, p. 247) mounted sections in water and flowed a drop or two of concentrated HNO_3 containing a small crystal of $NaNO_2$ under the cover glass. These reactions are all transient and capricious. At least two or three trials should be made before considering the reaction negative.

Lillie and Pizzolato (*J. Histochem. Cytochem.*, **15**: 600, 1967) reported an adaptation of Samuely's ("Abderhalden's Handbuch der biochemischen Arbeitsmethoden," vol. 2, Urban & Schwarzenberg, Berlin, 1910) chloroform bromine reaction. A 0.5% (v/v) solution of bromine in carbon tetrachloride applied to deparaffinized sections for 10 to 90 min produces a series of colors varying in individual bile casts or hematoidin crystals from dark violet through red-purple to rose-pink. Preparations may then be rinsed in xylene and mounted in cellulose caprate. The colors remain stable for some weeks but finally fade. The test is regarded chemically as equivalent to the Gmelin reaction.

Gmelin colors may sometimes be seen in the early stages of the argentaffin reaction and may be seen in desilvered preparations. Paraffin sections kept for 3 months after cutting before dewaxing and study also exhibit fair numbers of violet, purple, and rose-colored casts, a series of Gmelin colors resulting apparently from air oxidation. These Gmelin colors may be produced by oxidation at 60°C in 2.5% $K_2Cr_2O_7$ for 3 to 5 days, by 1 to 2 h at 24°C in 0.1% CrO_3 as well as with Br_2–CCl_4 and diammine silver. Prolongation of the CrO_3 oxidation to 4 h decolorizes completely.

Luna and Ishak (*Am. J. Med. Technol.*, **38:** 459, 1967) used an iron alum–celestin blue B, phosphotungstic acid, acidified acid fuchsin sequence technic for the demonstration of bile canaliculi in both normal and pathologic formol-fixed livers.

The celestin blue solution contains 600 mg dye, 5 g iron alum, 2 ml concentrated HCl, 15 ml glycerol, and 85 ml distilled water. Boiling the mixture for 3 to 5 min was prescribed following the method of Proescher and Arkush, but perfectly satisfactory iron alum–celestin blue is now prepared (Lillie et al., *Am. J. Clin. Pathol.*, **60:** 817, 1973) by simply dissolving the dye and the salt in separate portions of water and then mixing.

The technic prescribes:

1. Dewax, hydrate, and wash 5 min in running water and distilled water.
2. Stain 30 min in celestin blue.
3. Rinse in distilled water, and mordant 15 min in 1% phosphotungstic acid.
4. Wash 3 min in tap water (preferably not too heavily chlorinated).
5. Stain 5 min in 10 ml 1% acid fuchsin, 1 ml 1% oxalic acid, and 80 ml distilled water.
6. Dehydrate rapidly through alcohols, clear in xylene, and mount in synthetic resin.

RESULTS: Bile canaliculi are seen as pink to red tubules; liver cell membranes are otherwise unstained. Small bile duct cytoplasm is pink to unstained; nuclei are blue. Liver cell cytoplasm is pink to bluish. Bile usually is green to olive or even reddish brown; erythrocytes are red. Lipofuscin and hemosiderin are unstained, fibrin is red, smooth muscle is pink with blue nuclei.

T. Dunn (*Milit. Surg.*, **109:** 350, 1951), who questions the identity of hematoidin and bilirubin, describes hematoidin crystals as orange-red, yellow, or green, rhomboid, monoclinic, birefringent, readily soluble in chloroform and in pyridine, very slightly soluble in alcohol, and insoluble in water, 10% formalin, glycerol, ether, or xylene. Carbol-xylene dissolved them. These findings closely parallel Frey's 1860 description.

She reported that the spectrum in chloroform differed from that of bilirubin, that the diazo reaction failed to show a reddish violet, that application of nitrous acid to fresh tissues dissolved them rapidly without a play of colors. In paraffin sections she found many empty rhomboid spaces and perceived very little identifiable hematoidin. But in Carbowax sections the rhomboid form and yellow color of even very small crystals were well preserved. For sectioning in paraffin, we have had fair success with gasoline clearing and brief vacuum infiltration.

Spectroscopically a chloroform solution of bilirubin showed a strong maximum at 430 mμ and slowly increasing absorption from 340 to 250 mμ. The chloroform solution of hematoidin from human infant adrenals and renal papilla (of similar color density to the bilirubin solution) showed no peak at 430 mμ, but a rapidly rising absorption from 450 out to the ultraviolet end of the spectrum.

TABLE 11-1   SOME CHEMICAL REACTIONS
OF BILIRUBIN AND HEMATOIDIN

Test	Bilirubin	Hematoidin
Trichloroacetic reaction	Blue-green	Faint pink
Ehrlich's diazo reaction	Red-violet	Negative
Rosin's iodine test	Green	Negative

Source: T. Dunn, *Milit. Surg.*, **109:** 350, 1951.

Dunn reports the results of A. S. Mulay's chemical tests (Table 11-1).

It is to be observed that Glenner prescribed cryostat sections for his procedure, thus avoiding the losses of bilirubin suggested above and of hematoidin as detailed by T. Dunn. It does not appear to be necessary, however, to avoid the use of formalin, since Dunn found hematoidin well preserved after several weeks in that fluid. We have demonstrated hematoidin in material fixed 1 to 3 weeks in formalin and carried through an alcohol, petroleum ether, paraffin sequence with 30-min vacuum infiltration. The green of the bile in hepatic biliary ductules and capillaries in unstained preparations contrasted with the golden yellow of the hematoidin. Dichromate oxidation at pH 2.2 (below) did not obviously alter the color of bile pigment; perhaps the green was yellower and less pronounced. Nor does brief dithionite reduction (5%, 10 min) alter the green of the bile pigment of obstructive jaundice.

There are a considerable number of oxidative methods purporting to change yellowish brown bilirubin to green biliverdin, e.g., Stein's iodine, Glenner's dichromate, and certain peroxide procedures. We have long known that bile pigment in Van Gieson–stained preparations was often a brilliant grass green. An analysis of the effects of this stain, by leaving out one or another or all of the reagents that went into an iron chloride hematoxylin–Van Gieson sequence, led us to the finding that the pigment was green in the unstained paraffin section and required eosin to make it the familiar yellow-brown. Delvaux has confirmed this finding, using Biebrich scarlet as well as eosin.

That various oxidative procedures ($K_2Cr_2O_7$, H_2O_2, I_2) convert bilirubin to green biliverdin in vitro is, of course, established.

ALDEHYDE OXIDATION OF BILIRUBIN.   While it is a familiar fact that gallbladder mucosa and icteric liver turn green in a few days when fixed in formol, it is seldom mentioned in the textbooks. McManus (1960) found it difficult to understand how a reducing agent like formaldehyde could convert bilirubin to biliverdin. R. D. Baker ("Essential Pathology," Williams & Wilkins, Baltimore, 1961) simply mentions the fact without explanation.

Lillie and Pizzolato (*Virchows Arch. Pathol. Anat.*, **350:** 52, 1970) tested mixtures of fresh human gallbladder bile with distilled water (1 : 9), formol to make 10% (1.33 M), chloral hydrate as final 1.33 and 0.133 M concentrations, paraldehyde (1.5 M), 1.25 M glutaraldehyde, 0.95 M benzaldehyde, 0.1 M p-dimethylaminobenzaldehyde, 0.1 M vanillin, 1.03 M isobutyraldehyde, 1 M glyoxalic acid, and 1.03 M glyoxal, all expressed as molarities in the final mixture. p-Dimethylaminobenzaldehyde formed an immediate bulky red precipitate which separated in 24 h into a clear red supernatant and a dark green precipitate. The rest of the aldehydes promptly developed shades of green. At 24 h clear green solutions were present with vanillin, chloral hydrate, and formaldehyde. Green precipitates formed with isobutyraldehyde, paraldehyde, benzaldehyde, glyoxal, and glyoxalic acid, with green to yellow supernatants. Glutaraldehyde reverted to a brownish yellow fluid with a brown-orange precipitate.

Blocks of strongly icteric liver placed in 80% alcohol, in chloroform methanol, in acetone, in alcohol-ether remained golden brown; those placed in 10% formol developed a greenish tint in 2 to 3 h and by 2 to 3 days were deep green. Tissue blocks fixed in anhydrous acetone and preserved in that fluid remained golden brown for over 3 months; similarly fixed blocks washed in distilled water and transferred to 10% formol, exposed to air and under a 3-cm layer of mineral oil alike, turned green in 5 h, dark green in 24 h, and remained green without discoloration of the fluid for 3 months.

Dark green formol-fixed blocks, washed in water and transferred to 0.1 M $Na_2S_2O_5$, change color back to bright yellow in 18 h; 0.1 M $Na_2S_2O_3$ had a similar effect. A 2-h water washing and a return to 10% formol reproduced the green. This cycle could be repeated several times, but with a gradual loss in intensity of the colors. Ferrous sulfate solution, acetone, and 70% alcohol also reduced green formol-fixed tissue to a pale greenish yellow; washing and return to formol restored the deep green.

It was considered that most aldehydes were capable of oxidizing bilirubin to biliverdin and that the biliverdin could be readily reduced back to bilirubin. The oxidation-reduction cycle could be repeated several times with satisfactory results.

STEIN'S IODINE.   A number of oxidative procedures for converting golden brown bilirubin to green biliverdin in sections have been reported. Stein [*C. R. Soc. Biol.* (*Paris*), **120:** 1136, 1935] used an iodine technic; Nizet and Borac [*C. R. Soc. Biol.* (*Paris*), **126:** 11282, 1952], Gomori (1952) (H_2O_2, $K_2Cr_2O_7$), and Glenner (*Am. J. Clin. Pathol.*, **27:** 1, 1957) used $K_2Cr_2O_7$ on frozen sections; Kutlik [*Acta Histochem.* (*Jena*), **4:** 141, 1957] used several procedures; Okamoto et al. (cited by Gomori, p. 134) used nitrous acid. Hall (*Am. J. Clin. Pathol.*, **34:** 313, 1960) used a ferric chloride trichloroacetic acid method, claiming a high specificity. Pearse (1960, p. 659; 1972, p. 1073) found the results of Stein's reaction rather unconvincing, and Lillie and Pizzolato (*J. Histochem. Cytochem.*, **16:** 17, 1968) observed that formol-fixed cholestatic liver bile casts are often green or even greenish black without application of oxidizing agents. If, however, sections are first reduced in dithionite, thiosulfate, or acidified bisulfite until bile casts are olive or yellow-brown and oxidizing agents are then applied and the results compared with the reduced sections, quite convincing demonstrations of the oxidation to biliverdin are seen. Exposures of 2 to 4 h in 0.1 M $Na_2S_2O_5$–0.15 to 0.2 N HCl or to 0.2 M $Na_2S_2O_3$ were usually sufficient. Gmelin colors are not reduced, and sections should be freshly cut from previously unsectioned paraffin blocks to avoid this complication.

The Stein's iodine used by Lillie and Pizzolato was derived from Stein and the AFIP "Manual of Special Staining Technics" (edited by Ambrogi, 1960). Dissolve 11 g potassium iodide in 16 ml distilled water. Add and dissolve 10 g iodine. Add 100 ml 95% alcohol and distilled water to make 400 ml. Graded exposures of 6, 20, 90 min, 5, 16, and 30 h were used. The iodine effect was essentially maximal in 20 min. After the iodine followed a 30- to 90-s bath in 5% sodium thiosulfate (until white), 10 min in running water, and treatment with alcohols, xylene, and synthetic resin.

THE HALL METHOD.   This method has displaced Stein's iodine in the third edition (1968, edited by L. Luna) of the AFIP "Manual of Special Staining Technics." Bring formol-fixed paraffin sections to water as usual, and immerse in Fouchet's reagent for 5 min [25 g trichloroacetic acid in 100 ml distilled water, and add 1 g ferric chloride ($FeCl_3 \cdot 6H_2O$) in 10 ml distilled water]. Wash in distilled water for 5 min (two or three changes), stain with Van Gieoson's stain for 5 min, and treat with alcohols, xylene, and synthetic resin.

GLENNER'S DICHROMATE PROCEDURE.   In this procedure attached cryostat sections of fresh unfixed tissue were immersed for 15 min in a mixture of equal volumes of 2.94% $K_2Cr_2O_7$ (0.1 M) and 0.1 M phosphate HCl buffer of pH 2.2, postfixed 20 min in calcium acetate–formol, washed, counterstained, dehydrated, cleared, and mounted as usual. Control, unchromated sections treated with 0.05 M phosphate buffer of pH 2.2 should be compared. Bilirubin is converted to green biliverdin.

Lillie (1965) stated that hematoidin was not converted to biliverdin by any of these iodine and dichromate methods. Walker et al. (*J. Histochem. Cytochem.*, **18**: 367, 1970) reported that Stein's iodine (as above) with 1- and 2-day exposures produced no color change in crystalline hematoidin, and only a slightly greenish yellow in 14 of 114 sections in amorphous hematoidin. Bisulfite-reduced icteric liver sections used as controls showed restoration of green colors in 75% of the sections.

In regard to Stein's and Glenner's oxidation methods to turn bilirubin into green biliverdin, it may be noted that some bile pigment in obstructive jaundice is already quite dark green in formalin-fixed tissue and that neither iodine nor dichromate oxidation occasions any change of color in this pigment. It is also to be observed that bile pigments appearing deep green in Van Gieson's stains appear quite brown when hematoxylin-eosin or azure-eosin stains are used. The green may also be seen in uncounterstained Prussian blue and acid silver nitrate preparations, as well as in unreacted controls.

To prove that the pigment present is in fact bilirubin, it is necessary that it be brown in uncounterstained, unoxidized controls, and that its color be converted to green by the oxidation step. This requirement applies both to Stein's and to Glenner's methods.

THE ARGENTAFFIN REACTION.   We had already observed ferric ferricyanide, diammine silver, and hot methenamine silver reactions of bile casts in icteric livers and of hematoidin deposits in brain infarcts when Kutlik's [*Acta Histochem. (Jena)*, **5**: 213, 1958] report appeared. Our early notes indicated that 0.1 N (1.7%) $AgNO_3$ in pH 4 acetate buffer (Lillie, *J. Histochem. Cytochem.*, **5**: 325, 1957), which blackened cutaneous and ocular melanins in 10 to 30 min, had no effect on bile casts or hematoidin in 24 h, that 3 h at 60°C in methenamine silver blackened both, as did 30 min at 60°C in diammine silver. The latter agent took 5 h or more to give adequate and consistent blackening at 25°C. Kutlik used only neutral (and ammoniacal) silver nitrate solutions, and his findings substantially agreed with ours. On formol- and chloroform-methanol-fixed tissue 25°C treatment with diammine silver showed some reaction as early as 10 to 20 min, but the reaction was only reasonably complete in 3 to 5 h, and some specimens required 24 h for maximal reaction.

Primary silver fixation of fresh icteric liver in 10% silver nitrate in water (Cater's fluid) or in 50% alcohol (Barnett and Bourne's fluid), both acidified with 10% glacial acetic acid, and neutral 10% $AgNO_3$ gave the most complete and striking demonstration of bile pigment in the liver.

Fix 4 h in the dark, wash 2 h in four changes of distilled water, immerse overnight in 2% sodium thiosulfate, wash again in several changes of water, and harden for 1 to 3 weeks in 80% alcohol.

Brief iodination with Weigert's iodine for 20 to 60 min followed by 2 min in 0.2 M $Na_2S_2O_3$ (5% hypo) usually desilvered all the bile pigment, leaving no trace behind except for an occasional purple cast (Gmelin reaction).

This procedure with Barnett and Bourne's fluid should be applicable also to hemorrhagic infarcts. Ordinary tissue background stains are not interfered with. Melanin is of course blackened.

PERIODIC ACID SCHIFF REACTION.   On application of periodic acid Schiff (PAS) reaction, (10 min in HIO_4, wash, 10 min in Schiff, bisulfite, and water washes) the brown, green, and greenish black of bile casts and cytoplasmic granules are completely replaced by an orange-red Schiff reaction. Salivary digestion does not prevent the reaction. Use of the direct Schiff reaction, even when extended to 24 h or more, gives no red, although it may reduce biliverdin to brown bilirubin. Bile duct mucin colors a purplish red in contrast to the more orange-red of the bile casts. Saponification designed to cleave the propionic acid ester of the bilirubin with the α-carbon of the glucuronic acid did not prevent the PAS reaction, but did alter its color toward the usual purple-red of a carbohydrate reaction. These findings suggest loss of part of the bilirubin by alkaline solution and that the glucuronic acid is probably protein-bound, perhaps by its carboxyl uniting with a protein amino group (Lillie and Pizzolato, *J. Histochem. Cytochem.*, **16**: 17, 1968).

When the combined acid ferrocyanide–PAS procedure of Lillie and Geer (*Am. J. Pathol.*, **47**: 965, 1965) was applied to hemorrhagic infarcts, hemosiderin colored blue, and hematoidin retained its native orange to yellow, contrasting with the positive PAS reaction of the hepatic bile cast (Lillie and Pizzolato, *J. Histochem. Cytochem.*, **17**: 738, 1969).

METAL UPTAKE REACTIONS OF BILE PIGMENTS AND HEMATOIDIN. Lemberg and Legge ("Hematin Compounds and Bile Pigments: Their Constitution, Metabolism and Function," Interscience, New York, 1949) noted the existence of cyclic zinc and copper complexes with biliverdin and with biliviolinoid products, but none with bilirubin. With ("Bile Pigments: Chemical, Biological and Clinical Aspects," Academic, New York, 1968) mentions zinc complexes of biliverdin and bilipurpurin. Lillie (*Histochemie*, **11**: 332, 1967) reported acetic hematoxylin staining of bile casts after 60°C, 24-h $K_2Cr_2O_7$ chromation. Other workers prior to 1969 do not appear to have studied histochemical metal uptake by bilirubin.

Lillie and Pizzolato (*J. Histochem. Cytochem.*, **17**: 467, 1969) noted that CrO_3, which first produced Gmelin colors in 1 to 2 h and in 4 h or more completely bleached bile casts, at the same time deposited sufficient Cr(IV) (Lillie, *Histochemie*, **20**: 338, 1969) to induce acetic hematoxylin staining of bile pigment granules and casts and hematoidin. Potassium dichromate, 2.5% at 60°C, pH 3.5, takes 3 to 5 days to produce Gmelin colors, but 4 h at that temperature induces blue-black acetic hematoxylin staining of bile pigment, casts, and hematoidin (and myelin and erythrocytes). Reduction of the temperature to 24°C and extension of the time to 10 days still gave blue-black bile pigments and erythrocytes with a decreased amount of concurrent nuclear and cytoplasmic staining, and 4 to 6 weeks at 3°C eliminated background staining, with bile pigments, myelin, and erythrocytes still giving blue-black with acetic hematoxylin.

Mordanting with Cr(III), Fe(III), Fe(II), Cu(II) and Sn(II) salts at 1-, 10-, and 50-mM concentrations, usually for 2 h, induced hematoxylin staining of the same objects; Fe(III) and Sn(II) uptake was demonstrable with 0.1% hematoxylin at pH levels of 2.8 to 7, and black and red were produced, respectively. The other metals required hematoxylin at higher pH levels: Fe(II) and Cu(II) at pH 5 to 7 gave blue-black and blue-green, and Cr(III) gave blue-black with hematoxylin only at pH 6 to 7. Fe(II) and Sn(II) are bound equally well by bile casts previously reduced for 24 h with 0.1 M $NaHSO_3 \cdot HCl$ mixture. Fe(II) and Fe(III) uptake is shown in both cases by both ferricyanide and ferrocyanide HCl technics, indicating partial oxidation of Fe(II) and partial reduction of Fe(III).

AZO COUPLING OF BILE PIGMENTS AND HEMATOIDIN. First reported by Ehrlich (*Zentralbl. Klin. Med.*, **4**: 721, 1883) and adapted to clinical quantitation of serum bilirubin by Hijmans van den Bergh (*Dtsch. Arch. Klin. Med.*, **110**: 540, 1913; *Berl. Klin. Wochenschr.*, **51**: 1109, 1914, and **52**: 1081, 1915), the reaction was first applied histochemically by G. Daddi (*Riv. Clin. Med.*, **34**: 78, 1933) with p-nitrodiazobenzene on fresh frozen sections. He was not able to prepare stable permanent preparations.

Since bilirubin as presently formulated presents no open α or β positions on the pyrrole rings, the mechanism of its azo coupling was long obscure.

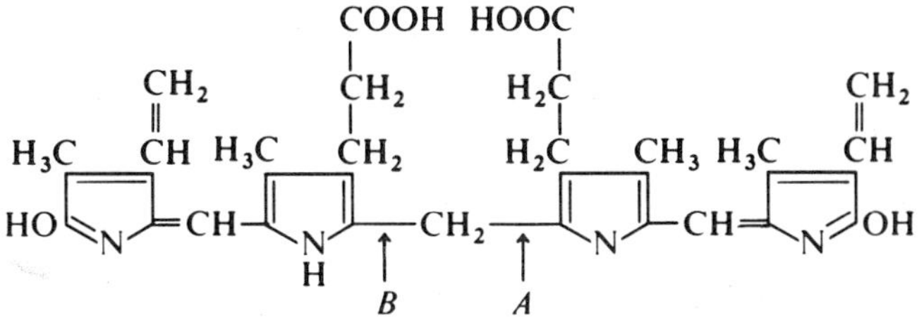

When bilirubin is attacked by a diazonium salt, a hydrolytic cleavage occurs at *A*, thus opening an α position on the third pyrrol ring above, where coupling promptly occurs. This is in accord with the general finding that when an aryl compound is linked by a methylene bridge to another aryl compound otherwise susceptible to azo coupling, the benzyl compound is ejected, forming a hydroxymethyl group on the ejected compound, $-CH_2OH$. It was long considered that this hydroxymethyl compound would not azo-couple (H. Fischer et al., *Hoppe Seylers Z. Physiol. Chem.*, **127**: 317, 1923, and **232**: 236, 1935), but Overbeek et al. (*Rec. Trav. Chim. Pays-Bas*, **74**: 81, 1955) found that the $-CH_2OH$ group was also hydrolyzed off as methylene glycol (formaldehyde) and that both dipyrroles coupled to form isomeric azobilirubins.

After Daddi's work, Nizet and Barac [*C. R. Soc. Biol.* (*Paris*) **146**: 1234, 1952] and Leibnitz [*Acta Histochem.* (*Jena*), **17**: 73, 1964] failed to attain histologically demonstrable results with *p*-diazobenzenesulfonic acid. The first really successful demonstrations were those of Raia [*Nature* (*Lond.*), **205**: 304, 1965] and Desmet (*J. Histochem. Cytochem.*, **16**: 418, 1968). Both these authors claimed specificity of one of their variants for conjugated as opposed to unconjugated bilirubin. Both used methods giving opportunity for loss of part of the diazo by triazene formation, and relied on frozen sections. Desmet et al. relied especially on a fresh aqueous diazotate of ethyl anthranilate at pH 2.5, which failed to couple with the unconjugated bilirubin of the Gunn rat renal papilla.

Lillie and Pizzolato (*J. Histochem. Cytochem.*, **17**: 788, 1969, and **18**: 75, 1970) made a more extended study using a dozen different diazonium salts at both acid and alkaline pH levels on variously fixed icteric livers and a variety of formol-fixed hemorrhagic infarcts for both conjugated bilirubin and unconjugated hematoidin pigment. As in the hands of Desmet et al. ethyl anthranilate diazo failed to couple on first trial not only with hematoidin, but also, in contrast to their findings, with hepatic bile casts.

The reason for this failure appeared to lie in the insolubility of ethyl anthranilate in 0.3 *N* aqueous HCl. Although the milky white suspension gradually cleared during the 20-min diazotization period, no color change was observed. This insolubility suggested a shift to another solvent, and Hilgetag and Martini (eds., "Weygand-Hilgetag Organisch-Chemische Experimentierkunst," third edition, p. 642, Barth-Verlag, Leipzig, 1964) offered a Claus procedure in which diazotization is done by adding sodium nitrite directly to a glacial acetic acid solution of the amine. A modified Claus procedure adopted by Lillie and Pizzolato (*J. Histochem. Cytochem.*, **17**: 738, 1969) is done as follows:

CLAUS DIAZOTIZATION PROCEDURE. Dissolve 1 mmol of the amine in glacial acetic acid to make a total volume of 9 ml, and cool to about 15°C. Then add 0.5 ml 20% $NaNO_2$ (100 mg). A little brown gas may evolve, and there may be some color change. After 20 min decompose excess HNO_2 by adding 50 mg urea in 0.5 ml distilled water. For coupling in acid solution dilute the 0.1 *M* solution with 1 *N* acetic acid for a pH of 2 to 2.3; 2 ml in 38 ml gives a 5 m*M* concentration, 1 ml in 50 ml gives 2 m*M*. For

TABLE 11-2 TRISODIUM PHOSPHATE BUFFER FOR 0.1 M CLAUS DIAZOTATES IN 90% ACETIC ACID

0.1 M diazo	Na$_3$PO$_4$ 1 M	H$_2$O	pH	0.1 M diazo	Na$_3$PO$_4$ 1 M	H$_2$O	pH
2.5	(in HAc) 97.5	....	2.3	2.5	50.0	47.5	6.2
2.5	....	97.5	3.1	2.5	52.5	45.0	6.4
2.5	0.5	97	3.3	2.5	55.0	42.5	6.5
2.5	1.0	96.5	3.4	2.5	57.5	40.0	6.6
2.5	1.5	96	3.6	2.5	60.0	37.5	6.7
2.5	2.0	95.5	3.7	2.5	62.5	35.0	6.8
2.5	2.5	95.0	3.8	2.5	65.0	32.5	6.84
2.5	3.0	94.5	3.9	2.5	67.5	30.0	6.9
2.5	3.5	94,0	4.0	2.5	70.0	27.5	7.0
2.5	4.0	93.5	4.1	2.5	71.0	26.5	7.2
2.5	4.5	93.0	4.2	2.5	72.0	25.5	7.3
2.5	5.0	92.5	4.3	2.5	73.0	24.5	7.5
2.5	7.5	90.0	4.4	2.5	74.0	23.5	7.6
2.5	10.0	87.5	4.5	2.5	75.0	22.5	7.7
2.5	12.5	85.0	4.6	2.5	77.5	20.0	7.8
2.5	15.0	82.5	4.7	2.5	80.0	17.5	7.85
2.5	17.5	80.0	4.75	2.5	82.5	15.0	7.9
2.5	20.0	77.5	4.8	2.5	85.0	12.5	8.0
2.5	22.5	75.0	4.9	2.5	85.5	12.0	8.1
2.5	25.0	72.5	4.95	2.5	86.0	11.5	8.2
2.5	27.5	70.0	5.0	2.5	86.5	11.0	8.3
2.5	30.0	67.5	5.2	2.5	87.0	10.5	8.4
2.5	32.5	65.0	5.3	2.5	87.5	10.0	8.6
2.5	35.0	62.5	5.4	2.5	88.0	9.5	8.7
2.5	37.5	60.0	5.5	2.5	88.5	9.0	8.8
2.5	40.0	57.5	5.6	2.5	89.0	8.5	8.9
2.5	42.5	55.0	5.7	2.5	89.5	8.0	9.0
2.5	45.0	52.5	5.9	2.5	90	7.5	9.1
2.5	47.5	50.0	6.0	2.5	92.5	5.0	9.5

Source: With interpolations, modified from Lillie and Pizzolato (*J. Histochem. Cytochem.*, **17**: 738, 1969).

alkaline coupling Lillie and Pizzolato offer Table 11-2 as a buffer table with graded amounts of 0.1 M trisodium phosphate added to 1 ml 0.1 M diazo in 90% acetic acid to yield graded pH levels from 3.8 to 9.5 with a fairly gradual progression up to 8.

On shifting to the above-described Claus diazotization, ethyl anthranilate now gave strong red reactions with both bile casts and hematoidin. This technic gave successful preparations also with anthranilic acid and transiently with aniline—here the red faded overnight, as in Daddi's early experiments.

A number of stable diazonium salts, fast red GG, fast black K, fast blue B, fast garnet GBC, and fast red B, all gave satisfactory reds both at acid and alkaline pH levels, but it was necessary that these salts be reasonably fresh, preferably under 1 year. Altogether the best results were obtained with freshly diazotized safranin O and methylene violet (dimethyl-phenosafranin) both in 1 N acetic acid and in disodium phosphate solutions (pH 2.5 and 7.8 to 8). These stains have been stable in cellulose caprate mounts for over 6 months with acid coupling, and 18 months with alkaline. Technics are given in Chap. 8.

In the demonstration of hematoidin with red diazo compounds a Prussian blue reaction before the azo coupling gives a satisfactory contrast of blue hemosiderin and red hematoidin.

After adequate azo coupling reactions azobilirubins and azohematoidins resist alcohol dehydration and mounting in resinous media. The azo colors resist a routine 30-min 0.1 N

HCl extraction. They are bleached at varying rates by fresh sodium dithionite solutions. Azo coupling of bile casts is prevented by acetylation in ethanol–acetic anhydride and restored by alcoholic potassium hydroxide saponification. A 4-h oxidation in 0.2% CrO_3 or by a diammine silver, iodine, thiosulfate sequence prevents the azo coupling reaction completely and irreversibly. Azo coupling is not prevented by 24-h 10% iodine methanol, 2-min 5% thiosulfate sequence or by a 24-h reduction in 0.1 M metabisulfite. Primary fixation in hot methanol chloroform and exhaustive extraction of formol-fixed sections in this solvent do not prevent the azo coupling reaction, although in both cases considerable amounts of yellow material appear in the solvent (Lillie and Pizzolato, *J. Histochem. Cytochem.*, **18:** 75, 1970).

DUBIN-JOHNSON PIGMENT. Ganter and Jolles (pp. 904–905) have tabulated the reactions in 24 cases of Dubin-Johnson disease. A dark brown fluorescence was present in 8 cases, none in 3. The Stein iodine and Perls Prussian blue reactions were negative (19 and 22 tests). An argentaffin reaction was reported in 11 cases; there were no negatives. Peroxide bleaching succeeded in 5, failed in 2. Oil red or Sudan stains succeeded in 5, failed in 5, and in 2 Sudan black succeeded while oil red failed. The PAS reaction was $+ +$ in 3, $+$ in 11, $\pm$ in 2, and $-$ in 5 cases. Thionin metachromasia was present once, absent in 3 cases. Basophilia was $+ +$ in 1, $+$ in 3 cases. A positive performic acid Schiff reaction occurred in 2 cases; it was negative in 1. There were one positive Feulgen reaction and negative tests for nucleic acids in 3 cases. Nile blue staining was reported in 2 cases; acid-fastness was moderate in 2, slight in 5, absent in 8. The Golodetz-Unna ferric ferricyanide reaction was positive in all 6 cases tested.

One presumes that overnight reactions with ammoniacal silver were used; exact data were not cited.

On liver pigments the Stein reaction is quite unreliable unless the section is first reduced by bisulfite or dithionite. It must be recalled that bilirubin is also argentaffin and reduces ferric ferricyanide, and in bile casts gives a positive PAS reaction. In regard to the lipid reactions it must be considered that hepatic lipofuscin may be present in addition to the Dubin-Johnson pigment. We must agree with Ganter and Jolles that much further study may be needed to elucidate the nature of this pigment.

The French school divides between a lipofuscin nature and a melanin. Ganter and Jolles seem to favor the latter concept. One recalls here that melanin is a far broader category in French writings than in German, British, or American usage. Pearse (1972) also remains uncertain about the nature of this pigment. Dubin and Johnson originally described the syndrome as a chronic icterus and thought of the pigment at the time as a mesobilirubin. Though Dubin later abandoned that view in favor of a lipid pigment, the fact that it is limited strictly to the liver lends weight to the speculation of a bilirubin origin.

Barone, Inferrero, and Carrozza (in M. Wolman, "Pigments in Pathology," chap. 10, Academic, New York, 1969) cover the foregoing reactions and note further that in the Dubin-Johnson syndrome normally conjugated bilirubin is not excreted. Pigment granule size ranges from 0.2 to 3 μ; it is almost restricted to the liver, chiefly in hepatocytes, but sometimes in Kupffer cells. It has rarely been found in spleen, regional lymph nodes, and bone marrow. The color ranges from ocher yellow to brownish black and yellow; orange and brown fluorescence is seen. Pigment is digested by pepsin and trypsin; a weak to moderate Millon reaction and a Ninhydrin Schiff reaction are reported; the Gibbs reaction is weak to moderate, azo coupling with fast red B is reported; and the ferric ion uptake reaction of Lillie is positive. The Golodetz-Unna ferric ferricyanide reaction, when ascribed to Chevremont and Frederic, is taken to denote presence of SH groups; when called *Schmorl's test*, it is taken to identify lipofuscins and melanins.

Pearse (1972, p. 1088) finds the identity of Dubin-Johnson pigment(s) still highly uncertain and refers to Barone et al.'s review cited above. Pearse further draws attention to an apparently related hepatic pigment in Cooriedale sheep with an associated phylloerythrin photosensitivity described by Cornelius et al., *J. Am. Vet. Med. Ass.,* **146:** 709, 1965.

UROBILINOID SUBSTANCES.   Kutlik [*Acta Histochem. (Jena),* **42:** 302, 1972] describes a green fluorescence, localizing particularly on the surfaces of intra- and extravascular erythrocytes in tissues fixed in acetone containing 5% $ZnCl_2$ for 24 h or more. Aqueous $ZnCl_2$ and $ZnAc_2$ give inferior results, and formol-fixed tissues posttreated with Zn salts are still worse.

Sections of tissue thus fixed are deparaffinized in xylene, dried, and covered by a coverslip without mountant of any kind. A mercury Höchstdrucklampe HBO 200 was used with two Zeiss filters ($\frac{3}{4}$ and $\frac{3}{2}$) or, if much heat developed, three $\frac{1}{2}$ filters and an NA 1.4 condenser, or for quantitative work a dark-field condenser was employed.

After $ZnCl_2$–acetone fixation the urobilin-urobilinogen remains soluble in alcohol and dioxan but resists acetone and benzene. Attempts to convert urobilinogen to urobilin with such oxidants as HIO_4, $K_2S_2O_8$, $NaBO_3 \cdot 4H_2O$, or 0.02% I_2-glycerol weakened or abolished the fluorescence. The procedure:

Fix for 24 h at 4°C in 5% $ZnCl_2$ in acetone, followed by six changes for 20 min each of acetone; acetone plus benzene for 20 min; benzene, 20 min; paraffin, 1 h and 18 h; and embed in fresh paraffin. Cut and float sections on slides on preboiled distilled water, warming to flatten. Use no adhesive. Deparaffinize in five changes of xylene for 1 h each at 22°C and 20 h at 45°C, then three quick rinses in fresh xylene. Dry, and cover with a coverslip without mounting medium. Examine on fluorescence microscope. Green, yellow-green, or occasionally orange-green fluorescence is localized on the surface of erythrocytes, which is nonfluorescent without zinc pretreatment.

Cohn (*J. Histochem. Cytochem.,* **3:** 342, 1955) described in the mouse Harderian gland lipid-filled epithelial cells whose lipid stained strongly by Sudan black and by Baker's acid hematein reaction, was apically red and basally blue with Nile blue, and was removed by pyridine extraction. The Liebermann-Burchard test revealed only a quickly fading pale blue-green. Little RNA was shown by toluidine blue; nuclear Feulgen reactions were light. No iron was shown by the Tirmann-Schmelzer ferricyanide, the Humphrey dinitrosoresorcinol, or the Macallum iron reaction. Gland secretion was foamy, inconstantly PAS–positive, direct Schiff–negative. The granular pigment grades from yellow when isolated to reddish brown in larger clumps and is present chiefly in the lumina of the glands. It reacts moderately to strongly with Sudan black and Baker's acid hematein, and loses these reactions, as well as part of its red-brown color, on pyridine extraction. It gives no iron reactions, darkens in OsO_4 and iron hematoxylin, and gives a weak ferric ferricyanide test, which is enhanced after peracetic acid oxidation. Peracetic acid produces a Nadi reaction. Protein tests—Millon, biuret, xanthoproteic, and Ninhydrin Schiff reactions—are negative. Pigment granules appear to arise in PAS–positive, peracetic acid Schiff–positive masses of material in gland lumina. Bromination prevents the latter reaction. The pigment appears to increase at the expense of and to replace the foamy PAS–, peracetic Schiff–positive material. Under ultraviolet light the pigment gives a bright red fluorescence.

The gland produces a partially oxidized unsaturated lipid, possibly partly phospholipid in nature, associated with a red fluorescent porphyrin pigment which seems to share most of the lipid reaction. The relation of the porphyrin to the lipid is unclear.

HEMATOPORPHYRIN.   This gives the same Gmelin reaction as bilirubin, and is insoluble in dilute acids and alkalies.

Lison (1953) recommends use of fluorescence microscopy for the detection of porphyrins in the tissues. With long-wave ultraviolet (365 mμ), orange to red fluorescence is observed.

Harderian gland of rodents, shell gland of the hen's oviduct, rodent placenta, and regressing corpora lutea form good test objects.

As already mentioned, Mallory refers to a dark brown or bluish black pigment occurring in old extravasations of blood as hematin, but does not further characterize it. We infer that it is iron-negative.

THE IRON-BEARING PIGMENTS

A golden brown granular pigment originating in sites of hemorrhage and congestion, darker in the lung and intestinal mucosa, soluble in sulfuric acid, and yielding red ash on ignition of the granules, which then yielded a Prussian blue reaction, was reported on at length by Rudolf Virchow (*Virchows Arch. Pathol. Anat.*, **1**: 379, 1847). Vogel in 1853 (Hueck, 1912) observed a black pigmentation of the intestinal mucosa which, unlike skin melanin, was removed by acid extraction and which he therefore named *pseudomelanosis coli*. Apparently it was to sections from a case of postmortem hemolysis that Grohe (*Virchows Arch. Pathol. Anat.*, **20**: 306, 1861) first applied the Prussian blue reaction and noted that if he applied acid first and then ferrocyanide, blue halos appeared about the granules, but that if the sequence was reversed, sharply stained blue granules were seen. It remained for Max Perls (*Virchows Arch. Pathol. Anat.*, **39**: 42, 1867) to work out a practical mixed ferrocyanide HCl reagent, which he applied systematically to a wide variety of tissues and lesions. Quincke added the ammonium sulfide reaction in 1880, although we think both Virchow and Vogel had used this reagent, but apparently on extracts or gross preparations only. Tirmann and Schmelzer (1895; Hueck, 1912) added a ferricyanide demonstration step after the ammonium sulfide. This is what Pearse rather loosely calls "Turnbull's blue reaction." Strictly speaking, that designation should apply to an acid ferricyanide reaction applied to tissue to demonstrate the presence of native Fe^{2+}. This occurs sometimes at iron absorption sites in the duodenum and jejunum and in cecal enterosiderosis in the guinea pig. We have occasionally seen it in corpora hemorrhagica of the ovary, and Bunting described it in iron and calcium incrustations of necrotic vessels. Otherwise, as Gomori said, it virtually does not occur. Gomori noted gas evolution in the application of the acid ferricyanide of the Tirmann-Schmelzer reaction and warned of the artifacts produced thereby.

It has been objected that ferricyanide is itself an oxidizing agent and would oxidize Fe^{2+} to Fe^{3+}. Since it would in this process be reduced itself to ferrocyanide, a blue precipitate of Prussian blue would result. However, ferricyanide exhibits this oxidizing capacity principally in alkaline solutions, while the ferricyanide test for Fe^{2+} is carried out in our laboratory at pH 1 or 2. At this pH we have not been able to demonstrate oxidation by ferricyanide, although ferric chloride at pH 1 to 2 oxidizes SH and diphenols readily.

In mammalian tissues we have not seen positive ferricyanide reactions except in loci where the ferrocyanide reaction was also positive.

There are a number of more recent organic colored chelate reactions for ferrous ions such as Humphrey's 1935 dinitrosoresorcinol reaction (dark green) and the Hukill and Putt recent highly sensitive bathophenanthroline reaction (deep red). The latter produces an alcohol-soluble pigment, and Hukill and Putt require blotting and drying in air before mounting in synthetic resin.

Granick's (*J. Biol. Chem.*, **147**: 91, 1943) experience with the reduction of ferric iron in ferritin by means of dithionite has been similar to ours with the reduction of hemosiderin iron with sulfide. It is very difficult to achieve complete reduction, and much Fe^{3+} remains demonstrable with ferrocyanide after the ammonium sulfide step of the Tirmann reaction.

For a long time hemosiderin was regarded simply as ferric oxide or hydroxide, though

Hueck (1912) spoke of a probable colorless protein carrier substance and drew attention to the concentrations of acids required to remove the iron from tissue sections.

In 1939 (*Am. J. Pathol.*, **15**: 225) Lillie noted enhancement of the resistance of hemosiderin iron to subsequent acid extraction by formaldehyde-containing fixatives, as well as the persistence of a brown iron-free pigment after certain acid-formalin fixations where only iron-positive pigment was present in parallel material fixed in neutral formalin. Considerable resistance against extraction by weaker acids was conferred on the iron content by prior fixation in formaldehyde at pH 7, but extraction by sulfuric or oxalic acid or sodium dithionite was not impeded (Lillie et al., *J. Histochem. Cytochem.*, **11**: 662, 1963).

APOSIDERIN.   In 1943 Lillie (*Public Health Rep.*, **58**: 30, 1943) reported the occurrence of a similar iron-negative brown renal pigment arising apparently by in vivo desiderization of intraepithelial hemosiderin in the later stages of chronic hemolytic intoxication. To this iron-free phase we applied the name *aposiderin* in 1948, and by extension the same term has been applied to artificially desiderized hemosiderin. The iron-free phase is quite resistant to strong acids and alkalies. Mallory's fuchsin stain is not retained. H_2S does not blacken it, and Fairhall's chromate reaction for lead is negative (Miller, *Public Health Rep.*, **56**: 1610, 1941).

In the same year McManus (*Stain Technol.*, **23**: 99, 1948) noted a positive reaction of hemosiderin by the periodic acid Schiff technic. After various isolated reports by Lillie, Goessner, and Gedigk, it became evident that a positive periodic acid Schiff reaction appears in new hemosiderins somewhat after the iron reaction has become strongly positive, thus accounting for the negative reaction in some specimens.

Gedigk (*Virchows Arch. Pathol. Anat.*, **324**: 373, 1953; **326**: 172, 1954) also reports on a positive tetrazonium reaction of the iron-bearing pigment and iron-free phases of the pigment, indicating, as of the early fifties, the presence of aromatic amino acids. Since Burstone (*J. Histochem. Cytochem.*, **3**: 32, 1955) has shown that cellulose, which contains no protein, also can give a postcoupled tetrazonium reaction, we must regard Gedigk's conclusion as only probable. However, Behrens and Asher (*Hoppe Seylers Z. Physiol. Chem.*, **220**: 97, 1933) isolated hemosiderin by centrifuging homogenized horse spleen over carbon tetrachloride (sp gr 1.594) and over 1,2-dibromoethane (sp gr 2.18) and mixtures thereof, thereby obtaining fractions below, between, and above those densities. The bulk of the material fell into the second fraction and could be fractionated further by use of mixtures of the two halogenated hydrocarbons. The fraction of 1.80 to 2.18 specific gravity contained 36.1% protein, 3% $CaHPO_4$, and 60.4% $Fe(OH)_3$. Ludewig (*Proc. Soc. Exp. Biol. Med.*, **95**: 514, 1957) got similar figures and also demonstrated hexosamine, galactose, mannose, and fucose chromatographically. These chemical findings agree well with what we presently know of the histochemistry of hemosiderin.

Granick (*Bull. NY Acad. Med.*, **25**: 403, 1949) conjectured that the protein of hemosiderin may be apoferritin. There are now acceptable amino acid analyses of this latter protein, but none for the glycoprotein remnant of hemosiderin, for which Goessner and Lillie have used the name *aposiderin*. Nor are there any analyses showing the presence of sugars in apoferritin. Dubin (*Am. J. Clin. Pathol.*, **25**: 514, 1955), however, seemed to regard the identity of apoferritin and the protein of hemosiderin as established, apparently from Granick's 1949 speculation.

From the published analyses of apoferritin it might be of interest to try histochemical tests for tyrosine and arginine. There is an approximate balance between the dicarboxylic and the basic amino acids which would agree with the shift from weak oxyphilia to basophilia at about the same pH levels as collagen and cytoplasms in hemosiderins. Levels of sulfur amino acids, tryptophan, and histidine are not remarkable.

Thus the concept has evolved that ionic iron is bound to one or more protein carrier substances. These carrier substances lose their reactive iron readily on acid extraction but give it up incompletely at pH 4.5 and scarcely at all at pH 7.6 in ethylenedi-aminetetraacetate solutions, even from 5- to 10-μ sections in 7 days at 25°C. They are also found iron-free in human and animal tissues under a variety of as yet undetermined circumstances. When naturally or artificially deprived of iron, they give a series of histo-chemical reactions which agree in all respects, except the specific iron reactions of the pigment to which they are related.

Histochemically we were able, in 1963, to recognize two pairs of such pigments. The one group, comprising the pigments of melanosis and pseudomelanosis coli and the villus core pigment of villus melanosis (*Zottenmelanose*), is localized almost strictly to the intestinal mucosa and has a suggested relationship to the phenomena of intestinal iron absorption and storage. The other group comprises the granular pigments resulting from phagocytosis of erythrocytes and resorption of hemoglobin by renal epithelium, and also manifests both iron-positive (hemosiderin) and iron-free (aposiderin) phases.

The relation of these histochemically demonstrable pigments to the biochemically extract-able and crystallizable protein ferritin and its iron-free phase apoferritin is uncertain. Physiologic and biochemical considerations almost demand the presence of these substances in the intestinal mucosa; much of the isolation work appears to have been done on spleen tissue, where hemosiderin is the characteristic histochemical iron pigment. Ferritin reputedly gives negative reactions for ionic iron (H_2S), which Granick imputes to its high degree of dispersion. His technic in his hands did demonstrate larger granules of hemosiderin in horse spleen as black masses, and it colored erythrocytes gray. The relation of ferritin to the diffuse ferrocyanide staining seen early in the evolution of hemosiderin pigmentation and called *protosiderin* in the 1948 and 1954 editions of this book is also uncertain, but a slight ferrocyanide reaction is easier to recognize than the sulfide test.

A parallel situation may exist in regard to the iron-bearing particles observed widely distributed in various rodent epithelial cells by Spicer (*J. Histochem. Cytochem.*, **10**: 528, 1962) and designated by him as *cytosiderin*. Spicer notes peracetic or permanganate aldehyde fuchsin staining of particles in the same locations, but he does not report whether the aldehyde fuchsin–staining particles persist after acid extraction of the iron.

Early in 1963 we found (*J. Histochem. Cytochem.*, **11**: 662, 1963) that short exposures to fresh 1% aqueous sodium dithionite[1] solutions completely removed demonstrable ferric and ferrous iron from tissues. Test tube experiments indicated that freshly precipitated ferric and ferrous hydroxides were readily dissolved by this reagent at pH 7, but not the sulfides, while at pH 4.5 both sulfides and hydroxides were dissolved. As little as 5 min in 1% $Na_2S_2O_4$ in 0.1 M, pH 4.5 acetate buffer removed all demonstrable iron in heavily pigmented liver, spleen, and lymph nodes (cytosiderin, hemosiderin) and experimental severe enterosiderosis of guinea pigs. Similarly 1% $Na_2S_2O_4$ in acetate and Veronal–HCl buffers at pH 5, 5.5, 6, 6.5, and 7 removed all iron from the same tissues in the shortest interval tested (15 min).[1]

It was promptly found that dithionite extraction did not impair the periodic acid Schiff reaction of hemosiderin or enterosiderin.

In general, the carrier substance(s) of the intestinal pigments presents stronger protein and lipid reactions, stronger reducing reactions, and a rather strong affinity for Nile blue, which after staining is not removed by acetone. These reactions, other than those for iron ions,

---

[1] Dithionite should be purchased in small bottles; it deteriorates slowly after opening. Solutions must be freshly prepared, at least daily.

appear identical in the iron-carrying and iron-free forms of the respective pigments and also remain unaltered when acid or dithionite extraction of iron salts is practiced.

PROTOSIDERIN (Lillie, 1948).   Diffuse and granular forms of iron-positive pigment may often be present in the same cell. The diffuse form is more easily lost on fixation in acid formalin. A similar diffuse Prussian blue reaction is sometimes seen in the contents of renal tubules in the presence of hemoglobinuria. This change is probably related to the positive Prussian blue reaction which may be induced in erythrocytes by fixation in 10% formalin buffered to pH 3.5 or 4, or by immersion of sections of material fixed by neutral formalin in solutions of mineral acids for varying periods.

HEART FAILURE CELL PIGMENT.   This usually iron-positive pigment, found in large phagocytic cells in pulmonary alveoli and in sputum in cases of chronic passive congestion of the lungs, is singly refractile and yellow to dark brown. In some cases it is more successfully demonstrated by the hydrochloric acid–ferrocyanide technic; in others this procedure produces blue halos about the granules. In some cases a positive periodic acid Schiff reaction is given, in others not. Ferric ferricyanide is not evidently reduced. The pigment is not acid-fast in the Ziehl-Neelsen technic. It does not retain iron hematoxylin in the myelin variant method for eosinophils or fuchsin in the Mallory hemofuscin method. Some of the darker brown examples of this pigment may be largely iron-negative with even the hydrochloric acid–ferrocyanide method. The iron reaction does not correlate with the periodic acid Schiff reaction. Usually the pigment granules are untinged by oil-soluble dyes.

Like some other iron-positive pigments, notably that of the involuting corpus hemorrhagicum of the ovary, this pigment may reduce methenamine silver at 60°C while nuclei and red cells are still unblackened. This reaction is probably similar to that of mucins and collagen, which blacken selectively with methenamine silver after treatment with ferric chloride. Diammine silver solutions as used for reticulum impregnation do not blacken the iron-positive pigments mentioned above.

Extraction of the iron by a 24-h bath in 10% sulfuric acid does not remove much of the darker brown material, though the iron reaction is negative. The periodic acid Schiff reaction of the granules is not decreased by the acid extraction. The amount of material which is dark brown to black after 3 to $3\frac{1}{2}$ h at 60°C in methenamine silver is much reduced by the acid extraction. This supports the thesis that the methenamine silver reaction is due to the iron itself.

OTHER IRON-POSITIVE PIGMENTS.   The iron-positive pigment in apocrine gland cells is of uncertain origin. That in renal epithelium is regarded as granular hemosiderin.

The iron-positive pigment in cutaneous xanthomas often appears to be related to the lipids.

In all these lipid-associated iron pigments the use of combined lipid and iron technics is needed to determine whether the lipid and the iron occur in separate granules or in the same granules.

The hemosiderins are by definition pigments which exhibit one or more of the reactions of ionic iron. Traditionally three principal reactions have been used: the formation of ferric ferrocyanide (Prussian blue) when the material is treated with acid solutions of ferrocyanides (Perls' reaction); the formation of black iron sulfides, probably both FeS and Fe_3S_4, perhaps also unreduced Fe_2S_3, when material is treated with ammonium sulfide (Quincke's reaction); and the formation of ferrous ferricyanide (Turnbull's blue) from the ferrous sulfide thus formed, by treatment with a ferricyanide and acid (the Tirmann-Schmelzer reaction).

Macallum [*J. Physiol. (Lond.)*, **22:** 92, 1897] used an 0.5% aqueous solution of pure hematoxylin to demonstrate inorganic iron in tissue. Only a short reaction time was required. Mallory apparently independently proposed a reaction with fresh unoxidized hematoxylin without mordant which colors iron salts blue-black and copper greenish blue. The two colors are comparable to those attained in Heidenhain-type hematoxylin procedures on tissues premordanted with $FeCl_3$ and $CuSO_4$, respectively. We have included also two more recent metal chelate reactions, both apparently for Fe^{2+} and used after reduction with sulfide or thioglycolate: Humphrey's and the reaction of Hukill and Putt.

It is often stated that the fixative of choice for demonstration of hemosiderin is alcohol. On comparative tests of the same material with various fixatives, Lillie (*Am. J. Pathol.*, **15:** 225, 1939) found that positive Prussian blue reactions are most often obtained when 10% formalin buffered to pH 7 is the fixing agent. This is definitely superior to alcohol, alcoholic formalin, Orth's fluid, and unbuffered 10% formalin. However, in more or less autolyzed human liver the iron-positive pigment in liver cells may be better preserved with alcohol.

According to Schmorl, Hall has proposed fixation of fresh tissue in alcoholic solutions of ammonium sulfide containing 70 parts of 100% alcohol. For liver, spleen, and bone marrow he made up the remaining 30 parts with strong ammonium sulfide solution; for other tissues he used 5 parts of ammonium sulfide and 25 of distilled water. The fixation interval was 24 h. This amounts to a Quincke reaction done in the block on tissue during fixation. Hall prescribed a ferrocyanide reaction to follow on the paraffin sections. A ferricyanide test could also be used, as in the Tirmann-Schmelzer reaction.

THE QUINCKE (*Dtsch. Arch. Klin. Med.*, **25:** 567, 1880) AND TIRMANN-SCHMELZER (Hueck; Schmorl; Mallory) REACTIONS. Formalin or alcohol fixation and celloidin or paraffin sections may be used. Quincke omits steps 4 and 5.

1. Bring sections to distilled water as usual.
2. Impregnate sections 1 to 2 h, or as long as 1 to 2 days in strong, slightly yellow ammonium sulfide solution. Mallory prefers to dilute this with 3 volumes of 95% alcohol, to avoid loss of sections. Otherwise it would seem indicated to soak sections first in 1% collodion for 5 to 10 min after deparaffinizing, drain 1 min, and harden 5 to 10 min in 80% alcohol before bringing to water. Highman (*Bull. Int. Assoc. Med. Mus.*, **32:** 97, 1951) impregnates instead for 24 h in saturated hydrogen sulfide water for hematite and other refractory iron ore dusts which fail to react to the usual procedure. Buffered Na_2S at pH 7 can be used to avoid the alkali detergent effect.
3. Wash thoroughly in distilled water.
4. Soak sections 15 min in equal volumes of 1% hydrochloric acid and 20% potassium ferricyanide, freshly mixed. This step is omitted in the Quincke method. (The 1% HCl is presumably 0.12 N, not 0.3 N).
5. Wash thoroughly with distilled water.
6. Counterstain with 0.5% basic fuchsin in 50% alcohol for 5 to 20 min, wash in water, differentiate in alcohol according to Mallory; or counterstain with alum carmine for 1 to 24 h, and wash in water according to Schmorl.
7. Dehydrate with alcohols, clear in xylene, mount in polystyrene, Depex, cellulose caprate, or other nonreducing resin.

RESULTS: Quincke's reaction gives a dark brown to black to the iron pigment; that of Tirmann and Schmelzer, a dark blue.

The brown of the sulfide is less readily distinguished from other brown-colored substances than the blues of the Prussian and Turnbull blue reactions. Silver, lead, and mercury also give dark brown to black deposits with this method. Other brown and black insoluble sulfides are noted in Chap. 12. Unless special precautions are taken, sections are often lost in the alkaline sulfide solution. The same objection applies also to the Tirmann-Schmelzer reaction. Further, if instead of potassium ferricyanide the ferrocyanide is applied to the sections treated with ammonium sulfide, a positive Prussian blue reaction is still obtainable. This indicates that only a portion of the ferric iron originally present was converted to ferrous sulfide, or perhaps it indicates a conversion to Fe_3S_4. It might be objected that Perls' reaction was open to the same criticism; i.e., that it demonstrated only ferric iron. However, with numerous tests of hemosiderin-containing material, Lillie only rarely obtained a direct positive Turnbull blue reaction on an intrinsic hematogenous pigment with acidulated potassium ferricyanide, when that salt was substituted for potassium ferrocyanide in the technic given later in this chapter. Bunting, however, reports positive diffuse ferricyanide staining as well as ferrocyanide staining in the mixed calcium-iron deposits occurring in necrotic areas (*J. Natl. Cancer Inst.*, **10:** 1368, 1950).

Lillie has occasionally seen ferricyanide-positive pigment in phagocytes in human ovaries, and part of the demonstrable iron in the epithelial cells of the distal extremities of the intestinal villi, especially in the duodenum, may be ferricyanide-reactive, though most of it is ferrocyanide-positive.

OTHER FERRICYANIDE REACTIONS.   Other insoluble ferricyanides include cobaltic and cobaltous—which are brown and red; cupric and cuprous—greenish yellow and brownish red; lead—red and soluble in hot water; nickel—brown; silver—orange; and stannous—white (Lange).

DINITROSORESORCINOL.   This gives a dark green color with iron salts. Humphrey (*Arch. Pathol.*, **20:** 256, 1935) substituted this reagent for the ferricyanide of the Tirmann-Schmelzer procedure, directing as follows:

After a 1-min bath in 30% ammonium sulfide, rinse in water and stain 6 to 20 h in dinitrosoresorcinol, either a saturated aqueous solution or a 3% solution in 50% alcohol. Wash in the same solvent, dehydrate, clear, and mount in Canada balsam. The dark green color is quite permanent in balsam.

BATHOPHENANTHROLINE.[1]   Hukill and Putt (*J. Histochem. Cytochem.*, **10:** 490, 1962) have reported a method utilizing the highly sensitive Fe^{2+} reagent bathophenanthroline (4,7-diphenyl-1,10-phenanthroline).

1. Fix in buffered 10% formalin (pH 7), embed in paraffin, section at 5 μ, and mount on slides using distilled water to float out. Deparaffinize and hydrate to distilled water as usual.
2. Stain sections 2 h in bathophenanthroline reagent. Dissolve 100 mg bathophenanthroline in 100 ml 3% acetic acid by heating to 60°C overnight, shaking well to suspend evenly at the start. Store at room temperature or in cold (stable for 4 weeks). For use, to 40 ml bathophenanthroline solution add 0.2 ml thioglycolic acid, and mix well. The used solution may be returned to stock. It is necessary to add more thioglycolic acid each day that it is used, since this reagent oxidizes readily in air.
3. After staining, rinse in distilled water, counterstain 3 min in 0.5% aqueous methylene blue, and wash in three changes of distilled water, for 1 min each.
4. Blot, dry thoroughly in oven at 60°C, and mount in Permount.

RESULTS: Iron stains red; nuclei are blue.

Note that this reagent demonstrates only Fe^{2+} and that if used without thioglycolic acid, it will serve as a specific reagent for Fe^{2+}. It does not permit separate demonstration of Fe^{3+} in distinction from Fe^{2+}, thus sharing the deficiency of the Quincke and Tirmann-Schmelzer methods. However, bathophenanthroline is claimed to be more sensitive than

---

[1] Bathophenanthroline was obtained from the Frederick Smith Chemical Co., Columbus, Ohio.

ferrocyanide in detecting minimal amounts of iron. Hukill and Putt do not recite the color reactions, if any, with other metal ions. The metal dye complex is readily soluble in alcohols and most lipid solvents, though essentially insoluble in water. The effect of gum syrup, glycerol gelatin, and similar mounting media was not reported.

THE FERROCYANIDE REACTION OF M. PERLS (*Virchows Arch. Pathol. Anat.*, **39:** 42, 1867). The following technic works well on the hemosiderins and most mineral iron. Hematite dust deposits may require more drastic treatment to obtain the reaction.

1. Fix 48 h or more in 10% formalin buffered with phosphates to pH 7.
2. Dehydrate, clear, embed in paraffin, and section as usual.
3. Make up fresh 2% solution of potassium ferrocyanide in distilled water and add an equal volume of 0.25 N (2% by volume of concentrated HCl) hydrochloric acid (Bunting) or 5% acetic acid (Highman). Heat sections 30 min at 60°C in this mixture, or, preferably (Bunting), let stand an hour at room temperature. Gomori used a 30-min bath at room temperature in a solution containing 2 g potassium ferrocyanide, 36 ml distilled water, and 4 ml concentrated hydrochloric acid (about 1.2 N).
4. Rinse in distilled water.
5. Counterstain 2 min in 0.2% safranin O or basic fuchsin in 1% acetic acid.
6. Wash in 1% acetic acid.
7. Dehydrate with 95 and 100% alcohols, 100% alcohol plus xylene; clear in two changes of xylene, and mount in polystyrene or other nonreducing resin. Cellulose caprate is satisfactory.

RESULTS: Reaction sites are blue or green; nuclei, red; background is pink. Freshly formed deposits of iron pigment react well with the acetic variant and are less likely to be dissolved out. Older deposits may require the stronger acid for adequate reaction. See also Chap. 12 for the demonstration of hematite and other highly insoluble iron. Heating to 80°C in the ferrocyanide, as in the Abbott variant, or even to 60°C is apt to produce a finely granular, blue deposit throughout the section. We have abandoned the practice of heating ferro- or ferricyanide reagents for iron. The Abbott variant was quoted in the eighth edition of Mallory and Wright, "Pathological Technique," Saunders, Philadelphia, 1924; the other references are Bunting, *Stain Technol.*, **24:** 109, 1949; Highman, *Arch. Pathol.*, **33:** 937, 1942; Gomori, *Am. J. Pathol.*, **12:** 655, 1936.

Recent experiments indicate that the optimal pH for the ferrocyanide reaction may be about pH 1.5. Cytosiderins of intestinal epithelia and of liver cells appear to be sensitive to more acid solutions. Such solutions as in Highman's methods for refractory iron ore dusts or Gomori's 1.2 N HCl solution or even 1 N HCl may completely remove these pigments, and even the 0.125 N HCl concentration traditionally used may give blurred borders to granules both of cytosiderin and of the phagocyte pigments hemosiderin and pseudomelanosis pigment. At high pH levels (7 to 5) all these pigments retain their native yellow-browns, even on prolonged exposures. At pH 3 to 4 a variable proportion reacts with green to blue.

The variant of this reaction currently in use in Lillie's laboratory in the study of hepatic and intestinal epithelial cytosiderins as well as the phagocyte pigments hemosiderin and aposiderin, pseudomelanosis and melanosis coli pigments, and natural and experimental enterosiderosis of guinea pigs conforms to the foregoing requirements.

LILLIE'S 1964 TECHNIC FOR THE PRUSSIAN BLUE AND TURNBULL BLUE REACTIONS FOR Fe^{3+} AND Fe^{2+} (*Am. J. Pathol.*, **47:** 965, 1965 revised in 1973).

1. Bring paraffin or frozen sections to distilled water as usual. Interpose desiderization or decalcification procedures at this point if required.

2. Dissolve 400 mg potassium ferrocyanide [$K_4Fe(CN)_6 \cdot 3H_2O$], yellow crystals, in 40 ml 0.06 N hydrochloric acid,[1] when testing for Fe^{3+} (ferric iron). For testing for Fe^{2+} (ferrous iron), substitute 400 mg of the red crystals of potassium ferricyanide $K_3Fe(CN)_6$. Make this solution fresh daily. Expose sections for 1 h. Positive reactions are obtained in 10 to 15 min, and the longer exposure gives no stronger reaction.
3. Wash in 1% acetic acid.
4. Variants:
   a. Stain 5 to 10 min in 0.5% basic fuchsin in 1% acetic acid; wash, dehydrate, and mount in polystyrene, Depex, cellulose tricaprate, or similar resin. Reducing resins tend to decolorize Prussian blue after a time.
   b. Stain by oil red O technic for concurrent demonstration of lipofuscin in paraffin sections or for lipids in frozen sections, wash, and mount in Apáthy's gum arabic medium.
   c. Traditionally, counterstain with alum carmine, nuclear fast red, or the like.
   d. For critical work omit counterstains, dehydrate in alcohols, clear in xylene, and mount in cellulose tricaprate.

RESULTS: Fe^{3+} is demonstrated as dark blue Prussian blue; Fe^{2+}, as dark blue Turnbull's blue. With low concentrations of iron, and with reactions done at pH 3 to 4, the color becomes less intense and appears more green, especially when the weak reaction is due to high pH action on considerable amounts of yellow-brown pigment. Variant 4b is particularly useful where iron-positive and lipid pigments are present in the same cell, to show whether separate or the same granules are giving the two reactions. If frozen sections are used, the various lipids, as well as iron pigment, may be shown in so-called xanthomas. Polarized light should be used on these tumors, as well, since they may contain birefringent lipids.

OTHER FERROCYANIDE REACTIONS. Uranium potassium ferrocyanide is dark brown; cupric and cuprous ferrocyanides are red-brown; cobaltous, gray-green; cobaltic, dark brownish red; mercuric, white; lead, yellowish white; nickel, greenish white; silver, yellow; barium, yellow; ferrous, bluish white; manganese, greenish white; and zinc, white. These are most of the insoluble ferrocyanides listed (Lange). Most of these reactions have not been used histologically.

MELANOSIS AND PSEUDOMELANOSIS PIGMENTS OF THE INTESTINE

These are coarsely granular pigments contained in large phagocytic cells of macrophage type occupying in greater or smaller numbers the stroma of the tips of the villi in the small intestine, especially the duodenum, and the stroma of the cecal, appendiceal, and colonic mucosa, particularly surrounding the mouths of the glands, thus giving rise to a tortoiseshell appearance when viewed from the surface.

The condition was apparently seen by Virchow (*Virchows Arch. Pathol. Anat.*, **1:** 379, 1847); it was described by Vogel about 1853 (Hueck, 1912) as pseudomelanosis coli, because the black pigment, unlike skin melanin, was soluble in sulfuric acid and contained iron. Solger (Inaug. Diss. Greifswald, 1898; Hueck, 1912) described a morphologically similar condition, in which the pigment lacked demonstrable iron, and named it *melanosis coli*. Pick (*Berl. Klin. Wochenschr.*, **48:** 840, 1911) and Hueck (1912) sharply distinguished pseudomelanosis and melanosis on the basis of the iron reaction of the former. Lubarsch described

---

[1] 1 ml 12 N (concentrated) HCl plus 199 ml H_2O.

the homologous condition in guinea pigs in 1917 (*Berl. Klin. Wochenschr.*, **54:** 65) and 1922 (*Virchows Arch. Pathol. Anat.*, **239:** 491). In this species the pigment is not infrequently partly iron-positive, a fact which led Lubarsch to the position that transitions existed between melanosis and pseudomelanosis. This was denied by most workers, Hueck (1922) taking the position that where iron-negative and iron-positive granules existed there were two distinct processes. As long as the iron reactions and their abolition by acid extraction were the sole criteria for identification of pseudomelanosis pigment, this position could not be successfully controverted.

While all these studies of the pigment of human melanosis and pseudomelanosis and of the homologous guinea pig pigment were going on and various theories were evolved as to its (or their) causation, involving such factors as chronic constipation, drug absorption (notably of cascara sagrada), and chlorophyll intake (particularly in the guinea pig), a few studies on intestinal iron absorption were made by physiologists and hematologists. Macallum [*J. Physiol. (Lond.)*, **16:** 268, 1894] fed certain ferrous salts to guinea pigs and observed pronounced uptake of granules of ferric iron in intestinal villus epithelium and in large phagocytic cells in the lamina propria of the mucosa. This condition appeared soonest in the duodenum, but on heavy feeding it extended through the jejunum and much of the ileum. Later studies by Gillman and Ivy (*Gastroenterology*, **9:** 162, 1940) and by Endicott, Gillman, et al. (*J. Lab. Clin. Med.*, **34:** 414, 1949) also demonstrated iron but failed to connect it definitely with the iron absorption mechanism. Their studies seem to have been limited to duodenum, cervical and mesenteric lymph nodes, spleen, and liver. Other parts of the intestine were not specifically mentioned.

There has been apparently no correlation between these physiologic studies and the studies of guinea pig melanosis, at least not up to 1954 (Hieronymi, *Zentralbl. Allg. Pathol.*, **91:** 428).

From 1954 to 1964 Lillie made a considerable number of studies of various reactions on human melanosis pigment, on a single case of pseudomelanosis encountered coincidentally with a carcinoid tumor of the appendix, and on guinea pig small and large intestine. Iron pigmentation appears to have been distinctly more frequent in guinea pigs studied at the National Institutes of Health than at Louisiana State University, perhaps because of the practice of giving green fodder in addition to pellets at the former institution. In New Orleans, where pellets only are fed, very small amounts of iron, limited to the duodenum and cecum, are found in most animals.

Lillie and Geer (*Am. J. Pathol.*, **47:** 965, 1965) concluded that human melanosis coli, pseudomelanosis coli, and villus melanosis pigments are homologous with the enterosiderosis pigment of the guinea pig. These pigments show a wide agreement in histochemical reactions apart from tests for ionic iron. Protein content is shown by the positive Morel-Sisley tyrosine reaction and by alkaline azo coupling with fast black K, diazosafranin, and diazomethylene violet. Tryptophan reactions are negative. The PAS reaction is positive in formol-fixed material, lost after Carnoy or chloroform-methanol fixation, and preserved through a 24-h extraction with acetone if followed by formol postfixation. Sudanophilia to Sudan black B and oil red O and fatty acid staining by Nile blue follows the same extraction patterns. Methylation of formol-fixed sections prevents Nile blue and PAS reactions but not Sudan staining. Ferric ferricyanide and diammine silver reductions are similar to those of the lipofuscins, requiring 18 to 24 h for silver reduction. Silver nitrate at pH 4 is reduced by skin and eye melanins in 10 to 30 min, but not by the enterosiderosis group pigments in 48 h. Azo coupling and PAS reactions are blocked by acetylation, benzoylation, and sulfation in like time intervals and relations to epithelial mucin reactions. Dithionite desiderization does not alter the above reactions.

The cytosiderin seen in villus epithelium of the duodenum lacks the foregoing reactions.

It is present in duodenal villi only in about 30% of normal guinea pigs. Its frequency in human beings is unknown. After 7 days of ferrous carbonate feeding it is present in duodenal villi of all guinea pigs and to a variable extent in those of jejunum and ileum as well. Iron-positive phagocytes are more often present in the cores of normal guinea pig duodenal villi. On iron feeding they become more numerous here and appear in jejunum and ileum. They are frequently present in cecal mucosa in the normal guinea pig, increasing in number on iron feeding and appearing also in other parts of the colon and in tributary lymph follicles and nodes. Iron-negative, PAS–positive phagocytes disappear early in the course of iron feeding and reappear in progressively increasing numbers after its cessation. Cytosiderin also disappears completely in some animals after cessation of iron feeding.

The iron reaction was modified in this study to preserve cytosiderin maximally, by reducing the acid content of the 1% potassium ferrocyanide to 0.5% concentrated HCl (0.06 N). At first a 1-h reaction time was used, but on detailed comparison by several observers without foreknowledge of the technic, it was found that a 15-min exposure was equal to—one observer said better than—the 1-h exposure.

Cytosiderin occurs as fine to very fine granules colored dark blue by acid ferrocyanide and black by 0.1% hematoxylin in 1% acetic acid, situated just beneath the striated border of the epithelium, occurring more on the tips of the villi, and extending a variable distance down their sides. Longitudinal sections through the pylorus show that cytosiderosis begins abruptly on the first duodenal villus, and the presence of coarsely granular iron phagocytes also starts abruptly in the tip, or core, of the first duodenal villus. The phagocyte granules are round or oval, up to 2 to 3 μ in diameter. That the same granules give both the PAS and the Prussian blue reaction was shown convincingly by staining a preparation by PAS, finding a villus crowded with PAS–positive phagocytes in a water mount, and running acid ferrocyanide under the coverslip while watching. As the fluid advanced across the field, the red granules suddenly turned blue.

The concept that black pigmentation of the colon was iron sulfide–produced by action of H_2S from intestine contents postmortem acting on a hemosiderin-like pigment dates back to Virchow (*Virchows Arch. Pathol. Anat.*, **1**: 379, 1847) and Vogel ("Pathologische Anatomie des menschlichen Körpers," L. Voss, Leipzig, 1845) and was widely accepted. Lillie, Geer, and Gutiérrez (*J. Histochem. Cytochem.*, **12**: 715, 1964) converted enterosiderosis pigment to sulfide by application of ammonium sulfide to sections. This caused the pigment to become insoluble in β-mercaptoethanol and sodium dithionite solutions. These extracting reagents completely remove untreated enterosiderosis pigment so that no iron reaction is obtainable with either ferrocyanide or ferricyanide. The unaltered pigment reacts only to ferrocyanide. After $(NH_4)_2S$ it reacts to both reagents, and after dithionite treatment of the sulfide it reacts only to ferricyanide. It was thus shown that iron sulfides were insoluble in sodium dithionite solution, which readily removes unaltered enterosiderin iron as well as hepatic hemochromatosis and other hemosiderin iron. Since all pseudomelanosis iron tested has been soluble in dithionite solutions (*J. Histochem. Cytochem.*, **11**: 662, 1963), it follows that none of this naturally occurring iron pigment was iron sulfide.

Ganter and Jolles give prominent consideration to the Lillie-Geer studies. Pearse (1972) completely ignores them. Ganter and Jolles discuss also the anthraquinone laxative theory Pick's melanin concept, Hueck's lipofuscin classification, and some late papers on a pigment of lipopigment and melanic characters citing Cabanne and Couderc [*Ann. Anat. Pathol. (Paris)*, **8**: 609, 1963], Kerisit and Weill-Bousson [*Ann. Anat. Pathol. (Paris)*, **13**: 355, 1968], and Debray et al. (*Sem. Hop. Paris*, **48**: 1897, 1967).

These authors described light and dark brown pigment granules, mixed even in the same histiocyte. They agree that ionic iron reactions are negative, but one infers that a positive reaction would be attributed to hemosiderin and not to the melanosis pigment. They agree that the pigment is basophil but not metachromatic, and set no pH limits. It is not evident that low-pH solutions were used. Cabanne noted positive Sudan black staining, Kerisit negative, and Debray et al. reported dark blue staining by Nile blue. All reported a minority of the pigment granules as argentaffin but used only ammoniacal and methenamine silver solutions and did not report on time or temperature conditions. The ferric ferricyanide reduction test was given by both the light and dark granules. Prolonged H_2O_2 treatment apparently bleached part of the pigment and prevented the argentaffin reaction. Protein reactions of weak grade were reported by Cabanne; Kerisit got negative reactions. The performic acid Schiff reaction was reported positive by Cabanne, negative by Kerisit; the periodic acid Schiff reaction was found weakly to moderately positive by Debray et al. Iron uptake was noted by Cabanne and Kerisit. All three conclude that most of the pigment is a lipofuscin with tendencies toward melanin which Cabanne differentiated from true cutaneous melanin, and Kerisit alone frankly calls it a melanin. None of the authors had fresh material to work with, all used paraffin sections of formol- or Bouin-fixed material, and none used experimental animals.

Spicer (*Am. J. Pathol.*, **36:** 457, 1960) found cytosiderin in epithelium of aging mice, in seminal vesicles and intercalated ducts of the parotid gland. Rawlinson and Pierce (*Science*, **117:** 38, 1968) reported similar cytosiderin in mammary gland epithelium of rats, mice, hamsters, and guinea pigs. The peracetic acid–aldehyde fuchsin sequence also showed fine red granules similarly located, but Spicer (*J. Histochem. Cytochem.*, **10:** 528, 1962) found aldehyde fuchsin–positive granules after oxidations by peracetic and performic acids, bromine, and (best) with an acidified permanganate–oxalic acid sequence. These granules did not accurately correspond in frequency or location with the cytosiderin. Besides the presence of cytosiderin in various epithelia in guinea pigs and rats—salivary glands, pancreas, adrenal, prostate, uterus, spleen, and choroid plexus—it was reported in human salivary, pyloric, and mammary glands, liver, kidney, epididymis and seminal vesicle, as well as phagocytized hemosiderin in uterus and spleen. This cytosiderin is not to be confused with the hemoglobin granules seen in proximal convoluted tubule epithelium in hemoglobinuria, which convert in a few days to iron-positive hemosiderin and in chronic cases to brown iron-negative aposiderin.

In hepatic cytosiderin Gedigk and Strauss (*Virchows Arch. Pathol. Anat.*, **324:** 373, 1953) noted the absence of the PAS reaction and Sudan black staining seen in granular hemosiderin.

The relation of Bantu intestinal siderosis of iron overfeeding to enterosiderosis seems highly probable. But the reports of Walker and Arvidsson [*Nature (Lond.)*, **166:** 438, 1950; *Trans. R. Soc. Trop. Med. Hyg.*, **47:** 536, 1953], Higginson, Gerritson, and Walker (*Am. J. Pathol.*, **29:** 779, 1953), and Bradlow, Dunn, and Higginson (*Am. J. Pathol.*, **39:** 221, 1961) mention only the Prussian blue reaction in their studies of the intestine, though PAS– and Sudan-stained pigments were seen in heart, liver, and arterial muscle. Without the other reactions of the enterosiderosis pigment, no definite identification can be made.

OCHRONOSIS

Ochronosis was first described by Virchow (*Virchows Arch. Pathol. Anat.*, **37:** 212, 1866). It is a hereditary metabolic defect in the catabolism of phenylalanine and tyrosine through 4-hydroxyphenyl pyruvic acid and homogentisic acid to fumaric acid and acetoacetic acid in which the destruction of homogentisic acid fails. It accumulates and is deposited in oxidized form in cartilages, coloring them black.

Phenylalanine $\longrightarrow$ tyrosine

$$\text{C}_6\text{H}_5\text{CH}_2-\overset{\overset{\displaystyle NH_2}{|}}{\text{CH}}\text{COOH} \longrightarrow \text{HO}-\text{C}_6\text{H}_4-\text{CH}_2-\overset{\overset{\displaystyle NH_2}{|}}{\text{CH}}-\text{COOH} \longrightarrow$$

p-hydroxyphenylpyruvic acid

$$\text{HO}-\text{C}_6\text{H}_4-\text{CH}_2-\text{CO}-\text{COOH}$$

$\longrightarrow$ homogentisic acid $\longrightarrow$ fumaric acid and acetoacetic acid

$$\text{HO}-\text{C}_6\text{H}_3(\text{OH})-\text{CH}_2-\text{COOH} \longrightarrow \overset{\displaystyle HOOC-CH}{\underset{\displaystyle HC-COOH}{\|}} \quad \text{and} \quad \text{CH}_3\text{COCH}_2\text{COOH}$$

Usually homogentisic acid appears in the urine (alkaptonuria).

The pigment occurs both in diffuse and granular forms, occurring in larger arterial walls, in cartilages, ligaments, dense fascia, and intervertebral disks, and sometimes as intracellular granules. It varies from yellow through browns to blue-black.

But few modern histochemical observations of this pigment have been reported. Friderich and Nikolowski (*Arch. Derm. Syph.*, **192**: 273, 1951) recorded gold yellow fluorescence with ultraviolet, basophilia to nuclear fast red and to cresyl violet and negative iron, silver and elastica reactions. Gomori thought the cresyl violet stain the most characteristic one and noted absence of an argentaffin reaction.

Oberndorfer's (*Erg. allg. Path.*, **19**: 47, 1921) review notes the occasional iron staining of cartilage as well as hemosiderosis in the bone marrow. In Moran's material (*Am. J. Path.*, **33**: 591, 1957) Lillie found diffuse ferrocyanide and ferricyanide reactions of the cartilage matrix.

Dr. L. B. Thomas at the Clinical Center, National Institutes of Health, supplied Lillie with material from a case studied there in 1957. In this case the pigment did not reduce diammine silver in 10 minutes to 16 hours at 25°C, even after dithionite reduction (2% $\text{Na}_2\text{S}_2\text{O}_4$ 2 hours), though it did moderately reduce ferric ferricyanide. It stained strongly with 0.05–0.1% Nile blue and azure A at pH 0.9–1, and the Nile blue stain was not removed by acetone. The Fe^{++} uptake reaction for melanin was negative. The pigment was unstained by Clara's (Mallory's) hematoxylin and yielded no Prussian blue or Turnbull's blue by treatment with HCl + potassium ferrocyanide or ferricyanide. Extraction with alcohol saturated with picric acid or ammonium picrate did not remove it. (Lillie, 1965.)

In addition Ganter and Jolles cite Lever's "Histopathology of the Skin" (3d ed., Pitman Medical Publishers, London, 1961) and Gomori as agreeing that the pigment is not argentaffin and Cooper and Moran's (*Arch. Pathol.*, **64**: 46, 1957) finding it both argentaffin and argyrophil (but see Lillie's note above). Pearse (1972) comments on the lack of precise histochemical data, refers to LaDu and Zannoni's review in M. Wolman's "Pigments in Pathology" (Academic, New York, 1969).

Thompson adds a negative Gmelin reaction as well as negative PAS, Sudan, acid-fast, Van Gieson, and elastin stains and no ash on microincineration.

LaDu and Zannovi (in Wolman, "Pigments in Pathology," Academic, New York, 1969) added the following step between homogentisic acid and the final fumaric, acetoacetic, maleylacetoacetic, and fumarylacetoacetic acids. They recited much the same distribution as others, noting pigmentation of derma and sweat glands, with black or brown sweat, corneoscleral and conjunctival pigmentation, general in fascia, deeper in tendons and

Maleylacetoacetic acid $\longrightarrow$ Fumarylacetoacetic acid

$$
\underset{HC}{\overset{HC}{\big|}}\overset{C}{\underset{\underset{COOH}{\big\backslash}}{\underset{\big|}{\big|}}} \overset{OH}{\underset{C=CH-COOH}{\overset{CH}{\big|}}} \qquad \underset{HC-CO-CH_2COCOOH}{\overset{HOOC-CH}{\big\|}}
$$

cartilages, heart valves, and endocardium, intimae of large arteries, pigmented renal and prostatic calculi, and black cerumen.

The pigment is apparently an oxidized and polymerized homogentisic acid, the first step of which is apparently α-benzoquinone acetic acid. The further steps are still obscure, and LaDu and Zannovi devote much space to discussing possible enzymatic pathways. A polyphenol oxidase which can act on homogentisic acid is stressed. Histochemical reactions receive little attention in this review.

LIPOFUSCINS

These pigments have been variously designated by the terms "wear-and-tear pigment," *Abnutzungspigment*, aging or waste pigment, lipochromes, chromolipoïdes, ceroid, yellow pigment, and probably others. The term "lipochrome" is more correctly applied to the readily ether-soluble carotinoid pigments which give blues with strong sulfuric acid. The term "chromolipoïdes," though of itself unequivocal, we consider too apt to be confused with "lipochrome." The terms "wear-and-tear pigment" and *Abnutzungspigment*, besides being cumbersome, imply a terminal pathologic physiologic process, which may not be so firmly established as was thought a generation ago. The latter objection applies also to "aging" or "waste pigment." The term "lipofuscin," despite its hybrid Greco-Latin etymology, is unequivocal in meaning and is widely used, both in English-speaking countries and in Europe and Latin America. The term "ceroid" was introduced to designate a usually, though not always, acid-fast sudanophil pigment occurring as granules up to several microns in diameter and as borders of large fat globules in experimental rat liver cirrhosis.

These yellow to brown pigments are generally characterized by sudanophilia, which characteristically persists in paraffin-embedded tissue, by a moderate basophilia manifest usually at pH levels above 3 in formalin-fixed tissue, and by staining blue with Nile blue at pH 1 and lower which is of oil-soluble-dye type and is promptly extracted by acetone applied immediately after staining. However, if a 5- to 10-min wash in tap water is interposed, the dissolved dye forms salt unions and acquires the same acetone resistance as usual basic dye stains at pH 4 to 7.

In addition to these fatty acid reactions, they usually show reactions for ethylene groups, blackening with osmium tetroxide and coloring red in performic and peracetic acid Schiff sequence stains.

In addition they usually reduce ferric salts in the ferric ferricyanide reduction test and blacken slowly in ammoniacal silver solutions without following photographic development; this occurs more rapidly in hot ammoniacal and methenamine silver solutions. Lipofuscins do not reduce acid silver nitrate. They would be expected to give the bromination silver reactions of Norton et al. and Mukherji et al. As recorded by Hueck, their basophilia and hence their Nile blue staining is not impaired by 24- to 48-h bleaching with hydrogen peroxide. It is often difficult to discern whether they lose any of their often light native color in this process. Their ready oxidation by organic peracids and by potassium permanganate has been referred to above in discussion of performic Schiff reaction.

These oxidants, applied beforehand, will prevent other reduction reactions, such as ferric ferricyanide and alkaline silver reductions. Peracetic acid for 2 h will induce the indophenol reaction for peroxides in some pigments (e.g., ovary).

In regard to diammine silver technics, most of those designed for the impregnation of reticulum commence with a permanganate or similar oxidation which tends to destroy the argentaffin reaction. In the original Maresch technic the exposure to diammine silver is probably too brief for an argentaffin reaction of lipofuscins, though melanins should be demonstrated. The diammine silver method to be used for lipofuscin is the Masson-Hamperl or Lillie's basically similar procedure.

Lipofuscins also usually give the periodic acid Schiff reaction. Since the oxidation of unsaturated fatty acids with peracids results predominantly in the production of glycolic fatty acids (oleic acid yields 9,10-dihydroxystearic acid, and polyene fatty acids yield corresponding polyglycolic acids), it is not necessary to postulate sugar conjugates to account for this reaction. Of course, galactolipids do exist, but the fatty acids in them are attached to the sphingosine group by an amide linkage, and such lipids would lack the free carboxyl which seems to be identified in the lipofuscins.

Methylation (about 6 h, 60°C, 0.1 N HCl–methanol) destroys the Nile blue reaction and the basophilia of lipofuscins; saponification (1 % KOH to 70 % alcohol, 20 min) restores it; the sudanophilia to oil red O and to Sudan black B is unimpaired by methylation, and methylated lipofuscin stains red to pink with Nile blue. These reactions seem to confirm the fatty acid nature of the lipofuscins.

The resistance to paraffin embedding mentioned above has long been known. However, it does not, as with the myelins and similar lipids, depend on prior fixation in dichromate, dichromate-formalin, or formalin. We have fixed fresh human autopsy tissues, brain, spinal cord, ganglia, liver, heart, scalp, and axillary sweat glands as well as the ceroid of experimental liver cirrhosis in hot methanol-chloroform and demonstrated lipofuscins with Nile blue and with Sudan black B in amounts not strikingly different from those in control formalin-fixed tissues. In other instances Carnoy's fluid or methanol-chloroform preserves distinctly less pigment than does formaldehyde. Hence for practical purposes the fat solvent fixations are not recommended except for the specific purpose of demonstrating insolubility of usually only a part of the total amount of pigment present in life.

In fluorescence microscopy, the lipofuscins of the liver, adrenal, testicular germinal epithelium, and heart muscle give a red-brown fluorescence. Hepatic lipofuscin fluoresces brown before extraction with alcohol, and red-brown after; and the extract gives the labile green fluorescence of vitamin A (cited from Popper, *Arch. Pathol.*, **41**: 766, 1941).

In human beings lipofuscins do not characteristically give reactions for iron ions, though sudanophilic granules may appear in the same cells together with hemosiderin granules in preparations stained to demonstrate lipofuscin with oil red O and hemosiderin with acidified ferrocyanide. This double staining is often readily demonstrable in macrophages in the margins of cerebral infarcts ("compound granule cells") and in involuting hemorrhagic corpora lutea. In such cases we do not consider that other reactions of granules in the cells in question can surely be imputed to one or the other of the two pigments.

We do not imply here that no sudanophil pigment can be an iron carrier at the same time, but that caution should be exercised in reaching the conclusion that a given pigment falls into that class.

Certain of these pigments, notably that of heart muscle and of the smooth muscle of the seminal vesicles, of the intestine, and of arteries, are distinctly less sudanophil than those of gland cells, nerve cells, and other epithelial structures. Here tinging by orange-red Sudan dyes may be difficult to discern, especially with browner pigments, and the use of blue, green, and black oil-soluble dyes is preferable, since green or blue discoloration is easier to discern in a yellow-brown pigment than orange.

On the other hand some lipofuscins are quite strongly sudanophil, notably that arising in involuting corpora lutea, in the reticular zone of the adrenal, and in neurons of the brain, spinal cord, and ganglia. Here it is possible that the clearer red staining with red Sudan dyes correlates with paler native color of the pigment.

In the brain three morphologic subtypes may be distinguished. In the olivary nucleus and in the subthalamic nucleus one finds single, rounded, mulberrylike masses lying more or less centrally in the cell, with the nucleus to one side, composed of small, rounded globules of dustlike particles of sudanophil material. In the major thalamic nuclei (anterior, lateral, median, and pulvinar) one finds about a dozen or two coarse sudanophil globules, perhaps the size of eosinophil leukocyte granules. In most other nerve cells lipofuscin granules are relatively fine, perhaps more accumulated to the central side of the somewhat eccentric nucleus but not forming any distinct globular mass.

Some histochemical peculiarities exist as well. The olivary granules tend to stain metachromatically with thionin and azure A at pH 3 to 3.5 after bromination, a peracetic acid oxidation of 10 to 120 min, or both in sequence. The coarse thalamic granules seem more likely than other lipofuscins to give a direct Schiff reaction for free aldehyde after relatively brief exposures. The nerve cells of the Gasserian ganglion contain sudanophil granules which stain with azure A at pH 1, rather than at a minimum of pH 3 as with most other nerve cell lipofuscins in formalin-fixed tissue. The acid-fastness of nerve cell lipofuscin noted by Wolf and Pappenheimer (*J. Neuropathol. Exp. Neurol.*, **4**: 402, 1945) has been a rather inconstant characteristic in our hands. Similarly, on extended experience the ceroid of dietary liver cirrhosis of rats has shown acid fastness in varying grade and frequency. Hence this characteristic for separating ceroid from the rest of the lipofuscins seems to break down, and we see no reason for continued use of the word except as a short, readily pronounced synonym for lipofuscin.

Pearse's concept of evolution of lysosome-contained unsaturated lipid through a series of oxidative polymerizations to a final lipofuscin seems eminently logical, though his sequences of development and suppression of the various reactions seems complicated and probably more subject to individual variation than his diagrammatic presentation indicates. For this reason we are adhering to our previous practice of presenting separately series of reactions observed in lipofuscins of various histologic localizations.

Adrenal lipofuscin pigment is seen as variably numerous, usually fine brown granules in the parenchyma cells of the reticular zone of the cortex and as coarser, usually darker brown granules staining similarly but more deeply and contained in scattered, small interstitial phagocytes in the reticular and inner fascicular zones of the cortex. This pigment is green to deep blue in the ferric ferricyanide reduction test, blackens with methenamine silver in the argentaffin method, and colors reddish brown to red-purple with the periodic acid Schiff method. The coarser globules, both in phagocytes and in parenchyma cells, are often peracetic acid Schiff–positive, and while some granules react even after Carnoy fixation, there appears to be more reaction after bichromate fixations.

The capacity to reduce ferric ferricyanide is abolished by bromination and by oxidation for 1 h in 5% chromic acid and for 20 min in 0.5% potassium permanganate, but not by the usual SH blocking reagents or by periodic acid. Conversely, the periodic acid Schiff reaction is unaffected by bromination, weakened by chromic acid and permanganate oxidations, and abolished by 24-h acetylation in 40% acetic anhydride–pyridine. It may often be colored with oil-soluble dyes after paraffin embedding. For this purpose the Sudan black method is recommended. Acid-fastness may be demonstrated in some cases. Here a night blue or Victoria blue technic may be more convincing than the usual fuchsin method because of the greater difficulty of distinguishing reddish brown from brown than blue or green from brown.

It is probable that the adrenal and testicular ceroid pigment described in mice by Firminger (*J. Natl. Cancer Inst.*, **13**: 225, 1952) is closely related to the foregoing pigment.

Two pigments are seen in the *seminal vesicles*, the one in the smooth muscle which is similar in most respects to that described in intestinal smooth muscle, the other occurring in the epithelial cells between the basally placed nuclei and the cell border against the lumen.

This epithelial pigment stains dark green with Sudan black B and dark blue with Nile blue. The sulfuric Nile blue stain is extracted by immediate application of acetone or alcohol, and the preparations may be restained by the same technic. Methylation for 4 to 8 h at 60°C in 0.1 N HCl–methanol prevents the staining by Nile blue, but 32 h has no effect on the sudanophilia. Slight staining is shown at pH 3, moderate at pH 4, with 0.1 to 0.05% solutions of thionin or azure A. Diazosafranin colors it dark red-brown to reddish black. Saturated iodine-methanol for 1 h at 3°C, acetylation (40% in pyridine, 20 h, 25°C), benzoylation (5% benzoyl chloride pyridine, 20 h, 25°C), and prolonged methylation (24 h, 60°C, but not 4 h) prevent the azo coupling reaction; performic acid (1 h) and 2,4-dinitro-fluorobenzene (24 h) do not. The periodic acid Schiff reaction is positive (red-purple), blocked by benzoylation or acetylation. Prolonged exposure (3 days) to Schiff reagent yields a direct (autooxidative) positive reaction; the usual 10-min exposure gives only a faint pink color. The peracetic Schiff reaction is positive.

Treatment for 1 to 2 h in 5% chromic acid at 25°C largely destroys seminal vesicle epithelial pigment, at least in some cases, so that the native brown color disappears and the granules largely or completely lose sudanophilia to Sudan black B. Partial blackening is seen in an hour in diammine silver at 60°C; this would indicate a 25°C reaction in 10 to 14 h. The ferric ferricyanide reaction is positive.

The Leydig cell lipofuscin is colored by Sudan black B and by oil red O in paraffin sections of formalin-fixed material. It reduces ferric ferricyanide. Part of it is acid-fast by the Ziehl-Neelsen method, and more retains the basic fuchsin in Mallory's hemofuscin stain. Methenamine silver is reduced in the argentaffin technic. A minority of the granules are Schiff-positive after peracetic acid and on 48-h exposure to Schiff reagent directly. The granules color red-purple in the periodic acid Schiff method, and the reaction is not impaired by a diastase digestion which removes all glycogen from germinal epithelium. In azure-eosin stains the pigment is brown to dark green, and the basophilia is destroyed by 1-day methylation at 25°C but not by methanol extraction for 7 days. Ferrocyanide reactions for iron are negative, and iron hematoxylin is not retained in the eosinophil myelin technic.

A quite similar pigment occurs in the epithelial cells of some tubules of the *epididymis*. It is basophilic, readily methylated, and periodic acid Schiff-positive with or without diastase digestion, as is the cuticular border of the epithelium. It is occasionally Schiff-positive on direct 48-h exposure and after peracetic acid oxidation. It is partly acid-fast and positive by the Mallory hemofuscin technic. It reduces methenamine silver and ferric ferricyanide. The ferrocyanide reaction for iron and the Weigert-Lillie myelin tests are negative.

Ovarian lipofuscin pigment occurs in phagocytes near involuting corpora lutea and perhaps in their walls. It colors light to dark green with thionin and with azure-eosin. It stains greenish black and greenish blue, respectively, with Sudan black and spirit blue in paraffin sections. The periodic acid Schiff reaction is moderately strongly positive. Part of the granules give the peracetic acid Schiff reaction as well. The granules are only partially acid-fast with Ziehl-Neelsen and Victoria blue technics. In myelin technics iron hematoxylin is retained longer than in nuclei, but not so well as in erythrocytes. The ferrocyanide test for ferric iron is negative. Ferric ferricyanide is sometimes reduced, the granules becoming green to deep blue, and the granules in part blacken with methenamine silver. With Mallory's hemofuscin method, retention of the dye is inconsistent.

In corpora hemorrhagica, iron-positive pigment is also seen, associated with strongly eosinophilic iron-negative granular material which is perhaps only partly degraded hemoglobin. It is not clearly evident that the iron-positive pigment is necessarily unrelated to the lipofuscin.

It is just in this situation that the ferrocyanide–oil red O combined stain is particularly valuable.

Occasionally in later stages some of this pigment reacts for Fe^{2+} as well as Fe^{3+}. One employs 1% $K_3Fe(CN)_6$ in 1% acetic or 0.125 N hydrochloric acid for 30 to 60 min at 25°C, washing in 1% acetic, then in alcohols and in xylenes, and mounting in cellulose caprate. A fuchsin counterstain may be employed; safranin gives brown crystalline precipitates after ferricyanide methods.

In some instances both the iron-positive and the iron-negative pigments blacken with methenamine silver. Sudan black B and oil red O clearly differentiate the two pigments, even when granules of both are present in the same cell. Likewise diammine silver blackens the lipofuscin pigment and not the iron-bearing granules. This last is a true argentaffin reaction, since the blackening is observed with no following reduction bath after as little as 2 min exposure to the silver solution.

Although the Winkler-Schultze indophenol synthesis is not usually effected by unaltered ovarian pigment, a preliminary 2-h oxidation by peracetic acid followed by thorough washing in water causes many ovarian pigment granules to stain a definite blue-green with the Winkler-Schultze procedure. This probably indicates the formation of (fatty acid?) peroxides from unsaturated compounds.

Apparently the *lutein* described by Schmorl was a mixture of lipofuscin and a carotenoid pigment.

Cardiac lipofuscin pigment often stains better than the other lipofuscin pigments with oil-soluble dyes after paraffin embedding. Sudan black or oil blue technics are preferred to Sudan IV or oil red O, because a green discoloration of the brown pigment is easier to discern than an orange discoloration. Ferric ferricyanide reduction is observed in some but not all cases. Part of the pigment gives the periodic acid Schiff reaction; a minor portion gives the peracetic Schiff. The pigment is not definitely acid-fast. It does not retain the myelin stain with iron hematoxylin. With ferrocyanide it does not react for iron. A minor portion of the granules reduce methenamine silver. When the permanganate oxidation is omitted, a minor portion of the granules blacken in the diammine silver reticulum technics; after potassium permanganate no silver reduction occurs. The pigment stains green with thionin at pH 3.

VON RECKLINGHAUSEN'S HEMOFUSCIN (*Versamml. Gesell. Natürf. Ärzte*, **62**: 324, 1889).   This was described as a finely granular yellow pigment not reacting to the iron methods. It occurred along with hemosiderin in the liver and certain other tissues in cases of hemochromatosis. Hueck classified it as a lipofuscin, we agree, and later Gillman and Gillman (*Arch. Pathol.*, **40**: 239, 1945) called it *cytolipochrome*. In frozen sections it may be stained with oil-soluble dyes, but according to Endicott and Lillie (*Am. J. Pathol.*, **20**: 149, 1944) it does not so stain after paraffin embedding and is not acid-fast. The material of Endicott and Lillie had been stored for several years in formalin; and it is possible that, like myelin, hemofuscin may lose part of its resistance to decolorization on long storage in formalin. However, the hemofuscin in this material still resisted decolorization with alcohol after prolonged staining with basic fuchsin according to Mallory's technic for the demonstration of this pigment.

MALLORY'S HEMOFUSCIN METHOD

1. Fix in Zenker's fluid, alcohol, or 10% formalin. Make paraffin or celloidin sections, and bring to water as usual, including an iodine–sodium thiosulfate sequence in the case of Zenker fixation.
2. Stain 5 to 10 min in Mallory's alum hematoxylin (other alum hematoxylins will serve).
3. Wash well in water.
4. Stain 5 to 20 min in 0.5% basic fuchsin solution in equal volumes of distilled water and 95% alcohol.
5. Wash in water.
6. Differentiate in 95% alcohol.
7. Dehydrate with 100% alcohol, clear in xylene, and mount in polystyrene.

RESULTS: Nuclei are blue; hemofuscin is bright red; melanin and hemosiderin, unstained in their natural browns.

Ceroid is also stained by this procedure, and it is not impossible that various other myelin-like substances would be.

Post, Benton, and Breakstone (*Arch. Pathol.*, **52**: 67, 1951) report a similar cytoplasmic pigment of normal, predominantly centrolobular *human* liver cells, which is insoluble in water, alcohols, aromatic and aliphatic solvents, acetone, acids, and alkalies. It is stained by the Mallory hemofuscin technic and resists alcohol decolorization but is not acid-fast. It gives negative reactions by the periodic acid Schiff method, by the ferric ferricyanide reduction test, by the Feulgen technic, by the acid ferrocyanide method for ferric iron, by Stein's methods for bile pigment, and by Gomori's alkaline phosphatase method. It resists digestion by ribonuclease, deoxyribonuclease, amylase, and trypsin. Unlike hemofuscin as characterized by Mallory, it is not dissolved by 5% hydrogen peroxide. It is considered to be otherwise similar to hemofuscin and is thought to be a functional metabolic complex.

In view of the high solubility of rosanilin base in fatty acids (oleic 20%, stearic 15%) recorded in the "Colour Index" as compared with its solubility in ethanol (0.3%), it is suspected that hydrolysis of the rosanilin chloride occurs in this procedure, and that we are probably seeing a differential solubility reaction. This would tend to support Hueck's views as to the lipofuscin nature of this pigment.

INTESTINAL SMOOTH MUSCLE PIGMENT.   This condition is seen occasionally as a dark brown coloration of the muscularis of the ileum, and appears to have been relatively commoner in the earlier German autopsy experience. Its causation is obscure.

The pigment occurs in formalin-fixed tissue as fine to coarse droplets in the muscularis and as coarser globules in macrophages in the submucosa. The muscularis mucosae is not involved.

The pigment colors moderate gray-green with Sudan black and moderately deep greenish blue with sulfuric Nile blue. The latter reaction is prevented by 6-h methylation in 0.1 N HCl–methanol and is restored by demethylation (saponification) in 1% potassium hydroxide in 70% alcohol for 20 min.

In azure A (0.7 mM), 1 : 5000, it stained strongly at pH 4 and above, moderately at pH 3, and remained almost unstained at pH 1 to 2. The pigment reacts slowly and lightly with Schiff reagent, attaining a fair pink color in 24 h, similar to that obtained after 10-min oxidation in peracetic acid. A 2-h peracetic acid oxidation followed by 10 min in Schiff reagent yields red. A similar sequence of 10 min each in 1% periodic acid Schiff reagent gives a dark red-purple.

With a myelin procedure a brown to black is retained when erythrocytes are still black and nuclei are decolorized.

Acid silver nitrate (pH 4) is not appreciably reduced in 2 h directly or after dithionite reduction. This reaction excludes melanin. Ammoniacal silver is strongly reduced in 24 h but is only slightly reduced in 3 h. Hot methenamine silver (60°C) is reduced in 3 h. The ferric ferricyanide test is positive.

The ferrous ion uptake reaction for melanins is very faint and is to be regarded as negative. Ferro- and ferricyanide reactions for Fe^{3+} and Fe^{2+} ions are negative, excluding the hemosiderins.

Oxidation by potassium dichromate at pH 2.2 gives no color reaction (bilirubin is said to give green).

Moderate acid-fastness to the Ziehl-Neelsen method is shown.

Only a yellow color is seen with diazotized p-nitroaniline; a deep red-purple is seen with diazosafranin and is prevented by prior acetylation. The Morel-Sisley reaction for tyrosine is completely negative (Paneth cells reacted strongly).

The postcoupled benzylidene reaction for tryptophan is negative.

The pigment remains strongly basophil to azure-eosin after 6-h peptic digestion, although basic protein is largely destroyed.

The reactions parallel closely those observed in the more familiar and much more frequent smooth muscle pigment seen in human seminal vesicles.

The carboxylic acid and aldehyde nature of the pigment, together with the presence of ethylenic and 1,2-glycol groupings, and the colorability with fat and fatty acid stains point to a lipofuscin pigment.

Hendy (*Histochemie*, **26**: 311, 1971) noted that sections stained with the Fontana or Schmorl methods for the identification of lipofuscin could be postfixed in osmium tetroxide and prepared for electron microscopy in the usual fashion. Lipofuscin is stated to be selectively identifiable.

CEROID

Since the original description of ceroid as an often, but inconstantly, acid-fast brown pigment occurring in experimental liver cirrhosis of rats (Lillie, Daft, and Sebrell, *Public Health Rep.*, **56**: 1255, 1941; Edwards and White, *J. Natl. Cancer Inst.*, **2**: 147, 1941) and mice (Lee, *J. Natl. Cancer Inst.*, **11**: 339, 1950), similar acid-fast pigments have been described in vitamin E deficiency of rats and monkeys, which apparently arise in muscle tissue. A similar pigment apparently evolves from adrenal and testicular Leydig cell lipofuscins (Firminger, *J. Natl. Cancer Inst.*, **13**: 225, 1952).

As characterized by Endicott and Lillie (*Am. J. Pathol.*, **20**: 149, 1944), ceroid possesses a bronze-brown color in gross preparations, and occurs microscopically as yellow globules 1 to 20 μ in diameter, located sometimes in liver cells, most often in large phagocytes, and also as rims of acid-fast material surrounding large fat globules. Fluorescence microscopy of frozen sections shows a greenish yellow fluorescence which fades to pale yellow. In paraffin sections the fluorescence is golden brown. It retains myelin stains, acid-fast stains, and Mallory's hemofuscin stain. It usually stains green in the azure-eosin technic, and is stained by oil-soluble dyes, both in frozen and paraffin sections. The oil-soluble dyes are readily removed by acetone, alcohol, and the like; and preparations may be restained and decolorized repeatedly. With the usual brief hemalum and iron chloride hematoxylin stains it does not color. Though it is gram-negative, if crystal violet staining is prolonged or accelerated by heat, the violet is retained with or without iodine treatment.

Ceroid is insoluble in dilute acids and alkalies but is saponified by boiling 10% NaOH (gross chemical procedure), and fatty acids are precipitated from the solution by acid. It is insoluble in alcohols, acetone, ether, aliphatic and aromatic hydrocarbons, chloroform, carbon tetrachloride, pyridine, acetic anhydride, glycols, and glycerol. It is not bleached by permanganate, chromic acid, hydrogen peroxide, or bromine or chlorine water.

It is blackened by osmium tetroxide, and blackens slowly with diammine silver carbonate. Most of it fails to reduce ferric ferricyanide, but definite deposits of blue pigments are formed in isolated globules and foamy masses. The Prussian blue (Perls) reaction is usually negative; but iron-positive pigment is sometimes associated with it, and necrotic foci containing ceroid exhibit both iron and calcium deposition.

Much of the pigment is colored red-purple in the periodic acid Schiff procedure, but often clear, unstained, small globules are seen enclosed in foamy red-purple masses. The peracetic and performic acid Schiff reactions are also positive in small globules, in part of the foamy masses, and in some of the large globules. In parallel preparations it appears as if part of the material gave the HIO_4 reaction, part the CH_3CO_3H reaction, and part both reactions.

Bromination prevents the peracetic Schiff reaction. Benzoylation or acetylation prevents the periodic Schiff reaction and retards but does not prevent the peracetic Schiff reaction. Ceroid is colored red by 2 to 3-day exposure to Schiff reagent, and yellow by similar long exposures to phenylhydrazine.

The basophilia of ceroid is abolished by benzoylation but not by bromination. Ferric ferricyanide reduction is prevented by benzoylation but not by bromination. Acid-fastness is unaffected by bromination, benzoylation, or strong halogen acids. Hydrochloric and hydriodic acids (16 h, sp gr 1.19 and 1.70, respectively) reverse the peracetic and periodic acid Schiff reactions but do not destroy acid-fastness or sudanophilia. Exposure to 5% phenylhydrazine or to molar aniline chloride for 24 h prevents both the peracetic Schiff reaction and the Schiff reaction, which occurs directly in 2 to 3 days. The periodic acid Schiff reaction is unaffected by these blockades.

SUDAN BLACK B METHODS FOR LIPOFUSCINS.   In this case any of the commonly used Sudan black technics may be used. Since the lipid substance under study is quite insoluble, the usual objection that the 70% alcohol and the propylene glycol technics remove much of the birefringent (steroid) lipid and perhaps part of the phospholipids as well does not apply.

We usually use the following Sudan black B method.[1]

Dilute 20 ml saturated Sudan black B in absolute alcohol with an equal volume of distilled water. Let stand 10 to 20 min and filter.

1.   Bring paraffin sections to water. Attached cryostat sections may be used if lipid solubility tests are contemplated. Lipofuscin is at least partly insoluble in an equal-volume methanol-chloroform mixture (ranging from 2 : 1 to 1 : 2) at 25 to 60°C for as long as 2 to 3 days.

2.   For a Feulgen nuclear counterstain, hydrolyze sections in 1 N HCl at 60°C. With routinely fixed formalin material in paraffin sections the optimal time varies between 10 and 20

---

[1] Note that a strong lipofuscin stain without precipitation on sections is produced by 30-min staining in Sudan black B, 500 mg; ethylene glycol, 30 ml; ethanol, 30 ml; and distilled water, 40 ml.

min; 3 min is enough for films or cryostat sections fixed a few minutes in methanol.

3. Transfer to Schiff reagent for 10 to 30 min; the 0.5% solutions take longer than the 1%, especially if they are not fresh.
4. Wash in three changes of 0.5% sodium metabisulfite ($Na_2S_2O_5$) and 10 min in running water. We use the bisulfite washes to avoid any oxidative recoloration of leucofuchsin by chlorinated city water.
5. Stain 10 to 20 min in the filtered Sudan black.
6. Wash in water, and mount in Apáthy's gum syrup or in glycerol gelatin.

To confirm the oil-soluble nature of the lipofuscin stain, decolorize in acetone (20 s to 5 min), examine in water or glycerol, wash, and restain (step 5). Lipofuscins decolorize by acetone and may be restained by the original technic. This decolorization and restaining may be repeated several times without evident loss in either stainability or decolorizability.

OTHER COMBINED PIGMENT STAINS

The oil red O stain for lipofuscins has also been successfully combined with the diammine silver method on central nervous tissues, to stain the neuromelanin of the substantia nigra and locus ceruleus in black or very dark brown and the nerve cell lipofuscins in nearby nuclei in red. The exposure to ammoniacal silver is purposely kept short to avoid the reduction of silver by lipofuscins which ensues on prolonged exposures.

We make the diammine silver by placing 2 ml 28% ammonia in a 100-ml beaker and then adding about 35 ml 5% silver nitrate, quickly at first, shaking between each addition to dissolve the dark brown silver oxide, and cautiously, in the last 5 ml, adding just enough silver to produce a faint but definite permanent turbidity.

1. Bring paraffin sections through xylene and alcohols to two changes of distilled water.
2. Immerse for 15 min in diammine silver at 20 to 25°C in the dark.
3. Wash in distilled water, and immerse for 5 min in 0.2% gold chloride ($HAuCl_4$). The gold toning is included to increase stability of the melanin stain.
4. Rinse in distilled water, and place in 0.2 M (5%) sodium thiosulfate for 2 min. Wash for 10 min in running water.
5. Stain with oil red O or by the Feulgen Sudan black B technic, as above in the normal technics. Mount in Apáthy's gum syrup, Farrants' gum arabic, or glycerol gelatin.

For cutaneous and ocular melanins the ammoniacal silver step may be decreased to 2 min, or it may be replaced by an hour in 2% silver nitrate in pH 4 acetate buffer. This makes the silver reaction more nearly specific for true melanins.

The Feulgen Sudan black silver variant gives red nuclei, black neuromelanins, dark green lipofuscin, and lighter green myelin and red corpuscles. In the oil red O technic the black neuromelanin contrasts well with the red lipofuscin. Although we have not specifically tested the use of an alum hematoxylin nuclear stain in this combined procedure, we believe it should work.

The more sensitive ferric ferricyanide reaction cannot thus be successfully combined with oil red O for lipofuscin. These reducing pigments also color blue with the ferric ferricyanide reaction, leaving the oil red O effect visible only as an orange staining of myelinated fibers.

Such technics as these should prove useful in helping to resolve the question of whether certain ganglion cell pigments are "lipomelanins" or a mixture of separate granules of lipofuscin and neuromelanin. The production or storage of multiple organic substances of a secretory nature in the same cells seems quite well documented in other locations, e.g., the presence of heparin, histamine, and serotonin in mast cells, along with a number of enzymes.

THE MELANINS

A variety of pigments have at one time or another been classed under this heading. The first, to which the name properly belongs, is the black to brown pigment found in epidermis, hair follicles, and hairs and in cutaneous melanoma. The iron-containing, acid-soluble pigment of human red hair is excluded, as is the yellow, readily alkali-soluble pigment of hair follicles and hairs seen in guinea pigs (trichoxanthin, *J. Histochem. Cytochem.*, **5**: 346, 1957) and possibly in mice (so-called "phaeomelanin," which is a misnomer, for this pigment is yellow, *xanthos*, not brown or dusky, *phaios*).

Second, we include the yellowish brown to black epithelial and connective tissue pigments of the iris, ciliary body, choroid, and retinal pigment epithelium. The last occurs in rod-shaped granules and may be chemically and functionally as well as morphologically distinct. It is sometimes called *fuscin*. The pigment of ocular melanoma appears to derive from that of the choroid and is histochemically similar to it.

Also quite similar histochemically to choroid pigment is the melanin occurring in patches in the piarachnoid in some human brains. It is often quite prominent in ruminant brains and is seen in other mammals. The pigment found in vertebrate chromatophores in the derma and elsewhere and in blue nevi is also classed as melanin.

The pigment found normally in the human substantia nigra, locus ceruleus, and nucleus dorsalis nervi vagi is histochemically distinct from ocular, pial, and cutaneous melanins and has been called *neuromelanin*. It occurs with less constancy and in smaller amounts in other primates, including Old World monkeys as well as the great apes, and has been reported also in Canidae and Felidae.

The brown to black pigment occurring in neurons of certain sensory nerve root ganglia and in some sympathetic ganglia is (or are) sometimes classed with lipofuscins, sometimes with melanins. The histochemical reactions applied to them have not been, perhaps, sufficiently critical to give a definitive answer to this question.

Finally the iron-free phase of the granular phagocyte pigment of human intestinal mucosa has also been classed as melanin by some workers, as lipofuscin by others, and as a substance closely related to the iron-bearing pseudomelanosis and villus melanosis pigments. This pigment has been discussed in full under the last-mentioned category.

Much chemical study has been devoted to the elementary analysis and destructive fractionation of melanins isolated by enzymatic or alkali digestion of black hair and wool, from epidermis of Negroes, from eyes of various species (mostly oxen), and from melanomas of human cutaneous and ocular origin, and from the common melanoma of white horses. These melanins were alkali-soluble acidic substances of undetermined molecular weight, containing 50 to 60% carbon, 4 to 6.5% hydrogen, 9 to 14% nitrogen, and a percentage of sulfur ranging from 1 to 12%. The balance was assigned to oxygen. That the sulfur was not assignable to contaminating keratins is shown by its presence in tumor and eye melanins. Amino acid analyses have been few; Serra [*Nature (Lond.)*, **157**: 771, 1946] identified chromatographically in hydrolysates of alkali extracted hair melanin the amino acids arginine, histidine, tyrosine, tryptophan, cystine, cysteine, and methionine, as well as an acid-insoluble black "melanoid" residue. Block et al. ("Amino Acid Handbook," Charles C Thomas, Springfield, Ill., 1956) do not list melanin.

The study of animal tyrosinase started around 1902; Otto von Fürth and his students were prominent in this work. In 1917 Bloch introduced the dopa reaction into the study of cutaneous melanogenesis. The dopa reaction is not, properly speaking, a reaction of any of the melanins. It is rather the enzyme histochemical reaction which produces a dark brown pigment from L-3,4-dihydroxyphenylalanine (dopa), the first oxidation product of tyrosine. The enzyme tyrosinase, in the presence of oxygen, oxidizes tyrosine slowly to dopa, and oxidizes that substance more rapidly to an indolic quinone or semiquinone-like substance which is deep brown and is considered to be the coloring matter of melanin.

Since decarboxylation is accomplished at stage VII, the acid nature of the melanoprotein is presumed to depend on coincident oxidation of cystine residues to protein-bound cysteic acid. Lorincz (in Rothman's "Physiology and Biochemistry of the Skin," chap. 22, pp. 515–563, University of Chicago Press, Chicago, 1954) adds step VIII to the original Raper (*Biochem. J.*, **21**: 89, 1927) sequence, considering the final state to be a quinone. Lillie (*J. Histochem. Cytochem.*, **5**: 325, 1957) added IX, to account for reducing capacity still present in melanogenic sites. Mild oxidations—I_2, H_5IO_6, $K_2Cr_2O_7$, $FeCl_3$, et al.—carry IX to VIII, while dithionite reduction carries melanin in stage VIII or IX to VII. Melanin in stage VII can be acetylated with consequent marked retardation of acid silver reduction. Acetylation is without effect on the reactions of native or oxidized melanins. Stage VII presents very rapid metal reduction reactions, like those of isolated catechols in vitro.

The dopa reaction occurs in skin, especially in stages of melanogenesis, as after radiation and in initial fetal and infantile pigment production. It is reported also in the fetal eye in

The Raper Cycle

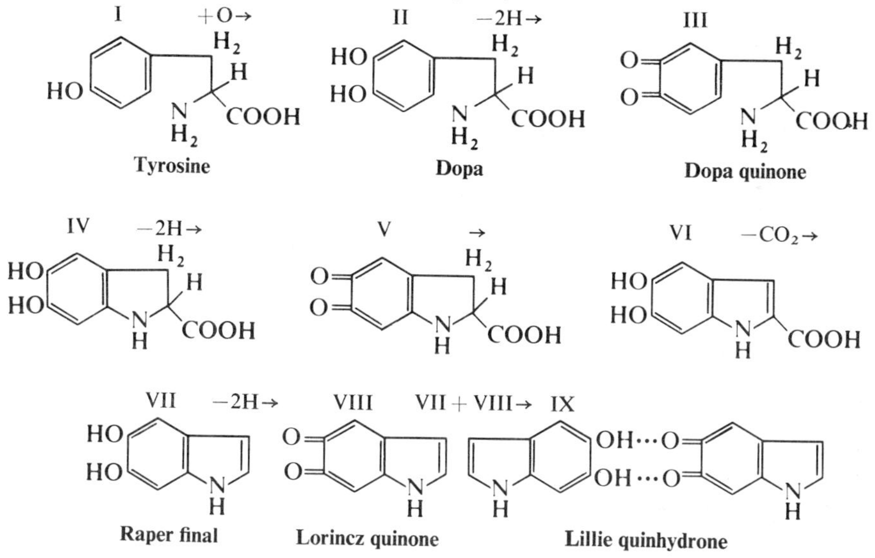

I $+O\rightarrow$ Tyrosine
II $-2H\rightarrow$ Dopa
III Dopa quinone
IV $-2H\rightarrow$
V $\rightarrow$
VI $-CO_2\rightarrow$
VII $-2H\rightarrow$ Raper final
VIII Lorincz quinone
VII + VIII $\rightarrow$ IX Lillie quinhydrone

early stages, and after irradiation. It occurs in cutaneous melanoma and probably also in melanoma of ocular origin, though we do not recall having found a report of the latter. Lillie made several trials of the dopa reaction on infant locus ceruleus and substantia nigra, within the period of pigment appearance, with consistently negative results.

Foster has reported the formation of the yellow pigment of mouse skin when tryptophan is used as substrate in place of tyrosine (*J. Exp. Zool.*, **117**: 214, 1951). Lillie found guinea pig hair follicles containing trichoxanthin dopa-negative, though Langerhans cells in the same area and black or brown pelage areas in the same animal gave positive dopa reactions (*J. Histochem. Cytochem.*, **5**: 346, 1957).

METAL REDUCTION REACTIONS.   These reactions depend on the reducing capacity of the melanins.

The crucial histochemical test for cutaneous, ocular, and pial melanins is the reduction of 0.1 M silver nitrate (1.7%) in pH 4 0.1 M acetate buffer in 1 h or less at 25°C in the dark. This is followed by 1- to 2-min washing in distilled water and in 0.2 M thiosulfate (5% hypo) and then in running water (10 min).

This procedure blackens guinea pig skin melanin in as little as 10 min. Monkey and human skin pigments require 30 to 60 min.

The trichoxanthin of yellow guinea pig skin is resistant to acid silver nitrate, requiring 18 to 24 h at 25°C to acquire a deep red-brown, and 24 h at 60°C for complete blackening. Neuromelanin of the human substantia nigra is similarly resistant, requiring 3 days in the dark at 25°C for blackening.

Masson's ammoniacal silver technic, which we simplify by placing 2 ml 28% ammonia in a 100-ml beaker and adding gradually 35 to 40 ml 5% $AgNO_3$, swirling to redissolve the dark brown silver oxide, and stopping at the point where the solution remains faintly turbid or opalescent, is more rapid and much less specific. Although it blackens the skin and eye melanins in about 2 min, it also blackens neuromelanin in 2 to 15 min; trichoxanthin may blacken in 2 min or may completely and irreversibly disappear, by reason of its alkali solubility. With longer exposures (18 to 24 h), lipofuscins, intestinal melanosis-pseudomelanosis pigment, hematoidin, bile pigment, enterochromaffin, and other substances also blacken.

The same is true of Gomori's methenamine silver used alone as an argentaffin reagent.

The ferric ferricyanide reaction is also sensitive but relatively nonspecific, and being an acid reagent it preserves trichoxanthin well.

BLOCKADES.   The foregoing metal reduction reactions are prevented by moderately active oxidants: 2% $FeCl_3$, 1 day at 24°C; 5% CrO_3, 3 to 30 min; 5% $K_2Cr_2O_7$, 2 to 3 days; 1% H_5IO_6, 3 to 6 h for ferric ferricyanide, less for diammine silver; 10% I_2 in methanol, 1 to 2 days. If after these oxidations reduction for 1 h in 1% (fresh) sodium dithionite ($Na_2S_2O_4$). is applied, the metal reduction reactions are restored. This is true for melanins, trichoxanthin, and neuromelanins.

After dithionite reduction, melanin, but not neuromelanin or trichoxanthin, reduces silver nitrate at pH 4 in 1 to 2 min. This behavior is observed with or without a preceding oxidation.

Acetylation is without effect on the metal reduction reactions of native melanin, but after dithionite reduction acetylation greatly retards (18 to 24 h) the pH 4 silver nitrate reaction. Acetylated catechol is similarly retarded in its reduction of acid silver nitrate.

One point about these oxidation studies is that dichromate or dichromate-sublimate fixations are apt to prevent the metal reduction reactions.

In combinations of the silver reaction with oil red staining or the periodic acid Schiff reaction, the silver reduction should be done first and followed by 10- to 15-min gold toning in 0.2% $HAuCl_4$ after thiosulfate removal of unreduced silver, to render the metal stain more permanent.

OTHER REACTIONS OF THE MELANINS.   Cutaneous and ocular melanins and trichoxanthin color dark green on 20-min staining in 0.05% azure A at pH 1. Methylation in 0.1 N HCl–methanol at 60°C is resisted 24 h or more, as with the cysteic acid of oxidized hair cortex. Neuromelanin does not stain below pH 3, where only a weak green appears. Strong staining occurs at pH 4 and above.

After prolonged (24 to 32 h at 25°C or 2 h at 60°C) exposure to weak (1 mM) solutions of ferrous salts at pH 4 to 5, melanin, trichoxanthin, and neuromelanin color blue-green, and enterochromaffin colors blue when reacted with ferricyanide at pH 1. Lipofuscins and heme pigments do not react (*Arch. Pathol.*, **64**: 100, 1957; *J. Histochem. Cytochem.*, **9**: 44, 1961).

Melanin and trichoxanthin give no tryptophan or tyrosine reactions. The Gibbs reaction for phenols is negative, as is Clara's staining with dilute unmordanted hematoxylin at pH 7 (1 to 2 days). No azo coupling is demonstrated with diazosafranin. Neuromelanin reacts moderately to the periodic acid Schiff reaction; cutaneous and ocular melanins and trichoxanthin do not.

Melanins are insoluble in water, alcohol, fat solvents, and dilute acids and alkalies. Though freed from the choroid and retinal epithelial cells and partly dispersed as granules and rods by peptic and tryptic digestions, the individual granules appear to resist digestion long after cytoplasms and nuclei have disappeared. Likewise, digestion with ribonuclease, with malt diastase, and with chondromucinase is resisted. Whether the alkali removal mentioned by Mallory is actually a solution of the pigment or is a dispersal from digestion of the cytoplasmic matrix we are still uncertain.

Melanin is slowly bleached by 10% hydrogen peroxide; it may require 1 to 2 days, and is not visibly affected by 3% peroxide in 24 h. Although slow bleaching with ferric chloride is reported, a 24-h exposure to a 1% solution has no evident effect. Similarly, no evident bleaching occurs with 24-h exposures to 1% sodium iodate in 0.3% nitric acid, to 3% potassium dichromate, to Weigert's iodine, to 0.5% sodium bisulfite, to 5% hydroquinone, to normal hydrochloric acid, to 5% formic acid, to 5% trichloroacetic acid, or to 5% acetic acid. Appreciable though often irregular and unpredictable bleaching occurs with 5% chromic acid (CrO_3) in 1 to 3 hours, and with 0.5% potassium permanganate in intervals of 20 min up to several hours. Chlorine water bleaches melanin, though treatment with 1 ml bromine in 39 ml carbon tetrachloride for an hour does not. The most pronounced bleaching occurs with performic and peracetic acids, even on 1- to 2-h exposures, and with 16- to 24-h exposures ocular melanin may be completely removed. This effect may be due to the large amount of hydrogen peroxide present in these reagents. Even 30-h exposure to 0.03 M periodic acid does not bleach ocular melanin. (See Table 11-3.)

Melanins are commonly said to reduce osmium tetroxide. Melanin remains the same shade of dark brown after the ammonium sulfide treatment used in the Gomori calcium-cobalt alkaline phosphatase method.

Ocular melanins do not evidently tinge with oil-soluble dyes. They do not retain the iron hematoxylin of the Weigert-Smith-Dietrich–type technics, or the basic fuchsin of the Mallory hemofuscin or Ziehl-Neelsen acid-fast methods. Lack of acid-fastness is also specifically

TABLE 11-3 BLEACHING OF MELANINS

	Skin melanin	Tricho- xanthin	Ocular melanins	Neuro- melanin
0.25% $KMnO_4$, 25°C	20 min	0–5 min	2–4 h	1–5 min
70% HNO_3, 25°C	6–8 h	30 s	3–4 h	4–6 min
5% CrO_3, 25°C	2 h	15 min	....	1 h (40°C)
1% H_5IO_6, 60°C	4 h	Unchanged, 4 h		
0.05 M HIO_4, 25°C	Partial, 4–24 h	Unchanged, 24 h		
Bromine water	8 h	3 h		
Peracetic acid, 25°C	2–12 h	1 h (3°C)		

recorded for cutaneous melanin and for that of the nerve cells of the substantia nigra. Perls' Prussian blue reaction with ferrocyanide is negative.

Lillie recorded (*J. Histochem. Cytochem.*, **5**: 346, 1957) bleaching of cutaneous and ocular melanins, neuromelanin, and trichoxanthin as detailed above. The experiments on eyes and brains were only fragmentary and were not reported previously.

It is of interest that prior fixation in dichromate fixatives prevents the prompt chromic acid bleaching.

Neuromelanin gives negative iron reactions with ferrocyanide and with the Tirmann-Schmelzer technic. Schiff reactions are negative on direct brief or prolonged exposure, after warm normal hydrochloric acid hydrolysis (Feulgen procedure) and after oxidation with potassium permanganate or chromic acid (Casella, Bauer procedures) or with peracetic acid (ethylene reaction). However, peracetic acid oxidation for 2 h causes the pigment to become oxidase-positive by the Winkler-Schultze indophenol method. This indicates probable formation of peroxides from unsaturated compounds. Neuromelanin is not tinged by Sudan black B or oil red O. Its color is not altered in myelin stains differentiated to the point where nuclei are decolorized but red corpuscles are still black. Osmium tetroxide is not evidently reduced in formalin-fixed tissue. The pigment does not retain fuchsin in the Mallory hemofuscin or Ziehl-Neelsen acid-fast stains.

Fe^{2+} ION UPTAKE REACTION OF MELANIN (Lillie). TECHNIC. Avoid all chromate fixatives. Other fixations, mercurial, formaldehyde, alcoholic, etc., are well tolerated.

1. Paraffin sections: deparaffinize and hydrate as usual.
2. Immerse for 1 h in 2.5% ferrous sulfate ($FeSO_4 \cdot 7H_2O$).
3. Wash 20 min in distilled water (four changes).
4. Immerse 30 min in 1% potassium ferricyanide [$K_3Fe(CN)_6$] in 1% acetic acid.
5. Wash in 1% acetic acid.
6. If desired, counterstain 5 min in Van Gieson's picric acid–acid fuchsin mixture (100 mg acid fuchsin in 100 ml saturated aqueous picric acid solution). Do not use hematoxylin.
7. Dehydrate in two changes each of 95 and 100% alcohol; clear in alcohol and xylene (50 : 50) and two changes of xylene.
8. Mount in synthetic resin. If not counterstained, use cellulose caprate.

RESULTS: Melanins of skin, eye, and pia, neuromelanin, and trichoxanthin are dark green; background is faint greenish or unstained; with Van Gieson, the usual red collagen and yellow and brown muscle and cytoplasm are demonstrated. Lipofuscins and heme pigments are unreactive.

FORMALDEHYDE-INDUCED FLUORESCENCE. In a series of papers [*Science*, **149**: 439, 1965; *Arch. Dermatol. Venereol.* (*Stockholm*), **46**: 65, 1966; ibid., **46**: 401 and **46**: 403, 1966; and *Arch. Dermatol.*, **94**: 363, 1966) Falck, Jacobsson, Olivecrona, Forsman, Olson and Rosengen describe a specific green to greenish yellow fluorescence of melanocytes, induced in freeze-dried tissue treated 1 to 3 h at 80°C in formaldehyde gas according to the Falck-Hillarp technic. Vacuum paraffin embedding follows, and 6- to 10-μ sections are studied by fluorescence microscopy.

The fluorescence was greater in malignant melanoma than in cutaneous nevi and greater in these than in normal melanocytes. It was found to be increased in normal melanocytes after therapeutic roentgen irradiation in human beings. It has been observed in the malignant melanoma in the Syrian golden hamster.

Analyses of seven human melanomas revealed appreciable amounts of dopa in three and small amounts of norepinephrine in four. Dopamine, epinephrine, and 5-hydroxytryptamine were not demonstrated.

Pearse (1972, p. 1052) notes agreement between emission spectra of melanin and dopa models (peaks at about 4400 and 4900 Å according to Rost and Polak, *Virchows Arch. Pathol. Anat.*, **347**: 321, 1969).

12

METALS, ANIONS, EXOGENOUS PIGMENTS

CAROTENES

The carotenoid pigments are defined by Lison (1960) as long-chain hydrocarbons with conjugated double bonds. They range from red to yellow.

They are insoluble in water, glycerol, dilute acids and alkalies, and formalin. They are soluble in cold alcohol, more so in chloroform, acetone, xylene, toluene, benzene, petroleum ether, and carbon disulfide. They give a green fluorescence in ultraviolet light.

They are quickly decolorized by oxygen and by dilute chromic acid or potassium dichromate. They yield a transient blue with concentrated sulfuric acid and a brown-green to violet or black with Weigert's iodine. Ferric chloride and hydrogen peroxide solutions decolorize them, but they are said not to reduce diammine silver.

Lison believes that the term *lipochrome* applied to these pigments should be abandoned. Inasmuch as it is also applied by some workers to the lipofuscins, we agree thoroughly. Its continued use produces confusion.

VITAMIN A

For demonstration of vitamin A, Popper (*Arch. Pathol.*, **31**: 766, 1941) recommends brief fixation in 10% formalin, preferably in the cold and not over 10 to 12 h. Blocks are cut 3 mm thick. Frozen sections are examined in water within 3 h of cutting, with ultraviolet light at 365 mμ in a fluorescence microscope. Vitamin A presents a brilliant green fluorescence which fades quite promptly in 10 to 60 s or more, depending upon the amount present.

Vitamin A is soluble in fat solvents and occurs dissolved in body fats as well as in liver cells, Kupffer cells, lutein cells, adrenal cortex, and other places. It resists treatment with 0.1 N hydrochloric acid, 0.1 N ammonium hydroxide, and saturated sodium dithionite (hydrosulfite) ($Na_2S_2O_4$) solution. The fluorescence of vitamin B_2 is destroyed by the last. Hydrogen peroxide produces a blue fluorescence of fats but does not destroy the green vitamin A fluorescence.

Vitamin A, like the carotenes, gives green to blue with sulfuric acid. It gives blue with antimony trichloride.

RIBOFLAVIN

Gomori quotes a method of Chèvremont and Comhaire. After reduction of riboflavin to leucoriboflavin (leucoflavin) it reoxidizes in air to red rhodoflavin.

Place frozen sections of formalin-fixed tissue in 1 to 2% hydrochloric acid containing enough zinc dust to give a steady evolution of hydrogen, and stir about gently for 30 min. Then wash in water and expose to air in a shallow vessel for several hours. Mount in glycerol gelatin. Flavoproteins are stained red. We have small faith in the efficiency of hydrogen reduction away from the immediate site of nascence. Gomori suggests replacing the hydrogen reduction with sodium dithionite.

VITAMIN C

For the demonstration of ascorbic acid, Deane and Morse (*Anat. Rec.*, **100**: 127, 1948) recommended immediate fixation for 30 min in the acetic alcohol–silver nitrate solution of Barnett and Bourne (*J. Anat.*, **75**: 251, 1940), followed by transfer directly to an acid fixing solution for 2 h, then overnight washing in running water.

The fluid of Barnett and Bourne was ambiguously stated to be a "saturated solution of silver nitrate in ethyl alcohol (5 parts), water (4 parts), and glacial acetic acid (1 part)" and to give a concentration of slightly less than 10% silver nitrate. Since silver nitrate is only slightly soluble in pure alcohol, it seems that saturation in the mixture was meant. The following emended formula is suggested: Dissolve 10 g silver nitrate in 40 ml distilled water, and add 10 ml glacial acetic acid and 50 ml 100% alcohol. This fluid does not keep.

The acid fixing solution contained sodium thiosulfate crystals ($Na_2S_2O_3 \cdot 5H_2O$), 5 g; sodium bisulfite ($NaHSO_3$), 1 g; distilled water, 100 ml.

After washing, the material is dehydrated, cleared, infiltrated, and embedded in paraffin as usual, and sectioned at 5 μ.

Deane and Morse deparaffinized and mounted for examination in the unstained state, or counterstained with paracarmine or hematoxylin and eosin.

Barnett and Bourne varied the procedure by simply washing in distilled water after fixation, embedding, and sectioning, and then toning sections for 4 to 10 min in "very dilute gold chloride solution," fixing in sodium thiosulfate solution for 4 to 10 min, followed by dehydration, clearing, and mounting.

Cater (*J. Pathol. Bacteriol.*, **63**: 269, 1951) claimed superior results by a 3-h fixation in aqueous 10% silver nitrate–10% acetic acid solution, followed by three changes of distilled water for 30 min each and a 90-min bath in 5% sodium thiosulfate, all in the dark. Washing in water is presumed. Dehydration, clearing, paraffin embedding, and sectioning at 5 μ were followed by a light counterstain with neutral red.

We have not used these procedures.

Clara (*Mikroskopie*, **7**: 387, 1952), though admitting the effectiveness of the acid silver nitrate method in demonstrating ascorbic acid, points out that a number of other substances also reduce silver under the prescribed conditions. He names melanin, pheochrome substance, enterochromaffin substance, pancreatic alpha granules, and the neurosecretory granules of the supraoptic and paraventricular nuclei of the thalamus.

Lillie and Pizzolato (*J. Histochem. Cytochem.*, **16**: 17, 1968) reported that these same acetic acid–silver fixations gave very complete demonstration of bile pigments in the icteric human liver.

Langeron cites Massa's (*Soc. Pharm. Montpelier*, **5**: 14, 1945) use of the ferric ferricyanide reduction test for the demonstration of ascorbic acid in plant tissues. Massa made up his potassium ferricyanide at 0.2% and his ferric chloride at 3.24%, both in 15%

acetic acid, and mixed equal volumes at the time of using. The reaction time should be restricted to 5 to 10 min. Because of the solubility and diffusibility of ascorbic acid, only fresh tissue can be used.

Ascorbic acid gave immediate blue coloration; phenols, tannoids, anthocyanins, and flavones gave greens or green-black precipitates. He noted the reaction of glutathione and cysteine and used the nitroprusside reaction as a control. The reaction of carotenoids, tocopherols, and vitamin A was noted as much slower than that of ascorbic acid.

The method should be applicable to Adamstone-Taylor or Linderström-Lang frozen sections of fresh animal tissue, transferred directly into the freshly mixed reagent from the microtome. Even brief drying should be avoided.

PNEUMONYSSUS PIGMENT

Sometimes difficult to distinguish from carbon is the pigment deposited in monkey lungs about cysts and remnants of the acarid parasite *Pneumonyssus foxi*. This includes particles of a rather deep brown which stain practically black with azures. However, examination of unstained preparations reveals angular black particles, brown granules, and doubly refractile needles.

EXOGENOUS PIGMENTS

Hueck lists the following, classified according to color. We have abbreviated, translated, and supplemented his list, chiefly from the mineral pigments in the "Colour Index" and from the colored ores noted in Lange. Not all these have been reported as pigments in tissues.

Black: Carbon as soot, coal, and graphite. Coal occurs as irregularly angular and jagged particles. Graphite crystallizes as hexagonal crystals.

Brownish, greenish, and grayish black: Oxides and sulfides of various metals, not specified by Hueck, but including iron, cobalt, nickel, lead, silver, copper, antimony, chromium, gold, iridium, manganese, mercury, molybdenum (MoS_2), palladium, platinum, rhodium, tin, tantalum, thallium, thorium, tungsten, uranium, and vanadium.

Brown: Bismarck brown used in tattooing; manganite, umber, and cupric ferrocyanide.

Brownish red: Iron and copper compounds, notably iron oxide.

Red: Cinnabar, carmine, and other dyes used in tattooing; alizarin and madder; to which we might add the native arsenic sulfides realgar and orpiment, the iron ore hematite (Fe_2O_3), and mercuric oxide.

Blue: Vivianite [$Fe_3(PO_4)_2 \cdot 8H_2O$], ultramarine (a mixed silicate and sulfide of aluminum and sodium), steatite (a colored talc), azurite [$Cu(OH)_2 \cdot CuCO_3$], azure blue ($CoAl_2O_4$), smalt (potassium cobaltous silicates), copper blue (CuS), blue dyes used in tattooing, etc.

Green: Casalis green (Cr_2O_3) and its hydrates; Kinmann's green and related products resulting from the fusion of varying proportions of zinc and cobaltous oxides; turquoise green (similarly made from chromium, cobaltous, and aluminum oxides); verdigris (basic copper acetates); Schweinfurt green (cupric acetoarsenite); Scheele's green (copper arsenite); and ultramarine green (a sodium aluminum silicate and sulfide); also green dyestuffs used in tattooing.

Yellow: Chromates of lead, barium, and zinc; cadmium and stannic sulfides; Naples yellow [$Pb_3(SbO_4)_2$].

Violet: Chiefly dyestuffs used in tattooing.

Gray: Chiefly various silicates.

White: Various lead, zinc, and bismuth pigments, as well as barite ($BaSO_4$) and titanium oxide (TiO_2).

CARBON

Carbon is one of the commonest extrinsic materials appearing as a pigment. Carbon is commonly deposited in the lungs and mediastinal lymph nodes, but may appear in axillary nodes as well, and sometimes in the skin as a result of tattooing or sterilization of hypodermic needles in sooty flames. It is distinguished by its black color and resistance to all solvents and bleaching agents.

The statements that carbon is insoluble in acids, alkalies, and nonpolar solvents and that it is not bleached by oxidizing agents should be considered in the light of Table 11-3 on melanin bleaching.

Metallic silver is readily removed by a 20-min exposure to 5% potassium cyanide solution or to a Gram-Weigert iodine–potassium iodide (1 : 2 : 100) solution followed by thiosulfate to remove the excess iodine. Most metallic sulfides are soluble in nitric acid even in moderate dilution.

HEMATOXYLIN AS A REAGENT FOR METALS

Under this head we do not refer to color changes with routine alum hematoxylin stains such as may be observed in calcified deposits, but to the use of pure hematoxylin in the absence of chelating reagents in the test solution. In a sense the reaction goes back to Böhmer (*Aertzl. Intell. Zentralbl. Muenchen.*, **12:** 539, 1865), who described colors produced in dichromate- and copper sulfate–fixed tissues, but the use of hematoxylin for the histochemical localization of tissue iron was introduced by Macallum (*J. Physiol.*, **22:** 92, 1897). This was reintroduced for iron and copper, and later lead by Mallory (1938). Lillie (*J. Histochem. Cytochem.*, **4:** 318, 1956) called attention to metal uptake by keratohyalin, reporting intense color reactions with Fe(III), Fe(II), Cr(VI), Cr(III), Cu(II), Sn(II), and Zn(II). Pizzolato and Lillie (*J. Histochem. Cytochem.*, **15:** 104, 1967) later made a comprehensive study of uptake of metals by human plantar skin keratohyalin. They reported dark blue to black with Al^{3+}, Cr^{4+}, Cr^{3+}, Ga, Hf, In, Fe^{3+}, Fe^{2+}, and Zr; blue, Be, Dy, Ho, Ir, Pb, Mn, Mo, Nd, Ni, Pt, Rh, Tb, U, Yb, and Zn; greenish blue, Cu; brown, Nb and Ti; brownish red, Ta; purplish red, Sn and Th; purple, Bi; and blue or greenish brown, Os. Lillie (*Histochemie*, **20:** 338, 1969) later studied the pH range of the hematoxylin solution required for a few metals. Cr(IV), Fe(III), and Sn(II) reacted at pH 2.5 to 7. Fe(II) reacted at pH 4 to 7, Cu at 5 to 7, Cr(III) at 6 to 7. This suggests that further exploration of pH ranges of reaction might make the test a useful screening procedure.

For cytosiderin the reaction seems more sensitive than the ferrocyanide reaction. Enterosiderin behaves similarly, but hemosiderin may completely fail to react when copiously present by the hydrochloric acid–ferrocyanide method, while reacting excellently in other cases with buffered neutral hematoxylin. Gomori's (1952, p. 39) suggestion that pretreatment with acid is necessary does not seem to account for the failure.

For this purpose 0.1 to 0.01% hematoxylin is diluted from a fresh 10% alcoholic stock solution (100 mg/ml absolute alcohol) with a sodium phosphate or acetate buffer of the desired pH at 0.1 to 0.01 M concentration made in double- or triple-glass-distilled water. Deionized water will serve but is less reliable for critical work. Ordinarily 1- to 2-h reaction times at 20 to 25°C are sufficient. For trace amounts we have used 0.01% for 24 to 48 h.

Avoid water washing; postchelation of adventitiously introduced metals is possible; dehydrate directly with 95% and absolute alcohol, clear in xylene, and mount preferably in a low-index-of-refraction resin such as cellulose tricaprate. This should leave the virtually unstained background visible with much recognizable morphologic detail.

IRON

Various ores may be black (magnetite, hematite), blue (vivianite), green (siderite), gray (siderite), red (hematite), and brown or yellow (siderite, goethite). Some react in acid solutions to ferrocyanides, others to ferricyanides, some to both. Iron ore dusts may be removed by treatment with oxalic or dilute nitric acids. Generally, however, petrographic examination may be necessary to identify precisely the dusts in question. See also below.

Application of specific extraction procedures before application of the hematoxylin reaction may aid in identification of the specific metal. Six-hour extractions with 0.5 M potassium oxalate or 1 M oxalic acid, with 1 to 2% fresh sodium dithionite, remove ionically reactive iron; prolonged extraction, up to several days in 1 N acetic acid, is resisted (basic ferric acetate is insoluble in water; iron forms soluble chelates with oxalates and dithionites). Iron sulfide, however, is reduced to insoluble ferrous sulfide by the dithionite reagent and becomes reactive only to ferricyanide, not ferrocyanide–hydrochloric acid mixtures. From Lange's data Fe(III) appears to be the only metal with both a soluble oxalate and an insoluble basic acetate. Neutral ferric acetate is very soluble in water.

A modified Clara hematoxylin, 0.1% in various buffers, has been used by Pizzolato and Lillie (*J. Histochem. Cytochem.*, **15**: 104, 1967) for the recognition of various metal ions in tissue as taken up on keratohyalin and other tissue elements, and at graded pH levels for separation of the action of Cr(IV) and Cr(III) in the study of the mechanism of the chromation hematoxylin (Smith-Dietrich) reaction (Lillie, *Histochemie*, **20**: 338, 1969) and Fe(II) and Fe(III), Cu, and Sn(II) (ibid.). Dark blues are produced by Al, Cr(III), Ga, Hf, In, Zr, and Fe(II) and Fe(III) at pH 7; Be, Dy, Ho, Ir, Pb, Mn, Mo, Nd, Ni, Pt, Rh, Tb, U, Yb, and Zn gave lighter blues; Bi gave purple; Ta, brownish red; Nb and Ti, brown; Cu, greenish blue; Sn and Th, purplish red; Os, blue to greenish brown. In these tests metals taken up by keratohyalin of human plantar skin resisted a 72-h distilled water extraction. Only limited studies have been made on the pH range of hematoxylin binding to metals. Lillie (*Histochemie*, **20**: 338, 1969) found that Cr(IV) bound from CrO_3 and $K_2Cr_2O_7$ solutions colored blue-black at pH 2.5 to 7, while Cr(III) from $CrAc_3$ or chrome alum bound at pH 6 to 7, Fe(III) at 2.5 to 7, Fe(II) at 5 to 7, Cu(II) at 6 to 7, and Sn(II) at 2.5 to 7. The usual acetate, phosphate, and barbital buffers are used. An 0.1% hematoxylin in 2 N acetic acid reads pH 2.35; in 1 N, 2.51; in 0.33 N (2%), 2.85; in 0.17 N (1%), 3.50. These figures are a little higher than those given by the solvent alone.

Kutlik [*Acta Histochem. (Jena)*, **37**: 259, 1970] uses a chlorate hematoxylin (1 g hematoxylin in 100 ml 7.3% $KClO_3$) as a reagent for tissue iron. The chocolate brown solution is filtered after each use and is good for 6 months. Sections of formol-fixed material are brought to water, stained for 2 h at 20 to 22°C, washed 10 min in running water, counterstained if desired a few seconds in 0.5% aqueous rose Bengal, rinsed in water, and dehydrated, cleared, and mounted through the alcohol-xylene series. Iron pigments are black with perhaps blue, violet, or brown tinges. Other pigments such as melanins, lipofuscins, and porphyrins are negative; copper gives a green-blue. Cobalt and nickel do not react; lead gives shades of violet. The method is somewhat less sensitive than the Prussian blue reaction, but photographs better.

Highman (*Bull. Int. Assoc. Med. Mus.*, **32**: 97, 1951) finds that some of the less soluble iron ore dusts (notably hematite, Fe_2O_3) fail to react to the usual ferrocyanide test. He notes that by increasing the concentration of the hydrochloric acid to 4 N or to even higher concentration (11 N), and heating to 60 to 80°C, positive Prussian blue reactions may be obtained even with the more refractory ores. For the Quincke and Tirmann-Schmelzer reactions a prolonged exposure (1 to 2 days) to saturated hydrogen sulfide water is substituted for the usual ammonium sulfide step.

Fenton et al. (*J. Histochem. Cytochem.*, **12**: 53, 1964) combined the use of micro-incineration and the Prussian blue reaction for a more sensitive histochemical demonstration of iron. They prescribe the use of three serial sections: one stained with hematoxylin and eosin, another stained with the usual Prussian blue reaction, and the third a paraffin-embedded section subjected to microincineration. (If a microincinerator is not available, the oxidizing area of a torch may be applied to a section on a slide on an asbestos support.) After incineration, the slide must be cooled slowly; otherwise it will crack.

The incinerated slide may be examined by dark-field illumination. Iron oxide appears yellow-orange to dark red. Be sure incineration is complete; remaining carbon also appears yellowish to red. By transmitted light carbon will appear black and iron will retain a reddish hue. The incinerated section may also be stained with the usual Prussian blue method provided the section is first collodionized; otherwise the ash is lost.

Snitzer et al. (*Tech. Bull. Med. Technol.*, **34**: 194, 1964) note that iron in tap water used to float sections can be adsorbed into tissue sections and give a positive iron stain.

These combinations of ashing and the Prussian blue reaction recall Virchow's use of the same procedures to demonstrate the iron of hemosiderin (*Virchows Arch. Pathol. Anat.*, **1**: 379, 1847).

ALUMINUM

Aluminum has been demonstrated after aluminum dust therapy of silicosis as globules and mulberrylike aggregates of dark red material lying in the center of the fibrous nodules by the aurine technic of P. C. Irwin (personal communication, 1948).

Fix in formalin, cut frozen or paraffin sections, and bring to water as usual.

1. Stain 5 min at 75°C in 2% aurine tricarboxylic acid (NH_4 or Na salt) in a pH 5.2 buffer composed of 3.8 parts 5 M (267.5 g/l) ammonium chloride, 3.8 parts 5 M (385.3 g/l) ammonium acetate, and 1 part 6 N (500 ml/l) hydrochloric acid.
2. Rinse for a few seconds in cool distilled water.
3. Decolorize 3 s in pH 7.2 buffer composed of 3.6 parts of above buffer and 10 parts 1.6 M (15.36%) ammonium carbonate.
4. Rinse quickly in distilled water.
5. Counterstain briefly in saturated aqueous picric acid solution or in 1 : 10,000 methylene blue.
6. Dehydrate, clear, and mount by alcohol, xylene, balsam (or synthetic resin) sequence.

Pearse [*Acta Histochem. (Jena)*, **4**: 95, 1957] has used a dye introduced as chromoxane pure blue B, C.I. 43830, and usually designated as pure blue B with various trade name prefixes, under the ICI trade name[1] Solochrome Azurine BS, for the demonstration of aluminum and beryllium in tissue.

A 0.2% solution of pure blue B is used, in distilled water for aluminum and beryllium, in 1 N NaOH for beryllium alone, the aluminum being dissolved at this pH level.

---

[1] ICI, Imperial Chemical Industries, Ltd.

Deparaffinize (remove mercury precipitate) and hydrate as usual. Stain 15 to 20 min, rinse in distilled water, dehydrate in alcohols, clear in xylene, and mount in synthetic resin.

Beryllium colors almost black in the alkaline solution; in the neutral bath, aluminum and beryllium are blue. (The "Colour Index" notes a quite stable dark blue barium lake used in printing.)

Alcoholic morin gives an intense green fluorescence in ultraviolet light tests. The fluorescence, after formation, is stable in 2 N hydrochloric acid, which dissolves out the morin compounds of beryllium, indium, gallium, thorium, and scandium. Although the aluminum salt is formed only in acetic acid or neutral conditions, the similar green fluorescence of *zirconium* is formed even in strong hydrochloric acid.

Feigl[1] used 0.001% morin in alcohol, 1 drop plus 1 drop test solution plus 5 drops concentrated hydrochloric acid, and he detected 0.1 μg Zr (1 : 500,000). For aluminum he used 1 drop saturated morin in methanol, 1 drop test solution, 1 drop 2 N acetic acid, detecting 0.2 μg (1 : 250,000).

ASBESTOS

Asbestos occurs in tissues as fine, white, doubly refractile fibers and as the so-called asbestosis bodies. The latter occur as beaded rods with large rounded ends, as fusiform bodies with beaded centers, as rosettes, and in other similar forms. They are golden yellow and give the Prussian blue reaction for ferric iron. Also some of the fine asbestos fibers are colored blue by acidulated potassium ferrocyanide solution. The asbestosis bodies and the fine fibers giving the Prussian blue reaction are dark under crossed Nicol prisms. (J. W. Miller, *Public Health Bull.* 241, pp. 96–101, GPO, 1938.)

TITANIUM

One drop (0.025 ml) saturated methanol solution of morin gives an intense brown spot with 1 drop 0.5 N hydrochloric acid solution containing titanium salts. The sensitivity is 10 ng or 1 : 5,000,000 (Feigl,[1] IST). Since titanium oxide is now being used extensively in white paints, the problem of its identification may arise. The morin test seems adaptable to histochemical use.

BERYLLIUM

Denz's (*Q. J. Microsc. Sci.*, **90**: 317, 1949) test for reactive beryllium compounds uses naphthochrome green G (CIBA) (C.I. 44530)[2] described as phenooxydinaphthofuchsondicarboxylate sodium. The dye forms a green beryllium lake, optimally at pH 5. The dark blue-green ferric lake and the yellowish green aluminum lakes are formed only at relatively high metal concentrations at this pH level.

DENZ'S BERYLLIUM METHOD

1. Fix tissues in formol–saline solution or alcoholic formalin. Dehydrate with alcohols, clear in xylene or cedar oil, embed in paraffin, section, deparaffinize with xylene, and hydrate through descending alcohols.

---

[1] References to Feigl's spot tests are abbreviated IST and OST for inorganic and organic volumes, respectively.

[2] According to the 1971 "Colour Index" there is now no known manufacturer of this dye.

2. Mix equal volumes of a phosphate buffer of pH 5 [*sic*] and of a freshly prepared 0.5 %
    aqueous solution of naphthochrome green B or G. Place slides in this mixture in a Coplin
    jar, and incubate 30 min at 37°C.
3. Wash in distilled water.
4. Differentiate 30 min in 100 % alcohol.
5. Wash in distilled water.
6. Counterstain 5 min in 1 % aqueous acridine red.
7. Wash in distilled water.
8. Differentiate rapidly in 100 % alcohol.
9. Clear in xylene and mount in Canada balsam.

RESULTS: Beryllium compounds are apple green; background is red.

The method demonstrates protein combinations of soluble beryllium salts, which
apparently remain in place for fairly long periods. Beryllium oxide and silicate dusts in
tissues fail to react.

The acridine red referred to may be acridine red 3B (C.I. 45000), which is a basic dye of
the pyronin class. Probably one of the latter could be substituted.

Beryllium gives a specific yellow-green fluorescence in ultraviolet when treated with 3
drops Versene solution plus 1 drop 0.02 % morin in acetone plus 1 drop concentrated
ammonia. The precipitate is washed successively with Versene, water, and acetone. The test
is sensitive to 70 ng beryllium and is selective in the presence of 200 parts aluminum, iron,
magnesium, or calcium. This spot test method appears readily adaptable to histochemical use.

Wyatt (*Stain Technol.*, **47**: 33, 1972) reports on the use of chromoxane pure blue
BLD (C.I. 43825) and chromoxane pure blue B (C.I. 43830). He gives specific technics for
beryllium in tissue sections. Reaction of the other metals is excluded by inclusion of
disodium ethylenediamine tetraacetic acid in the reagents. The first dye is used in an
acetate buffer at pH 4; the second is adjusted to pH 9. Both yield blues on a pink to
unstained background.

THE CHROMOXANE PURE BLUE BLD TECHNIC

1. Bring paraffin sections through xylene and alcohols to water.
2. Stain 2 h in 5 ml 1 % dye plus 9 ml 1 N hydrochloric acid plus 10 ml 1 N sodium
    acetate plus 27 ml distilled water plus 4.14 g Versene. Mix buffer, dissolve Versene just
    before using, then add dye solution.
3. Rinse in distilled water.
4. Immerse in 2 ml 28 % ammonia and 38 ml distilled water for 2 min.
5. Wash for 3 min in running water; rinse in distilled water.
6. Counterstain with 0.1 % nuclear fast red in 5 % ammonium oxalate.
7. Rinse in distilled water.
8. Stain for 1 to 2 min in 1 % naphthol yellow S.
9. Rinse, dehydrate, clear, and mount in DPX, polystyrene, or neutral resin.

RESULTS: Red nuclei, yellow cytoplasm, and blue beryllium are demonstrated.

THE CHROMOXANE PURE BLUE B TECHNIC. Dissolve 9 % EDTA (Na$_2$) in
1.0 N NaOH adjusted to pH 9 with 5 N HCl. To 45 ml freshly prepared pH 9 EDTA
solution add 5 ml 1 % dye. Bring sections to water, stain 1 h, rinse in distilled water,
counterstain 5 to 10 s in aqueous neutral red, rinse in distilled water, dehydrate, clear,
and mount in resin. When EDTA is omitted, Be, Cr, Cu, Gd, Ni, and Y are blue with
both dyes; Ba, Fe(II), and Fe(III) give blue with BLD only; Au, Mn, and Pb give blue
with the blue B only. Al gives pink with both; Bi, Li, Mo, and Zn give pink with blue B;

Mg is purple with B; Sn, green with BLD, Ag, As, Cs, Ca, K, Na, Sb, Ti, U, and W give no colors with either. EDTA inhibits all reactions except that of Be.

Beryllium Oxide

Wyatt (*Stain Technol.*, **47:** 33, 1972) states the following stain is relatively specific for beryllium oxide:

1. Deparaffinize formalin-fixed sections and bring them to water.
2. Stain for 1 h at 23°C in either chromoxane pure blue BLD (C.I. 43825) or chromoxane pure blue B (C.I. 43830), 1% in 1 M acetate–HCl buffer pH 9 containing 9% EDTA.
3. Rinse in distilled water.
4. Immerse sections in a solution of a 1 : 20 dilution of NH_4OH (28% NH_3) for 1 to 2 min until sections are colorless.
5. Rinse in distilled water.
6. Counterstain in nuclear fast red (0.1 g in 5% aqueous ammonium sulfate) for 3 min at 23°C.
7. Rinse in distilled water.
8. Stain for 1 to 2 min in 1% aqueous solution of naphthol yellow S (C.I. 10316).
9. Rinse in distilled water, dehydrate, and mount in synthetic medium.

RESULTS: Nuclei are red, cytoplasm is yellow, beryllium oxide is blue. Wyatt notes that omission of EDTA results in blue staining of FeO, Cu_2O, and $CaCl_2$ as well as beryllium oxide. A 1% neutral red solution may also be used as a counterstain.

BARIUM AND STRONTIUM

According to Waterhouse's [*Nature (Lond.)*, **167:** 358, 1951] method first fix tissue in neutral 10% formalin in 70% alcohol. Then soak the blocks for 30 to 60 min in a freshly prepared 1 to 2% solution of sodium rhodizonate in distilled water or pH 7 phosphate buffer. Then wash out excess rhodizonate in 50% alcohol. Tissues can then be sectioned as usual, and permanent preparations may be obtained. Barium and strontium compounds both give intensely red reaction products. Treatment with aqueous potassium chromate solution before the rhodizonate prevents formation of the color with barium salts, and chromate treatment after coloration removes the color from barium but not from strontium.

McGee-Russell (*J. Histochem. Cytochem.*, **6:** 22, 1958) fixes, embeds in paraffin, and sections at 8 to 10 μ, deparaffinizes, and brings through descending alcohols to 50%. Then rinse quickly in distilled water and immerse section in a saturated solution of sodium rhodizonate for 1 h or longer. Examine in distilled water. Calcium deposits containing Sr or Ba color reddish orange; background is unstained. Spectroscopically pure calcium salts do not react. If desired, sections may be counterstained 5 to 10 s in 0.5% toluidine blue, dehydrated, and cleared through acetone and xylene and mounted in a resinous mount.

McGee-Russell finds the K_2CrO_4 tests above equally applicable to sections; he notes further that dilute HCl removes the Sr rhodizonate but not the Ba salt.

CALCIUM

The procedure of feeding madder or injecting alizarin compounds to mark in red newly deposited calcium salts during life, and the recently observed intravital fluorochrome staining with tetracycline are considered in Chap. 18.

Allied to these procedures are the chelate staining with alizarin red S, purpurin, nuclear fast red, chloranilic acid, and similar mordant dyestuffs.

ALIZARIN RED S.   Langeron prescribed 80 to 90% alcohol fixation. We have found phosphate-buffered 10% formalin satisfactory. *Avoid calcium in fixatives.* Stain 1 h in 0.1% aqueous alizarin red S, rinse, and counterstain briefly in 0.1% thionin, azure A, toluidine blue, or methylene blue, rinse briefly in 1% acetic acid, dehydrate with acetone, clear in xylene, and mount in synthetic resin. Dahl (*Proc. Soc. Exp. Biol. Med.*, **80**: 479, 1952) added 1 : 1000 volume ammonia water to 1% alizarin red S to give pH 6.4, stained alcohol-fixed paraffin sections 2 min, washed with distilled water for 5 to 10 s, dehydrated in alcohols, cleared in xylene, and mounted in cedar oil. McGee-Russell's 1958 variation of these technics seems superior and is cited in detail.

McGEE-RUSSELL ALIZARIN RED S PROCEDURE FOR CALCIUM (*J. Histochem. Cytochem.*, **6**: 22, 1958).   Make a 2% aqueous solution of alizarin red S (C.I. 58005) and adjust to pH 4.2 with dilute ammonia. Fix material in alcoholic formalin. Bring paraffin sections through xylene and alcohols to 50% alcohol, rinse quickly in distilled water, cover section with the alizarin red S, and watch under microscope until red calcium lake forms over deposits (30 s to 5 min). Drain, blot with filter paper, dehydrate quickly (10 to 20 s) with acetone, acetone plus xylene, xylenes, balsam, or synthetic resin. Ca^{2+} deposits are orange-red, except oxalate. McGee-Russell did not consider sulfate or fluoride. The former should probably react; the latter probably will not.

The purpose of microscopic control is the avoidance of diffusion artifacts. Calcium deposits are orange-red on a colorless background. Toluidine blue counterstains tend to react with the deposits and are not recommended.

Alizarin red S forms a crimson lake with calcium, and scarlet lakes with aluminum and with barium. Magnesium gives a clear scarlet solution; mercury, a clear dark red solution.

Meloan et al. (*Arch. Pathol.*, **93**: 190, 1972) noted that metastatic calcifications are preserved very well in tissues fixed in 100% ethanol, Carnoy's, and methanol–chloroform–acetic acid (6 : 3 : 1), and that deposits stained optimally with alizarin red S at pH 8.8 to 9.4.

Puchtler et al. (*J. Histochem. Cytochem.*, **17**: 110, 1969) studied arteriosclerotic lesions and noted that alizarin (C.I. 58000) stained calcium selectively and intensively around pH 12, that alizarin red S (C.I. 58005) stained calcium selectively around pH 9, and that substantial diffusion of these stains occurs at neutral and acid pH levels.

PURPURIN OF GRANDIS AND MAININI.   Schmorl recommended a similar method employing purpurin (C.I. 58205) or anthrapurpurin (alizarin SX) (C.I. 58255) in saturated alcoholic solution (they are respectively very sparingly soluble in water and insoluble in water). He prescribed fixation with strong alcohol or neutral 10% formalin.

1. Paraffin sections are deparaffinized and brought through 100% alcohol into the saturated alcoholic stain for 5 to 10 min. The anthrapurpurin solution should contain also 1% sodium chloride and a trace of ammonia water.
2. Wash in 0.75% aqueous sodium chloride solution for 3 to 5 min.
3. Wash thoroughly in 70% alcohol until no more color clouds come out of sections.
4. Dehydrate, clear, and mount as usual.

RESULTS: Calcium deposits are stained red.

It is probable that sodium anthrapurpurin monosulfonate (C.I. 58260), which is water-soluble, can be substituted as an aqueous solution in this technic.

KERNECHTROT,[1] NUCLEAR FAST RED, CALCIUM RED (C.I. 60760), OR HELIOECHTRUBIN BBL.   This is an aminoanthraquinone sodium sulfonate which gives a red precipitate with soluble calcium salts. McGee-Russell (*J. Histochem. Cytochem.*, **6**: 22,

---

[1] Lillie has adopted the first two terms as the preferred common names for this dye in the next (ninth) edition of "Conn's Biological Stains" following prevailing usage, although the original name was "Helioechtrubin BBL" in the FIAT list.

1958) reports that lakes are formed with Pb^{2+}, Fe^{3+}, Cu^{2+}, K^+, Sn^{4+}, and Sr^{2+}, as well as with Ca^{2+}, but not with Ba^{2+} or Mg^{2+}. In tissue, selective staining of $CaCO_3$ and $-PO_4$ deposits is achieved, but oxalate remains unstained.

McGee-Russell directs purifying a sample of the dye by first washing in three small washes of distilled water and then making a saturated solution of the residue (about 0.25%) in distilled water. The pH should be about 4.9.

TECHNIC. The technic is simple. Bring paraffin, ester wax, celloidin, gelatin, or frozen sections (2 to 10 μ) to distilled water. Flood sections on slides with saturated aqueous kernechtrot. Stain with microscopic control until deposits are deep red (5 to 55 min). Drain and blot, dehydrate with acetone 20 s (2 to 3 changes or dropper bottle), acetone plus xylene, xylene, balsam, or other suitable resin.

Eisenstein et al. (*J. Histochem. Cytochem.*, **9**: 154, 1961) report a method producing yellow-brown rhomboid and needle-shaped microcrystals of calcium chloranilate at Ca^{2+} sites. Fe^{2+}, Fe^{3+}, and Mg^{2+} do not precipitate with neutral Na chloranilate. In large amounts Cu^{2+} reacts, and small amounts of heavy metals give more macrocrystalline deposits.

EISENSTEIN'S CHLORANILATE METHOD FOR CALCIUM. Dissolve 4 g sodium hydroxide in 600 ml distilled water. Add 11 g chloranilic acid (2,5-dichloro-3,6-dihydroxy-*p*-benzoquinone) (Eastman 4539, mw 209), and bring total volume to 1000 ml. Shake to dissolve. Adjust pH to 7 with additional chloranilic acid if necessary. Store in refrigerator. Filter before using. The reagent is stable for some 6 months.

Deparaffinize and hydrate sections as usual; wash in distilled water. Immerse sections for 1 h with constant vigorous agitation. Wash in two 30-min changes of 50% isopropyl alcohol. Counterstain with Giemsa or other suitable stain, dehydrate in alcohols, clear in xylene, and mount in balsam or synthetic resin.

The fine yellow-brown crystals are readily identified and are brilliantly birefringent under polarized light. Iron deposits do not react. Crystal size is such that very precise cytologic localization is not accomplished. The function of the agitation during the reaction is to reduce crystal size.

VON KÓSSA'S METHOD (*Beitr. Pathol.*, **29**: 163, 1901). Widely used for demonstration of calcification is the von Kóssa silver nitrate method, which actually demonstrates the presence of phosphates, soaps, and amorphous but not crystalline carbonates, rather than calcium itself.

McGee-Russell (*J. Histochem. Cytochem.*, **6**: 22, 1958) notes also the reaction of calcium oxalate, and Cogan et al. (*J. Histochem. Cytochem.*, **6**: 142, 1958) report intense blackening over gypsum ($CaSO_4 \cdot 2H_2O$) crystals. Pizzolato (*J. Histochem. Cytochem.*, **12**: 333, 1964) cites negative reactions of experimentally produced deposits of CaC_2O_4, CaF_2, and gypsum.

Alcohol fixation is often recommended for calcium methods, but neutral 10% formalin works well in practice. Calcium carbonate, calcium acetate, and calcium chloride formulas are, of course, to be avoided, because of the possibility of false-positive reactions.

1. Wash in several changes of distilled water.
2. Immerse in 5% silver nitrate for 10 to 60 min (von Kóssa: 5 min) and expose to bright daylight (Mallory, Romeis, Langeron), not direct sunlight. Schmorl used 1 to 5% silver nitrate for 30 to 60 min; Cowdry, 10% for 30 min or more.
3. Wash well in distilled water.
4. Treat for 2 to 3 min in 5% aqueous sodium thiosulfate solution.
5. Wash in water.
6. Counterstain 20 to 60 s in 0.5 to 0.1% aqueous safranin O.
7. Differentiate and dehydrate in 95 and 100% alcohol.

8. Clear with 100% alcohol and xylene and two changes of xylene. Mount in synthetic resin.

RESULTS: Calcium deposits are black; nuclei, red; other tissues, pink.

In place of safranin other counterstains may be used. In case of hard calcium deposits, ragged sections may require collodion treatment. With larger deposits it may be better to use the procedure of silvering before decalcification.

It should be recalled that melanin may also become black or very dark brown by this technic. Although ascorbic acid should also react, this material is not preserved in sections except with very special precautions.

The most widely used criterion for the detection of small calcareous deposits is their staining deep blue with alum hematoxylin or gray-violet with iron hematoxylin. This is not specific but is very useful.

Some calcareous deposits are encrusted or admixed with ferric salts, which may be detected by the usual ferrocyanide reaction.

Barker et al. (*Johns Hopkins Med. J.*, **127**: 2, 1970) reported that calcium phosphate as ordinarily prepared in the laboratory fails to stain with the von Kóssa method, whereas calcium phosphate precipitated in the presence of citric acid stains avidly with the von Kóssa method. Bills et al. (*Johns Hopkins Med. J.*, **128**: 194, 1971) precipitated calcium phosphate in the presence of several organic acids and concluded that the von Kóssa silver staining reaction indicates calcium phosphate molecularly integrated with certain organic acids. Typical reactive organic acids were open-chain compounds having at least one hydroxyl group and two or more carboxyls.

Heeley and Irving (*Calcif. Tissue Res.*, **12**: 169, 1972) employed ^{45}Ca and ^{32}P in a combined autoradiographic histochemical study in an attempt to determine which methods could first detect calcified tissue. ^{45}Ca labeling was detected simultaneously with a positive staining reaction with the GBHA method of Kashiwa and House (*Stain Technol.*, **39**: 359, 1964). ^{32}P detection and a positive stain with the von Kóssa method appeared within 6 h (in the heads of rat fetuses labeled by an IM injection of the mother on the fourteenth day of pregnancy).

Kaufman and Adams (*Lab. Invest.*, **6**: 275, 1957) and Hill (*Z. Wiss. Mikrosk.*, **68**: 65, 1967) advise the use of murexide as a selective stain for calcium in tissue sections. Fresh frozen or formalin- or ethanol-fixed paraffin-embedded sections are deparaffinized and brought to water. Sections are then exposed consecutively to 6.5% KCN for 1 to 3 min, 0.1% aqueous murexide for 2 to 15 min, rinsed, counterstained in hematoxylin, if desired, dehydrated, and mounted in synthetic medium. Sites of calcium are stained blue-purple. Hill states that the stained preparations are permanent.

Kashiwa and Atkinson (*J. Histochem. Cytochem.*, **11**: 258, 1963), Kashiwa and House (*Stain Technol.*, **39**: 359, 1964), and Kashiwa (*Stain Technol.*, **41**: 49, 1966) have developed several modifications of a stain which they believe identifies ionic calcium. They caution that strict observance of the prescribed method must be followed if accurate localization is to be achieved.

1. Immerse thin slices (1 mm) of fresh tissues overnight in a solution containing 5% glyoxal bis(2-hydroxyanil) (GBHA)[1] and 3.4% NaOH in 75% ethanol. The GBHA must be added immediately before use. Tissues are then rinsed, dehydrated in graded ethanols and xylene, embedded in paraffin, and sectioned.

---

[1] Fisher Scientific Company, St. Louis, Mo.

2. *Without removal of paraffin*, sections are immersed for 15 min at 23°C in a saturated solution of KCN and Na_2CO_3 in 90% ethanol.
3. Rinse in two changes of 95% ethanol.
4. Dehydrate and mount in synthetic resin.

RESULTS: Ionized calcium is stained red; calcium in apatite is unstained. Kashiwa states if the GBHA solution is applied for 5 min to tissue sections (fresh, paraffin-embedded, or previously treated with GBHA in the whole block), then insoluble calcium salts are also stained.

In control experiments, red-stained GBHA granules could not be obtained in specimens subjected to EDTA decalcification. Kashiwa has recently written a review article on this procedure (*Clin. Orthop.*, **70**: 200, 1970).

CRETIN'S REACTION.   Lison regarded this reaction as very sensitive and highly specific. Gomori agreed as to specificity but found the reaction so capricious as to be almost useless. Lillie always regarded it as too complicated for practical use and did not try it; it was given in detail in the second edition of this book (Lillie, 1954).

SPECIFIC CHEMICAL TESTS.   The most specific chemical tests are those with sulfuric and oxalic acids. In the unstained state the calcium deposits are granular, opaque, and white. By mounting the sections in water and running in acetic acid under one edge of the cover glass while drawing water out with a piece of filter paper from the other, the deposits may be seen to dissolve; if carbonates, with the formation of gas bubbles; if phosphate only, without gas bubbles. If sulfuric acid is used in place of acetic acid, the deposits dissolve as before, but monoclinic gypsum crystals are formed; but if 5% or 10% oxalic acid is used, the characteristic cubic calcium oxalate crystals appear.

Although it seems to be generally accepted that calcium oxalate deposits do not react to the organic colored chelate reactions for Ca^{2+}, reports on the von Kóssa reaction of oxalates and sulfates are conflicting. Silver oxalate is fairly insoluble, though less so than the calcium salt, and silver sulfate is moderately soluble (Lange: about 0.8%, 22°C). Hence pure deposits of calcium oxalate and of gypsum crystals could well be unreactive to the von Kóssa test. Although x-ray diffraction studies have assured the identification of the two crystal forms, we do not believe that they afford assurance of the absence of carbonate, phosphate, or other reactive salt. Since it is well known that $1 N$ solutions of acetic and formic acid are efficient decalcifying agents for bone and other calcified tissue and that calcium oxalate is promptly precipitated from these solutions, though not from hydrochloric acid above about 0.5 N, it would appear that if the positive von Kóssa reaction of a given oxalate deposit is due to the presence of contaminating carbonate or phosphate, for example, a 24-h extraction in 1 N acetic acid should render a following von Kóssa test negative. Pizzolato's H_2O_2 von Kóssa variant could then be applied to demonstrate the calcium oxalate, as usual.

The microincineration methods for calcium oxalate are presented in Chap. 19.

Stimulated by these successes in conversion of calcium oxalate to carbonate by 450°C microincineration, with the resultant development of reactivity to usual calcium tests, Pizzolato was led to search for an oxidant which would convert oxalate to carbonate *in situ*, without solution of the product. The chemical method of titration in acid permanganate solution was naturally inapplicable. After a number of trials, strong hydrogen peroxide used in the presence of silver nitrate was found to yield the desired results (*J. Histochem. Cytochem.*, **12**: 333, 1964).

PIZZOLATO'S PEROXIDE-SILVER METHOD FOR CALCIUM OXALATE   Routine paraffin sections of formalin-fixed tissue are usable. It is preferable to use a minimal amount of albumen fixative, since this is apt to occasion excessive bubble formation on the slides.

1. Deparaffinize in xylene; hydrate through graded alcohols as usual.
2. Mix equal volumes of 30% hydrogen peroxide and 5% silver nitrate, allowing 2 ml for each slide. The pH of this mixture is about 6.
3. Lay slides face up on two glass rods over a glass bowl, and deposit on each slide 2 ml fresh peroxide-silver mixture, rocking slightly to spread the reagent over the section.
4. Expose to a 60-W tungsten-filament electric lamp or a 25-W fluorescent bulb at a distance of 15 cm (6 in) above the slides for 15 to 30 min. If excessive bubbling develops, pour off reagent and replace with fresh.
5. Wash thoroughly in distilled water, counterstain for 2 to 3 min in 0.1% safranin in 1 to 2% acetic acid if desired.
6. Dehydrate in two changes each of 95% and absolute alcohol and absolute alcohol plus xylene, clear in xylene, mount in Permount or cellulose caprate.

RESULTS: Nuclei are red; calcium oxalate is black; calcium fluoride and barium sulfate did not react. In vitro tests with calcium sulfate were also negative.

Silver and Price (*Stain Technol.*, **44**: 257, 1969) report that Pizzolato's method for calcium oxalate (*J. Histochem. Cytochem.*, **12**: 333, 1964) effectively stains calcium oxalate in plant tissues.

Yarom and Chandler (*J. Histochem. Cytochem.*, **22**: 147, 1974) used x-ray microanalysis with the analytic electron microscope (EMMA 4) for studies on the distribution of calcium in frog sartorius muscles. Significant amounts were found in dense parts of nuclei, the triads, thin filaments, and the sarcolemma.

The following revised technic of Lillie's has been applied to the study of calcified atheroma and the lipids therein:

Select four blocks of formol-fixed atherosclerotic artery containing partly calcified plaques. Immerse block 1 (control) for 7 days in distilled water, changing a few times on the first day. Wash blocks 2, 3, and 4 in six changes of distilled water, for 20 min each, and then place them in 1.7% (0.1 M) silver nitrate and keep them in the dark. Use fresh silver nitrate at 2, 5, and 7 days. Remove block 2 at 5 days, block 3 at 7 days, and block 4 at 10 days. Then wash the blocks in six changes of distilled water, for 20 min each, and transfer them to 2 g sodium bromide in 94 ml distilled water for 3 h. Then add 6 ml glacial acetic acid. Change this 2% NaBr–6% HAc decalcifying fluid daily, testing 1 ml of the discarded fluid with 1 ml 2% sodium oxalate. Decalcification is regarded as complete when no turbidity develops within 2 min with the oxalate. Then divide each block by a vertical cut through the center of the lesion. Use one half for paraffin sections, one for frozen sections.

Frozen sections are stained with oil red O–hemalum, with Nile blue, with Schultz and Lewis-Lobban cholesterol technics, and with such other lipid methods as may be desired. The paraffin sections may be used for the usual collagen and elastin technics, for oil red O and Sudan black stains for insoluble lipids, as well as routine hematoxylin and eosin (H & E).

Silver may be removed from any individual section by a 20-min exposure to Weigert's iodine followed by 20 min in 2% sodium thiosulfate or by 20 min in 2% sodium or potassium cyanide solution.

Otherwise calcified areas are black.

The method has been used at intervals over the last decade, and the above description reflects several minor revisions.

Dempsey et al. (*J. Histochem. Cytochem.*, **21**: 580, 1973) recorded x-ray emission spectra from several tissues using a multichannel energy-dispersive analyzer with a retractable semiconductor detector coupled to a Cambridge Mark II scanning microscope. He was able to detect calcium, chlorine, silver, and phosphorus in quantities that occur in tissues. Trace amounts of copper, zinc, lead, sodium, iron, arsenic, osmium, and uranium were also detected in various tissue specimens.

Frazier and Nylen (Fourth International Congress of Histochemistry and Cytochemistry, 1972, Kyoto, Japan) use the Komnick and Komnick method (detailed below under sodium) for the histochemical detection of calcium. Tissues are treated as usual with the antimony reagent. Complexes of the antimony reagent with calcium are soluble in a neutral EDTA extraction fluid, whereas complexes with the sodium reagent remain insoluble in the tissues. Positive identification of the calcium-antimonate complexes may be made using a modern electron probe.

SODIUM

Komnick and Komnick (*Z. Zellforsch. Mikrosk. Anat.*, **60**: 163, 1963), Kaye et al. (*Science*, **150**: 1167, 1965), and Kaye et al. (*J. Cell Biol.*, **30**: 237, 1966) conclude that sodium is identified at the ultrastructural level in tissues fixed in 1% OsO_4 with 2% $KSb(OH)_6$ in 0.01 N acetate buffer of pH 7.6. An electron-opaque $NaSb(OH)_6$ deposit is assumed to be formed. Zadunaisky (*J. Cell Biol.*, **31**: C11, 1966) modified the medium slightly. He used 3% glutaraldehyde with 2% $KSb(OH)_6$ in 0.1 M phosphate buffer. The mixture was brought to the boiling point to bring about solution of the pyroantimonate. After cooling, the pH was adjusted to 7.4. After fixation, tissues were rinsed in 0.1 M phosphate buffer of pH 7.4 containing 10% sucrose. Postfixation with osmium tetroxide and embedding was then carried out in the usual fashion. Further studies on the specificity of the reaction need to be conducted.

Hardin and Spicer (*J. Ultrastruct. Res.*, **31**: 16, 1970) modified the Komnick and Komnick reagent (5% pyroantimonate and 2% OsO_4 were used) and noted three distinct types of antimonate deposits, designated fine, moderately fine, and coarse, in nucleoli of normal rat trigeminal ganglion cells. The deposits were located in the granular component, dense component, and foci of pars amorpha of nucleoii, respectively. Areas of nucleolus devoid of antimonate deposits corresponded to nucleolar vacuoles. Experiments by Hardin and Spicer provided no evidence to prove or disprove the claim that staining is due to the presence of Na^+ ions.

Bulger (*J. Cell Biol.*, **40**: 79, 1969) used the potassium pyroantimonate method of Komnick (*Protoplasma*, **55**: 414, 1962) and Komnick and Komnick (*Z. Zellforsch. Mikrosk. Anat.*, **60**: 163, 1963) for localization of sodium at the light and electron microscopic levels. Bulger was not convinced that the method provided specific localization of sodium ions. Diffusion and fixation problems persisted.

Sumi and Swanson (*J. Histochem. Cytochem.*, **19**: 605, 1971) also studied the Komnick potassium pyroantimonate method for sodium and concluded the histochemical method is unreliable because correlation could not be obtained with the chemical method.

Shiina et al. (*J. Histochem. Cytochem.*, **18**: 644, 1970) note that not only sodium but also calcium and magnesium ions complex and precipitate with potassium pyroantimonate.

Lane and Martin (*J. Histochem. Cytochem.*, **17**: 102, 1969) used an electron probe analysis of alternate sections of Epon-embedded sections of mouse vas deferens to confirm localization of sodium in the lamina propria and the absence of cations (Mg^{2+}, Ca^{2+}, and Zn^{2+}) that could interfere with the analysis of sodium in buffered potassium pyroantimonate–osmium tetroxide solutions.

MAGNESIUM

A blue magnesium compound is formed from alkaline solutions of *p*-nitrobenzene-azoresorcinol. An excess of potassium cyanide in the solution decreases interference of Cu, Zn, and other metals which form complex cyanides (McNary, *J. Histochem. Cytochem.*, **8:** 124, 1960).

Make a 1% solution of the dye in 0.4% (or 0.1 *N*) NaOH. Deposit a drop of this on the section or smear; add a drop of 1% potassium cyanide. Cover with a coverslip and seal with paraffin. The method detects 1.3 μg magnesium in spot tests.

The same author has also adapted Glick's (*J. Biol. Chem.*, **226:** 77, 1957) "titan yellow" (C.I. 19540, thiazol yellow G) method for the staining of leukocytes in blood smears.

Rinse smear in distilled water to lake red cells and remove plasma magnesium. Dry. Place on smear a drop of 0.1% thiazol yellow G in 1 *N* NaOH, cover with coverslip, and seal with paraffin.

Magnesium is demonstrated by a flame red color. Sensitivity is about 1 μg.

McNary also notes a green magnesium reaction with zincon.

Bowling and Wertlate (*Stain Technol.*, **41:** 329, 1966) report that sections may be microincinerated with a bunsen burner, cooled, coated with a few drops of 0.2% aqueous solution of thiazol yellow G (C.I. 19540), air-dried, alkalinized with 2 *N* NaOH for 30 to 60 min, rinsed very carefully with 2 *N* NaOH, mounted in 2 *N* NaOH, and sealed in with fingernail lacquer. Magnesium is stained orange-red. Calcium salts dissolved in serum failed to stain. The histochemical method appears to be a modification of an earlier biochemical method by Garner (*Biochem. J.*, **40:** 828, 1946) used for the detection of Mg^{2+} in plasma and serum.

SILVER

Appearing as dark brown to black granules which are turned black by soluble sulfides, silver deposits are removed by treatment with a solution containing 1 g potassium ferricyanide and 22.5 g sodium thiosulfate ($Na_2S_2O_3 \cdot 5H_2O$) per 100 ml or by treatment for 1 to 2 h with Weigert's iodine solution followed by rinsing and immersion in 5 to 10% sodium thiosulfate until white (Schmorl). According to Timm (*Virchows Arch. Pathol. Anat.*, **297:** 502, 1936), silver sulfide is also removed by potassium cyanide solution. It is not dissolved by ammonium hydroxide or by sodium sulfide solution.

It is here to be noted that a 20-min exposure to Weigert's iodine, followed by thiosulfate, or a simple 20-min treatment with 5% potassium cyanide will completely decolorize Masson-Hamperl preparations, even when they are quite dark. The presence of oxygen is, of course, necessary to convert metallic silver to the soluble double cyanide $KAg(CN)_2$ (*J. Histochem. Cytochem.*, **9:** 184, 1961).

COPPER

Copper (cupric?) sulfide is also a dark brown to black material which is soluble in potassium cyanide solution (Timm, *Virchows Arch. Pathol. Anat.*, **297:** 502, 1936), but is not removed by sodium sulfide or ammonium hydroxide. Copper ferrocyanide is red, and Lison quotes this as a specific reaction. According to Mallory copper compounds give a light to clear dark blue with unoxidized fresh aqueous hematoxylin solution.

MALLORY'S HEMATOXYLIN STAIN FOR IRON AND COPPER.   Fix in alcohol. After formalin fixation only yellow to brown are obtained with iron. Embed in paraffin or celloidin. Hydrate sections as usual.

1. Dissolve 5 to 10 mg hematoxylin in 0.5 to 1 ml 100% alcohol. Add 10 ml distilled water which has been boiled 5 min to drive off carbon dioxide. Stain sections 1 h or more in the hematoxylin solution.
2. Wash 1 h in several changes of tap water.
3. Dehydrate, clear, and mount with alcohol, xylene, and balsam.

RESULTS: Nuclei are bluish gray; hemosiderin is black; copper, light to clear dark blue. We have had no experience with this method, but the prolonged staining of tissues in Clara's 1 : 10,000 hematoxylin demonstrates enterosiderin and hemosiderin iron in 1 to 2 h, and tissue premordanted with 1 mM $CuSO_4$ gives beautiful blue-green nuclear staining.

Boyce and Herdmann (*Proc. R. Soc. Med.*, **62:** 30, 1898) used acid ferrocyanide to demonstrate reddish brown copper ferrocyanide and hematoxylin to show a blue combination.

Okamoto and Utamura used rubeanic acid (dithiooxamide) for the demonstration of copper. Romeis directs: Bring frozen, paraffin, or celloidin sections to water as usual. Add 2 to 5 ml 0.1% alcoholic rubeanic acid to 100 ml 10% aqueous sodium acetate solution. Incubate sections 12 to 24 h at 37°C in this solution in a tightly covered vessel. Wash in water; counterstain in alum carmine; treat with alcohol, xylene, and balsam as usual.

RESULTS: Copper rubeanate is greenish black; cobalt, yellowish brown; nickel, blue-violet. Silver and lead yield black sulfides. According to Feigl (1949) copper rubeanate is formed also from weak acid solutions; for cobalt and nickel the sodium acetate is required.

Poulson and Bowen quote (*Exp. Cell Res. Suppl.*, **2:** 161, 1952) Waterhouse's use of sodium diethyldithiocarbamate as a histochemical reagent for copper in blowflies. They applied equal volumes of a 0.2% solution of this reagent and of 2% HCl (0.25 N?) to freshly dissected *Drosophila* larvae. Copper stains yellow-brown. Waterhouse (*Counc. Sci. Indust. Res. Bull.* 191, 1945) used 0.1% in either acid or neutral solution, both on fresh tissue and on material coagulated by heat, alcohol, or formalin. Iron also forms a brown carbamate compound, but the sensitivity to iron is much lower than to copper. The blowfly material failed to react to the rubeanic acid method.

McNary (*J. Histochem. Cytochem.*, **8:** 124, 1960) reports a positive (blue) zincon reaction with copper. Posttreatment with a solution of Versene decolorizes the blue reactions of zinc but leaves the copper reaction unaltered.

Copper forms a red-brown dithizone complex but not from the zinc-specific dithizone complex forming buffer solution. Sensitivity is high, 0.03 μg, 1 : 660,000.

WILSON'S DISEASE. Schaffner, Sternlier, Barka, and Popper (*Am. J. Pathol.*, **41:** 315, 1962) reported the presence in (cirrhotic) liver cells of yellow-brown pigment granules which fluoresce golden brown in ultraviolet light and stain with oil red O and Nile blue in paraffin sections and in part red with periodic acid Schiff after diastase digestion. Acid phosphatase in formol-calcium frozen sections was much reduced in the pigmented liver cells and was almost normal in other cells. Acid phosphatase activity was increased in Kupffer cells, as was that of alkaline glycerophosphatase and adenosine triphosphatase. Intranuclear glycogen deposit was often prominent. (Most of the recited pigment reactions are those of the usual hepatic cell lipofuscin.—R.D.L.)

High assay figures for copper in hepatic tissue were reported, ranging in the 10 cases from 60 μg to 1.56 mg/g dry weight.

Howell (*J. Pathol. Bacteriol.*, **77:** 473, 1959) used both rubeanic acid and dimethylamino-benzalrhodanine (*p*-dimethylaminobenzylidinerhodanine) (DMABR) for copper. His technics are simple: 5 ml 0.1% alcoholic rubeanic acid in 100 ml distilled water; 3 ml 0.1% DMABR in absolute alcohol to 100 ml distilled water. Incubate sections for 12 to 24 h at 37°C in a tightly closed container, rinse in distilled water, and mount in Farrants' gum arabic. The first gives greenish black granules, the second purplish red. It is noted that DMABR is slightly soluble in hot alcohol, insoluble in water, but moderately soluble in acetone. Feigl

Rubeanic acid

$$H_2NC-C-NH_2$$
$$\parallel \quad \parallel$$
$$S \quad \quad S$$

Dimethylaminobenzalrhodanine

$$(Ag,Cu)\ HN\!\!-\!\!-\!\!-\!\!C\!\!=\!\!O$$

uses DMABR in acid solution as a specific reagent for Cu(I) and Ag. Cu(II) must be reduced with bisulfite to react to the acid solution. Alkaline solutions react to heavy metals generally.

Later writers have followed Howell. Lindquist (*Arch. Pathol.*, **87**: 470, 1969) and Barden (*J. Neuropath. Exp. Neurol.*, **30**: 650, 1971) relied chiefly on Howell's method with dimethylaminobenzylidinerhodanine and his and Uzman's usage of rubeanic acid.

COBALT

Despite the wide use of cobaltous sulfide to demonstrate the precipitation of phosphate ions in calcium phosphate, little other histochemical information is available concerning this metal. Its phosphate reacts with solutions of 1-nitroso-2-naphthol to form a red-brown deposit. Iron phosphate does not react. The more reactive iron salts form brown-black with this reagent. Copper salts give brown but may be rendered nonreactive by conversion to cuprous iodide by an iodide sulfite mixture. Cuprous iodide is soluble in potassium iodide solution.

Alcoholic thioglycolic acid anilide yields a red-brown, acid-insoluble cobalt compound from ammoniacal solution. Black precipitates are yielded by 4-methyl- or 4-chloro-1,2-mercaptobenzene with copper, cobalt, and nickel; red with tin, bismuth, and molybdenum; yellow with lead, silver, and antimony; and pale yellow with cadmium, mercury, and arsenic.

The foregoing account is gleaned largely from Feigl (1949).

McNary (*J. Histochem. Cytochem.*, **8**: 124, 1960) notes also a violet complex with dithizone, a blue-green zincon complex which is unstable in acid, and a yellow-brown rubeanic acid complex. Rubeanic acid detects 30 ng cobalt (1 : 660,000).

Edetic acid (EDTA) at about pH 10 produces blue to violet complexes with Co and Fe. Place a drop of 10% Versene in 1% Na_2CO_3 on the slide, add a drop of 1% H_2O_2, put on cover glass, and watch for color reaction.

NICKEL

According to Lison a fresh alcoholic pure hematoxylin solution stains nickel salts lilac, grading to blue in thicker sections. He prescribes fixation in 30 ml formalin, 100 ml saline solution (*sérum physiologique*), and 0.3 ml ammonium sulfide. Soak in ammonium phosphate solution to produce the insoluble double nickel-ammonium phosphate. Decalcify, embed, and section. Stain in fresh alcoholic hematoxylin.

Nickel sulfide (NiS) is black, soluble in nitric acid and nitrohydrochloric acid (aqua regia) and partly in hydrochloric acid, and insoluble in ammonia water and in sodium sulfide solution. Its ferrocyanide is greenish white; its ferricyanide, rusty brown (Lange). Since nickel phosphate is insoluble in water, the phosphate-buffered neutral formalin should serve as well as the sulfide for preservation of relatively soluble nickel salts.

Choman (*Stain Technol.*, **37**: 325, 1962) has adapted the dimethylglyoxime method for nickel from Feigl for the histochemical demonstration of this element.

1. Transfer cryostat sections directly to clean slides.
2. Expose face down for 10 s to fumes of concentrated ammonia water in small beaker.
3. Deposit on section, face up, a drop 1% dimethylglyoxime in 95% alcohol.
4. After 30 s wash off reagent gently with 70% alcohol.
5. Dehydrate and clear by dropping onto sections successively a few drops each of 95% and absolute alcohols and xylene. Mount in balsam. Superior tissue detail will be seen by use of cellulose caprate.

RESULTS: Nickel is demonstrated as red acidular crystals. Sensitivity is about 1 in 10,000,000 (0.1 ng).

LEAD

THE LEAD SULFIDE METHOD. For the detection of lead, Timm (*Virchows Arch. Pathol. Anat.*, **297**: 502, 1936) fixed tissues, especially bone, in absolute alcohol saturated with hydrogen sulfide gas. He decalcified in 30% formic acid in water, also saturated with hydrogen sulfide. Acid was removed by several changes of 5% sodium sulfate and of water, all saturated with hydrogen sulfide. Frozen sections were cut and treated with potassium cyanide solution to remove copper and silver sulfides and with yellow ammonium sulfide to remove tin. Sections were washed in water, mounted on slides, dehydrated in alcohols, cleared in bromobenzene, and mounted in bromobenzene-balsam. Lead remains as brown granules of sulfide. Lead sulfide is converted into white lead sulfate by hydrogen peroxide. The crucial test is the restoration of the brown at the same sites by a fresh 5- to 10-min application of dilute ammonium sulfide.

THE CHROMATE METHOD. Fairhall treated paraffin sections of formalin-fixed lung for several days in a solution of potassium chromate acidified with acetic acid. The lead salts are converted into yellow monoclinic lead chromate crystals (*Public Health Bull.* 253, pp. 22–24, GPO, 1940). Fixation in Orth's or Möller's (Regaud's) fluid yields the same result according to Lison, who attributes the general method to Frankenberger and to Cretin.

THE HEMATOXYLIN METHOD. Mallory (p. 143) prescribed fixation in alcohol or 10% formalin and embedding in celloidin or paraffin.

1. Stain 2 to 3 h at 54°C in fresh 0.05 to 0.1% solution of hematoxylin in water saturated previously with calcium carbonate.
2. Wash 10 to 20 min in running water (or several changes).
3. Dehydrate with 95% alcohol, clear in terpineol, and mount in terpineol-balsam.

RESULTS: Lead is stained bluish gray to black. Even slightly aged hematoxylin stains brown and is useless. Old xylene-balsam also immediately turns the stain brown. We suggest the use of one of the nonreducing synthetic resins.

RHODIZONATE TEST FOR LEAD. Following Molnar's suggestion (*Stain Technol.*, **27**: 221, 1952) for improving phosphatase technics by substituting a lead nitrate bath for the cobalt and a rhodizonate treatment for the ammonium sulfide, it would appear that a simple 40-min bath in 0.5% sodium or potassium rhodizonate solution should demonstrate lead deposits in dark brown.

According to Feigl (1949) neutral rhodizonate-metal salt mixtures yield precipitates as follows: lead, blue-violet; zinc, brown-violet; cadmium, bismuth, calcium, strontium, and barium, red-brown to brown-red; uranyl salts, brown; and thallium, dark brown. Silver is reduced to a black (metallic?) state. Iron gives a soluble blue-green and hence does not confuse. At pH 2.8 only silver (black), tin (violet), thallium (brown-black), cadmium and mercurous salts (brown-red), barium (red-brown), and lead (scarlet red) yield precipitates. Prior treatment with sulfuric acid should render barium nonreactive and should extract mercury, tin, cadmium, thallium, and silver. If the silver is, however, in the metallic state, application of the iodine-thiosulfate sequence or of cyanide would remove it. Under these conditions the test should be specific for lead.

ARSENIC

For arsenic fix in 10% formalin containing 2.5% copper sulfate ($CuSO_4 \cdot 5H_2O$) for 5 days. Wash 24 h in running water. Embed in paraffin. Deparaffinized sections present green granules of Scheele's green ($CuHAsO_3$) which, though insoluble in water, is dissolved by acids and by ammonium hydroxide. By substituting cupric acetate for the sulfate, the granular Paris green or green cupric acetoarsenite is produced. Its solubilities are similar (Castel's method, *Bull. Histol. Appliq.*, **13**: 106, 1936). We suggest a light safranin counterstain.

ANTIMONY

Duckert (*Helv. Chim. Acta*, **20**: 362, 1937) records that 9-methyl-2,3,7-trihydroxy-6-fluorone (-xenthenone in some catalogs) at about pH 4 gives a red precipitate with Sb(III) and Sb(V) at high dilutions, cerium(IV) and germanium(IV) in acid levels. Buffered solutions yield dark violet with Fe in a saturated alcoholic solution of the reagent; slightly acidified (pH 4) with HCl or H_2SO_4 reacts in a few seconds.

The reaction appears reasonably specific and should be applicable histochemically. The reagent is otherwise used in histochemistry to identify DNA (blue) and RNA (rose) (Turchini reaction).

BISMUTH

The Christeller-Komaya (Christeller, *Med. Klin.*, **22**: 619, 1926; Komaya, *Arch. Dermatol. Syph.*, **149**: 277, 1925) bismuth reaction uses Leger's reagents which are a 4% aqueous solution of potassium iodide and a 2% solution of quinine sulfate [3 g quinine sulfate, 1 ml concentrated nitric acid (30 drops), 150 ml distilled water]. Mix 5 ml of each, add 2 drops (0.067 ml) nitric acid for use. Cut frozen sections of formalin-fixed material; prestain with carmine, fuchsin, or crystal violet; wash in water; apply the mixed Leger's reagent for 1 min; rinse in water containing 2 drops nitric acid to each 10 ml. Mount, blot dry, treat with carbol xylene plus alcohol, carbol xylene, xylene, and balsam in sequence. Christeller describes the bismuth compound as bright yellow; according to Schmorl the needle-shaped crystals are orange-yellow. Cowdry refers to brown granules and cites a modification of Castel's (*Arch. Soc. Sci. Med. Biol. Montpelier*, **16**: 453, 1934–1935), which utilizes a few drops of sulfuric acid in place of the nitric acid to dissolve the quinine sulfate. This gives red quinine iodobismuthate.

Wachstein and Zak (*Am. J. Pathol.*, **22**: 603, 1946) modified the Leger-Castel method as follows: Fix in formol, prepare frozen or paraffin sections. Bring to water. Apply several drops of 30% hydrogen peroxide. The black bismuth sulfide is converted to the sulfate, which promptly hydrolyzes to the insoluble white hydroxide (Lange, 1961). Wash in water. Immerse for 1 h in Castel brucine reagent [250 mg brucine sulfate dissolved in 100 ml distilled water containing 2 or 3 drops concentrated sulfuric acid; add 2 g potassium iodide (KI); store in a brown glass bottle; and filter before use].

Wash sections in a 1 : 3 dilution of Castel reagent, agitating to dislodge precipitates. Counterstain 4 min if desired in 0.01% light green SF in Castel reagent (filter before use), blot, and mount in 60% fructose (w/w) syrup containing a drop of dilute Castel reagent.

Bismuth deposits appear in orange-red, and the color darkens slightly as the preparations age. In moist tropical and subtropical climates seal coverslips with xylene, cellulose caprate, or asphalt cement.

Bismuth deposits are blackened by hydrogen sulfide water or with ammonium sulfide solution.

GOLD

According to Elftman and Elftman (*Stain Technol.*, **20**: 59, 1945) gold may be demonstrated by simple incubation of paraffin sections of formalin-fixed tissue in 3% hydrogen peroxide at 37°C for 1 to 3 days. The hydrogen peroxide bleaches other interfering pigments, and gold appears as rose, purple, blue, and black deposits. For critical evaluation omit counterstains. An alum hematoxylin stain may be used to aid in topographic studies and interferes only slightly. Probably light green SF (C.I. 42095) interferes least of the cytoplasmic stains.

THE STANNOUS CHLORIDE REACTION. The same authors also recommend the following modification of Christeller's (*Verh. Dtsch. Pathol. Ges.*, **22**: 173, 1927) stannous chloride reaction, which produces the purple of Cassius. Make a stock aqueous solution of 5% stannous chloride ($SnCl_2 \cdot 2H_2O$) in which some pieces of metallic tin are placed to prevent oxidation to $SnOCl_2$. For use, mix 40 ml stannous chloride and 4 ml concentrated hydrochloric acid. Incubate paraffin sections of formalin-fixed tissue at 56°C for 24 h in a covered Coplin jar. Wash repeatedly in distilled water, dehydrate, clear, and mount. Gold is evident as purple to brown particles, and the red, blue, and black of colloidal gold may be present as well.

According to G. Brecher (personal communication) the Christeller reaction may also be applied to tissues fixed in Zenker's fluid, provided that the thiosulfate treatment is omitted after the customary iodizing to remove mercury precipitates.

MERCURY

For mercury Almkvist (Schmorl) prescribed fixation in 100 ml saturated aqueous picric acid solution to which is added 3 g 25% (0.75 ml 70%) nitric acid. The mixture is allowed to stand 1 day, shaken, filtered, and saturated with hydrogen sulfide. Fix for 8 to 24 h or up to 3 days for maximal effect. Embed in paraffin. Mercury appears as fine yellow to brown granules of mercury sulfide. This is soluble in sodium sulfide solution but not in sodium thiosulfate. Other heavy metal sufides are also precipitated by this fixative solution, such as iron, silver, and cobalt. The precipitate should be tested with acidulated potassium ferricyanide as in the Tirmann-Schmelzer reaction to exclude iron (blue) and cobalt (red). Mercuric ferricyanide is soluble, and potassium ferricyanide solution does not dissolve mercuric sulfide.

Christeller's fixative (Schmorl) consisted of 15 ml distilled water, 2 g tin chloride, and 1 g nitric acid, and produced a black granular precipitate.

Brandino's method (Lison, 1936, p. 102) tests sections of alcohol- or formalin-fixed tissue with 1% 1,5-diphenylcarbohydrazide ($C_6H_5NH \cdot NH)_2CO$ (in alcohol? very slightly soluble in water). A violet precipitate is formed.

Voigt [*Acta Histochem.* (*Jena*), **14**: 315, 1962] has modified the Timm silver sulfide method to make it specific for *mercury*. Fe, Zn, Cu, Pb, Co, and perhaps Cd are also demonstrable, but of all these sulfides HgS is the least easily oxidized. Hence paraffin sections of material fixed in H_2S alcohol are treated 15 min in 15% hydrogen peroxide.

The balance of the technic follows Timm's usual silver procedure for heavy metals (*Dtsch. Z. Ges. Gerichtl. Med.*, **46**: 706, 1958) as presented in the Timm zinc method.

Cafruny (*Biochem. Pharm.*, **9**: 15, 1962) localized mercury in the cytoplasm and membranes of proximal convoluted tubule cells of kidneys of dogs injected with chlormerodrin, $HgCl_2$ or *p*-chloromercuribenzoate. Paraffin-embedded sections fixed in either 10% formalin or 10% trichloracetic acid were employed. After hydration, sections are stained for 20 min at 22°C in a saturated solution of D-β-naphthylthiocarbazone (NTC), rinsed, decolorized in 0.1 N HCl, and mounted in an aqueous mount such as Kaiser's glycerogel. The NTC reagent is prepared by first dissolving 5 mg NTC in 15 ml *N,N*-dimethyl formamide. Deionized water is then added (approximately 45 ml) until a slight precipitate is obtained. Cafruny notes that both tissue sections and a complex of NTC-chlormerodrin dissolved in chloroform and dried in a slide revealed an absorption spectrum maximum at 510 mμ.

MANGANESE

From Lison's 1953 account the suggested methods for manganese appeared unsatisfactory. Pearse, 1960, notes in his Table 52 the color reactions of $Mn(NO_3)_2$ with diethyldithio-carbamate (brown), dithizone (light brown), and pure blue B (Solochrome Azurine) (red), but otherwise does not discuss histochemical demonstration of the metal. Barka and Anderson do not consider Mn.

The lower oxides are insoluble and colored: MnO, gray-green; Mn_2O_3, dark brown; and MnO_2, black. MnS is green or pink, MnS_2 is black, and there is also a pink native silicate.

Feigl (IST, pp. 175–177) uses the benzidine blue reaction for the detection of MnO_2, reporting a sensitivity of 0.15 μg (1 : 330,000). MnO_2 acts as the oxygen source to produce benzidine blue from solutions of benzidine or its halogen acid salts. Manganous salts in 0.05 N NaOH in the presence of air spontaneously oxidize to MnO_2, which is black and insoluble. Rinse and apply benzidine solution, and observe the formation of blue. The color fades on drying but may be redemonstrated by a fresh application of benzidine solution.

This spot test seems to be of such nature as to be readily used histochemically. MnO_2 promptly disappears in 2% oxalic acid.

Feigl also notes the rapid reduction of diammine silver by Mn^{2+}.

$$Mn(OH)_2 + 2(NH_3)_2AgOH \longrightarrow MnO_2 + 2Ag + 4NH_3 + 2H_2O$$

In view of the ready air oxidation of Mn^{2+}, this test seems less applicable to histochemistry, despite its high sensitivity (50 ng, or 1 : 1,000,000). Melanin also reacts promptly with diammine silver.

McNary (*J. Histochem. Cytochem.*, **8**: 124, 1960), however, has used it successfully on blood films. He prescribes adding 0.900 sp gr (28%) ammonia to saturated silver nitrate until the precipitate is redissolved, then adding an additional equal volume of ammonia. Saturated silver nitrate at 25°C takes about 25 g to 10 ml water. We suggest taking 5 ml 28% ammonia and adding this strong silver solution to it drop by drop until a slight turbidity remains on shaking. Then add an additional 5 ml ammonia water. Place a drop or two of this ammoniacal silver solution on the section or smear, put on a coverslip, seal with paraffin, and wait 10 min for formation of MnO_2 and metallic silver.

Eggert, Dean, and Heyden (*Stain Technol.*, **44**: 161, 1969) apply Feigl's (1958, pp. 175, 302) tetrabase periodate spot test to the histochemical demonstration of manganese in plant tissue. The reagent reacts with saturated KIO_4 in the presence of manganese to form a blue pigment, thus:

Freeze a drop of saturated aqueous KIO_4 (about 0.8%) on a clean glass slide. Place a 10-μ cryostat section on the frozen drop. Thaw and add 2 drops 1% tetrabase solution in chloroform. Cover with a watch glass and react 5 min, replenishing the chloroform solution as needed, so that the preparation does not dry. Then add a drop of 50% glycerol, cover with a coverslip, and examine. Feigl (IST) notes that chromates may also give blue.

The reagent tetrabase is a tetramethyldiaminodiphenylmethane and is found in organic chemical catalogs (Eastman 214).

POTASSIUM

Lison rejected Macallum's potassium method as of little value; Gomori considered it a usable reaction. He modified it as follows: Dissolve 2 g cobalt nitrate in 5 ml 20% acetic acid. Dissolve 6 g sodium nitrite in 10 ml distilled water. Mix. Allow most of fumes to escape. Chill to 5°C. Immerse small fragments of fresh tissue (or cryostat sections) in cold reagent for 2 min. Rinse in ice-cold distilled water, and wash in four or five changes of ice-cold 50 to 70% alcohol. Convert cobalt to sulfide by immersion in 0.5 to 1% yellow ammonium sulfide for 2 min. Wash in water; counterstain with safranin, hemalum, carmine, or the like; dehydrate with alcohol, clear in xylene, spread out, and mount in Canada balsam.

The specific precipitate of potassium cobaltinitrite is converted to dark brown cobalt sulfides, and these serve to localize potassium. It is uncertain whether this demonstrates ionic or bound potassium or both. The alleged nonspecific creatine reaction has been shown to be due to potassium contamination.

Ryder (*J. Histochem. Cytochem.*, **7**: 133, 1959) used freeze-dried sections for the method of Poppen et al. (ibid., **1**: 160, 1953), who used alcohol-formalin fixation and paraffin embedding. Poppen et al. modified Macallum's cobaltinitrite method as follows:

COBALTINITRITE REAGENT

1. Dissolve 25 g cobaltous nitrate in 50 ml distilled water; add 12.5 ml glacial acetic acid.
2. Dissolve 120 g sodium nitrite (potassium-free) in 180 ml distilled water. Of this solution take 210 ml and add to the acid cobaltous nitrate solution.
3. Bubble air through the mixture until no more brown fumes evolve. Store i: refrigerator and refilter before each use. The reagent remains potent for 3 to 4 months t 0 to 5°C.

TECHNIC

1. Cut paraffin sections at 7 μ. Press directly onto slides, or float on warm absolute alcohol to flatten; drain and dry. Or, according to Poppen, transfer ribbons directly into two changes of xylene and two of absolute alcohol. With attached paraffin sections we suggest Naidoo's method (*J. Histochem. Cytochem.*, **10**: 580, 1962) of deparaffinizing in petroleum ether, washing in isopentane, and drying instantly in air.
2. Transfer directly to cobaltinitrite reagent. Roll loose sections on a glass rod in the last alcohol bath, and immerse below the surface of the aqueous reagent until diffusion turbulence subsides. They will then float on the reagent smoothly. Leave in reagent 3 min or longer.
3. Wash quickly in two to four changes of ice-cold distilled water until yellow stops coming out.
4. Float loose sections onto slides, let dry, blot, dehydrate with two changes each of 95% and absolute alcohol, clear in xylene, and mount in synthetic resin. This leaves the potassium cobaltinitrite compound as birefringent chrome yellow crystals.

Alternatively, the cobalt may be converted to black cobalt sulfide, thus:

Immerse 20 to 30 min in 2% dilution of yellow ammonium sulfide, wash in several changes of distilled water (float onto slides), dehydrate, clear, and mount as above. As usual for cobalt sulfide preparations, Canada balsam may be a better mounting medium.

Control slides are immersed 2 to 5 min in distilled water before immersion in cobaltinitrite reagent. This removes potassium salts.

LITHIUM

For lithium Nelson, Bensch, Herman, and Barchas (*J. Histochem. Cytochem.*, **21**: 241, 1973) take advantage of the presence in ^{6}Li of α particles. ^{6}Li is a relatively stable isotope, constituting about 7.5% of the element, the other 92.5% being ^{7}Li. Atomic absorption spectroscopy and flame photometry are used to determine the presence of Li in tissues, but these methods do not permit histologic localization. X-ray microanalysis has been used for C, N, and O, but elements of lower atomic number have not been measurable by this method.

Radiographic track registration of ^{6}Li$(n,\alpha)^3$H tracks requires superimposition of α-particle-sensitive plastic film over the section and bombardment with thermal neutrons. The tracks made by the α particles are thus etched in the film, and localization is determined by superimposing the treated film on the section. Similar technics using neutron fission fragment (n, f) reactions have been used to localize uranium and plutonium in skeletal tissue (Becker and Johnson, *Science*, **167**: 1370, 1970; Hamilton, *Calcif. Tissue Res.*, **7**: 150, 1971). Nelson et al. reviewed theoretical aspects of this method (*J. Theor. Biol.*, **34**: 73, 1972).

Work with this method required thoroughly clean surroundings, benches, glassware, equipment; cleaning with borax-free soap, detergents, or surfactants; demineralized distilled water; and handling only with forceps after washing. Male rats of 200 g were injected intraperitoneally with 10 mmol/kg LiCl. The LiCl was prepared by dissolving ^{6}Li$_2$CO$_3$ with HCl and adjusting to pH 7 with NaOH. Rats were decapitated 30 min and 2 h after injection, the brains frozen on solid carbon dioxide and stored in plastic bags at $-20°$C until sectioned. Blocks were cut coronally 1 cm thick, sectioned at 20 μ in the cryostat, and mounted on cellulose acetate plates $31.5 \times 50.4 \times 3$ mm ($1.25 \times 2 \times 0.06$ in) and freeze-dried. Polymethylmethacrylate (Plexiglas)[1] plates are also suitable. Three kinds of dielectric films were used: cellulose triacetate (Triafol TN), cellulose acetate, butyrate (Triafol BN),[2] and cellulose nitrate. These are available in 12×12 sheets 0.2 mm thick.

These are cut to the size of the cellulose acetate plates and placed over the sections smooth side down and fastened by wrapping with plastic tape. Two small holes are drilled through the film into the plate carrying the tissue, to ensure accurate realignment after irradiation. Cellulose nitrate films were made by dipping glass slides into cellulose nitrate solution and drying at room temperature 30 min and then 20 h in the oven at 60°C. Dipping and drying were repeated twice to obtain a film about 50 μ thick. Films were separated from the glass by cutting along a long edge with a knife and then immersing in water. After being dried the films were placed over the tissue and fastened at one end with Eastman 910[3] adhesive or with adhesive tape.

IRRADIATION. The mounted sections are then placed in plastic bags and irradiated with slow neutrons in the thermal column of a reactor. The Stanford University 10-kW pool

---

[1] Cadillac Plastic & Chemical Co., 15841 Second Avenue, Detroit, Mich. 48232.

[2] The Triafols were supplied by Farbenfabriken Bayer, Leverkusen, Germany; the nitrate, by Hercules Powder Co., Wilmington, Del.

[3] Armstrong Cork Co., Lancaster, Pa.

reactor was used. The α particle range is about 5 μ. The exposure timing is computable from the predetermined profile of the thermal neutron flux.

TRACK FORMATION. Visible pits on tracks form in the dielectric film as a result of damage by penetrating α particles after etching the Triafol TN and BN films in 6 N KOH at 60°C for 50 and 120 min, respectively. The cellulose nitrate films were etched 25 min in 6 N NaOH at 40°C. After the etching, the films are washed well in water and dried.

During etching of the film the sections on the plates are protected by taping on a rigid plastic sheet. Sections are stained with cresyl violet and for the Triafol films mounted with a cover glass; for the cellulose nitrate, directly with immersion oil as a (n_D 1.640) mounting medium. Because of their matt surface the Triafol films and the sections must be studied in sequence.

Localization of lithium in the brain has been shown by this method. The method at present is restricted to specially equipped research laboratories. Boron is the principal interfering contaminant.

THALLIUM

Thallium deposits form yellow crystals of iodide (TlI) on fixation of tissues in alcohol colored with Lugol's solution (Barbaglia, from Lison, 1953). Gomori suggests 2.5 g iodine, 5 g KI in 50 to 100 ml 95% alcohol; USP Tincture of Iodine, diluted with an equal quantity of alcohol, should serve. The yellow thallous iodide is insoluble in alcohol, acetone, potassium iodide solution, or water but is soluble in nitrohydrochloric acid (aqua regia).

Giusti and Fiori (*Stain Technol.*, **44**: 263, 1969) propose the following method for thallium:

1. Tissues fixed in neutral formalin or 95% ethanol are paraffin-embedded, sectioned, and brought to water.
2. Sections are treated with gaseous H_2S in the plastic bag of a Kipp apparatus under a hood until the sections become gray.
3. Treat sections in a solution of 20% $(NH_4)_2S$ saturated with an excess of black powdered selenium for 10 min at 23°C or until the sections turn green.
4. Wash thoroughly to remove the sulfide completely.
5. Cover the sections with 2 to 3 ml of 20% H_2O_2 for 10 min or until sections are colorless.
6. Cover each slide placed on a flat rack with 2 to 3 ml of the following mixture from stock solutions immediately before use: (*a*) 25% gum acacia in 10 ml water; (*b*) 2% hydroquinone in an aqueous solution of 1 ml 5% citric acid and 0.1 ml 10% aqueous $AgNO_3$. Allow development for 20 to 30 min at 23°C *in darkness*. Terminate the development when sections are brown rather than black.
7. Wash in running water, dehydrate, and mount in synthetic medium. Controls included omission of H_2S, selenium, or the development procedure. Prolongation of the development procedure results in silver impregnation. Thallium deposits are identified as small black granules. Selenium is used in this procedure to prevent oxidation and dissolution by dilute acids and H_2O_2. Silver and mercury can interfere with the reaction. Of course, mercury and silver can be detected in sections not treated with selenium.

ZINC

Lison's 1953 text cited a method of Mendel and Bradley (*Am. J. Physiol.*, **14**: 313, 1905) for the demonstration of zinc. Treat paraffin sections 15 min at 50°C in 10% sodium nitroprusside solution. Wash in gently flowing water for 15 min. Cover section with a cover glass, and introduce at one side a drop of sodium or potassium sulfide solution. An intense purple is produced. Lison regarded the reaction as specific.

Mager, McNary, and Lionetti (*J. Histochem. Cytochem.*, **1**: 493, 1953) propose the following variation of the dithizone method for zinc:

REAGENTS

1. Stock dithizone: Dissolve 10 mg diphenylthiocarbazone (dithizone) in 100 ml reagent-grade anhydrous acetone. Store at 3°C in a brown glass bottle.
2. Complexing buffer: Dissolve 55 g sodium thiosulfate ($Na_2S_2O_3 \cdot 10H_2O$), 9 g sodium acetate ($NaCO_2CH_3 \cdot 3H_2O$), and 1 g potassium cyanide (KCN) in 100 ml distilled water. Shake in separatory funnel with several successive portions of dithizone in carbon tetrachloride, until the CCl_4 layer remains clear green, to remove traces of zinc.
3. Normal acetic acid (6%).
4. Rochelle salt: A 20% aqueous solution of sodium potassium tartrate [$NaK(CO_2CHOH)_2 \cdot 4H_2O$].

REACTION MIXTURE.   Mix 24 ml stock acetone dithizone with 18 ml distilled water, adjust with normal acetic acid to pH 3.7, add 5.8 ml complexing buffer and 0.2 ml Rochelle salt solution. Use at once.

PROCEDURE

1. Fix thin tissue fragments 1 h in two changes of cold 100% ethanol or, preferably, methanol (Rixon and Whitfield, *J. Histochem. Cytochem.*, **7**: 262, 1959).
2. Cut frozen sections at 15 μ, or clear in xylene, infiltrate 1.5 h in paraffin, embed, and section at 6 μ, or prepare paraffin sections by the freeze-dried procedure. Attached cryostat sections may be postfixed 10 min in cold alcohol.
3. Dry frozen sections on slides, deparaffinize paraffin sections in xylene, and let dry.
4. Flood sections with dithizone reaction mixture, stain 5 to 10 min, and drain.
5. Wash off excess dye by flooding with chloroform; drain.
6. Rinse quickly in distilled water and mount in Karo, fructose syrup, or Arlex gelatin. Rixon and Whitfield used glycerol; glycerol gelatin should serve.

RESULTS: Zinc is demonstrated as red to purple granules or diffuse red. The complexing buffer serves to render nonreactive the other metals which normally react with dithizone: Mn, Fe, Co, Ni, Cu, Ag, Pd, Pt, Cd, Sn, In, Au, Hg, Tl, Pb, Bi. Yellow crystals of dithizone resulting from evaporation of acetone are readily distinguished.

Haumont (*J. Histochem. Cytochem.*, **9**: 141, 1961) used this method on undecalcified sections of ossifying young rat bones, and on ground sections of adult bone, using 95% alcohol fixation in both cases. Longer exposures, up to 1 h in dithizone, were used for the 50-μ ground sections.

McNary (*J. Histochem. Cytochem.*, **8**: 124, 1960), besides several variants of the dithizone method, used zincon (2-carboxy-2-hydroxy-5-sulfoformazylbenzene).

ZINCON METHOD FOR ZINC, COPPER, COBALT, MAGNESIUM

1. Mix 2 ml 0.15% zincon in 0.1 N NaOH with 8 ml borate buffer of pH 8.8 (this should result in a pH about 9.1 to 9.2).
2. Flood slides with this mixture; let stand 3 min.
3. Rinse, and mount in Karo.

RESULTS: Zinc and copper yield deep blue complexes; cobalt, a blue-green; magnesium, green. The zinc complex is decolorized by 5 to 10% Versene; the cobalt, by dilute acid.

THE TIMM-GESSWEIN SILVER SULFIDE METHOD FOR ZINC.   Gesswein (*Vir-*

chows Arch. Pathol. Anat., **332**: 481, 1959) used a variant of the silver sulfide method of Timm (*Dtsch. Z. Ges. Gerichtl. Med.*, **47**: 428, 1958) and Voigt (*Acta Pathol. Microbiol. Scand.* **44**: 146, 1958).

METHOD.   Fix small pieces 10 to 20 h in H_2S–saturated 70% alcohol, then immerse for 10 h in 95% and 10 h in absolute alcohol, in two changes of xylene for 30 min each; embed in paraffin. Cut 4- to 6-μ sections, float on preboiled double-distilled water, and attach to cover glasses. Silver $1\frac{1}{2}$ to 2 h in artificial light under close microscopic control, wash in running water, stain nuclei with kernechtrot or hematoxylin; treat with ascending alcohols, xylenes; mount in resin.

The following solutions are stored in quartz glass containers:

Gum arabic: 40 g well-ground gum arabic suspended in 100 ml double-distilled water, with daily stirring with glass rod for 14 days.

Silver nitrate: 10 g dissolved in 100 ml double-distilled water and kept 14 days in the dark before using.

Reduction fluid: Dissolve 0.4 g hydroquinone and 0.9 g citric acid in 20 ml double-distilled water; shake hard for 5 min. Let stand in dark for 12 h before using.

Working silver solution: To 60 ml 40% gum arabic add 0.5 ml 10% $AgNO_3$, and shake well for 10 min. Then add 5 ml reduction solution, shake about 30 s, and pour at once over sections.

Ibata and Otsuka (*J. Histochem. Cytochem.*, **17**: 171, 1969) used Timm's sulfide-silver method in an electron microscopic investigation of zinc in the hippocampus of rabbits. Zinc was detected in synaptic vesicles in boutons terminaux (terminal boutons, or end feet) of the mossy fibers.

Smith, Jenkins, and Gough (*J. Histochem. Cytochem.*, **17**: 749, 1969) reported that 8-hydroxyquinoline forms complexes with 26 divalent and trivalent metals. Of these Ca, Co, Cu, Fe, Mg, Mn, and Zn are of biologic importance. Only Ca, Mg, and Zn form fluorescent complexes, and Zn has a considerably higher binding constant than the other two. For the reagent, use 3% 8-hydroxyquinoline in absolute alcohol, 0.1 ml in 25 ml pH 8 Michaelis universal buffer. For the buffer, dissolve 19.428 g sodium acetate ($3H_2O$?) plus 29.428 g sodium phenobarbital in 1000 ml distilled water. To 100 ml of this solution add 40 ml 0.85% sodium chloride and about 20 ml 0.1 N HCl to adjust to pH 8.

Blood smears are air-dried for 1 h or more. They are then stained directly in the above reagent for 15 min, drained, dipped twice in double-distilled water, dried, mounted in non-fluorescent immersion oil or other nonfluorescent mountant, and examined. Use 3850 A mercury-vapor lamp with exciter filter BG-88, BG-12, and BG-3 in a Zeiss photomicroscope or equivalent fluorescence microscopy.

Dithizone controls are prepared according to McNary (*Blood*, **12**: 644, 1957).

RESULTS: Greenish yellow fluorescent granules, large and discrete in eosinophil and basophil leukocytes, less strongly fluorescent fine granules in neutrophils. The fluorescence is stable for 2 weeks. Washing with 1% acetic acid before staining prevents the fluorescence. Alcohol fixation was not tried, and Giemsa controls were not used. The acetic acid may be removing the entire granule and not just the zinc. Sternberg, Cronin, and Philip (*Am. J. Pathol.*, **47**: 325, 1965) had previously used the green fluorescence with 8-hydroxyquinoline (oxine, 8-quinolinol) for demonstration of zinc in the rat prostate. They fixed in 5 ml 2% oxine–absolute alcohol plus 4 ml 0.1 M pH 5.6 acetate buffer plus 1 ml formol (38 to 40% formaldehyde).

An in vivo technic was also used. The oxine was pretreated with 1 meq HCl to convert to the hydrochloride, then diluted to 0.5% with isotonic saline solution and injected intravenously. Fresh frozen sections were used.

Similar technics were used with dithizone, yielding magenta granules in the same locations. Fresh frozen sections were stained by repeated dipping in 10 ml acetone and 1 mg dithizone plus 10 ml metal-free water. The metal-free water is prepared by shaking with repeated 1-ml portions of 1 mg dithizone to 100 ml chloroform until the chloroform is no longer colored by the water.

URANIUM

The several oxides of uranium are insoluble in water, as is uranyl phosphate (Lange). Hence, the phosphate-buffered formalin should be adequate for fixation. Lison cites Schneider's fixation in 50 ml saturated aqueous picric acid, 50 ml 5% potassium ferrocyanide, and 10 ml hydrochloric acid. This was followed by washing in 4% hydrochloric acid, then hydrochloric acid–alcohol. Dehydration and embedding as usual follow, and uranium salts appear as the deep brown double ferrocyanide of potassium and uranium. Gerard and Cordier [*Arch. Biol. (Liege)*, **43:** 367, 1932] used the Prussian blue reaction as for iron, obtaining the same dark brown double ferrocyanide.

The α particle autoradiographic technic (Lillie, 1954, p. 439) might be used to study the distribution of uranium salts in tissues.

THORIUM

Thorium was used at one time in roentgenography of blood vessels, and it has occasionally been demonstrated in tissues. The most probable form is the insoluble dioxide ThO_2. Thorium 232 (^{232}Th) is almost the only naturally occurring isotope, with a half-life of 1.39×10^{10} years. It is also an alpha and gamma emitter and should be localizable by the autoradiographic method for those emanations.

Feigl notes the reactivity of Th in the quinalizarin test for Mg (IST 225).

Th, along with Be, In, Ga, and Sc, forms fluorescent morinates in the morin test for Al which, in distinction from those of Al and Zr, are unstable in 2 N HCl.

According to Pavelka (*Mikrochemie*, **4:** 199, 1926; Feigl, IST 201), spot testing with alizarin and then exposing the spot to ammonia gives violet-red with Ti (sensitivity 0.18 μg, 1 : 166,000); purple-red (("raspberry")) with Zr (sensitivity 0.29 μg, 1 : 103,000); and violet with Th (sensitivity 0.24 μg, 1 : 125,000).

SULFATES

Gomori cites Macallum's ("Abderhaldens Handbuch der biologischen Arbeitsmethoden," vol. 2, pp. 1145–1146, 1912) sulfate method: Treat frozen sections (Adamstone-Taylor or Linderström-Lang technics should be more successful) of fresh tissue with 0.1 N lead acetate for 10 min or more. Wash thoroughly in water and then in 0.1 N nitric acid to remove lead phosphate, carbonate, and chloride. Wash in water, and treat with equal volumes of glycerol and ammonium sulfide, converting the lead sulfate to the brown sulfide. Molnar's suggestion of substitution of a 40-min bath in 0.5% sodium rhodizonate for the sulfide treatment might be tried. According to Feigl neutral sodium rhodizonate reacts with lead sulfate or sulfide to give an insoluble deep violet, while rhodizonic acid, at pH 2.8, gives a red precipitate.

Macallum's ammonium sulfide is produced by saturating 16% ammonia (sp gr 0.96) with hydrogen sulfide gas.

See also Chap. 8 for identification of sulfuric and sulfonic acid residues.

PHOSPHATES

For phosphates Bunting (*Arch. Pathol.*, **52**: 458, 1951) used a variant of the molybdenum blue reaction adapted by him from Feigl and from Serra and Queiroz-Lopes for paraffin sections of formalin-fixed tissues:

BUNTING'S MOLYBDENUM BLUE TECHNIC FOR PHOSPHATES

1. Deparaffinize and hydrate as usual.
2. Cover section with a few drops of 5% ammonium molybdate and add an equal volume of 1% nitric acid. Let stand for 5 min.
3. Wash thoroughly with water.
4. Cover with benzidine solution for 1 min. (Dissolve 50 mg benzidine base in 10 ml glacial acetic acid, and dilute to 100 ml with distilled water.)
5. Flood with 45% (saturated) sodium acetate solution. Apply cover glass and examine at once.

RESULTS: Sites of phosphate ion are blue. The color fades and diffuses in a few hours.

Serra and Queiroz-Lopes (*Portugaliae Acta Biol.*, **1**: 111, 1945) hydrolyzed small blocks of tissue for 2 to 3 weeks at 10 to 12°C in a solution containing 1 g ammonium molybdate in 100 ml 2 N HCl and then continued the hydrolysis for 2 to 3 days in the same fluid at 20 to 25°C. They then added an equal (small) volume of the same benzidine solution as Bunting (step 4), reacted for 3 min, and added 2 volumes saturated sodium acetate. By mounting in glycerol and sealing with Romeis' rosin lanolin they were able to preserve the reaction for some months.

Johansson et al. (*Nucl. Inst. Methods*, **84**: 141, 1970) and Jundt et al. (*J. Histochem. Cytochem.*, **22**: 1, 1974) used recently available high-resolution Si(Li) x-ray detectors to detect quantitatively individual elements throughout the periodic table by measuring characteristic x-rays emitted when thin samples are fluoresced by protons or other charged particles. Theoretically, detection as low as a few picograms (10^{-12} g) are possible. The method is believed to have approximate uniform sensitivity for all elements above Z (atomic number) 22. Jundt et al. analyzed tissue homogenates, frozen sections, and formalin-fixed paraffin-embedded sections of tissues obtained at surgery or autopsy.

13

LIPIDS

Lipids generally are natural products containing fatty acids that are usually saponifiable. They are generally insoluble in water and soluble in the common fat eluents. They are converted to water-soluble substances by saponification.

CLASSIFICATION

I. Triglycerides. Chemically, these have the general formula illustrated below. These natural fats contain five to twelve different fatty acids. The fatty acids may be either saturated or unsaturated.

$$CH_2-O-CO-R$$
$$|$$
$$CH-O-CO-R'$$
$$|$$
$$CH_2-O-CO-R''$$
Mixed triglyceride

II. Phospholipids. Phosphatides are esters of fatty acids whereby the alcoholic portion of the molecule contains a phosphate group. Most are glycerophosphatides and sphingomyelins which contain choline. They are notably absent from depot fat, but are generally constituents of all tissues—especially active ones such as brain and other nervous tissues. In this form fats are transported, undergo metabolic change, and may participate as structural elements of cells and tissues.

A. Glycerophosphatides. These are composed of α-glycerophosphoric acid esterified with fatty acid or other constituents. Several types are known:

1. Phosphatidic acids. These are the simplest of the glycerophosphatides. They may be regarded as triglycerides wherein one of the fatty acid residues has been replaced by phosphoric acid.

2. Phosphatidyl esters. These are glycerophosphatides wherein the phosphatidic acids are esterified with the hydroxyl groups of ethanolamine, choline, or serine as indicated below.

TABLE 13-1   CLASSIFICATION AND STRUCTURAL COMPONENTS OF SAPONIFIABLE LIPIDS

Classification of lipids*	Structural components (other than fatty acids)†		
	Alcohol	Nitrogenous base	Other
Triglycerides { 1. Fats	Glycerol	. . . .	
2. Oils	Glycerol	. . . .	(Higher proportion of unsaturated fatty acids than in fats)
Glycerophosphatides 1. Phosphatidic acids	Glycerol	. . . .	Phosphoric acid
2. Phosphatidyl esters { a. Phosphatidylcholines	Glycerol	Choline	Phosphoric acid
b. Phosphatidylethanolamines	Glycerol	Ethanolamine	Phosphoric acid
c. Phosphatidylserines	Glycerol	Serine	Phosphoric acid
3. Lysophosphatides	Glycerol	Choline / Ethanolamine / Serine	Phosphoric acid
4. Inositol phosphatides { mono- / di-	Glycerol / Glycerol	. . . .	Phosphoric acid, inositol / Phosphoric acid, inositol
5. Acetal phosphatides	Glycerol	(As for lysophosphatides)	Phosphoric acid, unknown long-chain alkyl group
Sphingolipids { 1. Sphingomyelins	Sphingosine	Choline	Phosphoric acid
2. Cerebrosides	Sphingosine	. . . .	Hexose, sulfate
3. Gangliosides	Sphingosine	. . . .	Hexose, neuraminic acid
Waxes { 1. True waxes	Long-chain aliphatic alcohols		
2. { Steryl esters / Vitamin A and D_3 esters	Complex cyclic alcohols		

* In Bloor's classification (*Chem. Rev.*, **2**: 243, 1925–1926) the triglycerides (neutral fats) and waxes together constitute the "simple lipids," while the remaining classes listed here are collectively designated "compound lipids" (phospholipids, cerebrosides, gangliosides). The phospholipids or phosphatides comprise the glycerophosphatides and sphingomyelins, grouped together by virtue of their sole common component, the phosphate group. The classification of the sphingomyelins with the other sphingolipids (derivatives of sphingosine) is more rational.

† All compounds yield fatty acids on hydrolysis, with the possible exception of some acetal phosphatides.

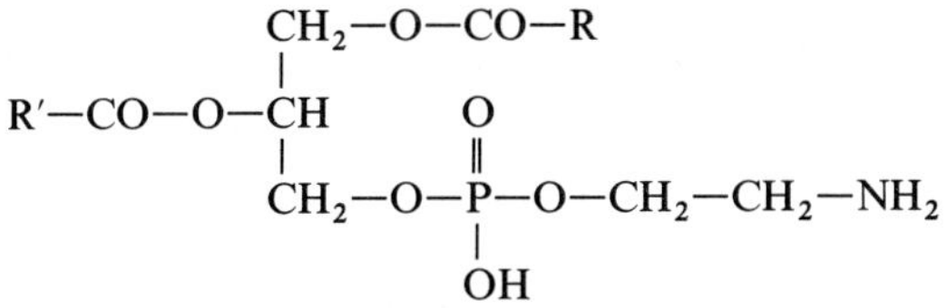

$$CH_2-O-CO-R$$
$$R'-CO-O-CH \qquad O$$
$$CH_2-O-\overset{\|}{\underset{OH}{P}}-O-CH_2-CH_2-NH_2$$

α-Phosphatidylethanolamines (cephalins)

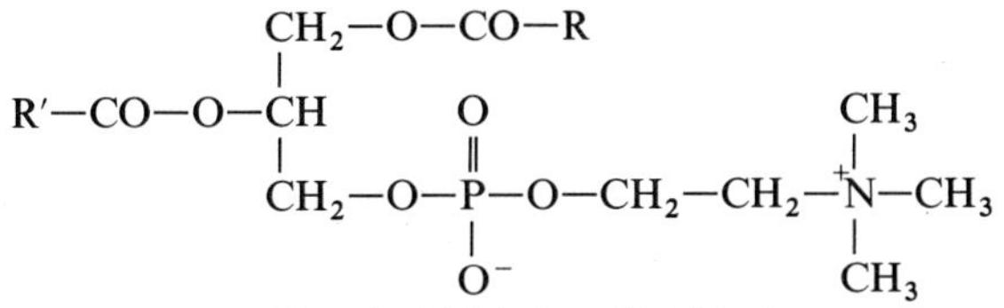

$$CH_2-O-CO-R$$
$$R'-CO-O-CH \qquad O \qquad\qquad CH_3$$
$$CH_2-O-\overset{\|}{\underset{O^-}{P}}-O-CH_2-CH_2-\overset{+}{\underset{CH_3}{N}}-CH_3$$

α-Phosphatidylcholines (lecithins)

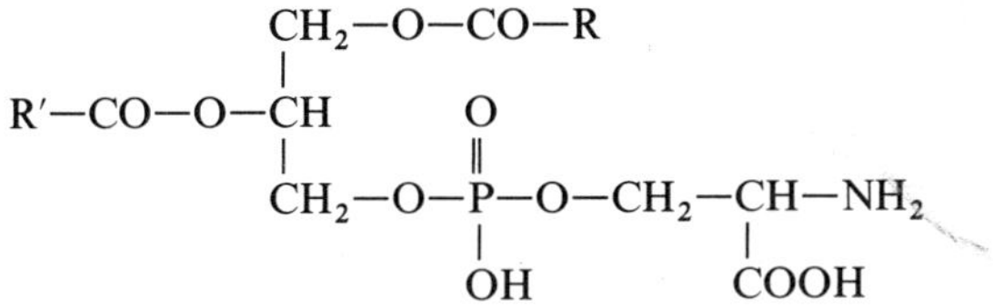

$$CH_2-O-CO-R$$
$$R'-CO-O-CH \qquad O$$
$$CH_2-O-\overset{\|}{\underset{OH}{P}}-O-CH_2-\underset{COOH}{CH}-NH_2$$

α-Phosphatidylserines

3. Lysophosphatides. Lysophosphatides are partially hydrolyzed glycerophosphatides.

4. α-Glycerophosphoryl compounds. These are α-glycerophosphatides which lack the fatty acids.

5. Phosphatidylinositides. Several inositides have been isolated and characterized by derivatives after hydrolysis. One or two phosphates may be esterified in the inositol ring. Phosphatidylinositides have been found in heart and liver tissues.

6. Acetal phosphatides (plasmalogens). The following structural formula illustrates the close similarity of plasmalogen to phosphatidyl esters:

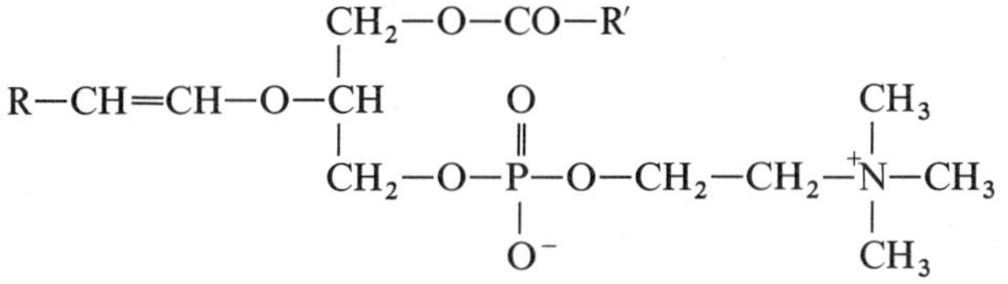

$$CH_2-O-CO-R'$$
$$R-CH=CH-O-CH \qquad O \qquad\qquad CH_3$$
$$CH_2-O-\overset{\|}{\underset{O^-}{P}}-O-CH_2-CH_2-\overset{+}{\underset{CH_3}{N}}-CH_3$$

Acetal phosphatides (plasmalogens)

B. Sphingolipids. In sphingolipids glycerol is replaced by the base sphingosine as indicated below. Rarely the base may be dehydrosphingosine or phytosphingosine.

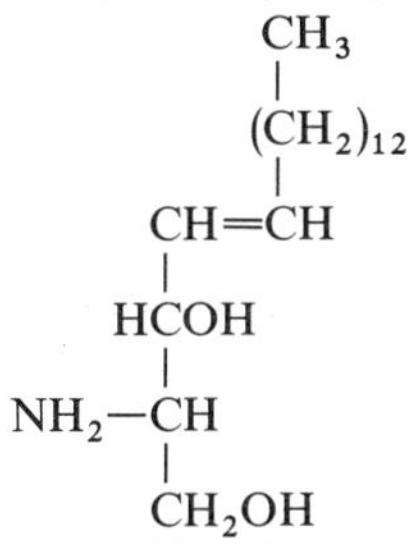

$$CH_3$$
$$(CH_2)_{12}$$
$$CH=CH$$
$$HCOH$$
$$NH_2-CH$$
$$CH_2OH$$

Sphingosine

The following sphingolipids are known:
1. Sphingomyelins. These resemble the glycerophosphatides, as illustrated below:

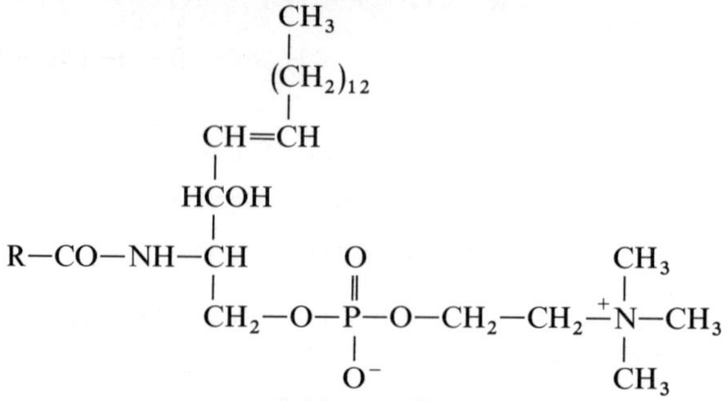

$$CH_3$$
$$|$$
$$(CH_2)_{12}$$
$$|$$
$$CH=CH$$
$$|$$
$$HCOH$$
$$|$$

$$\begin{array}{ccc} & & CH_3 \\ & O & | \\ R-CO-NH-CH & \| & + | \\ | & & \\ CH_2-O-P-O-CH_2-CH_2-N-CH_3 \\ | & | \\ O^- & CH_3 \end{array}$$

Sphingomyelins

Lignoceric acid $CH_3(CH_2)_{22}COOH$ is the predominant fatty acid in sphingomyelins.

2. Cerebrosides. Most cerebrosides contain sphingosine or dehydrosphingosine, a fatty acid, and a hexose. They are found predominantly in the brain. The cerebrosides include kerasin, phrenosin, nervon, and hydroxynervon.

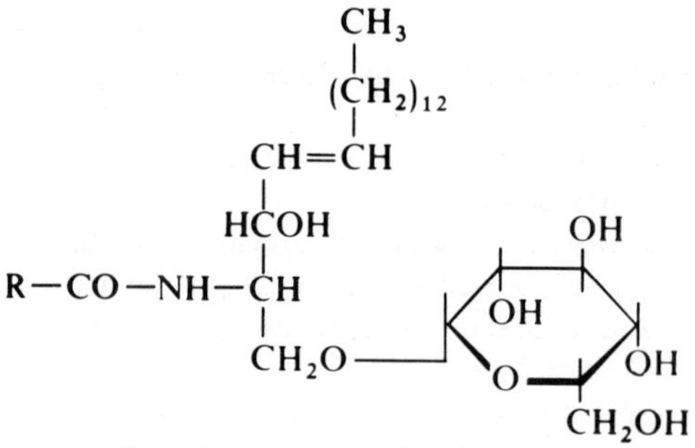

Probable structure of cerebrosides

3. Gangliosides. These complex structures present in many nervous and parenchymatous tissues yield sphingosine, long-chain fatty acids, hexose (usually galactose), and neuraminic acid.

III. Waxes. Saponifiable waxes may be categorized as either true waxes or sterols.
  A. True waxes. These are mixtures of fatty acid esters of the higher aliphatic straight-chain monohydric alcohols such as cetyl $[CH_3(CH_2)_{14}CH_2OH]$, octadecyl $[CH_3(CH_2)_{16}CH_2OH]$, and higher-chain alcohols.
  B. Sterols. Sterols have the 17-carbon cyclopentanophenanthrene ring illustrated below. They are widely distributed in tissues including the structure of vitamin D and certain hormones. Cholesterol is an important sterol. It is present in high concentrations in the adrenal cortex and is a precursor of adrenocortical hormones.

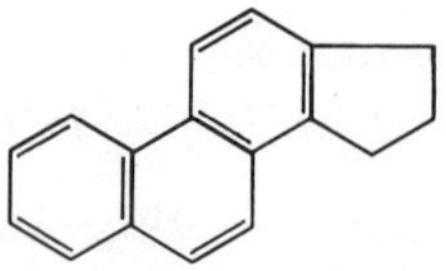

Cyclopentanophenanthrene ring

PHYSICAL PROPERTIES

Surface properties of lipids vary with the nature of the lipid. Penetration of dyes is affected by surface properties. As noted above, phospholipids possess highly polar phosphoryl and basic groups. These groups confer hydrophilia, whereas paraffin chains in the remainder of the molecules favor hydrophobia. Thus the structure of phospholipids is said to be "amphipathic." The hexose constituents of cerebrosides can be expected to confer slight hydrophilia.

In contrast to the above, the triglycerides, waxes, and cholesterol esters are hydrophobic lipids. Surface tension develops at the lipid-water interfaces, and a globular shape ensues to reduce volume-surface ratio.

Most lipids are mutually soluble in each other. Physical characteristics of lipids are greatly modified depending upon the degree of saturation of fatty acid esters and by associated proteins.

Birefringent Lipids

POLARIZED LIGHT.   When examined in the dark field produced by crossing polaroids or Nicol prisms, neutral fats ordinarily remain dark. Substances forming Lehmann's "liquid crystals," such as cholesteryl esters, phosphatides, and cerebrosides, may exhibit the black cross of polarization with luminous quadrants between the arms of the cross filling out a circle. This phenomenon is suppressed if the temperature is above that at which the liquid crystals in question can exist and the globules remain dark. Any fatty substance in solid crystalline form may be luminous under polarized light. Sections showing such crystals should be compared with paraffin sections of the same tissue, where the luminous crystals should be absent if they are fatty in nature. Or a control frozen section may be extracted 10 min in chloroform-methanol (2 : 1 v/v).

The use of polarized light has been suggested by Prickett and Stevens (*Am. J. Pathol.*, **15:** 241, 1939) as a means of differentiating between normal birefringent myelin and degenerating singly refractile myelin. In practice we have not found this method particularly helpful.

In cutaneous xanthomas, this means of examination reveals more or less numerous, fine, needle-shaped, doubly refractile crystals associated with more plentiful isotropic sudanophilic fat droplets. These tumors often contain iron-positive pigment as well, and the fat may be cholesterol-positive. Here it would be of interest to apply the combined ferrocyanide plus oil red O technic.

Adams (*Adv. Lipid Res.*, **7:** 1, 1969) is convinced that not all forms of cholesterol are revealed as Maltese crosses under polarized light. Also, one must be aware that nonlipid constituents in tissues such as formalin pigment and amyloid may be birefringent. Cholesterol is often seen as birefringent rhomboid needles.

Differential Lipid Solubility

Free lipids in tissues can be effectively extracted with chloroform-methanol (2 : 1, v/v). Lipids bound to proteins cannot be removed unless the tissues are subjected to proteases or possibly acid hydrolysis.

Preferential removal of hydrophobic lipids including cholesterol, cholesterol esters, triglycerides, and fatty acids from frozen sections, while leaving phospholipids, has been affirmed by several investigators including Keilig (*Virchows Arch. Pathol. Anat.*, **312**: 405, 1944), Wolfgram and Rose [*Neurology (Minneap.)*, **8**: 839, 1958], Bubis and Wolman [*Nature (Lond.)*, **195**: 299, 1962], Dunnigan (*J. Atheroscler. Res.*, **4**: 144, 1964), and Adams and Bayliss (*J. Histochem. Cytochem.*, **16**: 115, 1968). Adams and Bayliss noted selective removal of hydrophobic lipids by acetone in most tissues; however, phospholipid also appeared to be removed from sections of atherosclerotic plaques. Elleder and Lojda (*Histochemie*, **14**: 47, 1968) also noted a failure of stains for phospholipids after acetone extraction of sections of atherosclerotic plaques as determined by sections stained by fat red 7B, acetylated Sudan black in 70% ethanol, controlled chromation followed by Elftman's acid hematein, Luxol Fast Blue, and chromatography.

Fixation

Some loss of polar lipids may occur if tissues are fixed in aqueous formalin alone. Feustel and Geyer [*Acta Histochem. (Jena)*, **25**: 219, 1966] have advocated the use of acrolein as a fixative for maximum retention of phospholipid. Baker (*Q. J. Microsc. Sci.*, **87**: 441, 1946; "Principles of Biological Microtechnique," Methuen, London, 1958), Elbers et al. (*J. Cell Biol.*, **24**: 23, 1965), and Adams noted retention of lipids in tissues was favored by addition of calcium to solutions of formalin.

The mechanism of action is presumed to be one whereby cationic bridges form between the polar groups of phospholipids, calcium, and other tissue constituents. Roozemond (*J. Histochem. Cytochem.*, **15**: 526, 1967) employed thin-layer chromatography and demonstrated convincingly the superiority of calcium-formalin solutions for fixation and retention of lipids in tissue sections.

Hydrophilic phospholipids may also be retained in tissues treated with osmium tetroxide as advocated by many including Wigglesworth (*Proc. R. Soc. Lond. [Biol.]*, **147**: 185, 1957) and Baker (*J. Histochem. Cytochem.*, **6**: 303, 1958) and by potassium dichromate as recommended by Elftman (*J. Histochem. Cytochem.*, **2**: 1, 1954). Osmium and dichromate treatments probably modify lipids in many ways including oxidation. The effects of fixation on the structure of lipids are largely unknown. Both these oxidants produce cyclic metal esters bridging a former double bond (see Chap. 8, section on ethylenes).

Wolman (*Lab. Invest.*, **14**: 460, 1965) noted that fixatives containing NaCl resulted in ruptured cell membranes, glycogen displacement, and distortion of lipid droplets, whereas fixatives containing $CaCl_2$ appeared to preserve normal cell structures.

Elbers et al. (*J. Cell Biol.*, **24**: 23, 1965) recommend for the study of lipids in membranes the use of several solutions containing suitable cations and anions in order to effect a "tricomplex flocculation" whereby fixation is said to be achieved without alteration of lipid structure. The method needs to be tested on tissues.

Elleder and Lojda (*Histochemie*, **34**: 143, 1973) note significant amounts of sphingomyelin, cerebrosides, sulfatides, and gangliosides can be detected in tissue sections prepared in the usual fashion after formalin fixation, ethanol acetone and xylene dehydration and clearing, and paraffin embedding.

Histochemical Staining Methods

Histochemical staining methods are reasonably reliable for the identification of fatty acids, triglycerides, glycerophosphatides, sphingomyelins, cholesterol, cerebrosides, gangliosides, plasmalogens, and lipofuscins. These methods are not absolutely specific. Results of staining reactions must be used in conjunction with other data available. Although the above staining methods are not quantitative, they may be used in conjunction with autoradiography, microdissection, thin-layer chromatography, scintillation counting, and other quantitative methods, thereby achieving substantial quantitative data in relation to tissue and cellular localization.

NEUTRAL FATS

The most ancient method (Schultze, *Arch. Mikrosk. Anat.*, **1**: 299, 1865) for coloration of fatty substances in tissues is the reduction of osmium tetroxide (commonly called "osmic acid"). This reagent is reduced to a black substance by unsaturated fats and fatty acids and by a variety of other reducing agents, such as eleidin and tannin, and if osmium tetroxide is followed by 60 to 70% alcohol, stearin and palmitin are also blackened. Myelin is also blackened by osmium tetroxide, but if previously treated with chromate solutions, only degenerating myelin is so blackened.

Generally speaking, the oil-soluble dye methods are much less troublesome and more satisfactory for the demonstration of fatty substances.

While quinoline blue (C.I. 1st ed., 806) was used in 1875 as a lysochrome for staining of myelin and fat in the course of intestinal absorption by Ranvier, it faded badly with light and was quickly replaced on Daddi's (*Arch. Ital. Biol.*, **26**: 143, 1896) introduction of Sudan III (C.I. 26100).

Michaelis (*Virchows Arch. Pathol. Anat.*, **164**: 263, 1901) later introduced the use of Sudan IV, or Scharlach R, C.I. 26105, which has been the most popular of the oil-soluble dyes.

Following the long use of these two Sudan dyes, Sudan III and IV, the property of staining with oil-soluble dyes generally has come to be referred to as sudanophilia, even though the oil-soluble dye used may not have the word Sudan included in its name.

FAT STAINS WITH OIL-SOLUBLE DYES

The findings of Kay and Whitehead (*J. Pathol. Bacteriol.*, **53**: 279, 1941) indicate that stronger staining of fats is obtained when two or more homologs or isomers of the naphthol Sudans are saturated in the same stock staining solution. This they explain on the basis that each of the chemical individuals dissolves to its own saturation point, more or less independently of the other dyes present, both in the hydroalcoholic dye solvent and in the fats themselves. Since the absorption maxima of the red Sudans lie fairly close together in the 510- to 530-mμ range, denser coloration is thus achieved in that range. This principle seems equally applicable to the various solvents employed for Sudan staining, since it is the final optical density in the demonstrated oil or fat that is significant.

Since oil red O (C.I. 26125) is one of a group of homologs of Sudan III (C.I. 26100), Sudan IV (C.I. 26105), and oil red 4B (C.I. 26120), admixture of commercial samples of all four should yield more potent staining mixtures of very similar chemical behavior. For histochemical use it is not recommended that naphthol-type dyes of this sort be mixed with the chemically distinct naphthylamine and aminoanthraquinone oil-soluble dyes.

Staining with oil-soluble dyes is based on the greater solubility of the dye in lipid substances than in the usual hydroalcoholic (and other) dye solvents. Michaelis so described it in Ehrlich's "Encyklopädie," Lison (1936) so characterized it, and Cain and Harrison (*J. Anat.*, **84**: 196, 1950) still state that these substances operate "not as a dye, but only as an oil-soluble colorant." However, the "Colour Index," (1956, 1971), classifies them as "solvent dyes."

This staining has long been regarded as quite specific for lipids. Cain and Harrison make the important exception that fats of high melting point do not color with Sudan dyes unless heated to near or above their melting points. Thus cholesterol and its esters, carotenoids, tristearin, and high-melting-point paraffins may be stained when melted, but not at room temperature.

There is an important corollary to this theory of staining by oil-soluble dyes. The dye should be again extracted by an excess of a suitable dye solvent which does not dissolve the lipid, and the lipid should then be again restainable by the original technic. Most of the lipids which survive paraffin embedding can thus be completely decolorized in a matter of a few seconds to 1 or 2 min with acetone, and stained again with oil-soluble dyes. The decolorization and restaining can be reperformed repeatedly. Similarly, frozen sections dried onto slides and stained by Sudan black B, C.I. 26150, or oil red O, C.I. 26125, can be decolorized by 3- or 4-h extraction in diethylene glycol, restained, and decolorized repeatedly. It is found, however, that some substances, such as the sudanophil granules of leukocytes, cannot be decolorized by prolonged exposure to acetone, xylene, chloroform, and the like, even at elevated temperature. Erythrocytes stain intensely on prolonged heating with Sudan black B, but not with oil red O or Sudan IV, and the stain is not extracted by acetone or xylene.

It is evident that the Sudans are not inert chemically (Lillie and Burtner, *J. Histochem. Cytochem.*, **1**: 8, 1953), as had been supposed, but are capable of forming firm unions with certain tissue elements, some of which, at least, are probably not lipids. These combinations occur less promptly than with ordinary fat stains, and may require elevation of temperature for their accomplishment. Hence any Sudan or other oil-soluble-dye staining is suspect in which one dye is allegedly much better as an oil-soluble dye than another or in which resistance to immersion oil or to xylene or absolute alcohol has been noted; and the lipid nature of the stained substance needs confirmation by the extraction and restaining tests mentioned above.

Particularly dye bases of arylamine dyes are suspect in this regard (Table 6-8) and have actually been observed under some circumstances to stain chromosomes, cartilage matrix, and other acid tissue components.

It is perhaps noteworthy that acetylated and benzoylated Sudan dyes still stain ordinary fats quite effectively but fail to color leukocyte granules. It is possible that these ester dyes may be useful in discriminating true fats from chemically sudanophilic substances.

The original method of Daddi's (1896) called for a saturated solution of Sudan III in 70% alcohol. This takes a half hour or more to stain, gives a rather light orange color, and dissolves out an appreciable amount of fat, in some instances all the demonstrable fat that was present.

HERXHEIMER'S TECHNIC (*Zentralbl. Allg. Pathol.*, **14**: 891, 1903). Employing a saturated solution of Sudan IV (scarlet, or Scharlach, R) in a mixture of equal volumes of acetone and of 70% alcohol, this technic gave deep orange-red staining in a few minutes but also removed a considerable amount of fat, and in some cases all the fat present.

HERXHEIMER'S ALKALINE SUDAN IV (*Dtsch. Med. Wochenschr.*, **27**: 607, 1901). This saturated dye solution has been much used by neurohistologists. It contained 1% (Conn) or 2% (Mallory) sodium hydroxide in 70% alcohol. The solution is said to be unstable and must be discarded after two or three days. Globus ("Practical Neuroanatomy," p. 256, Wood, Baltimore, 1937) used 3.33% sodium hydroxide in 50% alcohol, again saturated with the dye.

TECHNIC

1. Place sections in a covered watch glass, heat until condensation droplets appear on the lid, and then allow to stand 15 min.
2. Wash 5 min in distilled water.
3. Counterstain in dilute Ehrlich's hematoxylin.
4. Wash in distilled water.
5. Mount in glycerol.

RESULTS: Fats are bright red.

This variant of Globus' method preserves some fats which are lost in the original Herxheimer solution, probably because of its lower (50%) alcohol content, as compared with the 70% of the original.

Romeis used a solution of Sudan IV in 40% alcohol and avoided fat losses but took some 18 to 24 h to stain.

Gross' (*Z. Wiss. Mikrosk.*, **47**: 64, 1930) technic, which used a saturated solution of Sudan IV in 50% diacetin, stained fats deep red promptly and did not dissolve out lipids. However, the solvent gradually decomposed and impaired the stain.

METHOD OF LILLIE AND ASHBURN (*Arch. Pathol.*, **36**: 432, 1943). Lillie and Ashburn developed the principle of using fresh aqueous dilutions of a saturated 99% isopropanol stock solution to 60 or 50% strength. These solutions were supersaturated with dye and stained vigorously and promptly in the first few hours after dilution, later becoming slow and relatively inefficient stains. Sudan IV, first used in this procedure, was soon replaced by dyes giving more stable supersaturated solutions and better color effects.

Oil red O, C.I. 26125, introduced by French (*Stain Technol.*, **1**: 79, 1926) in the Herxheimer technic, is found to be one of the best fat stains available in this supersaturated isopropanol method (*Stain Technol.*, **19**: 55, 1944). It gives a deep scarlet to fats. Oil red 4B, C.I. 26120, is similar.

Sudan II (C.I. 12140) gives a bright orange-yellow in this technic which contrasts well with a Weigert myelin stain. Sudan brown (C.I. 12020) gives a deep brownish red.

Coccinel red (*Stain Technol.*, **20**: 73, 1945) stains successfully from as low as 30% isopropanol, and also gives a deep scarlet. This dye is 1,5-bisamylaminoanthraquinone, and its 1,4 isomer oil blue N (C.I. 61555, *Stain Technol.*, **20**: 7, 1945) gives deep blue fats from 40% isopropanol solution.

SUPERSATURATED ISOPROPANOL METHOD. The following technic may be used with *Sudan brown*, *oil red 4B*, or *oil red O*, any of which is better than Sudan IV or Sudan III.

Prepare stock saturated dye solution in 99% isopropanol using 250 to 500 mg per 100 ml.

1. Dilute 6 ml stock solution with 4 ml water. Use 1% dextrin to stabilize and intensify the stain.
2. Let stand 10 to 15 min and then filter. The filtrate can be used for several hours.
3. Stain thin frozen sections for 10 min.
4. Wash in water.
5. Stain 5 min in an acid alum hematoxylin of about 0.1% hematoxylin content (undiluted Mayer's, 1 part of Lillie's to 4 of 2% acetic, or 1 of Ehrlich's to 5 of 2% acetic).
6. Blue in 1% disodium phosphate or in preboiled or unchlorinated tap water.
7. Float out in water, and mount on slides.
8. Drain and mount in gum syrup or glycerol gelatin.

RESULTS: Fats and lipofuscins are deeply stained red or brown according to the color of the dye. Birefringent lipids color less deeply but are conspicuously bright in polarized light.

Rinehart (*Arch. Pathol.*, **51**: 666, 1951) used oil red O, Sudan IV, Sudan black B, and coccinel red as supersaturated 60% ethanol solutions made in the same way. His staining interval was 5 min.

The amylaminoanthraquinone dyes oil blue N, coccinel red, and carycinel red were quite satisfactory in the supersaturated isopropanol technics. The first stained well from a final 40% isopropanol concentration; the other two, from as low as 30% isopropanol. Deep blue, scarlet, and deep crimson, respectively, were imparted to fats (*Stain Technol.*, **20**: 7, 73, 1945). For relatively alcohol-soluble birefringent lipids it would appear that dye solvents containing 40 to 50% water should be preferred and that Carbowax embedding should be avoided.

Chiffelle and Putt (*Stain Technol.*, **26**: 51, 1951) strongly recommend propylene or ethylene glycol as a solvent for fat stains. These solvents virtually do not attack aliphatic fatty acid esters but are said to have some solvent capacity for aromatic compounds, such as cholesterols and ketosteroids. The dyes recommended are Sudan IV and Sudan black B. These are soluble to about 0.5% in propylene glycol.

PROPYLENE GLYCOL METHOD OF CHIFFELLE AND PUTT.   Dissolve 0.7 g of Sudan IV or Sudan black B in 100 ml propylene glycol at 100 to 110°C. Do not exceed 110°C. Filter hot through Whatman No. 2 filter paper. Cool, and refilter with vacuum through a medium-porosity fritted glass filter.

TECHNIC

1.   Cut frozen sections, wash in water 2 to 5 min to remove formaldehyde.
2.   Dehydrate 3 to 5 min in pure propylene glycol, moving sections at intervals.
3.   Transfer to the dye solution for 5 to 7 min. Agitate occasionally.
4.   Differentiate in 85% propylene glycol for 2 to 3 min.
5.   Wash in distilled water 3 to 5 min.
6.   Counterstain if desired.
7.   Float onto slides, drain, and mount in glycerol gelatin.

RESULTS: Neutral fats, myelin, mitochondria, and other lipids are orange-red or greenish black; cytoplasm is unstained. No data are presented regarding preservation of birefringent lipids.

To avoid the dye precipitation often seen with supersaturated 50 to 60% alcohol solutions of oil-soluble dyes and the losses of birefringent crystalline lipids seen with glycol solutions, we have employed with encouraging results a solution of 500 mg Sudan black B in 30 ml glycol plus 30 ml ethanol plus 40 ml distilled water. The solution is approximately saturated and can be used with 15- to 30-min staining intervals for 2 to 3 days. Stains are very clean and limited to lipids while the solution is fresh, but more or less diffuse gray staining of muscle and cytoplasm appears in week-old solutions. Hence it has been our practice to make the solution fresh every second day (Gutiérrez and Lillie, *Stain Technol.*, **40**: 178, 1965).

M. E. BOTT'S OIL RED O PROPYLENE GLYCOL FAT STAIN   (1953). Substitute 0.5 g oil red O for Sudan IV or Sudan black B in the Chiffelle-Putt method. Stain frozen sections for 20 to 30 min in the solution, differentiate in 85% glycol, wash, stain in Harris' hematoxylin, blue, and mount in glycerol gelatin.

Adams, Abdulla, Bayliss, and Weller (*J. Histochem. Cytochem.*, **14**: 385, 1966) have recently developed a specific method for the triglycerides provided a pure enzyme is available and employed. In essence, pancreatic lipase is employed in the presence of calcium ions. Released fatty acids are precipitated as calcium soaps which are subsequently converted to lead soaps. A black lead sulfide stain is developed by exposure to dilute ammonium sulfide.

A crucial element in the development of the specific stain is the purity of the pancreatic lipase. However, this can be checked. Chromatographic and other methods may be used to check the purity of the enzyme to be employed. In addition, digestion products obtained after use of the enzyme can be checked to determine purity. Specific inhibitors and activators can also be used. Adams' (*Adv. Lipid Res.*, **7**: 1, 1969) experience suggests that waxes as well as triglycerides are revealed by the method.

THE ADAMS ET AL. LIPASE–LEAD SULFIDE TECHNIC FOR TRIGLYCERIDES
(*J. Histochem. Cytochem.*, **14**: 385, 1966)

1. Frozen sections are cut from tissue fixed in 1% calcium acetate–10% formalin.
2. Incubate free-floating sections in the reaction medium for 2 to 4 h at 37°C. The medium is prepared by adding 50 mg porcine pancreatic lipase to 10 ml 2% calcium chloride, 15 ml tris buffer (pH 8), and 25 ml distilled water. (The pancreatic lipase must be shown to be uncontaminated by other lipolytic enzymes. Purity can be established by testing the lipase against the individual spots obtained by previously separating a suitable lipid mixture on a thin-layer chromatoplate. The reaction products are then separated in the other dimension of the plate. Lipase from California Biochemical Corporation has been found free of contaminating enzymic activity. Other sources may also be satisfactory but require testing.) As with other enzymes, the lipase must be fresh and of known potency.
3. Wash well in several changes of distilled water for 15 min.
4. Treat with 1% lead nitrate for 15 min.
5. Wash well in several changes of distilled water.
6. Immerse in dilute ammonium sulfide (10 drops per 25 ml) for 1 min.
7. Wash well, counterstain with Mayer's alum hematoxylin, and mount in glycerin jelly.

Triglycerides and waxes are stained brownish black. Only the surface of large droplets is stained, because the reagents only slowly penetrate such droplets and lipolysis proceeds only at the water-fat interface.

A duplicate control section should be processed from step 3 onward in order to determine the extent of nonspecific lead binding such as that encountered with negatively charged lipid micelles (Rostgaard and Barrnett, *Anat. Rec.*, **152**: 325, 1965). Calcium deposits also give a false-positive reaction; they can be eliminated by preliminary treatment of the section with 20% EDTA at pH 6.9 or 1% acetic acid for 30 min.

Sudan Black B

Sudan black B has two secondary amines available that may permit it to act as a basic dye. The secondary amines are unavailable on acetylated Sudan black B. Puchtler and Sweat (*Histochemie*, **4**: 20, 1964) mordanted sections of Carnoy or ethanol-fixed tissues with 1% aqueous phosphomolybdic acid and thereafter immersed the sections in either of the above Sudan black B stains. Inasmuch as Sudan black B stained mordanted basement membranes, reticulum, and collagen fibers intensely and these structures failed to stain with acetylated Sudan black B, the conclusion was drawn that the basic groups of Sudan black B may be reactive with anionic groups. Likewise, oil red O and Sudan IV failed to stain phosphomolybdic mordanted membranes.

BAYLISS AND ADAMS BSB METHOD.   Bayliss and Adams (*Histochem. J.*, **4**: 505, 1972), noting that cholesterol crystals failed to stain with Sudan black and that considerable losses of phospholipids, notably phosphatidylcholine, occurred in the dye solvent, have largely remedied these defects with their bromine Sudan black (BSB) technic:

1. Cut fresh tissues on the cryostat or use frozen sections of formol-fixed tissue.
2. Brominate 1 h in 2.5% bromine water.
3. Wash, and dip into 0.5% sodium bisulfite to consume excess bromine.
4. Wash well in water, and stain 10 min in filtered saturated Sudan black B (C.I. 26150) in 70% alcohol.
5. Differentiate in 70% alcohol until a control predefatted stained section is colorless.
6. Wash in water (and counterstain if desired with kernechtrot) and mount in glycerol gelatin.

RESULTS: Lipids color blue-black. Nuclei are shown by the counterstain.
Controls may be carried out omitting bromination.

Cohen (*Stain Technol.*, **24:** 177, 1949) reports that nuclear material of blood smears immersed in acetic, citric, oxalic, or formic acids stains with Sudan black B.

Wigglesworth (*J. Cell Sci.*, **8:** 709, 1972) describes a method for "bound" lipid as follows:

1. Fix tissues in 2.5 or 5% glutaraldehyde in cacodylate buffer containing 2% sucrose for 2 to 3 h at 24°C, or, alternatively, in 1% osmium tetroxide in Ringer's solution for 2 to 3 h at 24°C.
2. Embed tissues in agar (2.5% for 1 h at 60°C).
3. Embed trimmed agar blocks in ester wax and cool rapidly (ester wax of Steedman 1947 formula and not the 1960 formula must be used. The wax of the 1960 formula will not impregnate the agar block.)
4. Cut 0.5- to 1-μ sections; float them on water. At this stage, the water bath may contain the hypochlorite (a stock solution of 10% chlorine is diluted 1:100, pH 11.5). Alternatively, the sections floated on water may be collected on coverslips and dried at room temperature, and a few drops of the diluted hypochlorite solution applied to the sections. The treatment with hypochlorite is not critical; 3 to 4 min is generally adequate. Wigglesworth indicates a time period of 15 s to 10 min may be employed.
5. Sections are rinsed, dried, and passed rapidly through xylene down to 70% ethanol.
6. Stain in Sudan black B (0.15% in 50% ethanol) for 10 min at 24°C.
7. Differentiate in 70% ethanol for 1/2 min.
8. Mount in Farrants' gum medium.

Wigglesworth notes that most, but not all, of the material stainable with Sudan black B may be extracted by chloroform-methanol (2:1) at 60°C prior to staining.

Schott et al. (*Histochemie*, **5:** 154, 1965) applied many lipids, proteins, and polysaccharides to filter papers and noted the avidity of Sudan III, Sudan IV, Sudan red B, Sudan red VIIB, Scharlach R, and Sudan black B staining. They recommend using Sudan black B in diacetin for greatest specificity, although some nonspecific staining was observed.

Irving (*Arch. Oral Biol.*, **1:** 89, 1959; **5:** 323, 1961) employs Sudan black B to identify lipid in sites undergoing calcification. He prescribes as follows:

1. Specimens are fixed in Baker's formol-calcium (1946), absolute alcohol, or Bouin's fluid.
2. Extract with pyridine according to Baker (1946): 1 h in 70% ethanol, 30 min in 50% ethanol, and 30 min in running water. Dehydrate in two changes of pyridine 20 to 25°C for 1 h each. Extract for 24 h in fresh pyridine at 60°C. Wash in running water for 2 h.
3. Decalcify with EDTA or nitric acid–formalin.
4. Dehydrate, embed in paraffin, and section.
5. Place sections briefly in 70% ethanol.

6. Stain in a saturated solution of Sudan black B in 70 % ethanol for 2 to 3 min.
7. Rinse briefly in 70 % ethanol.
8. Rinse in 50 % ethanol for 1 min.
9. Rinse several times in distilled water.
10. Mount in glycerin jelly.

RESULTS: Without pyridine extraction, no specific staining of bones, teeth, or other structures undergoing calcification is observed. With pyridine extraction, intense staining is observed at sites undergoing calcification, i.e., at the predentin-dentin junction, at sites of calcospherites, of enamel matrix at the junction where the matrix becomes acid-soluble, and at the sites where bone and cartilage are undergoing mineralization.

Dobrogorski and Braunstein (*Am. J. Clin. Pathol.*, **40**: 435, 1963) studied 223 human neoplasms with stains for lipids, acid mucopolysaccharides, and glycogen. Positive stains for lipid were noted consistently in estrogenic ovarian tumors, renal cell carcinomas, prostatic carcinomas, liposarcomas, and "a variety of stromal tumors."

Hopwood (*Histochemie*, **14**: 270, 1968) applied several stains for lipids to a study of the adrenals of the ox and sheep.

Broderson and Hayes (*Histochemie*, **16**: 97, 1968) compared results obtained with the Hayes histochemical method for plasmal (*Stain Technol.*, **24**: 19, 1949) with the biochemical method advocated by Williams et al. (*J. Lipid Res.*, **3**: 378, 1962) in a variety of tissues and noted close agreement. Using 10-μ sections Broderson and Hayes calculated 0.7 μmol/g tissue is required for detection with the histochemical procedure.

Hojdu et al. [*Acta Cytol.* (*Baltimore*), **15**: 31, 1971] are convinced that finding oil red O–stained cells in unfixed smears of urine sediment is a great aid in the diagnosis of renal cell carcinoma of the kidney. Smears from epidermoid carcinomas of the renal pelvis were lipid-negative in his small series.

STAINING OF LIPIDS WITH ESTERIFIED SUDAN DYES

Benzoylated oil red O was made for Lillie by the National Aniline Division, Allied Chemical & Dyestuffs Corp. Acetylated Sudan IV and Sudan black B can be made quite simply: Dissolve 2 to 2.5 g dye in 60 ml pyridine, then add 40 ml acetic anhydride, and let the solution stand overnight. Then pour it into 3 or 4 l distilled water, let it stand a few hours, and filter out the precipitated dyestuff on a Buchner funnel with vacuum. Dry the precipitate, the funnel, and the glassware used in the precipitation. Dissolve the precipitate from the glass, funnel, and filter paper with acetone, and evaporate to a constant weight at 60°C. Yields range from 75 % up to nearly the amount postulated. While acetylated oil red O may also be prepared, the resultant ester is a tarry red mass at room temperature.

Acetylated Sudan black B appears to give less background staining and just as intense lipid staining as the untreated dye. Acetylated and benzoylated oil red O and acetyl Sudan IV fail to stain human neutrophil leukocytes in 16 h at 37°C, but at 60°C there is apparently some hydrolysis of the esters, and some coloration results on prolonged staining.

By observing the precaution of staining at 37°C or lower and by using a control human blood film fixed either with formaldehyde gas or with 75 % alcohol for 10 min, these esterified dyes can be used to aid in discriminating lipid staining from stable sudanophilia.

The technics are those usual for fat stains. We have employed stock solutions at 550 to 600 mg per 100 ml in 100% ethanol, diluting at time of use to 60 or 50% alcohol content with distilled water. Filtration of the freshly diluted mixture reduces precipitation on sections, and washing after staining in 50 to 60% alcohol is helpful.

FLUORESCENCE MICROSCOPY   For the demonstration of fats in fluorescence microscopy, Popper (*Arch. Pathol.*, **31**: 766, 1941) used either a 10-s stain in 1% aqueous methylene blue (C.I. 52015) or a 3-min stain in 0.1% aqueous phosphine (C.I. 46045). The first gave a blue fluorescence which suppressed green fluorescences other than that of vitamin A. The second produced a silvery white fluorescence and demonstrated more fats than traditional strong alcoholic Sudan methods.

Both these dyes are basic, and both have been used also for demonstration of free or bound nucleic acids. Hence the specificity is questionable.

As with stains with Sudan dyes, the fluorescence color should be extractable by a dye solvent which does not dissolve the lipid, and the fatty substance should be restainable, if the fluorescence demonstration is to be regarded as evidence of lipid nature of the stained substance.

3,4-Benzpyrene is a colorless oil-soluble hydrocarbon which is highly fluorescent (blue) in ultraviolet light. Lison (3d ed., pp. 479–480) recites that it is the only lysochrome (oil-soluble dye) known which is certainly completely lacking in polar groups, that it is highly sensitive, and that it is the most specific colorant for lipids. He apparently considers it completely unreactive for forming other types of staining than simple oil-solubility.

However, it must not be forgotten that 3,4-benzpyrene is one of the most chemically reactive hydrocarbons known (Fieser and Fieser, "Organic Chemistry," 3d ed., Heath, Boston, 1956). It azo-couples readily at C-5 even with *p*-nitrophenyldiazonium chloride; it forms a 5-acetoxy derivative with lead tetraacetate in acetic acid and a 5-thiocyano derivative with thiocyanogen. Moreover, this carcinogenic hydrocarbon is readily metabolized in vivo to noncarcinogenic phenolic derivatives. It would seem to be a mistake to assume from its hydrocarbon structure that it is not capable of chemical reactivity under staining conditions.

According to Lison, Berg (*Acta Pathol. Microbiol. Scand. Suppl.* 90, 1951) used a 0.00075% solution in 0.75% caffeine in water. Apparently observation is done directly in water, and no permanent preparations result. The fluorescence is blue to bluish white.

In our judgment, if the preparations cannot be destained with an appropriate solvent and then restained as before, the lipid nature of the material stained is just as questionable as with phenolic and amine dyes.

Lison's explanation of stable sudanophilia as true oil-soluble dye staining, despite the nonextractability of the dye, directly controverts the whole theory of differential oil-solubility of dyes as the basis of fat staining. Many organic reactions are accomplished better in nonpolar than in polar solvents.

Phosphoglycerides

BAKER'S ACID HEMATEIN TEST   (*Q. J. Microsc. Sci.*, **87**: 441, 1946)

1. Fix 6 h in Baker's calcium chloride–formalin.
2. Transfer for 18 h to 5% potassium dichromate containing 1% calcium chloride ($CaCl_2$).
3. Transfer to a second bath of 1% $CaCl_2$–5% $K_2Cr_2O_7$ for 24 h at 60°C.
4. Wash 6 to 18 h in running water.

5. Cut frozen sections at 10 μ directly, or after Baker's gelatin embedding. Harden the gelatin block 18 h in calcium chloride–formalin, and wash the gelatin block in running water for 30 min before sectioning.
6. Incubate sections 1 h at 60°C in 1% $CaCl_2$–5% $K_2Cr_2O_7$ solution, and wash in several changes of distilled water (5 min total).
7. Incubate in acid hematein solution for 5 h at 37°C. [Acid hematein: Boil 50 mg hematoxylin with 10 mg sodium iodate (1 ml of 1%) in 49 ml distilled water. Cool and add 1 ml glacial acetic acid. Prepare fresh daily.] (The $NaIO_3$ can probably be decreased to 3 to 5 mg, and used cold.—R.D.L.)
8. Rinse in distilled water and leave 18 h at 37°C in borax ferricyanide [250 mg each of borax ($Na_2B_4O_7 \cdot 10H_2O$) and of potassium ferricyanide $K_3Fe(CN)_6$ in 100 ml distilled water].
9. Wash in distilled water (four or five changes, 10 min total).
10. Mount in Farrants' or Kaiser's medium, *or* dehydrate, clear, and mount in balsam.

RESULTS: Phospholipids—lecithin, cephalin, and sphingomyelin—are dark blue to blue-black; galactolipids from brain, blue-black to pale blue; gelatin,[1] black or brown; mucin, dark blue to pale blue or brown.

The Baker method is probably not specific for phospholipids (Adams, *Adv. Lipid Res.,* **7**: 1, 1969). Bourgeois and Hack [*Acta Histochem.* (*Jena*), **14**: 297, 1962] noted that neutral lipids stain equally well, and specificity was not enhanced by increasing the period of chromation. Bourgeois and Hubbard (*J. Histochem. Cytochem.,* **13**: 571, 1965) believe that only choline-containing phospholipids are convincingly stained by the method. Their conviction was based upon stained chromatograms. Choline-containing phospholipids stained, whereas nonphosphorylated lipids and non-choline-containing phospholipids failed to stain. Adams (*Adv. Lipid Res.,* **7**: 1, 1969) suggests that the presence of unsaturated fatty acid chains accentuates the staining reaction; however, he finds enhancement of the staining reaction by a bromination blockade inexplicable and puzzling. Adams also points out that protein-bound phospholipids cannot be removed by a pyridine extraction (such as recommended by Baker), although he suggests that a chloroform–methanol–hydrochloric acid extraction (66 : 33 : 1, v/v/v) may be successful.

BAKER'S PYRIDINE EXTRACTION TEST. Fix 20 h in dilute Bouin's fluid: saturated aqueous picric acid, 50 ml; formalin (40% formaldehyde), 10 ml; glacial acetic acid, 5 ml; water, 35 ml. Extract for 1 h in 70% alcohol, for 30 min in 50% alcohol, for 30 min in running water. Dehydrate in two changes of pyridine at 20 to 25°C, 1 h each, and extract for 24 h in fresh pyridine at 60°C. Wash for 2 h in running water. Transfer to 1% $CaCl_2$ 5% $K_2Cr_2O_7$ at step 2 of the acid hematein test, and proceed as before with that test.
RESULTS: Lecithin, cephalin, sphingomyelin, and galactolipids remain unstained. Mucin, gelatin, chromatins are stained black, blue-black, or dark brown. Erythrocytes stain black both with and without pyridine extraction. Mitochondria and myelin are positive without extraction, negative after pyridine. Nuclei stain after extraction but not before.

We regard sudanophilia to Sudan black B as a more critical test of the adequacy of the pyridine extraction than Baker's hematein reaction. Besides the lipofuscins, which remain sudanophil after primary hot methanol-chloroform fixation and in formalin-fixed tissue remain sudanophil after 24-h 60°C extraction with pyridine, rat liver ceroid resisted 3 days at 60°C; we have noted that formalin-fixed human brain and nerve root myelin quite

---

[1] Information about avoiding the gelatin staining artifact was given in an early chapter.

regularly stains almost as well with Sudan black B after 24-h, 60°C extraction in pyridine. In the same experiment the periodic acid Schiff (galactolipids) and peracetic acid Schiff (unsaturated lipids) reactions remained at most slightly impaired after similar pyridine extraction.

ELFTMAN'S CONTROLLED CHROMATION HEMATOXYLIN METHOD FOR PHOSPHOLIPIDS (*J. Histochem. Cytochem.*, **2**: 1, 1954). Elftman considerably simplified the rather complicated Baker acid hematein method and further controlled it by setting the pH of the dichromate solution at 3.5.

1. Place fresh tissue in 2.5% $K_2Cr_2O_7$ buffered to pH 3.5 (see Table 3-5, and substitute 5% $K_2Cr_2O_7$ for the 6% in the table), and heat for 18 h at 56°C.
2. Wash for 6 h in running water or in a siphonage device.
3. Dehydrate in alcohols, clear in xylene or other solvent, embed in paraffin.
4. Section at 5 to 25 μ, deparaffinize, and hydrate as usual.
5. To 50 ml 0.1 M acetate buffer pH 3 add 0.5 ml 0.5% potassium ferricyanide and 50 mg hematoxylin. Warm to 56°C. Stain sections in this solution for 2 h at 56°C.
6. Wash in distilled water, dehydrate in alcohols, clear in xylene, mount in synthetic resin.

RESULTS: Phospholipid sites—myelin sheaths, Golgi lipids—are dark blue.

The function of the ferricyanide in the staining bath is to prevent background staining and to replace the Weigert borax–ferricyanide differentiation. Acid ferricyanide is not an oxidant.

If the available oven or water bath is not regulated to 56°C, the chromation time should be adjusted. Since the speed of chemical reactions approximately doubles with each 10-degree Celsius rise in temperature, adjustments for lower or higher temperatures may be made by adding one-tenth of the logarithm of 2, or 0.030103, to the logarithm of 56 for each degree below that temperature, or subtracting similarly for each degree above it.

The applicability of such methods as this to material previously fixed in formalin needs further study. Hence for routine fixed material the more traditional methods are preferred.

Bevan et al. (*J. Chem. Soc.* [*Org.*], 1951, p. 841) used phosphomolybdic acid to identify choline-containing lipids in paper chromatograms, developing the molybdenum blue with stannous chloride. Levine and Chargaff (*J. Biol. Chem.*, **192**: 465, 1951) used a similar method. Landing et al. (*Lab. Invest.*, **1**: 456, 1952) adapted this procedure to tissue staining.

1. Frozen sections of formalin-fixed tissue are floated on gelatin-coated slides, drained, blotted, and exposed to formaldehyde vapor for 10 to 15 min. Fresh frozen sections are thawed onto slides and dried briefly. Paraffin sections are deparaffinized.
2. Dry sections thoroughly; dip in acetone and ether (50 : 50).
3. Immerse for 15 min in 1% phosphomolybdic acid in ethanol and chloroform (50 : 50). Rinse in ethanol chloroform and two changes of chloroform; dry.
4. Immerse for a few seconds in fresh 1% stannous chloride in 3 N hydrochloric acid (1 volume concentrated HCl, 3 volumes H_2O).
5. Wash in water, counterstain in aqueous eosin, dehydrate, clear, and mount in balsam or synthetic resin.

RESULTS: Gaucher lipid (kerasin) is deep blue; Niemann-Pick lipid (sphingomyelin), lighter blue-green; Tay-Sachs ganglion cells are deep blue; normal myelin, deep blue. Faint to moderate staining is seen in adrenal cortical cells, lymphocyte and leukocyte cytoplasm, some squamous epithelial cells, margins of striated and cardiac muscle fibers, "endothelium" of alveolar and glomerular capillaries, ganglion cells, cytoplasm of renal, hepatic, and basal epithelial cells. In cryostat sections, myelin stained strongly; the other structures stained less strongly. Paraffin sections showed strong staining of myelin, cytoplasm of liver, adrenal cortical and ganglion cells, some muscle fibers, erythrocytes, and collagen in many areas. Weak staining appeared in smooth muscle, mucous glands, lingual epithelium, pancreas, fat, lymph nodes, spleen, bladder epithelium, testicular, gastric and intestinal epithelia, and adrenal medulla.

The binding to known lipids is much decreased or prevented by prior extraction with fat solvents, but binding by other structures seems relatively increased.

This procedure cannot be regarded as a binding of molybdenum by intrinsic PO_4 groups, since the phosphoacid is used. Phosphomolybdic acid is known to bind many basic organic reagents and is used in the dye industry to convert soluble basic dyes into insoluble pigments. Kerasin, the Gaucher lipid, contains neither choline nor phosphate.

However, strong staining of sudanophilic materials may well be associated with phosphatide nature of the fat concerned.

The other reactions observed in myelinated fibers are probably attributable to specific groupings in certain of the foregoing lipids.

Reactions for ethylenes, the peracetic Schiff reaction and the direct osmium tetroxide acetic hematoxylin after cold dichromate reactions, and probably also the bromine-silver methods, can be assigned to unsaturated fatty acid groupings in the phosphatides, lysophosphatides, and ceramide groups. The sphingosine double bond should also prove as reactive as that in oleins.

Reactions for carbohydrates, notably the periodic acid Schiff reaction, should be expected in the galactoside and glucoside cerebrosides and in the sulfatides, since sulfation of galactose C-6 should not interfere with a Criegee cleavage in one of the vicinal glycol sites. The linkage, whether 1,3 or 1,4, in the tetrasaccharide residue of the gangliosides could determine whether or not these would react to periodic acid. Lison has no hesitation about assigning a positive periodic acid Schiff reaction to gangliosides.

For the Gaucher lipid kerasin, Morrison and Hack (*Am. J. Pathol.*, **25**: 597, 1949) have specifically reported a positive periodic acid Schiff reaction in the foam cells. Bauer and Casella reactions were negative in Lillie's hands, using Morrison's material.

The Niemann-Pick lipid sphingomyelin, which lacks hexose, is periodic acid Schiff–negative according to Morrison and Hack (ibid.). Wolman's conflicting report was based on a 24-h periodic acid treatment and is explained on the assumption that the ceramide bond was hydrolyzed, thus leaving a 2-amino-3-hydroxy grouping for a positive Nicolet-Shinn reaction. With the normal short periodic acid treatment the reaction is negative (Lison, 1960).

As noted before, diphosphoinositides should give a positive periodic acid Schiff reaction, if insolubilized by the fixation employed.

Basophilia, as Lison says, should be shown by gangliosides because of the sialic acid residue, and by sulfatides on account of the SO_4H residue on galactose C-6. The same behavior toward methylation and demethylation would be expected as with the sialic acid mucins and the sulfated aminopolysaccharides. But hot methanol controls must be performed.

The traditional myelin technics include fixations in dichromate alone, in formalin dichromate, in formalin followed by dichromate, and in formalin alone, without subsequent use of dichromate. Sometimes copper salts and trivalent chromium salts have also been interposed as "mordants." The hematoxylin used for staining has in some instances lacked any mordant metal, in some cases it follows treatment by a ferric salt, and in some it is mixed with ferric alum or ferric chloride.

In all cases, however, successful technics have required treatment by dichromate or by a ferric salt before extraction with fat solvents, and the hematoxylin, if uncombined with a ferric salt at the time of using, has to depend on prebound chromium or iron to form a blue to black lake complex.

Some myelin staining can be achieved with iron hematoxylin on material fixed in formalin and embedded in paraffin without prior chromation or ferric chloride oxidation, but the stain is generally weaker and of poorer quality.

Despite successes with sequence iron hematoxylins on frozen sections, the technics requiring chromation before embedding have been generally favored, either with unmordanted hematoxylins, depending on the chelated chromium, or with sequence or mixed iron hematoxylins.

It appears in our experience that primary fixation in formalin followed by cutting of thin blocks which are then chromated some days at 25°C in 2.5% $K_2Cr_2O_7$ before paraffin embedding is the preferable procedure. Primary dichromate-formalin fixations appear to give less even myelin preservation in brain blocks.

Hot dichromate treatments, as in the Baker and Elftman procedures, induce rather pronounced changes in staining with acid and basic dyes. Brain and root ganglion tissue which normally gives a satisfactory red-blue balance with azure A–eosin B at pH 4 or 4.5 after formalin or Kose's dichromate formalin fixation, after a 24-h chromation at 60°C in 5% $K_2Cr_2O_7$ requires a pH of 6.5 to 7 to get any nuclear staining. It is, of course, recognized that the lowering of tissue pK from the pH 6 which is useful for azure-eosin after simple methanol or ethanol fixation to 4 or 4.5 is due probably to the attachment of $=CH_2$ or $—CH_2OH$ groups to tissue amine residues, thereby decreasing their oxyphilia and consequently increasing their basophilia. It appears probable that their $=CH_2$ or $—CH_2OH$ groups are oxidized to formic acid or carbon dioxide by the hot dichromate, thereby restoring the basicity of the reliberated amine groups. Perhaps

$$RN{=}CH_2 + 2O \longrightarrow RNH_2 + CO_2 \quad \text{or}$$
$$RN{=}CH_2 + H_2O + O \longrightarrow RNH_2 + HCOOH$$
$$RNH{-}CH_2OH + 2O \longrightarrow RNH_2 + CO_2 + H_2O \quad \text{or}$$
$$RNH{-}CH_2OH + O \longrightarrow RNH_2 + HCOOH$$

We have not been too well satisfied with the density of the blue staining of myelin achieved by Baker and Elftman technics, as compared with more traditional myelin procedures. We have made some further alterations in the procedure first published by Lillie in 1944 (*Arch. Pathol.*, **37**: 392), and we are indebted to Windle and Rassmussen at the National Institutes of Health for their kindly evaluation of results at some stages in the differentiation procedure.

Holczinger (*Histochemie*, **4**: 120, 1964) applied several saturated and unsaturated fats on filter paper and tested their stainability with Baker's acid hematein method. Fatty acids with two or more double bonds were reactive.

Singh (*J. Histochem. Cytochem.*, **12**: 812, 1964) conducted a comparative histochemical study of Baker's acid hematein, Elftman's controlled chromatin, Pearse's copper phthalo-cyanin, and the phosphomolybdic acid method of Landing et al. (*Lab. Invest.*, **1**: 456, 1952) on avian central nervous tissues. Singh recommends the Elftman method for intraneuronal phospholipid identification.

DUNNIGAN'S NILE BLUE SULFATE METHOD FOR PHOSPHOLIPID
(*Stain Technol.*, **43**: 249, 1968)

1. Cut frozen sections at 10 μ from tissue blocks fixed in 10% formol-calcium. Gelatin-embedded blocks give the best results. Extract several sections with acetone for 1 h at room temperature (uncritical, 15 to 30°C).

2. Stain extracted and unextracted sections for 30 min in a 1% acid-hydrolyzed solution of Nile blue sulfate at 70°C. This solution is prepared by boiling 200 ml 1% solution of the commercial dye with 10 ml 1% H_2SO_4, with refluxing, for 4 h. The pH of this solution is almost exactly 2. It should be filtered before use.
3. Wash the stained sections thoroughly in water at 70°C, and mount in glycerol jelly.
4. Stain an extracted and an unextracted section in a saturated solution of Sudan black B in 70% alcohol for 30 min at room temperature. Rinse briefly in 70% alcohol and wash in water. Mount in glycerol jelly.

RESULTS: Phospholipids stain blue; lipids of other classes stain red or magenta. An overall picture is given by Sudan black B.

ADAMS OSMIUM TETROXIDE-α-NAPHTHYLAMINE (OTAN) TECHNIC FOR PHOSPHOLIPIDS (*J. Histochem. Cytochem.*, **8**: 262, 1960)

1. Cut 10- to 15-μ frozen sections from tissue fixed in 1% calcium acetate–10% formalin.
2. Treat free-floating sections for 18 h (overnight) with a mixture of 1 part 1% osmium tetroxide and 3 parts 1% potassium chlorate. The reaction vessel should be filled to the top and tightly stoppered to prevent vaporization of osmium tetroxide.
3. Wash the sections in distilled water, and then mount them on glass slides.
4. Treat the sections with a saturated aqueous solution of α-naphthylamine at 37°C for 20 min (10 to 15 min for 15-μ sections). The saturated solution of α-naphthylamine is prepared by adding an excess to distilled water, gently warming to 40°C, and then filtering. [α-Naphthylamine may contain the volatile carcinogenic β derivative. For this reason the reagent should not be handled; the solution should be prepared in a fume cupboard (hood); rubber gloves should be used throughout this stage; the reaction vessel should be sealed to prevent vaporization during incubation; and the vessel should be opened in a fume cupboard.]
5. Wash the sections in distilled water for 5 min.
6. Counterstain the sections with 2% Alcian blue in 2% acetic acid, usually for 15 to 60 s.
7. Wash in tap water and mount in glycerin jelly.

Unsaturated phospholipids are stained orange-red or orange-brown, while hydrophobic unsaturated lipid ester (triglycerides and cholesterol esters) and unsaturated fatty acids are stained black. Mixtures of hydrophilic and hydrophobic lipids may stain in intermediate shades. However, hydrophobic lipids may be removed by preliminary 2-h treatment with acetone at 4°C, leaving phospholipids intact. For reasons that are not at present understood, all phospholipid reactions of atherosclerotic plaques are nearly extinguished by preliminary acetone extraction (Adams, *Adv. Lipid Res.*, **7**: 1, 1969). The presence of protein-dispersed hydrophobic lipid may also be detected by prior brief trypsinization of tissue sections.

Interpretation of results observed with the OTAN method should be tempered with the following and perhaps other, as yet unknown, cautions:

1. Phosphoglycerides and sphingomyelin types of unsaturated phospholipids will give an orange-brown OTAN reaction; however, it is possible for hydrophobic dispersed lipids in lipoproteins to react in an analogous fashion.
2. Colors will appear darker in thick sections and in sections exposed to α-naphthylamine for periods longer than prescribed. Conversely, treatment of sections with α-naphthylamine for periods of time less than that prescribed may result in the orange-brown of lipids that stain black under prescribed conditions.
3. Quantitative estimations should be based on data other than the degree of color density or hue developed.

OTAN and Acid Hematein Stains

Elleder and Lojda (*Histochemie*, **24:** 21, 1970) advise caution in the interpretation of results for lipids derived by the use of the acid hematein and the OTAN methods. They convincingly demonstrated that hemoglobin in red blood cells can give a positive stain with both methods. They do not advance a theory of how hemoglobin A is stained and hemoglobin F is not. They conclude that the OTAN method is the most sensitive method for the detection of phospholipids; however, it is not specific and must be used in conjunction with extraction procedures. They also conclude that the acid hematein method is neither specific nor sensitive and that the gold-hydroxamate method detailed below is more specific than the OTAN method because it fails to stain erythrocytes.

Elleder and Lojda (*Histochemie*, **36:** 149, 1973) provide the following method for the histochemical identification of phospholipids:

Fresh frozen sections are used. One section is extracted in acetone for 10 min at 4°C and then dried in air. A comparable section is treated with chloroform-methanol (2 : 1) at 25°C for 10 min, washed briefly in acetone, and then washed in distilled water. Both sections are then fixed in formol-calcium, washed with distilled water, and stained in the following mixture (Lillie, 1965, p. 169):

Solution A:

Distilled water	298 ml
Concentrated HCl	2 ml
$FeCl_2 \cdot 6H_2O$	2.5 g
$FeSO_4 \cdot 7H_2O$	4.5 g

Solution B:

Distilled water	100 ml
Hematoxylin	1 g

Gentle heating may be needed to bring the hematoxylin into solution. For use, three portions of solution A are added to one portion of solution B, and the stain is used immediately. After 1 h or more the stain is less useful.

Sections are stained for 8 to 10 min, rinsed in distilled water, dipped several times in dilute (0.2%) HCl, washed thoroughly in tap water, and mounted either in Kaiser's glycerogel or in synthetic medium after acetone and xylene dehydration.

Phospholipids are stained blue in acetone-extracted sections. Sections extracted with chloroform-ethanol should be unstained. Elleder and Lojda recommend the method above that of any other for phospholipids. On the basis of tests with test compounds, they suggest that phosphate groups are responsible for the stain.

Elleder and Lojda (*Histochemie*, **37:** 371, 1973) assert the following stain is specific for sphingomyelin: Frozen sections cut at 12 to 16 μ from tissues fixed in 10% formalin are suitable. Fresh frozen sections subsequently fixed in 10% formalin for 10 min are also recommended.

The staining procedure is that prescribed for phospholipids by Elleder and Lojda using aqueous iron hematoxylin. Phospholipids are identified in sections that stain with iron hematoxylin after acetone extraction. Sphingolipids are identified in sections that stain with iron hematoxylin after acetone extraction *plus* alkaline hydrolysis in 1 N NaOH for 1 to 2 h at 25°C. Alkaline hydrolysis is believed to split ester bonds in phosphoglycerides but leaves amide bonds between sphingosine and fatty acids. Phosphoglycerides are thereby converted to water-soluble compounds, and the remaining sphingomyelin is stained with the iron hematoxylin.

NaOH–OTAN TECHNIC FOR SPHINGOMYELIN (modified from Adams and Bayliss, *J. Pathol. Bacteriol.*, **86**: 113, 1963).   Before step 2 of the OTAN method, hydrolyze free-floating sections with 2 *N* aqueous NaOH at 37°C for 1 h. Next gently wash the sections in water, rinse them in 1% acetic acid for 1 min, and again wash them with water. Then proceed with the OTAN method as detailed earlier.

RESULTS: Sphingomyelin and other alkali-resistant phospholipids are stained orange-red, while alkali-labile phospholipids (phosphoglycerides) are destroyed by hydrolysis. Cholesterol esters and triglycerides are only partly hydrolyzed, because NaOH is applied in aqueous and not alcoholic solution. Preliminary acetone extraction may be used to remove hydrophobic lipids.

THE GOLD-HYDROXAMIC ACID REACTION

Hydroxamic acids are formed by the action of alkaline hydroxylamine on ester bonds. Adams (*Adv. Lipid Res.*, **7**: 1, 1969) advocates the following method modified from Gallyas (*J. Neurochem.*, **10**: 125, 1963) for the histochemical detection of hydrophilic phosphoglycerides (lecithins, cephalins, phosphoinositides, and cardiolipin). Inasmuch as the alkaline hydrolysis is carried out in an aqueous medium, hydrophobic lipids are believed to be insusceptible to hydrolysis and thereby fail to develop hydroxamic acids needed for detection. Those who use the reaction are advised to remember that lipoprotein dispersal of hydrophobic esters may be reactive due to the changed environment.

The staining reaction can be abolished by prior ester bond hydrolysis. Blockade of aldehyde groups prior to treatment of sections with the silver reagent did not decrease the degree of silver reduction.

Adams notes that dispersed lipid in sections of atherosclerotic plaque are stained lightly by silver without prior alkaline hydroxylaminolysis. The nonspecific staining was not noted either in myelin or in the stratum germinativum of epidermis. Adams is also somewhat disturbed that the method does not consistently stain membranes of erythrocytes as might be expected.

GOLD-HYDROXAMIC ACID REACTION FOR PHOSPHOGLYCERIDES

(Adams et al., *J. Histochem. Cytochem.*, **11**: 560, 1963; Gallyas, *J. Neurochem.*, **10**: 125, 1963)

1. Cut frozen sections from tissues fixed in 1% calcium acetate–10% formalin.
2. Hydrolyze free-floating sections for 20 min in a mixture of equal parts of 12% sodium hydroxide and 5% hydroxylamine hydrochloride.
3. After hydroxylaminolysis, wash the sections with three changes of distilled water for 5 min each.
4. Next treat the sections for 1 to 2 h with an aqueous solution of 0.2% ammonium nitrate, 0.1% silver nitrate, and 0.025% NaOH; the silver reagent should be adjusted to pH 9.5 on the glass electrode. The reaction is faster in bright light (sunlight).
5. After washing in distilled water for 10 min, immerse the sections for 5 min in 1% acetic acid, wash them for 10 min in distilled water, and then tone them for 10 min with 0.2% yellow gold chloride.
6. Next briefly rinse the sections in distilled water, immerse them for 5 min in 5% sodium thiosulfate, and wash them for 10 min in distilled water.
7. Dry the sections onto slides, dehydrate, clear, and mount in DPX or Canada balsam. Sections may also be mounted in glycerin jelly.

RESULTS: Phosphoglycerides are stained a stable red-purple. A slight nonspecific reaction is encountered in some tissues; the extent of such staining is revealed by omitting step 2.

Böttcher and Boelsma-van Houte (*J. Atheroscler. Res.*, **4**: 109, 1964, and **7**: 269, 1967) have reported that *cis*-aconitic anhydride reacts with the choline-containing phospholipids (lecithin and sphingomyelin) to form a specific stain for these lipids. A handicap of the method is that quartz glassware must be employed. Elleder et al. (*Histochemie*, **16**: 294, 1968) studied the reliability of the *cis*-aconitic anhydride method of Böttcher and Boelsma-van Houte for choline-containing phospholipids (*J. Atheroscler. Res.*, **7**: 269, 1964). They found the prescribed oxidative polymerization with cobalt chloride and sodium periodate ineffective, and therefore do not recommend use of the method in practical histochemistry.

Erratic results on tissue sections have been obtained for phospholipids in the histochemical stains advocated by Landing et al. (*Lab. Invest.*, **1**: 456, 1952) and Landing and Freiman (*Am. J. Pathol.*, **33**: 1, 1957). Likewise, the specificity of methods for phospholipids which employ Luxol Fast Blue, bismuth iodide, mercuric nitrate–diphenylcarbazone, and hydroquinone-tetrazolium can be seriously questioned. Until further investigations reveal greater specificity, they can be advocated for disclosing the location of lipids only at the histologic level provided adequate controls are employed.

Schneider (*J. Lipid Res.*, **7**: 169, 1966) advocates the following stain for choline-containing phospholipids on thin-layer chromatograms:

1. Thin-layer chromatograms on silica gel prepared and developed in the usual manner are sprayed evenly with polyvinyl propionate (Neatan, E. G. Merck, Darmstadt, Germany).
2. After thorough drying, immerse the plate in water to aid in the release of the film.
3. Peel the film from the glass with the aid of a razor blade or sharp spatula.
4. Immerse the wet film in a 2% aqueous phosphomolybdic acid for 1 min at 24°C.
5. Wash in running water for 5 min.
6. Dip in 4% stannous chloride (40% stannous chloride in a concentrated HCl stock solution diluted 1 : 10) for 30 s.

RESULTS: Choline-containing phospholipids are stained blue on a white or faintly blue background. Inadequate washing gives more blue background. Table 13-2 provides Schneider's data on the staining of non-choline-containing lipids.

TABLE 13-2 APPROXIMATE LIMITS OF DETECTION FOR SOME LIPIDS NOT CONTAINING CHOLINE

Lipid	Micrograms detectable
Cholesterol	25
Cholesteryl palmitate	100
Palmitic acid	> 100
Dipalmitin	25
Tripalmitin	50
Phosphatidic acid	50
Phosphatidylethanolamine	10
Phosphatidylserine	10
Sphingosine	5

Source: Schneider, *J. Lipid Res.*, **7**: 169, 1966.

LUXOL FAST BLUE ARN. Lycette et al. (*Stain Technol.*, **45:** 155, 1970) noted that phosphatidylethanolamine, phosphatidylserine, and phosphatidylinositol bind Luxol Fast Blue ARN in aqueous solution apparently in a stoichiometric ratio of one dye molecule to two of lipid. Phosphatidylcholine, sphingomyelin, and palmitic acid reacted weakly.

PLASMALOGENS

Plasmalogens contain an unsaturated ether group which on hydrolysis forms an α,β-unsaturated alcohol. This in turn undergoes an enol-ketol shift to yield an aldehyde. They comprise both neutral lipids and phospholipids. Plasmalogens are the most abundant lipids of white matter and constitute 14% of total brain lipids, 33% of total phosphatides in ox brain. In this species 90% of the brain plasmalogen is phosphatidylethanolamine. The traditional methods for its histochemical demonstration are the application of Schiff reagent as in the plasmal reaction of Feulgen and Voit (*Hoppe Seylers Z. Physiol. Chem.*, **135:** 219, 1924) as standardized by Hayes (*Stain Technol.*, **21:** 19, 1949) and the use of phenylhydrazine as in Bennett's (*Am. J. Anat.*, **67:** 151, 1940) method or other hydrazines. More recent is the diphenylcarbazone method for localizing the specifically bound reactive mercury of Norton, Brotz, and Korey (*J. Neuropath. Exp. Neurol.*, **24:** 252, 1965).

Plasmalogens are readily soluble in alcohol and other fat solvents. They occur also in peripheral nerve myelin, adrenal cortex, corpus luteum, mammary and preputial glands, and fat cells. They are also found in some tissues showing no sudanophilic substances, such as muscle, kidney, liver, thyroid epithelium, and prostatic and seminal vesicle epithelium. The aldehyde reaction in elastic fibers was described by Feulgen and Voit as not prevented by fat solvents, not requiring acid or mercuric chloride treatment, and distinct from both the plasmal and nucleal reactions. It is discussed further under Elastic Fibers in Chap. 15 and noted in the discussion of aldehydes in Chap. 8. Nuclei react neither to the Feulgen plasmal method nor to the dinitrophenylhydrazine method.

The technics generally used employ Schiff reagent or a hydrazine to couple with the open carbonyl group. The first group includes the technic of Feulgen and its modifications—the use of frozen sections of unfixed tissue oxidized or hydrolyzed with mercuric chloride to release *plasmal* from *plasmalogens;* the technic of Verne—the use of tissue fixed in mercuric chloride or platinum chloride without controls; and the technic of Gerard—in which frozen sections of formaldehyde-fixed tissue are carefully washed in water and then treated with mercuric chloride. Untreated controls are used with the Feulgen and Gerard methods.

The second group includes the methods of Bennett (*Am. J. Anat.*, **67:** 151, 1940), employing phenylhydrazine or dinitrophenylhydrazine, which form yellow aldehyde complexes, and those of Seligman and Ashbel (*Cancer*, **4:** 579, 1951), in which 3-hydroxy-2-naphthoic acid hydrazide is first reacted with tissue carbonyls and then coupled with a stabilized diazonium salt to yield a colored compound.

For the dinitrophenylhydrazine reaction Albert and Leblond (*Endocrinology*, **39:** 386, 1946) prescribe 48-h fixation in 10% formalin (neutralized with magnesium carbonate), 24-h washing in running water, and frozen sections at 10 to 15 μ. Their technic is as follows: (1) Extract frozen sections for 4 h in 17% alcohol. (2) Soak overnight in saturated 30% alcohol solution of 2,4-dinitrophenylhydrazine, 17 ml, plus 0.2 N (1.64%) aqueous sodium acetate, 13 ml (emended; Albert and Leblond stated simply "enough to raise the pH to neutrality,"[1] noting a decrease of alcohol content to 17%). (3) Wash 20 min in 17% alcohol to remove

---

[1] Cf. Walpole's buffers.

excess stain. (4) Wash in distilled water and mount in glycerin gelatin. The positive reaction is yellow.

For the Feulgen plasmal reaction Albert and Leblond fix, wash, and section as above.

COMMENT. The use of such a long formaldehyde fixation without inclusion of a reducing agent should provide ready opportunity for spontaneous autooxidation of unsaturated fats, with formation of peroxides and aldehydes. Substitution of attached cryostat sections, treated with $HgCl_2$, as in the method of Ferrans and Hack for step 1 in the foregoing dinitrophenylhydrazine reaction, should operate to restrict it to plasmal lipids. A Seligmen-Ashbel technic using the 3-hydroxy-2-naphthoic acid hydrazide fast blue B sequence as the demonstration reaction could be similarly modified.

Hayes (*Stain Technol.*, **24:** 19, 1949) has redefined the plasmal reaction to make it more specific for acetal lipids.

THE PLASMAL REACTION OF FEULGEN AND VOIT, HAYES' 1949 MODIFICATION

1. Cut frozen sections of tissues at 15 μ either unfixed or fixed in neutral 10% formalin for *less than 6 h*. Or use unfixed smears or impression preparations. [Immerse unfixed tissue in (15%) aqueous gum arabic solution for 5 to 10 min, and freeze on a drop of the same to facilitate sectioning; or use Hack's polyethylene glycol method.] The use of attached cryostat sections of unfixed tissue is suggested.
2. Wash in several changes of 0.9% sodium chloride solution. (Use distilled water if sections are formalin-fixed.)
3. Place sections in 1% mercuric chloride solution for 2 to 10 min to allow complete penetration. For controls omit this step. Rinse glass section lifters before going into mercury solution.
4. Transfer mercury-treated and control sections directly to separate small, closed dishes of Schiff reagent, and let stand 5 to 15 min. Controls should remain negative. (Discard used Schiff reagent daily.)
5. Wash in three changes of 0.5% $NaHSO_3$ in 0.05 N HCl, for 2 min each.
6. Wash in several changes of water, float onto slides, and blot down.
7. Counterstain briefly in 0.5% methyl green in 0.5% acetic acid or in acetic hemalum (0.1% hematoxylin, 2% acetic acid) for 2 min. Wash in water. Blot.
8. Treat successively with 95% alcohol, 100% alcohol, 100% alcohol plus xylene (1:1), and two changes of xylene. Blot between changes if necessary. Mount in synthetic resin: polystyrene, Permount, HSR, etc.

RESULTS: Aldehyde sites (plasmal?) are shown in red. It is not clear how the alcohol-soluble aldehyde is insolubilized to resist step 8, and no alcohol extraction control before step 3 is prescribed.

Norton et al. (*J. Histochem. Cytochem.*, **10:** 375, 1962) fix ox adrenals and rat brains in Baker's calcium-cadmium-formalin with an excess of calcium carbonate at pH 5.7 to 5.8 or in calcium acrolein: 10 ml freshly distilled acrolein, 90 ml water, 1 g calcium chloride, 50 mg hydroquinone, calcium carbonate in excess (pH 6.4 to 6.8). The mixture is stable for several days at 3 to 5°C. Fixation intervals should be short, 1 to 3 h, and blocks should be only 3 to 4 mm thick. Norton and coworkers used the Hayes technic for their histologic studies. From concurrent chemical work it appeared that 90% of plasmalogen was phosphatidylethanolamine. Acrolein was so bound as to give much additional (artifactual) aldehyde reactivity. The use of acrolein fixation in plasmal studies seems contraindicated.

Ferrans, Hack, and Borowitz (*J. Histochem. Cytochem.*, **10:** 462, 1962) now identify the actual aldehyde-producing substance as an α,β-unsaturated glyceryl ether and consider that the acetal phosphatides are degradation products formed during isolation. They used cryostat sections or freeze-dried Carbowax sections of unfixed tissue.

1. Immerse sections for 2 min in 1.358% (50 mM) mercuric chloride.
2. Wash in distilled water for 1 min.
3. Treat with Schiff reagent for 10 min.
4. Wash in three changes of 0.05 M sulfurous acid.
5. Mount in glycerol and seal with lacquer.

CONTROLS: Omit step 1; this reveals native, including lipid, aldehydes. Extract with 2:1 chloroform-methanol before step 1; this reveals nonlipid or insoluble lipid aldehyde.

West and Todd formulate the plasmalogen as phosphatidylcholines or phosphatidyl-ethanolamines in which the γ fatty acid is replaced by an unsaturated ether group, which on hydrolysis undergoes an enol-ketol shift to form an alcohol-soluble aldehyde.

$$H_2C-O-CH=CH-R' + H_2O \longrightarrow HO-CH=CH-R'$$

Enol

$$R^2-CO-O-CH$$

$$H_2C-O-P-O- \left(\begin{array}{c} \text{Choline or} \\ \text{ethanolamine} \\ \text{N} \\ + \end{array} \right)$$

$$O=CH-CH_2-R'$$
Ketol
(aldehyde)

For their mercury reaction Norton, Brotz, and Korey (*J. Neuropathol. Exp. Neurol.,* **24:** 252, 1965) suggest a series of reactions in which the final aldehyde product still contains bound mercury:

$$RC(OH)=CHR' \rightleftharpoons R\overset{+}{O}=CH-CHR' \overset{+HgCl}{\rightleftharpoons} RO\overset{+}{C}H-\underset{HgCl}{CHR'} \overset{H_2O}{\rightleftharpoons}$$

$$\underset{HO}{\overset{RO}{>}}CH-\underset{HgCl}{CHR'} \rightleftharpoons ROH + HO\overset{+}{C}H-\underset{HgCl}{CHR'} \longrightarrow O=CH-\underset{HgCl}{CHR'}$$

This series of reactions is prevented by application before reacting with mercuric chloride of a 5-min iodination in 0.1 M iodine (KI₃) followed by decoloring in 0.1 M Na₂S₂O₃; by hydrolysis of the enol bond (0.1 N HCl for 1 h at 50°C); by lipid extraction with 2:1 chloroform-methanol or pyridine for 2 h at 25°C, by bromination, or by performic acid oxidation for 1 h at 25°C.

Iodination does not affect ordinary unsaturated lipids by this method; ICl or IBr is required for them.

Norton [*Nature (Lond.),* **184:** 144, 1959] observed that mercury was bound into the lipid in the plasmal reaction and conceived the idea that it might be localized by Brandino's reagent diphenylcarbazide. It was found that conversion of this reagent by oxidation to the carbazone was necessary for the reaction, and Norton, Brotz, and Korey (*J. Neuropathol. Exp. Neurol.,* **24:** 252, 1965) now prescribe Eastman's *sym*-diphenylcarbazone, which is a 1:1 mixture of diphenylcarbazone and diphenylcarbazide.

NORTON, BROTZ, AND KOREY'S MERCURY–DIPHENYLCARBAZONE PLASMAL METHOD

1. Fix overnight, 6 to 24 h, optimally about 20 h in 10% acrolein containing 1% CaCl, an excess of CaCO, and 0.05% hydroquinone. The acrolein should be freshly redistilled for making the solution.

2. Wash blocks thoroughly in water, section on freezing microtome at 20 μ. Handle sections on glass spatula.
3. Wash sections thoroughly in distilled water.
4. Hydrolyze for 30 min at 25°C in 1% mercuric chloride.
5. Wash in five changes of 1% NaCl and in five of distilled water.
6. React sections for 4 min in this solution:

> 0.1 g *sym*-diphenylcarbazone (Eastman)
> 70 ml 95% alcohol
> 25 ml distilled water
> 5 ml 2 *N* potassium hydroxide (final concentration 0.1 *N*)

7. Wash in three changes of distilled water, mount in glycerol gelatin. Study sections at once; the stain fades completely in a few days.

RESULTS: Plasmalogens are deep purple or violet.

In addition to the blockade procedures noted above for application before $HgCl_2$, the color reaction is prevented by iodine and thiosulfate or KCN applied after the $HgCl_2$ bath, by removal of bound Hg.

Application of SH blocking procedures is as follows: 30 min in 0.1% sodium *p*-chloromercuribenzoate; overnight in 0.1 *M* N-ethylmaleimide at pH 7.4 (0.1 *M* phosphate buffer) followed by a rinse in 1% acetic acid; or overnight in 0.1 *M* iodoacetamide in 0.1 *M* phosphate buffer of pH 8. This blocks SH reaction, but does not interfere with the reaction of plasmalogen.

Feulgen and Voit (*Pfluegers Arch.*, **206**: 389, 1924) developed the plasmal reaction for the detection of plasmalogen phospholipids. The α,β-unsaturated bond of plasmalogens appears to be specifically oxidized to aldehyde with the mercuric chloride reagent. A pseudoplasmal reaction must be excluded (aldehyde derived by atmospheric oxidation of other ethylene groups). Tissues should be fixed only briefly in formalin or acrolein, because these reagents rapidly condense with the aldehyde. Aldehydes formed by mercuric chloride oxidation may be blocked by several of the aldehyde-blocking reagents.

Adams (*Adv. Lipid Res.*, **7**: 1, 1969) has criticized the stain for plasmalogens advocated by Norton and Korey [in H. Jakob (ed.), "Proceedings of the Fourth International Congress of Neuropathology," vol. I, pp. 227–233, Thieme, Stuttgart, 1962] on the basis of lack of specificity. The method is stated to develop a purple chelate with diphenylcarbazone with Hg^{2+} attached to the α,β-unsaturated group.

Elleder and Lojda (*Histochemie*, **32**: 285, 1972) noted that in addition to reactivity with unsaturated bonds, bromination may block the pseudoplasmal reaction in unsaturated lipids (0.1 *N* bromine in water, 10 min), and the reaction of higher fatty aldehydes with Schiff's reagent (0.1 *N* bromine in water, 30 to 60 min). It may also oxidize a hydroxyl group of cholesterol to a keto group (0.1 *N* in water, 60 min) which is demonstrable with Schiff's reagent, and the intensity of the periodic acid Schiff stain is increased in areas containing monohexoside and dihexoside ceramides (0.1 *N* bromine in water, 1 min to 16 h).

Sphingolipids

CEREBROSIDES

Cerebrosides contain a hexose that may be detected with the periodic acid Schiff stain. The lipid may be distinguished from other carbohydrates, certain mucopolysaccharides, or sialic acid by removal in neutral or acidified chloroform-methanol according to Wolman (*J. Neurochem.*, **9**: 59, 1962).

Attention should be paid to the possible presence of lipid ethylene bonds that may be subject to periodic acid–engendered aldehyde (Adams and Bayliss, *J. Pathol. Bacteriol.*, **85**: 113, 1963) as well as direct atmospheric oxidation of these bonds. Other possible adulterating structures include the inositol ring in phosphoinositides, the 1-hydroxy-2-keto group in α-keto-corticosteroids, and the 1,2-hydroxyl groups in monoglycerides which are subject theoretically to developing a stain with the periodic acid Schiff method.

In addition to the lipid solubility of the cerebrosides, one may employ certain blocking procedures to obtain further information on the nature of the stained constituent. For example, 1-hydroxy-2-amino compounds and ethylene groups may be blocked by prior treatment of sections with chloramine-T–performic acid followed by 2,4-dinitrophenyl-hydrazine. Adams and Bayliss (ibid.) further recommend an alkaline hydrolysis to remove the possibility of reactivity of 1,2-hydroxyl groups in phosphoinositides, monogalactosyl diglyceride, and monoglycerides.

Cerebrosides with a sulfated 1,2-glycol group, known as *sulfatides*, are insusceptible to periodic acid oxidation to aldehyde under conditions employed histochemically. These sulfatides will be basophilic at low pH due to the strongly acidic sulfate group. Peiffer and Hirsch (*Excerpta Med. Sec. VIII*, **8**: 802, 1955) and Dayan (*J. Histochem. Cytochem.*, **15**: 421, 1967) have noted the development of a brown metachromasia with sulfatides in sections stained with an acetic acid–cresyl violet mixture. Dayan noted that the deposits subjected to polarized light develop a green dichroism. Mast cells do not display the green dichroism.

CRESYL VIOLET TECHNIQUE FOR SULFATIDE [Hirsch and Peiffer, in J. N. Cumings (ed.), "Cerebral Lipoidoses," pp. 68–76, Charles C Thomas, Springfield, Ill., 1957, and Dayan, *J. Histochem. Cytochem.*, **15**: 421, 1967]

1. Cut frozen sections from tissue fixed in 1% calcium acetate–10% formalin. Mount sections on glass slides and dry them in air.
2. Stain in 0.02 to 0.1% cresyl violet in 1% acetic acid for 10 to 30 min.
3. Wash in tap water.
4. Examine under water and then mount in glycerin jelly.

Deposits of sulfatide in metachromatic leukodystrophy are stained metachromatically brown. In polarized light these stained deposits exhibit green dichroism (Dayan, *J. Histochem. Cytochem.*, **15**: 421, 1967) which distinguishes them from the occasionally troublesome brown-tinted red reaction of mast cell granules.

Myelin stains lilac (orthochromatically) when sections are mounted in glycerin jelly. However, when sections are examined under water, central nervous system myelin is stained metachromatically brown, but peripheral nerve myelin remains orthochromatic violet.

To confirm the presence of sulfatide, the stained constituent should be extractable in lipid solvents. Mosen et al. (*Biochim. Biophys. Acta*, **116**: 146, 1966) has noted that cholesterol may be present in tissues as sulfate esters. Possible reactivity of such a possible contaminant should be considered.

Holländer (*J. Histochem. Cytochem.*, **11**: 118, 1963) proposes a method for demonstration of sulfuric esters of cerebrosides based on their metachromatic basophilia at low pH levels.

CEREBROSIDE SULFURIC ESTERS IN BRAIN

1. Fix in 10% formalin; cut frozen sections at 20 to 30 μ.
2. Stain 6 min in 1 : 20,000 (5 mg/100 ml) acriflavine in 0.1 M citrate HCl buffer of pH 2.5.
3. Wash 1 min in 70% isopropanol.

4. Transfer to a mixture of 15 ml 2% *p*-dimethylaminobenzaldehyde in 6 *N* HCl and 35 ml isopropyl alcohol for 75 s.
5. Wash in distilled water.
6. Counterstain 6 min in Mayer's acid alum hematoxylin.
7. Wash 15 min in running water.
8. Isopropyl alcohol, 70, 80, 90, 96, 100%.
9. Clear in xylene; mount in resin.

For fluorescence microscopy, omit steps 4 to 7 and use nonfluorescent resin for mounting. The full procedure yields carmine red sulfolipids and blue nuclei. The fluorescence procedure gives golden yellow to orange sulfolipids on light green myelin.

GANGLIOSIDES

Specific histochemical identification of gangliosides is precarious despite the hexose, hexosamine, and sialic acid content. The sialic acid constituents of gangliosides may be detected with R. Carlo's method for sialic acids detailed below.

METHOD FOR SIALIC ACIDS   (R. Carlo, *J. Histochem. Cytochem.*, **12:** 306, 1964)

1. Cut frozen sections from tissue fixed in 1% calcium acetate–10% formalin. Mount on glass slides.
2. Spray sections with Svennerholm-Bial reagent in a fume cupboard (hood). It is particularly important to use a very fine sprayer. The reagent solution is prepared as follows:

Orcinol	200 mg
Copper sulfate (0.1 *M*)	0.25 ml
Hydrochloric acid (12 *N*)	80 ml
Distilled water to make	100 ml

The solution should be allowed to "mature" for 4 h before use. It is important that the concentrated hydrochloric acid should be fresh.

3. Heat the sprayed sections in hydrochloric acid vapor at 70°C for 10 min. A thin layer of 12 *N* hydrochloric acid is poured into the bottom of a polyethylene Wilson jar; the whole container is brought to 70°C before the slides are inserted.
4. Rapidly dry sections in air, rinse in xylene, and mount in Canada balsam.

The deposits of ganglioside in Tay-Sachs disease are stained red, as are the sialomuco-polysaccharides in salivary gland acini.

Shear and Pearse [*Nature (Lond.)*, **198:** 1273, 1963] at first believed they had developed a specific histochemical method for the gangliosides; however, Pearse (1968, vol. 1, pp. 356–360) later indicated the test is unreliable.

Derry and Wolfe (*Exp. Brain Res.*, **5:** 32, 1968) used a fluorimetric method for the analysis of *N*-acetylneuraminic acid to determine the distribution of gangliosides in serially cut cryostat sections of the hippocampus and cerebellar folium area of ox brain. Wide differences were observed in relation to the varied structure as depicted in Fig. 13-1.

In gangliosides the ceramide residue is linked by its terminal alcohol group in glycosidic linkage to a tetrasaccharide consisting of a series of two hexose residues, *N*-acetylgalactos-amine, and sialic acid.

The class of proteolipids makes up about a tenth of the myelin lipids extracted from white substance. They are insoluble in water, soluble in fat solvents, and readily cleaved to

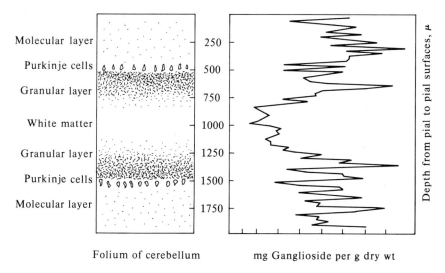

Molecular layer

Purkinje cells

Granular layer

White matter

Granular layer

Purkinje cells

Molecular layer

250
500
750
1000
1250
1500
1750

Depth from pial to pial surfaces, μ

Folium of cerebellum

mg Ganglioside per g dry wt

Fig. 13.1 Analyses of gangliosides in serial cryostat sections, 24 μ thick, through a folium of the ox cerebellum. On the left is shown the histological appearance after staining with eosin and methylene blue. In the center is shown the depth in microns from pial to pial surfaces. On the right the ganglioside content of individual sections in milligrams of ganglioside per gram dry weight at each depth is shown. The total number of sections analyzed was 83. (*From Derry and Wolfe, Exp. Brain Res.*, **5:** 32, 1968. *By permission of the publisher.*)

protein and lipid fractions. Proteolipid A is about 20% protein, 70% cerebroside, and a little phosphatide; B is about 50% protein, 20% cerebroside, 15% sphingomyelin, and 15% other phosphatides; C is 70 to 75% protein and 25% phosphatides.

The staining procedures applicable to the foregoing myelin lipids depend on a variety of factors, some common to the group, others restricted to certain members thereof.

Aside from sudanophilia in material untreated with fat solvents, which appears generally applicable to the whole group, the persistent sudanophilia in chromated material after paraffin or celloidin embedding and sectioning is probably the most generally applicable to the group, though specific data are lacking on some individual lipids. About half the total myelin lipids are thus rendered solvent-resistant, and our own experience indicates that there is also much binding of chromium to the soluble fraction coupled with the extensive oxidative changes which occur in the soluble as well as the insolubilized fractions. We have seen quite extensive sudanophilia of mesenteric fat in formalin-fixed human carcinoid material which was postchromed 24 h at 60°C in pH 3.5 3% $K_2Cr_2O_7$ before paraffin embedding through an alcohol-chloroform sequence.

Although chemical extraction studies do not indicate any extensive insolubilization of myelin lipids by formaldehyde fixation without chromates or other metallic oxidants, there is still quite extensive sudanophilia, particularly well demonstrated with Sudan black B after neutral or acid formalin fixation and routine paraffin embedding.

That this sudanophilia is due to myelin lipids is indicated by its total abolition when primary fixation is done in hot methanol–chloroform or other purely fat-solvent fixatives.

The glycolipoprotein vector substance of enterosiderin is demonstrated by its positive PAS reaction, its sudanophilia to Sudan black, its red staining by freshly diazotized safranin at pH 8. It is preserved by formaldehyde-containing fixatives against later treatment with fat solvents; it is removed by fat-solvent fixatives such as Carnoy and methanol-chloroform. A 24-h acetone extraction, if followed by a postfixation in formalin, does not remove it. Neither does acetone precipitation alone protect it from solution in fat solvents when no formol postfixation is employed (Lillie and Geer, *Am. J. Pathol.*, **47**: 965, 1965). The iron reactions of enterosiderin are not affected by fat-solvent fixations. This type of proof should be readily applicable to other, similar phospholipid and glycolipid deposits.

CHOLESTEROL

Cholesterol is often manifest in necrotic tissue and in the granulomatous tissue replacing it as long, rhomboid crystals which glow under polarized light and are extinguished and light up alternately once in each 90° of rotation of the stage.

THE SCHULTZ METHOD.   Cholesterol may be more definitely identified by Schultz' (*Zentralbl. Allg. Pathol.*, **35**: 314, 1924) adaptation to histology of the Liebermann-Burchardt sterol reaction. The technic as modified by Mallory follows:

1. Cut thin (10 to 15 μ) frozen sections of formalin-fixed tissue.
2. Mordant sections in a closely stoppered bottle for 3 days (Schmorl: 2 to 4 days) at 37°C in a 2.5% iron alum solution (or 1.4% $FeCl_3 \cdot 6H_2O$). (Hershberger, in our laboratory, found that 1 to 3 min in 3% hydrogen peroxide or in 1% sodium iodate will suffice in place of the iron alum.)
3. Rinse in distilled water, float onto slides, and blot dry.
4. Treat with a few drops of acetic sulfuric mixture made as follows: Place 2 to 5 ml glacial acetic acid in a small test tube, and immerse in ice water. Then add gradually the same volume of concentrated sulfuric acid while the tube is still in the ice water.
5. Cover with a cover glass and examine at once. The preparations may be kept for a few days if sealed with petrolatum.

RESULTS: A blue-green appears in a few seconds, becoming stronger in the first few minutes and often turning to brown in half an hour. Or the blue-green may persist as long as 24 h. Positive controls such as sections of previously tested adrenal cortex should always be used. At least two sections (three would be better) of the material under test should be treated with the acetic-sulfuric mixture and examined, before considering the test negative. The presence of glycerol inhibits the test; but if stearic acid is present as well, the inhibition is destroyed. Both cholesterol and cholesteryl esters react.

Lewis and Lobban (*J. Histochem. Cytochem.*, **9**: 2, 1961) modified the traditional Schultz test by using a sulfuric acid mixture containing 0.5% iron alum (or 0.28% $FeCl_3 \cdot 6H_2O$), 20 to 30% water, and no acetic acid. With this modified reagent cholesterol itself reacts weakly, yielding a weak red only after some 15 min, but other steroids react more promptly and with stronger colors. Premordanting 2 to 3 days in 2.5% ferric alum enhances the color.

METHOD OF LEWIS AND LOBBAN FOR STEROLS

1. Cut frozen or polyvinyl alcohol or gelatin sections as usual. Cryostat sections should be thawed onto slides and dried a few minutes to ensure adhesion.
2. Treat sections in 2.5% ferric ammonium alum (or 1.4% $FeCl_3 \cdot 6H_2O$) at 37°C for 1 to 3 days; at 25°C, 2 to 4 days may be tried.
3. Blot firmly onto slides, apply a drop of sulfuric acid–iron alum reagent, and cover at once with a coverslip. Examine.

REAGENT. Add cautiously 4 ml concentrated sulfuric acid (sp gr 1.84) to 1 ml 2.5% iron alum[1] in distilled water. Cool the mixture by immersing the test tube in cold water. The mixture is stable for a week or two. A 70% reagent (3 ml 1.7% iron alum, 7 ml concentrated sulfuric acid) results in a more rapid reaction.

RESULTS: Testosterone, Δ^4-androstene-3,17-dione, dihydrotestosterone, dihydroisoandrosterone, and androsterone yield blue-greens developing in 6 to 10 min. The androsterone color is the weakest, a brown-green. Pregnenolone, pregnanedione, methyltestosterone, and α-estradiol yield strong to moderate purple-red to pink, developing in 4 to 10 min. Pink to weak reds develop in 10 to 12 min with progesterone and 11-dehydro-17-hydroxycorticosterone; after 15 min cholesterol gives a weak red, which is weakened further as the water content of the reagent is increased.

CONTROLS. Unmordanted frozen sections are treated as above with iron-free 80% (v/v) sulfuric acid. If the reds seen in the test were due to steroids, they should fail to appear.

Apparently the Golodetz reaction (Schmorl, Romeis) was a procedure closely related to the foregoing. In it frozen sections were treated for 1 to 2 min in a mixture of 5 ml concentrated sulfuric acid and 2 ml 30% formalin, and "cholesterol" developed a brown-red. Romeis questioned the specificity of the reaction, since a blackish brown was given also by various phenols, fats, and oils.

WINDAUS' DIGITONIN REACTION (*Hoppe Seylers Z. Physiol. Chem.*, **65**: 110, 1910). This is a reaction for free sterols. All writers caution about the need for absolute cleanliness of slides and coverslips, since cholesterol is present in sweat and hence fingerprints give positive reactions. This is particularly important when the reaction is carried out on the slide under a cover glass on fresh or formalin-fixed material. Cowdry recommends a variant of Lison's which seems to avoid this difficulty: Fix in formalin and cut frozen sections. Immerse sections in a 0.5% solution of digitonin in 50% alcohol in a small covered dish for several hours. Rinse in 50% alcohol. Counterstain part of the sections only, by the usual hematoxylin Sudan IV or oil red O technic. Mount all sections as usual in Apáthy's gum syrup or a glycerol gelatin.

Examine the uncounterstained sections under polarized light with crossed Nicols. Needles or rosettes of complex cholesteryl digitonids are formed. In the counterstained preparations the cholesterol compound remains doubly refractile and does not stain, but the cholesteryl ester compound colors with the oil-soluble dye and loses its birefringence.

Digitonin precipitation is specific for steroids with a 3-hydroxy configuration, but an epi configuration of a methyl group at C-10 prevents the reaction. The digitonides are insoluble in water, cold alcohol, acetone, and ether; slightly soluble in methanol and hot ethanol; soluble in glacial acetic alcohol; very soluble in pyridine (Windaus, *Ber. Dtsch. Chem. Ges.*, **42**: 288, 1909; *Hoppe Seylers Z. Physiol. Chem.*, **65**: 110, 1910). Mallory (p. 225) added "very soluble in pyridine and chloral hydrate" at the end of the list of solvents, without further details.

Cholesterol is not blackened by osmium tetroxide. It is soluble in acetone, ether, benzene, xylene, and strong alcohol. There is some loss of doubly refractile material from adrenal cortex after treatment with 70% alcohol, more loss with 80%, and complete loss with 95 to 100%.

Adams [*Nature (Lond.)*, **192**: 331, 1961, and "Neurochemistry," pp. 6–66, Elsevier, Amsterdam, 1965] has developed the perchloric acid–naphthoquinone (PAN) method for

---

[1] An 0.84% solution of anhydrous ferric chloride equal to 1.4% $FeCl_3 \cdot 6H_2O$ (0.05 M) should serve, if ferric ammonium alum is not readily available.

cholesterol and related 3-hydroxy-$\Delta^{5,7}$-sterols, which has several advantages over the Schultz and Lewis and Lobban methods; namely, a more stable stain is developed, and fewer gas bubbles develop which distort and obscure the staining reaction. The stain is believed to develop by the formation of cholesta-3,5-diene by a condensation action by perchloric acid on cholesterol. The diene subsequently reacts with 1,2-naphthoquinone to form a dark blue pigment. The staining procedure follows:

PERCHLORIC ACID–NAPHTHOQUINONE TECHNIC FOR CHOLESTEROL ACCORDING TO ADAMS [*Nature (Lond.)*, **192:** 331, 1961]

1. Cut frozen sections from tissues fixed in 1 % calcium acetate–10 % formalin or in "routine" formalin.
2. Leave the sections free-floating in calcium-formalin for at least a week (preferably 3 to 4 weeks) in order to promote preliminary atmospheric oxidation of cholesterol. Then mount the sections on glass slides and dry them in air.
3. Paint the sections with a thin layer of reagent solution and heat them on a hot plate or bottom of an oven at 65 to 70°C for 5 to 10 min or until the red that first appears turns completely blue. Care must be taken not to overheat and burn the sections. Prepare the reagent solution fresh daily by adding 10 mg of 1,2-naphthoquinone-4-sulfonic acid to 10 ml of a mixture of ethanol–perchloric acid (60 %)–formalin (= 40 % formaldehyde-water) (2 : 1 : 0.1, v/v/v).
4. Place a drop of perchloric acid (60 %) on the reacted section, and gently lower a coverslip into position.

RESULTS: Cholesterol and related sterols are colored blue. The reaction product is stable for a variable number of hours and then turns grayish black; it is unstable in water, glycerin jelly, and other mounting media. Cholesterol esters do not react with digitonin, whereas free cholesterol forms a digitonide. Although prior reaction with digitonin is advocated to permit discrimination of cholesterol from cholesterol esters, the procedure has little practical value, because the alcohol solvent employed with the digitonin extracts cholesterol at a rate that may exceed the formation of digitonide.

Many histochemical methods for ketosteroids utilize the formation of aldehydes subsequently reacted with hydrazides for detection. Many investigators now conclude that these methods are nonspecific [Albert and Leblond, *Endocrinology*, **39:** 386, 1947; Wegmann, *Ann. Histochim.*, **1:** 116, 1956; Holczinger, *Acta Histochem. (Jena)*, **6:** 36, 1958; and Wolman, *Acta Histochem. (Jena) Suppl.* **2**, p. 140, 1961].

FATTY ACIDS

Fatty acids are readily soluble in ether and alcohol. They form calcium soaps when fixed in a 10 % formalin which has been saturated with calcium salicylate (about 1.5 %). These calcium soaps are insoluble in an ether and 100 % alcohol mixture and in dilute hydrochloric acid when tested with these reagents separately, but they dissolve in a hydrochloric acid solution in ether and alcohol. After mordanting with copper acetate, fatty acids and their calcium soaps form a black hematoxylin lake which is very resistant to decolorization with Weigert's borax-ferricyanide mixture. This forms the basis of Fischler's (*Zentralbl. Allg. Pathol.*, **15:** 913, 1904) method for fatty acids.

FISCHLER'S TECHNIC (modified slightly from Mallory)

1. Fix in 10% formalin and cut frozen sections.
2. To differentiate fatty acids from calcium soaps, extract some of the sections with two or three changes of 100% alcohol-ether mixture (50 : 50), first dehydrating through 95 and 100% alcohol, and afterwards rehydrating through 100, 95, and 80% alcohol. Only the soaps will remain in these. The specific colored metal chelate reactions for Ca^{2+} should be applicable.
3. Treat both extracted and unextracted sections with saturated (perhaps 10%?) aqueous cupric acetate solution for 2 to 24 h at 20 to 30°C.
4. Wash in distilled water.
5. Stain for 20 min in Weigert's lithium carbonate–hematoxylin (Table 13-4).
6. Differentiate in Weigert's borax-ferricyanide mixture (Table 13-4) "greatly diluted" with distilled water, until red corpuscles are decolorized.
7. Wash thoroughly in distilled water, and mount in glycerol, glycerol gelatin, or gum syrup. If desired, one may add a counterstain with a yellow or red oil-soluble dye before mounting.

Mallory notes that iron, hemoglobin, and calcium may stain as well. If differentiation is carried to the point where red corpuscles are thoroughly decolorized, hemoglobin should give no difficulty. Pretreatment of a few sections with dilute hydrochloric acid in place of step 2 should serve to eliminate calcium as a source of error, but adds the soap fatty acid to the free form. Similarly, overnight extraction in 5% oxalic acid or (15 min) sodium dithionite solution should serve to remove iron. Or a control section may be subjected to the ferrocyanide test.

Since Wigglesworth's iron sulfide method is said to demonstrate carboxylic acids as well as phosphoric acid complexes, this method might also be applicable to the demonstration of fatty acids.

Tandler's lead sulfide methods are said to demonstrate fatty acids [*Cienc. Invest. (Buenos Aires)*, **8:** 44, 1952] and phosphoric acids (*Arch. Histol. Norm. Pat.*, **4:** 275, 1951). He employs a primary 12-h fixation in 10% neutral lead acetate, with or without about 0.2% picric acid included, or in 5% lead acetate with a few drops up to 1% acetic acid. After this, tissues are washed thoroughly in water.

At this point tissues intended for demonstration of phosphoric esters are dehydrated, cleared, embedded in paraffin, sectioned, deparaffinized, and brought either into 70% alcohol containing 2% yellow ammonium sulfide or into saturated hydrogen sulfide water.

Tissues intended for demonstration of fatty acid esters are treated with very dilute (0.7 to 0.5%) ammonium sulfide without prior alcohol treatment and washed and mounted in glycerol.

Lead phosphate esters are insoluble in alcohol, chloroform, xylene, and alcohol ether, but soluble in dilute acetic acid. Even 0.5% will remove them in a short time. Unsaturated fatty acid lead soaps are soluble in alcohol (70 to 80%), while saturated fatty acid lead salts resist alcohol but are dissolved by xylene, chloroform, etc. Both types of lead soaps resist extraction in 40% acetic acid.

Substitution of Molnar's rhodizonate method should give a darker color in both technics and permit final dehydration and mounting in a resinous medium.

We question whether the differential solubilities of the lead soaps are sufficiently great to permit their histochemical differentiation.

Meyer-Brunot (*Z. Wiss. Mikrosk.*, **60:** 476, 1952) adapts the Gomori lead sulfide–lipase technic for the differentiation of fatty acids and neutral fats:

1. Fix smears by heating on an albumen-glycerol smeared slide over a small flame.
2. Immerse in 0.5% calcium chloride in a 0.5% Veronal sodium solution (pH 9.4) for 10 min, thereby converting free fatty acids to calcium soaps.
3. Wash thoroughly, and immerse for 10 min in a 2% lead nitrate solution, converting the calcium soaps to lead soaps.
4. Wash thoroughly in water, and immerse for 30 s in dilute ammonium sulfide—we suggest 1 : 50 dilution—thus converting lead soaps to sulfide; or in 0.5% potassium dichromate, forming the yellow lead chromate.
5. For neutral fats, counterstain with Sudan III as usual.

Undoubtedly the method can be applied to frozen sections of formalin-fixed material. Treatment with 1% acetic acid after step 3 to dissolve lead phosphate would tend, according to Tandler, to eliminate lead phosphate derived from cell nuclei. Use of rhodizonate in place of ammonium sulfide could give a darker color to the lead deposits.

Cain (*J. Anat.*, **84:** 196, 1950) again stresses the view (from which we strongly dissent, agreeing rather with Smith's original report) that Nile blue sulfate solutions containing the red oxazone which Lison calls "Nile red" behave simply as mixtures of a (rather inefficient) oil-soluble dye and a blue basic dye which colors nucleic acids and other acid-bearing tissue elements as well as the open carboxyl groups of free fatty acids. Staining with the red oxazone component does not indicate unsaturation, but appears to be governed by the staining temperature and the melting point of the specific fats and fatty acids concerned. Solid fats do not readily stain with oil-soluble dyes at temperatures far below their melting points. The melting points of the fatty acids are oleic, 14°C; palmitic, 63°C; and stearic, 70°C; of their glyceryl esters, respectively, 4, 65, and 71°C.

The technics are simple: Fix in formalin, cut frozen sections, stain for 20 to 30 min in 0.1 to 0.15% aqueous solution of Nile blue, differentiate for 1 to 20 min in 1% acetic acid, wash well in tap water, and mount in lukewarm glycerol gelatin, or in a buffered neutral gum syrup (modified from Romeis). To include the high-melting-point fats in the stained substances, raise the staining temperature to 70°C. We suggest buffering the Nile blue to pH 2.5 or 3 with acetates to eliminate if possible the necessity for regressive differentiation. Staining for 20 min in 0.1% Nile blue in 1% acetic acid, followed by washing in water and mounting in glycerol gelatin, has given sharply selective staining of certain lipofuscin pigments.

Further exploration (*J. Histochem. Cytochem.*, **4:** 377, 1956) disclosed that the method could be made more specific for fatty acids by doing it at a pH level below that at which carboxylic and phosphoric acid residues stain with basic dyes. It was found that lipofuscins, pepsinogen granules, and in vitro fatty acids stained dark blue when exposed to 0.05% Nile blue in 1% sulfuric acid by volume (pH about 0.9). If such sections are dehydrated directly with acetone, the blue staining is at once lost from the pigment and peptic zymogen granules, only mast cells and perhaps cartilage retaining the stain. But if, instead, mounting is done in glycerol gelatin or syrup, the fatty acid residues remain dark blue. The latter staining does not occur with the usual blue thiazin dyes of the thionin, azure, methylene blue group. However, if the pH level of staining is raised to 3 or 3.5 ($H_2SO_4-KH_2PO_4$ buffer), staining of carboxylic acid residues occurs alike with Nile blue and the thiazins, and preparations may be dehydrated and mounted in resins. The staining of lipofuscins is green, however, and easily distinguished from the dark blue of the glycerol gelatin mounts.

Liquid neutral fats, shaken with this sulfuric acid–Nile blue solution, color red. Addition of a drop or two of anhydrous acetic or propionic acid changes the color to dark blue; shaking with distilled water takes the acetic acid out of the oil and restores the red color. Methyl esters of fatty acids also color red when shaken with Nile blue, but the technic of methylation is applicable only to the histochemical study of such lipids as are insoluble in hot methanol.

Not all samples of Nile blue are satisfactory for this technic, even when certified. Naturally, the above procedure on staining of fatty acid materials which are then dehydrated applies only to the insoluble or relatively insoluble fatty acid complexes.

LILLIE'S SULFURIC NILE BLUE TECHNIC FOR FATTY ACIDS (*Stain Technol.*, **31**: 151, 1956).   Use frozen, attached cryostat or paraffin sections, according to the nature of the material under study. Bring sections to water.

1. Stain 20 min in 0.05% Nile blue in 1 ml concentrated sulfuric acid plus 99 ml distilled water. (Dissolve dye in water and add acid.)
2. Wash 10 min in running water. Mount in glycerol gelatin.

RESULTS: Fatty acids are dark blue; neutral fats, pink to red.

It is to be noted that all technics using acid Nile blue solutions will decompose calcium soaps and stain their fatty acid residues.

Holczinger [*Acta Histochem. (Jena)*, **8**: 167, 1959] has recommended a method for fatty acids (provided below) whereby the initial copper soap formed converted subsequently to a greenish black rubeanate.

COPPER RUBEANIC ACID TECHNIC FOR FATTY ACIDS [Holczinger, *Acta Histochem. (Jena)*, **8**: 167, 1959]

1. Cut unfixed tissues on the cryostat, or cut frozen sections from tissues fixed either in 1% calcium acetate–10% formalin or in "routine" formalin. Mount sections on glass slides and dry them in air.
2. Treat sections with 0.005% cupric acetate for 3 to 5 h.
3. Wash twice for 10 s with 0.1% disodium–EDTA at pH 7.1. Then wash for 10 min in distilled water.
4. Immerse sections in 0.1% rubeanic acid in 70% ethanol for 30 min. Dissolve the rubeanic acid in ethanol, warm slightly (avoiding a naked flame), and make up to volume with distilled water.
5. Wash for a few minutes in 70% ethanol and then wash in water. Mount in glycerin jelly, or dehydrate, clear, and mount in DPX or balsam.

RESULTS: Fatty acids are stained greenish black.

Elleder and Lojda (*Histochemie*, **32**: 301, 1972) noted that phospholipids, lipopigments, and even nonlipid substances may interfere with the staining reaction for fatty acids. To exclude these interferences, Elleder and Lojda recommend the use of parallel control sections pretreated with 2 N HCl for 1 h at 24°C, one of which is thereafter extracted with anhydrous acetone overnight at 24°C. Only substances positive in the section pretreated with HCl and negative after pretreatment with HCl followed by acetone extraction can be considered as fatty acids. This would exclude insoluble polyene fatty acids and protein carboxyls.

Unsaturated groups may be detected histochemically employing three types of methods, namely, oxidation of the unsaturated groups to aldehydes, bromination followed by silver reduction, and the reduction of osmium tetroxide.

Aldehyde Formation

Performic or peracetic acid will oxidize unsaturated groups with the formation of an aldehyde (Lillie, *Stain Technol.*, **27**: 37, 1952). The aldehyde may be detected by Schiff's reagent and blocked by bromination. Belt and Hayes (*Stain Technol.*, **31**: 117, 1956) have shown air oxidation with ultraviolet light to also effect aldehyde formation from unsaturated groups.

ULTRAVIOLET SCHIFF TECHNIC FOR UNSATURATED LIPIDS ACCORDING TO BELT AND HAYES (*Stain Technol.*, **31**: 117, 1956)

1. Cut unfixed tissues on the cryostat, or cut frozen sections from tissues fixed in 1% calcium acetate–10% formalin. Mount sections on glass slides and dry them in air.
2. Irradiate sections with ultraviolet light for 1 to 4 h.
3. Immerse in Schiff's reagent 15 min.
4. Rinse in three changes of sulfurous acid (3 min each).
5. Rinse in distilled water and mount in aqueous media.

RESULTS: Unsaturated lipids appear pink to red because of the formation of "pseudoplasmal" aldehydes from fatty acid ethylene bonds.

Bromination and Silver Reduction Method

Mukherji, Deb, and Sen (*J. Histochem. Cytochem.*, **8**: 189, 1960) and Norton, Korey, and Brotz (*J. Histochem. Cytochem.*, **10**: 83, 1962) brominated unsaturated groups of fatty acids. Subsequent treatment with silver nitrate resulted in the formation of silver bromide. Photographic developer was then applied to form a black silver stain. Adams (1965, pp. 6–66) is convinced that only hydrophobic unsaturated lipids are disclosed with this method.

BROMIDE–SILVER NITRATE TECHNIC FOR UNSATURATED LIPIDS (Norton et al., *J. Histochem. Cytochem.*, **10**: 83, 1962)

1. Frozen sections are cut from thin slices of tissue fixed in 1% calcium acetate–10% formalin. Mount sections on glass slides and dry them in air.
2. Immerse the sections in 2% potassium bromide–0.1 N bromine (1 ml Br_2 in 390 ml 2% KBr) or in saturated bromine water for 1 min.
3. Wash in water, rinse in 1% sodium bisulfite for 5 min, and wash 7 times in distilled water.
4. Treat with 1% silver nitrate in 1 N nitric acid for 18 h in the dark.
5. Wash 7 times in distilled water.
6. Develop for 10 min in Kodak Dektol developer (or equivalent) diluted with water (1 : 1, v/v).
7. Wash well in water and mount in glycerin jelly.

RESULTS: Some unsaturated lipids are stained brown-black. Cholesterol reacts when impregnated on paper, but crystals of the sterol in tissues are not stained.

Osmium Tetroxide Reduction

Korn (*Biochim. Biophys. Acta*, **116**: 317 and 325, 1966; *Science*, **153**: 1491, 1966; *J. Cell Biol.*, **34**: 627, 1967) has clearly shown that osmium reacts with unsaturated groups to form the structure illustrated below.

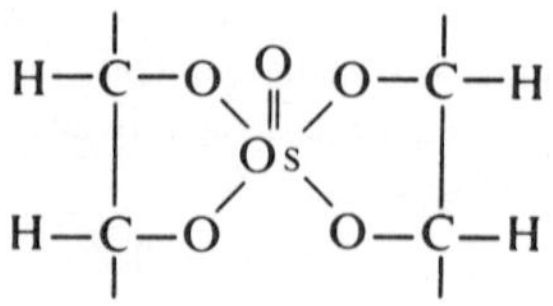

$$\text{H}-\overset{|}{\underset{|}{\text{C}}}-\text{O} \quad \overset{\text{O}}{\underset{/}{\underset{\text{Os}}{\overset{\backslash\,\|\,/}{}}}} \quad \text{O}-\overset{|}{\underset{|}{\text{C}}}-\text{H}$$
$$\text{H}-\overset{|}{\underset{|}{\text{C}}}-\text{O} \qquad \text{O}-\overset{|}{\underset{|}{\text{C}}}-\text{H}$$

Wigglesworth (*Proc. R. Soc. Lond. [Biol.]*, **147**: 185, 1957) and Baker (*J. Histochem. Cytochem.*, **6**: 303, 1958) had previously postulated such a diester formation. However, the actual structure formed in tissue sections with heterogeneous composition is unknown.

Korn (1966, 1967) noted that bisosmates did not react with the acidic or basic groups of phospholipids. For this reason, the osmiophilic-dense lines observed in electron micrographs may not be due to polar groups of phospholipids.

The specificity of osmium for the identification of unsaturated groups is in dispute. Porter and Kallman (*Exp. Cell Res.*, **4**: 127, 1953), Bahr (*Exp. Cell Res.*, **7**: 457, 1954), Wolman (*J. Neurochem.*, **1**: 270, 1957), and Rogers (*J. Ultrastruct. Res.*, **2**: 309, 1959) express the belief that osmium tetroxide is reduced by proteins or mucopolysaccharides, whereas Adams (*Adv. Lipid Res.*, **7**: 1, 1969) is unconvinced that structures other than ethylene bonds are responsible for osmium blackening in tissue sections.

Hayes et al. (*J. Cell Biol.*, **19**: 251, 1963) studied osmium uptake of many proteins and lipids and concluded that osmium could react with either unsaturated or carboxyl groups or both. Their data tended to favor reactivity with unsaturated groups predominantly. Saunders (*J. Cell Biol.*, **37**: 183, 1968) also concluded that osmium reacts with unsaturated but not saturated lipids. Lipids reactive with osmium were preserved in tissues prepared for microscopy to a greater extent than unreactive lipids. None of these workers considered the known reactivity of catechols.

MYELIN LIPIDS

CHEMICAL CONSIDERATIONS.   A number of papers have appeared recently bearing on the alteration of extractable brain lipids occasioned by exposures to formaldehyde for a varying period of time. The inference is fairly plain that those lipids which do not appear in extracts and are not converted into others which do must represent those retained in tissue and are demonstrated by various largely empirical histologic procedures.

Heslinga and Deierkauf (*J. Histochem. Cytochem.*, **9**: 572, 1961; **10**: 79, 704, 1962) have published a series of papers on human, ox, and rat brain lipids and their chromatographic distribution as modified by the action of formaldehyde and other fixatives. Fresh brain yielded seven fractions: F, near the front, of cholesterol, cholesterin esters, and glycerides; E, phosphatidylethanolamine; D, phosphatidylserine; C, lecithins; B, sphingomyelin and lysophosphatidylethanolamine; A, a double spot of lysolecithin and the larger half of phosphoinositides and cerebrosides. The starting spot was not characterized.

Formaldehyde fixation, even in 19 h, largely suppresses the phosphatidylethanolamine spot; in 93 h this change is further advanced.

Prolonged fixation, over a year, also decreases lecithin but increases the lyso compounds in spots A and B (lysolecithin and lysophosphatidylethanolamine).

Mercuric chloride apparently forms addition complexes with the more unsaturated lipids, so that spots E, C, and B also appear double (phosphatidylethanolamine, lecithin, and lysophosphatidylethanolamine).

Chromic acid and dichromate render 50% or more of the lipids insoluble and partly polymerize lipids bound to an insoluble aggregate of chromous hydroxides (Heslinga, thesis, Leiden, 1957). It is presumed that this chromous hydroxide serves to bind hematoxylin in the Weigert-Smith-Dietrich-Baker-Elftman myelin and phospholipid methods which do not employ an iron hematoxylin. The various fractions in the chromatograms exhibit oxidation "tailing" and a decrease of distinguishable fractions.

With osmium tetroxide the fusion and the tailing of the spots are even more pronounced and extend back to the starting spot. Spots D and E (phosphatidylserine and phosphatidylethanolamine) have disappeared.

Later studies indicated considerable diminution of both amount and mobility of phosphatidylethanolamine and its lyso derivative. About a third of the lecithin is changed to lysolecithin. Other lipids seem little altered by formaldehyde. Sulfatides are now recognized as present in the same fraction as cerebrosides. The front spot F is increased by an increment of free fatty acids almost corresponding to the losses in the two ethanolamine fractions.

Altogether, little if any lipid is insolubilized by formalin. In this regard the histochemical and chemical findings appear to conflict. But the chemical report nowhere relates the total amount of extractable lipid to fresh wet weights, to dry weights, or to total protein nitrogen, and it is not clear that the total amount of extractable lipid has not decreased.

The papers of Riemersma and Booij (*J. Histochem. Cytochem.*, **9**: 560, 1961; **10**: 89, 1962) establish spectrophotometrically and gravimetrically that lecithin binds osmium tetroxide in the proportion of 1 mol per double bond. This was true also for oleic acid, methyl oleate, decene, and octadecene.

MYELIN STAIN WITH SUDAN II AND IRON HEMATOXYLIN.   Use frozen sections at 10 to 15 μ of material fixed in formalin and chromated in 2.5% potassium dichromate for 2 to 4 days and then washed in water.

1. Make a fresh mixture of 5 ml fresh 1% aqueous hematoxylin and 5 ml 4% iron alum in a covered dish. Stain sections 45 min at 55 to 60°C, agitating the dish gently from time to time to ensure even staining.
2. Wash sections in water. Save out some sections in water.
3. Decolorize in 0.5% iron alum for 1 h, agitating from time to time.
4. Wash in water. (Steps 3 and 4 can probably be omitted.)
5. Treat with 1% borax–2.5% potassium ferricyanide solution for 10 min, agitating several times. (This is Lillie's variant of Weigert's borax-ferricyanide, Table 13-4, formula 15.) At the start of this step dilute the Sudan II for step 7. If step 3 is omitted, time variation is done in this step.
6. Wash in water.
7. Take 6 ml of a stock solution of Sudan II (C.I. 12140) saturated in 99% isopropanol, and dilute with 4 ml water, let stand 7 to 8 min, and filter. Stain sections for 10 min in the fresh filtrate.
8. Wash in water.
9. While a nuclear stain with a red, green, or brown dye might be useful at this point, we have had little success in attempting to insert one.
10. Float onto slides, drain, and mount in gum syrup.

RESULTS: Normal myelin is blue-black; nerve cells are gray; nuclei, deeper gray; red corpuscles, yellow to black; fats, orange-yellow.

The persistence of black in nerve cells indicates underdifferentiation; loss of color in myelin indicates overdifferentiation. If sections are not satisfactory, those saved at step 2 may be differentiated for a shorter time in 0.5% iron alum if the myelin was too pale, or an hour in 1% iron alum if other structures were too dark. With freshly fixed formalin or Orth material chromated for not over 4 days, this should not be necessary.

Feyrter developed a staining technic which is said to be specific for the so-called "onkocytes." Hamperl (*Arch. Pathol.*, **49**: 563, 1950) performs it as follows:

1. Fix in 10% formalin, and cut frozen sections at 10 to 15 μ.
2. Float from distilled water onto a clean slide.
3. Cover with 1% thionin or 1% cresyl violet in 0.5% tartaric acid solution. (Let stand a few days before using. Do not filter.)
4. Place cover glass on the stain drop, and seal with cellulose caprate, polystyrene, or lanolin rosin.

RESULTS: Nuclei are blue; myelin is red-purple; other lipids, onkocytes, and mucus are pink to red. Cardiolipids show blue metachromasia with cresyl violet. The onkocytes owe their metachromasia to poorly sudanophilic lipids in their cytoplasm. The full color of the staining develops only after some hours. Hamperl states that the preparations are stable for years. Romeis comments on their limited life.

Consideration of the special requirement of Feyrter (Romeis) that the thionin of only one German firm is suitable for his metachromatic staining of myelin and of the fact that some materials staining metachromatically by Feyrter's method are themselves removed from frozen sections by brief (2-h) alcohol extraction (Thorén, *Acta Soc. Med. Ups.*, **55**: 125, 1950) suggests that the dye concerned may contain a good deal of a relatively fat-soluble dye, such as Bernthsen's methylene violet. The solutions of this dye in chloroform are red. It would appear that the procedure may be closely related to the Ciaccio-type methods insofar as it applies to myelin.

TOTAL EXTRACTION OF LIPIDS. For total extraction of lipids we now use a 48-h fixation at 60°C in methanol-chloroform mixtures. The first 18- to 24-h period is divided between two changes of 2 : 1 methanol-chloroform; the second working day uses a 4- to 6-h bath in 1 : 1 methanol-chloroform; and the remaining period is divided between two changes of 1 : 2 methanol-chloroform. The extraction is done in tightly sealed screw-capped bottles with metal foil solvent-resistant inner caps. The fluid level is marked on the outside of each container with a grease pencil, and the bottles are inspected at intervals to see that no evaporation loss is occurring. Heating is normally done in a 60°C paraffin oven. At the end of the heating period blocks are transferred through two 30- to 60-min chloroform baths into paraffin directly, or after a sequence of two 1-h baths in absolute alcohol and alcohol-ether into a 1- to 3-day bath in 1% celloidin in alcohol-ether. The thin celloidin-infiltrated tissue is then carried through chloroform into paraffin as before. The double embedding permits thinner sections than the plain paraffin. We have cut 2-cm blocks of human cerebellum serially at 3 μ, without skipping sections.

CIACCIO-POSITIVE LIPIDS. The methods used for demonstration of sudanophilia or staining by oil-soluble dyes after paraffin (or celloidin) embedding and sectioning are all essentially variants of Ciaccio's original method.

It is perhaps of interest that when linseed oil or cod liver oil, which present iodine numbers of 175 to 202 and 137 to 166 respectively, are oxidized with dichromate in vitro, they yield insoluble, basophilic, acid-fast lipids. Cod liver oil contains no phosphorus. When these oils are parenterally administered, they provoke inflammatory reactions in which after a time the offending oils are identified in nonchromated, paraffin-embedded, formalin-fixed tissue as globules of sudanophil, acid-fast, basophilic material coloring blue to violet with Nile blue, black with osmium tetroxide, brown to black with ammoniacal silver carbonate, and *gray to black* with the Weil-Weigert myelin stain (Endicott, *Arch. Pathol.*, **37**: 49, 1944).

CIACCIO'S METHOD (*Anat. Anz.*, **35**: 17, 1909). This method depends on simultaneous formaldehyde and chromate treatment of fresh or formalin-fixed tissue, followed by prolonged chromation, dehydration, clearing, and paraffin embedding. Sections are stained in a super-saturated solution of Sudan III or Sudan IV, counterstained in hemalum, and mounted in gum syrup. This method stains myelin red; *Ciaccio-positive* lipids, orange; and nuclei, blue. By introducing an osmium tetroxide step just after chromation, Ciaccio stained neutral fats (probably oleins) black.

Simple chromation for 2 to 4 days after 2 days' or more fixation in 10% formalin seems as effective as Ciaccio's rather elaborate schedule, and nuclear staining is less impaired with the shorter chromation.

Though this and other variants of the classic Ciaccio procedure using Sudan III and Sudan IV are often quite satisfactory for uncolored insoluble lipids, in lipid pigments possessing natural browns the orange tingeing of the native color is often difficult to discern with any degree of certainty. With such lipids it is better to use a blue or dark green, oil-soluble dye. Blues to greens are easily distinguished from unmodified yellows and brown.

The following technic, devised by Lillie and Laskey (*Bull. Int. Assoc. Med. Mus.*, **32:** 77, 1951) for the demonstration of the rod acromeres in the retina, works very well on cardiac, ovarian, and other similar lipofuscins and ceroid pigments.

MODIFIED CIACCIO PROCEDURE OF LILLIE AND LASKEY WITH SUDAN BLACK B

1. Fix in calcium acetate–formalin or calcium chloride–formalin. For those lipids which require chromate oxidation, soak in 3% potassium dichromate for 4 to 8 days, changing every 2 days. Wash out chromate overnight in running water (or siphon washer). For many lipids chromation is unnecessary, and with some it renders the material useless for other histochemical procedures by oxidizing their reactive groups. Prepare paraffin sections, deparaffinize, and hydrate as usual. Frozen (or attached cryostat) sections may be used for comparison or for special multiple-solubility tests.
2. Hydrolyze 15 min at 60°C in preheated 1 N hydrochloric acid.
3. Transfer directly to Schiff reagent for 10 to 60 min.
4. Wash 5 min in 0.5% sodium metabisulfite, three changes of $\frac{1}{2}$ to 1, 2, and 2 min.
5. Wash in running water or in several changes of water.
6. Dilute 20 ml of a 1% solution of Sudan black B in 99% isopropanol with 20 ml of a 1% aqueous solution of borax (0.026 M). Let stand 10 to 15 min. Filter into a Coplin jar. Stain sections 5 min.
7. Wash 5 min in running water.
8. Mount in Arlex gelatin, glycerol gelatin, or other aqueous mountant.

RESULTS: Nuclei are red-purple; lipids, dark gray-green with brown pigments, definitely altered in color toward green or greenish black if lipid-positive; background is light greenish gray.

Several batches of slides may be stained successively in the same Coplin jar within the first hour or so after the dilution and filtration of the Sudan black B. Fresh dilutions must be made each day, as further precipitation occurs on longer standing and the solution loses its supersaturated state.

Longer staining in Sudan black B increases the density of background staining and decreases the contrast. The 50% alcohol solution turns brown in a few days and stains cytoplasm and various other structures gray-green but fails to stain fats.

Oil red O, similarly supersaturated in 50% alcohol, may be used for a red lipid stain in the above technic but requires 1 to 2 h for adequate staining. With this a *following* dilute alum hematoxylin counterstain is used in place of the *preceding* Feulgen procedure, steps 1 to 5.

If the alcoholic Sudan black solution and the aqueous diluent are preheated to 60°C, and then mixed and filtered in the paraffin oven as above, a 5-min stain at 60°C will color erythrocytes and the granules of eosinophil and neutrophil leukocytes and myelocytes dark gray-green, even after formalin fixation and routine paraffin embedding.

Thomas (*Q. J. Microsc. Sci.*, **89:** 333, 1948) gives a technic basically similar to the Lillie-Laskey, using 3-day chromation in saturated aqueous potassium dichromate (12% at 20°C) before embedding and a 7- to 10-min stain in a saturated solution of Sudan black B in 70% alcohol, a quick rinse in 50% alcohol, a 3- to 5-min counterstain in alum carmine, and mounting in Farrants' glycerol gum arabic.

THE KLÜVER-BARRERA LUXOL FAST BLUE MBS FOR MYELINATED FIBERS. Although at first sight this myelin stain with Luxol Fast Blue MBS or MBSN (Du Pont) (Azosol Fast Blue HLR, General Dyestuffs), C.I. Solvent Blue 38, might appear to be related to the Sudan staining of myelin, since the dye is classed as a solvent (oil-soluble) dye, it requires overnight staining at an elevated temperature and prolonged 70% alcohol extraction to remove it from other structures while leaving it in myelin.

Since the dye is a diarylguanidine derivative of a sulfonated copper phthalocyanin (Salthouse, *Stain Technol.*, **37**: 313, 1962) and the myelin stain is blocked by deamination, unaffected by bromination, and enhanced by methylation (Smith and Dunbar, *J. Pathol. Bacteriol.*, **91**: 117, 1966), it would appear to be acting as a typical acid dye. Prolonged formalin storage gradually weakens its staining capacity for myelin, concomitantly with the known gradual disappearance of phospholipids and free fatty acids and the fall in iodine number of extractable fat.

As a practical method for myelin it appears to be quite selective, and some writers (Pearse, 1960, 1967; Adams, 1965) have considered it specific for phospholipids.

The Armed Forces Institute of Pathology "Manual of Histologic and Special Staining Technics," second edition, cites a report by Klüver and Barrera (*J. Neuropathol. Exp. Neurol.*, **12**: 400, 1953), which we summarize as follows:

1. Use 15- to 30-μ celloidin or 15- to 20-μ paraffin sections of formalin-fixed tissue. Deparaffinize and wash in several changes of absolute (two) and 95% (three or four) alcohol.
2. Stain 18 to 24 h in 0.1% Luxol Fast Blue in 95% alcohol–0.05% acetic acid at 55 to 60°C.
3. Wash in 95% alcohol and in distilled water, and immerse briefly in 0.05% Li_2CO_3.
4. Differentiate in 70% alcohol until gray substance is paler than white, return again briefly (20 to 30 s) to Li_2CO_3, and resume 70% alcohol differentiation (several changes) until colorless gray substance contrasts sharply with blue-green-white substance.
5. Wash thoroughly in distilled water.
6. Variant A: Add 0.4 ml 10% acetic acid to 40 ml 0.1% cresyl echt violet (Chroma) in distilled water, warm (35 to 40°C?), and counterstain sections 6 min (celloidin sections 3 min). Differentiate in 95% alcohol, dehydrate in two changes of absolute alcohol, clear in xylene, and mount in synthetic resin.

   Variant B: Carry through periodic acid Schiff procedures: 0.5% periodic acid, 5 min; two changes of distilled water; Schiff's reagent, 10 to 30 min; bisulfite, three changes, for 2 min each; running water, 5 min; alum hematoxylin, 1 min; tap water (or 0.5% sodium acetate), 5 min; acid alcohol, 16 to 30 s; alcohols, xylene, Permount.

RESULTS: Myelin is blue-green. Variant A: Nissl substance is violet; variant B: fungi, basement membranes, other periodic acid Schiff–positive substances are red; nuclei, dark blue. THE TRADITIONAL WEIGERT HEMATOXYLIN METHODS FOR MYELIN. These depend on chromate oxidation producing polymerization and insolubilization of myelin lipids with concomitant binding of chromium in one of its lower valences. Although we cannot follow Lison's (1960, p. 469) exact formulation of the process, it appeared quite probable that a chemical process similar to that which he postulated had occurred, and that the end product was a chelated Cr^{2+} or Cr^{3+} ion bound to a complex lipid polymer. Elftman (*J. Histochem. Cytochem.*, **2**: 1, 1954) clearly identified the bound metal as Cr^{3+}, but did not indicate how it was bound. Empirically the reaction seems to be restricted to phospholipids, but no specific reaction of the phosphate group seems to be involved. It is questionable whether a cephalin or lecithin containing only saturated fatty acids would react. Of the reactivity of unchromated formalin-fixed tissue, further data appear in other sections of this text.

THE SMITH-DIETRICH PROCEDURE (*Verhl. Dtsch. Ges. Pathol.*, **14**: 263, 1910). With this procedure there is a sequence formalin-chromate treatment and an overstaining with ripened acetic hematoxylin followed by differentiation with Weigert's borax–potassium ferricyanide and mounting in fructose syrup. In this procedure frozen sections are made after the formalin fixation, and actually no test of solubility in fat solvents is made. Dietrich states that sections may be carried through alcohol and xylene into balsam, with loss of part of the lipids. Mallory notes that the procedure may also stain iron, hemoglobin, and blood pigments. Baker (*Q. J. Microsc. Sci.*, **85**: 1, 1944) used it as a reagent for demonstration of the Golgi substance, reducing the borax-ferricyanide time from 15 to 8 h, as follows:

1. Add 1% calcium chloride to 10% formalin, and neutralize it with suspended calcium carbonate. Fix tissue in this for 3 days.
2. Embed tissue in 25% gelatin, or if thinner sections are desired, evaporate it for 30 h in a desiccator over anhydrous calcium chloride at 37°C, stopping the evaporation while the gelatin solution is still liquid.
3. Cool in refrigerator (5°C), cut out tissue block, and harden 1 day in 1% calcium chloride–1% cadmium chloride–10% formalin solution.
4. Cut frozen sections at 15 μ from the 25% gelatin, or as thin as 5 μ if the concentration procedure has been used.
5. Attach sections by floating onto slides previously coated with 2.5% gelatin, drained, and dried.
6. Expose slides to the fumes of concentrated formalin for 10 min to harden the gelatin, and put back into calcium-cadmium-formalin until required.
7. Wash the slides to be stained by the Smith-Dietrich procedure for 3 min in water.
8. Place in cold 5% potassium dichromate solution in a Coplin jar, and then place the jar in an oven at 60°C (not 57°C) for 48 h.
9. Lift slides out of the solution a few times during the first few hours to get rid of air bubbles.
10. Then take out the jar and allow it to cool.
11. Wash slides in several changes of distilled water.
12. Stain 5 h in a modified Kultschitzky's hematoxylin (hematoxylin, 1 g; distilled water, 98 ml; sodium iodate, 0.2 g*; glacial acetic acid, 2 ml) at 37°C. The hematoxylin forms a resistant black lake with the chromium held by the lipids.
13. Normally one differentiates in Weigert's 1% borax–2.5% potassium ferricyanide solution for 15 h; for Golgi substance, only 8 h.
14. Then wash slides 5 min in running water, and mount in glycerol gelatin.

In vitro tests indicate that cephalin and sphingomyelin are stained black. Lecithin is black in the presence of other Smith-Dietrich–negative lipids, but not when alone. Galactolipids (cerebrosides) are gray (J. R. Baker, *Q. J. Microsc. Sci.*, **85**: 1, 1944).

Like Dietrich's method, the foregoing procedure makes no prestaining test of the solubility of the reacting lipids.

Ciaccio (*Bull. Microsc. Appl.*, **4**: 45, 1954) proposes the following modification of the Baker technic: Fix for 2 to 3 h in Zenker formol, wash (1 h) in running water, chromate in 5% $K_2Cr_2O_7$ at 60°C for 1 h, wash in water, and cut frozen sections. Or, alternatively, dehydrate in five 1-h changes of 50, 60, 70, 80, and 90% acetone containing 1% cadmium

---

* For usual American hematoxylin, the 200 mg $NaIO_3$ is excessive; use 50 mg.

chloride, then wash in two changes, for 1 h each, of anhydrous acetone, and then 1 to 2 h in petroleum ether (two changes?), and infiltrate and embed in paraffin.

Lipids thus preserved may be stained by the Smith-Dietrich hematoxylin procedure, differentiating in a 1 : 10 dilution of Weigert's borax-ferricyanide mixture, by the periodic acid Schiff procedure, directly with Schiff reagent, or by diammine silver hydroxide. They are also stained by basic aniline dyes.

The lecithins (Baker, *Q. J. Microsc. Sci.*, **87**: 441, 1946) react to the Smith-Dietrich-Baker type of procedure. The suggested mechanism is that the unsaturated fatty acid residues undergo oxidation by dichromate, with polymerization and consequent loss of solubility. Concurrently with this process Cr^{3+} is bound to the oxidation sites and serves to bind such mordant dyes as hematoxylin in a colored lake.

LILLIE'S MYELIN METHOD.   Fix routinely in buffered formol for 4 to 6 weeks for human brains, for smaller brains 3 to 4 weeks. Glutaraldehyde, 5%, is said to insolubilize about 20% of myelin lipids against 10% for formol. Selected blocks may be dehydrated and sectioned in paraffin before chromation or may be chromated in block in 3% $K_2Cr_2O_7$ buffered to pH 3.5 with acetates (Table 3-7). For strict selectivity for C=C double bonds chromate 4 to 6 weeks at 2 to 5°C. Chromation of sections in the same fluid at 3 to 5°C should be adequate in 4 weeks. For routine Baker-type Smith-Dietrich staining, 4 h at 60°C in 3% $K_2Cr_2O_7$ of pH 3.5 is adequate, or for superior results with plump myelin sheaths closely surrounding the axons mordant attached frozen sections for 2 to 6 h at 25°C in 0.1 to 0.2% CrO_3.

After any of these chromations, wash well in running water and then in distilled water, and stain 2 to 4 h in 0.1% hematoxylin in 1% acetic acid, wash in 80 and 95% alcohol, dehydrate with absolute ethanol, clear, and mount in synthetic resin. Regressive differentiation is generally unnecessary (*Stain Technol.*, **43**: 121, 1968; *Histochemie*, **20**: 338, 1969).

RESULTS: Myelin, erythrocytes, bile casts, keratohyalin, and eosinophil leukocyte granules are black. Other tissues are gray or unstained.

The chemistry of this process as reported in the above references is discussed in Chap. 8, in a discussion of ethylene. It is there noted that the use of heat extends the action of the dichromate to tissue amino and perhaps hydroxyl groups as well as ethylene groups. The aging of Kultschitzky's hematoxylin appears to be unnecessary for this technic. Lillie stresses that the bound metal demonstrated is Cr^{4+}, not Cr^{3+} or Cr^{2+}.

Augulis and Sepinwall (*Stain Technol.*, **46**: 137, 1971) report dark violet myelin staining with green nuclei and Nissl substance by a sequence iron alum, gallocyanin, methyl green technic. Formol-fixed tissue is required.

Make 1% *iron alum* fresh daily, dissolving only clean violet crystals in cold distilled water. *Gallocyanin solution:* Dissolve 90 mg sodium carbonate (Na_2CO_3) in 100 ml distilled water, add 250 mg gallocyanin (C.I. 51030), bring to a boil, simmer 2 to 3 min, let cool slowly, best overnight to 20 to 25°C with occasional shaking, and, last, add 5 ml absolute alcohol. This solution is stable and can be reused with weekly or semiweekly filtration for 2 to 3 months. It gradually changes from dark violet toward gray and may need replacement sooner if much used. The pH is about 7.4. *Methyl green:* Dissolve 500 mg methyl green (C.I. 42585 or 42590; certified by Biological Stain Commission) in 100 ml 0.1 *M* acetate buffer of pH 4.5 (60 ml 0.1 *M* acetic acid, 40 ml 0.1 *M* sodium acetate), and extract methyl violet by shaking in a separatory funnel with several changes of chloroform until no more violet appears in chloroform. The solution is stable for some months and can be used repeatedly.

TABLE 13–3 SCHEDULES FOR TRADITIONAL MYELIN STAINS

	Weigert	Pal	Kultschitzky	Wolters	Wright	Spielmeyer	Weil	Lillie, 1944
Fixation	Formalin or Orth 2–3 days	Formalin 2 days or Müller 2–3 wk	Formalin 2+ days	Formalin 2 days or Müller 2–3 wk	Formalin 2+ days	Formalin 3+ days	Formalin 2+ days	Formalin or Orth 2 days
Washing	.....	.....	.....	Müller 2–3 wk	.....	1 h	.....	Rinse
First mordant	Weigert 1st 4–6 days or 2–3 days 37°C	Müller 2–3 wk	Weigert 2d, 4–5 days					Dichromate 2–4 days
Washing	.....			Water				Rinse
Second mordant	Weigert 2d 1 day 37°C							
Section method	Graded alcohols to celloidin	Graded alcohols to celloidin	Graded alcohols to celloidin	Graded alcohols to celloidin	Frozen sections	Frozen sections	Graded alcohols to paraffin	Graded alcohols to paraffin or frozen sections
Washing	.....							
Hematoxylin method	Combined Weigert's neutral iron	Combined Weigert's lithium, Table 13–4 No. 5	Combined, Kultschitzky's acetic, Table 13–4 No. 8	Combined, Wolters' acetic, Table 13–4 No. 10	Sequence, 5 min 10% $FeCl_3$, and Wright's, Table 13–4 No. 11	Sequence, 6 min 2–5% iron alum and Spielmeyer's, Table 13–4 No. 12	Combined, Weil's iron	Combined, Lillie-Weil
Staining time	24 h	6–48 h	12–24 h	24 h	30 min	10–24 h	15 min	40 min
Temperature	15–20°C	15–20°C	15–20°C	"Warm"	20°C	20°C	50°C	58°C
Washing	30–60 min tap	Water and 2–3 drops sat. aq. Li_2CO_3	.....	Müller few seconds	Rinse	Rinse	Rinse	Rinse
First differentiator	Weigert's borax ferricyanide 15–30 min	0.25% $KMnO_4$ 15–20 s	Kultschitzky's decolorizer 3–4 changes 4–12 h	0.25% $KMnO_4$ 20–30 s	10% $FeCl_3$ brief	2.5% iron alum microscopic control	4% iron alum to gross differentiation	0.5% iron alum 1 h
Second differentiator	.....	Pal's bleach few seconds	.....	Pal's bleach 30 s to 3 min	.....	.....	Weil's borax-ferricyanide	Lillie's borax-ferricyanide 10 min
Washing	24-h tap	Thorough	Thorough	Thorough	Thorough	2 changes distilled 1–2 h tap	Water	Water
Counterstain	.....	(Carmine optional)	.....	.....				Acetic safranin, 5 min, wash
Mounting procedure*	Alcohol, aniline xylene, or carbol xylene, balsam	95% alcohol, carbol xylene, xylene, balsam	95% alcohol, terpineol, or origanum oil, xylene, balsam	95% alcohol, terpineol, or origanum oil, xylene, balsam	95% alcohol, terpineol, or origanum oil, xylene, balsam	95% alcohol, carbol xylene, or xylene (blot), balsam	Alcohols, xylene, balsam	Acetone, acetone, xylene, xylene, balsam

* Routine synthetic mounting media may be substituted for balsam.

Sources: Weigert's iron chloride–hematoxylin method is taken from his article in Ehrlich's "Encyklopädie," pp. 937–944, and is quoted essentially without change by Mallory, "Pathological Technic," Saunders, Philadelphia, 1938. Pal's method is taken from Weigert's article and from Schmorl, "Die pathologisch-histologischen Untersuchungsmethoden," 15th ed, Vogel, Leipzig, 1928, and from Mallory. Kultschitzky's and Wolters' technics are given by Romeis, "Mikroskopische Technik," Leibnitz, Munich, 1948, and by Schmorl and by Mallory. Wright's technic is as given in Mallory and Wright, "Pathological Technic," 8th ed., Saunders, Philadelphia, 1924, which varies somewhat from that in the 6th ed., 1918. Spielmeyer's technic is essentially as in Mallory, Romeis, and Schmorl; Weil's, from Arch. Neurol., 20: 392, 1938. Lillie's first appeared in Arch. Pathol., 37: 392, 1944; however, the 1944 Lillie method is replaced in the text by his 1968 and 1969 methods (Stain Technol., 73: 121, 1968; Histochemie, 20: 338, 1969).

THE TECHNIC

1. Bring paraffin sections through xylenes and alcohols to distilled water.
2. Mordant in 1% iron alum for 30 to 60 min.
3. Wash in distilled water for 3, 3, 3, and 3 min.
4. Stain 5- to 7-μ sections in gallocyanin for 1 h, 7- to 10-μ sections for 90 to 120 min, 12- to 20-μ sections for 2.5 h, or thicker sections for 15 to 30 min more for each 5 μ.
5. Wash in distilled water for 3, 3, 3, and 3 min until the water remains clear.
6. Counterstain for 5 to 10 min in methyl green.
7. Wash in distilled water for 2 min (two changes).
8. Immerse in 95% and absolute alcohol 2 times for 5 min each.
9. Treat with absolute alcohol and xylene, two or more changes of xylene, and synthetic resin.

RESULTS: Dark violet myelin, erythrocytes, and neuronal nucleoli, and green Nissl granules and nuclei are demonstrated.

Strangely enough, chromate fixatives not containing formaldehyde are ineffective in the usual 1- to 3-day fixation periods. It is to be noted that the older writers required 2 to 3 weeks' or months' hardening in Müller's fluid, and that Endicott found about 3 weeks'

TABLE 13-4 FORMULAS FOR REAGENTS USED IN VARIOUS MYELIN TECHNICS

1. The formalin is a 10% dilution in water of 40% formaldehyde solution.	9. Kultschitzky's decolorizer
	1% potassium ferricyanide — 10 ml
2. Weigert's first mordant	Saturated aqueous lithium carbonate — 100 ml
Potassium dichromate — 5 g	
Chromium fluoride — 2.5 g	10. Wolters' hematoxylin
Distilled water to make — 100 ml	Hematoxylin — 2 g
	Alcohol to dissolve — 10–20 ml
3. Weigert's second mordant	Glacial acetic acid — 2 ml
Cupric acetate — 5 g	Distilled water to make — 100 ml
Chromium fluoride — 2.5 g	
Glacial acetic acid — 5 ml	11. Wright's hematoxylin was an extempore solution of a few crystals in 15 ml distilled water.
Distilled water to make — 100 ml	
4. Weigert's borax-ferricyanide	
Borax ($Na_2B_4O_7 \cdot 10H_2O$) — 2 g	12. Spielmeyer's hematoxylin
Potassium ferricyanide — 2.5 g	5% alcoholic hematoxylin (aged) — 4 ml
Distilled water to make — 100 ml	Distilled water — 36 ml
5. Weigert's lithium-hematoxylin	
Hematoxylin — 0.75–1.0 g	13. Weil's borax-ferricyanide
Alcohol — 10 ml	Weigert's formula — 50 ml
Distilled water to make — 100 ml	Distilled water — 50 ml
Saturated aqueous lithium carbonate — 1–2 ml	14. Lillie's dichromate
6. Müller's fluid	Potassium dicromate — 5 g
Potassium dichromate — 2.5 g	Water — 100 ml
Sodium sulfate crystals — 1 g	
Distilled water to make — 100 ml	15. Lillie's borax-ferricyanide
7. Pal's sulfite oxalic bleach	Borax — 1 g
Potassium sulfite — 0.5 g	Potassium ferricyanide — 2.5 g
Oxalic acid — 0.5 g	Distilled water — 100 ml
Distilled water — 100 ml	
8. Kultschitzky's hematoxylin	16. Lillie's acetic safranin
10% alcoholic hematoxylin, aged 6 months* — 10 ml	Safranin O (C.I. 50240) — 100 mg
Glacial acetic acid — 2 ml	Glacial acetic acid — 1 ml
Distilled water to make — 100 ml	Distilled water to make — 100 ml

* Lillie (1969) finds the aging unnecessary.
Sources: Same as for Table 13-3.

TABLE 13-5   CONCENTRATION OF SATURATED LITHIUM CARBONATE SOLUTIONS
AT VARIOUS TEMPERATURES

Temperature, °C	0	5	10	15	20	25	30	40	50	60	70	80	90	100
g/100 ml	1.54	1.48	1.43	1.38	1.33	1.29	1.25	1.17	1.08	1.01	0.93	0.85	0.78	0.72
mM	208	200	193	187	180	175	169	158	146	137	126	115	105	97

Source: Primary data from N. A. Lange, "Handbook of Chemistry," rev. 10th ed., McGraw-Hill, New York, 1967. Data for 5, 15, 25, 70 and 90°C are interpolated.

treatment with dichromate necessary to render certain unsaturated fats insoluble and myelin-positive in vitro. It is further to be noted that Kaufmann and Lehmann insisted that the chromation in the Smith-Dietrich procedure be done at 60°C at least. This again follows a formaldehyde fixation.

In frozen sections, myelin of peripheral nerves is stained vividly red to red-purple by the periodic acid and peracetic acid Schiff procedures. These reactions are obtained (though less vividly) in material fixed with formaldehyde and chromated before dehydration and embedding, and the peracetic reaction is still strong in brain fixed 2 weeks in formalin and sectioned in paraffin without chromation.

THE MARCHI REACTION FOR DEGENERATING MYELIN.   Marchi (*Riv. Sper. Freniatr.*, **12:** 50, 1886) noted that osmium tetroxide blackened degenerating myelin in the presence of potassium dichromate; however, the lipids of normal myelin failed to blacken. Adams (*Adv. Lipid Res.*, **7:** 1, 1969) suggests that hydrophobic unsaturated lipids such as triglycerides, cholesterol esters, and free fatty acids are stained in degenerating myelin by the Marchi method whereas dichromate or perchlorate penetrates into the phospholipid or other hydrophilic molecules and is reduced instead of osmium tetroxide, thereby resulting in failure of a staining reaction.

Adams (*J. Histochem. Cytochem.*, **8:** 262, 1960) believes the lipid stainable with the Marchi method in areas of Wallerian degeneration of myelin is largely esterified cholesterol. The evidence is based upon the demonstration of elution of the Marchi substance in chromatograms in the esterified cholesterol fraction derived from degenerated human brain tissue at autopsy.

CARBONYL LIPIDS

This class probably includes ketosteroids, as well as plasmalogens and various ill-defined, readily soluble lipids which give aldehyde and ketone reactions. Verne derives them from ethylenic oxidation products. Since ethylenic linkages readily oxidize in air to epoxides and peroxides which hydrolyze respectively to 1,2-glycols and to aldehydes, this derivation is not improbable.

Cain (*Q. J. Microsc. Sci.*, **90:** 411, 1949) considers this latter group distinct from true plasmalogens, which become Schiff-positive promptly on treatment with mercuric chloride solutions, while ethylenic derivatives oxidize slowly to aldehydes, and the ethylenic linkages themselves react slowly with Schiff reagent. This last we question.

The lipids demonstrated by Liang (*Anat. Rec.*, **97:** 419, 1947) in axis cylinders of nerves are believed by Chu (*Anat. Rec.*, **108:** 723, 1950) to be unsaturated fatty acid lecithin esters, which he says react directly with Schiff reagent. In view of the slow reaction of ethylenic groups observed by Cain and by us, it seems more probable that Chu was dealing with intermediate peroxidation products of the unsaturated fat.

Seligman considers that his hydroxynaphthoic acid hydrazide method demonstrates ketones as well as aldehydes, and gave two supplemental tests for the distinction of the latter.

The following carbonyl method of Seligman and Ashbel (*Cancer*, **4**: 579, 1951) has been altered slightly to accord with the report of Herman and Dempsey (*Stain Technol.*, **26**: 185, 1951), with certain variations of our own and with Seligman and Ashbel (*Endocrinology*, **50**: 338, 1952).

M. L. Karnovsky and H. W. Deane (*J. Histochem. Cytochem.*, **3**: 85, 1955) have shown reasonably conclusively that the supposed histochemically demonstrable ketosteroids are actually autoxidation artifacts: aldehydes arising by oxidation of unsaturated lipids during formalin (and other) fixations. Nevertheless we retain the method as the only reasonably workable strongly chromogenic one which should show ketones as well as aldehydes. A comparison should always be made with the Schiff reaction and with the *m*-aminophenol fast black K technic, which should show aldehydes alone. The latter technic, unfortunately, is not applicable to readily soluble lipids as at present developed, since these are readily soluble in the glacial acetic acid in which the Schiff's base condensation reaction is effected. It is probable that if the amine were treated with a stoichiometric equivalent of concentrated HCl, a water-soluble chloride would be formed, as with aniline.

THE SELIGMAN-ASHBEL METHOD FOR ACTIVE CARBONYL GROUPS

1. Fix in 10% formalin, cut frozen sections at 10 to 20 μ, float onto slides, and dry for 10 min to make sections adhere.
2. Wash in water for 2 h to remove formaldehyde.
3. Stain controls by oil red O as usual. Perform solvent tests on experimental material, and bring sections back to water. Omit this step in plain demonstration technic.
4. Immerse extracted and unextracted sections for 2 h at 25°C in a fresh 0.1% solution of 2-hydroxy-3-naphthoic acid hydrazide. (Dissolve 40 mg in 2 ml hot glacial acetic acid, and add 38 ml freshly prepared 50% alcohol.)
5. Wash in four changes each of 5% acetic acid in 50% alcohol, of 50% alcohol, and of distilled water, for 2 min each (total 12 washes, 24 min); *or* (Seligman and Ashbel, 1952) six washes of 50% alcohol, 20 min each, and 1 of 30 min in 0.5 N hydrochloric acid, followed by rinses in water and in 1% sodium bicarbonate.
6. Mix 25 ml of 0.67 M pH 7.4 phosphate buffer with 25 ml 100% alcohol. Introduce slides, dissolve 50 mg fast black B or fast blue B, pour at once over slides in a Coplin jar, and let stand for 2 min.
7. Wash in 0.1% acetic acid and in water.
8. Counterstain in 0.1% safranin in 0.1% acetic acid for 1 to 2 min, if desired. Wash in water.
9. Mount in glycerol gelatin, Arlex gelatin, or the like; *or* dehydrate quickly with alcohol, clear with xylene, and mount in polystyrene or other synthetic resin.

RESULTS: "Active" carbonyl stains dark blue or greenish blue. Myelin and ceroid from frozen sections are stained. Compare with similar sections reacted for 20 min at step 4 with Schiff reagent and treated with bisulfite as usual. Ketones react only slowly with Schiff reagent.

Pearse (1960, p. 867) cites a method for "α-ketol groups of corticoids" from Khanolkar (*Indian J. Pathol. Bacteriol.*, **1**: 84, 1958), stating that the theoretical specificity is high and is supported by in vitro tests.

1. Attach cryostat sections of fresh tissue to slides, and immerse 20 min in aniline–glacial acetic acid (10 : 20) to block preexisting aldehydes.
2. Wash gently for 1 to 3 min in distilled water.
3. Oxidize 30 min at 50°C in 5% ferric chloride.
4. Rinse gently in distilled water.
5. Immerse 20 min in Schiff reagent.
6. Wash in three changes of bisulfite water as usual.
7. Mount in glycerol gelatin.

RESULTS: Corticosteroids stain magenta; androgens, estrogens, pregnanediol, and cholesterol do not react.

COMMENTS. Step 1 or a similar immersion in glacial acetic acid appears to remove all the birefringent crystals of adrenal cortex and all lipids stainable by Sudan black B except lipofuscins. Substitution of phenylhydrazine or its 2,4-dinitro derivative in this step blocks not only the direct Schiff reaction of adrenal cortex and atheroma but also the essentially equal one produced by ferric chloride. We suspect that the aldehydes demonstrable by the Schiff reaction, with or without ferric chloride oxidation, are the usual lipid aldehydes from oxidation of unsaturated fats, and that when these are blocked by *adequate aqueous* blockade methods, such as phenylhydrazine hydrochloride (0.5 M, 24 h) or 0.05% KBH_4 in 1% Na_2HPO_4 for 5 min, the amount of new aldehyde producible by ferric chloride oxidation will be too small to permit histochemical detection.

In any case it is to be noted that after a 24-h phenylhydrazine blockade of the periodic acid Schiff reaction of gastric mucosa, a similar ferric chloride treatment reengenders a partially positive Schiff reaction of the epithelial mucin. So the possibility of a positive lipid aldehyde reaction's arising by reoxidation or hydrolysis of previously blocked aldehyde is not to be disregarded.

ACID FASTNESS. The mechanism of acid fastness of mycobacteria has been much discussed. It has been attributed to mycolic acid, to a hypothetical capsule, to a peculiar deoxyribonucleic acid, or to an oil-solubility phenomenon similar to fat staining. The sudano-

TABLE 13-6 ACID FASTNESS AFTER VARIOUS BLOCKADE REACTIONS*

Blocking agent	End group	Hair cortex	Rat liver ceroid	*Mycobacterium tuberculosis*
5% phenylhydrazine hydrochloride	—CHO	24 h, 25°C Unimpaired	24 h, 25°C Unimpaired	24 h, 25°C Unimpaired
1 *M* Aniline chloride....................	—CHO	24 h, 25°C Unimpaired	24 h, 25°C Unimpaired	24 h, 25°C Unimpaired
40% Acetic anhydride–pyridine	—OH —NH$_2$	60 h, 60°C Unimpaired	4 h, 60°C Unimpaired	4 h, 25°C Lost
1 *N* NaNO$_2$–1 *N* HAc	—NH$_2$	4 h, 25°C Unimpaired	4 h, 25°C Unimpaired	4 h, 25°C Unimpaired
0.1 *N* HCl–MeOH.....................	—COOH —PO$_4$H —SO$_4$H	3–7 days, 25°C 24 h, 60°C Impaired	7 days, 25°C Lost	2 h, 25°C Lost
CCl$_4$	Lipid	24 h, 25°C Unimpaired	24 h, 25°C Unimpaired	24 h, 25°C Unimpaired
Pyridine	Lipid	6 h, 60°C Unimpaired	24 h, 60°C Unimpaired	2–4 h, 60°C Lost
Methanol	Lipid	7 days, 60°C Unimpaired	7 days, 60°C Unimpaired	7 days, 60°C Unimpaired

* The tissues used were fixed in formalin and embedded in paraffin.
Source: From Lillie and Bangle, *J. Histochem. Cytochem.,* **2:** 300, 1954.

philia of the organisms supports this. Lartigue and Fite (*J. Histochem. Cytochem.*, **10**: 611, 1962) support the last theory and stress the need for some phenol or phenol-like substances in the staining solution to facilitate the penetration of the dye. They also note the resistance to fat solvents shown in Table 13-6, but do not mention hot pyridine.

Our 1954 data indicate that the acid fastness of lipofuscins, as typified by ceroid, and of hair cortex and mycobacteria are not necessarily identical. However, in all three the reaction is impaired or abolished by methylation, while hot methanol is without effect.

It is perhaps worthy of note that the "Colour Index" records a number of amine dye bases as much more soluble in fatty acids than in neutral fats. Rosanilin base affords a particularly striking example. Its solubility in oleic and stearic acids is recorded as 20 and 15%; other solvents dissolve either none (linseed oil, paraffin, mineral oil, "white spirit," toluene) or quite small amounts (ethanol 0.3%, acetone 0.6%, benzene 0.2%, butyl acetate 0.5%; ethyl acetate is very slightly soluble).

PARAFFIN

Paraffin may occur in sections as a result of incomplete deparaffinization, as birefringent, often intranuclear crystals. The birefringence disappears at the melting point of the paraffin. In the cold these crystals do not stain appreciably with oil-soluble dyes; but if the staining solution is heated to above the melting point of paraffin, staining occurs (Nedzel, *Q. J. Microsc. Sci.*, **92**: 343, 1951). Xylene, alcohol, water, alcohol, xylene sequence treatments are more effective in removal of these birefringent crystals than a similar exposure to xylene alone, both at 23 to 25°C. Paraffin has occasionally been introduced into tissues, such as the breast, for cosmetic purposes.

LIMITATION OF SLIDE HISTOCHEMISTRY

Translocation of soluble substance occurs during processing of tissues for histochemical analyses. However, one must not be unmindful of the fact that this also occurs during the course of cell fractionation procedures. Despite the fact that some histochemical methods are less than absolutely specific for a particular constituent, they may still be helpful by providing some information on probable localization of a particular constituent. A combination of histochemical and one or more other methods may provide exquisite localization and qualitative and quantitative analysis.

ELECTRON MICROSCOPIC HISTOCHEMICAL METHODS

Detection of lipids at the electron microscopic level is difficult because lipids are soluble in reagents used in preparation of the tissues for microscopy. However, Korn and Weisman (*Biochim. Biophys. Acta*, **116**: 309, 1966) believe that some phospholipid is retained in specimens fixed in osmium or potassium permanganate. From Lillie's (*Histochemie*, **20**: 338, 1969) findings, binding of Cr(IV) by unsaturated lipids is probable. Methods have been advocated for the demonstration of unsaturated groups [Casley-Smith, *J. R. Microsc. Soc.*, **81**: 235, 1963 and **87**: 463, 1967; Jones et al., *Exp. Mol. Pathol.*, **2**: 14, 1963; and Seligman et al., *J. Cell Biol.*, **30**: 424, 1966], although they are relatively nonspecific. Likewise, the method Weller et al. (*J. Histochem. Cytochem.*, **13**: 690, 1965) advocated for the histochemical identification of phosphoglyceride could be improved.

Casley-Smith [*J. R. Microsc. Soc.*, **87**: 463, 1967] is convinced that there is no general fixative for lipids and that there is no generally satisfactory embedding medium for lipids (for electron microscopy). Likewise, there is no suitable stain for all lipids. Identification of lipid often involves observation of holes where presumed lipid was extracted.

Cope and Williams [*J. R. Microsc. Soc.*, **88**: 259, 1967] conducted a thorough quantitative study on the preservation of neutral lipids during the course of preparation of tissues for electron microscopy. Losses of neutral glycerides varied from 50 to 100%. Cope and Williams suggest use of embedding media containing a high percentage of water or the use of epoxy resins at low temperatures.

Saunders et al. (*J. Cell Biol.*, **37**: 183, 1968) believe osmium tetroxide fixes lipids by reacting with olefinic bonds to form relatively insoluble osmates. Substantial retention of oleate and linoleate was achieved in intestinal mucosa fixed in osmium and dehydrated in ethanol.

Cope and Williams [*J. R. Microsc. Soc.*, **90**: 31, 1969] conducted a quantitative study of the preservation of choline and ethanolamine phosphatides during processing of tissues for electron microscopy. Incorporation of ^{14}C-aminoethanol and ^{3}H-choline chloride into phosphatidylethanolamine and phosphatidylcholine was noted and followed. Column and thin-layer chromatography and double-label scintillation spectrometry were used in the quantitative study of rat livers.

After glutaraldehyde fixation alone, a complete loss of phosphatidylcholine and a loss of one-half of phosphatidylethanolamine were detected. Negligible loss of phosphatide occurred in specimens fixed either in glutaraldehyde followed by osmium or in osmium alone. No gain of phosphatide occurred by prefixation in glutaraldehyde. Correlated autoradiographic studies were carried out. Cope and Williams caution that retention of constituents may not always signify retention *in situ*.

COPE-WILLIAMS ELECTRON MICROSCOPIC METHODS FOR LIPID RETENTION

PROCEDURE 1:   Glutaraldehyde, osmium tetroxide, Araldite

1. 5% glutaraldehyde, 4 h, 0°C
2. Bicarbonate buffer washes, 4 times, 1 h, 0°C
3. 1% w/v osmium tetroxide, 1 h, 0°C
4. 70 and 95%, 3 times; 100% ethanol, 5 times, 10 min, 20°C
5. Epoxypropane, 30 min, 20°C
6. Epoxypropane-Araldite (1 : 1, v/v), overnight, 20°C
7. Araldite monomer, 8 h, 20°C
8. Encapsulated and cured with heat, 48 h, 60°C

PROCEDURE 2:   Osmium tetroxide, Araldite

1. 1% w/v osmium tetroxide, 1 h, 9°C
2. 70 and 95%, 3 times; 100% ethanol, 5 times, 10 min, 20°C
3. Epoxypropane, 30 min, 20°C
4. Epoxypropane-Araldite, overnight, 20°C
5. Araldite monomer, 8 h, 20°C
6. Encapsulated and cured with heat, 48 h, 60°C

LIPID AUTORADIOGRAPHY

Wilske and Ross (*J. Histochem. Cytochem.*, **13**: 38, 1965) employed ^{3}H-estradiol and ^{3}H-aspirin in an autoradiographic investigation to determine the best method for retention of lipid and water-soluble compounds. The following method was found highly efficacious:

1. Quick-freeze fresh tissue blocks in isopentane suspended in liquid nitrogen.
2. Freeze-dry at $-40°C$.
3. Bring tissue slowly to room temperature, and place in a sealed desiccator jar containing Drierite in a vapor of either osmium tetroxide (at 25°C) or paraformaldehyde (60°C) for 24 h.
4. Place tissue blocks in nonaccelerated Epon, and infiltrate under a vacuum (30 in Hg) at 25°C for 8 h. Tissues are infiltrated more rapidly if a low-viscosity resin is employed. Infiltrated blocks will settle to the bottom of the vessel.
5. Cut sections in the usual fashion.
6. Place each section in a drop of water on a glass slide, and dry on a hot plate.
7. After section adheres to the slide, coat section with either NTB-2 or NTB-3 (Eastman) nuclear track emulsion.
8. Following appropriate exposure, develop sections, fix, wash, and stain them with basic aniline dye.

Control tissues examined revealed quantitative retention of the isotope. The degree of retention and localization of labeled constituents obtained with the method was impressive.

Stumpf and Roth (*J. Histochem. Cytochem.*, **14**: 274, 1966) claim superior results with respect to retention of soluble constituents without translocation of activity with the following method:

1. A tissue sample (approximately 1 to 2 mm^3) is mounted on a brass tissue holder and quenched in liquid propane at $-180°C$.
2. Section at 0.75 to 1 μ with a cryostat. Detailed instructions for sectioning are provided by Stumpf and Roth [*Nature (Lond.)*, **205**: 712, 1965].
3. Sections are freeze-dried for 24 h.
4. The vacuum is broken by nitrogen, and sections are mounted directly on emulsion (Kodak NTB-3 or NTB-10), employing slight finger pressure under a safe light, preferably in a room with low humidity.
5. The slide with tissue and emulsion is stored in a desiccator over Drierite overnight at $-15°C$.
6. After exposure, the section is brought to room temperature and breathed on once or twice to enhance adherence to the slide.
7. The emulsion is processed by development with Kodak developer D19 for 2.5 min at 25°C.
8. Rinse in tap water briefly, and fix with Kodak fixer for 4 to 5 min.
9. Rinse in tap water for 5 min, and stain with hematoxylin and eosin.

The degree of localization of ^{3}H-estradiol and ^{3}H-mesobilirubinogen achieved by Stumpf and Roth was excellent.

Torvik and Sidman (*J. Neurochem.*, **12**: 555, 1965) have employed autoradiographic techniques using ^{3}H-acetate; Adams and Morgan [*Nature (Lond.)*, **210**: 175, 1966] and Kramsch et al. (*J. Atheroscler. Res.*, **7**: 501, 1967) have employed ^{3}H-cholesterol. Cholesterol-4-^{14}C was used by Schlant and Galambos (*Am. J. Pathol.*, **44**: 877, 1964), and Friedman et al. (*J. Clin. Invest.*, **38**: 539, 1959) used ^{131}I-labeled triolein.

The isotope favored by most investigators is tritium, because its β particle track is about 1.5 μ, which permits accurate detection of the source of the particle emitted. ^{32}P allows localization less precisely because of its 1.25-mm track. Those who employ tritium must remember, therefore, that close adherence of the emulsion to the section is essential to permit detection. ^{125}I has a track about 10 to 15 μ in length, whereas ^{14}C has a particle track of approximately 40 to 60 μ.

QUANTITATIVE LIPID HISTOCHEMISTRY

Linderstrøm-Lang (*Harvey Lect.*, **34**: 214, 1939) developed a method whereby serial sections were cut from an organ and monitored chemically and histologically. The method is particularly efficacious in organs that have a layered histologic pattern. Adjacent sections can be analyzed histologically and biochemically. Derry and Wolfe (*Exp. Brain Res.*, **5**: 32, 1968) and McDougal et al. (*J. Gen. Physiol.*, **44**: 487, 1961; **47**: 419, 1964) have employed this method for the study of lipids in the white matter of the brain.

Quantitative thin-layer chromatographic methods have been applied to a study of layers of aortic wall by Abdulla and Adams (*J. Atheroscler. Res.*, **5**: 504, 1965) and Adams et al. (*J. Atheroscler. Res.*, **8**: 679, 1968). Atherosclerotic plaques were studied by Liadsky and Woolf (*J. Atheroscler. Res.*, **7**: 718, 1967). Glick et al. (*J. Histochem. Cytochem.*, **3**: 6, 1954) studied lipid and potassium distribution in adrenal glands.

Posalaky et al. [*Acta Histochem. (Jena)*, **18**: 152, 1964] fixed cryostat-cut sections of adrenal glands in a solution containing 4% formalin, 1% $CaCl_2$, and 1% $CO(NO_3)_2$, rinsed in tap water, rinsed in 50% isopropyl alcohol, and stained in a saturated solution of Sudan black B in 50% isopropyl alcohol. Microspectrophotometry was employed to related sudanophilia with the structure of the adrenal gland.

Employment of the Linderstrøm-Lang method permits direct chromatography of sections; the results are not explicitly quantitative.

14

POLYSACCHARIDES; MUCINS

Spicer et al. (*J. Histochem. Cytochem.*, **13**: 599, 1965) have analyzed the problems arising from the usage by histochemists of terminology employed by biochemists for the designation of chemically defined carbohydrates and mucopolysaccharides. Histochemists must recognize that histochemical methods provide relatively little chemical information concerning the stained constituents in mucus components of cells and tissues. A general designation is needed to encompass a great variety of biochemically characterized and uncharacterized carbohydrate-rich constituents in tissues. The term *mucosubstance* appears appropriate for this purpose.

Spicer et al. (*J. Histochem. Cytochem.*, **13**: 599, 1965) recommend that mucosubstances demonstrable in tissue sections should still be named by (1) stating their localization and (2) characterizing them to the best degree possible as neutral mucosubstances, mucopolysaccharides, sulfomucins, and sialomucins. The latter designations would be derived on the basis of histochemical reactions for *vic*-glycols, sulfate groups, and sialocarboxyls. Customarily, the mucosubstances in connective tissues are designated *mucopolysaccharides;* those in epithelia are called *mucins*. The term *polysaccharide* excludes aminosugars. Aminosugars are essential constituents of mucosubstances. Spicer et al. (*J. Histochem. Cytochem.*, **13**: 599, 1965) have proposed the histochemical classification of mucosubstances presented in Table 14-1. They provide examples of the various mucosubstances as shown in Table 14-2.

Spicer (*Ann. N.Y. Acad. Sci.*, **103**: 379, 1963, slightly emended) lists histochemical methods for mucopolysaccharides as shown in Table 14-3.

According to Spicer (*J. Histochem. Cytochem.*, **8**: 18, 1960) the strongly acidic sulfated mucosubstances have extinction values for staining with 0.02% azure A at or below pH 0.5 (0.6 N HCl). These mucins display strong gamma metachromasia. They fail to stain with Alcian blue at pH 3, but they stain red with an Alcian blue (pH 3)–safranin O (pH 2) sequence. However, these mucins perhaps somewhat surprisingly will stain with Alcian blue at a pH level below 0.5 perhaps due to an incompletely dissociated polymer. These same mucins stain with aldehyde fuchsin in an aldehyde fuchsin–Alcian blue sequence (Spicer and Meyer, *Am. J. Clin. Pathol.*, **33**: 453, 1960). They also stain weakly or negatively with the periodic acid Schiff method; they are reactive with the iron diamine methods and are unaffected by periodate oxidation (Spicer, *Am. J. Clin. Pathol.*, **36**: 393, 1961; Spicer and Jarrels, *J. Histochem. Cytochem.*, **9**: 368, 1961).

Weakly acidic sulfomucins have extinction values for azurophilia in the range of pH 1.5 to 3 despite the presence of sulfate demonstrable by autoradiography. These mucins stain

611

TABLE 14-1 HISTOCHEMICAL CLASSIFICATION OF MUCOSUBSTANCES

I. Neutral mucosubstances—neutral glycoproteins, immunologically reactive glycoproteins, fucomucins, mannose-rich mucosubstances in epithelia and connective tissues (all periodate-reactive)

II. Acid mucosubstances

    A. Sulfated

        1. Connective tissue mucopolysaccharides (periodate-unreactive)

            a. Resistant to testicular hyaluronidase

                (1) Alcianophilic in the presence of 1 (or less) M MgCl$_2$—keratan sulfate, heparin

                (2) Alcianophilic in the presence of 0.7 (or less) M MgCl$_2$—dermatan sulfate

            b. Susceptible to testicular hyaluronidase

                (1) Alcohol-resistant affinity for 0.02% azure A at or above pH 2—chondroitin sulfates in cartilage*

                (2) Alcohol-resistant affinity for 0.02% azure A at or above pH 4—chondroitin sulfates in vascular tissues*

        2. Epithelial sulfomucins (testicular hyaluronidase resistant)

            a. Periodate-unreactive

                (1) Sulfate esters on *vic*-glycols

                (2) Sulfate esters not on *vic*-glycols

                    (*a*) Alcohol-resistant affinity for 0.02% azure A at or above pH 2

                    (*b*) Alcohol-resistant affinity for 0.02% azure A at or above pH 4.5

            b. Periodate-reactive—acid glycoproteins (?)

                (1) Alcohol-resistant affinity for 0.02% azure A at or above pH 2

                (2) Weak or negligible, alcohol-resistant affinity for 0.02% azure A at or above pH 4.5

    B. Nonsulfated

        1. Hexuronic acid-rich mucopolysaccharides—hyaluronic acid, chondroitin

        2. Sialic acid-rich mucosubstances

            a. Connective tissue mucopolysaccharides containing sialic acid (?)

            b. Epithelial sialomucins—acid glycoproteins

                (1) Highly susceptible to *Vibrio cholerae* sialidase, periodate-reactive; and stainable metachromatically with azure A

                (2) Slowly digestible with *Vibrio cholerae* sialidase

                    (*a*) Periodate-reactive

                    (*b*) Periodate-unreactive

                (3) Resistant to *Vibrio cholerae* sialidase

                    (*a*) Rendered metachromatic and susceptible to enzyme by prior saponification

                    (*b*) Sialidase-resistant after saponification

                      i. Periodate-reactive

                      ii. Periodate-unreactive

---

* These mucopolysaccharides show the same metachromasia in both wet and dehydrated sections, unlike epithelial sulfomucins, which often lose metachromasia after alcohol dehydration.

Source: Spicer et al., *J. Histochem. Cytochem.*, **13**: 599, 1965.

avidly with aldehyde fuchsin and Alcian blue at pH 3. They generally stain intensely with the periodic acid Schiff method.

According to Spicer (*Ann. N.Y. Acad. Sci.*, **103**: 379, 1963) nonsulfated acid mucosubstances manifest extinction of azure A staining below pH 2.5 in the range of pK values for strongly dissociated acid carboxyls. In an aldehyde fuchsin–Alcian blue sequence, these mucins favor Alcian blue uptake. On the basis of chemical sialic acid analysis, Spicer (*J. Histochem. Cytochem.*, **9**: 400, 1961) suspects the presence of substantial amounts of sialic acid in many sites of the nonsulfated acid mucosubstances. Other attributes such as strong staining with the periodic acid–phenylhydrazine–Schiff and the altered diamine staining after periodate oxidation or methylation suggest the presence of *vic*-hydroxyls in proximity to acid groups.

THE PERIODIC ACID SCHIFF METHOD

One of the most widely used stains for the histochemical detection of carbohydrates is the periodic acid Schiff method. The general chemistry and history of the method have been covered in Chap. 8.

TABLE 14-2 EXAMPLES OF HISTOLOGIC SITES CONTAINING EACH OF THE TYPES OF MUCOSUBSTANCES CLASSIFIED HISTOCHEMICALLY IN TABLE 14-1

I. Gastric surface epithelia and thyroid colloid of human beings, guinea pig, and rabbit; coagulating gland fluid in mouse and rat

II. *A.* 1. *a.* (1) Cornea and mast cells; cartilage?
     (2) Not demonstrated histochemically as yet—identified biochemically in mucosubstances isolated from skin, aorta, heart valves
   b. (1) Cartilage, ovarian follicular fluid
     (2) Aorta, heart valves, renal papilla, some areas of cartilage
  2. *a.* (1) Sublingual glands of Chinese hamster and glossal mucous glands and colonic goblets of rabbit
     (2) (*a*) Colon of guinea pig
       (*b*) Exorbital lacrimal gland of mouse
   b. (1) Glossal mucous glands and rectosigmoid colonic goblets in mouse and rat
     (2) Some duodenal goblets and pyloric and some laryngotracheal glands in mouse and rat

 B. 1. Ganglion cysts of synovial membranes and vitreous humor in man, uterine cervical stroma of estrogen-treated mouse, and cock's comb
  2. *a.* Cartilage?
   b. (1) Sublingual glands of mouse, Syrian hamster, and guinea pig
     (2) (*a*) Mucified vaginal epithelium of pregnant mouse
       (*b*) Rectosigmoid mucous cells of mouse
     (3) (*a*) Sublingual gland of rat
       (*b*) i. Sublingual glands of human being and monkey
         ii. Mammary gland secretion of lactating mouse

Source: Spicer et al., *J. Histochem. Cytochem.*, **13**: 599, 1965.

TABLE 14-3 HISTOCHEMICAL METHODS FOR MUCOPOLYSACCHARIDES

1. Autoradiography with $^{35}SO_4{}^{2-}$ or tritiated sugars
2. Visualization of acid groups with basic dyes (thiazins, azin, phthalocyanins, colloidal iron, aldehyde fuchsin, diamines)
 a. Determination of pH extinction values
 b. Demonstration of metachromasia
 c. Use of basic dye sequences (aldehyde fuchsin–Alcian blue, Alcian blue–safranin)
 d. Differentiation with acid or alkali, poststaining
3. Demonstration of aldehydes produced by periodate oxidation of *vic*-glycols (Schiff reagents, hydrazines, diamines)
4. Combination of basic dye and *vic*-glycol methods (Alcian blue–periodic acid Schiff, colloidal iron–periodic acid Schiff)
5. Blockage of staining by specific reactions (methylation, acetylation, phenylhydrazine condensation, saponification)
6. Induction of basophilia by sulfation
7. Alteration of staining by enzymatic digestion (hyaluronidase, sialidase)
8. Localization of specific mucoproteins with fluorescent labeled antibodies

Source: Slightly amended from Spicer, *Ann. N.Y. Acad. Sci.*, **103**: 379, 1963.

  In the use of the periodic acid Schiff procedure care should be taken that the Schiff reagent has not gradually lost in potency. A good test object is the tunica muscularis of a human appendix. The fine red meshwork of sarcolemma between individual smooth muscle fibers, especially in cross section, should be demonstrated. Prolonging the exposure compensates for the deterioration to some extent, but it is better to make small batches frequently, using the "cold Schiff" procedure.

  Traditionally the Schiff reaction is followed by several rinses in dilute sulfurous acid to remove excess reagent and prevent adventitious pigmentation from air oxidation of adsorbed leucofuchsin. Leuchtenberger (Danielli's "General Cytochemical Methods," vol. 1, 1958), Lison, Davenport, Gomori, McManus, Mowry, and Lillie have adhered to this practice; Pearse now uses simple water washing, as does the AFIP "Manual," 1960. With copiously chlorinated water supplies, we regard this practice as hazardous.

Since there is always a slight carryover of leucofuchsin into the sulfite rinses and since even considerably diluted leucofuchsin can produce appreciable staining on recoloration, it is necessary to discard these rinses at frequent intervals. We find it good practice to make up fresh dilutions from a stock of 10% $Na_2S_2O_5$ solution daily and to discard at least the first rinse more often, moving the second and third rinses up one, and supplying fresh bisulfite for the last rinse if many slides are being stained. The times allotted are entirely arbitrary.

The solution has usually been prescribed as 0.05 M $NaHSO_3$, or simply as sulfurous acid (McManus) or as 0.4% potassium metabisulfite ($K_2S_2O_5$) in a 1% dilution of concentrated hydrochloric acid (Hotchkiss). Apparently 0.5% sodium metabisulfite ($Na_2S_2O_5$), which is slightly stronger than 0.05 M $NaHSO_3$, is quite satisfactory.

THE PERIODIC ACID SCHIFF–SULFITE LEUCOFUCHSIN REACTION (PAS)

1. Deparaffinize and hydrate through xylene, alcohols, and distilled water as usual.
2. Oxidize 10 min in 1% (0.044 M) H_5IO_6.
3. Wash 5 min in running water.
4. Immerse for 10 min in Schiff reagent.
5. Transfer quickly and directly to three successive baths, 2, 2, and 2 min in 0.5% sodium metabisulfite ($Na_2S_2O_5$). Replace sulfite rinses daily or more often.
6. Wash 5 min in running water.
7. Counterstain nuclei and cytoplasm as desired; e.g., stain 2 min in a 2% acetic hemalum solution of about 0.1% hematoxylin content (Mayer's alum hematoxylin or a one-fifth dilution of Lillie's). Wash in water, and blue with 1 or 2 drops of 20% Na_2CO_3 in 200 ml water. *For critical histochemical work omit all counterstains.*
8. Dehydrate (and differentiate) in two changes each of 95% and 100% alcohol. Clear in one change of alcohol-xylene mixture (1 : 1) and two changes of xylene. Mount in suitable resin, such as polystyrene, HSR, Permount, ester gum, cellulose caprate, Depex, or the like.

It is to be noted that alum hematoxylin and freshly mixed Weigert's iron hematoxylin may stain nonnuclear structures excessively, sometimes partly converting the red-purple of Schiff reactions to gray-purple or violet. This may be overcome by keeping the staining period short and by allowing the Weigert mixture to age an hour or so before using it. Acid differentiation is useful, and this is one of the functions of the picric acid counterstain. A 20-s dilute Pal's bleach renders the iron hematoxylin stain almost purely nuclear.

When diastase or other enzymatic digestion tests are employed, the sections are likely to become detached during the oxidation or aldehyde demonstration procedures. Hence it is recommended that collodionization be interposed immediately *after* the enzymatic digestion. Collodion films almost totally inhibit enzyme penetration; therefore the collodionization should not be inserted before enzyme digestion. Since anhydrous methyl alcohol dissolves nitrocellulose, of necessity collodionization, if practiced at all, must follow a methylation or methanol extraction procedure. Similarly, since pyridine dissolves both paraffin and collodion, collodionization must follow pyridine extractions and pyridine acetylation technics.

On the other hand saponification procedures and various other chemical blockade procedures which also tend to detach sections may more conveniently follow collodionization.

For the Bauer reaction, substitute at step 2 a 1-h bath in 5% chromic anhydride (CrO_3), and for the Casella reaction, a 20-min immersion in 1% potassium permanganate. These are the usual directions, though we have reduced the chromic acid treatment to 20 or 30 min at the same concentration and the permanganate treatment to 0.5% for 10 min with essentially similar results and fewer section losses.

One may combine such procedures as the diammine silver reticulum methods or the ferric mannitol or dialyzed iron or Alcian blue mucin methods with the periodic acid Schiff by interposing them before the periodic acid oxidation, before step 2.

RESULTS: Oxidation Schiff reactions have been recorded as positive from a quite extensive list of substances (Lillie, *Stain Technol.*, **26:** 123, 1951). The substances reacting more strongly with the periodic acid method usually give positive Bauer and Casella reactions as well. These include the polysaccharides glycogen, starch, cellulose, and the rather slowly digestible polysaccharide of *Sarcocystis*. Similar Bauer-positive polysaccharides are present in *Eimeria stiedae*, in *Klossiella muris*, in the egg cytoplasm of *Capillaria hepatica*, and in *Toxoplasma* species, for which the periodic and permanganate reactions have not been recorded. Among the epithelial mucins those of the gastric surface epithelium, peptic gland neck cells, cardial and pyloric glands, uterine cervical glands, and rodent vaginal epithelium give all three reactions, as do also the prostatic gland secretion and corpora amylacea, the seminal vesicle contents, and renal hyalin casts. The ocular lens capsule and the membrane of Descemet and the colloids of the thyroid and anterior hypophysis also give strong periodic acid and positive Bauer and Casella reactions. Rossman noted the latter two reactions for ovarian pigment; it often reacts strongly to periodic acid.

With a second group of substances the Casella reaction is generally weak or negative; the Bauer, moderately strong; and the periodic Schiff, positive to strongly positive. This group includes the mucins of the salivary glands, of the conjunctiva, of the oropharyngeal glands of most of the species studied, of Brunner's duodenal glands, of intestinal goblet cells, of tracheobronchial glands and goblet cells, and of ovarian follicles and cysts. The cuticular borders of the epithelium of the villi of the small intestine and of the renal convoluted tubules react similarly, as does the zona pellucida of rodent ova. The Casella reaction of cartilage matrix is weak or negative; the Bauer and periodic acid Schiff reactions are positive, with or without antecedent diastase digestion. Agar used as an embedding matrix behaves similarly. Hypophyseal cyanophil or beta cells react similarly, as do cerebral corpora amylacea. Amyloid gives a rather light red coloration with the periodic acid Schiff, and weakly positive Bauer and Casella reactions. Positive Bauer and HIO_4 Schiff reactions, with negative Casella reaction, are given by the cell walls of various yeasts and molds, by the filamentous material in the granules of actinomycosis and botryomycosis, and by certain diphtheroid organisms. The coagulum in the vitreous humor and corneal collagen usually stain well with the HIO_4 Schiff method and weakly with the Bauer and Casella methods.

With a third group of substances the periodic acid Schiff reaction is positive, and the Bauer and Casella reactions vary from weakly positive to negative. This category includes the acromere lipid of the retinal rods, the ceroid pigment of the choline deficiency cirrhosis of rats, the pharyngeal gland mucin of the rabbit, the mucins in the bases of the colonic glands, chordoma mucin, Russell bodies, and the kerasin of Gaucher's disease.

Collagens from areolar connective tissue and sclera give positive HIO_4 Schiff reactions, negative Bauer reactions, and weak or negative Casella tests. Megakaryocyte granules color pink with the HIO_4 Schiff and vary according to species with Bauer and Casella reactions. Chromaffin in chromate-fixed adrenal gives a gray-red after HIO_4 only.

Vascular and certain epithelial basement membranes, and reticulum give good positive HIO_4 Schiff reactions, a negative Bauer reaction, and a dubious or negative Casella reaction. Human colonic melanosis pigment; a similar pigment in the guinea pig intestine; human, mouse, and guinea pig adrenal lipofuscin pigment; fibrin; often plasma and serum; salivary zymogen granules of human beings and various rodents; pancreatic zymogen granules of human beings, rabbits, and some mice, but not rats or guinea pigs; Paneth cell granules of rats, guinea pigs, and rabbits, but not regularly those of human beings and mice; granules of many mast cells in human beings, few in mice, none in rats; and the bacterial cell walls of some organisms (including anthrax, streptococcus, and various intestinal bacteria) give fairly good positive HIO_4 Schiff reactions and negative Bauer and Casella tests.

A relatively weak periodic acid Schiff reaction and negative Bauer and Casella reactions are recorded for bone matrix; myxoma mucin, the mucinous coagula in the lymphatic spaces of the umbilical cord matrix; nucleus pulposus mucin; *collacin*, or collagen in the state of "basophil degeneration"; and the lipofuscin of heart muscle.

Shanklin and Azzam (*Stain Technol.*, **40**: 117, 1965) note that substances that stain red with the PAS method appear blue when observed by phase microscopy.

Elftman (*Stain Technol.*, **38**: 127, 1963) advised a combined direct silver–periodic acid Schiff method for investigations of the Golgi apparatus such as in the pituitary and during spermiogenesis.

DIRECT SILVER–PAS

1. Immerse fresh tissue in a solution of 2% silver nitrate in 15% formalin for 2 to 3 h.
2. Decant the solution from the vial, but leave a small amount adhering to the tissue for its effects in step 3.
3. Add 2% hydroquinone in 15% formalin, and leave for 2 to 3 h for intensification.
4. Complete fixation by transfer into 15% formalin or Bouin's fixative overnight. Many other fixatives may be used, sometimes with bleaching of the silver but reappearance of the image after immersion in gold chloride. The quality of the fixation has, however, been essentially determined before this step is reached.
5. Run up and embed in paraffin, cut at suitable thickness, and run down to water in the usual fashion.
6. Immerse in 1% gold chloride for 10 min.
7. Rinse in distilled water.
8. Oxidize with 1% periodic acid for 10 to 15 min.
9. Rinse in three changes of distilled water, for 2 min each.
10. Stain in Schiff reagent for 20 min.
11. Rinse well. Nuclei may be stained with Harris' hematoxylin or with 0.2% toluidine blue O buffered at pH 5.2 with acetate.
12. Remove excess stain with 95% alcohol, run up, and mount.

The silver is especially useful for direct visualization of the Sertoli cells.

Petri (*Acta Pathol. Microbiol. Scand.*, **73**: 303, 1968) claims that celestin blue prepared according to Gray et al. (*Stain Technol.*, **31**: 141, 1956) is a better counterstain for PAS–stained preparations for studies of the kidney glomerulus in that contrast is greater than that provided by hematoxylin. His method follows:

1. After periodic acid Schiff staining including sulfite rinses, rinse the slide in water for 10 min.
2. Stain in celestin blue for 2 min.[1]

3. Rinse in water for 3 min.
4. Place in saturated aqueous picric acid for 2 min.
5. Dehydrate in absolute ethanol, three baths.
6. Clear, and mount in DPX or other synthetic resin.

Di Bella and Hashimoto (*J. Invest. Dermatol.*, **47**: 503, 1966) recommend the following PAS stain for Araldite-embedded specimens for electron microscopy. Thick tissue sections, approximately 2 mm square, are cut with glass knives on a Porter-Blum ultramicrotome, fixed to glass slides by heating them in 40% acetone until evaporation is completed, and stained for PAS reaction in the following manner:

1. Apply a few drops of 0.5% periodic acid for 10 min to the Araldite-embedded section which is fixed to a slide, and heat to 65°C on a hot plate. Avoid total evaporation of the periodic acid by adding it freshly once or twice during the heating. We prefer to place two thick sections at a time on an ordinary glass slide, for sometimes a single section is lost or damaged, as in rinsing. After the 10-min application of periodic acid, the sections are rinsed carefully with distilled water and then dried slowly on the hot plate. If a section is detached from the slide during heating or rinsing, it can be recovered or reattached to the slide by heating with a trace of distilled water until dry.
2. Then apply several drops of Schiff's reagent to the dry sections at a temperature of 65°C. At this temperature, it will be noted that the solution soon becomes intensely purple-red. Fresh solution must be added at intervals to prevent drying out. Five minutes or so of such treatment is adequate. The deep purple-red persists only on heating, and the usual faint pink returns to the solution upon cooling. Once imparted to the tissue, however, the staining is permanent. The specimens so treated reveal the classic distribution of PAS–positive stain of the perinuclear glycogen surrounding the "punched-out" spaces representing the PAS–negative nuclei. Sections thus stained are briefly rinsed with distilled water and dried. Then they may be permanently mounted.

Allen and Perrin (*J. Histochem. Cytochem.*, **22**: 919, 1974) provide the following method for the histochemical detection of aldehydes at the light and electron microscopic levels. The method is particularly suitable for the staining of glycogen and nuclei after a Feulgen-type hydrolysis.

Allen and Perrin synthesized ferrocenylmethylcarboxyhydrazide formulated below:

---

[1] Place 1 g celestin blue B in a 250-ml Pyrex beaker, tilting the beaker sideways so that the dye accumulates in a small area. Add 0.5 ml concentrated H_2SO_4, and rub the dye to a paste with the acid. This paste swells and effervesces but sets after a few minutes to a friable mass that can be reduced readily to coarse granules by the pressure of a glass rod.

Flood the granules, with constant stirring at a temperature of about 50°C, a solution of 2.5 g ferric alum in 100 ml water to which has been added 14 ml glycerol. Cool to room temperature, and adjust the pH of the solution to 0.8 with concentrated H_2SO_4.

Tissues are fixed for 3 h at 4°C in Karnovsky's formaldehyde-glutaraldehyde fixative buffered to pH 7.4 with 0.1 M sodium cacodylate buffer, washed overnight in the same buffer at 4°C, postfixed in 2% OsO_4 for 3 h at 4°C in the same buffer, dehydrated in ethanols, and embedded in Durcupan.

Thin sections mounted on either gold or stainless steel grids are oxidized for 1 h at 25°C with 1% periodic acid, rinsed *thoroughly* in three changes of distilled water (10 min each), stained in a 2% aqueous alcoholic solution of ferrocenylmethylcarboxyhydrazide for 1 h at 60°C, rinsed in the buffer solution at 40°C, rinsed in two changes of 80% ethanol, and examined by electron microscopy. For light microscopy, 1-μ epoxy sections were then mounted on glass slides and stained as indicated above except that after the buffer rinse which follows ferrocenylmethylcarboxyhydrazide staining, sections are treated with 2% silver nitrate in distilled water for 1 h in the dark, dehydrated, and mounted in synthetic medium.

The solvent for the hydrazide is prepared by dilution of acetate buffer of pH 5 (15.2 ml 1 M acetic acid and 10 ml 1 M KOH in 1 l distilled water) with an equal volume of ethanol followed by readjustment of the pH to 5. Slight warming may be necessary. The solution should be absolutely clear and golden yellow.

Glycogen is particularly well delineated and stained for both light and electron microscopy. Mucosubstances are also stained. The silver step is omitted for electron microscopy because the reaction products were too bulky for precise localizations. Failure of staining occurred in sections with omission of the oxidative step and in sections blocked with *m*-aminophenol after oxidation.

Crippa (*Boll. Soc. Ital. Biol. Sper.*, **27**: 599, 1951) reports on the use of lead tetraacetate as an oxidant in place of chromic or periodic acid in a procedure utilizing the oxidation-induced aldehyde reaction for the demonstration of mucin and mucoid substances.

The reagent, which is stable for an indefinite period, is a 1% solution of recrystallized lead tetraacetate in glacial acetic acid. The solution is clear and colorless. The technic follows. Shimizu and Kumamoto (*Stain Technol.*, **27**: 97, 1952) advocate a similar procedure.

CRIPPA'S LEAD TETRAACETATE METHOD FOR MUCINS AND MUCOIDS

1. Fix as usual. Embed in paraffin, section, and deparaffinize in xylene.
2. Wash in 50 : 50 xylene–glacial acetic acid and in glacial acetic acid.
3. Immerse for 30 to 60 min at room temperature in 1% lead tetraacetate in glacial acetic acid.
4. Wash in three changes of glacial acetic acid to remove lead acetates.
5. Pass through xylene–glacial acetic acid 50 : 50 mixture, xylene, and descending alcohols to water.
6. Immerse in Schiff reagent for 15 to 20 min.
7. Wash in three changes of 0.5% sodium bisulfite, for 1, 2, and 2 min.
8. Wash in running water for 10 min.
9. Counterstain if desired, as with the periodic acid Schiff method, or dehydrate, clear, and mount directly in balsam or synthetic resin.

RESULTS: Aldehyde deposits engendered by lead tetraacetate oxidation show in red-purple.

If desired, steps 7 to 13 of Gomori's methenamine silver method for glycogen and mucins may be substituted for steps 6 to 9 above. The mucins then appear in black.

McManus (February 1952) wrote Lillie that lead tetraacetate renders Schiff-positive essentially the same structures as does periodic acid. Jordan and McManus use a fresh mixture of 1 part of a strong to saturated solution of lead tetraacetate in glacial acetic acid with 3 parts of a mixture of 20 g anhydrous sodium acetate, 25 ml glacial acetic acid, and 50 ml water. They use a shorter oxidation time, 30 to 120 s. Lhotka (*Stain Technol.*, **27**: 213, 1952) uses 5 g potassium acetate dissolved in 100 ml glacial acetic acid saturated with lead tetraacetate, and the same short oxidation times as McManus. Leblond and coworkers (*Stain Technol.*, **27**: 277, 1952) observed that glycogen was not stained by room temperature oxidation but did react vigorously at higher temperatures. Our experience with the Lhotka reagent indicated prompt reaction of glycogen.

THE IRON REACTIONS.   The mucins, as well as collagen, reticulum, and various other tissue components, have the property of taking up in more or less selective fashion $FeOH^{2+}$ ions from colloidal solutions of ferric salts in 2 M acetic acid. These combined $FeOH^{2+}$ ions may then be demonstrated by the usual HCl ferrocyanide reaction.

Scott (*J. Histochem. Cytochem.*, **21**: 1084, 1973) could find no evidence that phosphotungstic acid cleaved glycols as proposed by Palladini et al. (*Histochemie*, **24**: 315, 1970). However, Scott noted that dried spots of phosphotungstic acid on filter paper processed through Schiff's reagent, bisulfite, and running water stained red. Inasmuch as protamine sulfate also stained under similar conditions, Scott believes that phosphotungstic acid combines via electrostatic bonds with cationic (or protonatable) groups, depending upon the pH.

Dahlqvist et al. (*J. Histochem. Cytochem.*, **13**, 423, 1965) conducted quantitative studies of the PAS staining reaction. The rate of the reaction, the amount of color produced, periodate consumption, and a variety of substrates were investigated. Table 14-4 reveals that oligosaccharides and heteropolysaccharides were either PAS–negative or weakly positive. Proteins and nucleic acids were negative.

Figure 14-1 and Table 14-5 indicate (in test tube experiments) no simple relation between the intensity of the PAS reaction and the amount of periodate consumed or the amount of aldehyde groups formed on oxidation by periodate.

Hardonk and van Duijn (*J. Histochem. Cytochem.*, **12**: 533, 748, 1964) synthesized cellulose films with (1) covalently bound amino groups, (2) sulfhydryl groups, and (3) phosphate-bound deoxyribonucleotides and stained them with the PAS method. On the basis of

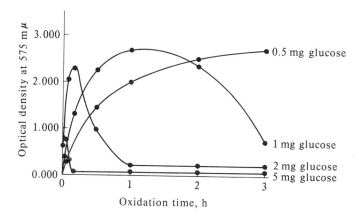

Fig. 14-1 Amount of color produced with Schiff reagent after the oxidation of varying amounts of glucose (0.5 to 5 mg) for times up to 3 h. (*From Dahlqvist et al., J. Histochem. Cytochem.*, **13**: 423, 1965. *By permission of the publisher.*)

TABLE 14-4   AMOUNT OF COLOR PRODUCED BY DIFFERENT
SUBSTANCES IN THE PERIODATE SCHIFF REACTION
AFTER REMOVAL OF EXCESS OXIDANT AND
DEVELOPMENT FOR 20 h WITH SCHIFF'S REAGENT

		Optical density at 560 mμ	
Substance	Amount of substance, mg	Oxidation time 5 min	Oxidation time 120 min
Monosaccharides:			
Glucose	0.5	0.940	2.790
Mannose	0.5	1.890	2.650
Galactose	0.5	1.160	3.672
Fructose	0.5	1.620	1.640
Glucosamine	0.5	0.500	2.500
Disaccharides:			
Maltose	1.0	0.080	0.140
Saccharose	1.0	0.010	0.030
Lactose	1.0	0.010	0.070
Trisaccharides:			
Raffinose	0.5	0.000	0.040
Polysaccharides:			
Homopolysaccharides:			
Glycogen	1.0	0.000	0.700
Starch	1.0	0.300	2.100
Dextran	1.0	0.060	1.100
Heteropolysaccharides:			
Heparin	1.0	0.000	0.008
Chondroitin sulfuric acid:			
Type A	1.0	0.000	0.000
Type B	1.0	0.000	0.000
Type C	1.0	0.005	0.020
Hyaluronic acid	1.0	0.050	0.100
Keratosulfate	1.0	0.030	0.040
Proteins:			
Serum albumin	1.0	0.000	0.000
Gelatin	1.0	0.000	0.000
Nucleic acids:			
Ribonucleic acid	1.0	0.000	0.000
Deoxyribonucleic acid	1.0	0.000	0.000

Source: Dahlqvist et al., *J. Histochem. Cytochem.*, **13:** 423, 1965.

absorption spectra of stained film with and without treatment with formaldehyde and sulfite, Hardonk and van Duijn concluded that the periodic acid Schiff reaction involves the formation of an aminoalkylsulfonic acid in agreement with previous studies by Hörmann et al. (*Justus Liebigs Ann. Chem.*, **616:** 215, 1958), Rumpf (*Ann. Chim.*, **3:** 327, 1935), and Nauman et al. (*Anal. Chem.*, **32:** 1307, 1960). Their studies found no evidence for the development of an aminosulfinic acid as proposed by Wieland and Scheuing (*Ber. Dent. Chem. Ges.*, **54:** 2527, 1921).

Heineman (*Stain Technol.*, **45:** 165, 1970) determined several physicochemical properties related to visible, ultraviolet, and infrared absorption spectra of basic fuchsin, leucofuchsin, and a formalin-Schiff reaction product.

TABLE 14-5 AMOUNT OF PERIODATE CONSUMED BY DIFFERENT SUBSTANCES ON APPARENTLY COMPLETE OXIDATION (2 TO 6 h)

Substance	Periodate consumed, μmol/mg
Glucose	18.0
Mannose	19.0
Galactose	22.0
Glucosamine	13.0
Maltose	9.0
Lactose	9.0
Sucrose	7.0
Glycogen	4.5
Starch	4.5
Dextran	7.5
Chondroitin sulfate A	<0.3
Chondroitin sulfate C	0.6
Keratosulfate	2.7
Gelatin	0.0
Serum albumin	0.7

Source: Dahlqvist et al., *J. Histochem. Cytochem.*, **13**: 423, 1965.

McManus (*Ann. Histochim.*, **7**: 57, 1962) notes as indicated by Table 14-6 that the intensity of the PAS staining reaction is increased sometimes in certain tissues when glacial acetic acid or diethyl ether is employed as a solvent. He is uncertain of the significance of the observation.

Horobin and Kevill-Davies (*Stain Technol.*, **46**: 53, 1971) advise substitution of the Schiff reagent by a 20-min stain in 0.5% v/v solution of basic fuchsin in ethanol (ethanol–water–concentrated HCl, 80 : 20 : 1) followed by a rinse in ethanol.

Culling et al. (*J. Histochem. Cytochem.*, **19**: 654, 1971) noted an increased intensity of stain with the PAS method in previously saponified sections of the intestine below the ileocecal valve. Saponification was conducted with 0.5% KOH in 70% ethanol at 24°C for 30 min. It is assumed that saponification increases the number of 1 : 2 glycol groups available for periodate oxidation.

TABLE 14-6 SOLVENT INFLUENCE ON OXIDATION

Material (human)	0.5% Periodic acid in		
	Water	Glacial acetic acid	Diethyl ether
Cartilage ground substance	*l* to *ll**	*lll*	*lll*
Cartilage glycogen	*l* to *ll*	*lll*	*llll*
Corpora amylacea, central nervous system	*ll*	*lll*	*llll*
Cardiac aging pigment	*ll*	*l*	*l*
Bronchial mucin	*lll*	*ll*	*ll*

Source: McManus, *Ann. Histochim.*, **7**: 57, 1962.

* The number of *l* are used to represent the intensity of stain.

Sawicki and Rowinski (*Histochemie*, **19:** 288, 1969) studied the effect of the PAS staining reaction on grain counts for autoradiography. They advise treatment of sections with periodic acid, processing for autoradiography, and staining thereafter with Schiff's reagent, whereupon no loss occurs.

Taylor and Clifford (*Stain Technol.*, **48:** 142, 1973) report that O-acetylated mucopoly-saccharides may be saponified with NaOH and then reacted by the PAS method. Their scheme for demonstrating both acetylated and unacetylated mucins in four serial sections is presented in the subjoined table. N-Acetyl groups resist the mild saponification used here.

	10% H_5IO_6, min	W*	pH 7.1 0.05– 0.075 NaBH$_4$ %, h	W	0.1 N NaOH, min	W	10% H_5IO_4, min	W	Schiff. min	Reactive	
										Nonacetyl mucin	Acetyl mucin
1.	5	W	1	W	..	..	..	..	10	−	−
2.	5	W	1	W	15	W	5	W	10	−	+
3.	..	..	..	..	..	..	5	W	10	+	−
4.	..	..	..	..	15	W	5	W	10	+	+

* W = water.

Sections are dewaxed, hydrated, and reacted as indicated with water (W) washes between steps, alcohol and water after the NaOH step. The customary bisulfite washes after Schiff reagent were replaced by a simple distilled water rinse. Counterstaining of nuclei with Mayer's alum hematoxylin, blueing, washing, alcohol dehydration, clearing, and mounting in DPX followed in all instances. Lillie and Pizzolato's 0.2% NaBH$_4$ in 1% Na$_2$HPO$_4$, 1 to 2 min, may be substituted for their 1-h 0.05 to 0.075% NaBH$_4$ in 0.05 M borate buffer of pH 7.1, which becomes more alkaline as the reaction progresses.

No great difference in distribution of O-acetylated and unacetylated mucins has so far been found in normal, adenomatous, and carcinomatous colonic mucosa.

In the same month Reid, Culling, and Dunn (*J. Histochem. Cytochem.*, **21:** 473, 1973) reported in a more elaborate study an almost identical periodic acid, borohydride, alkali hydrolysis, PAS method for showing O-acetylated mucins. They also studied colonic mucins in human beings, rabbits, rats, and mice. Their borohydride reduction was accompanied by adding gradually over a 30-min period 167 ml ice-cold water containing 1.89 g NaBH$_4$ to 100 ml ice-cold 2.45% boric acid in which the sections are immersed before starting the borohydride addition. Lillie and Pizzolato's borohydride technic was noted in the bibliography but not mentioned in the text. Alkali hydrolysis before the first periodic acid oxidation prevented appearance of the postreduction saponification effect, as might be expected. Effects of the procedures on pH 2.5 Alcian blue staining for carboxyls, on high-iron diamine for sulfate esters, mixed diamine for acid groups were studied, as well as the effects of a 4-h 60°C 1% HCl methanol and acetylation in the McManus-Cason acetic anhydride pyridine. The procedures were also tested on glycogen acetate where the presence of N-acetyl groups can be excluded. The glycogen acetate contained 4.15% acetyl residues (corresponding to slightly over two acetyl groups per hexose unit) and the saponification removed about one acetyl group per unit. If C-6 were preferentially and more firmly bound, removal of one acetyl group from C-2 or C-3 would serve to reactivate the glycogen to periodic acid oxidation by re-creating the 2,3-diol (Lillie). Reid, Culling, and Dunn did not arrive at what they considered an adequate explanation, and they seem to assume that they are dealing with a triacetyl glycogen, rather than the predominantly diacetyl compound indicated by their assay. Accepting the lactonization during methylation of Vilter and Sowari and Stoward, they also accept the latter group's claim that lactones are reduced by borohydride, which is disputed in the chemical literature (see Chap. 8). Since they do not record the final pH reached in their borohydride mixture, simple alkali hydrolysis is not excluded. Lillie and Jirge (*Histochemistry*, **41**: 249, 1975) reject the lactonization theory.

Stoward (*J. R. Microsc. Soc.*, **87**: 393, 407, and 437, 1967) recommends the demonstration of carbohydrate-containing mucosubstances with a formazan developed by condensing periodate-engendered aldehydes with an arylhydrazine which is subsequently coupled with a diazotized aromatic amine at high-alkaline pH. Stoward suggests the following may occur:

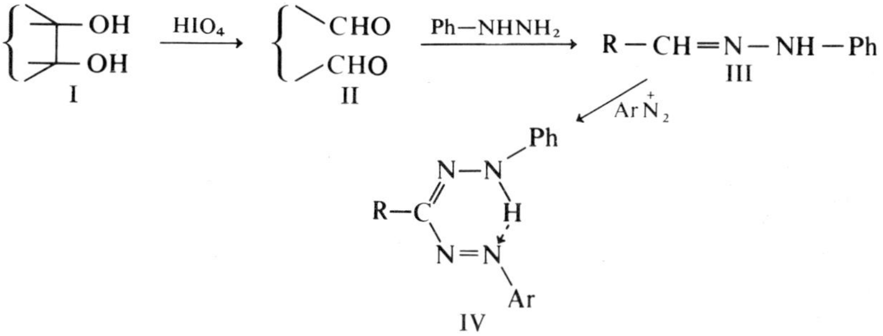

Stoward indicates several factors including the following which may affect the conversion of mucosubstances into formazans:

FACTORS AFFECTING THE CONVERSION OF MUCOSUBSTANCES INTO FORMAZANS. The yield, color, and properties of mucosubstance formazans varies from one tissue site to another and may be related to some or all of the following factors:

1. The nature of the periodate-engendered "dialdehyde" intermediates formed in step I
2. The stability of the corresponding "dialdehyde" arylhydrazone obtained in step II
3. The mechanistic course of the diazonium salt coupling reaction in step III
4. The presence of groups other than *vic*-glycols in the mucosubstance macromolecule (e.g., sulfate, carboxyl). Such groups may either interact within the formazan directly or react with arylhydrazine or with diazonium salt in side reactions to give colored products which would contribute to the final color observable in mucosubstance sites.

The relative influence of each factor depends on a number of experimental variables.

On the basis of many investigations, Stoward believes it is unreasonable to assume that all dialdehydes formed in mucosubstances will react with phenylhydrazine even under optimal conditions. Stoward notes that best results with phenylhydrazine are achieved in phosphate buffer of pH 4 to 6.

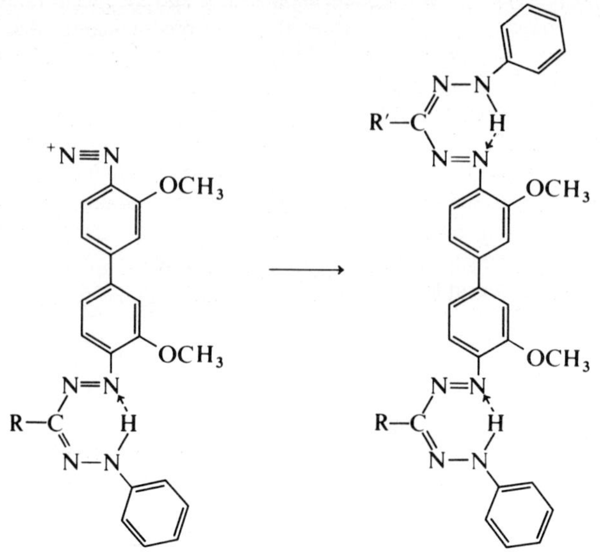

| Monoformazan VII | Diformazan VIII |

Stoward provides the above postulated monoformazans and diformazans that could theoretically form with the aldehyde-phenylhydrazine-arylamine. Diformazans are black, and black formazans are not observed. For this reason, Stoward believes that various types of monoformazans are formed.

Stoward (*J. R. Microsc. Soc.*, **87**: 407, 1967) prescribes the following procedure for the production of formazans from periodate-engendered aldehydes in mucosubstances:

Sections known to contain periodate-reactive mucosubstances are deparaffinized, taken to water, and then treated as follows (all reactions are carried out at room temperature unless stated otherwise).

1. Oxidize in 1% aqueous periodic acid (pH adjusted to be within a range of 3 to 5 with 1 *N* NaOH) for 10 min.
2. Rinse briefly in water.
3. Immerse in a freshly prepared 2.5% solution of phenylhydrazine hydrochloride in 5% mixed phosphate buffer [2.5 g each of monosodium dihydrogen phosphate ($NaH_2PO_4 \cdot 2H_2O$) and disodium monohydrogen phosphate ($Na_2HPO_4 \cdot 12H_2O$) in 100 ml aqueous solution] for 1 min upward (usually 3).
4. Rinse in at least two changes of water.
5. Flood with a fresh solution (concentration approximately 0.1 to 1%, and preferably less than 1 min old) of tetrazotized 3,3′-dimethoxybenzidine fluoroborate (TDMBF)[1] in 25% aqueous pyridine (1 : 3 pyridine–distilled water, pH 9.3 to 10.4) for 3 min.[2]

---

[1] *Warning:* Fluoroboric acid, TDMBF, benzene diazonium fluoroborate, and their solutions must not be allowed to come into contact with the skin. If they do, the skin may itch persistently (which may not start until several hours afterward) and should be scrubbed thoroughly with soap and hot water.

[2] Stoward stresses the point that TDMBF is not precisely a fast blue B stabilized with BF_4. Commercially available diazonium salts that he tried to substitute for TDMBF were unsuitable. Diazonium salts are unstable compounds. They need to be recently prepared.

6. Rinse briefly in water.
7. Take through alcohols, clear in xylene, and mount in a synthetic mountant such as Permount (Fisher Scientific Co.) or DPX (British Drug House Ltd., England).

Red, brown, and yellow formazans are developed. Stoward believes they are mostly mono-formazans with perhaps some interlinkages. Reasons for the color variations are unknown. Stoward states that all mucosubstances that are potentially periodic acid Schiff-positive form formazans using the method prescribed above. The one exception is mucin in the deeper mucous cells of the colon. Reasons for this failure are unknown.

Stoward (*J. R. Microsc. Soc.*, **87**: 437, 1967) believes he has excluded the possibility that the colored products observed are azo compounds, tetrazenes, or mixtures of these and that they are in fact formazans. He was able to treat the derivatives *in situ* first with *N*-bromosuccinimide which oxidizes acid-labile formazans into acid-stable colorless tetrazolium salts; second, azo derivatives and tetrazenes were destroyed with HCl in hot ethanol, which leaves tetrazolium salts unaffected; and third, the mucosubstance tetrazolium salts were restored to the original colored formazans by application of an aqueous solution of ammonium sulfide.

Stoward (*J. R. Microsc. Soc.*, **87**: 237, 1967) appears to have improved Kasten's (*Stain Technol.*, **33**: 39, 1958) fluorescent Schiff-type reagent method for the detection of periodate-engendered aldehydes in mucosubstances. Stoward prescribes the following method:

METHYLATION–PERIODIC ACID SO$_2$ DYE METHOD

1. Take deparaffinized sections through two changes of methanol, and methylate with a 2% solution of thionyl chloride in methanol (1 to 7 days old) for 4 h at room temperature. Afterward rinse in methanol and take to water.
2. Oxidize in 1% periodic acid (pH previously adjusted to be within a range of 3 to 5) for 10 min.
3. Rinse briefly in water.
4. Immerse in a freshly prepared solution of sulfur dioxide in water (0.25% v/v thionyl chloride per 100 ml distilled water; pH adjusted to appropriate value—usually 3—with sodium hydroxide) for 20 min. (It is possible that $Na_2S_2O_5$ would serve.)
5. Rinse briefly in water.
6. Immerse in a 0.01% (acridine orange, coriphosphine C.I. 46020) or 0.1% aqueous solution of dye at appropriate pH—usually 3—for 20 min.
7. Rinse briefly in water.
8. Blot dry, dehydrate in isopropyl alcohol, and mount in resin.
9. Examine for fluorescence.

The advantage of the above method over the original Kasten procedure is that methylation removes acid groupings which bind the fluorescent dye complex resulting in fluorescence in nonaldehyde sites.

Stoward (*J. R. Microsc. Soc.*, **87**: 247, 1967) has found that salicylhydrazide forms blue fluorescent derivatives with all potentially Schiff-reactive, periodate-oxidized mucosubstances in sections of fixed (Carnoy) tissues. Intense fluorescence is generally emitted by sulfomucins and sialomucins, whereas less fluorescence is emitted by neutral mucosubstances. Fluorescence is enhanced by aluminum salts and inhibited by zinc salts. The reaction has the advantage that stable salicylhydrazide fluorescent products are formed with aldehydes only. Stoward's method is detailed below.

PERIODIC ACID–SALICYLHYDRAZIDE TECHNIC. Potentially periodate-reactive mucosubstances were demonstrated as follows (all reactions being carried out at room temperature):

1. Take deparaffinized sections (of formalin-fixed or coagulant-fixed tissue, preferably Carnoy-fixed) to water.
2. Oxidize in 1% periodic acid (pH adjusted to a range of 3 to 5 with 1 N sodium hydroxide) for 10 min.
3. Rinse briefly in water.
4. Immerse in a filtered, 0.5% solution of salicylhydrazide (Koch-Light, Colnbrook; Eastman Kodak) in 5% acetic acid for 20 to 30 min. (A brown, resinous precipitate is slowly deposited in the salicylhydrazide solution after a few weeks, but this does not impair the effectiveness of the reagent.
5. Rinse in water.
6. Rinse in a freshly prepared, dilute (approximately 0.5%) solution of trisodium penta-cyanoammine ferroate (British Drug House, Ltd. or Hopkins & Williams Ltd.).
7. Rinse in two changes of water.
8. If desired, counterstain the section in a solution of a solochrome black–aluminum complex for 10 min. This is prepared by addition of a 1% aqueous solution of either Salicin black EAG (C.I. 15710) or diamond black PV (C.I. 16500) to twice the volume of a 5% aqueous solution of potassium–aluminum sulfate and then filtering. The dyes are obtainable from Imperial Chemicals Industry, England, and other sources.
9. Rinse in water.
10. Dehydrate through the alcohols, clear in xylene, mount in DPX, and examine for fluorescence using ultraviolet excitation (366 mμ).

Stoward (*J. R. Microsc. Soc.*, **88:** 571 and 487, 1968) studied the influence of trace metals on the fluorescence of periodate-oxidized mucosubstance salicylhydrazones. Salts of aluminum, iron, and copper were added to the salicylhydrazide solution and the rinse water. In general, cupric and ferric salts quenched the fluorescence even at 10^{-4} gram ions per liter. Fluorescence was enhanced by aluminum. Stoward recommends a rinse of tissues in 5% potassium alum after the pentacyanoammine ferroate rinse. Fluorescence of different mucin salicylhydrazones was noted to fade at different rates. Strongest fluorescence and least fading occurred in specimens prepared with fresh, pure salicylhydrazide solutions and exposed to 5% potassium alum as indicated above.

Burns and Neame (*Blood*, **26:** 674, 1966) applied the Stoward periodic acid–salicyl-hydrazide fluorescent method for the detection of carbohydrates in smears of blood and bone marrow. The method compared favorably with the periodic acid Schiff method.

GLYCOGEN DEMONSTRATION METHODS

The traditional methods for the demonstration of glycogen are the iodine method—which is applicable also to amyloid, corpora amylacea, starch, amylose, and other substances—and the alkaline carmine solution of Best. More recently we have added the aldehyde reaction with Schiff's reagent after oxidation with chromic acid or potassium permanganate, periodic acid, lead tetraacetate, and sodium bismuthate.

IODINE METHODS.   We omit the Langhans method cited in Lillie, 1948, since it is now demonstrated that ordinary aqueous formalin will preserve glycogen. We prefer to fix in Lillie's acetic alcohol–formalin, alcoholic formalin, or Carnoy's fluid. The method is as follows: Dehydrate completely with ascending alcohols; changes of 1 h each suffice for the 100% alcohol; the lower grades should have 8 to 16 h each. Clear in cedar oil and embed in paraffin as usual. Carry sections through two changes each of xylene and of 100% alcohol into 1% collodion in equal volumes of ether and 100% alcohol. Soak for

5 to 10 min, drain for 1 min, and harden in 80% alcohol for 5 to 10 min. Transfer to water and proceed.

1.   Treat with Gram's iodine for 5 to 10 min.
2.   Dehydrate in two changes of 2% iodine in 100% alcohol.
3.   Clear and mount in origanum oil. Seal with pyroxylin cement.

For the demonstration of newly formed amylose and glycogens Takeuchi immerses sections after incubation for a few minutes in a tenfold dilution of Gram's iodine and then mounts in iodine glycerol and seals with paraffin. Amylose and starch color blue, glycogens color red-purple to red-brown.

The staining solution is I–KI–H$_2$O, 1 : 2 : 3000; the iodine glycerol is 9 parts pure glycerol and 1 part Gram's iodine.

Lower's (*Stain Technol.*, **32:** 127, 1957) silver conversion procedure renders iodine preparations permanent. It was used chiefly for the study of insect cuticle and muscle but seems readily adaptable to other tissues.

1.   Deparaffinize and hydrate sections as usual, removing mercury deposits with iodine, thiosulfate, and water washes. This avoids carrying any mercury over to the silver bath.
2.   Immerse for 10 min in Weigert's iodine (I–KI–H$_2$O, 1 : 2 : 100). Wash thoroughly in distilled water.
3.   Immerse for 10 min in 1% silver nitrate. Wash thoroughly in distilled water.
4.   Develop in a fine-grain photographic developer or fresh 1% hydroquinone for 10 min. Wash thoroughly in distilled water.
5.   Tone for 15 min in 0.1% gold chloride in distilled water. Wash thoroughly in distilled water.
6.   Fix for 2 min in 0.2 *M* sodium thiosulfate (5% Na$_2$S$_2$O$_3$ · 5H$_2$O). Wash for 10 min in running water.
7.   Differentiate for 2 min in 1% K$_2$S$_2$O$_5$ in 0.125 *N* HCl. Wash, dehydrate, clear, and mount in balsam, Permount, cellulose caprate, or the like.

RESULTS: Iodophil material is purple to black.

Glassware should be cleaned with concentrated nitric acid to remove silver after each use for this method, and then washed thoroughly in distilled water.

Since the whole operation is conducted apparently in daylight it is probable that the silver iodine compounds formed are at least partly reduced without development. Since the silver exposure is brief, iodine is probably not liberated from aliphatic substitutions as in Norton's procedure, but only the loosely bound iodine in (oxidized) elementary form is shown. This is removed in the Norton procedure by the bisulfite bath which follows halogenation.

Takeuchi wrote Lillie (1963) that he found it better to use Gomori's CrO$_3$–methenamine silver technic when permanent preparations were required. In any case Lower's method does not preserve the differential color reactions with iodine.

BEST'S CARMINE SOLUTION.   This is prepared by gently boiling 2 g carmine, 1 g potassium carbonate, and 5 g potassium chloride in 60 ml distilled water for several minutes or until the color darkens. Cool and add 20 ml 28% ammonia water. Ripen for 24 h and store in refrigerator at 0 to 5°C. The carmine should be certified by the Biological Stain Commission. The stock solutions may be kept for several weeks only. The foregoing directions are derived from C. Bensley and from Mallory, and vary materially from Best's as cited in Ehrlich's "Encyklopädie."

The staining solution is composed of 8 ml of the stock solution above, 12 ml 28% ammonia water, and 24 ml methyl alcohol, a total of 44 ml. Mallory used 10, 15, and 15 ml respectively, total 40 ml, and directed filtration of the stock solution before mixing. C. Bensley directed thorough mixing and warned strictly against filtration. This dilute solution is good for 1 or 2 days only, and it is preferable to use it only once.

1. Mallory prescribes celloidin sections, Bensley paraffin, and we generally take the paraffin sections from 100% alcohol into 1% collodion for 5 to 10 min, drain for 1 min, harden for 5 min in 80% alcohol, and transfer to water.
2. Stain for 5 min in an acid alum hematoxylin (Ehrlich's, Lillie's, or other).
3. Wash briefly in tap water.
4. Stain for 20 min in the freshly diluted staining solution.
5. Paraffin sections:
    a. Rinse sections with three changes of fresh methyl alcohol (preferably using a dropper bottle).
    b. Dehydrate, and remove collodion with two or three changes of acetone.
    c. Pass through acetone and xylene into two changes of xylene, and mount in synthetic resin.
6. Celloidin sections:
    a. Wash sections in several changes of a mixture of 10 ml water, 8 ml 100% ethyl alcohol, and 4 ml methyl alcohol.
    b. Dehydrate in 80 and 95% alcohol.
    c. Clear in origanum oil and mount in balsam.

RESULTS: Glycogen is red; nuclei stain blue.

The periodic acid Schiff procedure, with diastase digestion controls, is far simpler and appears to be quite uniformly successful.

The Bauer and Casella oxidation Schiff methods and their results are discussed in an early chapter. The preferred method for demonstration of glycogen is the periodic acid Schiff–leucofuchsin technic. A useful method, giving red-purple glycogen and other periodic Schiff–positive substances (but blue collagen and reticulum), is the allochrome procedure (Chap. 15).

It may be noted here that prolonged bromination (1 ml bromine in 39 ml carbon tetrachloride for 3 to 6 h at 25°C) completely prevents the reaction of glycogen by the periodic acid Schiff method, without evidently affecting the reactivity of any other tissue substances. The reaction is probably analogous to the chlorination of starch, bromine replacing the hydrogen on C-1 and C-4, and removing both hydrogens, leaving O= on C-2 and C-3 (Kerr, "Chemistry and Industry of Starch," Academic, New York, 1950). This bromination effect is reversed by a 3-day immersion in 5% silver nitrate.

Mitchell and Wislocki (*Anat. Rec.,* **90:** 261, 1944, cited in *J. Tech. Methods,* **25:** 159, 1945) reported on the demonstration of glycogen by an *ammoniacal silver* nitrate technic. Gomori (*Am. J. Clin. Pathol.,* **10:** 177, 1946) explained this as an aldehyde reduction of ammine silver and extended the method to the demonstration of mucus. We dissent from this explanation.

Murgatroyd and Horobin (*Stain Technol.,* **44:** 59, 1969) find that hematoxylin, hematein, alizarin saphirol ES (C.I. 63000), anthracene blue SWR (C.I. 58605), alizarin red S (C.I. 58005), or gallein (C.I. 45445) may be substituted for carminic acid or carmine in Best's glycogen stain.

The stock solution is prepared by dissolving 1 g dye, 1 g K_2CO_3, and 5 g KCl in 60 ml boiling water. For use mix 20 ml stock solution with 15 ml 35% ammonia or 18.75 ml 28%

and an equal volume of absolute methanol (15 or 18.75 ml, respectively). With American reagents the staining solution is a little weaker, but the somewhat higher average American laboratory temperature should compensate for this so that the same staining time can be used. The staining solution is said to be stable for 24 h.

The staining directions differ somewhat from those for the traditional Best's carmine.

1. Hydrate as usual. Interpose enzyme digestion or control procedure if desired. A hemalum nuclear stain, or kernechtrot if red nuclei are desired, may be interposed. Wash as usual.
2. Stain in the specific glycogen stain for 2 to 5 min.
3. Wash off excess stain with absolute methanol.
4. Use absolute alcohol, xylene, Permount, Depex, or balsam as usual.

RESULTS: Nuclei stain blue with hemalum and red with kernechtrot, and are unstained without these. Glycogen stains blue with anthracene blue SWX or alizarin saphirol ES, purple with gallein, red with alizarin red S or carminic acid, red changing to brown on standing with hematoxylin or hematein.

Murgatroyd and Horobin invoke a hydrogen-bonding mechanism to account for the specific staining (Goldstein, *Q. J. Microsc. Sci.*, **103**: 477, 1962).

Murgatroyd (*J. Med. Lab. Technol.*, **27**: 512, 1970) and Horobin and Murgatroyd (*Histochem. J.*, **3**: 1, 1971) conducted studies to determine the mechanism of staining of Best's carmine with glycogen. The effects of sulfation, benzoylation, deamination, and methylation blocking procedures as well as variations in the salt, ethanol, and pH of the staining solutions were studied. Hydrogen bonding of the dye molecules with glycogen is postulated.

DIASTASE DIGESTION TEST. Make a 0.1% solution of malt diastase in 0.02 M phosphate buffer of pH 6. Sodium chloride up to 0.8% may be added if desired, but larger quantities should be avoided as they may inhibit the enzyme action. Filter before using, to remove starch granules. Include a known glycogen-positive section as a control on the potency of the enzyme. Amylases gradually deteriorate, even in cold storage.

TECHNIC

1. Bring paraffin sections to water as usual.
2. Digest for 30 to 60 min at 45 to 35°C in 1 : 1000 malt diastase.
3. Wash in water.
4. Dehydrate with alcohols or acetone.
5. Soak for 5 to 10 min in 1% collodion in ether-alcohol (50 : 50).
6. Drain for 1 min.
7. Immerse in 80% alcohol for 5 to 10 min to harden the collodion.
8. Proceed with the glycogen demonstration method chosen, using an undigested control section along with the digested one.

Goetsch, Reynolds, and Bunting (*Proc. Soc. Exp. Biol. Med.*, **80**: 71, 1952) report successful digestion of glycogen from liver cells by amylase, using undeparaffinized sections. This procedure could well avoid some of the section losses. In our hands diastase solutions required 6 to 12 h to effect fairly complete digestion of liver glycogen, compared with about 10 min for the usual deparaffinized ones. Heating of the sections to the melting point of the paraffin before digestion almost totally inhibits digestion, probably by forming a continuous paraffin film over the section.

Purified α- and β-amylases are now commercially available. Takeuchi (*J. Histochem. Cytochem.*, **6**: 208, 1958) uses them at 0.5% in distilled water, buffering with 4 mM acetates to pH 5.5 to 6 for α-amylase and to pH 4 to 5.7 for β-amylase. Digestion periods range from 30 min to 10 h at 37°C. The β-amylase promptly digests amylose, starch, and similar unbranched polysaccharides, but it attacks glycogen only slightly; α-amylase promptly digests both.

McManus and Saunders (*Science*, **111:** 204, 1950) reported on the use of *pectinase, pectinol O, pectin esterase, polygalacturonase,* and *β-glucuronase* as microdissection agents for the study of tissues fixed in cold acetone. The enzymes were used in 0.4% solution in acetate buffer of pH 4, probably Walpole's.

Pectinase and polygalacturonase remove periodic acid Schiff–positive materials: mucin, glycogen, splenic and lymphadenoid reticulum, cartilage matrix, hyalin, etc. Pectinol O gave qualitatively similar though less complete removals. Pectin esterase and β-glucuronase did not attack the periodic acid Schiff–positive materials. Incubation periods were 48 h at 37°C.

Paegle (*J. Histochem. Cytochem.*, **12:** 919, 1964) noted that most amylases and diastases contain peptidases and/or ribonucleases. On this basis caution is advised on interpretation of absence and/or failure of staining of constituents. See also Lillie (*Anat. Rec.*, **103:** 611, 1949).

Malinin (*J. Histochem. Cytochem.*, **18:** 834, 1970) proposed the following method for the demonstration of glycogen. A metachromatic staining reaction is observed after bisulfite is added to aldehyde engendered to glycogen by periodic acid as indicated below:

1. Oxidize hydrated sections or cell cultures with 0.5% aqueous periodic acid for at least 15 min.
2. Rinse in distilled water for 1 min.
3. Immerse in 50% $NaHSO_3$ for 1 h. (To ensure good contact between tissues and bisulfite solution slides are moved up and down during the first minute.)
4. Rinse in distilled water for 1 min.
5. Stain for 2 to 3 min with 0.5% toluidine blue at a 2 to 5 pH range.
6. Dehydrate in absolute ethanol.
7. Clear in xylene and mount in resin as usual.

THE METHENAMINE SILVER METHOD.   For this purpose Gomori found best a methenamine silver nitrate solution: Add 5 ml 5% silver nitrate to 100 ml 3% methenamine. A heavy white precipitate appears and easily redissolves on shaking. This stock solution may be kept for months in the refrigerator or for about 2 weeks at room temperature.

Gomori prescribed an alcoholic fixative for glycogen. Mitchell and Wislocki used an alcoholic picroformalin. Grocott prescribed 10% formalin. Grocott's (*Am. J. Clin. Pathol.*, **25:** 975, 1955) variant is much used for fungi in tissues. Gomori's technic follows:

1. Deparaffinize and collodionize in 0.5% collodion. For mucin omit collodion.
2. Bring sections to water.
3. Treat 1 to $1\frac{1}{2}$ h in 5% chromic acid (Grocott: 1 h).
4. Wash 10 min in running water (this can be reduced to 1 min).
5. Treat 1 min in 1 to 2% sodium bisulfite ($NaHSO_3$) to remove traces of chromic acid (Grocott: 1%).
6. Wash 5 min in running water.
7. Rinse in distilled water, three or four changes.
8. Silver at 37 to 45°C (Lillie, AFIP, 1960, 58 to 60°C) in 25 ml stock methenamine silver solution, 25 ml distilled water, 1 to 2 ml 5% borax (Grocott: 2 ml). Observe slides at intervals under the microscope until mucin, glycogen, and fungi are dark brown while background remains yellow. This may take 1 to 3 h (30 to 60 min at 60°C).
9. Rinse in several changes of distilled water.

10. Tone 5 min in 0.1% gold chloride.
11. Rinse thoroughly in 2 to 3% sodium thiosulfate solution to remove excess silver. (Grocott used 2%; we prefer 5%.)
12. Wash in running water.
13. Counterstain as desired (Grocott stained 40 s in 0.2% light green SF in 0.2% acetic acid), dehydrate, clear, and mount as usual. The collodion membrane sometimes stains rather intensely. It may be removed by acetone or by ether-alcohol mixture during dehydration.

RESULTS: Mucin, glycogen, and melanin color deep gray-brown to black. Insoluble calcium salts may also blacken. In comparison with the Bauer method this technic appears rather laborious, and on brief trial the visual contrasts appear inferior. We have had a number of failures with it on Bauer-positive material. It shares with the Bauer reaction the staining of mucin, which gives difficulties in the study of glycogen in muciparous cells.

Many investigators have used Grocott's variant with good results for the demonstration of fungi in tissues, notably *Histoplasma*. Grocott increased the temperature to 45 to 50°C and reduced the time to about 1 h. The AFIP "Manual," 1960, gives 58°C and 30 to 60 min as Grocott's method.

McLean and Cook state that staining for 15 to 20 min in a saturated solution of *chlorazol black E* in 70% alcohol demonstrated nuclei in black; cytoplasm and inclusions, gray; chitin, green; and glycogen, red. If sections are overstained, differentiate with terpineol.

We have tried this with American chlorazol black E. Chitin of cestodes is fairly well shown in green, but much of the connective tissue is also gray-green, and (to say the most for the method) it requires a certain amount of imagination to discern a reddish fluorescence where the glycogen ought to be. The carmine and periodic acid methods are much better for glycogen.

Leske and Mayersbach (*J. Histochem. Cytochem.*, **17**: 527, 1969) conducted correlative biochemical and histochemical studies in attempts to determine the best method for preservation of glycogen for histochemical assay. They were surprised to conclude that on the basis of detection with the periodic acid Schiff stain, the best preservation was observed in fresh frozen sections, and fixation of frozen sections with ethanol or Carnoy's fluid diminished the intensity of the staining reaction. Leske and Mayersbach conclude that the utilization of the "glycogen fixatives" and low temperatures does not result in a staining reaction which corresponds in intensity to the amount of glycogen measured chemically. These results are in accordance with those of Dahlqvist (see details of the PAS stain, above).

Sasse et al. [*Acta Histochem.* (*Jena*), **20**: 25, 1965] tested the suitability of several fixatives (Gendre, Friberg, Dubosq, Brasil and Pasteels, and Leonard) for the histochemical preservation of glycogen and concluded that Gendre's was most suitable for maximum retention and localization.

For the preservation of glycogen in bones and teeth Yoshiki et al. (*Histochemie*, **21**: 256, 1970) recommend 0.5% cyanuric chloride in methanol containing 1% *n*-methylmorpholine as advocated by Goland et al. (*Stain Technol.*, **42**: 41, 1967).

Rambourg et al. (*J. Cell Biol.*, **40**: 395, 1969) employed either a combined periodic acid–chromic acid–silver methenamine stain or chromic acid–phosphotungstic acid stain for the identification of carbohydrates in several rat tissues at the electron microscopic level. Comparable results were obtained with both methods and were similar to those observed by light microscopy with the periodic acid Schiff stain. Reactive carbohydrate was observed in the region of the Golgi apparatus, small vesicles and multivesicular bodies, cell coats, and basement membranes.

Vye and Fischman (*J. Ultrastruct. Res.*, **33**: 278, 1970) call attention to the fact that substantial alterations in tissues occur, including morphologic changes in glycogen, in blocks stained with unbuffered aqueous uranyl acetate.

Reynolds (*J. Cell Biol.*, **17**: 208, 1963) advises the use of lead citrate to stain cytoplasmic membranes, ribosomes, glycogen, and nuclear material for electron microscopy. The stain is suitable for embedding in Araldite or in Epon. The lead citrate stain is prepared as follows:

Place 1.33 g $Pb(NO_3)_2$, 1.76 g $Na_3(C_6H_5O_7) \cdot 2H_2O$, and 30 ml distilled water in a 50-ml volumetric flask. Shake the suspension vigorously for 1 min and allow to stand with intermittent shaking to complete conversion of lead nitrate into lead citrate. After 30 min, 8 ml 1 N NaOH is added, and the suspension is diluted to 50 ml with distilled water and mixed by inversion. Lead citrate dissolves, and the staining solution is ready for use. The pH of the staining solution is about 12 ± 0.1. Faint turbidity, if present, is usually readily removed by centrifugation.

The staining solution, stored in glass or polyethylene bottles, is stable for a period of up to 6 months. With "aged" staining solutions it is advisable to centrifuge before use.

STAINING PROCEDURE. Grids with sections are stained by floating on single drops of staining solution. A petri dish with a dental-wax–coated bottom is a suitable container, and as many as 30 or 40 grids can be accommodated at a single time. Staining times and stain concentrations vary depending upon fixation and embedment. Usually methacrylate-embedded tissue is stained for 5 to 10 min and that in Epon, Maraglas, and Araldite for 15 to 30 min. Tissues fixed in phosphate-buffered osmium tetroxide or glutaraldehyde stain so intensely, however, that staining times are reduced to 5 min and the stain solution diluted 1 : 5 to 1 : 1000 with 0.01 N NaOH to prevent overstaining. Digestion and leaching of tissue components have not been observed within the limits of the staining times employed. Following staining, grids are washed sequentially in jets of 0.02 N NaOH and distilled water from plastic wash bottles, allowed to dry, and, if desired, coated with carbon.

Although carbon coating is necessary in sections mounted on grids coated with celloidin films, stained silver-gold or thinner sections of Epon-embedded material mounted on bare grids may be routinely examined without coating. Fine granularity of staining patterns due to heating of the section in the electron beam rarely occurs, even at high magnification when the intensity of the beam is at or near crossover. When such disturbance occurs, it is often the result of overstaining and can be corrected by reduction of staining time for that material or by dilution of the stain with 0.01 N NaOH.

Reynolds speculates that divalent lead salts in alkali form compounds of the general type depicted below:

$$Pb(OH)_2PbX_2 \rightleftharpoons [Pb(OH)_2Pb]^{2+} + 2X^-$$

As alkalinization proceeds from pH 12 to 14, a progressive decrement of staining with lead citrate occurs. Staining is prevented in the presence of EDTA. Reynolds believes that under the prescribed conditions lead could bind to cysteine, orthophosphate, and pyrophosphate.

DEXTRAN

This highly water-soluble glucose polysaccharide may be demonstrated in the renal epithelium and in casts in the collecting tubules by using the periodic acid Schiff method. However, it is necessary to fix in strong alcoholic fixatives, to float paraffin sections on 95% alcohol, and to use alcoholic reagents at least through the periodic acid step, and preferably throughout (Mowry, *Am. J. Pathol.*, **29**: 523, 1953).

CHITINS

Chemical studies have been done largely on crustacean carapace material; it should be remembered that other phyla than arthropods and other classes within Arthropoda also elaborate exoskeletal material which bears the anatomic name of chitin. It appears quite clear that these chitins are not all chemically identical, and indeed it would be extremely surprising if they were. Those of most interest in pathology constitute the cell walls of some fungi and bacteria and the exoskeletons and eggshells of various parasitic and commensal worms and other less frequent parasites. Lobster shell is probably infrequent in human pathology.

Pearse (1960, p. 229) takes the view that there is a single chitin, composed of N-acetyl-glucosamine in pairs linked by β-1,4 linkages, and notes its presence in arthropods, annelids, and mollusks "and also in insects" and in certain lower plants.

Pearse (1960, p. 264) notes that all varieties of chitin on which he has tried his alkaline tetrazolium method for SH have a blue color. "Tanned chitin" (quinonized) does not react. He notes further that chitins may be sulfated by one of the Kramer-Windrum methods and are then strongly basophil.

Pearse further commented on the histochemical inapplicability of the alkaline p-dimethyl-aminobenzaldehyde reaction for N-acetylhexosamines, even when the preferred pH 9,8 borate buffer is used. Chitin is normally periodic acid Schiff–negative (ibid., p. 237); Gomori stated specifically (1952, p. 62) that chitin reacted to both the Bauer and the periodic acid Schiff reactions. Lison (1960, p. 403) comments that the N-acetylglucosamine polymer should *not* react. However, sclerosed insect chitin softened with NaOH is strongly positive, and that of mushroom (champignon) membrane reacts strongly. Lillie noted (1954, p. 279) a negative reaction in cestodes and adult schistosomes and in the egg walls of ascarids and trichurids, but a positive Bauer and periodic acid Schiff reaction in schistosome egg cuticle. The cell wall material of *Bacillus anthracis*, many other bacteria, fungi, and yeasts reacts.

Runham (*J. Histochem. Cytochem.*, **9:** 87, 1961; **10:** 504, 1962) reported strong periodic acid Schiff, Hale, Alcian blue, and metachromatic staining reactions in the radula of the mollusk *Patella vulgata* and newly formed puparia of *Calliphora*. However, alpha chitin from crab shells and beta chitin of the *Aphrodite chaete* and the *Aloteuthis* pen did not so react. Preliminary chromatographic studies yielded only N-acetylglucosamine; Runham suggests a highly branched structure with many terminal residues. In cockroaches in various stages of chitin formation, Salthouse (ibid., **10:** 109, 1962) reported negative reactions to the Hale, Alcian blue, and periodic acid Schiff methods.

The chitin of the exoskeleton of arthropods is insoluble in water, alcohol, ether, alkalies, dilute acids, and ammine copper hydroxide. It dissolves in hot concentrated hydrochloric acid or sulfuric acid.

When tested with the iodine and zinc chloride test it yields a violet color. Add 3 to 5 drops of concentrated iodine potassium iodide solution to 10 ml 33% zinc chloride solution.

Apply this to chitin which has previously been treated with potassium hydroxide and thoroughly washed. The chitin is colored brown on the surface, violet within.

A solution of 50 mg iodine, 500 mg potassium iodide, 16 g calcium chloride (probably $CaCl_2 \cdot 6H_2O$ was meant), and 4 ml distilled water stains chitin red-violet.

Zander directed as follows for sections: Mount (frozen or paraffin) section in water under a cover glass. Simultaneously draw water out from one side of the cover glass with filter paper and introduce Weigert's iodine solution on the other. Tissues are stained brown. Then draw out the iodine solution with filter paper, and at the same time introduce (33%) zinc chloride solution. The brown tissues are partly decolorized. In the same way replace the zinc chloride solution with distilled water. Chitin now assumes a violet color. The foregoing tests for chitin are derived from Ehrlich's "Encyklopädie."

CELLULOSE AND STARCH

Cellulose colors violet when treated with a solution composed of 25 parts zinc chloride, 8 parts potassium iodide, and 8.5 parts distilled water, saturated with iodine crystals (Behrens, from Ehrlich's "Encyklopädie").

If cellulose is treated with Lugol's solution and then with a mixture of 2 parts (by weight) concentrated sulfuric acid and 1 part water, it gives an intense blue (ibid.).

Cellulose fibers and vegetable tissue fragments in intestinal contents stain red to purple by the permanganate, chromic acid, and periodic acid Schiff technics. Some fibers are Schiff-positive without oxidation, and more, but not all, react after peracetic acid treatment. These perhaps contain lignin. They still rotate the plane of polarized light after application of these methods. With other stains these fibers may give striking color alterations when rotated under crossed Nicol prisms.

Cellulose resists ptyalin and malt diastase digestions; raw starch is digested quite slowly, but by both β- and α-amylases. When treated with Lugol iodine solutions, starch granules give the familiar dark blue. While this fades soon in usual mounts, Takeuchi's iodine glycerol helps, and Lower's iodine-silver method can be tried.

Starch granules remain almost unstained with most ordinary stains but are doubly refractile, exhibiting the black cross of polarization separating four bright quadrants. On rotation the orientation of the black cross follows the position of the plane of polarization, not that of the starch granules. Also on rotation it is seen that in certain positions alternate quadrants are respectively yellowish and bluish white, and that on rotation these colors fade in some granules but are maintained in others.

On Gram-Weigert staining, starch granules stain black and transmit no polarized light. After Casella or Bauer staining the anisotropy is maintained, the quadrants appearing bright red-purple on a dark background. On direct illumination with nonpolarized light these two methods give red-purple starch granules. After the periodic acid–leucofuchsin technic the granules are dark purple and may remain dark under crossed Nicol prisms, or give a dark red illumination. We assign the total extinction of polarized light noted above simply to the density of the stains, not to any alteration in molecular structure.

LIGNIN, INULIN, PECTIN

According to McLean and Cook (1941, p. 77) a basic fuchsin solution decolorized by addition of strong ammonia water, drop by drop, is a test reagent for aldehyde and may be used for the demonstration of lignin, which gives purple.

Plant cuticle and lignin, because of their aldehyde content, react also with Schiff's reagent.

Lignin (McLean and Cook, 1941, p. 78) also gives Maule's reaction. Treat first with aqueous potassium permanganate (we suggest 0.5 % for 10 min). Wash with dilute hydrochloric acid (10 % anhydrous, *ca.* 3 *N*) and then with dilute ammonia. Red develops in lignified tissue only.

McLean and Cook describe the orcinol test for *inulin*. Soak sections first in 0.5 % orcinol in 90 % alcohol. Then transfer to strong hydrochloric acid and warm. An orange-red is developed at sites of inulin deposit.

McLean and Cook (1941, p. 77) state that ruthenium red $(Ru(OH)Cl_2 \cdot 3NH_3 \cdot H_2O)$ (mw 258.75) colors pectin substances and pectic mucilage deep red.

Deparaffinize sections and soak 24 h in 1 part hydrochloric acid and 3 parts alcohol. Then place in dilute ammonia for some hours. Then place in 1 : 5000 aqueous ruthenium red in the dark until adequately colored.

MUCIN, CARTILAGE, MAST CELLS

Mucin is precipitated by dilute acetic acid and rendered insoluble in water. Pseudomucin and gastric mucin are not precipitated by acetic acid. When precipitated by alcohol, mucin will redissolve in water. Dilute alkalies readily dissolve it.

Consequently alcoholic and acid fixatives are preferred. Formalin is quite serviceable if not alkaline, as it would become from magnesium carbonate neutralization. Tellyesniczky's acetic formalin alcohol should be excellent. We have used most of the following methods on routine material fixed in formalin that was probably acid.

The most uniformly successful method for demonstration of epithelial mucins is the periodic acid Schiff method. Mucin in the gastric surface epithelium stains most intensely by this method; then that in intestinal goblet cells. The mucins of intestinal glands, trachea and bronchi, minor salivary and buccal glands, female genital tract, and prostatic gland and seminal vesicle contents are generally quite well stained. The mucins of the rabbit's pharyngeal glands, those of the basal portion of the colonic glands of some rodents, and connective tissue mucins may stain quite poorly or not at all. Cartilage matrix, however, usually stains quite well, and more densely in the pericellular capsular areas than between them.

Bauer and Casella methods are generally successful on those mucins which color intensely with the periodic acid Schiff method; they fail where this method colors them lightly.

Technics for these procedures appear in an early chapter. The allochrome procedure gives excellent contrasts for epithelial mucins—red-purple, contrasting with gray-green or greenish yellow epithelial cytoplasm and blue connective tissue.

METACHROMATIC DYES. Perhaps the simplest stains are those using simple aqueous solutions of certain basic dyes which stain mucin and cartilage matrix *metachromatically*. Of these perhaps the best are thionin (C.I. 52000), azure A (C.I. 52005), azure C (C.I. 52002), toluidine blue O (C.I. 52040), Bismarck brown Y (C.I. 21000), and safranin O (C.I. 30240). Highman (*Stain Technol.*, **20**: 85, 1945) recommends especially new methylene blue N (C.I. 52015) in his buffer technic.

With these, one stains perhaps for 30 to 60 s in a 1 : 1000 aqueous solution, dehydrates with acetone, and clears in xylene. The azures, thionin, and toluidine blue give deep blue nuclei, light blue cytoplasm, green erythrocytes, red-purple to violet mucus and cartilage matrix, and deep violet mast cell granules. With the Bismarck brown Y, nuclei are brown; mucin and cartilage matrix, brownish yellow. With safranin O, mucin and cartilage are orange; nuclei, red; mast cell granules, orange-red. However, gastric mucus is said to stain with these dyes only when freshly formed (Mallory). It is readily observed that the mucins in cardial and pyloric glands stain metachromatically while those of the surface epithelium do not.

BUFFERED THIONIN OR AZURE A STAIN

1. Deparaffinize and hydrate sections as usual.
2. Stain for 30 min in 0.05% thionin in 0.01 M acetate buffer.
3. Rinse, dehydrate in alcohol, clear with xylene, mount in polystyrene.

RESULTS: Connective tissue mucins are purplish red; mast cell granules, red-purple; nuclei, blue-violet; red corpuscles, pale yellow; basophil cytoplasm, blue; muscle and connective tissue, faint greenish or unstained.

Using a pH 5 buffer demonstrates muscle and connective tissue in light blue-greens, while pH levels of 3 and 2 cause red corpuscles to remain unstained. Cytoplasm stains more poorly. Mucins appear to be only partly demonstrated at pH 3, and at pH 1 to 2 metachromatic staining seems restricted to cartilage and mast cells. Nuclear basophilia of formalin-fixed tissue persists to about pH 1.5 to 1.2. (Note: Phosphates precipitate thionin; use acetate and citrate buffers with this dye.)

This method of staining with buffered thiazin dyes is based on that of Highman (*Stain Technol.*, **20**: 85, 1945), and a basically similar procedure has been extensively used by Dempsey et al. (*Anat. Rec.*, **98**: 417, 1947).

METACHROMASIA

Ehrlich (*Arch. Mikrosk. Anat.*, **13**: 263, 1877) appears to be the first to note metachromasia as a characteristic of basic aniline dyes. Polyanions appropriately positioned in a suitable density appear to be requisites for engendering metachromasia by certain basic dyes in tissue sections. The connective tissue acid mucopolysaccharides are readily identified by their metachromatic staining with basic dyes such as azure A. Carboxylic acid as well as sulfuric acid groupings may contribute to the development of metachromasia. Thus non-sulfated and sulfated acid mucopolysaccharides are revealed by metachromasia. Epithelial mucins vary widely in their ability to induce metachromasia. Some fail to induce meta-chromasia.

Schoenberg and Moore (*Biochim. Biophys. Acta*, **83**: 42, 1964) propose the model shown in Fig. 14-2 for a metachromatic complex. They indicate that theoretical and experimental considerations suggest that changes in the absorption spectra result from the formation of a complex having a linear arrangement of regularly spaced, extensively conjugated dye molecules aligned in such a fashion that their planes are parallel and approximately normal or inclined toward the long axis of the chain. The array is believed to be stabilized by Van der Waals–type forces between neighboring dye molecules. The attractive forces between dye molecules are long and additive (Kuhn, *J. Chem. Phys.*, **17**: 1198, 1949; Coulson and Davies, *Trans. Faraday Soc.*, **48**: 777, 1952).

Stone (*Biochim. Biophys. Acta*, **148**: 193, 1967) and Stone and Bradley (*Biochim. Biophys. Acta*, **148**: 172, 1967) conducted quantitative studies on the aggregation of cationic dyes on acid polysaccharides.

Stone (*Biopolymers*, **3**: 617, 1965) noted the induction of anomalous optical rotatory dispersion of symmetric dyes (acridine orange and methylene blue) with metachromatic complexes with chondroitin sulfates and keratosulfate suggesting the existence of a secondary structure of these biopolymers.

Ramalingam and Ravindranath (*Stain Technol.*, **46**: 221, 1971) studied the effects of ethanol dehydration on metachromasia. They concluded that ethanol does not abolish the metachromatic reaction of toluidine blue O attached to protein-polysaccharide complexes.

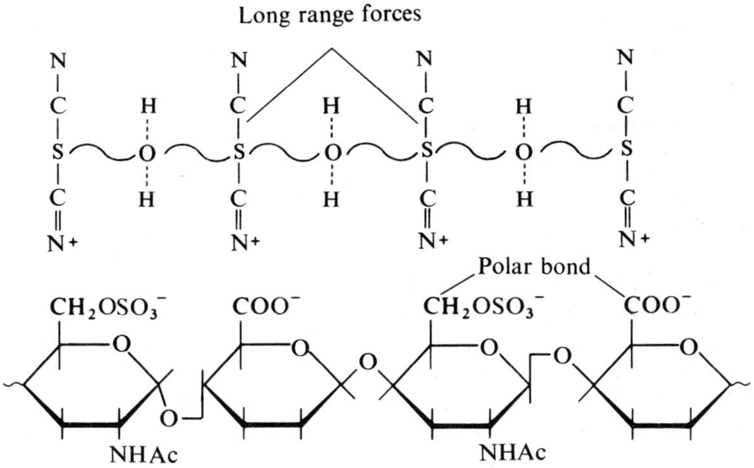

Fig. 14-2 Schematic illustration of a metachromatic complex. (*From Schoenburg and Moore, Biochim. Biophys. Acta,* **83:** 42, 1964. *By permission of the publisher.*)

Pal (*Histochemie,* **5:** 24, 1965) notes that metachromasia is reduced by dimethyl urea as well as by ethanol.

Lev and Spicer (*J. Histochem. Cytochem.,* **12:** 309, 1964) caution that in studies of basophilia at low pH (such as at 0.5) sections should be rinsed and dried in fluids at the same pH as the staining solution. Otherwise, dye may adhere to the section and stain during the water bath at an elevated pH.

Kelly and Chang (*J. Histochem. Cytochem.,* **17:** 658, 1969) studied the thermal effects on metachromasia and concluded that metachromasia in solutions and in solid systems is not a fundamentally different phenomenon.

Keutter et al. (*Histochemie,* **24:** 187, 1970) studied the "strength of metachromatropy" of several polysaccharides with dyes capable of manifesting metachromasia including dichloropseudoisocyanin. They state one advantage of the use of dichloropseudoisocyanin is that it may be photographed beautifully in black and white with the aid of a line filter (579 mμ).

Power et al. (*Int. J. Radiat. Biol.,* **20:** 111, 1971) observed similar enthalpy and entropy values for six dyes of markedly different chemical structure in combination with tracheal chondroitin 4-sulfate.

For retention of metachromasia in methacrylate-embedded sections, Izard (*J. Histochem. Cytochem.,* **12:** 487, 1964) advocates the following method if the specimen is to be previously dehydrated:

1. Rinse metachromatically stained sections in distilled water.
2. Dehydrate sections in ethanol.
3. Clear in xylene.
4. Mount specimens in prepolymerized butyl methacrylate.

Sections can also be dehydrated in acetone and mounted directly, as methacrylate and acetone are miscible. The mounting medium is a mixture of butyl methacrylate monomer and 1% Luperco C.D.B. prepolymerized at 57°C for 45 to 60 min and then cooled to 24°C.

Izard states that metachromatically stained specimens may also be mounted directly in prepolymerized butyl methacrylate without previous dehydration with retention of metachromasia for 2 to 3 weeks.

Green and Pastewa (*J. Histochem. Cytochem.*, **22**: 767 and 774, 1974) introduced a new dye for the histochemical identification of mucosubstances. Earlier Dahlberg et al. (*J. Mol. Biol.*, **41**: 139, 1969) used the cationic carbocyanin dye, namely, 1-ethyl-2-[3-(1-ethylnaphtho-[1,2-*d*]thiazolin-2-ylidene)-2-methylpropenyl] naphtho[1,2-*d*] thiazolium bromide as a sensitive stain for ribonucleic acid (bluish purple), deoxyribonucleic acid (blue), and protein (red) in electrophorectically separated polyacrylamide agarose composite gels. Earlier Bean et al. (*J. Phys. Chem.*, **69**: 4368, 1965) and Kay et al. (*J. Phys. Chem.*, **68**: 1896 and 1907, 1964) studied the same dye under the chemical name 4,5,4',5'-dibenzo-3,3'-diethyl-9-methylthiacarbocyanine bromide and referred to the dye as DBTC. Dahlberg et al. and Green and Pastewa refer to the dye as *Stains all*. The dye is available from Eastman Organic Chemicals, Rochester, New York. Lillie (*J. Histochem. Cytochem.*, **23**: 1169, 1974) has provided the name *carbocyanin DBTC* as that to be used in the next edition of Conn's "Biological Stains."

The stock solution of carbocyanin DBTC is made up as 0.1% solution in formamide. Tissues are stained either at pH 2.8 (190 ml, 1 : 20 dilution of Michaelis buffer of pH 2.6 in deionized water, 0.1 ml concentrated HCl, and 10 ml stock dye solution) or at pH 4.3 (190 ml, 1 : 20 dilution of Michaelis buffer of pH 3.6 in deionized water and 10 ml stock dye solution).

Tissues are fixed in Carnoy's fixative (6 : 3 : 1) and paraffin-embedded. Sections are deparaffinized, brought to water, stained for 1 h at 25°C *in the dark*, dehydrated with *tert*-butyl alcohol, and mounted in synthetic medium.

At either pH 2.8 or 4.3 mucous glands of mice stained blue-green, whereas cartilage and mast cells stained red-purple and purple. Nuclei were purple at both pH levels. Cytoplasm of mast cells was red at pH 4.3 and unstained at pH 2.8. Methylation (4 h, 37°C) prevented staining of mucous glands, and staining was restored by saponification (1% KOH in 70% ethanol, 20 min at 25°C). Staining of cartilage was abolished in sections digested with hyaluronidase. Sialidase-digested sections of mouse sublingual glands failed to stain.

ALCIAN BLUE. Steedman (*Q. J. Microsc. Sci.*, **91**: 477, 1950) reported a technic for staining chondroitin and mucoitin sulfuric acid mucins with the phthalocyanin dye Alcian Blue 8GS.[1] This dye stained mucin a clear blue-green. Mast cells were not stained. Prolonged staining colored almost all nonnuclear components of tissue. Treatment (after step 3) for 2 h or more with alkaline alcohol (pH 8 or higher) converted the dye into the insoluble pigment monastral fast blue, which is highly resistant to decolorization, and thus permits the subsequent use of various histologic reagents.

STEEDMAN'S ALCIAN BLUE 8GS FOR MUCINS.   Material fixed in Bouin, susa, or Zenker's fluid was recommended. Formalin-fixed tissue was considered unsuitable.

1. Deparaffinize and hydrate as usual.
2. Stain 10 to 40 s in 1% aqueous Alcian Blue 8GS.
3. Rinse in distilled water.
4. If required, treat 2 h in 80% alcohol containing 0.5% borax.
5. Counterstain with alum hematoxylin and eosin, or otherwise as desired.
6. Dehydrate, clear in xylene, and mount in synthetic resin or balsam.

---

[1] The Alcian Blue solution should contain thymol to prevent mold growth and should be filtered every 7 to 10 days.

Scott et al. (*Histochemie*, **4:** 73, 1964) provided the formula below (emended) as representative of a "family" of Alcian blues. They concluded that Alcian blues are a family of poly-

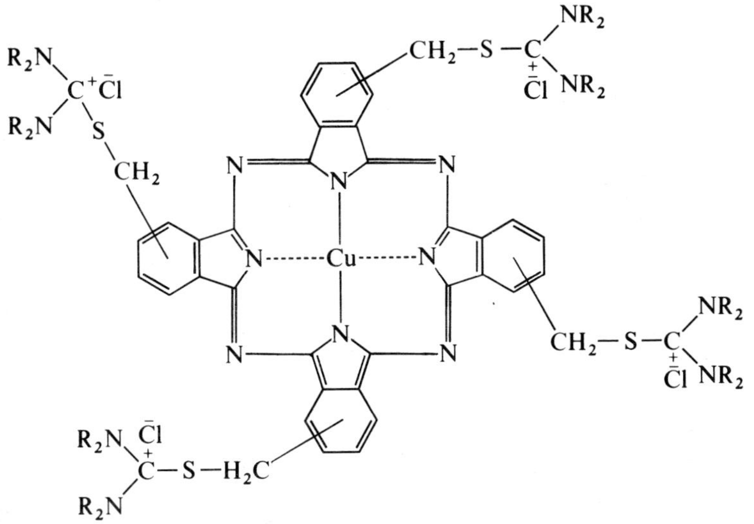

R_2N
R_2N $C^+\bar{C}l$
S
CH_2

CH_2-S-C NR_2 $\bar{C}l$ NR_2

N Cu N

CH_2-S-C NR_2 $\bar{C}l$ NR_2

R_2N $\bar{C}l$
R_2N $C-S-H_2C$

valent basic dyes with at least two and as many as four isothiouronium groups per phthalocyanin ring. The isothiouronium groups are more basic than ammonia but less basic than sodium hydroxide. In aqueous solutions Alcian blue reacts via a salt linkage with polyanions including heparin, DNA, and hyaluronate to give insoluble precipitates. Tests with model compounds failed to provide evidence for any unusual specificity for acid mucopolysaccharides. Increases in concentration of electrolyte such as magnesium chloride in the staining solution resulted in increased combination of Alcian blue molecules with heparin.

Quintarelli et al. (*Histochemie*, **4:** 86 and 99, 1964) noted enhanced staining of salivary gland mucins in sections stained with solutions of Alcian blue containing added electrolyte (NaCl, KCl, LiBr, MgCl, etc.).

Yamada [*Nature (Lond.)*, **198:** 799, 1963] tested the stainability with Alcian blue of several sulfated polysaccharides embedded in casein. All stained; however, inasmuch as acetylation impaired subsequent staining, Yamada concluded that factors other than the presence of sulfate groups influence staining with Alcian blue. Studies on the specificity of Alcian blue for acid mucopolysaccharides were conducted by Palladini et al. (*Histochemie*, **16:** 15, 1968).

Lev and Spicer (*J. Histochem. Cytochem.*, **12:** 309, 1964) recommend the following method for the selective demonstration of sulfated mucosubstances: Sections of fixed, paraffin-embedded tissues are brought to water and stained for 30 min at 23°C in 1% Alcian Blue 8GS at pH 1 (approximately 0.1 N HCl). Sections are blotted dry *immediately* after staining, dehydrated, and mounted in synthetic medium.[1]

Scott and Dorling (*Histochemie*, **5:** 221, 1965) applied the "critical" electrolyte concentration (CEC) concept to the differentiation of acidic glycosaminoglycans (Tables 14-7 and 14-8). With increasing concentrations of $MgCl_2$, Alcian Blue 8GS stains with increasing

---

[1] If sections are rinsed after the staining step, nonsulfated mucosubstances may stain, possibly due to elevation of pH during the rinse period.

TABLE 14-7 STAINING OF HUMAN NEWBORN LUNG IN ALCIAN BLUE, pH 5.8 $MgCl_2$, AND ALCIAN BLUE, pH 2.5 $MgCl_2$

		Molarity $MgCl_2$													
		0	0.05	0.1	0.2	0.3	0.4	0.5	0.6	0.7	0.8	0.9	1.0	1.2	1.5
Nuclei	pH 5.8	++	++	+	−	−	−	−	−	−	−	−	−	−	−
	pH 2.5	+	∓	−	−	−	−	−	−	−	−	−	−	−	−
Mucin	pH 5.8	++	++	++	++	++	−	−	±	±	±	±	±	±	±
	pH 2.5	++	++	++	+	++	+	+	±	±	±	±	±	±	±
Cartilage	pH 5.8	++	+++	+++	+++	+++	+++	+++	+++	∓	−	−	−	−	−
	pH 2.5	++	+++	+++	+++	+++	+++	+++	+++	+++	+	±	−	−	−
Mast cells	pH 5.8	?+	?+	+++	+++	+++	+++	+++	++	+	±	−	−	−	−
	pH 2.5	?∓	?∓	+	++	+++	+++	+++	+++	+++	+++	++	++	++	−
Collagen	pH 5.8	++	∓	+	+	++	++	++	++	++	++	++	++	−	±
	pH 2.5	+	+	+	+	++	++	++	++	++	++	++	+	+	±

Source: Scott and Dorling, *Histochemie*, **5**: 221, 1965.

TABLE 14-8 CRITICAL ELECTROLYTE CONCENTRATIONS (MOLARITIES OF $MgCl_2$) OF METHYL GREEN, PYRONIN, AND ALCIAN BLUE COMPLEXES ABSORBED ON FILTER PAPER

Polyanion	Methyl green	Pyronin	Alcian blue
RNA	<0.05	>2.0	0.3
DNA	>1.0	0.4	0.1
Hyaluronate	<0.05	<0.05	0.05
Chondroitin sulfate	<0.05	<0.05	0.45
Heparin	0.1	<0.05	0.65
Keratan sulfate	<0.05	<0.05	1.0

Source: From Scott and Dorling, *Histochemie*, **5:** 221, 1965; and Scott, *Histochemie*, **9:** 30, 1967.

selectivity. Binding of the dye to carboxylic acids or phosphoric acids was abolished at about $0.3\ M\ MgCl_2$, whereas sulfuric acids retained staining at 5 to 10 times that concentration.

Tissues may be fixed in neutral 10% formalin or in 95% ethanol at 4°C, dehydrated, paraffin-embedded, sectioned at 5 μ, and mounted on albuminized slides.

Deparaffinized sections are stained in 0.05% Alcian Blue 8GX (C.I. 74240) in 0.025 M acetate buffer of pH 5.8 at various concentrations of $MgCl_2$ for 18 h at 25°C, rinsed in a stream of distilled water, rinsed in a freshly distilled water bath, dehydrated in ethanols, cleared, and mounted in synthetic medium. Table 14-7, provided by Scott, reveals the concentrations of $MgCl_2$ required to abolish staining of nuclei, carboxylated mucins, and sulfated mucins. Scott and Dorling believe the light staining of collagen that persists in staining through 1 M $MgCl_2$ is probably due to a sulfated polyanion. Best discrimination is achieved at pH 5.8, because at a low pH such as 2.6 salt links may occur between proteins and polysaccharides, thereby masking the effect of the critical electrolyte concentration. Proteins and glycoproteins are only weakly negative at pH 5.8. Scott and Dorling caution that not all carboxyl polyanions and also not all sulfuric polyanions manifest the same critical electrolyte concentration. Molecular weight and the distance between charges (charge density) affect the CEC. The distinction between carboxyl, sulfuric, and phosphoric polyanions was more effectively revealed by $MgCl_2$ than by NaCl. Scott and Dorling noted comparable results in sections carboxylated, phosphorylated, or sulfated.

In Figure 14-3, Scott provides the spectra of 0.004% w/v Alcian blue in water and in 20, 40, 60, 80, and 100% ethanol. He notes that metachromasia of Alcian blue is favored by low concentrations in aqueous solvents.

Scott and Mowry (*J. Histochem. Cytochem.,* **18:** 842, 1970) have proposed that in the selection of an Alcian blue sample for critical histochemical purposes (1) it should be Alcian Blue 8GX (not GS or 5GX or 7GX), (2) to ensure freshness it should be obtained preferably from ICI, Ltd., (3) it should be soluble in water to at least 5%, (4) it should not precipitate for at least 24 h when mixed at 1% w/v concentration in 2 M $MgCl_2$ buffered to pH 5.7 with 0.025 M acetate buffer, (5) it should have the spectrum indicated in Fig. 14-3, and the age of the sample should not exceed 3 years.

Scott et al. (*J. Histochem. Cytochem.,* **16:** 383, 1968) suggest that one may be able to employ the "critical electrolyte concentration" principle to dissociate protein aggregates with tissue anions. They were able to demonstrate basophilia with Alcian blue in horse nasal cartilage,

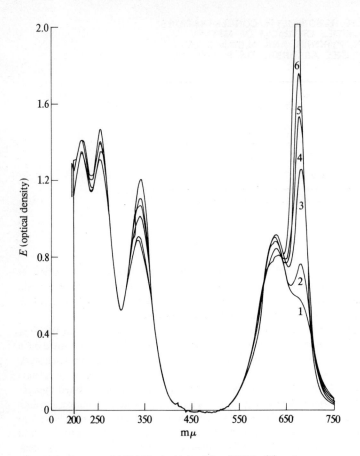

Fig. 14-3 Spectra of 0.004% w/v Alcian Blue 8GX in (1) water, (2) 20% v/v EtOH–H$_2$O, (3) 40% v/v EtOH–H$_2$O, (4) 60% v/v EtOH–H$_2$O, (5) 80% v/v EtOH–H$_2$O, (6) EtOH. (*From Scott, Histochemie,* **21:** 227, 1970. *By permission of the publisher.*)

provided electrolyte was added to the staining solution; without addition of electrolyte, no staining occurred at pH 2.8.

Sorvari and Näntö (*Histochem. J.,* **3:** 261 and 269, 1971) failed in their attempt to employ an Alcian yellow–Alcian blue sequence to discriminate between carboxyl and sulfated mucopolysaccharides, because displacement of the first dye by the second occurred during the staining reaction.

On the basis of (1) a staining pattern comparable to that of Alcian blue, (2) loss of staining after methylation, and (3) an appreciable restoration of staining after saponification, Yamada (*Histochemie,* **23:** 13, 1970) has postulated the following reaction of ruthenium red with carboxyl groups of acid mucopolysaccharides at pH 2.5:

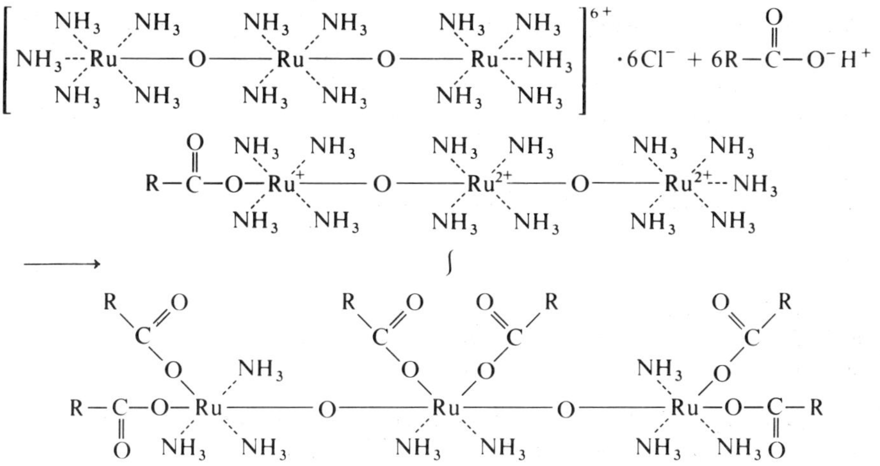

$$\left[\begin{array}{c} NH_3\ NH_3\quad NH_3\ NH_3\quad NH_3\ NH_3 \\ NH_3\text{-}Ru\text{---}O\text{---}Ru\text{---}O\text{---}Ru\text{-}NH_3 \\ NH_3\ NH_3\quad NH_3\ NH_3\quad NH_3\ NH_3 \end{array} \right]^{6+} \cdot 6Cl^- + 6R-\overset{O}{\underset{\parallel}{C}}-O^-H^+$$

$$R-\overset{O}{\underset{\parallel}{C}}-O-\overset{NH_3\ NH_3}{\underset{NH_3\ NH_3}{Ru^+}}\text{---}O\text{---}\overset{NH_3\ NH_3}{\underset{NH_3\ NH_3}{Ru^{2+}}}\text{---}O\text{---}\overset{NH_3\ NH_3}{\underset{NH_3\ NH_3}{Ru^{2+}}}\text{-}NH_3$$

$$\longrightarrow \qquad \int$$

$$R-\overset{O}{\underset{\parallel}{C}}-O-\overset{R\diagdown C\diagup O}{\underset{O\ NH_3}{\underset{NH_3\ NH_3}{Ru}}}\text{---}O\text{---}\overset{R\diagdown C\diagup O\quad O\diagdown C\diagup R}{\underset{NH_3\ NH_3}{Ru}}\text{---}O\text{---}\overset{NH_3\ O\diagdown C\diagup R}{\underset{NH_3\ NH_3O}{Ru}}-O-\overset{O}{\underset{\parallel}{C}}-R$$

Sulfate groups are also believed to be reactive. The above formulas are derived on a theoretical basis. Specific investigations are needed to determine the validity of the structures.

Luft (*J. Cell Biol.*, **23**: 54A, 1964) noted that ruthenium red tends to favor localization in highly polymerized acidic polysaccharides such as pectin. Epithelial mucins and dilute heparin solutions were also noted to precipitate with ruthenium red. Cell membranes were also noted to stain.

Groniowski and Biczyskowa (*Lab. Invest.*, **20**: 430, 1969) employed the Hale colloidal iron and Luft's ruthenium red methods to study the alveolar surface coat of lungs of rabbits at the electron microscopic level. A dense stain was observed lining the alveolar epithelium. The stain was abolished in sections subjected to sialidase.

Yamada [*Acta Histochem.* (*Jena*), **35**: 90, 1970] employs the following staining procedure for the detection of mucopolysaccharides with acriflavine:

1. Deparaffinize and hydrate sections through xylol, ethanol, and water.
2. Stain in acriflavine solution for 30 min.
3. Rinse briefly in water.
4. Place in 96% ethanol to remove excess dye.
5. Rehydrate through graded ethanols and water.
6. Counterstain nuclei briefly in Mayer's alum hematoxylin.
7. Wash in water.
8. Dehydrate through graded ethanols.
9. Clear in xylol and mount in balsam.

Saunders (*J. Histochem. Cytochem.*, **12**: 164, 1964) has attempted to discriminate between the acid mucopolysaccharides by employing a ribonuclease digestion and acridine orange staining with variable electrolyte concentration in the staining solution. He prescribes as follows:

SAUNDERS' MUCOPOLYSACCHARIDE IDENTIFICATION METHOD.   Three parallel slides are treated with 1% cetyltrimethylammonium chloride and washed for 10 min in running tap water. All three slides are treated with ribonuclease, 1 mg/ml, in glass-distilled water at 45°C for 2 h.

Slide 1 is treated again with cetyltrimethylammonium chloride, washed, and stained for 3 min in neutral 0.1% acridine orange and then washed again in tap water for 10 min, air-dried, and mounted in Gurr's fluor-free mounting medium.

Slide 2 is stained for 5 min in 0.1% acridine orange at pH 3.2, washed in running water, differentiated in 0.3 M NaCl in 0.01 M acetic acid, then washed in running water, air-dried, and mounted.

Slide 3 is treated as slide 2, but 0.6 M NaCl replaces 0.3 M NaCl.

According to Saunders,

Slide 1: Red fluorescence is due to hyaluronic acid.

Slide 2: Red fluorescence is due to chondroitin sulfuric acid and heparin.

Slide 3: Red fluorescence is due to heparin.

One must employ further methods to be certain of the specificity of the above staining reactions. Newcomer's fixation is preferred; avoid formalin.

Tice and Barrnett (*J. Histochem. Cytochem.*, **10**: 688, 1962) noted that ground substance between connective tissue cells can be distinctly observed by electron microscopy in Epon-embedded thin sections stained in 0.5% Alcian blue at pH 3 for 2 h at 24°C.

Lillie and Pizzolato (*Histochemie*, **17**: 138, 1969) developed a very useful stain for sulfate-containing mucins. The stain employs brilliant cresyl blue (C.I. 51010) 1% in 50 to 70% ethanol-water containing 1% concentrated HCl.

1. Deparaffinize sections cut at 5 to 6 μ, and bring them through xylenes and alcohols to 80% ethanol.
2. Stain 30 min in 1% brilliant cresyl blue in 0.12 N HCl–70% alcohol.
3. Dehydrate briefly in two 15-s washes in 70% alcohol, then through ascending alcohols and xylenes, and mount in a synthetic medium.

If a red nuclear stain is desired, precede step 2 by the Feulgen procedure allowing 15 min at 60°C in 1 N HCl for formol-fixed material. A modified Van Gieson stain may also be used to obtain red connective tissue and yellow muscle. After the 70% alcohol rinse of step 3, sections may be counterstained 1 min in 1.5% picric acid in 70% alcohol containing 50 to 100 mg acid fuchsin per 100 ml. The amount of acid fuchsin may need to be adjusted depending upon the dye batch being used. After staining, dehydrate and clear as above.

Without the counterstains mucins are blue-black on an almost unstained background. The specific stains of gastric epithelial and some intestinal goblet cell mucins are lost if staining is done using less than 50% or more than 70% alcohol as solvent. This specific staining is also lost at pH 2 and above.

A 16-h, 60°C 0.1 M methylsulfuric acid methylation is required to block the stain of gastric and some intestinal mucins, whereas 1 to 2 h will suffice for cartilage, mast cells, and laryngeal and salivary (submaxillary) gland mucins. The methylation block of gastric mucin is not deblocked by the usual 1% KOH–70% alcohol 30-min saponification.

The brilliant cresyl blue must be of recent manufacture; old samples do not work. Lillie noted in later studies (*Acta Histochem. Suppl.* 9, p. 669, 1971) that good stains were also obtained with neutral red (C.I. 50040), dimethylphenosafranin (C.I. 50205), Victoria pure blue BO (C.I. 42595), a recent sample of Victoria blue 4R (C.I. 42563), and two synthetic elastin stains. Thirty other basic dye samples failed to stain in the acid alcohol technic.

Spicer (*J. Histochem. Cytochem.*, **13**: 211, 1965) has developed diamine methods for partial differentiation of the mucosubstances histochemically. Mixtures of *N,N*-dimethyl-*m*-

phenylenediamine hydrochloride with the para isomer are employed. The mixed diamine, low-iron diamine–Alcian blue, and high-iron diamine–Alcian blue methods are detailed below:

MIXED DIAMINE STAIN

1. Deparaffinize duplicate slides in xylene, and rehydrate in graded alcohols.
2. Hydrolyze both sections for 10 min in preheated 60°C 1 N HCl (Feulgen hydrolysis to remove interfering staining of nucleic acids).
3. Wash in running water for 5 min.
4. Oxidize one section for 10 min in 1% aqueous periodic acid; rinse 5 min in running water.
5. Stain both sections for 20 to 48 h in mixed diamine solution freshly prepared as follows: Dissolve 30 mg N,N-dimethyl-m-phenylenediamine $(HCl)_2$ and 5 mg N,N-dimethyl-p-phenylenediamine HCl in 50 ml distilled water, then adjust the pH to 3.4 to 4 with 0.2 M Na_2HPO_4 (0.15 to 0.65 ml).
6. Omitting a water rinse, dehydrate directly through two changes of 95% and two of absolute alcohol; clear in 1 : 1 xylene–absolute alcohol and in xylene; and mount.

Diamines were obtained from Eastman Chemical Co., Rochester, New York; Schuchardt Chemical Co., Munich, Germany; Gallard-Schlesinger Co., Carle Place, Long Island, New York; or K and K Laboratories, Inc., Jamaica, Long Island, New York. With fresh uncontaminated preparations of the meta diamine, sections stain optimally in 24 h at pH 3.8 to 4, and inclusion of the para diamine is essential. Impure and perhaps aged batches of the reagent stain acid mucosaccharides more selectively at lower pH levels (3.4 to 3.8) and may give better staining—although perhaps slower—without the added para isomer.

LOW-IRON DIAMINE–ALCIAN BLUE METHOD

1. Deparaffinize one or two sections and rehydrate through graded alcohols.
2. Oxidize one section for 10 min in 1% H_5IO_6 if demonstration of neutral as well as acid mucosaccharides is desired.
3. Rinse for 5 min in running water.
4. Stain both sections for 4 to 24 h in a solution freshly prepared as follows: Add 30 mg N,N-dimethyl-m-phenylenediamine $(HCl)_2$ plus 5 mg N,N-dimethyl-p-phenylenediamine HCl simultaneously to 50 ml distilled water. When the reagents are dissolved, pour this solution immediately into a Coplin jar containing 0.5 ml of official $FeCl_3$ solution. (This National Formulary solution, available from Fisher Scientific Company, contains 10% Fe^{3+} w/w or 62% $FeCl_3 \cdot 6H_2O$ w/v and contributes 64 mg Fe^{3+} to the 50 ml of staining solution.) Unless stated otherwise, results reported here are for sections stained for 18 h. Staining of some sialomucins (e.g., some cells of mouse sublingual gland) may be weaker and of some sulfomucins (e.g., cornea) slightly stronger if the solution is prepared with tap water. Slides should be immersed in staining solution within 5 min of its preparation and should not be crowded in the staining jar.
5. Rinse quickly in and out of water before and following 30 min in Alcian Blue 8GX, 1% in 3% acetic acid. This Alcian blue step may be omitted if demonstration of the few low-iron-diamine–unreactive, Alcian blue–positive acid mucosubstances is unnecessary.
6. Dehydrate, clear, and mount as for mixed diamine method.

HIGH-IRON DIAMINE–ALCIAN BLUE METHOD. Steps of this procedure are the same as in the low-iron diamine–Alcian blue sequence, but the iron diamine solution differs. Dissolve 120 mg of the meta diamine and 20 mg of the para isomer simultaneously in 50 ml distilled or tap water; when dissolved, pour into a Coplin jar containing 1.4 ml (180 mg Fe^{3+}) of the National Formulary solution of $FeCl_3$ described in step 4 of the above procedure. Except where stated otherwise, results presented are for 18-h staining.

RESULTS: The mixed diamine method imparts purple to many sulfated and nonsulfated acid mucosubstances in epithelial and connective tissue sites. Periodate-reactive are distinguished from periodate-unreactive acid mucosubstances by loss of staining of the former in the periodate-oxidized section. Periodate-reactive acid mucosubstances can also be distinguished as those that lose affinity for azure A or Alcian blue after prior oxidation with periodate and exposure a few hours to aqueous 2% meta diamine adjusted to pH 5.

Ferric chloride added to a solution of both diamines permits selective demonstration of most acid mucosubstances. The low-iron diamine method followed by Alcian blue stains most sulfomucins and many sialomucins black, although some sialomucins stain blue. Some epithelia contain sialomucins stained both black and blue.

Employment of the high-iron diamine–Alcian blue method results in most sulfomucins staining black and sialomucins staining blue. Results of the stains appear to show that sulfomucins exist in some epithelial or connective tissue sites, sialomucins occur in others, and a mixture of the two occurs in other sites.

THE HALE COLLOIDAL FERRIC OXIDE, OR DIALYZED IRON, PROCEDURE FOR ACID MUCOPOLYSACCHARIDES. This technic has undergone considerable evolution since its introduction. Mowry (*Ann. N.Y. Acad. Sci.*, **106**: 402, 1963) has adapted Müller's preparation of colloidal iron and further controlled its behavior by a moderate dialysis to remove most of the free hydrochloric acid and, by adding acetic acid, to achieve a pH level of about 1.6 to 1.9 for optimal selectivity.

PREPARATION OF COLLOIDAL IRON SOLUTION. Boil 250 ml distilled water. Into the boiling water pour 4.4 ml of the official iron chloride solution (USP XI) and stir. (The equivalent in iron content of this solution would be 2.73 g $FeCl_3 \cdot 6H_2O$ dissolved in 4 to 5 ml distilled water, and such an extempore solution may be used if the official solution is not readily available.) Keep boiling during the addition of the iron solution. When the solution becomes dark red, let it cool.

Remove free acid and unhydrolyzed (ionizable) iron salts by dialysis. Transfer the red solution to 41-mm dialysis tubes containing a glass marble as a weight at the lower end in about 25-ml portions. Suspend each tube in a 250-ml cylinder, and add distilled water to fill the cylinder outside the tube. Dialyze 24 h, changing the water twice during the dialysis interval. Filter contents of dialysis tube through Whatman No. 50 or similar very fine paper, to remove particulate iron oxide. This filtrate is the *stock colloidal iron solution* and is stored at room temperature (20 to 25°C).

WORKING COLLOIDAL IRON SOLUTION. Use 18 ml distilled water, 12 ml glacial acetic acid, 10 ml stock colloidal iron. Prepare fresh daily, or oftener if many slides are to be stained.

Hale dialyzed iron staining is generally used in combination with the Van Gieson stain. The procedure follows:

HALE–VAN GIESON PROCEDURE

1. Deparaffinize and bring sections to water. Specimens may be fixed in neutral formalin or Carnoy's fluid.
2. Immerse in 12% acetic acid for 30 s.
3. Place sections in the dialyzed iron solution (prescription above).

4. Rinse in 12% acetic acid 3 times for 3 min each.
5. Treat sections with a freshly prepared solution of (*a*) 60 ml 1% HCl and (*b*) 30 ml 2% potassium ferrocyanide.
6. Rinse in running water for 5 min.
7. Stain with iron hematoxylin.
8. Wash in running water for 2 min.
9. Stain with the Van Gieson mixture: 5 ml 1% acid fuchsin, 95 ml saturated aqueous picric acid, 0.25 ml concentrated HCl.
10. Dehydrate and differentiate in two or three changes of 95 and 100% ethanol.
11. Clear through ethanol and xylene, and mount.

Advantages of the above stain are that the acidic mucosubstances stain blue, collagen stains red, muscle stains yellow, and nuclei are greenish black.

Bélanger (*Ann. N.Y. Acad. Sci.*, **106**: 364, 1963) compared autoradiographic and histochemical staining methods in sections of mucus-secreting cells. Hale-positive stains were observed in mucus cells that did and did not manifest labeled sulfur. The staining pattern for Alcian blue was the same as that observed for the Hale stain. Toluidine blue, safranin metachromasia, and labeled sulfate were located in the same places. No similarity in pattern was detected in the periodic acid Schiff staining and that for sulfated mucopolysaccharides.

Conklin (*Am. J. Anat.*, **112**: 259, 1963) believes five zones can be delineated in adult tracheal cartilage by utilization of Hale's colloidal iron, Alcian blue, azure A, methylene blue, and the periodic acid Schiff stains in concert with several blocking procedures.

Jirge (*Histochemie*, **22**: 82, 1970) reported a puzzling observation—a substance in the yolk sac of young fish larvae fails to stain with basic dyes including Alcian blue and azure A although it is stained blue with the dialyzed iron method of Hale.

De Moraes and Villa (*Ann. Histochim.*, **8**: 411, 1963) studied salivary glands of several animals and human beings with several stains for the acid mucopolysaccharides.

Colloidal iron has been employed in electron microscopy by Matukas (*J. Cell Biol.*, **39**: 29, 1968), Gasic and Berwick (*J. Cell Biol.*, **19**: 223, 1963), and Curran et al. (*J. Anat.*, **99**: 427, 1965).

Módis et al. (*Acta Morphol. Acad. Sci. Hung.*, **13**: 207, 1965) tested the efficacy of several metal colloidal solutions for the histochemical staining of acid mucopolysaccharides. Silver, copper, and molybdenum were found somewhat useful.

Lillie, Vacca, and Pizzolato's (*J. Histochem. Cytochem.*, **21**: 161, 1973) hydroxyferric ion methods for mucins follow:

REAGENT A. Prepare a 1 M solution of ferrous chloride. This can be made extempore by reacting 2 to 3 g coiled bright iron wire (analytical grade) for 2 to 3 days in 20 ml preboiled distilled water plus exactly 4 ml 12 N HCl (sp gr 1.19) under a 3-cm layer of mineral oil in a 25-ml cylinder, loosely stoppered to allow escape of hydrogen. The resulting 1 M $FeCl_2$ solution remains clear; its pH is about 4. The excess iron wire remains bright for 14 days.

To prepare the ferric hydroxychloride take 10 ml 1 M $FeCl_2$ and 5.1 ml 3% H_2O_2 (freshly diluted from 30%), dilute to 100 ml with boiled, cooled distilled water to yield 0.1 M $FeOHCl_2$. This stock solution is stable for some 4 months. For use dilute to 10 mM (5 ml plus 45 ml distilled water). The 10-mM solution remains usable for about 3 weeks and can be used repeatedly.

REAGENT B. For this variant, make a 0.2 M solution of ferric chloride $FeCl_3 \cdot 6H_2O$ (2.7%), and add to it an equal volume of 0.2 N NaOH. This stock 0.1 M solution remains stable for 2 to 3 months. For use dilute 5 ml with 45 ml distilled water to give 10 mM $FeOHCl_2$ plus NaCl.

TECHNIC

1. Dewax 5- to 6-μ paraffin sections and bring through alcohols to water as usual. Rinse in distilled water.
2. Treat sections for 2 h in 10 mM FeOHCl$_2$.
3. Wash sections in four 5-min changes of distilled water.
4. Develop color by immersion of sections for 15 min in 1% potassium ferrocyanide in 0.5% concentrated HCl (0.06 N).
5. Wash sections briefly, counterstain as desired, dehydrate, clear, and mount in synthetic resin (polystyrene, Depex, cellulose caprate).

RESULTS: Gastric surface epithelial mucoprotein, cartilage, rat mast cells, cardiac, pyloric, and Brunner gland, submaxillary and sublingual but not parotid gland, duodenal, intestinal, and colonic goblet cell and basiglandular mucins and experimentally sulfated collagen color blue. Methylation blocks the reaction; saponification restores it in some sites, not in others.

Reagent A appears to be somewhat superior to reagent B. Preparation of fresh FeCl$_2$ was resorted to because of the frequent partial oxidation of commercial ferrous salts in a warm, humid climate. Ferrous sulfate did not yield a satisfactory soluble hydroxy salt.

Martin and Spicer (*J. Histochem. Cytochem.*, **22**: 206, 1974) recommend the following method using concanavalin A[1] and iron-dextran for staining cell surface mucosubstances. They employed placental tissues. Fresh bits of placental tissues are fixed in 3% glutaraldehyde buffered with either 0.1 M cacodylate or phosphate-buffered saline solution at pH 7.4 for 1 h at 25°C, rinsed in three changes (5 to 15 min each) of phosphate-buffered saline solution, placed in concanavalin A (1 mg/ml in phosphate-buffered saline solution) for 1 h at 25°C, rinsed in three changes (5 to 15 min each) of phosphate-buffered saline solution at 25°C, placed in phosphate-buffered saline solution containing 5 to 50 mg iron dextran (Imferon, Lakeside Laboratories, Inc., Milwaukee, Wisconsin) for 30 min at 25°C, rinsed in three changes (5 to 15 min each) of phosphate-buffered saline solution at 25°C, postfixed in 1% OsO$_4$ in phosphate-buffered saline solution for 1 h, dehydrated, and embedded in Spurr. Sections were not stained with a heavy metal. Control sections either were placed in 0.2 M α-methyl-D-mannoside in phosphate-buffered saline solution for 60 min at 25°C prior to the iron-dextran step or were not exposed to concanavalin A.

Electron photomicrographs magnified 90,000× reveal localization of iron-dextran particles on cell surface coats external to the plasmalemma. Sections not exposed to concanavalin A or subjected to α-methyl-D-mannoside did not possess iron-dextran on cell surface coats.

Stoddart and Kiernan (*Histochemie*, **33**: 87, 1973) labeled concanavalin A with fluorescein isothiocyanate (FITC) and used the conjugate to stain gastric, intestinal, and salivary gland mucins, granules of mast cells, basement membranes, and cerebral gray matter. Collagen fibers stained moderately and cartilage stained weakly. Concanavalin A is a lectin derived from jack beans (*Canavalia ensiformis*) which is believed to attach specifically to sugars containing the α-D-arabinopyranoside configuration at C-3, C-4, and C-6 (Goldstein et al., *Biochemistry*, **4**: 876, 1965). These configurations include α-glucose, α-mannose, and derived glycosides, but not galactose, β isomers of the above, or sialic acids.

The conjugate was prepared in the usual manner, purified by chromatography on Sephadex G25 and dialyzed against Carbowax 15,000 to 20,000 to give a reagent of approximately 5 mg/ml protein in 0.1 M Na$_2$HPO$_4$–NaH$_2$PO$_4$ buffer of pH 7.2. The opalescent solution was cleared by a twofold dilution with distilled water and final adjustment of pH to 6. The reagent can be stored at -20°C for at least 6 months.

---

[1] Concanavalin A is a protein-containing extract from jack beans.

STAINING PROCEDURE. Suppress completely the autofluorescence by immersion of slides in 0.2% OsO_4 for 5 min, and then wash in running water for at least 4 h. Rinse sections in distilled water, and stain in the FITC–concanavalin A reagent for 30 to 60 min at 37°C. Agitate slides in three changes of distilled water for a total of 5 min, dehydrate, clear in xylene, and mount in Depex. Sections are examined with a fluorescence microscope at 350 to 450 mm with an exciter filter BS 12/6 mm and SP3 barrier filter.

Complete inhibition of staining was observed in sections of tissues incubated with 0.5 M sucrose with the FITC–concanavalin A reagent except mucous acini of sublingual and submaxillary salivary glands. Stoddart and Kiernan could not account for the disparate behavior of salivary glands. Complete inhibition of staining of all structures occurred in sections incubated with 100 mg/ml disodium EDTA along with the FITC–concanavalin A reagent. Complete blockage of staining occurred in sections previously acetylated in 40% acetic anhydride in pyridine for 18 h at 22°C.

Kiernan and Stoddart (*Histochemie*, **34**: 77, 1973) and Stoddart and Kiernan (*Histochemie*, **34**: 275, 1973) recommend the use of Trasylol (aprotinin), a polypeptide protease conjugated to FITC for the detection of acid mucosubstances. The conjugates retain both tryptic and chymotryptic inhibitory actions. The conjugate was shown to adhere to sialosyl or uronosyl groups. Formalin-fixed, dewaxed, paraffin-embedded sections were treated with 0.5% OsO_4 for 5 min and washed to suppress autofluorescence. Sections were then treated with a dilute solution of the FITC conjugate for 1 min, rinsed, dehydrated, and mounted in Depex. Fluorescence can be noted in structures that usually stain with Alcian blue. Glycogen does not stain. Trasylol is obtained from Bayer Pharmaceuticals Ltd., Haywards Heath, Sussex, England.

Dougherty and Lee (*J. Ultrastruct. Res.*, **31**: 1, 1970) treated myofibrils isolated from glycerol-extracted rabbit psoas muscles with dialyzed iron at pH 1.5 to 2 and noted a marked deposition in the I bands, with some staining in M bands. Methylation for 4 h at 60°C prevents staining. Trypsin digestion for 10 min removes M lines and Z disks; however, I band staining persists. Dougherty and Lee suggest the presence of a three-dimensional network of acid mucosubstances in I bands.

Kiernan and Stoddart (*Histochemie*, **34**: 77, 1973) conjugated FITC to Trasylol. The conjugate had an acid pH of 4.5 to 5.5. Aqueous solutions of the conjugate applied to a variety of formalin- or Carnoy-fixed, paraffin-embedded sections resulted in staining tissue sites containing acid mucosubstances. Further studies are needed to determine the mechanism of the staining reaction.

HEMATOXYLIN AND CARMINE METHODS

Two of the traditional methods for mucus are *Mayer's mucihematein* and *mucicarmine* stains. We have not used them to any great extent but cite them because the first is said by Mallory to be excellent for connective tissue mucin and the second for epithelial mucin. For both, 100% ethyl alcohol fixation is prescribed. Paraffin and celloidin (nitrocellulose) sections may be used.

MUCIHEMATEIN. For the mucihematein dissolve 0.2 g hematein (not hematoxylin) and 0.1 g aluminum chloride in 100 ml of either 40% glycerol or 70% alcohol. To the alcoholic solution add 1 or 2 drops of concentrated nitric acid. The latter is preferable when the mucus tends to swell in aqueous stains. With either solution stain sections 10 to 60 min or more, wash in several changes of distilled water for 5 to 10 min, dehydrate, clear, and mount with alcohols, xylene, and balsam or synthetic resin. Mucus stains violet-blue; other tissue elements remain unstained. Gastric mucin is not stained. Cowdry suggests prestaining with alum carmine to give red nuclei.

Laskey (*Stain Technol.*, **25**: 33, 1950) uses hematoxylin and ripens with 40 to 100 mg sodium iodate per g. Solutions are ready to use at once and remain useful for over 6 months. Her directions for preparation of the solution read: Dissolve 1 g hematoxylin in 100 ml 70% alcohol. Add 0.5 g aluminum chloride and 5 ml 1% aqueous sodium iodate solution, and make up to 500 ml with 70% alcohol.

LASKEY'S TECHNIC FOR MUCIHEMATEIN STAINING

1. Deparaffinize and hydrate as usual.
2. Lay slides face up on staining racks, and flush off several times with distilled water.
3. Deposit 2 ml of mucihematein on each slide, and stain for 5 to 10 min.
4. Wash 15 min in distilled water (three changes).
5. Dehydrate with 95 and 100% alcohols, two changes each, and clear with alcohol plus xylene and two changes of xylene. Mount in polystyrene or other resin.

RESULTS: Epithelial mucins stain deep blue-violet; connective tissue mucins and cartilage matrix, lighter violets. Gastric surface epithelial mucus is unstained.

MUCICARMINE. For mucicarmine heat a mixture of 1 g carmine (alum lake), 0.5 g aluminum chloride crystals, and 2 ml distilled water over a small flame, agitating constantly, until the color darkens, about 2 min. Then add with constant stirring 100 ml 50% alcohol. Let stand for 24 h and then filter. This stock solution is stable. According to Mallory, take 1 ml stock mucicarmine, dilute with 10 ml distilled water or 50 to 70% alcohol, stain 10 to 15 min or more, rinse in water, dehydrate with alcohols, clear with xylene, and mount in balsam. Cowdry first stained sections in alum hematoxylin, and washed and stained for 5 min in undiluted mucicarmine. Only mucus, including gastric mucus, is stained red; other structures are unstained or stained with hematoxylin, according to the variant chosen.

Acid methyl blue and aniline blue stains in the various collagen technics often stain epithelial mucus, including gastric mucus, selectively a light blue, contrasting well with counterstains and with the deep blue collagen. Mucus may remain deep violet by Gram-Weigert variants.

Laurén and Sorvari [*Acta Histochem.* (*Jena*), **34**: 263, 1969] studied the histochemical specificity of the mucicarmine stain along with eight other stains for mucins. Staining behavior was assessed in sections of epithelial mucins from rabbits and mice before and after sulfation, carboxylation, periodate oxidation, and digestion by sialidase. The staining pattern was observed to be comparable with that of Alcian blue, whereupon Laurén and Sorvari concluded reactivity of the stain with sulfate and carboxylic acid groups.

Stoward (*J. R. Microsc. Soc.*, **87**: 215, 1967) advocates the following fluorescent histochemical methods for the demonstration of sulfomucins:

A. FEULGEN METADIAMINE-CORIPHOSPHINE METHOD

1. Take sections to water.
2. Immerse in 5 N hydrochloric acid for 10 min (Feulgen hydrolysis).
3. Rinse in water.
4. Condense aldehydes with *m*-diamine by leaving the hydrolyzed sections overnight in a 0.1% solution of *m-N,N*-dimethylphenylene diamine dihydrochloride (Eastman Kodak, New York; ICI, Pharmaceutical Division, Wilmslow, Cheshire) in phosphate buffer (2.5 g $NaH_2PO_4 \cdot 2H_2O$ + 2.5 g $Na_2HPO_4 \cdot 12H_2O$ per 100 ml solution, pH 6.1).
5. Rinse in two changes of water.
6. Stain in 0.01% coriphosphine O, at pH 3, for 20 min.

7. Rinse in two changes of distilled water, and dehydrate and mount as described previously.
8. Examine for fluorescence using ultraviolet-light excitation.

RESULTS: Nuclei do not fluoresce whether excited by ultraviolet light or by blue light. Sulfomucins emit a characteristic red, orange-yellow, or yellow fluorescence (depending on the mucin) which can be easily distinguished from the green fluorescence of protoplasmic proteins. Some sulfomucins, such as those in Syrian hamster vaginal epithelia, which normally fluoresce only when excited with blue light, fluoresce here with ultraviolet-light excitation. Sialomucins emit an orange or reddish brown fluorescence and oxidized sulfo-proteins a yellow fluorescence, and therefore the method is not specific for sulfomucins.

B.   FERRIC ALUM–CORIPHOSPHINE METHOD

1. Take sections to water.
2. Immerse in an aqueous 4% solution of ferric ammonium alum for 10 min.
3. Wash in two changes of water.
4. Stain in a 0.01% solution of coriphosphine O (pH approximately 6) for 20 min.
5. Rinse in two changes of distilled water, and dehydrate and mount as described previously.
6. Examine for fluorescence first with ultraviolet-light and second with blue-light excitation.

RESULTS: Most, but not all, sulfomucins emit a characteristic green or a dull red fluorescence with ultraviolet-light excitation. The fluorescence of proteins and nuclei are completely quenched.

With blue-light excitation, sulfomucins emit a bright yellow fluorescence that can be distinguished easily from the green fluorescence of proteins and nuclei. Sialomucins, however, also emit a yellowish fluorescence, and therefore the method is not absolutely specific for sulfomucins. Oxidized sulfoproteins emit a green fluorescence similar to that emitted by most connective tissue proteins.

C.   CORIPHOSPHINE–THIAZOL YELLOW METHOD

1. Take sections to water.
2. Stain in 0.01% coriphosphine O (C.I. 46020), pH approximately 6, for 20 min.
3. Rinse briefly in two changes of distilled water.
4. Stain in 0.001% thiazol yellow G (C.I. 19540) (which from most suppliers need not be purified for this purpose), pH 2 to 2.1, for 1 min.
5. Rinse briefly in distilled water, and dehydrate and mount as described previously.
6. Examine for fluorescence first with ultraviolet-light excitation, and second with blue-light excitation.

RESULTS: Sulfomucins emit their characteristic red or orange-yellow fluorescence, which stands out clearly against the bright blue fluorescence emitted initially by most connective tissue proteins and the green (ultraviolet-light excitation) or greenish yellow (blue-light excitation) fluorescence of nuclei. The initial blue fluorescence emitted by proteins usually changes color irreversibly to dull green after 30 to 60 s exposure to ultraviolet light. Sialomucins do not fluoresce; but cells which secrete a mixture of sialomucins and sulfomucins fluoresce distinctly with a dull orange-brown.

D.   LITHIUM ALUMINUM HYDRIDE (LiAlH$_4$) REDUCTION METHOD FOR SULFOMUCINS

1. Take at least two deparaffinized sections to absolute ethanol and then through two changes of absolute methanol.

2. Methylate sections in a 2% solution[1] of thionyl chloride (Hopkin and Williams' purified grade) in absolute methanol for 4 h. The methanolic thionyl chloride solution should preferably be 1 to 7 days old.
3. Rinse in two changes of methanol, followed by two changes of dioxane.
4. Incubate in a 1% filtered solution of lithium aluminum hydride (British Drug House Ltd. or Hopkin and Williams reagent grade: preferably purified further in dioxane and stored over barium oxide) at 60°C for 48 h.
5. Wash briefly in dioxane and then in a 1% solution of concentrated hydrochloric acid in 70% alcohol for 3 to 5 min until any aluminum hydroxide deposit on the section has dissolved.
6. Rinse in three changes of 70% alcohol, and take to absolute alcohol and then to 50 : 50 ether–absolute alcohol.
7. Collodionize sections by immersing them in an 0.8% solution of nitrocellulose in 50 : 50 ether–absolute alcohol (British Drug House Ltd. "Stanvis" Necoloidine solution, diluted 1 : 10) for 5 min and then leaving to dry in air.
8. Saponify the collodionized sections in a 1% solution of potassium hydroxide in 80% alcohol for 20 min.
9. Rinse in two changes of 80% alcohol. Leave in 95% alcohol for 20 min to remove remaining nitrocellulose.
10. Take to water, and stain one section in 0.01% coriphosphine and a second at pH 4 (or for a conventional stain use azure A).
11. Rinse in distilled water, and dehydrate and mount as described previously.
12. Examine for fluorescence (ultraviolet-light excitation).

RESULTS: Normally only sulfomucins emit any significant fluorescence (dull brown or red). Nuclei and proteins emit a faint blue fluorescence. Exceptions to this generalization are usually due to desulfation during the methylation stage, overreduction if highly impure $LiAlH_4$ is used, or oxidation of the tissue by dioxane peroxides if the $LiAlH_4$ solvent is too "dry." The fixation of the tissue is also important.

Stoward indicates that his $LiAlH_4$ method requires further investigations to prove explicitly that the fluorescence is due to a sulfate–coriphosphine O complex and not to another conjugate such as coriphosphine O and a basophilic group such as uronic acids. Although the lithium–aluminum hydride method should have provided this proof, the variable results achieved do not promote faith in this proof.

Lev and Stoward (*Histochemie*, **20:** 363, 1969) note that eosin as employed in routine hematoxylin and eosin sections will induce a green fluorescence in mucins and other structures. The intensity of staining appears to be dependent upon the amount of amino groups available for reaction with acidic groups on eosin. Thus the protein rather than the carbohydrate components of mucins appear to be identified by eosin fluorescence.

Yamada (*Histochemie*, **20:** 271, 1969) employs the periodic acid Schiff (PAS) and acriflavine (C.I. 46000, 2,8-diamino-10-methylacridinium chloride) for the demonstration of sulfated mucins and periodate-reactive carbohydrates.

COMBINED STAINING IN SECTIONS OF PAS AND ACRIFLAVINE-REACTIVE MUCOPOLYSACCHARIDES.

1. Take sections to water.
2. Incubate 5 min in 0.5% periodic acid.

---

[1] Care should be exercised when preparing this solution. Thionyl chloride spits vigorously when it first comes into contact with methanol.

3. Wash well in distilled water.
4. Immerse in Schiff reagent, 15 min.
5. Rinse in sulfite water (three rinses).
6. Wash well in water.
7. Stain in a 0.5% acriflavine–hydrochloric acid solution of pH 1.5 at room temperature for 30 min.
8. Rinse briefly in distilled water.
9. Immerse in 95% ethanol to remove excess dye. If necessary, rehydrate through descending grades of ethanol, rinse in water, counterstain lightly with Mayer's alum hematoxylin for nuclei, and wash in water.
10. Dehydrate through ascending grades of ethanol.
11. Clear in two changes of xylene, and mount in balsam or in a synthetic resin.

RESULTS: The usual periodate-reactive constituents stain red. Sulfated acid mucopolysaccharides are believed to stain yellow. Yamada states that yellow staining is unaffected in sections digested with ribonuclease. However, yellow staining fails in methylated sections, and staining is not restored after saponification.

Puchtler et al. (*J. R. Microsc. Soc.*, **89**: 329, 1968) developed a combined PAS–myofibril stain for the demonstration of early lesions of striated muscle. The method follows. Carnoy fixation of human autopsy specimens is advised.

1. Bring sections to water, and wash in running tap water for 5 to 10 min.
2. Perform the PAS reaction, and wash in running tap water for 10 min.
3. Transfer to a 5% aqueous solution of tannic acid for 10 min.
4. Rinse in three changes of distilled water.
5. Place in a 1% aqueous solution of phosphomolybdic acid for 10 min.
6. Rinse in three changes of distilled water.
7. Stain in Sulfone Cyanin GR Ex (C.I. 26400)* solution for 5 min. Dissolve 2 g Sulfone Cyanin in 90 ml methanol, and let stand overnight. Add 10 ml glacial acetic acid shortly before use. Do not filter.
8. Rinse in two changes of methanol–glacial acetic acid, 9 : 1.
9. Rinse briefly in absolute alcohol, clear in xylene, and mount in Permount or similar synthetic resin.

RESULTS: With the above method, A and Z bands in normal cardiac and skeletal muscle, smooth muscle cells in the media, and myoepithelial cells of the intima of blood vessels were stained blue. I bands were colored yellow. Normal connective tissue fibers and amyloid stained yellowish orange. Areas of connective tissue fibrosis were stained pinkish red. Lipofuscin stained brownish red. Myofibrils in experimental myocardial infarctions manifested blurring of A bands and gradual disorganization of striations within 1 to 3 h after coronary artery ligation. Under various pathologic conditions, A and Z bands of human cardiac and skeletal muscle lost affinity for Sulfone Cyanin GR Ex Cyanin and thereafter stained yellow.

Puchtler et al. suggest that Sulfone Cyanin GR Ex is staining myosin inasmuch as A but not I bands are stained. Fibrin is also strongly stained by Sulfone Cyanin GR Ex. Yokoyama et al. (*Arch. Pathol.*, **59**: 347, 1955), Kilonsky (*Am. J. Pathol.*, **36**: 575, 1960), and the Puchtler group agree that a rapid loss of glycogen occurs within 5 min after occlusion of a

---

* Obtainable from Matheson, Coleman and Bell or through Roboz, Washington, D.C., from Chroma as "Levanol fast cyanine 5RN," Farbenfabriken Bayer.

coronary artery and the subsequent appearance of a PAS-positive substance in myocardial fibrils within 1 to 2 h after ligation.

Scott and Dorling (*Histochemie,* **19:** 205, 1969) claim to have developed the first *chemical* method for localization of the acid mucopolysaccharides. The method takes advantage of the fact that vicinal glycols on neutral mucosubstances oxidize quickly to aldehydes which are subsequently reduced by borohydride. Sections are thereafter oxidized again, whereupon acid mucopolysaccharides containing 1,4-linked uronic acids such as chondroitin sulfate are oxidized by $NaIO_4$ for longer periods at higher temperatures. The aldehydes engendered by the second oxidation are visualized by Schiff's reagent. The method follows:

1. Fix tissues in cold ethanol-formalin at $-20°C$ for 3 days with intermittent agitation.
2. Dehydrate tissues and embed in paraffin.
3. Cut sections at 5 μ, mount on slides, deparaffinize, and bring to water.
4. Oxidize sections for 1 h at 30°C with 2% aqueous sodium periodate ($NaIO_4$).
5. Rinse in running water for 10 min.
6. Reduce with 1% sodium borohydride ($NaHB_4$) in distilled water prepared immediately before use for 3 min (0.2% in 1% $NaHPO_4$ for 1 min should serve).
7. Rinse in running water for 10 min.
8. Oxidize with aqueous 2% sodium periodate ($NaIO_4$) for 24 h at 30°C.
9. Wash in running water for 10 min.
10. Immerse in Schiff's reagent for 30 min at 24°C.
11. Rinse in three solutions of 0.5% sodium metabisulfite ($Na_2S_2O_5$) for 1, 2, and 3 min respectively.
12. Rinse in running water 10 min.
13. Dehydrate and mount.

Scott and Dorling suggest that borohydride reduction occurs as follows:

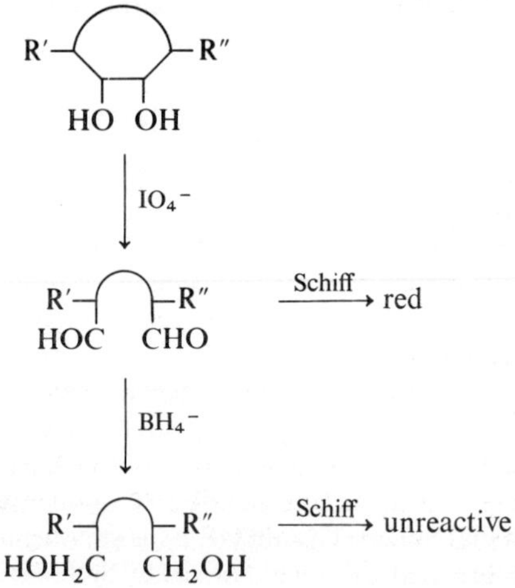

Pizzolato and Lillie (*Histochemie,* **27:** 335, 1971) noted that some salts of tin, titanium, zirconium, hafnium, niobium, tantalum, molybdenum, and tungsten may act as mordants which can combine with polysaccharides, whereupon they may be detected as color compounds after treatment of sections with an acid alcoholic solution of gallocyanin. They noted methylation prior to mordanting inhibited the staining reaction, and sulfation enhanced staining by titanium and molybdenum.

ALDEHYDE FUCHSIN. Gomori's (*Am. J. Clin. Pathol.*, **20**: 665, 1950) aldehyde fuchsin stain for elastin is a transient dye product formed as a product of basic fuchsin and paraldehyde at room temperature in acid alcoholic solution. Specifically, dissolve 0.5 g basic fuchsin (pararosanilin, C.I. 42500, or rosanilin, 42510) in 100 ml 70 % ethanol, add 1 ml concentrated HCl and 1 ml USP Paraldehyde. It is important for users of aldehyde fuchsin to keep in mind several points where difficulties may arise.

First, it is essential that the paraldehyde reagent be fresh and no older than 3 to 4 months. Next, aldehyde fuchsin is a transient dye. It is not made immediately in amounts sufficient for effective staining. It is best formed at room temperature in a stoppered bottle over a period of a few days. In our laboratory, it is prepared on a Friday for use the following Monday. Because it is a transient dye, discarding of the dye after 2 weeks is recommended. Apparently, other structures form after a week or two, resulting in insufficient aldehyde fuchsin available for effective staining qualities. It should always be filtered before use.

Aldehyde fuchsin stains many structures that contain mucins or acid mucopolysaccharides including umbilical cord, rooster comb, dental papillae of developing teeth, cartilage, mucin glands of the tongue, stomach, colon, goblet cells, beta cells of the anterior pituitary, mast cells, argentaffin cell granules, and thyroid colloid. Certain other structures are stained if sections are previously oxidized with periodic acid, namely, glycogen, gastric mucus, and striated and brush borders (Scott and Clayton, *J. Histochem. Cytochem.*, **1**: 336, 1953). After stronger oxidation with peracetic acid or permanganate, other structures, notably oxytalan fibers, will stain with aldehyde fuchsin (Fullmer and Lillie, *J. Histochem. Cytochem.*, **6**: 425, 1958). Aldehyde fuchsin–stained constituents appear to be selectively removed by β-glucuronidase in sections previously oxidized with peracetic acid (Fullmer, *J. Histochem. Cytochem.*, **8**: 113, 1960).

Hadler et al. [*Acta Histochem. (Jena.)*, **29**: 34, 1968; **30**: 41 and 54, 1968] conducted studies on selective staining of aldehyde fuchsin with neurosecretory substances.

Pease (*J. Ultrastruct. Res.*, **15**: 555, 1966; *J. Histochem. Cytochem.*, **18**: 455, 1970) advocates the use of phosphotungstic acid as a specific stain for the identification for "complex carbohydrates" at the electron microscopic level. Pease suggests that the specificity of phosphotungstic for the "complex carbohydrates" is pH-dependent. Staining is best at a pH of less than 1, and the specificity dwindles rapidly at a pH greater than 3. He notes that in weakly acidic, neutral, and alkaline solutions, phosphotungstic acid stains proteins avidly. Likewise, some organic solvents such as ethylene glycol favor staining with protein. Pease notes that Rambourg et al. (*J. Cell Biol.*, **40**: 395, 1969) obtained comparable results with Pease's method and another method (periodic acid–chromic acid–methenamine silver stain) for the histochemical detection of carbohydrates. Pease does not propose a definite mechanism of attachment of carbohydrate to phosphotungstic acid; however, he suggests that weak forces are operative inasmuch as phosphotungstic acid–carbohydrate stains are easily washed out if the pH is kept low.

Glick and Scott (*J. Histochem. Cytochem.*, **18**: 455, 1970) declare that phosphotungstic acid is not a stain for carbohydrate. They point out that not many model carbohydrates have been tested and the mechanism of staining reaction is unknown. They are concerned with the staining of many noncarbohydrate substances. They believe it may be possible for cationic polysaccharides to stain with phosphotungstic acid; however, very few natural cationic polysaccharides are known. They would like to have Pease explain how hydroxyl groups of complex carbohydrates differ from other hydroxyls. In other words, specific proof is requested for the contention.

Scott and Glick (*J. Histochem. Cytochem.*, **19**: 63, 1971) tested the efficacy of phosphotungstic acid in precipitation of several polyhydroxy compounds as indicated in Table 14-9. The results show that

TABLE 14-9 EFFECT OF ACIDITY ON PRECIPITATION OF POLYHYDROXY COMPOUNDS AND PROTEINS BY PHOSPHOTUNGSTIC ACID

Polyhydroxy compounds (1%, w/v)	Control (H$_2$SO$_4$, normality in last column)	Phosphotungstic acid (0.5% w/v)	Phosphotungstic acid (0.5% w/v) plus H$_2$SO$_4$ (normality at which precipitation observed)
Hyaluronate	−	−	7.5
Chondroitin sulfate	−	−	6.2
Alginate	−	−	7.5
Polyvinyl alcohol	−	−	0.5
Mannosan	−	−	9.0
Araban	−	−	4.4
Galactan	−	−	2.5
Glycogen	−	−	0.5
Fucoidin	−	−	9.5
Proteins			
Lab-Trol	−	+ +	+ + at 2 N
Human serum	−	+ +	+ + at 2 N
Glycoprotein from human amniotic fluid	−	+	+ + at 2 N

Source: Scott and Glick, *J. Histochem. Cytochem.*, **19:** 63, 1971.

1. Proteins *are* precipitated, with phosphotungstic acid either alone or in the presence of added strong acid.
2. Polyhydroxy compounds in general are precipitated by phosphotungstic acid only at very low pH.
3. The polyhydroxy compounds need not be polysaccharides (note polyvinyl alcohol).
4. The pH at which most of the polyhydroxy compounds can be precipitated is considerably lower than that currently used in phosphotungstic acid staining.

If the mechanism of the precipitation reaction involved hydrogen bonding between the participants, as proposed by Pease, it is not clear why the proposed hydrogen bonds should be formed only in strongly acid solutions.

Results by Scott and Glick would tend to suggest that not only hydroxyl groups are involved, since glycoproteins are precipitable at a pH higher than that in which many hydroxyl groups would be protonated. Peptide bonds and amide groups are also basic enough to become protonated in acid solution:

$$R'-\overset{\overset{\displaystyle O}{\|}}{C}-NH-R'' + H_3O^+$$

$$= R' + \overset{\overset{\displaystyle O-H^+}{\|}}{C}-NH-R'' + H_2O$$

and thus the protein moiety of a glycoprotein must also contribute to the total positive charge.

Neerbos and Vries-Lequin (*Clin. Chim. Acta*, **26:** 271, 1969) studied the composition of the oxidizing agent, the oxidation time and temperature, the rinsing solutions, and the

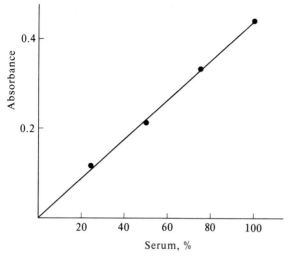

Fig. 14-4 The relation between the color intensity and the glycoprotein concentration by means of a serum dilution series. (*From Neerbos and Vries-Lequin, Clin. Chim. Acta,* **26:** 271, 1969. *By permission of the publisher.*)

staining times and temperatures on the efficacy of the PAS staining reaction for serum glycoproteins separated by electrophoresis. The following method was thereafter prescribed:

1. After electrophoresis allow the strips to dry for 10 min at 100°C and soak them in 96% alcohol for further 10 min.
2. Oxidize the strips by incubating for 30 min in a solution having the following composition:

   2 g periodic acid
   30 ml distilled water
   15 ml 0.2 M sodium acetate
   100 ml 96% alcohol

   This solution must be freshly prepared before use.
3. Remove the oxidation reagent by washing the strips twice with 0.001 N HCl, soaking them for 1 min in the first bath, and leaving them for 5 min in a second bath.
4. Incubate the strips for 50 min in a Schiff reagent, brought to room temperature before use.
5. Remove the Schiff reagent by washing 5 times successively in baths for 5 min each time, in a freshly prepared solution composed of 4 g $K_2S_2O_5$, 10 ml HCl, 38%, made up to 1000 ml with distilled water.
6. Finally, remove the remainder of the rinsing solution with distilled water.

The relation between color intensity and glycoprotein concentration was checked by means of serum dilution series. Figure 14-4 shows that with the method described above a linear correlation is found with 0.4% $K_2S_2O_5$ also in the rinsing solution.

SIALOMUCIN, SIALIDASE

In 1960 Spicer and Warren (*J. Histochem. Cytochem.,* **8:** 135) reported the identification of sialic acid–containing mucins in sublingual glands of mouse, rat, and guinea pig. The mucin was identified by digestion with sialidases prepared from cultures of *Clostridium* *perfringens* or of *Vibrio cholerae*. Commercial polyvalent (A + B) influenza vaccine also exhibited sialidase (neuraminidase) activity.

Sections were digested with 0.05 ml enzyme (0.1% protein) in 0.1 *M* phosphate buffer of pH 6 for unstated periods. The digestion fluid and washings were assayed for sialic acid, as were digested and undigested sections, thus demonstrating the liberation of sialic acid.

Drzeniek (*Histochem. J.*, **5**: 271, 1973) provides a review of substrate specificities of sialidases (neuraminidases):

Histologic preparations were stained for 30 min either in 0.02% azure A at pH 3 or by the Alcian blue–PAS technic. Sialomucins, strongly metachromatic before sialidase digestion, failed to stain afterward. Alcian blue–PAS staining of sialomucins is changed from a strong blue before digestion to purple-red after.

Quintarelli included the ox submaxillary gland as another locality for sialomucin (*J. Histochem. Cytochem.*, **9**: 176, 1961), identifying the constituents as sialic acid and *N*-acetylgalactosamine in 1 : 1 molar proportions.

Quintarelli used apparently a 24-h digestion in a dilution of influenza vaccine with 4 parts 5 m*M* phosphate-buffered 0.85% sodium chloride solution; the pH was not stated, but the composition of the buffer indicates perhaps pH 7.5 (2.72 g KH_2PO_4 plus 30 g Na_2HPO_4 plus 170 g NaCl per l, diluted 1 : 20 for use).

After digestion, preparations were stained by the iron hematoxylin, fast green FCF–safranin O technic; by the Alcian blue–PAS method; or by a simple PAS method. Metachromatic and Alcian blue staining were lost by sialomucins; the periodic acid Schiff reaction persisted.

In a later report by Warren and Spicer (*J. Histochem. Cytochem.*, **9**: 400, 1961) they used a 2500-fold sialidase concentrate from *V. cholerae* cultures [method of Ada and French, *Nature (Lond.)*, **183**: 1740, 1959]. Sialomucin is identified also in the vaginal epithelium of pregnant rats and mice, as well as during mucification periods of the estrus cycle. Sialidase failed to release sialic acid from the sialomucins of the vagina and salivary gland of the pregnant rat, though assays revealed the presence of copious sialic acid. Both *V. cholerae* and *C. perfringens* enzymes were used.

In methylation saponification sequences and in minimum pH levels of basic dye staining, the sialic acid mucins behave as carboxylic acid mucins.

Carlo (*J. Histochem. Cytochem.*, **12**: 306, 1964) recommends the following method for the histochemical detection of sialic (neuraminic) acids. In spot tests on filter paper, cerebroside, cephalin, hyaluronic acid, heparin, and ribonucleic acid were unstained.

1. Cut frozen sections, formalin-fixed, at 10 to 15 μ; attach them to slides and let dry.
2. Spray the sections with the Bial reagent as modified by Svennerholm:

Orcinol	200 mg
$CuSO_4$, 0.1 *M*	0.25 ml
HCl, 12 *N*	80 ml

Make up the volume to 100 ml with distilled water. (Let stand 4 h before use. If stored at 0°C, the solution is stable for a week.)
3. Place sections face down on a glass frame in a preheated container, which has on the bottom a thin layer of concentrated HCl, at 70°C for 5 to 10 min.
4. Dry the sections in air.
5. Clear in xylene; mount in Canada balsam.

RESULTS: Sites containing sialic acids are colored in various shades of red; the structure of tissues remains generally well preserved. The color fades within a day.

TECHNIC OF HYALURONIDASE DIGESTIONS

1. Fix in alcohol, alcohol acetic formalin, Carnoy, lead, or mercury salt mixtures, or the like. Dehydrate, embed in paraffin, and section as usual. Dichromate fixatives impair the digestibility of cartilage matrix.
2. Deparaffinize and hydrate through alcohols, treating with iodine and thiosulfate to remove mercury precipitates if necessary. Do not collodionize, for this greatly retards digestion. Fragile material may be digested without removal of paraffin, but the digestion may then require days rather than hours.
3. Digest for graded intervals in hyaluronidase or malt diastase buffered to pH 5.5 to 6.5 (0.1 M) with phosphates or acetates. We have used testicular hyaluronidase concentrations of 25 to 2 mg/100 ml. Digestion times vary from 1 to 18 h. Steps of 1, 2, 4, 6, and 18 h are convenient. Malt diastase is used at 1 : 1000 dilution in pH 6 buffer (0.1 M acetate or phosphate). Digest at 37 to 45°C.
4. After digestion, wash in water and stain with toluidine blue, thionin, the iron hematoxylin–fast green safranin method, or one of the iron methods, as usual. Dehydrate, clear, and mount by methods prescribed for the staining technic employed.

Hyaluronic acid and chondroitin sulfates A and C are digested.

Yamada (*J. Histochem. Cytochem.*, **21**: 794, 1973) compared the effects of *Streptomyces* hyaluronidase digestions with those of testicular hyaluronidase digestions on many tissues from several species. He noted that *Streptomyces* hyaluronidase digestions of tissues result in diminished staining in sites containing hyaluronic acid, but sites containing neutral and sulfated acid mucopolysaccharides stained as usual. Specific and selective removal of hyaluronic acid was noticed.

Tighe (*J. Pathol. Bacteriol.*, **86**: 141, 1963) employed azure A, PAS–mucicarmine, Van Gieson's, and Gomori's reticulum stains for a study of connective tissue tumors. No correlation was found between the type of staining reaction for mucin and malignancy.

Johnson and Helwig (*Ann. N.Y. Acad. Sci.*, **106**: 794, 1963) conducted a comparative study of the mucosubstances produced by primary tumor cells, metastatic tumor cells, and normal cells from which the tumor arose using Hale's colloidal iron, Alcian blue, aldehyde fuchsin, and the PAS methods with and without various digestions.

Mucins produced by carcinomas of the breast, bronchus, colon, rectum, and ovary were indistinguishable by histochemical methods employed from mucins produced by cells in the metastatic lesions and by normal cells at the site of the primary tumors.

Hou-Jensen (*Acta Pathol. Microbiol. Scand.*, **69**: 372, 1967) provides evidence that the PAS staining reaction may sometimes be helpful for differentiating Paget's disease of the skin from melanomas (Table 14-10). This study confirms an earlier report by Fisher and Beyer (*Arch. Pathol.*, **67**: 140, 1959) indicating the presence of a diastase-resistant PAS–reactive constituent in cells of Paget's disease of the skin. Further studies are needed.

Kvist and Finnegan (*J. Exp. Zool.*, **175**: 221, 1970) studied the distribution of glycosaminoglycans in the axial region of developing chicks employing the PAS and several basic stains.

Quintarelli (*Ann. N.Y. Acad. Sci.*, **106**: 339, 1963) employed several types of digestions, blocking procedures, and histochemical staining reactions to characterize the mucins of the major salivary glands of the sheep and dog and the submaxillary glands of fetal swine and monkeys.

Tan and Heptinstall (*Lab. Invest.*, **20**: 62, 1969) conducted a correlated light microscopic–electron microscopic–histochemical study of an initial buildup and then reduction of numbers of PAS–positive granules in the cytoplasm of interstitial cells of the kidney during the

TABLE 14-10 STAINING REACTIONS OF THE CYTOPLASM
OF LARGE INTRAEPIDERMAL NEOPLASTIC
CELLS IN PAGET'S DISEASE AND
MALIGNANT MELANOMAS

	PAS reaction		Alcian blue reaction	
No. of cases	Positive	Negative	Positive	Negative
36 Paget's disease	9	27	0	36
25 melanomas	0	25	0	25

Source: Hou-Jensen, *Acta Pathol. Microbiol. Scand.*, **69**: 372, 1967.

course of development and subsequent decline of experimental pyelonephritis in rats. Electron microscopy revealed the cytoplasmic PAS–stained granules were cytosomes in macrophage-type cells.

Manley and Kent (*Br. J. Exp. Pathol.*, **44**: 635, 1963) employed basic dyes and chemical analyses of glucosamine and galactosamine to study the mucopolysaccharides in relation to aortic aneurysms of man.

Dunn and Spicer (*J. Histochem. Cytochem.*, **17**: 668, 1969) provide the following results of their studies of granulocytes and platelets of adult human beings (Table 14-11). In addition, they noted that the mucopolysaccharide-containing granules of the three myeloid series also manifest acidophilia at a high pH indicative of the presence of strongly cationic proteins.

Olsson and Dahlqvist (*J. Histochem. Cytochem.*, **15**: 646, 1967) homogenized and fractionated human leukocytes. PAS–stained fractions revealed glycogen, nonglycogen, and possible glycolipid constituents.

Danes and Bearn (*Lancet*, **1**: 241, 1967) recommend the use of metachromatic staining procedures to cell cultures of skin fibroblasts for the study and detection of the mucopoly-saccharidoses (Table 14-12).

For the metachromatic staining of the acid mucopolysaccharide of Hurler's disease, Haust and Landing (*J. Histochem. Cytochem.*, **9**: 77, 1961) recommended fixation of fresh frozen sections for 20 to 30 min in equal volumes of acetone and tetrahydrofuran and staining for 2 min in 0.5% toluidine blue in 25% acetone. Rinse *directly* in acetone, *clear in xylene*, and mount.

Lagunoff, Rosa, and Benditt (*Am. J. Pathol.*, **41**: 273, 1962) applied the Falck-Torp technic of freeze-drying followed by formaldehyde gas fixation for 48 h at 50°C and subsequent embedding in paraffin. Sections of this material were brought through xylene and alcohols to 1% toluidine blue in 75% alcohol, 0.01 *M* acetates, at pH 3.5. With this technic "gargoyle" cells contained intensely metachromatic material, whereas with aqueous formalin fixation no toluidine blue metachromasia is preserved. These cells give a weak PAS reaction and light Sudan black B staining. A few cells contain intensely sudanophilic granules. Luxol Fast Blue staining paralleled the Sudan black results. With ordinary aqueous formalin fixation the lipid is not demonstrable in paraffin sections, only in frozen sections. Sudan black and Luxol Fast Blue gave strong staining in frozen sections; oil red O gave weak staining. Most "gargoyle" cells contain either metachromatic or sudanophilic material; a few possess both. Glutaraldehyde fixation may be tried. More lipid is fixed than with formol.

TABLE 14-11 REACTIVITY OF LEUKOCYTE GRANULES AND PLATELETS TOWARD METHODS FOR MUCOSUBSTANCE AND BASIC PROTEIN*

Histochemical technic	Predominant granule in					Granules in early and late basophils	Platelets
	Early neutrophil	Late neutrophil	Early eosinophil	Late eosinophil			
Aldehyde-fuchsin	+ +	0	+ + + +	+ + +		+ + + +	+ +
High-iron diamine	+ +	0	+ + +	+ +		+ + + +	+ +
Alcoholic PAS after diastase	+	0	+ + +	+ +		+ + +	+ + +
Aqueous PAS after diastase	0	0	+ + + +	+ +		0	+ +
Biebrich scarlet, pH 9.5	+ + +	0	+ + + +	+ + + +		+ + +	+ +

* The designated values indicate reactivity from unstained (0) to maximally stained (+ + + +). The values apply for all the specimens of bone marrow and buffy coat examined; marked individual variability was not encountered.

Source: Dunn and Spicer, *J. Histochem. Cytochem.*, **17**: 668, 1969.

TABLE 14-12   METACHROMASIA OF SKIN FIBROBLASTS GROWN IN CELL CULTURE FROM FAMILIES WITH THE GENETIC MUCOPOLYSACCHARIDOSES

Type of mucopolysaccharidosis	No. families studied	Affected individuals		Carriers (by pedigree)	
		No. investigated	Cultures positive for metachromasia, %	No. investigated	Cultures positive for metachromasia, %
Hurler	4	7	100	15	100
Hunter	7	9	100	12	100
Sanfilippo	2	2	100	4	100
Morquio (limited to skeletal systems)	6	6	0	4	0
Morquio (with clinical evidence of Hurler) .	2	2	100	6	100
Scheie	1	1	100	2	100

Source: Reprinted with permission from Danes and Bearn, *Lancet*, **1**: 241, 1967.

Tenconi et al. (*Experientia*, **26**: 1238, 1970) observed metachromatic granules in cultured leukocytes and fibroblasts of individuals with type II glycogenesis. Results are provided in Table 14-13.

Befalluy et al. (*Lancet*, **1**: 973, 1971) warn not to use heparinized syringes to obtain blood for use in cultures to search for metachromatic granules in cells in a marker for heterozygous carriers of certain diseases such as the mucopolysaccharidoses.

TABLE 14-13   SKIN FIBROBLAST METACHROMASIA AND LEUKOCYTE AND FIBROBLAST α-GLUCOSIDASE ACTIVITY IN PATIENTS WITH TYPE II GLYCOGENOSIS AND IN THEIR NEXT OF KIN

Subjects	Metachromasia, % positive cells	α-1,4-Glucosidase activity*	
		Leukocytes	Fibroblasts
Patient L.U.	40–80	0	4.5
Family of L.U.:			
Father	10–50	7.6	
Mother	30–70	4.8	16.5
Brother A . . .	40–80	5.9	23.1
Brother B . . .	40–80	6.1	16.0
Patient L.M.	40–90	0	
Family of L.M.:			
Father	40–90	6.0	
Mother	40–80	5.8	
Brother	40–90	5.2	
Controls 10	1–5		
Controls 9	. . . . . .	11.9 ± 3.6	
Controls 1	. . . . . .	. . . .	36.2
Controls 1	. . . . . .	. . . .	45.7

* Millimicromoles maltose hydrolyzed per minute per milligram protein.

Source: Tenconi et al., *Experientia*, **26**: 1238, 1970.

Khan and Overton (*J. Exp. Zool.*, **171**: 161, 1969) noted lanthanum-stainable material was absent from the surfaces of freshly isolated cells but present between cells in all stages of reassociation. They noted that the stainable material differed between cartilage and liver cells. The following method was employed:

1. Fix tissues for 2 h in 2.5% glutaraldehyde in cacodylate buffer of pH 7.2 containing 1% lanthanum nitrate.
2. Rinse in buffer 15 min.
3. Optional: Postfix in 1% OsO_4 in cacodylate buffer for 1 h.
4. Dehydrate and embed in Araldite 6005.

O'Brien (*J. Clin. Pathol.*, **23**: 784, 1970) noted inhibition of platelet aggregation if they were previously stained with dyes employed to detect acid mucopolysaccharides, namely, lanthanum salts, ruthenium salts, toluidine blue, brilliant cresyl blue, and Alcian blue (depending upon the concentration used). O'Brien suggests that acid mucopolysaccharides are operative during platelet adhesion and that attached dye molecules thereby interfere with aggregation.

AMYLOID

Amyloid appears to have been first described by Rokitansky ("Handbuch der pathologischen Anatomie," vol. 3, Braumuller und Seidel, Vienna, 1846). Present data indicate that amyloid is a peculiar fibrillar protein that is deposited in tissues under pathologic conditions. Interference with metabolic functions ensues which may become sufficiently severe to cause destruction of the organ or organs and death. Amyloid may be identified by Congo red birefringence (Missmahl and Hartwig, *Virchows Arch. Pathol. Anat.*, **324**: 480, 1953; Puchtler et al., *J. Histochem. Cytochem.*, **10**: 355, 1962), tryptophan staining (Cooper, *J. Clin. Pathol.*, **22**: 410, 1969), a peculiar fibrillar appearance revealed by electron microscopy [Cohen, in E. Mandema et al. (eds.), "Amyloidosis," p. 149, Excerpta Medica, Amsterdam, 1968; Cohen and Calkins, *Nature (Lond.)*, **183**: 1202, 1969] and a "cross-β" x-ray diffraction pattern when oriented (Eanes and Glenner, *J. Histochem. Cytochem.*, **16**: 673, 1968; Glenner et al., *Am. J. Med.*, **52**: 141, 1972). Harada et al. (*J. Histochem. Cytochem.*, **19**: 1, 1971) noted the morphologic similarity of amyloid derived from different individuals but noted occasional significant differences. All amyloids exhibited the "β-pleated sheet" conformation upon x-ray diffraction analysis, and those examined after orientation displayed a "cross-β" pattern. Harada et al. noted that the major amyloid components can be fractionated using 5 *M* guanidine-HCl in 1 *N* acetic acid on Sepharose 4B and Sephadex G100 or G75. The major amyloid components contained tryptophan and were deficient in hydroxylysine and hydroxyproline. The amyloid proteins from each individual differed from each other in molecular weight, amino acid composition, and the presence or absence of specific tryptic peptides. Amyloid proteins derived from the spleen and liver of the same individual were identical. Chemical differences distinguishing primary from secondary amyloidosis were not detected. Harada et al. noted a striking chemical similarity between amyloid proteins and the NH_2-terminal variable of the light and heavy chains of immunoglobulins.

Glenner et al. (*J. Histochem. Cytochem.*, **19**: 16, 1971) conducted a study of murine amyloid. Electron microscopy revealed morphologic similarity to human amyloid fibrils; namely, two parallel filaments approximately 35 to 40 Å in width are loosely twisted like a ribbon to form a fibril approximately 100 Å in width. X-ray diffraction patterns of murine amyloid revealed a "backbone" spacing at 4.75 Å and a "side chain" spacing at 11 Å, indicating a "β-pleated sheet" structure similar to human amyloid. The amyloid was identified as a unique glycoprotein with a molecular weight of 7200 and a high content of dicarboxylic acids and short-chain amino acids, substantial tryptophan, and an unreactive NH_2-terminal amino acid.

Pras et al. (*J. Exp. Med.*, **130**: 777, 1969) and Franklin and Pras (*J. Exp. Med.*, **130**: 797, 1969) isolated amyloid from tissues of nine patients with amyloidosis. On the basis of physical, chemical, and ultrastructural studies, they concluded that amyloids of different individuals resemble each other, but are not identical. Benditt and Eriksen (*Am. J. Pathol.*, **65**: 231, 1971) also noted similarities and differences of amyloids.

Linker et al. (*Biochim. Biophys. Acta*, **29**: 443, 1958) and Bitter and Muir (*Lancet*, **1**: 819, 1965) have isolated heparatin sulfate from amyloid.

Studies by Zucker-Franklin (*J. Ultrastruct. Res.*, **32**: 247, 1970) tend to indicate that amyloid fibrils resist digestion by human polymorphonuclear leukocytes except in the presence of specific antiserums to amyloid.

Westermark (*Histochemie*, **38**: 27, 1974) obtained staining for tryptophan in all amyloids except in pancreatic islets of Langerhans. Weak staining for tyrosine was also observed.

MALLORY'S IODINE REACTION (emended)

1. Stain paraffin or frozen sections of formalin- or alcohol-fixed material in 1 : 10,000 iodine solution for 10 to 15 min. (Dilute 1 ml Weigert's iodine solution with 99 ml water or 3 ml Gram's iodine solution with 97 ml water.)
2. Wash in water, and examine in water or glycerol. Or mount in Takeuchi's iodine glycerol.

RESULTS: Amyloid is mahogany brown. Postmortem tissues should be rinsed in 1% acetic acid before the iodine is applied.

Mallory (p. 133) gave a traditional iodine–sulfuric acid method for frozen sections of fresh unfixed tissue which appears readily applicable to cryostat sections:

1. Stain sections lightly with Gram's iodine ($I-KI-H_2O = 1 : 2 : 300$). Amyloid colors mahogany red-brown.
2. Flood slide with 2 to 5% H_2SO_4. Amyloid turns violet and then blue in a few minutes.

RESULTS: Amyloid stains mahogany brown at step 1, violet to blue at step 2. Cholesterol crystals are deep brown at step 1, brilliant blue at the edges in step 2. Corpora amylacea in brain and prostate stain brown with iodine. Starch granules color blue with iodine. Cellulose colors yellow with iodine; when washed and treated with strong H_2SO_4, it turns blue.

Brundelet (*Acta Pathol. Microbiol. Scand.*, **59**: 156, 1963) advises the following method for the identification and display of amyloid in gross museum specimens:

REAGENTS.   It is advisable to use freshly made-up reagents.

Iodine solution:

Iodine	10 g
Potassium iodide	250 g
Distilled water	1000 ml

This solution will stain the amyloid substance brown. (If freshly made-up Lugol's solution is available, the above reagent may be rapidly prepared by adding 230 g potassium iodide per l of Lugol's solution.)

Mounting fluid for the brown-stained amyloid organs:

Sodium chloride	25 g
Formaldehyde (40%)	250 ml
Sodium dihydrogen phosphate, monohydrate	6.1 g
Disodium hydrogen phosphate, anhydrous	3.9 g
Distilled water	750 ml

Mounting fluid to obtain the dark blue-green stained amyloid organs:

Sulfuric acid (concentrated)	50 ml
Sodium chloride	25 g
Formaldehyde (40%)	250 ml
Distilled water	700 ml

(Here, the sulfuric acid is incorporated into the mounting fluid.)

TECHNIC

1. Dip the fixed, sliced organs into the iodine solution for about 18 h. (The ratio of volume of staining fluid to volume of organs should be at least 5 : 1.)
2. Mount using the mounting fluid corresponding to the final staining desired. (The ratio of volume of mounting fluid to volume of organ in the mounting jar is not critical: it may be 2 to 1, or even 1 to 1.)

Brundelet notes that shortly after mounting, fluid in the jar will become stained with iodine. Although the stain will fade within a few weeks to months, amyloid will remain stained for several months. After fading, fluid around the specimen may be drained off, the specimen may be restained, and the mounting fluid may be replaced.

Shirahama and Cohen [*Nature* (*Lond.*), **206**: 737, 1965] observed negatively stained amyloid by electron microscopy.

According to Langhans (Ehrlich's "Encyklopädie") permanent mounts may be made thus:

1. Prestain 10 to 15 min with carmine.
2. Stain 5 to 10 min in Gram's iodine solution.
3. Dehydrate with 100% alcohol containing 1 to 2% of iodine crystals.
4. Clear and mount in origanum oil.

Lower's silver procedure for the iodine reaction of glycogen also seems applicable.

Bennhold's (*Munch. Med. Wochenschr.*, **2**: 1537, 1922) Congo red method for amyloid is given here as modified by Puchtler (*J. Histochem. Cytochem.*, **10**: 355, 1962). Alcoholic fixatives such as Carnoy's fluid and absolute alcohol permit better stains, but good results are obtainable after aqueous formalin, formol Zenker, or Kaiserling's fluid. Embed in paraffin as usual and section at 5 μ.

1. Hydrate through graded alcohols, removing formalin pigment and mercury precipitates as usual.
2. Stain for 10 min in Mayer's acid alum hematoxylin.
3. Wash in three changes of distilled water.
4. Transfer to alkaline NaCl alcohol for 20 min. To 40 ml saturated sodium chloride solution in 80% alcohol (*stable stock solution*) add 0.4 ml 1% NaOH just before using.
5. Stain for 20 min in freshly alkalized Congo red solution. To 40 ml stable stock saturated solution of Congo red in 80% alcohol saturated with NaCl add 0.4 ml 1% NaOH, filter at once, and use alkalinized solution within 15 min.

6.   Dehydrate quickly in three changes of absolute alcohol, clear in xylene, and mount in Permount.

RESULTS: Amyloid stains red to pink; elastic tissue, lighter red; nuclei stain blue; other structures are largely unstained.

The stock 80% alcohol saturated with NaCl is stable for at least some months. It is conveniently made by dissolving 2 g NaCl in 20 ml distilled water and adding 80 ml alcohol, when immediate precipitation of the excess NaCl occurs. The stock Congo red solution is readily made by saturating a quantity of the foregoing solvent with the dye. It also keeps for months. But the alkaline staining solution deteriorates promptly.

The dyeing of amyloid by sulfonated benzidine, tolidine, and dianisidine dyes appears to depend on a mechanism similar to that in the direct textile dyeing of cotton. The linearity of the dye configuration permits hydrogen bonding of the azo and amine (and naphthol) groups of the dye to similarly spaced carbohydrate hydroxyl radicals of cellulose and amyloid. It is to be noted that prior acetylation of tissue prevents Congo red staining of amyloid by blockade of the carbohydrate hydroxyls while deamination of tissue amine groups, which are the natural targets of the dye sulfonic acid groups, does not.

Permanganate and chromic acid oxidations have a greater suppressing effect on Congo red staining of amyloid than does periodic acid, since their action is not limited to 1,2-glycols, attacking primary alcohols as well.

Saturation of the dye solvent with sodium chloride tends to depress dye ionization and acid-base–type staining, as does also the high alcohol content. The presence of free alkali also acts to inhibit union of tissue amine groups with the sulfonic acid groups of the dye. The high alcohol percentage also protects sections against the detergent action of the alkali.

The function of the pretreatment with alkaline alcohol is to release native internal hydrogen bonding between adjacent polysaccharide chains and thus render more potential sites available for binding of dye. This is particularly effective on the more highly birefringent deposits of amyloid. The alkali pretreatment also has a greater enhancement effect on amyloid fixed in formalin, Zenker-formol, or Kaiserling's fluid, in contrast to that fixed in alcoholic fixatives.

Other dyes of the same series, Congo Corinth (C.I. 22145), benzopurpurin 4B (C.I. 23500), vital red (C.I. 23570), and trypan blue (C.I. 23850) are only slightly less effective than Congo red (C.I. 22120). Acid dyes of other classes could not be substituted. For a fuller discussion, see Puchtler's paper (ibid.).

Puchtler and Sweat (*J. Histochem. Cytochem.*, **13**: 693, 1965) noted that when sections were stained with the alkaline Congo red method and viewed with the aid of a mercury-vapor lamp employed in combination with exciter filter (BG 12/3 mm or BG 12/6 mm), bright-field condenser, and ultraviolet-blue absorption filter (GG 9/1 mm plus OG 1/1.5 mm), amyloid was colored bright red and other tissue structures were pale greenish gray. Puchtler indicates the above method is especially efficacious in detection of small amounts of amyloid.

According to Herzenberg (*Virchows Arch. Pathol. Anat.*, **253**: 656, 1924), amyloid may be stained intravitally by injection of 0.1 to 1 mg Congo red in 1 ml intravenously in mice.

THE CRYSTAL VIOLET METHOD.   Mallory cites a crystal violet method which we have modified by adding methyl violet 2B to give a redder color. Formalin or alcohol fixation, and frozen or paraffin sections may be used. From water:

1.   Stain sections 3 to 5 min in 1 g crystal violet (C.I. 42555), 0.5 g methyl violet 2B (C.I. 42535), 10 ml alcohol, and 90 ml distilled water.

2. Wash in 1% aqueous acetic acid.
3. Wash thoroughly in tap water.
4. Examine directly in water, or mount in glycerol, glycerol gelatin, fructose syrup or Apáthy's gum syrup.

RESULTS: Nuclei and cytoplasm are varying shades of blue-violet; amyloid and fibrinoid, red-purple. The preparations fade after a time (Mallory, 1938; Conn and Darrow, 1943). Water mounts may be sealed temporarily with petrolatum.

HIGHMAN'S CRYSTAL VIOLET METHOD (*Arch. Pathol.*, **41**: 559, 1946)

1. Stain 5 min in Weigert's acid iron hematoxylin.
2. Wash in water.
3. Stain 1 to 30 min in 0.5 to 0.1% crystal violet or methyl violet in 2.5% acetic acid (we suggest 5 min in 0.2% crystal violet).
4. Wash in water.
5. Mount in Highman's potassium acetate Apáthy gum syrup, glycerol gelatin, or Arlex gelatin.

RESULTS: Amyloid, cartilage, and certain types of mucus are red-purple in a bluish background with blue-black nuclei. The acetic acid in the crystal violet prevents overstaining of cytoplasm, and the salt added to the syrup prevents diffusion, or "bleeding," of the violet. Omission of the iron hematoxylin (step 1) renders the stain less dense.

Lieb (*Am. J. Clin. Pathol.*, **17**: 413, 1947) used about 0.3% crystal violet (10 ml saturated alcoholic solution in 1 ml hydrochloric acid and 300 ml water), stained for 5 min to 24 h, and mounted from water in a solution of 50 g Abopon in 25 ml water.

Fernando varies the crystal violet method by staining for 10 min in a 3% formic acid–1% crystal violet solution, differentiating amyloid to red with 1% formic acid in about 2 to 3 min, and washing and mounting in a dextrin–sucrose–sodium chloride solution of n_D 1.54 and pH 3.75. He prescribes blotting with filter paper after each step (*J. Inst. Sci. Technol.*, no. 2, 1961).

Iodine green is stated by Conn (1940) to color mucin and amyloid red, while it stains nuclei and basophil cytoplasms green. Mallory cited the following technic for frozen sections of fresh or formalin-fixed material: Stain 24 h in 0.33% aqueous iodine green, wash in water, and examine in water or glycerol.

The so-called paramyloid is characterized by inconstancy and irregularity of staining with Congo red and with crystal violet. Commonly colors are displayed between that of typical amyloid and that of serum or plasma, and considerable variability is evident even in the same section. The Van Gieson stain gives a yellow to paramyloid, whereas it stains typical amyloid more orange. Paramyloid bodies often present concentric lamination, like that of corpora amylacea in prostatic acini, and may show central calcification. Its distribution is not that of typical amyloidosis. It occurs in bone, fat, fascia, etc., rather than in the liver, spleen, pancreas, kidney, adrenal, and heart (Bauer and Kuzma, *Am. J. Clin. Pathol.*, **19**: 1097, 1949).

King (*Am. J. Pathol.*, **24**: 1108, 1948) has applied a Del Río-Hortega silver carbonate method for demonstration of both usual and atypical amyloid in frozen sections of formalin-fixed tissues.

1. Cut thin frozen sections; wash in distilled water.
2. Add concentrated (28%) ammonia water drop by drop to 5 ml 10% silver nitrate until the brown precipitate is just dissolved; then add 6 to 8 ml 3.5% sodium carbonate

(Na_2CO_3), and dilute to 75 ml with distilled water. Store surplus solution in a dark, cold place. To 10 ml of the stock diammine silver carbonate solution in a small beaker add a few drops of pyridine. Introduce the sections with a glass rod, and warm gently to about 45°C with continuous gentle agitation until the sections become a rather deep brown ("tobacco brown").

3. Transfer directly with a glass rod to (5%) sodium thiosulfate ($Na_2S_2O_3 \cdot 5H_2O$) solution for 2 to 4 min.
4. Wash in several changes of tap water.
5. Float onto slides, smooth out, and blot down with hard filter paper.
6. Counterstain lightly if desired.
7. Dehydrate and clear by the usual alcohol-xylene sequence, and mount in Canada balsam or an unsaturated synthetic resin.

RESULTS: Amyloid, nuclei, and lipochrome (lipofuscin?) pigment show dark brown to black. Some collagen may impregnate, and cytoplasm is often light brown, as are hyalin casts in the renal tubules.

In that no reduction step is used after silvering, this is an argentaffin method, and the positive reaction indicates the presence of reducing groups in these amyloids. Lillie has suspected for some time that carbonate methods may be more sensitive and less specific than silver oxide technics.

The periodic acid Schiff procedure colors amyloid light red, and in its allochrome variant the color remains predominantly red, though with a bluish cast. The pinacyanole technic of Humason and Lushbaugh gives carmine red amyloid.

Vassar and Culling (*Arch. Pathol.*, **68**: 487, 1959) stain tissues for 3 min in 1% aqueous thioflavine TCN (C.I. 49005) and differentiate for 10 min in 1% acetic acid to reduce background fluorescence. Wash and mount in Apáthy's medium. Nuclear fluorochrome staining sometimes occurs; this can be prevented by a 2-min prestain in alum hematoxylin.

Both primary and secondary amyloids fluoresce strongly in near ultraviolet illumination, and Vassar and Culling consider the stain superior to both Congo red and methyl violet.

Puchtler et al. (*J. Histochem. Cytochem.*, **12**: 900, 1964) noted satisfactory staining of amyloid by Sirius red 4B (C.I. 28160) could be accomplished without a differentiation step. Tissues may be fixed in buffered 10% formalin, absolute ethanol, Carnoy's fluid, Zenker formol, or Kaiserling's solution.

1. Bring sections to water.
2. Stain in 1 or 2% Sirius red 4B in 1% NaCl for 1 h at 60°C.
3. Wash in distilled water or 0.2 M borate buffer of pH 9.
4. Counterstain with alum hematoxylin.

RESULTS: Amyloid is stained selectively without differentiation. Collagen, reticulum fibers, basement membranes, muscle, and cytoplasms remain unstained; elastic fibers stain weakly to moderately. Nuclei are faintly stained, but fail to stain after deamination. Erythrocytes stain strongly.

Puchtler et al. list the following dyes as also strongly selective for amyloid: C.I. 29065, chlorantine fast red 6BLL, direct red 79; C.I. 27925, Sirius light blue F3R, direct blue 67; C.I. 35780, Sirius red F3B, direct red 80; C.I. 40270, Sirius light scarlet 2G, direct red 76; and the sulfonated copper phthalocyanin dye C.I. 74180, Sirius light turquoise blue GL, direct blue 86, which last should be interesting for permanence and electron density (heliogen blue SBL).

Wolman (*Lab. Invest.*, **25:** 104, 1971) prescribes the following stain for amyloid utilizing toluidine blue:

1. Bring sections to water.
2. Stain sections in 1% toluidine blue O in 50% isopropanol.
3. Blot sections carefully with filter paper.
4. Immerse sections in absolute isopropanol 1 min.
5. Blot again.
6. Clear in two changes of xylene and mount in Canada balsam.

Wolman believes the results compare favorably with other methods for the identification of amyloid.

Cooper (*J. Clin. Pathol.*, **22:** 410, 1969) evaluated several stains for the histochemical identification of amyloid. His studies suggest that amyloids can be reasonably expected to stain for tryptophan and to exhibit green birefringence after Congo red and/or Sirius red 4B (C.I. 28160) staining.

Braunstein and Buerger (*Am. J. Pathol.*, **35:** 791, 1959), Mowry and Megginson (*Am. J. Pathol.*, **43:** 38a, 1963), and Mowry and Scott (*Histochemie*, **10:** 8, 1967) are convinced that amyloid is not stained distinctly with crystal violet or methyl violet metachromasia. Hyalin may be particularly confusing at times. Why amyloid does not usually manifest metachromasia with the usual stains employed, e.g., toluidine blue or azure A, is puzzling. Mowry and Scott (ibid.) also studied the basophilia of amyloid.

Amyloid is stained yellow with the Van Gieson method (Dahlin et al., *Med. Clin. North Am.*, **34:** 1171, 1950; Lillie, *Stain Technol.*, **26:** 123, 1951) and red with the periodic acid Schiff method (Lillie, 1954, p. 127; Windrum and Kramer, *Arch. Pathol.*, **63:** 373, 1957).

Lendrum et al. (*J. Clin. Pathol.*, **25:** 373, 1972) on the basis of staining reactions concluded changes occur in amyloid deposits with age.

THE LANGHANS IODINE METHOD. For amyloid bodies the Langhans iodine method for glycogen may be used. Schmorl recommends a variant of Siegert's (no reference) in which Müller or alcohol fixation is recommended:

1. Wash sections well in water.
2. Stain deep brown with strong iodine–potassium iodide solution.
3. Decolorize with strong alcohol.
4. Immerse in 20% hydrochloric acid (2.4 N?) until the amyloid bodies reappear as darkly colored points.
5. Wash out acid quickly with water.
6. Dehydrate in tincture of iodine diluted with 4 volumes alcohol (1.4% I_2).
7. Mount in origanum oil.

RESULTS: Amyloid bodies in prostate, brain, lung, and urinary tract are stained deep brown; other tissues are colorless.

Toluidine blue O stains the amyloid bodies of the central nervous system metachromatically reddish purple. They are red to purple with the periodic acid and chromic acid leucofuchsin methods and resist diastase digestion.

The periodic acid Schiff method colors prostatic secretion pink to purplish red, and amyloid bodies appear in deeper purplish red. Diastase digestion does not alter these reactions. When prostate sections are incubated about $3\frac{1}{2}$ h in methenamine silver by the Gomori-Burtner technic, to the point where stroma reticulum and nuclei begin to blacken, prostatic secretion still stains pink by the safranin counterstain, and the amyloid bodies are peripherally or (more often) completely blackened.

Congo red colors prostatic amyloid bodies purplish to orange-red and the secretion gray-violet to pink. The crystal violet amyloid technic colors them red-purple, while the secretion remains violet. Lillie et al. (*Am. J. Clin. Pathol.*, **65**: 876, 1965) report dark purple staining of nervous system amyloid bodies by the phenocyanin Fe^{2+} method (C.I. 51145).

HYALIN

Mallory's "alcoholic hyalin" is said to give reactions for phosphates and to stain by Mallory's hematoxylin method for lead. After Zenker fixation it stains intensely red with Mallory's phloxine methylene blue method and blue with his phosphotungstic acid hematoxylin. With alcohol or formalin fixation it is stained deep blue by Mallory's (regressive sequence) iron chloride hematoxylin method.

MALLORY'S PHLOXINE METHOD. When overstained with phloxine and then decolorized with lithium carbonate, hyalin remains red: Fix in alcohol or formalin; section in paraffin or celloidin.

1. Stain nuclei in Mallory's or other alum hematoxylin.
2. Wash in water.
3. Stain 20 to 60 min in 0.5% phloxine B (C.I. 45410) in 20% alcohol.
4. Wash in tap water.
5. Decolorize 30 to 60 s in 0.1% aqueous lithium carbonate.
6. Wash in tap water.
7. Dehydrate and clear through an alcohol-xylene sequence and mount in balsam.

RESULTS: Nuclei appear blue; fresh hyalin appears as red droplets and threads; older hyalin, pink to colorless. Hyalin droplets in renal epithelium are also well shown, and hyalin casts color purplish red more than pink. Amyloid stains pale pink.

MALLORY'S THIONIN METHOD. Similar material is used.

1. Stain 5 to 10 min in 0.5% thionin (C.I. 52000) in 20% alcohol.
2. Differentiate several minutes in 80% alcohol.
3. Differentiate in 95% alcohol.
4. Clear in terpineol and mount in terpineol balsam.

RESULTS: Granules and networks of red to purple material appear; old hyalin and nuclei stain blue.

Laqueur (*Am. J. Clin. Pathol.*, **20**: 680, 1950) recommends a combination of Altmann's aniline acid fuchsin with a Masson trichrome procedure for formalin- or Zenker-fixed material.

1. Hydrate deparaffinized sections as usual, using the iodine thiosulfate sequence of Zenker-fixed tissue.
2. Stain 5 min in alum hematoxylin.
3. Wash in water.
4. Flood with 20% acid fuchsin in aniline water (shake 2 ml aniline thoroughly with 100 ml distilled water; filter) and heat with a small flame to fuming. Let stand 5 min.
5. Wash in water.

6. Differentiate in a mixture of 7 parts 20% alcohol and 1 part saturated alcoholic picric acid (about 1% picric acid in 30% alcohol) until only hyalin and red corpuscles remain red, and collagen is faint gray or unstained.
7. Wash thoroughly in water.
8. Mordant 4 to 18 h in 1% phosphomolybdic acid solution.
9. Transfer directly to 1% light green SF in 1% acetic acid for 1 h.
10. Wash in water.
11. Differentiate in 80% alcohol until collagen fibrils appear discrete. This occurs quickly, and so care is needed.
12. Clear through 95 and 100% alcohol, xylene; mount in synthetic resin.

RESULTS: Mallory's alcoholic hyalin is brilliant red; cytoplasm, pale brown; bile pigment, green; collagen, green; hemosiderin and hemofuscin, unstained or yellowish brown in their natural color.

Renal hyalin droplets and casts appear red-purple with the periodic acid–leucofuchsin technic. Casts also react to the Casella and Bauer methods, though less strongly. With the allochrome procedure the droplets and casts assume an orange color similar to that given by intravascular fibrin and that of pneumonic exudates. It contrasts well with the gray-green or gray-yellow of renal epithelial cytoplasm. Both droplets and casts often color red with Mallory's phloxine technic and may color violet with Weigert's fibrin method. They are eosinophilic with azure-eosin methods.

The hyalin of intimal degeneration of small renal arteries gives a positive Sudan IV and a Liebermann-Burchard cholesterol reaction in frozen sections. It is often periodic acid Schiff–positive (R. D. Baker, *Am. J. Pathol.*, **27**: 680, 1951).

Bavia (*Lab. Invest.*, **13**: 301, 1964), Flax and Tisdale (*Am. J. Pathol.*, **44**: 441, 1964), and Yokov et al. (*Am. J. Pathol.*, **69**: 25, 1972) employed electron microscopy to demonstrate a fibrillar appearance of alcoholic hyalin.

Wiggers et al. (*Lab. Invest.*, **29**: 652, 1973) conducted an electron microscopic investigation of Mallory's "alcoholic hyalin." After separation from cellular constituents by differential centrifugation, Mallory bodies were noted to be composed of branching tubular filaments measuring 5.5 ± 0.6 mμ in diameter *in situ* and 7.4 ± 0.2 mμ in diameter after separation.

MAST CELLS

The histochemistry of these cells depends on their content of heparin and their sulfation status, on their histamine content, and on their 5-hydroxytryptamine content as well as on other factors which determine the granule solubility in water. They were identified by Ehrlich as mast cells because of a concept that they were well fed (the word *mast* means nuts, such as beechnuts used as feed for animals; cf. the German *Mast*, food). He stained them with "Dahlia" in 1 to 2 N acetic acid. This dye is now thought to be Hofmann's violet (C.I. 42530).

Earlier workers considered alcohol fixation necessary for preservation of the granules, notably in rabbits, guinea pigs, hogs, and cats.

Holmgren (*Z. Wiss. Mikrosk.*, **55**: 419, 1938) showed that in a considerable number of species the granules were well preserved by basic lead acetate fixation though in many not by formol fixation. Neutral lead acetate, lead nitrate, and their mixtures with formol have served us well in hog and guinea pig. Mercuric chloride without formol also preserves guinea pig mast cells, provided that sections are not treated with iodine before toluidine blue staining (Donaldson and Lillie, *Stain Technol.*, **48**: 47, 1973). Rat, mouse, hamster, and Mongolian gerbil mast cells are well preserved by formol fixation, and they are often seen in formol-fixed human tissues though it is not yet established that fixation in human beings could not be improved by use of a heavy metal.

In accord with the presence of the sulfated polysaccharide heparin, mast cells stain well with thiazin dyes down to pH 0.5 to 1. Limits are not established for all species, and it might be thought that trisulfated heparin could be basophil at a lower pH than monosulfated. Thionin, toluidine blue, azure A, and safranin O have been much used for this purpose. Nile blue in the sulfuric Nile blue technic for fatty acids has been observed to stain human mast cells intensely blue, but in preparations mounted in Apáthy's gum syrup as usual for the lipofuscin stain the mast cells fade in a few days, just as Ranvier (1875) observed for nuclei when he stained fats with quinoline blue and mounted them in glycerol.

Mild methylation prevents the metachromatic staining of mast cell granules, and KOH–alcohol saponification does not restore it (Spicer and Lillie, *J. Histochem. Cytochem.*, 7: 123, 1959). It is usually stated that mast cell granules are not acid-fast. However, after staining with 0.1% safranin for 20 min at pH 1, rat mast cells are found to resist aqueous 0.1 N HCl decolorization for some hours but are decolorized promptly by HCl–alcohol (Lillie et al., *J. Histochem. Cytochem.*, 21: 161, 1973). The granules are gram-negative, Gram-Weigert–negative, and iron-free by the ferrocyanide and ferricyanide methods. The periodic acid Schiff reaction has been found negative in rats, variable in mice, and positive in human beings (*Anat. Rec.*, 108: 239, 1950), and a direct Schiff reaction has been observed in the rat (Pizzolato, unpublished). The argentaffin reactions with diammine and methenamine (Gomori-Burtner, Chap. 8) silver reactions as well as the ferric ferricyanide reaction have been observed to be positive in rats and mice, negative in human beings. These reactions need further study in other species to make correlations with 5-hydroxytryptamine and histamine content.

The granules do not ordinarily stain with alum or iron hematoxylin or by alum or borax carmine or by mucicarmine or mucihematein. The behavior toward Alcian blue has been somewhat variable; sometimes the stain fails; often excellent results are obtained at low pH levels (0.5 to 1.5); at pH 2 to 3 behavior seems variable and may show species variations (Spicer, *Ann. N.Y. Acad. Sci.*, 103: 322, 1963).

In deamination experiments we have often noted that azure A mast cell staining is suppressed by nitrous acid treatment, chiefly in rat tissue. We have made no systematic study of this phenomenon.

Mast cells contain histamine (Riley, *J. Pathol. Bacteriol.*, 65: 471, 1953; Riley and West, *J. Pathol. Bacteriol.*, 69: 269, 1955; and Benditt et al., *J. Histochem. Cytochem.*, 4: 419, 1956). Hedbom and Snellman (*Exp. Cell Res.*, 9: 148, 1955) employed differential centrifugation methods to obtain fractions of mast cells from ox liver capsules. Histamine was located in the granular fraction.

Tredway et al. (*J. Exp. Med.*, 102: 307, 1955) conducted correlated histochemical and biochemical assays of histamine in mast cells in several organs of dogs and cattle. Histamine amount per mast cell varied from 7 to 32 pg in beef liver capsule.

Histamine has been demonstrated in rat and other mast cells by Lagunoff, Phillips, and Benditt (*J. Histochem. Cytochem.*, 9: 534, 1961) using Falck-Torp freeze-dried, hot formaldehyde gas fixation and *p*-bromodiazobenzene in alkaline solution. *o*-Phthalaldehyde has been shown by Juhlin and Shelley (*J. Histochem. Cytochem.*, 14: 525, 1966) and Häkanson et al. (ibid., 18: 93, 1970) to give a quite specific fluorescence for histamine.

Van-Orden et al. (*Pharm. Exp. Ther.*, 158: 195, 1967) conducted a combined biochemical, histochemical, and electron microscopic study to identify 5-hydroxytryptamine in multivesicular bodies of murine neoplastic mast cells.

Barrnett et al. (*Biochem. J.,* **69:** 36. 1958) identified 5-hydroxytryptamine. histamine. and heparin in mast cell granules isolated from mouse mast cell tumors using a sucrose gradient.

5-Hydroxytryptamine was isolated and identified by chemical means from dog and mouse mastocytomas by Meier (*Proc. Soc. Exp. Biol. Med.,* **100:** 215, 1959). Rice and Mitchener's [*Nature (Lond.),* **189:** 767, 1962] identification of 5-hydroxytryptamine in a dog mastocytoma rested on demonstration of histochemical reactions similar to those of enterochromaffin which we now assign to a catechol and not to serotonin. Benditt et al. (*Proc. Soc. Exp. Biol. Med.,* **90:** 303, 1955) identified the serotonin isolated from a 90% mast cell suspension washed from the rat peritoneal cavity by pharmacologic tests and by chromatographic identity. Its presence in mast cells has been studied by the Falck-Torp freeze-dried, hot formaldehyde gas technic, utilizing its intense yellow fluorescence [Coupland and Riley, *Nature (Lond.),* **157:** 1128, 1960]. Adams-Ray et al. (*Experientia,* **20:** 80, 1964) showed that rat mast cells gave an intense yellow fluorescence by the Falck-Torp technic, hamster and rabbit only a weak green. After parenteral administration of L-dopa (a tenfold dose of D-dopa was required) but not dopamine, rabbit mast cells acquire an intense green (dopamine) fluorescence, dischargeable by reserpine; guinea pig, hamster, and cat mast cells did not respond to dopa administration. Lillie et al. (*J. Histochem. Cytochem.,* **21:** 161, 1973) reported that diazosafranin colored most rat, mouse, dog, and gerbil and some armadillo mast cells red, both at pH 8 and at pH 3; human and dog mast cells only exceptionally reacted, and monkey, cat, and guinea pig mast cells did not react.

Radden (*J. Histochem. Cytochem.,* **9:** 165, 1961) noted the presence of metallic cations in the staining solution delays acquisition of metachromatic staining.

Spicer (*Ann. N.Y. Acad. Sci.,* **103:** 322, 1963) and Combs et al. (*J. Cell Biol.,* **25:** 577, 1965) correlated the embryonic development and maturation of mast cells from loose connective tissues of the rat with Alcian blue–safranin staining, and ^{35}S and tritiated thymidine autoradiography. Figure 14-5 (provided by Combs) depicts mast cells and mast

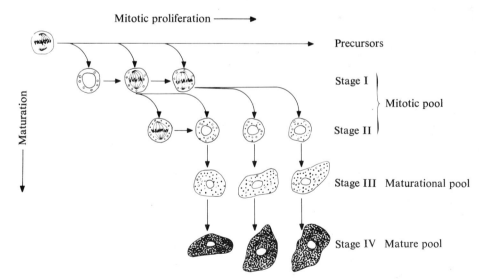

Fig. 14-5 Schematic representation of mast cell proliferation and maturation. Immature granules are represented by circles; mature granules by dots. (*From Combs et al., J. Cell Biol.,* **25:** 577, 1965. *By permission of the publisher.*)

cell granule development in four stages. Also portrayed is the staining of granules observed with an Alcian blue–safranin sequence.

Combs et al. also employed a nitrous acid hydrolysis (*Arch. Biochem. Biophys.*, **99**: 396, 1962) which is believed to be selective in removing sulfamido sulfate but not ester sulfate from acid mucopolysaccharides. In control sections they noted that the degree of safranin staining of cartilage was largely unaffected after the nitrous acid (0.05% $NaNO_2$ plus 0.33% acetic acid, 1 : 1) treatment. Cartilage is known to be *N*-sulfate–poor. Combs et al. propose that the metachromasia of mast cell granules is due to heparin content. Heparin contains both *N*- and ester sulfate. On this basis Combs et al. suggest that mast cell granules stain red in an Alcian blue–safranin sequence to the extent of mature *N*-sulfated heparin content and that blue-stained granules identify a polysaccharide deficient in *N*-sulfate—perhaps a heparin precursor. They also speculate that the same blue-stained constituent may be responsible for reactivity with the periodic acid Schiff staining. Combs (*J. Cell Biol.*, **31**: 563, 1966) correlated electron microscopic with histochemical and autoradiographic observations of mast cell granule development. Table 14-14 is Combs' postulated sequence of events in rat mast cell granule formation developed on the basis of combined histochemical, autoradiographic, and electron microscopic studies.

Pretlow and Cassady (*Am. J. Pathol.*, **61**: 323, 1970) employed gradients of Ficoll (poly-sucrose) in rate-zonal sedimentation studies to separate rat peritoneal mast cells in their four stages of development. Pretlow's study was able to determine that stage I mast cell granules actually comprise two sharply identifiable subgroups.

Parekh and Glick (*J. Biol. Chem.*, **237**: 280, 1962) conducted cell fractionation studies of mast cells from rats and could find evidence for no sulfated acid mucopolysaccharide except heparin, and it was present in the granules. Kelly and Bloom (*Exp. Cell Res.*, **16**: 538, 1959) conducted quantitative spectrophotometric studies of mast cells stained metachromatically. One must be cautious in the identification of mast cells on the basis of the presence of metachromatic granules in cells only. Higginbotham et al. (*Proc. Soc. Exp. Biol. Med.*, **92**: 256, 1956) noted degranulation of mast cells in the skin of mice following mild trauma. Shortly thereafter metachromatic granules were detected in cytoplasm of fibroblasts.

One must also be cautious in the interpretation of staining results after the application of proteins, enzymes, etc. For example, Morris and Krikos (*Proc. Soc. Exp. Biol. Med.*, **97**: 527, 1958) noted that commercial preparations of hyaluronidase contain constituents that may prevent mast cell metachromasia at pH 2 or 3, thereby simulating enzymatic hydrolysis of mucopolysaccharides. The possibility that hyaluronidase itself may be responsible for this action was not excluded.

Spicer (*Ann. N.Y. Acad. Sci.*, **103**: 322, 1963) studied the distribution, prevalence, and basophilia of mast cells in five species. Table 14-15 indicates the results of several staining reactions.

Wegelius and Hjelmman (*Acta Pathol. Microbiol. Scand.*, **36**: 304, 1955) noted that toluidine blue O in 1 : 10,000 to 1 : 100,000 concentrations stains hamster mast cells in vivo. They also noted (*Acta Pathol. Microbiol. Scand.*, **36**: 309, 1955) that administration of anti-hamster serum caused mast cell degranulation. Degranulation of mast cell granules was accomplished by compound 48/80 (a condensation product of *p*-methoxyphenethylmethyl-amine with formaldehyde consisting of a mixture of dimers and trimers) by Fawcett (*Anat. Rec.*, **121**: 29, 1955; *J. Exp. Med.*, **100**: 217, 1954).

The following stain for mast cells is advised by Bloom and Kelly (*Histochemie*, **2**: 48, 1960). As a fixative they used formaldehyde 40%, 10 parts; absolute ethanol, 80 parts; phosphate buffer, 10 parts. The final pH was adjusted to 6.8 to 7.

TABLE 14-14   A POSTULATED SEQUENCE OF EVENTS IN RAT MAST CELL GRANULE FORMATION

Morphologic steps		Hypothetical chemical correlates
Step 1: Progranule formation		Formation of heparin precursor; *O*-sulfation
Step 2: Aggregation of progranules		Accumulation of heparin precursors, forming a nidus for the future mast cell granule
Step 3: Addition of finely granular material and continued aggregation of progranules		Accretion of basic proteins including mast cell chymase
Step 4: Fusion of progranules and continued addition of finely granular material		*N*-sulfation
Step 5: Reorganization to form strands with disappearance of the finely granular and dense components		
Step 6: Compaction		Maximum ionic binding between heparin and basic proteins

Source: Combs, *J. Cell Biol.*, **31**: 563, 1966.

TABLE 14-15 COMPARATIVE STAINING OF MAST CELLS IN FIVE SPECIES WITH TWO METHODS OF FIXATION

| | Fixation | Azure A | | | AF | AB | AB saf. | AB*-PAS | AB-PAS after MeOH | Meta diamine | BS pH 9.5 |
		pH 0.5	pH 1.5	pH 4.0							
Mouse	Formalin	4P	4P	4P	4P	0-B	4R†	0-B	1-3B	4P	0
Mouse	Carnoy	1-3P	0-P	3-RP‡	0-P	1-3B	4R-B	1-3B	4B	4P	4 OrR
Rat	Formalin	4P	4P	4P	4P	0-B	4R†	0-2PB	1-3B	4P	0
Rat	Carnoy	1-4P‡	0-P	3RP	0-P	3B	4B	3B	4B	4P	4 OrR
Hamster	Formalin	4P	4P	4P	4P	0-B	4R	0-B	2-3B	4P	0
Hamster	Carnoy	2-3P	0-P	2-3RP	0-P	1-3B	0-3R	1-3B	4B	4P	4 OrR
G. pig	Formalin	0	0-P	0-2P	0-3P	0-2B	0-2B	0-B	1-2B	0-3P	0
G. pig	Carnoy	2-3P	0	1-3RP	0-P	2-4B	2-4B	2-4B	2-4B	4P	4 OrR
Rabbit	Formalin	0	0	0-B§	0	0	0	0	0	0	0
Rabbit	Carnoy	1-2P	0	3P	0	3-4B	2-4B	3B	3-4B	0-P	3 OrR¶

* In sections stained with the PAS method alone weak reactivity was observed in some of the rat and guinea pig mast cells, none in those of the other species.
† Except for numerous blue cells in cervical lymph nodes of mouse and rat and uterus of mouse.
‡ Redder and smaller in cervical lymph nodes.
§ Identification uncertain.
¶ Difficult to distinguish but stained in duodenum at least.

Note: AB = Alcian blue; AF = aldehyde fuchsin; BS = Biebrich scarlet; PAS = periodic acid Schiff; saf. = safranin; MeOH = methylation 1 h at 60°C in acid methanol; B = blue; Or = orange; P = purple; R = red. Numbers indicate relative intensity of color of mast cells from 0 = unstained to 4 = maximal staining.

Source: Spicer, *Ann. N.Y. Acad. Sci.*, **103**, 322, 1963.

1. Bring paraffin-embedded sections to water.
2. Stain for 10 min in 0.1% Astrablau adjusted with concentrated HCl to pH 0.2 to 0.3 (0.6 to 0.7 N HCl).
3. Rinse in 0.6 to 0.7 N HCl for 10 min. This eliminates nonspecific staining.
4. Dehydrate and mount in synthetic resin.

Juraj and Weiskopf (*Arch. Pathol.*, **82**: 430, 1966) describe the following fluorescent method for staining mast cells in formalin-, ethanol-, or Zenker-fixed, paraffin-embedded tissues.

1. Bring sections to water.
2. Stain in Weigert's hematoxylin for 5 min.
3. Wash in tap water for 3 to 5 min.
4. Stain in acridine orange 1 : 1000 aqueous solution for 5 to 6 min.
5. Rinse in tap water for 1 min.
6. Dehydrate, and mount in synthetic medium.

Gustafson and Pihl [*Nature (Lond.)*, **216**: 697, 1967] claim mast cell granules react selectively with ruthenium red, thereby permitting observation by electron microscopy. Pihl (*Histochemie*, **22**: 302, 1970) believes that ruthenium red binds stoichiometrically to heparin.

Shea (*J. Cell Biol.*, **51**: 611, 1971) advises the use of lanthanum in combination with either Alcian Blue 8GX or cetylpyridinium chloride for excellent demonstration of muco-polysaccharide complexes at cell surfaces by electron microscopy. Shea advises fixation of blocks of tissue for 2 h at 24°C in a formaldehyde-glutaraldehyde mixture as advocated by Karnovsky (*J. Cell Biol.*, **27**: 137A, 1965). The solution is as follows: 2 g paraformaldehyde powder is dissolved in 25 ml water by heating to 60 to 70°C and stirring. One to three drops of 1 N NaOH are added with stirring until the solution clears. A slight milkiness may persist. The solution is cooled, 5 ml 50% glutaraldehyde (Biological Grade, Union Carbide Company) is added, and the volume is made to 50 ml with 0.2 M cacodylate or phosphate buffer, pH 7.4 to 7.6. The final pH is 7.2. If cacodylate is used, 25 mg anhydrous $CaCl_2$ is added. The fixative is diluted 1 : 1 with 0.1 M cacodylate buffer at pH 7.2, or in a 3% glutaraldehyde solution in 0.1 M cacodylate buffer at pH 7.2. Alcian blue to make 0.1 or 0.5%, or cetylpyridium chloride to make 0.1%, is added. After a brief rinse in 0.1 M cacodylate buffer of pH 7.2, tissues are postfixed for 2 h at 24°C in osmium tetroxide–lanthanum nitrate (1% of each) in 0.1 M *sym*-collidine buffered to pH 8. Tissues are dehydrated in ethanol and propylene oxide and embedded in Epon 812. Sections are cut with glass knives and examined with an electron microscope.

Glenner and Cohen [*Nature (Lond.)*, **185**: 846, 1960] demonstrated the existence of a trypsin-like enzyme in mast cells of dog and man. Sodium benzoyl-DL-arginine-β-naphthyl-amide was employed as a substrate.

Benditt et al. (*J. Exp. Med.*, **110**: 451, 1959) and Lagunoff and Benditt [*Nature (Lond.)*, **192**: 1198, 1962] demonstrate the existence in mast cells of several species of an enzyme resembling α-chymotrypsin. The substrate employed was 3-chloroacetyl-2-naphthoic acid anilide. The enzyme was also demonstrated by Perera and Mongar (*Immunology*, **6**: 472 and 478, 1963).

Lagunoff et al. (*Exp. Cell Res.*, **61**: 129, 1970) demonstrate N-acetyl-β-glucosaminidase in mast cell granules of rats.

Hall (*Arch. Oral Biol.*, **11**: 1325, 1966) believes that the most reliable method for the demonstration of all mast cells is the esterase method of Lagunoff, Benditt, and Watts (*J. Histochem. Cytochem.*, **10**: 672, 1962). Further studies are required to demonstrate conclusively that all reactive cells are, in fact, mast cells.

Glick (*Exp. Cell Res.*, **65**: 23, 1971) used centrifugal elutriation to separate mast cells from peritoneal washings of rats.

Nalavade and Varute [*Acta Histochem. (Jena)*, **47**: 70, 1972] conducted a comparative histochemical study of mast cells in tongues of 20 vertebrates.

15

CONNECTIVE TISSUE FIBERS AND MEMBRANES

COLLAGEN

Collagen, the most prevalent protein in the body and the major constituent of most connective tissues, is a white fibrous intercellular substance that swells in dilute acids and becomes gelatin when denatured. It is a protein composed of three polypeptide (alpha) chains wound in a tight coil. Each alpha chain is in a left-handed helical configuration, and the three alpha-chain helices wind around each other to form a right-handed superhelix. The two most characteristic features of the wide-angle diffraction pattern of native collagen are a principal axial repeat of 2.9 Å and an equatorial periodicity of 11 Å which varies somewhat with the degree of hydration of the specimen. The wide-angle diffraction pattern is believed to derive from the triple helical structure made possible by the presence of glycine in every third position along the chains. The 2.9-Å x-ray reflection represents the distance of one amino acid residue along the helical axis direction and is a measure of the inclination of that residue with respect to the helical axis. The 11-Å line is related to the distance between two laterally adjacent molecules.

The amino acid composition of collagen and elastin is given in Table 15-1. Collagen is a protein characterized by a high content of glycine (one-third of the total amino acid residues), a high content of proline and hydroxyproline, and an absence of tryptophan and cystine.

Molecules of collagen are approximately 3000 Å in length and 14 Å in diameter and have a molecular weight of approximately 300,000. Normally, they align themselves into fibrils with an axial periodicity of approximately 680 Å (in the wet state). The peculiar 680-Å periodicity is a consequence of an actual gap of approximately 430 Å plus an axial displacement of individually aligned collagen molecules. Characteristic collagen fibrils will form spontaneously from a buffered solution of collagen warmed to 37°C. Collagen molecules in acid solution precipitated with adenosine triphosphate (ATP) will align themselves without an axial stagger (Gross et al., *Proc. Natl. Acad. Sci. U.S.A.*, **40**: 679, 1954). Structures formed by ends of each molecule aligned in precisely the same position relative to each other are called *segment-long-spacing* (SLS) fibrils as observed by electron microscopy. This is a very useful form inasmuch as the preparation can be used to determine the effects of enzymatic cleavage, etc.

Bouteille and Pease (*J. Ultrastruct. Res.*, **35**: 314, 1971) conducted an electron microscopic study of the tridimensional structure of collagen filaments and fibrils. In specimens of rat aorta adventitia "inertly dehydrated," embedded in hydroxypropyl methacrylate, sectioned, and stained with uranyl acetate or lead citrate, they observed collagen filaments

679

TABLE 15-1 AMINO ACID COMPOSITION OF ELASTIN, COLLAGEN, AND BASEMENT MEMBRANES

	Ox ligamentum nuchae*	Day-old chick aorta†	Year-old chick aorta†	Human skin†	Rat skin‡	Dog Descemet's membrane§	Human anterior lens capsule§	Human glomeruli§
3- and 4-hydroxyproline	7.1	22	23	94	92.9	84.0	106.3	65.0
Aspartic acid	7.0	2	2	45	46.0	58.8	56.9	70.0
Threonine	10.1	4	5	18	19.6	36.0	31.1	40.3
Serine	9.9	3	4	36	43.0	42.0	43.4	54.2
Glutamic acid	17.4	12	12	73	71.0	94.8	94.4	101.3
Proline	125.4	124	124	128	121.0	98.6	67.3	64.1
Glycine	316.2	351	352	330	331.0	221.0	260.0	225.2
Alanine	213.3	180	177	110	106.0	53.4	40.6	58.6
Valine	134.0	177	174	24	24.0	45.2	33.2	36.0
Methionine	....	0	0	6	7.8	8.0	5.0	7.0
Isoleucine	26.6	19	20	10	10.8	29.7	32.1	28.6
Leucine	64.7	56	58	24	23.8	75.0	57.6	60.3
Tyrosine	6.1	12	12	3	2.4	21.2	13.1	20.5
Phenylalanine	33.6	22	22	12	11.3	25.6	29.7	28.3
Isodesmosine	1.1⎫							
Desmosine	1.7⎭	7	11					
Hydroxylysine		0	0	6	5.7	21.0	34.5	24.5
Lysine	3.6	4	2	27	28.1	25.0	19.4	26.4
Histidine	0.5	0	0	5	4.9	11.0	15.2	18.7
Lysinonorleucine	1.8							
Half cystine	0¶	0	0	0		11.0	21.0	22.0
Tryptophan	0¶							
Arginine	6.6	6	5	51	51.0	39.0	39.5	48.3
Amide N	....	....	....	....	(41)			

* Franzblau et al, *Nature (Lond.),* **205:** 802, 1965.
† White, Handler, Smith, "Principles of Biochemistry," 4th ed., p. 872, McGraw-Hill, New York, 1968.
‡ Piez et al., *Biochemistry,* **2:** 58, 1963.
§ Kefalides, *Arthritis Rheum.,* **12:** 627, 1969.
¶ Block and Weiss, "Amino Acid Handbook," p. 274, Charles C Thomas, Springfield, Ill., 1956.

30 to 35 Å in diameter which were multiples of tropocollagen molecules 10 to 15 Å in diameter. A helical pattern within fibrils was observed with a gyrus length slightly exceeding 1 μ. A matrix between filaments was observed.

Except for a short, uncoiled portion of the collagen molecule at the NH_2-terminal end, undenatured collagen is not enzymatically digested by nonspecific proteases. Specific collagenases have been detected from gingivae [Fullmer and Gibson, *Nature (Lond.)*, **209**: 728, 1966; Fullmer et al., *J. Dent. Res.*, **48**: 646, 1969], polymorphonuclear leukocytes (Lazarus et al., *J. Clin. Invest.*, **12**: 2622, 1968; Robertson et al., *Science*, **177**: 64, 1972), synovial tissues (Lazarus et al., *N. Engl. J. Med.*, **279**: 914, 1968; Evanson et al., *Science.*, **158**: 499, 1967), skin (Eisen et al., *Biochim. Biophys. Acta*, **151**: 647, 1968; Lazarus and Fullmer, *J. Invest. Dermatol.*, **52**: 545, 1969), and bone (Fullmer and Lazarus, *J. Histochem. Cytochem.*, **17**: 793, 1969). All the above collagenases cleave the collagen molecule once approximately three-fourths of the distance from the NH_2-terminal end of the molecule. The two pieces thereafter behave differently biologically from the original total molecule and become susceptible to other protease actions. Thus collagen degradation is controlled by specific enzyme activity.

RETICULUM AND BASEMENT MEMBRANES

His ("Die Haute und Holen des Korpers," Schweighauser, Basel, 1865) pointed out a membrane between epithelial and connective tissues of skin. Kupfer (*Arch. Mikrosk. Anat.*, **12**: 353, 1876) and Mall (*Abhandl. Math. Phys. Kl. Königl Sächs. Ges. Wiss.*, **17**: 295, 1891; *Johns Hopkins Med. J.*, **1**: 171, 1896) used enzymes, acids, and alkalis to separate collagenous from reticular fibers.

The nature of reticulin has been in dispute for several reasons: Reticulin is generally identified by silver impregnation staining methods which are capricious. Wide variations in staining patterns occur after apparently slight modifications in staining procedures. The staining reactions are nonspecific and depend upon unknown, apparently complex physical and chemical factors. Furthermore, data are consistent with the view that not all reticulins have an identical composition, whereupon differences in staining reaction may result. Table 15-1 provides the amino acid composition of elastin, collagens, and certain basement membranes.

Argyrophilic staining reactions selectively distinguish precollagenous reticulins, stromal reticulins, and basement membranes.

PREFIBROUS COLLAGEN RETICULIN. Newly formed collagenous fibers blacken with silver impregnation methods and are called *reticulin*. Bundles of collagen later become refractory to silver stains. Jackson and Williams [*Nature (Lond.)*, **178**: 915, 1956] and Robb-Smith [in G. Stainsby (ed.), "Recent Advances in Gelatin and Glue Research," p. 38, Pergamon, New York, 1958] noted a loss of argyrophilia of this type of reticulin after 0.2 *M* NaCl extraction with simultaneous acquisition of stainability with the Van Gieson method for collagen. These reticular fibers are solubilized in citrate buffer of pH 3.6. Stromal and basement membrane reticulins are not solubilized in citrate buffer of pH 3.6.

STROMAL RETICULINS. Typical reticular fibers of this type are those of lymph nodes and spleen. They stain red with the periodic acid Schiff method and variably blue to red in allochromatically stained sections. They are metachromatic only after sulfation, and they fail to stain with aldehyde fuchsin after peracetic acid or Oxone oxidation. They are more resistant to peptic digestion than are basement membranes (Lillie, *Lab. Invest.*, **1**: 30, 1952).

BASEMENT MEMBRANES. These may derive embryologically from mesodermal or epidermal tissues. In the adult stage they may be associated with epithelial cells, endothelial cells, or both.

On the basis of amino acid and carbohydrate composition, various basement membranes are similar qualitatively, but differ quantitatively. All basement membranes examined to date (Kefalides, *Arthritis Rheum.*, **12:** 427, 1969) have at least two carbohydrate-rich proteins, one being collagen and the other a noncollagen glycoprotein. Basement membranes of epithelial origin such as the lens capsule contain a greater proportion of collagen than do those of endothelial origin such as Descemet's membrane. Collagen from Descemet's membrane differs from other mammalian collagens in that it contains larger amounts of hydroxyproline, hydroxylysine, and hexose. It has significant amounts of half-cystine.

Acid-insoluble and acid-soluble fractions of basement membranes are obtained after limited pronase digestion. The acid-insoluble fraction represents collagen remaining in association with noncollagen peptides. Immunochemical studies reveal the antigenic determinants of basement membranes reside in peptides with an amino acid composition unlike collagen. Antigenic determinants are shared by both epithelial and mesenchymally derived basement membranes.

Typical fibers of the basement membrane type are those surrounding kidney tubular epithelial cells and between dermis and epidermis. These reticular fibers stain purplish red with the periodic acid Schiff method and retain the red through the allochrome procedure. They stain purple with the peracetic acid (or Oxone)–aldehyde fuchsin–Halmi method and become metachromatic with the sulfation-metachromasia procedure of Kramer and Windrum (*J. Histochem. Cytochem.*, **2:** 196, 1954). They stain moderately with the Barrnett-Seligman (*J. Biophys. Biochem. Cytol.*, **4:** 169, 1958) method for α-acylamido groups. They are digested readily by pepsin and less readily by trypsin.

Romhánya et al. (*Histochemie*, **36:** 123, 1973) noted the induction of strong basophilia of basement membranes oxidized with periodic acid and then sulfated. They attribute the basophilia to selective sulfation of OH groups of collagen. Further work is needed to confirm this suggestion. See also sulfation in Chap. 8.

Although the methods for collagen, reticulum, basement membranes, and elastin are different in some respects, it is profitable to discuss them together, since it is often desirable to stain two types of fibers simultaneously.

Four basic processes exist for the selective demonstration of basement membranes and collagen and reticulum fibers: the silver impregnation from ammoniacal solutions, the staining with acid aniline dyes from strongly acid solution, the phosphotungstic and phosphomolybdic acid hematoxylin methods, and the periodic acid–leucofuchsin method.

THE SILVER METHODS

The variations of the Bielschowsky-Maresch diammine silver technic for reticulum impregnation are far more numerous than the few samples quoted here. Maresch apparently got quite acceptable preparations in his original study, but since the process depends on a silver impregnation usually done in diffuse daylight at varying temperatures, variations in results are to be expected. Various pretreatments have been introduced to attempt to control the variability of these results. Each has its devotees, and the tendency has been to add more steps and make the process increasingly complex.

The use of preoxidants which oxidize tissue elements tending to give spontaneous argentaffin reactions probably does serve a useful purpose. But temperature control has generally been very crude, and usually little attention is paid to the amount and quality of light present during the silvering process. We have noted a material difference in optimal impregnation time when the working area was moved to another, less well lighted room.

It is questionable whether the various methods of preparing diammine silver oxide have any influence, provided the same final point of approximate saturation is reached. Carbonate technics do appear to be slightly different, and a few attempts to use diammine silver chloride which we made at one time never reached a measurable degree of success.

The most important factor appears to be experience in appraising the proper grade of section color change during impregnation. This experience is often acquired painfully.

It seems dubious that any considerable amount of reducible silver remains in the tissue by the time the final thiosulfate step is reached, but this step is traditional, and no one omits it.

The silver methods purport to differentiate between reticulum and collagen fibers. The former are colored black; the latter brown, lavender, and gray in varying shades and tones. These methods are often uncertain in their action. One learns by experience to discern whether or not a given preparation has afforded a reasonably complete demonstration.

On account of the high alkalinity of the silver solutions, paraffin sections are often partially or completely loosened from the slides. To combat this tendency Masson used a gelatin glue which we have found no more uniformly successful in keeping sections on than was Mayer's albumen glycerol. Mallory, however, recommended it highly.

Masson (*J. Tech. Methods.*, **12**: 75, 1929) dissolved 50 mg gelatin in 25 ml distilled water and floated out sections on a large drop of this placed on the slide on a warm plate. When the sections were smooth, the excess gelatin water was drained off, and the section was blotted dry with filter paper and placed in a large, closed, moist chamber with formaldehyde vapor at 40 to 50°C for several hours to overnight. Shorter periods are said to be adequate for other than alkaline silver methods.

With the Foot silver carbonate method Mallory recommended alcohol as the final diluent of the silver solution in place of water, to prevent sections from floating off the slides in the alkaline silver solution.

We prefer to use the routine paraffin sections which are deparaffinized with xylene and transferred to 100% alcohol. They are then soaked 5 to 10 min in 1% collodion. The collodion is then drained for 1 to 2 min, and the preparations are next immersed in 80% alcohol for 5 min or more and then transferred to water.

The Bielschowsky-Maresch method, the Perdrau–Da Fano, one Foot variant, the Wilder method, the Gomori variant, and a method of ours depend on silver oxide (or hydroxide) dissolved in ammoniacal solution. The Del Río-Hortega method and Foot's and Laidlaw's variants use ammoniacal solutions of silver carbonate. Except Foot's two technics, which specify Zenker material but work well after formalin fixation, and Laidlaw's, which specifies Bouin's fluid as an alternative method, all methods specify formalin fixation. The details of manufacture of the ammoniacal silver solution vary somewhat, but in all except our oxide method one produces a precipitate from silver nitrate with sodium or potassium hydroxide, with ammonium hydroxide and then sodium hydroxide, or with lithium or sodium carbonate. One then barely dissolves the precipitate with ammonium hydroxide, except in Gomori's, in which one back-titrates with more silver nitrate. In all, the re-solution in ammonium hydroxide is done in quite concentrated solution; and, except in Laidlaw's method, the solution is considerably diluted for use.

In the following formulas solutions are expressed in grams per 100 ml solution, except in the Bielschowsky-Maresch and Del Río-Hortega formulas, where it is presumed that the continental European custom of expressing solutions in percentage by weight was followed. Thus the 10% silver nitrate solution with a specific gravity of 1.088 contained 10.88 g/100 ml in Bielschowsky's laboratory, while according to American custom 100 ml 10% silver nitrate solution contains 10 g.

MARESCH'S BIELSCHOWSKY SILVER SOLUTION (*Zentralbl. Allg. Pathol.*, **16:** 641, 1905). Add 5 drops of 40% sodium hydroxide (4 g NaOH, 6 g H_2O) (this would be about 0.3 ml and would contain about 172 mg NaOH, the equivalent of 731 mg $AgNO_3$) to 10 ml 10% (10 g $AgNO_3$, 90 g H_2O) silver nitrate solution (1088 mg $AgNO_3$). The brown precipitate is then redissolved by constant shaking while 28% ammonia water is being added drop by drop (perhaps 1 ml). It is best to leave the last few granules undissolved, as an excess of ammonia inhibits the impregnation. Dilute to 25 ml with distilled water for use. Use once.

GRIDLEY AND COUCEIRO AND FREIRE. Gridley (*Am. J. Clin. Pathol.*, **21:** 207, 1951) and Couceiro and Freire (*Acad. Brasil. Cien.*, **24:** 11, 1952) use the Maresch solution diluted to 60 ml instead of 25 ml.

PERDRAU–DA FANO SILVER SOLUTION Perdrau (*J. Pathol. Bacteriol.*, **24:** 117, 1921) used Da Fano's silver solution, which we have cited from Bailey and Hiller (*J. Nerv. Ment. Dis.*, **59:** 337, 1924). Add 2 drops of 40% sodium hydroxide (40 g NaOH in 100 ml solution) to 5 ml 20% (20 g in 100 ml solution) silver nitrate and 28% ammonia drop by drop (about 1.1 ml) with constant shaking until precipitate is almost or just dissolved. Dilute to 50 ml. Use once.

FOOT'S SILVER OXIDE SOLUTION (*J. Lab. Clin. Med.*, **9:** 777, 1924). Add 20 drops of 40% sodium hydroxide to 20 ml 10% silver nitrate. Dissolve the brown precipitate by adding strong (28%) ammonia water drop by drop with constant shaking until only a few granules remain. About 2 ml (theoretically 1.7 ml) is required. Dilute to 80 ml with distilled water. Use once. Assuming 17 drops per ml for the sodium hydroxide solution, an exact equivalent of the silver nitrate is used in this variant.

WILDER'S SILVER OXIDE SOLUTION (*Am. J. Pathol.*, **11:** 817, 1935). Add 28% ammonia water drop by drop (about 0.5 to 0.6 ml) to 5 ml 10.2% silver nitrate until the precipitate is dissolved. Then add 5 ml 3.1% sodium hydroxide, and redissolve the precipitate with a few drops of ammonia water. Make up to 50 ml with distilled water.

GOMORI'S SILVER OXIDE SOLUTION (*Am. J. Pathol.*, **13:** 993, 1937). To 20 ml 10% silver nitrate add 4 ml (3.3 to 5 according to Gomori) 10% potassium hydroxide. (This amount precipitates 1 to 1.5 g of the total 2 g silver nitrate as oxide.) Then add 28% ammonia water drop by drop to dissolve the precipitate completely. Then again add 10% silver nitrate drop by drop until the precipitate formed dissolves easily on shaking. Dilute the solution with an equal volume of distilled water. It can be kept stoppered for 2 days.

The end point in the addition of silver nitrate is not clear in the above directions, which are as Gomori originally published them, and as they were copied by Mallory. Gomori wrote Lillie (1945) that he added silver until the precipitate dissolved only on vigorous shaking or even to faint permanent opalescence. In the last case he again added 1 drop of about 2% ammonia water to clear.

LILLIE'S SILVER OXIDE METHOD (*Stain Technol.*, **21:** 69, 1946, modified 1954). Place 1 volume 28% ammonia water in a small flask and add rapidly 14 to 16 volumes 5% silver nitrate solution, shaking the flask constantly. Then cautiously add more silver nitrate, shaking until the brown precipitate completely dissolves between each addition. Continue the addition of silver nitrate until a faint permanent opalescence is produced. This takes a total of 19 to 20 volumes silver nitrate solution. Use the solution by placing 2 ml on each slide, and discard. The solution is good for 1 to 2 days after mixing.

Krajian employed the silver solution produced by adding concentrated ammonia water to 10% silver nitrate solution until the precipitate is almost completely dissolved. This should yield essentially the same mixture as the preceding 10% silver solutions.

THE DEL RÍO-HORTEGA SILVER CARBONATE SOLUTION (cited from Romeis). To 5 ml 10% silver nitrate add 15 ml 5% sodium carbonate. Then carefully add (28%) ammonia water drop by drop (about 0.4 ml) with constant shaking until the precipitate is just dissolved. Then add 55 ml distilled water. The solution may be kept a long time in brown glass.

FOOT-MENARD SILVER CARBONATE SOLUTION (*Arch. Pathol.,* **4:** 211, 1927). To 10 ml 10% silver nitrate add 10 ml saturated (about 1.3%) aqueous lithium carbonate solution. (The solubility of lithium carbonate is less in hot water than in cold; do not heat to saturate.) Wash the precipitate 3 times with distilled water by decantation. Add 25 ml distilled water, and then add 28% ammonia water drop by drop (about 0.8 ml) until the precipitate is almost dissolved, shaking vigorously the while. Make up to 100 ml with distilled water. Mallory modified this by making the final dilution to 100 ml with 95% alcohol instead of water and adding a few more drops of ammonia water to dissolve the resultant precipitate.

LAIDLAW'S CONCENTRATED SILVER CARBONATE SOLUTION (*Am. J. Pathol.,* **5:** 239, 1929). Dissolve 12 g silver nitrate in 20 ml distilled water in a 250-ml graduate. Add 230 ml saturated (1.33%) aqueous lithium carbonate solution. Shake well and let precipitate settle to the 70-ml mark. Carefully decant supernatant fluid, refill with distilled water, shake, and again let settle to 70 ml and decant, repeating to a total of three or four washes. Again let settle to 70 ml, decant, and add 28 % ammonia water with constant shaking until almost clear (about 9.5 ml). Dilute to total volume of 120 ml, filter through a Whatman No. 42 or 44 filter, and store in stock bottle. This solution may be kept for months and may be filtered and reused a dozen times or more.

A modification of Foot's silver carbonate solution used for years in our laboratory was prepared thus: To 5 ml 10% silver nitrate add 5 ml saturated (16 to 20%) aqueous sodium carbonate. Let settle and wash 5 times with 35- to 40-ml portions of distilled water by decantation. Then add 15 ml distilled water and 28% ammonia water drop by drop (about 0.4 ml) with constant shaking until the precipitate is almost dissolved. Dilute to 50 ml with distilled water.

The relative virtues of sodium, potassium, and ammonium hydroxides as precipitants are difficult to evaluate, and all solutions should contain diammine silver, hydroxyl, nitrate, and ammonium ions as well as the potassium or sodium ions that may have been added. The alkalinity is controlled with fair precision by limitation of the excess ammonia to the amount just sufficient to form the diammine silver radical.

The use of lithium or sodium carbonate in the Foot, the Laidlaw, and our variants of the Del Río-Hortega method seems to be a matter of indifference, since in all three of these methods the silver carbonate is carefully washed free of soluble salts before the diammine compound is formed.

GENERAL CONSIDERATIONS OF THE TECHNICS. The technics follow a quite uniform general pattern, with quite pronounced variations in duration of individual steps and in concentration of some of the reagents. (See Tables 15-2 and 15-3.)

THE IODINE SODIUM THIOSULFATE SEQUENCE. This sequence in the two Foot technics is to be regarded simply as the usual treatment for mercury-fixed tissues, and we have always omitted it with the carbonate method on formalin-fixed material. The function of this iodine thiosulfate treatment in Laidlaw's variant is not clear, but Laidlaw seems to have regarded it as an integral part of the Mallory bleach. The thiosulfate in any case should be thoroughly washed out before proceeding with the next step.

TABLE 15-2 SCHEDULES FOR DIAMMINE SILVER HYDROXIDE RETICULUM METHODS

	Maresch	Perdrau	Foot	Wilder
Fixation	Formalin or alcohol	Formalin	Zenker or formalin	Formalin, Zenker, or Helly
Section method	Wash several hours, cut frozen sections	Wash 1 day tap, 1 day dist. water. Frozen or paraffin sect.	Paraffin sections	Paraffin or loose or mounted frozen or celloidin sections
Iodine	....	....	0.5% alc., 5 min	....
Wash	....	....	Water	....
Thiosulfate	....	....	0.5% aq., 5 min	....
Wash	....	....	Tap water	....
Oxidant	....	0.25% $KMnO_4$ 10 min	0.25% $KMnO_4$ 5 min	0.25% $KMnO_4$ 1 min or 10% phosphomolybdic acid 1 min
Wash	....	Dist. water	Tap water	....
Reducer	....	Pal's bleach until white	5% oxalic acid 15–30 min	10% HBr 1 min or none
Wash	Dist. water	Dist. water 5 min	Dist. water	Dist. water
Sensitizer	2% $AgNO_3$ 24 h	2% $AgNO_3$ 24 h	2% $AgNO_3$ 48 h	1% uranyl nitrate 5 s
Wash	Dist. water 1–3 s	Dist. water less than 5 min	Dist. water brief	....
Silver	2–30 min 20°C until yellow-brown	40–60 min 25°C	30 min 25°C	1 min 25°C
Wash	Dist. water 2–3 s	Dist. water quick	Dist. water quick	95% alcohol quick
Developer	20% formalin 5–30 min	20% formalin 30 min	5% formalin 2 changes 30 min	1% formalin 0.03% uranyl nitrate, 1 min
Wash	Dist. water	Dist. water	Tap, rinse	Dist. water
Gold chloride	5 drops 1% 1 drop glacial acetic in 10 ml water, until violet; 10 min	0.2% until violet	1%, 1 h	0.2%, 1 min
Wash	Dist. water	Dist. water, rinse	Tap, rinse	Dist. water
Sodium thiosulfate*	5%, 15–60 s	5%, 2 min	5%, 2 min	5%, 1–2 min
Wash	Running water	Thorough, dist.	Tap, several h	Tap water
Counterstain	....	....	Hematoxylin, Van Gieson	Hematoxylin and eosin, or Van Gieson
Mounting	Alcohols to 95 or 100%, carbol-xylene or xylene, balsam	Alcohols, carbol-xylene or creo-sote-xylene, balsam	Alcohols to 100%, xylene, balsam	Alcohols, xylene, balsam

Gomori	Lillie	Gridley	Krajian	Couceiro and Freire
Formalin	Formalin or Orth	Formalin, Zenker, Bouin, alcohol, Carnoy	Formalin (?)	Formalin, alc. form., Helly, Orth, Carnoy, Gendre
Paraffin sections; use gelatin glue	Paraffin sections collodionized	Paraffin sect. well dried	Frozen sect. 7–10 μ	Frozen (12–15 μ) or paraffin 6 μ
. . . .	. . . .	Only for Zenker		
. . . .	. . . .	Only for Zenker		
. . . .	. . . .	Only for Zenker		
Tap water	Tap water	Tap and dist. water	Distilled water	Dist. water 3–5 min, 2–3 changes
0.5–1.0% $KMnO_4$ 1–2 min	0.5% $KMnO_4$ 2 min	0.5% H_5IO_6 aq., 15 min	15 ml 0.3% $KMnO_4$ + 5 ml 28% NH_3, 5 min	4% CrO_3, 40 min
Rinse tap	Tap water	Distilled water	Dist. water, few sec	Dist. water 3–5 min, 4–5 changes
1–3% $K_2S_2O_5$ 1 min	5% oxalic acid 2 min	None	2% oxalic to decolorize	None
Tap, several min	Tap water	. . . .	2 changes dist. water	
2% iron alum 1 min	3% H_2O_2 2 min or 1.24% (46 mM) $FeCl_3 \cdot 6H_2O$[†] 2 min	2% $AgNO_3$ 30 min	None	None
Tap, few min dist. water, 2 changes 1 min 25°C	Tap, 3 min dist. water, 2 changes 3 min 25°C	Dist. water, 2 changes 15 min 25°C	15 min 60°C	20 min 25°C
Dist. water 5–10 s	Dist. water 5–10 s	Dist. water, quickly	Dist. water, 2 changes	Dist. water 10–15 s, 2 changes
10 to 20% formalin, 3 min	10% formalin, 2 min	30% formalin, 3 min	20% formalin, 55°C, 2 min	30% formalin, 5 min
Tap, few min	Tap, 3 min	Wash well dist. water	Rinse, dist. water	Distilled water
0.1–0.2% 10 min, rinse dist. water. Reduce 1 min 1–3% $K_2S_2O_5$	0.2%, 2 min	0.5%, 5 min	1 : 300, 2 min	0.1%, 3–4 min
. . . .	Tap, rinse	Dist. water	Rinse, dist. water	Distilled water
1–2%, 1 min	5%, 2 min	5%, 3 min.	5%, 2 min	5%, 2 min
Tap water	Tap water	Running water	Wash	Prolonged. water
. . . .	Hematoxylin and eosin, or Van Gieson, acid-fast, or others			
Alcohols, xylene, balsam	95% alcohol, acetones, acetone-xylene, xylene, balsam	Alcohols, xylene, balsam	Glycerol gelatin	Alcohols, xylene, balsam

* Crystals, $Na_2S_2O_3 \cdot 5H_2O$.

† Original-$\frac{1}{50}$ USP Liq. Ferr. Chlor.

Sources: Maresch, *Zentralbl. Allg. Pathol.*, **16**: 641, 1905; Perdrau, *J. Pathol. Bacteriol.*, **24**: 117, 1921; Foot, *J. Lab. Clin. Med.*, **9**: 777, 1924; Wilder, *Ann. J. Pathol.*, **11**: 817, 1935; Gomori, *Am. J. Pathol.*, **13**: 993, 1937; Lillie, *Stain Technol.*, **21**: 69, 1946; Gridley, *Am. J. Clin. Pathol.*, **21**: 207, 1951; Couceiro and Freine. *Acad. Brazil. Cien.* **24**: 11, 1952.

TABLE 15-3 SCHEDULES FOR DIAMMINE SILVER CARBONATE RETICULUM METHODS

	Del Río-Hortega	Foot-Menard	Laidlaw	Lillie
Fixation	Formalin	Formalin or Zenker	Bouin or formalin	Formalin
Section method	Frozen sections	Paraffin sections	Paraffin sections. Wash Bouin 20 min, formalin 5 min	Paraffin sections collodionized
Iodine............	....	0.5% alc., 5 min	1% alc., 3 min	
Wash				
Thiosulfate*.......	....	0.5% aq., 5 min	5% aq., 3 min	
Wash	....	Tap water	Tap water	Tap water
Oxidant	....	0.25% $KMnO_4$, 5 min	0.25% $KMnO_4$, 5 min	0.25% $KMnO_4$, 5 min
Wash	....	....	....	Tap water
Reducer	....	5% oxalic acid, 10 min	5% oxalic acid, 3 min	5% oxalic acid, 10 min
Wash	....	Tap water	Tap water, 10 min	Tap water
Wash	Dist. water	Dist. water	Dist. water, 3 changes	Dist. water
Silver	1–2 min 45–50°C	10–15 min 37°C	5 min, 50°C	5 min, 60–70°C, until golden brown
Wash	....	Dist. water, rinse	Dist. water, rinse	Dist. water, rinse
Developer	1% formalin until yellow	20% formalin, 5 min	1% formalin several changes, 3 min	20% formalin, 2 min
Wash	....	Thorough, tap	Dist. water, rinse	Tap water
Gold chloride	0.2%, 30 s	0.2%, 5 min	0.2%, 10 min	0.2%, 2 min
Wash	....	Tap water	Dist. water	Tap water
Sodium thiosulfate*	5%, brief	5%, 2 min	5%, 10 min several changes	5%, 2 min
Wash	Tap water	Tap water	Tap water	Tap water
Counterstain	Van Gieson	Hematoxylin Van Gieson	As desired	Hematoxylin Van Gieson
Mounting†	95% alcohol, carbol-creosote-xylene, xylene, balsam	Alcohols, xylene, balsam	Alcohols, xylene, balsam	95% alcohol, acetones, acetone-xylene, xylene, salicylic balsam

* $Na_2S_2O_3 \cdot 5H_2O$.

† Usual synthetic resins are substituted for balsam.

Sources: Del Río-Hortega, cited from B. Romeis; Foot-Menard, *Arch. Pathol.*, **4**: 211, 1927: Laidlaw, *Ann. J. Pathol.*, **5**: 239. 1929; Lillie, "Histopathologic Technic and Practical Histochemistry," 3d ed., McGraw-Hill, New York, 1965.

WEIGERT'S BLEACH. The sequence treatment with potassium permanganate and oxalic acid, or a substitute for it, appears in all the variants after the Bielschowsky-Maresch and the Del Río-Hortega. Its introduction into diammine silver methods is due to Perdrau (*J. Pathol. Bacteriol.*, **24**: 117, 1921) and appears to be purely pragmatic. Omission of the permanganate step results in failures of impregnation, and omission of the oxalic acid step causes irregularities of impregnation, so that both excellent impregnations and failures are obtained on similar material. Perdrau, Foot, Laidlaw, and Wilder prescribed 0.25% potassium permanganate for 10, 5, 3, and 1 min, respectively; Gomori used 0.5 to 1.0% for 2 to 1 min; and we have used 0.5, 0.33, and 0.25% solution for 2 to 5 min. In any case the sections are colored fairly deep brown, and the reaction is probably completed in a few seconds. Wilder's substitution of a 1-min 10% phosphomolybdic acid treatment probably serves the same purpose, as the hexavalent molybdenum of the molybdic acid is easily reduced to a lower valence. The washing after this step serves to remove the excess reagent. The reducing agent in Foot's methods and in Laidlaw's and our variants has been 5% oxalic acid, varying in time thus: 15 to 30 min in Foot's oxide method, 10 min in his and our carbonate methods, 3 min in Laidlaw's, and 2 min in our oxide method. Actually the brown permanganate stain disappears in a few seconds in this reagent, and what benefit may accrue from further exposure is difficult to discern. Gomori used a 1 to 3% potassium metabisulfite ($K_2S_2O_5$, which hydrates to $KHSO_3$ in water) solution for 1 min, and Perdrau used Pal's 0.5% oxalic acid, 0.5% acid potassium sulfite ($KHSO_3$) solution until the sections were white. Wilder prescribed a 1-min bath in a one-fourth dilution of concentrated hydrobromic acid to follow the permanganate step and stated that this could be omitted if phosphomolybdic acid was substituted for permanganate. The hydrobromic acid removes the brown permanganate stain instantly. Couceiro and Freire oxidized 40 minutes in 4% chromic acid solution, Gridley used a 15-min bath in 0.5% aqueous periodic acid (H_5IO_6). The latter should lead to greater constancy of results, since periodic acid carries oxidation of 1,2-glycols and 1,2-amino alcohols only to the aldehyde stage, while permanganate and chromic acid both produce aldehyde and oxidize it further.

SENSITIZERS.   Maresch employed a 24-h bath in silver nitrate solution, with no antecedent oxidation. Wilder substituted 1% uranyl nitrate for 5 s. Gomori prescribed 1 min in 2% iron alum as preferable to several other metal salt impregnations tried by him. We have substituted a 2-min bath in a 1 : 50 dilution of USP ferric chloride solution (*ca.* 46 mM, or 1.24% $FeCl_3 \cdot 6H_2O$) as equivalent and have found a number of oxidant solutions also successful: 3% H_2O_2, 1% $NaIO_3$, 1% iodine, or a 1% acetic acid–5% potassium dichromate mixture. After Orth fixation this sensitizing step is better omitted. Since ferric chloride treatment also sensitizes connective tissue fibrils so that they more or less specifically reduce the methenamine silver of the argentaffin cell method, it would appear that there may be actual metal organic compounds formed in which the silver later replaces the metal. Aldehyde sites do not appear to reduce ammine or methenamine silver complexes selectively, perhaps because the aldehyde is bound with amino groups under alkaline conditions.

However, aldehyde sites, engendered by permanganate or periodic acid, may well operate to initiate silver deposition from the 2% silver nitrate of the Maresch, Perdrau-Da Fano, Foot, and Gridley methods and thus afford nuclei for further deposit in the alkaline ammine silver solutions. Gomori found that the same effect could be achieved in 10 min with 10% silver nitrate and noted that thorough washing in distilled water after sensitization was preferable to the brief rinses prescribed by the Bielschowsky-Maresch and the Foot silver oxide technics. Bailey and Hiller washed not over 5 min with distilled water, using the Perdrau-Da Fano method.

Sensitizing agents have not been commonly employed in diammine silver carbonate technics. All technics prescribe washing with distilled water before the silver bath.

SILVER BATH.   Times in the silver oxide baths vary thus: 1 min at room temperature prescribed by Gomori and Wilder, the 2- to 30-min interval of the Bielschowsky-Maresch technic, Foot's 30 min, and Perdrau's 40 to 60 min. These methods vary in concentration in grams of silver nitrate used to make 100 ml solution thus: 4 g for the Bielschowsky-Maresch; 4 to 5 g for our oxide method and the Gomori; 2.5 g for the Foot; 2 g for the Perdrau-Da Fano; and 1 g for the Wilder. With carbonate methods concentrations are 10 g for Laidlaw's; 1 g for Foot's and ours; and $\frac{2}{3}$ g for Del Río-Hortega's. Times and temperatures are respectively 5 min at 50°C, 10 to 15 min at 37°C, 3 to 5 min at 60 to 70°C, and 1 to 2 min at 45 to 50°C. Sections turn yellow to brown during silvering by the carbonate methods and in the original Bielschowsky-Maresch oxide technic. With the other oxide methods they remain unchanged in color until developed. Though little correlation is seen among time, temperature, and concentration, nevertheless, when the same variant is employed, increase in temperature accelerates the process, and variation of dilution alters the time requirement in the expected direction.

Most technics prescribe a quick rinse of 1 to 10 s in distilled water between the silver bath and the reduction bath. Wilder used alcohol, and Del Río-Hortega carried sections directly from his weak silver bath to weak formalin.

REDUCTION.   All methods use formaldehyde in concentrations varying from a 30% dilution of the concentrated 37 to 40% formaldehyde down to 1%. Gridley used 30% formalin for 3 min; Couceiro and Freire used the same for 5 min; Bailey and Hiller, 20% formalin for 30 min for the Da Fano technic used by Perdrau; Maresch, for 5 to 30 min; Foot's carbonate method, for 5 min; ours, for 2 min; Foot's oxide method used two changes of 5% over a 30-min period; Gomori required 10 to 20% for 3 min; Laidlaw, 1% for 3 min; Del Río-Hortega, 1% until yellow; and Wilder, a 0.03% uranyl nitrate–1% formalin solution for 1 min. Actually reduction occurs visibly in the first few seconds after adding formalin, and no further darkening is evident after 30 seconds or so. We have used a 2-min period in 10% formalin for the oxide variant, and in 20% formalin for the carbonate.

Washing after the formalin is usually in tap water, though Wilder and Laidlaw prescribe distilled water. This is especially advisable when tap water is heavily chlorinated.

TONING.   All technics use a toning in yellow auric chloride (which is 1 : 500), thus: until violet according to Bailey and Hiller in Perdrau's method, 1 min in Wilder's method, 30 s in Del Río-Hortega's, 5 min in Foot's carbonate method, 10 min in Laidlaw's, 2 min in our carbonate and oxide variants. Foot's silver oxide method used 1% for 1 h, while the Bielschowsky-Maresch technic called for 5 drops of 1% solution with 1 drop of glacial acetic acid in 10 ml distilled water until the sections turned violet, and Gomori used either 1 : 500 or 1 : 1000 for 10 min. Couceiro and Freire used 0.1% for 3 to 4 min, controlling microscopically; and Gridley employed 0.5% for 5 min and insisted that golden brown tones be replaced by pale yellowish gray (taupe) or lavender. Actually the sections change color in a few seconds, and probably even Wilder's 1-min interval is longer than necessary.

Following gold toning usually a brief wash or rinse in tap or distilled water is recommended. Gomori prescribes a second reduction with 1 to 3% potassium metabisulfite for 1 min, followed by another wash.

FIXING.   Then all methods remove unreduced silver by means of sodium thiosulfate. Gomori prescribes 1 to 2% for 1 min; the rest use 5% for 15 to 60 s (Bielschowsky-Maresch), brief (Del Río-Hortega), 1 to 2 min in Wilder's, 2 min in Perdrau's and Foot's two cited methods and our carbonate and oxide variants and Couceiro and Freire's method, 5 min in Gridley's technic, and 10 min with Laidlaw's strong silver carbonate method.

COUNTERSTAINS, ETC.   Thorough washing follows in all technics, and then counterstains are inserted by Foot, Wilder, Del Río-Hortega, and our carbonate and oxide variants. Usually a brief hemalum stain, tap water blueing, and a 45- to 60-s Van Gieson technic are recommended; they have been quite successful in our hands. If Gomori's iron alum sensitizer or our iron chloride bath is used before the silver oxide bath, an iron hematoxylin effect is produced. This we find too dense and prefer to omit the hematoxylin with this variant. We have had very interesting results by following the reticulum method with a Ziehl-Neelsen acid-fast stain or by a periodic acid Schiff reaction.

Alcohol dehydration, clearing with xylene, carbol xylene, or carbol creosote xylene and inclusion in balsam are recommended. With our collodion film technics we find it advisable to complete the dehydration with acetone (thus removing the collodion and any precipitate which may have formed on it) and then pass through acetone and xylene into two or three changes of xylene.

RETICULUM TECHNIC. The following reticulum technic is based on *Wilder's, Gomori's,* and *our variants* of the *Bielschowsky-Maresch method.*

Place 1 ml 28% ammonia water in a small flask. Add 5% silver nitrate, the first 14 to 16 ml fairly rapidly, the rest of the 19 to 20 ml total cautiously, shaking between each addition to clear the brown clouds of silver oxide until a faint permanent turbidity remains. This diammine silver hydroxide solution can be used for 1 or 2 days and is discarded after one use.

The procedure of impregnation is as follows:

1. Fix in formalin or Orth's fluid. Dry 5-μ paraffin sections several hours in 58°C oven, then take through two changes of xylene and two of 100% alcohol into step 2.
2. 1% collodion in ether and absolute alcohol (equal volumes) for 5 to 10 min.
3. Drain for 1 to 2 min.
4. 80% alcohol for 5 min or more.
5. Rinse in tap water.
6. 0.5% potassium permanganate for 2 min.
7. Wash in water.
8. 5% oxalic acid for 2 min.
9. Wash in water.
10. Wilder's 1% uranyl nitrate for 5 to 10 s, or Gomori's 2% iron alum for 1 min, or liquor ferri chloridi, 1:50 (46 mM FeCl$_3$) in distilled water for 2 min, or 3% hydrogen peroxide for 2 min. With Orth-fixed material omit this step.
11. Wash for 3 min in running water, and then rinse in two changes of distilled water. Wilder omitted this wash, but it does not interfere even after uranyl nitrate.
12. Lay slides face up on glass rods over a large pan, and deposit on each about 1.5 to 2 ml diammine silver hydroxide. Let stand 3 min, and decant.
13. Rinse quickly in distilled water.
14. Reduce for 2 min in 10% formalin.
15. Wash for 3 min in running water.
16. Tone for 2 min in 0.2% acid gold chloride (HAuCl$_4$).
17. Rinse in tap water.
18. Fix for 2 min in 5% sodium thiosulfate (0.2 M).
19. Wash in tap water.
20. Counterstain as desired. For example, acetic alum hematoxylin for 2 min, tap water for 2 min, Van Gieson's picrofuchsin for 1 min. Differentiate in two or three changes of 95% alcohol. Or stain by Ziehl-Neelsen for acid-fast organisms. Or stain 5 min in 0.1% safranin, thionin, or toluidine blue, differentiate 1 min in 5% acetic acid, wash well in tap water. Or use the periodic acid Schiff for red basement membranes.
21. Complete dehydration and decollodionization in three changes of acetone, clear with one change of acetone and xylene (50:50) and two or more changes of xylene. Mount in synthetic resin. The acetone removes the collodion film; if this is to be preserved, use isopropanol.

Gottlob and Hoff (*Histochemie,* **13:** 70, 1968) conducted histochemical studies in an effort to determine the mechanism of action of silver impregnation methods. Further studies are required in order to understand this mechanism. Velican and Velican (*Biochim. Biol. Spermentale,* **7:** 223, 1968) reported that permanganate– or chromic acid–induced argyrophilia of reticulin and basement membranes is prevented in sections acetylated prior to the oxidative step.

Reticulin is stained by a bromine-silver sequence. Sections are exposed to bromine vapor for 10 min at 4°C, washed thoroughly in distilled water, and then treated with the diammine silver reagent. Mukherji and Sen (*Acta Anat.*, **66:** 365, 1967) were unable to prevent the bromine-silver stain by acetylation; however, silver staining was abolished by lipid solvent extraction prior to the staining reaction.

Cooper et al. (*Science*, **141:** 526, 1963) noted that the administration of reserpine to dogs in doses sufficient (2.5 mg/kg) to deplete myocardial catecholamines resulted in reduction of silver staining, particularly of the perimysial plexus. The catecholamine content of the heart is related to sympathetic motor innervation. The means whereby reserpine brings about this effect is unknown.

Lynch (*Stain Technol.*, **39:** 19, 1964) advocates the following stain for reticulum. The method does not employ alkaline silver solutions. For this reason, disconcerting detachment of sections from slides is stated to be avoided.

Tissues fixed in buffered neutral formalin are recommended.

1. Dewax and bring sections to distilled water.
2. Immerse for 1 h in a solution consisting of liquid bromine, 2 ml; potassium bromide, 100 mg; and distilled water, 1000 ml.
3. Rinse through three changes of distilled water, for 3 to 5 s each.
4. Place for 5 min in Weigert's iodine, I–KI–H$_2$O, 1 : 2 : 100.
5. Rinse through three changes of distilled water, for 3 to 5 s each.
6. Place for 5 min in a 1% w/v solution of chloroauric acid (HAuCl$_4$, British Drug House, Ltd.) in distilled water.
7. Rinse through three changes of distilled water, for 3 to 5 s each.
8. Place for 2 to 4 h at 37°C in freshly made 3% H$_2$O$_2$ (30% diluted 1 : 10), or for 1 to 3 h in 2% w/v aqueous oxalic acid.
9. Dehydrate, clear, and mount in a synthetic resin, e.g., Permount.

RESULTS: If peroxide is used in step 8, fine reticular fibers are black, and coarse collagen fibers are golden brown. Background staining varies from unstained to light brown. If oxalic acid is used in step 8, reticulin stains light blue, background is unstained to gray, and sarcolemma is distinctly stained.

ACID AND BASIC FLUOROCHROME METHOD

Vassar and Culling (*Arch. Pathol.*, **68:** 487, 1959) found good differential fluorochrome staining of collagen, reticulum, and basement membranes by staining 3 min in equal volumes of 1% alcoholic solutions of rhodamine B and fluorescein, differentiating in 1% acetic alcohol, dehydrating, clearing, and mounting in Harleco resin (HSR). The photomicrograph shows moderate (red?) fluorescence of proximal tubule cytoplasm, with strong fluorescence of glomerular stroma, presumably yellow. Vassar and Culling neglect to give the colors.

THE HEMATOXYLIN METHODS

Mallory's phosphotungstic acid hematoxylin stains nuclei, centrioles, spindles, mitochondria, fibrin, fibrils of neuroglia, and the so-called myoglia and fibroglia and the contractile elements of striated muscle blue; collagen, reticulum, elastin, cartilage, and bone matrix are yellowish to brownish red.

For staining of mitochondria Mallory prescribed a prolonged oxidation with ferric chloride before dehydration of the formalin-fixed blocks, and prolonged the staining interval. For other elements the following general technic is prescribed. The solution is the same in either case and is prepared by dissolving 1 g hematoxylin and 20 g phosphotungstic acid in 1000 ml distilled water. This ripens in several weeks and the naturally ripened product is thought to be best. However, Mallory states that the addition of 177 mg potassium permanganate will ripen it at once (50 mg $NaIO_3$ might serve better).

Zenker fixation was prescribed by Mallory. Earle (personal communication) had satisfactory staining of material fixed in buffered formalin, without subsequent mercury treatment. Peers (*Arch. Pathol.*, **32:** 446, 1941) stated that sections of formalin-fixed material should be brought to water as usual and then mordanted 3 h at 57°C (paraffin oven temperature) in saturated aqueous mercuric chloride solution. Sections thus mordanted are then treated just as though Zenker fixation had been employed, thus:

THE MALLORY PHOSPHOTUNGSTIC ACID HEMATOXYLIN (PTAH) TECHNIC

1. Iodine in 95% alcohol (0.5%), 5 min.
2. 0.5% sodium thiosulfate, 5 min (or 5% for 1 min).
3. Wash in tap water.
4. 0.25% potassium permanganate, 5 min.
5. Wash in water.
6. 5% oxalic acid, 5 min (Mallory 10 to 20 min).
7. Wash in running water 1 to 2 min.
8. Stain in phosphotungstic acid hematoxylin overnight (12 to 24 h).
9. Dehydrate rapidly in 95 and 100% alcohol or in acetone, clear with a 50% mixture of the dehydrating agent and xylene, then two changes of xylene. Mount in balsam or synthetic resin.

Lieb (*Arch. Pathol.*, **45:** 559, 1948) introduces mordanting in ferric ammonium alum after step 7 thus: Rinse in distilled water, mordant 1 h in 4% iron alum, rinse in tap water and in distilled water, and stain in the hematoxylin solution 2 to 24 h. The color of normally blue staining elements is intensified by this treatment.

Linder (*Q. J. Microsc. Sci.*, **90:** 427, 1949) has described a hematoxylin method for collagen, reticulum, and capillary basement membranes. It appears to be limited in its usefulness to material fixed with mercuric chloride fixatives. The hematoxylin solution requires 10 weeks' ripening to stain basement membranes well, though collagen and reticulum are stained by the fresh solution. The lake retains its staining potency up to at least 7 months. We have had no experience with this method.

Thomas' phosphomolybdic acid hematoxylin was made by Linder as follows:

Dissolve 2.5 g hematoxylin in 49 ml dioxane and 1 ml aerated water. Dissolve 16.5 g phosphomolybdic acid in 44 ml distilled water and 11 ml ethylene glycol or glycerol, and filter. Mix equal volumes of the two solutions, and ripen the mixture at (British) room temperatures (18 to 20°C) for 10 weeks. The similarity of the solution to Mallory's phosphotungstic acid hematoxylin is noteworthy.

THE LINDER-THOMAS PHOSPHOMOLYBDIC ACID HEMATOXYLIN TECHNIC

1. Fix in 9 volumes saturated aqueous mercuric chloride and 1 volume commercial formalin.
2. Embed in paraffin by usual technics, section at 3 μ, deparaffinize, and hydrate as usual.

3. Treat with Weigert's iodine solution for 1 min.
4. Treat with 5% sodium thiosulfate (0.2 *M*) for 1 min.
5. Wash in distilled water.
6. Barely cover section with phosphomolybdic acid hematoxylin, and stain for 10 min.
7. Wash with distilled water.
8. Dehydrate, clear, and mount.

RESULTS: Collagen, reticulin, and basement membranes of pulmonary capillaries stain deep violet; red corpuscles, purple; nuclei, blue but lightly stained; cytoplasm is bluish. A counterstain with 0.02% acriflavine in 1% acetic acid for 30 s gives a greenish background that contrasts well.

THE ACID ANILINE DYE METHODS: PICRIC ACID MIXTURES

The general subject of collagen staining with acid aniline dyes Lillie reviewed (*J. Tech. Methods*, **25**: 1, 1945), and the reader may find there many details and methods for which we do not have space in this account.

The selective collagen stains with acid aniline dyes apparently depend on the selectivity of collagen for certain acid dyestuffs from fairly strongly acid solutions. With one group of methods a mixture of aniline blue, methyl blue, indigocarmine, acid fuchsin, or other dyestuffs is made in appropriate proportion with picric acid, which acts both as the acidifying agent and as a counterstain for muscle, cytoplasm, and other materials. In another group the counterstain precedes the fiber stain, and the latter is mixed with or preceded by an acid such as phosphotungstic, phosphomolybdic, or picric acid, or even hydrochloric acid.

Stains in the first group are the simplest in application and are widely used also as general stains.

Picroindigocarmine was the first differential connective tissue stain reported (by Jullien in 1872). In Ramón y Cajal's (*Rev. Cien. Med. Barcel.*, **22**: 97, 1896; "Histology," 1933) modification it is still widely used. In this technic, first stain nuclei red with carmine, then wash in water, and stain for 5 to 10 min with a solution of 250 mg (originally 333 mg) indigocarmine (C.I. 73015) in 100 ml saturated (1.2%) aqueous picric acid solution. Rinse in weak (perhaps 0.5%) acetic acid, and dehydrate and differentiate in absolute alcohol. Connective tissue is blue-green; muscle, greenish yellow; nuclei are red.

An iron hematoxylin stain, such as Weigert's, for 3 to 6 min can be substituted for the carmine, and a sequence of 95% and absolute alcohol is just as satisfactory for differentiation and dehydration. This method on the whole is less selective and less complete for fine fibers than are its methyl blue or acid fuchsin counterparts.

Van Gieson's mixture consists of acid fuchsin (C.I. 42685) and picric acid. (See *J. Tech. Methods*, **25**: 1, 1945, for literature.) The usual proportion is 5 ml 1% acid fuchsin (50 mg dye) to 95 ml saturated aqueous picric acid solution. The proportion of acid fuchsin can profitably be raised to 100 mg/100 ml picric acid solution (*Weigert*). Addition of 0.25 ml concentrated hydrochloric acid to the last mixture sharpens the differentiation, so that muscle is purer yellow and collagen is deeper red (Lillie, *J. Tech. Methods*, **25**: 1, 1945). *Unna's variant*, containing 250 mg acid fuchsin, 1.5 g nitric acid (the P.G., 1884, was 30%, sp gr 1.185, and the 1.5 g would equal 0.5 ml of present-day concentrated, 70% acid, sp gr 1.42), 10 ml glycerol, 90 ml water, and picric acid to saturation, gives deep

TABLE 15-4 VAN GIESON PICRO ACID FUCHSIN MIXTURES

	Van Gieson	Weigert	Unna	Lillie	Freeborn–Van Gieson glia*
Acid fuchsin, mg/100 ml	50	100	250	100	150
Picric acid saturation ..	95%	Saturated	Saturated	Saturated	Half saturated
Other additives	. . . . . .	. . . .	1.5 HNO₃ 10 glycerol	0.25 ml conc. HCl	

* Mallory uses 15 ml 1% acid fuchsin, 50 ml 1.22% picric acid, and 50 ml water for Freeborn's glia mixture.
Source: Lillie, *J. Tech. Methods*, **25**: 1, 1945.

crimson collagen and bright yellow muscle. Table 15-4 lists all of these picric acid fuchsin mixtures.

TECHNIC.   Bring sections to water as usual.

1.   Stain for 5 min in Weigert's acid iron chloride or other, similar hematoxylin.
2.   Wash in water.
3.   Stain 5 min in the picrofuchsin mixture.
4.   Dehydrate, and differentiate with two or three changes each of 95 and 100% alcohol.
5.   Clear with a mixture of 100% alcohol and xylene followed by two or three changes of pure xylene. Mount in polystyrene.

Kattine (*Stain Technol.*, **37**: 193, 1962) substitutes alum hematoxylin in step 1. She stains for 7 to 10 minutes, washes in water, acid alcohol (1 ml concentrated HCl to 99 ml 70% alcohol), and again in water. Then she immerses sections 30 to 60 s in phosphotungstic phosphomolybdic acid mixture (1.25% each) and washes with water before proceeding with the Van Gieson step. Nuclei are then stained black.

Picromethyl blue mixtures were used by Dubreuil, Curtis, Ohmori, and others. We find that 100 mg methyl blue (C.I. 42780) or aniline blue (C.I. 42755) per 100 ml saturated aqueous picric acid solution is a suitable proportion (*J. Tech. Methods*, **25**: 1, 1945).

1.   Stain first with iron hematoxylin (Weigert's acid iron chloride) for 5 min.
2.   Wash with water.
3.   Stain in picromethyl blue or picroaniline blue for 5 min.
4.   Differentiate briefly in 1% acetic acid or directly in 95% alcohol.
5.   Then dehydrate and clear in 100% alcohol and xylene, two changes of xylene, and mount in polystyrene, Permount, or similar resin.

RESULTS: Connective tissue, including much reticulum and mucosal and renal glomerular basement membranes, is deep blue; muscle and cytoplasms are varying shades of yellowish green to gray; mucus is pale blue; nuclei are black.

Violamine R (C.I. 45190) is a good substitute for acid fuchsin in the Van Gieson methods, yielding a similar color picture, with less tendency to fade than with acid fuchsin. Amido black 10B (C.I. 20470) can be used in place of methyl blue, giving a very precise dark bluish green to collagen. It is one of the most selective and precise collagen stains that we have encountered. For the best contrasts, use 20 to 40 mg of this dye to the 100 ml of saturated picric acid solution.

LILLIE-GUTIÉRREZ-PALMER ACID FUCHSIN–IRON FLAVIANIC ACID STAIN FOR COLLAGEN (*Anat. Rec.*, **159**: 365, 1967). This method too is useful:

1. Deparaffinize and hydrate sections as usual.
2. Stain 10 min at 25°C in alum hematoxylin of 0.5% hematoxylin content; for increased density of nuclear staining, stain 5 to 10 min at 60°C.
3. Wash briefly in water.
4. Stain 3 min in flavianic acid, 1.75 g; acid fuchsin, 200 mg; ferric chloride ($FeCl_3 \cdot 6H_2O$), 2.7 g; and distilled water to make 100 ml. The solution keeps.
5. Dehydrate and differentiate in two changes of 95% alcohol.
6. Completely dehydrate with absolute alcohol, clear with xylene, and mount in synthetic resin (Permount, Harleco, cellulose caprate, etc.). Naphthol yellow S, 2 g, may be substituted for the flavianic acid, but a stoichiometric equivalent ($Na_2C_{10}H_4O_8N_2S$, mw 412.25) of concentrated hydrochloric acid (0.8 ml) should be added to bring the pH to about 1.2.

Lillie et al. note that the substitution of flavianic acid (50 mM) or of a stoichiometric mixture of naphthol yellow S and hydrochloric acid in place of picric acid in the Van Gieson–type mixtures gives deeper yellows to cytoplasm, muscle, and erythrocytes. Higher concentrations of acid fuchsin can be used with consequent greater density of collagen fiber staining and improved contrast. Retention of excellent nuclear staining is obtained by inclusion of 0.1 M ferric chloride in the Van Gieson mixture. Nuclear staining with alum hematoxylin is thereby permitted.

MASSON'S VAN GIESON VARIANT (*J. Tech. Methods*, **12**: 75, 1929). This method introduced a plasma stain with 2 ml saturated aqueous metanil yellow (C.I. 13065) diluted to 100 ml in 1% acetic acid between the iron hematoxylin and the picrofuchsin, thus:

1. Stain nuclei black with Regaud's iron hematoxylin.
2. Wash.
3. Differentiate in two-thirds saturated alcoholic picric acid.
4. Wash 15 min in running water.
5. Stain 5 min in the acetic metanil yellow.
6. Rinse in distilled water.
7. Mordant 5 min in 3% potassium dichromate.
8. Stain 30 to 60 s in 1% acid fuchsin in saturated aqueous picric acid.
9. Rinse quickly in distilled water.
10. Differentiate 5 min in 1% acetic acid.

We see no particular advantage to this method over an acidified Van Gieson stain such as Unna's, in which there are fewer steps.

BIEBRICH SCARLET PICROANILINE BLUE. In 1940 Lillie reported a similar method (*Arch. Pathol.*, **29**: 705, 1940):

1. Weigert's acid iron chloride hematoxylin for 5 min in place of Regaud's.
2. Wash.
3. 0.2% Biebrich scarlet (C.I. 26905) in 1% acetic acid for 4 min in place of the metanil yellow.
4. Wash, omitting the chromation.
5. Stain for 4 to 5 min in 0.1% aniline blue in saturated aqueous picric acid, instead of in picro acid fuchsin.

6. Transfer directly to 1% acetic for 3 min.
7. Dehydrate, clear, and mount in polystyrene.

RESULTS: Connective tissue was stained blue, including renal glomerular stroma, basement membranes, and much reticulum; erythrocytes were orange-scarlet; muscle was pink; cytoplasm, pink to gray; nuclei, gray to black; mucus, light blue.

Further experimentation with this method corrected some of its defects and apparently increased its completeness of staining connective tissue (*J. Tech. Methods*, **25**: 1, 1945), thus:

PICRO-AMIDO BLACK 10B.[1]   Bring sections to 80% alcohol as usual.
1. Stain for 6 min in acid iron hematoxylin (Weigert's or similar formula).
2. Wash in tap water.
3. Stain for 5 min in brilliant purpurin R (C.I. 23510), 0.6 g; azofuchsin G (C.I. 16540), 0.4 g; glacial acetic acid, 1 ml; and distilled water to make 100 ml.
4. Rinse in water.
5. Stain for 1 to 5 min in amido black 10B (C.I. 20470), 50 to 100 mg; saturated aqueous picric acid, 100 ml.
6. Differentiate directly in 1% acetic acid for 2 min.
7. Dehydrate, clear, and mount by an alcohol, alcohol xylene, xylene, synthetic resin sequence.

RESULTS: Dark green demonstrates collagen, reticulum, basement membranes, and renal glomerular stroma; pale greenish blue, mucus; brown to greenish brown, epithelial cytoplasm; light brown, muscle; red, erythrocytes.

Since with fresh samples of brilliant purpurin R and azofuchsin G it may be necessary to adjust the proportions so as to maintain the brown-red balance between muscle and erythrocytes, it is well to prepare these two dyes in 1% solutions in 1% acetic acid and mix them in 6:4 proportion for use. This permits ready variation to increase or decrease the red. Aniline blue or methyl blue at the same concentration can be used in place of amido black 10B, with only slightly inferior results and somewhat better demonstration of mucus.

FAST GREEN VAN GIESON.   A similar method giving red connective tissue, gray-green cytoplasm and muscle, green erythrocytes, and brown nuclei is done thus (Lillie, *J. Tech. Methods*, **25**: 1, 1945). We have found it a useful method.

TECHNIC.   Bring sections to water as usual.

1. Stain 6 min in acid alum hematoxylin.
2. Wash in tap water.
3. Stain 4 min in 0.1% fast green FCF (C.I. 42053) in aqueous 1% acetic acid.
4. Wash in 1% acetic.
5. Stain 10 to 15 min in 0.2% acid fuchsin in saturated aqueous picric acid.
6. Wash 2 min in 1% acetic acid.
7. Dehydrate, clear, and mount through alcohols, alcohol and xylene, and xylene to polystyrene or cellulose tricaprate.

RESULTS: Nuclei are brown; collagen, reticulum, and basement membranes, deep purplish red; ocular lens capsule and Descemet's membrane, a deeper red-purple; muscle is grayish

---

[1] Called naphthol blue black in Lillie, 1954, and in the original publication.

or yellowish green after formalin fixatives and a deeper and purer green with nonformalin fixatives; cytoplasm, ellipsoids of retina, and red corpuscles are green; Paneth cell granules, sometimes red; hypophyseal alpha granules, light green; beta granules, brownish red; adrenal cortex cells, green; medulla cells, pink to gray-pink except after chromate fixatives, when they are a darker and grayer green than cortex cells. Mucus is almost unstained.

In step 3 we have substituted 0.3% wool green S (C.I. 44090) with good results. The greens are bluer. In step 4, 0.2% violamine R (C.I. 45190) or 0.1 to 0.5% ponceau S (C.I. 27195) may be substituted for the acid fuchsin with good results.

The mechanism of the foregoing group of stains remains obscure. The dyes usable in picric and related acid mixtures are generally di- to tetrasulfonic acids; they often, but not regularly, contain one or more primary, secondary, or even tertiary amino groups; and they belong to several distinct classes of dyes, the disazo, triphenylmethane, rhodamine, and indigo groups all being represented.

Unlike what happens with many other structures stained by acid dyes, the staining of collagen fibers by Van Gieson mixtures is not prevented by deamination and is only poorly prevented by pyridine acetylation. Sulfation in acetic anhydride acetic acid solvent mixtures prevents acid dye staining of the usual type and also prevents the selective collagen stains by Van Gieson stains. This blockade is resolved by methanolysis in the same manner as (and concurrently with) the destruction of sulfation-induced metachromatic basophilia at pH 1 to 2. Since both O and N acylations appear to occur in such mixtures, this does not aid greatly in elucidating the staining mechanism.

The selectivity of the Van Gieson–type stains for collagen fibers appears to be enhanced by addition of mineral acids to the stain mixtures (see also Lillie, *J. Histochem. Cytochem.*, **12:** 821, 1964).

PHOSPHOMOLYBDIC AND PHOSPHOTUNGSTIC ACID METHODS

It appears perhaps indicated by Puchtler's work (*J. Histochem. Cytochem.*, **6:** 265, 1958) that the selectivity of the following group of methods may be due to specific uptake of molybdenum or tungsten acids by the fibers and subsequent binding of the usual sulfonated aminotriphenylmethane dyes by their NH_2 groups. Puchtler indicates that the same structures are demonstrated if, after the phosphomolybdic acid bath, sections are reduced with stannous chloride to produce molybdenum blue, and she finds further that basic dyes may be used in place of the usual water-soluble aniline blue or light green SF. If this mechanism is accepted, it relates this group of stains more to the phosphotungstic acid hematoxylin group than to the Van Gieson stains, to which they have sometimes been regarded as equivalent.

These methods include technics in which the acid and the fiber stain are used in sequence, as in the 1900 Mallory method and in the Heidenhain and Masson variants, and those in which the acid and the fiber stain are used together, as in the 1905 and 1936 Mallory methods. Often Zenker, Helly, Regaud, Bouin, or other mercurial or picric acid fixations are prescribed for these methods, and if the material has been fixed with formalin, mordanting in Zenker's or Bouin's fluid or in saturated mercuric chloride or picric acid solution is prescribed, either of blocks of unembedded tissue or of sections before staining. We have found saturated mercuric chloride in water and saturated picric acid in alcohol acceptable means of accomplishing this end. The effect of these fixations and aftertreatments is to enhance the staining of cytoplasm and to render connective tissue somewhat more difficult to stain fully.

Although there has been considerable dispute both as to the relative efficiency of phosphomolybdic and phosphotungstic acids and as to the proper concentration and exposure time, it seems to make little difference whether one uses one, the other, a combination of both, or neither, as long as sections are treated with acid before and during the fiber stain. We have used picric acid, ferric chloride, and other reagents successfully in place of phosphomolybdic acid (Lillie, *J. Tech. Methods*, **25**: 1, 1945).

Numerous variants of the Mallory technics have been published, and the tendency of recent years seems to have been in the direction of greater complexity of method and introduction of more visually controlled differentiations. Although we do not question the efficacy of such methods in the hands of those who have had long experience with them, they are difficult for the new or occasional user to execute and require a familiarity with the results to be attained which is not general among other than specially trained technicians.

Of the group, the most widely used have been Mallory's 1905 and 1936 variants, Heidenhain's 1916 "azan" method, and Masson's 1928–1929 "trichrome" methods.

MALLORY'S 1905 METHOD (*J. Med. Res.*, **13**: 113, 1905)

1. Zenker fixation.
2. Usual iodine and thiosulfate sequence.
3. Wash in water.
4. Stain 5 min in 0.5% acid fuchsin.
5. Drain and transfer to aniline blue (C.I. 42755), 0.5 g; orange G (C.I. 16230), 2 g; phosphomolybdic acid, 1 g; distilled water, 100 ml. Stain 10 to 20 min.
6. Dehydrate and differentiate in several changes of 95% alcohol, then 100% alcohol, xylene, and balsam.

MALLORY'S 1936 METHOD (*Stain Technol.*, **11**: 101, 1936).

As above, but use 1 g phosphotungstic acid in place of the phosphomolybdic in the aniline blue stain. Stain with this for 20 to 60 min or longer. In 1936 Mallory recommended 0.25% acid fuchsin for 30 min, but in his 1938 text he reverted to a 1- to 5-min stain in 0.5%, as in the 1905 technic.

RESULTS: With both Mallory's variants collagen fibrils are deep blue; cartilage and bone matrix, mucus, amyloid, and some hyalin materials, varying lighter shades of blue; fibrin, nuclei, glia fibrils, "fibroglia and myoglia," and axis cylinders, red; erythrocytes and myelin, yellow.

HEIDENHAIN'S "AZAN"[1] VARIANT (*Z. Wiss. Mikrosk.*, **32**: 361, 1916).

This technic gave red nuclei and erythrocytes, orange muscle, reddish glia fibrils, blue mucin, and dark blue collagen and reticulum, including glomerular stroma.

TECHNIC. Fix in Zenker's Helly's, Bouin's, or Carnoy's fluid.

1. Stain 30 to 60 min in a covered dish at 50 to 55°C and then 1 to 2 h at 37°C in 0.25 to 1 g azocarmine B (C.I. 50090) in 100 ml cold water and 1 ml glacial acetic; or 100 ml water saturated by boiling with 0.1 g azocarmine G (C.I. 50085), cooled, and acidified with 1 ml glacial acetic acid.
2. Wash in distilled water.

---

[1] The term *azan* is derived from the first syllables of *Azokarmin B* and of *Anilinblau W,* the names of the principal dyes used.

3. McGregor (*Am. J. Pathol.*, **5**: 545, 1929) inserted the step of differentiating in 0.1% aniline in 95% alcohol and rinsing in 1% acetic acid in 95% alcohol.
4. Heidenhain then mordanted 30 min to 3 h in 5% phosphotungstic acid.
5. Rinse in distilled water.
6. Stain 1 to 3 h in a 50 to 33% (Mallory: 25%) dilution of a stock solution of aniline blue, 0.5 g; orange G, 2 g; glacial acetic acid, 8 ml; distilled water, 100 ml.
7. Rinse in water.
8. Dehydrate and differentiate in 95% alcohol; then 100% alcohol, xylene, and balsam as usual.

We have not used this method, regarding a 3- to 9-h technic as too cumbersome for frequent use, and having found some of the picric acid and hydrochloric acid–methyl blue methods fully effective. The periodic acid–leucofuchsin technic and its allochrome variant are suggested for kidney and lung.

MASSON'S TRICHROME STAIN (*J. Tech. Methods*, **12**: 75, 1929). This also followed a sequence procedure: Fix in Bouin's or Möller's (Regaud's) fluid or premordant formalin or alcohol material with Bouin's or Möller's fluids, respectively. Paraffin sections are fastened to slides with Masson's gelatin.

1. Stain with Regaud's iron hematoxylin.
2. Differentiate to a pure nuclear stain in two-thirds saturated alcoholic picric acid solution (6%).
3. Wash 15 min in running water (3 min is adequate to remove the picric acid).
4. Stain 5 min in 2 parts 1% "xylidene ponceau" (ponceau 2R, C.I. 16150) and 1 part 1% acid fuchsin (C.I. 42685), both in 1% acetic acid.
5. Rinse in distilled water.
6. Mordant 5 min in 1% aqueous phosphomolybdic acid.
7. Drain, and stain 5 min in 2.5% aniline blue (C.I. 42755) in 2.5% acetic acid or in 2% light green SF (C.I. 42095) in 1% acetic acid.
8. Return aniline blue stains to the phosphomolybdic acid for another 5-min period. (We find this step unnecessary.)
9. Take the aniline blue stains from the phosphomolybdic acid and the light green stains directly from the stain, and differentiate 2 min in 1% acetic acid. Dehydrate, clear, and mount by an alcohol, xylene, balsam sequence. Masson recommends Curtis' salicylic acid balsam or natural acid balsam. Polystyrene, Permount, or other resin should serve well.

RESULTS: Results are similar to those of other Mallory stains, depending on the colors of the plasma and collagen stains used.

Other writers have substituted the simpler Weigert's acid iron chloride hematoxylin for Regaud's. Fast green FCF (C.I. 42053) and wool green S (C.I. 44090) have been found (Lillie, *J. Tech. Methods*, **25**: 1, 1945) to be good substitutes for light green, giving green and blue-green collagen, respectively, and staining mucus in paler tones of the same colors. Biebrich scarlet (C.I. 26905), Bordeaux red (C.I. 16180), chromotrope 2R (C.I. 16570) in the reds and brilliant purpurin R (C.I. 23510) as a brown are good substitutes for the ponceau 2R and acid fuchsin mixture. Saturated alcoholic picric acid and saturated aqueous mercuric chloride are good premordants for formalin material.

A single solution stain giving muscle and cytoplasms in pink, with blue collagen, reticulum, and hypophyseal beta granules, is our buffered Mann stain.

Gomori (*Am. J. Clin. Pathol.,* **20:** 661, 1950) found that one-solution trichrome mixtures can be made from blue or green acid triphenylmethane or diphenylnaphthylmethane dyes as collagen stains with red sulfonated azo or disazo dyes as plasma stains, and acetic and phosphotungstic or phosphomolybdic acids. Phosphotungstic acid tends to intensify the plasma stain; phosphomolybdic, the fiber stain; and alcohol weakens the plasma stain. Pretreatment with hot Bouin's fluid intensifies staining of muscle and plasma. Water washing extracts plasma stains more than fiber stains, whereas acetic rinsing makes the preparation more transparent without altering the color balance. Gomori recommended the following technic:

GOMORI'S TRICHROME STAIN

1. Fix smears in alcohol or alcohol ether. Embed tissue blocks in paraffin, and section at 3 to 5 μ.
2. Bring smears or sections to water as usual.
3. Stain nuclei 5 min with alum hematoxylin (Delafield, Harris, Lillie-Mayer, or Ehrlich).
4. Wash in water.
5. Stain 5 to 20 min in the following mixture:

Chromotrope 2R	0.6 g
Fast green FCF	0.3 g
Phosphotungstic acid	0.6 g
Glacial acetic acid	1 ml
Distilled water	100 ml

6. Rinse in 0.2% acetic acid.
7. Dehydrate and clear by the usual alcohol-xylene sequence. Mount in polystyrene, Permount, HSR, or other synthetic resin.

RESULTS: Connective tissue is green; muscle and cytoplasm stain red; nuclei, gray-blue.

Like other fast green methods of the Masson type, it gives a rather diffuse and incomplete picture of the more finely fibrillar stroma. Much of the reticulum of liver and spleen is difficult to discern.

Engel and Cunningham [*Neurology (Minneap.*), **13:** 919, 1963] recommend the following trichrome method for muscle biopsies. Normal muscle myofibrils are green with distinct A and I bands. Intermyofibrillar material is red. Nuclei are reddish purple, and interstitial connective tissues are green. Engel and Cunningham state that the stain is particularly efficacious as an aid in the diagnosis of neuromuscular diseases.

1. Striated muscle may be obtained by biopsy or autopsy.
2. Rapidly freeze the muscle, and fix it to a chuck with 5% gum tragacanth by plunging it into isopentane cooled to about $-160°C$ with liquid nitrogen.
3. Cut sections 2 to 10 μ thick in a cryostat at $-25°C$, pick them up on cool coverslips, and let them dry at room temperature for 2 to 20 min.
4. Omit fixation.
5. Stain in Harris' hematoxylin for 5 min.
6. Rinse briefly in three changes of distilled water.
7. Stain for 10 min in Gomori's trichrome mixture raised to pH 3.4 with 1 N NaOH: chromotrope 2R, 0.6 g; fast green FCF, 0.3 g; phosphotungstic acid, 0.6 g; glacial

acetic acid, 1 ml; and distilled water to make 100 ml. The trichrome mixture should be freshly made each week.

8. Differentiate by a few dips in 0.2% acetic acid.
9. Dehydrate and mount in Permount.

THE HYDROCHLORIC ACID METHODS

A hydrochloric acid–orange G–methyl blue method of Lillie (*J. Tech. Methods*, **25**: 1, 1945) omits all so-called mordants and yields deep blue reticulum, fine collagen fibrils, renal glomerular stroma, and basement membranes, blue to brown coarse collagen bundles, lighter blue epithelial mucus, orange-pink erythrocytes, orange to gray cytoplasm, orange-yellow muscle, and black nuclei. Brush borders of renal epithelium sometimes stain differentially blue. Aniline blue can be used instead of methyl blue.

Formalin material is used. Paraffin sections are brought to water.

1. Stain 6 min in Weigert's or similar acid iron hematoxylin.
2. Wash in water, and stain 10 min in 1% aqueous phloxine B (C.I. 45410).
3. Wash 2 min in 1% acetic acid, and stain 10 min in methyl blue, 100 mg; orange G, 600 mg; concentrated hydrochloric acid (37%), 0.25 ml; distilled water, 100 ml.
4. Wash in 1% acetic acid 5 min, dehydrate, clear, and mount by an acetone, acetone and xylene, xylene sequence in synthetic resin.

The hydrochloric acid–Biebrich scarlet–methyl blue variant of this technic, giving a denser overall stain with more brilliant colors, may be done as follows (*Arch. Pathol.*, **29**: 705, 1940):

1. Stain paraffin sections of formalin-fixed material 6 min in Weigert's acid iron chloride hematoxylin.
2. Wash in water, and stain 5 min in 1% Biebrich scarlet in 1% acetic acid.
3. Wash 2 min in 1% acetic acid, and stain 5 min in a 0.5% dilution of concentrated hydrochloric acid containing 0.5% methyl blue or aniline blue or 0.2% amido black 10B. Satisfactory results can also be attained by reducing the methyl blue to 0.1% or the amido black 10B to 0.05% in the same hydrochloric acid solution if the staining time is prolonged to 20 to 40 min.
4. Dehydrate with 95 and 100% alcohol, clear with 100% alcohol and xylene mixture and two changes of xylene, and mount in synthetic resin.

RESULTS: Muscle is red with darker cross striae; erythrocytes are scarlet; cytoplasm, gray-pink; brush borders of renal epithelium, sometimes blue or green; mucus, pale blue or blue-green; collagen, reticulum, basement membranes, and renal glomerular stroma, deep blue or blue-black.

Sweat and Puchtler (*Stain Technol.*, **43**: 227, 1968) recommend the following one-step trichrome method particularly for the demonstration of fine connective tissue fibers:

1. Bring sections to water. If necessary, remove hematogenous formalin pigment with saturated alcoholic picric acid, or mercurial precipitates with the iodine–sodium thiosulfate sequence. Wash in running tap water for 5 min or until sections are decolorized.
2. Place in freshly prepared Bouin's fluid for 1 h at approximately 56°C (paraffin oven).
3. Wash in running tap water until sections are decolorized.

4. Stain 1 min in trichrome solution. Formula: chromotrope 2R, 0.6 g; aniline blue WS, 0.6 g; phosphomolybdic acid, 1.0 g; dissolve in 100 ml distilled water, then add 1 ml concentrated hydrochloric acid. Allow the solution to stand 24 h in the refrigerator before use; store in the refrigerator and use only the cold solution. Do not filter.
5. Rinse in a 1% aqueous acetic acid solution for 30 s.
6. Dehydrate rapidly in one change of 95% alcohol and three changes of absolute alcohol, clear in xylene, and mount in Permount or similar synthetic resin.

RESULTS: Collagen, reticulum fibers, basement membranes, intercalated disks in cardiac muscle, and cartilage stain blue. Nuclei, cytoplasm, fibrin, muscle, and elastic fibers stain red.

MISCELLANEOUS RECENT COLLAGEN METHODS

Sweat et al. (*Arch. Pathol.*, **78:** 69, 1964) advocate the following stain for connective tissues. The stain is said not to fade for at least 15 months.

1. Bring sections to water and hydrate at least 10 min.
2. Stain in picro-Sirius red F3B (C.I. 35780) solution for 30 min (0.1% Sirius red F3B in saturated aqueous picric acid).
3. Dehydrate rapidly in three changes of absolute ethanol, clear in xylene, and mount in synthetic resin.

Puchtler and Leblond (*Am. J. Anat.*, **102:** 1, 1958) used as a mordant 5% aqueous tannic acid for 5 min at 24°C or 1% phosphomolybdic acid (10 min, 24°C) or both prior to immersion of sections in several basic or amphoteric dyes, or in amido black. Cell membranes on the lateral surfaces of epithelial cells of intestinal mucosa of rats, basement membranes, and collagen fibers were noted to stain intensely with amido black or basic dyes after a tannic acid mordant.

Kallenbach (*Exp. Cell Res.*, **33:** 581, 1964) noted that the cell wall in the area of the cleavage furrow of dividing cells stained with the Puchtler and Leblond tannic acid–phosphomolybdic acid–amido black stain. Selective staining was observed in ameloblasts and human mammary carcinoma. Collagen, reticulum fibers, basement membranes, sarcolemma, and capillary stroma are stained intensely.

Constantine and Mowry (*J. Invest. Dermatol.*, **50:** 414 and 419, 1968) tested several stains for collagen and noted two distinctive features, namely, selective staining with picrofuchsin and strong yellow birefringence after staining with picro-Sirius red F3B.

Tonna [*J. R. Microsc. Soc.*, **83:** 307, 1964] noted enhanced birefringence of collagenous and other tissues in sections stained with basic dyes that exhibit metachromasia.

Olsen et al. (*Science*, **182:** 825, 1973) conjugated purified antibody to proline hydroxylase to ferritin with glutaraldehyde using a method of Kishida et al. (*J. Cell Biol.*, **64:** 340, 1975) and demonstrated prolyl hydroxylase in the cisternae of the rough endoplasmic reticulum of chick embryonic tendon cells.

PERIODIC ACID OXIDATION METHODS

Another method for the demonstration of collagen, reticulum, and basement membranes is the periodic acid–leucofuchsin method. This procedure is apparently unrelated chemically to the silver methods or to the acid-aniline dye collagen methods. Basically it is supposed to depend on the presence of hydroxyl groups or one hydroxyl and one primary or secondary amine group on adjacent carbon atoms. It now appears that this reaction in the case of collagen and reticulum (and gelatin) is due in small measure to the presence of hydroxylysine in the protein molecules, and, more largely, to the presence of a closely bound, nonglucosamine saccharide (Bangle and Alford, *J. Histochem. Cytochem.*, **2**: 62, 1954).

The periodic acid Schiff reaction of collagen and reticulum is quite readily blocked by sulfite treatment after periodic acid oxidation; that of basement membranes, somewhat less so. Immersion after periodic acid for a 2-h period in 0.05 M NaHSO$_3$ followed by a 10-min bath in Schiff reagent results in positive reactions of glycogen, starch, cellulose, most mucins, renal casts, yeasts, lens capsule, Descemet's membrane, rod acromeres, and vitreous coagulum. The connective tissues including collagen, reticulum, basement membrane, bone and cartilage matrix, zymogen granules, and thyroid colloid fail to react.

Acetylation with a 40% acetic anhydride–60% pyridine mixture before periodic acid oxidation prevents the aldehyde formation in most substances, if adequately prolonged. Cartilage matrix and starch resist acetylation the longest, 18 to 24 h at 25°C being required to block this reaction. Contrariwise, much shorter periods are adequate to prevent the reaction in collagen, reticulum, and basement membranes (2 h). Glycogen and the mucins are intermediate in their requirements.

The reaction of renal reticulum and glomerular basement membranes with the periodic acid–leucofuchsin technic is not abolished by 16 h digestion in 1 : 4000 bull testis hyaluronidase at pH 5 and 37°C, but the metachromatic staining of umbilical cord matrix was completely destroyed by the same solution in 2 h.

The usual periodic acid Schiff procedure used after the above chemical treatments was given in an early chapter. A useful variant specifically valuable for the differentiation of the connective tissues is the allochrome procedure (*Am. J. Clin. Pathol.*, **21**: 484, 1951). This method colors most collagen and reticulum bright blue; lens capsule, suspensory ligament, Descemet's membrane, renal tubule basement membranes and many other epithelial basement membranes, capillary basement membranes, and part of the medial stroma of arteries, deep red. It depends on the use of a picromethyl blue counterstain after a usual periodic acid Schiff sequence. The technic follows:

LILLIE'S ALLOCHROME CONNECTIVE TISSUE METHOD

1. Deparaffinize and hydrate paraffin sections through xylene, alcohols, and water as usual. Treat with iodine and thiosulfate if required for removal of mercury precipitates. Wash frozen sections with two changes of water to remove excess formaldehyde.
2. Oxidize 10 min in 1% aqueous H$_5$IO$_6$.
3. Wash 5 min in running water.
4. Immerse 10 min in Schiff reagent.
5. Wash 2, 2, and 2 min in three changes of 0.5% sodium metabisulfite.
6. Wash 5 min in running water.
7. Stain 2 min in Weigert's acid iron chloride hematoxylin.
8. Wash 4 min in running water.
9. Stain 6 min in saturated aqueous picric acid containing 40 mg methyl blue (C.I. 42780) per 100 ml.
10. Dehydrate and differentiate in two changes each of 95 and 100% alcohol. Wash in 50 : 50 alcohol-xylene mixture and clear in two changes of xylene. Mount in synthetic resin: polystyrene, Permount, HSR, cellulose caprate, Depex, or the like.

Kattine (*Stain Technol.*, **37**: 192, 1962) substitutes alum hematoxylin in step 7, staining for 10 min, washing in water, differentiating briefly in acid alcohol (1 ml concentrated

HCl, 99 ml 70% alcohol), washing again, and converting the hematoxylin stain to a black lake by immersing for 30 to 60 s in phosphotungstic-phosphomolybdic acid mixture (1.2% each). Preparations are again washed in water.

RESULTS: Nuclei are black, gray, or brown; cytoplasms and muscle cells, gray-green to greenish yellow; most reticulum, blue; ring fibers of splenic reticulum, typically red-purple; basement membranes of glomerular, renal, and many other capillaries, red-purple; medial stroma of arteries, red with fewer blue fibrils; epithelial basement membranes, characteristically red-purple in the kidney, where they are relatively thick, perhaps deep violet in other organs where they may be quite thin; but usually one can find at least small areas where an inner red lamina is slightly separated from an adjacent blue reticulum fibril. Amyloid varies from fairly deep to quite pale purplish red to lavender. Arterial hyalin colors orange-red to purple-red. Fibrin is red to orange. Bone matrix colors gray-orange; cartilage, red-purple. The collagen of areolar tissues is colored bright blue; that of denser masses may retain some coarse pink or greenish yellow fibers among the predominant blue or greenish blue fibers. Glycogen, starch, cellulose, Descemet's membrane; lens capsule and suspensory ligament; bacterial, yeast, and mold chitins; epithelial mucins and that of acute torulosis; thyroid and hypophyseal colloids; follicle fluid and zona pellucida in the ovary, prostatic secretion and corpora amylacea; intestinal melanosis and adrenal and ovarian lipofuscin pigments—all retain the red-purple colors of the Schiff reaction. Renal brush borders remain purplish pink, hyalin or colloid droplets color orange, and casts color deep red-purple.

The "black periodic acid procedure" is useful for demonstration of the stroma of smooth and striated muscle, mucosal basement membranes, and reticulum, as well as glycogen, starch, periodate-reactive mucins, and other substances.

Sweat and Puchtler (*Arch. Pathol.*, **78**: 73, 1964) claim that the aniline blue WS in the above allochrome procedure fades fairly rapidly. Sweat and Puchtler believe they have overcome this feature by substitution of a 0.07% solution of Sirius light blue GL (C.I. 23160) in aqueous saturated picric acid for 5 min at 24°C.

Sweat et al. (*Arch. Pathol.*, **86**: 33, 1968) advise the following stain particularly for disclosing fibrosis in glomeruli:

1. Bring sections to water and hydrate for at least 5 min.
2. Oxidize with 1% periodic acid (H_5IO_6) for 10 min at 24°C.
3. Wash in running water for 5 min.
4. Immerse in Schiff's reagent for 10 min at 24°C.
5. Transfer to three successive baths in 0.5% sodium metabisulfite ($Na_2S_2O_5$) for 2, 2, and 2 min at 24°C.
6. Wash in running water for 5 min.
7. Place in 1% phosphomolybdic acid for 5 min.
8. Rinse briefly in distilled water 3 times for 10 s each.
9. Stain in Sirius light blue GL solution for 5 min (2 g Sirius light blue GL in 100 ml distilled water with 2 ml glacial acetic acid added).
10. Rinse in two changes of distilled water for 10 s each.
11. Dehydrate rapidly in three changes of absolute ethanol, clear in xylene, and mount in synthetic resin.

ELASTIN

Elastic tissues contain a characteristic fibrous protein called *elastin*. Table 15-1 provides comparative amino acid analyses of collagen, elastin, and basement membranes. As contrasted with collagen, elastin contains substantially more nonpolar amino acids such as alanine, valine, isoleucine, leucine, and phenylalanine. It is further characterized by the exclusive content of desmosine and isodesmosine and a lesser content of the polar amino acids, hydroxyproline, and aspartic and glutamic acids as contrasted with collagen.

It is important to recognize that biochemical data on elastin were derived on the basis of the presumed near-immutability of elastin to weak alkalies (Lowry et al., *J. Biol. Chem.*, **139:** 795, 1941), to guanidine hydrochloride (Greenlee et al., *J. Cell Biol.*, **30:** 59, 1966), to autoclaving (Partridge, *Fed. Proc.*, **25:** 1023, 1966), and to certain enzymes (Fitzpatrick and Hospelhorn, *J. Lab. Clin. Med.*, **56:** 812, 1960; Hospelhorn and Fitzpatrick, *Biochem. Biophys. Res. Commun.*, **6:** 191, 1961).

Elastic tissues appear to contain a small amount of carbohydrate. Moret et al. (*J. Atheroscler. Res.*, **4:** 184, 1964) detected galactose, mannose, xylose, ribose, arabinose, fucose, glucosamine, galactosamine, and sialic acid in an NaCl-extracted autoclaved sample of ox ligamentum nuchae. Others who have detected carbohydrate include Banga and Balo [*Nature (Lond.)*, **178:** 310, 1956], Gotte et al. (*J. Atheroscler. Res.*, **3:** 244, 1963), Walford et al. (*Lab. Invest.*, **8:** 948, 1959), and Walford et al. (*Arch. Pathol.*, **72:** 158, 1961). Jensen and Berstelsen (*Acta Pathol. Microbiol. Scand.*, **51:** 241, 1961), Paule (*J. Ultrastruct. Res.*, **8:** 219, 1963), Yu and Lai [*J. Electron Microsc. (Tokyo)*, **19:** 362, 1970], and many others have shown the presence of acidic mucopolysaccharides surrounding elastic fibers.

Some investigators have provided evidence for lipid in elastic tissue (Lansing et al., *Anat. Rec.*, **114:** 555, 1962).

It is of interest to note the existence of two new amino acids (desmosine and isodesmosine) in elastic tissue demonstrated by Partridge et al. [*Nature (Lond.)*, **197:** 1297, 1963; ibid., **200:** 651, 1963]. The biosynthetic steps in the formation of desmosine and isodesmosine and their development into cross-links in elastin were demonstrated by Miller et al. (*J. Biol. Chem.*, **240:** 3623, 1965).

Inhibition of cross-linking occurs during the course of copper deficiency. Copper deficiency in growing swine results in the formation of altered elastic tissue (Carnes et al., *Fed. Proc.*, **20:** 118, 1961; Shields et al., *Am. J. Pathol.*, **41:** 603, 1962; Kimball et al., *Exp. Mol. Pathol.*, **3:** 10, 1964; Carnes et al., *Ann. N.Y. Acad. Sci.*, **121:** 800, 1965; Weissman et al., *Lab. Invest.*, **14:** 372, 1965). Increase in formic acid solubility was noted concomitant with progressive copper deficiency.

Sandberg et al. (*Biochemistry*, **8:** 2940, 1969) have prepared an elastin gel from aortas of copper-deficient swine. The gel appears to be a better preparation than that prepared earlier by Grant and Robbins [*Nature (Lond.)*, **185:** 320, 1960].

Bucciante and Gotte [*Arch. Sci. Biol. (Bologna)*, **42:** 490, 1958] noted elastin from aged individuals or from calcified regions of arteriosclerotic lesions was less resistant to alkaline or acid hydrolysis.

Labella and Lindsay (*J. Gerontol.*, **18:** 111, 1963) noted that elastin purified by defatting, autoclaving, and treatment with hot 0.1 N NaOH is progressively more resistant to the action of pancreatic elastase as a function of increasing age. They were impressed with the progressive increase of a fluorescent substance in elastic tissue with increasing age of the individual.

Elastic tissue appears to become increasingly subject to calcification with increasing age. Urry (*Proc. Natl. Acad. Sci.*, **68:** 810, 1971) has proposed that this propensity may be engendered in neutral uncharged binding sites by the high glycine content of elastin with beta turns with associated conformations that are known (from studies on ion-transporting antibiotics) to interact with cations. After deposition of calcium occurs, negatively charged phosphate and carbonate follow and mineralization proceeds.

Elastic fibers generally appear yellow by light microscopy. They may branch or become fenestrated in structures such as vessels. They are birefringent only when stretched.

Electron microscopic studies of elastogenesis were undertaken by Haust et al. (*Exp. Mol. Pathol.*, **4**: 508, 1965), Fahrenbach et al. (*Anat. Rec.*, **155**: 563, 1966), Yu and Lai [*J. Electron Microsc.* (*Tokyo*), **19**: 362, 1970], and others. Pollicard et al. (*Bull. Microsc. Appl.*, **4**: 139, 1954) in an electron microscopic study noted the interdigitation of elastic and collagen fibers in the lung permitting function during respiration.

On the basis of ruthenium red–stained preparations, Yu and Lai (*J. Electron Microsc.*, **19**: 362, 1970) suggest the presence of a network of microfibrils in elastin. The presence of microfibrils has also been suggested by Gotte et al. (Structure and Function of Connective and Skeletal Tissue, *Proc. NATO Conf. 1964*, Butterworth) and by Greenlee et al. (*J. Cell Biol.*, **30**: 59, 1966).

Winkleman and Spicer (*Stain Technol.*, **37**: 303, 1962) noted a selective uptake of tetraphenylporphine sulfonate into elastic fibers after administration of the reagent in vivo. Subsequently, Albert and Fleischer (*J. Histochem. Cytochem.*, **18**: 697, 1970) complexed silver or gold to tetraphenylporphine sulfonate and noted selective staining of aortic elastic fibers of young and old mice. Aortas were fixed in 3% glutaraldehyde and/or 2% OsO_4 in phosphate buffer of pH 7.2, dehydrated, embedded in Vestopal *W*, and sectioned. For the silver preparation, sections were stained for 45 min in the silver tetraphenylporphine sulfonate (available from Electron Microscopy Sciences, Box 251, Fort Washington, Pennsylvania 19304), and then with uranyl acetate and lead citrate. The gold preparation could be applied only after uranyl acetate and lead citrate due to formation of nonspecific precipitates. Selective dense staining of elastic fibers was observed.

Sirius rose BB (C.I. 25380, Roboz Surgical Instrument Co., Washington D.C.) and Chicago blue 6B (C.I. 24410; E. Gurr's Microme No. 91), which may be used as vital stains to mark sites undergoing calcification, have been noted to localize in elastic fibers in all tissues (Moss, *Stain Technol.*, **29**: 247, 1954; Storey, *Stain Technol.*, **43**: 101, 1968).

ELASTIC FIBERS

At alkaline pH levels elastic fibers may be stained quite selectively with a number of acid dyes, and in acid alcohol solution by certain basic dyes. Spicer and Lillie (*Stain Technol.*, **36**: 365, 1961) used eosin B and Biebrich scarlet at pH levels up to 10 for basic proteins and demonstrated elastin staining. This appears assignable to arginine content, and Lillie et al. (*J. Histochem. Cytochem.*, **19**: 487, 1971) demonstrated arginine in rat arterial elastica with the 1,2-naphthoquinone-4-SO_3Na (NQS) method. The native aldehyde reaction of elastin was reported first by Feulgen and Voit (*Pfluegers Arch.*, **206**: 389, 1924) and may be demonstrated by Schiff's reagent in elastin of young animals and children. It is generally absent in adult human beings (Pizzolato and Lillie, *Arch. Pathol.*, **88**: 588, 1969) and in the probably old horse nuchal ligament (Dempsey et al., *Anat. Rec.*, **113**: 197, 1952). The aldehyde reaction appears to be largely responsible for acid alcohol staining with pararosanilin, Bismarck brown R and Y, enianil azurin J (Horobin and James direct blue 152) and certain other experimental dyes used by Lillie et al. [*Acta Histochem.* (*Jena*), **44**: 215, 1972], but not for staining with orcein, resorcein, resorcin fuchsin, or amidol black I. Elastic fibers color pink to red with the Morel-Sisley diazotization coupling tyrosine reaction of Lillie (*J. Histochem. Cytochem.*, **5**: 528, 1957) and less certainly by the Millon reaction; they give a positive azo coupling reaction by the diazosulfanilic acid–pH 1 azure A procedure (Lillie and Donaldson, *J. Histochem. Cytochem.*, **20**: 929, 1972) which is blocked by nitration but rather enhanced by zinc acetate blockade of histidine (Lillie, unpublished studies, 1973). The stain extraction studies followed the procedure of Fullmer and Lillie (*J. Histochem. Cytochem.*, **4**: 64, 1956) on a number of other dyes. Lillie et al. [*Acta Histochem.* (*Jena*), **44**: 215, 1972] found that all "extraction procedures" failed on Bismarck brown and diazosafranin, where covalent bonding is presumed to exist with acid dyes; partial to complete extraction occurred with alkalis. Most of the elastin stains resisted attempts at acid extraction. Formamide weakened or removed staining by enianil azurin J, amidol black I, resorcin fuchsin, and other experimental dyes. The conclusion reached was, in contrast with Horobin and James, that there is no single mechanism covering all elastin staining. Horobin and James (*Histochemie*, **22**: 324, 1970) had proposed that elastin staining, generally, occurred by large dye molecule adhesion by van der Waals forces; Lillie et al. [*Acta Histochem. (Jena)*, **44**: 163, 1972] reported that a number of monoazo dyes consisting of only two benzene rings performed effectively. The mode of elastin staining by unmordanted hematoxylin in young animals and children remains obscure. This is the Clara hematoxylin method as used for enterochromaffin.

The direct Schiff reaction of young elastin noted above complicated other procedures in which Schiff reagent is the demonstrating agent: the Feulgen nucleal and plasmal reactions; the Bauer, Casella, and PAS–glycol reactions; the performic and peracetic Schiff stains. The elastin reaction with Schiff reagent is readily prevented by phenylhydrazine or borohydride (Chap. 8).

The part played by iron ions in such procedures as the Gallego iron fuchsin method, the resorcin fuchsin stain, and Verhoeff's method, the premixed iron hematoxylin stain which sometimes brilliantly demonstrates elastin, remains uncertain.

Verhoeff's Iodine–Iron Hematoxylin

Verhoeff's procedure is an overstaining with an iodine plus ferric chloride plus hematoxylin mixture, followed by a ferric chloride differentiation (Mallory). It seems to be quite permanent. We have seen a section stained by Verhoeff nearly 50 years before which was still excellent. The elastic tissue was deep violet, nuclei were blue-violet, the acromeres of the rods and cones were a good violet.

VERHOEFF'S METHOD, ACCORDING TO MALLORY. Fix in Zenker or in formalin, embed in paraffin or celloidin, bring sections to 80% alcohol. Do not iodize before staining.

Dissolve 1 g hematoxylin in 20 ml hot 100% alcohol. Add 8 ml 10% ferric chloride solution. Mix and add 8 ml Lugol's solution containing 2% iodine and 4% potassium iodide. This solution is best fresh but can be used for 2 to 3 weeks. Immerse sections in this mixture for 15 min or more, until quite black.

Differentiate a few seconds in 2% ferric chloride, observing microscopically in water. If overdifferentiated, restain at once. Wash in water; then in 95% alcohol to remove the excess iodine; then again in water. Counterstain in 0.5% aqueous eosin, dehydrate in alcohols, clear in origanum oil, mount in balsam.

RESULTS: Elastin is black; nuclei are blue-black; collagen, fibrin, glia, and myelin, pink; erythrocytes, orange-red. The acromere staining noted above is probably a lipid stain.

Brissie et al. (*J. Histochem. Cytochem.*, **22**: 895, 1974) conducted an ultrastructural study of Verhoeff's iron hematoxylin stain for elastic fibers. Ribosomes, heterochromatin, mucus in goblet cells, and granules in Paneth cells and acinar cells of rat parotid glands were also stained. Brissie et al. conclude that ionic binding is responsible for the staining of mucus inasmuch as staining is prohibited in solutions saturated with NaCl. Hydrogen bonding is assumed to be responsible for staining of elastic tissues and granules in acinar cells of the parotid, since staining was prevented in solutions saturated with urea.

No single mechanism can be invoked which accounts for all elastin staining [Lillie et al., *Acta Histochem. (Jena)*, **51**: 109, 1974]. Using nigrosin base, C.I. 50415B, solvent black 7, highly selective black elastin staining is obtainable with a 1% solution in 60% isopropanol or 70% ethanol acidified with 1% concentrated hydrochloric acid. Spirit-soluble nigrosin, C.I. 50415, solvent black 5, is not effective with this technic. But both dyes in neutral 60% isopropanol selectively stain arterial elastin in elderly human normal and arteriosclerotic arteries, but not that in young human or animal arteries. The findings suggest the appearance in old arteries of a lipoprotein component.

Orcein

THE TAENZER-UNNA ACID ORCEIN METHOD. Bring paraffin sections to 70% alcohol by the usual sequence.

1. Stain for 1 to 18 h at room temperature or for 10 to 30 min at 30 to 37°C in a freshly filtered solution of 1 g orcein (natural or synthetic) in 100 ml alcohol of 65% (Stutzer), 70% (Mallory, Romeis), 75% (Taenzer), 80% (Merk), 90% (Kornhauser, from Conn and Darrow), 100% (Cowdry, Schmorl, Lee) to which 1 ml of official PG or concentrated hydrochloric acid is added. (Stutzer, Taenzer, and Merk are cited from Ehrlich's "Encyklopädie.")
2. Wash in alcohol (70 to 100%) and in water, either first.
3. Counterstain briefly in polychrome methylene blue (Unna), 1 : 1000 azure A or toluidine blue for 1 min, or alum hematoxylin for 3 to 6 min, if desired.
4. Dehydrate and clear by an alcohol- or acetone-xylene sequence and mount in balsam or synthetic resin.

One may further add a connective tissue stain of Van Gieson or Mallory type, or a picroindigocarmine. The foregoing technic is a composite from the various modern texts.

Fullmer and Lillie (*J. Histochem. Cytochem.*, **4**: 64, 1956) were unable to prevent staining of elastic fibers with the Taenzer-Unna orcein method in formalin-fixed acetylated, benzoylated, deaminated, methylated, or phenylhydrazine-treated sections. Braun-Falco (*Arch. Klin. Exp. Dermatol.*, **203**: 256, 1956) conducted comparable studies on the mechanism of action of aldehyde fuchsin and Weigert's resorcin fuchsin stain for elastic fibers. Neither stain could be prevented by aldehyde, hydroxyl, amine, or carboxyl group blocking reagents. Braun-Falco also noted that phloroglucin behaves like resorcin and orcinol with ferric chloride in acid ethanolic solution to form a dye selective for elastic fibers.

Lillie (*Histochemie*, **19**: 1, 1969) conducted studies on the Taenzer-Unna orcein type of staining of elastic tissues. He noted that the dyes and indicators lacmoid, resorcin blue MLB (C.I. 51020), part of the Musso (Musso et al., *Angew. Chem.*, **73**: 434, 1961) resorcin blue products and the two *m*-aminophenol oxidation products, namely, elastin purple FP and elastin violet PR, selectively stain elastic fibers in an acid (0.12 N HCl) ethanolic (70%) solution. On the basis of failure of azolitmin and resorufin (both 7-hydroxy-2-phenoxazones) and the success of resorcin blue MLB (7-*N*,*N*-dimethylamino-2-phenoxazone) in staining elastic fibers, he suggests that a determining characteristic may be the aminophenoxazone structure. The view is supported by success with *m*-aminophenol substitution in the Musso air oxidation NH_3 synthesis. Trials with triphenylmethane dyes to replace orcein were generally unsuccessful.

Lillie et al. (*Stain Technol.*, **43**: 203, 1968) noted that sections stained with the usual Taenzer-Unna orcein stain may be transferred to a 70% ethanolic solution containing 0.02% ferric chloride ($FeCl_3 \cdot 6H_2O$) or 0.02% copper sulfate ($CuSO_4 \cdot 5H_2O$) for 1 min, thereby converting brownish elastic fibers and nuclei to reddish black. Use of the Van Gieson counterstain thereafter is not recommended.

For staining collagen and elastin Salthouse (*J. Histochem. Cytochem.*, **13:** 133, 1965) brought formol-fixed sections to absolute methanol after any desired pretreatments and stained for 3 min in a saturated solution of Luxol Fast Blue G in methanol (about 0.9%), rinsed in methanol, and if desired counterstained for 5 min with nuclear fast red 0.1% in 5% aluminum sulfate for nuclei and 0.1% aqueous tartrazine for 1 min.

Methylation enhanced background as well as collagen and elastin staining; methylation followed by permanganate oxidation prevented elastin but not collagen staining; methylation and saponification permitted both elastin and collagen staining. A Van Slyke deamination for 8 h at 25°C prevented elastin staining but not collagen. Tannin (5%) treatment if prolonged to 6 h blocked both elastin and collagen staining; acetylation blocked neither; benzoylation produced collagen and elastin fragmentation and partial lysis. Sulfation induced some background staining, prevented elastin stains, and did not impair collagen stains.

Luxol fast blue G forms a water-soluble complex with gelatin.

Lillie, Pizzolato, and Donaldson [*Acta Histochem. (Jena) Suppl.* **9**, p. 625, 1971] later reported on a Wurster synthesis of a new dye resorcein, by hydrogen peroxide oxidation of resorcinol in the presence of ammonia.

Resorcein in a 1% solution in 0.12 N HCl–70% alcohol gives in a 6-h Taenzer-Unna technic black elastin, dark red nuclei, pink to pale pink collagen, muscle, and gland cytoplasm, orange-yellow to pale yellow erythrocytes.

The dye is an indicator, red in acid, blue in alkali, changing at pH 4.3 to 5.5. The same synthesis has recently been applied to the manufacture of orcein with excellent results in one of the commercial laboratories.

The performance of the dye is further discussed in the report of Lillie et al. [*Acta Histochem. (Jena)*, **44:** 215, 1972] on mechanisms of elastin staining.

Shellow and Klingman (*Arch. Dermatol.*, **95:** 221, 1967) studied elastic fiber networks of skin in three dimensions utilizing thick sections. They recommend the following method:

1. Cut biopsy or necropsy sections horizontally and transversely at 100 μ on a freezing microtome.
2. Place sections on albuminized slides, and air-dry for several hours.
3. Stain with acid orcein (1% in 70% ethanol and 0.6% HCl).
4. Decolorize in 0.5% acid ethanol for 5 min.
5. Wash in running water for 10 min.
6. Dehydrate thoroughly; clear and mount in synthetic medium.

FRAENKEL'S METHOD (Schmorl)

1. Stain nuclei red with (Orth's) lithium carmine. Differentiate in hydrochloric acid alcohol (1%).
2. Stain 24 h: Of a stock solution of 1.5 g orcein, 120 ml 95% alcohol, 60 ml distilled water, and 6 ml nitric acid, add sufficient to a 3% alcoholic hydrochloric acid solution to give a dark brown color. This is the staining solution.
3. Differentiate in 80% alcohol.
4. Stain 10 to 15 min in 0.25% indigocarmine in saturated aqueous picric acid solution.
5. Rinse in 3.5% acetic acid.
6. Dehydrate quickly in 95 and 100% alcohol, clear in oil or xylene—we suggest the 100% alcohol plus xylene 50 : 50 mixture first, then two changes of xylene. Mount in balsam (or synthetic resin).

RESULTS: This method gives red nuclei, dark brown elastin, blue-green collagen, greenish yellow muscle.

ROMEIS' METHOD. Bring paraffin sections to 70% alcohol.

1. Stain 1 h in 1% orcein in 0.125 *N* hydrochloric acid, 70% alcohol.
2. Wash thoroughly in two changes of distilled water.
3. Stain heavily with Ehrlich's acid hematoxylin, acid hemalum, or for 3 min in Hansen's iron alum hematoxylin.
4. Wash for 10 min in distilled, tap, and distilled water.
5. Stain in 60 ml 0.1% acid fuchsin in saturated aqueous picric acid solution to which is added 0.25 to 0.3 ml 2% acetic acid (5 min should suffice).
6. Rinse in 60 ml distilled water containing 2.5 ml of the acidified picrofuchsin solution above, for not more than 2 to 4 s.
7. Blot dry.
8. Wash in 95% alcohol for 1 min.
9. Dehydrate in 100% alcohol for 3 min, then clear with xylene and mount in balsam.

RESULTS: Black to red-brown elastica, yellow muscle, bright red collagen, dark brown nuclei are shown.

Weigert's acid iron chloride hematoxylin can probably be used in place of Hansen's formula, and any alum hematoxylin would undoubtedly serve. The addition of the few drops of 2% acetic acid (pH probably above 2) to the Van Gieson stain already at pH 1.95 seems rather futile, and it seems that the technic could be simplified by direct transfer from the Van Gieson mixture to alcohol.

Resorcin Fuchsin

Another important group of elastin stains are the iron resorcin lakes of basic fuchsin, crystal violet, and other basic dyes. Weigert (*Zentralbl. Allg. Pathol.*, **9**: 289, 1898; Ehrlich's "Encyklopädie") directed thus: To 200 ml 1% basic fuchsin solution add 4 g resorcinol and boil until dissolved. Then add 25 ml PG Liquor Ferri Sesquichlorati* (the modern United States official solution of iron chloride is the same), and boil 2 to 5 min longer. Cool and collect the precipitate on a filter. Take up the precipitate from the sides and bottom of the original vessel as well as from the filter paper with 200 ml 95% alcohol (by boiling as necessary). Add 4 ml concentrated hydrochloric acid and filter, washing the filter through afterward with enough fresh alcohol to restore the total volume to 200 ml. Commercially prepared resorcin fuchsin is available.

Weigert directed 15- to 30-min staining in this solution, followed by alcohol differentiation and clearing in xylene. This gave black elastic fibrils on a pale violet background.

One may use 2 g crystal violet in 200 ml water in place of the basic fuchsin. In this case 0.5 to 2 g dextrin should be added, according to French (*Stain Technol.*, **4**: 11, 1928; also personal communications). The resulting stain is green. Or if 1 g basic fuchsin and 1 g crystal violet are used, a deep blue-green is achieved. If 2 g safranin is used, elastic fibrils are stained brownish red. The resorcinol used should be fresh and crystalline. When heating the alcohol to dissolve the precipitate, a closed electric hot plate or a steam table is preferred

---

* 25 ml PG or US = 62 ml British 1932 = 29 ml France 1905 = 27 ml Spain 1905 = 16.5 ml Netherlands 1905. USP Liquor Ferri Chloridi is about 2.3 *M* $FeCl_3$.

because of the fire hazard. Romeis specified the hydrochloric acid as the PG official 25%, sp gr 1.126; Ehrlich specified it as concentrated. Most writers have followed the latter, and some even add more hydrochloric acid for staining.

Usually this procedure is combined with other stains which, because of the high acidity and alcohol content of the resorcin fuchsin stain, must follow it rather than precede. Silver impregnations for reticulum, however, should precede the elastic tissue stain. In this case the crystal violet compound is preferable to resorcin fuchsin, because the green contrasts better than blue-black with the black silver deposit. The more common combinations are with fibrin or collagen stains. Some simply precede the elastin stain with a carmine stain for nuclei.

Histochemical studies on the staining behavior of resorcin fuchsin with normal and abnormal renal tissues were conducted by Jackson et al. (*J. R. Microsc. Soc.*, **88**: 473, 1967) and Joiner et al. (*J. R. Microsc. Soc.*, **88**: 461, 1967).

For combination with staining of tubercle bacilli, first stain as usual with hot (1 h at 55 to 60°C) carbol fuchsin, wash off in water, and then stain for 20 to 30 min in the acid alcoholic resorcin fuchsin solution, which decolorizes cells and other structures at the same time. Then differentiate in alcohol, counterstain for 5 min in 0.1% methylene blue in 1% acetic acid, dehydrate with alcohol or acetone, and clear with a 50% xylene mixture with 100% alcohol or acetone followed by two changes of xylene. Mount in synthetic resin.

For Hornowski's (*Z. Wiss. Mikrosk.*, **26**: 128, 1909) Weigert's elastica Van Gieson's collagen stain we have substituted variations of Hart (*Zentralbl. Allg. Pathol.*, **19**: 1, 1908) and Kattine (*Stain Technol.*, **37**: 193, 1962):

1. Deparaffinize and bring to 70% alcohol. Rinse in acid alcohol (1 ml concentrated HCl, 99 ml 70% alcohol).
2. Stain 24 h in Weigert's resorcin fuchsin solution diluted with 9 volumes acid alcohol.
3. Rinse in acid alcohol, and place for 1 to 3 min in 95% alcohol.
4. Stain 7 to 10 min in alum hematoxylin (Ehrlich, Harris, Delafield, or other).
5. Rinse in water, differentiate briefly in acid alcohol, rinse 1 min in running water.
6. Form black dye lake by immersing 30 to 60 s in phosphomolybdic phosphotungstic acid mixture (1 g each in 80 ml distilled water).
7. Rinse in distilled water, and counterstain 2 to 5 min in Lillie's HCl–Van Gieson (0.25 ml concentrated HCl, 100 ml saturated aqueous picric acid, 100 mg acid fuchsin).
8. Rinse briefly in distilled water, dehydrate, and differentiate in two changes each of 95% and absolute alcohol, clear in xylene, mount in synthetic resin.

Puchtler and Sweat (*Stain Technol.*, **35**: 347, 1960) use commercial prepared resorcin fuchsin obtained from Chroma, staining for 4 to 5 h at 20 to 25°C in 0.2% resorcin fuchsin in 70% alcohol containing 1% by volume of concentrated hydrochloric acid. Distilled water rinsing and a Van Gieson counterstain follow, then alcohols, xylene, and synthetic resin.

Higher concentrations gave less selective staining; lower concentrations required longer staining, 24, 48, or more hours for 0.1, 0.05, or 0.02%.

Puchtler and Sweat (*Stain Technol.*, **39**: 164, 1964) recommend the following as a selective stain for renal basement membranes. The method is selective and useful only for Carnoy-fixed tissues.

1. Bring sections to water. Treat with 0.5% aqueous HIO_4 for 5 min.
2. Rinse in two changes of distilled water.
3. Place in $NaHSO_3$ solution overnight (approximately 15 h). Formula: Dissolve 20 g $NaHSO_3$ in 40 ml distilled water; add 10 ml absolute alcohol.

4. Wash in five or six changes of distilled water.
5. Stain in resorcin fuchsin solution for 4 h. Formula: Resorcin fuchsin (Chroma), 0.2 g; 70% alcohol, 100 ml; concentrated HCl, 1 ml.
6. Rinse in distilled water.
7. Counterstain:
   a. For red nuclei, counterstain with kernechtrot (nuclear fast red, Chroma) for 1 to 5 min; rinse in distilled water. Formula: Dissolve 0.1 g kernechtrot in a 5% solution of aluminum sulfate by aid of heat, cool, and filter. *Or*
   b. Counterstain with Van Gieson's picrofuchsin for Van Gieson background.
8. Dehydrate in alcohol, clear in xylene, mount in Permount or similar synthetic resin.

RESULTS: Basement membranes and elastic fibers are stained black; collagen fibers and reticulum fibers are stained red; nuclei are pink.

THE VOLKMAN-STRAUSS METHOD. Others have combined the resorcin fuchsin with one of the aniline blue methods, such as the Volkman-Strauss (*Z. Wiss. Mikrosk.*, **51:** 244, 1934) combination with the Mallory-Heidenhain procedure; and no doubt picromethyl blue and similar procedures, with or without a red plasma stain, could be used.

An elastic tissue fibrin combination is sometimes useful, and such a method is quoted by Schmorl. A variant of this which we have found useful follows:

ELASTIN FIBRIN TECHNIC. Bring paraffin sections to 95% alcohol.

1. Stain 20 to 40 min in resorcin fuchsin or the resorcin fuchsin crystal violet. Mercury-fixed tissue should be stained 1 to 2 h.
2. Wash in 95 and 80% alcohol.
3. Stain 20 min in carmine, 5 min in acid iron hematoxylin, or 10 min in 1% Bismarck brown R in 1% acetic acid.
4. Wash in water.
5. Stain 2 to 3 min in aniline methyl violet (saturated methyl violet in 9 ml distilled water and 1 ml saturated methyl violet in 20% aniline–80% alcohol mixed at time of using). Any methyl violet or crystal violet solution will serve.
6. Rinse quickly in 0.9% aqueous sodium chloride solution.
7. Flood with Weigert's iodine, pour off, and again flood, giving total exposure of 20 to 30 s.
8. Rinse in water.
9. Blot dry.
10. Clear and decolorize in equal volumes of aniline and xylene as long as a fresh drop of mixture is colored violet by the section.
11. Wash in two or three changes of xylene. Mount in balsam.

RESULTS: Nuclei are red, gray, or brown; elastin is blue-black; fibrin, violet; gram-positive bacteria are blue-black.

RESORCIN FUCHSIN OIL RED O. The resorcin fuchsin method may also be combined with a fat stain if a suitable quantity of the stock stain is diluted with water to bring its alcohol content down to 60%. The frozen sections are stained 1 to 2 h in a covered dish, then rinsed with 60% alcohol, and stained with oil red O for 5 min (dilute 3 ml stock saturated isopropanol solution with 2 ml distilled water; let stand 7 to 8 min, filter, and use at once). Then wash in water, counterstain 3 to 5 min in 0.1% aqueous Janus green B (C.I. 11050), and wash 1 min in 5% acetic acid. Wash in water, float onto slides, and mount in gum syrup. Or use an alum hematoxylin nuclear stain for 5 min, wash, and mount.

Orcinol New Fuchsin (Fullmer and Lillie, *Stain Technol.*, **31:** 27, 1956)

This reagent appears to be more selective for elastic fibers than the preceding ones. The reagent is prepared in a manner similar to that for Weigert's resorcin fuchsin.

Add 2 g new fuchsin (magenta III, C.I. 42520) and 4 g orcinol (reagent grade) to 200 ml distilled water, boil 5 min, and add 25 ml USP IX Liquor Ferri Chloridi and boil 5 min more. (Or use 15.5 g $FeCl_3 \cdot 6H_2O$ and water to make 25 ml, if the USP solution is not available.) Cool, collect precipitate on a filter, and dissolve it in 100 ml 95% alcohol. This is the staining solution.

Sections are deparaffinized, brought to absolute alcohol, stained for 15 min at 37°C in the above solution, differentiated in three changes of 70% alcohol, for 5 min each, dehydrated, cleared, and mounted in resin as usual. Only elastic tissue and organic enamel matrix stain deep violet.

Various counterstains may be combined with it. A Van Gieson stain may be interposed either before or after the alcohol differentiation.

Staining in a 50 : 50 mixture of redox-buffered iron hematoxylin and orcinol new fuchsin gives excellent results.

GALLEGO'S IRON FUCHSIN METHOD.   This method was included in Lillie, 1954, pp. 363–364, in German (*Am. J. Clin. Pathol.*, **3:** 13, 1939), and in Langeron, and was attributed by them to Gallego. A modification of it by Lillie is in Conn and Darrow.

Aldehyde Fuchsin

GOMORI'S ALDEHYDE FUCHSIN METHOD FOR ELASTIC FIBERS (*Am. J. Clin. Pathol.*, **20:** 665, 1950). Dissolve 0.5 g basic fuchsin (C.I. 42510) in 100 ml 70% alcohol. Add 1 ml concentrated hydrochloric acid and 1 ml USP Paraldehyde. In 24 h the mixture becomes a deep violet and is ready to use. Store at 0 to 5°C. (The molar ratio of acetaldehyde to rosanilin is nearly 15 : 1.)

Avoid chromate fixations. Formalin and Bouin's fluid give colorless backgrounds; mercury fixatives give a pale lilac.

1. Deparaffinize and hydrate paraffin sections as usual.
2. Treat 10 to 60 min with 0.5% iodine.
3. Decolorize 30 s with 0.5% sodium bisulfite.
4. Wash 2 min in water.
5. Transfer to 70% alcohol.
6. Stain in aldehyde fuchsin: for elastic fibers, 5 to 10 min; for pancreatic islet cells, 15 to 30 min; for hypophysis, $\frac{1}{2}$ to 2 h. Rinse in alcohol (70%) and inspect microscopically from time to time.
7. Wash in several changes of 70% alcohol.
8. Counterstain with hematoxylin and orange G or with the Masson trichrome or Mallory-Heidenhain method. The latter two are preferred for hypophysis. Fast green FCF or light green SF should be substituted for the aniline blue in these methods.
9. Dehydrate, clear, and mount through an alcohol, xylene, synthetic resin sequence.

RESULTS: Elastic fibers, mast cell granules, gastric chief cells, beta cells of pancreatic islets, and certain of the hypophyseal beta granules stain violet to purple. Alpha granules of the hypophysis stain orange-red, delta granules stain green to greenish blue, and chromophobe cells present pale gray-green cytoplasm. Collagen is stained green, using the fast green Mallory-Heidenhain variant.

According to Vassar and Culling (*Arch. Pathol.*, **68:** 487, 1959), elastic fibers show quite a strong autofluorescence in unstained sections or in hematoxylin eosin–stained preparations. The fluorescence is enhanced by a 3-min stain in a 1% alcoholic solution of acriflavine, followed by dehydration, clearing, and mounting in Harleco resin (HSR).

Sumner (*J. R. Microsc. Soc.*, **84:** 329, 1965) conducted histochemical studies of the mechanism of action of aldehyde fuchsin staining with various proteins and other constituents with and without previous oxidation of tissues. Decisive results were not achieved; however, tissue basophilia appears to be involved.

Kwaan and Hopkins (*Stain Technol.*, **39:** 123, 1964) advise the following stain for simultaneous demonstration of elastic fibers and lipids.

1. Rinse section in distilled water to remove all residual formalin and then in 60% isopropanol.
2. Stain in a tightly covered container with a freshly filtered oil red O solution for 6 to 10 min. (Formula: oil red O, 2 g; 60% isopropanol, 50 ml.)
3. Differentiate for 5 to 10 s in 60% isopropanol, which must be kept covered or made up as needed.
4. Stain in aldehyde fuchsin for 4 to 10 min, depending on the shade of purple desired. (Formula: basic fuchsin, 1 g; concentrated HCl, 2 ml; paraldehyde, 2 ml; 70% isopropanol, 200 ml.) Keep at room temperature in a closed container for about 48 h or until the stain turns to a deep purple.
5. Rinse briefly in 60% isopropanol; transfer to water.
6. Mount in glycerol jelly.

RESULTS: Neutral fat is stained orange to bright red; elastic tissue and mast cells, light to deep purple. A good contrast is obtained between the lipid-laden intima and the elastic media in specimens with early atheromatous lesions. The staining of mast cells is an added advantage with this method.

Horobin and James (*Histochemie*, **22:** 325, 1970) recommend the following stain for elastic fibers. The method differs from others in that staining occurs at high pH (9) without iron and naphthols.

1. Fix selected tissue in 10% neutral formalin, and prepare paraffin-embedded sections in routine manner. Dewax in xylene and alcohol, and pass into water.
2. Stain sections in dye bath overnight at room temperature. [Dye-bath formula: Prepare 0.5% solution of enianil azurin J, direct blue 152 (C.I. 24360), in 2-methoxyethanol or dimethyl sulfoxide. Mix the dye solution with an equal volume of barbiturate or "universal" buffer at pH 9. Use within 1 week.]
3. Wash in tap water.
4. Dehydrate in alcohol, pass through xylene, and mount in a synthetic medium.

RESULTS: Elastic fibers stained red. With older solutions background may be stained blue.

This dye apparently stains by a combination of condensation of its amine group with the lysinal aldehyde and anionic dye binding to basic groups, probably arginine in elastin. The stain is weakened but not blocked by prior borohydride reduction of aldehydes, is partially extracted by borax solution but not completely, and when borohydride pretreatment is combined with borax extraction after staining, complete loss of staining occurs [Lillie et al., *Acta Histochem. (Jena)*, **44:** 215, 1972].

Menzies (*Stain Technol.*, **38:** 245, 1963) and Menzies and Roberts [*Nature (Lond.)*, **198:** 1006, 1963] noted the acidophilia of elastic fibers is modified by several factors. Acidophilia was best demonstrated (1) in tissues fixed in formol sublimate, (2) by staining with bromophenol blue for at least 30 min, and (3) at an alkaline pH up to 11. He noted that for reasons unknown, the bromophenol blue readily washes out of elastic tissues of elderly subjects but only slowly from those of infants.

Cooper (*J. Histochem. Cytochem.*, **17**: 539, 1969) attempted to determine the nature of the reducing material in elastic tissue responsible for the blue ferric ferricyanide stain. He concluded that phenolic groups probably in tyrosine were responsible. Small contributions from lipochrome or polytyrosines could not be excluded. He suggests that this reducing property may be useful for analysis of suspected pathologic material such as "elastotic" material.

Cooper (*J. Histochem. Cytochem.*, **19**: 564, 1971) studied the Casella (permanganate-Schiff) staining of elastic fibers. He noted that brief (5 min) oxidation with acid (0.1% aqueous sulfuric) permanganate resulted in subsequent excellent staining either with Schiff's reagent or with basic dyes (Alcian blue and Nile blue A). The permanganate-induced basophilia is subject to a critical electrolyte concentration of about 0.15 M with the Alcian blue–$MgCl_2$ system and is readily blocked by postoxidative methylation and greatly diminished by preoxidative acetylation, dinitrophenylation, and methylation.

Pizzolato and Lillie (*Arch. Pathol.*, **88**: 581, 1969) noted that arterial elastic tissues of newborn infants stain readily with unmordanted hematoxylin and with Schiff's reagent in sections not previously oxidized. Staining with unmordanted hematoxylin gradually diminished with age until staining could not be demonstrated in persons over age twenty. Staining with Schiff's reagent was prevented in sections blocked by aldehyde blocking reagents; however, the blockade was somewhat ineffective in the prevention of staining with unmordanted hematoxylin. The presence of an aldehyde-quinoid structure in young arterial elastic tissue is postulated.

Maher (*Arch. Pathol.*, **67**: 175, 1959) noted a reduction or failure of staining of elastic lamina of postpartum uterine vessels with Weigert's resorcin fuchsin and Verhoeff's stains for elastic tissues for approximately 30 days postpartum. Subsequently, he notes regeneration of the elastic tissue and the acquisition of properties reactive for usual stains. He proposes that enzymatic activity is responsible for the altered staining during the postpartum period.

The abnormal elastic tissues in pseudoxanthoma elasticum have been studied histochemically by Fisher et al. (*Am. J. Pathol.*, **34**: 977, 1958) and by Moran and Lansing (*Arch. Pathol.*, **65**: 688, 1958). Parallel results obtained from abnormal and normal elastic fibers of the dermis prompted those writers to conclude that the abnormal tissue was indeed elastic.

Histochemical studies by Wechsler and Fisher (*Arch. Pathol.*, **77**: 613, 1964) failed to reveal any qualitative differences from normal in elastic fibers from individuals with Ehlers-Danlos syndrome.

Elastic fibers are stainable with several elastic tissue stains even in Egyptian mummies [Sandison, *Nature (Lond.)*, **198**: 597, 1963].

Elastase

Recrystallized trypsin (Armour and Co.) has no effect on elastin, according to Lansing et al. (*Anat. Rec.*, **114**: 555, 1952), but commercial trypsin does digest elastin on overnight exposure at 37°C. This difference is due to the presence of an elastase in the cruder product, which is extractable by pH 6, 0.1 M phosphate buffer and is salted out by 0.4 saturated ammonium sulfate. Lansing's group used digestion at pH 9 for 1 h at room temperature or for less time at 37°C. Baló and Banga (*Biochem. J.*, **46:** 384, 1950) prepared an elastase concentrate from defatted and powdered pancreas.

Recrystallized elastase is now available from a number of commercial sources.

Fullmer (*J. Histochem. Cytochem.*, **6:** 425, 1958) prescribed 6-h digestion at 37°C in 0.015% "elastase" in 0.1 *M* Veronal HCl at pH 8.8, staining afterward with orcein, resorcin fuchsin, orcinol new fuchsin, or aldehyde fuchsin.

The Worthington preparation available at that time exhibited some proteolytic action. Fullmer later (*J. Histochem. Cytochem.*, **8:** 290, 1960) reported quite satisfactory results with the twice crystallized Worthington product at 8 mg/30 ml 0.2 *M* borate buffer pH 9, using a 6-h 37°C digestion.

Trowbridge and Moon (*Lab. Invest.*, **21:** 288, 1969) prepared antiporcine and antihuman elastase by immunizing rabbits with porcine or human elastase. The antiserum to porcine elastase reacted with human elastase, and the converse also occurred. Fluorescein-conjugated sheep antirabbit globulin was employed to demonstrate nicely the presence of elastase in acinar cells and exocrine secretions of human pancreas.

Banga (*Acta Physiol. Hung.*, **24:** 1, 1963) employed orcein-impregnated elastin for a biochemical assay for elastase and an elastase inhibitor. Presence of the inhibitor in pancreatic extracts was proved by Loeven (*Acta Physiol. Pharm. Neerland.*, **10:** 228, 1962), who separated the inhibitor by employing starch column electrophoresis.

Oxytalan Fibers

Oxytalan fibers were first described in human periodontal membranes (Fullmer, *Science*, **127:** 1240, 1958; Fullmer and Lillie, *J. Histochem. Cytochem.*, **6:** 425, 1958; Fullmer, *J. Histochem. Cytochem.*, **8:** 290, 1960). The name oxytalan is derived from Greek words meaning acid, enduring, and to bear or resist in recognition of the resistance of these fibers to acid solution in contrast to collagen. The fibers stain with aldehyde fuchsin only after a strong oxidation such as with peracetic acid, performic acid, potassium persulfate (Oxone) permanganate, and less adequately with bromine. Best performance of the stain for identification is achieved with aldehyde fuchsin. Fairly unsatisfactory results may be obtained with resorcin fuchsin and orcein. The other elastic tissue stains (Verhoeff's and orcinol new fuchsin) fail to stain the oxidized or unoxidized fibers.

Table 15-5 provides data for distinguishing oxytalan from fibers of other types. Several factors permit the development of the thesis that oxytalan fibers are related to elastic fibers. They are stained with varying degrees of success with three of five elastic tissue stains after oxidation. Ordinarily, oxidized fibers remain unstained with basic aniline dyes. Rannie (*Trans. Eur. Orthodont. Soc.*, **39:** 127, 1963) is able to stain oxidized fibers with basic dyes in Newcastle, England; however, we have been unable to do so in the United States. Oxytalan fibers exhibit the same morphologic features and array in human periodontal membranes as do elastic fibers in several animals such as cattle. The absence of elastic fibers in human periodontal membranes and the array of human oxytalan fibers analogous to elastic fiber array in animals invites further speculation of comparability. Although elastic fibers are not normally detected in human periodontal membranes, periodontal structures around teeth involved with scleroderma contain fibers that manifest all criteria typical of elastic fibers. It is suggested that perhaps scleroderma promotes changes in oxytalan fibers to convert or alter them to elastic-type fibers (Fullmer and Witte, *Arch. Pathol.*, **73:** 184, 1962).

TABLE 15-5 COMPARISON OF OXYTALAN FIBERS WITH OTHER FIBER TYPES

Fiber type	Characteristics	Oxytalan fiber characteristics
Collagen	Birefringent	Nonbirefringent
	Acid-soluble	Resists acid solution
	Stains with methods for collagen	Not identified with stains for collagen
Reticulum	Stains with methods for reticulum	Unstained with methods for reticulum
Nerve	Stains:	
	Silver	Unstained by methods for nerve
	Myelin	Unstained by methods for myelin
	Cholinesterase	Unstained by methods for cholinesterase
		Morphologic features dissimilar to nerve
		Distribution dissimilar to nerve
Elastic	Animal periodontal membrane distribution	Distribution in human and monkey periodontal membrane similar to that of elastic fibers in other animals
	Usual stains for elastic fibers	Unstained with usual stains for elastic fibers
	Stains with aldehyde fuchsin after oxidation	Stains with aldehyde fuchsin after oxidation
		Dissimilar to elastic fibers at electron microscopic level
Preelastic	Stains with aldehyde fuchsin oxidation	Stains with aldehyde fuchsin after oxidation
	Stainable component removed by β-glucuronidase after oxidation	Stainable component removed by β-glucuronidase after oxidation
		Similarities and dissimilarities at the electron microscopic level

Unoxidized oxytalan fibers resist digestion by crude preparations of elastase, although oxidized fibers are readily digested (Fullmer, *J. Histochem. Cytochem.*, **8:** 113, 1960). The stainable component is readily removed by β-glucuronidase from oxidized fibers. Lysozyme digestion is less effective. Oxytalan fibers have a characteristic structure revealed by electron microscopy (Carmichael and Fullmer, *J. Cell Biol.*, **28:** 33, 1966). Filaments 150 to 200 Å in diameter are interspersed with a nonfilamentous amorphous material occupying approximately the same amount of space. A definite periodicity has not been observed.

The presence of oxytalan fibers in sites other than periodontal membranes is less well established. This is because developing preelastic fibers are demonstrable with the oxytalan fiber stain prior to their identification with the usual stains for elastic fibers (Fullmer, *J. Histochem. Cytochem.*, **8:** 290, 1960). Present histochemical methods cannot discriminate between oxytalan fibers and preelastic fibers. By electron microscopy, oxytalan fibers may be discerned among preelastic and elastic fibers. Oxytalan fibers develop in reparative tissue in sites where they are normally present (Fullmer, *Arch. Pathol.*, **70:** 59, 1960). For a review, see Fullmer et al., *J. Oral Pathol.*, **3:** 291, 1975.

Mander et al. (*J. Histochem. Cytochem.*, **16:** 480, 1968) noted that oxytalan fibers in mice differ from those in human beings in that they are stained with aldehyde fuchsin after periodic acid oxidation. Strong oxidation such as peracetic is required to effect stainability in human oxytalan fibers.

16

SMEAR PREPARATIONS, BACTERIA, PROTOZOA, AND OTHER PARASITES

For the study and demonstration of smear preparations, bacteria, protozoa, and other parasites, a variety of special methods are prescribed. Some of them use the usual fixed material, but some perhaps function better with special fixations.

PREPARATION OF SMEARS AND FILMS

In place of sections or to supplement them it is often desirable to utilize spread films or smears of tissues, blood, and exudates. Films are preferable for the study of the cytology of the blood, the red bone marrow, the spleen pulp, and various inflammatory exudates. For the demonstration of bacteria, protozoa, and rickettsiae they are often preferable to sections. For the demonstration of scarce blood protozoa, thick blood films are made and hemolyzed before or during staining so that a relatively large volume of blood can be scanned quickly for the presence of malarial plasmodia, trypanosomes, leishmaniae, and even microfilariae.

ENDICOTT'S MARROW SMEAR METHOD. For the staining of spleen and marrow Endicott's (*Stain Technol.*, **20:** 25, 1945) technic for the preparation of films is recommended. Dip a capillary pipette into a tube of human serum or plasma, and take up a column about 10 mm in length. Then immediately place the point of the pipette in the red marrow or spleen pulp, and aspirate about 2 mm of tissue. Blow the tissue and serum onto a clean slide near one end, and mix thoroughly by repeated aspiration and expulsion with the same pipette. Finally, leaving the drop on the slide, smear it in the usual manner by placing another slide against the first at an angle of 30 to 45° over the drop and between it and the center of the first slide. Then draw the slide back toward the drop until it makes contact and the fluid spreads along the acute angle between the two slides. Then push the second slide along the first away from the original site of the drop so that the fluid film follows in the acute angle. The thickness of the film can be regulated by varying the angle of contact; the more acute the angle, the thinner the film.

IMPRESSION FILMS. Useful tissue films can also be prepared by simply pressing a clean slide lightly on a freshly cut surface of the tissue in question. These films are called *impression films*. The method is often used for brain.

Thin blood films are simply prepared by depositing a small drop of blood near one end of the slide and drawing it along in the acute angle made by a second slide, as above.

Thick blood films are made by depositing several large drops of blood near one end of a clean slide and spreading them with a glass rod, the corner of a clean slide, or a match into a circular area about 10 to 15 mm in diameter. It is often useful to make a thick film on one end of a slide and a thin film on the remaining two-thirds. Since the value of thick films depends on the removal of the hemoglobin from the red corpuscles by hemolysis in distilled water or some other agent, one should be careful with such combined films not to allow the fixative to come in contact with the thick film before hemolysis.

Smears of thick pus may perhaps require dilution with serum (as for spleen and marrow smears). Thinner exudates can be smeared as is blood. Relatively clear or even turbid fluids may require centrifugation and resuspension of the sediment in a small drop of serum or serous exudate for the preparation of satisfactory films. With serous inflammatory exudates of relatively high protein content the supernatant fluid is satisfactory for resuspension of the sediment, but with urine and spinal fluid, serum is a more satisfactory diluent.

Freeman [*Acta Cytolog. (Baltimore)*, **13:** 416, 1969] and Kaplow (*Stain Technol.*, **46:** 177, 1971) advocate the use of Aqua Net hair spray (Rayette Inc., 1290 Avenue of the Americas, New York, New York 10019) as a fixative for vaginal, oral, or blood smears. The aerosol is stated to be less expensive than the usual fixatives, and specimens stained with the Papanicolaou, hematoxylin and eosin, PAS, Sudan black, methyl green–pyronin, peroxidase, M-Nadi oxidase, and alkaline phosphatase methods were stated to be excellent.

CENTRIFUGATES. According to Arcadi (*J. Urol.*, **61:** 814, 1949) centrifuged urinary sediments may be smeared on slides, air-dried, and then fixed by immersion for 1 min in 99% isopropyl alcohol. After this fixation he stained for 2 min in alum hematoxylin (Harris'), washed in water, counterstained for 1 min in 2% aqueous eosin Y, washed for 15 s in water, dehydrated with isopropyl alcohol, cleared in xylene, and mounted in Permount, dammar, or other suitable resin.

Undoubtedly other stains can be applied after this fixation technic.

Aspirated material and washings from hollow viscera may be fixed by mixing with an equal volume of 15% formalin and then centrifuging. Addition of alcohol to precipitate mucus may be necessary with gastric washings. The sediment can then be dehydrated after some hours in formalin, dealcoholized, and embedded in paraffin. Paraffin sections may be prepared as usual and stained in a variety of ways. The method, according to Wollum et al. (*J. Natl. Cancer Inst.*, **12:** 715, 1952), is superior to smears in cancer diagnosis, in avoiding thick areas, in preserving cell relationships of small tissue fragments, and in permitting multiple stains.

FIXATION. Thin blood films and marrow films, tissue smears, and exudate smears should be fixed at once in methyl alcohol for 3 to 5 min and then allowed to dry until it is convenient to stain them. Such films can be stained successfully for 2 to 3 weeks, whereas films not fixed at the time of taking soon deteriorate, so that staining becomes inferior.

Sometimes special procedures are better served by fixing still moist films by the vapor of osmium tetroxide or by gaseous formaldehyde (films are placed face down over a shallow vessel containing a 1 to 2% osmium tetroxide solution or a little concentrated formalin) or, perhaps more conveniently, by depositing 1 to 2 ml strong formalin in a Coplin jar and then standing the films in it, film end up, and putting on the lid.

In the case of quite thin smears, where focusing may be difficult, it may be useful to draw a line, or more than one, on the slide with a grease pencil or, if the preparation is to be covered with a cover glass, with black marking ink to establish a readily visible plane of focus.

Thick blood films should ordinarily be dried in a place protected from dust, flies, and roaches for 18 to 24 h before staining. Staining should not be delayed any longer after drying than necessary. If staining cannot be carried out within 24 h, it is preferable to hemolyze with distilled or even tap water for 5 to 10 min and then fix 5 min in methyl alcohol.

Umlas and Fallon (*Am. J. Trop. Med. Hyg.*, **20:** 527, 1971) hemolyze thick films by applying 3 drops of 0.5 to 1% saponin to the air-dried film and gently agitating for about 5 s. The fluid is then drained off and the film dried for about 10 min before staining as usual.

Such films may be stained by a variety of methods, including simple solutions of basic aniline dyes, Gram's stain, the acid-fast stain, Macchiavello's stain, Goodpasture's, Giemsa's, Wright's, Leishman's, and many others. After staining they may be allowed to dry or mounted in synthetic resin.

In place of the ether-alcohol mixture recommended by Papanicolaou ("Diagnosis of Uterine Cancer by the Vaginal Smear," Commonwealth Fund, New York, 1945), Davidson, Clyman, and Winston (*Stain Technol.*, **24:** 145, 1949) highly recommend 3 : 1 mixtures of *tert*-butyl alcohol with ethyl alcohol and with ethyl phosphate, and report excellent results with Papanicolaou's stain.

PAPANICOLAOU STAIN FOR CANCER AND PRECANCER DIAGNOSIS. There has appeared a considerable number of single-solution methods giving the general color effects of the Mallory aniline blue connective tissue stain. Several of them have been applied, mainly to the staining of smears of desquamated epithelia and notably in the study of the variations of vaginal contents during the estrus cycle, in the diagnosis of uterine cervical carcinoma, and the like. Some are applicable also to paraffin and celloidin sections. They have also been widely applied to smear and exudate material from other organs, to millipore filtrates of ureteral urine, spinal fluid, and thin exudates, gastric washings, etc.

In the Papanicolaou (*J. Lab. Clin. Med.*, **26:** 1200, 1941; *Science*, **95:** 438, 1942; *J. Natl. Cancer Inst.*, **7:** 357, 1947) technics for staining smears for cancer diagnosis, ether alcohol (50 : 50 *or* 1 : 2) or isopropyl alcohol fixation (2 min) is usually recommended. Smears are then brought through descending alcohols to water and stained for 5 to 10 min in an alum hematoxylin with or without added acetic acid; washed, blued, and dehydrated through ascending alcohols; counterstained 30 to 100 s in a phosphotungstic acid–orange G solution in alcohol (OG-5, 6, or 8); washed in three changes of alcohol; stained in an alcoholic solution of light green SF, Bismarck brown, and eosin Y for $1\frac{1}{2}$ min [designated as EA 25, 31, 36, 50, or 65 (Table 16-1)]; and washed 3 times in 95% alcohol, then 100% alcohol, alcohol xylene, two changes of xylene, and balsam.

Some of these technics are presented in quite elaborate detail. Gates and Warren ("Handbook for Diagnosis of Cancer of the Uterus," Harvard University Press, Cambridge, Mass., 1947), for example, prescribe dipping 5 times into each of the descending and ascending graded alcohols.

The subjoined technics (Papanicolaou and millipore) now used at the Charity Hospital in New Orleans were kindly furnished us by Dr. Nelson D. Holmquist. Part of the timings are given in "dips," which take about 1 s each, and in "slow dips," which take about 5 s. These are not translated into seconds since the motion implied in the word "dip" is an essential part of the process.

Gynecologic smears (vaginal, cervical, or combined scrapes) are received in the laboratory in specimen bottles containing fixative (95% ethyl alcohol, or Carnoy's solution if the specimen appears bloody). Each specimen is accompanied by a cytopathology request form containing all pertinent clinical information. An accession number is given to each specimen and is written on both the slide and the cytopathology form. After the specimen has been logged, it is stained using one of the modified Papanicolaou techniques.

TABLE 16-1 COMPOSITION OF PAPANICOLAOU EA STAINS

	EA 25	EA 31	EA 50*	EA 65†
Light green SF 0.5% alcoholic	44 ml	50 ml	45 ml	4.5 ml
Bismarck brown 0.5% alcoholic	12 ml	8 ml	10 ml	10 ml
Eosin Y 0.5% alcoholic	44 ml	42 ml	45 ml	45 ml
Phosphotungstic acid	170 mg	170 mg	200 mg	200 mg
Li_2CO_3 saturated aqueous	1 drop	1 drop	1 drop	1 drop
95% alcohol	....	....	....	40.5 ml

* EA 50 was originally described as EA 36.

† The amount of light green in EA 65 has been further reduced since the second edition.

The staining procedure for the hand staining of gynecologic smears is as follows:

1. 95% ethyl alcohol or Carnoy's solution (fixative)
2. Tap water, 10 dips.
3. Tap water, 10 dips.
4. Alum hematoxylin,[1] 2 min.
5. Tap water, 10 dips.
6. Tap water, 10 dips.
7. Scott's tap water; substitute bluing agent, 1 min.
8. Tap water, 10 dips.
9. Tap water, 10 dips.
10. 95% ethyl alcohol, 10 dips.
11. 95% ethyl alcohol, 10 dips.
12. OG, $1\frac{1}{2}$ min.
13. 95% ethyl alcohol, 10 dips.
14. 95% ethyl alcohol, 10 dips.
15. EA 50, $1\frac{1}{2}$ min.
16. 95% ethyl alcohol, 10 dips.
17. 95% ethyl alcohol, 10 dips.
18. Absolute ethyl alcohol, 30 s.
19. Absolute ethyl alcohol and xylene (1 : 1), 1 min.
20. Xylene, 2 min.
21. Xylene, 4 min.
22. Mount in synthetic medium.

Most of the gynecologic smears are stained using the 23-stage Shandon-Elliot Automatic Stainer. Times cut on the staining disk are as follows:

1. Tap water, 30 s.
2. Tap water, 30 s.
3. Alum hematoxylin,[1] 2 min.
4. Tap water, 30 s.
5. Tap water, 30 s.
6. Scott's tap water, 1 min.
7. Tap water, 30 s.
8. Tap water, 30 s.
9. 95% ethyl alcohol, 30 s.
10. 95% ethyl alcohol, 30 s.
11. OG-6, $1\frac{1}{2}$ min.
12. 95% ethyl alcohol, 30 s.
13. 95% ethyl alcohol, 30 s.
14. EA 50, $1\frac{1}{2}$ min.
15. 95% ethyl alcohol, 30 s.
16. 95% ethyl alcohol, 30 s.
17. Absolute ethyl alcohol, 1 min.
18. Absolute ethyl alcohol, 1 min.
19. Absolute ethyl alcohol and xylene (1 : 1), 1 min.
20. Xylene, 2 min.
21. Xylene, 4 min.
22. Mount in synthetic medium.

---

[1] Lillie et al. (*Am. J. Clin. Pathol.*, **60:** 817, 1973) found that iron alum–celestin blue B could be used to replace Harris' hematoxylin.

Millipore technics offer a convenient method of concentrating the cells from sparsely cellular body fluids and washings. Millipore is a cellulose plastic material of unstated composition, furnished in filter disks 150 μ thick and of varying diameters. Disks of 15 mm diameter have been found convenient for filtrations of quantities of about 10 ml. Slight suction is used to expedite filtration. An open manometer attached to the suction line by a T tube should read not over 10 mm mercury.

The millipore disks are conveniently labeled with a single typewriter character which serves both to identify the case and to mark the upper surface bearing the cell film. They can then be attached to a carrier assembly by small spring hemostats strung by the spring end on the carrier and holding the millipore disks in their jaws. Larger, 40- to 50-mm disks can be similarly carried through the stain baths, but they require larger staining vessels and are mounted finally on 50 × 75 mm (2 × 3 in) slides.

Apparatus, instructions, and millipore disks are obtainable from the Millipore Filter Corp., Bedford, Massachusetts. Similar material is available from the Gelman Corp., Chelsea, Michigan.

MILLIPORE STAIN TECHNIC

1. 95% ethyl alcohol, 30 s.
2. Tap water, 30 s.
3. Tap water, 30 s.
4. Alum hematoxylin, 1 min.
5. Tap water, 30 s.
6. Tap water, 30 s.
7. 0.05% HCl, 30 s.
8. Tap water, 30 s.
9. Tap water, 30 s.
10. Scott's water, 1 min.
11. Tap water, 30 s.
12. Tap water, 30 s.
13. 95% ethyl alcohol, 30 s.
14. 95% ethyl alcohol, 30 s.
15. OG-6, 2 min.
16. 95% ethyl alcohol, 30 s.
17. 95% ethyl alcohol, 30 s.
18. EA 50, 2 min.
19. 95% ethyl alcohol, 30 s.
20. 95% ethyl alcohol, 30 s.
21. Absolute propanol, 2 min.
22. Absolute propanol and xylene (1 : 1), 2 min.
23. Xylene, 4 min.
24. Xylene, 4 min.
25. Xylene and Eukitt[1] (1 : 1), 6 min.
26. Mount in synthetic medium (Eukitt).

The orange G solutions as used are supersaturated in 95% alcohol and must be freshly diluted. For OG-5 take 1 ml 10% aqueous orange G and 19 ml absolute alcohol and add 5 mg phosphotungstic acid. For OG-6 reduce the phosphotungstic acid to 3 mg; for OG-8, to 2 mg. Except for the phosphotungstic acid the solutions are identical.

The Shorr (*Science*, **94**: 545, 1941) stains are also used principally for smears. Shorr also fixed in ether-alcohol for 2 min, then stained 1 min in "S-3," washed (10 dips each) in 70, 95, and 100% alcohol, cleared in xylene, and mounted. Foot (in "Pathology in Surgery," p. 11, Lippincott, Philadelphia, 1945) brought smears (or sections) to water, stained in alum hematoxylin, washed, blued, and counterstained 5 min in his modified Shorr stain, and carried through ascending alcohols to xylene and balsam or synthetic resin. See Table 16-2.

Dart and Turner (*Lab. Invest.*, **8**: 1513, 1959) use an acridine orange fluorochrome technic for exfoliative cytology.

---

[1] Eukitt may be purchased from O. Kindler, Freiburg, West Germany, Silberbachstrasse, 25.

TABLE 16-2 COMPOSITION OF SHORR STAINS

Ingredient	Shorr, 1941	Foot, 1945
Biebrich scarlet, C.I. 26905	500 mg	300 mg
Aniline blue WS, C.I. 42755	....	75 mg
Orange G, C.I. 16230	250 mg	125 mg
Fast green FCF, C.I. 42053	75 mg	25 mg
Glacial acetic acid	1 ml	1 ml
Phosphotungstic acid	500 mg	25 mg
Phosphomolybdic acid	500 mg	250 mg
Alcohol	100 ml	50 ml
Water	....	50 ml

Sources: Shorr, *Science*, **94**: 545, 1941; Foot, "Pathology in Surgery," Lippincott, Philadelphia, 1945.

Fix smears in equal parts of diethyl ether and 95% ethyl alcohol for 1 h or longer. Stain as follows:

1. Five dips each in 80, 70, and 50% alcohol and in distilled water.
2. Four dips in 1% acetic acid.
3. Distilled water, 2 min.
4. pH 3.8 buffer, 3 min [Walpole (Table 20-8) or McIlvaine (Table 20-10)].
5. Buffered 0.01% acridine orange, 3 min, with initial agitation. Mix 1 ml 0.1% acridine orange in 9 ml distilled water containing 0.2% "Tween 80" in pH 3.8 buffer. The working stain is refrigerated when not in use. The stock 0.1% acridine orange is stable at room temperatures of 20 to 25°C.
6. Differentiate 4 min in pH 3.8 buffer.
7. Wipe ends of the slides, then gently blot with coarse blotting paper to remove excess buffer.
8. Mount under a coverslip in pH 3.8 buffer and examine. Let each smear stand 2 min after application of coverslip.

Immediately before screening blot excess buffer from the slide with a towel. It is convenient to store the buffer used as a mounting medium in a small dropper bottle that is cleaned frequently. If the slides dry during examination, add more buffer. After screening remove the coverslips gently. The slides may be reexamined within 3 months by fluorescence microscopy after remounting with buffer. Beyond that time, decolorize and restain in acridine orange. Or slides may be decolorized with 50% alcohol and restained by the usual Papanicolaou method.

RESULTS: Normal superficial squamous cervical epithelial cells show fluorescent green nuclei and gray-green transparent cytoplasm; basal and intermediate cells present green or yellow-green nuclei and greenish or reddish cytoplasm; endocervical cells present green nuclei, blepharoplasts, and cilia, and reddish cytoplasm.

Bacteria, yeasts, yeast spores, and monilia stain red; trichomonads, red with a yellow nucleus. Leukocytes are bright green; mucus is pale dull green.

Atypical and dyskaryotic cells present increased green nuclear staining. Atypical hyperplastic and hyperchromatic cells present brilliant green to orange-red nuclear fluorescence and gray-green to red cytoplasm. Atypical brilliant red nucleoli are often present.

The Gram Stain

The mechanism of action of the Gram stain in bacteria remains unknown despite many efforts to clarify the problem. Several investigators conclude permeation factors by various reagents employed are responsible for the differential staining reaction (Benians, *J. Pathol. Bacteriol.*, **23**: 401, 1920; Burke and Barnes, *J. Bacteriol.*, **18**: 69, 1929; Bartholomew et al., *Stain Technol.*, **34**: 147, 1959, and *J. Gen. Microbiol.*, **36**: 257, 1964; Salton, *J. Gen. Microbiol.*, **30**: 223, 1963). Stearn and Stearn (*J. Bacteriol.*, **9**: 463, 1924), Barbo and Kennedy (*J. Bacteriol.*, **67**: 603, 1954), and Lamanna and Mallette (*J. Bacteriol.*, **68**: 509, 1954) believe differentiation is due to differences in basic dye uptake. Some believe a particular constituent of cells is responsible for the differentiation, including proteins (Deussen, *Z. Hyg. Infekt. Kr.*, **93**: 512, 1921), carbohydrate (Webb, *J. Gen. Microbiol.*, **2**: 260, 1948), nucleic acids [Henry and Stacey, *Nature (Lond.)*, **151**: 671, 1943], sulfhydryl groups (Fischer and Larose, *J. Bacteriol.*, **64**: 435, 1952), glycerophosphate and teichoic acids [Mitchell and Moyle, *Nature (Lond.)*, **166**: 218, 1950, and *J. Gen. Microbiol.*, **10**: 533, 1954] and lipids (Schumacher, *Zentralbl. Bakteriol. [Orig.]*, **109**: 181, 1928; Basu and Biswas, *Histochemie*, **16**: 150, 1968).

Differences also are observed depending upon whether dry slides or wet slides are used for the Gram stain. Chelton and Jones (*J. Gen. Microbiol.*, **21**: 652, 1959) noted that ruptured cells of gram-positive bacteria and yeast stained gram-negative if wet slides were used and positive if dry slides were used. Nonruptured cells stained gram-positive with both methods. Bartholomew et al. (*J. Gen. Microbiol.*, **36**: 257, 1964) repeated experiments by Chelton and Jones and noted that Gram positivity and negativity of the preparations examined were a function of the degree of decolorization time. All ruptured cells became gram-negative if decolorization time was extended to 30 min instead of 2 min. They also noted that all organisms were decolorized more rapidly when wet preparations were used.

It seems unlikely that a particular gram-positive substrate-dye-iodine complex is responsible for Gram differentiation if cells can be gram-differentiated by the use of iodine alone (Bartholomew et al., *Stain Technol.*, **34**: 147, 1959; *J. Gen. Microbiol.*, **36**: 257, 1964) or dye alone (Benians, *J. Pathol. Bacteriol.*, **23**: 401, 1920; Bartholomew and Mitwer, *Stain Technol.*, **25**: 103, 1950) provided care is exercised with the decolorization step, although the "stability" of that complex may be a factor (Biswas and Basu, *Int. Rev. Cytol.*, **29**: 1, 1970). The data permit the suggestion that permeation factors are involved with Gram differentiation. Further investigations are needed to provide convincing proof.

Basu and Biswas (*Histochemie*, **16**: 150, 1968) report that after *Escherichia coli* are treated with absolute ethanol for 10 min at 24°C and then with a 1 : 1 mixture of chloroform and methanol for 30 min, they become gram-positive. They suggest that the Gram staining of bacteria is related to the manner in which lipids are structurally related to other constituents of the cell wall rather than to the presence of lipid per se.

Bergh et al. (*Can. J. Biochem.*, **42**: 1141, 1964) could find no correlation between Gram staining of bacteria and the kinds of fatty acids in lipids.

GRAM-POSITIVE BACTERIA

Fibrin, certain *hyalin* droplets seen in degeneration of renal epithelium, and *keratohyalin* granules share to a greater or lesser extent the property of *gram-positive bacteria* of retaining the dye complex formed by the action of iodine upon crystal violet when certain solvents are applied. Gram-positive bacteria, however, are more resistant to solvent extraction than are fibrin and hyalin droplets. Thus ethyl alcohol, acetone, and their mixtures usually leave gram-positive bacteria blue-black and decolorize fibrin and hyalin droplets. Keratohyalin is intermediate in its resistance. Aniline and aniline-xylene mixtures leave fibrin as well as bacteria stained violet.

Various other substances are usually added to crystal violet or methyl violet solutions in mixtures of water and alcohol. The dye concentrations are usually high, and some solutions are apparently supersaturated when made. The function of the added phenol or aniline is not clear. We are not inclined to credit the alleged mordant action of these substances, since the Hucker-Conn ammonium oxalate variant seems as good as any and is considerably more stable than the aniline water solutions, or even the phenol water solutions. Many of these solutions were formerly described as gentian violet solutions. Inasmuch as gentian violet seems to have been a variable mixture of dextrin, crystal violet (hexamethylpararosanilin), and methyl violet (its tetra- and pentamethyl homologs), and crystal violet alone serves the same purpose better (Conn and Darrow), this dye is here prescribed in all formulas.

EHRLICH'S ANILINE CRYSTAL VIOLET AS EMENDED BY CONN (Conn and Darrow). Dissolve 1.2 g crystal violet (C.I. 42555) in 12 ml 95% alcohol, and add 100 ml freshly prepared and filtered aniline water made by shaking 2 ml aniline in 100 ml water. This keeps about 2 weeks. Weigert (Ehrlich's "Encyklopädie") used methyl violet (C.I. 42535) in a similar formula.

STIRLING'S ANILINE CRYSTAL VIOLET, AS EMENDED BY CONN (Conn and Darrow). Use crystal violet, 5 g; 100% alcohol, 10 ml; aniline, 2 ml; and distilled water, 88 ml. This solution is quite stable.

NICOLLE'S CARBOL CRYSTAL VIOLET, AS EMENDED BY CONN (Conn and Darrow). Use crystal violet, 1 g; 95% alcohol, 10 ml; phenol, 1 g; and distilled water, 100 ml. Dissolve the dye in alcohol, the phenol in water, and mix; or, perhaps easier, dissolve both in alcohol and add the water. Schmorl cites a carbol crystal violet containing 2.5 g phenol, and Mallory cites one with 3 g, both attributed to Nicolle. Both these authors give the amount of gentian or crystal violet as 10 ml saturated alcoholic solutions. According to Conn this would be about 1 g of most commercial samples.

THE HUCKER-CONN AMMONIUM OXALATE CRYSTAL VIOLET, MODIFIED (*Arch. Pathol.,* **5**: 828, 1928). Use crystal violet, 2 g; 95% alcohol, 20 ml; ammonium oxalate, 800 mg; and distilled water, 80 ml. Dissolve the dye in the alcohol, the oxalate in the water, and mix. The solution keeps at least 2 or 3 years.

GRAM'S IODINE. Dissolve 2 g potassium iodide in 2 to 3 ml distilled water; dissolve 1 g iodine crystals in this solution. Dilute with distilled water to 300 ml for Gram's solution, or to 100 ml for Weigert's.

GRAM STAINING OF SMEARS FOR BACTERIA

TECHNIC

1. Fix by quickly passing the smear face down through the blue flame of a Bunsen burner or alcohol lamp 3 times.
2. Stain with crystal violet for 20 to 60 s. Acid-fast bacilli require at least 1 to 2 min, and perhaps it is safer to heat to 60 to 80°C as well.
3. Wash in water. Conn blots off excess dye and does not wash.
4. Cover with Gram's iodine for 1 to 2 min or with Weigert's for 20 to 30 s.
5. Decolorize for 30 to 60 s by dropping alcohol on the film or by agitation in two or three changes of alcohol until color clouds no longer come out and the film, if of an

exudate, is largely decolorized. Instead of alcohol, one may use acetone from the dropper bottle. With this reagent decolorizing is complete in 5 to 10 s.

6. Wash in water.
7. Counterstain for 30 to 60 s in a 0.1 to 0.5% solution of safranin O (C.I. 50240), basic fuchsin (C.I. 42510), Bismarck brown Y or R (C.I. 21000 or 21010), or pyronin Y or B (C.I. 45005 or 45010).
8. Wash in water, dry, and examine in immersion oil.

Lillie preferred the Hucker-Conn crystal violet formula, the Weigert iodine-acetone decolorization, and 0.2 to 0.5% safranin as counterstain.

RESULTS: Gram-positive organisms are blue-black; gram-negative, red or brown, according to the counterstain used.

Bartholomew's (*Stain Technol.*, **37:** 139, 1962) directions, designed for staining of thin culture suspension smears, are perhaps not directly applicable to tissue smears or sections. He used the Hucker-Conn ammonium oxalate–crystal violet formula, the Burke formula for iodine, which is identical with Weigert's, and 0.25% safranin in 10% alcohol. He compared 99% ethanol, acetone, methanol, *n*-propanol, *n*-butanol, and *n*-amyl alcohol as decolorizers, and appraised the effect of water dilution on them.

Lower concentrations of crystal violet and of iodine were less effective in achieving proper Gram differentiation. Overwashing in water after crystal violet and iodine steps is to be avoided. Washing after safranin counterstains also needs to be brief.

Methanol is too rapid and unselective as a decolorizer, and butyl and amyl alcohols are too slow and ineffective for practical use. *n*-Propanol gave the widest time range between under- and overdecolorization; although slower than acetone it had a wider margin of safety. Addition of small amounts of water (5 to 10%) to the decolorizers accelerated destaining. At 95% *n*-propanol gave fully adequate decolorization over a far wider time range than ethanol; though slower than acetone, its time range was moderately wider.

The recommended procedure follows. It should be usable for tissue and exudate smears and can probably be adapted for sections.

Heat fixation is recommended.

1. Flood slide with crystal violet 1 min.
2. Wash 5 s in running water.
3. Weigert's iodine, two changes, 5 s and 1 min.
4. Wash 5 s in running water.
5. Decolorize in three changes of *n*-propanol, 1 min each. (Discard propanol No. 1 after each 10 slides, putting in fresh at No. 3 and moving the others forward.)
6. Wash 5 s in running water.
7. Two changes of safranin, 5 s and 1 min.
8. Wash 5 s, dry, and examine.

RESULTS: Blue-black gram-positive organisms; red gram-negative organisms.

GRAM STAIN FOR SECTIONS.   The acetone technic may be applied to sections (*Arch. Pathol.*, **5:** 828, 1928), thus:

1. Bring paraffin sections through xylene and alcohols to water as usual.
2. Stain 30 s in Hucker-Conn crystal violet.
3. Rinse briefly in water.
4. Treat 20 to 30 s with Weigert's iodine.

5. Decolorize with acetone 10 to 15 s.
6. Wash in water.
7. Counterstain 30 s in 0.5% safranin.
8. Differentiate and dehydrate (10 to 15 s) with acetone from a dropper bottle.
9. Clear by acetone and xylene and two changes of xylene. Mount in polystyrene, Depex, or other resin.

RESULTS: Gram-positive bacteria are blue-black; nuclei, deep red; gram-negative bacteria and fibrin, red; cytoplasm is pink.

Glynn's method (*Arch. Pathol.*, **20**: 896, 1935) differs from the foregoing in using Nicolle's carbol crystal violet for 2 min and in counterstaining (step 7) first with 1 : 2000 basic fuchsin in 0.002 *N* hydrochloric acid for 3 min and then for 30 to 60 s in saturated aqueous picric acid. Cytoplasm and red corpuscles are yellow; serum, fibrin, and collagen are pale pink; myelin is violet; and gram-positive bacteria are blue-black.

The *Kopeloff-Beerman* formula for crystal violet has given more intense gram-positive reactions than the ammonium oxalate formula (Bartholomew and Mittwer, *Stain Technol.*, **25**: 103, 1950). We cite the method from Conn and Darrow.

1. Hydrate sections or use air-dried films.
2. Stain 5 min or more in a fresh mixture of 4 parts 5% sodium bicarbonate and 15 parts 1% aqueous crystal violet solution.
3. Wash off with sodium hypoiodite. (Dissolve 2 g iodine crystals in 10 ml 1 *N* NaOH, and dilute to 100 ml.)
4. Immerse in or cover with hypoiodite for 2 min.
5. Wash in water; blot lightly and at once.
6. Decolorize with acetone or 30% ether acetone, dropping onto slide until color stops coming out (10 s or less). A higher proportion of ether (50%) slows the differentiation.
7. Counterstain 5 to 10 s in 2% safranin or 0.1% basic fuchsin.
8. Wash in water, dry, and examine. Sections can probably be dehydrated by dripping on acetone and xylene in sequence as in the Gram acetone technic (above).

THE BROWN-BRENN PROCEDURE FOR GRAM-POSITIVE AND GRAM-NEGATIVE BACTERIA IN TISSUES (from the Armed Forces Institute of Pathology, "Manual of Histologic and Special Staining Techniques," 2d ed., McGraw-Hill, New York, 1960).   Paraffin sections of formalin-fixed tissues (4 to 6 μ) are brought through xylene and alcohols to water as usual.

1. Stain flat, face up, for 1 min by depositing on each slide 1 to 1.5 ml fresh mixture of 4 ml 1% crystal violet and 1 ml 5% sodium bicarbonate.
2. Wash in water, and flood with Gram's iodine for 1 min.
3. Rinse in water and blot dry.
4. Decolorize individually with equal-volume ether-acetone mixture from a dropper bottle until no more color comes out.
5. Stain 1 min in a 0.1% dilution of saturated aqueous basic fuchsin (25 mg/1000 ml final dilution).
6. Wash and blot gently.
7. Dip in acetone, and then differentiate to yellowish pink with 0.1% picric acid in acetone.
8. Treat with acetone, acetone and xylene, three changes of xylene, Permount.

RESULTS: This procedure gives blue-black gram-positive bacteria, red gram-negative bacteria, red nuclei and mast cell granules, and yellow background.

Perhaps the most frequently used Gram technic for tissues has been the Gram-Weigert method for fibrin and bacteria.

FUNGI

Mallory's variant of Weigert's fibrin stain for the demonstration of actinomyces seems worthy of note:

1. Stain paraffin sections of formalin- or alcohol-fixed material for 3 to 5 min in alum hematoxylin.
2. Wash in water.
3. Stain 15 min at 57°C in 2.5% phloxine B (C.I. 45410) or 5% eosin Y (C.I. 45380).
4. Wash in water.
5. Stain in Ehrlich's or Stirling's aniline crystal violet for 5 to 10 min.
6. Wash in water.
7. Treat with Gram's iodine for 1 min (or Weigert's for 20 to 30 s).
8. Wash in water and blot dry with filter paper.
9. Differentiate with several changes of aniline until no more color is removed.
10. Wash in three or four changes of xylene and mount in balsam (or synthetic resin).

RESULTS: Mycelia are blue; clubs, red.

GENERAL INFORMATION ON FUNGI.   The same method is useful for other ray fungi in mycetoma, Madura foot, and allied conditions. Mycelia are often well brought out in light blue by azure-eosin methods, which are much better for study of tissue cellular reactions. The Bauer chromic acid leucofuchsin and the periodic acid Schiff methods can be recommended for mycelial fungi and yeasts in tissues. With the latter, it is desirable to suppress the background staining of collagen and reticulum by use of a sulfite blockade procedure, by use of highly diluted (20 to 50 mg fuchsin per 100 ml) Schiff reagent, or by use of a triphenylmethane sulfonic acid dye counterstain as recommended by Kligman, Mescon, and DeLamater (*Am. J. Clin. Pathol.*, **21:** 88, 1951), who used a brief counterstain with light green.

The allochrome method has given us brilliant results, particularly with *Cryptococcus neoformans* (*Torula histolytica*). *Coccidioides immitis*, *Histoplasma capsulatum*, the organisms of cutaneous blastomycoses, and various mycelial fungi have been successfully demonstrated by this method. Anthrax bacilli are well stained. The granules of mycetoma, actinomycosis, and botryomycosis are densely stained by this method; quite thin sections are required if any internal detail is to be discerned. All these organisms retain the red-purple in their cell walls.

Contrasts are better with the Bauer method or with the Kligman-Mescon-DeLamater periodic acid Schiff variant, but tissue structure is much better shown by the allochrome method. In the Bauer method a brief nuclear stain with Weigert's acid iron hematoxylin is preferable to alum hematoxylin, as it seems less apt to overpower the red of the smaller fungi such as *Histoplasma*.

Cryptococcus neoformans is also well shown by simple brief stains with metachromatic dyes, such as 1 : 1000 toluidine blue or thionin for 30 s. Organisms are blue-violet, the copious mucin of acute torulosis is red-purple; cell nuclei and tigroid granules are deep blue; and mast cell granules, violet to purple.

Use of safranin O in the Gram acetone technic gives red nuclei, pink cytoplasm, orange-red and blue-black yeast cells, orange mucus, and orange-red cartilage and mast cell granules.

Kligman, Mescon, and DeLamater (*Am. J. Clin. Pathol.*, **21**: 88, 1951) prepare skin scrapings for examination for fungi by first applying a drop of Mayer's glycerol albumen to the surface of the lesion and then scraping with a blunt knife. The scales are then smeared on a slide, and the smear is fixed 30 s in 95% alcohol. Then they are passed through periodic acid (1% H_5IO_6) for 1 min and washed in water; then Schiff reagent is used for 5 min. Wash for 5 to 10 min in 0.5% $K_2S_2O_5$ in 0.05 N HCl, using two or three changes. Wash in water, dry, and examine. The fungi appear red-purple.

Fetter and Tindall (*Arch. Pathol.*, **78**: 613, 1964) note that *Sporotrichum schenckii* may be identified in formalin-fixed, paraffin-embedded sections stained with the periodic acid Schiff hematoxylin method. Diastase digestions were helpful.

The internal filamentous structure of botryomyces granules should be well shown by this technic, as it is by the Bauer method. The staphylococcal component is better demonstrated by Gram-Weigert variants, which do not, however, stain the filaments (*J. Lab. Clin. Med.*, **32**: 76, 1947).

Kligman's periodic acid fuchsin stain is probably derived from Arzac's and may well depend on the aldehyde binding capacity of the arylamine of the basic fuchsin. The Schiff bases formed by the complexing of aldehyde with fuchsin are highly resistant to acid and alcohol decolorization. Kligman's technic has variants for smears and for sections (letter from Dr. Kligman, January 31, 1952).

KLIGMAN'S PERIODIC ACID FUCHSIN METHOD. Fix smears 1 min in 95% alcohol. Deparaffinize sections and immerse briefly in 100% alcohol.

		Smears	Sections
1.	Rinse in distilled water.	No	Few seconds
2.	Immerse in aqueous periodic acid.	5%, 1 min	1%, 10 min
3.	Wash in running water.	No	5–10 min
4.	Stain in 0.1% fuchsin in 5% alcohol.	2 min	2 min
5.	Wash in tap water.	Rinse	30 s
6.	Immerse in a 0.5% tartaric acid–1% zinc dithionite solution.	10 min Sputum 1 min	10 min and 30 min to 3 h
7.	Wash in tap water.	Rinse	3–5 min
8.	Saturated aqueous picric acid solution.	2 min	6 min
9.	Dehydrate in 95 and 100% alcohol, 10 s and 1 min respectively, clear in two changes of xylene, 1 min each, and mount in HSR or other synthetic resin or in Canada balsam.		

The color of the positive reaction is somewhat more purple than a plain basic fuchsin stain. In a similar procedure utilizing 0.5 N alcoholic hydrochloric acid as the decolorizing reagent in place of the tartaric acid plus zinc dithionite mixture, we have interposed also a 5-min stain in Weigert's acid iron chloride hematoxylin before the picric acid step.

Gridley (*Am. J. Clin. Pathol.*, **23**: 303, 1953) has modified the Bauer stain for demonstration of fungi in tissues by adding counterstains with Gomori's aldehyde fuchsin and metanil yellow.

GRIDLEY'S TECHNIC FOR FUNGI IN TISSUE SECTIONS

1. Deparaffinize paraffin sections at 6 μ, and hydrate as usual. Rinse in distilled water.
2. Oxidize 1 h in 4% chromic acid.
3. Wash 5 min in running water.
4. Immerse in Schiff reagent (0.5% fuchsin) for 15 min.

5. Rinse in three changes of 0.5% sodium metabisulfite $Na_2S_2O_5$ in 0.05 N hydrochloric acid, 2 min each.
6. Wash 15 min in running water.
7. Stain 15 to 20 min in Gomori's aldehyde fuchsin.
8. Rinse in 95% alcohol and wash well in water.
9. Counterstain 2 to 5 min in 0.25% metanil yellow in 0.25% acetic acid.
10. Wash in water, dehydrate, clear, and mount in Permount.

RESULTS: Hyphae are deep blue; conidia, rose to purple; elastin and mucin, deep blue; yeast capsules, deep purple; general background is yellow.

The simple Bauer method, with or without an iron or alum hematoxylin counterstain, is to be preferred for histochemical study of fungus cell walls.

Filaments of *Actinomyces* or *Nocardia* are not demonstrated by the Gridley method according to Luna (*Am. J. Med. Technol.*, **30**: 139, 1964).

Perhaps the most popular of the carbohydrate methods adapted for finding of fungi in tissues has been Gomori's chromic acid–methenamine silver sequence as used by Grocott.

Grocott (*Am. J. Clin. Pathol.*, **25**: 975, 1955) provides a very useful stain for many kinds of fungi including filaments of *Actinomyces bovis* and *Nocardia asteroides*. Sutter and Roulet (*Stain Technol.*, **40**: 49, 1965) have also noted that the method is particularly useful for the detection of *Mycobacterium leprae* even in sections cut from paraffin blocks 10 years old. In addition, Lopez and Grocott (*Am. J. Clin. Pathol.*, **50**: 692, 1968) were able to use the method to demonstrate *H. capsulatum* in a blood smear from an individual with disseminated disease. The Grocott method below is taken from Luna (*Am. J. Med. Technol.*, **43**: 101, 1968). Formalin-fixed tissues are satisfactory.

1. Deparaffinize through two changes each of xylene, absolute alcohol, and 95% alcohol.
2. Rinse in distilled water. (Slides previously stained with most other stains may be used by removing cover glasses in xylene and hydrating through alcohols to water. Subsequent chromic acid treatment will remove any remaining stain.)
3. Place in 5% chromic acid solution for 1 h.
4. Rinse in 1% sodium bisulfite for 1 min. (This removes any residual chromic acid.)
5. Wash in tap water for 5 to 10 min.
6. Wash with three or four changes of distilled water.
7. Place in methenamine–silver nitrate solution. Heat in oven at 58 to 60°C for 30 to 60 min, or until section turns yellowish brown. (Use paraffin-coated forceps to remove slides from this solution.)
8. Dip each slide in distilled water to stop reaction, and check under microscope for adequate silver impregnation. (Fungi should be a dark brown at this stage.)
9. Rinse in six changes of distilled water.
10. Tone in 0.1% gold chloride solution for 2 to 5 min.
11. Rinse in distilled water.
12. Remove unreduced silver with 2% sodium thiosulfate (hypo) solution for 2 to 5 min.
13. Wash thoroughly in tap water.
14. Counterstain with 0.2% light green solution for 30 to 45 s.
15. Dehydrate with two changes each of 95% alcohol and absolute alcohol.
16. Clear with two changes of xylene and mount in Permount.

RESULTS: Fungi are black; mucins, taupe to dark gray; inner parts of mycelia, rose; background is pale green. If *Nocardia* is suspected, it may be necessary to increase the time in step 7 to 90 min.

THE BLACK BAUER METHOD. Another chromic acid variant which seems useful is the "black Bauer" method reported by Lillie, Gilmer, and Welsh (*Stain Technol.*, **36:** 361, 1961).

1. Bring formalin-fixed tissue, 5-μ paraffin sections, through xylenes and alcohols to water as usual.
2. Oxidize 15 to 60 min (usually 40) in 4% chromic acid.
3. Wash in running water 10 min and in glacial acetic acid 1 min.
4. Immerse $1\frac{1}{2}$ h in 11% (1 M) *m*-aminophenol in glacial acetic acid.
5. Wash 10 min in running water; rinse in distilled water.
6. Azo-couple 2 min at 3°C in a freshly prepared solution of fast black K (3 mg/1 ml) in ice-cold 0.1 M Michaelis Veronal–HCl buffer of pH 8. (Fast black K should be less then 1 year old.)
7. Wash 15 min in 0.1 N HCl (three changes, 5 min each) and 10 min in running water.
8. Dehydrate in alcohols, clear in xylene, mount in cellulose tricaprate or Permount.

RESULTS: Cytoplasm and nuclei are pink to red; glycogen, epithelial mucins, starch, cellulose, fungus cell walls, black. Organisms are less conspicuous than with the chromic methenamine–silver technic, but they show more internal structure. *Candida albicans, Blastomyces dermatitidis, Coccidioides immitis, Cryptococcus neoformans, Toxoplasma gandii,* and *Histoplasma capsulatum* were demonstrated by this method; *Endamoeba histolytica* is also well shown by reason of its glycogen content (Gilmer).

For staining nuclear chromatin in certain yeastlike fungi, DeLamater (*Stain Technol.*, **23:** 161, 1948) recommends a procedure of acid hydrolysis and aldehyde mordanting followed by staining in basic fuchsin.

DELAMATER'S FORMALDEHYDE FUCHSIN METHOD

1. Fix (cultures) in Schaudinn's fluid.
2. Hydrolyze 5 min at 30°C, 5 min at 60°C, and then 5 min at 30°C in 1 N hydrochloric acid.
3. Wash in one to three changes of distilled water.
4. Mordant in 2% formalin 4 min (2 ml 40% HCHO plus 98 ml H_2O).
5. Wash in distilled water.
6. Stain 15 min with 0.5% basic fuchsin in 0.04 N hydrochloric acid.
7. Wash in distilled water.
8. Dehydrate and decolorize in graded alcohols.
9. Clear in xylene and mount in balsam or synthetic resin.

RESULTS: Nuclei stain an intense magenta red; cytoplasm stains light pink.

The *streptothrices* are usually gram-positive and may be acid-fast; hence the technics for gram-positive or acid-fast organisms may be used. Their reaction to the formaldehyde fuchsin methods of Goodpasture and of Fite we do not know.

BAKER AND SMITH'S PICROINDIGOCARMINE. Baker (*Am. J. Pathol.*, **32:** 287–307, 1956) recommends a previously unpublished picroindigocarmine stain of Baker and Smith for *Mucor* in tissues:

1. Deparaffinize and hydrate as usual; rinse in 0.5% acetic acid.
2. Stain 5 to 10 min in Goodpasture's aniline carbol fuchsin.[1]

---

[1] Goodpasture's aniline carbol fuchsin: To 100 ml 30% alcohol add 0.59 g basic fuchsin, 1 ml aniline, 1 g phenol.

3. Wash in running water and in 0.5% acetic acid.
4. Stain 5 to 10 min in Ramón y Cajal's picroindigocarmine.
5. Wash in 0.5% acetic acid.
6. Dehydrate rapidly in 95% and two changes of 100% alcohol, clear in xylene, and mount in Permount.

RESULTS: Nuclei stain red; fungi and collagen, blue-green; other tissues, in varying colors.

Other mycelial fungi in tissues may be studied with Gram-Weigert, with azure-eosin technics, and with the Bauer method.

THE ALKALI METHOD. A useful quick method for epidermal fungi is to scrape off material from the suspected area and macerate on the slide under a cover glass in 20% (Mallory) or 15% (Schmorl) sodium (or potassium) hydroxide solution. The epidermal cells are dissolved or cleared, leaving the fungus mycelia as refractile, perhaps branching, and often septate filaments and spores. Reduced illumination is often desirable for study of details by this method. This technic may be used for identification of the fungi of ringworm, favus, epidermophytosis, and the like.

Tkacz et al. (*J. Bacteriol.*, **105:** 1, 1971) conjugated fluorescein isothiocyanate to concanavalin A. The conjugate stained the yeast *Saccharomyces cerevisiae*, but not *Schizosaccharomyces pombe* or *Rhodotorula glutinis*. The conjugate appears to react with α-mannosan in cell walls.

Lindegren and Miller (*Can. J. Genet. Cytol.*, **11:** 987, 1969) noted that cobalt provided to yeast during growth localizes in mitochondria which may be observed as increased density by electron microscopy.

DIPHTHERIA ORGANISMS

Christensen (*Stain Technol.*, **24:** 165, 1949) recommends a method with sequence of acid toluidine blue, iodine, and safranin to replace Albert's method for diphtheria organisms. We have not tried Christensen's method.

CHRISTENSEN'S STAIN FOR *CORYNEBACTERIUM DIPHTHERIAE*. Use air-dried, heat-fixed smears of the usual Loeffler medium cultures.

1. Stain 1 min in 0.15% toluidine blue (52% dye content),[1] 5 ml glacial acetic acid, 2 ml ethyl alcohol, and 100 ml distilled water.
2. Wash with water, and apply Albert's iodine solution (iodine, 2 g; potassium iodide, 3 g; water, 300 ml) for 1 min.
3. Wash with water and counterstain with safranin.

RESULTS: Cell bodies are seen in light pink; protoplasmic striations, in red or brownish red; metachromatic granules, in black.

Excellent results may also be attained by simple, brief (1 to 2 min) staining in polychrome methylene blue. Loeffler's solution when aged several years performs excellently. For a freshly prepared solution we suggest 0.1% azure A in the same solvent. The orthochromatic color is blue-violet; the metachromatic, red-purple. Loeffler's methylene blue consists of 0.3 g methylene blue in 30 ml alcohol plus 1 ml 1% potassium hydroxide and 90 ml distilled water.

---

[1] For zinc-free toluidine blue of 80 to 90% dye content, 0.1% should serve.

INFLUENZA BACILLI AND ENCEPHALITOZOA

The *Goodpasture-Perrin* (*Arch. Pathol.*, **36**: 568, 1943) method for influenza organisms, encephalitozoa, and toxoplasmas is as follows:

1. Zenker or Orth fixation is preferred, but Perrin found 10% formalin material post-chromated with 2.5% potassium dichromate for 2 days quite satisfactory.
2. Bring paraffin sections to water as usual, including iodine and thiosulfate sequence if Zenker material is used.
3. Stain in Goodpasture's carbol aniline fuchsin for 5 min at 70°C—steaming on a hot plate.
4. Rinse quickly in tap water.
5. Decolorize with strong formalin (40% formaldehyde), a few drops at a time, until no more color is removed, 15 to 20 min.
6. Rinse in tap water.
7. Counterstain 1 min in saturated aqueous picric acid solution.
8. Dehydrate with two changes of 95% and two of 100% alcohol. Clear with 100% alcohol and xylene followed by two changes of xylene. Mount in synthetic resin.

RESULTS: Encephalitozoa are blue-black; the chromatin of toxoplasmas is brownish red; cell nuclei are light red; cytoplasm is pinkish yellow; erythrocytes are bright yellow; influenza bacilli are blue.

The *Wright and Craighead method* as modified by Perrin (*Arch. Pathol.*, **36**: 568, 1943) for encephalitozoa and toxoplasmas is as follows: formalin fixation with 48-h postchromation in 2.5% potassium dichromate, or Orth fixation; paraffin sections brought to water as usual.

1. Stain 10 min at 70°C in carbol fuchsin.
2. Rinse in tap water.
3. Differentiate with concentrated formalin (40% HCHO) from a dropper bottle until no more color is removed.
4. Rinse in tap water.
5. Stain 4 min in methylene blue, 1 g; alcohol, 20 ml; glacial acetic acid, 0.5 ml; distilled water, 80 ml.
6. Dehydrate with acetone. Clear with 50 : 50 acetone xylene mixture followed by two changes of xylene. Mount in synthetic resin.

RESULTS: Encephalitozoa are stained deep bluish red; cell nuclei and toxoplasmas are blue; cytoplasm is light blue to pink.

ACID-FAST STAINS

By acid fastness we refer to the property of retaining stains when other stained elements of tissues are decolorized by treatment with dilute solutions of mineral acids in water or alcohol. This property is shown by *Mycobacteria* and by certain other organisms to a lesser extent, by bacterial spores, by hair cortex and sometimes keratohyalin, and by some of the lipofuscin pigments, notably the coarsely granular one called *ceroid*. The dyes used have generally been basic aniline dyes, and usually the term "acid fastness" is understood to mean retention of these dyes. Although the coupling of active diazonium salts with tissue phenols yields acid-resistant dye tissue compounds, this reaction is not generally included in the meaning of the term "acid fastness."

The dyes used have generally been basic triphenyl or diphenylnaphthylmethane dyes, with pararosanilin (C.I. 42500), rosanilin (C.I. 42510, magenta I, basic fuchsin), and new fuchsin

(C.I. 42520) being the most used. Night blue (C.I. 44085) and auramine O (C.I. 41000) have also been used considerably, and Koch first used methylene blue. The basic nature of these dyes has naturally directed attention to the possibly acid nature of the substance demonstrated.

Lillie and Bangle (*J. Histochem. Cytochem.*, **2**: 30, 1954) found that acid-fast staining of ceroid, hair cortex, and tubercle bacilli was not prevented by 24-h exposure to 5% phenylhydrazine or to 1 *M* aniline hydrochloride. Tubercle bacilli lost their acid fastness on 2- to 4-h extraction in pyridine at 60°C and on 4-h acetylation in acetic anhydride-pyridine (40 : 60) but resisted methanol, 7 days at 25, 37, or 60°C, and carbon tetrachloride, 6 h at 60°C; ceroid and hair cortex resisted all these. Methylation at 60°C in 0.1 *N* HCl–methanol destroyed the acid fastness of tubercle bacilli in 2 h, ceroid lost its acid fastness in a week at 25°C, and hair cortex showed moderate impairment on 24-h, 60°C, or 7-day, 25°C, methylation. Reversal of methylation effects by saponification was not known at that time. Nitrosation in 1 *N* NaNO$_2$–1 *N* acetic acid for 4 h had no effect on acid fastness.

Lartigue and Fite (*J. Histochem. Cytochem.*, **10**: 611, 1962) failed to obtain successful acid-fast stains with new fuchsin in 9.5% alcohol but succeeded when various other phenols were substituted for phenol or aniline. On the basis of these studies and on the known sudanophilia of tubercle and lepra bacilli they felt that the oil-solubility theory of Lamanna's (*J. Bacteriol.*, **52**: 99, 1956) was the most probable.

The "Colour Index" records that rosanilin base exhibits solubilities of 20 and 15% in oleic and stearic acids, but only 0.5% in butyl acetate, no solubility in linseed oil and mineral oil, and 0.3 and 0.6% in ethanol and in acetone. But rosanilin chloride has a 0.39% solubility in water and 8.16% solubility in alcohol (Conn).

The oil-solubility theory would seem to require hydrolysis of the chloride (or acetate) during the staining procedure and solution of the free base in the fatty acids whose presence was indicated by the successful methylation blockade.

Nyka and O'Neill (*Ann. N.Y. Acad. Sci.*, **174**: 862, 1970) noted that certain *Mycobacterium tuberculosis* which fail to stain readily with acid-fast stains conducted in the usual fashion may be made "acid-fast" by immersion of sections or smears in 10% periodic acid for 24 h at 24°C prior to conduct of the acid-fast stain.

Harada (*Stain Technol.*, **48**: 269, 1973) noted that potassium permanganate (1% for 20 min), performic acid, or peracetic acid (60 min) oxidation of smears of mycobacteria resulted in enhanced acid-fast staining.

Murohashi and Yoshida (*Am. Rev. Resp. Dis.*, **92**: 817, 1965; *Ann. N.Y. Acad. Sci.*, **154**: 58, 1968) note that acid fastness of mycobacteria is lost in smears subjected to ultraviolet irradiation for variable periods of time. The period of irradiation required for loss of acid fastness varies with different mycobacteria. Reasons for the loss are unknown.

Segal (*Am. Rev. Resp. Dis.*, **91**: 285, 1965) concludes that *M. tuberculosis* grown in the host differs significantly in lipid composition from the same strain grown in an artificial culture medium. Bacilli grown in vitro characteristically fix neutral red and are *not* acetone-fast after staining with Sudan black B, whereas comparable bacilli from mouse lungs do not fix neutral red and *are* acetone-fast after Sudan black B staining.

CARBOL FUCHSIN–METHYLENE BLUE METHOD. In this method for acid-fast organisms, the stock carbol fuchsin is traditionally composed of saturated alcoholic solution of basic fuchsin (C.I. 42510), 10 ml, and 5% aqueous phenol solution, 90 ml. This solution may keep for years. Rosanilin chloride is soluble to about 6% in alcohol; pararosanilin chloride is soluble to about 3.5% of commercial samples and to 8.16 and 5.93% of the pure substances according to Conn. Neelsen [*Zentralbl. Med. Wiss.*, **21**: 497,

1883; *Fortschr. Med.*, **3**: 200 (footnote), 1885] prescribed first 0.75, later 1 g fuchsin in 100 g 5% phenol, adding "a little" alcohol, which he later specified as 10 g. Conn prescribes 300 mg fuchsin, 10 ml alcohol, 5 g phenol, and 95 ml distilled water. Kinyoun's (*Am. J. Public Health*, **5**: 867, 1915) formula—fuchsin, 4 g; phenol, 8 g; alcohol, 20 ml; heat to dissolve and then add water, 100 ml—is far stronger and is said to be a more energetic stain. This may well be true when the stain is freshly prepared, but it soon deposits a considerable quantity of excess dye, and thereafter is probably no better than other formulas.

Mallory recommended Verhoeff's formula. This is kept as a stock solution which is diluted at the moment of use. Dissolve 26.8 g phenol (25 ml melted crystals) in 50 ml 100% alcohol. Add 2 g fuchsin and heat at 37°C with occasional shaking for 18 to 24 h. Filter and store. For use dilute 1 ml stock solution with 6 ml water. This represents a final dilution of 0.281% fuchsin, 9.53% alcohol, and 4.76% phenol. The main advantage seems to be in the permanence of the stock solution. In view of the usual stability of the ordinary formula, this advantage may be outweighed by the disadvantage of having to dilute it for use.

Fite (*Am. J. Pathol.*, **14**: 491, 1938) recommended new fuchsin and prescribed thus: 1 g dye, 5 g phenol, 10 ml methyl alcohol; dissolve completely, and then add gradually, with shaking, enough distilled water to make 100 ml. Later he reduced the dye to 0.5 g and used ethyl or methyl alcohol.

Lillie's 1954 directions read: Dissolve 25 g phenol in 50 ml alcohol, add and dissolve 5 g fuchsin, and then dilute to 500 ml with distilled water.

Carbol fuchsin solutions gradually form a dark red, caked deposit which fails to redissolve on warming and shaking. This deposition results in progressive weakening of the solution. Consequently positive controls should be used at frequent intervals to avoid false-negative reactions from stain failure. Although some batches of solution have remained effective for years, others become useless in as little as a year.

With any of the carbol fuchsin or carbol new fuchsin solutions the sections are heavily stained either by means of heat or by prolonged exposure, and then decolorized with acids, alcohol, or usually both. The heating methods occasion somewhat more section shrinkage and definitely more dye precipitation from evaporation. Lillie tried adding glycerol to carbol fuchsins to prevent drying while heating but abandoned it because bacilli appeared less well stained. Fite's observation that previously heated and cooling or cooled carbol fuchsin stains more brilliantly than unheated, just as does hot carbol fuchsin, appears to indicate that supersaturation plays an important part in brilliancy of staining. This would also account for the greater vigor of staining obtained with the freshly prepared (and supersaturated) Kinyoun's solution.

TECHNIC FOR CARBOL FUCHSIN AND METHYLENE BLUE

1. Bring paraffin sections to water as usual. (Smears are prepared as usual and heat-fixed.)
2. Stain 10 min at 70°C, 30 min at 55°C, 2 h at 37°C, or 4 to 16 h at 25 to 20°C with any of the carbol fuchsins.
3. Wash in water.
4. Decolorize with 2 ml concentrated hydrochloric acid plus 98 ml 95 to 70% alcohol. This ordinarily takes 20 s or more and may be extended to several minutes without harm.
5. Wash 2 to 3 min in running water.
6. Counterstain with acid hemalum for 2 to 5 min or with 1% methylene blue or 1% Janus green B in 1% acetic acid–20% alcohol for 3 min. (The latter two counterstains are suitable for smears.)

7. Wash in water. (At this point smears are dried in a warm airstream and examined directly in immersion oil.)
8. Dehydrate and clear with an acetone-xylene sequence, and mount in a suitable non-reducing resin, such as polystyrene, Depex, or cellulose caprate.

RESULTS: Acid-fast bacilli are red; ceroid is red; nuclei are blue or green; mast cell granules, blue-violet with methylene blue but unstained by hematoxylin. Red corpuscles are often pink in formalin material, and hair shafts and keratohyalin may retain more or less red.

By acidifying the counterstain the dense staining seen with ordinary methylene blue counterstains is avoided. This was at least part of the value of the Gabbett (*Lancet*, **1**: 757, 1887) solution (2 g methylene blue in 25 ml concentrated sulfuric acid and 75 ml distilled water–48.8 g sulfuric acid per 100 ml), which was recommended for simultaneous decolorization and counterstaining. Gabbett did not specify the amount of methylene blue. This was supplied in the 1901 and later editions of Mallory and Wright. The omission of alcohol was the most serious defect of the Gabbett solution. Picric acid counterstains have been used by some workers, with good contrast, but give no tissue detail, fail to show leukocytes and non-acid-fast organisms, and, according to Fite, cause fading of the fuchsin stain.

In clearing of sections, carbol xylene should be avoided. It decolorizes nearly all the previously well stained tubercle bacilli and even impairs the acid-fast staining of ceroid.

Similar acid-fast stains may be achieved, with opposite color effects, by substituting night blue (C.I. 44085) or Victoria blue R (C.I. 44010) for fuchsin in the carbol fuchsin formula (given above) and using 0.1% safranin in 1% acetic acid as a counterstain. The technic follows that above. Acid-fast bacilli appear dark blue, cell nuclei in red.

HAGEMANN'S PHENOL AURAMINE. Hagemann's (*Munch. Med. Wochenschr.*, **85**: 1066, 1938) technic of staining smears with phenol auramine and examining with fluorescence microscopy has been widely used. The principal advantages claimed are that organisms are readily discerned at lower magnifications than with Ziehl-Neelsen technics and hence larger areas can be scanned in the same time. Use of magnifications as low as 180× is claimed for scanning; but to determine morphologic characteristics 600× was needed. Lempert (*Lancet*, **247**: 818, 1944) used a 16-mm ($\frac{2}{3}$ in) objective for focusing the lamp and condenser system, and a 6-mm ($\frac{1}{4}$ in) objective for identification. The use of immersion objectives is unnecessary, but should one desire to use them for higher magnification, Lempert advises the use of glycerol for immersion, because of the fluorescence of cedar oil. Our white modified mineral oils (Shillaber's, Crown) are designated "low fluorescence" and "very low fluorescence" for high and low viscosities, and should be usable.

Hagemann dissolved 1 g auramine (C.I. 41000) in 100 ml 5% phenol; Lempert used 300 mg in 100 ml 3% phenol and filtered after vigorous shaking and warming to 40°C. The technic follows:

1. Stain heat-fixed films 8 to 10 min at room temperature in Lempert's phenol auramine.
2. Wash well in tap water.
3. Decolorize in two changes of 2 min each of hydrochloric acid alcohol. Lempert used 0.5 ml concentrated hydrochloric acid, 0.5 g sodium chloride, 25 ml distilled water, and 75 ml methanol. Fite used his usual 2 ml concentrated hydrochloric acid and 98 ml 95% alcohol mixture.
4. Wash well in tap water.
5. Treat for 20 s with 0.1% potassium permanganate solution.
6. Wash in water, dry at room temperature in an air current, and examine.

RESULTS: The bacilli appear as bright yellow rods on a very dark red background. The method has not been found adaptable for histologic study. Part of the alleged superiority of the method disappears when one uses acetic methylene blue as a counterstain for Ziehl-Neelsen stains, either of smears or of tissues. We have often readily found tubercle bacilli by this latter technic with an 8-mm objective, and the tissues and other non-acid-fast organisms are seen as well.

Gilkerson and Kanner (*J. Bacteriol.*, **86**: 890, 1963) modified Hagemann's phenol auramine stain so that ferric chloride is used to suppress background fluorescence. The stain is composed of 3 g auramine O (C.I. 41000), 40 ml phenol, and 60 ml glycerol in 900 ml distilled water. Dry smears fixed with heat and formalin-fixed tissues may be used.

Deparaffinized sections or smears are brought to water, stained with the phenol auramine reagent for 10 min at 24°C, washed in tap water, reacted with 10% ferric chloride for 10 min at 24°C, washed, air-dried, mounted in a nonfluorescing medium, and examined microscopically using a blue-light exciter filter and an orange-yellow barrier filter. Contrast between bacilli and background is said to be great. Under low magnification golden yellow bacilli were clearly visualized in smears or in tissues.

Gilkerson and Kanner compared the sensitivity and specificity of the stain to the acid-fast stain. More positives were found with the phenol auramine stain than with the acid-fast stain. Disease was later confirmed in the positives. No false-positive results were obtained, and a false-negative reaction was rare.

Koch and Cote (*Am. Rev. Resp. Dis.*, **91**: 283, 1965) are also convinced of the superiority of the fluorescent method for detection of acid-fast bacilli. Wang (*Am. J. Clin. Pathol.*, **51**: 71, 1969) concludes that the fluorescent method is extremely sensitive. He was able to detect in tissue sections contamination from saprophytic acid-fast bacilli in the water bath, in the gelatin, and in the paraffin dispenser. The fact that Koch and Cote found acid-fast bacilli in 6 of 21 cases of confirmed sarcoidosis remains bewildering, or perhaps reminiscent of the earlier era when certain cutaneous tuberculids were called "sarcoids."

Mansfield (*Am. J. Clin. Pathol.*, **53**: 394, 1970) and other authors recommend the fluorescent method of Kuper and May (*J. Pathol. Bacteriol.*, **79**: 59, 1960) modified from Matthaei (*J. Gen. Microbiol.*, **4**: 393, 1950) for a rapid selective and sensitive method for acid-fast bacilli. Heat-fixed smears or formalin-fixed, paraffin-embedded sections are used.

1. Deparaffinize sections (two changes, 6 min each). Mansfield notes that staining is enhanced in paraffin-embedded tissues if sections are deparaffinized in a 2:1 xylene–peanut oil or mineral oil mixture. Sections are rinsed briefly in tap water (45 s). Not all the oil should be removed. (See also Fite's Oil Fuchsin Method, below.)
2. Stain sections in the following *preheated* mixture for 10 min at 60°C: auramine O (C.I. 41000), 1.5 g; rhodamine B (C.I. 45170), 0.75 g; glycerol, 75.0 ml; liquefied phenol crystals at 50°C, 10 ml; distilled water, 50 ml.
3. Rinse slides for 2 min.
4. Decolorize for 1 to 3 min with 0.5% acid alcohol.
5. Rinse briefly, and counterstain with 0.5% potassium permanganate for 1 min.
6. Rinse briefly, dehydrate, and mount in a nonfluorescent synthetic medium (Permount is satisfactory).
7. Examine stained sections with a fluorescence microscope with appropriate heat and barrier filters and a BG12 excitation filter. As an alternate, a regular light microscope with a high-intensity light source, BG12 excitation filter, and appropriate barrier filter may be used.

RESULTS: Individual bacilli may often be seen at low magnification, whereas many bacilli are usually required for detection with the Ziehl-Neelsen technics. Fluorochrome-stained slides may be readily restained with carbol fuchsin; however, the converse may not be done.

For step 5 a 10-min immersion in 10% ferric chloride (fresh) may be substituted with omission of the acid alcohol step (step 4). If a delay occurs between the time the smear is made and staining, the fluorochrome staining method is advised for greater reliability in detection.

FITE'S NEW FUCHSIN FORMALDEHYDE. Except for the acid alcohol step contained therein, Fite's (*Am. J. Pathol.*, **14**: 491, 1938) method for acid-fast bacilli was quite similar to DeLamater's chromatin method and the Wright-Craighead procedure described earlier.

1. Stain paraffin sections 30 to 60 min or more at room temperature (22 to 25°C) in phenol, 5 g; ethyl or methyl alcohol, 10 ml; new fuchsin, 1 g; dissolve, and add distilled water to make 100 ml.
2. Immerse in concentrated formalin (40% HCHO) for 5 min.
3. Decolorize nearly completely with acid alcohol (2 ml concentrated hydrochloric acid, 98 ml 95% alcohol).
4. Immerse again in concentrated formalin for a few seconds.
5. Counterstain with hematoxylin and Van Gieson's picrofuchsin. Dehydrate, clear, and mount through 95 and 100% alcohol, 100% alcohol and xylene, and two changes of xylene to polystyrene or Permount.

RESULTS: Acid-fast bacilli are stained violet; hair shafts, keratohyalin, and mast cell granules are decolorized. The behavior of ceroid with this stain has not been reported.

THE FITE OIL FUCHSIN METHOD. Lepra bacilli sometimes fail to stain by the foregoing procedures. Fite recommended a procedure based on protection of organisms from oil solvents, which Wade (letter, May 21, 1956) regarded as an entirely new procedure, perhaps suggested by Faraco's oil restoration step but actually completely original. Wade (*Stain Technol.*, **32**: 287, 1957) has further modified this oil fuchsin method of Fite's by combining it with the formaldehyde fuchsin method, thus achieving effects which Fite considers very superior for lepra bacilli.

THE WADE-FITE OIL FORMALDEHYDE NEW FUCHSIN PROCEDURE FOR LEPRA BACILLI. Fix preferably in Zenker's fluid or in formalin. Dehydrate with alcohols and thin cedar oil, and embed in paraffin as usual. Section at 6 to 8 μ, and mount on slides with Mayer's glycerol albumen. Dry overnight at 37°C.

1. Deparaffinize in 2 parts rectified turpentine plus 1 part heavy liquid petrolatum (paraffin oil), two changes in 5 min.
2. Drain, wipe back and edges of slide, blot with filter paper until section appears opaque. Let stand in water until rest of slides are ready.
3. Stain 16 to 24 h in Fite's carbol new fuchsin (later formula). Wash in water.
4. Immerse 5 min in reagent-grade 37 to 40% formaldehyde until bacilli appear blue. Section color remains red or may turn blue.
5. Extract 5 min in 5% (v/v) sulfuric acid. Wash in water. No obvious color change occurs.
6. Treat with 1% potassium permanganate, 3 min.
7. Bleach individually by agitation in 2% oxalic acid, preferably less than 30 s, not more than 60 s. Use 5% oxalic acid if sections do not decolorize readily.
8. Stain 3 min in a dilute Van Gieson stain: 10 mg acid fuchsin, 100 mg picric acid, 100 ml distilled water (a 10% dilution of Weigert's formula in distilled water should serve).
9. Dehydrate directly and rapidly without aqueous rinse, in 95 and 100% alcohol. Clear in xylene.
10. Mount in synthetic resin: Permount, HSR, or the like.

RESULTS: Acid-fast bacilli and free granules are dark blue; connective tissue is red; background, yellowish.

In some cases it may be necessary to soak sections several hours in the turpentine-oil mixture; 6 h should suffice, 10 may occasionally be needed, to "refat" "decrepit" bacilli.

Blanco and Fite (*Int. J. Lepr.*, **16**: 367, 1948) found that aqueous formalin fixation gave fair to good staining of lepra bacilli in biopsy material from human skin, using the Ziehl-Neelsen and oil fuchsin (above) technics. Excellent results were obtained with Zenker fixation, poor with Bouin, and fair with alcoholic 20% formalin.

For preservation of the acid-fast material in lepra bacilli, Wade (*Stain Technol.*, **27**: 71, 1952) recommends Carbowax embedding and sectioning and one of the oil fuchsin technics.

Blanco and Fite (*Arch. Pathol.*, **46**: 542, 1948) have used a modified Jahnel procedure for the demonstration of lepra bacilli in tissues. Although much too prolonged for regular diagnostic use, the procedure is said to afford a truer picture of the relations of the organisms to the tissues than most of the commoner methods.

BLANCO-FITE SILVER METHOD FOR *MYCOBACTERIUM LEPRAE*

1. Fix in 10% formalin for 2 or more weeks.
2. Soak blocks in pyridine 1 to 3 days.
3. Wash 24 h in several changes of distilled water.
4. Immerse in 10% formalin 4 days.
5. Wash 24 h in several changes of distilled water.
6. Treat with 95% alcohol 3 to 8 days, changing alcohol daily.
7. Transfer to distilled water until blocks sink.
8. Incubate at 37°C in the dark in 0.5% silver nitrate for 5 to 8 days.
9. Wash 10 min in distilled water.
10. Reduce 2 days in pyrogallol, 4 g; formalin, 5 ml; distilled water, 95 ml.
11. Dehydrate, clear, and embed in paraffin as usual.
12. Section, deparaffinize, and mount.

RESULTS: Most lepra bacilli are black, some are brown. Melanin and keratohyalin are also blackened. Tubercle bacilli and spirochetes should also blacken by this method.

BACTERIAL SPORES

Bartholomew and Mittwer (*Stain Technol.*, **25**: 153, 1950) modified the bacterial spore stain of Schaeffer and Fulton (*Science*, **77**: 194, 1933) with some simplification, thus:

	Schaeffer and Fulton	Bartholomew and Mittwer
1. Fix smears on slides by passing through flame.	3 times	20 times
2. Stain in malachite green.	5%, 4 or 5 times in 1 min	Sat. sol. (7.6%), 10-min, cold
3. Rinse in water.	30 s	Rinse
4. Stain in safranin.	0.5%, 30 s	0.25%, 15 s
5. Rinse, blot dry, and examine.		

RESULTS: Bacterial bodies are red to pink; spores, green.

Bartholomew et al. (*J. Bacteriol.*, **90**: 1146, 1965) recommend the following simple stain for spores: Heat-fixed smears are stained with auramine O (C.I. 41000) for 2 min, rinsed briefly, differentiated with 0.25% safranin O (C.I. 50240) for 1 min, and rinsed. Safranin replaces auramine O in vegetative cells, but not in spores. *Bacillus subtilis* was tested.

In studies of phagocytosis, it is sometimes useful to identify vegetative cells, nongerminating spores, germinating spores, and cytologic details of phagocytes. Booth et al. (*Stain Technol.*, **46**: 23, 1971) recommend the following staining procedure:

1. Spread a small drop of the exudate or other sample evenly on a slide and air-dry for 2 to 4 h at 25 to 37°C. Air drying for longer than 4 h did not impair the staining quality of the sample.
2. Flood with Wright's stain for 2 min.
3. Add an equal volume of distilled water and let stand for 2 min.
4. Rinse with distilled water.
5. Stain for 30 s in 0.005% methylene blue.
6. Rinse with distilled water.
7. Flood with Ziehl-Neelsen's stain for 3 to 4 min.
8. Destain with a 1 : 1 acetone–95% ethanol mixture for 10 s, allowing it to flow over the smear.
9. Quickly rinse with distilled water.
10. Flood for 1 min with Wright's stain.
11. Add an equal volume of distilled water, 2 to 3 min.
12. Pour slowly about 5 ml phosphate buffer of pH 6.5 directly over the stained smear, and allow it to drain off.
13. Rinse with distilled water, blot, and allow the preparation to air-dry.
14. Examine directly in oil, or apply a cover glass with a resinous medium.

RESULTS: Spores are light red with a red wall. Germinating spores are deep red throughout; vegetative cells are blue, and leukocytes have a light blue cytoplasm with a dark purple nucleus. The color of leukocytes varies with the degree of exposure to carbol fuchsin. Thin smears are important; thick smears tend to be overstained with carbol fuchsin.

BACTERIAL CELL ENVELOPE

Hale (*Lab. Pract.*, **2**: 115, 1953) stains bacterial cell envelopes by first mordanting unfixed smears for 5 to 10 min in 1% phosphomolybdic acid and then staining for a few seconds in 1% methyl green or 0.1% Janus green.

BACTERIAL MITOSIS

DeLamater and Mudd (*Exp. Cell Res.*, **2**: 499, 1951) demonstrate bacterial mitosis by the following procedure:

1. Cut a small slab of agar from a 2-h-old 37°C culture. Place on a small glass plate (slide), stand upright in a Coplin jar containing a little 0.5% osmium tetroxide, and cover the jar for 5 min.
2. Cut small fragments of the agar, and press face down on cover glasses or slides to make impression preparations.
3. Hydrolyze 6 min in 1 *N* hydrochloric acid at 60°C (preheated).
4. Rinse in distilled water.
5. To 10 ml 0.25% aqueous thionin add 1 drop of thionyl chloride, mix well, and stain preparations in this for 2 h or more.

6. Rinse once in distilled water to remove stain particles; drain on filter paper.
7. Immerse in 100% alcohol in jars surrounded by solid carbon dioxide to maintain a temperature below $-50°C$. Let stand 12 h to complete dehydration.
8. Dip in fresh 100% alcohol at 25°C. Clear in xylene, drain on filter paper, and mount in synthetic resin.

The reaction of thionin (or azure A, which may also be used in the foregoing technic) in the presence of the HCl and H_2SO_3 formed by hydrolytic decomposition of the thionyl chloride $SOCl_2$ is considered by DeLamater and Mudd to be an aldehyde reaction. In a sense, the procedure parallels the Feulgen method, but it uses an undecolorized thionin sulfurous acid complex in place of Schiff reagent.

GRAM-NEGATIVE BACTERIA AND RICKETTSIAE

Gram-negative bacteria and rickettsiae in tissues are generally best studied with stains of the azure-eosin type, such as Mallory's phloxine plus borax–methylene blue, Maximow's hematoxylin azure II–eosin, or our buffered azure eosinate and azure A–eosin B variants. Giemsa's blood stain in 1 : 40 to 1 : 50 dilution has been much used. Wolbach added 5 drops of 0.5% sodium carbonate to 100 ml final stain mixture and differentiated after staining with colophonium alcohol. We prefer to buffer to a relatively acid level, say pH 4 for formalin material, and thus obviate differentiation. Except for Wolbach's variant, these methods have been given under general methods in an earlier chapter.

WOLBACH'S GIEMSA VARIANT ("The Etiology and Pathology of Typhus," Harvard University Press, Cambridge, Mass., 1922, pp. 13–14). Fix thin slices of tissue for 24 to 48 h in Zenker's or Möller's (Regaud's) fluid. Cut thin paraffin sections and take to water as usual.

1. Stain 1 h in Giemsa's stain, 1 ml; methyl alcohol, 1.25 ml; 0.5% sodium carbonate solution, 0.1 ml (2 drops); distilled water, 40 ml.
2. Pour off and replace with two further changes of the same mixture during the first hour, and leave in the third change overnight.
3. Differentiate in 95% alcohol containing a few drops of 10% colophonium alcohol.
4. Dehydrate with 100% alcohol, clear in xylene, and mount in cedar oil.

RESULTS: Rickettsiae stain an intense reddish purple; nuclei, dark blue to violet; cytoplasm stains varying lighter blue shades; collagen and muscle, pale pink; erythrocytes, gray to yellow or pink. Further differentiation occurs after mounting in cedar oil, and exposure to sunlight for prolonged periods has been used to bring sections to the proper point. We have not used this method extensively but would caution against identifying as rickettsiae the metachromatic granules of tissue mast cells, which also stain in redder shades than nuclear chromatin.

There are a number of methods used for the identification of rickettsiae in smears. Among them one of the most useful has been the Attilio Macchiavello technic, which is not separately published but appears in the article by Hans Zinsser, Florence Fitzpatrick, and Hsi Wei (*J. Exp. Med.*, **69**: 179, 1939). Macchiavello states "Rapid examination of *Rickettsia* in agar tissue cultures is now made as a routine by the staining technic first worked out in our (Harvard) laboratory."

MACCHIAVELLO'S STAIN FOR RICKETTSIAE

A 0.25% solution of basic fuchsin is made either in a phosphate solution, buffered at pH 7.4, or in distilled water brought to pH 7.2 to 7.4 with sodium hydroxide. Preparations are made by smearing a bit of tissue on the slide and drying gently by heat after drying in the air, and the fuchsin solution is filtered over the preparation through a coarse filter paper in a funnel. The fuchsin is left on the slide for 4 min and is washed off very rapidly with 0.5% citric acid solution. The citric acid solution is poured on and off the slide and very rapidly washed with tap water. It is then stained for about 10 s with a 1% aqueous solution of methylene blue. With a little practice and adjustment to individual laboratory materials this method gives excellent contrast stain, the rickettsiae, intra- and extracellular, being stained red, the cellular elements blue. This method, though excellent for culture smears, is not successful in tissue sections of typhus animals, for which the Giemsa method seems to be the only reliable technique at the present time.

BENGSTON'S VARIANT OF THE MACCHIAVELLO METHOD. The following is Bengston's variant, as used at the National Institute of Health:

1. Fix thin films by passing quickly, face down, through a blue flame 3 times.
2. Stain 5 min in basic fuchsin, 0.5 g (saturated); 0.1 M disodium phosphate, 3.6 ml; 0.1 M sodium acid phosphate ($NaH_2PO_4 \cdot H_2O$), 1.4 ml; distilled water, 95 ml. Filter. This buffer corresponds to pH 7.2.
3. Rinse rapidly with 0.5% aqueous citric acid solution.
4. Wash thoroughly in tap water.
5. Counterstain 1 to 2 min in a 0.1% aqueous methylene blue or 10 s in a 1% solution.
6. Rinse with water, dry, and examine in immersion oil.

RESULTS: Rickettsiae are red; cells and bacteria, varying shades of blue. Thus far this method has not been adapted to tissue sections.

The other method most often used is the Giemsa stain. With this one may stain at pH 7 to 7.2 and then differentiate with faintly acid water until red corpuscles are pink, or one may stain at pH 6.5 to 6 and simply rinse and dry smears. Smears of blood, marrow, pus, or exudates should be spread thin, so that in most areas cells lie separated from one another. These smears should be fixed at once with methyl alcohol or 100% ethyl alcohol for 2 or 5 min, respectively.

Nyka (*J. Pathol. Bacteriol.*, **67**: 317, 1945) recommends a relatively simple methyl violet–metanil yellow technic for typhus rickettsiae in mouse lungs:

Fix in 10% neutral formalin. The usual embedding and sectioning procedures are presumed, and sections are deparaffinized and hydrated as usual.

1. Stain 30 to 60 min in 1 : 10,000 aqueous methyl violet.
2. Differentiate in weak acetic acid (0.03 to 0.04%: 2 drops of glacial acetic acid in 100 ml distilled water) until cell cytoplasm is decolorized.
3. Counterstain a few seconds in 1 : 10,000 aqueous metanil yellow.
4. Dehydrate and clear with an acetone-xylene sequence.
5. Mount in synthetic resin. Gurr's Xam and the Media Manufacturing Centre's DPX4 are mentioned by Nyka. Our American synthetic resins such as Permount, polystyrene, etc., will probably serve.

In the few preparations seen by Lillie that were stained by this method, rickettsiae were still difficult to distinguish from other basophilic granules and mast cell granules.

Anderson and Greiff (*J. Histochem. Cytochem.*, **12**: 194, 1964) provide the following method for rickettsiae (*R. mooseri*):

1. Fix smears of infected tissues in anhydrous methanol or absolute ethanol and chloroform (1 : 1). Tissues may be fixed in either 10% neutral buffered formalin or Zenker's, Bouin's, or Carnoy's fluid.

2. Stain deparaffinized sections or smears with the Mowry Alcian blue–periodic acid Schiff method.
3. Rinse briefly 3 times in distilled water.
4. Immerse in McIlvaine's buffer of pH 5.4 (0.1 M citric acid, 0.2 M Na_2HPO_4). The prescribed buffer concentrations and pH are believed critical.
5. Immerse specimens in acridine orange (0.05 g acridine orange, 50 ml McIlvaine's buffer, 50 ml distilled water at pH 5.4) for 5 min.
6. Rinse specimens in McIlvaine's buffer of pH 5.4, 3 times for 1 min each.
7. Dehydrate and mount in synthetic medium.

RESULTS: Rickettsiae are revealed by orange fluorescence at 4150 Å. Anderson and Greiff state that background fluorescence is greatly reduced by use of the Alcian blue–periodic acid Schiff stain.

Fogh and Fogh (*Proc. Soc. Exp. Biol. Med.*, **117**: 899, 1964) recommend acetic orcein (2% orcein is boiled in 60% glacial acetic acid until dissolved; the original volume is thereafter restored by addition of water) as a stain for pleuropneumonia-like (*Mycoplasma*) organisms in tissue-cultured cells. Monolayers on slides or coverslips are rinsed with 0.6% sodium citrate, treated with 0.45% sodium citrate for 10 min, fixed in Carnoy's fluid for 10 min, air-dried, stained with acetic orcein for 5 min, dehydrated in ethanols, and mounted.

BLOOD, TISSUE, AND PROTOZOA

Giemsa's stain is also widely used for the morphology of blood, spleen, and marrow cells and for the identification of protozoan parasites such as trypanosomes, leishmaniae, plasmodia, and bartonellae.

The stain is best purchased from a reputable dye manufacturer. Insist on certification by the Biological Stain Commission. Either the prepared glycerol-methanol solution or the dry mixed powder may be obtained. The latter should be dissolved in equal volumes of glycerol and methanol (methyl alcohol) at 800 mg/100 ml of the mixed solvent. Although a small undissolved residue is found on dissolving Giemsa stains in amounts over 300 mg/100 ml, the solution increases in strength, staining capacity, and optical density with further addition of dye—the last in proportion to the total amount of dye added, up to 1.1 g/100 ml. Consequently the presence of a small residue is not to be taken as evidence of saturation of the glycerol-methanol solvent. Quantities in excess of 1.1 g/100 ml occasion no further increase in staining power or optical density.

The best solvent appears to be an equal-volume mixture of C.P. methanol and neutral C.P. glycerol. Special treatment of the methanol to render it acetone-free is expensive and of no discernible value. Glycerol may be of either 95 or 98% strength without affecting the quality of the stain. More important than traces of acid in the reagents are traces of alkali. A trace of acid actually acts more as a stabilizer of the azures and methylene blue against the alteration which occurs spontaneously in glycerol-methanol solutions, and is readily overcome in staining by the use of the appropriate buffers. Traces of alkali, on the other hand, fairly rapidly convert methylene blue and azure B into lower azures and methylene violet and alter the staining effect profoundly.

The older German texts required equal weights of glycerol and methanol. This corresponds nearly exactly to 60 volumes methanol and 40 volumes glycerol. Giemsa stain deteriorates somewhat more rapidly in this mixture than in the equal-volume mixture. The

75 : 25 mixture used for MacNeal's stain appears to be little or no better for preserving the stain than plain methanol.

An equal-volume mixture of 100% ethyl alcohol and 98% glycerol appears to be as good a solvent for Giemsa stain as the corresponding methanol mixture, and to preserve the stain at least as well.

The recommended composition of Giemsa stain, using American dyes, is as follows: methylene blue eosinate, 4 g; azure B eosinate, 5 g; azure A eosinate, 1 g; and methylene blue chloride (85 to 88% dye content), 2 g. The mixture should be kept in a cool, dry place, tightly stoppered.

It is not recommended that the eosinates be made in the laboratory, particularly not from commercial azure B or from methylene blue, unless special precautions are taken.

Commercial azure B is apparently a variable substance, containing varying proportions of methylene blue and azure A, often with an adequate amount of one or the other to modify considerably the character of the Giemsa stain. Further, eosinates of azure B and of methylene blue are quite susceptible to partial demethylation ("polychroming") on drying even at moderate (55 to 60°C) temperatures. (See Lillie, 1954, pp. 386–388.)

THE GIEMSA STAIN FOR FILMS

1. Thin films should be fixed as soon as taken in methyl alcohol for 3 to 5 min. Thick films are first hemolyzed and then fixed or stained without fixation.
2. Stain 40 to 120 min in 1 ml Giemsa stain, 2 ml stock buffer, and 47 ml distilled water. For marrow a phosphate buffer of pH 5.8 to 6 is preferable; for blood cytology, pH 6.4 or 6.5; for both blood and protozoa some prescribe pH 6.8; and for malaria survey work on thick films pH 7 to 7.2 is prescribed. For most things we use a pH 6.5 buffer.
3. Rinse in distilled water, dry, and examine. For thick-film malaria staining Wilcox used a pH 7.2 stain and prolonged the distilled water washing for some minutes. This takes out some of the excess basic dye deposited at the higher pH level. Use of a pH 6.5 to 6.8 buffer obviates this differentiation.

The 40-min stain is adequate for blood work and for tertian and quartan parasites. For staining of Schüffner's granules or undulating membranes, longer staining, up to 2 h, is often desirable. Platelets are often well demonstrated by these technics (better than by Wright's stain), with purple central granule and pale blue periphery.

ACCELERATED GIEMSA STAIN FOR THICK FILMS. Stain unfixed thick films, after 1-h drying, in Giemsa stain, 4 ml; acetone, 3 ml; pH 6.5 buffer, 2 ml; distilled water, 31 ml. Stain 5 to 10 min, rinse in distilled water, dry, and examine. By this brief staining one avoids the loss of thick films so often suffered with brief drying periods.

Some malariologists find it advantageous to cover positive thick films with coverslips after staining, to protect them from roaches. Polystyrene appears to be excellent for this purpose. Dr. J. A. (Johnnie) Walker wrote Lillie about 1958 that he had found good color preservation in films so mounted after a 10-year interval. It is presumed that the film collection was kept in the dark under ordinary tropical temperature conditions.

Methods for protozoa in sections are discussed elsewhere.

Some investigators such as Yam (*Am. J. Clin. Pathol.*, **47:** 797, 1967) advocate the use of Riu's stain (*J. Niigata Med. Assoc.*, **70:** 635, 1956) for rapid preparation of air-dried smears from effusions and aspirates.

The stain is prepared as follows:

Solution A:

Eosin Y	0.18 g
Methylene blue	0.07 g
Absolute methanol	100 ml

Solution B:

Methylene blue	0.7 g
Azure I (B?)	0.6 g
KH_2PO_4	6.25 g
$Na_2HPO_4 \cdot 12H_2O$	12.6 g
Distilled water	500 ml

The solutions are filtered separately 24 h after preparation. Solution B is then adjusted to pH 6 to 6.6 by the addition of saturated solution of KH_2PO_4 as desired. The more acid pH imparts a reddish color to the cells, and the more alkaline pH a bluish color.

STAINING PROCEDURE.   Air-dried smears are covered with solution A for from 30 s to 1 min. Solution B is then added to the slide and mixed well with solution A by gentle blowing. The amount of solution B used is approximately twice that of A. After from 90 s to 2 min, the slides are rinsed, first with distilled water and then with tap water.

RESULTS: Cells are stained similarly to those obtained with the Giemsa method. Stated advantages are that Riu's method requires only 2 to 3 min, red blood cells are less red than with Giemsa or Wright, and nuclei are more blue. Lillie regards this as not especially novel and the reported results as inferior to those with a good Giemsa.

Steedman (*Stain Technol.*, **45**: 247, 1970) states the following one-solution triacid stain differentiates oxygenated from nonoxygenated erythrocytes:

1. Fix tissues for 5 to 24 h at 23°C in the following fixative:

40% formaldehyde	10 ml
p-Toluene sulfonic acid	5 g
Distilled water	85 ml

2. Dehydrate, paraffin-embed, and section in the usual manner.
3. Bring sections to water, and stain for 10 min or longer at 23°C in the following mixture (solution A, 30 ml; solution B, 40 ml; solution C, 30 ml, pH 3.4):

    Solution A:

Chlorantine fast blue 2RLL, 200%*	0.5 g
Glacial acetic acid	0.5 ml
Propylene glycol monophenyl ether	0.5 ml
Distilled water	98.5 ml

    * Ciba, C.I. direct blue 80, a metallized disazo dye of unrevealed constitution.

    Solution B:

Cibacron turquoise blue G-E*	0.5 g
Glacial acetic acid	0.5 ml
Propylene glycol monomethyl ether	0.5 ml
Distilled water	98.5 ml

    * Ciba, C.I. 74460, reactive blue 7, a copper phtalo-cyanin dye.

Solution C:

Procion red M-G*	0.5 g
Glacial acetic acid	0.5 ml
Propylene glycol monomethyl ether	0.5 ml
Distilled water	98.5 ml

* ICI, C.I. reactive red 5, a dichlorotriazinyl monoazo dye.

4. Rinse sections in distilled water 2 min.
5. Differentiate in 70% ethanol about 2 min.
6. Dehydrate and mount.

RESULTS: Mammalian oxygenated erythrocytes are bright red; deoxygenated red blood cells are blue; nuclei are red or blue; cytoplasms are blue to red depending upon the tissue; mucin is green; and connective tissue is green to blue.

This is an example of the use of reactive dyes in histology. It is regretted that Steedman used two dyes of unrevealed constitution. For further elucidation consult the original.

Wright's and *Leishman's stains* are compounds of eosin Y with methylene blue altered by the action of alkalies so as to contain a variable amount of azure B, azure A, and methylene violet. The stains are best procured as dry powders from commercial manufacturers, and should bear the certificate of the Biological Stain Commission.

As so certified, these stains usually consist chiefly of the eosinates of azure B and of methylene blue, regardless of the method of manufacture. Spectroscopically they should present two absorption bands of nearly equal density, the one at about 517 mμ (eosin) in a proportion of about 0.9 : 1 of the other, which represents the blue component, lying preferably between 652 and 658 mμ. A satisfactory stain with these characteristics may be made by mixing equal parts of the methylene blue and azure B eosinates as prepared for the Giemsa stain.

The solvent recommended is C.P. methyl alcohol (methanol), which meets American Chemical Society specifications. Redistillation from silver oxide and sodium hydroxide to prepare a neutral, acetone-free, and aldehyde-free methyl alcohol has been found to be unnecessary.

Lillie prefers Wright's original prescription of 0.5 g/100 ml methanol to the weaker solutions in vogue. Although some residue remains undissolved when as little as 150 mg/100 ml is used, the solution increases in optical density and staining power up to 600 or 700 mg/100 ml, and it is probable that the solubility lies in that range.

One of the advantages usually urged for Wright's and Leishman's stains is that the undiluted stain may be used for fixation and water may then be added for the staining period. We prefer to have films fixed at once at the bedside in methanol, and then later, on return to the laboratory, we mix stain and water in a test tube in a proportion of 1 ml stain to 3 ml water, allowing 2 ml total for each slide to be stained. The water should be buffered by addition of 1 ml 0.1 *M* phosphates at pH 6.5 for each 30 to 40 ml water. If it is necessary to use other than distilled water or rainwater, a larger amount of stock buffer may be needed, or a solid buffer (ibid.) may be added in such quantity as is found necessary under local conditions (Sörensen's phosphates, p. 878).

THE TECHNIC FOR WRIGHT STAIN

1. Fix films in methanol for 2 to 3 min (or deposit 0.5 ml stock stain on each slide and let stand 2 min).
2. Deposit 2 ml 25% dilution with buffered water of the stock stain on each slide (or add 1.5 ml water to the stock stain on the slide). Let stand 3 to 5 min.

3. Rinse in water, dry in an airstream (or with compressed air), and examine in immersion oil. Modified mineral oils preserve the stain if left on; cedar oil decolorizes the blue and purple elements.

LILLIE'S WRIGHT STAIN TECHNIC. Superior results in regard to the sharpness of nuclear and parasite chromatin staining may be achieved by using a 1% solution in equal volumes of glycerol and methanol, thus: Fix films 2 to 3 min in methanol. Stain 5 min in 4 ml stock stain, 3 ml acetone, 2 ml 0.1 M pH 6.5 phosphate buffer, and 31 ml distilled water in a Coplin jar. Rinse, dry, and examine. Such a mixture may be used *at once* for a second group of 10 slides with only slightly inferior results.

A slower, similar method uses 2 ml stock 1% stain, 2 ml pH 6.5 buffer, and 46 ml water, and requires 20 to 30 min.

GENERAL RESULTS OF ROMANOVSKY STAINS ON FILMS. With all these Giemsa, Wright, Leishman, and similar Romanovsky stains, the cytoplasm of lymphocytes should be a clear medium blue; their nuclei, a deep purple to violet; chromatin of malaria parasites and of trophonuclei of trypanosomes, red-purple; of blepharoplasts, a darker, perhaps more violet, purple; undulating membranes, pink; cytoplasm of plasmodia and trypanosomes, light blue; nuclei of monocytes, lighter red-purple; their cytoplasm, an opaque, faintly bluish gray; central granules of blood platelets, red-purple; periphery, light blue; the granules of mast leukocytes are deep blue-violet; eosinophil granules are orange-pink; neutrophil granules, purple to violet. Bartonellae appear faintly blue with red-purple chromatin dots and are best discerned when stained with Giemsa stain at pH levels between 6.2 and 6.8. Bacteria stain deep blue to violet. Diphtheroids may exhibit deeply stained bars and polar granules in a light blue body. The color of erythrocytes varies with the pH of the diluting water from pink at pH 6 through yellowish pink at 6.5, pinkish or grayish yellow at 6.8, grayish yellow to greenish yellow or even gray-blue at 7 to 7.2. Differentiation with distilled water, which is often faintly acid, or with very dilute (0.1 to 0.05%) acetic acid displaces the color of erythrocytes from the gray-blue toward the pink limit of the color series above. Eosinophil granules are often best stained when chromatin is much understained and pale blue.

SUDANOPHIL GRANULES OF LEUKOCYTES

These granules occur in neutrophil and eosinophil leukocytes, monocytes, and probably basophil granulocytes as well, in general in much the same cells as show peroxidase and indophenol oxidase activities. They do not stain in the cold with the usual oil-soluble dye methods which suffice for the staining of true fats and lipids. At temperatures of 20 to 25°C they require prolonged exposures to the naphthol-type oil-soluble dyes such as oil red O, Sudan III, and Sudan IV. Naphthylamine and aminoanthraquinone dyes act more rapidly but still do not stain in the usual 5- to 10-min intervals used for lipids. Reasonably rapid staining, 1 to 2 h for oil red O, for example, may be attained at 37°C, but elevation of temperature to 55 to 60°C or higher appears to be positively deleterious in comparison with 37°C staining.

After Savini's and Sehrt's papers [*Wien. Med. Wochenschr.*, **46:** 1964, 1921; *C. R. Soc. Biol. (Paris)*, **87:** 744, 1922; *Munch. Med. Wochenschr.*, **74:** 139, 1927] these granules were generally accepted as lipid in nature, though a number of writers commented on the resistance of Sudan black stains to extraction with 100% alcohol or with xylene. Lillie and Burtner (*J. Histochem. Cytochem.*, **1:** 8, 1953) demonstrated their water-soluble nature, their resistance to extraction with a variety of fat solvents (aromatic and aliphatic hydrocarbons, pyridine, carbon tetrachloride, etc.), their unstainability with esterified Sudan dyes, and their capacity to react with alcohols and phenols, and concluded that they were not lipid. Since they are readily rendered unstainable by alcohol of 60% concentration or higher at 60°C, staining is preferably done at 50% alcohol concentration. Indeed we have seen decolorization of previously well-stained preparations in the customary saturated 70% alcohol solutions of Sudan dyes.

LILLIE-BURTNER TECHNIC FOR STAINING SUDANOPHIL GRANULES IN LEUKOCYTES

1. Fix air-dried blood films 10 min in 75% alcohol, in 75% alcohol containing 10% formalin, or with formaldehyde gas over a little strong formalin. Wash briefly in water to remove excess formaldehyde.
2. Mix 0.5% oil red O in 100% alcohol or, better, 99% isopropanol with an equal volume of distilled water, and place in oven or water bath at 37°C. Stain smears 2 to 4 h. (Fewer but larger granules are shown by staining 30 to 60 min in dye mixture pre-heated to and kept at 60°C.)
3. Wash in water.
4. Counterstain 5 min in Lillie's acetic hemalum or other alum hematoxylin.
5. Wash 5 min in running water.
6. Blow dry with airstream. Wash 1 to 2 min in xylene or acetone to remove dye precipitate. Mount in polystyrene with cover glass, or again blow dry and examine in immersion oil.

RESULTS: Nuclei are blue; numerous fine to moderately numerous medium bright red granules are seen in unstained to pale yellowish cytoplasm of neutrophil leukocytes; fewer similar granules are thus seen in monocytes; granules of eosinophils are largely pale gray-green to brownish yellow, usually with a few to moderately numerous bright red to orange-red, medium to coarse granules. Erythrocytes vary from greenish gray to very pale gray. Often erythrocytes and plasma show globules to fine granules of red deposit, which is also highly resistant to solvents. This deposit varies from none to copious from one slide to another, and in different areas of the same slide. It seems to be uninfluenced by filtration of the stain mixture and is extracted by alcohol of 60 to 90% in much the same way as the coloration of the leukocyte granules.

By substituting Sudan black B for oil red O in the foregoing technic the color of granules in the leukocytes is changed to greenish black; the granules of the eosinophils stain quite uniformly in dark gray-green, often appearing more lightly colored centrally. The erythrocytes assume varying shades of gray-green, becoming dark gray to black if staining is prolonged, especially at 60°C. Staining time with Sudan black B is shorter than with oil red O. A half hour at room temperature, 5 to 10 min at 60°C (preheated), or 15 min at 37°C suffice. Counterstains for nuclei are less satisfactory than with the hemalum of the oil red stain. A 2-min stain in 0.5% safranin in 1% acetic acid is fairly satisfactory. Feulgen staining, as practiced on tissue sections, gives rather pale colors.

Blood films fixed with formaldehyde (either gaseous or in aqueous solution) and stained with oil-soluble dyes for the demonstration of leukocyte granules are quite apt to show areas of partial detachment, wrinkles running in mosaic fashion, and irregular precipitates both of oil-soluble dye and of the hematoxylin counterstain. These precipitates and the droplets stained in red corpuscles may be quite difficult to remove, whether with alcohol, acetone, or xylene, after staining. They seem to be prevented in considerable measure by prior treatment with 60 to 80% alcohol, and are much less conspicuous in films fixed for 10 min in 75% alcohol.

TABLE 16-3   RESULTS OF NONSPECIFIC ESTERASE, CHLOROACETATE ESTERASE, PEROXIDASE, AND METACHROMASIA STAINS OF BLOOD AND MARROW CELLS

	Nonspecific esterase	Chloroacetate esterase	Peroxidase Without cyanide	Peroxidase With cyanide	Metachromasia
Myeloblasts	$0-\pm$	$0-+++$	$0-+++$	0	0
Promyelocytes	$0-\pm$	$++++$	$++++$	0	0
Neutrophilic granulocytes	$0-\pm$	$++++$	$++++$	0	0
Eosinophilic granulocytes	$0-+$	0	$++++$	$+++$	0
Basophilic granulocytes	0	$0-+$	$0-++$	0	$++-++++$
Megakaryocytes	$++++$	0	0	0	0
Erythroblasts	$0-+$	0	0	0	0
Lymphocytes	$0-+$	0	0	0	0
Plasma cells	$0-++$	0	0	0	0
Monocytes	$++++$	$0-+$	$0-+++$	0	0
Reticulum cells (histiocytes)	$++++$	$0-+$	0	0	0
Mast cells	?	$++++$	0	0	$++++$

Key: 0 = no activity, $\pm$ = questionable, + = weak, ? = unknown, + + = moderate, + + + = strong, + + + + = very strong activity.

Source: Yam et al., *Am. J. Clin. Pathol.*, **55**: 283, 1971.

The sudanophil granules of polymorphonuclear leukocytes, though quickly destroyed by 60 to 100% alcohol at 60°C, resist 90% and lower alcohol concentrations for quite long periods at room temperatures and lower. They are water-soluble at 60°C after brief 75% alcohol fixation but are rendered insoluble in water by mercuric chloride, lead nitrate, and formaldehyde.

Their sudanophilia is destroyed by treatment with ferric chloride, ferrous chloride, potassium dichromate, potassium permanganate, periodic acid, and hydrogen peroxide.

Though it is uncertain whether Sudan staining demonstrates preexisting granules or whether the granules simply represent deposition of dye complex at sites of reaction, it is convenient to refer to these sites as the *sudanophil granules*.

Yam et al. (*Am. J. Clin. Pathol.*, **55**: 283, 1971) provide Table 16-3 as an aid in the cytochemical identification of monocytes and granulocytes. Blood and bone marrow smears were employed. Smears were stored unfixed at 23°C for at least 2 weeks without noticeable change in enzyme activity. Before staining for enzymes, smears were fixed for $\frac{1}{2}$ min at 4 to 10°C at pH 6.6 in 20 mg Na_2HPO_4, 100 mg KH_2PO_4, 30 ml H_2O, 45 ml acetone, and 25 ml 40% formaldehyde.

For a detailed extensive review on the cytochemistry of monocytes and macrophages the reader is referred to Schmalzl and Braunsteiner (*Ser. Haematol.*, **3**: 93, 1970).

Booyse (*J. Histochem. Cytochem.*, **19**: 540, 1971) employed soluble horseradish peroxidase–antihorseradish peroxidase complex to demonstrate that the actomyosin-like contractile protein thrombostenin is both situated in the cytoplasm and associated with the fluffy membrane coat of platelets. Brief exposure of platelets to pronase resulted in the removal of all protein, including the thrombostenin from platelet membranes.

Lycette et al. (*Am. J. Clin. Pathol.*, **54**: 692, 1970) noted a direct correlation between platelet uptake of either Luxol Fast Blue ARN (C.I. solvent blue 37) or rhodamine 6G (C.I. 45160), their aggregation, and their phospholipid content. They suggest the participation of surface phospholipids during aggregation.

Balogh [*Nature* (*Lond.*), **199**: 1196, 1963] demonstrated aminopeptidase activity in platelets.

Carstairs (*J. Pathol. Bacteriol.*, **90**: 225, 1965) used the Coons indirect fluorescent immunohistochemical method to demonstrate platelets in tissue sections containing thrombi.

While blood platelets are readily recognized in Giemsa- or Wright-stained blood smears, their recognition in formol-fixed tissue sections is more difficult. Immunofluorescence technics have been invoked successfully, but this puts the question into a special research category beyond the scope of the usual diagnostic laboratory. Success has been claimed for modifications of Lendrum's picro-Mallory method.

For example, Carstairs (*J. Pathol. Bacteriol.*, **90**: 225, 1965) claims consistent results in differentiating platelets from fibrin by the following modified picro-Mallory method:

Fix at least 48 h in neutral formol-saline solution. Prepare paraffin sections as usual, dewax, and hydrate through graded alcohols.

1. Mordant 5 min in 5% iron alum.
2. Wash in running water.
3. Stain 5 min in Mayer's alum hematoxylin (which formula not specified).
4. Wash in running water.
5. Stain 30 to 60 s in orange G, 200 mg; saturated aqueous picric acid, 20 ml; saturated isopropanol picric acid, 80 ml.
6. Dip quickly in distilled water.
7. Stain 1 to 5 min in ponceau 2R, 500 mg; acid fuchsin, 500 mg; glacial acetic acid, 1 ml plus 99 ml distilled water.
8. Rinse in distilled water.
9. Differentiate in 1% phosphomolybdic acid until muscle is red and background pale pink.
10. Rinse in distilled water.
11. Stain 30 s to 2 min in 1% soluble blue in 1% acetic acid.
12. Wash in tap water, dehydrate, clear, and mount.

The soluble blue is probably either water blue I, C.I. 42755, or methyl blue, C.I. 42780. The aniline blue of the classic Mallory stain was an approximately equal mixture of these two dyes.

By this technic platelets color a dark slate blue, fibrin red, and erythrocytes yellow. The technic is long, involved, and with variable steps depending on the technician's experience. We have not used the method.

PERIODIC ACID SCHIFF REACTION

After alcohol fixations, a 10-min oxidation in 1% H_5IO_6, 5-min washing, 10 min in Schiff reagent, 5-min washing in three changes of 0.5% sodium metabisulfite, 10-min washing, 5 min in acetic hemalum as counterstain, and 5-min washing in tap water, films are blown dry in an airstream and examined in immersion oil or mounted in synthetic resin.

Nuclei stain deep blue; erythrocytes, pale yellow to gray; cytoplasm of neutrophils, purplish red; granules of eosinophils, clear and unstained in foamy pink cytoplasm. Dark red-purple glycogen granules occur in lymphoid cells and sometimes in neutrophil leukocytes. Platelets contain purplish red oval granules.

Pretreatment with diastase solutions weakens the cytoplasmic staining of neutrophil leukocytes but does not destroy it. Glycogen is removed by diastase digestion. The staining of platelets is unaffected. Treatment of films with water or pH 6 buffer weakens cytoplasmic staining of leukocytes only slightly.

For glycogen digestion tests, films fixed in formaldehyde gas or alcoholic formaldehyde solutions should be employed, since after simple alcoholic fixations, the proteolytic enzymes in the polymorphonuclear leukocytes selectively autolyze these cells, destroying first their cytoplasm and then their nuclei. This leukocyte protease is quite active at 60°C and is not destroyed by 10 min in boiling acetone, benzene, toluene, or xylene.

Acetylation for 4 h at 60°C in 40% acetic anhydride–pyridine mixture renders leukocytes and platelets periodic acid Schiff–negative, and a 16-h deacetylation in 20% ammonia water–80% alcohol solution restores the reactivity of platelets and in part that of neutrophils. Benzoylation in 1 : 19 benzoyl chloride–pyridine is less successful at 25 or 37°C, and at 60°C the reagent completely destroys the capacity of nuclei to stain with alum hematoxylin or Giemsa stain. Deamination in 2 N NaNO$_2$–HAc, sufficient to destroy oxyphilia of erythrocytes, does not impair this reaction of leukocytes or platelets.

FLUORESCENCE MICROSCOPY.   Primulin, berberin, and rivanol have been used for the demonstration of protozoan parasites in fluorescence microscopy. They give respectively blue, bright yellow, and yellowish green fluorescence to leukocyte nuclei; yellow and bright yellow to leukocyte cytoplasm; and blue-white, golden yellow, and yellowish green to malaria parasites (*Haemoproteus*, *Plasmodium nucleophilum*, and *P. vivax*). One stains methyl alcohol–fixed smears 2 to 5 min in saturated aqueous or alcoholic solutions of the fluorochromes. Parasites and leukocytes stand out as brilliantly fluorescent objects against a dark field and are readily discerned at 200 diameters under dry lens systems. Nothing is said about the behavior of blood platelets with this method, and most of Patton and Metcalf's work was done on the two avian parasites (*Science*, **98**: 184, 1943). We have had no experience with this method.

CATECHOL MORDANT DYES.   When Gomori's variant of Clara's method for enterochromaffin cells using very dilute aqueous solutions of hematoxylin, brazilin, gallocyanin, or celestin blue B is applied to alcohol-fixed or formaldehyde-fixed human blood films, the granules of the eosinophil leukocytes are selectively stained. Gallocyanin gives brilliant purplish violet eosinophil granules with 48-h staining in a 1 : 20,000 solution. The red corpuscles appear in pale yellowish gray; nothing else stains. With celestin blue at 1 : 20,000 for 2 days, white cell nuclei appear pink; red cells, faint yellow; eosinophil granules, blue-green; and neutrophils, unstained. Hematoxylin gives blue-gray, brazilin and alizarin pink, but the contrasts are inferior to those with the oxazin dyes. In sections, however, 0.01% hematoxylin, 32 to 48 h, gives blue-black eosinophil granules.

Perhaps the most contrastful results are obtained with a 1 : 100,000 aqueous solution of celestin blue, staining 6 h at 60°C. This yields light red-purple nuclei, blue-green eosinophil granules, unstained neutrophil cytoplasm, and pale yellow erythrocytes. It is necessary to employ alcoholic formaldehyde or formaldehyde-vapor fixation to avoid the autolytic digestion of the neutrophils which occurs in distilled water at 60°C.

In tissue sections a safranin counterstain may be used after 1 : 20,000 gallocyanin (2 days, 37°C), but the staining is variable and generally less successful than in blood films. Lillie, Pizzolato, and Donaldson [*Acta Histochem. (Jena)*, **49**: 204, 1974] assign this reaction to the arginine content of the granules.

RETICULOCYTES

Brecher (*Am. J. Clin. Pathol.*, **19**: 895, 1949) prefers *new methylene blue N* (C.I. 52030) to the more commonly used brilliant cresyl blue for the staining of reticulocytes in blood. Various lots of brilliant cresyl blue vary considerably in staining properties and spectroscopic characteristics, indicating differences in composition. As Conn states, brilliant cresyl blue is not used in industry and must be specially manufactured in small lots for biologic staining. This fact undoubtedly explains the variations. Brilliant cresyl blue now includes four chemically distinct oxazin dyes. New methylene blue, on the other hand, is manufactured by several manufacturers for textile dyeing and appears to be quite constant both in its absorption spectrum ($\lambda = 630$–632.5 mμ) and in staining performance.

BRECHER'S NEW METHYLENE BLUE TECHNIC FOR RETICULOCYTES

1. Dissolve 0.5 g new methylene blue and 1.6 g potassium oxalate in 100 ml distilled water.
2. Mix approximately equal drops of stain and of fresh or oxalated blood on a slide.
3. Draw up the mixed drop into a capillary pipette, and let stand for 10 min.
4. Expel the mixture in small drops on several slides, and make thin smears as usual. Dry in air and examine with oil immersion.

RESULTS: Erythrocytes are light greenish blue; reticulum is a deep blue and sharply outlined.
   Deeper blue staining of erythrocytes indicates that an excess of dye was used. Generally the blood drop should equal or slightly exceed the stain drop in size.

   As an example of a brilliant cresyl blue dry smear technic we quote the Cunningham-Isaacs technic from Conn and Darrow, 1948 (ID3-8); it agrees well with Clark (ed.), "Staining Procedures," 3d ed., p. 19, Williams and Wilkins, Baltimore, 1973.

BRILLIANT CRESYL BLUE METHOD FOR RETICULOCYTES

1. Dry a drop of 0.3% alcoholic solution of brilliant cresyl blue (C.I. 51010) on a cleaned and polished slide or cover glass.
2. Deposit a drop of fresh blood 2 to 3 mm in diameter on a similarly cleaned slide or cover glass.
3. Appose the stain-covered area on the first slide or cover glass to the blood drop, and move this slide up and down hinge fashion until the dye film is all dissolved and the blood appears blue-black.
4. Allow the two slides or cover glasses to cohere in parallel position and spread the blood drop.
5. Draw apart along the plane of contact, and allow the two films to dry.

RESULTS: Sharply stained blue reticulum and pale blue erythrocytes are seen. If desired the films may be counterstained with Wright's or Giemsa's stain by one of the usual technics. In this case the background is the usual one with these stains, and the reticulum appears deep blue.

   Phase microscopy reveals a number of erythrocytes containing rods and granules which appear dark under dark-contrast phase illumination, and which Brecher (*Bull. Int. Assoc. Med. Mus.*, **30:** 99, 1949) finds to be associated with, but not identical with, the stainable reticulum in the same cells.
   Since the presence of the hemoglobin largely obscures these rods and granules, hemolysis is a necessary part of the technic for their phase contrast demonstration. Phenylhydrazine intoxication renders reticulocytes more numerous and more conspicuous under phase microscopy, perhaps because of in vivo hemolysis, so that under these conditions they may be found without artificial hemolysis.

BRECHER'S TECHNIC FOR PHASE MICROSCOPY OF RETICULOCYTES

1. Mix a drop of blood with a drop of hypotonic ammonium oxalate solution (1.2%) on a clean cover glass.

2. Pick up the blood and cover glass by bringing a clean slide down on top of it.
3. Turn over and apply Vaseline (USP Petrolatum, BP Paraffinum Molle) along the edges of the cover glass to seal the preparation. This is done conveniently from a 5-ml syringe fitted with a large-bore needle.

INCLUSION BODIES

Oxyphil inclusion bodies such as Negri bodies, Guarnieri bodies, herpes, varicella, and molluscum inclusions and the like, are often quite well shown by azure-eosin and phloxine-azure sequence methods. They were excellently demonstrated in hematoxylin-eosin preparations which had been mounted in Canada balsam for 2 or 3 years. Such old preparations appear to be distinctly superior to fresh stains for this purpose. The safranin O–eriocyanin A method has given good results.

Most of the so-called specific Negri body methods depend on balanced mixtures of two basic or two acid dyes, or on sequence procedures using an acid and a basic dye. The preferred material for Negri bodies is hippocampus and cerebellar cortex. The major lesions of rabies, however, are found in the brainstem.

STOVALL-BLACK METHOD (*Am. J. Clin. Pathol.*, **10:** 1–8, 1940). Stovall and Black used acetone fixation and a sequence stain:

1. Stain 2 min in a 1% alcoholic solution of ethyl eosin (C.I. 45386, sodium ethyl eosinate) adjusted to pH 3 with 0.1 N hydrochloric acid.
2. Rinse in water.
3. Stain 30 s in 10 ml 1% methylene blue in 95% alcohol, 10 ml 0.2 M acetate buffer of pH 5.5, and 20 ml water.
4. Then differentiate in 0.38% acetic acid (0.063 N) in water (13 drops in 60 ml) until sections are brownish red.
5. Wash, dehydrate, and clear.
6. Mount in balsam.

RESULTS: Negri bodies are brownish to pure red; nucleoli, pale blue; other structures, pink.

A variant of this method which we have used successfully is the following:

1. In 90 ml 100% alcohol or 94 ml 95.5% alcohol, 3.25 ml 1% acetic acid, and water to make 100 ml, dissolve 950 mg ethyl eosin. Stain in this for 2 min.
2. Then wash in alcohol and in water.
3. Counterstain formalin material in 0.5% methylene blue in 25% alcohol; alcohol-fixed material, in alum hematoxylin.
4. Differentiate in 0.25% acetic acid for 2 to 5 min.
5. Wash, dehydrate, and clear.
6. Mount in balsam.

GERLACH'S METHOD (Kraus, Gerlach, and Schweinburg)

1. Stain paraffin sections of formalin-fixed material in a fresh mixture of 3 ml carbol fuchsin, 6 ml Loeffler's methylene blue (about 0.35% in 22% alcohol containing 1 : 10,000 potassium hydroxide), and 50 ml distilled water. Heat sections to steaming 4 times in four changes of this mixture.
2. Wash in water.
3., Differentiate and dehydrate in alcohols and clear in xylene.
4. Mount in synthetic resin or mineral oil; the stain fades rapidly in balsam.

RESULTS: Negri bodies stain red; chromatin and nucleoli, rather light blue.

MANN'S METHYL BLUE–EOSIN TECHNIC (Kraus, Gerlach, and Schweinburg). This method is classic: Zenker fixation was prescribed. Paraffin sections are carried through 0.5% iodine and 5% sodium thiosulfate as usual and washed in water.

1. Stain 24 h in 6 ml 1% aqueous eosin Y (C.I. 45380), 6 ml 1% aqueous methyl blue (C.I. 42780), and 28 ml distilled water.
2. Wash in water, and differentiate in 100% alcohol containing 4 mg sodium hydroxide per 100 ml (0.001 N; add 0.1 ml 1% NaOH in alcohol to 25 ml 100% alcohol).
3. Wash in 100% alcohol.
4. Then wash in water containing a few drops of acetic acid—say 0.1%.
5. Dehydrate and clear through alcohols and xylene, and mount in polystyrene or cellulose caprate.

RESULTS: Negri bodies and erythrocytes are stained red; nuclei and inner granules of the inclusions, blue.

SCHLEIFSTEIN'S RAPID METHOD (*Am. J. Public Health,* **27:** 1283, 1937, emended). Schleifstein prescribed 4-h fixation in Zenker's fluid at 37°C, 30-min washing in water, dehydration 1 h at 37°C in dioxane over anhydrous calcium chloride, infiltration 1 h in dioxane-paraffin 50% mixture at 56°C and in pure paraffin for 1 h. In place of the dioxane schedule we suggest substitution of the rapid acetone-benzene schedule.

Schleifstein's stain consists of 1.8 g basic fuchsin (rosanilin chloride) and 1 g methylene blue, dissolved in 100 ml glycerol and 100 ml methyl alcohol. For use it is diluted 1 : 80 (Mallory, 1 drop to 2 ml) with 1 : 40,000 potassium hydroxide.

1. Bring paraffin sections to water as usual.
2. Steam gently for 5 min (70°C?) in the diluted stain.
3. Then wash in tap water.
4. Differentiate to a faint violet in 90% alcohol.
5. Dehydrate rapidly with 95 and 100% alcohol, and clear through 100% alcohol and xylene (50 : 50) and two changes of xylene. Mount in Permount or other resin.

RESULTS: Negri bodies are a deep magenta red; erythrocytes, coppery red; nucleoli, blue-black; cytoplasm, blue-violet.

ZLOTNIK'S METHOD FOR NEGRI BODIES IN FIXED TISSUE [*Nature (Lond.),* **172:** 962, 1953]

1. Bring paraffin sections to water as usual.
2. Stain 5 min in Ehrlich's hematoxylin.
3. Blue 2 min in tap water.
4. Counterstain 1 min in saturated aqueous picric acid containing 0.5% orange G (C.I. 16230).
5. Wash in water until only erythrocytes remain yellow. Rinse in distilled water.
6. Stain 10 min in 0.5 g acid fuchsin, 0.5 g phosphotungstic acid, 100 ml 1% acetic acid.
7. Rinse in distilled water.
8. Differentiate 5 min in 2 g phosphotungstic acid, 2 g phosphomolybdic acid, 30 ml 100% alcohol, and 70 ml saturated aqueous picric acid.
9. Rinse in distilled water and in 1% acetic acid.
10. Stain 15 min in 1% aniline blue in 2% acetic acid.
11. Rinse in 1% acetic acid, dehydrate, and clear by the alcohol-xylene sequence; mount in synthetic resin.

RESULTS: Negri bodies are purplish red with blue inner granules; nerve cell cytoplasm is bluish; nucleoli are dark purple; erythrocytes, yellow.

This is essentially a somewhat complicated variant of the Masson-Mallory trichrome procedure and should yield blue connective tissue as well. We have not had occasion to use it.

For various oxyphil inclusion bodies, notably Guarnieri and Kurloff bodies, the inclusions of infantile giant cell pneumonia, and others (but not Negri bodies), A. C. Lendrum (*J. Pathol. Bacteriol.*, **59**: 399, 1947) recommended his phloxine-tartrazine stain. Fluorane dyes are required, and of these phloxine B (C.I. 45410) and rose Bengal (C.I. 45440) appear to be the best. Eosin Y and erythrosin are too readily extracted. Calcium chloride is used as an intensifier for the fluorane staining. As a differentiator Lendrum specifies the tartrazine NS of Imperial Chemical Industries (ICI), dissolved in Cellosolve (ethylene glycol monoethyl ether). It is presumed that this is C.I. 19140 and that the tartrazines of other manufacturers will serve. ICI lists only tartrazine N in the 1971 "Colour Index." Tartrazine NS is supplied by NSR (Sumitomo in Osaka).

LENDRUMS'S PHLOXINE-TARTRAZINE METHOD (emended)

1. Fix preferably in mercuric chloride formalin (9 parts saturated aqueous mercuric chloride solution and 1 part formalin) for 24 h, dehydrate with iodized 70% alcohol and ascending alcohols, clear, infiltrate, and embed in paraffin as usual.
2. Bring paraffin sections to water as usual, including iodine and sodium thiosulfate steps if tissue was not iodized before embedding.
3. Stain as usual in Mayer's alum hematoxylin or in Weigert's acid iron chloride hematoxylin.
4. Blue and wash as usual.
5. Stain 30 min in 0.5% phloxine B or rose Bengal in 0.5% (0.045 M) aqueous calcium chloride solution.
6. Rinse in water.
7. Differentiate with saturated tartrazine solution in Cellosolve, either briefly from a dropper bottle, or with strongly phloxinophil objects for as long as several hours in a Coplin jar.
8. Rinse in 60% alcohol, dehydrate with 95 and 100% alcohol, then 100% alcohol and xylene, and clear in two changes of xylene. Mount in balsam or other suitable resin.

RESULTS: Kurloff bodies in guinea pig lung, Guarnieri bodies, inclusions of infantile giant cell pneumonia, and others (but not Negri bodies) are stained (red) by the phloxine; collagen is stained yellow; nuclei, according to the hematoxylin stain selected.

We have not tried this method. It has been used by a number of British workers.

According to Wolman (*Proc. Soc. Exp. Biol. Med.*, **74**: 85, 1950) the elementary bodies of smallpox may be demonstrated by fixing smears of scrapings from incised recent papules or vesicles in ether and alcohol (50:50) for a few minutes, drying in air, and staining by the Feulgen procedure.

The strongly acidophil intranuclear inclusion bodies seen in some cases of chronic lead and bismuth poisonings may be acid-fast when stained by the Ziehl-Neelsen technic. Sections were stained in carbol fuchsin for 3 h at 56°C, washed in water, and decolorized for 3 to 5 min in 3 ml concentrated HCl in 97 ml 70% alcohol and counterstained with Harris' alum hematoxylin. Those seen in lead poisoning failed to react to ammonium sulfide. (M. Wachstein, *Am. J. Clin. Pathol.*, **19**: 608, 1949.)

Perrin and Littlejohn (*J. Clin. Pathol.*, **3**: 40, 1950) detail a rapid method which may have some value in the cytologic examination of fresh sputum for tumor cells. Their carbol fuchsin–methylene blue stain is composed of 4 volumes of a 0.5% methylene blue solution in 20% glycerol (by volume), mixed extempore with 1 volume of a 1% fuchsin solution in 10 parts alcohol to 90 parts 5% aqueous phenol solution. Purulent or bloody particles are picked out of the morning sputum, mixed thoroughly with a drop of the stain, and warmed gently over a Bunsen burner for 30 s. The uniformly stained, sticky specimen is then covered with a large cover glass and spread by gentle pressure. Specimens may be examined at once or allowed to stand a while. Staining continues and in 5 or 6 h becomes so intense as to obscure nuclear detail.

Improved nuclear detail was obtained by substituting 2% iodine green (C.I. 42556) for the fuchsin–methylene blue mixture. It is probable that the commoner, closely related dyes ethyl and methyl green can be used in the same way.

With the fuchsin–methylene blue, cytoplasm of squamous and columnar cells stains bright pink; nuclei and bacteria stain violet to purple. Leukocyte and lymphocyte cytoplasm is pale pink or green; that of macrophages, purplish red; and of plasma cells, violet to purple. Tumor cell cytoplasm is red when well keratinized, and violet to blue-green in more anaplastic cells.

Goldberg (*Blood*, **24:** 305, 1964) demonstrated acid phosphatase in Auer bodies using the Gomori lead method.

Wicker and Avrameas (*J. Gen. Virol.*, **4:** 465, 1969) used horseradish peroxidase, alkaline phosphatase, or glucose oxidase to label both T antigens and structural antigens of three viruses, namely, SV_{40}, adenovirus 12, and rat K virus in infected cell cultures. Specific localization was achieved, and different colors could be used to localize different antigens.

Siverd and Sharon (*Proc. Soc. Exp. Biol. Med.*, **131:** 939, 1969) used the enzyme-conjugated antibody technic to detect vaccinia virus in tissue culture cells.

Lafora Bodies

These were first described by Lafora (*Virchows Arch. Pathol. Anat.*, **205:** 295, 1911) under the name of *amyloid bodies* as globules and granules in the cytoplasm of nerve cells in a fatal case of myoclonic epilepsy. The affected cells were numerous in various parts of the cerebral cortex, thalamus, pons, medulla, and spinal cord. One to as many as seven rounded bodies could occur in a cell, centrally, peripherally, or in a cell process. Tigroid granules were displaced or lysed; pigment granules and nuclei were displaced. The globules colored purple with Delafield's hematoxylin, brown, gray, or black by Heidenhain's iron hematoxylin, rose with alum carmine, red-brown with Lugol's iodine, brown with iodine–sulfuric acid, red with Best's carmine, metachromatically red-purple with toluidine blue, red by the Unna-Pappenheim (methyl green–pyronin) method, inconstantly red-purple with methyl green (from contaminating crystal violet), pale green with Bismarck brown, green by the Russell method, bluish with gentian violet, and brown with various silver impregnations (Ramón y Cajal, Bielschowski, Levaditi). With the silver methods concentric lamellation and radial striation were sometimes seen. Black crystals on a brown background appeared centrally with iron hematoxylin; these were seen also as red central crystals in peripherally blue bodies with the Alzheimer-Mann (methyl blue–eosin) stain. Lafora did not note the fixation or section methods employed; one presumes formol and paraffin, just as Lillie found in the same hospital (St. Elizabeths in Washington, D.C.) 25 years later.

Collins, Cowden, and Nevis (*Arch. Pathol.*, **86:** 239, 1968) contributed later data on histologic, phase microscopic, histochemical, and electron microscopic examination of another case of myoclonic epilepsy. They described globular bodies often displacing and compressing the nucleus to crescentic form. Their size varies, and occasionally two or even three may be seen in one neuron. They are hyalin and faintly osmiophilic under phase microscopy or centrally denser and more osmiophilic. They are strongly positive by the PAS reaction, resisting amylase, hyaluronidase, acetylation, mild hydrolysis, and chloroform-methanol extraction in formaldehyde-fixed tissue. Proteolytic enzyme digestion did not affect either structure or PAS reactivity. They give no Deitch-Sakaguchi arginine reaction, no Adams DMAB-nitrite indole reaction, and no Bennett mercury orange sulfhydryl reaction. There is a weak Morel-Sisley tyrosine reaction. They stain by methylene blue at pH 8 but not at pH 4, and lightly with fast green at pH 2.3. Hot 5% trichloroacetic acid does not affect fast green staining; acetylation somewhat enhances the methylene blue stain. Peptic digestion after acetylation decreases the methylene blue reaction; papain has no effect; chymotrypsin prevents the pH 8 methylene blue stain. The centers and margins react alike to most of the foregoing tests; the centers stain selectively with Luxol Fast Blue, paralleling the osmiophilia. Electron microscopically the bodies are composed of fine granules and fibrils, sometimes with a concentric lamellation; some include cytoplasmic organelles such as mitochondria and endoplasmic reticulum; denser bodies do not. Lamellation may be closely spaced—5 mμ; fibrils measure 5 to 10 mμ in diameter and are composed of fine, dense granules. Denser aggregates appear to correspond to the denser centers of phase and optical microscopy.

Abundant lipofuscin is concurrently present in other neurons. A glycoprotein nature of the Lafora body is indicated, and a possible relation to the lipofuscin is suggested.

Russell Bodies

Russell bodies are colorless spheres, polyhedra, rods, or ovoids occurring singly and in clusters in the cytoplasm of cells of the plasma cell type, ranging upward in size to spheres twice the diameter of surrounding plasma cells. In fixed tissue they are strongly oxyphil on staining with mixtures of the azure-eosin type, but, according to Kindred (*Stain Technol.*, **10:** 7, 1935), they differ from hemoglobin in not staining with eosin at pH levels of 6.2 to 6.6 when in the unfixed state.

Russell (Ehrlich's "Encyklopädie") prescribed Müller fixation and stained 10 to 30 min in saturated fuchsin solution in 2% aqueous phenol, then washed 3 to 5 min in water and 30 s in 100% alcohol, counterstained in 1% iodine green (C.I. 42556) in 2% aqueous phenol for 5 min, dehydrated with 100% alcohol, cleared in xylene, and mounted in balsam. Nuclei stained green; the fuchsin bodies, red. Klien (ibid.) prestained with alum hematoxylin, stained in warm carbol fuchsin, and differentiated in a strong fluorescein solution in alcohol. The usual alcohol, xylene, balsam sequence followed. Schmorl quotes Russell's original method without change; Cowdry refers to Kindred's paper (above); other, more recent authors make no reference to these bodies. Askanazy, writing in Aschoff's "Pathologische Anatomie" (Fischer, Jena, 1936), noted that they are gram-positive. This is true with the Gram-Weigert method, but with acetone differentiation they are gram-negative. With hemalum and oil red O they do not take the fat stain in frozen sections. On application of the Weigert myelin technic they become deep gray while red corpuscles are still black; and they retain some gray for a little while after the erythrocytes decolorize. The periodic acid Schiff–leucofuchsin technic colors them red, with or without antecedent diastase digestion; but after oxidation with chromic anhydride or with 1% potassium permanganate, only a pink or gray-pink is produced by the leucofuchsin. They do not blacken with diammine silver and are iron-negative.

Welsh (*Am. J. Pathol.*, **40:** 285, 1962) notes also the occurrence of periodic Schiff–negative forms and finds a correlation between the degrees of reactivity and of intracisternal electron density. The grade of eosinophilia also varies from quite strong to inappreciable.

Using 0.2 mM methylene blue and 0.1 mM eosin at pH ranges from 3 to 9.5, Goldberg and Deane (*J. Histochem. Cytochem.*, **8:** 327, 1960) found the pK of formalin-fixed Russell bodies to lie at about 6 to 7.5. Congo red and crystal violet amyloid stains were negative; the periodic acid Schiff stain was positive.

For Russell bodies of cancer cells (and plasma cells) Bangle (*Am. J. Pathol.*, **43:** 437, 1963) records slight to strong acid fastness (Ziehl-Neelsen) unaltered by prior 2-h, 60°C, 2 : 1 chloroform-methanol extraction or HCl hydrolysis (1 N, 60°C, 20 min); a positive Gram-Weigert and negative Gram acetone stain; positive periodic acid Schiff reaction, also after diastase and in the allochrome method (orange to red); orthochromatic basophilia above pH 4 with 0.01 to 0.05% thiazin dye solutions; oxyphilia to eosin–acid fuchsin and Biebrich scarlet only in the acid range (pH 5, 6); a positive Millon reaction but negative SH and SS reactions with ferric ferricyanide and the peracetic thionin method; negative fat stains with oil red O and Sudan black B; negative peracetic Schiff and direct 72-h Schiff reactions; a negative benzidine nitroprusside test for hemoglobin; and a negative phosphotungstic acid hematoxylin stain for so-called alcoholic hyalin.

Döhle Bodies

Oski et al. (*Blood*, **20:** 657, 1962) described the Döhle bodies in neutrophil leukocytes of peripheral blood smears from a case of familial May-Hegglin anomaly. These bodies are round and stain blue with Wright and Giemsa stains. They occur in the cytoplasm of myelocytes, metamyelocytes, band forms, and polymorphonuclear neutrophil and eosinophil leukocytes. Feulgen and periodic acid Schiff reactions are negative, and the inclusions are not stained by Sudan black. They stain red with methyl green–pyronin, losing this reaction after ribonuclease digestion (Petz et al., *Clin. Res.*, **8:** 215, 1960).

Döhle [*Cbl. Bakt. Paras Knde. Infekt.* (*Orig.*), **61:** 63, 1912] referred to round or oval inclusion bodies in leukocytes in cases of scarlet fever. They stained lighter blue with Giemsa, red with methyl green–pyronin, and lighter blue than nuclei with "Michaelis's azure blue."

SPIROCHETES

Spirochetes are generally demonstrated in tissues by one of the numerous reported silver methods. For fresh clinical material the dark-field technic is preferable, and this may be applied also to fresh autopsy material. For smears, various methods using Wright's and Giemsa's stains as well as silver methods and "negative" methods have been used.

The methods employing Wright's and Giemsa's stains either add a small quantity of an alkali carbonate or heat the staining solution or both. Heated stains are replaced 3 to 5 times and allowed to act for 15 s to 3 or 4 min each time.

GIEMSA'S METHOD (modified slightly from *Dtsch. Med. Wochenschr.*, **31:** 1026, 1905)
1. Fix films 15 min in 100% alcohol.
2. Dilute 1 ml Giemsa stain with 38 ml distilled water and 2 ml 0.1% potassium carbonate. Stain smears 10 to 30 min.
3. Rinse, blow dry with compressed air, or blot dry with filter paper, and examine in immersion oil.

WRIGHT'S (?) RAPID METHOD. Fix films 15 min in 100% alcohol or by passing 3 times through a blue flame. Make a 1 : 40 dilution of Giemsa stain in distilled water. Flood slide with 2 to 3 ml of this, heat to steaming, let stand 15 s, decant, and repeat flooding and heating 5 times more. Let cool for 1 min the last time, rinse with distilled water, dry, and examine. This technic appeared in the 1908 edition of "Pathological Technique," by Mallory and Wright, apparently as an original modification of Wright's. With both the foregoing methods spirochetes are stained dark purplish red.

Negative methods are derived from Burri's india ink method, which itself, according to Conn and Darrow, has largely been abandoned on account of difficulties in obtaining ink free of bacteria. Dorner (*Stain Technol.*, **5**: 25, 1930) introduced nigrosin ws for the same purpose, and Harrison (*Br. Med. J.*, **2**: 1547, 1912) used collargol. Congo red posttreated with hydrochloric acid to turn it blue has been similarly employed, but Cumley (*Stain Technol.*, **10**: 53, 1935) notes that after the slides are made they may again turn red or fade altogether.

Mallory recommended a 10 to 25% suspension of india ink in distilled water, which is to be autoclaved before use. Dorner prescribed dissolving 10 g water-soluble nigrosin (C.I. 50420) in 100 ml distilled water by heating 30 min in a bath of boiling water. Harrison prepared a 5% suspension of collargol in distilled water. This is good for months.

TECHNIC.   Mix a bacteriologic loopful of exudate on a clean slide with a loopful of one of the above fluids and spread into a thin smear. Dry and examine. Spirochetes and bacteria appear unstained in a dark brown, gray, or reddish brown background respectively for the ink, nigrosin, and collargol.

Silver Methods

FONTANA'S METHOD.   This is the traditional silver method for smears:

1. Conn and Darrow prescribe heat fixation, Mallory a 1-min treatment with Ruge's 1% acetic acid–2% formalin solution using several changes, followed by rinsing in water.
2. Steam 30 s in phenol, 1 g; tannic acid, 5 g; distilled water, 100 ml. Use only a few drops.
3. Rinse 30 s in distilled water.
4. Steam (70 to 80°C) 30 s in Fontana's ammoniacal silver hydroxide: To 5% silver nitrate add 28% ammonia water drop by drop until the dark brown precipitate just dissolves (about 0.5 ml for 10 ml 5% silver nitrate); then add more silver nitrate, shaking to dissolve the brown clouds of silver oxide between drops, until a faint permanent turbidity is attained. This is identical with our usual diammine silver solution, except that we put in the 0.5 ml ammonia first. Conn and Darrow state that this solution is good for several months, but judging from experience with similar solutions for reticulum, it seems preferable to prepare a small quantity fresh each time.
5. Wash, dry, and mount in balsam.

Blenden (*J. Invest. Dermatol.*, **45**: 68, 1965) and Blenden and Goldberg (*J. Bacteriol.*, **89**: 899, 1965) recommend the following stain for spirochetes. It is stated to be particularly useful for smears of suspected *Treponema pallidum*. Reagent A is 5 g tannic acid, 1.5 g ferric chloride, 2.0 ml 15% formalin, 1 ml 1% NaOH in 100 ml distilled water. Reagent B is ammoniated silver nitrate solution. A thin film of exudate is smeared on alcohol-cleaned slides, air-dried, stained with solution A for 2 to 4 min, rinsed briefly with distilled water, stained with solution B for about 30 s, air-dried, and examined under oil immersion. Spirochetes are stained dark brown to black on a light to golden background. Other structures may be stained black, but are distinguished by their morphologic features.

LEVADITI'S METHOD.   This is the traditional method for spirochetes in blocks of tissue (Conn and Darrow). Except for preparation of teaching material it is not recommended. It is slow and cumbersome, as well as capricious and uncertain. Slide methods are preferred for diagnosis.

1. Fix blocks about 1 to 2 mm thick in 10% formalin for 1 to 2 days. Older material stored in formalin can be used.
2. Rinse in water and soak in 95% alcohol for 1 day.
3. Place in distilled water until the tissue sinks.
4. Impregnate in 2% silver nitrate at 37°C for 4 days, changing the solution daily.
5. Wash in distilled water.
6. Reduce for 48 h in 3 g pyrogallol, 5 ml 40% formaldehyde, and 100 ml distilled water.
7. Wash in several changes of distilled water.
8. Dehydrate with graded alcohols, clear in cedar oil, and embed in paraffin. Section at 5 μ. Deparaffinize with xylene and mount in balsam or synthetic resin.

RESULTS: Tissues are varying shades of yellow and brown; spirochetes, black.

THE WARTHIN TECHNIC. Of the numerous proposed single-paraffin-section methods, we finally selected a *modification of Warthin's* (*Am. J. Syph.*, **4**: 97, 1920) technic which has given us more consistent results than any previously tried. *Faulkner and Lillie* (*Stain Technol.*, **20**: 81, 1945) substituted an acetate buffer for Kerr's (*Am. J. Clin. Pathol.*, **8**: 63, 1938) dilute citric acid solution, which seems to control the process better. The technic is as follows:

1. Bring paraffin sections of formalin-fixed material to water as usual.
2. Wash with 0.01 M pH 3.6 to 3.8 acetate buffer.
3. Impregnate for 45 min at 55 to 60°C in a Coplin jar filled with 1% silver nitrate in water buffered to pH 3.6 to 3.8 as above.
4. While slides are incubating, heat and mix the developer solution. Make in advance a stock gelatin solution by dissolving 10 g gelatin in 200 ml distilled water buffered as above (pH 3.6 to 3.8), heating in a paraffin oven for 1 h. Add 2 ml 1 : 10,000 Merthiolate (sodium ethylmercurithiosalicylate) as preservative. Cool and store. Melt the stock gelatin solution; take 15 ml and heat it to 60°C. Heat 3 ml 2% silver nitrate buffered as above at pH 3.6 to 3.8 to 60°C, and add to the gelatin. Then add 1 ml freshly prepared 3% hydroquinone in the same buffered water. Use the mixed developer at once.
5. Place slides face up on glass rods, and pour on the warm developer. When sections become golden brown to grayish yellow and developer begins to turn brownish black, pour off and rinse with warm (55 to 60°C) tap water, then with distilled water.
6. Tone 3 min in 0.2% gold chloride (for greater permanence).
7. Wash in water and counterstain with hemalum and eosin or by other methods. This step may be omitted.
8. Then dehydrate, clear, and mount as usual.

RESULTS: Underdevelopment gives a pale background and slender or pale spirochetes; overdevelopment gives thick spirochetes, and dark background and precipitates; optimal development gives pale yellowish brown tissue and black spirochetes. At the lower pH level, tissue staining is less; at the higher, organisms are denser. This method has performed well on syphilitic fetal liver, on yaws lesions, on Weil's disease, on Vincent's fusospirillosis mouth lesions, and on other spirochetoses.

TRICHINAE

THE UNFIXED MUSCLE SPREAD. The usual technic for examination for trichinae in muscle is taken from *Nolan and Bozicevich* (*Public Health Rep.*, **53**: 652, 1938). Take about 1 g fresh muscle in the aggregate, as small fragments snipped with scissors from various areas, best from the diaphragm near the tendinous portions. Lay these in rows on one or more 50 × 75 mm slides. Place a second slide on top, and press the fragments out flat. Heavy spring paper clips may be used to supply the pressure, one on each side, or heavy rubber bands. The muscle fragments should be pressed out to a state of transparency. Examine directly with a dissecting microscope or with the low power of the microscope with reduced illumination. The preparations do not keep.

THE FORMIC ACID METHOD. Even more satisfactory crush preparations can be made after preliminary fixation in 20% formic acid for 12 to 24 h. This method has the advantage of allowing removal of the material to the laboratory and working up at leisure. The muscle swells greatly. Small snips are cut off, teased and crushed out to transparency. Stains may be applied, and permanent preparations mounted. Lillie's technic was this:

Cut small pieces of muscle from various locations at autopsy, especially diaphragm and intercostal muscles. Pectoralis and rectus abdominis muscles can be used. Deltoid and gastrocnemius are favorable sites for biopsy. Masseter is also a good muscle for biopsy, if cosmetic considerations can be disregarded. Immerse these at once in a 1 : 3 or 1 : 4 dilution of concentrated (90%) formic acid in distilled water and leave overnight. Wash 30 to 60 min in running water, and transfer to a mixture of equal volumes of glycerol and of 50% alcohol for at least 18 h. The material can be kept in this mixture.

When desired, snip off small fragments of muscle with scissors, taking at least a dozen from various pieces of muscle. Drop these into 30 ml 1% acetic acid to which 0.3 to 0.6 ml Lillie's acid hemalum has been added. Other hematoxylins can be used, if diluted in 1% acetic acid to give a final concentration of about 5 to 10 mg hematoxylin per 100 ml. Stain overnight. Place the fragments on cover glasses in two rows of two or three pieces each, and tease and crush gently with needles. Add a few drops of Apáthy's gum syrup. Lay the cover glass on a blotter on a flat surface. Put a slide down on top of the preparations, and press down firmly with a slight rotatory motion, so as to crush and spread the muscle fragments. The excess gum syrup is squeezed out into the blotter. The preparations may be examined at once, using a 32-mm objective for finding and an 8-mm objective for detailed study; but they clear further on standing.

Muscle nuclei and cross striae are well shown; the nuclei of the giant cells surrounding recently encysted parasites are readily identified.

If the preparations acquire air bubbles after release of the pressure, run in a little more syrup from one edge, but do not press on the cover glass directly. It is likely to break. The syrup dries hard. In hot humid climates, seal with xylene cellulose caprate.

Use of very dilute safranin in dilute acetic acid solution in place of the hematoxylin yields more transparent preparations with a purer nuclear stain. Use a 1 : 1000 to 1 : 10,000 safranin O (best, about 1 : 2000) in 1% acetic acid, and stain overnight as with the hematoxylin.

Muscle previously fixed in formalin does not yield satisfactory crush preparations.

The *digestion technic*, quoted from Bozicevich (*Public Health Rep.*, **53:** 430, 1938), is more sensitive but requires more apparatus. Set up a 3-l glass funnel with a large rubber tube on the stem. Into the open end of this tube fit a 15-ml centrifuge tube with conical bottom. Between the bottom of the funnel and the centrifuge tube attach a screw clamp for closing the tube when the centrifuge tube is to be removed after the digestion is complete. Support the whole in a suitable circular hole in an inch plank. Lay in the funnel a 15-cm (6-in) perforated porcelain plate (Fig. 16-1).

Make up the digestion fluid by dissolving 15 g pepsin in 3000 ml warm (40°C) tap water. Add 21 ml concentrated hydrochloric acid. Place this solution in the funnel with attached centrifuge tube.

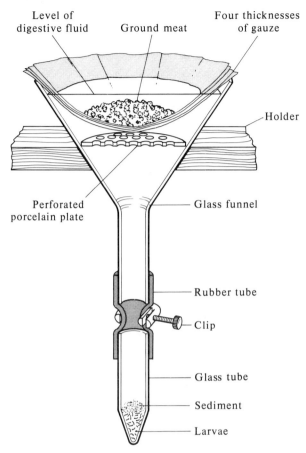

Level of digestive fluid   Ground meat   Four thicknesses of gauze

Fig. 16-1 Modified Baermann apparatus. (*Courtesy J. Bozicevich, Public Health Reports,* **53:** 430, 1938, *Washington, D.C., Government Printing Office.*)

Holder

Perforated porcelain plate

Glass funnel

Rubber tube

Clip

Glass tube

Sediment

Larvae

Grind 70 g fresh muscle, preferably from the diaphragm; place the ground muscle a little at a time on a quadruple thickness of 40-mesh gauze or cheesecloth which has been previously laid in the funnel over the porcelain plate. Place the whole apparatus in a 37°C incubator, and digest for 15 to 18 h (overnight). During the digestion the living larvae are liberated from the muscle and pass through the gauze, settling into the bottom of the centrifuge tube. Then clamp off the rubber tube, remove the centrifuge tube, and take up larvae with a pipette from the bottom of the tube. Most of the fluid may be decanted off first.

17

GLIA AND NERVE CELLS AND FIBERS

Valuable as the silver methods may be in the study of reactive gliosis, it is often of even more value to study brain tissues stained by general oversight methods such as the azure-eosin technics, which reveal cellular gliosis, perivascular infiltration, hemorrhages, necroses, bacteria, inclusion bodies, and tigrolysis. Use of diammine silver technics for reticulum may aid in establishing the relation of focal gliosis to blood vessels. We have successfully used the picric acid– and hydrochloric acid–methyl blue methods for this purpose. The periodic acid Schiff–leucofuchsin and allochrome methods seem worthy of trial. Fat stains and myelin technics have their place. Perivascular and interstitial macrophages can well be studied with iron reactions as well as with fat stains, separately or in the same preparation.

Ependyma is quite well shown with the above general methods for cells and fibers. Some of the special glia procedures can be of value in the study of ependymitides. Choroid plexus and their epithelium are well shown by the azure-eosin methods. The usual stroma methods of general histology are applicable to these essentially vascular and epithelial structures. Fat stains may be of value, especially for the plexal epithelial lipid.

The meninges may require almost any of the methods applicable to the general tissues. The azure-eosin methods are useful for the cytology of exudates; the bacterial methods, collagen methods, pigment methods, fibrin and other methods may all have their specific applications.

Generally the amateur has indifferent success with the metallic impregnation methods, but many workers have used them and secured useful preparations. Unfortunately special fixations and very fresh material are often required, and this limits their usefulness. We have endeavored to select methods which are adaptable to routinely formalin-fixed material when possible and have relied to a considerable extent on the selection made by Davenport, Windle, and Rhines for Conn and Darrow. These authors have supplied a number of valuable modifications.

Block impregnations have been used with considerable success by many neuroanatomists. Since these often prevent the use of other, perhaps unexpectedly more significant methods on adjacent sections, we have omitted them here. Davenport discusses them to some extent.

Generally collodionization of paraffin sections should be employed when alkaline silver solutions are to be used. Frozen or loose celloidin or nitrocellulose sections should be handled with glass needles during the metal impregnation stages. Paraffin-coated forceps [dipped in smoking hot (100 to 120°C) paraffin] may be used for handling tissue blocks.

GLIA CELLS AND GLIA TUMOR CELLS
General

WEIL AND DAVENPORT (quoted from Conn and Darrow). Weil and Davenport prescribe 10% formalin fixation for several days and paraffin sections at 10 μ.

1. Deparaffinize sections, and bring through 100% alcohol into a 1.5% celloidin solution in ether-alcohol mixture for 2 to 3 min.
2. Drain partially, and then hold horizontal face down with slight tilting movements, both lateral and longitudinal, until the remaining film congeals.
3. Then harden the film in 80% alcohol.
4. Impregnate for 6 to 48 h at 37°C in 8 g silver nitrate dissolved in 10 ml distilled water and diluted with 90 ml 95% alcohol.
5. Rinse quickly in 95% alcohol.
6. Reduce in 5 g pyrogallol, 5 ml 40% formaldehyde, and 95 ml 95% alcohol for about 1 min, more or less according to desired intensity.
7. Wash thoroughly in running water.
8. Tone for 5 to 10 min in 0.2% aqueous gold chloride solution.
9. Wash in distilled water.
10. Fix in 10% sodium thiosulfate (0.4 M) for 1 min.
11. Dehydrate with alcohols; dissolve the celloidin with 100% alcohol and ether, and clear in xylene. We suggest substitution of an acetone and xylene sequence; methanol also dissolves celloidin.

RESULTS: Pathologic glia stains gray to black; axis cylinders, black or gray; background, gray-violet.

Addition of 0.25 to 0.5 ml 1 N nitric acid to the silver bath may be required to inhibit staining of normal glia cells.

Weil and Davenport (ibid.) also cite a modification of *Stern's* diammine silver hydroxide method for *microglia* and *oligodendroglia*. Fixation in 10% formalin and celloidin sections at 15 μ are prescribed.

1. Wash sections in distilled water. If it is desired to increase the relative density of impregnation of oligodendroglia, add 0.5 ml 28% ammonia water to 100 ml distilled water, and soak for 3 min.
2. Impregnate for 10 to 20 s (the longer intervals favor the impregnation of oligodendrocytes) in a diammine silver hydroxide prepared by adding 10% silver nitrate to 2 ml 28% ammonia water until a faint permanent opalescence is produced. This requires 18 to 20 ml. Dilute to 40 ml with distilled water. This solution is essentially identical with that used in Lillie's argentaffin and reticulum technics.
3. Transfer sections directly to fresh 15% formalin, and agitate until deep brown. For oligodendrocytes use 10% formalin, and commence agitation only after celloidin is blackened and section begins to turn brown.
4. Wash in tap water, and dehydrate with two changes of isopropyl alcohol; clear in one change of a 50% mixture of isopropyl alcohol with xylene and two changes of xylene. Mount in balsam or synthetic resin. Introduction of gold toning at step 4 tends to prevent fading in xylene (Davenport).

Oligodendrocytes

The following methods are designed especially for oligodendrocytes but often impregnate microglia as well.

DEL RÍO-HORTEGA'S SILVER CARBONATE METHOD.   According to this method, slightly altered from Conn and Darrow, fix in Ramón y Cajal's formalin–ammonium bromide for 12 to 48 h.

1. Heat the block in fresh formalin–ammonium bromide at 45 to 50°C for 10 min. Ramón y Cajal heated at 50 to 55°C.
2. Cut frozen sections at 15 to 20 μ.
3. Wash in 1 : 100 dilution of 28% ammonia water and then in distilled water.
4. Impregnate 1 to 5 min in an ammoniacal silver carbonate solution prepared as follows: To 5 ml 10% silver nitrate add 20 ml 5% sodium carbonate solution (a three-fold excess), and then drop by drop add 28% ammonia water to barely dissolve the precipitate (about 0.4 ml should be required). Add about 20 ml distilled water, and filter, bringing volume to 45 ml. Keep in a brown bottle.
5. Wash sections 15 s in distilled water.
6. Reduce 30 s in a 1 : 100 dilution of strong formalin (Davenport: 2 to 3 min).
7. Wash thoroughly in tap water.
8. Tone in 0.2% gold chloride until gray. Ramón y Cajal specified 10 to 15 min in cold (15°C?) solution, less time if solution is warmer.
9. Fix in 5 to 10% sodium thiosulfate (2 to 5 min). Ramón y Cajal specified 5%.
10. Wash thoroughly in tap water, float onto slides, blot down with filter paper, dehydrate and clear by a 100% alcohol–xylene sequence, and mount in balsam or synthetic resin.

RESULTS: With this method cytoplasm and processes of oligodendrocytes should be black; cell nuclei, unstained; and general background, gray. Since the exposure to the diammine silver carbonate is brief and at a lower temperature than in the same author's reticulum method and a considerably greater excess of sodium carbonate is present, reticulum should not be impregnated. The strongly acid formol NH_4Br should also tend to hydrolyze collagen.

Longer fixation than that prescribed is said to make the impregnation less selective and to favor impregnation of astrocytes.

PENFIELD'S VARIANT (ibid.).   In Penfield's variant this method gives both oligo-dendrocytes and microglia. By increasing the volume of the silver solution above from 45 to 75 ml with distilled water and by taking out sections after 20, 45, and 120 s in the silver the rest of the technic being the same, microglia and processes are black, other neuroglia cells are dark gray to black, and background is pale.

Astrocytes

Large protoplasmic astrocytes are often well demonstrated with azure-eosin stains, and with Mallory's aniline blue or his phosphotungstic acid hematoxylin both astrocytes and neuro-glia fibrils may be well shown. The classic method, however, is the one that follows:

S. RAMÓN Y CAJAL'S GOLD SUBLIMATE METHOD.   This technic follows Conn and Darrow for the most part.

Fix about 5 days (not less than 2 or more than 25) in Ramón y Cajal's formalin–ammonium bromide. Cut frozen sections at 15 to 30 μ, and store in the same fixative.

1. Wash in two changes of distilled water.

2. Impregnate well-spread-out sections for 3 to 4 h in 5 ml 1% yellow or brown gold chloride, 5 ml 5% mercuric chloride, and 30 ml distilled water. Ramón y Cajal used 5 ml 1% gold chloride, 4 to 5 g mercuric chloride (saturated solution?), and 20 to 25 ml water. Temperatures around 25°C are preferable; 18 to 40°C is permissible. Check impregnation from time to time by examination of a wet section under the microscope. Astrocytes should appear dark with a relatively light background, and sections acquire an overall purple coloration.
3. When impregnation is satisfactory, wash in distilled water.
4. Fix in 5 to 10% sodium thiosulfate solution ($Na_2S_2O_3 \cdot 5H_2O$) for 5 to 10 min. Ramón y Cajal preferred a saturated solution with an added 2% of a normal sodium bisulfite solution.
5. Wash thoroughly in several changes of tap water, float onto slides, blot down with filter paper, dehydrate with 100% alcohol, and clear with xylene, blotting between changes. Mount in balsam or synthetic resin.

According to Globus (*Arch. Neurol.*, **18**: 263, 1937) material stored in formalin for a long time may be used if one soaks frozen sections 24 h in a 10% dilution of 28% ammonia water, rinses twice in distilled water, and then immerses for 2 to 4 h in a 10% dilution of concentrated (40%) hydrobromic acid. Following this treatment rinse in two changes of a 1 : 2000 dilution of strong ammonia water, and proceed as above.

ACHÚCARRO'S TANNIN SILVER METHOD. This method, according to Del Río-Hortega, emended slightly from Conn and Darrow, prescribes as follows: Fix for 10 days or more in 10% formalin (alkalinized to litmus with ammonia, according to Ramón y Cajal). Cut frozen sections (Ramón y Cajal: not over 10 μ).

1. Wash in distilled water.
2. Use 3% aqueous tannic acid at 50°C for 5 min. Ramón y Cajal used 10% tannin and cites Del Rio-Hortega as using 3% tannin rather than tannic acid.
3. Wash in a 1% dilution of 28% ammonia water until sections become pliable.
4. Impregnate sections a few at a time in three changes of diammine silver hydroxide. The recommended solution is made by precipitating silver oxide from 30 ml 10% silver nitrate by the addition of 2 ml (40 drops) 40% sodium hydroxide drop by drop (a slight excess). The precipitate is then filtered out on hard filter paper and washed free of excess alkali with 10 or more washes of distilled water. (Ramón y Cajal omitted this filtration and washing.) Then transfer the precipitate to a 250-ml flask with 50 ml distilled water, and dissolve by adding 28% ammonia water drop by drop [2 to 2.38 ml; the latter is the theoretical quantity for $Ag(NH_3)_2OH$, presuming no loss of silver], avoiding any excess of ammonia and preferably leaving a few granules undissolved. Dilute to 150 ml with distilled water, and store this stock solution in a brown glass bottle. For use, dilute 5 ml with 45 ml distilled water. (This would correspond to perhaps a 1 : 25 dilution of Lillie's diammine silver solution.) *Caution:* Dry silver oxide may be explosive.

   Sections become yellowish brown when sufficiently impregnated. The weak silver solution requires frequent renewal.
5. Wash in three changes of distilled water.
6. Tone for 20 min at 40 to 45°C in 0.2% gold chloride solution.
7. Wash in water.
8. Fix in 5 to 10% sodium thiosulfate (for 2 to 5 min). (Ramón y Cajal specified 5%.)
9. Wash in (several changes of) water. Float onto slides and blot down. Dehydrate and clear with a 100% alcohol-xylene sequence, blotting between changes to keep sections in place and to accelerate clearing. Mount in balsam or synthetic resin.

RESULTS: Protoplasmic astrocytes should be dark gray to violet; fibrils, black; connective tissue, pale; other tissue elements, reddish to violet. This method is considered less selective than the gold sublimate method.

NERVE CELLS AND THEIR PROCESSES

Procedures which randomly select individual nerve cells and display them with all (or most of) their cell processes are generally derived from the chromate silver procedure introduced by Camillo Golgi in 1873–1878. They are of necessity block impregnation methods and have rather little use in pathology. Variants have been numerous. We have selected Fox's zinc chromate method for presentation here because it can be applied to the routinely formalin-fixed material which is left over after gross sectioning and selection of blocks for regular histologic procedures have been done.

FOX'S ZINC CHROMATE GOLGI TECHNIC FOR NERVE CELLS AND THEIR PROCESSES

1. Cut thin (2 to 5 mm) slices of brain which has been fixed in formalin for 2 months to 2 years. Longer fixations, over 4 months, are preferable.
2. Soak blocks for 2 days in zinc chromate, 6 g; 90% formic acid, 4 ml; distilled water to make 100 ml. Davenport reduces the time to 18 h at 25°C.
3. Blot free of excess chromate; agitate in 0.75% silver nitrate until all surfaces are deep red.
4. Pass a thread through each block, and suspend in 0.75% $AgNO_3$, two changes, 24 h each. Brush off silver chromate precipitate with camel's hair brush after each change.
5. Dehydrate with two changes each of 95% and absolute alcohol, for 15 min each, clear in xylene for 10 min, and infiltrate for 10 min in paraffin. Here the use of vacuum should give much better infiltration, but this has apparently not been tried, or perhaps thorough infiltration is not desired (see step 7).
6. Section entire block serially at 90 to 100 μ, and store sections serially in 95% alcohol.
7. Dehydrate sections with absolute alcohol, several changes; clear with alcohol-xylene and several changes of xylene, washing off loose silver chromate in the process. Mount on slides.
8. Coat the sections with several applications of Permount, and let dry 5 to 7 days. Davenport omits step 7 and drys in Permount 3 to 7 days.
9. Moisten surface of Permount with toluene, apply cover glass, and place slides in a warm place for a few hours with small lead weights on the cover glasses.

RESULTS: Background is pale yellow; isolated cells and processes are black.

Klatzo (*Lab. Invest.*, **1**: 345, 1952) recommends a shorter Golgi variant for neuroglia cells and processes.

1. Fix in fresh chloral hydrate–formalin dichromate: 5 g $K_2Cr_2O_7$ in 90 ml distilled water; add 10 ml formalin and then 3 g chloral hydrate. For fibrillary astrocytes fix for 18 to 48 h; for protoplasmic astrocytes, 12 to 17 h; for oligodendroglia, 18 to 24 h; for microglia, 12 to 18 h. At 24 h, rinse blocks in 10% formalin, and place them in fresh fixing solution if a second day is required.
2. Wash rapidly in three or four changes of distilled water.
3. Pass blocks individually through four baths of 25 ml 1% silver nitrate, allowing about 2 min in each, handling with paraffin-coated forceps.
4. Impregnate for 18 to 24 h in 1.5% silver nitrate.
5. Rinse in distilled water and brush off surface precipitate.

6. Cut frozen sections at 60 to 100 μ. Collect in 30% alcohol (two changes).
7. Dehydrate in 95 and 100% alcohol, and clear in toluene. We suggest use of alcohol-toluene mixtures to facilitate the clearing.
8. Float onto slides, blot, cover liberally with toluene Permount, and apply cover glass.

Fregerslev et al. (*Histochemie*, **25**: 63, 1971) made x-ray powder diagrams of dissected tissues containing Golgi stains prepared by treatment of formalin-fixed brain and liver with potassium dichromate and silver nitrate. Crystalline silver chromate only (Ag_2CrO_4) was detected. No other silver or chrome compounds were detected.

NEUROFIBRILS, AXONS, AND NERVE ENDINGS

These methods generally require special fixations or fresh tissue and hence are often not applicable in pathologic studies. Davenport's method and Foley's variant of Bodian's protargol method utilize material fixed in 10% formalin.

DAVENPORT (*Arch. Neurol.*, **24**: 690–695, 1930).   Davenport preferred nitrocellulose sections at 15 to 30 μ but stated that paraffin sections could be used.

1. Spread celloidin or nitrocellulose sections on slides, and then immerse in 2% collodion and drain. Pass paraffin sections through xylene and 100% alcohol into 2% collodion in ether-alcohol mixture, and drain.
2. When the collodion film has set, immerse in 80% alcohol for 5 to 10 min or as much longer as convenient.
3. Immerse sections in 10 g silver nitrate, 0.5 ml 1 N nitric acid, 10 ml distilled water, and 90 ml 95% alcohol. Let stand until sections are light brown at 37 to 40°C. Formalin-fixed material usually requires about an hour. With other fixations longer intervals may be necessary. Avoid exposure to bright light.
4. Rinse in 95% alcohol.
5. Reduce in 5 g pyrogallol, 5 ml strong formalin, 100 ml 95% alcohol, watching under the microscope. This may take as little as 2 min. The reduction may be slowed by adding more alcohol. To avoid precipitates, the preparation should be kept moving. Addition of 100 mg glucose in 0.5 ml water (0.125 ml corn syrup, 0.375 ml water) also aids in preventing precipitates. One 50-ml portion of reducer is sufficient for at least 20 slides.
6. Rinse in two or three changes of 95% alcohol, agitating constantly.
7. Dehydrate with 100% alcohol, remove collodion with ether plus 100% alcohol mixture, clear in xylene, and mount in balsam.

RESULTS: Nerve cells are yellow to brown; fibers, dark brown to black.

For demonstration of peripheral nerve endings in frozen sections the following variant of the *Bielschowski-Gros method* appears useful (Garven and Gairns, *Q. J. Exp. Physiol.*, **37**: 134, 1952; Davenport, p. 252). It is probable that material fixed routinely in sodium acetate formalin or phosphate-buffered formalin can be used, but Davenport prescribes an $MgCO_3$ formalin.

Shake 25 to 30 ml formalin with 1 to 2 g $MgCO_3$, filter off 24 ml, and dilute to 200 ml with distilled water. Fix blocks 5 to 7 mm thick in three daily changes of this 12% formalin.

1. Wash blocks 5 min in running water. Cut frozen sections at 15 μ. Wash sections individually in three or four changes of distilled water, 5 to 10 s each.
2. Float sections individually on 25% silver nitrate, in small, flat staining dishes, spreading

well with glass needles to avoid wrinkles. Cover with tin can or cardboard box to exclude light, and let stand 10 to 120 min, inspecting periodically for darkening to a shade of brownish yellow which must be learned by experience. On first trial of this method, inspect at 10-min intervals up to 2 h, taking out one or two sections each time after some color change appears.

3. Wash in four changes of formalin diluted to 20% with distilled water containing 1 mg NaCl to each 200 ml. Davenport suggests 5, 10, 20, and 60 s duration for the four baths. No AgCl cloud should form around the section in the last bath.

4. Treat with ammoniacal silver solution, watching under a microscope for proper blackening of nerve fibers: To 5 ml 20% silver nitrate add 28% ammonia (sp gr 0.890 to 0.900) drop by drop until only one or two black granules of silver oxide remain. Use 1 ml of this solution for each section, picked up on a slide, adding 1 drop 28% ammonia to each 1 ml.

5. When impregnation is adequately developed, transfer section to a 20% dilution of 28% ammonia to stop the impregnation.

6. Rinse in distilled water, and tone 2 to 10 min in 0.2% gold chloride, if desired.

7. Dehydrate, clear, and mount in Permount or cellulose tricaprate.

BODIAN COPPER PROTARGOL METHOD, FOLEY'S VARIANT (*Stain Technol.*, **18:** 27, 1943). Dehydrate and infiltrate with celloidin or low-viscosity nitrocellulose as usual, and section at 15 to 25 μ, or infiltrate for 24 h at 58°C in 5% agar, and cut frozen sections at 20 to 40 μ.

1. Soak for 24 h in 1 ml 28% ammonia and 99 ml 50% alcohol.

2. Drain and place in 1% protargol (silver albumose) in distilled water for 6 to 8 h at 37°C. In the protargol solution, but not in contact with the sections, place a 200- to 300-mg piece of 0.002-in (50 μ) electrolytic copper foil which has been thinly coated with collodion. Use no metallic instruments.

3. Drain and transfer sections to a mixture of 50 ml 1% aqueous protargol, 50 ml 95% alcohol, and 0.5 ml pyridine (0.1 to 2 ml: higher quantities of pyridine accentuate impregnation of thin fibers; lower quantities accentuate that of cell bodies and dendrites). Put in another piece of copper foil, very thinly coated with collodion. Incubate for 24 to 48 h at 37°C.

4. Rinse for 5 s in 50% alcohol.

5. Reduce for 10 min in the following reducer: Dissolve 1.4 g boric acid in 85 ml distilled water; then add and dissolve 2 g anhydrous sodium sulfite; then add and dissolve 0.3 g hydroquinone; then add 15 ml C.P. acetone and mix well.

6. Wash in several changes of distilled water.

7. Tone for 10 min in 0.2% gold chloride in about 0.3% acetic acid (20 drops of glacial acetic acid to 100 ml).

8. Wash in several changes of distilled water.

9. Reduce for 1 to 3 min in 2% aqueous oxalic acid solution.

10. Rinse in distilled water.

11. Fix for 3 to 5 min in 5% sodium thiosulfate (0.2 *M*) solution.

12. Wash in distilled water.

13. Stain for 18 to 24 h in Einarson's chrome alum gallocyanin.

14. Wash thoroughly in distilled water.

15. Mordant for 30 min in 5% aqueous phosphotungstic acid.

16. Transfer directly to the following stain: aniline blue, 10 mg; fast green FCF, 500 mg;

orange G, 2 g; glacial acetic acid, 8 ml; distilled water to make 250 ml. (We fail to see the function of the minute amount of aniline blue.) Stain for 1 h.

17. Dehydrate and differentiate with 70 to 95% alcohol.

18. Complete dehydration with normal butyl alcohol, and clear in cedar oil. (Or use the isopropyl alcohol-xylene sequence for dehydration and clearing.)

RESULTS: Nerve fibers and neurofibrils are blue-black; nuclei, variable—blue-black if high pyridine concentration was used; tigroid is light blue; collagen, blue to green; myelin, yellow.

The essential part of the foregoing method consists of steps 1 to 12. The counterstains can well be varied according to taste, as with other silver methods where gold toning and thiosulfate fixation are employed. For instance, an acid picrofuchsin stain, giving red collagen and yellow cytoplasm, might well provide better contrast between collagen and nerve fibers than the blue-green prescribed.

ROMANES' METHOD   (*J. Anat.*, **84**: 104, 1950).   Romanes recommends a very dilute silver chloride ammonia procedure for staining nerve fibers in paraffin sections. Fix in alcohol, Bouin's fluid, Carnoy's fluid, acetic alcohol, or acetic alcohol–formalin. Treat blocks for a few hours in 2 ml "0.880 ammonia" (35.3%) in 98 ml 70% alcohol (= 2.5 ml 28% NH_3 plus 97.5 ml 70% alcohol). Decalcify if necessary, dehydrate, clear, and embed in paraffin. Section and attach to slides with albumen fixative. Deparaffinize and hydrate as usual. The following directions are as emended by Powers (*J. Dent. Res.*, **31**: 383, 1952, and in letters):

1. Incubate at 58°C for 16 h in the dark in a fresh ammoniacal silver chloride. Mix with 2.9 ml 0.1% silver nitrate with 100 ml distilled water, add 1 ml 0.1% sodium chloride, and adjust to pH 7.8 with 0.7 ml 1% dilution of 28% ammonia water. This fluid does not keep.

2. Drain, and transfer without washing to 1% hydroquinone in 10% sodium sulfite (crystals, $Na_2SO_3 \cdot 7H_2O$ or 5% of the anhydrous salt) for 5 min at 20°C.

3. Wash well in running water, and rinse in distilled water.

4. Tone in 0.5% gold chloride for 2 min or (Powers) 0.2% for 10 min. Wash 1 min in two changes of distilled water. Reduce not more than 3 min in 2% oxalic acid. Wash in running water. Fix for 3 to 5 min in 5% sodium thiosulfate. Dehydrate, clear, and mount in unsaturated synthetic resin or balsam.

RESULTS: Nerve fibers appear purple to black; nuclei, red; neurofibrils, purple; keratin, yellow; bone cells, black. Raising the pH to 8.3 and introducing 5 g fine copper wire give black nerve fibers throughout. Powers replaces this step by premordanting for 18 h at 37°C in 0.5% cupric nitrate [*ca.* 24 m*M* $Cu(NO_3)_2 \cdot 3H_2O$].

Powers (ibid.) recommends use of both Romanes and Ungewitter silver variants for the demonstration of nerve fibers in dentine. For the Romanes method she prefers fixation for 1 to 3 days in 10% formalin containing 10% chloral hydrate; for the Ungewitter method, 3 or more days' fixation in a Bouin fluid containing 1 g trichloroacetic acid, 75 ml saturated aqueous picric acid, and 25 ml 40% formaldehyde (w/v).

After either fixation, wash for 16 to 24 h in running water, and decalcify in the Evans-Krajian formic acid citrate mixture or in 5% trichloroacetic acid in 50% alcohol. Wash for 24 h in running water, dehydrate in successive 24-h changes of 30, 50, and 70% ethyl alcohol, in a mixture of 95% alcohol and *n*-butanol, and finally in two 24-h changes of *n*-butanol. Infiltrate for 12 to 24 h in hard paraffin (56 to 58°C mp), embed, and section serially at 10 to 15 μ.

For the Romanes procedure, deparaffinize, hydrate, and mordant for 18 h at 37°C in 0.5% cupric nitrate [crystals, $Cu(NO_3)_2 \cdot 3H_2O$, ca. 20 mM]. Wash in four changes of distilled water, and carry through the Romanes-Powers procedure (above). The copper mordanting reduces the amount of background staining and obviates the copper wire in the silver solution.

For the Ungewitter method, deparaffinize and bring to 80% alcohol. Transfer directly to cold 1% aqueous silver nitrate containing 20 to 30% urea and 1 to 3 drops (0.1 ml) per 100 ml of a 1% picric acid–1% mercuric cyanide solution. Incubate in a paraffin oven for 60 to 90 min. Rinse quickly in two changes of distilled water, and reduce for 3 to 5 min in a urea-hydroquinone solution containing 1 to 2 g hydroquinone, 20 to 30 g urea, and 10 g anhydrous sodium sulfite in 100 ml distilled water, agitating gently for the first 2 min. Wash in five changes of distilled water and examine microscopically in 80% alcohol. If sections are not adequately silvered, repeat silvering and reduction as before. Finally dehydrate, clear, and mount.

THE NONIDEZ METHOD (*Am. J. Anat.*, **65:** 361, 1939). This method is highly recommended by Davenport, Windle, and Rhines in Conn and Darrow:

1. Fix 1 to 3 days in 25 g chloral hydrate in 100 ml 50% alcohol.
2. Drain and blot, and place in ammoniated alcohol (95% alcohol, 30 ml; 28% ammonia water, 0.1 ml or 2 drops) for 24 h, changing once or more if there is much fat.
3. Wash in distilled water 5 min.
4. Place in 2% silver nitrate (aqueous) at 37°C for 5 to 6 days, changing after 1 to 2 days or sooner if the solution becomes yellowish brown.
5. Wash for 2 to 3 min in distilled water.
6. Reduce for 24 h in pyrogallol, 2.5 to 3 g; 40% formaldehyde, 8 ml; distilled water, 100 ml.
7. Wash in six changes for 20 to 30 min each of distilled water. Embed, section, and mount.

Nonidez prescribed graded alcohols, amyl acetate clearing, and paraffin embedding. Other methods can undoubtedly be used. This method is used for nerve endings, both peripheral and central, and for neurofibrils and axis cylinders generally. They appear in shades of brown.

Nauta and Gygax (*Stain Technol.*, **26:** 5, 1951), observing that increasing amounts of NaOH in ammonia–silver nitrate mixtures increase the sensitivity and decrease the selectivity of silver impregnation, explored the effect of varying mixtures on impregnation of degenerating axons.

For axon degenerations in the rat brain they recommend the following modification of the Glees method:

BIELSCHOWSKY-GLEES METHOD FOR DEGENERATING AXONS (Nauta and Gygax)

1. Fix in neutral 10% formalin for 14 days to 6 months.
2. Cut frozen sections at 15 to 20 μ.
3. Soak sections 6 to 12 h or more in 50% alcohol containing 1% of 25% ammonia water. Handle sections only with glass rod from steps 3 to 10.
4. Wash in three changes of distilled water.
5. Soak for 24 h in 1.5% silver nitrate solution in 5% aqueous pyridine solution (a 10-ml lot is adequate). Transfer directly to step 6.

6. Impregnate 2 to 5 min in a covered vessel in diammine silver: $AgNO_3$ 490 mg in 20 ml distilled water plus 10 ml 95% alcohol. Let cool, and add 2 ml 27 to 28% ammonia water and 1.5 ml 1 N NaOH (emended slightly for American reagents).

7. Transfer directly to a mixture of 45 ml 10% alcohol, 2 ml 10% unbuffered formalin, and 1.5 ml 1% aqueous citric acid (crystals) solution. The sections spread on the surface and turn golden brown rapidly.

8. Fix for 1 to 2 min in 2.5% sodium thiosulfate.

9. Wash in three to five changes of fresh distilled water.

10. Dehydrate with graded alcohols, clear with xylene, mount in synthetic resin—Caedax, Depex, polystyrene, etc.

Loewy [*Acta Neuropathol. (Berl.)*, **14:** 226, 1969] was able to stain selectively products of fiber degeneration with ammoniacal silver nitrate and citric acid–formalin reduction alone, and lipid solvents (acetone, carbon tetrachloride, and chloroform-methanol) did not abolish selective staining of the degenerating products. Selective impregnation of degenerating fibers *can* be achieved without various pretreatments including phosphomolybdic acid, potassium permanganate, hydroquinone, oxalic acid, and aqueous silver nitrate.

FINK-HEIMER METHODS. Heimer (*Brain Res.*, **5:** 86, 1967) conducted a comparative evaluation of silver impregnation methods for terminal degeneration in the forebrain. Heimer concluded that the Nauta-Gygax method failed to show the quantity and mode of termination in all fiber systems. The method was particularly deficient in the lateral olfactory tract and in the two commissural systems. No method was completely satisfactory. The Fink and Heimer method (*Brain Res.*, **4:** 369, 1967) was favored for reliability and is provided below:

1. Fixation should be initiated, if possible, by perfusion with saline solution followed by 10% formalin or formalin–saline solution. Material to be studied is carefully removed and stored in 10% formalin for 1 to 2 weeks.

2. Blocks to be sectioned are soaked for 2 to 3 days (or longer if needed) in a 30% sucrose solution. Formalin-fixed material initially floats in this heavy liquid.

3. Blocks are frozen without previous rinsing, and sections 15 to 30 μ thick are collected in dilute (1 to 2%) formalin. The sugar-soaked tissue requires more thorough freezing than is the case when blocks are frozen directly from formalin, water, or 20% alcohol. Therefore, blocks are best kept surrounded by pulverized dry ice on the freezing stage.

STAINING PROCEDURE I

1. After a brief rinse in distilled water, soak sections for 5 to 15 min in 0.05% potassium permanganate.

2. Rinse, and bleach sections in a mixture of equal parts of 1% oxalic acid and 1% hydroquinone for 30 to 60 s.

3. Rinse thoroughly, and transfer sections for 30 to 60 min to a mixture of the following:

0.5% uranyl nitrate	10 ml
2.5% silver nitrate	12 ml
Distilled water	28 ml

4. Transfer sections for 30 to 40 min to a mixture of:

0.5% uranyl nitrate	20 ml
2.5% silver nitrate	30 ml

5. Rinse thoroughly and transfer sections for 1 to 5 min to a freshly prepared ammoniacal silver nitrate solution composed of:

2.5% silver nitrate	30 ml
Strong ammonia water	1.0 ml
2.5% sodium hydroxide	1.8 ml

6. Transfer sections, without rinsing, to a Nauta-Gygax reducing solution, 2 volumes, for 1 to 2 min each, made up of:

Distilled water	910 ml
95% alcohol	90 ml
10% formalin	27 ml
1% citric acid	27 ml

7. Rinse, and transfer sections for 1 min to 0.5% sodium thiosulfate, rinse again, transfer sections to slides, dehydrate, and mount.

RESULTS: This procedure results in selective silver impregnation of degenerating axons and their synaptic endings in the central nervous system. Normal fiber impregnation is stated to be minimal.

Fink and Heimer (*Brain Res.*, **4**: 369, 1967) and Heimer and Peters (*Brain Res.*, **8**: 337, 1968) advise staining procedure II, below, for electron microscopic investigations, although they indicate it may also be employed for light microscopy. Animals are perfused with glutaraldehyde-formalin fixative buffered to pH 7.2, and, after sacrifice, brains are fixed in the same solution for 1 to 2 days more. Small hand-cut slices are stained with the Fink-Heimer method II:

STAINING PROCEDURE II

1. After a brief rinse in distilled water, soak sections for 5 to 10 min in 0.025% potassium permanganate.
2. Rinse, and bleach sections in a mixture of equal parts of 1% oxalic acid and 1% hydroquinone.
3. Rinse thoroughly, and transfer sections for 5 to 10 min to a 2.5% uranyl nitrate solution.
4. Rinse thoroughly, and transfer sections to a 0.2% silver nitrate solution for 1 to 2 h. The addition of 0.2 ml pyridine to 10 ml of the silver nitrate solution may improve the result.
5. Without washing, the sections are transferred for 2 to 5 min to a freshly prepared ammoniacal silver nitrate solution composed of:

1.5% silver nitrate	20 ml
95% ethyl alcohol	12 ml
Strong ammonia water	2 ml
2.5% sodium hydroxide	1.6–1.8 ml

6. Transfer sections, without rinsing, to the reducing solution described in procedure I (step 6).
7. Rinse, and transfer sections for 1 min to 0.5% sodium thiosulfate.
8. Rinse, transfer sections to slides, dehydrate, and mount.
9. If desired, mounted sections can be counterstained by immersion in cresyl echt violet, preferably after coating with a celloidin film.

After step 7, specimens may be postfixed in 2% OsO_4, dehydrated in ethanols, embedded in Epon-Araldite, sectioned, and examined by electron microscopy.

Hjorth-Simonsen (*Stain Technol.*, **45**: 199, 1970) modified the Fink-Heimer method to permit the use of frozen sections attached to slides. Fresh frozen sections are attached to slides using chrome alum–gelatin in a water bath (0.3 g chrome alum in 600 ml 0.5% aqueous gelatin), air-dried at 24°C for at least 2 h, fixed in neutralized aqueous phosphate buffered formalin overnight, and stained with Fink-Heimer staining procedure II.

Walberg (*Brain Res.*, **31**: 47, 1971) also conducted a comparative evaluation of several silver impregnation methods (Glees, Nauta-Laidlaw, and Fink-Heimer). Evaluations included electron photomicrographs. On the basis of observations of the inferior olivary nucleus, Walberg states one should be extremely careful concerning interpretations of the nature of open rings and clubs (boutons) observed with silver stains. Walberg is convinced that the capriciousness of silver stains is such that general conclusions should not be made.

After the rinse and thiosulfate treatment (step 7) of the Fink-Heimer-Peters method above, Kawamura and Niimi (*Stain Technol.*, **47**: 1, 1972) suggest it may be helpful to use a cresyl violet counterstain as follows:

1. Soak sections for 10 to 20 s in a working solution made by diluting 1 part stock solution (filtered before use) with 4 parts distilled water. Stock ferricyanide: borax, $Na_2B_4O_7 \cdot 10H_2O$, 2 g; potassium ferricyanide, 2.5 g; distilled water, 100 ml. Agitate the section gently, and observe progressive bleaching of the yellow background. Stop the bleaching before the solution has had time to act on the black silver deposits associated with nerve fibers, and proceed immediately to step 2.
2. Wash well through at least three changes of distilled water.
3. Soak for 2 min in 1% $Na_2S_2O_3 \cdot 5H_2O$ and follow with another wash in three changes of distilled water.

Before counterstaining, the sections are attached to albuminized slides in the usual manner with Mayer's 1 : 1 albumen-glycerol.

4. Immerse sections for about 5 min in Carnoy's alcohol–chloroform–acetic acid 6 : 3 : 1 mixture to remove lipid and to effect dehydration.
5. Draw the floating section from the Carnoy's fluid onto the albuminized side of a slide held at a low angle in the fluid. A small artist's brush may be used to manipulate the section (as with aqueous flotation fluids).
6. Allow the slide to drain in a nearly vertical position, and then dry in air, 5 min or longer.
7. Immerse the slides in 80% alcohol for 3 to 5 min, and then wash for 5 to 10 s in distilled water.
8. Stain 30 min in a 0.1% cresyl violet solution acidified to pH 3.5 to 3.8 by addition of about 0.7 ml of 10% acetic acid to each 100 ml solution.
9. Differentiate and dehydrate through 70% alcohol and two changes of 95% alcohol.
10. Complete the dehydration with two changes of absolute alcohol, and clear in two or three changes of xylene.
11. Apply a cover glass with a resinous mounting medium.

RESULTS: Cells and Nissl substance are delineated by the counterstain. It is important to note that the borax ferricyanide will also destroy finely stained silver on nerve fibers. The bleaching must be watched and discontinued at an appropriate time.

Heimer (*Brain Res.*, **12**: 246, 1969) modified the Nauta-Gygax method for light microscopic examination of semithin sections cut from Epon-Araldite–embedded material for electron microscopy. Brains from rats perfused with a glutaraldehyde-formaldehyde fixative are fixed in glutaraldehyde-formaldehyde overnight. The following day appropriate areas of the brain are dissected out using a dissecting microscope. Pieces of brain are rinsed for 2 h in phosphate buffer of pH 7.4, postfixed in 2% OsO_4, dehydrated, and embedded in an Epon-Araldite mixture (Epon 812 = 12.4 ml; dodecenyl succinic anhydride = 27.6 ml; Araldite resin 502 = 7.6 ml; dibutyl phthalate = 1.5 ml; dimethylaminomethylphenol = 1 ml) for electron microscopy, and sectioned at 5 to 10 μ.

1. Transfer the sections with jeweler's forceps to a silver-pyridine solution for 18 to 24 h (3 to 5 ml pyridine to each 100 ml 0.5% silver nitrate solution).
2. Without washing, transfer the sections with a brush for about 10 min to a freshly prepared ammoniacal silver solution composed of concentrated ammonia water, 2 ml; 2.5% sodium hydroxide, 2 ml; 1.5% silver nitrate, 20 ml; and 95% ethyl alcohol, 12 ml.
3. Transfer sections, without rinsing, for 5 to 10 min to a Nauta-Gygax reducing solution made up of distilled water, 910 ml; 95% alcohol, 90 ml; 10% formalin, 27 ml; and 1% citric acid, 27 ml.
4. Rinse and transfer to gelatin–chrome alum–coated slides. (Dissolve 300 mg chrome alum in 600 ml 0.5% aqueous gelatin.) After the section has dried onto the slide, rinse in xylene and mount.

Johnstone and Bowsher (*Brain Res.*, **12**: 47, 1969) state that the following silver carbonate-neurofibrillar impregnation stain is unusually reliable and reproducible:

1. Fix material by perfusion with neutral formol, and store in 10% formalin.
2. Cut frozen sections at 25 μ (or less), and assemble in 10% formalin.
3. Rinse sections individually in three changes of distilled water.
4. Immerse in 1% phosphotungstic acid for 60 min.
5. Transfer, without washing, to 0.05% potassium permanganate for 5 to 10 min.
6. Rinse rapidly in distilled water.
7. Decolorize for 1 min in a mixture of equal parts of 1% oxalic acid and 1% hydroquinone.
8. Wash thoroughly in five changes of distilled water (3 min each).
9. Stain for 15 to 20 min in 20% silver nitrate.
10. Rinse in three changes of distilled water.
11. Place sections in individual dishes, for 5 to 10 min, in the following solution, pre-warmed to 45°C: 20% silver nitrate, 7.5 ml, plus 5% sodium carbonate, 12 ml, plus distilled water, 80 ml; add 28% ammonia until the solution is clear (usually about 10 to 12 ml); then add 1 ml pyridine.
12. Reduce in two successive dishes of 10% formalin.
13. Wash in four changes of distilled water.
14. Tone for 5 to 15 min in 1% brown gold chloride, prewarmed to 45°C.
15. Rinse in distilled water.
16. Fix for 5 min in 5% sodium thiosulfate.
17. Rinse in five changes of distilled water.
18. Place sections in gelatin-alcohol (1% aqueous gelatin and 80% ethanol, 1 : 1) warmed to 45°C which has been filtered before use.
19. Blot slides and mount in Albrecht's medium.

RESULTS: Degenerating axon terminals are described as particularly well delineated.

THE GLEES METHOD FOR AXONS IN PARAFFIN SECTIONS (Marsland, Glees, and Erikson, *J. Neuropathol. Exp. Neurol.*, **13**: 587, 1954)

1. Fix in 10% formalin or formalin–saline solution, dehydrate in alcohols, clear in xylene, and embed in paraffin. Section at 6 to 8 μ.
2. Deparaffinize in xylene; bring through 100% alcohol.
3. Collodionize for 1 min in 0.5% celloidin; drain off most of celloidin, and allow surface to *gel* but not *dry*.
4. Immerse for 1 min in 70% alcohol.
5. Wash in distilled water.
6. Incubate for 15 to 30 min in 20% silver nitrate at 37°C in an incubator, until the section turns amber.
7. Wash off directly with 10% formalin in tap water from a dropper bottle until the solution comes off clean. Flood with formalin for another 10 to 15 s, until a fine white precipitate rises to the surface. The section should now be yellow.
8. Wash off directly with the Glees ammonia silver solution. Flood with ammonia silver, and let stand 30 s. *Glees ammonia silver solution:* Mix 30 ml 20% aqueous silver nitrate solution and 20 ml 95% alcohol. Add concentrated (28%) ammonia water drop by drop until precipitate just dissolves, and add 3 or 4 more drops of ammonia (use dropper bottle).
9. Drain, and wash off with 10% formalin (tap water) for 1 min or more. The section becomes brownish yellow; the solution turns black. Check impregnation microscopically. If the section is underimpregnated, repeat steps 8 and 9 using shorter intervals in each.
10. Wash in distilled water, fix for 30 to 60 s in 5% sodium thiosulfate, and wash for 10 min in running water.
11. Dehydrate, clear, and mount by alcohol, xylene, balsam sequence.

RESULTS: Axons and degenerate boutons terminaux are black. For intracellular and intraaxon detail, reduce concentration of formalin and silver solutions.

Proceeding from studies on acid phosphatase (*J. Neurophysiol.*, **9**: 121, 1946), Lassek evolved a method (*Stain Technol.*, **22**: 133, 1947) for the demonstration of normal axons in myelinated tracts (letter, March 29, 1951).

Brain tissue fixed in 10% formalin for 2 to 3 days may be used.

LASSEK'S LEAD METHOD FOR AXONS

1. Embed in paraffin, section, and hydrate sections as usual.
2. Immerse in buffered lead nitrate at 37°C for 1 to 96 h.

Molar acetate buffer pH 4.7	5 ml
5% aqueous lead nitrate solution	2 ml
Distilled water	36 ml

At time of using add 50 mg ascorbic acid.

3. Wash in distilled water.
4. Immerse in H_2S water or in 1% ammonium sulfide for 1 to 5 min.
5. Wash in water, and counterstain as desired; 0.1% safranin or thionin in 0.1% acetic acid for 5 min is suggested. Rinse in 1% acetic acid.
6. Dehydrate, clear, and mount in balsam through alcohol-xylene sequence. Axons should appear in brown.

METHENAMINE SILVER. In the use of the Burtner argentaffin cell variant for the study of nerve cell pigments we have seen in oversilvered preparations excellent demonstration of axons, along with generally blackened nerve cell bodies. This occurs with formalin material after about 3- to $3\frac{1}{2}$-h incubation at 60°C or in about $\frac{1}{2}$ h less in material chromated before embedding.

Prior methylation of formalin-fixed material greatly accelerates the silvering of axons. We have seen excellent, perfectly black axons and brown to black tigroid substance after a 24-h methylation at 60°C in anhydrous methanol containing 0.1 N hydrochloric acid. The silvering period in the same Gomori methenamine silver solution as above should be $1\frac{1}{2}$ to $2\frac{1}{2}$ h at 60°C. This methylation completely inhibits the usual safranin counterstain, as it does all other basic aniline dye stains. If any counterstain is desired, it would be necessary to use the Feulgen procedure or alum hematoxylin for nuclei, since these nuclear stains are not prevented by methylation.

De Martino and Zamboni (*J. Ultrastruct. Res.*, **19:** 273, 1967) tried to apply the silver methenamine stain to tissues prepared for electron microscopy. In some cases, it appeared that the precipitation of silver was related to staining time.

Ha (*Anat. Rec.*, **155:** 59, 1964) advocates the following method (emended from Golgi-Cox) for staining synapses.

1. Perfuse anesthetized animals with gum acacia–formalin as described by Koenig et al. (*Stain Technol.*, **20:** 13, 1945).
2. Fix tissues in 10% formalin in 0.9% NaCl for 2 weeks to 1 year at room temperature.
3. Slices of brain 3 mm thick are washed overnight in running water followed by several changes of distilled water.
4. Immerse tissues for 6 to 8 weeks in the Golgi-Cox impregnation mixture at room temperature. The mixture is prepared by adding 450 ml 5% $HgCl_2$ to 450 ml 5% $K_2Cr_2O_7$ and 100 ml 3% $K_2Cr_2O_7$. The last 100 ml is added with constant stirring.
5. Dehydrate impregnated blocks in ascending ethanols and embed in Parlodion. The blocks may be stored in 70% ethanol indefinitely.
6. Rinse sections of 80 to 100 μ briefly in distilled water and transfer them to 5% osmium tetroxide until they blacken (3 to 5 min).
7. Wash in distilled water.
8. Counterstain in 0.25% cresyl violet for 2 to 3 min at 23°C (5 drops of 10% acetic acid is added to 30 ml of 0.25% cresyl violet and filtered).
9. Wash, dehydrate in graded butyl alcohols, clear in cedarwood oil, and mount in a neutral medium.

RESULTS: Neurons are selectively stained. Nerve processes and cell bodies are counterstained with cresyl fast violet. Nissl substance is stained except granules are obscured. Boutons terminaux with parent fibers are disclosed in intimate contact with counterstained cell bodies and their processes. Sections of spinal cord are also stained with good results. Endings of mossy fibers and climbing fibers are clearly demonstrated.

Smit and Colon (*Brain Res.*, **13:** 485, 1969) conducted a quantitative analysis of Golgi-Cox–stained material in the cerebral cortex of rabbits. They conclude that approximately 1% of neurons are stained by the Golgi-Cox method. Good agreement was obtained between cell body volume of Golgi-Cox–stained neurons and Nissl-stained neurons.

Gwyn and Heardman (*Stain Technol.*, **40:** 15, 1965) recommend the following method for the simultaneous demonstration of nerves and motor end plates:

1. The fresh mammalian muscle is placed in 10% formol–saline solutions for 6 h at 20 to 25°C. Specimens should not be more than 5 mm thick.

2. Wash in distilled water for at least 16 h.
3. Cut sections 50 μ thick on a freezing microtome, collect them in a 10% sucrose solution, and wash them in distilled water for 1 h.
4. Incubate in the following solution for 15 min at 37°C:

   a. $CuSO_4 \cdot 5H_2O$, 24.9 g/l (0.1 M), 0.2 ml
   b. 3.7% aqueous aminoacetic acid (0.5 M), 0.2 ml
   c. Acetate buffer, pH 4.7, 0.1 M, 5 ml
   d. 2% aqueous acetylthiocholine iodide
   e. The supernatant fluid obtained by mixing 1 volume of a with 3 volumes of d and centrifuging until clear (5 to 10 min), 0.8 ml

   These volumes are suitable for the incubation of four sections of rat tibialis anterior muscle 50 μ thick.
5. Wash in distilled water for 5 min.
6. Place in 5% ammonium sulfide for 2 min.
7. Wash in distilled water with three changes for 1 h.
8. Soak the sections in 10% neutral formalin containing 2% pyridine in normal saline solution (neutralized with $MgCO_3$) for at least 15 days.
9. Wash in tap water with two or three changes for 1 h.
10. Stain in 10% $AgNO_3$ for 1.5 h at 37°C in the dark.
11. Wash in 10% formalin in tap water until the white precipitate no longer forms (about six changes).
12. Place in 10 ml ammoniacal $AgNO_3$. To 10 ml 10% $AgNO_3$ add NH_4OH, sp gr 0.880 (35% NH_3), drop by drop until the brown precipitate disappears, and then add 3 drops in excess.
13. Wash sections as follows: 1% ammoniated water (1 ml 35% ammonia and 99 ml H_2O) for 2 min; 1% acetic acid, 2 min; distilled water, 2 min; 3% $Na_2S_2O_3$, 5 min; and distilled water, 30 min.
14. Dehydrate, and mount in synthetic medium.

RESULTS: Copper sulfide deposits turn black. Motor nerves can be followed into end plates even at low magnification. The stain is particularly useful in studies of nerve regeneration and reinnervation of mammalian skeletal muscle. Incubation time for the demonstration of cholinesterase is critical. If it is too short, definition of the motor end plate is obscured; if too long, end plates are overstained, and definition is again lost. Likewise, the time of soaking in formol-pyridine is important. At least 15 days are required. If less time is used, small nerve fibers cannot be traced into the motor end plates. Deterioration of staining was not observed if specimens were treated in excess of 15 days.

RANVIER'S METHOD. The classic method for *motor end plates* and other *peripheral nerve endings* is that of Ranvier. We have had some excellent preparations and some utter failures in a rather brief experience with this method. Carey has used this method with quite consistent success, and we quote here his variant (*Am. J. Pathol.*, **18**: 237–289, especially p. 242, 1942):

RANVIER'S GOLD CHLORIDE METHOD FOR PERIPHERAL NERVE ENDINGS (Carey)

1. Cut tissue (muscle) in 3- to 5-mm slices, and immerse in filtered fresh lemon juice for 10 to 15 min (they become transparent in 5 to 10 min, according to Ranvier), handling tissues with glass or paraffin-coated instruments.

2. Pour off lemon juice, and add 1% gold chloride (HAuCl$_4$) without rinsing. (Ranvier rinsed in water.) Let stand for 10 to 60 min until a uniform golden yellow tone is reached, not *brown*. (Ranvier used 20-min immersion and then washed in water.)
3. Transfer directly to a 25% dilution of concentrated (90%) formic acid, and reduce in the dark for 8 to 12 h (Ranvier used 20% for 1 day).
4. Wash in water, and store in a 50:50 mixture of C.P. neutral glycerol and 50% alcohol.
5. Snip out a small fragment and tease gently in glycerol, or crush out under a cover glass. Mount in glycerol sealed with cellulose caprate or in glycerol gelatin.

RESULTS: Nerve fibers and endings appear black; other tissues, in varying shades of red, purple, and violet.

Other writers have used a 20% dilution of formic acid in distilled water in place of the lemon juice and reduced in a fresh portion of the same fluid for 24 h. Tissues after spreading can be dehydrated, cleared, and mounted in balsam or synthetic resin. We have tried formic, acetic, and citric acids, and the last seems to give the cleanest impregnations. With these, 30 to 60 min in the gold chloride seems to be necessary. We have used Apáthy's syrup for mounting these preparations.

Various attempts have been made to apply this method to material previously fixed in formalin, but with poor success.

PETERS' PROTEIN SILVER FOR NERVE FIBERS. Peters (*Stain Technol.*, **33**: 47, 1958) gives methods for nerve fiber impregnation said to be applicable to routine paraffin sections of material fixed in aqueous formalin, acetic alcohol–formalin, or Bouin's or Huber's fluid (omit the usual iodine treatment of sections). These methods avoid the inconstancies of manufactured silver proteinates by the direct use of proteins with silver nitrate.

1. Attach paraffin sections with albumen fixative, dry, deparaffinize, and hydrate through xylene and graded alcohols as usual.
2. Impregnate sections 16 to 24 h at 37°C or 56°C in the dark in albumen-silver or casein-silver, made as follows: Dissolve 300 mg dried egg albumen or casein by sprinkling powder onto surface of 60 ml distilled water. Filter through Whatman No. 1 paper. To 50 ml filtrate add 1.8 ml 2% silver nitrate, and adjust to pH 8.2 to 8.5 by adding drop by drop a 5% dilution of 28% ammonia (instead one may use a boric acid sodium borate buffer of pH 8.2 or 8.5 as the solvent for the albumen or casein). This amount of silver solution suffices for five slides only.
3. Wash slides in two or three changes of distilled water.
4. Develop in 1% hydroquinone in 10% sodium sulfite (Na$_2$SO$_3$) for 5 to 10 min.
5. Wash for 10 min in running tap water.
6. Wash in distilled water, and tone for 10 min in 0.2% gold chloride.
7. Rinse in distilled water, and reduce for 10 min in 2% oxalic acid.
8. Wash for 5 min in running water.
9. Fix for 10 min in 5% sodium thiosulfate (0.2 *M*).
10. Wash for 10 min in running water.
11. Dehydrate in graded alcohols; clear in alcohol-xylene and two or three changes of xylene. Mount in synthetic resin.

For embryonic tissue increase 2% silver nitrate to 2.5 ml; in adult tissue greater intensity is attained with 2 ml. With casein-silver use only 1 ml 2% silver nitrate.

The borate buffer may be replaced by 0.2% ammonium hydrogen tetraborate (NH$_4$HB$_4$O$_7$·3H$_2$O).

Impregnation time may be prolonged to 48 h without harm.

For the method of Rogers, Pappenheimer, and Goetsch (*J. Exp. Med.*, **54:** 167, 1931), see Lillie, 1954, p. 417, or the original reference.

EHRLICH'S SUPRAVITAL METHYLENE BLUE METHOD.   This is the other classic method for study of peripheral nerve endings. We have had no personal experience with this method, but have relied largely on Romeis, Lee, and Mallory for the following: The solution should be made from medicinal, zinc-free methylene blue. The dye should be dissolved in 0.9% sodium chloride of C.P. grade (0.6% for amphibia), not in Ringer's or Locke's or similar fluids, as these produce precipitates. The methylene blue is introduced into the tissues by intravascular perfusion, by instillation into body cavities, by injection into loose connective tissues, or simply by immersion of small fragments. Within the tissues the methylene blue is reduced to leucomethylene blue and then reoxidized by atmospheric oxygen. Tissues must be taken as soon after cessation of circulation as possible. Human autopsy tissues are ordinarily not usable. Blood must first be washed out with saline solution before the perfusion methods are used, though some workers use an intravital perfusion in living animals. The strength of methylene blue solution varies between 0.5 and 0.06%.

PERFUSION AND INJECTION TECHNICS.   Use about 0.2% methylene blue at 37°C, and perfuse with sufficient volume to replace the salt solution used to wash out the blood, or inject with enough to distend the tissues moderately. After 30 min remove the tissues to be studied, and place on glass wool moistened with 0.06 to 0.1% methylene blue in a loosely covered container in a 37°C incubator. Tissue should be taken in small fragments or thin slices. Leave in the incubator until tissues become bluish—perhaps 15 to 120 min.

IMMERSION TECHNIC.   Immerse thin slices or small fragments of tissue in 0.1 to 0.25% methylene blue at 37°C. Romeis prefers to lay them on glass wool moistened with 0.05 to 0.1% methylene blue and to drip onto the surface of the fragments (which may measure 5 to 20 mm in thickness) a 0.1 to 0.25% dye solution to flood the surface. Keep the preparations at 37°C for 1 to 2 h, observing them under low magnification from time to time until nerves are stained blue.

SECTIONING AND FIXATION.   After staining by the variants above has reached its optimum, Romeis prescribes cutting fresh sections with a razor and immediate study, or fixation with ammonium molybdate according to Dogiel—a freshly prepared 5 to 8% solution, which is to be filtered if not clear. Thin membranes are adequately fixed in 30 to 60 min. Thicker tissue sections may require up to 24 h. This requires 50 to 100 volumes solution. Then wash for 1 to 6 h in several changes of water according to size and thickness. Preparations are to be dehydrated rapidly in several changes of 95 and 100% alcohol during a total period of 10 min to 2 h—again according to thickness. Alcohol tends to extract the stain. Clear in terpineol, changing once. Wash out in xylene and mount in balsam (or Permount).

Romeis also suggests the preparation of thick frozen sections after washing out the ammonium molybdate, with or without gelatin embedding. Paraffin embedding extracts the color badly.

Dogiel also proposed a fixation in saturated aqueous ammonium picrate solution for 2 to 24 h, followed by mounting in equal volumes of glycerol and ammonium picrate solution.

Schabadasch (*Z. Zellforsch. Mikrosk. Anat.*, **10:** 221–385, 1930) proposed addition of 50 to 700 mg resorcinol (or in some instances *p*-aminophenol) per 1 methylene blue, and buffering of the methylene blue solution with phosphates to various pH levels ranging from 5.7 for colon to 7.2 for isolated nerve. The methylene blue concentrations ranged from 150 to 350 mg/l for the most part. He fixed in a mixture of ammonium iodide or ammonium thiocyanate ("rhodanid") in ammonium picrate. The latter is to be impure and contaminated with enough "ammonium picraminicum" (picramate?) so that the crystals are large, orange needles, not the fine, yellow needles of the pure salt. Either a 2.11% solution of ammonium iodide or a 1.12% solution of ammonium thiocyanate is saturated with this ammonium picrate at 60°C and then allowed to stand several days at 15 to 20°C. One fixes the stained preparations for 3 to 6 h at 35°C, and mounts directly in a mixture of equal volumes of C.P. glycerol saturated with ammonium picrate and of 10% gelatin solution. Thicker pieces may require preliminary clearing in saturated ammonium picrate–glycerol.

Apáthy uses his gum syrup as a mounting medium for crushed or spread whole mounts.

Bethe proposes fixation in 5% ammonium molybdate containing either 0.25% osmium tetroxide or 1% chromic acid (CrO_3). In either case add 1 drop (0.05 ml) of concentrated hydrochloric acid to 20 ml fixative. Fix for 4 to 12 h in the osmium tetroxide variant, 45 to 60 min in the chromic acid; then wash, dehydrate quickly in alcohols, clear in benzene, and embed in paraffin; or clear in xylene for direct mounting in balsam or Permount (cited from Romeis).

The acetylcholinesterase methods are also much used now for the study of nerve endings.

Richardson (*Anat. Rec.*, **164**: 359, 1969) vitally stained autonomic nerves in rodents using a methylene blue perfusion. After fixation in potassium permanganate and embedding, sections were examined by electron microscopy. By controlling the pH and saturating with oxygen during perfusion staining, only cholinergic axons were stained at pH 6.5 to 7, whereas at pH 5 to 5.3, both cholinergic and adrenergic axons stained.

Ehinger et al. (*Histochemie*, **13**: 105, 1968) combined the methylene blue and fluorescent adrenergic nerve stains to illustrate in the same preparation adrenergic and nonadrenergic nerve fibers. Tissues (such as irides) are incubated for 15 min at 37°C in the following: NaCl, 6 g; glucose, 1.7 g; $MgCl_2$, 1 g; $CaCl_2$, 1 g; KCl, 1.0 g; 1/3 M phosphate buffer of pH 5.9, 17 ml; and 0.2 g methylene blue (C.I. 52015). The solution is heated to 50°C (but not above) to effect solution. After cooling, the solution is diluted to 1000 ml with distilled water. Addition of 1 μg adrenaline per ml incubation medium resulted in a superior and constantly reproducible demonstration of axons.

Incubated irides are then quick-frozen, freeze-dried, treated with formaldehyde vapor, vacuum-embedded in paraffin, sectioned, and examined by fluorescence and other microscopy.

LIANG'S SULFUROUS ACID LEUCOFUCHSIN METHOD. H. M. Liang's (*Anat. Rec.*, **97**: 419, 1947 [Proceedings]) method utilizing Schiff reagent to demonstrate nerve endings is interesting but requires further study as to its significance.

Liang prescribed 1-h fixation in 1% acetic or formic acid, followed by two rinses in sulfurous acid of unspecified strength and 1- to 3-h immersion in Schiff reagent until nerves are deep purple. He then washed in several changes of sulfurous acid, rinsed with distilled water, counterstained 15 min in 1% aqueous methyl green solution, dehydrated, cleared, and mounted as whole mounts. Or after the washing following the second sulfurous acid bath, he dehydrated, cleared, and embedded in paraffin, sectioned, brought sections to water, stained 5 min in 1% methyl green, dehydrated quickly with alcohols, cleared, and mounted.

For the paraffin sections, an acid hemalum or iron hematoxylin–picric acid counterstain can be used and renders other structures more visible. Elastic tissue of blood vessels sometimes stains red-purple by this method, and myelin shows red-purple vacuolation figures around the solid red-purple axon.

Since the Schiff reaction follows directly an acid treatment, one suspects that this is related to Feulgen's plasmal reaction.

For methods of separation of neurons and neuroglia, the reader is referred to Sinha and Rose (*Brain Res.*, **33**: 205, 1971). Three methods were compared.

Flanagen et al. (*J. Histochem. Cytochem.*, **22**: 952, 1974) noted that Procion yellow M-4

RS binds covalently to subcellular organelles of rat brain under physiologic conditions. Binding to primary amines is assumed.

The studies of Tetzloff, Peterson, and Ringer (*Stain Technol.*, **40:** 313, 1965) improve the demonstration of nerves but leave the mechanisms of the reaction unresolved. They direct as follows: Fix for 1 h in 1% formic acid, transfer through two 5-min baths in 6% sulfuric acid, drain, and immerse for 30 min in 5% phenylhydrazine hydrochloride, wash in running water for 10 min, immerse for 5 min in distilled water, and place in "cold Schiff" reagent for 4 h. Wash in four changes of 6% sulfuric acid, for 5 min each, then in three changes of distilled water, for 5 min each. Dehydrate in four changes of acetone for 10 min each. Clear in two changes of methyl benzoate, for 1 h each; wash in xylene; and mount in balsam or synthetic resin. The times here are for chicken skin; for turkey skin increase times 50%.

The phenylhydrazine treatment prevents the staining of elastic tissue. Axis cylinders and neurilemma are stained deep red; myelin remains uncolored.

RETINA

Apparently three lipid substances are demonstrable (Lillie, *Anat. Rec.*, **112:** 477, 1952) in the rods and cones. One is relatively insoluble, is located in the acromeres, is apparently partly protein-bound, and is preserved in paraffin sections even after some partly alcoholic fixations; but it is removed by alcohol-ether and alcohol-chloroform extraction from fresh unfixed tissue. It is best demonstrated by the periodic acid Schiff reaction after various aqueous fixations. It retains iron hematoxylin poorly in myelin-type technics, and retains it best after formalin-dichromate sequence fixations, and may be demonstrated with oil-soluble dyes in Ciaccio-type technics. It was thought that this lipid might be kerasin.

The second lipid substance, also located in acromeres, required aqueous formaldehyde or dichromate fixations for its preservation in paraffin sections, and was best demonstrated by peracetic acid or performic acid Schiff technics after formalin-dichromate sequence fixations. Conditions which produce Kolmer's droplets give also fine droplets between the rod acromeres which are performic acid Schiff–positive. It is probably largely responsible for the positive myelin stains which are obtained after formalin-dichromate combination or sequence fixations. It is assumed that this lipid also stains with fat stains in paraffin sections. That it is not identical with the periodic acid Schiff–positive material is indicated by its presence in Kolmer's droplets and by its loss from sections first fixed in lead or mercury salts without chromates or formaldehyde and then dehydrated and embedded by an alcohol-gasoline-paraffin sequence. This lipid is perhaps an unsaturated fatty acid lecithin.

The third lipid is demonstrated by oil-soluble dyes and by iron hematoxylin methods after aqueous formaldehyde or chromate fixations, but not by the periodic, performic, or peracetic acid Schiff methods. It is localized in the ellipsoids and is associated with a strongly eosinophilic basic protein, which is especially well shown after Carnoy and alcoholic formalin fixations.

Glycogen is demonstrated in the retina by the periodic acid Schiff method, removed by diastase digestion, and localized chiefly in the myoid segments of rods and cones and in the zones of synapses. Sometimes entire granule cells seem to be outlined by heavy glycogen deposits.

Myelin seems more difficult to demonstrate in the nerve fiber layer of the retina than in the optic nerve or in the small oculomotor branches seen outside the bulb. It reacts with the performic acid Schiff as well as with hematoxylin methods and oil-soluble dyes.

For the demonstration of Kolmer's droplets the preferably dark-adapted retina should be fixed for 12 to 24 h in Held's fluid. This is a vague mixture composed of 4 volumes each of saturated aqueous potassium dichromate solution and of 10% formalin, with the addition of 1 volume of glacial acetic acid. By increasing the amount of acetic acid, droplets can be shown in relation to cones as well as to rods. After fixation, the tissues are washed overnight in running water, dehydrated through alcohols, and embedded in paraffin, not celloidin. Staining is done with an iron hematoxylin procedure of the Heidenhain type, and preparations are differentiated, according to Kolmer, to the point where rod acromeres become colorless. At this point ellipsoids as well as the Kolmer droplets should still be rather darkly stained. According to Walls (*Anat. Rec.*, **73:** 373, 1939) these lipid droplets are artifactual, produced by high concentrations of acetic acid from acromere and pigment epithelial substances and oxidized to an insoluble state by chromation.

For the demonstration of rhodopsin (visual purple) fix retina in 2.5% platinic chloride solution for 12 to 24 h (Romeis), dehydrate in alcohols, and embed in paraffin as usual. The rod acromeres are colored an intense orange. This is essentially the method of Stern (*Arch. Ophthalmol.*, **61:** 561, 1905), who prescribed the use of animals dark adapted for at least 2 h and the use of red light only, during dissection, blocking, and fixation.

LIPID DROPLETS IN HUMAN RETINAL PIGMENT EPITHELIUM. Ciaccio positive lipid is almost regularly demonstrable with Sudan black B in routine paraffin section of formalin-fixed surgical specimens removed for a wide variety of conditions, as well as in normal eyes. Only in retinal detachments with atrophy and of considerable apparent duration and in ophthalmophthisis (phthisis bulbi) have we noted pronounced diminution or absence of the lipid.

According to Streeten (*Arch. Ophthalmol.*, **66:** 391, 1961) these droplets first appear in children about 16 months of age and gradually increase in number. They vary in size from barely visible to 0.5 μ or more. Faint staining with Sudan IV and oil red O is recorded; staining is rapid and intense with Sudan black B. The Sudan black may be extracted with acetone or alcohol and the preparations restained with little or no loss of intensity, even after repeated decolorization and restaining. Schultz cholesterol reactions are negative; Baker's acid hematein test is essentially negative, as was the Luxol Fast Blue myelin stain. With Nile blue the granules stain violet, or more reddish in older eyes, and a purer blue after permanganate bleaching of pigment.

The lipid is slowly extracted by hot methanol-ether, after formalin fixation, but no data are available about solubility on primary alcoholic fixations. The direct Schiff reaction was negative even in 48 h, but peracetic and performic acids, though no' periodic acid, engendered a definitely positive Schiff reaction (purple). The lipid is not acid-fast. Methenamine silver and ferric ferricyanide are not reduced by the lipid, though the pigment reacts.

Lillie recorded lipid droplets up to 4 or 5 times the diameter of the retinal fuscin granules and in one series of cases found the lipid still strongly stainable with Sudan black after a 24-h extraction in pyridine at 60°C.

It is of interest also that Sudan black demonstrates quite well in paraffin sections of formalin-fixed eyes the lipid in the more or less numerous macrophages which appear in and near the nerve head in chronic glaucoma.

OTHER OCULAR STRUCTURES

The vitreous humor contains a webwork of coagulum staining red by the periodic acid Schiff procedure and blue by the allochrome method. It colors blue with spirit blue in the Ciaccio-type method using this dye but not with Sudan black B. It does not stain with basic thiazin dyes, either ortho- or metachromatically. It sometimes colors blue by the Lillie-Mowry ferric mannitol method. Its staining evidently is not affected by bovine hyaluronidase chondromucinase preparations.

The suspensory ligament of the lens merges into this vitreous web in periodic acid Schiff preparations but, like lens capsule and Descemet's membrane, remains red in allochrome preparations.

Corneal connective tissue differs from that of the sclera and various ocular areolar tissues in its metachromasia to thionin, which may be demonstrable after lead nitrate or mercuric chloride fixatives, and in its pure deeper red with periodic acid Schiff procedures when a picric acid counterstain is used.

Under polarized light corneal and scleral collagen are strongly birefringent; lens capsule is moderately so and requires rotation for its demonstration; lens fibers are scarcely visible.

Conjunctival epithelium may show quite pronounced alkaline phosphatase activity as well as moderate amounts of glycogen and some melanin. The conjunctival goblet cell mucin is rather weakly stained by metachromatic dyes and Alcian blue; it is stained quite well by HIO_4 and CrO_3 Schiff procedures.

The allochrome method often fails to distinguish a distinct red basement membrane, though one is often shown under the corneal epithelium and the retinal epithelium over choroid and ciliary.

Ocular pigments are considered in an earlier chapter.

18

HARD TISSUES; DECALCIFICATION

Bones, teeth, and other calcified tissues are sometimes studied by the preparation of ground sections of the macerated specimen, but usually the removal of most of the calcium salts is required to permit the preparation of sections thin enough for the study of cellular detail.

Adequate fixation of hard bone is facilitated by sawing thin slices with a fine-toothed bone saw or hacksaw. This procedure produces a narrow zone of mechanical distortion adjacent to the saw cut. Mallory discarded the first half dozen sections; we prefer to slice off this zone with a razor blade after decalcification and then section from the freshly cut surface. Making such a cut through the center of a block further serves as an adequate test for completion of decalcification.

When the marrow of the cancellous bone is the primary objective of study, time may be saved and the acid damage to marrow cells reduced by cutting off the cortical bone as soon as the cancellous portion is soft enough.

Fixatives for bone and marrow should be chosen with the primary objective of study in mind, and are much the same as for soft tissues. Chromate fixatives such as Orth's, Kose's, and Möller's potassium dichromate–formalin mixtures and formalin-Zenker variants such as the Spuler-Maximow fluid are preferred by many for the study of marrow cells. They render the nucleic acids less susceptible to hydrolysis either by ribonuclease or by acids and engender moderate resistance even to normal hydrochloric or nitric acid.

When preservation of iron-bearing pigments is important and for routine diagnostic purposes, formalin is the fixative of choice. The calcium phosphate of the bone substance serves as a fully adequate buffer to keep the pH level of the fixative probably above 6 within the tissue. If the action of the formaldehyde is not prolonged beyond the 2 to 3 days recommended for adequate fixation, nucleic acids are left quite susceptible to ribonuclease digestion or hydrolysis by mineral acids. Hence formalin-fixed bone is preferably decalcified with 1 N acetic or formic acid or with buffer mixtures of pH 2 or higher.

For general diagnostic purposes, 5 to 7.5% nitric acid has been found effective over many years of trial. Recently the 8% hydrochloric–10% formic acid fluid of Richman, Gelfand, and Hill (*Arch. Pathol.*, **44**: 92, 1947) has had considerable vogue. This fluid is 1 N HCl containing 2.4 N HCOOH. It is a quite active and effective decalcifying agent. It will decalcify 1- to 2-g pieces of compact bone in 12 to 24 h at 24°C and in 6 to 8 h at 37°C.

The effect of the electrolytic bath recommended by these authors for use with this fluid seems to be purely that of the heat produced by the passage of the current, since equally rapid decalcification occurs with external heating to the same temperature or in bones suspended in the electrolyte at the maximum possible distance from either electrode.

Decalcification at 55 to 60°C is hazardous. Loss of calcium salts occurs rapidly and is followed promptly by swelling and hydrolysis of the bone matrix collagen, which soon results in complete digestion. This occurs in as little as 24 h in the 8% hydrochloric acid–10% formic acid mixture and in 2 to 3 days in 5% formic acid.

Decalcification at 37°C impairs alum hematoxylin staining and Weigert's iron hematoxylin staining of nuclei to some extent. Eosin stains of cytoplasm are well preserved. Feulgen staining of nuclei is unsuccessful, and azure-eosin stains give pink cytoplasms and nuclei. Both the formic acid–hydrochloric acid mixture and 5% formic acid produce these results. The latter also impairs Van Gieson and Masson staining of collagen and bone matrix. Mineral acid decalcification is 2 or 3 times faster at 37°C than at 20°C; formic acid is actually somewhat slower.

With decalcification at 15 to 25°C, either in mineral acids or in formic acid or in mixtures of formic and hydrochloric acids, satisfactory hematoxylin-eosin, Van Gieson, Masson, and azure-eosin stains may be obtained, if exposure to the mineral acids is not prolonged. Feulgen staining of nuclei is well preserved after formic acid decalcification at 24°C, but with mineral acids at 20°C or even at 3°C it is impaired or destroyed.

Altogether, for other than enzyme demonstration, 1 N acetic and formic acids seem the best general decalcifying agents. They are more economical reagents than the Kristensen and other buffer mixtures and seem equally effective. The preferred procedure is to use two changes daily of about 25 ml/g bone, testing the removed fluid at each change to determine the presence of calcium ion by mixing 5 ml of it with 1 ml 5% sodium or ammonium oxalate. When a negative test is obtained—wait several minutes for the development of delayed turbidity—the specimens may be considered decalcified. This test seems more objective and less liable to underdecalcification than the common practice of testing with a pin. It coincides well with the cessation of weight loss. It is equally applicable to decalcification with mineral acids, but it must be recalled that calcium oxalate will not precipitate from hydrochloric or nitric acid above about 0.7 N or 0.8 N final dilution. The concentration of stronger acid must be reduced by dilution or by addition of an amount of 2 N NaOH which will not quite neutralize; a drop of Congo red solution will give a color change at about pH 4 to 5.

Following this procedure 1- to 2-g blocks of compact bone 3 to 4 mm thick decalcify in 4 to 5 days in 5% formic acid; similar blocks of cancellous bone, in 2 to 3 days. In 5% nitric acid it takes about half that time, and in the 8% hydrochloric–10% formic acid mixture, even less: some 4 to 8 h for cancellous bone and 12 to 24 h for compact bone.

C. E. Jenkins (*J. Pathol. Bacteriol.*, **24:** 166, 1921) recommended an alcoholic acid solution which fixes and decalcifies human rib in 48 h, permits normal blue nuclear staining with Ehrlich's hematoxylin, and occasions no swelling of collagen fibers in Van Gieson preparations. A 17 to 20°C temperature in Glasgow is presumed.

JENKINS'S DECALCIFYING FLUID

Concentrated hydrochloric acid	4 ml
Glacial acetic acid	3 ml
Distilled water	10 ml
Alcohol, 100%*	73 ml
Chloroform	10 ml

* If 95% alcohol is used, the proportions are 76 ml alcohol and 7 ml distilled water.

Fix bones until soft; wash in two changes absolute alcohol, for 4 h each; treat with chloroform to remove alcohol, in two changes, for 30 min each; embed in paraffin.

A. J. Schmidt (*J. Exp. Zool.*, **149:** 171, 1962) used this solution after freeze substitution for salamander legs, for 5 days at −70°C in absolute alcohol saturated with picric acid. His decalcification interval was 5 days at 20 to 25°C (*J. Histochem. Cytochem.*, **11:** 443, 1963). He reported demonstration of glycogen by the periodic acid Schiff method equal to that in similarly freeze-substituted but undecalcified tissue.

In not exposing tissue to water before or during decalcification this procedure meets the requirements which we have specified for the preservation of glycogen through decalcification (*J. Histochem. Cytochem.*, **10:** 763, 1962).

Our findings (*Am. J. Clin. Pathol.*, **21:** 711, 1951) of very slow decalcification of dog femur by 0.5 N HCl in 80% alcohol, for 8 days at 25°C or 12 weeks at −23°C, indicate that decalcification of compact bone may take considerably longer than stated above by Jenkins for human rib.

Trichloroacetic acid is also a fully satisfactory decalcifying agent which permits just as satisfactory nuclear and marrow cell staining as does formic acid. However, on account of its high molecular weight (163) as compared with that of formic acid (46), much larger quantities (3.55 times) are needed for efficient decalcification.

Sulfurous acid, available as a 5 to 6% solution of sulfur dioxide in water, is also a prompt decalcifying agent, but if its action is prolonged more than 48 h, nuclear staining is seriously impaired.

After decalcification with nitric, formic, trichloroacetic, or sulfurous acid, tissues should be washed 18 to 24 h in gently running water to remove acid, then hardened in 80% alcohol, dehydrated, and embedded as usual in paraffin or nitrocellulose. Frozen sections of decalcified bone for the study of fats and myelinated nerves are readily prepared on the freezing microtome. In this case, naturally, all contact with alcohol or other fat solvents is to be scrupulously avoided.

Alcoholic solutions of acids are quite inefficient decalcifying agents. Solutions in 70 or 80% alcohol are slow: 0.5 N HCl took 8 days at 25°C, compared with 4 days for 0.5 N HCl in 40% alcohol and 2 days for aqueous 0.5 N HCl. With formic acid 5% solutions took 3 to 5 days in water and about 4 months in 30% alcohol, and failed to decalcify in several months in 80% alcohol. This occurs probably more because of the suppression of ionization by alcohol than because of insolubility of certain calcium salts in alcohol. Calcium nitrate, calcium chloride, and calcium bromide are listed as soluble to very soluble in alcohol; calcium acetate is slightly soluble; the formate is insoluble; but addition of 5% of either glacial acetic acid or 90% formic acid to a saturated 70% alcohol solution of calcium chloride occasions no precipitation, though small quantities of oxalic or sulfuric acids give copious precipitates.

In the study of fractures, periosteal hemorrhages, and similar lesions where preservation of precise anatomic relation is of more importance than cytologic preservation, it is often desirable to embed in nitrocellulose either after or preferably before decalcification. Embed before decalcifying to make sure that soft tissues are not displaced from hard.

For the gross opening of long bones in scurvy, rickets, and similar conditions it may be desirable to embed the bones temporarily in a plaster of paris matrix before sawing in two lengthwise. Procedural fractures at the epiphyseal line are thus avoided. This embedding may be done either before or after fixation.

When embedding the bone in plaster, its position may be marked by inserting toothpicks in the soft plaster to mark the plane of sawing. Allow the plaster to set for about an hour. Saw with a thin-bladed hacksaw. Then break off the plaster, and wipe the surface with gauze or with a brush moistened in 5% acetic acid. The bones are usually greasy enough to separate readily from the plaster. If difficulty is encountered in this respect, the bones may be first dipped in light mineral oil before embedding in plaster. We are indebted to Dr. J. H. Peers and Dr. T. H. Tomlinson for this note.

Mechanical agitation appears to have no great effect on the speed of decalcification, so that apparatus designed for this purpose seems scarcely worth the trouble or expense.

The use of vacuum during decalcification does not materially expedite the process, but it does prevent the formation of large gas bubbles in the marrow in large pieces of bone. With small bones it has no evident value.

Prolonged, slow decalcification with fluids of relatively high pH level (5), such as 5 to 10% solution of ammonium chloride, ammonium nitrate, or potassium or sodium acid phosphate, gives excellent preservation of marrow cells after 2 to 3 weeks; but cortical bone is still hard at the end of 3 weeks.

Frank and Deluzarche (*Bull. Histol. Appl.*, **27**: 35, 1950) recommended for the decalcification of enamel and dentin in the crowns of human teeth first a primary fixation in 10% formalin followed by a 12-h refixation in Bouin's fluid. Then the crown is rinsed in water, in alcohol, and in ether, and is coated with three layers of collodion on the enamel surface. Decalcification follows in a fluid composed of 2 g nitric acid (sp gr 1.38) (1.23 ml of our 69 to 70% acid, sp gr 1.42), 8 g trichloroacetic acid, 20 ml Bouin's fluid, and 70 ml water. Maintenance of partial vacuum (about 100 mm) during decalcification is recommended. Dehydration, clearing, and infiltration through graded alcohols, cedar oil, toluene, and paraffin should be fairly rapid.

Hahn and Reygadas (*Science*, **114**: 462, 1951) report the use of a solution of sodium ethylenediaminetetraacetate for demineralization of hard tissues. This solution is alkaline, and the evolution of carbon dioxide is thereby prevented. Complex soluble electronegative calcium-containing ions are formed. If the edetic acid (ethylenediaminetetraacetic acid, or EDTA) is neutralized with quaternary sodium pyrophosphate instead of hydroxide, the demineralization effect for iron salts is increased. No further details as to concentration of solutions have been published. Since iron salts were also removed along with calcium salts, this process was not recommended in those cases where demonstration of iron-containing pigments might be important.

RETENTION OF HEMOSIDERIN IRON IN DECALCIFIED TISSUE. Various expedients have been adopted in the past to retain in the tissues as much iron as possible. Some of these seem particularly applicable in the case of tissue which is to be decalcified, either with acids or with Versene. It is true that much hemosiderin remains demonstrable in marrow littoral cells when formic acid decalcification is practiced. But both of the iron formates are soluble in water; the ferric is somewhat more soluble than the ferrous, which is only slightly soluble. It is probable that protein and polysaccharide chelate bonding of the iron protects them to some extent, at least in the pH 2.5 to 4.5 range of acid treatment. The question of iron retention is usually disregarded by the Versene decalcification enthusiasts.

However, use of a soluble sulfide, buffered to neutrality to prevent deleterious alkali effects on tissue, as a component of the fixing fluid should act to keep iron undissolved. $Ca(SH)_2$ is very soluble in water. FeS has a solubility of 6.16 mg/l; Fe_3S_4 is also quite insoluble, though both dissolve in strong mineral acids. Hence nitric and hydrochloric acids should be avoided as decalcifying agents under these circumstances.

It is probable that much of the tissue hemosiderin, normally ferric in state, is reduced to Fe_3S_4 by H_2S. In the Tirmann-Schmelzer method after sulfide reduction it is known to react with $K_3Fe(CN)_6$. The existence of a ferrocyanide reaction in sulfide-treated tissue is evidence of the preexistence of Fe^{3+}. Ferrous salts are comparatively uncommon in tissue, and their prior existence cannot be proved in sulfide-treated tissue or in tissue subjected to other reducing agents.

To apply this method, fix bone with red marrow in 10% formalin to which 2% yellow ammonium sulfide is added or 1% NaSH. Na_2S yields strongly alkaline solutions which should be adjusted electrometrically to pH 7, or saturated with hydrogen sulfide gas. After usual fixation intervals decalcify with 5% formic acid or by Christensen's solution, or perhaps with Versene.

1 *N* ACETIC ACID.   This decalcifies effectively, though perhaps not quite so rapidly as 1 *N* formic acid, according to A. Gutiérrez in Lillie's laboratory. Hemosiderin iron resists 1 to 2 *N* acetic acid for prolonged periods (*Am. J. Pathol.*, **15**: 225, 1939) and is readily shown by ferrocyanide after acetic acid decalcification. The decalcification technic is the same as with 5% formic acid.

EDETIC ACID (ETHYLENEDIAMINETETRAACETIC ACID, EDTA, VERSENE, SEQUESTRENE).   Generally 5 to 10% Versene solutions are used. Alkaline solutions tend to hydrolyze protein to some extent (Cook and Ezra-Cohn, *J. Histochem. Cytochem.*, **10**: 560, 1962); hence solutions adjusted to between pH 6.5 and 7.5 are recommended. We (*J. Histochem. Cytochem.*, **11**: 662, 1963) have shown that in 5-μ sections there is no appreciable loss of the iron in cytosiderin, hemosiderin, or enterosiderin during 7 days' exposure to 6% Versene at pH 7.5, and that the loss is scarcely appreciable at 4.5. Hence, neutral Versene decalcification would seem applicable to the study of iron-bearing pigments in bone marrow.

Cook and Ezra-Cohn (*J. Histochem. Cytochem.*, **10**: 560, 1962) reported a somewhat less complete removal of calcium from ground unfixed bone by Versene than by hydrochloric or formic acid. But loss of protein nitrogen did not occur to any appreciable extent with Versene. Such loss was considerable with hydrochloric and formic acids from unfixed ground bone, but was reduced to a relatively minor amount in bone prefixed in formalin, especially if acid exposure was limited to the period of active calcium removal. In this connection the analyses of Cook and Ezra-Cohn are not closely enough spaced in relation to the end of the active decalcification period.

Trott (*J. Histochem. Cytochem.*, **9**: 699, 1961) noted use of Versene at pH 7.2 by Schajowicz and Cabrini, and he himself used a Versene solution at pH 7.5 for 7 days in the study of glycogen in rat jaw tissues.

For the demonstration of nicotinamide-adenine dinucleotide and nicotinamide-adenine dinucleotide phosphate (NAD, NADP)–linked dehydrogenases and the diaphorases and of succinic dehydrogenase in bones, Balogh (*J. Histochem. Cytochem.*, **10**: 232, 1962) decalcified fresh rat and mouse jaws and other bones, as well as $5 \times 5 \times 1$ mm blocks of cancellous and compact human bone, by a 3-, 4-, or 5-day immersion at 4 to 10°C in 10% Versene in 0.1 *M* phosphate buffer, pH 7, readjusting to pH 7 with NaOH (or HCl) after addition.

On the other hand Cabrini and Rosner (*J. Histochem. Cytochem.*, **11**: 119, 1963) found it necessary to do these dehydrogenase formazan reactions in thin unfixed blocks, postfix in formalin, and then decalcify and section by usual procedures. Among the fluids used for decalcifying was a 5% Versene at pH 7.

Tonna et al. (*J. Histochem. Cytochem.*, **11**: 720, 1963) used 10% Versenate (disodium edetate) for decalcification of mouse femora.

GENERAL REFERENCES FOR DECALCIFICATION.   Useful references for decalcification are R. D. Lillie, *Am. J. Pathol.*, **20**: 291, 1944; R. D. Lillie, A. Laskey, J. Greco, H. J. Burtner, and P. Jones, *Am. J. Clin. Pathol.*, **21**: 711, 1951.

FORMULAS FOR DECALCIFYING FLUIDS

1. 5% NITRIC ACID

Concentrated (68–70%) nitric acid (sp gr 1.41)	50 ml
Distilled water	950 ml

Older directions called for tenfold dilution of the pharmacopeial acid, sp gr 1.25, assay about 500 g HNO_3 per l (Lange). Later some of the German writers prescribed 75 ml of the 68% acid (specific gravity, 1.41) per l, apparently neglecting the specific gravity in their computations. (Schmorl, 1907, 1928; Roulet, 1948; Romeis, 1932, but corrected in 1948.)

2. VON EBNER'S HYDROCHLORIC ACID–SODIUM CHLORIDE MIXTURE

Concentrated hydrochloric acid (sp gr 1.19)	15 ml
Sodium chloride	175 g
Distilled water	1000 ml

During decalcification add 1 ml concentrated hydrochloric acid daily to each 200 ml of the above mixture, until decalcification is complete.

3. RICHMAN-GELFAND-HILL FORMIC ACID–HYDROCHLORIC ACID MIXTURE

Concentrated formic acid (90%)	100 ml
Concentrated hydrochloric acid (38.8%, sp gr 1.19)	80 ml
Distilled water	820 ml

4. 5% FORMIC ACID (ESSENTIALLY 1 *N*)

Concentrated formic acid (90–92%)	50 ml
Distilled water	950 ml

5. KRISTENSEN'S FLUID

1 *N* sodium formate (6.8%)	500 ml
8 *N* formic acid (see Table 20-26)	500 ml

6. FORMALIN FORMIC ACID, ACCORDING TO SCHMORL

Formic acid (90%)	500 ml
Formaldehyde (40%)	50 ml
Distilled water	450 ml

Schmorl recommended this as working rapidly, without producing swelling. Nuclear staining was said to be well preserved.

7. EVANS AND KRAJIAN'S FLUID (*Arch. Pathol.*, **10**: 447, 1930, emended)

Sodium citrate crystals	10 g
90% formic acid	25 ml
Distilled water	75 ml

8.   EVANS AND KRAJIAN'S FLUID, KRAJIAN'S VARIANT

85% formic acid	100 ml
95% ethanol or 99% isopropanol	100 ml
Sodium citrate crystals	20 g
Trichloroacetic acid	1 g
Distilled water	100 ml

9.   MOLAR HCl CITRATE BUFFER OF pH 4.5

1 N hydrochloric acid	540 ml
1 M sodium citrate solution	460 ml

(29.4% of the dihydrate, or 35.7% of the $Na_3(CO_2)_3C_3H_4OH \cdot 5\frac{1}{2} H_2O$)

10.   LORCH'S CITRATE HCl BUFFER OF pH 4.4

Citric acid crystals	14.7 g
0.2 N sodium hydroxide	700 ml
0.1 N hydrochloric acid	300 ml
1% zinc sulfate	2 ml
Chloroform	0.1 ml

11.   NORMAL ACETATE BUFFER OF pH 4.5

Normal acetic acid	520 ml
Normal sodium acetate (8.2% anhydrous; 13.6% crystalline)	480 ml

Add 2 ml 1% zinc sulfate and 0.1 ml chloroform.

12.   NORMAL AMMONIUM CITRATE CITRIC ACID BUFFER OF pH 4.5

Normal citric acid (monohydrate 7%)	50 ml
Normal ammonium citrate (anhydrous 7.54%)	950 ml

Add 2 ml 1% zinc sulfate and 0.1 ml chloroform.

Fullmer and Link (*Stain Technol.*, **39**: 387, 1964) recommend the following demineralization procedure for enzyme histochemical studies:

1.   Slices of bones or teeth made as thin as possible are suspended in a 10% solution of EDTA in 0.1 M tris buffer pH 6.95. KOH pellets are added to bring the EDTA to pH 6.95. The beaker of decalcifying fluid is kept in a cold room at 4°C and slowly and constantly stirred with a magnetic stirrer. Thin sections of bone decalcify within 1 day. Dense bone requires slightly longer. Thin (1 to 2 mm) sections of teeth may be decalcified within 2 to 3 days. Thicker sections of hard teeth require longer periods.
2.   Slices are quick-frozen, mounted on the chuck of a freezing microtome, sectioned, and mounted directly on slides. The knife must be colder than usual. A block of CO_2 ice may be placed on the head of the microtome to increase cooling.

Sections may be used for qualitative studies of many enzymes including acid and alkaline phosphatases, esterases, and many dehydrogenases. Substantial succinic dehydrogenase appears to remain in specimens demineralized as long as 7 days. Balogh (*J. Histochem.*

TABLE 18-1 PERCENTAGE OF CALCIUM REMOVED FROM TIBIAS OF MICE BY DIFFERENT INTENSITIES OF ALTERNATING CURRENT IN DIFFERENT DECALCIFYING SOLUTIONS

Decalcifying fluids	Electric current (ac)				Controls (without current)
	50 mA	100 mA	200 mA	400 mA	
Trichloroacetic acid	26°C 91.3%	29°C 91.9%	30°C 89.1%	40°C 98.5%	25°C 47.4%
Nitric acid	26°C 76.3%	28°C 64.0%	30°C 77.2%	49°C 73.7%	25°C 64.3%
Formic acid	27°C 66.5%	35°C 64.5%	60°C 77.2%	64°C 73.7%	25°C 68.9%
Picric acid	27°C 60.4%	30°C 55.5%	36°C 47.4%	40°C 44.0%	25°C 32.7%
EDTA disodium salt	27°C 29.5%	30°C 43.2%	37°C 33.6%	55°C 56.1%	25°C 29.1%

Source: Goncalves and Olivêrio, *Mikroskopie*, **20**: 154, 1965.

Cytochem., **10**: 232, 1962) obtained good results also using EDTA with phosphate buffer in the cold. For reasons unknown, more rapid demineralization appears to occur with the tris buffer advised above. In other studies, Balogh (*J. Histochem. Cytochem.*, **12**: 485, 1964) demonstrated the presence of several other enzymes in mineralized tissues including cytochrome oxidase, glyceraldehyde 3-phosphate dehydrogenase, dihydroorotate dehydrogenase, and dihydrolipoic dehydrogenase.

Goncalves and Oliverio (*Mikroskopie*, **20**: 154, 1965) decalcified tibia of mice for 2 h in 1 *M* solutions of trichloroacetic acid, nitric acid, formic acid, picric acid, or the disodium salt of EDTA at various temperatures as indicated in Table 18-1. Alternating current supplying 50, 100, 200, and 400 mA was applied to determine the effect on speed of decalcification. The table gives the percentage of calcium removed, which was determined by subtracting the amount of calcium remaining in the bone after decalcification from the amount existent initially; this value was obtained statistically through a regression analysis. It is assumed that the amount of calcium is a function of weight of the tibia. Ten tibias of 10 mice of comparable age were individually weighed on a torsion balance, and the amount of calcium in each one was determined after incineration. Within the range of intensities used, no direct relation between the intensity of the current and speed of decalcification could be discerned. Likewise, temperature did not have a uniform effect on all decalcifying fluids tested.

MATRIX STRUCTURE OF BONES AND TEETH

GROUND SECTIONS OF BONES AND TEETH. These are sometimes required for the study of bone lamellae and canaliculi, and dentin tubules and enamel prisms. They are made from macerated and dried bone.

To macerate bones and teeth, saw into fairly thin pieces and soak in several changes of water over a period of several months. Then wash out and dry thoroughly in air.

Then grind sections until nearly transparent between two pieces of pumice stone with water or between two pieces of plate glass with powdered pumice and water. When sections are thin enough, wash in water and dry.

White (cited from Lee) recommended cutting moderately thin slices of bone or tooth, presumably previously macerated and dried, and soaking in ether for 1 or 2 days. Then soak in collodion colored red with basic fuchsin for 2 or 3 days. Then transfer to 80% alcohol for another 2 or 3 days, and finally grind down nearly to transparency by rubbing between two pieces of ground glass with water and pumice powder. Dry the surfaces and mount.

To mount ground sections, place on a slide a small fragment of solid balsam or other resin, and heat until it just melts. Do the same with a cover glass. Place the ground section in the resin on the cover glass, and press down the resin area on the slide on top.

RESULTS: Lacunae and canaliculi are filled with air or with the colored collodion. Enamel requires preliminary etching with 0.6% hydrochloric acid–alcohol or weak aqueous picric acid and mounting in solid camsal,[1] which has a lower refractive index (1.478).

Matrajt and Hioco (*Stain Technol.*, **41**: 97, 1966) believe that chromoxane cyanin R (C.I. 43820) may be useful as an indicator dye for staining undecalcified bones and teeth, particularly for the portrayal of changes occurring at the "calcification front." Bones are fixed in Lillie's aqueous neutral formalin, dehydrated, embedded in polyester resin, and sectioned at 6 μ on a Jung microtome without decalcification. Without removal of plastic, sections are immersed for 10 min at 24°C in 1% chromoxane cyanin R (E. Gurr) containing 2% glacial acetic acid. Sections are differentiated in warm (30°C) running water until sections turn blue and the water is no longer red. Sections are dehydrated and mounted in synthetic resin.

RESULTS: Osteoid is orange, and a succession of alternate layers of blue and orange may be observed at various calcification fronts. Areas where the first mineral is deposited stain blue and may appear granular.

Andrus and Donikian (*Stain Technol.*, **39**: 56, 1964) state they have been able to cut sections 4 to 14 μ in thickness from unfixed, unembedded frozen tibias and femurs from 5-week-old rats. Legs with the attached muscles were frozen on a block of CO_2 ice and sectioned with a standard International Harris cryostat without injury to the knife.

Some investigations require analysis of mineral content of bones or teeth in nondecalcified sections at the ultrastructural level. Boothroyd (*J. Cell Biol.*, **20**: 165, 1965) called attention to the loss of bone mineral during sectioning and in the water bath while employing the usual methods for electron microscopy. Viscous embedding media do not penetrate bone thoroughly (particularly compact bone), thereby permitting displacement of mineral during cutting and in the water bath. Demineralization and dislocation in 600-Å-thick sections occurred in the water bath within $6\frac{1}{2}$ to 8 min. Although Boothroyd recommended the use of a saturated (calcium and phosphate) solution for the water bath, more work may need to be done to ensure that exchange and dislocation of mineral does not occur.

BONE MATRIX IN PARAFFIN OR NITROCELLULOSE SECTIONS OF DECALCIFIED BONE. This stains pink with hematoxylin-eosin and azure-eosin methods. Remnants of cartilage in newly ossified bone are stained a deep purple with the latter. With Van Gieson's picrofuchsin bone matrix is a deep red. Osteoid tissue is similarly

---

[1] Camsal is an ordinarily liquid mixture of phenyl salicylate (salol) and camphor, in such proportion that a little camphor remains undissolved.

stained, contrasting with the pink to yellow of cartilage in fracture callus. Aniline blue and methyl blue collagen methods stain bone matrix only lightly.

SCHMORL'S THIONIN METHOD. This is sometimes employed for the demonstration of lamellae and canaliculi in compact bone. Formalin fixation is recommended and should be followed by 6 to 8 weeks at 20°C or 3 to 4 weeks at 37°C in Müller's fluid. Then wash for 24 h in running water and decalcify according to preference. Schmorl recommended direct transfer from Müller's fluid to hydrochloric acid–alcohol: hydrochloric acid, 2.5 ml; sodium chloride, 2.5 g; distilled water, 100 ml; 95% alcohol, 500 ml. When decalcification is completed, wash in several changes of 80% alcohol. Then dehydrate and embed in paraffin or celloidin. Cut thin sections.

1. Bring to water as usual.
2. Stain for 5 min in half-saturated aqueous (0.125%) thionin. Staining may be accelerated and intensified by addition of a drop or two of ammonia water to the thionin.
3. Rinse in water.
4. Differentiate for 1 to 2 min in 95% alcohol.
5. Again rinse in water.
6. Transfer to concentrated aqueous phosphotungstic acid. Differentiation is completed in a few seconds, but longer exposure is harmless.
7. Wash for 5 to 10 min in water until sections are sky blue. Longer washing is harmless.
8. Fix the color in a mixture of equal parts of concentrated formalin and water for 1 to 2 h.
9. Dehydrate in two changes of 95% alcohol and two of 100% alcohol, and clear in one change of equal parts 100% alcohol and xylene and two changes of xylene. For celloidin sections substitute 99% isopropyl alcohol for the 100% ethyl alcohol in the three steps in which it is used.

RESULTS: The walls of bone cavities and their processes are blue-black; cells are diffuse blue; nuclei, only little darker than cytoplasm. Nuclear staining may be intensified by insertion of an alum hematoxylin stain after the formalin step. The method is excellent for teeth. Bone matrix is light blue. Cement lines between lamellae are readily seen. Fibrillar structure is distinct.

Powers, Rasmussen, and Clark (*Anat. Rec.*, **111:** 17, 1951) have adapted the Romanes method for the demonstration of dentin tubules. They recommend fixation in a modified Bouin fluid composed of 75 ml saturated aqueous picric acid, 25 ml formalin (37% HCHO), and 1 g trichloroacetic acid; decalcification in 5% trichloroacetic acid in 50% alcohol; and paraffin embedding. Sections are deparaffinized, hydrated, and mordanted for 18 h in 1% (41.4 mM) aqueous cupric nitrate solution [$Cu(NO_3)_2 \cdot 3H_2O$] at 37°C, washed 3 times in distilled water, and put through the Romanes procedure. The solution used in step 1 is adjusted by addition of 16 drops (1 ml) 1% ammonium hydroxide (a 1% dilution of 28% ammonia water?).

G. J. ROMANES' DIAMMINE SILVER CHLORIDE METHOD (*J. Anat.*, **84:** 104, 1950). Dilute 2.9 ml 0.1% $AgNO_3$ with 100 ml distilled water; add 1 ml 0.1% sodium chloride. Adjust to pH 7.8 with a few drops of very dilute ammonia. Incubate sections in this solution 16 h at 58°C in the dark. Drain and transfer directly to 5 g anhydrous sodium sulfite, 1 g hydroquinone, 100 ml distilled water for 5 min at 20°C. Wash well, rinse in distilled water; tone for 10 min in 0.5% gold chloride; distilled water, two changes, for a total of 1 min; reduce for 2 to 3 min in 2% oxalic acid; wash and fix for 3 to 5 min in 5% sodium thiosulfate. Wash, dehydrate, clear, and mount in balsam.

RESULTS: Nerve fibers are black to purple.

Bélanger [*Calcif. Tissue Res.*, **4** (*Suppl.*): 16, 1970] advises the following method for impregnation and demonstration of bone canaliculi: Slices of bone fixed in acetic alcoholic formalin are demineralized in EDTA, paraffin-embedded, and sectioned at 7 μ. Sections are soaked in 5% water-soluble starch (Fischer Scientific Co.) for 30 min, dried, exposed for 1 h to iodine vapor in a closed jar containing Lugol's solution, cleared in toluene, and mounted in synthetic medium. The stain is good for several months and then fades. Canaliculi, lacunar borders, osteoid borders, osteocyte cytoplasms, cartilage matrix, and mucoids in mucous glands are stained. Canaliculi are particularly well delineated.

According to Lillie, mounting in iodine, 0.1% KI, 0.2% Farrants' or Apáthy's media, could conserve the stain longer.

LILLIE'S METHOD OF SILVER IMPREGNATION.   Lillie's method (*Z. Wiss. Mikrosk.* **45:** 380, 1928) with subsequent decalcification distinguishes osteoid tissue and new, uncalcified bone lamellae from older calcified lamellae, and in denser bone often gives a very sharp definition of lacunae and canaliculi. Gomori has used a basically similar method with similar results, but his is somewhat more complicated (*Am. J. Pathol.*, **9:** 253, 1933). Lillie's method follows:

1. Fix in 10% formalin for 2 days or more.
2. Wash thoroughly in several changes of distilled water.
3. Silver for 4 to 5 days at 37°C in 2 to 2.5% silver nitrate solution.
4. Wash thoroughly in distilled water.
5. Decalcify in Ebner's sodium chloride–hydrochloric acid mixture: concentrated hydrochloric acid, 3 ml; saturated aqueous (35%) sodium chloride solution, 100 ml; distilled water, 100 ml. To this add 1 ml hydrochloric acid daily until decalcification is complete.
6. Wash out acid in 2 or 3 changes of half-saturated sodium chloride solution over 4 to 5 days. If the salt solution becomes acid to litmus, add a few drops of diluted ammonia water to neutralize.
7. Wash for 18 to 24 h in running water, harden for 1 to 2 days in 80% alcohol, embed in paraffin or nitrocellulose. Section and counterstain with hemalum and eosin or, better, with Weigert's iron hematoxylin and Van Gieson's picrofuchsin.

RESULTS: Thin trabeculae of bone are completely blackened; thicker bone shows black outer lamellae, with perhaps an unsilvered lamella or two adjacent to the periosteum or endosteum, and deep lamellae respectively pink or red according to counterstain, with black lacunae and canaliculi. Calcified areas in cartilage and other calcium phosphate deposits also blacken. Dentin is blackened; enamel is partly blackened.

Actually, like the von Kóssa method, this demonstrates phosphates rather than calcium salts, but since soluble phosphates are first washed out, it is essentially calcium phosphate that is demonstrated by both methods.

The following revised technic of Lillie's has been applied to the study of calcified atheroma and lipids therein: Four blocks of formol-fixed atherosclerotic artery containing partly calcified plaques are selected. Block 1 (control) is immersed 7 days in distilled water, changing a few times on the first day. Blocks 2, 3, and 4 are washed in six changes of distilled water, for 20 min each, and then placed in 1.7% (0.1 *M*) silver nitrate and kept in the dark. Fresh silver nitrate is used at 2, 5, and 7 days. Block 2 is removed at 5 days, block 3 at 7 days, and block 4 at 10 days. The blocks are then washed in six changes of distilled water, for 20 min each, and transferred to 2 g sodium bromide in 94 ml distilled water for 3 h. Then 6 ml glacial acetic acid is added. This 2% sodium bromide–6% glacial acetic acid decalcifying fluid is changed daily, and 1 ml of the discarded fluid is tested with 1 ml 2%

sodium oxalate. Decalcification is regarded as complete when no turbidity develops within 2 min with the oxalate. Each block is then divided by a vertical cut through the center of the lesion. One half is used for paraffin sections, one for frozen sections.

Frozen sections are stained with oil red hemalum, with Nile blue, with Schulz and Lewis-Lobban cholesterol technics, and such other lipid methods as may be desired. The paraffin sections may be used for the usual collagen and elastin technics, for oil red and Sudan black stains for insoluble lipids, as well as routine hematoxylin and eosin.

Silver may be removed from any individual section by a 20-min exposure to Weigert's iodine followed by 20 min in 2% sodium thiosulfate or by 20 min in 2% sodium or potassium cyanide solution.

Otherwise calcified areas are black.

Lansdown (*Histochemie*, **13:** 192, 1968) recommends the following stain for detection of sites of calcification and acid mucopolysaccharides in developing cartilage and bone:

Dewaxed and hydrated sections are treated with 1% $AgNO_3$ in bright sunlight for 60 min, thoroughly washed in distilled water, and rinsed for 10 min in Kolthroff's buffer (sodium tetraborate–succinic acid, pH 3.3). Sections are then drained and stained in 0.25% toluidine blue for 1 to 5 min at 24°C, rinsed in the above buffer, dehydrated, and mounted in synthetic medium.

RESULTS: Nuclei are blue, cytoplasms are light blue, cartilage matrix is lilac to deep mauve, and calcium deposits are intensely black.

LABELING OF BONES AND TEETH

For many years it has been known that the bones of animals fed madder (an herb containing alizarin) become colored. Alizarin red S (C.I. 58005) is widely used, but many dyes of various colors may be employed to mark sites undergoing mineralization at the time of injection. Some dyes (such as the alizarin group and the tetracyclines) are lost during demineralization. Others such as the procions remain after decalcification.

Alizarin red S (C.I. 58005) may be used to stain a fetal skeleton. Jensh and Brent (*Stain Technol.*, **41:** 179, 1966) conducted a survey of the literature of previous methods and also a laboratory study of several methods for the preparation of stained rat skeletons. Ideal results include a well-stained skeleton with other tissues transparent. Table 18-2 from Jensh and Brent details several methods advocated. They recommend the following method for fetal or newborn rats after conducting many studies.

1. Remove skin, eyes, subcutaneous fat, and viscera.
2. Fix for 7 to 14 days in nonbuffered 10% formalin.
3. Dehydrate in acetone (two changes of 30 min and 12 h). Use at least 8 ml fluid per g specimen here and in steps 4 and 5.
4. Transfer to 1% aqueous KOH to which alizarin red S (6 mg/1) has been added for 3 days 24°C.
5. Place specimens in 10% KOH for 3 days 24°C.
6. Wash in cold water for 2 min.
7. Place specimens in benzyl alcohol, ethanol, and glycerol, 1 : 2 : 2 by volume, for 12 h at 24°C. (Use 4 ml fluid per g specimen.)
8. Transfer to pure glycerol for storage. (Use 4 ml glycerol per g specimen.)

RESULTS: Bones are red and tissues are transparent.

Hong et al. (*Calcif. Tissue Res.*, **2:** 286, 1968) injected several dyes to follow bone growth in rabbits. Alizarin red S at 10 mg/kg animal weight reduced the rate of bone growth and at

TABLE 18-2 PREVIOUSLY PUBLISHED STAINING PROCEDURES FOR FETAL BONE

Author	Fixation: chemical and time (days)	Maceration: KOH% and time (days)	Staining: alizarin red; concentration time	Clearing: fluids and time (days)	Total time, days
Schultze (*Anat. Ges.*, **11**:3, 1897)	95% ETOH(*)	3 × (*)	(*), (*)	(*), (*)	?
Mall (*Am. J. Anat.* **5**: 433, 1906)	Form. (30)	10 (*)	(*), (*)	95% ETOH-glyc. (4–60)	35–95
Dawson (*Stain Technol.*, **1**: 123, 1926)	95% ETOH (2–3)	1 (2–3)	1:10,000 in 1% KOH until stained	95% ETOH-glyc. (12–19)	16–25
Lipman (*Stain Technol.*, **10**: 61, 1935)	95% ETOH (*)	2 (until clear)	(*), 1 day	95% ETOH-glyc. (8)	?
Richmond and Bennett (*Stain Technol.*, **13**: 77, 1938)	95% ETOH (14) then 1% K$_2$CO$_3$ (28)	1 (10), then 20% glyc. and 1% KOH	0.1, 7–14 days	95% ETOH-glyc. H$_2$O (7)	67–74
Cumley et al. (*Stain Technol.*, **14**: 7, 1939)	95% ETOH (2–3)	1 (7)	(8), 3–12 h	1% KOH (7); 95% ETOH (4); toluene (4); naphthalene (4); anise oil (4)	33
Williams (*Stain Technol.*, **16**: 23, 1941)	10% form. (7)	2 (5–7)	drops in 2% KOH (24 h)	Cellosolve (0.75), to methyl-salicylate (3)	17–19
Noback and Noback (*Stain Technol.*, **19**: 51, 1944)	95% ETOH (14–60) (†)	2–4 with alizarin red drops (3–21)	....	Increasing conc. of glyc.	17–81
Hood and Neill (*Stain Technol.*, **23**: 209, 1948)	95% ETOH (3) or form. (*)	2 (3) 5 (*)	(*), 6–12 h	2% KOH (1), to glyc. (4–5)	12–13
Sedra (*Stain Technol.*, **25**: 223, 1950)	10% form. (7)	2% NaOH (3–7)	(*), 12 h	NaOH-glyc. (6)	17–20
St. Amand and St. Amand (*Stain Technol.*, **26**: 271, 1950)	95% ETOH or form. (2)	Boiling, 10 min	(*) (*)	(*) (*)	?
Green (*Ohio J. Sci.*, **52**: 31, 1952)	No fixation	1 (5)	1% KOH: 0.5% alizarin red, 5 days	Glyc. (storage)	10
Crary (*Stain Technol.*, **37**: 124, 1962)	No fixation	1% KOH: drops 5% alizarin red (3): repeat (3)	....	Alc.; ETOH; glyc. (1:2:2), 1 day	7
Staples and Schnell (*Stain Technol.*, **39**: 61, 1964)	(*), then acetone (0.5)	1 (1.5) with 6 mg/l alizarin red	....	Benzyl alcohol; ETOH; glyc. (1:2:2) 12 h	2.5
Burdi (*Stain Technol.*, **40**: 45, 1965)	Form. (†)	1 (3) with 2–3 acetoalcohol 40 min	....	ETOH; glyc. (1:1) until clear	8–12

* Data not given.
† Toluidine blue portion of procedure omitted.
Note: Abbreviations: form., formalin; glyc., glycerol; ETOH, ethyl alcohol
Source: Jensh and Brent, *Stain Technol.*, **41** : 179, 1966.

TABLE 18-3 COMPARISON OF EFFECTIVENESS OF DYES STUDIED

Staining	Blue H-GRS	Blue M-RS	Green H-7GS	Green M-3GS	Yellow H-4GS	Yellow M-4GS
Gross staining of animal	+	+	+	+	+	0
Gross fluorescence of animal	0	0	0	0	0	0
Gross staining of bone	+	+	+	+	+	+
Gross fluorescence of bone	0	0	0	0	0	0
Microscopic soft tissue staining	0	0	0	0	0	0
Microscopic soft tissue fluorescence	0	0	0	0	0	0
Microscopic bone staining	+	+	+	+	0	0
Microscopic fluorescent bone staining	0	0	0	0	0	0
Microscopic staining of dentin	0	0	0	0	0	0
Microscopic fluorescent dentin staining	0	0	0	0	0	0

Source: Prescott et al., *Am. J. Phys. Anthrop.*, **29** : 219, 1968.

Note: In the third edition of the "Colour Index" the final S of all the above suffixed letter symbols is dropped, but the prefix Procion is always retained.

150 mg/kg caused complete cessation of bone growth. Trypan blue at 100 mg/kg and chlortetracycline at 50 mg/kg caused a decrease of osteogenesis, and trypan blue at 350 mg/kg caused a complete arrest of bone formation.

Toxicity develops in rabbits given higher doses of vital stains. Yen et al. (*J. Dent. Res.*, **50:** 1666, 1971) noted reduction in the rate of dentin formation with alizarin red S administered at 10 mg/kg rabbit weight. Trypan blue could be given at 100 mg/kg without adverse effect, but dentinogenesis is arrested completely at a dose of 350 mg/kg.

Adkins (*Stain Technol.*, **40:** 69, 1965) notes that sometimes alizarin red S uptake in bones and teeth may be insufficient to be observed by the usual light microscopy; however, the uptake can be readily observed under ultraviolet light.

Goland (*J. Dent. Res.*, **43:** 70, 1965) in an abstract noted that certain Procion dyes had the ability to be deposited in sites undergoing calcification and to persist in bones and teeth after demineralization. The various tetracyclines are known to be localized in sites undergoing calcification, and these sites can be observed in mineralized sections under ultraviolet light. Prescott et al. (*Am. J. Phys. Anthrop.*, **29:** 219, 1968) tested several procion dyes indicated in Table 18-3 for their toxicity and efficacy of localization to sites undergoing calcification. They recommend Red H-8BS[1] and Scarlet M-GS[2]. Administered as 100 mg/kg doses intraperitoneally to rats, cats, dogs, and monkeys, the dyes caused minimal, if any, toxicity. The dyes do not cross the placental barrier in cats, mice, and rats. Discrete localization in sites undergoing calcification was observed, and the dyes were retained in demineralized sections. The dye could be visualized by either light or ultraviolet microscopy. Pink hematoporphyrin was fairly useful except that it was fairly toxic to animals. The procion dyes are reactive mono- and dichloro-S-triazinyl compounds obtained from ICI (Imperial Chemicals Industries, 55 Canal Street, Providence, Rhode Island 02901).

[1] C.I. Reactive Red 31, constitution unrevealed.
[2] C.I. 17908, Reactive Red 8.

Procion scarlet MG, C.I. 17908

Red M-8BS	Scarlet H-RS	Yellow tetracycline	Blue H-5GS	Purple H-3RS	Red H-8BS	Scarlet M-GS	Pink hematoporphyrin
0	+	0	++	++	++	+	0
0	0	0	0	0	+	+	++
+	+	0	++	++	++	++	0
+	+	++	0	0	+	+	+
+	0	0	++	+	++	+	0
+	0	0	0	0	++	+	+
++	0	0	++	++	++	+	0
+	0	++	0	0	++	+	++
0	0	0	++	++	++	++	0
0	0	++	0	0	++	++	++

Goland and Grand (*Am. J. Phys. Anthrop.*, **29**: 201, 1968) prepared several Procion M and Remazol dyes as 2% solutions in 5% dextrose and thereafter administered them intravenously to rabbits at a concentration of 1 ml/100 g animal weight. The dyes were concentrated at tissue sites undergoing calcification, and dyes were retained in tissues after decalcification. The triazinal dyes were obtained from ICI and Organics Inc., 151 South Street, Stamford, Connecticut 06904. Seiton and Engel (*Am. J. Anat.*, **126**: 373, 1969) conducted similar studies. Sometimes dyes with different colors are injected at different times (Cleall et al., *Arch. Oral Biol.*, **9**: 627, 1964).

Rahn and Perren (*Stain Technol.*, **46**: 125, 1971) provide the following formula for xylenol orange (available from Siegfried Ltd., Zofingen, Switzerland). Rahn and Perren advise an intraperitoneal or intravenous injection of a 3% aqueous solution at a dose of 90 mg/kg to rabbits and rats. The dye fluoresces orange at sites undergoing calcification. It is therefore useful if multiple injections are made. Rahn et al. (*Eur. Surg. Res.*, **2**: 137, 1970) conclude that xylenol orange is less toxic than alizarin red S, hematoporphyrin, and tetracycline.

Xylenol orange

Lead acetate can be used as a vital stain to mark sites undergoing calcification at the time of administration of the lead. Lead remains after demineralization and is converted to a sulfide. Okada and Mimura (*Jap. J. Med. Sci. Pharm.*, **11**: 116, 1933) appear to be the first to have employed lead acetate as a vital stain. Several methods have been advocated since that time. The following method by Schneider (*Am. J. Phys. Anthrop.*, **29**: 197, 1968) is an example.

1. Inject rabbits or rats with lead acetate, 1 to 4 mg/kg body weight.
2. Fix tissues in 10% formalin for 48 h.

3. Decalcify in 1% HCl into which hydrogen sulfide gas is bubbled before and during decalcification. The HCl should be saturated with H_2S.
4. Embed decalcified specimens in 10% gelatin for 24 h, 20% gelatin for 24 h, and 30% gelatin for 24 h. Harden 30% gelatin blocks in 10% formalin at 4°C for 24 h.
5. Section on a freezing microtome at 20 to 30 μ.
6. Place sections in distilled water for 5 min.
7. Place sections in 0.1% aqueous gold chloride for 10 min.
8. Wash briefly, and place sections in 5% aqueous sodium bisulfate for 10 min at 24°C.
9. Transfer sections to distilled water and observe for 1 to 3 h until lines become the desired degree of brown to purplish black. Bone becomes light pink.
10. Place sections in 5% aqueous sodium thiosulfate for 5 min.
11. Wash sections, and mount in an aqueous medium such as Kaiser's.

RESULTS: Sites undergoing mineralization at the time of the injection of lead acetate are stained. With suitable timing, multiple injections can be made. With multiple injections the investigator must be mindful of the release of lead during normal bone turnover.

Pizzolato and Lillie (*Histochemie*, **16:** 333, 1968) tested 225 metal salts and determined that 34 of them can enter the calcium carbonate-phosphate-protein complex of bone. Of these only mercurous nitrate and perchlorate and the water-soluble salts of silver blackened areas of decalcified, paraffin-embedded sections of bone.

TETRACYCLINE FLUOROCHROME STAINING IN VIVO.   Milch, Tobie, and Robinson (*J. Histochem. Cytochem.*, **9:** 261, 1961) explain the specific golden yellow fluorescence of newly deposited calcium in bone, cartilage, and calcifying neoplasms in tetracycline treatment as a chelation with Ca^{2+} ions which they formulate as binding 2 mol of the drug in vitro, but in tissue they consider the Ca^{2+} as bound covalently to two phosphate groups which are bound to collagen protein chains and chelate with the quinone oxygen of the drug by secondary valences. One might speculate that a Ca^{2+} ion bound to only one phosphate radical could form a more effective ring chelate complex.

TECHNIC.   The technic is simple. Tissue from tetracycline-treated patients is quickly frozen in petroleum ether with solid CO_2 or dry-ice–alcohol mixture and kept at -20°C in a refrigerator until sectioned. Cryostat sections are prepared on a rotary microtome at 2 to 8 μ, cutting at -15°C. Sections are transferred directly to glass slides, caused to adhere by thawing with a finger on the back of the slide, removed from the cryostat, and dried for 30 min in an airstream. They are then mounted in glycerol and examined in 365 mμ light using a Corning 5840 2.25-mm filter in the light source and a Wratten 2A eyepiece filter to transmit blue and yellow light. Control hematoxylin-eosin and alizarin red S stains are recommended.

Newly deposited calcium fluoresces golden yellow on a blue autofluorescence background (collagen).

Malek et al. (*Chemotherapia*, **5:** 269, 1962), Tapp et al. (*Experientia*, **20:** 393, 1964), and Tapp et al. (*Stain Technol.*, **40:** 199, 1965) note that tetracyclines localize not only at sites undergoing calcification, but also in sites of necrosis.

Frost (*Can. J. Biochem.*, **41:** 31 and 1307, 1963; *J. Bone Joint Surg.* [*Am.*], **48:** 1192, 1966; *Am. J. Phys. Anthrop.*, **29:** 183, 1968) has used tetracycline labeling to measure the rate of bone formation at the level of the osteon in man or animals. The method involves the administration of tetracycline, acquisition of a sample at a known time interval, and preparation of an undecalcified ground section. A measurement of the amount of new bone formed in individual osteons is made and multiplied by the number of osteons actively producing new bone per unit area to provide a figure for rate of bone formation. Differences were observed between normal individuals and those with bone diseases, such as senile osteoporosis, osteogenesis imperfecta, and vitamin D–resistant rickets.

ENZYME DETECTION   For the qualitative histochemical demonstration of a wide variety of hydrolytic and oxidative enzymes, EDTA solutions in the cold with agitation are far superior to all others available today. Unfixed sections are employed for staining.

Fullmer et al. (*Stain Technol.*, **39:** 71, 1964) recommend the following stain for the simultaneous demonstration of bone formation (alkaline phosphatase) and bone resorption (succinic dehydrogenase in osteoclasts):

1. Bones or teeth may be demineralized with the EDTA method according to Fullmer and Link or the Balogh (*J. Histochem. Cytochem.*, **10:** 232, 1962) method. The Balogh method uses 0.1 *M* phosphate buffer. Unfixed sections cut with the cryostat are used.
2. Place slides directly in the succinic dehydrogenase incubating solution for 30 min at 37°C. The incubating solution is made by mixing equal volumes of solutions A and B. Solution A is a 1 : 1 mixture of 0.2 *M* sodium succinate and 0.15 *M* phosphate-phosphate buffer of pH 7.6. Solution B is a solution of either nitro BT or tetranitro BT (Nutritional Biochemical Corp., Cleveland, Ohio) at a concentration of 1 mg/ml in distilled water.
3. Rinse in distilled water.
4. Incubate for 20 min at 25°C in the naphthol AS-MX phosphate mixture according to the method of Burstone for alkaline phosphatase. Prepare this mixture as follows: Dissolve 5 mg naphthol AS-MX phosphate in 0.2 ml dimethylformamide. Add 25 ml 0.2 *M* tris buffer at pH 8.3, 25 ml distilled water, and 30 mg Red Violet LB salt (Sigma Chemical Co., St. Louis, Missouri). Stir the mixture until the reagents dissolve; then filter. The incubation period for alkaline phosphatase may be varied to suit the activity of individual specimens and the investigator's requirements.
5. Rinse sections in distilled water, and mount in a water-soluble medium such as Kaiser's. The sections may be dehydrated; however, both stains are slightly to moderately soluble in alcohol and xylene; therefore, prolonged dehydration must be avoided.

RESULTS: When tetranitro BT is used, osteoclasts stain intensely brownish black at sites of succinic dehydrogenase activity. When nitro BT is used, osteoclasts are stained purple. Muscle and a number of other tissues stain intensely for succinic dehydrogenase. However, in bone, osteoclasts are discretely stained purple or black against a red background. The red stain indicates alkaline phosphatase activity, displayed principally by osteoblasts. A glance at low-power magnification permits the investigator to visualize areas of both bone resorption and formation.

LORCH'S METHOD FOR ALKALINE PHOSPHATASE AND PREFORMED PHOSPHATES (*Q. J. Microsc. Sci.*, **88:** 367, 1947)

1. Fix for 12 to 24 h in 80% alcohol, dehydrate in ascending alcohols, and clear and infiltrate in paraffin. We suggest the petroleum ether–vacuum infiltration sequence. Section at 8 to 10 μ. The method is applicable only to developing bones which may be sectioned successfully without decalcification.
2. Deparaffinize and hydrate as usual.
3. Immerse in 2% cobalt nitrate for 5 min.

4. Wash in three to five changes of distilled water, for 1 to 2 min each.
5. Immerse for 30 s in freshly diluted, 1% ammonium sulfide.
6. Wash for 5 min only in tap water.
7. Incubate for 20 min to 15 h at 37°C in this substrate:

2% calcium nitrate [Ca(NO$_3$)$_2$·4H$_2$O]	10 ml
2% magnesium chloride (MgCl$_2$·6H$_2$O)	10 ml
4% sodium β-glycerophosphate·5H$_2$O	10 ml
1% sodium barbital	70 ml
Ammonium sulfide	1 drop

8. Wash in 1% calcium nitrate.
9. Immerse for 10 min in saturated aqueous gallamin blue (C.I. 51045)[1] buffered to pH 7 (we suggest Michaelis' Veronal sodium).
10. Rinse for 10 to 12 s in 0.5% sodium hydroxide.
11. Wash in water, dehydrate, clear, and mount as usual. Interpose counterstain if desired: safranin, methyl green, neutral red, eosin, or orange G.

ALBERT-LINDER BONE SECTIONING FOR ALKALINE PHOSPHATASE. (*Stain Technol.*, **35**: 277, 1960). Fix for 2 h in 80% alcohol, and dehydrate for 56 h in several changes of absolute alcohol, all at 4°C. Clear for 24 h at 4°C in two changes of benzene. Infiltrate for 2 h in vacuo at 48 to 50°C in tropical ester wax (British Drug Houses, Inc.).

Cut blocks on a base sledge microtome using a Jung extrahard steel knife with a tool-edged profile. Knife slant may vary between 45 and 90° to direction of cutting; tilt should be about 10°. Cut at normal speed.

Press a piece of Scotch tape larger than the section onto the block face before each cut. Lay pieces of tape sticky side up on glass plate. Coat each section with 2% celloidin and return it to its place. Dry for 30 min. Number sections with "brushing cellulose paint"; staple together in sequence and store at 4°C.

When ready to use, dissolve off Scotch tape adhesive with several changes of chloroform. This hardens the celloidin, and the sections separate readily. The wax is removed at the same time. The sections are now transferred to 70% alcohol and can be handled as ordinary celloidin sections. Albert and Linder used Gomori's 1952 azo method; more recent methods are undoubtedly applicable. The Burstone and Barka variants seem applicable; and the procedure can probably also be used with the cobalt sulfide technics.

Although no such procedure appears to have been tested as a whole, current trends in soft tissue alkaline phosphatase technics would indicate the advisability of trying alcohol or acetone freeze substitution at −70°C, followed by pH 7 Versene or pH 5 formic acid–sodium citrate decalcification, according to Schajowicz and Cabrini, and frozen sections according to the more recent soft tissue technics (Chap. 10).

Göthlin and Ericsson (*Histochemie*, **36**: 225, 1973; *Isr. J. Med. Sci.*, **7**: 488, 1971) studied the localization of alkaline phosphatase and ATPase (*Histochemie*, **35**: 111, 1973) in fracture callus of rats.

ACID PHOSPHATASE IN BONE. For acid phosphatase in developing bone and cartilage Schajowicz and Cabrini (*Stain Technol.*, **34**: 59, 1959) compared fixations in slightly acid and neutral 10% formalin, Baker's chloral formalin, in acetone and 80% alcohol, with decalcification in 5% Versene at pH 7, and in buffer mixtures of acetic acid and sodium hydroxide at pH 4.2.

---

[1] Celestin blue B, C.I. 51050, can probably be substituted, if gallamin blue is not available.

PHOSPHORYLASE

Cobb (*Arch. Pathol.*, **55:** 496, 1953) reports a technic for the histochemical localization of phosphorylase activity in young rat tibiae. The bones from freshly killed rats were split lengthwise and frozen in isopentane at −150°C. They were then desiccated in vacuo at −30°C for at least 3 days. Next followed embedding in vacuo for 15 min in the chloroform-soluble fraction of carnauba wax[1] and sectioning in this very hard wax without decalcification. Sections were dewaxed in chloroform.

COBB'S PHOSPHORYLASE TECHNIC

1. Dewax for 10 min in chloroform, thus also partially inactivating the bone alkaline phosphatase.
2. Digest for 90 min in α amylase, saliva, or diastase at 37°C to remove glycogen.
3. Soak overnight in distilled water to dissolve out all soluble bone salts.
4. Inactivate phosphoglucomutase by heating in a few drops of water in a moist chamber. (In vitro this enzyme resists heating at 65°C.)
5. Incubate for 6 h at 37°C in pH 5.9 acetate buffer containing a "high" concentration of glucose 1-phosphate and an unstated amount of barium chloride to precipitate liberated phosphate ions.
6. Fix glycogen overnight in 100% alcohol. Stain glycogen by the periodic acid Schiff procedure. Cobb used the Hotchkiss A (?) variant. For controls omit steps 3 to 5 above. The excess over the controls represents glycogen newly formed as a result of phosphorylase activity.

Although we have not tried this method and as published it lacks some necessary details, the rationale seems good and the procedure is worthy of trial. (See also Takeuchi, whose method is described earlier.)

GLYCOGEN

Glycogen also presents special problems in its preservation *in situ* during decalcification procedures. Total losses have been reported during acid decalcification, and even Versene demineralization does not fully protect it (Trott, *J. Histochem. Cytochem.*, **9:** 699, 1961; Schajowicz and Cabrini, ibid., **3:** 122, 1955).

Although 24-h, 25°C fixations in such fluids as acetic alcohol–formalin, Rossman's and Gendre's fluids, and the like seem to give about as full preservation of soft tissue glycogen as does freeze-drying, the glycogen so precipitated remains relatively susceptible to aqueous extraction in the presence of acids. However, it was observed that in material primarily fixed in aqueous formalin deliberately acidified with acetic or formic acid the glycogen in muscle and marrow cells withstood decalcification much better.

This suggested (*J. Histochem. Cytochem.*, **10:** 763, 1962) that glycogen could be preserved in tissue attached to bone during decalcification by achieving a more thorough fixation of the protein surrounding it or by prior embedding in celloidin. Both these procedures should create semipermeable membranes surrounding the precipitated glycogen particles and thereby hinder re-solution and diffusion. In practice these procedures proved successful in preserving soft tissue glycogen in virtually undiminished amount during formic acid decalcification.

---

[1] Carnauba wax is used industrially in furniture and floor waxes and should be procurable through the makers of these products.

DECALCIFICATION PROCEDURES FOR THE SUBSEQUENT DEMONSTRATION OF LABILE GLYCOGEN (Lillie)

CELLOIDIN PROCEDURE

1. Fix for 24 h at 25°C or 3 to 4 days at 5°C in acetic alcohol–formalin or for 6 h at 60°C in methanol-chloroform, 65 : 35.
2. Complete dehydration with 95% and absolute alcohol and infiltrate for 3 days in 1% celloidin in alcohol plus ether (50 : 50).
3. Transfer to 80% alcohol overnight to harden celloidin.
4. Decalcify with daily changes of 5% formic acid in water, testing discarded fluid with 1% sodium oxalate for the presence of calcium ions.
5. As soon as a negative test is obtained, wash for 6 to 8 h in running water, dehydrate, clear, embed, and section in paraffin as usual.

HARD PROTEIN FIXATION PROCEDURE

1. Fix as above in acetic alcohol–formalin or in Gendre's fluid.
2. Transfer to Bouin's fluid for 3 days at 25°C.
3. Decalcify in daily changes of 10% formalin containing 5% formic acid, testing as above.
4. When a negative oxalate test is obtained, wash for 8 h in running water, dehydrate, clear, embed, and section in paraffin as usual.

CARTILAGE

Dixit (*Calcif. Tissue Res.*, **10**: 49, 1972) conducted a quantitative study of alkaline phosphatase and glucose 6-phosphate dehydrogenase in the growth plates of rachitic and normal rat tibiae. Pertinent regions of freeze-dried sections were weighed on a quartz fiber fish pole balance, and microassays were conducted. Alkaline phosphatase activity in the normal hypertrophic zone of cartilage is about 3 times greater than that in the proliferative zone, and about 4 times greater than in the rachitic rat. Following a single dose of vitamin D (8000 IU), a significant elevation of alkaline phosphatase occurs in the proliferative zone, and a decrease of the enzyme in the central region of the hypertrophic zone; however, the effects of vitamin D were minimal in rats fasted for 48 h. In normal rats, glucose 6-phosphate dehydrogenase is greater in the proliferative than in the hypertrophic zone, and in rachitic rats enzyme content is nearly the same in both regions. Although a reduction in glucose 6-phosphate dehydrogenase activity occurred in both zones in rachitic rats, no effect was observed in fasted rats. Dixit does not attempt an explanation of the disparity in the effects of fasting on the quantity of enzyme detected.

Yoshiki et al. (*Histochemie*, **29**: 296, 1972) note that incubation of demineralized sections with 100 mM $MgCl_2$ in tris maleate or cacodylate buffer for 1 h results in a reactivation of alkaline phosphatase (α-glycerophosphate) far in excess of that observed if the Mg is added to the enzyme incubation medium.

Thyberg (*J. Ultrastruct. Res.*, **38**: 332, 1972) used Goldfischer's (*J. Histochem. Cytochem.*, **13**: 520, 1965) method to localize aryl sulfatase intracellularly in dense bodies and in portions of the Golgi apparatus in all cartilage cells of growth plates of guinea pigs. Extracellularly, aryl sulfatase activity was found in type I vesicles (location not specified).

Hirschman and McCabe (*Calcif. Tissue Res.*, **4**: 260, 1960) note that basic dyes such as azure A stain intracellular granules either orthochromatically or metachromatically in cartilage cells of the proliferative zone in fresh, unfixed, nonfrozen, nondecalcified, hand-cut sections at 0.1 to 4 mm in thickness. The granules cannot be detected in frozen or demineralized sections or in sections prepared by usual histologic methods.

Hall et al. [*J. Microsc. (Oxf.)*, **99:** 177, 1973] used quantitative x-ray microanalysis of calcium in ultrathin sections of cartilage with vesicles in areas undergoing calcification.

Anderson and Sajdera (*J. Cell Biol.*, **49:** 650, 1971) were able to extract as much as 85% of total proteoglycan from 0.5-mm-thick slices of bovine nasal cartilage (as determined by hexuronic acid assay) with either 1.9 M CaCl$_2$ for 48 h at 24°C or 4 M guanidinium chloride for 24 h at 24°C. Both were buffered to pH 5.8 with 0.05 M N-morpholinoethane sulfonic acid–NaOH.

19

VARIOUS SPECIAL PROCEDURES

VASCULAR INJECTIONS

The practice of vascular injection with colored masses has had wide application in the study of the circulatory system of various organs. The Berlin blue and carmine gelatin masses are classic. They serve well for gross preparations and those intended for clearing and study under low magnification. With both of these some color loss occurs during paraffin embedding, and a carbon mass has been found more satisfactory for strictly histologic purposes.

CARBON GELATIN MASS. *Ashburn and Endicott*, in Lillie's laboratory, devised a carbon gelatin based on Mall's statement that he used such a mass for study of the circulation of the liver. This mass must be freshly prepared each time, though it might be kept frozen for a few days if desired. The carbon particles are apparently held back by the capillaries so that either the afferent or efferent vessels may be injected without filling the other. This technic has been used in the study of lobular relationships of destructive and cirrhotic processes in rat livers. The technic follows:

Dissolve 8 g gelatin in 100 ml warm water. Sift animal charcoal through a 100-mesh sieve, and suspend 20 g of the sifted charcoal in the gelatin solution in a conical glass of perhaps 200 ml capacity immersed in a water bath at 45 to 50°C. While preparing the animal for injection, keep the gelatin mass agitated with a small propeller-type glass stirrer driven by a small electric motor. The propeller is directed downward and is kept fairly near the bottom of the conical glass.

Cannulae are made of glass with a slight bulbous enlargement on the end so that they can be tied in. It is preferable to have the opening on the side of the bulb, so as to avoid tears in the veins when introducing the cannula. At the same time, the opening must be distal to the constricted area about which the tie is made.

For injection of the hepatic veins, remove the thoracic and abdominal walls to expose the liver fully, tie off the vena cava below the liver, sever the portal vein, and insert the cannula into the right atrium and thence downward into the inferior vena cava. Then pass a ligature around the vena cava, and tie proximal to the bulbous end of the cannula. Then wash out the blood from the liver with warm saline solution from a 20- to 50-ml Luer syringe attached by a 15-cm (6 in) length of thin-walled rubber tubing to the cannula, carefully avoiding introduction of air. When clear saline solution flows from the portal vein, substitute a 10- to 20-ml Luer syringe filled with the warm carbon gelatin, and inject. In changing the syringes, carefully avoid air bubbles. The surface of the liver blackens quickly.

Gentle digital massage of the liver surface during injection facilitates both removal of blood and penetration by the mass.

Injection of the portal system is done in a similar manner, by placing the cannula in the portal vein below the liver and opening the inferior vena cava above it.

With larger animals or human organs, some type of gravity perfusion apparatus with a heating arrangement to maintain temperature is required. A manometer can readily be attached to the injection system with a T tube, and the pressure can be regulated by varying the height of the reservoir of the gravity apparatus.

Some workers use 0.5% ammonium oxalate and 0.75% sodium chloride in the fluid used for washing out the blood before injection. This tends to prevent clotting. The inclusion of 0.2% sodium nitrite, in either the washing or injection fluid, will promote vascular dilation.

Ashton (*Brit. J. Ophthalmol.*, **34**: 38, 1950) injects india ink into the retinal vessels. The long nerve end of the freshly enucleated eye is dipped in formalin for a few minutes. Then, under a dissecting microscope, the optic nerve is cut across, and a fine glass cannula is inserted into the isolated central artery. The vessels are first perfused for a few minutes with distilled water at about 500 mm mercury pressure. This perfusion is followed by one with 10% formalin, and then by the india ink. The eyeball is then immersed in formalin for 12 h before opening. Although Ashton's procedure was intended for gross display of the retina, Ashburn's experience with the liver indicates that the material can be used for histologic purposes.

FISCHER'S MILK METHOD (*Zentralbl. Allg. Pathol.*, **13**: 977, 1902).   Since these carbon masses do not penetrate the capillaries, and this is perhaps their greatest virtue, it is necessary to resort to other methods for demonstration of the capillaries. One may wash out with salt solution as above and then inject milk or cream, tie off both afferent and efferent vessels, and fix in 7.5 ml formalin, 1.5 ml glacial acetic acid, and 100 ml distilled water. Cut frozen sections and stain with Sudan III or IV (Fischer) or oil red O as usual, and mount in Apáthy's gum syrup. Or one may use one of the carmine or Berlin blue gelatin masses.

CARMINE GELATIN MASSES.   These are made by dissolving carmine in water with heat and ammonia water, and then either first mixing with the gelatin solution and then neutralizing, or the converse. Overacidification with acetic acid is likely to occur and ruins the mass by producing a granular red precipitate. Mallory avoids this difficulty by driving off the excess ammonia by heat. We suggest the combination of gentle heat and blowing an air current over the surface of the carmine or carmine gelatin until ammonia is no longer detectable by odor or with moist litmus or nitrazine paper. Most of the directions for carmine gelatin masses are very vague as to the amounts of gelatin and of carmine in the final mixture. Robin alone among the authors cited in Lee prescribed 1 part gelatin and 7 to 10 parts of water for an aqueous gelatin mass, and 50 g gelatin, 300 ml water, and 150 ml glycerol for a glycerol gelatin mass. He dissolved carmine in ammonia water and diluted it in glycerol and neutralized with acetic acid in glycerol, attaining perhaps a 3 to 4% stock solution of carmine, which was diluted with 3 or 4 volumes of either of the above masses.

BERLIN BLUE MASSES.   These we have generally found too pale for histologic use. Robin prescribed (solution A) 90 ml (?) saturated aqueous (28 to 30%) potassium ferrocyanide solution plus 50 ml glycerol and (solution B) 3 ml (30 ml?) pharmacopoeial (French) ferric chloride solution (about 26%) plus 50 ml glycerol. These were mixed slowly, a few drops of hydrochloric acid were added, and the mixture was then combined with 3 volumes of one of his gelatin vehicles at 45 to 50°C. This is nearly a fifteenfold excess of ferrocyanide. We suggest 17 g crystalline potassium ferrocyanide, 85 ml distilled water, and 50 ml glycerol for solution A above and 30 ml USP or PG ferric chloride solution in B.

Or one may make a saturated aqueous "solution" of "soluble Berlin blue" at 60°C and mix with 3 volumes 12% gelatin solution at the same temperature.

Since glycerol masses especially tend to cause contraction of vessels, it is well to add 0.1 to 0.2% sodium nitrite either to the preceding saline solution or to the mass itself, or to both.

After injection of any gelatin mass, quickly cool the organs and fix in 10% formalin. With Berlin blue masses, avoid alkali, since this tends to convert the ferric ferrocyanide into ferric hydroxide and soluble alkali ferrocyanide. The inclusion of a little ferric chloride in the dehydrating alcohol is suggested by Romeis.

CORROSION TECHNICS

By substituting some such substance as rubber or neoprene latex for the injection mass and then digesting off the tissues with hydrochloric acid or artificial gastric juice, interesting casts of the vascular system may be obtained.

Duff and More (*J. Tech. Methods*, **24**: 1, 1944) used such a method on adult human kidneys. Kidneys as soon as possible postmortem are washed with tap water at 75 mm mercury pressure through a cannula tied into the renal artery. After a 30-min washing, the pressure is raised to 150 mm, and washing is continued for several hours (until the kidney becomes uniformly pale) but not over 12 to 15 hours. Then, leaving the cannula in place, disconnect from the water faucet, and place the kidney in a covered dish at 4°C for 6 to 12 h to allow as much water to escape as possible. Then keep at 20 to 25°C for about 5 h before injecting neoprene. Completely fill the tubing with the neoprene before connecting to the cannula. Inject with air pressure at 150 mm for $3\frac{1}{2}$ min for normal kidneys. Injection into sclerotic kidneys may require a 5-min injection, or raising the pressure to 250 mm, or warming the kidney to 60°C for 30 min before injection. Then disconnect and immerse the whole kidney in commercial hydrochloric acid at 56°C for 24 to 36 h, agitating gently from time to time until all the renal tissue can be removed by gently washing the cast in warm water.

The cast maintains its form floating in water but collapses when removed. Small fragments of the vascular tree may be teased apart and cut off for microscopic study. Duff and More suggest Farrants' medium for examination under a cover glass of such fragments of the vascular tree. According to Lieb, neoprene casts should be stored in some mold-inhibiting fluid (*J. Tech. Methods*, **20**: 48, 1940).

Similar corrosion methods have been used for injection into lungs through the bronchi, and neoprene latex of a different color can simultaneously be injected into the vascular system. Rigid plastics may be substituted for the latex.

McClenahan and Vogel (*Am. J. Roentgenol. Radium Ther. Nucl. Med.*, **68**: 406, 1952) use an alloy of bismuth (44.7), lead (22.6), indium (19.1), tin (8.3), and cadmium (5.3), designated as "Cerrolow 117," which melts at 47.2°C.

Arteries of organs to be injected are ligated before removal from the body to prevent access of air. The injection system is cleared of air, and the cannula is inserted into the artery and ligated in place under water. The organ is perfused with cool kerosene at 120 to 150 mm mercury pressure until the return flow is free of blood clots. The organ and the alloy reservoir are then warmed for about 30 min in a water bath at about 50°C. The organ is so oriented that the metal enters from the lowest point, so as to avoid entrapping kerosene, and perfusion with metal is carried on until no more kerosene emerges from the efferent veins by maintaining a pressure nearly equal to the antemortem

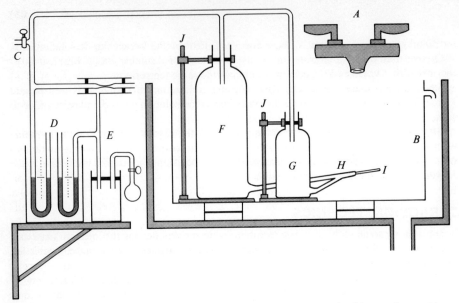

Fig. 19-1 Injection apparatus for metal corrosion preparations. Hot and cold water faucet *A* is over large sink *B* with a water bath containing kerosene reservoir *F*, metal reservoir *G*, Y tube *H* with clamps (not shown) on each arm, cannula *I*, and ring stands *J* to support reservoirs. Air pressure enters from inlet *C*, is measured by manometers *D*, and controlled by a maximum-pressure Starling escape valve *E*.

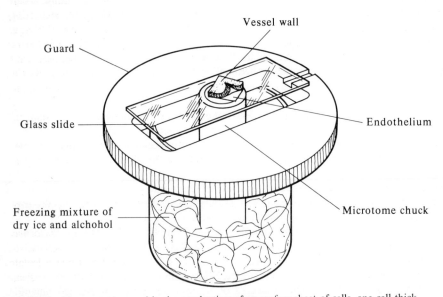

Fig. 19-2 The apparatus used in the production of an *en face* sheet of cells, one cell thick, from a vessel. A small segment of dry vessel wall, endothelial surface downward, is shown in place on the glass slide. The slide is cooled by means of a microtome chuck standing in a freezing mixture of alcohol and dry ice. When the endothelium is judged to be frozen to the slide the remainder of the vessel wall is rolled back with forceps. After this procedure sheets of endothelial cells are left on the slide. A metal guard serves to protect the specimen from stray drops of alcohol from the freezing mixture.

maximum. The organ is then chilled with cold water while pressure is still maintained. When cold, the organ is transferred to 20% potassium hydroxide (15% NaOH should serve), and the soft tissue is corroded off over a period of 24 h, with two changes of alkali and periodic gentle washing in water. See Fig. 19-1.

Large casts tend to sag at room temperature. Embedding in clear plastic and cold storage are suggested as means of keeping casts undistorted.

Warren (*J. R. Microsc. Soc.*, **84**: 407, 1965) recommends the following method for *en face* preparations of one-cell thickness. The method is stated to be particularly useful for observation of the endothelial cell layer of blood vessels. An example follows:

The area of the aorta to be observed is dissected out, rinsed with saline solution, blotted with facial tissues such as Kleenex, and applied to a clean slide with the endothelial surface downward as in Fig. 19-2. Care must be taken to prevent air bubbles or moisture between endothelium and the glass. A microtome chuck standing in an ethanol-CO_2 ice mixture provides the flat freezing source. A metal guard fits on top of a beaker. The clean, dry glass slide is then placed on the chuck with the tissue in contact. Within a few seconds the tissue freezes. With correct timing, the aorta may be pulled off with a forceps, leaving a single layer of cells on the slide for subsequent processing and staining.

AUTORADIOGRAPHY

Autoradiography permits detection and localization of radioactive isotopes in sections of tissues. Becquerel [*C. R. Acad. Sci. [D] (Paris)*, **122**: 420, 1896] may have been the first to have used a photographic film to detect radioactivity. A crystal of potassium uranyl sulfate was exposed to a photographic plate. Later, London (*Arch. Elec. Med.*, **12**: 363, 1904) obtained an autoradiograph of a frog previously exposed to radium. For an exhaustive treatment of autoradiography, readers are referred to Pelc [in J. F. Danielli (ed.), "General Cytochemical Methods," vol. 1, Academic, New York, 1958], Baserga (in H. Busch (ed.), "Methods in Cancer Research," Academic, New York, 1967), Baserga and Malamud ("Modern Methods in Experimental Pathology: Autoradiography, Techniques and Applications," Hoeber-Harper, New York, 1969), Rogers ("Techniques of Autoradiography," Elsevier, Amsterdam, 1967), Roth and Stumpf ("Autoradiography of Diffusible Substances," Academic, New York, 1969).

Stumpf (*J. Histochem. Cytochem.*, **18**: 21, 1970) calls attention to several factors concerned with accurate autoradiographic localization of isotopes. The isotope should be pure, and all molecules should be labeled. This may be unattainable; however, the purity and specific activity should be as high as possible.

Maximal autoradiographic resolution is achieved with ^3H and ^{125}I. Tritium emits pure beta particles with a maximum energy of 0.018 MeV. According to Fitzgerald et al. (*Science*, **114**: 494, 1951) and Leblond et al. (*Lab. Invest.*, **8**: 296, 1959) beta particles may travel a maximum of 6 μ in tissues. Approximately one-half of the particles are believed to travel less than 1 μ. Eidinoff et al. (*Proc. Soc. Exp. Biol. Med.*, **77**: 225, 1951) calculate the half-life of tritium to be 12.26 years. Hughes (Autoradiography with Tritium, *Proc. Symp. Tritium Tracer Applications*, p. 38, New England Nuclear Corp., New York, 1957) calculates that all silver grains exposed to beta particles of tritium are within 1 μ of the source and that most beta particles expose only one grain of silver bromide.

^{14}C also emits beta particles with a maximum energy of 0.016 MeV. It has a half-life of 5500 years. Lajtha and Oliver (*Lab. Invest.*, **8**: 214, 1959) calculate the average penetration of beta particles from ^{14}C in an emulsion to be 60 μ and a maximum of 300 μ. Thus, resolution of autoradiographs made with ^{14}C is less than that obtained with tritium if all other parameters are equivalent. It is important to recognize that beta particles have a low mass which is easily deflected, resulting in a circuitous path. In contrast, paths of alpha particles are straight with abrupt endings. Thus, the origin of the isotope sometimes may be determined with some difficulty on the basis of beta emission on an autoradiograph. [125]I has a half-life of 60 days, and Auger, gamma, and x-rays emitted have a maximum energy of 0.034 MeV.

For maximum resolution, sections should be as thin as possible, and the space between section and film should be minimal. Doniach and Pelc (*Br. J. Radiol.*, **23:** 184, 1950), Gross et al. (*Am. J. Roentgenol. Radium Ther. Nucl. Med.*, **65:** 420, 1951), and Stevens (*Br. J. Radiol.*, **23:** 728, 1950) have calculated that a reduction of resolution from 17 to 1.5 μ occurred upon a reduction of interspace between section and film from 3 to 0.01 μ. Stumpf (*J. Histochem. Cytochem.*, **18:** 21, 1970) advises that tissue sections not exceed 2 to 3 μ in thickness.

Emulsions should be as thin as possible to enhance resolution. The nature of the silver bromide crystals also affects resolution. Large crystals provide a more rapid emulsion speed, which requires less exposure time; however, large grains also give less resolution. Gross et al. (*Am. J. Roentgenol. Radium Ther. Nucl. Med.*, **65:** 420, 1951) have also called attention to the desirability of a uniform grain size in the emulsion, which results in increased contrast. Stevens (*Br. J. Radiol.*, **23:** 728, 1950) and many others have noted that increasing exposure times result in an increase in the diameter of the image with consequent decreased resolution. Stevens also noted loss in resolution with overdevelopment. It is important to use fresh film (background from cosmic and other radiation is reduced); exposure should be minimal and in the cold; and minimal optimal development time should be employed.

The quality of autoradiographs is affected greatly by the dose of isotope, the solvent employed, the route of administration, and the time interval between administration of the isotope and the time when the specimen is taken. No general prescription can be provided, because each isotope is peculiar and different information is desired by the various investigators. One can gain general impressions by a review of the literature in the area under investigation. Some helpful advice may be obtained from Pelc (Autoradiography as a Cytochemical Method, in J. F. Danielli (ed.), "General Cytochemical Methods," vol. 1, p. 279, Academic, New York, 1958).

Care needs to be exercised with respect to retention of the isotope at the site occupied in vivo. This is particularly true of diffusible substances. Losses may also occur during tissue processing and preparation of the autoradiogram. Finally, it is important to identify the nature of the isotope at the emitting site. It may be a derivative of the injected isotope from which different interpretations may be justified.

Fixatives containing mercury should be avoided, because they may cause fogging of the emulsion. Formalin, ethanol, acetone, and Bouin's solution are recommended fixatives if translocation and loss of isotope are not problems (Fitzgerald et al., *Lab. Invest.*, **8:** 319, 1959).

Other problems that may be encountered may be enumerated: The dark room must be spotless to prevent dirt, etc., from causing contamination. Only clean and filtered solutions should be used. The emulsion must not have direct contact with fingers. Excessive pressure should not be applied to film and sections. Sections of comparable nonradioactive tissues should be concomitantly processed as controls. An accumulation of air bubbles between the section and emulsion may cause a problem in interpretation. Specimens should be agitated during development to prevent uneven development. Processing of the film should occur in solutions of constant temperature to prevent reticulation. Be alert to the possibility of electrostatic discharges especially when working with the stripping film technic.

Autoradiographic Methods

CONTACT METHOD.   Several methods have been described. The following method was described by Gross et al. (*Am. J. Roentgenol. Radium Ther. Nucl. Med.*, **65**: 420, 1951):

1. Fixed paraffin-embedded or fresh frozen sections are affixed to chemically clean slides. Increasing resolution is achieved with decreasing thickness of sections.
2. An appropriate emulsion is applied to the surface of the section and slide in the darkroom after the slide and section have been thoroughly examined and found to be entirely free of dust, etc. Firm, even pressure is applied, possibly by using a lightproof x-ray pressure cassette.
3. The specimen is exposed at 1 to 2°C for the desired period.
4. After separation, the film is processed according to specifications of the manufacturer. The tissue section is dewaxed (if paraffin-embedded), stained, and mounted in the usual fashion.

Some authors recommend removal of paraffin prior to film exposure. If this is contemplated, one should consider the possibility of loss or translocation of the isotope. Due to the relatively poor apposition of the film to the specimen, the resolution achieved by contact autoradiographs is poor. The method may be used for a pilot study or a screening process. Short exposure time may be employed if high-speed x-ray film is used. A comparison microscope may be used in situations where film and tissue sections have been separated for processing.

LIQUID EMULSION COATING TECHNIC (according to Kopriwa and Leblond, *J. Histochem. Cytochem.*, **10**: 269, 1962). This technic is generally credited to Leblond, the earliest reference cited by him and his associates being Bélanger and Leblond, *Endocrinology*, **39**: 8, 1946.

Fixation of tissue should be in accord with the special requirements for the specific tissue elements under study. Most ordinary fixations are suitable. Very thorough removal of mercury with iodine and of the iodine with cysteine should be practiced.

Routine paraffin sectioning is done; sections are deparaffinized and stained with alum hematoxylin if desired, or left unstained.

COLLODIONIZATION.   This step is perhaps preferably omitted for routine purposes. Messier and Leblond prescribed two dips into 1% collodion and drying overnight in a vertical position. The collodion film prevents 15 to 35% of the tritium beta particles from reaching the emulsion. Some stains, however, cause emulsion fogging if not covered.

EMULSION.   Eastman NTB 2 bulk emulsion seems generally the most suitable. NTB 3 and the British Ilford K 5, which latter must be diluted after melting with an equal volume of triple-distilled water to give the same consistency as NTB 2, also seem quite serviceable.

COATING.   This is done in a photographic darkroom maintained at about 28°C (82.4°F) and 80% humidity, at about 1 m (3 ft) from a Wratten No. 2 safelight. Solid emulsion, stored in the refrigerator in the darkroom at 5°C, is scooped out with a porcelain spoon and deposited in a cylindrical staining jar which is standing in a 40 to 45°C water bath. In about an hour the emulsion is completely melted and free of bubbles. Slides bearing sections are warmed on a warm plate set at 40°C, and dipped in emulsion for 1 to 2 s. The excess is drained back into the jar and the slides dried in a vertical position for about 90 to 100 min with the safelight turned off.

EXPOSURE.   Transfer the slides to black plastic boxes containing a perforated metal capsule loaded with Drierite. Seal the box with black adhesive tape, mark the box with an orange grease pencil, identifying the slides and dates. Put the box in a cold room, at 5°C, standing it on end so that the slides are horizontal with the emulsion side down.

Exposure times vary with the isotope used, its state of decay if short-lived, the amount present in the tissue, and the substance in which it is incorporated. Previous published work with the isotope in use can be a fairly good guide, but in early experiments of an investigation it is well to try a series of exposure times designed to bracket the expected optimum.

DEVELOPING. This should be done in a separate photographic darkroom regulated at 17 to 18°C (about 63°F). Develop for 2 min in Dektol (D-72), or 6 min in Dolmi (D-170), or 3 min in D-178. Fix in acid fixer with hardener or in 24% hypo for 10 min. Wash for 15 min in running water at 17 to 18°C. Dehydrate in 95% alcohol and two changes of absolute alcohol, clear in cedar oil, wash in xylene, and mount in balsam or synthetic resin.

It is to be noted that a considerable variety of stains may be applied to sections before coating, including hematoxylin-eosin, the periodic acid Schiff method, and Van Gieson's stain. Harris' hematoxylin and Bullard's are to be avoided. Apparently the mercuric oxide used in ripening is deleterious. Instead use a naturally ripened formula such as Boehmer's, Delafield's, or Ehrlich's or an iodate formula, e.g., Mayer's or Lillie's.

THE STRIPPING FILM METHOD. Stripping film Kodak England AR.10 seems to be preferred by some workers. Sections are prepared as for the coating technic, deparaffinized, stained as needed, and brought to distilled water.

Film is cut in the darkroom to a size usually somewhat larger than the slide to be covered and placed face down on fresh distilled water for 2 to 3 min to swell the plastic layer beneath. The slide is then dipped into the water bath under the cut film, and the latter is picked up on it. Dry with a fan at 20 to 25°C, and store in black boxes as for the coating method (Pearse, 1960).

For ^{65}Zn and other highly water-soluble substances the method of Millar, Vincent, and Mawson (*J. Histochem. Cytochem.*, **9:** 111, 1961) seems applicable. Tissues were fixed in 9 : 1 methanol-formalin, 4- to 5-μ serial (paraffin) sections were cut, Kodak autoradiographic permeable-base 35-mm stripping film was used. Sections were dewaxed in dry chloroform and mounted on slides in the same.

A piece of film 10 cm long was cut from the roll, removed from the base, and allowed to swell emulsion side down on distilled water. It was then transferred through two dry chloroform baths and mounted on the slide in the second bath with the aid of a stainless steel wire-mesh fork holder with a space slightly wider than a 25-mm slide between the two lateral arms. The slide was brought up beneath the film between the fork arms of the film holder, and the film was withdrawn from the holder on the slide. The slide and film were then washed in absolute alcohol, dried, stored, developed, and mounted according to Pelc's method (*Int. J. Appl. Radiat. Isot.*, **1:** 172, 1956). Instead the developer and fixer prescribed by the maker of the film should be used.

Fitzgerald (*J. Histochem. Cytochem.*, **9:** 597, 1961) has avoided diffusion of water-soluble materials in the preparation of radioautographs by using vacuum-freeze-desiccated cryostat sections or freeze-substituted cryostat sections of immediately frozen experimental material. Stripping film is stripped and pressed down on the dry sections without prior wetting and exposed for the requisite period at −20°C. The preparations are then rehydrated either by exposure for several hours in the dark to an atmosphere of 80 to 90% humidity or by 5-min immersion in formalin solution, followed by redrying before development. This improved adhesion of the stripping film to the glass slide. Development was done as usual with DK-50 for 3 min. Cytidine ^{3}H and thymidine ^{3}H were thus localized.

BASERGA'S TWO-EMULSION PROCEDURE FOR ^{3}H AND ^{14}C. Utilizing the known shorter range of beta particles from ^{3}H (1.5 μ) as compared with that of the higher-energy beta particles from ^{14}C (60 μ), Baserga (*J. Histochem. Cytochem.*, **9**: 586, 1961), in studies of the incorporation of thymidine ^{3}H into deoxyribonucleic acid and of uridine-2-^{14}C into ribonucleic acid, adapted the following technic from Joftes (*Lab. Invest.*, **8**: 131, 1959):

1. Deparaffinize and hydrate 4-μ sections.
2. In the darkroom dip in Eastman NTB-2 emulsion at 40°C.
3. Here it seems preferable to follow the latest method of Kopriwa and Leblond: After 15 days, develop, fix, wash, and stain with Mayer's hematoxylin.
4. Then dip 3 s in a diluted celloidin made by diluting Randolph Products celloidin-embedding solution A-4700 with 2 volumes absolute alcohol plus ether (50 : 50). Drain and dry. This is probably equivalent to 2 to 4% celloidin.
5. Dip again as before in Eastman NTB-2 emulsion.
6. Expose in dark for another 15 days, develop, fix, wash, dehydrate, and mount following the general directions of Kopriwa and Leblond.

The lower emulsion, immediately over the section, registers beta-particle tracks both from the tritiated thymidine and from the radiocarbon-labeled uridine; the upper emulsion contains tracks only from ^{14}C.

Baserga regards this procedure of interposing a layer of celloidin between the two emulsion layers as an improvement over that of Krause and Plaut [*Nature (Lond.)*, **188**: 511, 1960], who used two layers of stripping film immediately superimposed.

Although other brands of nitrocellulose could undoubtedly be used, it would be necessary to determine experimentally the concentration which would yield a film of the required thickness to stop tritium beta particles and not materially impair the passage of the ^{14}C particles.

Baserga's second report (*J. Histochem. Cytochem.*, **10**: 628, 1962) indicated that substitution of AR.10 stripping film in either or both of the stages gave poorer results and that the intermediary celloidin film was necessary and was superior to Permount, Formvar, polyvinyl alcohol, and other substances. Additional ^{3}H and ^{14}C compounds were used, with similar results.

Endicott and Yagoda (*Proc. Soc. Exp. Biol. Med.*, **64**: 170, 1947) and Evans (ibid., **64**: 313, 1947) sectioned tissues, floated them on a water bath, and transferred them to the darkroom, where the sections were placed and dried directly on the photographic plate. A fan may aid in drying. After exposure at 4°C in the dark with a desiccant, tissue is deparaffinized, passed through graded alcohols to water, developed, fixed, and washed. Tissue sections may then be stained and mounted in Permount. The authors indicate that blood and tissue smears may be processed in the same manner.

A note of caution may be interposed here. Board (*J. Cell Physiol.*, **38**: 377, 1951), Boyd and Board (*Science*, **110**: 586, 1949), and Fitzgerald and Engstrom (*Cancer*, **5**: 643, 1952) have called attention to the fact that fresh nonradioactive bone marrow, spleen, liver, heart, lung, and muscle tissues of rats and guinea pigs produce a blackening of photographic film. Everett and Simmons (*Anat. Rec.*, **117**: 25, 1953) and later Tonna (*Stain Technol.*, **31**: 255, 1958) demonstrated the blackening was not due to SH groups as suggested by the previous authors. Board noted that the blackening effect could be prevented by brief collodionization of the surface between the fresh tissue and film.

Stumpf and Roth (*J. Histochem. Cytochem.*, **14**: 274, 1966) conducted a comparative autoradiographic study of six methods using two diffusible compounds, namely, ^{3}H-estradiol and ^{3}H-mesobilirubinogen. They compared (1) frozen sections subsequently lyophilized and then dry-mounted on emulsion-coated slides, (2) frozen sections thawed on emulsion-coated slides, (3) frozen sections thawed on glass slides and dipped in liquid emulsion, (4) freeze-dried tissue, vapor-fixed (OsO_4 at 25°C or paraformaldehyde at 25°C), epoxy-embedded, sectioned, and dipped in liquid emulsion, (5) freeze-dried tissue, paraffin-embedded, sectioned, deparaffinized, and dipped in liquid emulsion, and (6) formalin-fixed tissue, paraffin-embedded, sectioned, deparaffinized, and dipped in liquid emulsion. Stumpf and Roth noted varying degrees of diffusion with methods 2 to 6. Neither fixative was superior to the other. Falck and Ourman (*Acta Univ. Lund.*, sec. II, no. 7, p. 5, 1965) noted that even gaseous fixation may permit translocation of certain substances. The following method (method 1 above) was highly recommended by Stumpf and Roth:

1. Tissue samples (1 to 2 mm³) mounted on a brass tissue holder are quenched in liquid propane at −180°C.
2. Tissues are sectioned at 0.75 to 1 μ at −70°C. Stumpf and Roth used an International Minot Custom Rotary Microtome with an ultrathin sectioning attachment mounted in a two-stage cryostat (Harris Model No. 3L-2-075). Cutting is observed with the aid of a dissecting microscope, and sections are handled with a fine brush. Serial sections as thin as 0.5 μ are attainable.
3. *Without thawing*, sections are lyophilized for 24 h. The vacuum is broken with nitrogen, and the section is mounted directly on the emulsion as follows. (At this point, sections may be stored over Drierite at 25°C if needed.) In the darkroom with low humidity at 25°C, the lyophilized section is placed on a small square of clean, smooth Teflon under the vision of a dissecting microscope. Using the safelight, an emulsion-coated microscopic slide (Kodak NTB-3 or NTB-10) previously stored over Drierite is carefully placed over the tissue section supported by the Teflon. Slight pressure from the thumb and forefinger is applied. Clamps are not used due to formation of pressure artifacts. Teflon is lifted off either with fingers or by inversion of the slide.
4. Exposure is conducted at −15°C over Drierite.
5. Bring the specimen to 25°C, breathe on the section once or twice to enhance adherence to the emulsion, and develop with Kodak Developer D19 for 2.5 min at 25°C, rinse in tap water, fix with Kodak Fixer for 4.5 min, and rinse in tap water for 5 min.
6. Stain section with hematoxylin and eosin, dehydrate, and mount in synthetic resin.

RESULTS: Results published by Stumpf and Roth are excellent. Translocation of substances is minimized. During dry mounting, a slight pressure is applied which may induce slight translocation; however, this is minimal and can be detected by the variable distribution of isotope observed if several sections are processed. The method has been employed by Ferin et al. (*Endocrinology*, **83:** 565, 1968) and Stumpf [*Science*, **162:** 1001, 1968; *Z. Zellforsch. Mikrosk. Anat.*, **92:** 23, 1968; *Endocrinology*, **83:** 777, 1968; *Science*, **163:** 958, 1969; *Endocrinology*, **85:** 31, 1969; *Nature* (*Lond.*), **205:** 712, 1965; *J. Histochem. Cytochem.*, **18:** 21, 1970] for localization of diffusible hormones.

Sams and Davies (*Stain Technol.*, **42:** 269, 1967) note that sometimes hematoxylin staining interferes with grain counting. They advise the following stain using nuclear fast red:

1. Fix tissues in 15% formalin in 0.9% NaCl, decalcify with EDTA, if necessary, embed in paraffin, and section at 5 μ.
2. Deparaffinize and bring sections to water, coat with emulsion, expose, develop, and fix as usual.
3. Rinse thoroughly in distilled water.
4. Stain section for 15 to 20 min at 37°C in 1% Alcian blue in 1% acetic acid.

5. Rinse in distilled water.
6. Stain sections for 1 to 5 min at 37°C in 0.1% nuclear fast red (C.I. 60760, G. T. Gurr) in saturated aqueous potassium aluminum sulfate. (The staining solution is prepared by boiling for 5 min and then filtering.)
7. Rinse in distilled water.
8. Stain for 1 to 4 min at 37°C in a 50% saturated solution of tartrazine (C.I. 19140) in Cellosolve.
9. Rinse in 90% ethanol, dehydrate, and mount in synthetic medium.

RESULTS: Sams and Davies note that good histologic detail results, and the emulsion is only slightly stained.

Roberts and Hunt (*Stain Technol.*, **42:** 7, 1967) recommend fixation in Bouin-Hollande's solution and paraffin embedment. Tissues are then sectioned, coated with Kodak NTB-3 liquid emulsion, developed, stained with an aqueous 1 : 1 mixture of 0.5% eosin Y–orange G mixture at pH 4.4 for 10 min at 25°C, differentiated in 0.5% acid alcohol, and counterstained with 0.25% toluidine blue O in 50% ethanol. Roberts and Hunt praise the method as an aid to differentiation of cells in lymph nodes.

Schmeer (*Stain Technol.*, **46:** 27, 1971) took tissues of animals injected with ^{3}H-thymidine, fixed them for 24 to 48 h in acetic acid–ethanol (1 : 3) and stained them *en bloc* with the Feulgen method. Stained preparations could be stored for at least 4 months at -25°C in 45% acetic acid, in 45% acetic acid in dimethyl sulfoxide, or in glycerol–acetic acid (85 : 15, v/v) without noticeable loss of grain counts.

Stenram (*Stain Technol.*, **37:** 231, 1962) noted a loss of silver granules in specimens stained with chrome alum gallocyanin. Many investigators have noted that eosin tends to overstain the emulsion. Thurston and Joftes (*Stain Technol.*, **38:** 231, 1963) tested 14 methods for staining ^{3}H-thymidine autoradiographs in formalin-fixed, paraffin-embedded sections of 13 different mouse tissues. They recommended the use of Feulgen fast green, Gomori's chrome alum hematoxylin–phloxine, and the aldehyde fuchsin–periodic acid Schiff stains *provided* sections are stained *prior* to being dipped in emulsion.

Thurston and Joftes further recommend celestin blue B–Mayer's alum hematoxylin; metanil yellow (C.I. 13065)–iron hematoxylin according to the method of Simmel et al. (*Stain Technol.*, **26:** 25, 1951); lithium carmine–picric acid according to the method of Witten and Holmstrom (*Lab. Invest.*, **2:** 368, 1953); Weigert acid–iron hematoxylin; methyl green–pyronin; and indigocarmine–picric acid provided the stains are applied after development of the emulsion. For sections to be stained with hematoxylin and eosin, Thurston and Joftes recommend, in sequence, deparaffinization, absolute ethanol, staining with eosin, hydrating rapidly, dipping into emulsion, exposing, developing, staining with hematoxylin, washing thoroughly in running water, dehydrating, and mounting in synthetic medium.

Sawicki and Rowinski (*Histochemie*, **19:** 288, 1969) state that the PAS reaction can be successfully conducted on sections prepared for autoradiography. They recommend oxidation of sections with periodic acid prior to coverage by the emulsion. After exposure and development, sections are treated with Schiff's reagent.

Wolberg (*Stain Technol.*, **40:** 90, 1965) recommends that emulsion-coated sections for autoradiographs be stained for 10 min in Darrow red (25 mg/100 ml in 0.2 M acetic acid), rinsed, counterstained in 0.2% aqueous light green SF yellowish, dehydrated, and mounted in synthetic medium.

Kennedy and Little (*J. Histochem. Cytochem.*, **22:** 361, 1974) used the Stumpf freeze-dried autoradiographic method for localization of labeled carcinogens in the lung with slight modifications; namely, the entire lung was frozen and smaller pieces were cut from it, sections were cut at 2 to 4 μ at a higher ($-35°$C) temperature, and the use of the thicker sections required that they be dipped into 1% gelatin to prevent slipping from the emulsion-precoated slides.

Salpeter et al. (*J. Histochem. Cytochem.*, **22**: 217, 1974) conducted resolution studies on light microscopic autoradiographic specimens of 0.5 to 1 μ in thickness coated with Ilford L4 and Kodak NTB-2 and AR.10 emulsions. They used their radioactive line source (*J. Cell Biol.*, **41**: 1, 1969; ibid., **50**: 324, 1971) and noted that with tritium labeling the half-distance (HD) was approximately 3500 Å. The half-distance resolution measure employed is the distance from the line source within which one-half of the developed grains fell. The value was not altered significantly with either section thickness beyond 0.5 μ or emulsion thickness beyond 2000 Å. A 0.5-μ section labeled with ^{14}C had a half-distance value of 7000 Å when coated with a monolayer of either Ilford L4 or Kodak NTB-2 emulsion. This value increased with an increase of either section or emulsion thickness. Salpeter et al. also noted that the density distributions, normalized in units of their own half-distance, gave a good fit to the set of universal curves derived from electron microscopic autoradiographic specimens.

Autoradiography for Electron Microscopy

Autoradiography at the electron microscopic level is finding increasing use. Methods have been developed to observe developed grains in focus simultaneously with the tissues, thereby permitting grain counts with relative ease. Quantitative studies are legitimate provided adequate controls are employed. Quantitative studies require knowledge of the efficiency of the method employed (sensitivity). This is obtained by a determination of how many silver grains are produced by one radioactive decay. It is also a requisite to know the expected grain distribution around a point source (resolution), and a sufficient number of developed grains are required for a statistical analysis. Inasmuch as layers of emulsion and sections of tissue now employed are thinner than the range of the radiation, variations in quantitative results are readily achieved by variations in section thickness. Bachmann and Salpeter (*Lab. Invest.*, **14**: 1041, 1965) note that photographic (E_p) and geometric (E_g) factors influence resolution independently. They calculate the total effect (E_t) in the following formula:

$$E_t = \pm[(E_p)^2 + (E_g)^2]^{1/2}$$

Thus section thickness (E_g) and emulsion thickness markedly influence results. Very little scattering of electrons from either ^{14}C or ^{3}H is believed to occur in tissue sections; however, in emulsions, substantial scatter from ^{3}H, though little scatter from ^{14}C, is noted.

For a determination of resolution, readers are referred to papers by Caro (*J. Cell Biol.*, **15**: 189, 1962; *Science*, **41**: 918, 1969), Caro and Schnos (*Science*, **149**: 60, 1965), Bachmann et al. (*Histochemie*, **15**: 234, 1968), Salpeter et al. (*J. Cell Biol.*, **41**: 1, 1969), and Salpeter and Salpeter (*J. Cell Biol.*, **50**: 324, 1971).

The efficiency (sensitivity) of autoradiographic methods may be expressed as grains times decay times 100%. Factors that influence efficiency include absorption and scattering in the specimen, in the emulsion, or anywhere; any possible chemical interaction between specimen and emulsion; physical and chemical properties of the emulsion; the possible formation of latent images during periods of long exposure; the type of radiation; and the type of developing technic employed. Salpeter and Szabo (*J. Histochem. Cytochem.*, **20**: 425, 1972) noted that sensitivity in electron microscopic autoradiography using Ilford L4 emulsion was affected by dose of radiation.

According to Peachey (*J. Biophys. Biochem. Cytol.*, **4**: 233, 1958), Bachmann and Sitte (*Mikroskopie*, **13**: 289, 1958), and Williams and Meek (*J. R. Microsc. Soc.*, **85**: 337, 1966), section thickness may be assessed by a correlation with interference colors observed in sections floating on water. Most authors believe the method is accurate, although Williams and Meek note some problems. Salpeter and Bachmann (in Hayat, vol. II, 1972) have noted that a Normanski-type interferometer attachment is available for attachment to a Reichert Zetopan microscope for critical studies.

FLAT SUBSTRATE METHOD (Salpeter and Bachmann, *J. Cell Biol.*, **22**: 469, 1964, and in Hayat, vol. II, 1972)

1. Place ribbons of sections on cleaned slides coated with collodion, and allow to dry.
2. Stain sections with appropriate method using drops of stain on horizontally oriented slides placed in a tightly closed chamber. Do not allow sections to air-dry. Rinse stain with distilled water.
3. Vacuum-coat stained sections with a layer of carbon.
4. Drop liquid emulsion (60°C) on the section, drain, dry in a vertical position. The slide may also be dipped into the emulsion.
5. Expose and develop specimens.
6. On the surface of a water bath, separate the collodion-section carbon emulsion from the slide, and float it on the water surface.
7. Place grids on the sections.
8. Examine sections by electron microscope.

Budd and Pelc (*Stain Technol.*, **39**: 295, 1964) offer another "Flat Substrate Method":

1. Sections are placed directly on microscope grids previously attached to a glass slide.
2. The pregelled emulsion (60°C) is applied over the section with a wire loop (Caro, *Science*, **41**: 918, 1969). The slide may also be dipped in emulsion (Hay and Revel, *Dev. Biol.*, **7**: 152, 1963).
3. Expose, process, and stain. The grid remains attached to the microscope slide during exposure and processing.

It is recommended that slides be cleaned using a detergent followed by rinsing in distilled water and drying with facial tissues, such as Kleenex. Do not use lens tissues or KimWipes, and do not air-dry. The collodion is prepared by making up a 2% (w/v) solution in amyl acetate the day before use. Before use, the dilution to 0.7% is made. Sections may be stained with either uranyl acetate or lead. Lead staining may result in nonspecific adherence and cause confusion with developed grains unless appropriate control sections are processed.

Kopriwa (*J. Histochem. Cytochem.*, **14**: 923, 1967) developed a semiautomatic device for application of liquid nuclear track emulsion. Subsequently, Williamson and Van Den Bosch (*J. Histochem. Cytochem.*, **19**: 304, 1971) developed an inexpensive mechanical device (with four pulleys 3.25, 2.25, 1.25, and 0.8 cm in diameter) that provides uniform monolayers of emulsion on up to 30 grids simultaneously. Both sides of the grids can be coated with emulsion, permitting a potential doubling of sensitivity with no loss of resolution, theoretically.

Attramadal (*Histochemie*, **19**: 64, 75 and 110, 1969) conducted several studies to determine maximum preservation of labeled soluble hormones (^{3}H-estradiol, progesterone, and testosterone) during several autoradiographic procedures. During formalin fixation of liver tissues (10% in 0.15 M phosphate buffer of pH 7, 20 h, 25°C) 30% of labeled material diffused into the fixative, and 19% more was subsequently extracted during dehydration prior to paraffin embedding. The loss was slightly less in tissues sacrificed 1 h rather than 15 min after injection of the isotope. From uterine tissue, 20.5% of radioactivity was lost during formalin fixation, and 58.1% more was lost during dehydration prior to paraffin embedding.

In tissues fixed with OsO_4 (1% in 0.15 M phosphate buffer of pH 7, 4 h, 4°C) only 3.9% diffused into the fixative. Liver tissues lost less than 2% during fixation. During dehydration, 72.8% was lost from uterine tissues, and 19 to 35% was lost from liver tissues. Embedding (24 h, 25°C) in glycol methacrylate resulted in a further 13 to 33% loss of radioactivity. Glutaraldehyde-perfused (2.5% in 0.15 M phosphate buffer of pH 7) tissues later postfixed in OsO_4 retained nearly all radioactivity; however, during ethanol and propylene oxide dehydration and Epon embedding, 93% of radioactivity was lost.

In tissues freeze-dried, vapor-fixed with either OsO_4 (18 h, 25°C) or formaldehyde (48 h, 25°C), and subsequently embedded in Epon or methacrylate, very little (0 to 3%) radioactivity was lost. A small amount (2 to 3%) was lost while sections were floated on a water bath. Attramadal is convinced freeze-drying, osmium-vapor fixation, and Epon embedding provide a valid method for accurate intracellular localization of steroid sex hormones. A preliminary study by Wilske and Ross (*J. Histochem. Cytochem.*, **13**: 38, 1965) on freeze-dried osmium- or paraformaldehyde-vapor-fixed, Epon-embedded tissues obtained results analogous to those by Attramadal. Merriam (*J. Histochem. Cytochem.*, **6**: 43, 1958) and Droz and Warshawsky (*J. Histochem. Cytochem.*, **11**: 426, 1963) have also conducted studies of losses of isotopes used for labeling during various fixations. A review article by Moses (*J. Histochem. Cytochem*, **12**: 115, 1964) covers many theoretical and practical aspects of autoradiography at the light and electron microscopic levels.

MICRORADIOGRAPHY

The reader is referred to works by Beneke (in G. Wied, "Introduction to Quantitative Cytochemistry," Academic, New York, 1966), Neumann ("Handbuch der Histochemie," **1**: 339, 1958), Lindstrom [*Acta Radiol. (Stockh.) Suppl.* 125, 1955], and Lindstrom and Philipson (*Histochemie*, **17**: 194, 1969). Philipson and Lindstrom (*Histochemie*, **17**: 201, 1969) recommend freeze-drying sections for microradiographic analysis, although defects of the method were observed, namely, shrinkage and incomplete drying of the specimens. Shrinkage varied from one tissue to the next and had to be determined individually. Philipson and Lindstrom note that results of microradiographic dry weight determinations are obtained as mass per unit area and that all methods employed for thickness determinations of freeze-dried sections are subject to large error.

QUANTITATIVE MICROCHEMICAL ANALYSES

Quantitative microchemical methods developed from elegant and critical investigations particularly in the laboratories of Oliver, Lowry, and Linderstrøm-Lang. A few exemplary papers include Linderstrøm-Lang et al., *C. R. Trav. Lab. Carlsberg*, **20**: 66, 1935; Linderstrøm-Lang and Lanz, ibid., **21**: 315, 1938; Linderstrøm-Lang and Møgensen, ibid., **23**: 27, 1938; Lowry, *J. Histochem. Cytochem.*, **1**: 420, 1953, *J. Biol. Chem.*, **236**: 2746, 1961, ibid., **239**: 18, 1964; Matschinsky et al., *J. Histochem. Cytochem.*, **16**: 29, 1968; Passonneau et al., *Anal. Biochem.*, **19**: 315, 1967; Giacobini, *J. Histochem. Cytochem.*, **17**: 139, 1969, *Acta Physiol. Scand.*, **36**: 276, 1956, *J. Neurochem.*, **1**: 234, 1957; *Acta Physiol. Scand.*, **66**: 49, 1966; and Giacobini and Holmstedt, *Acta Physiol. Scand.*, **42**: 12, 1958.

Lowry and coauthors have applied many of their technics to studies of the nervous system. Lowry and Passonneau (*Biochem. Pharmacol.*, **9**: 173, 1962) note three basic requirements for direct histochemical analysis: (1) The structure to be analyzed must be isolated without *loss* or *alteration* of chemical composition. (2) The sample size must be measured. (3) Analytical methods of appropriate sensitivity must be employed. The appropriate steps in a quantitative analysis are listed below:

1. Quick-freeze tissues at $-150°C$.
2. Section tissues at 5 to 25 μ at -15 to $-25°C$.
3. Dry sections under a vacuum.
4. Bring sections to 25°C under a vacuum, and dissect out appropriate cells (etc.) with the aid of a dissecting microscope.
5. Weigh samples with a quartz fiber balance.
6. Subject samples to appropriate chemical analysis.

Lowry and Passonneau note that samples as small as nuclei of nerve cells may be readily isolated by dissection. Dissections down to 10 μ are feasible and not difficult. In some parts of the nervous system, such as the retina and cerebellum, cells of a uniform type are layered. Sections can be made parallel to the layers, permitting samples of 0.1 to 2 μg dry weight to be made. These are pure samples of one type of cell and matrix.

Samples to be analyzed must be weighed. The quartz fiber fish-pole balance was developed by Lowry (*J. Biol. Chem.*, **140**: 183, 1941). Oliver also developed a quartz fiber torsion balance (*J. Biol. Chem.*, **152**: 293, 1944) with a sensitivity of 5 ng and a reproducibility of 20 ng. The fish-pole balance has a sensitivity of 10 ng. Bonting and Mayron (*Microchem. J.*, **5**: 31, 1961), Burt (*Microchem. J.*, **11**: 18, 1968), McCann (*Microchem. J.*, **11**: 255, 1966), Nelson (*Anal. Biochem.*, **35**: 542, 1970), and Moss (*J. Histochem. Cytochem.*, **20**: 545, 1972) have made various modifications of the quartz fiber fish-pole balance. Oliver recommends simply the use of a 4-mm-long quartz fiber of 0.3 μ diameter mounted horizontally in a tuberculin syringe. This type has a sensitivity of 2 pg (picograms) (capable of weighing one red cell with 10% accuracy).

Lowry and Passonneau note that most of the enzymes they have studied produce in 1 h between 0.1 and 100 mol product per kg dry weight of brain. This amounts to 2×10^{-10} to 2×10^{-7} mol for a 2-μg sample and 10^{-14} to 10^{-11} mol for a 0.1-ng sample. Substrate concentrations studied are much lower (10^{-5} to 10^{-1} mol/kg dry weight).

The method can utilize NAD and NADP as analytical tools. This provides a substantial advantage, because many biologically important substances can be channeled through an enzyme system to reduce or oxidize NAD or NADP. If reduced coenzyme is formed, the unchanged excess oxidized coenzyme is readily destroyed with weak alkali without loss of the reduced form. If oxidized coenzyme is formed, excess reduced coenzyme is destroyed with weak acid without loss of oxidized form. The remaining nucleotide may be measured fluorometrically or otherwise.

In the event the amount of substance to be measured is less than 10^{-11} to 10^{-12} mol, sensitivity can be increased almost without limit by use of the final pyridine nucleotide product to catalyze an enzymatic dismutation of two substrates. Thereafter, one of the products can again be measured with an NAD or NADP system. Sensitivity is thereby increased 10,000-fold. Lowry and Passonneau state that the process may be repeated to give a 100,000,000-fold increase in sensitivity permitting measurement of 10^{-19} mol of most biologic substances.

Lowry and Passonneau provide the following example. The problem is to measure the amount of glucose (such as 10^{-14} mol) in a sample of freeze-dried axoplasm weighing 0.5 ng.

1. Sample is weighed and treated with weak acid to destroy enzymes present.
2. Hexokinase, ATP, glucose 6-phosphate dehydrogenase, and 5×10^{-14} mol $NADP^+$ are added. During brief incubation NADPH is formed in an amount equivalent to the amount of glucose present, i.e., 10^{-14} mol.
3. The excess $NADP^+$ is destroyed by heating at 60°C in 0.01 N NaOH.
4. The entire sample is now added to a solution containing the following two enzyme systems:

    a. Glucose 6-phosphate, glucose 6-phosphate dehydrogenase
    b. α-Ketoglutarate, NH_4^+, and glutamic dehydrogenase

    During subsequent incubation the following cyclic process occurs:

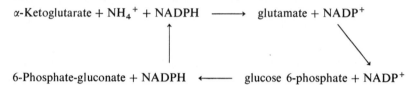

$$\alpha\text{-Ketoglutarate} + NH_4^+ + NADPH \longrightarrow \text{glutamate} + NADP^+$$

$$\text{6-Phosphate-gluconate} + NADPH \longleftarrow \text{glucose 6-phosphate} + NADP^+$$

   After 60 min, the cycle has been repeated 10,000 times or more with the formation of at least 10^{-9} mol 6-phosphate-gluconate from the 10^{-14} mol NADP present.

5. After heating to 100°C to destroy the enzymes, the sample is added to 1 ml reagent containing 6-phosphate-gluconate dehydrogenase and 10^{-8} mol $NADP^+$. After incubation, 10^{-9} mol NADPH are formed, and this is measured by its fluorescence.

Lowry and Passonneau note that if more sensitivity is needed after step 5, excess $NADP^+$ may be destroyed with alkali, and remaining NADPH could be recycled again. Success is claimed with such recycling.

In working with very small samples (less than 1 μg dry weight), it is preferable to work with small aqueous droplets *under oil*. The method was developed by Lowry (*Harvey Lect.*, **58**: 1, 1963). Matschinsky et al. (*J. Histochem. Cytochem.*, **16**: 29, 1968) provide a detailed description of the oil well technic. Giacobini (*J. Histochem. Cytochem.*, **17**: 139, 1969) conducted a critical evaluation of quantitative neurochemical analyses.

As a comparative note, using the Warburg apparatus, sensitivity may be attained down to 10^{-6} to 10^{-7} mol; using colorimetric methods with microcells, sensitivity may reach 10^{-9} to 10^{-10} mol; and using the Cartesian diver refined by Giacobini (*J. Histochem. Cytochem.*, **17**: 139, 1969) sensitivity is increased to 10^{-14} mol. Fluorometry may be employed for assay of pyridine nucleotides down to 10^{-11} to 10^{-12} mol depending upon the apparatus employed. The reader is referred to current and other papers by the above authors for details of these procedures.

Silverman and Glick (*J. Cell Biol.*, **40**: 761, 1969) reacted model amino acids, lipids, carbohydrates, chondromucoprotein, and several proteins with glutaraldehyde, stained them with an aqueous solution of phosphotungstic acid, and examined them by electron microscopy. Under certain conditions, the stain density reflected the concentration of protein based on the quantitative reaction of phosphotungstic acid with the positively charged groups, although the stoichiometry of the reaction between phosphotungstic acid and protein varied with the type of protein.

ZYMOGRAMS

The zymogram technic was developed by Hunter and Markert (*Science*, **125**: 1294, 1957). The method is essentially a starch-gel electrophoresis of enzymes. After completion of the separation, enzyme histochemical stains are employed to stain the enzymes in the gel. Markert and Hunter (*J. Histochem. Cytochem.*, **8**: 58, 1960, and **7**: 42, 1959) used the method to separate esterase isozymes. Many authors have since used the method to separate many enzymes. The reader is referred to a review article by Shaw and Prasad (*Biochem. Genet.*, **4**: 297, 1970) for detailed methods for the separation of numerous isozymes.

SCANNING ELECTRON MICROSCOPY

Initial studies by Knoll (*Z. Tech. Phys.*, **16**: 467, 1935) and Knoll and Theile (*Z. Phys.*, **113**: 260, 1939) resulted in the development of the scanning electron microscope with a resolution of approximately 100 μ. Von Ardenne (*Z. Tech. Phys.*, **109**: 553, 1938) constructed an instrument with an improved and reduced electron beam diameter. Zworykin et al. (*ASTM Bull.*, **117**: 15, 1942) developed an instrument with the possibility of 500 Å resolution. The first commercial scanning electron microscope was developed in Cambridge, England, and was available in 1965. For a current review on the potential usefulness of the scanning electron microscope for biologists, the reader is referred to Hollenberg and Erickson (*J. Histochem. Cytochem.*, **21**: 109, 1973) and Hayes and Pease (*Adv. Biol. Med. Phys.*, **12**: 85, 1968). The basic theory of the scanning electron microscope and its application to biology is only briefly detailed below.

Figure 19-3 illustrates the basic design of a scanning electron microscope. Electrons emitted from a heated tungsten cathode of an electron gun, possibly made of lanthanum hexaboride (LaB_6), usually from a 10- to 20-kV below-ground potential, pass through a column in a vacuum maintained at 10^{-5} torr. Lenses in the column serve to reduce the final electron beam to 100 Å in diameter with a 10^{-10}- to 10^{-12}-A current focused on the

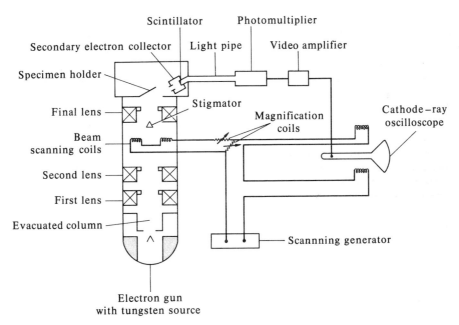

Fig. 19-3 Schematic representation of the scanning electron microscope. (*After Oatley et al., Adv. Electronics Electron Phys.*, **21**: 181, 1965. *By permission of the publisher.*)

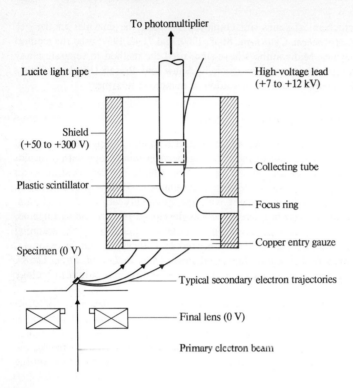

To photomultiplier

Lucite light pipe ——————————————— High-voltage lead
(+7 to +12 kV)

Shield ——
(+50 to +300 V)

Collecting tube

Plastic scintillator ——————————————

Focus ring

Specimen (0 V)

Copper entry gauze

Typical secondary electron trajectories

Final lens (0 V)

Primary electron beam

Fig. 19-4 Schematic representation of a secondary electron detector.
(*After Hayes and Pease, Adv. Biol. Med. Phys.,* **12:** 85, 1968. *By
permission of the publisher.*)

surface of the specimen. Deflected primary electrons and emitted secondary electrons from
the surface of the specimen are directed to a photomultiplier and video amplifier as indicated
in Fig. 19-4. Output from the video amplifier controls the potential of the modulating
electrode of the cathode-ray oscilloscope and thereby the intensity of the image on the screen.
Current originating from the scanning generator and passing through the beam scanning
coils (Fig. 19-3) causes coordinated deflection of the primary electron beam (Fig. 19-4) and
cathode-ray oscilloscope spot in a raster pattern, thereby scanning the specimen.

Inasmuch as the amount of current reaching the electron collector depends upon the
number of electrons emitted from a particular point on a specimen at any moment, brightness of the cathode-ray oscilloscope will vary accordingly, whereupon a light-dark image is
developed on a screen. The cathode-ray oscilloscope image portrays the specimen as it would
appear if the observer viewed it in the direction of the primary beam. The reader is referred
to articles by Oatley et al. (*Adv. Electronics Electron Phys.,* **21:** 181, 1965), Kimoto and
Hashimoto (*SEM,* **68:** 63, 1968), Kimoto et al. (*SEM,* **69:** 81, 1969), Broers (*SEM,* **70:** 1,
1970), Nixon (*SEM,* **69:** 1, 1969), Pease (*SEM,* **71:** 9, 1971), and Everhart (*Stereoscan,* **70:** 1,
1970) for special and detailed aspects of scanning electron microscopy not provided herein.

It is important to recognize that scanning electron microscopy is a relatively new venture.
In many cases, inadequate data are available to determine if the morphologic features
observed are those existent in vivo. A major problem is that generally water must be

removed from biologic samples before they are subjected to the high vacuum. Removal of water generally results in distortion.

Surface features are detailed by scanning electron microscopy. Natural surfaces such as chitin, enamel, etc., can be examined after only cleansing treatments. Isolated cells and smears may be examined provided excessive mucus or blood is removed to expose the cell surface. Surfaces may be examined after freeze-fracturing, freeze-etching, or even after slicing, ripping, or tearing of tissues. In tissues thoroughly embedded, some of the embedding material must be removed before surface topography can be revealed.

Most biologic specimens must be dried before examination can be conducted by scanning electron microscopy; otherwise, the vacuum employed will cause boiling and distortion of the specimen. Various dehydration methods may be employed including freeze-drying (from water or organic solvents), air drying, and critical point drying. Large defects frequently occur in relation to air drying. Freeze-drying frequently provides excellent preservation. The principles of critical point drying are detailed by Anderson [*Trans. N.Y. Acad. Sci.*, (2)**13**: 130, 1951]. In essence, water in the specimen is successively replaced by ethanol, amyl acetate, and liquid CO_2. The temperature of the specimen is then elevated above the critical point for liquid CO_2 (31°C) in an enclosed space, whereupon the liquid becomes a gas which evaporates leaving minimal surface distortion. Of course, a disadvantage of critical point drying is that ethanol and amyl acetate are fixatives causing ill-defined effects. Tissues also become very hard and brittle. In a similar effort, Water and Buck [*J. Microsc. (Oxf.)*, **94**: 185, 1971] vacuum-dried from camphene.

After dehydration, the specimen is coated with a thin film of carbon or metal rendering it conductive. After replication, the specimen may be examined by scanning electron microscopy. Further, the tissue may be digested (for example with sodium hypochlorite) and the replica examined by transmission electron microscopy. Boyde and Wood [*J. Microsc. (Oxf.)*, **90**: 221, 1969] provide further details on the preparation of soft tissue for examination by scanning electron microscopy.

A few of the recent advances in instrumentation will be briefly mentioned. Pawley and Hawes (*SEM*, **71**: 105, 1971) have developed a micromanipulator for use with the scanning electron microscope which permits microdissection of the specimen in any direction. Lane (*SEM*, **70**: 41, 1970) developed an environmental control stage whereby water vapor may be constantly supplied to a specimen. Vapor is then constantly drawn off by the vacuum. Theoretically, this should minimize the need for dehydration of the sample. Echlin et al. (*SEM*, **70**: 49, 1970) have developed a controlled-temperature stage from 25°C to −180°C.

Knox and Heslop-Harrison [*Nature (Lond.)*, **223**: 92, 1969] and Scarpelli and Lim (*Stereoscan*, **70**: 143, 1970) used the scanning electron microscope to localize enzyme activity on cell surfaces. Certain insects can be examined alive in the scanning electron microscope (Pease et al., *Science*, **154**: 1185, 1966; Humphreys et al., *Stereoscan*, **70**: 177, 1970). Lo Buglio et al. (*SEM*, **72**: 313, 1972) used latex particles in an immunologic study to identify cell surface antigens.

Freeze-Etching

Steere (*J. Biophys. Biochem. Cytol.*, **3**: 45, 1957) is credited with development of principles of freeze-etching. In essence, an object is quickly frozen and briefly lyophilized. A replica is made and examined by electron microscopy. Moor and coworkers (*J. Biophys. Biochem. Cytol.*, **10**: 1, 1961; *Z. Zellforsch. Mikrosk. Anat.*, **62**: 546, 1964; *Int. Rev. Exp. Pathol.*, **5**: 179, 1966; *Int. Rev. Cytol.*, **25**: 391, 1969) have made several refinements of the methods that are described below.

In order to maintain integrity of structures in a state which most nearly approximates that in vivo, tissues must be supercooled as rapidly as possible. Studies by Riehle (*Chem. Ing. Tech.*, **40**: 213, 1968) indicate that ice crystals grow rapidly between 0 and $-100°C$, and that the speed of crystal growth is much less at $-100°C$ than at $-15°C$. In order to obtain an amorphous crystalline state, it is essential for the temperature to fall from 0 to $-100°C$ as rapidly as possible. Damage to cells by ice crystals results from both mechanical injury and chemical injury due to a dehydration in areas of cells exclusive of ice crystals. During ice-crystal formation, a separation is believed to occur between pure water and a more concentrated remaining solution or content.

Moor (*Int. Rev. Cytol.*, **25**: 391, 1969) believes better structural preservation is obtained in tissues immersed briefly (10 to 30 min) in 20 to 30% glycerol in buffer or saline solution prior to quenching. Moor also recommends the use of Freon-22 ($CHClF_2$, bp $-41°C$, mp $-160°C$) rather than isopentane, propane, or other explosive gas. Moor and Mühlethaler (*J. Cell Biol.*, **17**: 609, 1963) and others recommend that thin tissue samples be quick-frozen on the surface of a small copper plate cooled in Freon-22 with liquid nitrogen.

Methods for fracturing cells vary widely. A commercial apparatus such as Balzers Model BA360M (Balzers High Vacuum Co., Furstentum Liechtenstein) employs a microtome with a remote-control mechanical advance similar to early models of the Porter-Blum ultramicrotome. It appears likely that a coarse advance is suitable. Some authors use a good-quality stainless steel razor blade such as the Eversharp Schick. The temperature of the specimen should be kept around $-100°C$. Maintenance of this temperature prevents minute condensations on the surface of the specimen. Knife marks always appear on sections; however, these are readily recognized and pose no problem for interpretation.

Etching is accomplished by sublimation of the ice. Moor et al. (*J. Biophys. Biochem. Cytol.*, **10**: 1, 1961) recommend a high vacuum of 10^{-6} torr with a distance of only 2 mm between the trap ($-196°C$) and the specimen ($-100°C$). Fluctuations of temperature (even $1°C$) are regarded as deleterious and causing distortions of surface architecture. It is believed important to keep the object temperature constant (i.e., $-100°C$) during cutting, etching, and coating. If sublimation described in this paragraph is not employed, the process represents the *freeze-fracturing* method. Mühlethaler (*Int. Rev. Cytol.*, **31**: 1, 1971) has described a device that permits retention and examination of both surfaces after fracture.

Coating is performed to obtain a replica of a surface exposed by the fracture and subsequent freeze-drying. The replica is obtained by shadowing the object with a heavy metal followed immediately with a carbon-backing film. The carbon provides mechanical stability to the replica. Replicas are made in many ways (Bradley, in D. H. Kay (ed.), "Technics for Electron Microscopy," Davis, Philadelphia, Pennsylvania, 1965); however, the carbon-platinum film is regarded as best for freeze-etching studies. Accurate replication down to 50 Å appears to be achieved in a grainless background. The use of elements such as iridium, tantalum, rhenium, and tungsten with higher melting points may be expected to increase resolution in future studies.

After formation, the replica must be removed from the freeze-etching apparatus. Within a few seconds, the specimen thaws. Most specimens may be dissolved in 40% chromic acid (CrO_3) within 2 h without interference with the replica. Other mineral acids or alkali may sometimes be needed. If the shadowing material is even slightly soluble in the solvent, saturation of the solvent with the corresponding metal will minimize changes in the replica. The freed replica is cleansed with distilled water, attached to an electron microscopic screen that has been covered with Formvar, and examined in the usual fashion. Surface structures and structures that may be opened to the surface are particularly well delineated.

NEGATIVE STAINING

Negative staining may be used in either light or electron microscopy. The process may be employed if the object under investigation does not admit conventional stains. For light microscopy, the specimen is placed in a black dye such as nigrosin or india ink. The object appears transparent against a black background.

Hall (*J. Biophys. Biochem. Cytol.*, **1**: 1, 1955), Huxley [*Proc. Eur. R. Conf. Electron Microsc. (Stockh.)*, p. 260, 1956], and Brenner and Horne (*Biochim. Biophys. Acta*, **34**: 103, 1959) developed the method for electron microscopy. The method consists of embedding an electron-transparent object in a structureless electron-dense medium. Potassium phosphotungstate is most frequently employed. In practice, either a suspension of the specimen in water or ammonium acetate mixed with phosphotungstate is sprayed onto carbon-coated grids (Brenner and Horne, ibid.), or a thin film is spread on the grid and allowed to dry (Bradley and Kay, *J. Gen. Microbiol.*, **23**: 553, 1960; Huxley and Zubay, *J. Mol. Biol.*, **2**: 10, 1960).

The reader is referred to several papers for details of the negative staining process as applied to electron microscopy. Bradley (*J. Gen. Microbiol.*, **29**: 503, 1962) studied the effectiveness of several materials for negative staining of viruses. Horne [*J. R. Microsc. Soc.*, **83**: 169, 1964; *Lab. Invest.*, **14**: 316, 1965] applied negative staining methods to the study of bacteria, collagen, viruses, subcellular particles, and membranes. Glauert (*Lab. Invest.*, **14**: 331, 1965) examined several biologic constituents with negative staining and emphasized the need for the use of several stains in order to retain accurate information on the structure under investigation. Gilëv (*Biochim. Biophys. Acta*, **79**: 364, 1964) applied negative staining to a study of muscle myofilaments. Muscatello and Horne (*J. Ultrastruct. Res.*, **25**: 73, 1968) studied the effect of tonicity of negative staining solutions on the structural appearance of membranes on red blood cells, mitochondria, and microsomes. Prezbindowski et al. (*Exp. Cell Res.*, **50**: 241, 1968) applied the phosphotungstate negative staining technic to a study of microsomal membranes. Monroe and Brandt (*Appl. Microbiol.*, **20**: 259, 1970) applied the negative staining technic to develop a rapid semiquantitative method to screen large numbers of virus samples. Müller and Meyerhoff [*Nature (Lond.)*, **201**: 590, 1964] note that anomalous contrast may play a role in negative staining. Harris and Agutter (*J. Ultrastruct. Res.*, **33**: 219, 1970) used negative staining in a study of human erythrocytes and liver cell nuclei. Stoeckenius (*J. Cell Biol.*, **17**: 443, 1963) used negative staining to study a spherical particle approximately 85 Å in diameter on the surface of mitochondrial cristae.

Callahan and Horner (*J. Cell Biol.*, **20**: 350, 1964) recommend the use of either vanadyl sulfate ($VOSO_4 \cdot 2H_2O$) or vanadatomolybdate [$3(NH_4)_2O \cdot 2V_2O_5 \cdot 4\ MoO_3 \cdot nH_2O$] as a stain for electron microscopy. They describe the stain as providing (1) uniform and effective contrast enhancement, (2) simplicity of preparation, (3) short and uncomplicated application, and (4) freedom from precipitate. All membranes, including Golgi, cytoplasmic, reticular, mitochondrial, and nuclear, stain uniformly. Ribosomes are well stained, but slightly less so than with lead hydroxide. Glycogen stains intensely. Secretion granules of pituitary cells stain with an intensity that appears to vary with stages of the secretory process. Background cytoplasmic material is not stained very intensely. This may contribute to the apparent selective contrast observed.

VANADYL SULFATE. A 1% solution of vanadyl sulfate ($VOSO_4 \cdot 2H_2O$) is prepared (pH 3.6) and is satisfactory for staining for about 2 weeks. Crystals of vanadium dioxide then form, whereupon the stain is ineffective.

VANADATOMOLYBDATE. The staining solution is prepared by mixing 20 ml 1% vanadyl sulfate ($VOSO_4 \cdot 2H_2O$) with 80 ml 1% ammonium heptamolybdate [$(NH_4)_6Mo_7O_{24}$]. A dark purple solution forms initially which oxidizes further to form a stable, clear yellow staining solution with a pH of 3.2. The staining solution is effective for 6 to 12 months. The stain is filtered through a 0.01-μ millipore filter prior to use.

In practice, sections of osmic acid–fixed, Araldite- or Epon-embedded sections are placed on a grid. A drop of stain is placed in a porcelain spot plate, and grids are placed on the drop with the tissue side down. A large glass slide is placed over the grids to eliminate dust accumulation and reduce evaporation. Sections are stained for varying periods of time from 5 to 30 min. Ten minutes frequently provided a good stain. After staining, sections are rinsed by holding the grid in a stream of distilled water from a wash bottle. Grids are then dried and viewed. Callahan and Horner favor slightly the vanadatomolybdate stain over vanadyl sulfate.

Although Huxley and Zubay (*J. Biophys. Biochem. Cytol.*, **11**: 273, 1961), Stoeckenius (*J. Biophys. Biochem. Cytol.*, **11**: 297, 1961), and Zobel and Beer (*Int. Rev. Cytol.*, **18**: 363, 1965) provide evidence for the selective uptake of uranyl acetate with nucleic acids (possible phosphate), Lombardi et al. (*J. Histochem. Cytochem.*, **19**: 161, 1971) employed deamination blocking procedures to formaldehyde-fixed pancreas and concluded that free amino groups may also participate in the reaction. [Deamination adequate?—EDITOR.]

Sternberger et al. (*J. Histochem. Cytochem.*, **11**: 48, 1963) were able to obtain specific staining of *Bordetella bronchiseptica* by using uranium-labeled antibody. Briefly, the steps involved are (1) removal of antibody from antiserum by agglutination with bacteria, (2) exposure of washed agglutinate to relatively large amounts of uranium (the specific combining sites of the antibody in the agglutinate are now covered by antigen), and (3) dissociation of antibody from antigen by brief treatment with alkali in the cold, yielding specifically purified antibody containing uranium in nonspecific areas, but preservation of crucial combining sites. In additional studies Sternberger et al. (*J. Histochem. Cytochem.*, **14**: 711, 1966) further increased contrast of the uranium by using thiocarbohydrazide as a bidentate ligand to bridge uranium to osmium.

Schreil (*J. Cell Biol.*, **22**: 1, 1964) added solutions of various types of DNA and nucleohistones to several osmic acid–, uranyl acetate–, or indium chloride–containing fixatives. The samples were then either viewed with polarized light or embedded and examined by electron microscopy. Schreil concluded that fixatives that cause gelation of DNA [OsO_4 at pH 6 in the presence of amino acids (tryptone) and calcium; uranyl acetate; indium chloride] result in the most accurate preservation of DNA.

ELECTRON PROBE MICROANALYZER

The electron probe developed from works by Mosley (*Phil. Mag.*, **26**: 1024, 1913, and **27**: 703, 1914), von Ardenne ("Elektronenubermikroskopie," Springer-Verlag, Berlin, 1940), Zworykin et al. ("Electron Optics and the Electron Microscope," Wiley, New York, 1945), Hillier (United States Patent 2,418,029; 1947), and Castaing and Guinier (*Proc. Int. Conf. Electron Microscopy (Delft)*, p. 60, 1949). For a current review of the instrument and methods for qualitative and quantitative analysis, the reader is referred to Anderson (see General References).

Under appropriate conditions, the electron probe microanalyzer makes possible a qualitative and quantitative chemical analysis of elements in thin tissue sections. Components of the electron probe microanalyzer include an electron optical system, an electron beam scanning system, an x-ray detection system, a light microscopic optical system, a mechanical stage, data recording and display systems, power and vacuum systems, and electronic circuitry. Elements above beryllium can be analyzed with good sensitivity in a volume of a few cubic microns. Limits of detection range in the order of 10^{-15} g, and relative weight fraction detection limits are of the order of 0.01 to 0.10% for most elements in biologic specimens. A light microscope or a scanning electron microscope is used to select the area to be analyzed. Several elements can be analyzed simultaneously and rapidly. The localized spectrographic analysis is accomplished by the generation of a characteristic x-ray spectrum from elements in the microscopic area which are excited by a finely focused beam of high-energy electrons. Either qualitative or quantitative analyses may be conducted. Qualitative analyses are unequivocal and absolute due to emission of characteristic x-ray spectra from each element. Quantitative analyses are obtained by comparison of the spectrographic intensity of the unknown to a standard.

TISSUE PREPARATION. Tissues are prepared in various ways depending upon the nature of the substance to be identified and whether it is diffusible or not. Tissues prepared as for ordinary light microscopy are sometimes employed. Tissues embedded as for electron microscopy may be used for certain purposes. Ingram and Hogben (1968) and Ingram et al. (*J. Histochem. Cytochem.*, **20:** 716, 1972) advise the use of freeze-dried, osmic acid–fixed, and Epon-embedded tissues. Ingram et al. stress the need for a tissue preparation that (1) presents a flat surface to the electron beam in a vacuum, (2) can withstand a vacuum, (3) can withstand electron bombardment, and (4) retains the in vivo localization of the material under investigation. Ingram et al. report that with a 10-kV, 50-nA electron beam, spatial resolution is comparable to that of light microscopy with detection limits for K, Cl, and Na of a few milliequivalents per liter. In a study of frog skeletal muscle fibers, Ingram et al. detected 122.4 ± 2.8 (standard error) meq/l for K; 10.5 ± 0.7 meq/l for Cl, and 9.2 ± 1.0 meq/l for Na as an average for nine fibers from one muscle.

Lehrer and Berkley (*J. Histochem. Cytochem.*, **20:** 710, 1972) described a procedure for the preparation of gelatin standards for x-ray electron probe microanalysis. Gelatin solutions (10 to 15%) containing varying concentrations of the elements to be measured are quick-frozen in capsules, sectioned with a cryostat at a thickness comparable to the tissues, freeze-dried, mounted with each tissue section, carbon-coated, and analyzed simultaneously with tissues. Lehrer and Berkley report good accuracy and reproducibility are achieved only with sections thicker than 10 μ; that is, in volumes 10^{-12} l (1000 μ^3) or greater. Concentrations of elements may be determined directly from a regression curve or an equation derived from the standards.

Langer et al. (*J. Histochem. Cytochem.*, **20:** 723 and 735, 1972) used electron probe microanalytical methods to study asbestos bodies from lungs of individuals with asbestosis.

Weavers (*Histochem. J.*, **5:** 173, 1973) provides a short review on the potentiality of the combined transmission electron microscope x-ray microanalyzer.

LASER MICROPROBE ANALYSIS

Policard (*Harvey Lect.*, **27:** 204, 1932) and Gerlach and Gerlach ("Die chemische Emissionsspektralanalyse," vol. II, "Anwendung in Medizin, Chemie, und Mineralogie," Voss, Leipzig, 1933; translated by J. H. Twymann, "Clinical and Pathological Application of Spectrum Analysis," Hilger, London, 1934) were early workers in the development of laser microprobe analysis. Glick (*J. Histochem. Cytochem.*, **14:** 862, 1966) describes the use of a ruby microprobe apparatus. In essence, the area to be analyzed in a tissue section is brought into focus with a microscope. The laser beam is flashed through the same objective to vaporize the sample. A Q-switch device is used whereby a main spike flash [5 MW (megawatts)/pulse] is followed within 200 ns by two weaker spikes lasting about 50 ns each. At the focus of the flash, a temperature of about 10,000 K develops, and a crater forms due to vaporization of tissue without an intermediate liquid phase. The light emitted directly due to tissue vaporization is too low for spectrographic analysis. However, when cross excitation provided by the electric discharge is applied, light intensity of the emission is sufficiently high for photographic recording with a recording microdensitometer. Sensitivity of the technic depends upon the efficiency of light collection, spectrographic resolution, and recording sensitivity. The method has been applied by Rosan et al. (*Science*, **142**: 236, 1963) in a study of freeze-dried sections of brain. Rosan et al. (*Fed. Proc.*, **24** (*Suppl.* 14): S126, 1965) also used the method to study sections of teeth. One can expect the method to be applied to histochemical staining reactions whereby tagged products are deposited specifically in a stoichiometric fashion.

Kaufmann et al. (*J. Histochem. Cytochem.*, **19**: 469, 1971) used argon and cadmium lasers to excite fluorescence in fluorescein isothiocyanate (FITC)–tagged protein conjugates in attempts to develop high-resolution scanning microscope applications. Laser beams induced at least 100 times more fluorescence than did light from conventional lamps. Futhermore, nearly complete recovery of initial fluorescence was observed in samples allowed to rest in the dark for 48 h. Also, laser excitation caused a change in color of FITC conjugates.

FLUORESCENCE MICROSCOPY

After irradiation with ultraviolet, violet, blue, or green light, certain substances have the property of emitting radiation of their own at a wavelength longer than that of the exciting source. Such emission is called *luminescence*. If the luminescence has the capability of persisting for an appreciable period of time, the luminescence is called *phosphorescence*. If substantial emission persists only while the exciting source acts on the substance, the luminescence is called *fluorescence*. If fluorescence is observed in specimens that have not been treated with a fluorochrome (dye), the emission is called *autofluorescence*. Fluorescence acquired after treatment of a specimen with a fluorochrome is called *secondary fluorescence*.

Well-known fluorochromes include coriphosphine O (C.I. 46020), acridine orange (C.I. 46005), aurimine O (C.I. 41000), rhodamine B (C.I. 45170), and phosphine 3R (C.I. 46045). Fluorescent compounds may also be obtained by treatment of specimens with various chemical reagents. For example, biogenic amines such as catecholamine and hydroxytryptamine condense with formaldehyde to form a fluorescent end product.

Special fluorescence microscopes are becoming more useful and popular. Fluorescence microscopes now permit transmitted fluorescence microscopy, incident light excitation fluorescence microscopy, and a fluorometer useful for quantitative studies.

Figure 19-5 is a diagram of a microscope setup for transmission fluorescence microscopy. A suitable exciting light source (1) is employed which passes through an exciting (primary) filter (2) generally situated in the lamp housing. These selected primary filters permit transmission of exciting radiation only, which passes through the condenser (3) to excite fluorescence of the desired object. Incident light excitation acts in the same fashion. Fluorescent light then passes through the objective (4) and another specifically selected secondary filter (5) matched for the appropriate wavelength to exclude radiation that may damage the eye. The filter may absorb (glass filter) or reflect (interference filter), resulting in a dark background as it appears through the eyepiece (6).

Figure 19-6 provides details on several kinds of filters. In addition to the designations of primary and secondary filters provided above, filters are characterized in other functional

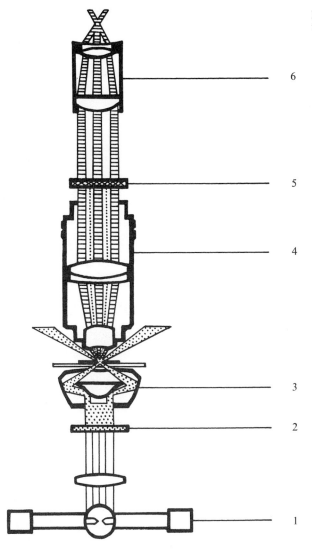

Fig. 19-5 Basic construction of the fluorescence microscope.

terms. Absorption filters act to disperse certain wavelengths. Interference filters reflect certain wavelengths and permit passage of others. Absorption filters are more widely used because (1) they are relatively inexpensive, (2) they are more efficient, though less selective, than interference filters in absorbance of undesired excitation light, and (3) they permit efficient transmission.

Interference filters are of two types: dielectric and metallic. Dielectric interference filters that are especially designed for fluorescein isothiocyanate (FITC) excitation comprise layers of substances such as sublimed zinc sulfate and magnesium fluoride on glass. The layers are selected to permit transmission of desired wavelengths only. They have a sharp cutoff. Light that is transmitted is called the *passband*. Such a filter is then called a *short-wave*

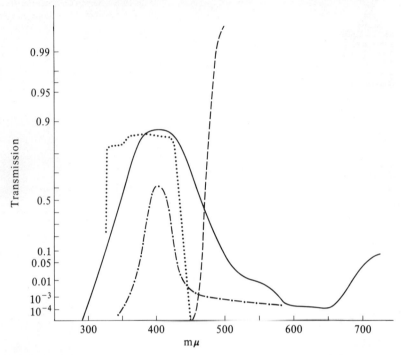

Fig. 19-6   Schematic presentation of characteristics of different types of filters. — Wide passband absorption filter; ---- long-wave pass absorption filter with sharp cut-on; -------- narrow passband interference filter; .... short-wave pass interference filter with sharp cutoff. Note that in this scheme the transmittance is not given in percentages, and that the wavelengths for each transmission spectrum are arbitrary. Wide passband filters have a half-band width of more than 50 mμ, and narrow passband filters have a half-band width of less than 25 mμ. (*From Faulk and Hijmans, Prog. Allergy,* **16**: 9, 1972. *By permission of the publisher.*)

pass interference filter. The term short-wave pass means that the filter has a cutoff for the longer wavelengths. A *long-wave pass* filter has cutoff for shorter wavelengths. Presently, all short-wave pass filters are interference filters, and all long-wave pass filters are absorption filters. Figure 19-7 illustrates some of these terms.

Inasmuch as stimulation of fluorescence of presently desired conjugates must be accomplished at a wavelength other than the most efficient, a powerful light source must be employed. The high-pressure mercury lamp is most commonly used. It provides powerful ultraviolet-blue light with principal peaks at 365 and 435 mμ. Strong light is also provided at 546 mμ (green) and 578 mμ (yellow). The strong mercury lines are very efficient for the excitation of several fluorochromes, and for this reason they will find use in epi-illumination systems. Table 19-1 provides details on several illuminating and filter systems. Faulk and Hijmans indicate that the ordinary low-voltage tungsten lamps and more especially the halogen lamps are sufficiently bright for many immunofluorescence studies. Maximum fluorescence is generally achieved by providing light of wavelengths maximally absorbed by the fluorochrome. Since the conjugates of fluorescein and rhodamine B absorb at 495 and 575 mμ respectively and emit at 520 and 595 respectively, light at maximal absorbance

TABLE 19-1 SEVERAL OPTICAL SYSTEMS WITH EXAMPLES OF THEIR APPLICATION

System of illumination	Numerical aperture*	Light source†	Primary filters	Interference dividing plate‡	Secondary filters§	Fluoro-chromes	Remarks on application
Transmitted light	Low (<0.70)	Halogen 100 W	Short-wave pass interference filter		515	FITC Rhodamine FITC + rhodamine	Tissue sections, in which autofluorescence does not interfere with assessment.
Epi-illumination	High (>0.70)	Halogen 100 W or mercury 50 W	Two short-wave pass interference filters	495	510	FITC	As above, especially where high magnification is required (e.g., cells, microorganisms).
		Mercury 100 W or xenon 75 W	Same	495	510	FITC	As above, especially when weak fluorescence can be anticipated (e.g., membrane fluorescence).
		Mercury 100 W or xenon 75 W	Same, plus a long-wave pass filter with 50% T‡ at 455 or 475 mμ; or two narrow passband interference filters with T_{max} at 480–485 mμ	495	510 or fluorescence selection interference filter with T_{max} at 525–530 mμ	FITC	Narrow-band excitation eliminates short-wavelength excitation, which causes autofluorescence; the use of the fluorescence selection filter is mandatory if the fluorescein dye is combined with rhodamine.
		Mercury 100 W	Narrow passband interference filter with high T at 546 mμ plus absorption filter with low T at 578 mμ	580	590	Rhodamine	This system is extremely efficient.

* Objectives with low magnification generally have a low numerical aperture (NA). Objectives with low magnification in combination with a high numerical aperture are being developed for use in epi-illumination.

† Both mercury and xenon lamps are of the high-pressure type.

‡ The numbers refer to the wavelengths in millimicrons (nanometers) at which the transmittance (T) of the filter is 50% of the peak transmittance (T_{max}). This is called λH.

§ Absorption-type barrier or secondary filters with a 50% transmittance at the indicated wavelength in millimicrons (nanometers) are inserted in the emission beam.

Note: These schemes are based on equipment available in 1972. They should be adapted to new developments. Improved schemes can be expected with a high numerical aperture for epi-illumination become available.

A heat-protective filter and a red-suppression filter should always be inserted in the excitation beam next to the light source.

Source: From Faulk and Hijmans, *Prog. Allergy*, **16**: 9, 1972.

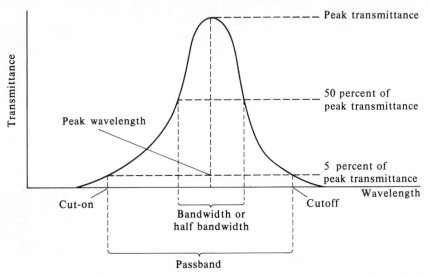

Fig. 19-7 Some graphic definitions of filters. (*From Faulk and Hijmans, Prog. Allergy,* **16:** 9, 1972. *By permission of the publisher.*)

peaks cannot be provided, because adequate selective filtering cannot be provided. Filters that prevent penetration of primary illumination also excessively reduce transmission of the desired fluorescence, although Tomlinson (*Immunology,* **13:** 323, 1967) has experimented with an iodine-quartz lamp and filter system in an attempt to overcome this difficulty. The Tomlinson system appears to lack the desired sensitivity and versatility. Attempts are being made to develop better filters (Rygaard and Olsen, *Acta Pathol. Microbiol. Scand.,* **76:** 146, 1969; Faulk and Hijmans, *Prog. Allergy,* **16:** 9, 1972).

EPI-ILLUMINATION

Faulk and Hijmans (*Prog. Allergy,* **16:** 9, 1972) predict a greater use of water immersion objectives and epi-illumination in the future. The water used for lens immersion soon evaporates, leaving a clean surface. The Ploem type of vertical illumination (*Z. Wiss. Mikrosk.,* **68:** 129, 1967) is illustrated in Fig. 19-8. The Ploem type of vertical illuminator is provided with an interference dividing plate (dichroic mirror) mounted above the objective at an angle of 45° to the illuminating beam. The objective thereby acts as a condenser. The interference plate can be matched for excitation and transmission to the fluorochrome being used. With the aid of several filters almost any light source can be employed. Other advantages of epi-illumination are elimination of the need for careful focusing and centering of the condenser and the fact that it may be readily combined with transmitted illumination. It permits a combination of fluorescence microscopy with epi-illumination and tungsten illumination using phase contrast, interference contrast, or polarization condensers. If microfluorometry is undertaken, transmitted tungsten illumination may be used to focus on cells and epi-illumination used for microfluorometric measurements. This permits reduction of exposure of the specimen to destructive wavelengths which cause fluorescence to fade.

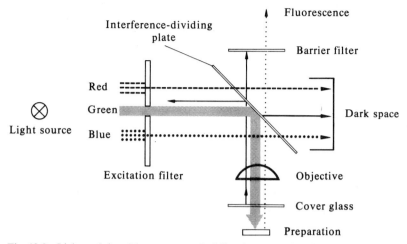

Fig. 19-8 Light path in a Ploem-type vertical illuminator. An interference dividing plate is mounted above the objective. The filter systems can be adapted to the fluorochrome. In this example, the illuminator is equipped for rhodamine with green excitation light. (*From Faulk and Hijmans, Prog. Allergy,* **16:** 9, 1972. *By permission of the publisher.*)

IMMUNOFLUORESCENCE

Fluorochromes such as fluorescein isothiocyanate (FITC) and tetramethyl rhodamine isothiocyanate (TRITC) may be coupled to antibodies by covalent bonds and subsequently reacted with a specific antigen. The antigen-antibody complex is identified by a specific fluorescence after excitation with the appropriate wavelength of light.

There is no uniform immunofluorescence method. Methods of preparation of antiserums and antigens vary. The following method for the preparation of antiserums is advocated by Nairn (*Clin. Exp. Immunol.,* **3:** 465, 1968). The reader is referred to this paper for further details.

Preparation of Antiserums

Laboratories that use large volumes of antiserums should attempt to attain large batches to last for a year or for the duration of a project. For example, it is advantageous to have antihuman globulin continuously available from the same batch. Comparisons can be made over a long period of time. Likewise, it is better to have antiserum from a single large animal, if possible, rather than a pooled batch from several small animals. The antigen should be as pure as possible, and an attempt should be made to obtain antiserums of high titer and narrow specificity. Prior to immunization, it is important to obtain a large sample of preimmune serum for control purposes.

For anti-IgG, a total dose of 0.5 mg/kg body weight of IgG obtained by diethylaminoethyl (DEAE) cellulose fractionation of human serum is injected intramuscularly in four sites in a goat. The globulin in 0.5% solution in phosphate-buffered saline solution (0.145 M NaCl, 0.01 M phosphate of pH 7.1) is first emulsified with an equal volume of Freund's adjuvant (Progressive Laboratories, Baltimore, Maryland). The dose is repeated 6 weeks later. Three weeks later 500 ml blood is removed from the jugular vein. At this stage, serum normally contains not less than 10 mg antibody protein per ml against IgG with minor activity against other immunoglobulins. One booster injection of the same antigenic preparation is permitted 6 to 12 weeks after the last immunization dose, and the animal is then bled to obtain 2.5 l blood. It is imprudent to use the goat for further booster immunizations because of increased nonspecific antibody produced. Antigen preparations must be in appropriate titer. Concentrations less than ideal may fail to provoke an adequate immune response, and concentrations too high may result in immunologic paralysis.

Serums should be characterized as definitively as possible. For precipitating systems, immunoelectrophoresis is generally employed (Grabar and Burtin, "Immuno-electrophoretic Analysis," Elsevier, Amsterdam, 1964). Protein assays of precipitates may also be employed (Kabat and Mayer, "Experimental Immunochemistry," Charles C Thomas, Springfield, Ill., 1964). Fluorescence polarization may be a useful method in the future (Dandliker et al., *Immunochemistry*, **1**: 165, 1964) for a quantitative determination of nonspecific antibody in a serum. In addition, whenever necessary, double dilution titrations of serums should be made instead of absolute estimations of antibody content of serums. They may be identified as immunofluorescence titers against specified antigens in a microscopic preparation or a specified solution of antigen. When international standards become available, potencies should be stated.

Characterized, titrated, and sterilized (if necessary) serums should be stored at $-20°C$ or lower in sealed containers of 1- to 5-ml volumes. Freezing and thawing should be accomplished rapidly. For freezing, CO_2–ethanol may be used. For thawing, a 37°C water bath is recommended. Lyophilization usually results in a small reduction in titer, but it may be necessary for prolonged storage or for shipping. Commercially produced antiserums may be satisfactory; however, the reliability of the samples should be verified. If commercial preparations are used, it is important to have sufficient material from the same batch for completion of a project. Standard reference serums for human IgG, IgA, and IgM are available (Rowe et al., *Bull. WHO*, **42**: 535, 1970). Faulk and Hijmans (*Prog. Allergy*, **16**: 9, 1972) provide the following basic information required of antiserums:

1. The species of animal immunized
2. The source and method of preparation of the antigen
3. The immunizing schedule, route, and adjuvant
4. The absorbent and absorption procedure, and evidence of specificity before and after absorption
5. The immunoglobulin separation process or other chemical treatment, particularly the preservative used and any other dilutions or additions
6. Specificity as determined by standard, reproducible tests
7. Some expression of the potency of the antiserum

Conjugation with Fluorochrome

Fluorescein isothiocyanate (FITC) appears to be the label of choice by many investigators. FITC has two isomers with varying impurities. Corey and McKinney (*Anal. Biochem.*, **4**: 57, 1962) noted that conjugation is effected more efficiently if pure samples of the fluorochrome are employed. Nairn (*Clin. Exp. Immunol.*, **3**: 465, 1968) advises the use of isomer I of FITC for optimal staining. The pure crystalline form may be obtained from Baltimore Biological Laboratories. Less pure reagents may be obtained more cheaply; however, purification is more difficult, and increased nonspecific staining results. A loss of FITC is noted in commercial batches of fluorochrome containing substantial amounts of HCl (up to 4.9%) (McKinney et al., *Anal. Biochem.*, **8**: 525, 1964). FITC should be stored in the dark over a desiccant.

Rhodamine B sulfonyl chloride is also a satisfactory fluorochrome and is frequently employed for color contrast if two antigens are to be identified in a single preparation. Rhodamine conjugates fade less rapidly than do FITC conjugates when exposed to ultraviolet light.

Tetramethylrhodamine isothiocyanate (TRITC) and rhodamine B sulfonyl chloride emit a red fluorescence. TRITC is more labile than FITC and should be stored over a desiccant in a freezer. TRITC is available from Baltimore Biological Laboratories, Baltimore, Maryland.

The structures of FITC and rhodamine B conjugates to proteins are depicted below:

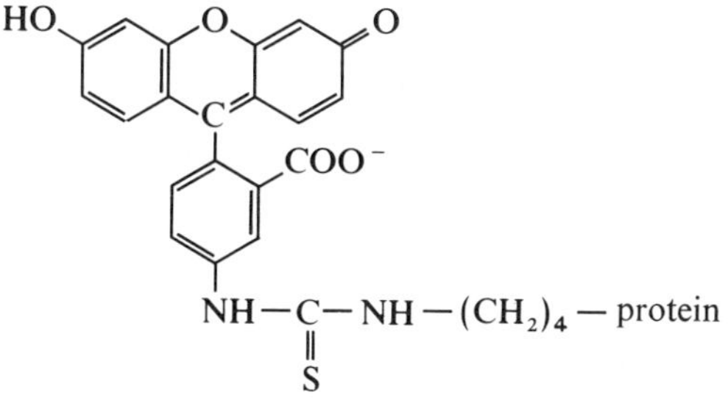

FITC conjugate with ε-amino group of a lysine residue in a protein

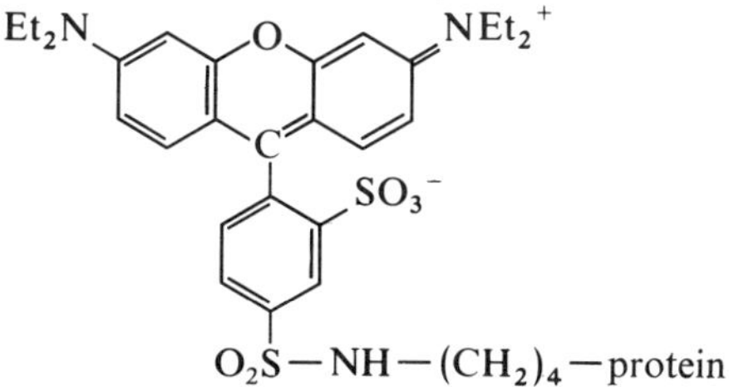

RB 200 conjugate with ε-amino group of a lysine residue in a protein

As a consequence of conjugation with either fluorochrome, an increase of two units of negative charge at pH 7 results for each molecule of fluorochrome added. This results in an increase in nonspecific attraction between conjugates and objects positively charged in microscopic preparations. To obtain a compromise between specific and nonspecific effects many workers attempt to achieve 0.5 to 1.3% fluorochrome conjugation. This represents about 2 to 5 molecules of fluorochrome per globulin molecule. It is important to employ suitable dye-protein ratios because (1) immunofluorescence is blocked by unconjugated antiserum, (2) antiserum insufficiently labeled will provide inadequate staining, and (3) excessively labeled antiserum results in substantial nonspecific staining.

For routine use Nairn recommends conjugated immunoglobulins containing 1% of the fluorescein thiocarbamyl portion. Nairn's (*Clin. Exp. Immunol.*, **3:** 465, 1968) procedure, which is a modification of McKinney et al. (*J. Immunol.*, **93:** 232, 1964), follows:

1. A pure γ-globulin fraction of serum is obtained by precipitation with 40% saturated ammonium sulfate.
2. Place all reagents in a water bath at 25°C.
3. Use 12.5 μg FITC per mg globulin. For example, a 4-ml sample of 2% globulin solution would require 1 mg FITC which is completely dissolved in 2 ml 0.1 M Na$_2$HPO$_4$, pH 9, by crushing and stirring with a glass rod for 5 to 10 min. This solution is unstable,

and after equilibration at 25°C it should be used within 2 h. TRITC is less soluble than FITC; however, it will dissolve if shaken vigorously or ground in a mortar with carbonate buffer (Faulk and Hijmans, *Prog. Allergy*, **16**: 9, 1972).

4. Add dropwise over 2 to 3 min while stirring 1 ml 0.2 M Na_2HPO_4 to the 4-ml globulin sample in a 15- to 20-ml glass stoppered container.
5. Add FITC solution in the same manner.
6. Rapidly adjust the pH to 9.5 with drops of 0.1 M Na_3PO_4.
7. Add sufficient 0.145 M NaCl to make a total volume of 8 ml, stopper, and mix reactants.
8. Place in a water bath at 25°C for 30 min without agitation.
9. Remove to an ice bath, swirl to cool, and remove the precipitate by centrifugation.

White [Fluorochromes and Labelling, in Holborow (ed.), "Standardization in Immunofluorescence," Blackwell, Oxford, 1970] has noted that the thiocyanates conjugate more efficiently to α- and β-globulins than to γ-globulins. More uniform labeling may be obtained if conjugates are prepared from purified IgG obtained after chromatography or the use of immunoabsorbents.

Although Wood et al. (*J. Immunol.*, **95**: 225, 1965) have been able to purify conjugates with the aid of DEAE cellulose fractionation, Nairn has obtained poor yields using this method. For routine purposes Nairn advises the use of less elegant nonspecific absorptions with homogenates of fresh tissues of frozen stored tissues. The most common absorption technic is to stir, without frothing, 1 volume tissue homogenate with 2 volumes serum for 2 h at 23°C. The serum is recovered by centrifugation at 10,000 g/20 min. It is important that parallel absorptions of different serums in immunofluorescence control studies are made identically.

Satisfactory labeling of globulin-enriched antiserum may be prepared by salt fractionation according to Stelos [in Weir (ed.), "Handbook of Experimental Immunology," Blackwell, Oxford, 1967] expecially if chromatography is subsequently used to remove unreacted dye. Salted out antiserum must be dialyzed to remove the salt, which interferes with labeling.

Purification of the Conjugate

The conjugate will always contain some unreacted fluorochrome which must be removed. This is best accomplished by gel filtration (Sephadex G25 at 2 to 4°C). The column volume should be 6 times the sample volume, and standard phosphate-buffered saline solution is used for elution. The first colored fraction to elute is the conjugated globulin which can be concentrated by negative-pressure dialysis in the cold against buffered saline solution. Fluorochrome-protein ratios should be determined to assess the degree of labeling [Fothergill, in Nairn (ed.), "Fluorescent Protein Tracing," 2d ed., p. 34, Williams & Wilkins, Baltimore, 1964]. White [Fluorochromes and Labelling, in Holborow (ed.), "Standardization in Immunofluorescence," Blackwell, Oxford, 1970] does not advise the use of DEAE cellulose chromatography to remove unreacted dye from antiserum excessively labeled with fluorochrome. The conjugates tend to remain on the column.

Arnold and Mayersbach (*J. Histochem. Cytochem.*, **20**: 975, 1972) recommend the following method for purification of conjugates by precipitation. Nonspecific staining is stated to be abolished or reduced maximally.

1. Conjugation of rabbit IgG with 0.8 or 1% fluorescein isothiocyanate (FITC).
2. Dialysis against 0.01 M sodium phosphate buffer of pH 6.2 to 6.8 for 3 days at 4°C.
3. The conjugates are centrifuged at 10,000 × gravity for 20 min at 4°C.

4. Readjust the pH to 7.5 by passage through a Sephadex G25 column equilibrated with 0.01 M sodium phosphate buffer of pH 7.5. The proportion of Sephadex bed to protein volumes is 3 : 1.

5. Using an Amicon Ultrafiltration cell, conjugates are concentrated to 10 mg protein per ml for immunologic studies and to 3 to 5 mg protein per ml for immunohisto-chemical studies.

6. Check for nonspecific staining in an immunologically unrelated IgG tissue system.

Arnold and Mayersbach note that IgG from varied sources may react differently with FITC and thereby result in differences in acid-solubility. It may be necessary to lower the pH to precipitate more of the nonspecific stainable material or to raise the pH to prevent heavy loss of protein. Nonspecific staining appears to be related to electrostatic adsorption of labeled IgG to positively charged components in tissues (Mayersbach, *J. R. Microsc. Soc.*, **87**: 295, 1967). According to Curtain [*Nature (Lond.)*, **182**: 1305, 1958], Goldstein et al. (*J. Exp. Med.*, **114**: 89, 1961), and McDevitt et al. (*J. Immunol.*, **90**: 634, 1963) the reduction of nonspecific staining in conjugates chosen after DEAE cellulose chromatography is due to the selection of fractions with low FITC-protein ratios which manifest limited electrostatic affinity for tissue constituents. Despite this reduction of nonspecific staining, DEAE cellulose chromatography is not generally employed because of a major loss of IgG during chromatography.

Preparation of Tissues for Immunofluorescence

Unfixed frozen sections are generally appropriate. Tissues quick-frozen in isopentane cooled with liquid nitrogen may be stored below $-60°C$ for long periods of time. Sections are cut with a cryostat in the usual fashion. If sections must be stored before staining, they should be kept below $0°C$ and in a sealed container containing a desiccant such as silica gel. Artifacts such as nuclear staining may occur in sections due to storage (Nairn, *Clin. Exp. Immunol.*, **2**: 697, 1967). Some antigens are preserved in sections stored in 2-octanol (Maxwell et al., *Stain Technol.*, **41**: 305, 1966).

Smears or monolayers of cells may also be stained with immunofluorescence methods. Cells may be disaggregated from tissues using 8 ml 5% EDTA, 20 ml 20% bovine serum albumin, 60 ml phosphate-buffered saline solution (0.145 M NaCl, 0.01 M phosphate of pH 7.1), 10,000 units penicillin G, and distilled water to make 100 ml. Dry $NaHCO_3$ is added to bring the pH to 6.8. After washing, cells may be concentrated using a cytocentri-fuge. Dead but intact cells permit conjugated antibody to enter, and intracellular identifica-tion of antigens is permitted.

Some antigens in tissues are not markedly affected by fixatives. For example, Yasuda and Coons (*J. Histochem. Cytochem.*, **14**: 303, 1966), Mazurkiewicz and Nakane (*J. Histo-chem. Cytochem.*, **20**: 969, 1972), and Poole et al. (*J. Histochem. Cytochem.*, **20**: 261, 1972) used fixatives containing formaldehyde and picric acid. Each antigen needs to be tested to determine if fixation is permissible. It is interesting to note that glutaraldehyde is fluores-cent and may interfere with fluorescence microscopy. Certain antigens may not withstand other procedures such as paraffin or other embedding methods. Depending upon the nature of the antigen and the tissue, fresh frozen, freeze-dried, freeze-substituted, or ethanol- or methanol-fixed tissues may be employed.

The staining procedure will vary depending upon the immunologic method selected. In general, sections equilibrated in phosphate-buffered (pH 8.6) saline are thereafter subjected to the specific conjugate for $\frac{1}{2}$ to 3 h at room temperature in a moist chamber. Non-reactive conjugate is removed by washes with phosphate-buffered saline solution. The

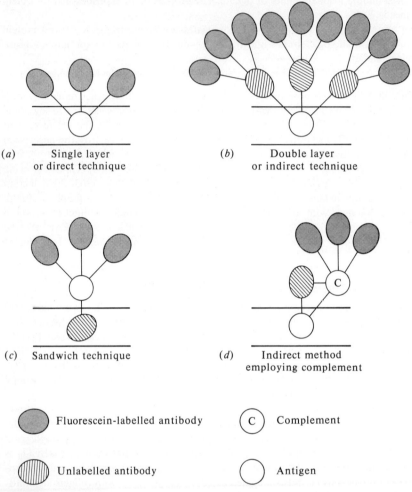

(*a*)   Single layer
or direct technique

(*b*)   Double layer
or indirect technique

(*c*)   Sandwich technique

(*d*)   Indirect method
employing complement

Fluorescein-labelled antibody

C   Complement

Unlabelled antibody

Antigen

Fig. 19-9   Diagrammatic representation of the various modifications of the fluorescent-antibody method. (*a*) Direct method for demonstrating antigen by the use of a single layer of fluorescein-labeled specific antibody. (*b*) Indirect or double layer technique: unlabeled specific antibody is used first and the material is subsequently treated with fluorescent antibody against γ-globulin. Thus if the first layer employed rabbit antibody, fluorescent antibody against rabbit γ-globulin would be used in the second layer. (*c*) Sandwich technique for the detection of antibody. The section is first treated with a dilute solution of antigen. After a wash to get rid of the excess antigen, the section is exposed to fluorescent antibody. (*d*) Indirect method for detection of antigen employing complement (fresh guinea-pig serum) and rabbit antibody to guinea-pig globulin. (*From Humphrey and White, "Immunology for Students of Medicine," 3d ed., Blackwell, Oxford, 1971. By permission of the publisher.*)

immunohistochemical methods for fluorescent, ferritin, and enzyme conjugates are very similar. A few selected fluorescent antibody methods in common use today are provided in Fig. 19-9. The single-layer, direct technic is advocated if it is possible to use it.

Nairn (1969) hesitates to recommend a permanent mounting procedure. He has had success with rapid dehydrations in ethanol with subsequent mounting in Fluormount or DPX. In some cases he notes excellent preservation of fluorescence for as long as 10 years. However, other preparations lost fluorescence for unknown reasons. Nairn suggests that specimens be air-dried after a buffered saline solution rinse followed by mounting in Fluormount or DPX.

Schenk and Churukian (*J. Histochem. Cytochem.*, **22**: 962, 1974) recommend the following counterstain for sections stained with immunofluorescent methods:

After sections are stained with either a direct or indirect immunofluorescence method, sections are stained for 5 min at 25°C in methyl green (C.I. 42590). The staining solution is 4 ml 0.1% solution of methyl green added to 36 ml phosphate-buffered saline solution. Sections are then washed for 5 min in phosphate-buffered saline solution and mounted in glycerol–phosphate-buffered (pH 8.6) saline solution.

Flannagen et al. also recommend staining sections as above followed by staining with Chromogen Black ETOO (C.I. 14645) for 10 s, rinsed in phosphate-buffered saline solution, and mounted in glycerol–phosphate-buffered saline solution.

Methyl green has maximal absorption at 420 mμ and emission at 630 mμ resulting in a red fluorescence of all nuclei. It does not affect background fluorescence, but excellent cellular detail is achieved. Elbadawi and Schenk (*J. Histochem. Cytochem.*, **15**: 580, 1967) first used methyl green as a counterstain in their studies of catecholamines. The combined methyl green–Chromogen Black ETOO method provides suppression of background fluorescence and good cellular detail; however, fading under ultraviolet light occurs more rapidly than with methyl green alone.

According to Faulk and Hijmans (*Prog. Allergy*, **16**: 9, 1972) factors that need to be considered in the control of nonspecific staining include quality of the dye, specificity of the antiserum, total serum or serum fraction conjugated, method of conjugation, presence of unreacted dye, dye-protein ratio, dilution of the conjugate, absorption with tissue powder, counterstains, specificity controls for staining, and method of tissue preparation.

Most nonspecific staining is related to factors that can be controlled. Nonspecific staining is reduced if high-quality dyes are used, if specific antiserums are used, and if labeling technics are carefully employed. If dye-protein ratios are high, increased electronegativity of the conjugate occurs with consequent increased attraction to cells producing staining artifacts. DEAE cellulose chromatography may be employed to remove the highly charged molecules (McDevitt et al., *J. Immunol.*, **90**: 634, 1963; Cebra and Goldstein, *J. Immunol.*, **95**: 230, 1965). Conjugates may also be absorbed with tissue homogenates to reduce non-specific staining [Holborow and Johnson, in Weir (ed.), "Handbook of Experimental Immunology," Blackwell, Oxford, 1967].

Faulk (*Prog. Allergy*, **16**: 9, 1972) has noted artifactual positive staining of normal human kidney with FITC-labeled antihuman IgG only if sections have been washed with phosphate-buffered saline solution and not if sections have been washed with nonionized solutions such as isotonic sucrose.

The most frequently employed mounting fluid is glycerol buffered with phosphate–saline solution. The pH should be about 8.6, and it should not be acid, because the intensity of fluorescence is reduced in acid solutions.

Kihm (*Histochemie*, **35**: 273, 1973) conjugated anticasein and anti-β-lactoglobulin to FITC and observed staining of secretory epithelial cells, particularly in the apical portions, in bovine mammary glands.

Immunohistochemical Detection of Cell Surface Receptors

During the past few years, it has been observed that mononuclear cells which participate in immune responses are functionally and morphologically heterogeneous. Several receptors and differentiation antigens have been identified on cells with different immunologic competences. Lymphocytes derived from bone marrow (B cells) may be identified by a surface immunoglobulin (Raff, *Immunology*, **19**: 637, 1970), and they possess a receptor for the Fc portion of the immunoglobulin molecule which is demonstrable with radioactive-labeled soluble antigen-antibody complexes (*J. Exp. Med.*, **135**: 610, 1972). Fluorescein-labeled aggregated human IgG may also be used (Dickler and Kunkel, *J. Exp. Med.*, **136**: 191, 1972). Most B cells also have receptors for the third component of complement (C′3) and are thereby identifiable by their ability to bind erythrocytes coated with antibody and complement (EAC) (Bianco et al., *J. Exp. Med.*, **132**: 702, 1970). Although most monocytes, macrophages, and histiocytes also possess C′3 receptors, they are distinguishable from B lymphocytes by additional receptors for cytophilic antibodies [Berken and Benacerraf, *J. Exp. Med.*, **123**: 119, 1966; Jaffe et al., *Br. J. Cancer*, **31**: (Suppl. II) 107, 1975]. As a consequence of this receptor, macrophages, histiocytes, and monocytes will also bind erythrocytes coated with IgG (IgGEA) [Huber et al., *Immunology*, **17**: 7, 1969; Jaffe et al., *Br. J. Cancer*, **31**: (Suppl. II) 107, 1975]. Although B lymphocytes also possess receptors for IgG (Basten et al., *J. Exp. Med.*, **135**: 610, 1972; Dickler and Kunkel, *J. Exp. Med.*, **136**: 191, 1972), the nature of the receptor on these cells is qualitatively or quantitatively different in such a way that they do not find the IgGEA reagent.

Fortunately, receptors for IgG and C′3 can be identified on cells in frozen sections and on cells in suspension [Dukor et al., *Proc. Natl. Acad. Sci. U.S.A.*, **67**: 991, 1970; Shevach et al., *Transplant. Rev.*, **16**: 3, 1973; Jaffe et al., *Br. J. Cancer*, **31**: (Suppl. II) 107, 1975]. Although thymus-derived lymphocytes (T cells) do not have these receptors, they may be positively identified by their capacity to induce nonimmune rosettes with sheep erythrocytes [Lay et al., *Nature (Lond.)*, **230**: 531, 1971; Jondal et al., *J. Exp. Med.*, **136**: 207, 1972; Jaffe et al., *New Engl. J. Med.*, **280**: 813, 1974, and *Br. J. Cancer*, **31**: (Suppl. II) 107, 1975]. On the basis of studies by Jaffe et al., it appears permissible to conclude that one can determine the origin of certain neoplastic cells on the basis of presence or absence of the receptors described above.

In the series of cases reported by Jaffe et al., the membrane receptors of the neoplastic cells in six patients with nodular lymphoma were similar to those of B lymphocytes; in two patients with leukemic reticuloendotheliosis the cells had receptors characteristic of histiocytes; and in three patients with diffuse lymphocytic lymphoma the cells formed spontaneous rosettes with sheep red blood cells, characteristic of T cells.

The following methods, rosette assay for binding of IgMEAC (immunoglobulin-erythrocyte-antibody-complement) and IgGEA (immunoglobulin-erythrocyte-antibody), in cell suspensions and in frozen sections are provided by Jaffe et al. [*New Engl. J. Med.*, **280**: 813, 1974; *Br. J. Cancer*, **31**: (Suppl. II) 107, 1975].

ROSETTE ASSAY FOR BINDING OF IgMEAC OR IgGEA USING FROZEN SECTIONS

1. Unfixed frozen sections are cut at 8 μ and air-dried.
2. Sections are layered with IgGEA, IgMEAC, or IgMEA (1×10^8 cells per ml) and incubated at 25°C for 30 min.
3. Slides are washed 3 times in phosphate-buffered saline solution.
4. Sections are fixed for 15 min in Perfix[1] (Applied Bioscience, Patterson, New Jersey), stained with hematoxylin and eosin, dehydrated, and mounted in synthetic medium.

---

[1] Perfix is a fixative with unknown composition.

Sensitized sheep erythrocytes (IgGEA or IgMEA) are prepared with either IgG or IgM antibodies to boiled sheep red cell stroma according to the method of Frank and Gaither (*Immunology*, **19**: 967, 1970). Normal mouse serum is used as a source of complement, and IgMEAC complexes are prepared according to the method of Shevach et al. (*J. Clin. Invest.*, **51**: 1933, 1972). Inasmuch as a receptor for IgM has not been demonstrated on human cells (Huber and Fudenberg, *Int. Arch. Allergy Appl. Immunol.*, **34**: 18, 1968), IgMEA should not bind to any cells, thereby serving as a control for nonspecific binding of red blood cells. It is important to remember that all cells in the spleen which possess surface immunoglobulin are probably also complement receptor lymphocytes, which contrasts with the status of lymphocytes in the peripheral blood (Ross et al., *J. Clin. Invest.*, **52**: 377, 1973).

Jaffe et al. noted that in frozen sections of normal lymph nodes, spleen, and tonsils treated with IgMEAC the reagent red cells were always localized to lymphoid follicles. The reaction was most intense in the germinal centers and less intense in relation to small lymphocytes in the surrounding cuff. B lymphocytes in the far cortex and in medullary cords failed to react with IgMEAC despite the belief that these areas are populated by B cells. Thymic-dependent portions of lymph nodes and spleens failed to react.

In frozen sections treated with IgGEA, the reagent reacted with histiocytes of splenic red pulp and the sinusoidal histiocytes of lymph nodes. Sinusoidal histiocytes of lymph nodes also formed rosettes with IgMEAC. Sections reacted with IgMEA were always negative. In frozen sections of tissues infiltrated with nodular lymphoma treated with IgMEAC, neoplastic nodules were always reactive in the cases studied by Jaffe et al.

ROSETTE ASSAY FOR BINDING OF IgMEAC AND IgGEA IN CELL SUSPENSIONS. Immediately after surgical removal, small bits of spleen or lymph nodes are finely minced in RPM1-1640 (Grand Island Biological Company, 3175 Staley Road, New York, New York 14072) with 10% fetal calf serum. After filtration through a stainless steel wire mesh (holes approximately $\frac{1}{4}$ mm in diameter) a pellet is obtained by centrifugation. To the pellet is added at 0°C approximately 25 ml cold NH_4Cl lysate solution (0.155 M NH_4Cl, 0.01 M $KHCO_3$, and 0.1 M EDTA), mixed, and allowed to incubate for 5 to 10 min. The remaining white cells are washed 3 times in serum-free medium and diluted to a concentration of 2 to 3×10^6/ml. The pH is adjusted to 7.4 at 0°C with either 0.1 N NaOH or HCl.

Peripheral blood lymphocytes are separated on a Ficoll-Hypaque[1] gradient and washed 3 times before use. The Ficoll is obtained from Pharmacia Fine Chemicals, Piscataway, New Jersey, and Hypaque-M, 90%, is obtained from Winthrop Laboratories, New York, New York (Boyum, *Scand. J. Clin. Lab. Invest.*, vol. 21, *Suppl.* 97, 1968). Jaffe notes that suspensions from spleen and lymph nodes may be purified in a similar manner.

Sensitized sheep erythrocytes (IgGEA or IgMEA) are prepared with either IgG or IgM antibodies to boiled sheep red cell stroma according to the method of Frank and Gaither (*Immunology*, **19**: 967, 1970). Normal mouse serum is used as a source of complement, and IgMEAC complexes are prepared according to the method of Shevach et al. (*J. Clin. Invest.*, **51**: 1933, 1972). Inasmuch as a receptor for IgM has not been demonstrated on human cells (Huber and Fundenberg, *Int. Arch. Allergy Appl. Immunol.*, **34**: 18, 1968), IgMEA should not bind to any cells, thereby serving as a control for nonspecific binding of red blood cells.

Equal volumes of the white cell suspensions (2 to 3×10^6/ml) and the IgGEA, IgMEAC, or IgMEA suspensions (1×10^8/ml) are mixed in plastic tubes and gently rotated at 37°C for 30 min. An aliquot of the suspension is then mixed with an equal volume of trypan

---

[1] Hypaque is sodium meglumine diatrizoate. Ficoll is a polysucrose with an average molecular weight of 400,000.

blue and examined in a hemocytometer. The stock solution of trypan blue is made up as a 0.2% solution in distilled water. The working solution is made up 4 : 1 trypan blue–4.5% saline solution. For use, 1 : 1 diluted saline trypan blue–cell suspension is prescribed. Only viable cells are counted (those that do *not* stain with trypan blue). Jaffe et al. report that use of this method generally results in a viability in excess of 90% of cells. A minimum of 200 cells is counted. Any cell with more than three adherent red cells is considered positive.

For cytologic identification of individual cells, rosetted cell suspensions may be fixed in Perfix (contents unknown), filtered on a millipore filter, stained with hematoxylin and eosin, dehydrated, and mounted in synthetic medium.

Jaffe et al. noted that in cell suspensions from normal spleen, the number of complement receptor lymphocytes (B cells) and sheep erythrocyte rosette-forming lymphocytes (T cells) was approximately equal. Cells morphologically identifiable as histiocytes on millipore filter preparations were not observed to form IgMEAC rosettes (Jaffe et al., *Am. J. Med.*, **57**: 108, 1974). Cell suspensions from lymph nodes displayed greater variations, thereby more nearly simulating lymphocytes of the peripheral blood; i.e., approximately 60% were T cells.

Immunofluorescence Quantitation

Two types of standards are employed in microfluorometry. One is employed in the calibration of light sources and electronic equipment. The other is a standard used to measure fluorescence (Ploem, in Holborow, "Standardization in Immunofluorescence," p. 63, Blackwell, Oxford, 1970). The amount of fluorescence in a cell stained with an FITC conjugate may be determined by comparison with a defined microvolume of a standard fluorochrome. Fluorescent phosphorus (Goldman, *J. Histochem. Cytochem.*, **15**: 38, 1967) or uranyl glass [Rigler, *Acta Physiol. Scand.*, vol. 67, *Supp.* 267, 1966; Ploem, in Holborow, "Standardization in Immunofluorescence," Blackwell, Oxford, 1970] is employed in the fluorometers. Jongsma et al. (*Histochemie*, **25**: 329, 1971) used microdroplets of a fluorochrome in a microchamber in an artificial resin mixture to serve as a fluorochrome. A standard unit for fluorescein has been defined (Jongsma et al., *Histochemie*, **25**: 329, 1971) which can be used in different laboratories that employ various optical systems, excitation filters, and light sources. It must be remembered that the above standards are optical standards and are standards to be employed for antiserum, etc.

Cherry et al. (*Stain Technol.*, **44**: 179, 1969), Goldman (*J. Histochem. Cytochem.*, **15**: 38, 1967), and Hijmans et al. (in Holborow, "Standardization in Immunofluorescence," Blackwell, Oxford, 1970) state that commercial photometers are available and suitable for quantitation of microfluorescence. They are reputed to be capable of measuring less than 1 μm^2 in a microscopic specimen. Background light must be subdued (Killander et al., *Immunology*, **19**: 151, 1970).

Enerbäck and Johansson (*Histochem. J.*, **5**: 351, 1973) studied rates of photodecomposition of several fluorophores during short (milliseconds) and longer (minutes) incident illumination periods with an Osram X BO 75 xenon burner. Fluorophores studied included the formaldehyde-induced fluorescent product from 5-hydroxytryptamine in mast cells, Berberin sulfate (C.I. 75160) bound to mast cells, Feulgen-stained DNA, and FITC conjugated IgG in an anti-nuclear-factor test. Significant fading occurred with all fluorophores during illumination for 3 min. FITC conjugates faded most rapidly with 50% fading after 2-s illumination. No fluorophore faded significantly during illumination lasting a few milliseconds. Enerbäck and Johansson constructed an inexpensive device containing a peak reader (with a response time of 2 ms) and memory circuit triggered by the flash synchronization tap of a camera shutter positioned in the activation beam. The device permits measurement of fluorescence intensity without fading.

FERRITIN-LABELED ANTIBODY METHOD. Singer [*Nature (Lond.)*, **183**: 1523, 1959] and Singer and Schick (*J. Biophys. Biochem. Cytol.*, **9**: 519, 1961) were first to localize macromolecules in biologic specimens using ferritin-conjugated antibody. Morgan et al. (*J. Exp. Med.*, **114**: 825, 1961), Easton et al. (*J. Exp. Med.*, **115**: 275, 1962), Baker and Loosli (*Lab. Invest.*, **15**: 716, 1966), Shahrabadi and Yamamoto (*J. Cell Biol.*, **50**: 246, 1971), Sternberger (1974), and others have employed the method for the localization of antigens on the surface and within cells. Ferritin is a spherical molecule with a diameter of 12 mμ having a molecular weight of 650,000 and an iron content of 23% by weight. It is particularly beneficial that the iron is concentrated in the center of the molecule forming a tetrad unit with a diameter of 55 Å. The characteristic structure permits recognition of a single molecule. Actually there are six gëthite units in ferritin situated at the six apexes of a bipyramidal octahedron. See Lillie and Geer, *Am. J. Pathol.*, **47**: 465, 1965 for literature. Getting the tetrad is a matter of orientation of the octahedron to the focal plane.

CONJUGATION METHOD. The procedure involves (1) removal of apoferritin present in commercial batches of ferritin (apoferritin is ferritin without the heme component) employing recrystallization methods and (2) conjugation of antibody (protein) to ferritin in *two stages*, as formulated in Fig. 19-10. The procedure below is advocated by Sternberger (1974) and is based substantially on the initial studies by Singer and Schick (*J. Biophys. Biochem. Cytol.*, **9**: 519, 1961) and Singer and McLean (*Lab. Invest.*, **12**: 1002, 1963).

To 100 ml 2% solution of ammonium sulfate add 1 g crude ferritin. Add 33.3 ml 20% solution of cadmium sulfate. Allow crystallization to occur for 1 day at 4°C, centrifuge, dissolve the precipitate in 100 ml 2% ammonium sulfate, recentrifuge, and to the supernate add again 33.3 ml 20% cadmium sulfate. Allow crystallization to develop again at 4°C for 1 day, and examine for noncrystalline material in the suspension. If present, repeat the above procedure until the suspension is free of noncrystalline material.

Centrifuge and dissolve the crystalline precipitate in 75 ml 2% ammonium sulfate, add 75 ml saturated ammonium sulfate, allow to stand for 1 h at 4°C, centrifuge, and dissolve the precipitate in 75 ml distilled water. Reprecipitate in ammonium sulfate 2 times, suspend in a small volume of distilled water, dialyze against running tap water in the cold overnight, and then against 0.05 M phosphate buffer of pH 7.5 in order to remove cadmium ions. Centrifuge the retentate at 35,000 r/min for 2 h and remove the colorless top portion of the solution. The pellet may be dissolved in 0.05 M phosphate buffer containing 0.15 M NaCl of pH 7.5 and used immediately or centrifuged at 4000 r/min for 30 min at 4°C, sterilized by millipore filtration, and stored at 4°C.

STEP 1 CONJUGATION. The procedure is carried out at 4°C. To 6 ml 0.05 M phosphate buffer of pH 7.5 containing 0.15 M sodium chloride is added 150 mg purified ferritin. While the mixture is continuously stirred, 0.15 ml toluene 2,4-diisocyanate is added. The pH of the stirred solution should be continuously monitored and maintained at 7.5 for 20 min by addition of 0.3 M sodium carbonate of pH 9.5, if necessary. The clear supernate is pipetted into a graduated cylinder and allowed to settle for 10 min, and the volume is recorded.

STEP 2 CONJUGATION. A volume of the supernate containing 125 mg ferritin is mixed with an equal volume of immunoglobulin containing 125 mg protein and an equal volume of 0.3 M sodium carbonate buffer, pH 9.5. The mixture is stirred for 48 h, dialyzed thoroughly first against 15 l 0.1 M ammonium carbonate of pH 7.5 and then against 15 l 0.05 M phosphate buffer of pH 7.5 containing 0.15 M sodium chloride, after which it is centrifuged at 15,000 r/min at 4°C for 5 min. The supernate is used.

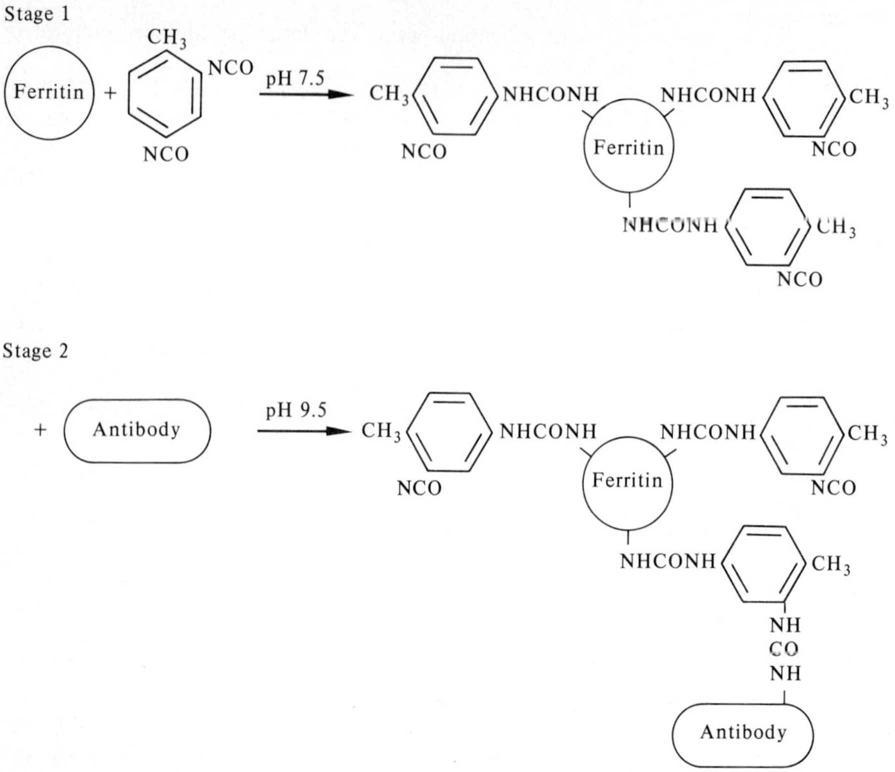

Stage 1

Stage 2

Fig. 19-10 A schematic representation of the two-stage reaction for the coupling of ferritin and antibody molecules by toluene 2,4-diisocyanate. The reaction of isocyanate groups with proteins is probably predominantly, but not exclusively, at the ε-NH$_2$ groups of lysine residues. Only three toluene diisocyanate groups are shown coupled to a ferritin molecule, but this number is actually much larger under the reaction conditions used. Residual ferritin bound isocyanate groups at the end of the second stage of reaction are destroyed by the subsequent addition of NH$_3$. (*From Singer and McLean, Lab. Invest.,* **12:** 1002, 1963. *By permission of the publisher.*)

Marinis et al. (*Immunology,* **17:** 77, 1969) and Sternberger (1974) advocate the following procedure for purification of the conjugate. Supernate from above is placed in Spinco tubes containing about 1 ml of a thick suspension of Sephadex G25, and centrifugation is conducted at 40,000 r/min for 1 h using Spinco Rotor 40. The sediment is dissolved in a small amount of the phosphate–NaCl buffer.

If removal of all unconjugated antibody is necessary, the solution is placed on a discontinuous sucrose gradient (60, 45, and 30%) and then centrifuged at 25,600 r/min for 4 to 5 h with Spinco Rotor SW25. OD$_{440}$ and OD$_{280}$ are recorded for each fraction. The peak fraction at 440 and the heavier fractions are collected. Fractions lighter than the peak are discarded, and the pellet is discarded as it forms. The fractions are dialyzed against 0.05 M phosphate buffer of pH 7.5 containing 0.15 M NaCl, concentrated using an Amicon membrane, sterilized by passage through a millipore filter, and stored at 4°C.

Staining procedures using a ferritin conjugate are the same as those employed for enzyme-labeled conjugates discussed above. Direct or indirect procedures may be employed. Controls are also identical. The steps concerned with the development of the enzyme stain are omitted. A general direct procedure is as follows. For indirect methods, see procedures by Avrameas discussed later in this chapter.

1. Sections may or may not be fixed, depending upon the nature of the antigen. A fixative that alters the reactivity of the antigen must not be employed.
2. If the sections are small, one may wish to mark the back of the slide for identification.
3. A drop or two of the specific ferritin conjugate is applied to the section for 15 to 30 min at 25°C. For controls, equal dilutions of nonimmune conjugate or immune conjugated antiserums against other antigens may be employed. The slides should be incubated in a moist chamber, and drying should be avoided.
4. Slides are rinsed with phosphate-buffered saline solution using a wash bottle. To ensure adequate rinsing, some authors advise flooding slides with buffer and using a mechanical shaker for 5 min at the rate of 30 horizontal excursions per min. The mechanical shaking may be repeated.

Ainsworth and Karnovsky (*J. Histochem. Cytochem.*, **20:** 225, 1972) were able to enhance substantially the contrast staining of ferritin at the electron microscopic level. They prescribe as follows: To 10 ml of 2 N NaOH is added 400 mg sodium tartrate. The alkaline tartrate solution is added drop by drop to 200 mg bismuth subnitrate while the solution is continually stirred with a magnetic bar. The solution will clear gradually until all tartrate is added. The final solution should be clear. The working solution is diluted 1 : 50 with distilled water. Both solutions keep for 1 month or more when stored at 4°C.

For use, sections on grids containing ferritin are floated on drops of the bismuth mixture on dental wax for 30 to 60 min at 25°C. The grids are then washed in several changes of distilled water with 10 to 20 rapid strokes and dried by touching with filter paper. Substantial enhancement of electron opacity and contrast of ferritin is observed. No loss of specificity of staining was observed. Sections may also be stained thereafter with either uranyl or lead salts or both without interference.

Several problems are encountered by those using ferritin conjugates. In the event one wishes to localize antigens on the surface of cells, it may be useful to keep the cells alive during the first application of antibody.

Another problem relates to the size of the ferritin conjugate. It is too large to penetrate into the interior of cells. Some investigators including Dales et al. (*Virology*, **25:** 193, 1965) advocate the use of digitonin. Cells suspended in phosphate-buffered saline solution containing 4×10^{-5} M digitonin for 12 h are then fixed in 1% glutaraldehyde and subsequently processed for electron microscopy.

Yet another problem relates to the propensity of ferritin to bind to plastic embedding media (Striker et al., *Exp. Mol. Pathol. Suppl.*, **3:** 52, 1966). To overcome this problem, McLean and Singer (*J. Cell Biol.*, **20:** 518, 1964) recommend the use of cross-linked polyampholyte. Later, the same authors used cross-linked bovine serum albumin (*Proc. Natl. Acad. Sci. U.S.A.*, **65:** 122, 1970). The McLean-Singer method for preparation of cross-linked bovine serum albumin containing cells follows:

Cells to be embedded are dispersed in 0.5 ml 30% bovine serum albumin in 0.15 M NaCl (0.877%).

This mixture is then concentrated by placing it in a short glass tube; one end of the tube is attached to a collodion dialysis membrane and immersed in the dehydrating agent Aquacide II (Calbiochem). When the bovine serum albumin has attained the consistency of a gel, the membrane-covered end of the tube is freed of adhering Aquacide and is immersed in a solution of either 2% glutaraldehyde in phosphate–saline solution buffer, pH 7.5, or 2% formaldehyde in the same buffer. These fixatives diffuse through the gel and cross-link the bovine serum albumin in about 3 h. The cross-linked bovine serum albumin is cut into narrow strips, washed briefly with water, drained, and dried in a desiccator above silica gel. The dried strips are then cemented onto an epoxy resin block and sectioned with a diamond knife. The sections are collected on water and mounted on carbon-coated Formvar or collodion grids which are first treated with a 4% solution of bovine serum albumin in phosphate buffer. (The bovine serum albumin treatment largely eliminates non-specific ferritin-conjugated antibody staining of the support film.) Staining of the sections with specific ferritin conjugate antibody is carried out by placing a drop of conjugate on the mounted section. After 5 min, the grid is floated face down on a succession of phosphate buffer solutions to remove excess antibodies, washed with water, and dried. Other post-staining of the sections with uranyl acetate or phosphotungstate may be used to increase contrast within the cell ultrastructure.

Shahrabadi and Yamamoto (*J. Cell Biol.*, **50**: 246, 1971) claim success in prevention of adherence of ferritin conjugates to embedding media. Cells were fixed in 4% formaldehyde in 0.1 *M* phosphate buffer of pH 7.2 at 4°C for 30 min, washed in the same buffer for 16 h, and embedded in glycol methacrylate according to Leduc and Bernhard (*J. Ultrastruct. Res.*, **19**: 196, 1967). Shahrabadi and Yamamoto noted a tendency for spreading of sections when spread on buffer or water during washing. The problem was reduced by addition of low concentrations of bovine serum albumin in the washing solutions.

Nicolson and Singer (*Proc. Natl. Acad. Sci. U.S.A.*, **68**: 942, 1971) conjugated ferritin to concanavalin A (a plant agglutinin that binds specifically to oligosaccharides with terminal D-glucose, D-mannose, or sterically related sugar residues) in a study of red cell membranes. Specific binding to the outer but not the inner surfaces of membranes was noted.

Olsen et al. (*Science*, **182**: 825, 1973; *J. Cell Biol.*, **64**: 340, 1975) and Olsen and Prockop (*Proc. Natl. Acad. Sci. U.S.A.*, **71**: 2033, 1974) provide two improved methods for the preparation of ferritin-protein conjugates for electron microscopy. Olsen and coworkers employed the methods to study collagen synthesis. Procollagen was detected in cisternae of endoplasmic reticulum and in large Golgi vacuoles of fibroblasts derived from chick tendon.

PROCEDURE I: PREPARATION OF Fab-FERRITIN OR IgG-FERRITIN WITH A WATER-SOLUBLE CAR-BODIIMIDE AND *N*-HYDROXYSUCCINIMIDE.   In the first step, the amino groups in apoferritin are acylated with succinic anhydride. In a typical experiment 101 mg ferritin in 1.1 ml water is mixed with 6 ml water saturated with sodium succinate at room temperature. The mixture is cooled to 4°C and reacted with 0.53 g succinic anhydride, an amount calculated to be 130 equiv per NH_2 group in apoferritin. The reaction is allowed to proceed at 4°C for 1 h and then at about 22°C for 1 h. The sample is then passed through a 2.5 × 90 cm column of 6% agarose (A-5m, 200- to 400-mesh, Bio-Rad Corp.), equilibrated and eluted with 0.01 *M* sodium phosphate buffer, pH 7, at 4°C. The peak fractions containing mono-meric ferritin are pooled and concentrated at 4°C to about 30 mg ferritin per ml in an Amicon ultrafiltration cell with a PM-30 membrane.

In a second step the succinyl ferritin is activated with the water-soluble 1-ethyl-3-(3-di-methylaminopropyl) carbodiimide hydrochloride (Calbiochem) and *N*-hydroxysuccinimide (Sigma) with a minor modification of the method of Cuatrecasas and Pakrikh (*Biochemistry*, **11**: 2291, 1972). Succinylated ferritin (15.2 mg in 0.5 ml 0.01 *M* sodium phosphate buffer at pH 7) is cooled on an ice bath, and 37 mg water-soluble carbodiimide and 23.4 mg *N*-hydroxysuccinimide are added. After 3 h, the sample is passed through a gel filtration column (Sephadex G25, coarse, Pharmacia Corp.) which is 0.9 × 80 cm and which has been equilibrated and eluted with 0.01 M sodium phosphate buffer, pH 7. The gel filtration step removes the excess water-soluble carbodiimide and N-hydroxysuccinimide from the activated succinyl ferritin. One must remember that the active ester in the succinyl ferritin which elutes in the void volume is subject to hydrolysis. For this reason it is used for conjugation in the third step as soon as it is recovered from the column.

In the third step the active ester of the succinyl ferritin is linked to either Fab or IgG. Fab is obtained from a papain digest of purified goat IgG as described by Porter (*Biochem. J.*, **73:** 119, 1959). For the conjugation with Fab the two peak fractions from the Sephadex G25 column (above), containing about 10 mg activated succinyl ferritin in 5.6 ml, are mixed with 1.7 ml 0.1 M sodium phosphate buffer, pH 7.3, containing 3.4 mg [^{125}I] iodine-labeled Fab. The reaction is allowed to proceed at room temperature for 24 h and the product isolated by gel filtration on a 6% agarose column (2.5 × 95 cm, equilibrated and eluted with 0.1 M tris-HCl buffer, pH 7.5). The peak fractions containing the Fab-ferritin are concentrated to 2.1 mg ferritin per ml using an Amicon ultrafiltration cell with a PM-30 membrane. The conjugation of ferritin with IgG is carried out under similar conditions, but Olsen et al. found it necessary to reduce the ratio of activated succinyl ferritin to antibody in order to minimize formation of polymers. In a typical experiment, 1 mg activated succinyl ferritin in 2 ml buffer is mixed with 4.4 mg rabbit [^{125}I]–IgG in 3 ml 0.01 M sodium phosphate buffer, pH 7. The reaction is allowed to proceed for 18 h at 4°C. Because the solution is relatively dilute, the reaction is stopped by the addition of 400 mg glycine, and thereafter concentrated to 0.8 ml in the Amicon ultrafiltration cell with an Amicon PM-30 membrane. The product is then isolated by gel filtration on a 6% agarose column (0.9 × 60 cm, equilibrated, and eluted with 0.1 M tris-HCl buffer, pH 7.5).

PROCEDURE II: PREPARATION OF Fab-FERRITIN IgG-FERRITIN WITH GLUTARALDEHYDE. In the first step, ferritin is activated with a 1200-fold molar excess of glutaraldehyde per amino group in apoferritin. In a typical experiment 10 mg ferritin in 2 ml 0.1 M sodium phosphate buffer, pH 7.3, is mixed with 1 ml 50% glutaraldehyde. The reaction is carried out with stirring at room temperature for 30 min. After centrifugation at 20,000 × gravity for 10 min to remove a small amount of precipitate, the unreacted glutaraldehyde is separated from the activated ferritin by gel filtration on a coarse Sephadex G25 column (2.5 × 90 cm, Pharmacia Corp.), equilibrated, and rapidly eluted at 4°C with 0.1 M sodium phosphate buffer, pH 7.3. The activated ferritin is used immediately to prepare the conjugate.

In the second step, activated ferritin is coupled to Fab or rabbit IgG. Usually, the reaction is carried out with about 1.5 mg activated ferritin and about 0.4 mg Fab or IgG per ml. These concentrations are achieved either by concentrating the reactants before they are mixed or by concentrating the mixture of the two reactants. In a typical experiment with Fab, the peak fractions of activated ferritin from the G25 column are concentrated at 4°C to about 3 mg ferritin per ml in an Amicon ultrafiltration cell with a PM-30 membrane. The concentrated, activated ferritin (11.4 mg in 4.4 ml) is mixed with 3 mg [^{125}I] iodine-labeled Fab dissolved in 2 ml 0.1 M sodium phosphate buffer, pH 7.3. After gently stirring at room temperature for 3 h, the solution is cooled to 4°C, and the reaction is allowed to proceed for an additional 65 h at 4°C. The conjugate is then separated from nonconjugated Fab by gel filtration on a 6% agarose column (A-5m, 200 to 400 mesh, Bio-Rad Corp.) which is 2.5 × 95 cm. In order to inactivate residual aldehyde groups on the ferritin, the column is equilibrated and eluted with 0.1 M tris-HCl buffer, pH 7.5, at 4°C. The product is concentrated to about 2 mg ferritin per ml with an Amicon ultra-filtration cell and a PM-10 membrane and stored at 4°C. In a typical experiment with IgG, the activated ferritin from the Sephadex G25 column (see above), 12.2 mg in 20 ml, is mixed directly at 4°C with 1 ml 0.1 M sodium phosphate buffer, pH 7.3, containing 3 mg rabbit [^{125}I] iodine-labeled IgG. The mixture is then concentrated to 8.8 ml using the Amicon PM-30 filter and kept at 4°C for 67 h. The product is chromatographed on a 6% agarose column as described for the isolation of Fab-ferritin.

TISSUE PREPARATION

Olsen and Prockop employed the following method for detection of procollagen in fibroblasts (*Proc. Natl. Acad. Sci. U.S.A.*, **71**: 2033, 1974).

Fibroblasts obtained from tendons of 17-day-old chick embryos are incubated for 3 h at 37°C in a modified Krebs medium containing 10% fetal calf serum (Dehm and Prockop, *Biochim. Biophys. Acta.*, **240**: 358, 1971), fixed for 3 h or 15 h at 4°C in 1% formaldehyde in 60 mM sodium phosphate buffer of pH 7.3 containing 0.14 M sucrose, centrifuged for 6 min at 600 × gravity, resuspended in 0.1 M sodium phosphate buffer of pH 7.3, and stored at 4°C for 15 to 24 h.

Cells are homogenized with 50 to 100 strokes in a Teflon and glass homogenizer and centrifuged at 20,000 × gravity for 10 min, and the pellet is incubated with the ferritin-conjugated antibodies (prepared as indicated above) for 24 to 48 h at 4°C. Fragments of 50 million to 100 million cells are incubated with 100 μl 0.1 M tris-HCl containing ferritin-labeled specific antibodies (1 mg ferritin and 0.3 mg antibody per ml).

Cell fragments are washed 2 times with 5 ml 0.1 M sodium phosphate buffer of pH 7.3, fixed with 3% glutaraldehyde in 60 mM sodium phosphate buffer of pH 7.3 containing 0.16 M sucrose, stained with 0.5% magnesium uranyl acetate in the block for 30 min at 25°C, dehydrated, and embedded in Araldite.

Olsen and Prockop located procollagen in the cisternae of endoplasmic reticulum. Olsen et al. (*Science*, **182**: 825, 1973) used the same general method for detection of prolyl hydroxylase in the rough endoplasmic reticulum of chick embryonic tendon cells. Since the method uses lightly homogenized cells, penetration of ferritin-conjugated antibodies through cell membranes is not a problem.

Labeling of molecules with other heavy elements in order to achieve sufficient electron density has been without substantial success. Kendall (*Biochim. Biophys. Acta*, **97**: 174, 1965) labeled antibody with mercury after thiolation of many amino groups with N-acetylhomocysteine thiolactone. Kendall claimed that virtues of the method include stability of the binding of the scattering material, chemical inertness of the newly introduced groups, retention of immunospecific activity, homogeneity of labeling, ease of removal of excess reagents from antibody after labeling, in addition to the endowment of increased electron density. Kendall calculated that each antibody molecule acquired 26 mercury atoms.

Zhdanov et al. (*J. Histochem. Cytochem.*, **13**: 684, 1965) used p-(aminophenyl)-mercuric acetate to label antibodies to Sendai virus, which resulted in increased electron density of the virus. Mekler et al. [*Nature (Lond.)*, **203**: 717, 1964] have employed iodine to label proteins to enhance electron density.

Tanaka (*Acta Haematol. Japonica*, **31**: 125, 1968) provides an explanation of why labeling with the above heavy metals has not met with great success. He states that to provide contrast the value of $NZ^{4/3}/S$ for a specified stain area should exceed $4700/m\mu^2$ where N is the number of heavy metal atoms of atomic number Z added as a label to an area of size S. Thus, Kendall's labeling of mercury above is very inadequate. However, due to the peculiar structure of ferritin, the calculation $NZ^{4/3}/S$ for the tetrad is $6000/m\mu$. The size of an antibody molecule is assumed to be 76 $m\mu^2$.

Marucci et al. (*J. Histochem. Cytochem.*, **22**: 35, 1974) prepared and purified soluble immune complexes of ferritin and chicken antiferritin. They were used as follows for the detection of avian leukosis virus in chick fibroblasts:

Virus-infected chicken fibroblast cell cultures were grown on 18-mm^2 coverslips, washed with neutral 0.01 M tris-buffered Earle's balanced salt solution, fixed with 2% glutaraldehyde in Millonig's buffer (*J. Appl. Physics*, **32**: 1637, 1916), and washed in 0.15 M NaCl containing 0.05 M tris pH 7.6 (3 times). Cells on coverslips are placed in a moist chamber and treated at 24°C with (1) chicken antiviral serum (1 : 10 in the tris–NaCl buffer) for 10 min, (2) buffer (three changes), (3) undiluted rabbit antichicken serum, 10 min, (4) buffer (three changes), (5) undiluted ferritin-antiferritin, 10 min, and (6) buffer (three changes). Cells were then osmicated, dehydrated, and prepared for electron microscopy.

Theoretically, the above method has an advantage over methods that require chemical conjugation of antibody to either ferritin or to an enzyme. Conjugation methods result in the production of variable amounts of complexes of nonantibody protein-ferritin, nonantibody protein-enzyme, or antibody-antibody which may not be completely removed by purification methods. In addition, not all antibody is labeled. Unlabeled antibody may compete with labeled antibody for antigenic sites, thereby reducing intensity of staining. Reduction of staining could also occur due to modification of specific reactive sites on the antibody molecule due to conjugation.

In the above method, the antibody is not labeled, and sensitivity is increased because the method permits an increase of antigenic determinants above those of the original antigen by deposition of different layers of antibody.

IMMUNOENZYME TECHNICS

An enzyme can be used to label an antigen or an antibody. Nakane and Pierce (*J. Histochem. Cytochem.*, **14**: 929, 1966; *Int. Cong. Electron Microsc. Kyoto*, p. 51, 1966), Ram et al. (*Fed. Proc.*, **25**: 732, 1966), and Avrameas and Uriel (*C. R.* **262**: 2543, 1966) appear to have simultaneously but independently developed methods for coupling of proteins with enzymes and their use in the detection of antigen-antibody–reactive sites.

Figures 19-11, 19-12, and 19-13 reveal several ways this may be accomplished. With the

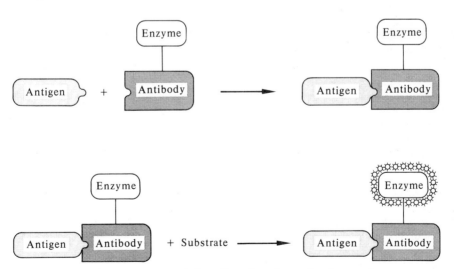

Fig. 19-11 Direct immunoenzyme technique. Detection of an antigen by use of an enzyme-labeled antibody. ☼ represents chromogenic or electron-dense substances produced by the action of the enzyme on its specific substrate. (*From Avrameas, Int. Rev. Cytol.*, **27**: 349, 1970. *By permission of the publisher.*)

direct method, an enzyme may be covalently linked to the antibody provided the corresponding homologous antibody is available (Fig. 19-11). After the enzyme-labeled antibody is reacted with the antigen, enzyme activity is detected by employment of histochemical methods, whereupon the antigen also is identified.

DIRECT IMMUNOENZYME TECHNIC FOR THE LOCALIZATION OF ANTIGENS AND ANTIBODIES (Avrameas, *Immunochemistry*, **6**: 43, 1969)

1. Fresh frozen sections or suspensions of cells placed on slides are fixed for 45 min at 22°C in a 60 : 40 mixture of ethanol-ether. Good results are also achieved in specimens fixed for 10 min at 4°C with acetone, absolute methanol, and 95% ethanol.
2. Wash for 15 min at 22°C in 0.1 M phosphate-buffered saline solutions of pH 7.2. At this stage, slides may be stored at 4°C.
3. To detect cells containing γ-globulins, incubate preparations for 3 h at 22°C with peroxidase-labeled antibody at a concentration of 0.5 mg antibody per ml (the stock solution of peroxidase-labeled antibody is diluted 10 times).
4. Rinse slides twice for 3 min each in the above phosphate-buffered saline.
5. Incubate preparations for 5 to 30 min in the reagent to detect peroxidase (75 mg 3,3-diaminobenzidine and 0.001% hydrogen peroxide in 100 ml 0.05 M tris buffer of pH 7.6).
6. Rinse in distilled water.
7. Preparations may be dehydrated and mounted in a synthetic medium.

Positive results are indicated by a yellowish brown stain. Experimental controls for the above method should include incubation of tissues with solutions of (1) the enzyme-labeled antibody adsorbed with its homologous antigen, (2) an unrelated antibody coupled with the enzyme, and (3) the enzyme alone. In general, the more impure the enzyme preparation used and the higher the proportion of nonspecific proteins in the antibody preparation, the greater will be the degree of nonspecific staining.

ANTISERA PREPARATION (Avrameas, *Immunochemistry*, **6**: 43, 1969; Avrameas and Ternynck, *Immunochemistry*, **6**: 53, 1969)

1. Using 1-year-old rabbits, 2 mg antigen in 0.7 ml of 0.15 M NaCl and 0.3 ml complete Freund's adjuvant is injected intradermally in both hind footpads.
2. Two months later 2 mg antigen is injected daily for 3 consecutive days. The first day the injection is given intramuscularly. The injections are given intravenously on the other 2 days.
3. Eight days later, step 2 is repeated.
4. Between the fifth and seventh day after the last injection, animals are bled by cannulation of the internal carotid artery, and serum is stored at -20°C until used.

Although a γ-globulin fraction from ammonium sulfate-precipitated immune serum has been employed (Nakane and Pierce, *J. Cell Biol.*, **33**: 307, 1967), Avrameas could obtain reproducible results only when purified antibody in high titer from hyperimmunized animals was used. The purification procedure employing specific and stable immunoadsorbents is detailed below.

PREPARATION OF INSOLUBLE PROTEIN DERIVATIVES (Avrameas and Ternynck, *Immunochemistry*, **6**: 53, 1967). Human IgG is used as an example. Into 5 ml 0.1 M phosphate buffer of pH 7 is dissolved 250 mg human IgG. With stirring is added dropwise

1 ml of a 2.5% aqueous solution of glutaraldehyde. A gel forms almost instantly. The mixture is allowed to stand for 3 h at 22°C. Completeness of the reaction can be determined by an assay for protein in the supernate.

PREPARATION OF IMMUNOADSORBENTS, BATCHWISE PROCEDURE (Avrameas and Ternynck, *Immunochemistry*, **6**: 53, 1969). Into 200 ml 0.2 M phosphate buffer of pH 7.2 to 7.4 is dispersed 500 mg insolubilized protein prepared as above. The protein is homogenized in small portions with a loose-fitting Potter homogenizer. The suspension is centrifuged for 15 min at 3000 r/min at 4°C. Repeat the homogenization and centrifugation steps 2 or 3 times. The insoluble protein is then suspended in 200 ml of the eluting fluid to be used (0.1 M glycine–HCl buffer of pH 2.8 and 2.5 M NaI in 0.05 M tris buffer of pH 9 have been found satisfactory) and centrifuged. Repeat the last step. The immunoadsorbent is then washed by centrifugation with phosphate buffer until the supernate has an optical density of 0 at 280 mμ. This generally occurs after two washings. If 2.5 M and 5 M MgCl$_2$ in 0.05 M tris buffer of pH 7.5 is to be used for elution, unbuffered saline solution must be used throughout the entire procedure.

USE OF THE IMMUNOADSORBENT (Avrameas and Ternynck, *Immunochemistry*, **6**: 53, 1969). Appropriate volumes of immune serum are mixed with the insoluble protein in centrifuge tubes. The mixture is stirred gently for 30 min and then centrifuged at 3000 r/min for 15 min at 4°C. Keep the supernate in order to measure nonadsorbed antibody. Suspend the precipitate in buffered physiologic saline solution in a volume 3 or 4 times the volume of antiserum added, and centrifuge as above. Wash with buffered physiologic saline solution 3 or 4 times until the supernate has an optical density of less than 0.04 at 280 mμ. To elute proteins 0.1 M glycine–HCl buffer of pH 2.8, or 2 M and 5 M NaI in tris buffer of pH 9, or 2.5 M and 5 M MgCl$_2$ in tris buffer of pH 7.5 is employed.

The washed immunoadsorbent is added in 20 ml NaI or MgCl$_2$ solutions. The suspension is stirred with a magnetic stirrer for 5 min at 22°C and centrifuged at 10,000 r/min for 15 min at 4°C. Repeat the last procedure 2 or 3 times. If 0.1 M glycine–HCl buffer is used, the immunoadsorbent is suspended in a minimum of volume (possibly 4 ml), and the resuspension and centrifugation steps are repeated. The supernates are passed through a 0.45-μ millipore filter and dialyzed against several changes of cold saline solutions. After the immunoadsorbent has been washed again with the elution fluid and several times again with saline solution, it is stored at 4°C in the presence of 10^{-4} M merthiolate and DFP (diisopropylfluorophosphate). This regenerates the immunoadsorbent for use again.

A column may also be employed. If so, columns 1 to 2 cm in diameter and 2 to 4 cm in height are advised. Whole immune serum is passed over the column at a rate of 30 to 40 ml/h at room temperature. The column is then transferred to a cold room at 4°C and antibody eluted with the glycine–HCl, MgCl$_2$, or NaI buffers detailed above.

The adsorption capacity of insoluble derivatives depends upon the antigens used. Avrameas and Ternynck noted that 50 mg insolubilized IgG could adsorb approximately 50 mg antibody whereas 50 mg insolubilized Bence Jones (K) protein could adsorb only 10 mg antibody. Best elution (65 to 98%) was obtained in both batch and column procedures with the glycine–HCl buffer above.

GLUTARALDEHYDE ENZYME-ANTIBODY CONJUGATION METHOD. Avrameas (*Immunochemistry*, **6**: 43, 1969) coupled peroxidase, glucose oxidase, tyrosinase, or alkaline phosphatase to human IgG, human serum albumin, sheep antibody, or rabbit antibody with glutaraldehyde. He states that glutaraldehyde is by far the most effective and suitable reagent for producing enzyme-protein complexes with retention of substantial enzymatic activity and immunologic specificity. The method follows:

In 1 ml 0.1 M phosphate buffer of pH 6.8 containing 5 mg antibody, 12 mg peroxidase is dissolved. During gentle stirring 0.05 ml 1% aqueous solution of glutaraldehyde is added dropwise. The mixture is allowed to stand for 2 h at 22°C and then dialyzed against two changes of 5 l buffered physiologic saline solution overnight at 4°C. The precipitate formed is removed by centrifugation for 20 min at 4°C at 20,000 r/min in a rotor 40 of a Spinco ultracentrifuge. This is the stock solution of peroxidase-labeled antibody which may be kept at 4°C until used. Deterioration of enzymatic activity or immunologic specificity has not been detectable for as long as 3 months. Avrameas has employed the same method for coupling of other enzymes detailed above with equal success (Table 19-2). Retention of enzyme activity and immunologic specificity of the conjugates in Table 19-2 would tend to suggest that these qualities are not substantially dependent upon amino groups.

Avrameas provides Table 19-3 of sedimentation coefficients of rabbit IgG and three enzymes before and after conjugation. He points out that the use of (1) glutaraldehyde, (2) a purified antibody, and (3) the purest commercially available enzymes permits the reliable preparation of protein-enzyme complexes with subsequent accurate and reproducible detection of intracellular antigens.

Avrameas (*Immunochemistry*, **6**: 43, 1969) obtained insoluble antigen and antibody antiderivatives when glutaraldehyde was added to solutions of antigen or antibody. The

TABLE 19-2 QUANTITIES OF REACTANTS GIVING PROTEIN-ENZYME COMPLEXES IMMUNOLOGICALLY AND ENZYMATICALLY ACTIVE

Protein	Quantity, mg	Enzyme*	Quantity, mg	Volume† of enzyme-protein solution before the addition of glutaraldehyde, ml	Quantity‡ of glutaraldehyde added, ml
Human IgG	5	Peroxidase	12	1	0.05
Human IgG	5	Phosphatase	10	1.3	0.10
Human IgG	12	Phosphatase*a*	50	2	0.40
Human IgG	12	Glucose oxidase*b*	50	2	0.40
Human serum albumin	5	Glucose oxidase*b*	50	1	0.80
Glucose oxidase*b*	5	Phosphatase	10	1.3	0.15
Sheep antirabbit γ-globulins	5	Phosphatase	10	2	0.15
Sheep antirabbit γ-globulins	5	Glucose oxidase	10	1	0.15
Sheep antirabbit γ-globulins	5	Tyrosinase	25	1	0.15
Sheep antirabbit γ-globulins	5	Peroxidase	12	1	0.05
Rabbit antihuman IgG	5	Peroxidase	12	1	0.05

* The different enzyme preparations employed were peroxidase = peroxidase RZ 3; phosphatase = alkaline phosphatase BAPSF, 10 units/mg; phosphatase*a* = alkaline phosphatase PC, 0.9 units/mg; glucose oxidase = glucose oxidase, 90 units/mg "analytical reagent grade"; glucose oxidase*b* = crude preparation of *Aspergillus niger* glucose oxidase type II; tyrosinase: mushroom tyrosinase TY 500 units/mg.

† Phosphate buffer 0.1 M, pH 6.8.

‡ 1% (w/v) of aqueous solution of glutaraldehyde.

Source: Avrameas, *Immunochemistry*, **6**: 43, 1969.

TABLE 19-3 SEDIMENTATION COEFFICIENTS OF RABBIT IgG AND
ENZYMES BEFORE AND AFTER THE COUPLING REACTION*

IgG and enzyme	Sedimentation coefficient (S) before coupling	Sedimentation coefficient (S) after coupling
IgG + peroxidase	6.25 3.13	3.38
IgG + glucose oxidase	6.25 4.24–6.87–9.3	4.72–7.58–10.12
IgG + alkaline phosphatase	6.25 1.3–5.58	2.09–6.2

* Coupling of IgG to peroxidase (RZ 3), phosphatase (10 units/mg), and glucose oxidase (90 units/mg) were carried out under the conditions reported in Table 19-2.

Source: Avrameas, *Immunochemistry*, **6**: 43, 1969.

insoluble derivatives were found to be efficient, specific, and stable immunoadsorbents that could be used either in a batchwise or column procedure for the isolation of antigens or antibodies. Certain protein antigens such as glucose oxidase and peroxidase appear to contain insufficient amino groups to effect insoluble derivatives. Alternatively, it is possible that after linkage of one aldehyde of glutaraldehyde, the structure is such that conjugation of the other aldehyde is physically hindered or prevented.

To conjugate peroxidase or acid phosphatase to antibody, Ram et al. (*J. Cell Biol.*, **17**: 673, 1963) used *p,p'*-difluoro-*m,m'*-dinitrodiphenyl sulfone (FNPS, a reagent that had been used to link ferritin to protein), and 1-ethyl-3-(3-dimethylaminopropyl)carbodiimide (CDI) has been used by Nakane and Pierce (*J. Cell Biol.*, **33**: 307, 1967), Ram et al. (*Fed. Proc.*, **25**: 732, 1966), and Goodfriend et al. (*Science*, **144**: 1344, 1964).

INDIRECT IMMUNOENZYME TECHNIC (Fig. 19-12) FOR THE INTRACELLULAR DETECTION OF ANTIGENS (Avrameas, *Immunochemistry*, **6**: 825, 1969). Avrameas uses as an example the detection of immunoglobulin in spleen cells of hyperimmunized rabbits using an anti-immunoglobulin and antiglucose oxidase or peroxidase antibody (mixed antibody method).

1. Fresh frozen sections or suspensions of cells placed on slides are fixed for 45 min at 23°C in a 60 : 40 mixture of ethanol-ether. Good results may also be achieved in specimens fixed for 10 min at 4°C in acetone, absolute methanol, or 95% ethanol.
2. Wash slides for 15 min in 0.1 M phosphate-buffered saline solution of pH 7.2. At this stage, slides may be stored at 4°C until use.
3. Incubate slides for 30 min at 22°C in a solution containing 2 to 6 mg/ml of sheep antirabbit immunoglobulin antibody. For controls, slides are incubated in a solution of normal γ-globulins at the same concentration.
4. Wash slides thoroughly for 3 min in 0.1 M phosphate-buffered saline solution of pH 7.2.
5. Incubate sections in a solution containing 0.2 to 0.4 mg/ml rabbit anti-glucose oxidase (or peroxidase) antibody for 45 min at 22°C.
6. Wash slides for 3 min in 0.1 M phosphate-buffered saline solution of pH 7.2.

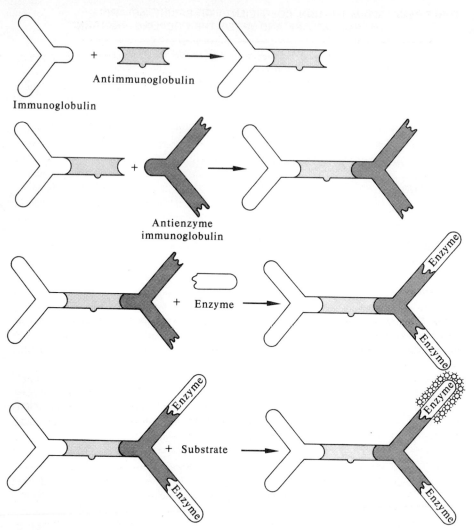

Fig. 19-12   Indirect immunoenzyme technique: mixed antibody method. Detection of an immunoglobulin by use of an anti-immunoglobulin and an antienzyme antibody. Anti-immunoglobulin antibody reacts with immunoglobulin of species A present in a cell. Because the immunoglobulin in the cell exists in an insoluble form, the antibody reacts with only one of its active sites leaving the other available. This site can then react with the subsequently added antienzyme antibody of species A (antienzyme immunoglobulin). The final addition of the enzyme allows the localization of the immunoglobulin. (*From Avrameas, Int. Rev. Cytol.,* **27**: 349, 1970. *By permission of the publisher.*)

7. Incubate slides in a solution containing 1 mg glucose oxidase (or peroxidase) per ml for 30 min at 22°C.
8. Wash sections in 0.1 M phosphate-buffered saline solution of pH 7.2 for 3 min.
9. Stain sections for peroxidase (5 to 30 min at 22°C in 75 mg 3,3′-diaminobenzidine and 0.001% hydrogen peroxide in 100 ml 0.05 M tris buffer of pH 7.6) or for glucose oxidase *in the absence of light* (15 to 30 min at 22°C in 20 ml 0.1 M phosphate buffer of pH containing 150 mg D-glucose, 10 mg MTT (the tetrazolium salt 3-(4,5-dimethylthiazolyl)-2,5-diphenylmonotetrazolium bromide), and 2 mg phenazine methosulfate. We see no reason why the tetrazolium salt nitro BT cannot be used. If so, specimens may be dehydrated and permanently mounted in a synthetic medium. According to Pearse (personal communication) MTT may be difficult to obtain.

    If peroxidase labeling is used, a yellowish brown stain is developed. The color that develops with glucose oxidase depends upon the tetrazolium salt employed. Likewise, if acid or alkaline phosphatase enzymes are employed as the detection system, the color that develops will depend upon the substrate and diazonium salts used.
10. Wash preparations for 3 to 5 min in 0.1 M phosphate-buffered saline solution of pH 7.2.
11. Wash in distilled water, and mount in a water-soluble mount such as Kaiser's glycerogel.

Figure 19-12 depicts how an antienzyme antibody and an anti-immunoglobulin can be used to detect any antigen at either light or electron microscopic levels. The methods can also be employed to detect the antigens in biologic fluids by employment of gel immunodiffusion technics.

Avrameas (*Int. Rev. Cytol.*, **27**: 349, 1970) in another variant of the indirect immunoenzyme method notes that if the quantity of, for example, peroxidase-labeled antibody fixed in a cell is very low, the preparation can be incubated in a solution of antiperoxidase antibody. The preparation may then be washed, incubated in peroxidase, washed again, and stained. In this manner, low levels of antigen may be detected. Avrameas refers to this as the *amplification antibody method.*

In still another variant of the indirect immunoenzyme method Avrameas cites the hybrid antibody method depicted in Fig. 19-13.

Fudenberg et al. (*J. Exp. Med.*, **119**: 151, 1964) were the first to prepare "hybrid" antibodies by treatment of rabbit 7S antibody with pepsin resulting in the removal of an immunologically inactive fragment leaving a bivalent residue with an average molecular weight of 106,000 and a sedimentation coefficient of approximately 5S. Subsequent reduction of one labile disulfide bond splits the residue into two univalent, nonprecipitating fragments that migrate as a single peak at 5S peak in the ultracentrifuge.

An example in this case is the use of isolate antiperoxidase and anti-immunoglobulin which are first digested with pepsin. Ten parts (by weight) of the digested antiperoxidase and two parts of the digested anti-immunoglobulin antibody are mixed, reduced with β-mercaptoethylamine, and then allowed to reoxidize. Yields of hybrid antibody vary with different preparations from 20 to 50%.

For use, dilute hybrid antibody, with respect to the anti-immunoglobulin antibody, to 0.1 to 0.2 mg protein per ml, and apply it for 2 h to cell preparations. Wash and incubate in peroxidase (1 mg/ml) in saline solution for 1 h at 22°C. Wash and stain for peroxidase.

Bretton et al. (*Exp. Cell Res.*, **71**: 145, 1972) compared the efficacy of peroxidase and ferritin labeling of cell surface antigens. A solid tumor (mouse plasmacytoma) and cultured

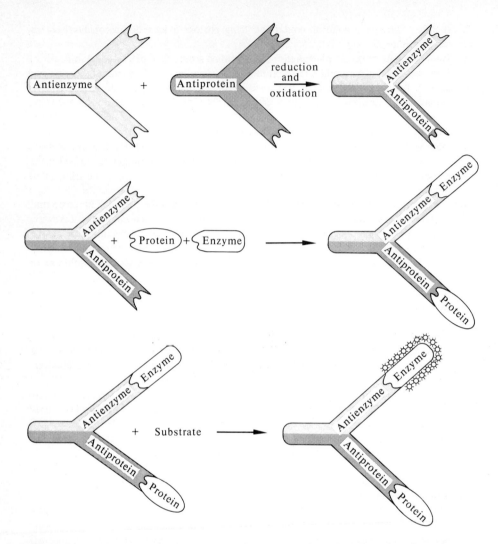

Fig. 19-13 Indirect immunoenzyme technique: hybrid antibody method. Detection of protein-antigen by use of a hybrid antienzyme antiprotein antibody. See legend for Fig. 19.11 for explanation of symbols. (*From Avrameas, Int. Rev. Cytol.,* **27:** 349, 1970. *By permission of the publisher.*)

green monkey kidney epithelial cells were used as test objects. Peroxidase labeled more antigenic sites than did ferritin, and discontinuities of ferritin labeling were noted. An advantage of peroxidase labeling over that of ferritin is believed to be smaller molecular size permitting better penetration into cells and tissues. The molecular weight of ferritin exceeds 750,000.

UNLABELED ANTIBODY ENZYME METHOD FOR IMMUNOHISTOCHEMISTRY. Sternberger (*Mikroskopie,* **25:** 346, 1969), Sternberger and Cuculus (*J. Histochem. Cytochem.,* **17:** 190, 1969), and Sternberger et al. (ibid., **14:** 711, 1966; **18:** 315, 1970) developed an unlabeled antibody enzyme method for immunochemistry. They demonstrated that the localization of antigens in tissues can be identified and the staining reaction intensified by the use of enzymes. Conjugated antibodies need not be employed. The procedure prescribes the sequential application to the tissue of (1) rabbit antiserum to the antigen to be localized, (2) sheep antiserum to rabbit immunoglobulin G added in sufficient excess to leave one free combining site after reaction with the rabbit antibody, (3) rabbit antihorseradish peroxidase, which reacts as an antigen with the free combining sites of anti-IgG, (4) horseradish peroxidase, (5) 3,3'-diaminobenzidine tetrahydrochloride and hydrogen peroxide, and (6) osmium tetroxide. Sternberger and his coworkers note that it is better to employ immunospecifically purified antihorseradish peroxidase rather than total serum IgG, because the portion of serum IgG devoid of antihorseradish peroxidase interfered with the next step of the procedure.

Moriarty and Halmi (*J. Histochem. Cytochem.*, **20:** 590, 1972) employed the method of Sternberger and his associates to identify ACTH-secreting cells of the anterior pituitary glands of rats. The method follows:

1. Tissues are satisfactorily fixed in 2 to 4 h at 23°C in 8% paraformaldehyde in 0.1 *M* phosphate buffer of pH 7.4, 2.5% glutaraldehyde in 0.1 *M* phosphate buffer of pH 7.4, or paraformaldehyde–picric acid as advocated by Zamboni and De Martino (*J. Cell Biol.*, **35:** 148A, 1967).

2. Tissues are embedded in either methacrylate (75% ethyl, 25% methyl) polymerized at 40°C or Araldite 6005 (Ladd Research Industries, Inc., Burlington, Vermont).

3. Thin sections are cut, mounted on Formvar-coated nickel grids, and stained immediately after sectioning. Methacrylate-embedded sections do not require etching; however, staining of Araldite-embedded sections may be increased if they are etched by flotation on a drop of 10% aqueous hydrogen peroxide for 20 min immediately prior to staining.

4. To cover nonspecific sites of protein absorption, grids are first floated on normal goat serum diluted 1 : 30 with 0.5 *M* tris–HCl buffer of pH 7.6 diluted 1 : 9 with 0.9% NaCl for 3 min.

5. Rinse in saline 0.5 *M* tris–HCl buffer of pH 7.6 for 3 min.

6. Place sections on a drop of rabbit anti-ACTH antiserum diluted 1 : 20 or 1 : 100 with the tris–HCl–saline solution buffer for 3 min.

7. Rinse in saline 0.5 *M* tris–HCl buffer of pH 7.6 as above for 3 min.

8. Place sections on a drop of goat antirabbit IgG diluted 1 : 10 (or 1 : 5 for Araldite-embedded sections) with the tris HCl–saline solution buffer above for 3 min.

9. Rinse sections on a drop of the tris–HCl–saline solution buffer for 3 min.

10. Place sections on a drop of the horseradish peroxidase–antihorseradish peroxidase (PAP) complex (the antiperoxidase is made in rabbits) diluted 1 : 10 with the tris–HCl–saline solution buffer above for 3 min. The horseradish peroxidase–antihorseradish peroxidase complex is kept frozen in 0.1-ml amounts and diluted just before use. Frozen-diluted horseradish peroxidase–antihorseradish peroxidase complex keeps no longer than 2 weeks.

11. Rinse section on a drop of the tris–HCl–saline solution buffer for 3 min.

12. Stain sections for 3 to 4 min with the 3,3'-diaminobenzidine (DAB) reagent with hydrogen peroxide prepared as follows: Place 22 mg DAB in a beaker containing 175 ml tris (0.5 *M* tris brought to pH 7.6 by addition of 1 *N* HCl) and 1.5 ml 0.3% hydrogen peroxide. Filter the solution. The solution must be kept moving during staining. The present authors agitated the incubation fluid with a magnetic stirrer. Kawarai and Nakane (*J. Histochem. Cytochem.*, **18:** 161, 1970) developed a "steady flow" apparatus to effect circulation.

13. Rinse grids in three consecutive changes of distilled water.
14. Stain sections in aqueous 2% osmium tetroxide for 60 min under a hood. No visible stain occurs without osmication. Osmium serves to stain the reaction product and to counterstain.

Moriarty and Halmi note the two cell types that stain for ACTH in the rat pituitary, namely, a distinctive cell type in the anterior lobe and all cells of the intermediate lobe. The normal ACTH cell in the anterior pituitary is small, stellate, or sometimes angular with a few processes extending to vessels and often around other cells. Secretory granules (approximately 200 to 300 mμ in diameter) are normally in a single row near the plasma membrane. Golgi complex generally stained intensely.

Moriarty et al. (*J. Histochem. Cytochem.*, **21**: 825, 1973) conclude that the sensitivity of the unlabeled antibody peroxidase-antiperoxidase complex immunocytochemical technic greatly exceeds that of radioimmunoassay. Poor-quality [17-19]ACTH antiserum which did not bind a significant amount of antigen beyond a 1 : 30 dilution could be employed at the ultrastructural level in the immunocytochemical staining method at antiserum dilutions as high as 1 : 1500.

Petrali et al. (*J. Histochem. Cytochem.*, **22**: 782, 1974) conducted a comparative quantitative assay of sensitivities of immunohistochemical stains using or not using the peroxidase-antiperoxidase complex. With the use of peroxidase-antiperoxidase complex the stains were 4 or 5 times more sensitive than with the use of purified antiperoxidase and 20 times more sensitive than by using antiserum to peroxidase. Petrali et al. ascribe the increase in sensitivity as due to the high binding affinity of peroxidase to antiperoxidase in the cyclic peroxidase-antiperoxidase molecule, thereby preventing loss of peroxidase that occurs during washing of sections when antiperoxidase and peroxidase are added in sequence. Petrali et al. calculate that with the use of anti-[17-19]ACTH and the intermedial lobe of the rat pituitary gland, the unlabeled antibody enzyme method is 16,000 to 100,000 times more sensitive than the present radioimmunoassay.

Parsons and Erlandsen (*J. Histochem. Cytochem.*, **22**: 340, 1974) used the unlabeled antibody enzyme method for specific identification of prolactin in anterior pituitary cells of rats.

Erlandsen et al. (*J. Histochem. Cytochem.*, **22**: 401, 1974) used the unlabeled antibody enzyme method to demonstrate lysozyme at the ultrastructural level in human Paneth cells. Specific staining was observed in secretion granules in the apical cytoplasm within the region of the Golgi complex and in some but not all lysosomes.

Masur et al. (*J. Histochem. Cytochem.*, **22**: 385, 1974) cultured rat pituitary cells known to synthesize growth hormone, fixed for 1 h at 25°C, then with Zamboni's fluid (2% formaldehyde generated from paraformaldehyde, 0.2% picric acid, and phosphate buffer of pH 7.3 to 7.4 at 330 mOsm), washed overnight in cold phosphate-buffered saline solution, air-dried, exposed to anti-growth hormone antibody conjugated to horseradish peroxidase for 1 h at 25°C, rinsed, and stained for peroxidase. Reaction product was localized in the endoplasmic reticulum and nuclear envelope. Nuclei stained intensely at the light microscopic level.

Baker and Yu (*Am. J. Anat.*, **131**: 55, 1971) employed the Nakane (*J. Histochem. Cytochem.*, **16**: 557, 1968) peroxidase-labeled antibody method to relate thyrotropic cells of the rat pituitary to staining characteristics. Thyrotropic cells were noted to stain with aldehyde fuchsin and with the periodic acid Schiff procedure.

Nayak et al. (*J. Histochem. Cytochem.*, **22**: 414, 1974) used the immunoglobulin enzyme bridge (sandwich) method developed in Spicer's laboratory (Mason et al., *J. Histochem.*

Cytochem., **17:** 563, 1969) for the demonstration of α-fetoprotein in human fetal liver cells. Specific staining was observed in hepatocytes surrounding efferent veins and randomly throughout the lobule.

The immunoglobulin peroxidase bridge procedure of Mason et al. (*J. Histochem. Cytochem.,* **17:** 563, 1969) was employed by Phifer and Spicer (*J. Clin. Endocrinol. Metab.,* **36:** 1210, 1973) to localize thyroid-stimulating hormone in thyrotropic cells of the human hypophysis. Thyrotropic cells were located mostly in the anteromedial area of the pars distalis. The same method was employed to demonstrate ACTH-productive cells in the pars intermedia of the hypophysis (*Lab. Invest.,* **23:** 543, 1970; *J. Clin. Endocrinol. Metab.,* **31:** 347, 1970). Likewise, Gavin et al. used the method to demonstrate IgG-productive cells in Hodgkin's disease (*J. Exp. Med.,* **139:** 1077, 1974), and Willingham et al. demonstrated antigens on human lymphocytes with the method (*Lab. Invest.,* **25:** 211, 1971).

Klockers and Osserman (*J. Histochem. Cytochem.,* **22:** 139, 1974) used the immunoperoxidase-immunoglobulin bridge technic as developed by Mason et al. (*J. Histochem. Cytochem.,* **17:** 563, 1969) for the demonstration of lysozyme in several rat tissues. Staining was particularly notable in proximal tubules of the kidney, Paneth cells, lung alveolar macrophages, and macrophages in perifollicular sinusoids of activated lymph nodes.

Gonatas et al. (*Lab. Invest.,* **26:** 253, 1972) used peroxidase-labeled antirat immunoglobulin G and IgM to determine the distribution of these immunoglobulins on normal rat thymus and lymph node cells. Light and electron microscopy were employed. Between 26 and 46% of lymph node cells were positive for IgG and 21 to 25% for IgM. Insignificant numbers of thymus cells were labeled. At the electron microscopic level, positive reaction product was noted in lymphocytes and plasma cells to be a thin line, approximately 100 Å surrounding the plasma membrane or as a dense patch (0.2 to 0.3 μ) which occasionally covered about one-third of the cell perimeter in a caplike fashion.

Nakane and Pierce (*J. Cell Biol.,* **33:** 307, 1967) conjugated either acid phosphatase or horseradish peroxidase to antibodies and noted retention of immunologic and enzymatic activity. The Gomori lead method for acid phosphatase and the Karnovsky method for peroxidase were employed with antiserums directed to epithelial basement membrane. Distinct advantages of the enzyme-labeled antibody (in contrast to fluorescein used in the older Coons method) are permanence of the preparations and that it can be used for both light and electron microscopic observations. Nakane and Pierce detected localization of antibody both in epithelial basement membrane cells and in the endoplasmic reticulum of cells known to produce basement membrane. Mazurkiewicz and Nakane (*J. Histochem. Cytochem.,* **20:** 969, 1972) obtained excellent localization of peroxidase-labeled antigen-antibody complexes at the light and electron microscopic levels in' tissues embedded in polyethylene glycol.

Streefkerk and van der Ploeg (*J. Histochem. Cytochem.,* **21:** 715, 1973) used a model system composed of horseradish peroxidase incorporated in polyacrylamide gels to determine if the quantities of dyes formed during the cytochemical horseradish peroxidase reaction can be used as a quantitative measure of enzyme activity. They achieved a linear relation between the amount of reaction product using 3,3'-amino-9-ethylcarbazole as the hydrogen donor, but a similar relation could not be obtained using benzidine as a substrate.

Yokota (*J. Histochem. Cytochem.,* **21:** 779, 1973) noted an inhibition of 35 and 40% respectively of horseradish peroxidase activity by 2.5 and 4% formaldehyde. The peroxidase was incorporated in polyacrylamide films and fixed at pH 7.4 for 90 min at 0 to 4°C.

Weir et al. (*J. Histochem. Cytochem.,* **22:** 51, 1974) recommend treatment of tissue with 0.074% HCl in ethanol (0.2 ml concentrated HCl in 100 ml ethanol) for 15 min at 24°C to abolish endogenous peroxidase activity.

Aronson et al. (*J. Histochem. Cytochem.*, **21**: 1047, 1973) coupled a relatively small molecule (hemin, mw 652) to sucrose, glucuronic acid, or several types of dextrans which were injected (IV) into mice after purification for use as a tracer at the ultrastructural level. Sections were stained with the Graham-Karnovsky method for peroxidase. The concentration of diaminobenzidine was increased to 1 mg/ml, the incubation medium was sometimes warmed to 60°C, and sections were treated for as long as 1 h in order to obtain sufficient peroxidase activity for disclosure. Heme has one five-thousandth the degree of peroxidase activity as does horseradish peroxidase.

Karnovsky and Rice (*J. Histochem. Cytochem.*, **17**: 751, 1969) used cytochrome c as a protein ultrastructural tracer in the diaphragm and kidney of mice. Karnovsky's method for peroxidase was used for detection. Cytochrome c with a molecular weight of 12,000 may be more useful sometimes than the usual peroxidases that have molecular weights from 40,000 to 160,000.

Avrameas (*C. R. Acad. Sci.* [*D*] (*Paris*), **270**: 2205, 1970), Bernhard and Avrameas (*Exp. Cell Res.*, **64**: 232, 1971), and Parmley et al. (*J. Histochem. Cytochem.*, **21**: 912, 1973) studied cell surfaces at the electron microscopic level using a lectin–horseradish peroxidase method. Parmley et al. used either concanavalin A (Con A) or *Lens culinaris* hemagglutinins (LcH) conjugated to horseradish peroxidase to identify either concanavalin A binding sites (hydroxyl groups in terminal nonreducing α-D-glucopyranosyl, α-D-mannopyranosyl, and β-D-fructopyranosyl residues according to Goldstein et al., *Biochemistry*, **4**: 876, 1965) or to LcH binding sites (α-D-N-acetylglucosamine in addition to mannosides and glucosides according to Young et al., *J. Biol. Chem.*, **246**: 1596, 1971). The Avrameas method as used by Parmley et al. is provided below:

1. Prepare buffy coats from heparinized blood, and fix in 3% glutaraldehyde in phosphate-buffered saline solution for 45 min at 24°C.
2. Centrifuge fixed cells into a pellet, and rinse in three changes of phosphate-buffered saline solution (5 min each).
3. Suspend cells in either 0.015 to 1.5 mg/ml concanavalin A (Sigma) or 1.5 mg/ml LcH (obtained from Dr. Harvey Sage, Duke University School of Medicine) in phosphate-buffered saline solution for 45 min at 24°C.
4. Rinse in phosphate-buffered saline solution (3 times).
5. Suspend in horseradish peroxidase (Sigma type II) at a concentration of 1 mg/100 ml phosphate-buffered saline solution for 30 min at 24°C.
6. Rinse in phosphate-buffered saline solution (3 times).
7. Treat cells with 3,3'-diaminobenzidine (75 mg in 0.001% hydrogen peroxide in 100 ml 0.05 M tris buffer of pH 7.6) for 30 min at 24°C.
8. Rinse cells in phosphate-buffered saline solution (2 times), and place in 1.5% bovine fibrinogen in phosphate-buffered saline solution.
9. Clot the pellet with 1% bovine thrombin, rinse in phosphate-buffered saline solution, postfix in osmium, dehydrate, and embed in Epon for electron microscopic examination.

RESULTS: Surface coats of platelets, granulocytes, monocytes, lymphocytes, and erythrocytes from human and rabbit peripheral blood, myeloblasts and lymphoblasts from peripheral blood of persons with leukemia, and erythroid precursors and granulocyte precursors in bone marrow of rabbits generally stained intensely with both concanavalin A and LcH-A and LcH-B. The surface of human erythrocytes stained less intensely than did rabbit erythrocytes.

Barbotin and Thomas (*J. Histochem. Cytochem.*, **22**: 1048, 1974) used the method of Broun et al. (*Biotechnol. Bioeng.*, **15**: 359, 1973) to construct a 50-μ-thick membrane developed from a solution of 0.5 to 10 mg horseradish peroxidase per ml, 5% serum albumin, and 0.75% glutaraldehyde in 0.02 M phosphate buffer of pH 6.8. Direct spectro-photometric assays revealed that the amount of insoluble polymer produced did not increase in a linear fashion with the quantity of enzyme in the membrane. Furthermore, geometric similarity in distribution of the enzyme product and the distribution of the enzyme in the membrane was not observed.

Harada et al. (*J. Histochem. Cytochem.*, **21**: 804, 1973) purified lactoperoxidase from fresh cow's milk which was used as an antigen for the production of specific antiserum in rabbits. Purified rabbit γ-globulins were conjugated to fluorescein isothiocyanate (FITC), and some were conjugated to tetramethylrhodamine isothiocyanate (TRITC). Purified conjugates were used to stain frozen sections of bovine mammary gland; sublingual, parotid, and lacrimal glands; liver; kidney; spleen; pancreas; thyroid; and thymus. Peripheral blood smears also were stained. Lactoperoxidase was found mainly in the cytoplasm of alveolar cells of the mammary gland, serous cells of the sublingual gland, acinar cells of the lacrimal gland, a few splenic cells of the red pulp, and certain nonspecified leukocytes of the peripheral blood.

A recent review of ultrastructural cytochemistry of the adenohypophysis has been written by Moriarty (*J. Histochem. Cytochem.*, **21**: 855, 1973).

Gonatas et al. (*J. Histochem. Cytochem.*, **22**: 999, 1974) demonstrated immunoglobulins intracellularly in paraformaldehyde-fixed cells from lymph nodes of rats at the ultrastructural level using ^{125}I Fab of sheep antirat antibody. ^{125}I Fab fragments penetrated fixed cells to permit excellent specific localization; however, the larger conjugate of either peroxidase and Fab antibody or ^{125}I sheep antirat IgG could not penetrate into cells. Gonatas et al. also were able to obtain simultaneous identification of plasma membrane and intracellular immunoglobulins using iodinated Fab of sheep antirat Ig and peroxidase-labeled Fab of sheep antirat Ig with subsequent processing of cells for peroxidase and autoradiography.

MICROINCINERATION

The practice of microincineration, according to Lison, dates back over a century to Raspail in 1833, and has been reintroduced several times *de novo*. It was used by Virchow in 1847 for identification of iron in hemosiderin.

It is necessary to utilize some method of preparation of tissues which neither adds extrinsic mineral elements, as do chromates and mercuric salts, nor removes those mineral elements naturally present. Some have tried using frozen sections of unfixed fresh tissue, but these were difficult to prepare or to flatten on slides without some flotation procedure before the introduction of cryostat methods. The procedure of rapid freezing with liquid air (or, easier, with one of the acetone carbon dioxide technics used for freezing blood plasma) followed by desiccation in vacuo at -35 to $-50°C$ and infiltration with paraffin directly after desiccation offers theoretically the best tissue preparation. Paraffin sections can then be cut and attached to appropriate slides as usual.

Fixation with 100% alcohol conserves most mineral elements well but gives indifferent to poor tissue fixation. The use of 10% formalin in alcohol is perhaps the best method. With 100% alcohol dehydration, clearing in benzene, and paraffin infiltration it affords losses of water-soluble salts perhaps as small as those with any other method.

The paraffin sections for incineration are to be floated on freshly filtered 95 to 100% alcohol on clean glass slides without albumen or other adhesive. Alternate sections should be prepared as usual for ordinary histologic examination and topographic comparison with the ashed sections. Sections should preferably be cut at 3 to 5 μ.

Place the slides on quartz plates or on slides in an electric furnace. Raise the temperature gradually at first, allowing 10 min for it to reach 100°C; then raise it more rapidly to 650°C, reaching this temperature in another 25 min. These are Scott's directions. Lison recommends the Schultz-Braun procedure of heating in flowing nitrogen until 500 to 530°C is reached and then admitting air for a few minutes to burn off the carbon. By this method, chlorides are said to be better preserved than at the higher temperatures, and the formation of not easily combustible partial oxidation products at lower temperatures is avoided.

When the oven is opened after incineration, remove the quartz plate and slides with heated forceps, to avoid rapid chilling and cracking of the glass. Then place the slides on an asbestos plate to cool. Cover the ashed section, or *spodogram* (σπόδοσ, ashes; γράμμα, a drawing, a writing, or a picture), with a cover glass and fasten down the edges with paraffin, petrolatum, sealing wax, or a pyroxylin cement. Some sections may be sprayed with thin collodion (0.5 to 1% nitrocellulose in ether-alcohol) for the purpose of making solubility tests and chemical reactions on the spodograms. This is not advised as the routine procedure, but see Fenton, Johnson, and Zimmerman later in this section.

The first examination is to be made with oblique illumination from a strong light. Dark-field condensers may also be used, and study under reflected light as practiced in mineralogic microscopy may be of value. Low magnifications are generally most profitable. Cell topography is often well maintained, but nuclei may not be evident.

Sodium and potassium chlorides and tricalcium and magnesium phosphates are preserved as such; carbonates are converted into oxides such as sodium, calcium, and magnesium oxides. Dipotassium phosphate (K_2HPO_4) is converted into the pyrophosphate ($K_4P_2O_7$). Iron compounds appear as yellow to red ferric oxide (Fe_2O_3). Silica generally combines with calcium to form various silicates. Sulfur is either volatilized or converted into (or remains as) sulfates.

Identification of elements still remains difficult. Certain birefringent silica crystals remain identifiable. Silica appears as a white, doubly refractile crystalline mineral. It is insoluble in water. Calcium is seen as a white, singly refractile ash, almost insoluble in water. It responds to the gypsum and oxalate tests. Magnesium is difficult to identify in the presence of calcium. Here the collodion technic, with some of the more recent color reactions (Chap. 13), could be useful.

According to Alexander and Myerson, iron is present as yellow to red ferric oxide. It is to be distinguished from remaining carbon. The latter is black in direct transmitted light. Application of the HCl–potassium ferrocyanide reagent to the collodionized ash gives the Prussian blue reaction, though the use of this test is not mentioned by these authors.

The papers of Scott (*Protoplasma*, **20**: 133, 1933; *Am. J. Anat.*, **53**: 243, 1933; *Anat. Rec.*, **55**: 75, 1933), of Cowdry (*Am. J. Pathol.*, **9**: 149, 1933), and of Alexander and Myerson (*Am. J. Pathol.*, **13**: 405, 1937) may be consulted. The last contains a bibliography of 124 titles, many of which deal with microincineration.

Erich Hintzsche's monograph "Das Aschenbild tierischer Gewebe und Organe," Springer-Verlag, Berlin, 1956, should prove a valuable and comprehensive reference of more recent date.

CALCIUM OXALATE. Johnson (*J. Histochem. Cytochem.*, **4**: 404, 1956) demonstrated that by microincineration in a muffle furnace at 425 to 475°C calcium oxalate was converted to calcium carbonate. The carbonate could then be identified by gently applying dilute acid to the ash while it is under microscopic observation. Bubbles of carbon dioxide were evolved. If the preparations were heated to 600°C, calcium carbonate broke down to calcium oxide, and no gas was produced on addition of acid. As a further control it would be demonstrable that unheated preparations did not evolve gas on acid treatment.

Johnson later noted (*J. Histochem. Cytochem.*, **6**: 405, 1958) that calcium succinate, citrate, malonate, and maleate also yield calcium carbonate on microincineration at 450°C, and that this may be demonstrated by evolution of carbon dioxide on acid treatment of the ash. However, pretreatment of sections with 2 *N* acetic acid removes such salts other than oxalate. Strontium and barium oxalates are also converted to carbonate by 450°C microincineration. Here Johnson noted that the calcium ash forms a red lake with alizarin red S and that those of barium and strontium do not. It should be noted that 450°C microincineration leaves a considerable amount of unburned carbon.

Wolman and Goldring (*J. Histochem. Cytochem.*, **10**: 505, 1962) applied the von Kóssa reaction to the ash of preparations made by Johnson's method, thereby obtaining permanent preparations. It was necessary to compare serial sections, one not incinerated and one incinerated at 450°C and reacted by von Kóssa's technic, and a third, incinerated but unsilvered. Only black deposits appearing in the silvered spodogram alone were considered as calcium oxalate. Wolman and Goldring state categorically that unashed calcium oxalate does not react to the von Kóssa test.

Johnson's method (*J. Histochem. Cytochem.*, **12**: 153, 1964) of microincineration, though less precise in temperature control than muffle furnace technics, can afford quite usable preparations without the more elaborate equipment required for the latter.

1. Cut 6- to 8-μ paraffin sections of material fixed in formalin, alcohol, or other non-metal-containing fixatives, and mount them on glass slides. Stain control preparations by usual appropriate histologic or histochemical methods.
2. Place the slide in middle of an asbestos board, and heat it cautiously with the blue flame of a gas burner (Fisher or Meeker type), directing the flame downward on the slide and keeping it in motion to heat evenly. The section turns brown and then white. Continue heating until it has almost disappeared.
3. Then cover the slide with an inverted metal pan resting on the asbestos board. The pan heats up and slows the cooling of the slide, so that breakage of the glass from too rapid cooling is avoided.

Fenton, Johnson, and Zimmerman (*J. Histochem. Cytochem.*, **12**: 153, 1964) have modified the foregoing technic by adding a specific histochemical reaction for iron after microincineration: The microincinerated section, when cold, is gently dipped into a very thin parlodion or celloidin solution (dilute 1.5 ml 1% celloidin to 100 ml with an equal-volume absolute alcohol–ether mixture); the slide is removed at once and allowed to dry in air. Fenton, Johnson, and Zimmerman then apply a mixture of equal volumes 5% potassium ferrocyanide and 0.6 *N* hydrochloric acid and allow it to act for 15 min. They then wash in distilled water, dry in air or in an oven, and mount in Permount.

The reaction can also be watched in wet preparations.

Boothroyd [*J. Microsc. (Oxf.)*, **88**: 529, 1967] adapted microincineration technics to electron microscopy. Mineral salts must be avoided throughout all procedures for specimens prepared for either light or electron microscopy. Boothroyd recommends fixation in either 5% glutaraldehyde or 10% formalin in 0.1 *M* sodium cacodylate buffer of pH 7.2. Tissues are rinsed, processed through graded alcohols, and embedded in Araldite. Sections cut at 1100 Å are mounted on nickel grids, coated with silicon monoxide film, incinerated at 600°C for 15 to 20 min, and examined by electron microscopy. Readers will recognize the opportunity for loss and shifting of minerals during the recommended fixation and washing periods. Also, sodium and arsenic are added in the buffer. An acetic acid–ammonium acetate buffer would avoid that difficulty. A tri- or tetramethylammonium salt would give a higher pH range.

20

BUFFERS AND
BUFFER TABLES;
NORMAL ACIDS
AND ALKALIES

Since the previous edition we have expanded considerably our use of buffers, not only for staining procedures but also during fixation methods, in enzyme localization technics, and in some decalcification procedures. Additional buffers in the acid and alkaline ranges have been found useful and are included here. Since the pH levels of buffer mixtures vary with their dilution, in a few cases figures for more than one dilution are included.

It is to be remembered that buffer mixtures are comparatively ineffective in preventing pH displacement when one of the salts is present in nearly pure form. The same phenomenon is observed with phosphates and citrates when the mixture is such that a monohydrogen or dihydrogen salt would be crystallized in nearly pure form if the mixture were evaporated. Hence, where adjacent pH figures in any buffer series are relatively widely separated and no other contraindications exist, it is preferable to use a different buffer series if the desired pH level is approximated by a number of consecutive readings in the tables. Therefore, closely spaced pH values in the tables are printed in boldface.

Due regard must be given to the effect of the buffer ions on the other ingredients of the solution in which they are used. For example, acid phosphate buffers precipitate thionin. Tartrates precipitate calcium salts, as do the less acid phosphate buffers. Acetates and citrates can be used in acid silver solutions. Formates and citrates reduce permanganate and chromate solutions, but acetates and phosphates can be used. (Acetic acid reduces permanganates and chromates slowly at 95 to 100°C.)

In the alkaline range buffers are less satisfactory because of their avidity for carbon dioxide from the laboratory air.

The salts and acids used should be of reagent grade—those specially made for buffer use when available. Particular note should be taken of the molecular weight specified on bottle labels. That weight should be used in preparing solutions, since some of the salts are available in a variety of states of hydration. It is often a matter of indifference whether a sodium or a potassium salt is used in a buffer mixture, provided that due account is taken of the change in molecular weight.

Distilled water is usually acid as prepared from block tin stills, and redistillation from glass does not seem to remedy this fault. Even freshly distilled or doubly distilled water boiled to expel carbon dioxide and cooled often gives pH levels between 5 and 6. However, the amount of acid present seems too small to affect seriously even hundredth-molar buffers, and we usually ignore the fact that fresh distilled water may measure pH 5.5. For most purposes singly distilled water of that pH level is quite satisfactory in buffers.

TABLE 20-1 FORMULAS AND MOLECULAR WEIGHTS OF COMMON BUFFER INGREDIENTS AND REAGENTS

	Formulas	Molecular weights
Acetic acid*	CH_3COOH	60.05
Ammediol (2-amino-2-methyl-1,3-propanediol	$\begin{matrix} CH_2OH \\ \mid \\ H_2N-C-CH_3 \\ \mid \\ CH_2OH \end{matrix}$	105.140
Ammonia*	NH_3	17.032
Barbital sodium (sodium diethyl barbiturate)	$C_8H_{11}O_3N_2Na$	206.18
Borax	$Na_2B_4O_7 \cdot 10H_2O$	381.43
Boric acid	$B(OH)_3$	61.84
Cacodylic acid	$(CH_3)_2AsO_2H$	137.99
Citric acid, anhydrous	$C_3H_4(OH)(COOH)_3$	192.12
Citric acid, crystals	$C_3H_4(OH)(COOH)_3 \cdot H_2O$	210.14
γ-Collidine	$2,4,6(CH_3)_3C_5H_2N$	121.18
Ferric chloride anhydrous	$FeCl_3$	162.22
Ferric chloride crystals	$FeCl_3 \cdot 6H_2O$	270.32
Formic acid*	$HCOOH$	46.03
Glycine	NH_2CH_2COOH	75.07
Hydrochloric acid*	HCl	36.465
Maleic acid	$HOOCCH=CHCOOH$	116.07
Nitric acid*	HNO_3	63.016
Oxalic acid	$(COOH)_2$	90.038
Potassium acid phosphate	KH_2PO_4	136.09
Potassium chloride	KCl	74.553
Potassium hydroxide	KOH	56.104
Potassium phosphate tribasic	K_3PO_4	212.275
Sodium acetate	CH_3COONa	82.04
Sodium acetate, crystals	$CH_3COONa \cdot 3H_2O$	136.09
Sodium acid phosphate	$NaH_2PO_4 \cdot H_2O$	138.01
Sodium chloride	$NaCl$	58.448
Sodium citrate, crystals	$C_3H_4OH(COONa)_3 \cdot 5\frac{1}{2}H_2O$	357.18
Sodium citrate, granular	$C_3H_4OH(COONa)_3 \cdot 2H_2O$	294.12
Sodium formate	$HCOONa$	68.015
Sodium hydroxide	$NaOH$	40.005
Sodium oxalate	$Na_2C_2O_4$	134.004
Sodium phosphate, dibasic	Na_2HPO_4	141.98
Sulfuric acid*	H_2SO_4	98.082
Trihydroxymethylaminomethane (tris)	$H_2NC(CH_2OH)_3$	121.14
Veronal sodium, Medinal (sodium 5,5-diethylbarbiturate)	$C_8H_{11}O_3N_2Na$	206.18

* See Tables 20-25 and 20-26 for preparation of normal solutions.

Source: Emended slightly from Lange, 1949.

TABLE 20-2 pH STANDARD BUFFER SOLUTIONS AT 15 TO 30°C

Molarity	Name	15°C	20°C	25°C	30°C	Grams per liter
0.05	Potassium tetraoxalate	1.67	1.68	1.68	1.69	12.71 $KH_3(C_2O_4)_2 \cdot 2H_2O$
	Potassium acid tartrate			3.56	3.55	Shake excess salt with water at 25°C, decant
0.05	Potassium acid phthalate	4.00	4.00	4.01	4.01	10.21 $KHC_8H_4O_4(o)$
$\left.\begin{matrix}0.025 \\ 0.025\end{matrix}\right\}$	$KH_2PO_4 + Na_2HPO_4$	6.90	6.88	6.86	6.85	3.40 KH_2PO_4 3.55 Na_2HPO_4
0.01	Borax ($Na_2B_4O_7 \cdot 10H_2O$)	9.27	9.22	9.18	9.14	3.81

Source: From N. A. Lange, "Handbook of Chemistry," 10th ed., McGraw-Hill, New York, 1961.

TABLE 20-3 ACID PHOSPHATES

HCl, ml	KH$_2$PO$_4$, ml	pH values	
		1 M	0.1 M
30	20	0.69	1.53
29	21	0.75	1.56
28	22	0.81	1.63
27	23	0.90	1.66
26	24	0.98	1.71
25	25	1.05	1.77
24	26	1.12	1.80
23	27	1.23	1.84
22	28	1.35	1.89
21	29	1.43	1.94
20	30	1.52	2.00
19	31	1.61	2.05
18	32	1.70	2.10
17	33	1.76	2.15
16	34	1.82	2.23
15	35	1.92	2.27
14	36	2.02	2.31
13	37	2.11	2.38
12	38	2.20	2.44
11	39	2.29	2.50
10	40	2.38	2.56
9	41	2.45	2.63
8	42	2.52	2.70
7	43	2.62	2.79
6	44	2.72	2.89
5	45	2.85	3.02
4	46	3.00	3.14
3	47	3.17	3.23
2	48	3.32	3.44

Note: Use 1 N and 0.1 N HCl for acid component. See Table 20-25, KH$_2$PO$_4$, mw 136.09. Use 1 mol and 0.1 mol/l, respectively.

Buffers containing sodium citrate are particularly liable to mold growth. Strong stock buffers should be made up in 20 to 25% alcohol to inhibit this growth. When diluted 1 : 20 or 1 : 25 for use, the alcohol concentration falls to an insignificant 1% or so.

In the acid range we have added a sulfuric acid, acid sodium phosphate series for Nile blue stains, Pearse's[1] HCl sodium acetate series, an additional 0.01 M dilution series for the Walpole acetates, a cacodylate buffer from Long's "Biochemist's Handbook," Gomori's tris HCl buffer from Pearse (1960), and Gancevici's K$_3$PO$_4$ HCl buffer in the 9.5 to 12.4 pH range, and we have expanded the useful Michaelis Veronal HCl series by interpolation, to give closer readings. Holmes' borate series and Gomori's collidine hydrochloric acid series are retained from previous editions, for we have found occasional use for them. The McIlvaine series as given varies to some extent from that given by Clark, whose figures are based on the undiluted aqueous mixtures. The figures here are based on a 1 : 25 dilution of the stock 25% methanol solutions in distilled water, the concentration at which this buffer is used in azure-eosin stains.

---

[1] Full references for works cited in this chapter are given at the end of the chapter.

TABLE 20-4 HCl SODIUM CITRATE BUFFER SERIES AT 1 M, 0.1 M, 0.01 M

HCl, ml	Sodium citrate, ml	pH values 1 M	0.1 M	0.01 M	HCl, ml	Sodium citrate, ml	pH values 1 M	0.1 M	0.01 M
50	0	0.25	1.11	2.10	25	25	4.69	5.10	5.42
49	1	0.27	1.14	2.15	24	26	4.78	5.20	5.52
48	2	0.30	1.19	2.20	23	27	4.85	5.28	5.61
47	3	0.35	1.24	2.25	22	28	4.92	5.35	5.70
46	4	0.40	1.29	2.30	21	29	5.00	5.41	5.79
45	5	0.45	1.34	2.35	20	30	5.05	5.48	5.87
44	6	0.51	1.38	2.40	19	31	5.10	5.54	5.93
43	7	0.58	1.42	2.45	18	32	5.17	5.60	5.97
42	8	0.67	1.50	2.50	17	33	5.22	5.65	6.01
41	9	0.80	1.65	2.60	16	34	5.28	5.70	6.06
40	10	0.97	1.85	2.70	15	35	5.33	5.75	6.10
39	11	1.30	2.12	2.85	14	36	5.37	5.79	6.14
38	12	2.00	2.52	3.02	13	37	5.42	5.83	6.19
37	13	2.60	2.90	3.24	12	38	5.49	5.87	6.25
36	14	2.95	3.24	3.50	11	39	5.56	5.93	6.31
35	15	3.25	3.58	3.80	10	40	5.63	6.00	6.36
34	16	3.50	3.82	4.03	9	41	5.70	6.06	6.42
33	17	3.67	4.00	4.22	8	42	5.79	6.14	6.50
32	18	3.84	4.18	4.40	7	43	5.88	6.21	6.60
31	19	4.01	4.36	4.60	6	44	5.97	6.30	6.69
30	20	4.18	4.52	4.75	5	45	6.06	6.39	6.78
29	21	4.30	4.67	4.91	4	46	6.16	6.48	6.88
28	22	4.40	4.79	5.05	3	47	6.27	6.57	6.98
27	23	4.50	4.90	5.20	2	48	6.42	6.73	7.11
26	24	4.59	5.00	5.31	1	49	6.66	6.97	7.25
					0	50	7.17	7.51	7.82

Note: Sodium citrate is used at 357.18, 35.718, and 3.572 g crystals per l for 1 M, 0.1 M, and 0.01 M, respectively. Granular sodium citrate, mw 294.12, contains less water; use 294.12, 29.412, and 2.941 g/l for 1 M, 0.1 M, and 0.01 M.

TABLE 20-5 SULFURIC ACID, ACID SODIUM PHOSPHATE BUFFER SERIES FOR NILE BLUE STAINS

0.1 M solutions pH	H_2SO_4	NaH_2PO_4	0.1 M solutions pH	H_2SO_4	NaH_2PO_4	0.1 M solutions pH	H_2SO_4	NaH_2PO_4
1.1	50	0	2.1	36	14	3.1	13.5	36.5
1.2	48.5	1.5	2.2	34	16	3.2	11.5	38.5
1.3	47	3	2.3	31.5	18.5	3.3	10	40
1.4	45.5	4.5	2.4	30	20	3.4	8	42
1.5	44.5	5.5	2.5	28	22	3.5	6	44
1.6	44	6	2.6	25	25	3.6	5	45
1.7	42	8	2.7	23	27	3.7	3.5	46.5
1.8	40.5	9.5	2.8	20	30	3.8	2.5	47.5
1.9	39	11	2.9	18	32	3.9	1.5	48.5
2.0	37	13	3.0	15	35	4.0	0.5	49.5

Note: 0.2 M H_2SO_4, pH 0.87. 1% H_2SO_4 v/v, pH 0.9. See Table 20-25 for sulfuric acid; 0.1 M = 0.2 N. For 0.1 M NaH_2PO_4, use 13.801 g disodium phosphate per l.

Source: From *Stain Technol.*, **31**: 151, 1956.

Data were determined on a Beckman pH meter, corrected to 25°C, and slightly smoothed. We are indebted to P. Jones for these measurements.

Tables based on Lange's data for density and grams per liter of the significant constituent are presented for the ready preparation of normal acid and ammonia solutions. For chemical titration procedures, these solutions must be standardized as usual, but for most histologic purposes they may be used as prepared.

TABLE 20-6   0.125 N MALEIC ACID PLUS SODIUM HYDROXIDE

1 N maleic acid	0.1 N NaOH	Distilled water	pH
5	6	39	**1.83**
5	7	38	**1.87**
5	8	37	**1.89**
5	9	36	**1.90**
5	10	35	**1.94**
5	11	34	**1.98**
5	12	33	**2.00**
5	13	32	**2.02**
5	14	31	**2.08**
5	15	30	**2.10**
5	16	29	**2.15**
5	17	28	**2.20**
5	18	27	**2.22**
5	19	26	**2.28**
5	20	25	**2.30**
5	21	24	**2.38**
5	22	23	**2.42**
5	23	22	**2.49**
5	24	21	**2.55**
5	25	20	**2.62**
5	26	19	**2.70**
5	27	18	**2.75**
5	28	17	**3.05**
5	29	16	3.31
5	30	15	3.62
5	31	14	4.40
5	32	13	4.71
5	33	12	**4.91**
5	34	11	**5.00**
5	35	10	**5.15**
5	36	9	**5.27**
5	37	8	**5.34**
5	38	7	**5.45**
5	39	6	**5.50**
5	40	5	**5.59**
5	41	4	**5.62**
5	42	3	**5.68**
5	43	2	**5.73**
5	44	1	**5.75**
5	45	0	**5.83**

Note: 1 N maleic acid = 116.07 g/l. 0.1 N sodium hydroxide = 4 g/l.

Source: Laskey, unpublished data, 1951.

MALEATE BUFFER (J. W. Temple, *J. Am. Chem. Soc.*, **51**: 1754, 1929)

STOCK SOLUTIONS

A: 0.2 M solution of acid sodium maleate (8 g NaOH + 23.2 g maleic acid or 19.6 g maleic anhydride in 1000 ml)
B: 0.2 M NaOH (50 ml A + x ml B, diluted to a total of 200 ml)

x	pH	x	pH
7.2	5.2	33.0	6.2
10.5	5.4	38.0	6.4
15.3	5.6	41.6	6.6
20.8	5.8	44.4	6.8
26.9	6.0		

CARBONATE-BICARBONATE BUFFER (G. E. Delory and E. J. King, *Biochem. J.*, **39**: 245, 1945)

STOCK SOLUTIONS

A: 0.2 M solution of anhydrous sodium carbonate (21.2 g/l)
B: 0.2 M solution of sodium bicarbonate (16.8 g/l) (x ml A + y ml B, diluted to a total of 200 ml)

x	y	pH
4.0	46.0	9.2
7.5	42.5	9.3
9.5	40.5	9.4
13.0	37.0	9.5
16.0	34.0	9.6
19.5	30.5	9.7
22.0	28.0	9.8
25.0	25.0	9.9
27.5	22.5	10.0
30.0	20.0	10.1
33.0	17.0	10.2
35.5	14.5	10.3
38.5	11.5	10.4
40.5	9.5	10.5
42.5	7.5	10.6
45.0	5.0	10.7

HYDROCHLORIC ACID–POTASSIUM CHLORIDE BUFFER (W. M. Clark and H. A. Lubs, *J. Bacteriol.*, **2**: 1, 1917)

STOCK SOLUTIONS

A: 0.2 M solution of KCl (14.91 g/l)
B: 0.2 M HCl (50 ml A + x ml B, diluted to a total of 200 ml)

x	pH
97.0	1.0
78.0	1.1
64.5	1.2
51.0	1.3
41.5	1.4
33.3	1.5
26.3	1.6
20.6	1.7
16.6	1.8
13.2	1.9
10.6	2.0
8.4	2.1
6.7	2.2

TABLE 20-7 WALPOLE BUFFER, 1914, MODIFIED FROM PEARSE, 1960

1 N HCl	1 M NaAc	Distilled water	pH
20	10	20	**0.65**
18	10	22	**0.75**
16	10	24	**0.91**
14	10	26	**1.07**
13	10	27	**1.24**
12	10	28	**1.42**
11	10	29	**1.71**
10.7	10	29.3	1.85
10.5	10	29.5	1.99
10.2	10	29.8	2.32
10.0	10	30	2.64
9.95	10	30.05	2.72
9.7	10	30.3	3.09
9.5	10	30.5	3.29
9.25	10	30.75	3.49
9.0	10	31	**3.61**
8.5	10	31.5	**3.79**
8	10	32	**3.95**
7	10	33	**4.19**
6	10	34	**4.39**
5	10	35	**4.58**
4	10	36	**4.76**
3	10	37	**4.92**
2	10	38	**5.20**

Note: In using this buffer other solutions added must replace part of the water, so that the 50-ml volume is not increased. 1 M sodium acetate = 82.04 g/l. For 1 N HCl, see Table 20-25.

Source: G. S. Walpole, *J. Chem. Soc.* [*Org.*], **105:** 2501, 1914, modified from A. G. E. Pearse, "Histochemistry," 2d ed., Little, Brown, Boston, 1960.

TABLE 20-8 WALPOLE ACETATE BUFFER

Acetic acid, ml	Sodium acetate, ml	0.2 M pH	0.01 M pH
20.0	0.0	2.696	3.373
19.9	0.1	2.804	3.420
19.8	0.2	2.913	3.477
19.7	0.3	2.994	3.503
19.6	0.4	3.081	3.523
19.5	0.5	**3.147**	**3.543**
19.4	0.6	**3.202**	**3.590**
19.2	0.8	**3.315**	**3.593**
19.0	1.0	**3.416**	**3.647**
18.5	1.5	**3.592**	**3.737**
18.0	2.0	**3.723**	**3.863**
17.0	3.0	**3.900**	**3.980**
16.0	4.0	**4.047**	**4.110**
15.0	5.0	**4.160**	**4.223**
14.0	6.0	**4.270**	**4.337**
13.0	7.0	**4.360**	**4.473**
12.0	8.0	**4.454**	**4.527**
11.0	9.0	**4.530**	**4.600**
10.0	10.0	**4.626**	**4.717**
9.0	11.0	**4.710**	**4.807**
8.0	12.0	**4.802**	**4.910**
7.0	13.0	**4.900**	**5.000**
6.0	14.0	**4.990**	**5.077**
5.0	15.0	**5.110**	**5.183**
4.0	16.0	**5.227**	**5.373**
3.0	17.0	**5.380**	**5.500**
2.0	18.0	**5.574**	**5.713**
1.0	19.0	5.894	6.003
0.5	19.5	6.211	6.227
0.0	20.0	6.518	6.777

Note: 1 N acetic acid, pH 2.2. 2 N acetic acid, pH 2. 0.2 M sodium acetate = 16.408 g/l; 0.01 M = 820 mg/l. For acetic acid, see Table 20-26.

Source: G. S. Walpole, *J. Chem. Soc.* [*Org.*], **105:** 2501, 1914.

TABLE 20-9 HORECKER-LILLIE BUFFER TABLE (1 : 25 DILUTION OF SOLUTIONS IN 25% ALCOHOL)

pH	40 mM citric acid	40 mM sodium citrate	pH	40 mM citric acid	40 mM sodium citrate
2.86	20	0	**4.84**	9	11
2.95	19	1	**5.04**	8	12
3.08	18	2	**5.26**	7	13
3.23	17	3	**5.50**	6	14
3.40	16	4	**5.74**	5	15
3.58	15	5	**5.97**	4	16
3.80	14	6	**6.17**	3	17
4.01	13	7	6.42	2	18
4.27	12	8	6.77	1	19
4.50	11	9	7.60	0	20
4.65	10	10	$M/10$ 21.0		$M/10$ 29.4

Note: Stock buffer is made at 0.1 M: thus 21.014 g citric acid, 35.718 g sodium citrate crystals = 29.412 g granular in 250 ml ethanol + 750 ml distilled water. Final dilutions contain 1% alcohol.

TABLE 20-10 McILVAINE-LILLIE BUFFER TABLE (1 : 25 DILUTION OF STOCK 25% CH$_3$OH SOLUTION)

pH	0.04 M citric acid	0.08 M disodium phosphate	pH	0.04 M citric acid	0.08 M disodium phosphate
2.5	20.0	0	**5.3**	9.5	10.5
2.6	19.5	0.5	**5.5**	9.0	11.0
2.65	19.0	1.0	**5.7**	8.5	11.5
2.7	18.5	1.5	**6.0**	8.0	12.0
2.75	18.0	2.0	**6.1**	7.5	12.5
2.8	17.5	2.5	**6.3**	7.0	13.0
2.9	17.0	3.0	**6.4**	6.5	13.5
3.0	16.5	3.5	**6.5**	6.0	14.0
3.05	16.0	4.0	**6.6**	5.5	14.5
3.1	15.5	4.5	**6.8**	5.0	15.0
3.2	15.0	5.0	**6.9**	4.5	15.5
3.3	14.5	5.5	**7.0**	4.0	16.0
3.45	14.0	6.0	**7.1**	3.5	16.5
3.6	13.5	6.5	**7.2**	3.0	17.0
3.75	13.0	7.0	**7.3**	2.5	17.5
3.95	12.5	7.5	**7.4**	2.0	18.0
4.1	12.0	8.0	**7.5**	1.5	18.5
4.3	11.5	8.5	**7.7**	1.0	19.0
4.5	11.0	9.0	8.0	0.5	19.5
4.75	10.5	9.5	8.3	0	20.0
4.95	10.0	10.0			

Note: The stock buffer is made at 0.1 M citric acid and 0.2 M Na$_2$HPO$_4$ in 25% methanol. 21.014 g citric acid 28.396 g Na$_2$HPO$_4$ in 250 ml methanol + 750 ml distilled water. Use 2 ml for 50-ml stain.

TABLE 20-11   CACODYLATE BUFFER, 0°C AND 25°C

	0°C				25°C		
pH	1 M cacodylic acid, ml	1 N NaOH, ml	Distilled water, ml	pH	1 M cacodylic acid, ml	1 N NaOH, ml	H_2O, ml
5.2	513	50	437	**5.2**	522	50	428
5.4	342	50	608	**5.4**	348	50	602
5.6	235	50	715	**5.6**	238	50	712
5.8	166	50	784	**5.8**	169	50	781
6.0	124	50	826	**6.0**	125	50	825
6.2	96.3	50	853.7	**6.2**	97.2	50	852.8
6.4	79.2	50	870.8	**6.4**	79.8	50	870.2
6.6	68.5	50	881.5	**6.6**	68.8	50	881.2
6.8	61.6	50	888.4	**6.8**	61.9	50	888.1
7.0	57.4	50	892.6	**7.0**	57.5	50	892.5
7.2	54.6	50	895.4	**7.2**	54.7	50	895.3

Note: Ionic strength, 0.05. Cacodylic acid 137.99 g/l = 1 M $(CH_3)_2AsO_2H$ recrystallized from warm alcohol (from Long's "Biochemists' Handbook." Sodium hydroxide, mw 39.999, or essentially 40 g/l.

CACODYLATE BUFFER (M. Plumel, *Bull. Soc. Chim. Biol.*, **30:** 129, 1949)

STOCK SOLUTIONS

A: 0.2 M solution sodium cacodylate (42.8 g $Na(CH_3)_2AsO_2-3H_2O$ per l)
B: 0.2 M HCl (50 ml A + x ml B, diluted to a total of 200 ml)

x	pH	x	pH
2.7	7.4	29.6	6.0
4.2	7.2	34.8	5.8
6.3	7.0	39.2	5.6
9.3	6.8	43.0	5.4
13.3	6.6	45.0	5.2
18.3	6.4	47.0	5.0
23.8	6.2		

SUCCINATE BUFFER (G. Gomori, unpublished)

STOCK SOLUTIONS

A: 0.2 M solution of succinic acid (23.6 g/l)
B: 0.2 M NaOH (25 ml A + x ml B, diluted to a total of 100 ml)

x	pH	x	pH
7.5	3.8	26.7	5.0
10.0	4.0	30.3	5.2
13.3	4.2	34.2	5.4
16.7	4.4	37.5	5.6
20.0	4.6	40.7	5.8
23.5	4.8	43.5	6.0

TABLE 20-12 SØRENSEN'S PHOSPHATES*

	Dry salt mixtures† for field use				Phosphates at 0.1 M, 0.067 M, 5 mM				
pH	mg Na$_2$H PO$_4$	mg NaH$_2$ + PO$_4$ ·H$_2$O	mg Na$_2$H PO$_4$	mg + KH$_2$ PO$_4$	KH$_2$ PO$_4$ or Na H$_2$PO$_4$ ·H$_2$O, ml	Na$_2$H PO$_4$, ml	pH values at specified dilutions		
							0.1 M	0.067 M	5mM
5.3	27	973	27	973	50	0	4.41	4.47	4.77
5.4	33	967	33	967	48	2	5.31	5.42	5.63
5.5	41	959	42	958	47	3	5.53	5.60	5.81
5.6	52	948	53	947	46	4	5.67	5.74	5.95
5.7	66	934	67	933	45	5	5.78	5.83	6.06
5.8	82	918	84	916	44	6	5.86	5.91	6.14
5.9	102	898	103	897	43	7	5.94	5.99	6.22
6.0	126	874	128	872	42	8	6.02	6.07	6.30
6.1	155	845	156	844	41	9	6.08	6.14	6.36
6.2	190	810	193	807	40	10	6.12	6.19	6.42
6.3	230	770	232	768	39	11	6.17	6.24	6.47
6.4	273	727	275	725	38	12	6.23	6.28	6.51
6.5	323	677	326	674	37	13	6.28	6.32	6.56
6.6	382	618	385	615	36	14	6.33	6.37	6.61
6.7	439	561	442	558	35	15	6.37	6.41	6.65
6.8	498	502	502	498	34	16	6.41	6.45	6.68
6.9	558	442	561	439	33	17	6.45	6.49	6.72
7.0	618	382	621	379	32	18	6.49	6.53	6.75
7.1	672	328	676	324	31	19	6.53	6.56	6.79
7.2	726	274	728	272	30	20	6.55	6.59	6.82
7.3	773	227	775	225	29	21	6.58	6.63	6.86
7.4	812	188	815	185	28	22	6.61	6.68	6.91
7.5	845	155	847	153	27	23	6.65	6.72	6.95
7.6	873	127	875	125	26	24	6.70	6.76	7.00
7.7	897	103	898	102	25	25	6.76	6.81	7.05
7.8	917	83	918	82	24	26	6.81	6.86	7.09
7.9	934	66	935	65	23	27	6.84	6.91	7.13
8.0	949	51	949	51	22	28	6.87	6.94	7.16
					21	29	6.89	6.96	7.18
					20	30	6.91	6.98	7.20
					19	31	6.94	7.01	7.22
					18	32	6.97	7.03	7.25
					17	33	7.00	7.05	7.28
					16	34	7.02	7.07	7.31
					15	35	7.06	7.11	7.34
					14	36	7.10	7.15	7.38
					13	37	7.14	7.20	7.41
					12	38	7.19	7.24	7.45
					11	39	7.24	7.28	7.51
					10	40	7.30	7.33	7.59
					9	41	7.36	7.40	7.65
					8	42	7.42	7.47	7.70
					7	43	7.49	7.54	7.75
					6	44	7.57	7.61	7.80
					5	45	7.65	7.69	7.87
					4	46	7.73	7.77	7.96
					3	47	7.81	7.85	8.05
					2	48	7.92	7.97	8.15
					0	50	8.98	8.93	8.73

* The mixtures on this page were made by P. Jones, and read electrometrically on a Beckman pH meter. Readings are corrected to 25°C and slightly smoothed. Phosphate buffers above 8 and below 5.3 are considered unreliable for histologic use, and readings are omitted.

TABLE 20-13 OXALATE BUFFER*

0.1 *M* oxalic acid	0.1 *M* sodium oxalate	pH
25	0	**1.34**
24	1	**1.40**
23	2	**1.45**
22	3	**1.50**
21	4	**1.55**
20	5	**1.65**
19	6	**1.71**
18	7	**1.80**
17	8	**1.89**
16	9	**1.99**
15	10	**2.12**
14	11	**2.36**
13	12	**2.59**
12	13	**2.88**
11	14	**3.22**
10	15	**3.35**
9	16	**3.51**
8	17	**3.61**
7	18	**3.73**
6	19	**3.92**
5	20	**4.05**
4	21	**4.18**
3	22	**4.37**
2	23	4.58
1	24	4.95
0	25	7.02

* Oxalic acid, mw 90.038. 0.1 *M* = 9,004 g/l. Sodium oxalate, mw 134.004. 0.1 *M* = 13.4 g/l.

TABLE 20-14 OXALATE PLUS $FeCl_3$*

0.1 *M* Na oxalate	0.1 *M* $FeCl_3$	pH
25	0	7.0
24	1	4.92
23	2	4.6
22	3	4.27
21	4	3.7
20	5	2.3
19	6	**1.9**
18	7	**1.88**
17	8	**1.80**
16	9	**1.75**
15	10	**1.70**
14	11	**1.60**
13	12	**1.59**
12	13	**1.55**
11	14	**1.51**
10	15	**1.49**
9	16	**1.48**
8	17	**1.44**
7	18	**1.40**
6	19	**1.38**
5	20	**1.36**
4	21	**1.33**
3	22	**1.32**
2	23	**1.30**
1	24	**1.28**
0	25	**1.24**

* Ferric chloride, $FeCl_3 \cdot 6H_2O$, mw 270.32. Use 27.032 g/l for 0.1 *M*.

† The dry salt mixtures calculated from Sørensen 0.067 *M* data. They are to be dissolved in rainwater at 1 % concentration. *ca.* 0.070 *M*; if higher dilutions are used, pH values may be approximated from the following table:

	Grams per liter		Milligrams per liter
	0.1 *M*	0.067 *M*	5 m*M*
KH_2PO_4	13.509	9.006	675
$NaH_2PO_4H_2O$	13.801	9.201	690
Na_2HPO_4	14.198	9.465	710

TABLE 20-15   LASKEY III : 3 : 51*

1 N NaOH	Distilled water	1 N glycine	pH 0.5 N	pH 0.1 N
	Glycine plus NaOH			
0.5	19.5	20	8.30	8.15
1	19	20	8.60	8.50
2	18	20	8.85	8.75
3	17	20	9.10	8.95
4	16	20	**9.40**	**9.30**
5	15	20	**9.52**	**9.40**
6	14	20	**9.64**	**9.55**
7	13	20	**9.73**	**9.65**
8	12	20	**9.80**	**9.78**
9	11	20	**9.90**	**9.88**
10	10	20	**9.98**	**9.99**
11	9	20	**10.04**	**10.00**
12	8	20	**10.12**	**10.09**
13	7	20	**10.22**	**10.18**
14	6	20	**10.28**	**10.22**
15	5	20	**10.40**	**10.33**
16	4	20	**10.50**	**10.48**
17	3	20	**10.65**	**10.60**
18	2	20	10.90	10.85
19	1	20	11.20	11.12
19.5	0.5	20	11.48	11.30

* Glycine, mw 75.07; use 75.07 g/l. NaOH, mw 39.999; use 40 g/l.

Note: These are made as stock 1 N solutions from which dilutions are made as required.

TABLE 20-16   SØRENSEN'S GLYCINE NaCl NaOH*

Glycine NaCl (0.1 N each)	N/10 NaOH	pH levels at		
		18°C	24°C	30°C
95	5	8.58	8.45	8.32
90	10	8.93	8.79	8.67
80	20	9.36	9.22	9.08
70	30	9.71	9.56	9.42
60	40	10.14	9.98	9.83
55	45	10.48	10.32	10.17
51	49	11.07	10.90	10.74
50	50	11.31	11.14	10.97
49	51	11.57	11.39	11.22
45	55	12.10	11.92	11.74
40	60	12.40	12.21	12.03
30	70	12.67	12.48	12.29
20	80	12.86	12.66	12.47
10	90	12.97	12.77	12.57

* For 0.1 N use glycine, mw 75.07, 7.507 g/l; NaCl, mw 58.448, 5.845 g/l; NaOH, mw 39.999, 4.00 g/l.

Source: Cited from W. M. Clark, "The Determination of Hydrogen Ions," 3d ed., Williams & Wilkins, Baltimore, 1928.

PHTHALATE–SODIUM HYDROXIDE BUFFER (W. M. Clark and H. A. Lubs, *J. Bacteriol.*, **2**: 1, 1917)

STOCK SOLUTIONS

A: 0.2 *M* solution of potassium acid phthalate (40.84 g/l)
B: 0.2 *M* NaOH (50 ml A + *x* ml B, diluted to a total of 200 ml)

x	pH	*x*	pH
3.7	4.2	30.0	5.2
7.5	4.4	35.5	5.4
12.2	4.6	39.8	5.6
17.7	4.8	43.0	5.8
23.9	5.0	45.5	6.0

GLYCINE–HCl BUFFER (S. P. L. Sørensen, *Biochem. Z.*, **21**: 131, 1909; **22**: 352, 1909)

STOCK SOLUTIONS

A: 0.2 *M* solution of glycine (15.01 g/l)
B: 0.2 *M* HCl (50 ml A + *x* ml B, diluted to a total of 200 ml)

x	pH	*x*	pH
5.0	3.6	16.8	2.8
6.4	3.4	24.2	2.6
8.2	3.2	32.4	2.4
11.4	3.0	44.0	2.2

PHTHALATE–HYDROCHLORIC ACID BUFFER (W. M. Clark and H. A. Lubs, *J. Bacteriol.*, **2**: 1, 1917)

STOCK SOLUTIONS

A: 0.2 *M* solution of potassium acid phthalate (40.84 g/l)
B: 0.2 *M* HCl (50 ml A + *x* ml B, diluted to a total of 200 ml)

x	pH	*x*	pH
46.7	2.2	14.7	3.2
39.6	2.4	9.9	3.4
33.0	2.6	6.0	3.6
26.4	2.8	2.63	3.8
20.3	3.0		

ACONITATE BUFFER (G. Gomori, unpublished)

STOCK SOLUTIONS

A: 0.5 M solution of aconitic acid (87.05 g/l)
B: 0.2 M NaOH (20 ml A + x ml B, diluted to a total of 200 ml)

x	pH	x	pH
15.0	2.5	83.0	4.3
21.0	2.7	90.0	4.5
28.0	2.9	97.0	4.7
36.0	3.1	103.0	4.9
44.0	3.3	108.0	5.1
52.0	3.5	113.0	5.3
60.0	3.7	119.0	5.5
68.0	3.9	126.0	5.7
76.0	4.1		

TABLE 20-17 GOMORI'S "TRIS MALEATE" BUFFER*

1 M maleic acid	1 M $(CH_2OH)_3$ $C-NH_2$	0.5 N NaOH	Water	pH
5	5	1	39	**5.08**
5	5	2	38	**5.30**
5	5	3	37	**5.52**
5	5	4	36	**5.70**
5	5	5	35	**5.88**
5	5	6	34	**6.05**
5	5	7	33	**6.27**
5	5	8	32	**6.50**
5	5	9	31	**6.86**
5	5	10	30	**7.20**
5	5	11	29	**7.50**
5	5	12	28	**7.75**
5	5	13	27	**7.97**
5	5	14	26	**8.15**
5	5	15	25	**8.30**
5	5	16	24	**8.45**

* Tris maleate = trihydroxymethylaminomethane. 1 M = 121.14 g/l. Maleic acid, 1 M = 116.07 g/l. Sodium hydroxide, 0.5 N = 20 g/l.

Source: *Proc. Soc. Exp. Biol. Med.*, **68**: 354, 1948.

TABLE 20-18 GOMORI'S TRIHYDROXYMETHYLAMINOMETHANE PLUS HYDROCHLORIC ACID: "TRIS" HYDROCHLORIC ACID (0.05 M)

pH	Tris 0.2 M*	HCl 0.1 N†	Distilled water
7.19	10	18	12
7.36	10	17	13
7.54	10	16	14
7.66	10	15	15
7.77	10	14	16
7.87	10	13	17
7.96	10	12	18
8.05	10	11	19
8.14	10	10	20
8.23	10	9	21
8.32	10	8	22
8.41	10	7	23
8.51	10	6	24
8.62	10	5	25
8.74	10	4	26
8.92	10	3	27
9.10	10	2	28

  * For 0.2 M tris, use 24.228 g/l.
  † For 0.1 N HCl, see Table 20-25.
  Source: Modified from A. G. E. Pearse, "Histochemistry," 2d ed., Little, Brown, Boston, 1960.

TABLE 20-19 SØRENSEN'S BORATE BUFFERS

0.1 N HCl, ml	Borate,* ml	0.1 N NaOH, ml	pH levels	
			20°C	30°C
475	525		**7.61**	**7.58**
450	550		**7.93**	**7.89**
425	575		**8.13**	**8.09**
400	600		**8.27**	**8.23**
350	650		**8.49**	**8.44**
300	700		**8.67**	**8.61**
250	750		**8.79**	**8.72**
200	800		**8.89**	**8.83**
150	850		**8.99**	**8.92**
100	900		**9.07**	**9.01**
50	950		**9.15**	**9.08**
	1000		9.23	**9.18**
			22°C	30°C
	1000		**9.21**	**9.15**
	900	100	**9.33**	**9.29**
	800	200	**9.46**	**9.43**
	700	300	**9.63**	**9.59**
	600	400	**9.91**	**9.86**
	500	500	**10.99**	**10.91**
	400	600	12.25	12.13

  * Borate is 12.404 g boric acid and 100 ml 1 N sodium hydroxide diluted to 1 l with distilled water.

TABLE 20-20   GOMORI'S γ-COLLIDINE HYDROCHLORIC ACID BUFFER

pH at 25°C*	100-ml portions			40-ml portions		
	Collidine + HCl†	0.1 N HCl‡	Distilled water	Collidine + HCl†	0.1 N HCl‡	Distilled water
6.45	30	40.0	30.0	12	16	12
6.62	30	37.5	32.5	12	15	13
6.80	30	35.0	35.0	12	14	14
6.92	30	32.5	37.5	12	13	15
7.03	30	30.0	40.0	12	12	16
7.13	30	27.5	42.5	12	11	17
7.22	30	25.0	45.0	12	10	18
7.31	30	22.5	47.5	12	9	19
7.40	30	20.0	50.0	12	8	20
7.49	30	17.5	52.5	12	7	21
7.57	30	15.0	55.0	12	6	22
7.67	30	12.5	57.5	12	5	23
7.77	30	10.0	60.0	12	4	24
7.88	30	7.5	62.5	12	3	25
8.0	30	5.0	65.0	12	2	26
8.18	30	2.5	67.5	12	1	27
8.35	30	0.0	70.0	12	0	28

* At 37°C subtract 0.08 from recorded pH values for 25°C.

† Collidine HCl: Dissolve 24.236 g γ-collidine in 200 ml 0.1 N HCl + 1000 ml distilled water. This gives the equivalent of the usually prescribed 25 ml 0.2 M collidine and 5 ml 0.1 N HCl to help dissolve it in each 30-ml portion above. The values for 0.1 N HCl are correspondingly reduced by 5 ml in the above table.

‡ For 0.1 N HCl, see Table 20-25.

Source: Gomori, personal communication.

TABLE 20-21   MICHAELIS VERONAL SODIUM HCl (40 ml)*

pH	0.1 N HCl†	0.1 M Veronal sodium‡	pH	0.1 N HCl†	0.1 M Veronal sodium‡
6.4	19.6	20.4	**8.1**	10.3	29.7
6.5	19.5	20.5	**8.2**	9.2	30.8
6.6	19.4	20.6	**8.3**	8.2	31.8
6.7	19.3	20.7	**8.4**	7.1	32.9
6.8	19.1	20.9	**8.5**	6.1	33.9
6.9	18.8	21.2	**8.6**	5.2	34.8
7.0	18.6	21.4	**8.7**	4.4	35.6
7.1	18.2	21.8	**8.8**	3.7	36.3
7.2	17.8	22.2	**8.9**	3.1	36.9
7.3	17.3	22.7	**9.0**	2.6	37.4
7.4	16.8	23.2	**9.1**	2.2	37.8
7.5	16.1	23.9	**9.2**	1.9	38.1
7.6	15.4	24.6	**9.3**	1.5	38.5
7.7	14.5	25.5	**9.4**	1.0	39.0
7.8	13.5	26.5	**9.5**	0.8	39.2
7.9	12.4	27.6	**9.6**	0.6	39.4
8.0	11.4	28.6	**9.7**	0.4	39.6

* Recalculated on a 40-ml volume and interpolated arithmetically from the table on p. 454 of Lillie, "Histopathologic Technic and Practical Histochemistry." 2d ed., Blakiston, Philadelphia, 1954.
† For 0.1 N HCl, see Table 20-25.
‡ For 0.1 M Veronal sodium, use 20.618 g/l.

TABLE 20-22   HOLMES' ALKALINE BUFFER FOR SILVER SALTS

pH	0.2 M H_3BO_3*	0.05 M $Na_2B_4O_7 \cdot 10H_2O$†
7.4	18	2
7.6	17	3
7.8	16	4
8.0	14	6
8.2	13	7
8.4	11	9
8.7	8	12
9.0	4	16

* 0.2 M boric acid = 12.4 g/l.
† 0.05 M borax = 19.0 g/l. 1% borax is pH 9.11 at 25°C.

Source: W. Holmes, *Anat. Rec.,* **86:** 163, 1943.

TABLE 20-23 HCl + 2-AMINO-2-METHYLPROPANE-1,3-DIOL (AMMEDIOL) BUFFER*

	Ionic strength 0.05				Ionic strength 0.1			Ionic strength 0.2		
pH	1 N HCl	1 M Ammediol‡	Water	pH	1 N HCl†	1 M Ammediol‡	Water	1 N HCl†	1 M Ammediol‡	Water
8.2	50	61.7	888.3	8.2	100	120.9	779.1	200	239	561
8.4	50	68.2	881.2	8.4	100	133.7	766.3	200	260	540
8.6	50	79.0	871	8.6	100	154	746	200	294	506
8.8	50	95.9	854.1	8.8	100	183	717	200	348	452
9.0	50	122.1	827.9	9.0	100	229	671	200	441	359
9.2	50	164	786	9.2	100	303	597	200	606	194
9.4	50	232	718	9.4	100	424	476	200	2 M 437	363
9.6	50	338	612	9.6	100	626	274	200	2 M 652	148
9.8	50	511	439	9.8	100	2 M 478	422			
10.0	50	790	160	10.0	100	2 M 764	136			

* Ammediol, mw 105.140. Ammediol HCl, mw 141.605. The solubility of Ammediol is 250 g/100 ml distilled water, according to "The Merck Index."

† For 1 N HCl, see Table 20-25.

‡ For 1 M Ammediol, use 105.14 g/l.

Source: Long, "Biochemists' Handbook," p. 35, Van Nostrand, Princeton, N.J., 1961.

TABLE 20-24 TRIPHOSPHATE
HCl BUFFER,
pH 9.5 TO 12.4

pH	K_3PO_4 1 M*	HCl 1 N	Water
9.5	40	24.5	35.5
10.0	40	23.5	36.5
10.5	40	22.0	38.0
11.0	40	19.0	41.0
11.5	40	12.0	48.0
12.0	40	5.5	54.5
12.4	40	0.75	59.25

* 1 M K_3PO_4 = 212.28 g/l. Solubility at 25°C, 1931 g/l. The sodium salt $Na_3PO_4 \cdot 12H_2O$ requires 380.14 g/l for 1 M and would approach saturation at 45°C; hence it cannot be substituted.

Source: Gancevici, *Arch. Roum. Pathol. Exp. Microbiol.*, **21**: 191, 1962.

TABLE 20-25 TABLES FOR PREPARATION OF NORMAL SOLUTIONS OF SULFURIC, HYDROCHLORIC, AND NITRIC ACIDS FROM THE USUAL CONCENTRATED ACIDS

Sulfuric acid

Specific gravity	% H$_2$SO$_4$, w/w	Grams H$_2$SO$_4$ per liter	Normality	Vol. = 49.04 g for 1 N sol.,† ml
1.8337	95	1742	35.51	28.2
1.8355	96	1762	35.93	27.86
1.8364	97	1781	36.31	27.6
1.8361	98	1799	36.68	27.3
1.8342	99	1816	37.03	27.1
1.8305	100	1831	37.34	26.8

Hydrochloric acid

Specific gravity	% HCl, w/w	Grams HCl per liter	Normality	Vol. = 36.47 g for 1 N sol.,† ml
1.1789	36	424.4	11.64	86.0
1.1837	37*	438.0	12.01	83.3
1.1885	38	451.6	12.38	80.8
1.1932	39*	465.4	12.75	78.4
1.1980	40	479.2	13.14	76.2

Nitric acid

Specific gravity	% HNO$_3$, w/w	Grams HNO$_3$ per liter	Normality	Vol. = 63.02 g for 1 N sol.,† ml
1.4048	68	955.3	15.16	66.0
1.4091	69	972.3	15.43	64.9
1.4134	70	989.4	15.70	63.8
1.4176	71	1006	15.96	62.7
1.4218	72	1024	16.25	61.6

* Figures for 37 and 39% HCl are interpolations.

† Figures in column 5 are calculated from Lange's data (N. A. Lange, "Handbook of Chemistry," 10th ed., McGraw-Hill, New York, 1961) as given in the first three columns. Where assay figures are stamped on printed concentrated acid labels, this figure should be used. Grams per liter is at 20°C.

To prepare normal acid solutions measure the amount specified in column 5 into a 1-liter volumetric flask, and then add distilled water up to the liter mark. These instructions are intended to be sufficiently accurate for histologic purposes.

If the normal solutions are to be used for quantitative chemical work, they must be standardized as usual for that purpose.

TABLE 20-26 FORMIC ACID, ACETIC ACID, AND AMMONIA

Formic acid, mw 46.027

Assay, %	Sp gr 20°C	Grams acid per liter	M	Vol. 1 mol ml
80	1.1860	948.8	20.614	48.517
81	1.1876	962.0	20.900	47.841
82	1.1896	975.5	21.097	47.400
83	1.1914	988.9	21.485	46.543
84	1.1929	1002	21.770	45.934
85	1.1953	1016	22.052	45.346
86	1.1976	1030	22.378	44.686
87	1.1994	1045	22.704	44.044
88	1.2012	1057	22.965	43.545
89	1.2028	1070	23.247	43.016
90	1.2044	1084	23.551	42.460
91	1.2059	1097	23.834	41.956
92	1.2078	1111	24.104	41.429
93	1.2099	1125	24.444	40.908
94	1.2117	1139	24.746	40.410
95	1.2140	1153	25.051	39.919
96	1.2158	1167	25.355	39.350
97	1.2170	1180	25.633	39.009
98	1.2183	1194	25.941	38.549
99	1.2202	1208	26.246	38.102
100	1.2212	1221	26.528	37.609

Acetic acid, mw 60.054

Assay, %	Sp gr	Grams acid per liter	M	Vol. 1 mol ml
80	1.0700	856.0	14.254	70.157
81	1.0699	866.6	14.429	69.298
82	1.0698	877.2	14.607	68.460
83	1.0696	887.8	14.784	67.643
84	1.0693	898.2	14.957	66.860
85	1.0689	908.6	15.130	66.095
86	1.0685	918.9	15.302	65.353
87	1.0680	929.2	15.473	64.630
88	1.0675	939.4	15.643	63.928
89	1.0668	949.5	15.811	63.249
90	1.0661	959.5	15.977	62.590
91	1.0652	969.3	16.141	61.956
92	1.0643	979.2	16.305	61.330
93	1.0632	988.8	16.463	60.734
94	1.0619	998.2	16.620	60.169
95	1.0605	1007	16.799	59.636
96	1.0588	1016	16.918	59.109
97	1.0570	1025	17.068	58.590
98	1.0549	1034	17.218	58.079
99	1.0524	1042	17.351	57.633
100	1.0498	1050	17.484	57.194

Ammonia, mw 17.032

Assay, %	Sp gr	Grams NH$_3$ per liter	M	Vol. 1 mol ml
16	0.9362	149.8	8.797	111.09
17	0.9328	158.5	9.309	107.45
18	0.9295	167.3	9.823	101.80
19	0.9262	175.9	10.331	96.890
20	0.9229	184.6	10.831	92.327
21	0.9196	193.1	11.338	88.254
22	0.9164	201.6	11.837	84.980
23	0.9132	210.0	12.328	80.912
24	0.9101	218.4	12.823	77.886
25	0.9070	226.7	13.311	75.128
26	0.9040	235.0	13.798	72.473
27	0.9010	243.2	14.279	70.032
28	0.8980	251.4	14.761	67.745
29	0.8950	259.5	15.230	65.659
30	0.8920	267.6	15.714	63.644
30.38	0.8946	271.8	15.959	62.661
31.36	0.8917	279.7	16.422	60.892
32.34	0.8889	287.5	16.881	59.239
33.32	0.8861	295.2	17.333	57.694
34.30	0.8833	303.0	17.790	56.210
35.28	0.8805	310.6	18.234	54.834

Source: Data for specific gravity, and in part for grams per liter, derived from N. A. Lange, "Handbook of Chemistry," 10th ed., McGraw-Hill, New York, 1961, with interpolations in NH$_3$ series. Molarities and molar volumes recomputed to agree with 1957 atomic weights. M is equal to millimoles per milliliter.

References

W. M. Clark: "The Determination of Hydrogen Ions," 3d ed., Williams & Wilkins, Baltimore, 1928.

G. Gomori: Personal communication.

W. Holmes: *Anat. Rec.*, **86:** 163, 1943.

B. L. Horecker and R. D. Lillie: Unpublished data, Sept., 1943.

N. A. Lange: "Handbook of Chemistry," 10th ed., McGraw-Hill, New York, 1961.

C. Long. "Biochemists' Handbook," Van Nostrand, Princeton, N.J., 1961.

L. Michaelis: *J. Biol. Chem.*, **87:** 34, 1930.

A. G. E. Pearse: Histochemistry, 2d ed., Little, Brown, Boston, 1960.

G. S. Walpole: *J. Chem. Soc. [Org.]*, **105:** 2501, 1914.

INDEX

Abbé test plate, 2
Abbott's variant of Perls' reaction, 507
Abbreviations, dye companies, 124
Abnützungspigment, 513
Abopon, 117, 122, 192
AB-PAS and AB stains for mucins, 639–642, 676
Absorption spectrophotometry, 13–14
Absorption spectroscopy of tissues, 13–14
Accelerated Giemsa stain, 745–746
Acetal phosphatides (plasmalogens), 560–561
Acetic acid:
  directions for 1 M solution, 889
  glacial: azo coupling medium, 240, 259
    diluent of indole reagents, 259–261
    diphenamine Schiff base reactions, 309–310
    lipid solvent, 606
Acetic acid decalcification, 791
Acetic acid silver nitrate fixations, 495, 580
Acetic acid-sodium acetate buffer series:
  normal, pH 4.5, 793
  pH 2.7–6.5, azure eosin staining, 195–196
  table 0.2 M, 0.01 M, 875
Acetic alcohol formalin, 34, 45
  glycogen fixation, 630
  nucleic acids, 177
Acetic $HgCl_2$ formalin, 52
Acetic orcein for chromosomes, 187
  mounting, 122
Acetic sublimate, 51
Acetone, as fixative, 47–48
Acetone dehydration, 80, 82–83
Acetone-paraffin preparations for enzymes, 360, 380, 407, 415
Acetone sublimate, 52

Acetylation:
  and acid dye stains, 289
    aldehydes, 315
    amyloid, 666
    basic dye nuclear stain, 212
    basic nucleoprotein, 186
    hydroxyls, 226–227
    Paneth granules, 350–351
  cartilage, 226
  connective tissues, 226
  mast cells, 226
  mucins, 226
  phenols, 254
  reduced melanin, 524
  saponification effects on blood cell periodic acid Schiff, 752
Acetylcholinesterase, 415–425
  (See also Cholinesterases)
N-Acetylglucosamine reaction of chitins, 633
N-Acetyl-β-glucosaminidase, 396–398, 400
N-Acetylglucosaminolactone inhibition of N-acetylglucosaminidase, 396–398
Acetyl Sudan IV, 147, 571
Acetyl Sudan black B, 145, 147, 571
Acetyltransferase, 372
Achúcarro's tannin silver method for astrocytes, 768–769
Acid anthrapupurin, 127
Acid azo coupling, 249–251
Acid deoxyribonuclease lead method, 441–442
Acid dye reactions with amines, 286–287
Acid extraction:
  of hemosiderin iron, 489–490, 502
  of nucleic acids, 168–171
Acid fast basophilia, 166, 606–607
  ceroid, 513, 515, 519–521

Acid fast basophilia:
  hair cortex, 221, 606
  lipofuscins, 513−519
  Russell bodies, 758−759
Acid fast stain for bacteria, 734−740
Acid fastness, 734−740
  after blockades, 606−607
Acid formaldehyde hematin, 488 489
Acid fuchsin, 136, 139
  diphenamine Schiff base reaction, 311
  picric acid collagen stains, 694−698
  stain for mitochondria, 328−330
Acid groups, 315−326
Acid hematins, 488−490
Acid orcein elastin stain, 709−710
Acid phosphatase (AcPase), 359−365, 402
  azo dye EM method, 363
  Barka azo dye method, 362
  bone and cartilage, 804
  Burstone azo dye method, 362
  conjugates for immunohistochemistry, 857,
    863
  5-iodoindoxyl phosphate, 361
  lead EM method, 359
  lysosomes, 364
  osteoclasts, 363
Acid phosphate for decalcifying, 790
Acid resistance of oxytalan fibers, 717−718
Acid Rhodamine B, 135−136, 150, 832, 835,
  837−840
Acidophilia, 125
Ackerman, aminopeptidase method,
  433−434
Aconitate-HCl buffer pH 2.5−5.7, 882
AcPase (*see* Acid phosphatase)
Acridine orange, 135−136, 181−182
  exfoliative cytology, 723−724
  nucleic acids, 181−182
Acridine red 3B, 129, 535−536
Acridine yellow G, 135−136
  Schiff reagent, 308−309
Acriflavine, 135
  elastin fluorochroming, 715
  Schiff reagent, 308−309
  stain for acid mucosubstance, 643
Acrolein fixation, 42−43
  Schiff reaction, 42
Acromere lipids in retina, 598, 615
ACTH productive cells:
  Mason *et al.*, immunoglobulin-peroxidase
    bridge method, 863
  Sternburger PAP stain identification,
    861−862
Actinomyces granules, 731

Acylation:
  and acid dye stains, 287−288
  phenols, 254
Acylation blockade, 254
  of amines, 287−289
  azo coupling blockade, 254
Adams DMAB reaction for indoles, 260
Adams and Glenner, nerve cathepsin method,
  435−436
Adams perchloric acid-napthoquinone (PAN)
  stain for cholesterol, 589, 599
Adams rosindole method for tryptophan,
  260−261
Adams-Tuqan gelatin silver film protease
  method, 437−438
Adamstone-Taylor sections, 71, 338, 361, 477,
  531
Adenine, 165−167
Adenosinediphosphatase (ADPase, pyrophos-
  phatase), 376−377
Adenosine-3′-phosphatase, 376−377, 442
Adenosine-5′-phosphatase:
  at pH 5.9, 442
  at pH 7.2, 368−369
Adenosinetriphosphatase (ATPase), 368−370
  electron probe analysis, 369
  quantitative Cs EM method, 370
Adenylcyclase method, 370
ADN, 165
ADPase, 376−377
Adrenal aryl sulfatase, 406
Adrenal cortex, lipids, quantitative his-
  tochemistry, 610
Adrenal lipofuscin, 232, 235, 237, 515−516
Adrenal medulla, 235, 237, 262−273
Adrenalin, 262−273, 464
  azo coupling reaction, 252, 268
  ferric ferricyanide reaction, 235−238
  osmic acid reaction, 217
  oxidation to pigment, 262
  periodic acid Schiff reaction, 265−268
  reaction with keratohyalin, 210
  Vulpian reaction, 264
Adrenochrome, 262, 265
Affixing of paraffin section to slides, 87−89
Ag⁺ inhibition of enzymes, 443
Agar, 615
  embedding, 80
Aging elastica of arteries, 709
Aging pigment, 513
Air under coverslips, 108, 110−111, 116
Al⁺⁺⁺ ion demonstration, 127−129, 534−535,
  538
Alanyl peptidase substrate, 433

Albert-Leblond phenylhydrazine method for plasmal, 581
Albert-Linder method for alkaline phosphatase in bone, 804
Alcian Blue 8GS:
  "Alcian Blue 8GX-300," 641
  critical electrolyte concentration, 639–641
  formula, 639
  ground substance stain, EM, 644
  mast cell staining, 676
  polysaccharide staining, 638–642, 659
  stain for collagen, cartilage, nuclei, mucin and mast cells, 640
  stain specifications, 641
  Steedman's mucin stain, 638
Alcian Blue 2GX, 134
Alcian Blue 5GX, 134
Alcian Blue 7GX, 134
Alcian Blue-PAS for rickettsiae, 744
Alcian Green 2GX, 134
Alcian Green 3BX, 134
Alcian Yellow GX, 134, 642
Alcohol dehydration, 80, 82, 107, 109–110
Alcohol dehydrogenase, 477
Alcohol fixation, 45–47
  paraffin methods for alkaline phosphatase in bone, 803–804
Alcohol formalin, 34
Alcohol groups, 224
Alcohol inactivation of enzymes, 433, 437–438
  inhibition of branching enzyme, 391
Alcoholic decalcifying fluids, 789
Alcoholic HCl acetic chloroform decalcification, 788–789
Alcoholic picroformalin, 61
Alcoholic sublimate (HgCl₂), 51
Alcohols, 224–226
Aldehyde fixatives, 29–45
Aldehyde fuchsin, 343, 655, 659, 714–716
Aldehyde reactions:
  arylamines, 309–310
  blockade, 312–315
  blockade after peracetic acid, 220
  carbonyls, 305–315
  hydrazines, 309
  reaction of rodent elastin, 707
  Sawicki reaction, 311–312
  Schiff's reagent, 305–309
Aldolase, 386
Alfert and Geschwind method for nuclear protein, 185
Aliesterases, 406–425
  by thioacetic substrate, 408–410

Alizarin, 126–129, 536–538
  ammonia method for Th, Ti, Zr, 556
Alizarin black P, 127
Alizarin black S, 127
Alizarin blue, 127
Alizarin blue S, 154
Alizarin cyanin BBS or 3RS, 127
Alizarin cyanin black G, 127
Alizarin cyanin R, 127
Alizarin green S, 127
Alizarin red S, 127, 129, 538
  phosphomolybdic acid hemoglobin stain, 486
  spodograms, 866
  stains for developing bone, 798–799
Alizarin SA, SX, 129
Alkali examination for fungi, 733
Alkaline formalin hematin, 489
Alkaline phosphatase (AlPase), 364–370
  Allen, 367
  bone, 803–804
  Burgos et al., 367
  calcium-cobalt sulfide methods, 365–367
  conjunctiva, 786
  cysteine inhibition, 380
  Feigin and Wolf, 367
  Freiman, 367
  Gomori, 367
  inhibitors, 366
  lead citrate LM and EM method, 364
  Padykula, 367
  substrates, 362, 367
  Wachstein, 367–368
Alkaline polyphosphatase, 384–385
Alkaline Sudan IV, 566
Alkyl halide alkylation of amines, 321
Alkylation of acid residues, 318–325
Allen and Bourne's zymohexase method, 386
Allen 1961 alkaline phosphatase substrates, 367, 376–377
  biochemical phosphate substrates, 376–377
Allen's B-13 Bouin variant, 61
Allochrome stain, 635, 668, 671, 704–705, 786
  fibrin, 297
  fungi, 729
  reticulum and basement membranes, 704–705
  vitreous, 729–730
Alloxan reaction, 282
Alloxan Schiff reaction, 282–283
AlPase, 364–370
Altmann's aniline acid fuchsin, 328–330, 333
Alum hematoxylins, 204–210
Aluminum, 534–535, 538
Aluminum mordant dyes, 126
Amann's Viscol, 118

Ameba, 202
Amethyst violet, 132
Amido black 10B, 139, 141
  collagen stains, 697, 702
Amine oxidase, 465−471
Amines, 282−289
  acid dye methods, 286−287
  acylation, 287−289
  Biebrich scarlet method, 287
  dansyl chloride method, 285
  Deitch's naphthol yellow S method, 286−287
  o-diacetylbenzene method, 285
  2,3-hydroxynaphthaldehyde method, 283, 285
  mercuric bromophenol blue method, 286
  sulfation basophilia, 225−226, 288−289
Amino dyes, as fat stains, 145
1-Aminoanthraquinone, 149
Aminoanthraquinone oil soluble dyes, 146−147
p-Aminodimethylaniline and α-naphthol, M-Nadi
    reaction, 451−452
4-Amino-N,N-dimethyl-α-naphthylamine, cy-
    tochrome oxidase substrate, 453−454
4-Aminodiphenylamine in cytochrome oxidase
    substrate, 452−453
3-Amino-9-ethylcarbazole:
  aminopeptidase method, 435
  cytochrome oxidase substrate, 453
2-Amino-2-methylpropane-1,3-diol HCl buffer,
    367, 886
8-Amino-1-naphthol-5-sulfonic acid, 255, 259
Aminopeptidase, 433−436
m-Aminophenol aldehyde condensation:
  acid, 310, 313−314
  borax/50% acetone, 311
  fast black K coupling, 310
  Schiff blockade, 313−314
N-(4-aminophenyl)maleimide method for SH,
    242
8-Amino-1,2,3,4-tetrahydroquinoline in cy-
    tochrome oxidase substrate, 453
Ammediol HCl buffer series pH 8.2−10.6, 886
Ammine silver carbonate methods for oligoden-
    droglia, 767
Ammonia, directions for 1 M solution, 889
Ammonia alcohol saponification, 228−229
  demethylation, 320−322
Ammonia leucofuchsin, 634
Ammoniacal silver after nuclear HCl hydrolysis,
    175
Ammoniacal silver reticulin technics, 683−692
  periodic acid Schiff combination, 691
  reaction of melanins, 524
  (See also Argentaffin reaction)
Ammonium bromide formol, 32, 34

Ammonium chloride, nitrate for decalcifying, 790
Ammonium citrate, citric acid 1 N pH 4.5, 793
Ammonium molybdate stabilization of in-
    dophenol blue, 450−451
Ammonium picrate, 137
Ammonium sulfamate in Morel-Sisley reaction,
    255
Ammonium sulfide, 364−365
Amplification antibody method of Avrameas, 859
Amylase digestion, 395−396
β-, α-Amylases:
  action on amylose and glycogen, 391−392,
    629−630
  action on cellulose and starch, 634
  localization by starch film method, 437−440
Amylbenzene, secondary, 116
Amyloid, 196, 615, 663−670
  Bennhold's Congo red stain, 665
  composition, 663
  crystal violet stains, 666−667
  EM appearance, 663
  fluorescence, 668
  heparatin sulfate isolation, 664
  Langhans' iodine stain, 669
  Mallory's iodine stain, 664
  pinacyanole staining, 192
  Puchtler et al., Sirius red 4B stain, 668
  Puchtler-Sweat alkaline Congo red, 666−667
  silver carbonate stain, 667
  stain for museum jar, 664−665
  structure, 663
  toluidin blue O stains, 669
  tryptophan, 664
Amylophosphorylase, 391−395
Amylose, 630
  amylase digestion, 392
  iodine reaction, 391, 626−627
Amylo-1,4→1,6 transglucosidase, 391−392
Anderson-Grief stain for rickettsiae, 743−744
Aniline, aniline hydrochloride blockade of alde-
    hydes, 220, 313−314
Aniline−acid fuchsin:
  Altmann's, 328−330
  Lacqueur's trichrome for hyalin, 670−671
  methyl green for islet cells, 342
  methyl green for mitochondria, 328
Aniline blue, alcohol soluble, 133
Aniline blue-orange G method for chromaffin,
    265
Aniline blue WS, 139, 141
  HCl collagen stain, 702−703
  Mallory collagen stains, 650, 699
  Masson trichrome stains, 696, 700
  picric acid collagen stains, 694−698

Aniline blue WS:
  Shorr stains, 723 – 724
  zinc leuco, hemoglobin peroxidase, 477
Aniline carbol fuchsin, Goodpasture's, 734
Aniline glycerol to improve paraffin sectioning, 89
Aniline pyronin, Graham's, 451
Anionic dye reactions with amines, 286 – 287
Anthocyanidins, 215
Anthocyanins, 215
Anthracene blue SWR, 126
  and eosin, 210
Anthrapurpurin, 127, 129
Anthrax bacilli, 616
Antibody localization, 303 – 305, 460 – 461, 837 – 865
Antimony (stains for), 548
Antimony trichloride reaction of vitamin A, 529
Antiserum conjugate purification, 840 – 841
Antiserum conjugation with fluorochrome, 838 – 840
Antiserum preparation for immunofluorescence, 837 – 838, 854 – 855
Aoyama silver variant for Golgi, 332 – 333
A-5′-Pase, 368, 377 – 378, 442
Apáthy's alum hematoxylin, 206
Apáthy's gum syrup, 19
  Lillie-Ashburn variant, 119 – 120
Apocrine cells, 344
  lipofuscin, 514
Apoferritin, 502
Aposiderin, 502
Aqueous mounting media, 117 – 122
Araldite embedding medium, 100, 102
Arcadi's urinary sediment smear technic, 720
Argentaffin cells, 262
Argentaffin reaction, 276 – 279
  ascorbic acid, 530 – 531
  carotid body, 281
  hematoidin, 495
  iron pigments, 504, 509
  lipofuscins, 513 – 519
  mast cells, 672
  melanin, 524
  melanosis pigment, 509
  neuromelanin, 522
  urates, 290 – 291
Arginine, 291 – 295
  Baker α-naphthol method, 294
  benzil blockade, 291
  1,2-cyclohexanedione blockade, 291
  deamination, $FeSO_4$, hematoxylin reaction, 294
  deamination effect on, 289, 294 – 295

Arginine:
  Deitch method, 291, 294
  fibrinoid, 299 – 300
  glyoxal blockade, 291
  β-naphthoquinone-4-sulfonate-Na-reaction, 291 – 294
  nuclei, 166
  Paneth granules, 350 – 351
  9,10-phenanthrenequinone blockade, 291
  Sakaguchi reaction, 291
  trichohyalin granules, 295 – 296
Argyrophilia, 276, 278 – 279
  islet α cells, 341
Arlex gelatin, 120 – 121
  sorbitol syrup, 121
Aromatic solvents of resins, 111
Aronson et al.:
  hemin ultrastructural tracer method, 864
  method for deoxyribonuclease II, 441 – 442
Arsanilate inhibition of esterase, 408, 411, 418
  other enzymes, 428
Arsenic, 548
Arsenic trioxide inhibition of enzymes, 442
Arteries, small, embedding, 78, 86
Artifacts:
  gelatin embedding mass staining, 78 – 79
  mercurial fixation precipitates, 50, 106 – 107
  paraffin in nuclei, 105
Aryl amines:
  azo coupling, 214, 259 – 260
  reaction with aldehydes, 309 – 310
Aryl sulfatase, 402 – 406
Arzac's HIO_4 fuchsin method, 310
Asbestos, 535, 831
Ascorbic acid, 530 – 531
Ashburn and Endicott's carbon gelatin injection in liver cirrhosis, 809
Ashton's method for retinal vessel injection, 810
Astra Blue 4R, 134
Astracyanin B, 134
Astrocytes, 767 – 769
Atabrine, 135
  fluorescence, 12
  stain for chromosomes, 189 – 191
Atherosclerosis:
  Lillie's silver impregnation stain for lipid and calcification, 797 – 798
  quantitative lipid histochemistry, 610
Atoxyl inhibition of esterase, 408, 411 – 412, 418
ATPase, 360, 368, 377
Attachment of celloidin sections, 93 – 94
Au^{+++} enzyme inhibitor, 443

Auramine O, 135–136
  acid fast stain, 737
  Schiff reagent, 308–309
Aurine tricarboxylic acid, 534
Autoradiographic method counterstains:
  aldehyde fuchsin-PAS, 819
  celestin blue B, 819
  Feulgen, 819
  Feulgen-fast green, 819
  Gomori's chrome alum hematoxylin-phloxine, 819
  hematoxylin and eosin, 819
  indigocarmine-picric acid, 819
  methyl green-pyronin, 819
  nuclear fast red, 818–819
  PAS, 819
  toluidin blue, 819
  Weigert's iron hematoxylin, 819
Autoradiographic methods:
  Baserga's 2 emulsion method, 816–817
  Budd and Pelc flat substrate method for quantitative EM, 821
  contact method (Gross *et al.*), 815
  counterstains, 818–819
  Fitzgerald method for water soluble substances, 816
  LeBlond's liquid emulsion method, 815–816
  Millar *et al.*, method for water soluble substances, 816
  Salpeter and Bachmann flat substrate method for quantitative EM, 821
  stripping film method, 816
Autoradiography:
  ³H-acetate, 609
  acetylcholinesterase, 424
  acid mucosubstance, 613, 647
  ⁴⁵Ca, 540
  ³H-cholesterol, 609
  ³H-estradiol, 608–609, 817–818
  fixatives, 814
  lipids, 608–609
  ³H-mesobilirubinogen, 609, 817–818
  noradrenalin I¹³¹, 263
  nucleic acid turnover, 165
  ³²P, 540
  PAS stain, 622
  preservation of isotope during processing, 821–822
  principles, 813–814, 820–822
  ³H-progesterone, 821–822
  resolution, 814, 820
  ³H-testosterone, 821–822
  uranium and thorium, 556
Autoxidation of SH, 234–235

Avrameas:
  amplification antibody method, 859
  direct immunoenzyme technic, 853–857
  glutaraldehyde enzyme-antibody conjugation method, 855–857
  immunoadsorbent preparation, 855
  indirect immunoenzyme method, 857–860
  insoluble protein derivative preparation, 854–855
  lectin peroxidase method, 864
Axons, 769–784
Azan aniline blue method, 699–700
  enterochromaffin cells, 279
Azide (NaN₃) inhibition of enzymes, 443, 453, 462
Azo coupling, amines, 240, 259–260
Azo coupling reaction, 246–253
  enterochromaffin, 246
  fresh diazos, 249–251
  mast cells, 252, 257, 673
  nuclei, 175–177
  pigments, 496–499
  prevention by tosylation, 228
  purines, 247
  pyrimidines, 247
  stable diazos, 246–249
Azo dye phosphatase methods, 358–359, 361–368
Azocarmine B, 154, 699–700
Azocarmine G, 154, 699–700
Azofuchsin G, 138
  erythrocyte stain, 697
Azofuchsin GN, S, 138
Azomethine method for amino groups, 309–310
Azure A, 130, 132–133, 154, 317
  mucins, 635–637
  nuclei with thionyl chloride, 741–742
  stain for DNA, RNA, 173
Azure A eosin B (1963), 195
  for inclusions, 754
Azure B, 132–133, 154
  stain for RNA, 181
Azure C, 132–133, 154, 635–636
Azure eosin stains, 193–197
  after decalcification, 788
  Nissl staining, 83, 193–197

B-4 sublimate sodium acetate, 51
B-5 sublimate sodium acetate formalin, 52–53
Ba⁺⁺ ion demonstration, 129, 537–538, 548
Bacteria, 193–194, 196, 725–744
  cell envelope, 741
  cell walls, 616

Bacteria:
  mitosis, 741—742
  negative stain, 829
  spores, 740—741
Bacteriologic procedures at autopsy, 25
Baker, J. R.:
  acid hematein phospholipid test, 572—573,
    576, 578
  bichromate formalin, 334
  calcium cadmium formol, 34
  chrome hematoxylin phospholipid methods,
    595—604
  formalum, 334
  formol calcium, 34
  gelatin embedding for lipids, 78
  α-naphthol method for arginine, 293—294
  pyridine extraction test, 573—574
  Sudan black B for Golgi substance, 334
Baker J. R., II, chloral formalin, 804
Baker, R. D.:
  arterial hyalin reactions, 670
  picroindigocarmine for *Mucor*, 732
BAL decolorization of mercurial SH stains, 239
Balantidium, 203
Balsam, Canada, 107—108, 111—112, 115
  salicylic acid, 107, 111—112
Barium, 537—538, 548
  ion in NQS arginine technic, 291—294
Barka acid phosphatase lead method (1962),
    359—360
Barka azo dye phosphatase methods, 362
Barnett-Bourne alcoholic acetic silver fixation,
    495, 530
Barr body stain, 190—191
Barrnett-Seligman's DDD reaction for SH,
    243—244
Barrnett-Seligman's 2-hydroxy-3-naphthoic acid
    hydrazide method for COOH, 323—324
Barrnett-Seligman's indoxyl esterase method,
    413—415
Barroso-Moguel silver method for carotid body
    tumors, 281
Bartholomew-Mittwer spore stain, 740
Bartholomew's Gram stain for smears, 727
Bartonella bacilliformis, 197
Basement membranes, 681—682
  allochrome, 703—705
  Bauer reaction, 230
  immunoenzyme stain, 863
  Lillie's Biebrich scarlet picroaniline blue,
    696—697
  Lillie's fast green Van Gieson, 697—698
  Lillie's hydrochloric acid variants, 702—703
  Lillie's picroamido black 10B, 697

Basement membranes:
  negative EM stain, 829
  oxytalan fiber stain, 717
  phosphomolybdic acid hematoxylin, 693—694
  phosphotungstic acid hematoxylin, 693
  picromethyl blue, 695
  Puchtler and Sweat, 712—713
  rhodamine B-fluorescein, 692
  sulfation, 226
  Sweat's modified allochrome, 705
  Sweat's picro-Sirius red F3B stain, 703
Baserga's two emulsion radioautography,
    816—817
  stripping film variant, 816—817
Basic aniline dyes, 129—134
  chromaffin stains, 264—266
  reactions of acid groups, 286—287
Basic fuchsin, 131, 133
  acid fast stains, 606—607, 734—740
  diphenamine Schiff base reaction, 309—310
  Schiff reagents, 306—309
Basic lead acetate fixation, 48—50
Basophil leukocytes, 747—748
Basophilia, 316
  brain lipids, 584—586
  conjunctival mucin, 786
  erythrocytes at pH 7.0, 485
  lipofuscins, 516—517, 519
  melanins at pH 1.0, 524—525
  metachromasia of saccharides, 636—637
  methylation and methanolysis, 226, 229, 288,
    318—319, 514, 516—519, 524, 658
  nucleic acids, 167, 177—185
  ochronosis pigment, 512
  pH effect on, 316
  sulfation induced, 225—227, 289
  sulfonic acids, 245
Bathophenanthroline reaction for iron, 506
Bauer reaction, 229—230, 615, 634—635,
    728—729
  fungi, 729—732
  polysaccharides, 627, 671
Be^{++} ion demonstration, 127, 129, 535—537,
    556
Beeswax in paraffin, 85
Belanger's bone canaliculi stain, 796
Bencosme's formalin Zenker, 55
Benda's iron hematoxylin, 199
Benditt and Arase's substrate for mast cell pro-
    tease, 427—428
Bengtson's Macchiavello stain, 743
Bennett mercury orange SH method, 238—239
  phenylhydrazine plasmal reaction, 581
Bennhold's Congo red for amyloid, 665—666

Bensley, R. R.:
  acetic osmic bichromate fixative, 329
  aniline acid fuchsin-methyl green for islets,
    328–329
    cell granules, 342–343
  aniline blue-orange G for islets, 340
  safranin-acid violet islet cells, 337–338
  stain for mitochondria, 329
Benzene, clearing agent, 80–84
Benzidine:
  myeloperoxidase methods, 448–451,
    853–854, 857–865
  myoglobin peroxidase method, 448
  nitroprusside hemoglobin peroxidase,
    447–448
Benzidine blue reaction of Mn, 550–551
Benzidine diazo, 248
Benzo eosin BL, 142
Benzo light eosin BL, 142–144
Benzo pure blue, 138
  +eosin, hypophyseal α and β cells, 346–347
Benzoflavine, 135–136
Benzopurpurin 4B for amyloid, 154, 666
Benzoyl-L-arginine methyl ester, trypsin inhibitor, 428
Benzoylation:
  blockade of aldehyde, 315
  blockade of nuclear hematoxylin stain, 227
  ceroid, 520
  hydroxyls, 226–227
  (See also Acylation)
6-Benzoyl-2-naphthyl sulfate for sulfatase,
    402–403
Benzoyl oil red O, 145, 147, 571
3,4-Benzopyrene:
  chemical reactivity, 572
  fluorochrome for fats, 572
Berberine, 135–136
  fluorochrome for blood protozoa, 752
Bergner's cholinesterase inactivators, reactivators, 419–420
Berg's inorganic polyphosphatase methods,
    382–385
Berlin blue gelatin injection masses, 810
Beryllium, 535–537, 556
Beryllium oxide, 537
Best's carmine, 627–628
−BF₄ diazo inhibition of enzymes, 433
Bichromate chloral formalin fixative, 769
Bichromate effect on OsO₄ reduction, 217
Bichromate fixatives, 54–60, 334
  mercury variants, 54–55
Biebrich scarlet, 154
  basic protein method, 287
  chromosome stain, 190–191

Biebrich scarlet:
  Masson trichrome stains, 700
  picroaniline blue, 696
  Shorr stains, 723–724
Bielschowsky silver methods, 262–263,
    683–684
Bilirubin:
  and biliverdin, 490–500
  reaction with diazosafranin, 252
Biological Stain Commission, 124
Bioloid resin, 113–116
Birefringence:
  acid hematins, 488–490
  cellulose and starch, 634
  corneoscleral collagen, 786
  formalin pigment, 8–9, 488–489
  lipids, 563
  paraffin, 105
  striated and smooth muscle, 8–10
Bismarck brown R, 154
Bismarck brown Y, 129–130, 635
  Papanicolaou EA stains, 724
Bismuth, 546–547, 554
  reaction with nucleic acids, 173
  stains for, 548–549
Bismuth poisoning inclusions, 756
Black B, BS, 148–149
Black Bauer method for fungi, 732
Black K, NK, 148–149
Black periodic acid technic and aldehyde,
    310–311, 705
  demonstration of stroma, 705
Blanco-Fite silver impregnation for M. leprae,
    740
Blank and McCarthy:
  Carbowax technic, 76
  chrome gelatin adhesive, 77
Blastomyces dermatitidis, 729, 732
Blazing red, 145
Bleaching:
  of hematins and formalin pigment, 488–489
  malaria pigment, 489
  melanins, 525
"Bleeding" of basic dyes in mounting media, 118
Blenden method for spirochetes, 760
Blepharoplasts, 197–198, 748
Bloch's Dopa reaction, 461–462
Blockade reactions:
  and acid fastness, 606–607, 735
  in periodic Schiff for connective tissue, 698
  (See also specific end groups and procedures)
Blocking tissue for fixation, 25–26
Blood platelets, 750
Blood smears, preparation, 719–720
Blue B, BNS, 148–149

Blue 2B, DB, 148–149
Blue naevus melanin, 522
Blue RR, 148–149
Blue tetrazolium (BT), 151, 442, 472–473
Blue V, 148–149
Bodian's copper protargol method, 771–772
Bodian's formol acetic alcohol, 34
Boehmer's alum hematoxylin, 206
Boiling saline, 67
Boiling water, 67
Bone, 616
  decalcification, 787–794
  enzymes, 477, 793–794, 803–806
  fixation, 787
  glycogen, 806
  ground sections, 794–795
  hydrolysis of collagen, protein, 788, 790–791
  iron preservation, 790–791
  matrix, 616, 699, 794–798
  and tooth stains, 795
    alizarin red S stains for sites undergoing
      mineralization, 788–789
    Belanger's bone canaliculi stain, 797
    Lansdown silver-toluidin blue stain for cal-
      cification, 798
    lead acetate for sites undergoing mineraliza-
      tion, 801–802
    Lillie's silver impregnation method, 797
    Powers *et al.*, stain for dentinal tubules, 796
    Procion dyes for sites undergoing minerali-
      zation, 800–801
    Romanes' diammine silver stain, 796
    Schmorl's thionin, 796
    tetracycline for sites undergoing mineraliza-
      tion, 802–803
    van Gieson, 795
  zinc demonstration, 554
Boothroyd EM microincineration method, 867
Borax, 870
Borax-ferricyanide, 575–576, 603
Bordeaux GP, 148–149
Bordeaux R, 154, 700
  plasma stain in Masson trichrome, 700
Boric acid blockade of *cis*-glycols, 233
Boric acid borax buffer series pH 7.4–9.0, 885
Borohydride aldehyde reduction, 314–315
Boron trifluoride, acid group esterification, 320
Botryomycosis granules, 615, 729
Bouin-Hollande copper picroformol, 61
Bouin's picroformol-acetic fixative, 61
  decalcification, 790
  Masson trichrome stains, 698
  postfixation, 106–107, 699
Brain:
  cytochrome oxidase, 453

Brain:
  fixation for histologic study, 26–27
  α-D-glucosidase, 390
  lipofuscins, 514
  phosphamidase, 381
  thiamine pyrophosphatase, 378
Brandino's method for mercury, 549
Brazilin, 126, 154
  keratohyalin stain, 354
  (*See also* Catechol dyes)
Breast, isocitric dehydrogenase, 477
Brecher's new methylene blue N for re-
    ticulocytes, 753
Brilliant cresyl blue, 130, 154
  reticulocyte stain, 753
Brilliant crocein, 138
Brilliant purpurin R, 138
  plasma strain in collagen methods, 697, 700
Bromcresol green inhibition of enzymes, 429
Bromide-silver nitrate stain for unsaturated
    lipids, 594
Bromination:
  blockade of ethylene, 219–223
  cystine-cysteic acid cleavage, 245
  glycogen, 222, 628
  oxidation of aldehyde, 314
  silver methods for ethylenes, 219–223
p-Bromoaniline diazo, 257
5-Bromo-4-chloroindoxyl acetate esterase sub-
    strate, 413–415
p-Bromodiazobenzene, 246
5-Bromoindoxyl acetate esterase method,
    413–415
6-Bromo-2-naphthol azo coupling method for β-
    glucuronidase, 399
Bromosulfalein inhibition of enzymes, 429
Bronchial gland mucin, 615, 635
Brown-Brenn Gram technic for sections, 728
Brown RR, 148–149
Brunner gland mucin, 615
Brush border, renal epithelium, 615
BT, 142, 151
Budd and Pelc flat substrate autoradiographic
    quantitative EM method, 821
Buffered azure A staining, 316–317
Buffered bichromate mixtures, 58
Buffered formalin, 32–33
Buffered Orth fluid, 58
Buffered osmic fluids, 62
Buffered sodium sulfide, 365, 415–417
Buffered thiazin dye mucin stains pH 2–5,
    635–636
Buffers:
  azo coupling, 250
  general considerations, 869

Bulb pipets, 21
Bullard's alum hematoxylin, 206
Bunsen burners, 22
Bunting's technic for Perls' reaction, 507
Bunting's technic for phosphates, 557
Burets, 22
Burgos, Deane and Karnovsky, alkaline phos-
    phatase method (1955), 367
Burri's India ink method for spirochaetes, 760
Burstone, M. S.:
    azo dye esterase methods, 411
    azo dye phosphatase methods, 362
    cytochrome oxidase methods, 453
    freeze-dried paraffin methods for enzymes,
        362, 411
Burstone and Folk, aminopeptidase method, 434
Burstone-Weisburger oxidative aminocarbazole
    method for aminopeptidase, 435
Burtner-Lillie methenamine silver argentaffin
    method, 277 – 278

Ca^{++} ion demonstration, 128 – 129, 547, 556,
    866 – 867
Cacodylate-HCl buffer, 877
Cacodylic acid NaOH buffer series pH 5.2–7.2,
    877
Cadmium, 546 – 547, 549, 554
Cadmium formalin fixatives, 32, 34, 332
Caedax, 114 – 115
Cain:
    aniline acid fuchsin for mitochondria, 329
    Nile blue for fatty acids, 592
Cajal (*see* Ramón y Cajal)
Calcium, 537 – 543
Calcium acetate formalin, 32 – 34
Calcium acrolein fixative, 582
Calcium cadmium formalin, 32, 34
Calcium copper hematoxylin for fatty acids, 590
Calcium formol, Baker's, 34
Calcium mordant dyes, 126
Calcium oxalate, 537 – 538, 866 – 867
Calcium red, 128 – 129, 537 – 538
Calcium soaps, conversion to lead soaps,
    317 – 318
Calcium stains:
    alizarin red S, 538
    ^{45}Ca, ^{32}P, 540
    Eisenstein's chloranilate, 539
    glyoxal bis(2-hydroxyanil) (GBHA), 540 – 541
    Grandis and Mainini (purpurin), 538
    hematoxylin, 540
    Kernechtrot, 538 – 539
    Komnick and Komnick, 543

Calcium stains:
    Lillie, 542
    McGee-Russell, 538
    microincineration, 866
    murexide, 540
    Pizzolato's peroxide-silver stain, 541 – 542
    Puchtler, 538
    von Kóssa, 539 – 540
    x-ray microanalysis, 542, 807
Calomel precipitation in mercurial fixatives,
    51 – 54
Canada balsam, 107 – 109, 112 – 113, 115
Candida albicans, 732
Capillaria hepatica, 615
Carazzi's hematoxylin, 207
Carbazole substrates for aminopeptidase, 434
Carbohydrate reactions of brain lipids, 575
Carbol aniline fuchsin, Goodpasture's, 734
Carbol auramine, 737
Carbol fuchsin, 735 – 737
Carbon, 532
    radioautography, 813 – 817
Carbon bisulfide, as clearing agent, 81
Carbon gelatin injection masses, 809 – 810
Carbon tetrachloride:
    clearing agent, 80 – 81
    radioautography, 813 – 817
Carbonate-bicarbonate buffer, 874
Carbonates, 539 – 540
Carbonic anhydrase, 443 – 445
Carbonyl lipids, 604 – 607
Carbowax sections, 76 – 78, 740
Carboxylic acid residues, 315
    methylation, demethylation, 318 – 323
Carcinoid tumors, 273, 281
Cardiac lipofuscin, 517
    fluorescence, 12
Cardiac muscle in polarized light, 9
Cardial gland mucus, 615
Carey's Ranvier technic for motor end plates,
    780 – 781
Carmine, carminic acid, 126
    Best's method for glycogen, 627 – 628
    mucicarmine, 650
    nuclear stains, 126 – 127, 190, 214 – 215
    Schultz's, 290
Carmine gelatin injection masses, 810
Carnoy fluids, 45, 46
Carnoy-Lebrun fluid, 52
Carotenes, 529
    fluorescence, 12
Carotid body tumors, 281
Cartilage:
    basic dye reactivity, 806

Cartilage:
  enzymes, 806
  proteoglycan, 807
  x-ray microanalysis of calcium, 807
Cartilage matrix, 193, 285, 615, 635–636,
  641–642
Carycinel red, 147, 568
Casella KMnO₄ Schiff reaction, 230, 232, 615,
  634–635, 670
Catalase, 453, 456–460
  substrate film method, 459–460
Catechol dyes:
  on blood films, 752
  without mordants, 128
  (*See also* Clara's hematoxylin)
Catechol reaction with keratohyalin, 354
Cater's acetic silver fixation, 495, 580
Cathepsin or aminopeptidase of nerves,
  435–436
Cathepsin Bl, 431
Cathepsin D, 432
Cd⁺⁺ inhibition of enzymes, 433
Cedar oil as clearing agent, 92–93, 113, 115
  mounting medium, 113, 115
Celestin blue, 126, 132, 154
  enterochromaffin, 246
  eosinophils, 752
  iron lakes, 211
Cell membrane:
  concanavalin A-FITC, 648–649
  concanavalin A-iron dextran, 648
  ruthenium red stain, 643
Cell surface receptors, immunohistochemical
  stain, 844–846
Celloidin embedding, 91–94
  before decalcification, 789
Celloidin-paraffin sectioning, 93–94
Celloidin section mounting, 93–94
Cellulase, 440–441
Cellulose, 634
Cellulose tricaprate, 115–117
Cellulose tridecanoate, 115–117
Centrifuge smear preparation, 720
Cephalins, 561, 595–596
Ceramides, 560, 562, 584, 586
Cerebrosides, 560, 562, 584–586
Ceroid, 513, 515, 519–521
  fluorescence, 12
  mouse adrenal, testis, uterus, 615–616
  peracetic Schiff reaction, 221
  periodic acid Schiff, 615
  peroxide reactions, 222–223
  rat liver, 615
Cerrolow 117 alloy, as corrosion mass, 811–812

Certified stains, 124
Cervix uteri, gland mucin, 615
Cetylpyridinium chloride in fixative, 46
Chang, J.:
  cryostat, 20
  freeze dried, aniline–acid fuchsin for mi-
    tochondria, 329–330
  section freeze substitution, 47
Charcoal in Schiff reagent preparation, 306
Charity Hospital Papanicolaou technic, 721
Chemical models, azo coupling reaction,
  247–248, 252
Chemical reagents, ferric ferricyanide test,
  235–238
Chemistry and biochemistry of melanin,
  522–527
Chief cells of peptic glands, 349
Chiffelle and Putt's propylene glycol Sudan stain,
  568
Chiquoine's method for glucose-6-phosphatase,
  373
Chitin, 633–634
Chloral bichromate formalin fixative, 769
Chloral formalin, 398
Chloral hydrate (in fixative), 43
Chloramine T deamination, 287
Chloranilic acid test for Ca⁺⁺, 539
Chlorazol black E, 155, 631
Chloride, electron probe analysis, 831
Chloroacetyl esterase, 426–428
p-Chloroanilidophosphonic acid, 380–382
p-Chloroanilidophosphoric acid, 381
Chloroform as clearing agent, 80–81, 84
  hardening of celloidin, 92–94
Chloroform methanol fixation, 47
4-Chloro-1,2-mercaptobenzene for Co, 546
p-Chloromercuribenzoate inhibition of enzymes,
  429, 436, 442–443
1-Chloromercuriferrocene method for SH, 241
p-Chloromercuriphenylazo-*β*-naphthol SH reagent, 238
Chlorophyll, fluorescence, 12
Chloropicrin, SH blockade, 236
Chlorous acid, oxidation of aldehyde, 314
Cholesterol, 560, 563
  and sterols, 588–590
Cholesterol esters, 589–590
Choline in phospholipids, 560, 578–581, 584,
  588
Choline acetyltransferase, 425–426
Cholinesterases, 415–425
  inhibitors and reactivators, 419–420
  substrates, 415–425
  thioacetic substrate technic, 408–410

Choman's method for Ni, 547
Chordoma mucin, 615
Choroidal melanin, 522
Christeller's method for mercury, 549
  stannous chloride method for gold, 549
Christensen's stain for diphtheria organisms, 733
Chromaffin, 235, 237, 262–273, 616
Chromaffin reaction, 262, 264, 266
Chromate effect on brain lipids, 595
Chromate fixations, 54–60
Chromate method for lead, 547
Chromate-Sudan black B for Golgi, 334
Chromatin:
  acetic orcein staining, 187
  pinacyanol staining, 187
Chromation of polyene fats *in vitro*, 597–598
Chromatophore melanin, 522
Chrome alum carmine, 215
Chrome alum gallocyanin, 212–213
Chrome alum hematoxylin, 209–210
  for islet cells, 341
Chrome alum mordant dyes, 126
Chrome cyanin R, 128, 155
Chrome gelatin adhesives for Carbowax sec-
    tions, 77
Chrome silver block methods for neurons, 769
Chrome violet CG, 128–129
Chromic acid glycol methods, 229–230, 306,
    615, 634–635, 728–729
Chromic acid methenamine silver method,
    630–631
Chromic acid oxidation of methylene blue, 133
Chromic acid Schiff reaction, 615
Chromium, 126–127, 129, 531
Chromolipoide pigment, 513
Chromosomes, stains for: acetic orcein, 187,
    189, 192
  Atabrine, 189
  Biebrich scarlet, 190–191
  carmine, 190
  Feulgen or basic dye, 188
  hematoxylin, 191, 204
  silver hexamethylenetetramine, 190
  Tjio and Whang, 187–188
  zinc after chromic acid oxidation, 190
Chromotrope 2R, 138
  Masson trichrome stain, 700
Chromoxane cyanin R, 128, 211
Chromoxane pure blue B, 128, 536–537
Chymotrypsin, 426–428
Chymotrypsin-like enzyme, 426–428
Ciaccio C:
  fixing fluids, 59
  insolubilized lipids, 597–598

Ciaccio C:
  reaction of Golgi lipid, 334
  variant of Baker acid hematein method, 598
Cibacron Brilliant Red BA, 160–161
Ciliary melanin, 522
Cinnabar, 531
Citric acid:
  disodium phosphate buffer series pH 2.6–7.5,
    876
  sodium citrate buffer series pH 2.8–6.8, 872
Citrulline, 295–297
  in hair follicles, 295–297
  Holmes-Fearon (α-nitrosoethyl methyl ketone)
    method, 296–297
  Rogers' method, 295–296
Clara's hematoxylin:
  blood films, 752
  enterochromaffin, 280–281
  eosinophils, 280, 752
  keratohyalin, 280, 354–355
  metal chelates, 202–203
  methylation, 320
Clarite, Clarite X, 114–115
Clark and Lubs:
  phthalate-HCl buffer pH 2.2–3.8, 881
  phthalate-NaOH buffer pH 4.2–6.0, 881
  potassium chloride-HCl buffer pH 1.0–2.2,
    874–875
Clayton yellow, 135
Clearax, 114–115
Clearcol, 118
Clearing:
  of celloidin sections, 110
  before paraffin, 80–84
Clorox in Sakaguchi methods, 291
Clostridium perfringens sialidase, 657–658
Cloudy resin mounts, 108–109
Coal, 129, 531
Cobalt, 531, 546, 549, 554
  formalin fixation, 231
  localization in yeast mitochondria, 733
  silver method for Golgi, 231–232
  stains for, 546
  sulfide preparations, mounting, 365
Cobaltinitrite method for potassium, 551–552
Cobb's method for bone phosphorylase, 805
Cocarboxylase, 380
Coccidioides immitus, 729, 732
Coccinel red, 155, 566–568
Cocheneel, 152
Cold acetone fixation, 47
Cold alcoholic fixatives, 45
Cold fixation, 25, 29–30, 37, 42, 47
Cold knife frozen section, 71, 338, 361, 477, 531

Cold Schiff reagent, 306–308
Collacin, 616
Collagen, 285, 298, 615, 679–681
 allochrome, 703–705
 amino acid composition, 680
 birefringence, 8–10
 iron uptake, 619
 Lillie's acid fuchsin–iron flavianic acid, 696
 Lillie's Biebrich scarlet picroaniline blue,
  696–697
 Lillie's fast green Van Gieson, 697–698
 Lillie's hydrochloric acid methods, 702–703
 Lillie's picroamido black 10B, 697
 Masson's Van Gieson, 696
 negative EM stain, 829
 phosphomolybdic acid hematoxylin, 693–694
 phosphotungstic acid hematoxylin, 693
 picromethyl blue, 695
 proline hydroxylase immunofluorescence, 703
 rhodamine B-fluorescein, 692
 selective stains, 138–141
 structure, 679
 sulfation basophilia, 225
 Sweat's modified allochrome, 705
 Sweat's picro-Sirius red F3B stain, 703
 trichrome stains, 698–702
 Van Gieson's stain, 694–695
 violamine R, 695
Collagenase digestion, 301–302
Collargol negative film method for spirochetes,
  760
Collidine HCl buffer series pH 6.8–8.3, 884
Collodion film solubilities, 105–106
Collodionization:
 of blocks during sectioning, 89
 of sections, 105
  ammoniacal silver technics, 683
  enzyme digestion of sections, 752
  periodic acid Schiff, 614
  radioautographs, 815–816
  spodograms for metal ion tests, 866–867
Colloidal iron, 646–648
Colloidal iron preparation, 646
Colonic gland mucin, 310, 615, 635
Colophony, 113
Color contrast sequence Schiff reactions,
  308–309
Colored minerals, 531
"Colour Index," 2nd and 3rd editions, 124
Combined iron hematoxylin methods, 202–203
Combined oil red O diamine silver method, 521
Composite nitro blue tetrazolium technic for
  dehydrogenase, 473–474
Compound granule cells, 514

Concanavalin A-FITC stain, 648–649
Concanavalin A-iron dextran stain (LM and
  EM), 648
Condenser centration, 4–5
Condensers, optical, 4
Congo Corinth G, 138
 amyloid stain, 666–667
Congo red, 139, 143
 amyloid stain, 665–668
Conjugation, protein, 847–865
Conjunctiva, 615, 786
Connective tissue mucins, 635
"Conn's Biological Stains," 8th and 9th editions,
  124
Conn's carbol fuchsin, 735
Coons' fluorescent antibody method, 303–305
Coplin jars, 20–21
Copolymer resin, Groat's, 114–115
Copper, 128, 531, 544–546, 549, 554–555
 mordant dyes, 125–126
 reaction with dilute hematoxylin, 280
 stains for, 544–546
 thiocholine method for cholinesterase,
  416–417, 420–421
Corinth LB, 144
Corinth V, 144
Coriphosphine O, 135–136
 Schiff reagent, 308
Corneal stromal mucin, 615, 786
Corpora amylacea:
 brain, 615
 seminal vesicle, 615
Corrosion procedures, 811–813
Corynebacterium diphtheriae, 733
Couceiro-Freire CrO_3 oxidation in reticulum
  methods, 687
Coujard models, phenols, 252
Counterstains, 166–167
Coupled tetrazonium reaction, 247
Coverslip cleaning, 22
Coverslipping or mounting of sections, 107–109
Cowdry's aniline–acid fuchsin for mitochondria,
  328
Cowdry's Janus green for mitochondria, 327
Cox's acetic sublimate formalin, 53
Cr^{+++} ion demonstration or mordanting,
  126–128
Cr^{+++} uptake by Clara's hematoxylin, 203
Creatine kinase, 385
Creatine phosphokinase isozymes, 483
Cresyl fast violet, 130, 132, 155
 counterstain for Luxol Fast Blue MBS, 599
Criegee oxidation cleavage reactions, 229–234,
  305–306, 615–616, 628–632

Crippa's lead tetraacetate method for mucins, 233

Critical illumination, 5 – 6

Cryostat sectioning, 19 – 20

Cryostat sections for enzymes, 71, 338, 361, 477, 531

Cryostats, 19 – 20

Crystal violet, 131, 154
  amyloid stains, 118, 666 – 667
  fibrin stain, 298 – 299
  Gram stains, 725 – 729

Crystallization artifacts in azo dye enzyme preparations, 435

Crystalloids of Leydig cells, 349

Cryptococcus neoformans, 729, 732

Cu^{++} inhibition of enzymes, 429, 433, 442

Cu^{++} ion demonstration, 128, 531, 544 – 546, 549, 554 – 555

Cu^{++} uptake by Clara's hematoxylin, 203

Cunningham-Isaacs' reticulocyte stain, 753

Curtis' picromethyl blue, 695

Curtis' salicylic acid balsam, 107, 112, 700

Cutaneous melanin, 522 – 527

Cuticular border, villus epithelium, 310 – 311, 615

Cyanide in dehydrogenase substrates, 471

Cyanide blockade of aldehydes, 313
  reversal by HIO$_4$, 313

Cyanide inhibition of enzymes, 429, 433, 436, 442, 453, 462, 466

Cyanin R, 128
  phosphoric acid method for nuclei, 211

Cyanol FF, 141, 154

Cyanol zinc leuco for hemoglobin peroxidase, 447

Cyanuric chloride, 48, 152

Cysteic acid method for protein sulfur, 245 – 246

Cysteine desulfurase, 442

Cysteine effect on enzyme activity, 435
  inhibition of alkaline glycerophosphatase, 380
  inhibition of thiamine monophosphatase, 380

Cysteine-cystine in Paneth granules, 350

Cystine, Swift's silver methenamine EM method, 353

Cystine-cysteic acid CH$_3$CO$_3$H and HIO$_4$ oxidations, 230

Cytochrome addition in oxidase methods, 453

Cytochrome oxidase:
  Graham and Karnovsky, 454
  Kerpel-Fronus and Hajos, 457, 474
  M- and G-Nadi oxidases, 451 – 452
  Plapinger *et al.*, 456
  substrates, 452 – 454, 456 – 457
  Tsou *et al.*, 457 – 458
  Winkler-Schultze, 451 – 452

Cytolipochrome, 517

Cytophotometry:
  chromosomes, 192
  nuclei, 182, 213

Cytoplasmic necrobiosis, 196 – 197

Cytosiderin:
  hepatic, 503
  of rodents, 503
  villus epithelium, 509 – 519

Cytosine, 165

Daddi's Sudan III method, 566

Da Fano's cobalt silver for Golgi, 331 – 332

Dahlqvist's disaccharase method, 386

Dahl's Viscol substitute, 119 – 120

Dammar, damar, 112

Dansyl chloride, reaction with histone, 186

Daoust's nuclease substrate film methods, 438 – 440

Dark field illumination, 8

Darrow red, 155

Dart and Turner's fluorochrome smear method, 723 – 724

Davenport's nerve fiber method for sections, 770

Dawson and Friedgood's HgCl$_2$ formalin, 53

DDD reaction for SH, 235, 243 – 244, 354

Deamination, 287 – 289
  amyloid, 282, 286, 665 – 666

Decalcifying fluids:
  acetate-acetic acid, 793
  1 N acetic, 791
  alcoholic citrate formic acid, 793
  ammonium citrate-citric acid, 793
  Balogh's EDTA for enzymes, 791
  EDTA, 791
  formaldehyde-formic acid, 792
  5% formic, 792, 806
  Frank and Deluzarche, 790
  Fullmer-Link EDTA for enzymes, 793 – 794
  Goncalves and Olivêro acid mixture, 794
  Hahn and Reygades alkaline EDTA, 790
  HCl-citrate, 793
  Jenkins' alcoholic acid, 788 – 789
  Kristensen's formate-formic acid, 792
  Lorch's citrate-HCl, 793
  5% nitric acid, 792
  Richman-Gelfand formic-HCl, 792
  sodium citrate-formic acid, 792
  sulfurous acid, 789
  trichloracetic acid, 789
  Von Ebner's HCl-NaCl, 792

Defatting in enzyme artifact control, 433, 473

De Galantha method for urates, 291

Dehydrant-clearing agent mixtures, water tolerance, 80

Dehydration, 82–87
  clearing: of celloidin sections, 109
    paraffin, schedules, 82–83
  of stained sections, 107
Dehydrogenases, 471–480
  composite nitro blue tetrazolium method,
    473–474
  diaphorase: NADH₂, 476–477
    NADPH₂, 476–477
  ethanol, 479
  folate, 480
  glutamic, 475, 479
  glyceraldehyde 3-phosphate, 477
  α-glycerophosphate, 479
  β-hydroxybutyrate, 479
  3-β-hydroxysteroid, 476
  17-β-hydroxysteroid, 476
  isocitric, 477, 479
  lactic, 475–476, 479, 481–482
  L-malate, 479
  6-phosphogluconate, 479
  succinic, 473–476, 479
  tellurite reactions, 479
  tetrahydrofolate, 480
  Versene decalcification, 791, 793–794,
    803
Deitch-Sakaguchi variant for arginine, 289
Deitch's basic protein method, 286–287
Delafield's alum hematoxylin, 206
DeLamater-Mudd method for bacterial mitosis,
  741
DeLamater's formaldehyde fuchsin, 732
Del Río-Hortega, Pío, Achúcarro technic,
  768–769
  oligodendrocyte method, 767
  silver reticulum method, 685–688
Demethylation by saponification, 320–324
  methylated cysteic acid, 245
  methylated pepsinogen granules, 349
Dempsey's pH signature method, 182–183,
  316–317
  protein pH signature, 182–183, 316–317
Dentin:
  nerve fibers, 772
  tubule stain, 796
Denz's naphthochrome green B for Be⁺⁺,
  535–536
2-Deoxyaldose liberation, 166
Deoxyribonuclease (DNase, DNAse), 135,
  168–169, 212, 441–442
  localization by substrate film, 437, 439–440
Deoxyribonucleic acid (DNA), 165–167
Deoxyribonucleoprotein (DNP), 165–166,
  185–186
2-Deoxyribose, 165–166

Deparaffinization:
  of sections, 105
  of tissue carriers, 18
Depex, 115–116
Depex resin, 114–115
Descemet's membrane, 615, 786
Destaining of celloidin, 110
Developing in silver reticulum methods, 690
Dewaxing of sections, 105
Dextran, 632
DFP cholinesterase inhibitor, 409, 412, 416,
  418–419, 421, 424
DFP chymotrypsin inhibitor, 426, 429
Diacetin Sudan solution, 567
Diacetone alcohol fixation for enzymes, 361
o-Diacetylbenzene, 285
Diamine purpurin 3B, 141
Diammine silver argentaffin methods, 276–278
Diammine silver oil red O stain, 521
Diammine silver reaction:
  of chromaffin, 262
  of Mn⁺⁺, 550
Diammine silver reticulum methods, 682–692
Diammine stains for mucosubstances,
  644–646
Diamox inhibition of enzymes, 443
Dianil blue RR, 141–142
Di-o-anisidine, diazo or tetrazo, 248
Diaphane, 113, 115
Diaphorases, 472, 476–477
Diastase digestion, 629
Diastase effect on periodic Schiff of blood cells,
  752
Diatom media, 114–115
Diazinon, cholinesterase inhibitor, 419
Diazo (see under Fast)
Diazo blue B, 149
Diazo blue BB, 149
Diazo Corinth LB, 149
Diazo garnet GC, 149
Diazo red AL, 149
Diazo red G, 149
Diazo red 3GL, 149
Diazo red RC, 149
Diazo scarlet R, 149
Diazonium bases, 150
Diazonium salts, stabilized, 150, 248
Diazosafranin, 251–254, 516, 519
  reaction for 5-hydroxytryptophan and 5-HT,
    275
Diazosulfanilic acid, 248–249
Diazosulfanilic acid-azure A method, 253, 275,
  281
  for histidine, 256–257
Diazotization of safranins, 251–253

Diazotization coupling tyrosine reaction, 255–256

Diazotization technic, 249–253

2,6-Dibromoquinone-4-chloroimide, 253–254

Dicarboxylic acid content of pepsin, 349–350

2,4-Dichloro-1-naphthol, as arginine reagent, 291

Diethylbenzene, as resin solvent, 111, 116

Diethyldithiocarbamate, sodium, 128–129, 545, 550

copper reagent, 545

Diethylene glycol distearate, 90–91

Diethylene glycol tissue storage, 35

Diethyl-p-nitrophenyl phosphate, cholinesterase inhibitor, 418

Diethylsafranin, 129, 132, 251–253

mitochondrial stain, 327

Diffusion artifact in M-Nadi reaction, 452

Digestion:

of fresh muscle for trichinae, 762–763

of oxytalan fibers, 718

Digestion tests for glycogen, 629

Digitonin reaction for cholesterol, 589

Dihydrofolate reductase, 480

Dihydronicotinamide adenine nucleotide, 472

phosphate, 472

Dihydroxydinaphthyl disulfide (DDD) SH reaction, 243–244

Diisopropylfluorophosphate (DFP):

as esterase inhibitor, 411, 416, 418–419, 421, 424

other enzyme inhibitor, 416–421, 423–424

Dimedone blockade of aldehydes, 313

2,3-Dimercaptopropanol decolorization of mercurial SH stains, 239

Dimethyl adipimate (fixative), 41–42

p-Dimethylaminobenzaldehyde:

citrulline stain, 295–296

indole reaction, 259–260

4-(p-Dimethylaminobenzene azo) phenylmercuric acetate method for SH, 242

p-N,N-Dimethylaminophenylmercuric acetate for SH, 240

Dimethylglyoxime for Ni, 546–547

Dimethyl malonimidate (fixative), 41–42

Dimethylol urea embedding medium, 99

Dimethylsafranin, 251–253

Dimethyl suberimidate (fixative), 41–42

Dimethylthiazolyldiphenyltetrazolium (MTT), 150–152, 476–477

2,4-Dinitrophenylhydrazine reaction, 309, 581

2,4-Dinitrosoresorcinol, 128–129

reaction for iron, 506

Dioxane, water and paraffin solvent, 80, 82

Diphenamine Schiff's base reactions, 309–310

o-Diphenol groups, 125

1,5-Diphenylcarbohydrazide for mercury, 549

Diphenylthiocarbazone, 128

method for zinc, 554

Diphtheria organisms, 733

Diphtheroids, 615, 733

Direct blue 3B, 141

Direct pure blue BF (Fran), 141

Direct Schiff reaction of lignin, 634–635

lipofuscins, 515–517, 519–521

Disaccharases, 386

Distrene resin, 114

Disulfide demonstration, 234–235

Dithionite extraction of iron, 503

Dithionite inhibition of enzymes, 462

Dithiooxamide, 128–129

reaction for cobalt, 546

reaction for copper, 545

Dithizone, 128, 545, 550, 554

DNA, DNP, DNS, 165

DNase, DNAase. 168–169, 212, 437, 439–442

Dohle bodies, 759

Dopa reaction, 461–462

leukocyte oxidase, 462

melanogenesis, 523–524

neuromelanin, 523–524

trichoxanthin, 524

Dopamine reactions, 270

HCHO fluorescence, 270–271

OsO₄ reaction, 63–64, 217–218

Dopamine-β-hydroxylase, 462–464

DPNH diaphorase, 472, 476–477

Drews and Engel's benzidine method for myoglobin peroxidase, 448

Drimarene Red Z-2B(S), 160–161

Dropping bottles, 21

Drying of paraffin sections, 89

Dubreuil's picromethyl blue, 695

Duff and More's neoprene corrosion mass, 811

Dunn, R. C.:

patent blue V and cyanol hemoglobin peroxidase methods, 447

and Thompson hemoglobin stain, 486

Dunn, T., reaction of hematoidin, 492–493

Dye concentration effect on basophilia, 316

Dye solubilities, table, 154–159

E600 esterase inhibitor, 418

EA Papanicolaou stains, 721–724

EDTA decalcification, 477, 791, 793–794, 803, 805

Ehlers-Danlos syndrome, 716
Ehrlich, P.:
  acid alum hematoxylin, 206
  aniline crystal violet, 726
  p-dimethylaminobenzaldehyde reagent: for ci-
    trulline, 295
    for indoles, 259−261
  supravital methylene blue for nerve endings,
    782
Ehringer et al., combined adrenergic-
    nonadrenergic nerve fiber stain, 783
Eimeria stiedae polysaccharide, 616
Einarson's chrome alum gallocyanin, 212−213
Einarson's sublimate formalin, 212
Eisenstein's chloranilate Ca method, 539
Elastase, 302−303, 716−717
  basement membrane digestion, 303
  oxytalan fiber digestion, 718
Elastic tissue:
  Biebrich scarlet stain, 287
  black PAS stain, 310
  composition, 680, 705−707
  Ehlers-Danlos syndrome, 716
  methylation, effect of, 320
  mummies, 716
  pinacyanol stain, 192−193
  rodents and children, native aldehyde, 172,
    616
  selective EM stain, 707
  selective vital stain, 707
  stain for tryptophan, 260
  stain for tyrosine, 255, 355
  Weiss et al., stain for tissue amines, 285
Elastic tissue stains, 708−716
  aging, solvent blacks 5 and 7, 709
  aldehyde fuchsin, 714−715
  alkaline enianil azurin J, 715
  bromophenol blue, 716
  Luxol fast blue G, 710
  nigrosin base acid alcohol, 709
  orcein methods, 710−711
  orcinol new fuchsin, 714
  permanganate-Schiff, 716
  resorcin fuchsin stains, 711−713, 716
  Taenzer-Unna orcein, 709−710
  unmordanted hematoxylin, 716
  Verhoeff's, 708−709, 716
Elastin, reactions, 143, 255, 705−709
Electricity, static during paraffin sectioning,
    89
Electrolytic decalcification, 787
Electron probe microanalyzer, 830−831
Eleidin, 354−355
  osmic reaction, 217

Elftman, H.:
  Aoyama silver method for Golgi, 332−333
  chromation, hematoxylin for phospholipid, 574
  direct silver Golgi method, 332−333
Ellipsoid lipid in retina, 784
Embalming before autopsy, 25
Embedding:
  Agar, 80
  Araldite, 100
  Carbowax, 76
  celloidin, 91−94
  dimethylol urea, 99
  for EM, 94−104
  epoxy resins, 101−104
  ester wax, 90−91
  gelatin, 78
  glycol methylacrylate, 96−98
  2-hydroxypropyl methacrylate, 98−99
  for immunofluorescence, 90
  Maraglas, 103−104, 655
  methods, 76
  methylacrylates, 100
  miscellaneous mixtures for EM, 104
  paraffin, 80−87
  polyampholyte, 99
  Rigolae, 101
  sectioning methods, 76−104
  urea formaldehyde, 99
  Vestopal W, 101
Encephalitozoa, 734
Endamoeba histolytica, 198, 203, 732
Endicott's marrow smear technic, 719
Enterochromaffin:
  azo coupling, 252, 279−281
  ferric ferricyanide reaction, 235, 278−279
Enterochromaffin cells, reactions, 211, 217, 228,
    246, 254−255, 260, 262, 264, 273−281,
    335−336, 530
Enterosiderin, 278, 508, 511, 587−588
Enzymatic digestion tests, 300−303
  amylases, 629, 634, 657−658, 668
  elastases, 302−303, 718
  mucopolysaccharides, 655, 657−659
  nucleases, 168−169, 181
  proteases, 298, 300−303, 519
Enzyme-antibody conjugation, 855−857
Enzyme histochemistry:
  azo dye principles, 358−359
  diazonium salt inhibition, 361
  enzyme diffusion, 401
  esterase histochemical problems, 425
  glycol-methacrylate embedding, 361
  kinetic assay, 441
  preparation of tissues, 357−359

Enzyme histochemistry:
  quantitative: bone cells, 482
    glycosidases, 482
  soluble enzymes, 481–482
  substrate film problems, 437, 440
Eosin, blockade by tosylation, 228
Eosin B, 139, 154
  +azure A, 195, 196, 754
  +methyl blue for hypophysis, 343
Eosin Y, 136, 139, 208–209, 485, 754, 788, 795
  in Papanicolaou EA stains, 721–723
Eosinophil granule stains with catechol dyes, 752
Eosinophil leukocytes, 186, 196, 211, 255, 260,
    285, 288, 343–344, 616, 748–752
Eosinophil myelin stain, hypophyseal α cells, 345
Eosinophilia, 125
Epidermis, 243–244, 352–355
Epididymis, 349
  lipofuscin, 516
  tubule fluid, 285
Epinephrine, 217, 236, 252, 262–273
Epithelial enzyme activity, 301, 388–389
Epoxy-ether method for COOH, 324
Epoxy resin embedding media, 101–104
Eränkö, adrenal medulla reactions, 267–269
Eriocyanine A, 154, 338–339
Erythrocytes, 196, 211, 248, 285, 694, 697,
    742–759
  carbonic anhydrase, 443
  cholinesterase, 406, 408
  concanavalin A-ferritin stain, 850
  hemoglobin peroxidase, 447–448
  negative stain, 829
  Sudan black B staining, 566
  Zenker fixed, birefringence, 9
  (See also Hemoglobin)
Erythrosin B, 139
Eserine inhibition:
  acetylcholinesterase, 411, 416, 418
  other enzymes, 428
Esophageal gland mucin, 285
Ester gum mountants, 113–115
Ester wax, 90–91
Esterases, 406–425
  acetylcholinesterase, 409
  acetylthiocholine, 416
  azo dye methods, 410–413
  Burstone's azo dye method, 411
  chloracetyl esters, 426–428
  cholinesterase, 415–425
  EM method, 410–415, 420
  Gomori, 407–408, 411, 415–416
  indoxyl methods, 413–415
  inhibitors, 418–420

Esterases:
  Karnovsky-Roots method, 420–421
  Koelle et al., methods, 421–424
  Koelle-Friedenwald method, 416–417
  lipase, 407
  pancreatic lipase, 409
  reactivators, 420
  thioacetic acid method, 408–410
  thiocholine, 425
  Tween method, 407
Esterified Sudans, 571–572
Ethanol dehydration, 69, 82–83
Ethanolamine in phospholipids, 560–561
Ethyl chloride alkylation of amines, 321
Ethyl maleimide, blockade of SH, 243
Ethylene glycol fat stains, 568
Ethylene groups, 217–223
  lipids, 560–562, 595, 604
  lipofuscins, 513, 515–517, 519–520
  myelin, 575
Ethyl eosin + methylene blue Negri body
    methods, 754
Ethyl green, 113, 131
Ethyl maleimide blockade of SH reaction,
    243–244
Euchrysine GGNX, 197
Euparal, 113, 115
Euparal vert, 113
Evans and Krajian's decalcifying fluids, 793
Evans blue, 142
Exfoliative cytology methods, 721–724
Eyepieces, microscope, 3
Eyes:
  celloidin schedule, 91–92
  fixation, 27
  gelatin embedding, 78
  postchroming, 78

Fading of stains in balsam, 112
Fairhall's chromate method for lead, 547
Falck, fluorescent formaldehyde reaction, 267,
    269–270, 274, 281
Farrants' glycerol gum arabic, 118, 120
Fast acid violet 10B, 138
Fast black B, 148, 155
Fast black G, 149
Fast black K, 148, 155, 248, 250, 257, 310, 433,
    732
Fast blue B, 148, 155, 248, 250, 390, 396, 399,
    403, 428, 434
Fast blue BB, 148, 155, 250, 362, 411
Fast blue BM, MB, 149
Fast blue RR, 148, 155, 411

Fast blue VB, 148
Fast Bordeaux GP, 148
Fast brown RR, 148
Fast brown V, 148
Fast brown VA, 149
Fast Corinth LB, 148
Fast Corinth V, 148, 428
Fast dark blue R, 148
Fast garnet AC, 149
Fast garnet GBC, 148, 155, 248, 250, 257, 411,
    427 − 428, 433 − 434, 436
Fast garnet GC, 148
Fast green FCF, 155, 697 − 698
    Masson trichrome variant, 700
    Shorr stain, 723 − 724
Fast green Van Gieson, 697 − 698, 701
    after acylation, 226, 288
Fast orange GR, 148
Fast red AL, 149
Fast red B, 149, 155, 248, 250, 433, 436
    enzyme inhibition, 436
Fast red 4GA, 149
Fast red GG, 149, 155, 248, 250, 280
Fast red GL, 149
Fast red 3GL, 149
Fast red ITR, 149, 155
Fast red 5NA, BN, 148
Fast red NRL, 149
Fast red RC, 149, 155, 396
Fast red RL, 149
Fast red TR, 149, 155
Fast red TRN, 149
Fast red violet LB, 149, 155, 362, 411
Fast scarlet GG, GGS, GSN, 149
Fast scarlet 4NA, 149
Fast scarlet R, 149
Fast violet B, 149
Fat staining by tetrazoles, 472
Fat stains, 144 − 150
Fats, lipids, 559 − 610
Fatty acids, 590 − 595
Faulkner's Warthin-Starry silver variant for
    spirochetes, 761
Fe in spodograms, 866
Fe⁺⁺ uptake of melanins, 526
Fe⁺⁺, Fe⁺⁺⁺ uptake, Clara's hematoxylin, 203
Fearon's reagent for citrulline, 295
Feder-Sidman freeze substitution, 62 − 63
Feder's polyvinyl alcohol embedding, 76
    after freeze substitution, for enzymes, 360
Feigin-Wolf 1957 alkaline phosphatase substrate,
    368
Fekete's acetic alcohol formalin, 34
Fenton et al., Fe⁺⁺⁺ reaction in spodograms, 867

Fernando's crystal violet amyloid technic, 667
Fernando's salt sucrose dextrin, 121
Ferrans-Hack-Borowitz 1962 Feulgen plasmal
    reaction, 582 − 583
Ferric chloride reaction, chromaffin, 264
Ferric chloride sodium oxalate buffer series pH
    1.3–1.9, 879
Ferric ferricyanide for reduction sites, 235 − 238
    ascorbic acid, 236, 531
    chromaffin, 265
    enterochromaffin, 278 − 279
    mast cells, 236, 257, 672
    melanins, 524
    pigments, 509, 513, 515, 517, 519 − 520
    sulfhydryls, 234, 354
Ferric ferricyanide stain:
    keratin, 236
    keratohyalin granules, 236
Ferric ion reaction, 495, 503 − 504, 507 − 508
Ferricyanide ferrocyanide buffer in esterase
    methods, 411
Ferricyanide reaction:
    for Fe⁺⁺, 501, 507 − 508
    of mast cells, 672
    other metals, 505 − 506, 544
Ferricyanide safranin artifacts, 132, 507, 517
Ferritin labeled antibody method, 847 − 853
    comparison to peroxidase, 859 − 860
    ferritin conjugation methods, 847 − 848
    staining methods, 849 − 850
Ferrocene-carbohydrazide method for nuclei,
    174
Ferrocenylmethylcarboxyhydrazide aldehyde re-
    agent for LM and EM, 617 − 618
Ferrocyanide reaction:
    for Fe⁺⁺⁺, 501, 503 − 504, 507 − 508
    of mast cells, 672
    other metals, 508, 544, 546, 556
Ferrous ion reaction, 506 − 508
Ferrous ion uptake, melanins, enterochromaffin,
    525 − 526
Feulgen reaction, 171 − 174
    decalcification, 788
    Guarnieri bodies, 756
    methylation, 779
    peracetic acid Schiff, 221
    ribonucleic acid extraction, 170 − 171
Feyrter's onkocyte stain, 596
Fibrin, 196, 211, 297 − 299, 616, 699
    elastin stain, 712
Fibrinoid, 211, 298 − 300, 667
Fibrinolysin, substrate film localization, 437
Fink-Heimer methods for degenerating axons,
    774 − 776

Fire red, 145
Fischer's milk method for vascular injection, 810
Fischler's fatty acid method, 591
Fisher-Lillie methylation technic, 318
Fishman β-glucuronidase methods, 398−402
Fite, G. L.:
  carbol new fuchsin, 736
  new fuchsin-formaldehyde method, 739
  oil fuchsin for *Mcb. leprae*, 739
Fitzgerald, freeze dry radioautography, 816
Fitzpatrick and Lerner's Dopa oxidase technic, 461
Fixation, 25−68
  acrolein, 42−43
  alcian blue (in fixative), 41
  alcohols, 45−47
  aldehyde fixatives, 29−45
  arginine demonstration, 293
  for autoradiography, 821−822
  Baker's formol calcium, 34
  Barnett-Bourne alcoholic acetic silver, 495, 580
  Benscome's formalin Zenker, 54−56
  bone, 787
  Bouin-Hollande copper picroformol, 61
  Bouin's, 60−62
  Bouin's picroformol-acetic, 60
  brain, 26
  Cater's acetic acid silver nitrate, 495, 580
  cetylpyridium chloride (in fixative), 46
  chloral hydrate, 43
  chrome salts, 54−60
  Ciaccio's fluid, 59
  cyanuric chloride, 48
  diethylpyrocarbonate, 43−45
  dimethyl adipimate, 41−42
  dimethyl malonimidate, 41−42
  dimethyl suberimidate, 41−42
  effect of buffer, 30
  effect on membrane size, 27
  eye, 27
  ferric ferricyanide reaction, 237
  Flemming's strong solution, 62
  Gendre's acetic alcohol picroformalin, 61
  glycogen demonstration, 806
  Guthrie's formic acid Zenker, 55
  heat, 67
  Heidenhain's susa, 52
  Helly's fluid, 54−55
  Hermann's, 62
  Huber's bichromate formol, 59
  Huber's trichloracetic mercury, 52
  Kolmer's, 60
  Kose's fluid, 59
Fixation:
  lanthanum nitrate, 41
  lead, 48−50
  lead citrate for EM, 49−50
  Lillie's acetic alcohol formalin, 34
  Lillie's alcoholic lead nitrate formalin, 48
  Lillie's salt Zenker, 56−57
  Mann's osmic sublimate, 53, 62
  Mann's tannin picrosublimate, 53
  Marchi's fluid, 62
  marrow *in situ*, 787
  mercury salts, 50−57
  N-methyl morpholine, 48
  microincineration, 865
  Möller's, 57−59
  mucins, 635
  Müller's, 58
  Newcomer's, 45−46
  Ohlmacher's sublimate alcohol, 52
  Orth's, 57−59
  osmium tetroxide, 27−68
  osmolality, 27−28
  Paneth granules, 350−351
  penetration, 27
  peptic glands, 349−350
  picric acid, 60−62
  pyroantimony, 66
  Ramón y Cajal's formol-ammonium bromide, 34
  Regaud's, 57−59
  Romeis' Orth fluid, 53, 57−59
  Rossman's alcoholic picroformalin, 61
  ruthenium red, 40
  Seligman OTO method, 64−66
  SH reactions, 235, 237−239, 241−246
  Solcia's GPA, 61
  Spuler's acetic sublimate formalin, 53
  Spuler's formalin Zenker, 55
  tannic acid, 40, 53
  Tellyesniczky (acetic-bichromate), 59
  Tellyesniczky (acetic-ethanol-formalin), 34
  thin films and smears, 720−721
  trichloroacetic acid, 52−53
  tris(1-aziridinyl)phosphine oxide, 40
  Zenker's, 54−57
  zymogen granules, 336−337, 349−351
Fixation effect on staining:
  acridine orange, 182
  azure eosin, 196
  pH of basic dye staining, 316
Fixed tissue mitochondrial stains, 328−331
Flaming red, 145
Flavianic acid, 156
  replacing picric acid in collagen stain, 696

Flavophosphine N, 156
  Schiff reagent, 308
Flemming's strong solution, 62
Fluorescein, 136, 156
  fluorochroming of collagen, 692
  isocyanate, isothiocyanate, 304, 838–839
Fluorescein mercuric acetate stain for SH, 241
Fluorescence:
  acid fast stain, 738
  adrenal lipofuscin, 515
  Anderson-Greiff stain for rickettsiae, 744
  3,4-benzpyrene, 572
  carotenes, 529
  chromaffin cells, 267–281
  concanavalin A-FITC for fungal cell walls,
    733
  concanavalin A-FITC stain, 648–649
  COOH method, 324
  Coons' immunofluorescent method, 303–305
  creatine kinase, 385
  Deitch fluorescent method for arginine, 294
  dopamine, 262–273
  Ehringer et al., combined adrenergic-
    nonadrenergic nerve fiber stain, 783
  enterochromaffin, 274
  glyoxylic acid, 271–272
  histamine, 257–258
  immunofluorescent stain for fibrin, 299
  juxtaglomerular granules, 352
  lipid stain, 572
  lipofuscins, 349, 514, 518
  mast cells, 673
  morin stains, 535–536, 556
  naphthol AS esters, 428
  native, 12
  noradrenalin, 262–273
  9,10-phenanthrenequinone fluorescent method
    for arginine, 291
  o-phthalaldehyde, stain for α-cells of pancreas,
    342–343
  Previero method for tryptophan, 260–261
  protozoan parasites, 752
  quantitation, 846–847
  rhodamine B-fluorescein for collagen, reticulin
    and basement membrane, 692
  Schiff reagents, 308–309
  sialic acid, 586
  Stoward carbohydrate stain, 625–626
  Stoward sulfomucin stains, 650–652
  sulfatide lipids, 585–586
  tetracycline label for calcium, 802–803
  Trasylol-FITC stain, 649
  vitamins A, B₂, 529
Fluorescence microscopy, 10–12, 832–847
  epiillumination, 835–836
Fluorescence microscopy:
  filters, 833–836
  immunofluorescence, 837–865
  microscopy setup, 832–833, 835
Fluorescent-tagged protein purification,
    304–305, 840–841
Fluorescent-tagged protein reconcentration, 305
Fluoride, calcium, 538–539
Fluoride effect on enzymes, 429, 433
Fluoride inhibition:
  of esterase, 408, 411, 418
  other enzymes, 428, 442
Fluorochrome stains, 134–136
  blood films, 752
  collagen, reticulum, basement membrane, 715
  fast stains, 572
  mucin stain, 636
  Schiff reagents, 308–309
  tubercle bacilli, 737
Focusing, 6–7
Foley's Bodian protargol variant, 771
Fontana's diammine silver for spirochetes, 760
Foot's silver carbonate reticulum method,
    683–691
Foot's silver oxide reticulum method, 683–691
Formal, 30
Formaldehyde blocking of amine reactions, 285
Formaldehyde fixation, 30–35
  effect on brain lipids, 589
  freeze dry, vapor for histamine, 257
  vapor for films, 720–721
Formaldehyde fuchsin methods, 732, 734, 739
Formaldehyde pigment, birefringence, 9, 488
Formalin bichromate chloral fixative, 769
Formalin pigment, 488–489
Formalum, Baker's, 334
Formazans, 150–152, 469–482
  formation from aldehyde (Stoward), 623–625
Formic acid, directions for 1 M solution, 889
Formic acid Bouin fluid, 60–61
Formic acid decalcification, 787, 792
Formic acid method for trichinae, 762, 792–793
Formic acid Zenker fluid, 54–56
Formol, 30
Formol bromide fixation, 32, 34
Formol calcium, 33–34
  for mitochondria, 330
Formol saline, 33
Formyl violet S4B, 139
Fox's zinc chromate Golgi method, 769
Fractures, special decalcifying procedures, 790
Fraenkel's orcein elastin method, 710–711
Frank and Deluzarche's procedure for decalcify-
    ing teeth, 790
Frankenberger chromate method for lead, 547

Free fatty acids, brain lipids, 595—596
Freeborn's Van Gieson stain, 695
Freehand sectioning, 69
Freeze dry methods, 68, 74—75, 357, 822—823
  carnauba wax for undecalcified bone, 805
  enzymes, 360, 362, 407, 409—410, 434, 805
  formaldehyde gas for histamine, 257
  mitochondria, 329—330
  paraffin, 68, 362, 407, 409—410, 434
Freeze-etching, 827—828
Freeze substitution:
  acetone, 384, 435
  alcohol, for mitochondria, 329—330
  alcoholic picric acid, 60—61
  alcoholic sublimate, 51
  alcohols for sulfhydryl, 243
  embedding methods for enzymes, 360
  ethanol, 73
  ethylene glycol for EM, 73
  glycerol for EM, 73
  osmic acetone, 62—63
  OsO_4, 73
  picric acid in ethanol, 73
  section methods for enzymes, 47, 360, 409, 435, 443
Freezing:
  and eye fixation, 27
  quick, 74
Freezing artifacts, 25
Freiman et al., 1962 alkaline phosphatase substrates, 368
Fresh muscle spreads for trichinae, 761
Friedenwald-Kaiser glycerol gelatin, 120
Friedland's boiling agar infiltration, 80
Frost's tetracycline quantitative bone assay, 802—803
Frozen section staining, 110—111
Frozen sections, 69—76
Fructose syrup, 119—120
Fructose-1,6-diphosphatase, 376—377
Fructose-1-phosphatase, 376—377
Fructose-6-phosphatase, 376—377
Fuchsin basic, 131, 133
  acid fast stains, 735—736
  diphenamine Schiff base reaction, 309—310
  Schiff reagents, 306—308
Fullmer-Lillie hematoxylin method for nucleoprotein, 186
Fullmer-Lillie orcinol new fuchsin elastic fiber stain, 714
Fullmer-Link decalcification method for enzymes, 793—794
Fullmer-Link stain for bone formation and resorption, 803

Fungi, 729—733
  cell walls, 615
  chromatin, 175, 732
  stains, 729—733

Gabbett's sulfuric methylene blue, 737
Galactose-1-phosphatase, 376—377
Galactose-6-phosphatase, 376—377
α-D-Galactosidase, 389
β-D-Galactosidase, 388—389
Galactosides, brain lipids, 560, 562, 584—586
Gallamine blue, 132, 156
  iron lake stain, 212
  lead reaction in phosphatase technic, 365
Gallego's iron fuchsin method, 714
Gallein-iron, mordant stain for nuclei, 202
Gallium, 535, 556
Gallocyanin, 126—127, 156
  chrome alum stain, 212
  iron lake stain, 212
  unmordanted, for eosinophils, 752
Gallocyanin-metal stain for polysaccharides, 654
Gancevici's triphosphate buffer pH 9.5-12.4, 887
Ganglion cell lipofuscin, 514—515
Gangliosides, 560, 562, 586—588
Gargoylism mucopolysaccharide, 660
Garnet AN, GC, GCD, 144
Garnet GBC, 144
Garven-Gairns silver method for nerve endings, 770—771
Gasoline clearing agent, 80, 83
Gasserian ganglion pigment, 515
Gastric cardial gland mucin, 615, 635—636
Gastric chief cells, 196, 317, 349, 714
Gastric epithelial mucin, 310, 615, 636, 648
Gastric parietal cells, 196, 349
Gastric washings, smear and section technics, 720
Gaucher lipid, 575, 615
Gelatin, 615
Gelatin adhesives for Carbowax sections, 77
Gelatin glue, 683
Gelatin embedding, 78—80
Gelatin paper models, phenols, 252
Gendre's acetic alcohol picroformalin, 61
Gendre's classification of stains, 123—124
General methods on nerve tissues, 765
Gentian violet in Gram stains, 726
Gerard-Cordier ferrocyanide method for uranium, 556
Gerlach's Negri body method, 754
Gesswein-Timm silver sulfide method for zinc, 555

Giant cell pneumonia inclusions, 755–756
Gibbs' reaction for phenols, 253–254, 279
Giemsa stain:
  for blood marrow and protozoa, 744–747
  for tissues, 195
  for spirochetes, 759
Giglioni's gelatin embedding variant, 77
Giovacchini's gelatin glycerol adhesive, 77
Giroud-Bulliard nitroprusside SH method, 238
Glaucoma, lipid macrophages in nerve head, 785
Glavind's fatty peroxide technic, 224
Glees' ammonia silver solution, 778
Glees-Bielschowsky method for degenerating
  axons, 773–774
Glees method for axons in paraffin sections, 778
Glenner, G. G.:
  amine oxidase technic, 465–466
  aminopeptidase variants, 436
  bichromate bilirubin method, 495, 519
  Fischer rosindole reaction for indoles, 260
  γ-glutamyl transpeptidase, 429–431
  postcoupled benzylidene indole reaction, 259
  rosindole reaction for tryptophan, 260
  section freeze substitution, 435
  trypsin-like peptidase, 428–429
Glia fibrils, 693
Glia methods, 765–769
Globus' HBr postmordanting of formalin fixed
  tissue, 768–769
Glucokinase, 387
Glucosamine-6-phosphatase, 376–377
Glucose oxidase conjugates for immunohistochemistry, 857–860
Glucose-1-phosphatase, 376–377
Glucose-6-phosphatase (G-6-Pase), 368–369,
  373–374
Glucose-6-phosphate in UDPG glycogen synthesis, 395–396
Glucose-6-phosphate dehydrogenase, 374, 824
Glucose substrate in disaccharase method, 386
Glucose syrup, 119
α-D-Glucosidase, 390
β-D-Glucosidase, 390–391
β-Glucuronidase, 398–402
β-Glucuronidase digestion, 655
  oxidized oxytalan and elastin, 717–718
Glutamic dehydrogenase, 477, 479
γ-Glutamyl transpeptidase, 429–431
Glutaraldehyde, 29, 35–42, 53
  effects on glycosidases, 39
  enzyme-antibody conjugation method,
    855–857
  for Feulgen stain, 172
  Schiff reaction, 37–38

Glutaraldehyde fixation, 35–42
Glutathione extraction of SH stains, 239, 244
Glyceraldehyde-3-phosphate dehydrogenase,
  376
Glycerides, 595
Glycerol:
  fluorescence, 12
  as mountant, 110
Glycerol gelatin, 118–120
Glycerol gum arabic, 118–120
Glycerol gum syrup, 120
Glycerol immersion microscopy, 737
Glycerol 3-phosphate dehydrogenase, 376
Glycerophosphatase, acid:
  pH 4.7-5.0 Pb methods, 359–360
  at pH 6.0, 7.2, 365–367
α-Glycerophosphatase dehydrogenase,
  468–471
Glycerophosphate substrates for alkaline phos-
  phatase, 367
Glycerophosphatides, 559–561
Glychrogel, 119–120
Glycine-HCl buffer pH 2.2-3.6, 881
Glycine NaCl:
  NaOH buffer series pH 8.8-12.8, 880
  NaOH buffer series pH 9.4-10, 880
Glycine reactivation of alkaline phosphatase in
  bone, 804
Glycogen, 310–311, 615, 617–618, 626–632,
  784
  amylase digestion, 392, 629–630
  in blood smears, 752
  bromination, 222–223
  chromic acid-methenamine silver, 630–631
  decalcified tissue, 805–806
  EM stain, 617–618, 631–632
  fixation, 32, 34, 45–46, 52–53, 61
  frozen sections, 631
  iodine reaction, 626–627
  lead citrate (Reynolds' EM), 632
  negative EM stain, 829
  periodic acid-toluidin blue, 630
  preservation, 631
  synthesis, enzymatic, 391–396
Glycogen synthetase, 392
Glycogen transferase, 392, 395–396
Glycol methacrylate, 96–98
Glycol solubility of sterols, 568
Glycols, 229–234
  in polysaccharides and glycolic fatty acids,
    229–234
Glycoprotein of hemosiderin, 502
Glyoxylic acid fluorescence, 271–272
G-Nadi reaction, 451–452

Gold, 531, 549, 554−555
  stains for, 549
Gold-hydroxamic acid stain for hydrophilic glycerides, 579−580
Gold sublimate method for astrocytes, 767−768
Goldman's iron hematoxylin, 198
Golgi substance, 331−335, 615
  enzyme localization, 335, 378
Golgi stains:
  Baker's Sudan black B, 334
  Da Fano's cobalt-silver, 331−332
  Elftman's Aoyama silver, 332
  Elftman's direct silver, 332−333
  osmium tetroxide, 333
  peracetic acid-Schiff, 334
  periodic acid-Schiff, 335
  potassium permanganate, 334
  Ramón y Cajal's uranium silver, 331
  Smith-Dietrich, 334−335
Golgi's chrome silver block methods, 769
Golodetz's sulfuric formalin sterol reaction, 589
Golodetz-Unna ferric ferricyanide reduction test, 235−238
Gomori, G.:
  acid phosphatase, 1950 method, 359−360
  aldehyde fuchsin for elastin, 714−716
    hypophysis, 343
  alkaline phosphatase, 1952 method, 367
  aniline blue-orange G method for chromaffin, 265
  borate buffers, pH 7.8−8.2, 277−278
  chloroacetyl esterase substrate, 426−427
  cholinesterase method, 415−416
  chrome alum hematoxylin for islet cells, 341
  chromic acid methenamine silver, 630−631
  collidine HCl buffer pH 6.8−8.3, 884
  lead sulfide technic for calcium soaps, 317−318
  methenamine silver argentaffin method, 272, 278
    urates, 291
  myristoylcholine for cholinesterase, 415−416
  phosphamidase method, 380−382
  silver method for reticulum, 681−693
  succinate buffer pH 3.8−6.0, 877
  tellurite for succinic dehydrogenase, 479
  trichrome stain, 701
  tris, tris-HCl buffer, pH 7.5−8.7, 883
  tris maleate buffer, pH 5.1−8.4, 882
  Tween esterase lipase methods, 407
Gomori-Burtner methenamine silver method, 277−278
Gomori-Coujard models, phenols, 252
Gomori-Perls ferrocyanide iron method, 507
Goodpasture-Perrin method for encephalitozoa, 734

Goodpasture's carbol aniline fuchsin, 734
G-6-Pase, 368−369, 376−377
Graduate cylinders, 22
Graham and Karnovsky monoamine oxidase method, 466−467
Graham and Karnovsky peroxidase method, 454−455, 864
Graham's benzidine peroxidase method, 450
Graham's α-naphthol myeloperoxidase method, 451
Gram, C.:
  iodine solution, 153, 726
  negative bacteria, 742−744
  stains for bacteria, 725−735
  stains for smears: Bartholomew's, 727
    traditional, 727
  stains for tissue sections, 727−729
Gram-Weigert technic for sections, 728−730
  mucins, 650
  Russell bodies, 758−759
  starch, 634
  (See also Weigert, C., fibrin stain)
Granados and Dam's peroxide reaction, 224
Grandis and Mainini's purpurin, 538
Graphite, 531
Graupner and Weissberger's dioxane paraffin, 80
Green diaphane, 113
Green euparal, 113
Greenspan, preparation of peracids, 220
Grenacher's alum carmine, 215
Gridley ammoniacal silver solution, 684
  Bauer variant for fungi, 730−731
  periodic acid in reticulum stains, 689
Grimley, tribasic stain for plastic-embedded sections, 183−184
Groat's styrene isobutyl methacrylate, 114−115
Grocott-Gomori CrO_3 methenamine silver for fungi, 630−631, 731
Gros-Bielschowsky method for nerve endings, 770−771
Gros-Schultze silver for islet cells, 341−342
Gross' contact autoradiographic method, 815
Gross' diacetin Sudan method, 567
Ground sections, 794−795
Growth hormone, identification with Sternberger PAP stain, 862
Guanine, 165−167
  murexide reaction, 283
Guarnieri bodies, 754, 756
Gum syrup, Apáthy's, 119−120
Gurr's mounting media, 115−116
Guthrie's formic acid Zenker, 55
Gutiérrez's Sudan black, 520−521, 568−569
Gwyn-Heardman stain for nerves and motor end plates, 779−780
Gypsum, 538−543, 866

H acid, 248, 255
Ha, modified Golgi-Cox stain for synapses, 779
Hack's freeze dry Carbowax method, 76
Hagemann's phenol auramine, 737
Hahn-Reygadas EDTA decalcification, 790
Hair, 352 – 355
Hair cortex, 221, 280, 298, 352 – 355, 615
  acid fastness, 606 – 607, 734 – 735
Hair follicle cytoplasm, 352 – 355
Hair sheath cells, 243 – 244, 255
Hale dialyzed iron method, 633, 646 – 648, 659
Hale-Mowry dialyzed iron technic, 646 – 647
Hale's method for bacterial cell envelopes, 741
Halowax in paraffin, 84
Hamperl, H.:
  argentaffin method, 277
  argyrophil technic, 278 – 279
Hansen's chrome-alum hematoxylin, 209 – 210
Hansen's iron hematoxylin, 201
Harderian gland stains, 500 – 501
Harleco synthetic resin, 114 – 115
Harris' alum hematoxylin, 206
Hart's Weigert elastin Van Gieson, 712
Hassall bodies of thymus, 196, 244
Hausler's Kurata variant for carbonic anhydrase, 443
Hayes' Feulgen plasmal reaction, 582
HBr postmordanting of sections, 107
HCl:
  acid phosphate buffer series pH 0.7 – 3.3, 871
  barbital buffer series pH 7.0 – 9.5, 885
  Biebrich scarlet, methyl blue collagen stain, 702
  directions for 1 N solution, 888
  formic decalcification of Richman et al., 787, 792
  NaCl decalcification, von Ebner's, 792, 797
  orange G, methyl blue collagen stain, 702
  sodium acetafe buffer series pH 0.6 – 5.2, 875
  sodium citrate buffer series pH 0.25 – 7.0, 872
  trisodium phosphate buffer series pH 9.5 – 12.4, 887
  Veronal sodium buffer series pH 7.0–9.5, 885
Heart failure cell pigment, 504
Heart muscle:
  cytochrome oxidase, 453
  lipofuscin, 514, 517
Heat effect in decalcification, 788
Heat fixation, 67
Heavy metals by Timm silver method, 550, 554 – 555
Heidenhain, M.:
  azan Mallory aniline blue method, 699
  iron hematoxylin, 199
  susa fixative, 52

Heidenhain, R., chromaffin reaction of enterochromaffin, 262, 265
Heimer's modified Nauta-Gygax degenerating axon stain for EM, 776 – 777
Helaktyn Blue F – 2R, 160, 162
Helaktyn Red F4BAN, 160, 162
Held's fixative fluid, 785
Heliotrope B, 2B, 130, 132
Hellweg's protargol for pancreatic α cells, 342
Helly's fluid, 54 – 55
Hemalum nuclear stain, benzoylation, 227
Hematein in hematoxylins, 204, 207
Hematite, 533
Hematoidin, 490 – 500
  reaction with diazosafranin, 252
Hematoporphyrin, 500 – 501
Hematoxylin, 154, 165
  basic nucleoprotein, 186
  chromosome, 191
  elastin, 280
  eleidin, 280, 354
  enterochromaffin, 280
  eosinophil leukocytes, 280, 752
  iron pigments, 280, 504 – 505, 508 – 509, 544 – 545
  keratohyalin, trichohyalin, 280, 354
  lead, 340
Hematoxylin, acetic: Smith-Dietrich-Baker method, 600
  stain for myelin, 217
Hematoxylin (alum):
  azure II eosin, 194
  benzoylation blockade, 227
  after decalcification, 788 – 789, 796
  eosin stains, 208 – 210
  erythrocyte staining, 485
  fading, 113 – 117
  formulae, 206 – 207
  for inclusion bodies, 754
  for mitochondria, 328
  for osmium-fixed tissues, 209
  phloxin for hypophysis, 343
  post-methylation staining, 779
Hematoxylin colors with metal ions, 532 – 533, 544 – 547
  calcium, 540
  chromium in myelin stains, 595
  iron and copper, 504, 510
  iron technics, 197 – 203
Hematoxylin substitutes:
  anthracene blue SWR, 203, 210
  brazilin, 203
  celestin blue B, 203
  chromoxane cyanin R, 211
  gallamin blue, 211
  gallein, 202

Hematoxylin substitutes:
  gallocyanin, 203
  phenocyanin TC, 214
Hemin-leucodichloroindophenol peroxide reaction, 224
Hemofuscin, 517—518
Hemoglobin:
  altered, 488
  Dunn-Thompson stain, 486
  general reactions, 485—488
  metal uptake reactions, 487
  Okajima stain, 486
  9-phenyl-2,3,7-trihydroxy-6-fluorone, 487
  Puchtler-Sweat stain, 487—488
Hemoglobin peroxidase, 447
Hemolysis, prevention in acetic fixatives, 57
Hemosiderin, 501—507
  pinacyanol stain, 192—193
Heparin in mast cells, 671—677
Heparin action on enzyme activity, 428
Hermann's fluid, 62
Herpes inclusions, 754
Herring sperm sibstrate for deoxyribonuclease II, 441
Herxheimer's Sudan IV methods, 566—567
Hetherington's pinacyanol neutral red, 327—328
Hexazonium pararosanilin, 113, 362, 397, 400, 402
Hexokinase, 387, 824
L-Hexonate dehydrogenase, 387—388
Hg^{++} inhibition of enzymes, 391, 433, 438, 443
HgCl$_2$ postmordant for phosphotungstic acid hematoxylin, 693
Highman, B.:
  Apáthy gum syrup, 118—119, 187, 435
  crystal violet amyloid method, 667
  hemosiderin method, 507
  iron ore dusts, 533
  mucin stains in buffered thiazins, 636
HIO$_4$, sulfite leucofuscin reaction, 229—234
Histamine:
  in mast cells, 257, 672—673
  o-phthalaldehyde method, 258—259
Histidine, 256—257
  azo coupling, 246
Histiocytes, 197
Histoplasma, 631
Histoplasma capsulatum, 631, 729, 731—732
HNO$_3$ decalcification, 788, 792
HNO$_3$ directions for 1 N solution, 888
Holländer's method for cerebroside sulfate esters, 585—586
Holmes' buffer for silver salts pH 7.4—9.0, 885
Holt's bromoindoxyl esterase method, 413—415

Holt's gum sucrose, 33
Hooke, logwood and cocheneel, 152
Horecker-Lillie citrate buffer series pH 2.8—6.8, 876
Hortega (*see* Del Río Hortega, Pío)
Hot formalin, 32
Hot plates, 22
Hotchkiss reaction, 230—232
  reducing rinse, 232
Hršel's stain for tryptophan residues, 261
Hruschovetz and Harder's Abopon, 117
  acetic orcein, 187
H$_2$SO$_4$ acid phosphate buffer series pH 1.1—4.0, 872
H$_2$SO$_4$ directions for 1 N solution, 888
HSR resin, 114—115
Huber's bichromate formol acetic, 59
Huber's trichloroacetic sublimate, 52
Hucker and Conn's ammonium oxalate crystal violet, 726
Hukill and Putt's reaction for iron, 501, 506
Humason-Lushbaugh pinacyanol stain, 192—193, 520, 668
Humphrey's reaction for iron, 501, 506
Hurler's disease, 660
Hyalin, 670—671
Hyaline intimal degeneration of small arteries, 671
Hyaluronic acid, 612, 641, 643—644, 659
Hyaluronidase digestion, 659
  resistance of vitreous, 786
Hybrid antibodies, 859
Hydration of paraffin sections, 106
Hydrazine reactions, 309
Hydrochloric acid hematin, 488—490
Hydrochloric acid sodium citrate, 1 M pH 4.5, 793
Hydrolysis of thick blood smears, 719—720
Hydroxamic method for COOH, 324
Hydroxyadipaldehyde fixation, 29, 39, 42
o-Hydroxybenzoic acid group, 125
β-Hydroxybutyric dehydrogenase, 477, 479
Hydroxyferric ion mucosubstance stain, 645
Hydroxylamine blockade of aldehyde, 313
Hydroxylysine, 137, 680
2,3-Hydroxynaphthaldehyde amine reagent, 283—285
2-Hydroxy-3-naphthoic acid hydrazide for —CHO, 309
Hydroxyproline role in sulfation basophilia, 225
2-Hydroxypropyl methacrylate, 98—99
8-Hydroxyquinoline:
  as arginine reagent, 291
  iron method for β-glucuronidase, 398
3-β-Hydroxysterol dehydrogenase, 476

5-Hydroxytryptamine, 274—275
  azo coupling, 275
  formaldehyde fluorescence, 274
  reaction with diazosafranin, 251
5-Hydroxytryptophan:
  azo coupling, 275
  reaction with p-nitrodiazobenzene, 248, 251
Hypoiodite, 728
Hypophysis, 343—349
  α cells, 211, 714
  β cells, 299, 429, 615
  chromophobe cells, 714
  colloid, 255, 615
  δ cells, 714
Hypophysis stains:
  alcian blue-PAS-celestin blue, 347—348
  aldehyde fuchsin, 343, 347
  aniline blue, 343
  benzo sky blue, 346—347
  coupled tetrazonium, 346
  eosin B-methyl blue, 343—344
  Gram, 345
  Halmi, 347
  immunofluorescent, 348—349
  iron hematoxylin, 345—346
  McLetchie-Lendrum method, 344
  methyl blue-eosin, 343
  PAS-orange G-methyl blue, 348
  Pearse's PAS multiple stain, 345
  periodic acid-Schiff, 343
Hyrax, 114—115

I[131] binding by noradrenaline, 263
Illumination, 4—8
Imferon in dextran localization, 632
Immersed condenser, 7
Immersion microscopy, 7
Immunoadsorbent preparation, 854—857
Immunoenzyme technics, 853—865
  p,p'-difluoro-m,m'-dinitrophenyl sulfone
    (FNPS) conjugates (Ram $et\ al.$), 857
  direct immunoenzyme method (Avrameas),
    853—857
  1-ethyl-3-(dimethylaminopropyl) carbodiimide
    (CDI) conjugates (Nakane and Pearse),
    857
  indirect immunoenzyme method (Avrameas),
    857—860
  Mason $et\ al.$, immunoglobulin peroxidase
    bridge method, 863
  Sternberger PAP method, 860—862
  unlabeled antibody enzyme methods,
    860—865

Immunofluorescence:
  antiserum conjugation, 838—840
  antiserum preparation, 837—838
  conjugate purification, 840—841
  diethylpyrocarbonate as fixative, 43—45
  counterstains, 843
  methanol as fixative, 46—47
  nonspecific staining artifacts, 843
  quantitation, 846—847
  tissue preparation, 841—843
Immunoglobulin, immunohistochemical EM de-
    tection, 865
Immunoglobulin peroxidase bridge method of
    Mason $et\ al.$, 863
Immunohistochemical method:
  ACTH, rat pituitary, 460—461
  aminopeptidase, 436
  carbonic anhydrase, 445
  cathepsin Bl, 431
  cathepsin D, 432
  cell surface receptors, 844—846
  counterstains, 843
  DNA-histone, 174
  dopa decarboxylase, 463
  dopamine-β-hydroxylase, 462—464
  elastase, 717
  glycerol 3-phosphate dehydrogenase, 376
  hormones of pituitary, 348—349
  immunoglobulin (EM), 865
  phenylethanolamine-N-methyl transferase,
    463
  phosphorylase, 393—394
  platelets, 751
  proline hydroxylase, 703
  quantitative assays, 846—847
  Sternberger, 460—461, 860—865
  thrombostenin in platelets, 750
  tissue preparation, 841—843
  uranium labeled specific antibody, 830
  uricase, 446
Impression films, 719
Inclusion bodies, 754—759
Incomplete dehydration, 86
Incubators, 18
Indazole blue, 129, 131
India ink injection of retinal vessels, 809
Indifferent examination media, 121—122
Indigocarmine, 154
  picric acid collagen stain, 694, 732—733
Indium, 554, 556
Indophenol reaction for phenols, 253—254
Indoxyl esterases, 413—415
  acetate, butyrate for esterases, 407, 412
Influenza organisms, 734

Influenza vaccine sialidase, 657–658
Inhibition of enzymes (*see* specific enzymes and inhibitors)
Injection apparatus for metal corrosion preparations, 808
Inorganic polyphosphatase, 382–385
Instruments, 20
INT, INPT, 150–152, 469–482
Intestine:
  embedding of small animal, 86
  iron pigments, 500–503
  mucins, 310–311, 615, 635
  muscle lipofuscin, 514, 518–519
  striated border, 386
Inulin, 634–635
Invert sugar syrup mountant, 121
Invertase, 386
In vivo oxidation of exogenous polyene oils, 597
Iodate reaction of noradrenalin, 262–263
  ripening of alum hematoxylin, 205
  solubilities of Na, K salts at 0-100° C, 263
Iodination in Morel-Sisley reaction, 255
Iodine, 152
  amyloid reactions, 663–667
  blockade of SH reactions, 244
  effect on azure eosin stains, 196
  glycerol mounting medium, 391, 396, 626–627, 664–665
  radioautography, 813
  reactions of saccharides, 391–395, 626–632
  sulfuric reaction of cellulose, 634
  test for carotenes, 485
Iodine green, 131, 133, 757
  amyloid stain, 667
Iodine inactivation of esterases, 408
  glucose-6-phosphatase, 373
  mast cell protease, 428–429
  phosphamidase, 380–382
Iodine iron hematoxylin elastin stains, 708
Iodine stabilization of indophenol blue, 452
Iodine-thiosulfate in silver reticulum methods, 685
Iodization for mercury removal, 106
Iodoacetamide SH blockade, 238
Iodoacetic acid:
  enzyme inhibitor, 429, 433, 436
  SH blockade reagent, 236, 238
Iodonitrotetrazolium (INT, INPT), 151, 469–482
Iridium, 531
Iris melanin, 522
Iron, 129, 531, 533–534, 539, 545–546, 548–549, 554
  absorption deposits, 508
  bearing pigments, 501–508

Iron:
  demonstration, 128–129, 505–508, 533–534
  hematoxylin reaction, 280, 509
  microincineration, 866
  preservation in bone, 787, 790–791
  spodograms, 865–867
  unmasking in hemoglobin, 488
Iron (tissue):
  catechol dye sequence stains, 202–203
  celestin blue, gallamine blue, gallocyanin lakes, 211–212
  mordant dyes, 125–126
  orcinol, new fuchsin for elastin, 714
  pyrogallol or hydroxyhydroquinone stain, 202–203
Iron dextran localization, 632
Iron hematoxylin methods, 197–204
  fast green safranin for mucins, 658
  hypophyseal α cells, 345
  iron toning of alum hematoxylin, 202
  mitochondria, 328
  myelin methods, 599–604
Iron methods for acid groups, 325–326
Iron salt equivalents of official solutions, 200
Iron uptake reactions:
  enterochromaffin, 524–526
  hematoxylin reaction, 203
  melanins, 524–526
  mucopolysaccharides, 646–647, 786
  sulfide reaction, 203
Irwin's Al method, 534
Isocitric dehydrogenase, 477, 479
Isoelectric point, 182–183
Isopropanol dehydration, 80
Isopropanol supersaturated Sudan solutions, 567–568
Isothiocyanate protein conjugation, 303–305, 838–841

Jaffe *et al.*:
  cell suspension rosette assay for binding Ig-MEAC or IgGEA, 844–845
  frozen section rosette assay for binding Ig-MEAC or IgGEA, 844–845
Jahnel silver method, lepra and spirochetes, 740
Janigan and Lillie's aniline blue leuco for hemoglobin peroxidase, 447
Janssens' iron hematoxylin, 200
Janus black I, 156
  mitochondrial stain, 327
Janus blue B, 130, 156
  mitochondrial stain, 327
Janus green, 130, 156

Janus green B, 130, 156
  mitochondrial stain, 327
Jarrett's cysteine desulfurase method, 442
Jenkins' decalcifying fluid, 788 – 789
Johnson, F. B.:
  calcium oxalate studies, 866 – 867
  microincineration without muffle furnace, 867
Johnstone-Bowsher silver carbonate stain for
    degenerating axons, 777
Jullien's picroindigocarmine, 694
Juxtaglomerular granules, 351 – 352

Kaiser's glycerol gelatin, 119 – 120
Karo, as mountant, 121
Kasten's fluorochrome Schiff reagents, 308 – 309
Kattine, V.:
  allochrome variant, 704 – 705
  black toning of alum hematoxylin, 202
  Hart elastin collagen method, 712
  hematoxylin Van Gieson, 695
Kawamura-Niimi stain for degenerating axons,
    776
Kay-Whitehead multiple Sudan staining, 565
KCl in spodograms, 866
Kephalins, 560, 584 – 588, 595
Kerasin, 584 – 588, 597, 615
  in retina, 784
Keratin:
  cutaneous, 193, 211, 236, 248, 280, 352 – 355,
    616
  rodent forestomach, 193, 236, 248, 255, 260,
    280, 352 – 355
Keratogenic zone of hair, 236, 243 – 244
Keratohyalin, 203
  acid fastness, 734
  cutaneous, 186, 236, 280, 287, 352 – 355,
    461 – 462, 616
  esophagus, 280, 352 – 355
  permanganate hematoxylin stain, 354
  rodent forestomach, 280, 352 – 355
Kerenyi-Taylor method for hypophysis,
    346 – 347
Kernechtrot, 156, 511, 538
Ketone blue 4BN (cyanol), 139
Ketose reductase, 368
Ketosteroids, artifacts, 605
Khanolkar FeCl$_3$ α-ketol method, 605 – 606
Kidney, enzymes, 403, 408, 429, 443
King's Del Río Hortega variant for amyloid,
    667 – 668
Kinyoun's carbol fuchsin, 736
Klatzo:
  chloral formol bichromate fixative, 769
  Golgi variant for glia, 769 – 770

Kligman-Mescon-DeLamater fungus method,
    730
Kligman's HIO$_4$:
  fuchsin method, 310
  for fungi, 730
Klossiella muris polysaccharide, 615
Klüver and Barrera's Luxol Fast Blue MBS, 599
Knife sharpening, 17
Koelle and Friedenwald's acetylthiocholines-
    terase, 416 – 417
KOH alcohol saponification, 228, 320 – 321
  ribonucleic acid extraction, 170 – 171
Köhler illumination, 4 – 6
Kolmer's droplets in retina, 784 – 785
  fixative, uranium acid formol bichromate, 60
Koneff and Lyons' nitrocellulose schedule, 93
Kopriwa-Leblond liquid emulsion radioau-
    tography, 815
Kósa on malaria pigment, 489 – 490
Kose's formol bichromate, 59
Krajian's silver solution for reticulum, 687
Krebs cycle dehydrogenases, 471 – 480
Kresylechtviolett, 132
Kristensen's buffered formic acid, 788, 792
Kultschitzky, N.:
  acetic hematoxylin, 602
  cells, 273 – 281
  lithium carbonate ferricyanide, 603
Kurata's carbonic anhydrase technic, 443
Kurloff bodies, 287, 616, 756
Kurnick's methyl green pyronin, 180

Labeling:
  paraffin blocks, 86
  reagents, stains, 24
  stained slides, specimen bottles, 22, 24
Lacrimal gland cytoplasmic granules (rat), 616
Lactic dehydrogenase, 479 – 482
Lactonization of uronic acid, 321 – 324
Lactoperoxidase in tissues, 865
Lactose, as disaccharase substrate, 386
Lafora bodies, 757
Lagunoff and Benditt's substrate for mast cell
    protease, 427 – 428
Laidlaw's silver reticulum method, 685 – 688
Lamanna's theory of bacterial acid fastness, 735
Lamps, microscope, 1 – 2
  alignment, 4 – 5
Landing-Hall phosphorylation reaction, 225
  tetrazonium method for hypophysis, 346
Landing's histochemical molybdenum blue
    method for phospholipid, 574
Langeron's alizarin red S for Ca^{++}, 538
Langhans' iodine for amyloid, 664, 669

Lanolin rosin, 110

Lansdown silver toluidine blue stain for calcification, 798

Lanthanum nitrate fixative, 41, 653, 677

Lanthanum staining, 663, 677

Laqueur's method for alcoholic hyalin, 670 – 671

Lartigue and Fite, mechanism of acid fast stain, 735

Larynx, stratified squamous epithelium, 354

Laser microprobe analysis, 831 – 832

Laskey, A.:
  frozen section Nissl method, 183
  glycine NaOH buffer, pH 9.4 – 10.6, 880
  maleate buffer series, pH 1.8 – 5.8, 873
  mucihematin formula, 650

Lassek lead method for axons, 778 – 779

Levine and Chargaff's molybdenum blue method for phospholipid, 574

Lead, 531, 546 – 548, 550, 554
  gallamine blue and rhodizonate reactions, 366
  stains for, 547 – 548

Lead acetate stain for sites undergoing mineralization, 801 – 802

Lead citrate staining for EM, 49 – 50

Lead nitrate pH 4.7 method for axons, 778 – 779

Lead rhodizonate demonstration, phosphatase methods, 366

Lead salts in fixation, 48 – 50, 671

Lead soap solubilities, 318

Lead tetraacetate Schiff reaction, 229, 618 – 619

Leblond et al., liquid emulsion coating technics, 815 – 816

Lebrun's Carnoy variant, 52

Lecithin, 560, 589 – 590, 595 – 605

Leishmania, 197, 744 – 745, 748

Leishman's stain for blood films, 747 – 748

Lempert's carbol auramine technic, 737

Lendrum's aniline glycerol, 89

Lendrum's phloxine tartrazine for inclusions, 756

Lens capsule (eye), 615, 785 – 786

Lens fibers, 785 – 786

Lepehne-Pickworth hemoglobin peroxidase method, 447 – 448

Lepra bacilli, 734 – 740

Leschke's method for urea, 289

Leucine aminopeptidase, 433 – 436

Leukocyte granules:
  basic protein stains, 661
  mucosubstance stains, 661

Leukocyte protease, properties, 751

Leukocyte stable sudanophilia, 288, 566, 748 – 751

Leukocytes:
  comparative enzyme activity, 750
  eosinophil, enzymes, 449 – 456

Leukocytes:
  neutrophil, enzymes, 389, 428, 438, 448 – 456

Leukodichloroindophenol, Glavind method for peroxide, 224

Levaditi block method for spirochetes, 760 – 761

Lewis and Lobban's sterol method, 588 – 589

Leydig cells, 349
  fluorescence, 12
  lipofuscin, 349, 513 -- 519

Lhotka's sodium bismuthate Schiff, 234

Liang's SO_2 Schiff for nerve endings, 783

Lieb, E.:
  Abopon mountant, 117
  crystal violet amyloid technic, 667
  phosphotungstic acid hematoxylin variant, 693

Light:
  for microscopy, 1
  in silver reticulum methods, 679

Light green SF, 132, 139, 156
  Masson trichrome stain, 700
  Papanicolaou EA stains, 721 – 724

Lignin, 634 – 635

Lignoceric acid, 562

Lillie, R. D.:
  acetic alcohol formalin (AAF), 34
  acetic anhydride acetic sulfation, 225 – 226
  alcoholic lead nitrate formalin, 48
  allochrome stain, 704 – 705
  azure A eosin B (1963), 195 – 196
  buffered sublimate, 51
  B-5 buffered sublimate formalin, 52
  calcium acetate formalin, 31 – 34
  carbol fuchsin, 735 – 736
  corollary to Michaelis fat stain theory, 565 – 566
  diammine silver method, 277
  diazotization coupling tyrosine reaction, 255
  elastin fibrin stain, 713
  elastin Sudan stain, 713
  Fe^{+++} and Fe^{++} 1964 technic, hemosiderin, 507 – 508
  formic and acetic acid decalcifications, 788
  glycogen in hard tissues, 805 – 806
  Gram stain for sections, 727 – 728
  hydroxyferric ion stain for mucosubstances, 648
  iron hematoxylin method for hypophysis, 345 – 346
  lead nitrate formalin, 48
  Mayer hemalum variants, 205 – 207
  methyl sulfuric methylation, 289
  1944 myelin technic, 601 – 602
  1964 myelin technic, 601, 603
  1,2-naphthoquinone-4-sodium sulfonate (NQS) method for arginine, 291 – 294

Lillie, R. D.:
  peracetic or performic acid Schiff reaction, 221
  periodic acid Schiff method, 230—232
  permanganate-Schiff, 230
  picroformic formalin, 60—61
  postsilvering decalcification for bone and
    callus, 796—797
  predecalcification Kóssa silver for bone,
    796—797
  Prussian and Turnbull's Blue stain for Fe^{+++}
    and Fe^{++}, 507—508
  safranin O-eriocyanin A, 338—339
  salt-Zenker fixative, 56—57
  silver carbonate reticulum technic, 685—692
  silver impregnation method for lipid and cal-
    cification, 797—798
  silver oxide reticulum technic, 683—692
  stable Weigert iron hematoxylin, 201
  sulfuric Nile blue for fatty acids, 317—318, 593
  Van Gieson stain, 694—698
  Wright blood stain, 747—748
  xanthydrol reaction for indoles, 261
Lillie and Ashburn, L. L.:
  Apáthy gum syrup, 119
  isopropanol fat stain technic, 567—568
Lillie and Burtner:
  leukocyte protease, 438
  M-Nadi reaction technic, 451—452
  stable sudanophilia of leukocytes, 566,
    749—750
Lillie, Gilmer, and Welsh:
  black Bauer stain, 732
  black periodic sequence, 310—311, 705
Lillie-Glenner mercurial SH reaction, 240
Lillie and Henson:
  cellulose caprate, 115—116
  1% celloidin paraffin, 94
  sucrose formalin method, 94
Lillie-Laskey Sudan black B Ciaccio method,
  598
Lillie-Mayer alum hematoxylin, 206
Lillie-Pasternack buffered azure eosinate stain,
  195
Lillie and Pizzolato:
  borohydride reduction of aldehydes, 622
  brilliant cresyl blue stain for sulfated mucins,
    644
Linder's hematoxylin method, 693—694
Lipase with thioacetic method, 409
  Tween methods, 407—408
Lipase-lead sulfide stain for triglycerides, 569
Lipid:
  Adams *et al.*, stain for triglycerides, 568—569
  in adrenal, 571, 574—575
  basophilia, 575

Lipid:
  birefringence, 563
  carbohydrate reactions, 575
  classification, 559—562
  EM methods, 607—611
  ethylenes, 575
  extraction, 597
  fixation, 564
  Gaucher, 574—575
  limitations of slide histochemistry, 607
  myelin, 575
  neutral, 565
  Niemann-Pick, 574—575
  physical properties, 563
  propylene glycol (Sudan B or IV), 568
  quantitative histochemistry, 610
  solubilities, 563—564
  Sudan black B for lipid-associated calcifica-
    tion, 570—571
  Sudan black B stains, 569—571
  supersaturated isopropanol stain, 567
  Tay-Sachs, 574, 586
  in tumors, 571
  in urine sediment, 571
  Wigglesworth bound lipid stain, 570
Lipid removal by fixatives, 45—48, 67
Lipids of retina, 784—786
Lipochrome pigment, 513—519
Lipofuscin, 221, 235, 237, 317, 349, 507—508,
  513—519, 592
  acid fastness, 735—736
  combined stains, 507, 521
  ferric ferricyanide reaction, 235
  fluorescence, 12, 349
  Leydig cells, 349, 516
  Mallory fuchsin method, 518
  Nile blue, 513—515
  peracetic Schiff reaction, 221
  peroxide reactions, 223—224
  Sudan black B, 520—521
Lipp's preparation of fluorescent protein, 305
Liquid emulsion coating technics, 815—816
Liquid fatty acids, 592
Lison's zinc leuco method for hemoglobin perox-
  idase, 447
Lissamine Rhodamine B-200, 135, 303—305,
  837—841
Lithium, stains for, 552—553
Lithium carbonate, ferricyanide myelin differen-
  tiator, 576
Liver cells, 196
  enzymes, 403, 408, 429
  lipofuscin, 518—520
Location of focal plane in thin films, 720
Locke's solution, 121
Locus coeruleus neuromelanin, 522

Loeffler's methylene blue, 733–734
Logwood, 152
Longley, J. B., illumination, 4–8
Loose celloidin section staining, 109–110
Loose frozen section staining and mounting, 110–111
Lorch's alkaline phosphatase method, 803–804
Lorch's HCl citrate decalcifying fluid pH 4.4, 793
Love-Rabotti nuclease digestions, 169
Low-temperature defatting in tetrazole methods, 472–473
Low-viscosity nitrocellulose, 92–93
Lower's iodine silver sequence for saccharides, 627
Lowry and Passonneau quantitative enzyme methods, 822–824
Lowry quantitative microchemical methods, 822–824
Ludford's osmic method for Golgi, 333
Lugol effect on pH of azure eosin stain, 196
Lugol's iodine, 153
Lutein, 517
Luxol Fast Blue AR, 131
Luxol Fast Blue G, 145, 156
Luxol Fast Blue MBSN, 131
  myelinated fibers, 599
Lymphocytes, 197, 748, 751
Lysine, 137
Lysochromes, 572
Lysolecithin, 575, 595–605
Lysophosphatides, 560–561
Lysophosphatidylethanolamine, 560–561, 575, 595–605
Lysosomes, 364
Lysozyme:
  identification with Mason *et al.*, peroxidase-immunoglobulin bridge method, 863
  identification with Sternberger PAP stain, 862

Macallum's cobaltinitrite reaction for K⁺, 551–552
  lead sulfide method for sulfate, 556
Macchiavello stain for rickettsiae, 742–743
McClenahan-Vogel metal corrosion mass, 811
McDonald 1950 acid phosphatase method, 359
Maceration of bone for ground sections, 794–795
McGee-Russell's calcium methods, 538–540
McIlvaine Lillie buffer series pH 2.6–7.5, 876
McLetchie's carbacid fuchsin for hypophysis, 344
McManus reaction, 230
  periodic acid method, 230–232, 615–616
  saponification after acetylation, 228–229

McNary's ammine silver for Mn, 550
  dithizone technic for zinc, 554
  method for magnesium, 544
Macrophages, 197
Madder feeding for new-formed bone, 537
Madura foot, 729
Magdala red, 136, 150
Magnesium, 544, 554–555
  microincineration, 866
Mahady's Micromount, 114–115
Malachite green, 131
Malaprade reaction, 229–231
Malaria parasites, 197–198, 747
Malaria pigment, 488–491
Malaria thick film stains, 745
Maleate buffer (Temple), 874
Maleic acid:
  NaOH buffer series pH 1.8–5.8, 873
  pH 5.6, 381
  pH 7.1, 378
Maleimide effect on enzymes, 433
Malic dehydrogenase, 471–480
Mall's vascular injection technic for liver, 809
Mallory, F. B.:
  alum hematoxylin, 206
  aniline blue method: for chromaffin, 265
    for collagen, 699
    for hypophysis, 343
    1905 technic, 699
    1936 technic, 699
    for pancreatic islets, 340
  bleach in silver reticulum stains, 688–689
  crystal violet amyloid technic, 666–667
  eosin methylene blue, 193
  fuchsin stain for pigments, 518
  hematoxylin: for iron and copper, 544–545
    for lead, 547
  hemofuscin stain, 518
  iodine for amyloid, 664
  iodine green for amyloid, 666–667
  iron hematoxylin, 199
  Kaiser glycerol gelatin formula, 119
  phloxine hematoxylin for hyalin, 670
  phloxine methylene blue, 193–194
  phosphotungstic acid hematoxylin (PTAH), 692–693, 699
  thionin for hyalin, 670
  Weigert method for fungi, 729
Malt diastase digestion, 629, 752
Maltose, isomaltose as disaccharase substrates, 376–377
Maltose syrup, 121
Manganese, 531, 550–551, 554
  stains for, 550–551
Mannose-1-phosphatase, 376–377

Mannose-6-phosphatase, 376–377
Mann's methyl blue eosin, 700
  for Negri bodies, 755
  for pituitary cells, 343–344
Mann's osmic sublimate, 53, 62
Mann's tannin picrosublimate, 53
Maraglas, 103, 655
Marchi method for myelin, 217–218, 604
Maresch silver method for reticulum, 684,
  686–687
Marsilid inhibition of enzymes, 466
Marsland *et al.*, axon method for paraffin section,
  778
Masek and Birn's polyvinyl alcohol mountant, 122
Mason *et al.*, immunoglobulin peroxidase bridge
  method, 863
Masson, P.:
  gelatin glue, 683
  iron hematoxylin, 199
  methyl green pyronin, 179
  section argentaffin method, 276–277
  trichrome stain, 700
  Van Gieson trichrome stains, 700
Mast cells, 193, 196, 236, 257, 277–278, 616,
  635–636, 638, 655, 671–678, 714–716,
  737, 742
  acid diazosafranin, 252, 275, 673
  alcian blue-PAS, 676
  basic stains, 672, 674, 676–677
  blocking studies, 672, 674
  carbohydrate content, 672
  cell fractionation studies, 674, 678
  α-chymotrypsin-like enzyme, 677
  degranulation, 674
  development, 673, 675
  esterase, 677
  β-glucosaminidase, 677
  heparin content, 672
  histamine content, 672
  5-hydroxytryptamine, 672–673
  proteases, 426–429
  ruthenium red, 677
  ^{35}S autoradiography, 673
  trypsin-like enzyme, 677
Matukas, colloidal iron for EM, 647
Maule's lignin reaction, 635
Maximow's azure II eosin, 194
Mayer, P.:
  albumen glycerol, 87
  carmalum, 215
  hemalum, 206
  mucicarmine, mucihematein, 650
  paracarmine, 215
Mazia's mercuric bromphenol blue, 286
M and B 1767 (tetrazole), 151

Mechanical stages, 15–16
Mechanical tissue changer, 18
Mechanisms of staining, 123–124
Megakaryocyte granules, 615
Meirowsky reaction of skin, 354
Melanins, 522–527, 631
  argentaffin reaction, 524
  bleaching, 525–526
  formaldehyde fluorescence, 526
  iron uptake, 526
  OsO$_4$ reaction, 217–218
Melanoma melanins, 522–527
Melanosis coli pigment, 503, 508–511
Melibiose, as disaccharase substrate, 386
Menadione, 472–473
Mendel-Bradley method for zinc, 553
Meningeal melanin, 522
Merbromin stain for SH, 240–241
Mercaptide azo coupling SH method, 240
β-Mercaptoethanol decolorization of mercurial
  SH stains, 239–240
Mercurial postfixation of sections, 106
Mercurial sulfhydryl reagents, 239–240
Mercuric bromphenol blue, 286
Mercuric chloride for SH blockade, 238
  reduction by acetone, 51
Mercuric oxide, 531
  ripening of hematoxylin, 204–208
Mercuric salt fixatives, 50–57
  bichromate mixtures, 54–57
  fixation artifacts, iodization, 50, 106, 108–109
  reaction on brain lipids, 589
Mercury, 531, 546–547, 549–550
  stains for, 549–550
Mercury labeled antibodies, 852
Mercury orange SH method, 238–239
Mercury vapor lamps, 10–11, 832–836
Metachromasia:
  mechanism, 636–637
  mucins, 635–636
  proteoglycans, 636–637
  in sections for EM, 637
Metachromatic basophilia, 130–133, 635–636
  fibrinoid, 299–300
  mast cells, 671–678
  mucopolysaccharides, 611–613, 635–638
  sulfation basophilia, 225, 288
Metal ion capture phosphatase methods, 359,
  365–370
Metal-gallocyanin stain for polysaccharides, 654
Metal uptake reactions:
  bile and hematoidin, 496
  Clara's hematoxylin, 203
Metallic impregnations, postbromuration for, 106
Metallic oxides and sulfides, 531

Metallophilia, 124 – 125

Metanil yellow, 138
  picrofuchsin, 700

Methacrylate embedding media, 100

Methanol chloroform fixatives, 47

Methanol dehydration, 80

Methanolysis of sulfate esters, 226, 318 – 325
  after sulfation of amines and alcohols, 289

Methemoglobin, 488

Methenamine silver:
  argentaffin technic, 630 – 631
    axons, 778 – 779
    mast cells, 257
    sulfhydryl and disulfide sites, 245
  Grocott-Gomori technic, 731
  reaction for uric acid, 291

Methionine, 236

Methyl blue, 141, 157
  HCl collagen stains, 702 – 703
  picric acid collagen stains, 694 – 698

Methyl green, 131, 133, 157
  pyronin for nucleic acids, 177 – 180

N-Methyl morpholine (in fixative), 48

Methyl salicylate, clearing agent, 80

Methyl sulfate, sulfuric acid, 319

Methyl violet, 131, 157

Methylation:
  acid fastness, 606 – 607
  acid readicles, 318 – 324
  amines, 320
  cysteic acid residues, 245
  deamination, 320
  demethylation, 245, 320 – 324
  Feulgen reaction, 779
  hematoxylin staining of nuclei, 779
  iron uptake, sulfide method, 203
  lipofuscins, 514, 516 – 517, 519
  melanin, 524
  pepsinogen granules, 349
  silver staining of axons, 779
  sulfonic acid residues, 318 – 324

Methylation saponification of mast cell granules,
    319, 672
  sulfate and carboxyl mucins, 671 – 672

3-Methyl-2-benzothiazolone hydrazone reaction
    for aldehyde, 311

Methylene blue, 130, 132 – 133, 150, 157, 212
  fluorochrome for fat, 572
  supravital for nerve endings, 782

Methylene green, 157

Methylene violet (Bernthsen), 130, 132

Methylene violet RR, 130, 157

Methyltrihydroxyfluorone for nucleic acids, 175

Meyer-Brunot lead method for fatty acids,
    591 – 592

Meyer and Weinman's phosphamidase method,
    380 – 382

Mg and Ca in spodograms, 865 – 866

Michaelis, L.:
  theory of oil soluble dye staining, 565 – 566, 572
  Veronal sodium HCl buffer, pH 7.0 – 9.5, 885

Microchemical assays, 822 – 824
  Lowry methods, 822 – 824
  quantitative principles, 822 – 824

Microglia, 766, 770 – 771

Microincineration, 865 – 867
  iron, 534

Micrometry, 15

Micromount, 114 – 115

Microperoxidase, 458 – 459

Microradiography, 822

Microreaction chamber, 19

Microscope, 2 – 16

Microtome knives, 17

Microtomy, 69 – 73, 87

Millar et al., stripping film method, 816

Millipore filtrates, 723

Millon reaction for tyrosine, 254

Mineral acid extraction of nucleic acids,
    170 – 171

Mineral oil mounting, 117

Mitchell-Wislocki silver method for glycogen,
    628

Mitochondria, 327 – 331, 455
  enzymes, 455, 471 – 480
  tetracycline fluorochroming in vivo, 12

Mitochondrial stains:
  acid fuchsin, 328 – 330, 333
  chrome alum hematoxylin phloxine, 341
  chromotrope 2R, 331
  cytochrome oxidase, 458
  diaminobenzidine, 455
  Janus green, 329
  Luxol Fast Blue MBS, 330
  phosphotungstic acid hematoxylin, 330

Mitoses, 187, 198, 202, 317, 741 – 742

Mixed antibody method (Avrameas), 857 – 860

M. leprae:
  fixation and sectioning, 740
  stains, 737 – 741

M-Nadi oxidase, 451 – 452
  Carbowax effect, 76 – 77
  peroxide reaction, 223 – 224

Model gelatin paper azo coupling, 252

Möller-Regaud fixation, 328 – 329

Möller's fluid, 57 – 59

Molluscum inclusions, 754

Molnar's rhodizonate for lead, 547

Moloney's smear method for mast cell protease,
    428

Molybdate fixation of supravital methylene blue, 783
Molybdenum, 531, 547
Molybdenum blue:
  at collagen sites, 698
  phosphate demonstration, 557
Molybdic acid mordant dyes, 125
Monilia, 724
Monoamine oxidase, 465 – 471
Monocytes, 197, 745, 749
Mordant dyes, 125 – 129, 197 – 215
  nuclei, nucleoproteins, 167, 185 – 186, 197 – 216
Morel-Sisley tyrosine reaction, 255
Morin, 535 – 536, 556
Motor end plates, 780 – 783
  (See also Cholinesterases)
Mounting of paraffin blocks for microtomy, 87
Mounting media, 111 – 117
Mounting technic, 107 – 109
Mowry, R. W.:
  colloidal iron periodic acid Schiff, 646 – 647
  ether H₂SO₄ sulfation technic, 225
MTT, 150 – 151, 469 – 482
  dehydrogenases, 469 – 482
Mucicarmine, 650
Mucihematein, 649 – 650
Mucins, 214, 635 – 636
Mucor, 732 – 733
Mucosubstances:
  blockades, 644
  carbocyanin DBTC, 638
  classification, 611 – 613
  concanavalin A-FITC stain, 648 – 649
  concanavalin A-iron dextran stain (LM and EM), 648
  diamine methods, 644 – 646
  digestions, classification, 612 – 613
  gargoyle cells, 660, 662
  Hale EM method, 647
  Hunter cells, 662
  Hurler's disease, 660, 662
  leukocyte granules, 661
  Lillie *et al.*, hydroxyferric ion stain, 647 – 648
  metachromasia, 635 – 637
  Morquio cells, 662
  platelets, 661
  Sanfilippo cells, 662
  Scheie cells, 662
  stain classification, 613
  "stains all," 638
  Trasylol-FITC stain, 649
  type II glycogenesis, 622
Mukherji *et al.*, bromine silver ethylene reaction, 222

Müller-Mowry colloidal iron method, 646 – 647
Müller's bichromate fluid, 58
  in myelin methods, 603
Murexide reaction, 283
Muscle, 192 – 193, 208, 211, 244, 248, 255, 317, 693 – 705, 710 – 716
  birefringence, 8, 10
  enzymes, 388, 392, 409, 417, 428, 447, 451, 468 – 471
  fluorescence, 12
  lipofuscins, 513 – 519
  negative stain, 829
Muscle adenylic acid (A5′P), 368, 391 – 396
Mycobacteria, 731, 734 – 735
  acid fastness, 606 – 607, 734 – 740
Mycoplasma acetic orcein stain, 744
Myelin, 211, 390, 595 – 605, 737 – 738
  lipids, 595 – 605
  phagocyte lipofuscin, 514
  retina, 784 – 785
  stains, 516 – 520, 572, 596, 599 – 604
Myelin technics for hemoglobin and methemoglobin, 488
Myeloperoxidase, 448 – 452
Myoepithelial cells, 352
Myoglobin peroxidase, 448
Myristoylcholine substrate for cholinesterase, 416
Myxoma mucin, 635

Nachlas *et al.*:
  cytochrome oxidase method, 452 – 453
  leucine aminopeptidase methods, 434
NaCl in spodograms, 866
NAD, NADH, NADP, NADPH, 472
Naidoo's thiamine pyrophosphatase methods, 378 – 379
Nakane, J.:
  peroxidase-labeled antibody method, 863
  stain for thyrotropic cells, 862
NaOH–OTAN stain for sphingomyelin, 579
Naphrax, 114 – 115
Naphthochrome green B, 128
Naphthochrome green G, 128 – 129
α-Naphthol, as arginine reagent, 291
Naphthol AS-LC glucuronide method for β-glucuronidase, 400
Naphthol AS phosphatase (BI, TR, LC, MX, E, CL, AN, KB, BR), 358, 361 – 362, 401 – 402
Naphthol blue black, 141, 157
  HCl collagen stain, 702
  picric acid collagen stains, 696 – 697
Naphthol dyes as fat stains, 145

Naphthol green Y, 128 – 129
α-Naphthol peroxidase method, 451
Naphthol yellow S, 138
    for basic protein, 286 – 287
1,2-Naphthoquinone-4-sodium sulfonate (NQS)
        method for arginine, 291 – 294
α-Naphthyl-N-acetyl-β-glucosaminide,
        396 – 397
α-Naphthylamine diazo, 149, 410
β-Naphthylamine substrates for aminopeptidase,
        434
Naphthyl esterases, 410 – 413
Naphthyl phosphatases, 362
α-Naphthyl phosphate, phosphatase substrate,
        362 – 363
Na₂SO₄ inhibition of enzymes, 442
Native fluorescence:
    of elastin, 715
    of various substances, 12
Natural resins, 112 – 113
Natural ripening, alum hematoxylin, 205 – 208
Nauta-Gygax variant of Glees axon method,
        773 – 774
Navy blue RN, 149
NBT, 151, 469
Necrosis of cytoplasm, 193, 197
Neelsen's carbol fuchsin acid-fast stain,
        735 – 736
Negative staining:
    principles, 829 – 830
    for spirochetes, 760
Negri bodies, 338 – 339, 754 – 756
Neoprene latex corrosion masses, 811
Neotetrazolium (NT), 151, 469
Nerve cathepsin (Adams), 435
Nerve endings, 770 – 784
Neumayer's anthracene blue eosin, 210
Neuraminidase digestion of mucin, 657 – 658
Neurofibrils, nerve fibers, 769 – 784
Neuromelanin, 522
    lipofuscin bicolor method, 521
Neurosecretory substance, 531
Neutral fats, 559 – 560, 565, 596
Neutral formalin, 31, 33
Neutral polysaccharides, 612 – 613
Neutral red, 130, 135, 157
    indicator for KCN solutions, 313
    Schiff reagent, 308 – 309
Neutral stains for basic proteins, 286
Neutral triphosphatase, 384 – 385
Neutralization of KCN solution, 313
Neutrophil leukocytes, 749 – 751
    enzymes, 389, 402 – 406, 428, 438, 448 – 456,
        460, 750
    sudanophilia, 288, 565 – 566, 749 – 750

New fuchsin, 130, 157
    acid-fast stains, 735 – 737, 739
    formaldehyde for mycobacteria, 739
    iron, orcinol elastin stain, 714
New methylene blue N, 130, 157, 540, 753
    reticulocytes, 753 – 754
Newcomer's fluid, 45 – 46
Niagara Blue 4B for hypophyseal beta cells, 346
Nickel, 129, 531, 545 – 547, 554
    stains for, 546 – 547
Nicolas' gelatin section of eyes, 78 – 79
Nicolet-Shinn reaction, 229 – 230
Nicolle's carbol crystal violet, 726
Nicolson and Singer ferritin-concanavalin A con-
        jugates, 850
Nicotinamide adenine nucleotide, 472
    phosphate, 472
Niemann-Pick lipid, 574 – 575
Night blue, 157
    acid-fast stains, 735 – 737
Nigrosin, 139 – 141
    alcohol soluble, 709
Nigrosin base, 709
Nigrosin negative film method for spirochetes,
        760
Nile blue, 130, 157, 317 – 318
    fatty acids, 592 – 593
    pepsinogen granules, 349 – 350
    pigments, 513 – 515, 518
Ninhydrin reaction, 282
Ninhydrin Schiff reaction, 282 – 283
Nissl stains, 182, 196, 211, 214
Nitration of tyrosine residues, 255 – 256
Nitrazol CF, 149
Nitric acid, normal solutions, 888
Nitric acid decalcification, 787, 791
Nitrite in injection masses, 811
Nitro blue tetrazolium (NBT; Nitro BT),
        150 – 151, 432, 469
Nitro BT, 150 – 151, 469
p-Nitroaniline diazo, 248 – 249, 251, 280
p-Nitrobenzeneazoresorcinol, 544
Nitrocellulose embedding, 91 – 94
Nitrocellulose section staining and mounting,
        109 – 110
Nitroneotetrazolium, 150 – 151
3-p-Nitroneotetrazolium (3p-NNT, Nachlas),
        150 – 151
5-m-Nitroneotetrazolium (NNT, Pearson),
        150 – 151
Nitroprusside method for SH, 238
Nitrosation of Paneth granules, 350 – 351
1-Nitroso-2-naphthol, 129
    cobalt reagent, 546
Nitrous acid reactions of amines, 287 – 289

3-*p*-NNT, 150–151
5-*m*-NNT, 150–151
Nonidez's method for neurofibrils and axons, 773
Noradrenalin:
    adrenal medulla, 262–273, 463
    azo coupling, 253, 268
    carotid body tumors, 281
    fluorescence, 269
    glutaraldehyde fixation, 37–38
    osmic reaction, 217–218
    oxidation to pigment, 262–263
Norepinephrine, 217–218, 253, 262–263, 281
Normal acid solutions, 888–889
Norton bromine silver ethylene reaction, 222
NQS arginine method, 291–294
NT, 150–151
Nuclear fast red, 157, 538–539
Nuclear inclusions, vas deferens, 616
Nuclear paraffin artifact, 105
Nuclear stains, 124–134, 165–216, 317,
    692–695
    deoxyribonuclease, 438–439, 441–442
Nucleic acids, 165–216
    cytophotometry, 182
    fluorochroming, 130, 134–136
    methylation and demethylation, 318–324
Nucleohistone methods, 185–186
Nucleoli, 165, 181, 211
Nucleoprotein, Fullmer-Clara hematoxylin, 186
Nucleoside phosphorylase, 371–372
5-Nucleotidase (A-5′-Pase), 368–369, 371
Nucleus dorsalis vagi neuromelanin, 522
Nucleus pulposus mucin, 615
Nyka's methyl violet, metanil yellow for rickett-
    siae, 743

Objectives, microscope, 3–4
Oblique illumination, 8
Ochronosis pigments, 511–513
Ocular melanins, 522–527
Ocular tissue stains, 785–786
Oculars, microscope, 3–4
OG-6, OG-5, OG-8, 721–723
Ogata silver reaction for chromaffin, 262
Ohlmacher's sublimate alcohol, 52
Oil blue N, 147, 567–568
Oil brown D, 145–146
Oil red 4B, 145–146, 565–567
Oil red EGN, 145–146
Oil red O, 145–147, 157, 565, 567–568
    + ferrocyanide for mixed pigments, 507–508,
        514, 517
    resorcin fuchsin combined stain, 711–712
    stable sudanophilia of neutrophils, 748–749

Oil soluble dyes, 144–150
    mounting, 107–109
    staining methods, 565–572
Okajima hemoglobin stain, 486
Okamoto's method for copper, 545
Oligodendroglia, 767, 769
Olivary nucleus lipofuscin, 515
Olsen *et al.*, method for protein-ferritin con-
    jugates, 850
Olsen and Prockop ferritin conjugate method for
    procollagen, 852
Onkocyte stain of Feyrter, 596
Ophüls' Orth fluid, 57–60
Opie and Lavin's acetic alcohol formalin, 34
Optical alignment, 4–7
Orange I, 157
Orange II, 157
Orange IV, 157
Orange G, 157
    Mallory aniline blue stains, 699
    Papanicolaou stains, 721–724
    Shorr stains, 724
Orcein, 144, 157
    acetic orcein, 187
    elastin stains, 709–711
Orcin reaction of inulin, 635
Orcinol, 144
    new fuchsin elastin stain, 714
Oregon fir balsam, 112, 115
Oropharyngeal gland mucins, 615, 635
Orth's fluid, 57–59
Orth's lithium carmine, 214–215
Osmic sublimate, Mann's, 53
Osmium tetroxide:
    brain lipid alterations, 595
    chromaffin, 262
    elastin, 680, 707–708
    ethylenes, 217, 223–224, 594–595, 608
    fixation, 62–67, 594–595
    Golgi substance, 333
    increased stain with *p*-phenylenediamine, 66
    lipid methods, 180–224, 591–592, 594–595
    loss of protein during fixation, 66–67
    nuclear loss of basophilia, 177
    pigment reactions, 217, 513, 519–520
    sterols, 589
    unsaturated fats, 144–145, 217, 222–223,
        592
Osmium vapor fixation of films, 62, 720
OTAN stain (Adams) for phospholipids,
    577–579
Ovary:
    cyst mucin, 615
    enzymes, 397, 429
    follicle mucin, 615

Ovary:
lipofuscin, 235, 513—517, 519, 615
macrophage pigments in corpora hemorr-
hagica, 501, 507—508, 516—517
zona pellucida of ovum, 615
Oxalate, 538—543, 865
Oxalate testing during decalcification, 788
Oxalate-FeCl₃ buffer pH 7.0–1.4, 879
Oxalic acid sodium oxalate buffer series pH
1.4–4.4, 879
Oxazin dyes, 129—132
iron lakes, 211
Oxidation:
of carotenes, 529
of lipofuscin, 513, 517
and reduction of melanin, 524
Oxidation basophilia of lipofuscin, 515
Oxidative dealkylation of methyl esters,
320—321
Oxidative enzymes, 446—482
Oxine, as arginine reagent, 291
Oxone oxidation (potassium monopersulfate), 66
Oxyphil inclusion bodies, 753—759
Oxyphilia, 124—125
fibrin, 297—299
gastric parietal cells, 349—350
guinea pig gastric zymogen granules, 349—350
hemoglobin, 485
Paneth granules, 287, 350—351
zymogen granules, 287, 336—337, 350—351
Oxytalan fibers, 717—718

Padykula-Herman 1955 alkaline phosphatase
method, 367, 377
Palade's buffered osmium, 62
Palladium, 531, 554
Pal's myelin method, 602—603
sulfite oxalic bleach, 603, 689
2-PAM, cholinesterase reactivator, 419—420
Pancreatic acinar cytoplasm, 196, 244—245
duct goblet cells, 615
enzymes, 403, 407—408, 429, 437, 443
islet cell granules, 260, 338—343, 531
zymogen granules, 196, 255, 260, 285, 337,
389, 439, 615
Pancreatic islet stains:
aldehyde fuchsin (insulin), 343
aniline blue-orange G, 340
Bensley's aniline-acid fuchsin-methyl green,
342
Bensley's neutral basic dyes, 337—338
chrome alum hematoxylin-phloxine, 341
N,N-diethylpseudocyanin, 339
eriocyanin A, 338—339
lead hematoxylin, 340

Pancreatic islet stains:
phosphotungstic acid hematoxylin, 339—340
o-phthalaldehyde, 342—343
safranin-acid violet, 337
silver, 341—342
Paneth cells, 350—351
granules, 196, 211, 244, 255, 260, 287, 439,
615
Pannett and Compton's saline, 121
Papanicolaou smear technics, 721—724
Para red, 149
Paracarmine, 214—215
Paraffin:
artifacts, 105
birefringence, 105
Paraffin embedding and sectioning, 80—90
infiltration, 84
ovens, 17—18
section hydration, 106
Paragon mountant, 118
Paramyloid, 667
Paraphenylenediamine stain for osmium fixed tis-
sues, 333
Pararosanilin, 130, 157
acid-fast stains, 735—737
hexazonium salt, 113, 359, 362, 397, 400
Schiff reagents, 306—308
Parathion, cholinesterase inhibitor, 419
Parietal cell reactions, 196, 349—350
Parotid zymogen granules, 196, 255, 260, 336,
615
M-Nadi reaction, 451—452
PAS acriflavine, 652—653
PAS alcian blue, 676
PAS mucicarmine, 659
PAS reaction, 229—234, 310, 612—623,
626—633, 635
PAS sulfone cyanin Gr Ex, 653
Patent blue V, 138
leuco for hemoglobin peroxidase, 447
Pathological glia, 766
Pauly reaction, 246, 259
Pb⁺⁺ enzyme inhibitor, 428—429, 433
Pb⁺⁺ ion demonstration, 126, 128—129,
546—549, 554
gallamine blue reaction, 365
rhodizonate reaction, 365
sulfide reaction, 359, 368
Pearse, A. G. E.:
Al⁺⁺⁺ and Be⁺⁺ method, 534
cyanin R, H₂PO₄ for nuclei, 211
α-D-glucosidase method, 390
orange G, HIO₄ Schiff for hypophysis, 343
Pectin, 634—635
Pectinase, pectinol O, pectinesterase digestion,
630

Pectinase digestion of fibrinoid, 300
Peers' phosphotungstic acid hematoxylin, 693
Penetration of fixatives, 25 – 29
Penfield's variant for oligodendroglia, 767
Penicillin fluorescence, 12
Pepsinogen granules, 255, 260, 317, 350
Peptic digestion of pigments, 518 – 519
Peptic digestion technic, 301
Peptic glands, 349 – 350
  neck cell mucin, 615, 635
Peracetic acid preparation, 220
  azure A for protein sulfur, 245 – 246
  indophenol peroxide reaction, 223 – 224,
    513 – 514, 517 – 518
Peracetic acid Schiff for ethylenes, 221 – 222
  Golgi lipid, 334 – 335
  lipofuscins, 513 – 520
  plant skeletal material, 634 – 635
  retinal lipids, 784 – 785
Perchloric acid extraction of nucleic acids,
  170 – 171
Perdrau-da Fano silver reticulum method, 684,
  686
Perdrau's fluid for postchroming, 78
Performic acid alcian blue for protein sulfur, 246
Performic acid preparation, 220
Performic acid-Schiff reaction, 219 – 223
Periodate ripening of alum hematoxylin, 205
Periodic acid-borohydride-PAS method for acid
  mucopolysaccharides, 654
Periodic acid-formazan, 623 – 625
Periodic acid-Schiff reaction, 229 – 234,
  612 – 623, 751 – 752
  acetic acid solvent, 621
  allochrome method, 704 – 705
  borohydride reduction, 622
  digestions, effect of, 629 – 630
  EM stain, 617 – 618
  glycoprotein stain after electrophoresis, 657
  Luxol Fast Blue MBS method, 599
  periodic acid concentrations, table, 231
  quantitative studies, 619 – 621
  saponification, 621 – 622
  silver reticulum combined method, 690 – 692
  stain for radioautography, 622
  of tissues, 232
    basement membranes, 615, 631
    blood cells, 751 – 752
    collagen, reticulum, 615
    chitins, 633
    chromaffin, 265
    conjunctival mucin, 786
    corneal stroma, 785
    fibrin, 297
    fibrinoid, 300
    fungi, 729

Periodic acid-Schiff reaction:
  of tissues:
    gargoyle cells, 660
    gastric glands, 349
    gastric surface epithelium, 635
    Golgi lipid, 334 – 335
    hypophyseal β cells, 343 – 349
    lipofuscins, 513 – 520
    neuromelanin, 524 – 526
    Paneth granules, 350
    pigments, 502
    retina, 784 – 785
    Russell bodies, 758 – 759
    saccharides, 615 – 616, 660, 667 – 668
    vitreous coagulum, 784 – 785
    zymogen granules, 336 – 337, 350
    Weigert fibrin stain, 298 – 299
Peripheral ganglion pigments, 522
Peripheral nerve myelin, 603
Perls' reaction for Fe^{+++}, 501, 507 – 508, 865
Permanent red R, 145 – 146
Permanganate-oxalic in silver reticulum
  methods, 688
Permanganate oxidation of ethylenes, 219
Permanganate ripening of alum hematoxylin, 205
Permanganate Schiff reaction, 219, 614 – 616
  for glycols, 614 – 616, 634 – 636, 660
Permount, 114 – 115
Peroxidase-labeled antibody methods, 853 – 854,
  857 – 865
  quantitative assay, 863
Peroxidases:
  benzidine peroxidase method, 450
  hemin as ultrastructural tracer, 864
  hemoglobin, 447
  horseradish, 454
  lactoperoxidase, 865
  leukocyte, 449
  myeloperoxidase, 448 – 449
  myoglobulin, 448
  α-naphthol, 451
  quantitative assay, 864 – 865
  submaxillary gland, 456
Peroxidation of fats, 604 – 605
Peroxide von Kóssa for calcium oxalate, 541
Peroxides, 223 – 224
Peroxisomes, 455 – 456
Perrin, T., method for encephalitizoa, 734
Perrin and Littlejohn's fuchsin methylene blue,
  756 – 757
Person and Fine, M- and G-Nadi oxidase reactions, 451 – 452
Peters' protein silver for nerve endings, 781
Peters' protein silver stain for nerve fibers, 781
Petroleum ether, clearing agent, 80 – 81
PFF fixative, Lillie's, 61

pH:
  of aqueous mountants, 117–118, 120
  of azure eosin, fixation effect, 196
  of HgCl₂ solutions, 51–56
  of nitrite solutions, 287
pH 1 basophilia of pigments, 513, 524
pH effect:
  on acid dye stains, 287
  on azure eosin staining, 182–183, 193–197
  on basic dye staining, 182–183, 316
  on basophilia of mucins, 635
  on neutral stains, 286
pH signature of Dempsey et al., 316–317
pH 4 silver reduction, 524
pH standard buffer solutions, 879
Phaeomelanin, 522
Pharyngeal gland mucin, rabbit, 615
Pharynx epithelium, 354
Phase microscopy, 14
  reticulocytes, 753
Phemerol, cholinesterase inhibitor, 418
9,10-phenanthrenequinone reaction for arginine,
  292
Phenazine methosulfate, 157, 386, 472, 477,
  480–481
Phenocyanin TC, 214
Phenols, 246–280
  and aniline in acid-fast staining, 734
Phenosafranin, 130, 132, 135
  Schiff reagent, 308–309
Phenylalanine, xanthoproteic reaction, 283
L-Phenylalanine, intestinal alkaline phosphatase
  inhibitor, 365
p-Phenylenediamine stain for OsO₄ embedded
  sections, 185
Phenylhydrazine blockade of aldehydes, 220,
  312–313
Phenylhydrazine inhibition of enzymes, 466
Pheochrome substance, 530
Phloxine B, 157, 208, 702
  azure A for inclusions, 754
  Mallory's alcoholic hyalin, 670
  methylene blue stain, 193
Phosphamidase, 380–382
Phosphatases, 357–370, 373–374, 376–385,
  402
  bone, 803–804
  inorganic, 382–385
Phosphate buffers in azure eosin stains, 196
  table, 871
Phosphates:
  stains for, 539–540, 557, 573
  Tandler's method, 591
Phosphatidyl esters, 559–561
Phosphatidylcholine, 560, 589–590
Phosphatidylethanolamine, 560–561, 575

Phosphatidylinositides, 560–561
Phosphatidylserine, 560–561, 575, 589–590
Phosphine, 135–136, 145
  fluorochrome, 572
  Schiff reagent, 308–309
Phosphocholine phosphatase, 376–377
Phosphocreatine phosphatase, 376–377
Phosphodiesterase (cyclic 3',5'-nucleotide), 371
Phosphoenolpyruvate phosphatase, 375
Phosphoglucomutase, 374–375
6-Phosphogluconate phosphatase, 376–377
Phosphoglycerides, 572–584, 593
  Adam's OTAN, 577–578
  Baker's acid hematein, 572–573
  choline-containing lipids, 580
  Dunnigan's stain, 576–577
  Elftman's controlled chromation, 574
  EM method, 608
  gold hydroxamic acid, 579–580
  hematoxylin, 578
  Luxol fast blue ARN, 580–581
  phosphomolybdic acid, 574
Phosphoinositides, 560–561, 589–590
Phospholipid nature of Golgi substance,
  334–335
Phospholipids, 559–560, 593
Phosphomolybdic acid, aniline blue methods,
  698–699
Phosphomolybdic acid hematoxylin, 693–694
Phosphomolybdic acid methyl green for bacterial
  cell envelope, 741
Phosphomolybdic acid mordant dyes, 126,
  698–702
Phosphopyridine nucleotide linked dehydrogen-
  ases, 471
Phosphoric acid, 165, 167
Phosphoric acid cyanin R for nuclei, 211
Phosphoric acid residues, 315
Phosphorus, 543
Phosphorylase, 391–395
  in bone, 805
  lead method, 394
Phosphorylation of hydroxyls, 224–225
O-Phosphoserine phosphatase, 376–377
O-Phosphothreonine phosphate, 376–377
Phosphotungstic acid:
  negative stain, 829
  quantitative stain for protein, carbohydrate and
    chondromucoprotein, 824
  stain for carbohydrate, 619, 655–656
Phosphotungstic acid aniline blue methods,
  698–702
Phosphotungstic acid hematoxylin, 693
  fibrin, 297–298
  fibrinoid, 299–300
  mitochondria, 330

Phosphotungstic acid mordant dyes, 698—702
Phosphotungstic orange G, 721—724
Phthalaldehyde method for histamine, 258
Phthalate buffer for pH 4 azure eosin stain, 196
Phthalate-HCl buffer pH 2.2–3.8, 881
Phthalate-NaOH (Clark and Lubs) buffer pH 4.2–6.0, 881
Phthalic anhydride, toluidin blue for OH groups, 227
Phthalic anhydride acylation of hydroxyls, 227
Phthalocyanin dyes, 131, 134
Phthalogen Blue 1B, 134
Phthalogen Blue Black IVN, 134
Phthalogen Brilliant Blue IF3GM, 134
Phthalogen Brilliant Green IFFB, 134
Physiological phosphates, as phosphatase substrates, 376—377
Physiological saline, 121
Picric acid, 60—62, 138, 157
    dye mixtures for collagen stains, 694—698, 704—705
    fixatives, 60—62
    postfixation, 106, 700—701
Picric fixation of methylene blue nerve preparations, 782
Picro-amido black 10B, 697
Picrofuchsin, 696—698
Picro-indigocarmine, 694, 732—733
    + orcein elastin stain, 710—711
Picro-methyl blue, 695, 704
Picro-naphthol blue black, 697
Picrosublimate, 61
Pigment epithelium of retina, 784—785
Pigments, exogenous, 531—532
Pinacyanol, 158, 192—193
    amyloid stain, 192—193, 667—668
    chromatin stain, 192
    frozen section stain, 192—193
    mitochondria, 327
Pipets, 21
Pituitary gland, 343—349, 865
Pizzolato's peroxide silver for calcium oxalate, 541—542
pK changes on nitrosation, 288
Plasma cells, 196—197
    pinacyanol stain, 192—193
Plasma stains, 137—144
Plasmalogens, 220, 581—584
Plasmodia, 197—198, 748
Platelets:
    aminopeptidase, 751
    immunofluorescent stain, 751
    Luxol fast blue stain, 750
    picro-Mallory stain, 751
    thrombostenin in, 750
Platinum, 531, 554

Pneumonyssus foxi pigment, 531
Poirier fluid for degenerating myelin, 218
Polariscopy, 8—10
Polyampholyte embedding media, 99
Polyethylene glycol embedding, 76
Polygalacturonidase digestion, 630
Polyglycol esters (Tweens) esterase substrates, 407
Polysaccharide:
    phthalic anhydride, 227
    sulfation, 225—226
Polystyrene mountants, 110, 116
Polyvinyl acetate, as mountant, 117
Polyvinyl alcohol:
    embedding, 76—78
    as mounting medium, 122
Polyvinyl pyrrolidone in dehydrogenase media, 472
    for mounting sections, 411, 433
Ponceau 2G, 158
Ponceau 2R, 158
    Masson trichrome stain, 700
Ponceau 5R, 158
Ponceau 6R, 158
Ponceau S, 139—140
"Ponceau S," 139—140
Ponceau S (C.I. 15635), 139, 158
Ponceau S (C.I. 27195, disazo), 139—140
Ponceau scarlet (C.I. 16140), 139—140
Ponceau SX (C.I. 14700), 139—140
Porphyrins, fluorescence, 12
Postbromination for silver technics, 107
Postchroming of eyes, 78—79
Postcoupled benzylidene tryptophan reaction, 259
Postcoupled tetrazonium reaction, 247, 346—347
Postfixation of sections, 106
Postmethylation saponification, 228—229, 320
Postmordanting of sections, 106
Potash alcohol saponification demethylation, 320
Potassium, stains for, 551—552
    electron probe analysis, 831
    microincineration, 866
Potassium acid phthalate, 871
Potassium acid tartrate, 871
Potassium chlorate, use in stain for degenerating myelin, 218
Potassium chloride-HCl buffer, 874—875
Potassium cyanide, aldehyde blockade, 313
Potassium dichromate-hematoxylin, oxidation of ethylenes, 217
Potassium dihydrogen and disodium phosphates, 871
Potassium hydroxide alcohol saponification, 228—229, 320

Potassium permanganate:
  oxidation of cystine, 245
  oxidation of ethylenes, 219
  oxidation of glycols, 230, 232
  stain for Golgi, 334
Potassium tetraoxalate, 871
Powers, M.:
  Romanes technic for nerve fibers, 772
  silver stain, 772
  Ungewitter technic for nerve fibers, 772
Preferred common names of diazonium salts, 149
Preoxidants in silver reticulum methods,
  682−683, 685−689
Pretlow and Cassady mast cell granule separation, 674
Preveiro et al., fluorescent method for tryptophan, 260−261
Prevention of formalin pigment, 489
Primary aryl amines:
  and diphenamine Schiff's bases, 309−310
  for Schiff reagents, 308−309
Primulin, 136
  fluorochrome for blood protozoa, 752
Procion Black H-G, 160−161
Procion Blue H-B, 160, 163
Procion Brilliant Blue H-GR, 160, 163
Procion Brilliant Blue MR, 160, 163
Procion Brilliant Red H-3B, 160, 162
Procion Brilliant Red H-7B, 160−161
Procion Brilliant Red M-2B, 160, 162
Procion dyes, 152−153, 160, 162−163
  for mineralization sites, 800−801
Procion Orange Brown HG, 160, 162
Procion Rubine H-2B, 160−161
Procion Rubine M-B, 160−161
Procion Scarlet H-3G, 160−161
Procion Scarlet M-G, 160−161
Procion Turquoise H-G, 160, 163
Procion Violet H-2R, 160, 162
Procion Yellow M-4RS, 164
  stain for nerve cells, 783−784
Procollagen localization in fibroblasts, 852
Proescher and Arkush, iron oxazin stains, 211
Prolactin identification in pituitary with Sternberger PAP method, 862
Proline hydroxylase, EM, 703
Prompt fixation, 25
Prontosil, fluorescence, 12
Propylene glycol fat stains, 568−569
Prostatic mucin, 615, 635
Prostigmine inhibition of cholinesterase,
  416−418, 421, 424
Protargol:
  Bodian copper method for nerve fibers,
  771−772

Protargol:
  Hellweg method for pancreatic α cells, 342
Proteases:
  gelatin, 437−438
  leukocyte, 438
  and peptidases, 426−438
  sperm, 438
  substrate-film localization, 437−438
Protein, conjugates, 847−865
Protein diazotization coupling reaction,
  255−256
Protein moiety of hemosiderin, 501
Protein pK shifts from fixations, 316
Protein silver method for nerve endings, 781
Protein tagging by fluorochromes, 135,
  303−305, 837−841
Proteins, 282−305
Proteolytic action of ribonuclease preparations,
  169
Protosiderin, 503−504
Protozoa, Feulgen reactions, 172
Prussian blue reaction, 507−508, 867
Pseudoplasmal, 519−520
  (See also Direct Schiff reaction)
Pseudomelanosis coli pigment, 503, 508−511
PTAH stain, 692−693, 699
Ptyalin digestion, 629
Puchtler, H., Congo red technic for amyloid, 666
Puchtler et al., Sirius red 4B amyloid stain, 668
Puchtler-Sweat:
  elastin technic with commercial resorcin fuchsin, 712
  hemoglobin stain, 487
  hemosiderin stain, 487
Pugh and Walker:
  N-acetyl-β-glucosaminidase, 396−397, 400
  β-glucuronidase, 396−397, 400
Pure blue B, 128, 534
Purine bases, murexide reaction, 283
Purines, 165−167
Purpurin, 126−127, 129, 157
  for calcium, 538−539
Pus smears, 719−720
PVP, 472
Pyloric gland mucin, 310, 635
Pyridine extraction of lipids, 573
Pyridine solvent acylations of hydroxyls, 227
Pyridoxine-5-phosphate activator for cysteine
  desulfurase, 442
Pyrimidines, 165−167
Pyrogallol reaction with keratohyalin, 354
Pyronins B and G, Y, 130, 132, 158
  Gram stain, 727
  methyl green stain, 177−180
  α-naphthol peroxidase, 451

Pyrrolidinone in dehydrogenase media, 472
Pyruvic dehydrogenase, 477

Q-10, 408, 472
Quinacrine, 12, 135—136, 158, 636
Quinalizarin, 127
Quincke's reaction for iron, 500, 505, 534
Quinine inhibition of lipase, 408, 417—418
Quinoline yellow (C.I. 47005), 139
Quinoline yellow (C.I. 47010), 139

Radioautography (*see* Autoradiography)
Radioimmunoassay, 862
Ramón y Cajal, S.:
   formol ammonium bromide, 34
   gold sublimate method, 767—768
   methyl green pyronin, 178
   picroindigocarmine, 694, 732—733
   uranium silver for Golgi, 331
Ranvier's gold method for motor end plates,
   780—781
Raper cycle, 523
Raw frozen sections for enzymes, 360, 386, 391,
   395, 407, 433, 438
Ray fungi, 729—730
Reagents, 24
Receptors, cell surface: Fc immunoglobin im-
      munohistochemical detection, 844
   IgMEAC or IgGEA immunohistochemical
      assays, 844—846
Red AL, 149
Red B, V, 149
Red G, 149
Red 2G, GG, 149
Red 3GR, 3GL, 3G, 149
Red ITR, 149
Red RC, 149
Red RL, 149
Red TR, TA, TRS, 149
Red violet LB, 149
Redox dyes, 139, 141
Redox stabilized iron chloride hematoxylin, 201
Reduction of Fe^{+++}, 501
   in silver reticulum method, 690
   of SS to SH, 244
Reduction reactions of bile pigments, 494
Reflecting microscope, 13
Reflux extraction, 19
Refractive indices of mounting media, 111, 115
   tables, 115, 120
Refrax, 114—115
Regaud:
   fixing fluid, 57—59
   iron hematoxylin, 197—202
   method for mitochondria, 328

Remazol Brilliant Blue R, 160, 163
Renal epithelium:
   brush border, 615
   iron pigment, 501—508
   protein (hyalin) droplets, 671
Renal hyaline casts, 615, 671
Reserpine effect on enterochromaffin, 274,
   278—279
Resinous mounting media, 111
   table, 115
Resolution criteria, optical, 2—4
Resorcin green, 158, 502, 506
Resorcin-fuchsin, 711—712
Resorcinol crystal violet elastin stain, 711—712
   fuchsin elastin stain (Weigert), 711—712
Restaining of mounted sections, 109
Retention of protein in EDTA decalcification,
   791
Reticulin:
   allochrome, 703—705
   bromine-chloroauric acid-peroxide stain, 692
   bromine-silver stain, 692
   rhodamine B-fluorescein stain, 692
   Sweat's modified allochrome, 705
Reticulin fibers, 681—692
Reticulocyte stains, 752—754
Reticulum, 211, 312, 615, 681—692
Retina, 784—785
   fuscin or melanin, 522
   rod acromere lipid, 221, 615
Reversal of cyanide blockade of aldehydes, 313
Rhenohistol, 114—115
Rhodamine B, 135—136, 150, 832, 835,
   837—840
   fluorochrome for connective tissue, 692
   isocyanate, isothiocyanate, 839
   mitochondria, 327
Rhodamine 3G, 135—136
   Schiff reagent, 308—309
Rhodindine, 135—136
Rhodin's dextran osmic, 62
Rhodium, 531
Rhodizonate, di-K, 129, 537, 547
Rhodoflavin, 529
Rhodopsin, 785
Riboflavin, 530
   fluorescence, 12
Ribonuclease (RNase, RNAase), 168—169, 212,
   350
   picric acid fixation inhibition, 60
   substrate film localization technic, 437,
      439—440
Ribonucleic acid (RNA), 168—182, 193—197
   acid extraction, 168—169
   alcoholic KOH extraction, 228, 320

Ribonucleoprotein (RNP), 165 – 167, 185 – 186, 196

D-Ribose, 165 – 166

Ribose-5-phosphatase, 376 – 377

Richardson methylene blue vital stain for autonomic nerves, 783

Richman, Gelfand and Hill's decalcifying fluid, 788, 792

Rickets lesions, preservation in decalcification, 789

Rickettsiae, 193 – 196, 742 – 744

Rigolac embedding medium, 101

Rinehart-Abul Haj colloidal iron Van Gieson, 646 – 647

Rinehart's supersaturated ethanol fat stains, 568

Ringer's solution, 121

Ripening of hematoxylin, 198, 203 – 208

Rivanol, 136
    fluorochrome for blood protozoa, 752

RNA, RNP, 165

RNase, RNAase, 168 – 169

Robin's gelatin injection masses, 810

Rogers, G.E., method for citrulline, 295 – 296

Romanes' silver:
    for dentin tubules, 772 – 773, 796
    nerve fibers, ammonia AgCl method, 772

Romanovsky stains, 744 – 748
    resin mounts, 112 – 117

Romeis, B.:
    40% alcohol Sudan method, 567
    lanolin rosin, 110
    orcein Van Gieson stain, 711
    Orth fluid, 57 – 59

Romieu reaction for tryptophan, 259

Rosanilin, 158
    acid-fast stains, 735 – 737
    Schiff base reactions, 311
    Schiff reagents, 305 – 309

Rose Bengal, 158
    tartrazine inclusion stain, 756

Rosette assay:
    for IgMEAC or IgGEA for cell suspensions, 845 – 846
    for IgMEAC or IgGEA for sections, 844 – 845

Rosin, 113

Rosindole reaction for indoles, 260

Rossman-Casella reaction, 230, 232, 615 – 617, 633, 635

Rossman's alcoholic picroformalin, 61

Roy's aryl sulfatase variants, 402 – 403

Rubeanic acid, 158
    cobalt reaction, 546
    copper reaction, 545

Russell bodies, 211, 615, 758 – 759

Russell's fuchsin, iodine green for inclusions, 758

Rutenburg, A. M., et al,:
    β-D-galactosidase method, 388 – 389
    α-D-glucosidase, 389

Rutenburg, Cohen and Seligman, aryl sulfatase, 402

Ruthenium red, 40, 642 – 643, 663, 677, 707

Ryder-Poppen cobaltinitrite technic, 551

S acid, 248, 255

Saathoff's methyl green pyronin, 179

Safety razor blades, 17

Saffron, 158

Safranin O, A, T, 129 – 132, 135
    acid violet, Bensley's, 337
    diazo, 247 – 248, 250, 254, 516, 519
        pH3, bilirubin, murine mast cells, 498, 673
    eriocyanine A, 338 – 339
    ferricyanide precipitate, 132, 507
    iron hematoxylin, fast green stain, 658
    mucin stain, 635
    nuclear stain, 117
    Schiff reagent, 308 – 309

Safranin B extra, 132

Sakaguchi reaction for arginine, 291

Salicylate (Na) fluorescence, 12

Salicylhydrazide method for COOH, 324 – 325

Salicylic acid balsam, 107, 112, 700

Saline formalin, 32 – 33

Saliva digestion, 629

Salivary glands:
    enzymes, 397, 429, 436
    mucins, 615, 635, 657 – 658
    secretory duct granules, 260
    zymogen granules, 336 – 337

Salpeter and Bachmann flat substrate autoradiographic quantitative EM method, 821

Salt:
    in acetic acid fixatives, 57
    in aqueous mountants, 118 – 119

Sandwich (immunoglobulin enzyme bridge) method, 863

Sanfelice's chromic acid acetic formol, 59

Saponification:
    of acetylated amines, 228, 288 – 289
    acetylated glycols, 228
    methylated carboxylic acids, 320 – 324
    methylated lipofuscins, 514, 519
    methylated nucleic acids, 318 – 319
    methylated pepsinogen granules, 349
    methylated sulfonic acids, 319
    removal of RNA, 228

Sarcocystis polysaccharide, 615

Sarcosporidia, 197 – 198

Sarin, cholinesterase inhibitor, 419

Sato's benzidine peroxidase method, 450

Scandium, 535, 556
Scanning electron microscopy, 825–827
Scarlet GG, 149
Scarlet R (stable diazo), 149
Scarlet R (Sudan IV), 145–146, 565–566, 568
Scarpelli *et al.*, MTT Co^{++} for dehydrogenases, 476–477
Schaeffer-Fulton spore stains, 740
Schajowicz and Cabrini:
  acid phosphatase in bone, 804
  alkaline phosphatase in bone, 803–804
Scharlach R, 565
Schaudinn's sublimate alcohol, 51
Schedules for reticulum technics, 685–688
Schiff's base reaction of aldehydes, 309–310
  with basic dyes, 166, 316
Schiff's sulfite leucofuchsin reagent, 306–308
  "Cold Schiff," 307
  other dyes and fluorochromes, 308–309
  fluorochrome reagents, 308–309
  for peracetic acid reaction, 220
  for performic acid reaction, 220
  for periodic acid reaction, 612–616
  Spicer's, 175
  traditional reagent, 306–307
Schistosoma egg shells, 615
Schleifstein's Negri body stain, 755
Schmorl, G.:
  Boehmer's hematoxylin, 205–206
  formalin formic acid, 792
  thionin for teeth and bone matrix, 796
Schüffner's granules, 745
Schultz method for urates, 290
  carmine solution, 290
Schultz's Liebermann Burchardt sterol reaction, 588
Scleral collagen, 615
Scott's microincineration procedure, 865
Scurvy lesions, preservation in decalcification, 789
Sealing of fluid mounts, 110
Sebaceous glands, 244, 355
Secondary aryl amines and diphenamine Schiff's bases, 309–310
Section flotation, 87
Section freeze substitution, 47
  for enzymes, 360, 409, 433, 443
Sectioning:
  Adamstone-Taylor, 71
  freehand, 69
  fresh tissue slicers, 69–70
  frozen, 70–71
  frozen-ultrathin for EM, 72–73
  in paraffin, 84–90
Sehrt, fatty peroxide M-Nadi reaction, 223
Seligman-Ashbel carbonyl lipid technic, 605

Seligman *et al.*:
  methods for β-glucuronidase, 399
  OTO method, 64–66
Semicarbazide blockade of aldehyde, 220, 312
Seminal vesicle:
  glutamyl transpeptidase, 429
  lipofuscins, 513, 516, 519
  mucins, 615
Semisynthetic resins, 113, 115
Sen's urease method, 445
Sensitizers in silver oxide reticulum methods, 689
Sequence iron hematoxylins, 197–201
Serine in phospholipids, 560–561
Serra and Queiroz-Lopes method for phosphates, 557
Sertoli cells, 616
Serum, as adhesive for sections, 87–88
Serum phenol models, 252
SH blockade reactions, 235, 238, 241, 243–244
Shackleford's phthalic anhydride acylation, 227
Shimizu and Kumamoto, lead tetraacetate Schiff, 233
Shnitka and Seligman method for inhibitor sensitive and resistant esterases, 411
Shorr stains, 723–724
Sialic acid in gangliosides, 560, 586
Sialic acid stain, 586, 658
Sialidase digestion, 643, 657–658
Sialomucins, 317, 657–658
Siderophilia, 125
Silica:
  birefringence, 9
  microincineration, 866
Silicates and silica in spodograms, 865–866
Silver, 531–532, 544, 546–548, 550, 554–555
  stains for, 544
Silver methods for spirochetes, 740, 760–761
Silver nitrate acetic acid primary fixation:
  ascorbic acid, 580
  bilirubin, 495
Silver oxide reticulum technic, 691
Silver solutions, for reticulum technics, 682–692
Silver technics (*see* Amyloid; Argentaffin reaction; Argyrophilia; Glia methods; Golgi substance; Reticulum)
Silvering after iodine stains, 627
Silver-PAS (Elftman), 616
Single-cell layer preparation, 813
Siphon tissue washer, 56
Sirius light blue F3R, 158
Sirius light blue G, 158
Sirius red 4B, 158
Skin enzymes, 442, 461–462
Slide carriers, 20–22
Slide cleaning, 22

Slider and Downey's methyl green pyronin, 179
Smear and film preparation, 719–721
Smith-Dietrich-Baker method for myelin, 219
Smith-Dietrich reaction:
  for Golgi, 334–335
  for phospholipids in myelin, 600–601
Smooth muscle:
  birefringence, 10
  lipofuscin, 513, 518
Sodium:
  electron probe analysis, 831
  microincineration, 866
Sodium acetate lead nitrate, 48–49
Sodium acetate sublimate (B-4), 51
Sodium acetate sublimate formalin (B-5), 52
Sodium bismuthate Schiff reaction, 229, 234
Sodium dithionite (hydrosulfite):
  extraction of iron, 503
  inhibition of enzymes, 462
  reduction of fuchsin for Schiff reagent, 306
Sodium iodate, 205–208, 263
Sodium plumbite SH and SS reaction, 234, 244, 354
Sodium stains, 543
Sodium sulfide, buffered, 366, 416–417, 505–506
Sodium thiosulfate for Schiff reagent, 306
Solcia:
  endocrine granules staining, 335–336
  GPA fixative, 61
Solochrome Azurine BS, 128, 534
Solochrome Cyanine R, 128
Solubilities:
  of cholesterol, 597
  of dyes, table, 154–159
  of Golgi lipid, 334
Solutions:
  Locke-Lewis, 121
  Pannett and Compton buffered saline, 121
  physiologic saline, 121
  Ringer's, 121
  Tyrode's, 121
Solvent resistance of lipofuscins, 514
Solvent table, 23
Sorbitol gelatin, 120–121
Sörensen's borate buffer pH 7.6–11.0, 883
  glycine-HCl buffer pH 2.2–3.6, 881
  glycine NaCl NaOH buffer pH 8.8–12.8, 880
  phosphates, buffer series pH 5.3–7.9, 878
Soybean trypsin inhibitor, 429
Specific gravity of acids and ammonia, 888–889
Specimen bottles, 24
Spectra of Wright stains, 747
Spectroscopic absorption of hemoglobin at 411 mμ, 486

Spectroscopy absorption of tissues, 13
Spermatozoa, 349
  acrosomes and headcaps, 287, 615
Sphingolipids, 560–561, 584–588
Sphingomyelin, 560, 562, 564, 578–579, 584–588
Sphingosine, 560–561
Spicer, S.S.:
  Bouin-Feulgen-azure A method, 175
  classification of mucosubstances, 611–613
  cytosiderin in rodents, 503
  high-iron diamine-Alcian blue stain, 646
  immunoglobulin enzyme bridge (sandwich) method, 863
  low-iron diamine-Alcian blue stain, 645–646
  "mild methylation" technic, 319
  mixed diamine stain, 644–645
  Schiff reagent, 175
  sulfuric acetic sulfation reagent, 226
Spicer-Lillie method for basic protein, 287
Spielmeyer's hematoxylin, 603
Spielmeyer's myelin method, 603
Spinal cord lipofuscin, 515
Spirit blue, 133, 147, 158
Spirochetes, 740, 759–761
Spleen, enzymes, 397
Spodography, 865–867
Spore stains, 740–741
Sporotrichum schenckii, 730
Spuler's acetic sublimate formalin, 53
  formalin Zenker fluid, 55
Squash chromosome preparation, 187–188
Sr^{++} ion demonstration, 123, 537, 547
Stability of Zenker mixtures, 53–54
Stable sudanophilia, 288, 566, 749–751
Stain solubilities, table, 154–159
Staining dishes, 20
Staining racks, 20–22
Staining sinks, 21
Stannous chloride method for gold, 549
Starch, 310–311, 615, 633–634
  amylase digestion, 392, 395–396, 629
  halogenation, 222
Starch film method for amylase localization, 437–439
Static electricity during paraffin sectioning, 89
Steedman:
  Alcian Blue, 638
  ester waxes, 90–91
  triacid stain for oxygenated red blood cells, 746–747
Stein's iodine test for bilirubin, 618
Sternberger PAP method, 460–461, 860–865
Stern's microglia and oligodendroglia method, 766

Stern's rhodopsin method, 785
Steroid solubility in Carbowax, 76
Sterols, cholesterol, 588—590
    in arterial intimal hyalin, 671
Stieve's acetic sublimate formalin, 53
Stirling's aniline crystal violet, 726
Stomach, enzymes, 397, 443
Storage:
    of paraffin blocks, 90
    of stained malaria smears, 745—746
Stovall-Black Negri body method, 754
Stoward:
    coriphosphine-thiazol yellow stain for sul-
        fomucins, 651
    ferric alum coriphosphine stain for sul-
        fomucins, 651
    Feulgen metadiamine-coriphosphine stain for
        sulfomucins, 650—651
    fluorescent stain for carbohydrate, 625—626
    lithium aluminum hydride reduction stain for
        sulfomucins, 651—653
    periodic acid-formazan stain for carbohydrate,
        623—625
Strandin, 560—561
Stratum lucidum, 287, 352—355
Streptothrix, 732
Striated border, intestinal epithelium, 310, 615
Striated muscle, birefringence, 8—10
Stripping film technic, 816
Stroma of muscle, 310
Strontium, 129, 537, 547
    ion in NQS arginine method, 291—294
Stumpf and Roth autoradiographic method,
    817—818
Sublimate, aqueous, alcoholic, 51
Sublimate acetic picric fluid (vom Rath), 61
Sublimate alcohol formalin, Einarson's, 212
Substantia nigra pigment, 522
Substrate film methods for enzymes, 436—441,
    459—460
Succinic acid-NaOH buffer pH 3.8–6.0, 877
Succinic dehydrogenase, 471—479
Sucrose:
    in dehydrogenase media, 472
    as disaccharase substrate, 386
    formalin, 33
    glycerol gelatin, 120
    osmic fixatives, 62
    syrup mounting medium, 118—120
Sudan I (orange R), 158
Sudan II, 145—146, 158, 567
    +Weigert myelin stain, 596
Sudan III, 146, 158, 565—569
Sudan IV, 146, 158, 566—567
    IV, neutrophil leukocytes, 748—751

Sudan black B, 146, 158, 566, 569—571
    Ciaccio procedures, 520—521, 597—599
    Golgi lipid, 334—335
    leukocytes and erythrocytes, 749—750
    lipofuscin, 513—521
    peptic gland parietal cells, 349—350
Sudan blue, 147
Sudan blue BZL, 158
Sudan brown, 145—146, 567
Sudan brown 5B, 158
Sudan Corinth 3B, 145—146
Sudan green, 147
Sudan R, 145—146
Sudan red (E. Gurr), 145—146
Sudan red 4BA, 145—146
Sudan red VII-B, 145—146, 158
Sudan violet, 147
Sudanophil droplets in gargoyle cells, 660
Sudanophil granules of leukocytes, 748—750
Sudanophilia, 125, 565
    arterial intimal hyalin, 671
    chromated fats, 585—586
    mycobacteria, 735
    pigments, 507—509, 513—521
Sugar content of hemosiderin, 502
Sulfa drug fluorescences, 12
Sulfanilic acid, diazo, 248—249
Sulfatases, 402—406
Sulfates, 556, 865
    stains for, 556
Sulfatides, 585—586
Sulfation, 225—226
    of amines, 226, 289
    blockade by methylation, 226
    blockade of periodic acid Schiff, 225—226
    hydroxyls, 225
    pH 1.0 basophilia of sulfation sites, 226
Sulfhemoglobin, 488
Sulfhydryl reactions, 236—244
Sulfide inhibition of enzymes, 453, 462
Sulfide method:
    for lead, 547
    for mercury, 549—550
Sulfide reaction for iron, 501—505
Sulfides in iron immobilization during decalcifica-
    tion, 790—791
Sulfite blockade of aldehydes, 220—221, 312,
    705
p-Sulfodiazobenzene coupling, 252
    demonstration of histidine, 252
Sulfomucins, 650—653
Sulfonamide inhibition of enzymes, 443
Sulfone cyanin GR, 159
Sulfonic acid methylation and demethylation,
    318

Sulforhodamine B, 134–136, 304–305, 837–841
    protein conjugation, 837–841, 847–865
Sulfur in hair cortex, 352–355
Sulfur amino acids, 234–235
    radioautography, 815
Sulfuric acid reaction:
    of carotenes, 529
    of vitamin A, 529
Sulfuric acid residues, 315
Sulfuric acid-sodium phosphate buffer, 872
Sulfuric Nile blue technic for fatty acids, 317–318
Sulfurous acid decalcification, 317–319
Sulfurous acid Schiff method for nerve endings, 783
Supersaturated isopropanol stain for lipid, 567–568
Supravital stains:
    methylene blue for nerve endings, 782
    for mitochondria, 327–328
Surviving cell study media, 121–122
Susa fixative, 50, 52
Swank-Davenport method for degenerating myelin, 218
Sweat gland lipofuscin, 514
    fluorescence, 12
Sylvén's lead fixatives, 48
Synthetic resins, 114–117

Tabun, cholinesterase inhibitor, 419
Taenzer-Unna elastin stain, 709–710
Takeuchi, T.:
    amylophosphorylase branching enzyme, 391–395
    amylose and glycogen stains, 391–395
    iodine glycerol mounting medium, 391
    UDPG glycogen synthesis, 395–396
Tandler's lead sulfide methods for fatty acids, 591
Tanned chitin, 633
Tannenberg's dioxane paraffin schedule, 80
Tannic acid:
    fixative with glutaraldehyde, 40
    fixative with mercury, 53
Tannin, OsO_4 reaction, 217
Tannin picrosublimate, 53
Tantalum, 531
d-Tartrate inhibition of acid phosphatase, 373
Tartrazine, 159
Taurocholate:
    activation of pancreatic lipase, 408, 418
    inhibition of enzymes, 408, 418, 428

Tautomerism of picric acid, 137
Tautomers of tetrazoles and formazans, 150–151
Technicon dehydration paraffin schedules, 83, 85
Technicon resin, 115
Technicon tissue changers, 18
Tellurite method for succinic dehydrogenase, 479
Tellyesniczky's acetic alcohol formalin, 34
Tellyesniczky's acetic bichromate fixative, 59
Temperature:
    and $HgCl_2$ saturation, 51
    in silver reticulum methods, 682–683
TEPP, cholinesterase inhibitor, 419
Terry's polychrome methylene blue, 69
    surface block stain, 69
Testis, 349
    enzymes, 397, 429
    fluorescence, 12
    lipofuscins, 513–516
Tetrabutylammonium iodide, trypsin inhibitor, 428
Tetracycline, mordant fluorochrome, 135–136, 159
    calcium (*in vivo*), 13, 538, 802–803
    mitochondria, 13
Tetraethyl pyrophosphate, AChE inhibitor, 419
Tetraethylsafranin, 132
Tetrahydrofolate reductase, 480
Tetramethylrhodamine isothiocyanate (TRITC), 839
Tetranitro blue tetrazolium (TNBT), 150–151, 468–471
Tetranitromethane:
    tryptophan blockade, 256
    tyrosine blockade, 256
Tetranitroneotetrazolium (TNNT), 150–151, 468–471
Tetrazoles, 150–152, 469–482
Tetrazolium formazan reaction, 151, 468–471
Tetrazolium salts:
    desirable characteristics, 469
    osmates, 469, 471
    quantitative histochemistry, 469
    tables, 151, 469
Tetrazolium SH method, chitin reaction, 633
Tetrazolium violet (TV), 150–151
Tetzloff *et al.*, stain for axis cylinders and neurilemma, 784
Thalamic lipofuscin, 515
Thallium, 531
    stains for, 553
Thiamine diphosphatase, 378–380
Thiamine pyrophosphatase (diphosphatase), 377–380

Thiazin dyes, 129–132
Thiazin red R, 136, 138, 145
Thiazol yellow G, 136, 145, 159
Thick blood film preparation, 720
Thin blood films, 720
Thioacetic esterase, 408–410
Thiocarbohydrazide, 401, 430–431
Thiochrome (oxidized thiamine) fluorescence, 13
Thioflavine S, 136
Thioflavine TCN, 159
    fluorochroming of amyloid, 668
Thioglycolic acid anilide for Co^{++}, 546
Thiol decolorization of mercurial SH stains, 239, 244
Thionin, 130, 159
    hyalin, 670
    mucins, 635
    Schiff reagent, 308, 740
    teeth and bone matrix, 796
Thionyl chloride:
    for methylation, 319–320
    for Schiff reagents, 308, 740
Thomas' Ciaccio procedure, 596
Thomas' phosphomolybdic acid hematoxylin, 693–694
Thorium, 531, 535, 556
    stains for, 556
Thymine, 165–166
Thymonucleic acid, 165
Thymus, 196, 397
Thymus deoxyribonucleic acid substrate for DNase II, 441–442
Thyroid:
    colloid, 196, 236–237, 255, 285, 615
    enzymes, 396, 429, 451
Thyroxin I^{131}, 813 ff.
Tigroid stains, 182, 211
Timm's method for lead, 547
Timm's silver method for metals, 550, 554
Tin, 531, 546, 554
Tirmann-Schmelzer iron reaction, 505
Titan yellow, 135, 145, 544
Titanium, 535, 556
Tjio and Whang chromosome stains, 187–188
TMB-4 cholinesterase reactivator, 417, 420
TNBT, 150–151, 468–471
TNNT, 150–151
Toluene, as clearing agent, 81
p-Toluenesulfonyl chloride acylation, 228
p-Toluidine blockade of aldehydes, 313–314
Toluidine blue, 130, 154
    mucins, 635
    after phthalic anhydride, 227
    stain for DNA, RNA, 173

Toluidine blue:
    after sulfation, 225–226
Tosyl-L-arginine methyl ester, trypsin inhibitor, 428
Tosylation, 228
    enterochromaffin, 228
    glycols, 228
"Total lipid extraction" of brain lipids, 597
Toxoplasma gandii, 732
    sp., polysaccharide, 615
TPN, TPNH, 468–471
TPNH diaphorase, 468–471
TPT, TTC, 150–151
Tracheal gland mucin, 615
Tracheal perfusion fixation, 26
Trehalose as disaccharase substrate, 386
Tremblay's amylase localization by starch film method, 439
Trevan and Sharrock's methyl green pyronin, 179
Triazinyl dyes, 152, 161–163
Trichinae, 761–763
Trichloroacetic acid:
    decalcification, 789
    fixation for sulfhydryl, 234–235
    fixative, 52–53
    inhibition of monoamine oxidase, 466
Trichohyalin, 203, 288, 296, 352–356, 461–462
Trichomonas, 725–726
Trichoxanthin, 522, 524
Triglycerides, 559
Trihydroxymethylaminomethane HCl buffer, 883
Trimetaphosphatase, 382–385
Trimethylbenzene as resin solvent, 111, 116
Triphenylmethane dyes, 130–133
Triphenyltetrazolium (TTC, TPT), 150–151
Triphosphate-HCl buffer pH 9.5–12.4, 887
Triphosphates, monohydrogen phosphates in spodograms, 865
Triphosphopyridine nucleotide, 468–471
Triphosphopyridine nucleotide phosphatase, 376–377
Tris, tris-HCl buffer series pH 7.5–8.7, 883
Tris(1-aziridinyl)phosphine oxide fixative, 40
Tris maleate buffer series pH 5.1–8.4, 882
Tritium labeled nuclear constituents, 165
Tropaeolin O, 159
Trophonuclei, 197–198, 748
Tropical ester wax, 90
    sectioning undecalcified bone, alkaline phosphatase, 804
Trypan blue, 142, 159
    amyloid stain, 666–667

Trypan red, 142–143, 159
Trypanosoma cruzi, 197–198
Trypanosomes, 744, 748
Trypsin, 426–429
Trypsin-like enzyme of mast cells, 428–429, 433
Tryptic digestion technic, 300–301
  collagen and fibrin, 298
  fibrinoid, 298, 300
Tryptophan, 259–261, 283
  amyloid, 664
  azo coupling, 259–261
  fibrin, 259–261
  hypophyseal β cells, 346
  pancreatic zymogen and α cells, 342
  Paneth granules, 350–351
  pepsinogen granules, 349–350
  salivary zymogen and duct cell granules, 336–337
  skin, 352–355
TTC, TPT, 150–151
Tubercle bacilli, 606, 734–740
Tumor cells in sputum, 757
Tungsten, 531
Turchini's method for nucleic acids, 176–177
Turgitol-7 for Carbowax section flotation, 77
Turnbull's blue reaction, 506–507
TV, 150–151
Tween 80 and 85 for lipase, 406
Tween esterase and lipase methods, 407
Two emulsion procedure for H^3 and C^{14}, 816
Tyrode solution, 121
Tyrosinase, 461–462
Tyrosine, 254–256
  aminotransferase, 432
  diazotization coupling reaction, 255–256
  hair and skin, 255
  hydroxylase, 464
  hypophyseal α cells, 346
  mast cells, 257
  Millon reaction, 254
  Morel-Sisley reaction, 255–256
  nitration of, in histidine azo coupling, 256
  Paneth granules, 255, 350–351
  pepsinogen granules, 349–350
  vascular elastic tissue, 255
  xanthoproteic reaction, 283
  zymogen granules, 255, 336–337, 349–351

UDP, UDPG, UTP, 395–396
Ultramarine, 531
Ultraviolet absorption by nucleic acids, 167
Ultraviolet absorption spectroscopy, 13

Umbilical cord stroma mucin, 615
Undecalcified sections:
  chromoxane cyanin R stain, 795
  EM preparation, 795
  preparation, 795
Undeparaffinized sections, 106
Unfixed tissues, 68
Unlabeled antibody enzyme (PAP) method of Sternberger, 860–862
Unna's P.G. methyl green pyronin, 179
  Van Gieson variant, 694–695
Unstained sections, 123
Uracil, 165–166
Uranium, 531, 547, 556
  stains for, 556
Uranium silver method for Golgi, 331
Uranyl acetate:
  reaction with nuclei, 830
  thiamine pyrophosphatase inhibitor, 380
Uranyl acid formalin bichromate, 60
Uranyl formalin fixation, 331
Urates, 290
Urea, 289–290
  in Morel-Sisley reaction, 256
Urea denaturation of fibrinoid, 300
Urea formaldehyde embedding medium, 99
Urease, 445
Uric acid, 290–291
Uricase, 445–446
Uridine-2 C^{14} radioautography, 865
Uridine diphosphate (UDP), 395
Uridine diphosphate glucose glycogen transferase, 395–396
Uridine triphosphate UTP, 395
Urobilinoid stains, 500–501
Uronic acid mucins, 659
Uterus, enzymes, 396, 433

Vacuum:
  during decalcification, 790
  desiccation, 68
  freeze dry for radioautography, 817–822
  oven, 17–18
  paraffin infiltration, 85
Vaginal epithelial mucin (rodent), 615, 657–658
Vanadatomolybdate negative stain, 829–830
Vanadium, 531
  V^{+++} enzyme inhibition, 443
Vanadyl sulfate negative stain, 829
van Duijn, P.:
  acrolein fixation, 42–43
  acrolein Schiff reaction, 42–43
  benzidine H_2O_2 solution, 448
  thionin Schiff reagent, 308

Van Gieson type stains, 137, 694 – 698
  elastin stains, 709 – 711
  mucins, 704 – 705
  reticulum methods, 691
Van Gieson's picrofuchsin, 694 – 698, 795
  acylation blockades, 698
  decalcification effect, 788, 794 – 795
  mounting media for, 107 – 108, 112, 114 – 115
  nitrous acid resistance, 698
  sulfation blockade, 226
Van Slyke's nitrous acid mixture, 287 – 289
Variamine blue B, BA, BD, 149
Varicella inclusions, 754
Variola inclusions, 754 – 756
Vascular injections, 809 – 813
Vascular perfusion fixation, 26
Vas deferens nuclear inclusions, 615
Vassar and Culling:
  elastin fluorescence, 715
  fluorochrome: for amyloid, 668
    for connective tissue, 715
Verdoperoxidase:
  Carbowax effect, 77
  demonstration technics, 448 – 452
Verhoeff's carbol fuchsin, 735 – 736
  elastin stain, 708
Veronal sodium-HCl buffer pH 6.4–9.7, 885
Versene decalcification, 477, 791, 793 – 794,
    803, 805
  decalcification for enzymes, 477, 803 – 804
  enzyme activity alteration, 428 – 429
  iron resistance to Versene, 790 – 791
Vestopal embedding medium, 101
Vibrio cholerae sialidase, 657 – 658
Victoria blue B, 130, 159
Victoria blue R, 130, 159, 737
Victoria blue 4R, 130, 159
Villamil and Mancini's benzidine peroxidase for
    tissue, 450
Villus epithelium, striated border, 310 – 311, 615
Villus melanosis pigment, 508 – 509
V+++ inhibition of enzymes, 443
Violamine R, 138
  picric acid collagen stains, 695, 698
Virus:
  immunoferritin stain, 852 – 853
  mercury-antibody conjugate, 852
  negative stain, 829
Viscol, Amann's, 118
Visual purple, 785
Vital new red, 141 – 142
Vital red,141 – 142, 159
  amyloid stain, 667
Vitamin A, 529
  fluorescence, 12

Vitamin B, 485
Vitamin C, 530 – 531
Vitreous humor, 615
  hyaluronic acid, 659
  staining reactions, 789
Vivianite, 531
Voigt-Timm silver method for mercury,
    549 – 550
Voisenet-Furth reaction for tryptophan, 259
Volkman-Straus elastin azan technic, 713
Voltage, lamp bulbs, 1
Vom Rath's acetic picrosublimate, 61
von Bertalanffy and Bickis, acridine orange,
    181 – 182
Von Ebner's decalcifying fluid, 792, 797
von Kóssa's silver method, 539 – 540, 867
  bone before decalcification, 797
  phosphatase demonstration, 365
  spodograms, 866
von Recklinghausen's hemofuschin, 517
Vorbrodt's method for deoxyribonuclease II, 441
Vulpian reaction of chromaffin, 264

Wachstein, M.:
  1952–1954 Ca++ CoS alkaline phosphatase
    method, 367 – 368
  1958 Gomori acid phosphatase method, 360
  thioacetic esterase technics, 408 – 410
Wachstein-Meisel pH 6.0, 7.2 phosphatase
    methods, 367 – 368
Wade-Fite oil fuchsin formaldehyde method,
    739 – 740
Walls' fixation of eyes, 60
Walpole acetate buffer series pH 2.7–6.5, 875
Walpole-Pearse HCl acetate buffer series pH
    0.6–5.2, 875
Warming plates, 22
Warren's single-cell layer preparation, 813
Warthin's silver spirochete methods, 761
Washburn-Lillie technic for myeloperoxidase,
    449
Washing after Zenker fixatives, 57
Waste pigment, 513
Water baths, 19
Water blue I, 139, 141, 159
Water in resin mounts, 109
Waterhouse's method:
  for copper, 545
  strontium and barium, 537
Waxes, 562
Wear and tear pigments, 513
Weigert, C.:
  acid iron chloride hematoxylin, 199 – 200
  bleach in silver reticulum technic, 688

Weigert, C.:
  borax-ferricyanide, 603
  elastin stains with oil red O, 713
  fibrin stain, 298–299
    renal hyaline droplets and casts, 670
  iodine solution, 152–153, 726
  lithium hematoxylin, 591, 603
  myelin methods, 595, 599–600, 602–603
  myelin mordants, 603
  resorcin fuchsin elastin stain, 711–712
  Van Gieson stain, 694
Weil and Davenport's glia cell methods, 766
Weil's iron alum hematoxylin, 200
  myelin technic, 602–603
Weill's methyl green pyronin, 179
Weir *et al.*, acid ethanol inactivation of en-
    dogenous peroxidase, 863
Weiss-Tsou-Seligman amino group method,
    283–285
White's ground section technic, 795
Wigglesworth's iron sulfide method for fatty
    acids, 591
  iron uptake reaction, 203
Wilder silver method for reticulum, 684–691
Wildi's alcoholic $AgNO_3$ for *o*-and *p*-diphenols,
    262
Wilson's disease, 545
Windaus' digitonin reaction for sterols, 589
Windle's thionin Nissl stain, 183
Winkler Schultze oxidase method, 451–452
  peroxide reaction, 223–224
Wolbach, S. B., Giemsa stain, 193, 195, 742
Wolfgram, localization of glucosidase and galac-
    tosidase, 390
Wolman spodography of calcium oxalate, 867
Wolters' hematoxylin, 603
  myelin method, 602–603
Wool green S, 159
  Masson trichrome variant, 700
Wool yellow G, 159
Wright, J. H.:
  blood stain, 747–748
  Giemsa variant for spirochetes, 759–760
  hematoxylin, 603
  myelin method, 602–603
Wright-Craighead method for encephalitozoa,
    734

Xam, 114–115
Xanthoma lipids, 563
  pigment, 504, 508
Xanthroproteic reaction, 283
Xanthydrol reaction:
  for indoles, 261

Xanthydrol reaction:
  for urates, 290
  for urea, 289–290
X-ray analysis, chromaffin reaction, 266
Xylene, as clearing agent, 81–84, 107
Xylidene ponceau in trichrome stain, 700

Yasuda-Montagna method for monoamine ox-
    idase, 465
Yasuma-Ichikawa ninhydrin Schiff reaction, 282
Yeast adenylic acid (A3' P), 376–377, 442
Yeast nucleic acid, 168
Yeasts, cell wall polysaccharide, 615
Yellow pigment, 513

Zander's $I_2 + ZnCl_2$ method for chitin, 633
Zenker, F. A., degeneration of muscle, 196
Zenker. K.:
  fixation for Masson trichrome stain, 700
  fixation for phosphotungstic acid hematoxylin,
    693
  fixation Technicon paraffin schedule, 83
  fixative fluid, 54–57
Zephiran action on enzymes, 428
Ziehl's carbol fuchsin, 735
Zinc, 126, 128, 531, 545, 547, 549, 553–556
  relation to carbonic anhydrase, 443
  stains for, 553–556
Zinc chloride, iodine reaction of cellulose, 634
Zinc chromate Golgi variant, 769
Zinc leuco method for hemoglobin peroxidase,
    447, 488
Zinc mordant dyes, 126
Zinc 65 radioautography (*see* Autoradiography)
Zincon, 544, 546, 554–555
Zirconium, 535, 556
Zlotnick's Negri body method, 755–756
[65]Zn autoradiography, 816
Zn^{++} inhibition of enzymes, 428–429, 433, 443
Zn^{++} ion demonstration, 126, 129, 545, 547, 549,
    553–556
Zn^{++} uptake, Clara's hematoxylin, 203
Zottenmelanosepigment, 508
Zugibe, F. T.:
  Carbowax technic, 76
  chrome gelatin adhesive, 76
Zwemer's gelatin embedding, 78
  glychrogel, 119–120
Zymogen granules, 261, 350–351
Zymograms, 824–825
Zymohexase, 386
Zymonucleic acid (RNA), 168